ANATOMY
Regional and Applied

R. J. LAST

M.B., B.S. (Adel.), F.R.C.S.

Visiting Professor in Anatomy to The University of California, Los Angeles; lately Professor of Applied Anatomy, Royal College of Surgeons of England; past Examiner in Anatomy for the Primary Fellowship Examination of the College; past Examiner in Anatomy for the Licence in Dental Surgery and for the Fellowship in Dental Surgery of the College; former Superintendent of Dissections and past Examiner in Anatomy to the Examining Board in England; former Examiner in Anatomy to the Worshipful Society of Apothecaries of London.

SIXTH EDITION

WITH 526 ILLUSTRATIONS,
46 IN COLOUR

CHURCHILL LIVINGSTONE
EDINBURGH LONDON AND NEW YORK 1978

CHURCHILL LIVINGSTONE
Medical Division of Longman Group Limited

Distributed in the United States of America by
Churchill Livingstone Inc., 19 West 44th Street,
New York, N.Y. 10036 and by associated companies,
branches and representatives throughout
the world

First Edition 1954
Second Edition 1959
Third Edition 1963
Fourth Edition 1966
ELBS Edition first published 1970
Fifth Edition 1972
ELBS Edition of Fifth Edition 1972
ELBS Edition reprinted 1974
Sixth Edition 1978
Reprinted 1978
ELBS Edition of Sixth Edition 1978
ELBS Edition reprinted 1978
Sixth Edition reprinted 1979
ELBS Edition reprinted 1979
Sixth Edition reprinted 1981

ISBN 0 443 01454 X (cased)
ISBN 0 443 01317 9 (limp)

British Library Cataloguing in Publications Data
Last, Raymond Jack
Anatomy, regional and applied—6th ed.
1. Anatomy, Human
I. Title
611 QM23.2 77-30085

Printed in Singapore by Ban Wah Co

PREFACE TO THE SIXTH EDITION

Those who already know this book will, I feel sure, recognize it in its new guise. The pages are larger and the type is smaller, in the now fashionable two columns. This has reduced the number of pages by one third. Each page takes longer to read but there are fewer pages! The total length of the text is no greater than in the last edition.

The basic plan and substance of the text remain the same as in all former editions. Emphasis continues to fall on the understanding of function. I have taken advantage of the type resetting to improve the accounts of the sympathetic nervous system in general and of its limb distribution in particular. Accounts of the anal canal have been rewritten, as also of the rhinencephalon and limbic system to bring them into line with current beliefs. There are innumerable minor adjustments throughout, to secure greater clarity of description and better understanding. I continue to use the old-fashioned diphthongs.

Many illustrations have been redrawn, in addition to which there are 56 new black and white figures. All 46 colour illustrations are now grouped in one section. This has some advantages over the former scheme of scattering them through the book, since multiple references appear in the text to most of them. It also lowers the cost of printing and therefore of the cost of the book, an important consideration in these days of mounting prices.

Having retired after 24 years of postgraduate teaching at the Royal College of Surgeons of England I am indeed favoured by the appointment of Visiting Professor in Anatomy to the University of California, Los Angeles. Continuing contact with both freshmen and graduate students in North America provides a welcome stimulus to continue the search for a fuller understanding of functional anatomy.

I send greetings to my former students (all 13,000 of you) and good wishes to the many strangers who have written to me. I continue to be comforted by the number of clinicians who find the book helpful in their medical practice.

1977 R. J. LAST

FROM THE PREFACE TO THE THIRD EDITION

Only material that has practical application or that illustrates a general principle is considered appropriate for inclusion. Consequently a great deal of osteology already existed in the book, studied as part of the region concerned. *This osteology remains in the text as before.*

Full accounts of each bone are now included in response to innumerable requests, some amounting almost to demands, from many parts of the world.

To the student who must 'learn the bones' the following advice is offered. Never attempt to learn the whole bone until the anatomy of the 'soft parts' has been thoroughly mastered; most *practical* details of the bones will have already been noted during the study of the region. *Never read osteology unless the bone is in front of you.* Even though the bone may need reversing for inspection (e.g. base of skull), wherever possible you should study osteology with the bone *strictly in its correct anatomical position.*

FROM THE PREFACE TO THE FIRST EDITION

I have attempted to include in the text all those parts of human anatomy which should occupy a place in the knowledge *and in the understanding* of the student or the general clinician, and I have sought to exclude details that have neither practical application nor value in illustrating a general principle. Often a structure insignificant in itself and of no importance in clinical anatomy possesses great interest because it illustrates a basic principle. Nobody would be any worse off without his rectus sternalis or his coraco-brachialis muscles, yet much may be learned from understanding their significance, and for that reason they and many structures like them are included in the text.

Finally, I want to say to the student: 'I sincerely hope that your reading of the following pages may not only prove profitable to you but will stimulate your permanent interest in a fascinating subject, much of which is still not fully understood.'

ACKNOWLEDGMENTS

It is very gratifying to be able to reproduce twenty-eight of Dr. Tompsett's drawings of his own prosections in the Museum of the Royal College of Surgeons; for this privilege I thank the President of the College. I am indebted to Messrs. H. K. Lewis & Co. for their permission to reproduce Figs. 6.15, 6.52 and 6.53 from *The Anatomy of the Eye and Orbit*. Fig. 6.51 is adapted, by permission of Messrs. John Wright & Sons, from French's *Differential Diagnosis of Main Symptoms.* I am grateful to the proprietors of the *Journal of Bone and Joint Surgery* for permission to reproduce from my articles Figs. 1.29, 2.44, 3.18, 3.46, 3.52, 6.73 and 6.74 and to the proprietors of the *British Journal of Surgery* for permission to reproduce Figs. 7.10 to 7.14 inclusive and Fig. 7.40. Fig. 2.40 has been drawn after Testut's *Anatomie Humaine.*

I am indebted to Miss J. Fairfax Whiteside for the care and skill with which she has drawn Figs. 1.4, 3.15, 4.26, 5.47, 5.54, 6.11 and 6.30 and Plates 4 and 37. I thank Dr. Seymour J. Reynolds for the X-ray films from which Figs. 4.6, 4.7, 5.31 and 6.66 have been made, and Mr. Frederic Mancini for modelling the face and ear on the skull photographed in Fig. 6.13. Dr. Douglas Silva kindly provided the photomicrographs for Fig. 1.23, and Dr. Frances S. Grover supplied the information from which Fig. 3.62 was drawn.

Finally it is with much gratitude that I acknowledge the biographical notes that continue to be provided by Miss Jessie Dobson, who is now a Hunterian Trustee at the Royal College of Surgeons of England.

CONTENTS

SECTION 3. THE LOWER LIMB

SECTION 4. THE THORAX

SECTION 5. THE ABDOMEN

SECTION 6. THE HEAD AND NECK

SECTION 7. THE CENTRAL NERVOUS SYSTEM

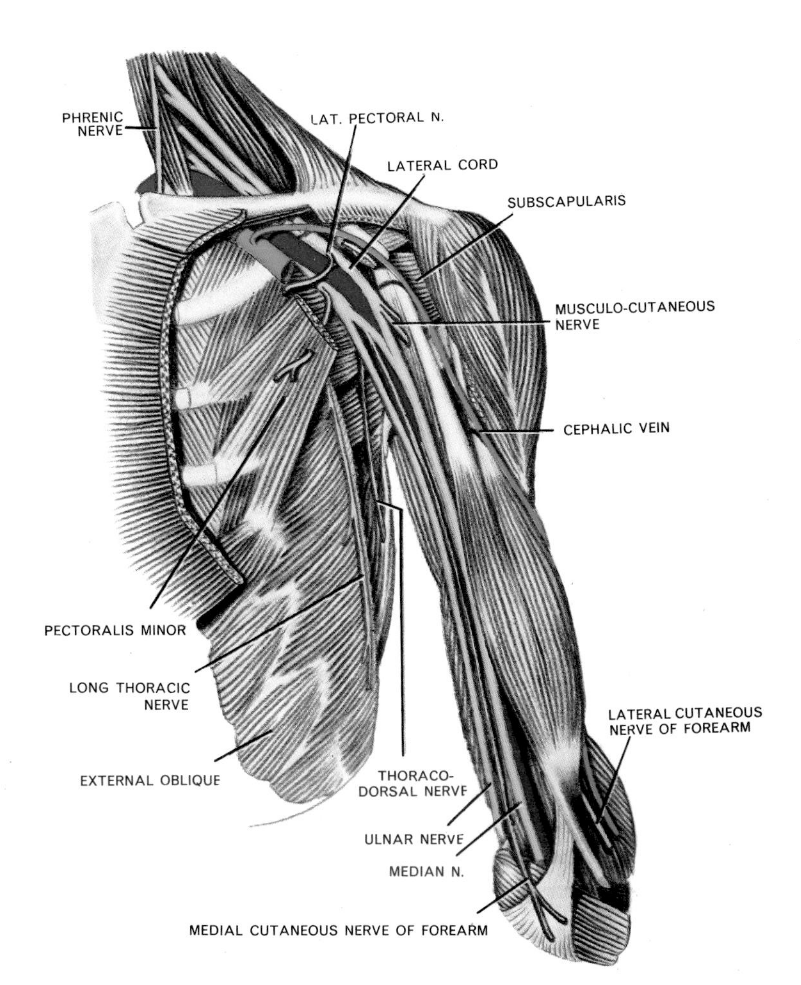

PLATE 1. THE LEFT AXILLA AND ARM.

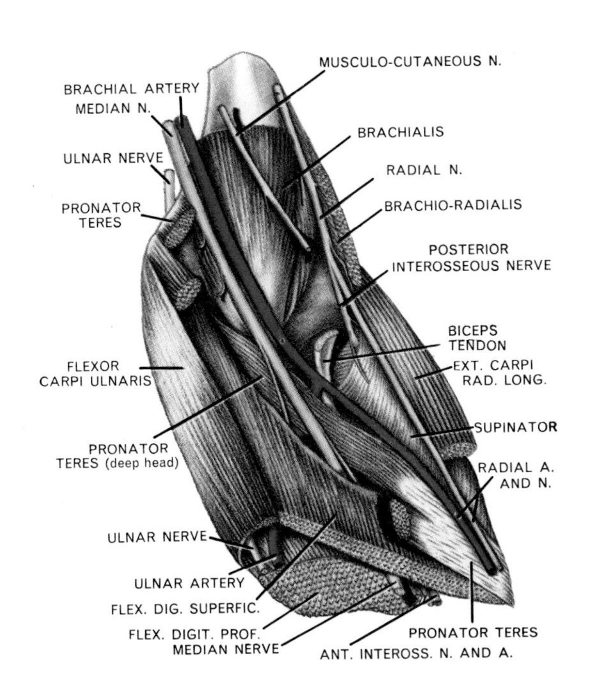

PLATE 2. THE FLOOR OF THE LEFT CUBITAL FOSSA.

PLATE 3. THE SUPERFICIAL PALMAR ARCH AND DIGITAL NERVES OF THE LEFT HAND.

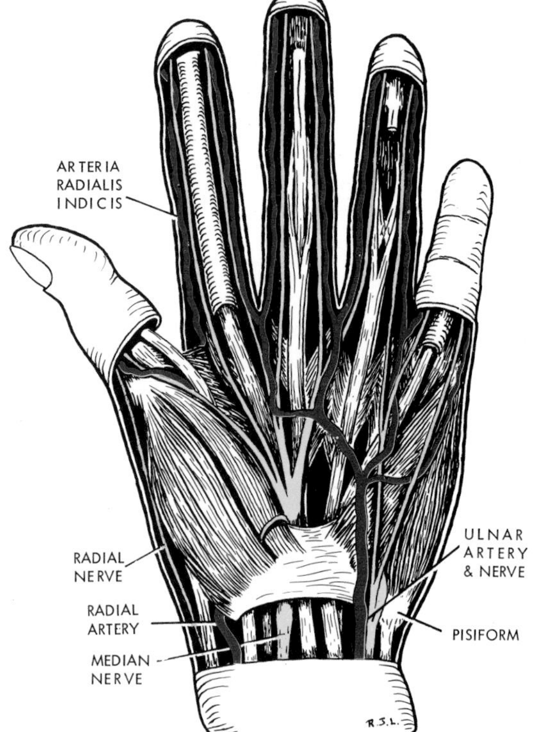

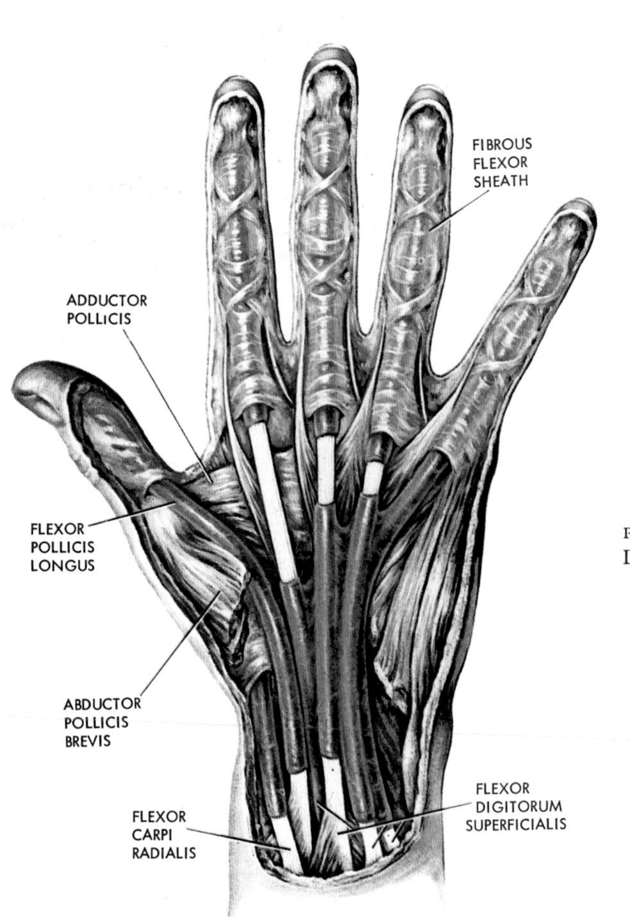

PLATE 4. THE SYNOVIAL FLEXOR SHEATHS OF THE LEFT HAND.
Drawn from Specimen S.51 in R.C.S. Museum.

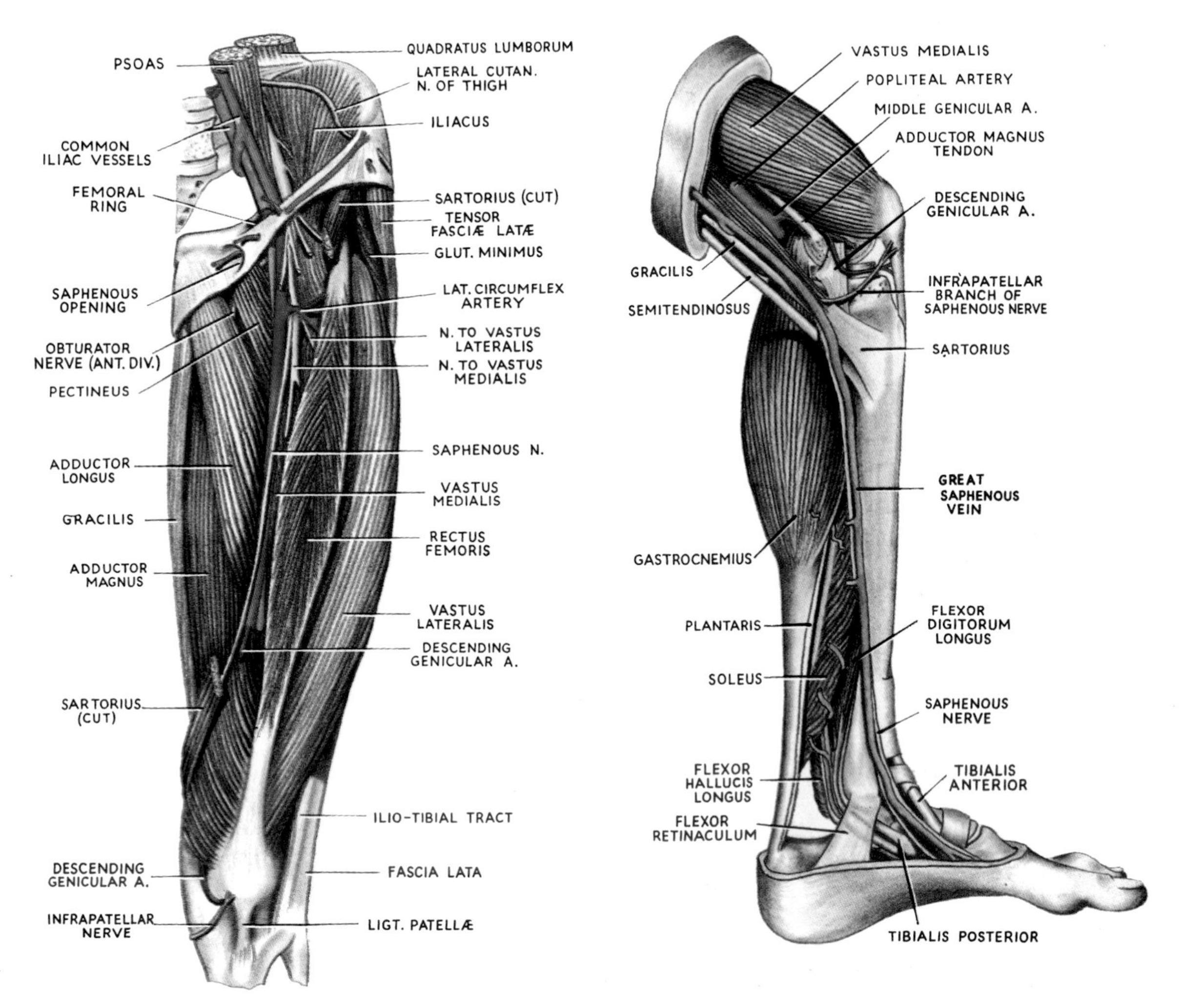

PLATE 5. THE LEFT THIGH, WITH THE SUBSARTORIAL CANAL EXPOSED.

PLATE 6. THE MEDIAL SIDE OF THE LEFT LEG.

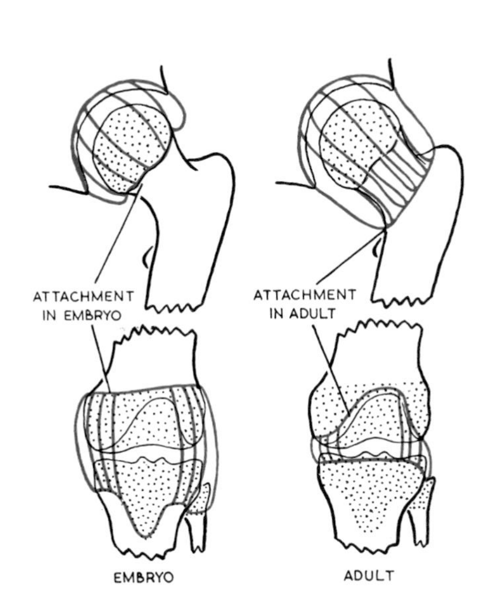

PLATE 7. THE MIGRATION OF JOINT CAPSULES FROM THE EPIPHYSEAL LINE. The epiphysis of the head of the femur becomes intracapsular, the lower epiphysis of the femur becomes extracapsular. There are many such examples.

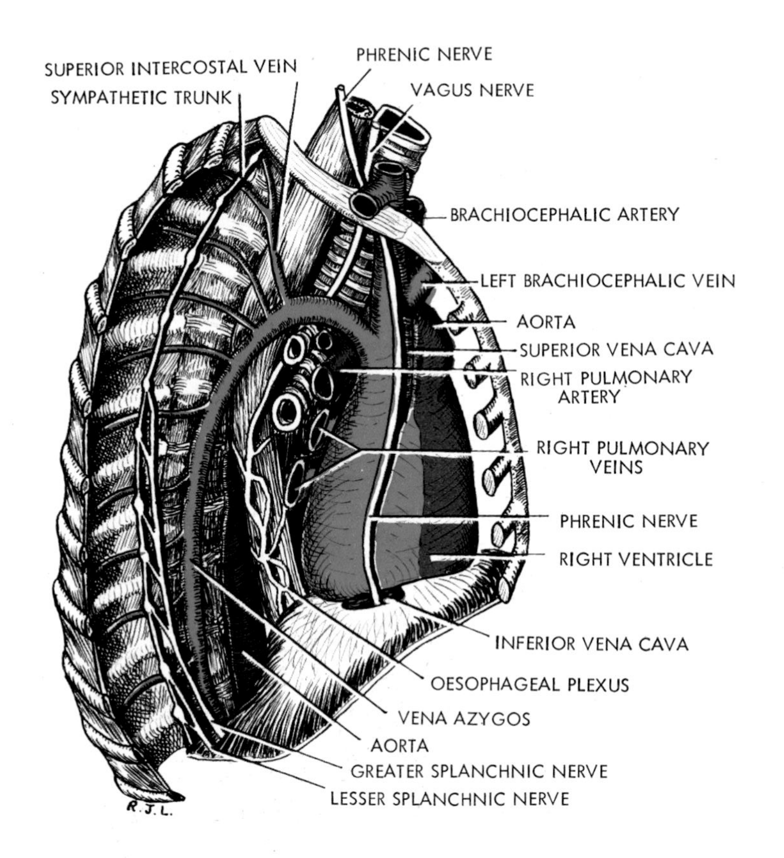

PLATE 8. THE MEDIASTINUM FROM THE RIGHT SIDE. The atria and the systemic *and pulmonary* veins are coloured blue. They are prominent on this side of the mediastinum. The phrenic nerve lies on venous structures. The vagus nerve touches the trachea. The apex of the lung is in contact with the trachea.

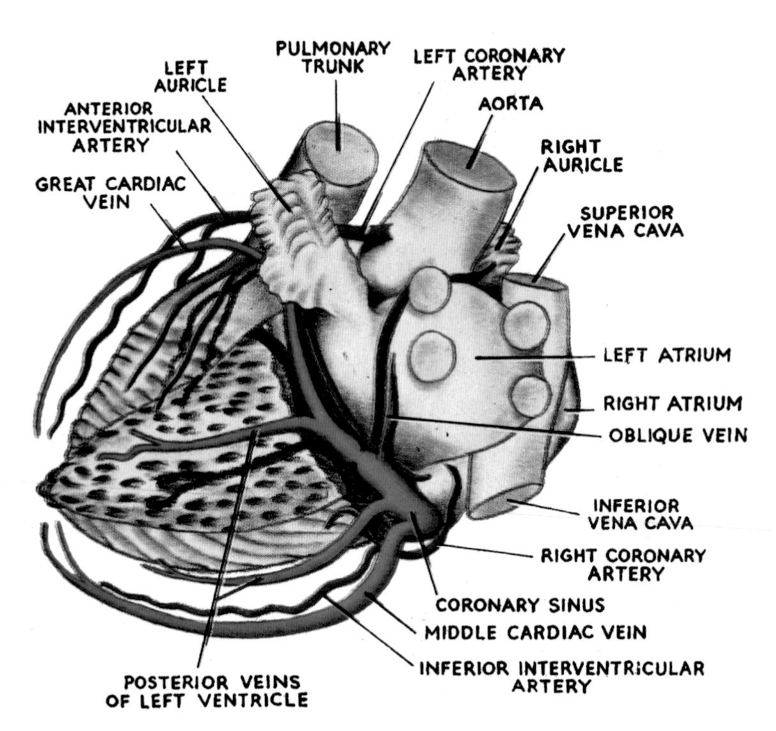

PLATE 9. POSTERIOR VIEW OF A CORROSION CAST OF THE VESSELS OF THE HEART (a resin injection in which the heart cavities and great vessels are also filled). Drawn from Specimen S.264 in R.C.S. Museum.

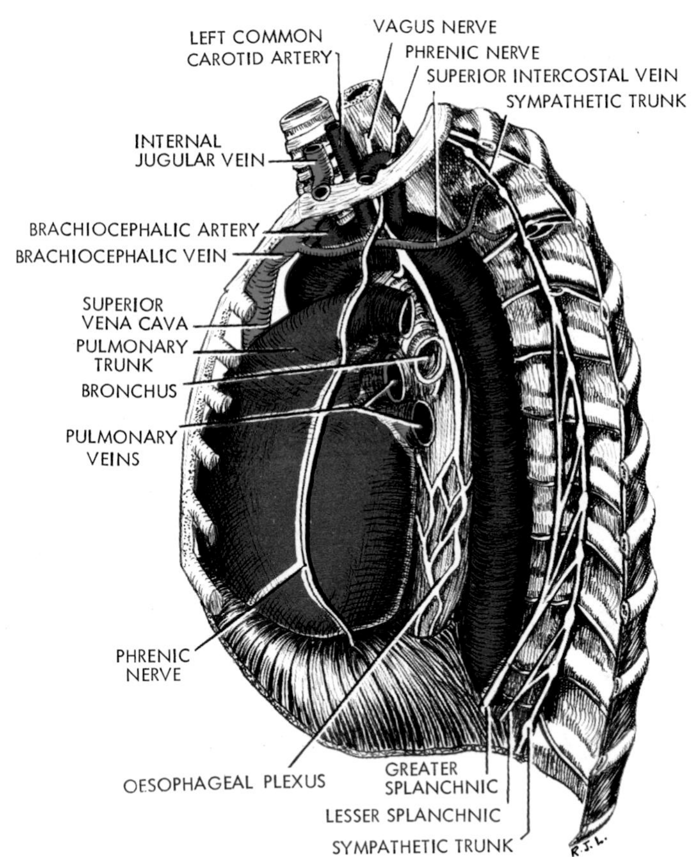

PLATE 10. THE MEDIASTINUM FROM THE LEFT SIDE. The ventricles and the systemic *and pulmonary* arteries are coloured red. They are prominent on this side of the mediastinum. The phrenic nerve lies on arterial structures. The vagus nerve and the apex of the lung are held away from the trachea by the arch of the aorta and its branches.

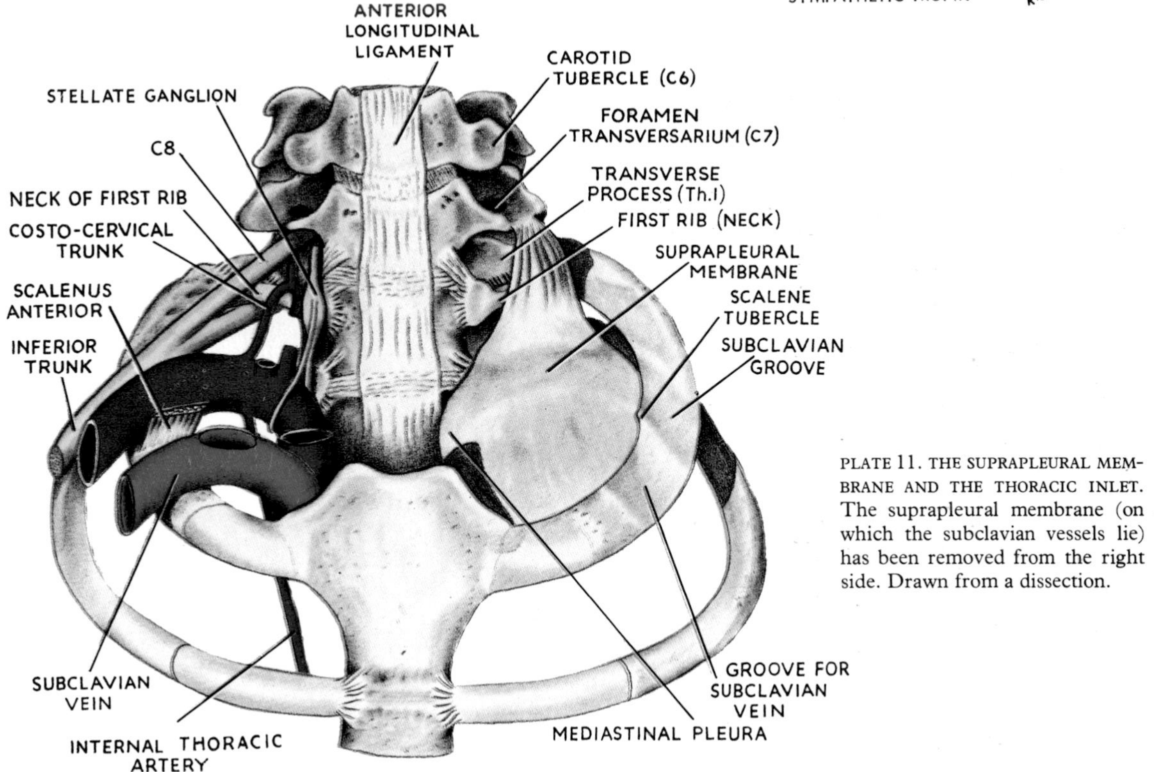

PLATE 11. THE SUPRAPLEURAL MEMBRANE AND THE THORACIC INLET. The suprapleural membrane (on which the subclavian vessels lie) has been removed from the right side. Drawn from a dissection.

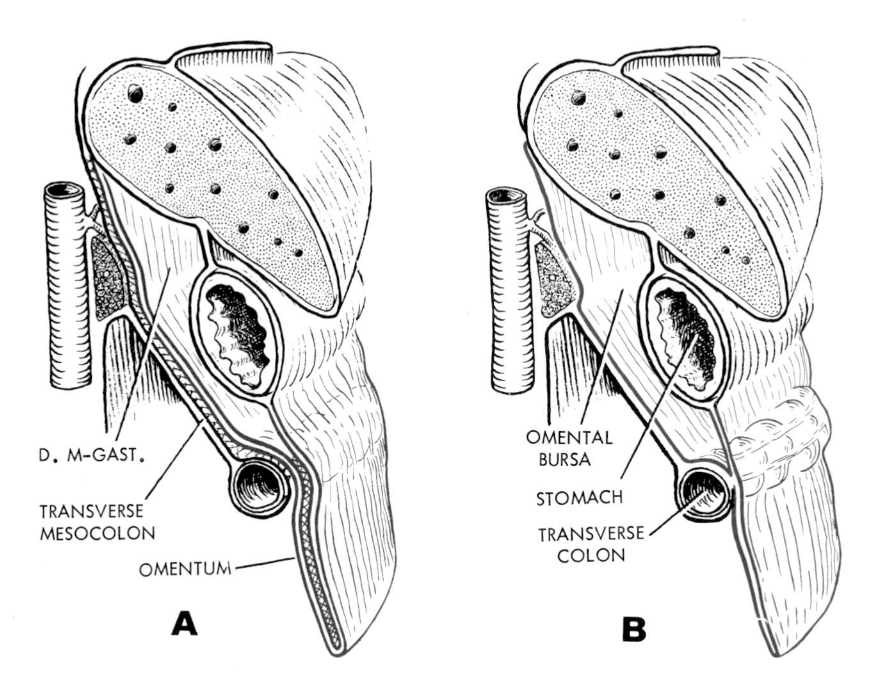

PLATE 12. THE FATE OF THE DORSAL MESOGASTRIUM, HERE SHOWN IN RED.

A. It sweeps to the left across the peritoneum of the posterior abdominal wall (D. M-GAST) as in Fig. 5.23, p. 278 and across the transverse mesocolon (which is itself a secondary fusion of midgut mesentery to the peritoneum across the pancreas, Figs. 5.20, 5.21, p. 277). Adjacent surfaces (indicated by cross-hatching) fuse and disappear. The dorsal mesogastrium continues as an empty fold, free across the transverse colon and slung to the greater curvature of the stomach. This is 'the omentum', and in it adjacent surfaces fuse and disappear (see cross-hatching).

B. The final result. Note that the posterior wall of the lesser sac, including the anterior layer of the transverse mesocolon, is the original right leaf of the dorsal mesogastrium (Fig. 5.18, p. 274).

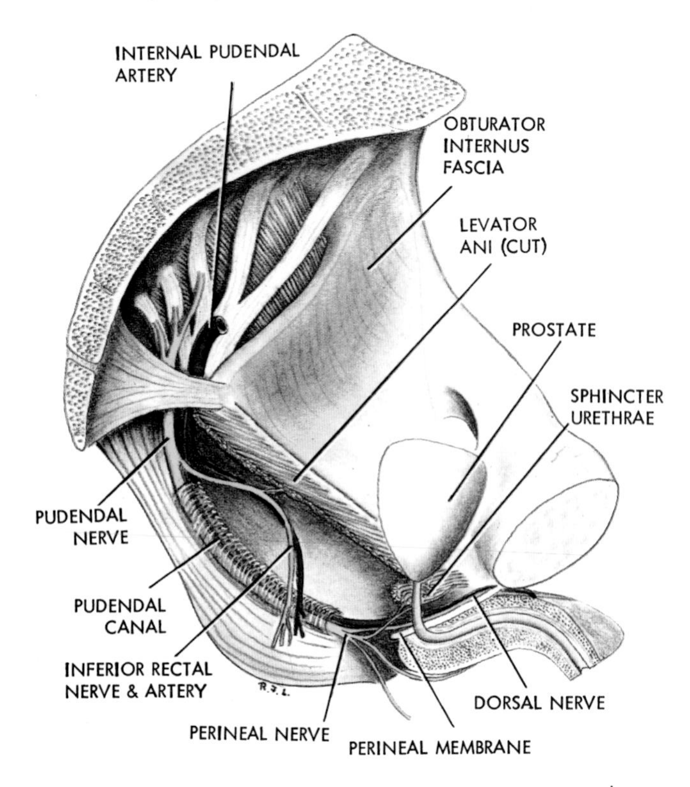

PLATE 13. THE LATERAL WALL OF THE ISCHIO-RECTAL FOSSA, EXPOSED BY REMOVAL OF MUCH OF LEVATOR ANI. The pudendal canal leads to the perineal membrane, and the ischio-rectal fossa extends forward deep to this (cf. Fig. 5.56, p. 323).

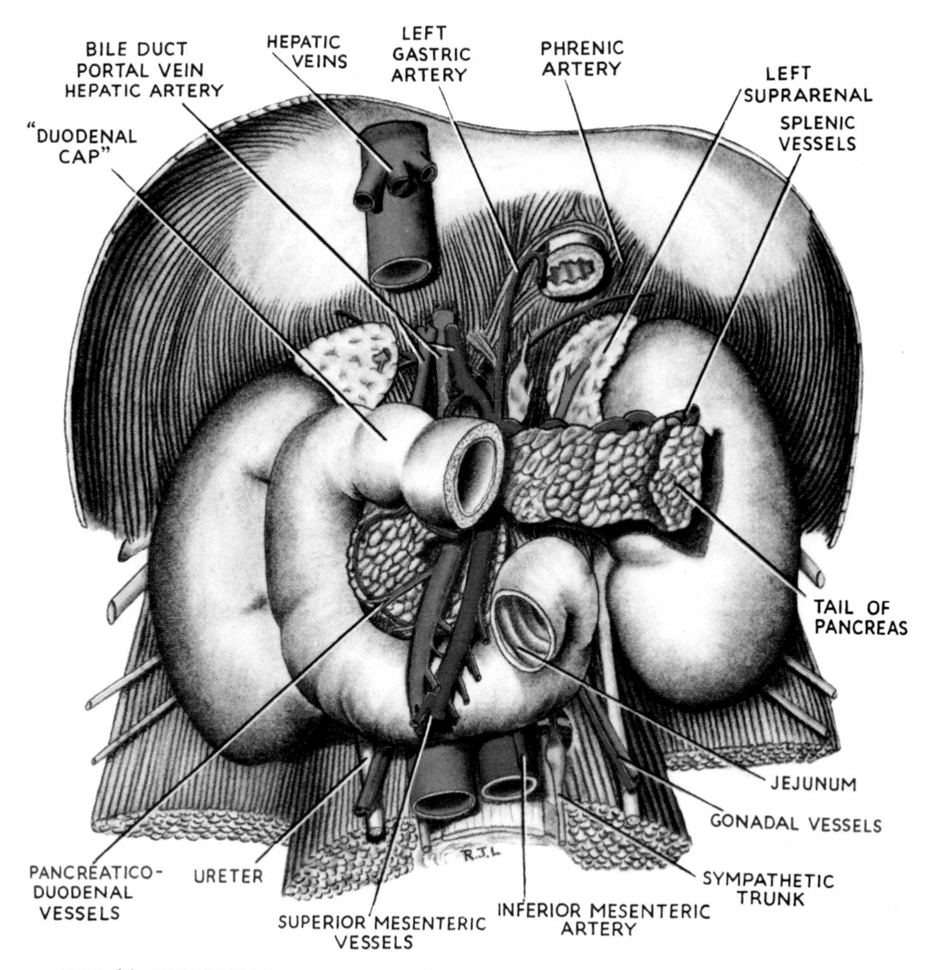

PLATE 14. THE DUODENUM AND PANCREAS IN POSITION ON THE POSTERIOR ABDOMINAL WALL.

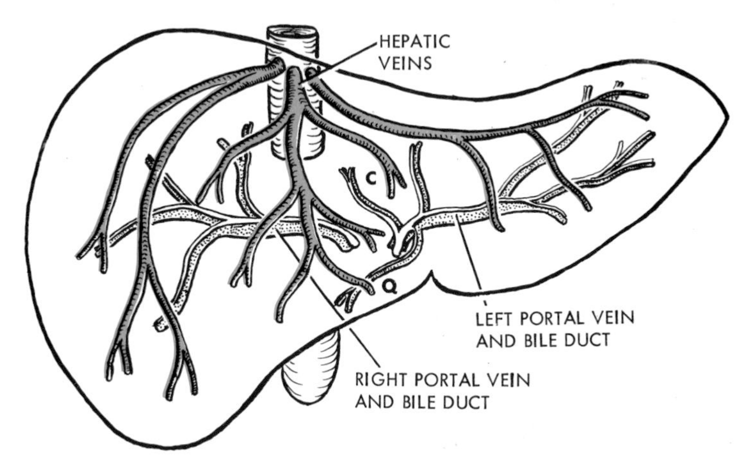

PLATE 15. THE VESSELS OF THE LIVER. The right and left **portal veins** enter the visceral surface and supply equal *halves* of the liver. The **bile ducts** always cross in front of the portal veins. The **hepatic artery** (not shown) is generally between the two; the order is V.A.D. even within the liver itself. The caudate (**C**) and quadrate (**Q**) lobes are supplied from the left side and drain into the middle hepatic vein. The three main **hepatic veins** lie high up near the diaphragmatic surface of the liver. Small accessory hepatic veins reaching the inferior vena cava at a lower level are not shown. Drawn from corrosion casts in R.C.S. Museum.

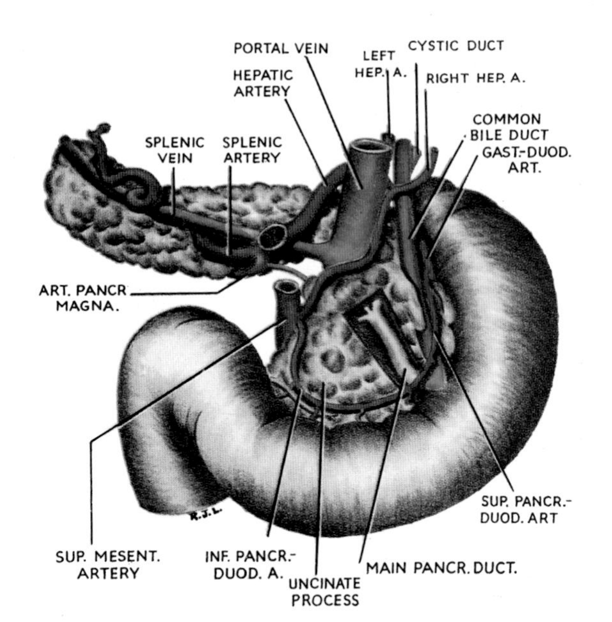

PLATE 16. THE PANCREAS AND DUODENUM SEEN FROM BEHIND. HERE THE RIGHT HEPATIC ARTERY ARISES FROM THE SUPERIOR MESENTERIC ARTERY, A COMMON VARIANT. Drawn from Specimen S.277 in R.C.S. Museum.

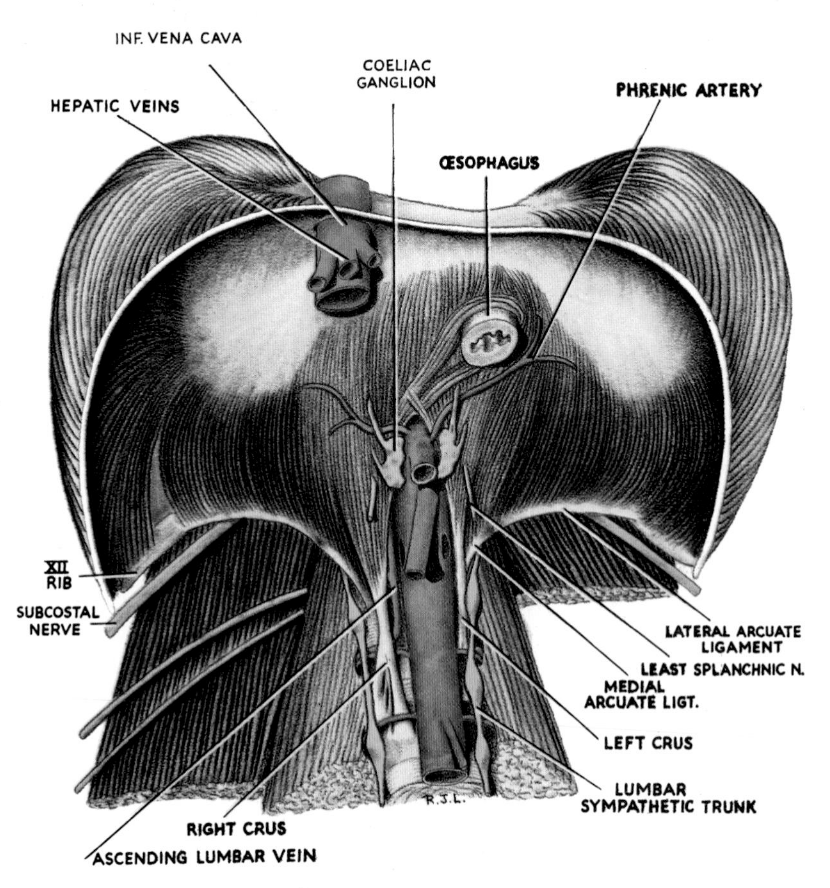

PLATE 17. THE CRURA AND POSTERIOR ATTACHMENTS OF THE DIAPHRAGM.

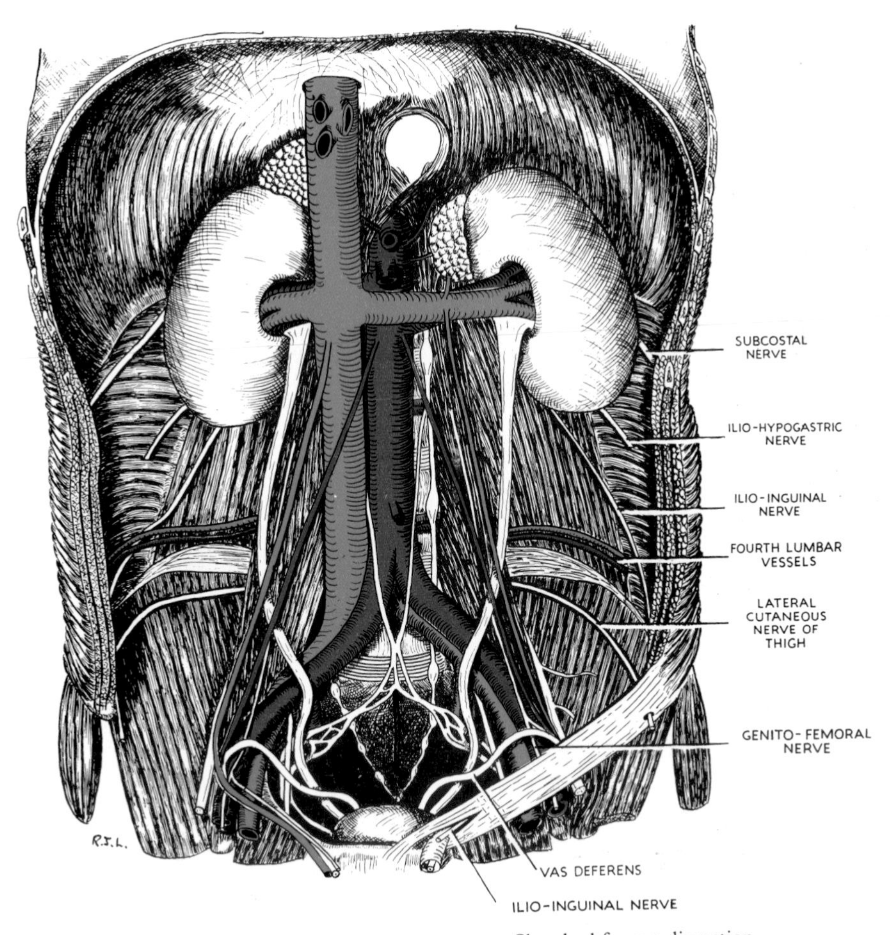

PLATE 18. THE POSTERIOR ABDOMINAL WALL. Sketched from a dissection.

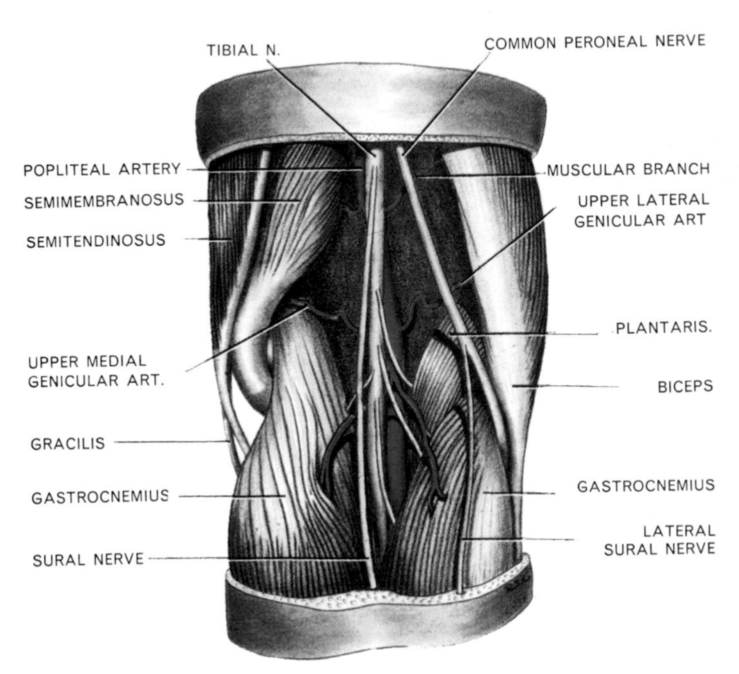

PLATE 19. THE RIGHT POPLITEAL FOSSA, THE SIDE IDENTIFIED BY THE COMMON PERONEAL NERVE. Drawn from Specimen S.133 in R.C.S. Museum.

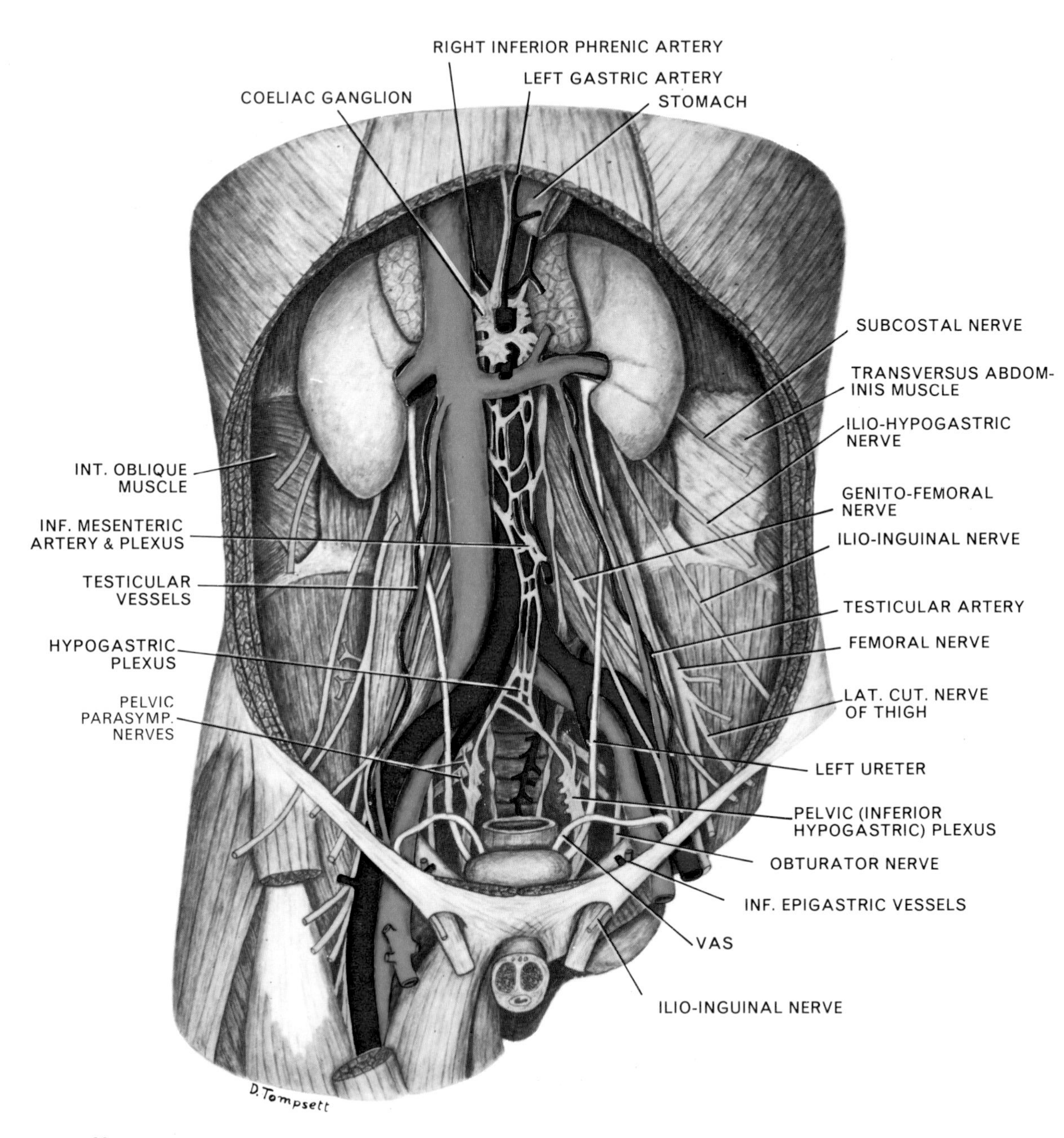

PLATE 20. A DISSECTION OF THE POSTERIOR ABDOMINAL WALL. THE TESTICULAR ARTERIES HERE ARISE FROM THE RENAL ARTERIES, A NOT UNCOMMON VARIATION. Illustration of Specimen S.354 in R.C.S. Museum.

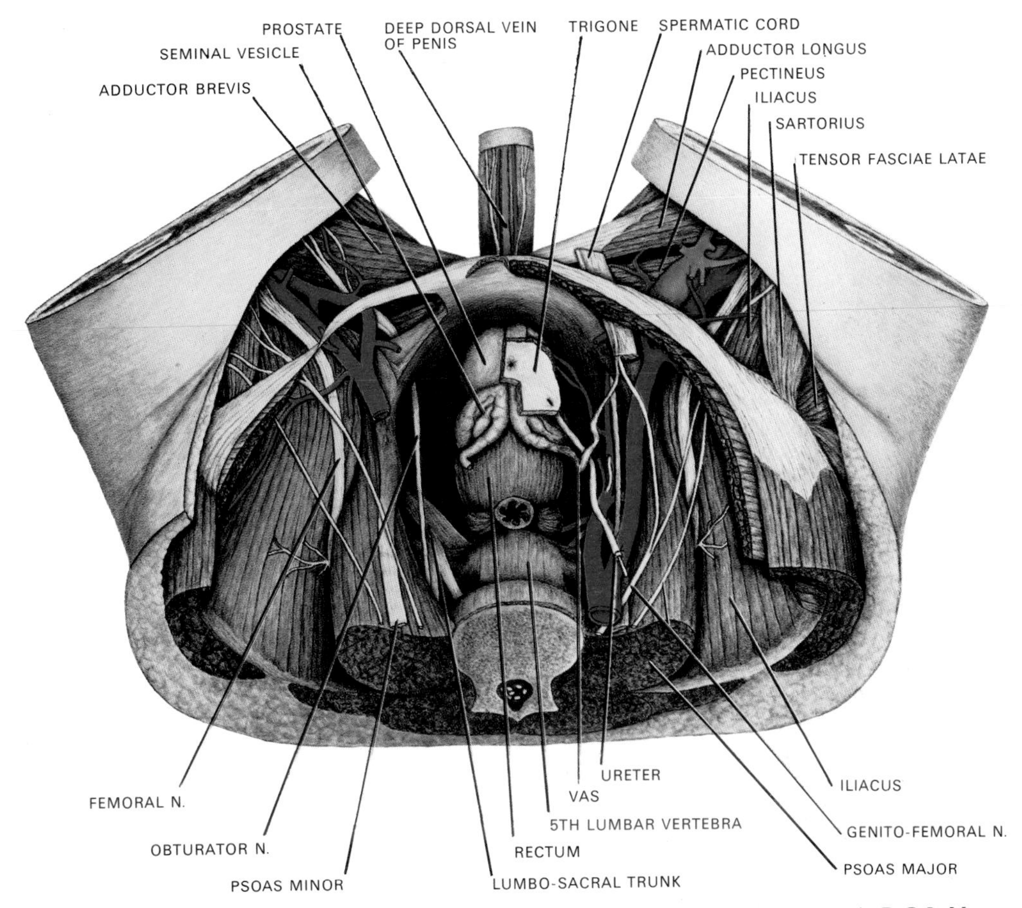

PLATE 21. A DISSECTION OF THE MALE PELVIS, VIEWED FROM ABOVE. Illustration of Specimen S.353 in R.C.S. Museum.

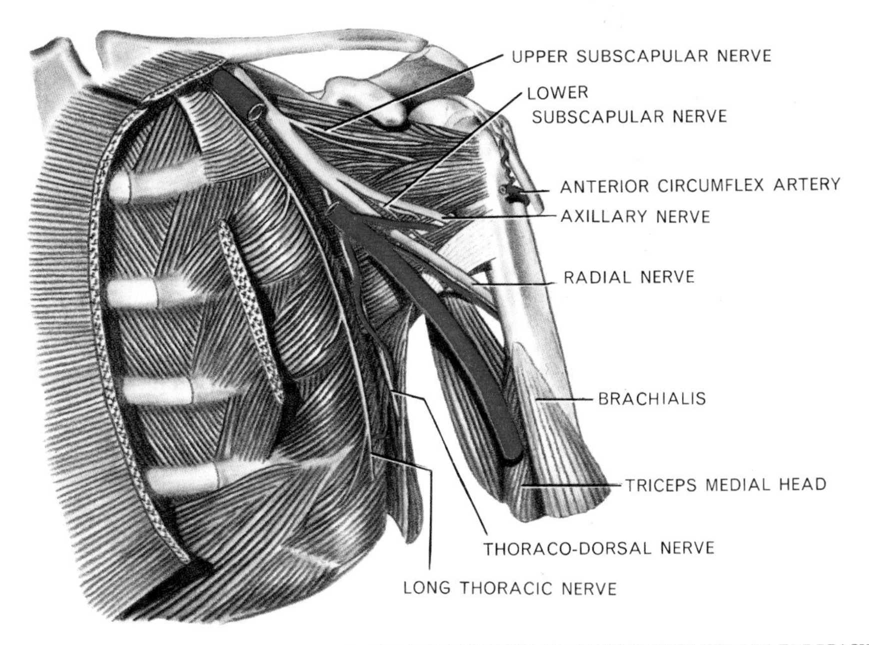

PLATE 22. THE POSTERIOR WALL OF THE LEFT AXILLA, AND THE BRANCHES OF THE POSTERIOR CORD OF THE BRACHIAL PLEXUS.

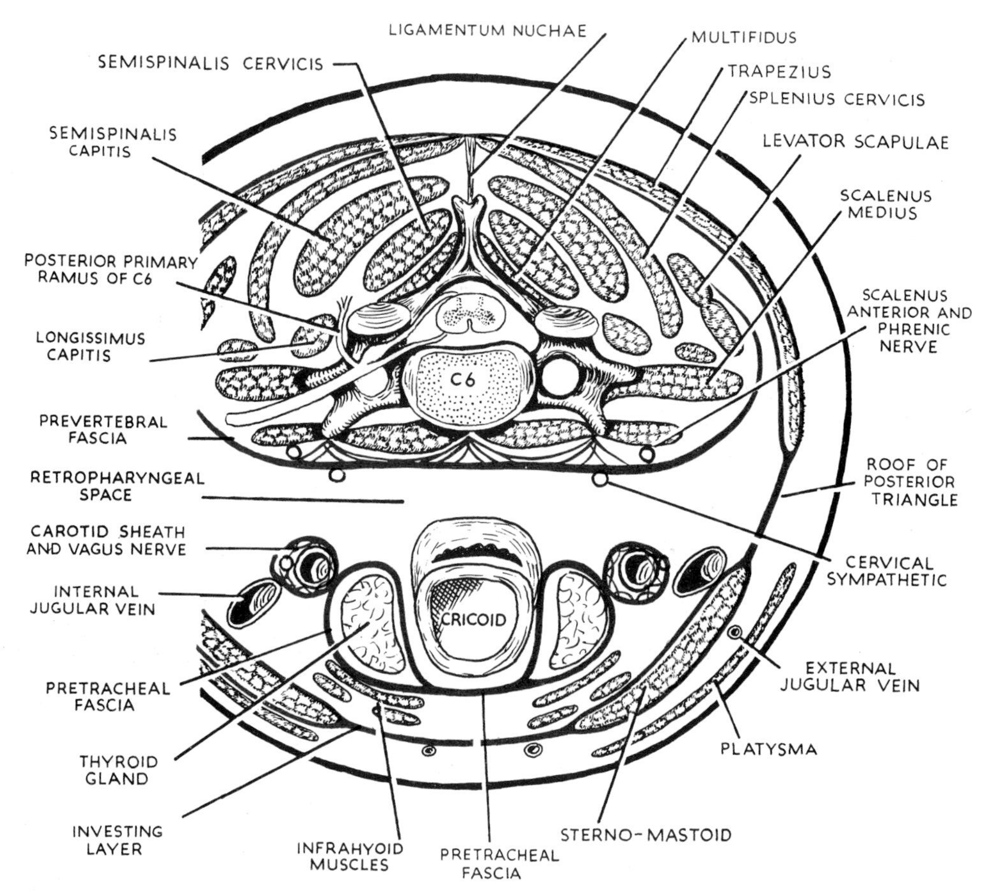

PLATE 23. THE FASCIAL PLANES AND TISSUE SPACES OF THE NECK, IN HORIZONTAL SECTION THROUGH C6. The deep fasciæ and the carotid sheath are shown in red. *The tissue spaces are artificially opened up*. The hypopharynx behind the cricoid actually lies in sliding contact with prevertebral fascia; the retropharyngeal space is only potential. The carotid sheath lies in sliding contact with prevertebral fascia posteriorly, and is adherent anteriorly to the investing fascia on the deep surface of sterno-mastoid. The roof of the posterior triangle lies for the most part in contact with prevertebral fascia over scalenus medius (cf. PLATE 25).

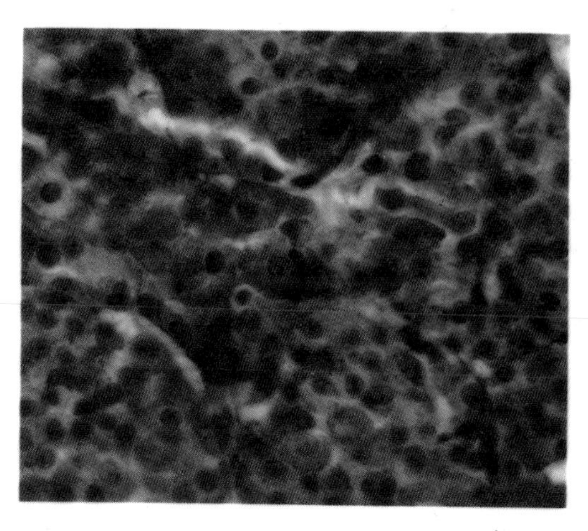

PLATE 24. THE ANTERIOR LOBE OF THE PITUITARY GLAND (HYPOPHYSIS), SHOWING EOSINOPHILS, BASOPHILS AND CHROMOPHOBES. (Photograph, × 400.)

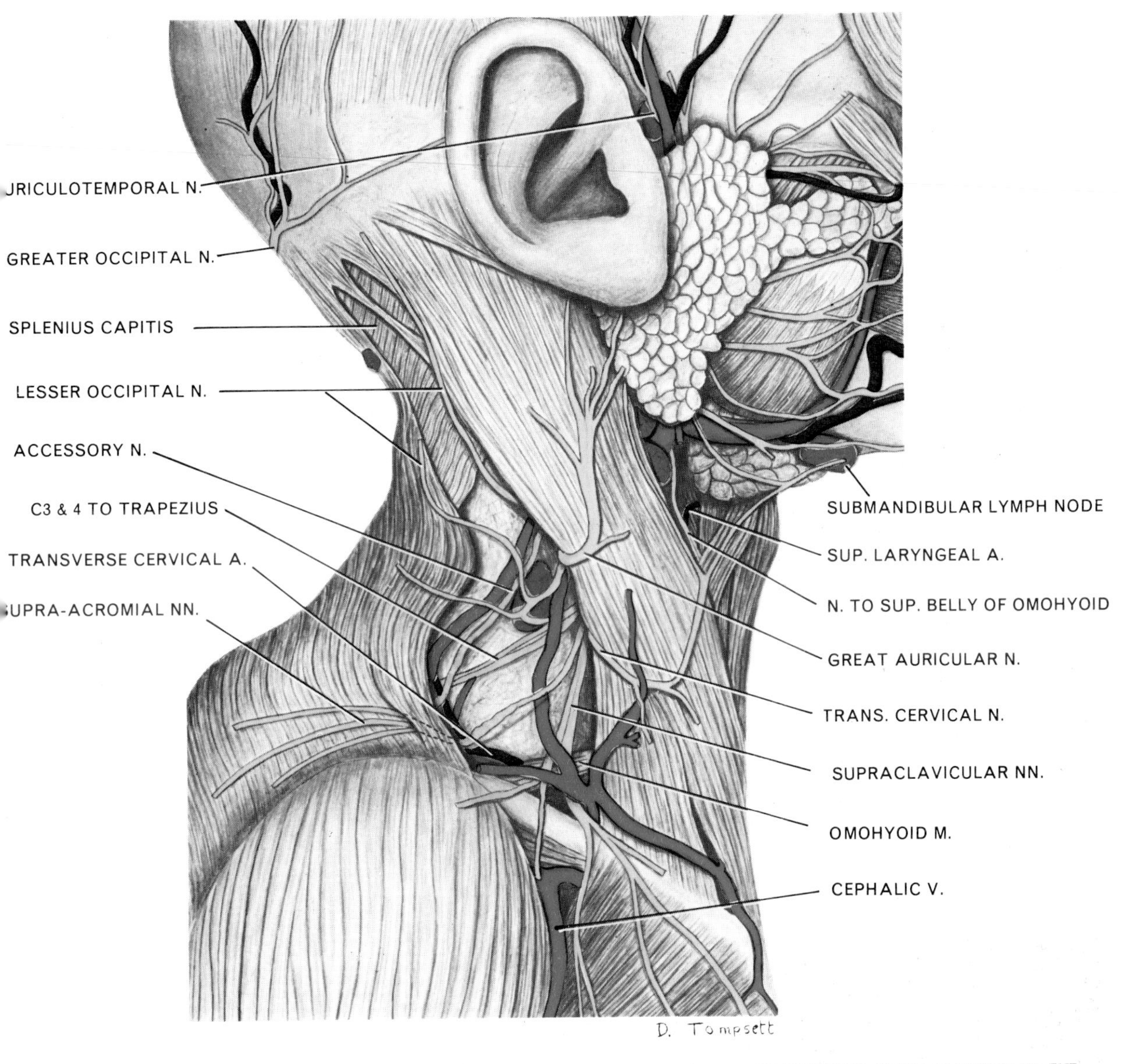

PLATE 25. A SUPERFICIAL DISSECTION OF THE RIGHT SIDE OF THE NECK AND FACE. THE PREVERTEBRAL FASCIA IS INTACT IN THE LOWER PART OF THE POSTERIOR TRIANGLE. Illustration of Specimen S.350 in R.C.S. Museum.

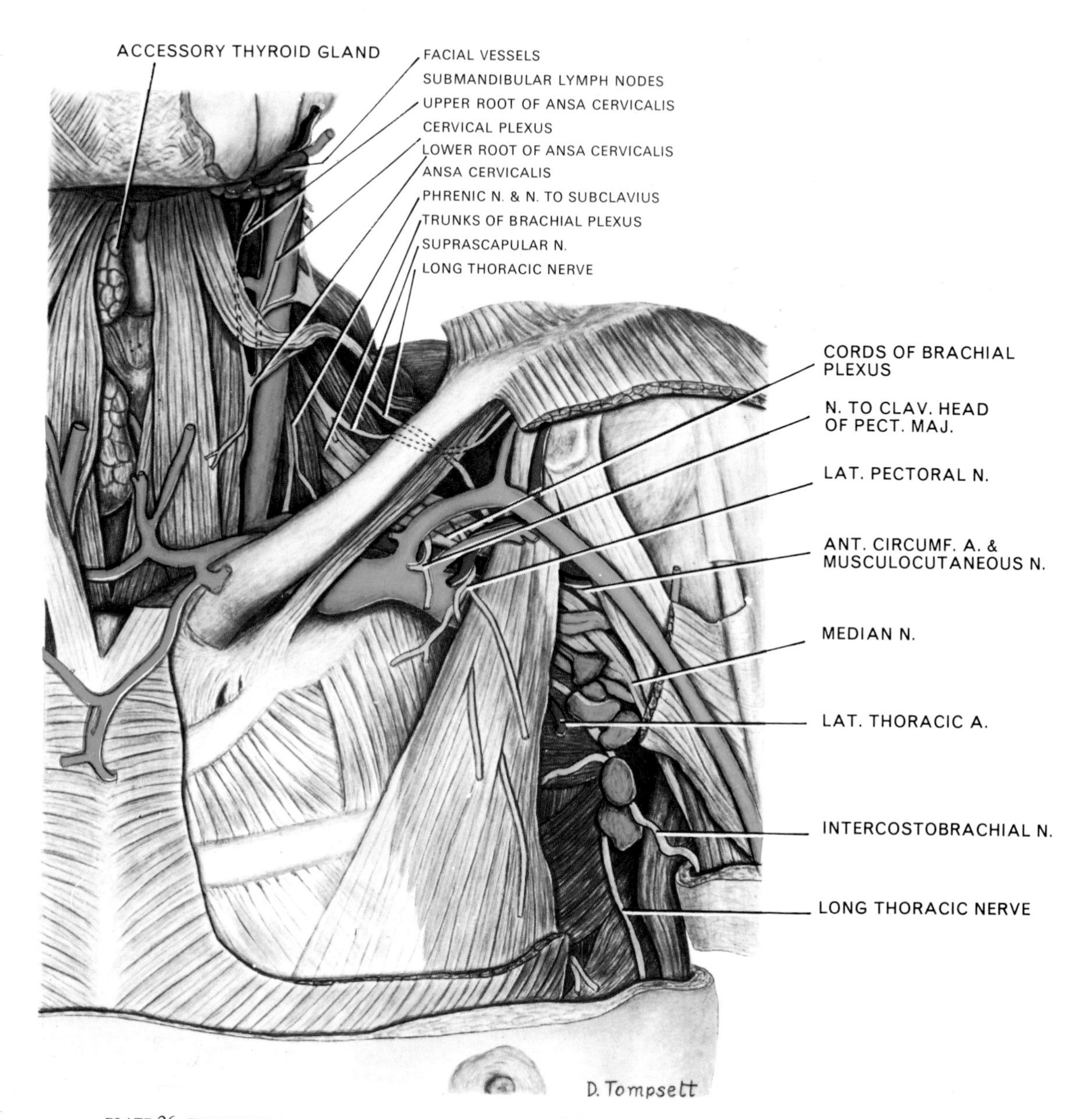

PLATE 26. THE FRONT OF THE NECK AND THE LEFT AXILLA. Illustration of Specimen S.350 in R.C.S. Museum.

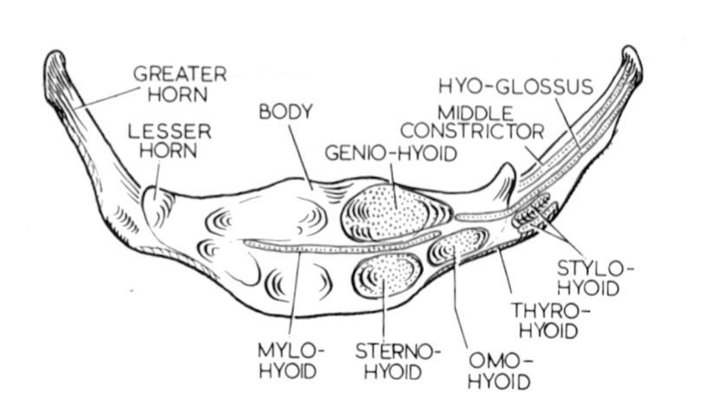

PLATE 27. THE HYOID BONE (ANTERIOR VIEW).

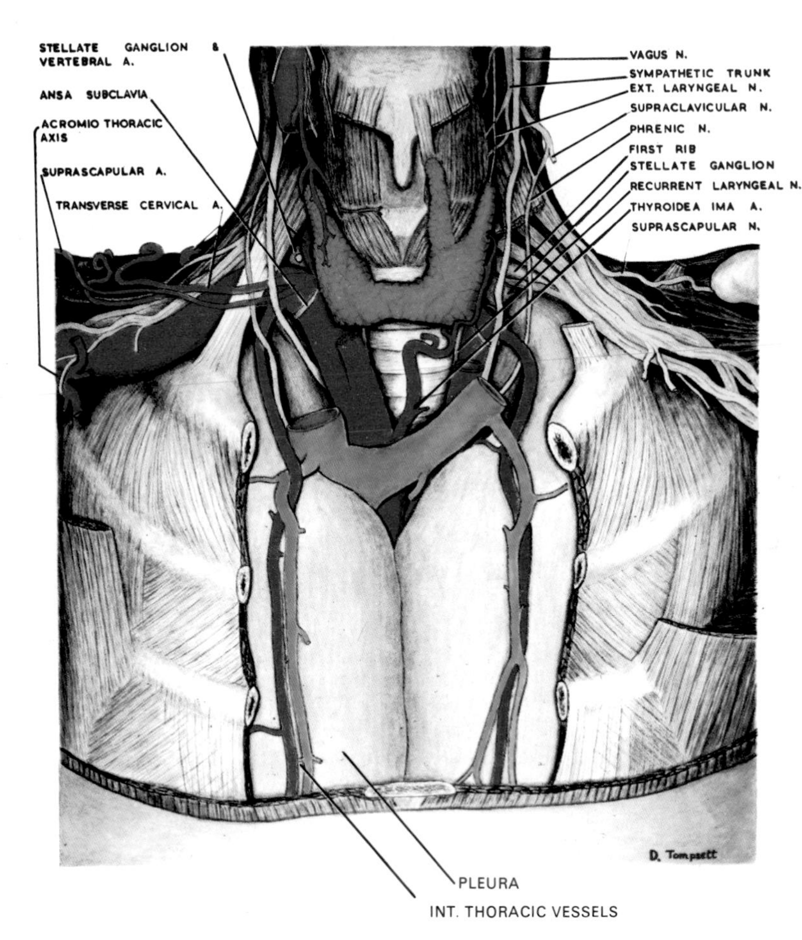

PLATE 28. THE THYROID GLAND AND THE ANTERIOR MEDIASTINUM. THE PLEURAL CAVITIES HAVE BEEN DISTENDED BY INJECTING GELATINE. Illustration of Specimen S.352 in R.C.S. Museum.

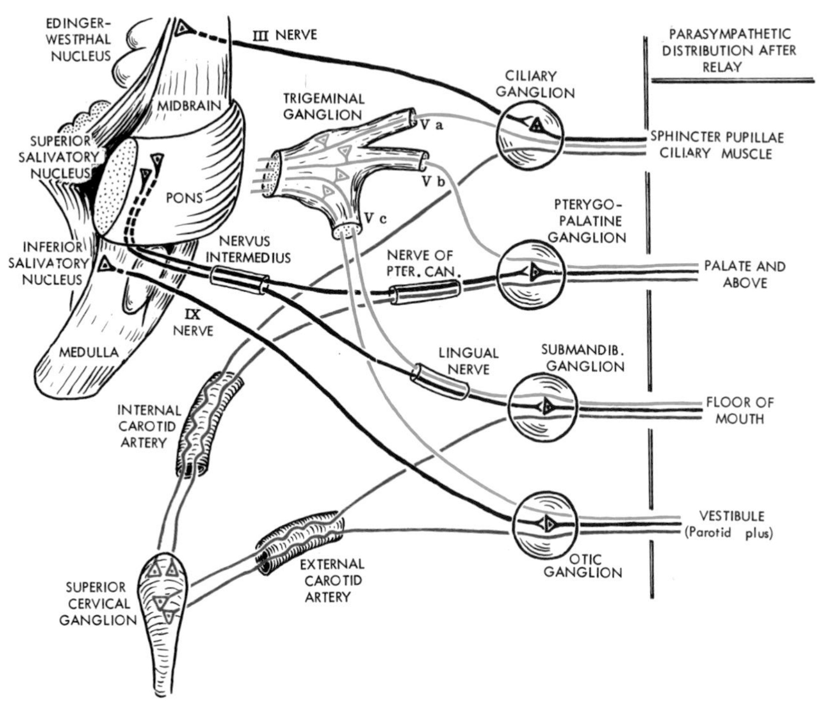

PLATE 29. THE CONNEXIONS OF THE CRANIAL AUTONOMIC GANGLIA. Do not be dismayed by these complex connexions. The otic ganglion lies high up near the skull base, but *morphologically* and by its connexions it is the most caudal of the four. (1) Only the **parasympathetic roots** (black) *relay* in the ganglia. They come from three nuclei, one each in midbrain, pons and medulla. The pontine nucleus (superior salivatory nucleus) relays in the middle two ganglia. (2) The **sensory roots** (blue) come from the trigeminal ganglion; the third division (V c) sends branches through the last two ganglia. (3) The **sympathetic roots** (red) come from cell bodies in the superior cervical ganglion. They hitch-hike along the internal and external carotid arteries, two on each artery, to reach the four ganglia.

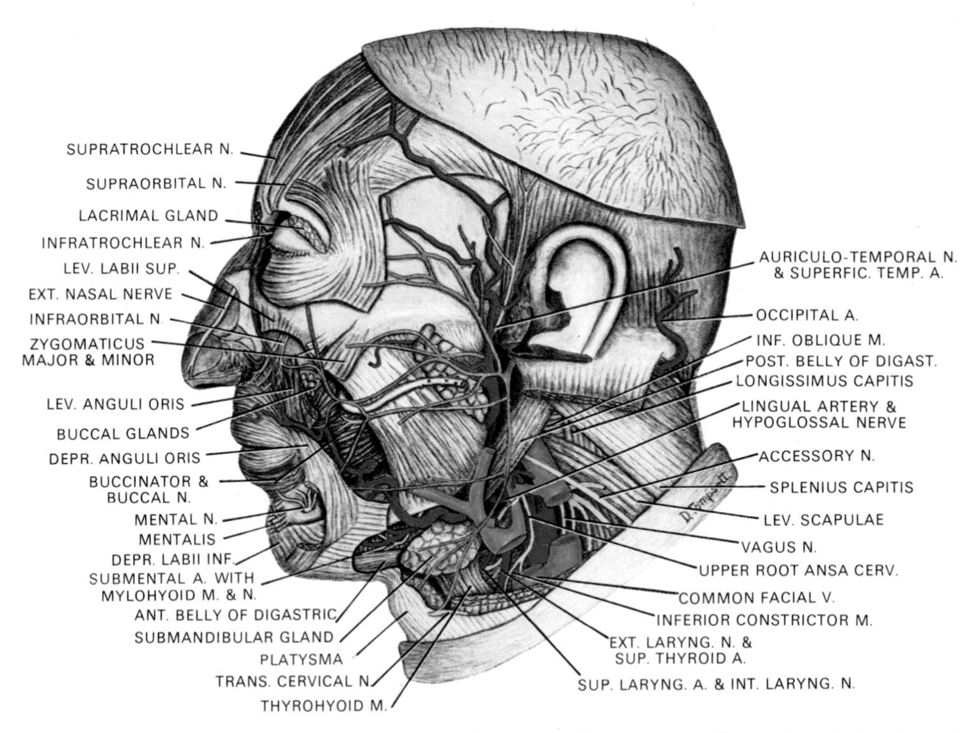

PLATE 30. A DISSECTION OF THE LEFT SIDE OF THE FACE. *Facial nerve is coloured orange*. Illustration of Specimen S.152 A, in R.C.S. Museum.

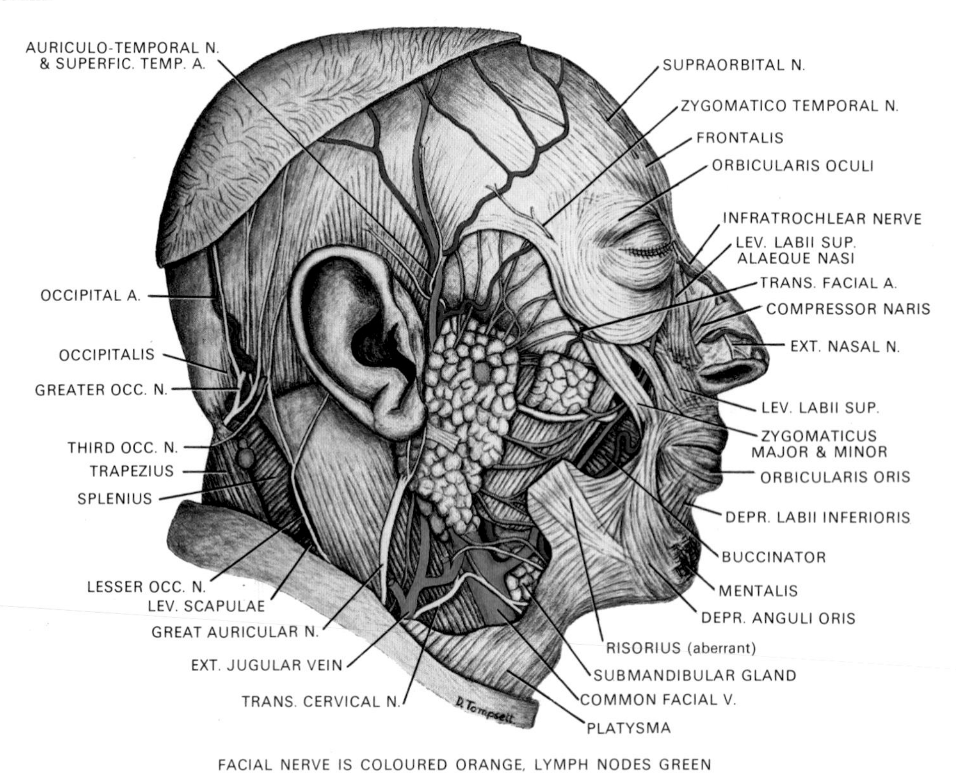

PLATE 31. THE SUPERFICIAL STRUCTURES OF THE RIGHT SIDE OF THE FACE AND NECK. Illustration of Specimen S.152 A, in R.C.S. Museum.

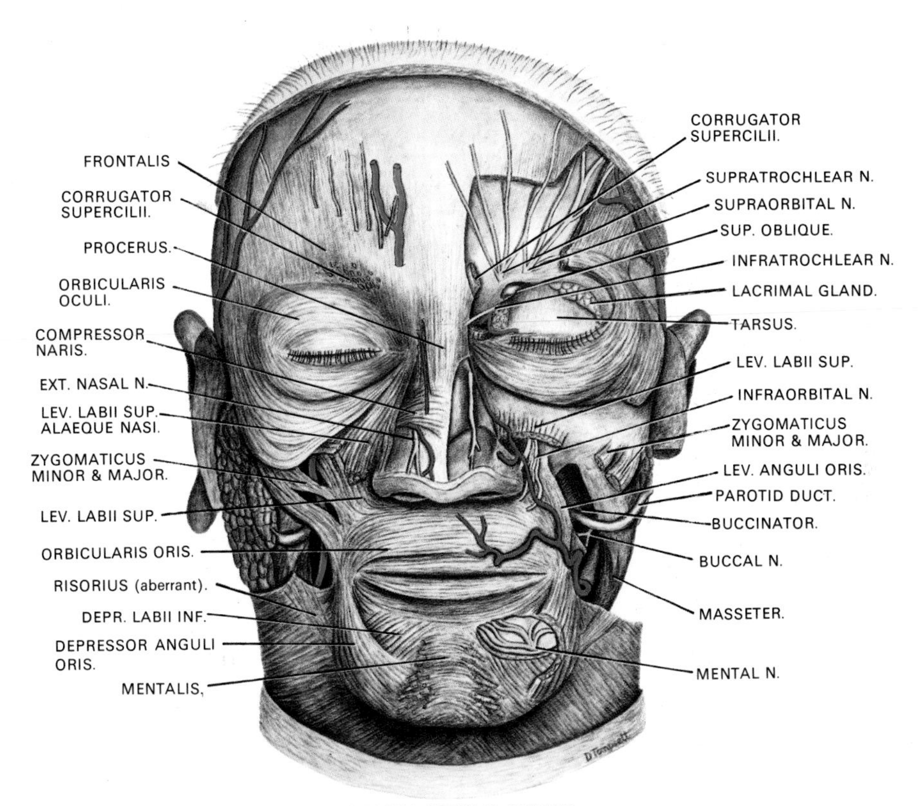

PLATE 32. A DISSECTION OF THE MUSCLES OF THE FACE. Illustration of Specimen S.152 A, in R.C.S. Museum.

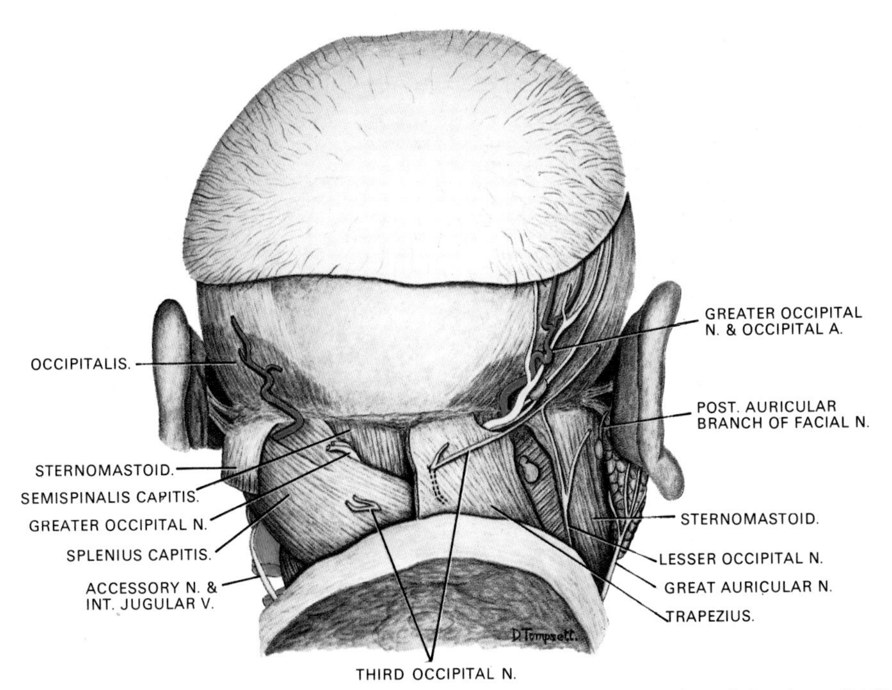

PLATE 33. THE SUPERFICIAL STRUCTURES OF THE BACK OF THE NECK AND SCALP. Illustration of Specimen S.152 A, in R.C.S. Museum.

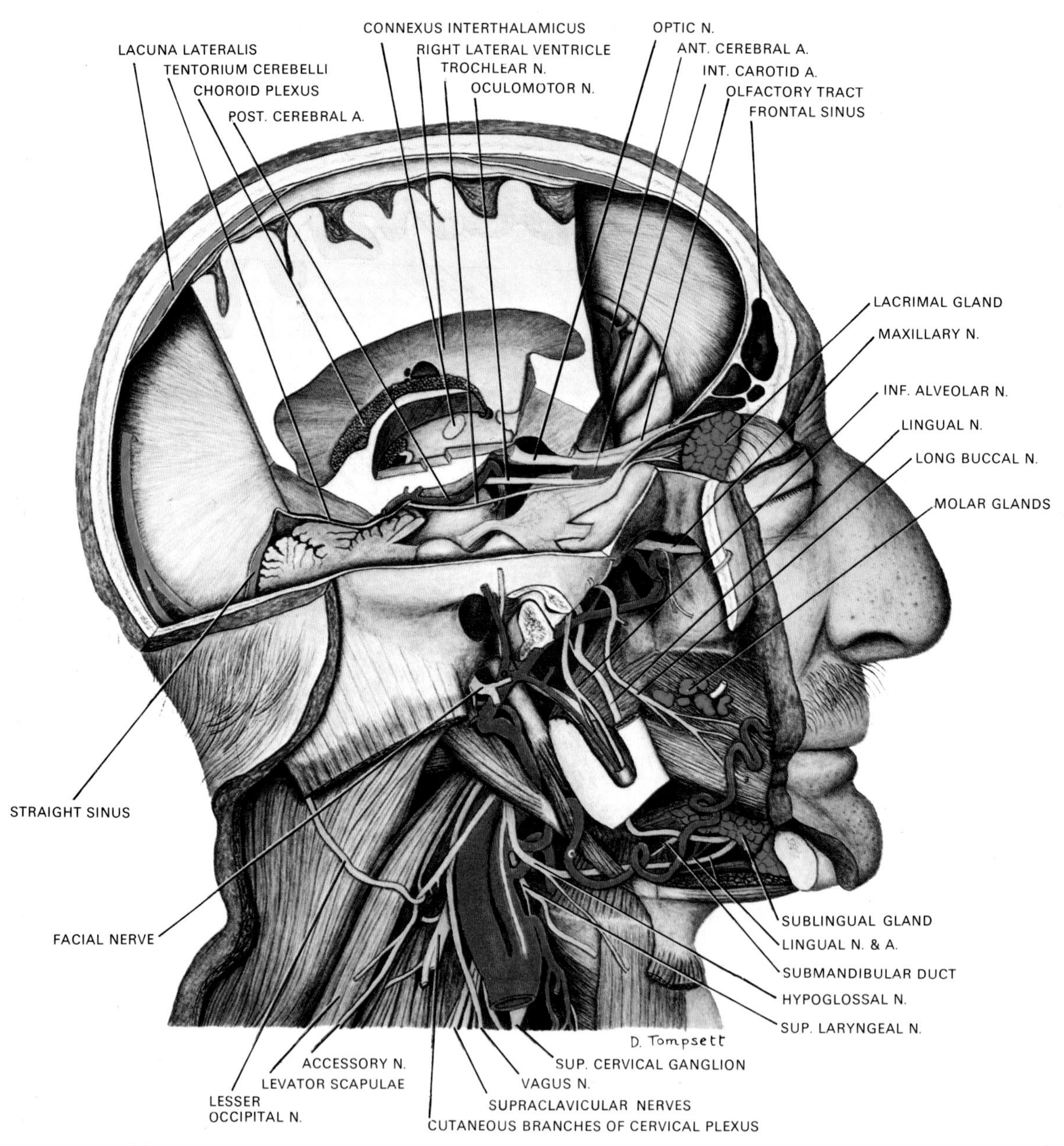

PLATE 34. A DEEP DISSECTION OF THE FACE, NECK AND BRAIN. Illustration of Specimen S.352 in R.C.S. Museum.

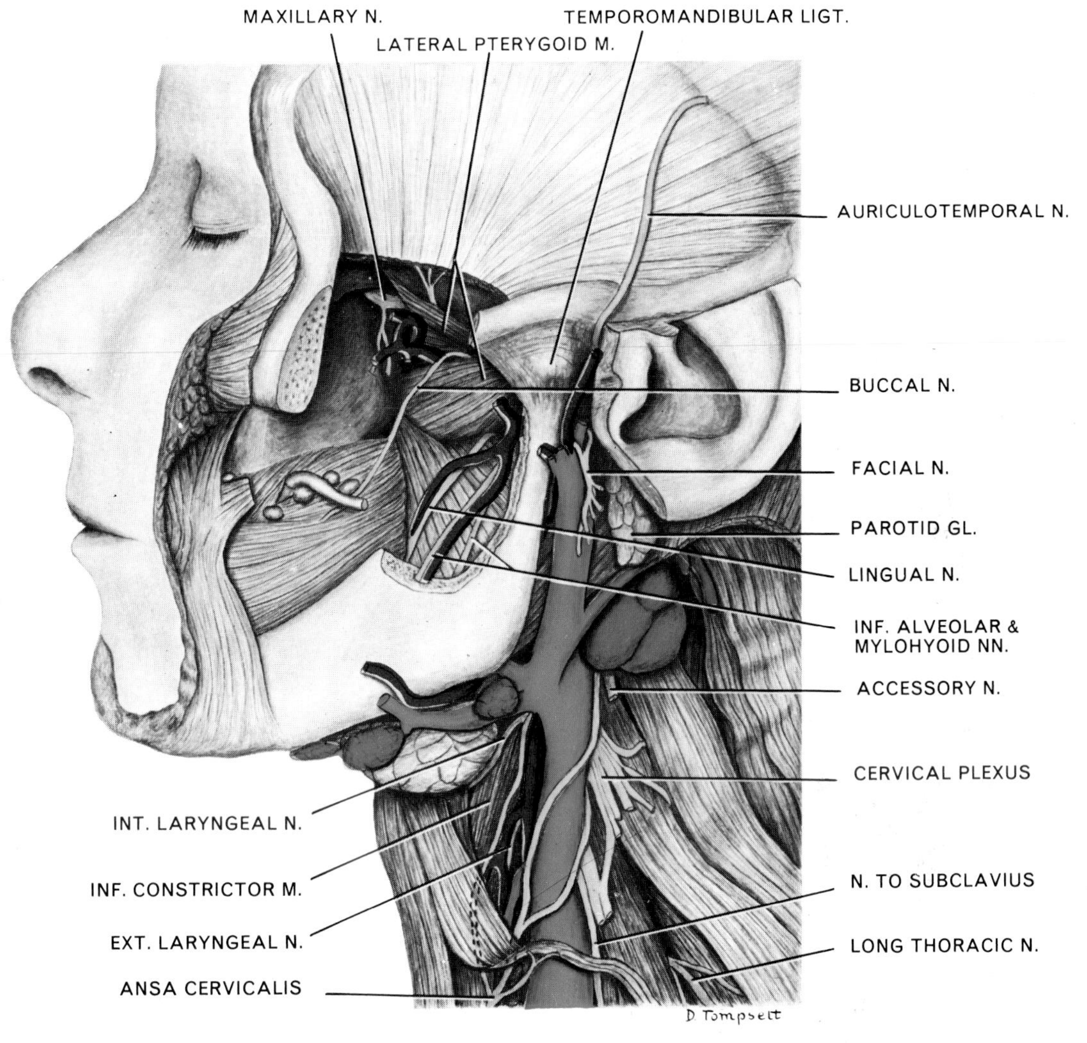

PLATE 35. THE LEFT INFRATEMPORAL FOSSA AND THE SIDE OF THE NECK. Illustration of Specimen S.350 in R.C.S. Museum.

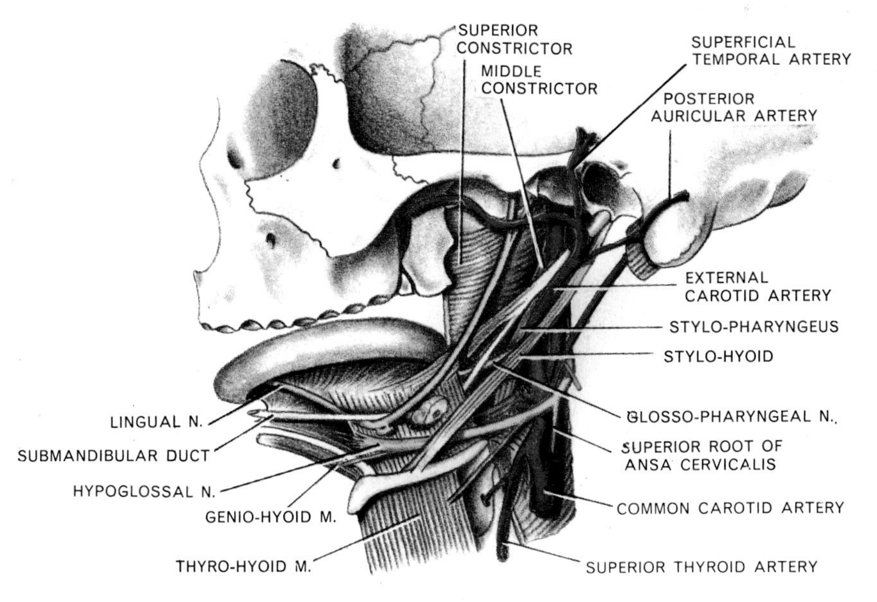

PLATE 36. THE STYLOID APPARATUS AND THE CAROTID ARTERIES. The three styloid muscles diverge widely, to tongue, hyoid bone and thyroid cartilage. The external carotid artery passes deep to stylo-hyoid but superficial to the styloid process. The internal carotid artery lies on the pharynx deep to the styloid process.

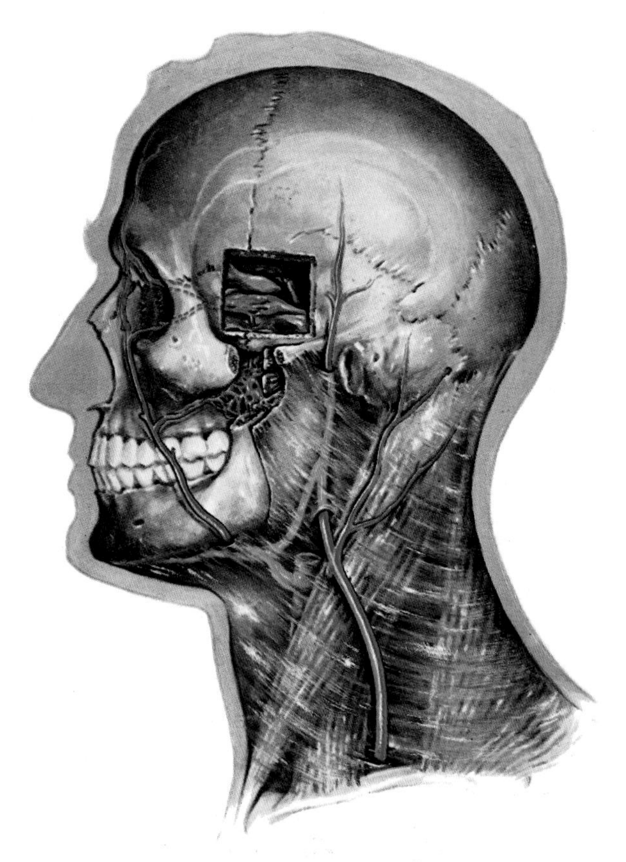

PLATE 37. VEINS OF THE FACE. **On the Face.** THE (ANTERIOR) FACIAL VEIN COMMUNICATES WITH THE CAVERNOUS SINUS AND WITH THE PTERYGOID PLEXUS. THE PTERYGOID PLEXUS ITSELF RECEIVES A VEIN FROM THE CAVERNOUS SINUS. **In the Neck.** THE RELATIONSHIPS OF THE FACIAL AND EXTERNAL JUGULAR VEINS TO THE DEEP FASCIA ARE SHOWN.

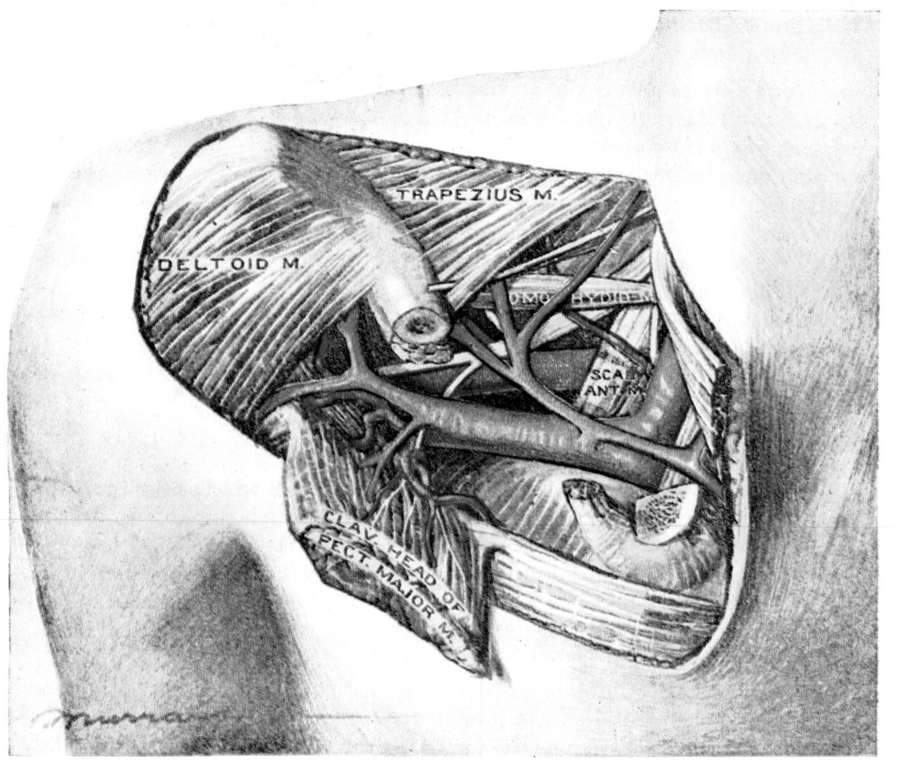

PLATE 38. THE BASE OF THE RIGHT POSTERIOR TRIANGLE. The medial two-thirds of the clavicle and the subclavius muscle are divided, and the clavicular head of pectoralis major is turned aside. Note its individual nerve supply (lateral pectoral) from the lateral cord of the brachial plexus. The clavicular head of sterno-mastoid is turned medially to expose the scalenus anterior, on which lie the phrenic nerve and the transverse cervical artery. The suprascapular artery here comes off the third part of the subclavian, a common variant.

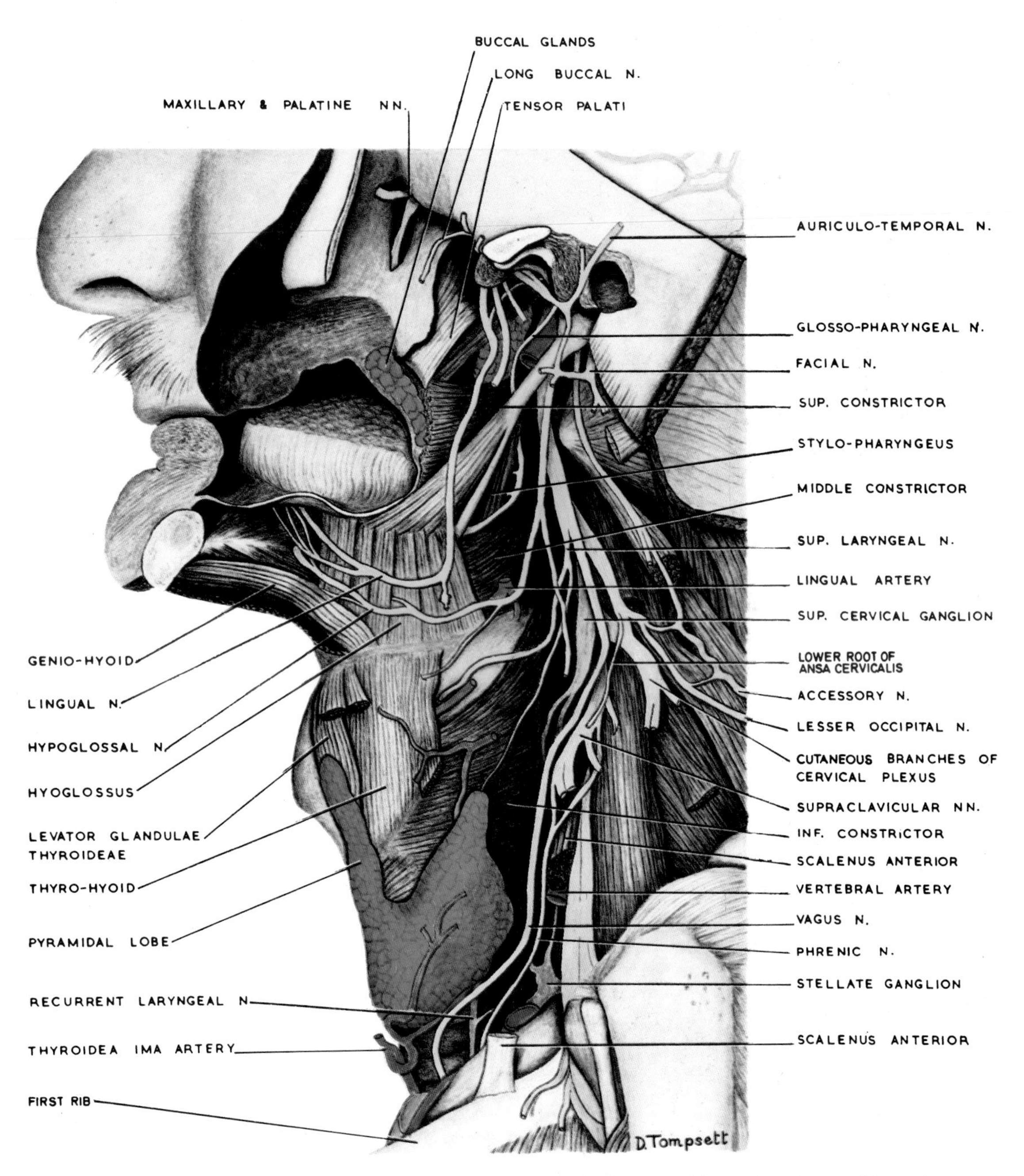

PLATE 39. THE INFRATEMPORAL FOSSA AND THE SIDE OF THE PHARYNX. Both pterygoid muscles, the carotid arteries and the internal jugular vein have been removed. Illustration of Specimen S.352 in R.C.S. Museum.

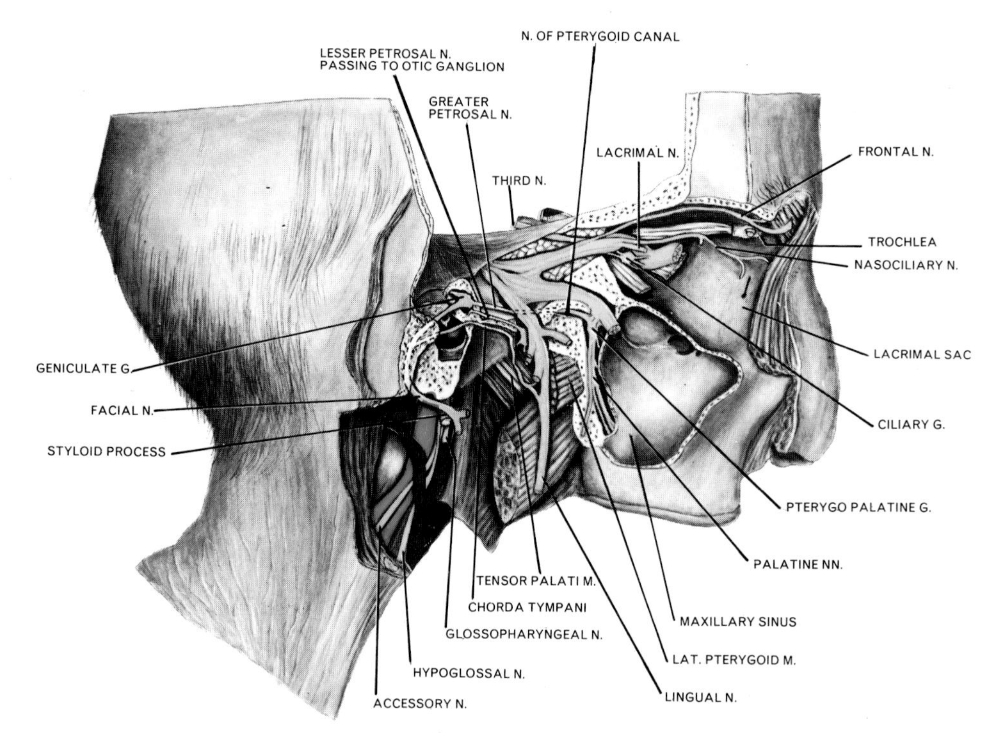

PLATE 40. THE TRIGEMINAL GANGLION, THE PETROSAL NERVES AND THE PTERYGO-PALATINE AND OTIC GANGLIA OF THE RIGHT SIDE. Lateral view. Illustration of Specimen S.165 C, in R.C.S. Museum. Compare PLATE 41.

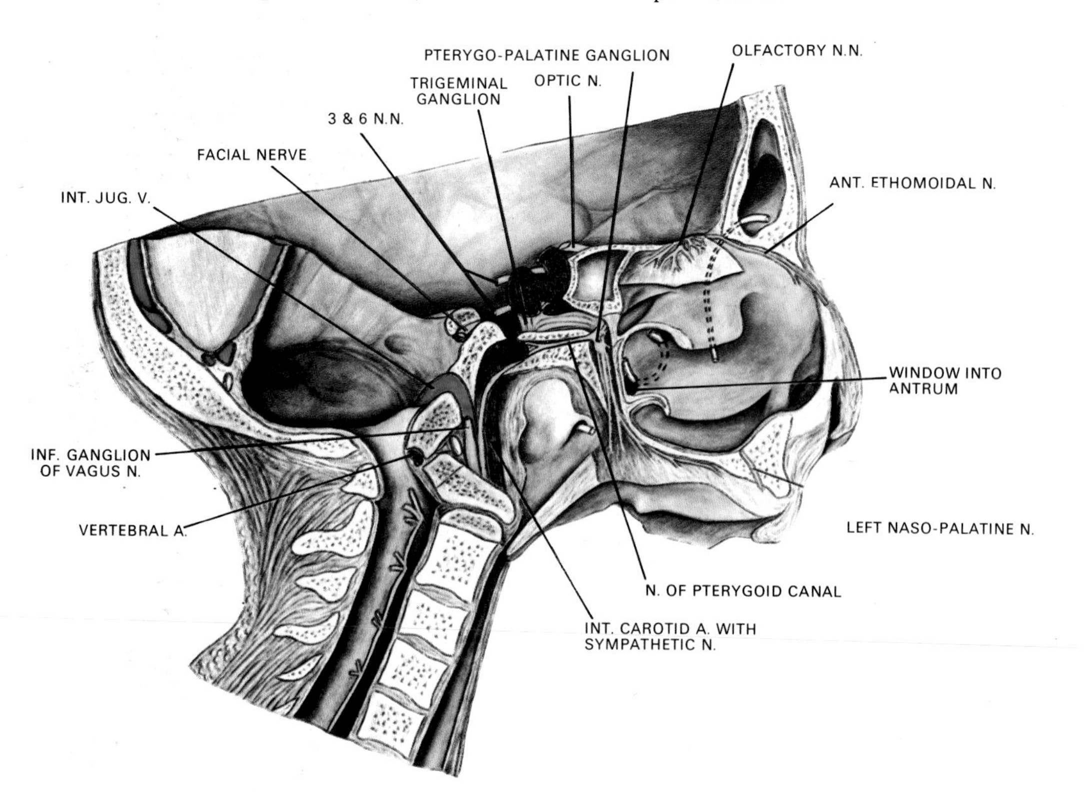

PLATE 41. THE LEFT PTERYGO-PALATINE (SPHENO-PALATINE) GANGLION, VIEWED FROM ITS MEDIAL ASPECT. Illustration of Specimen S.166 1, in R.C.S. Museum.

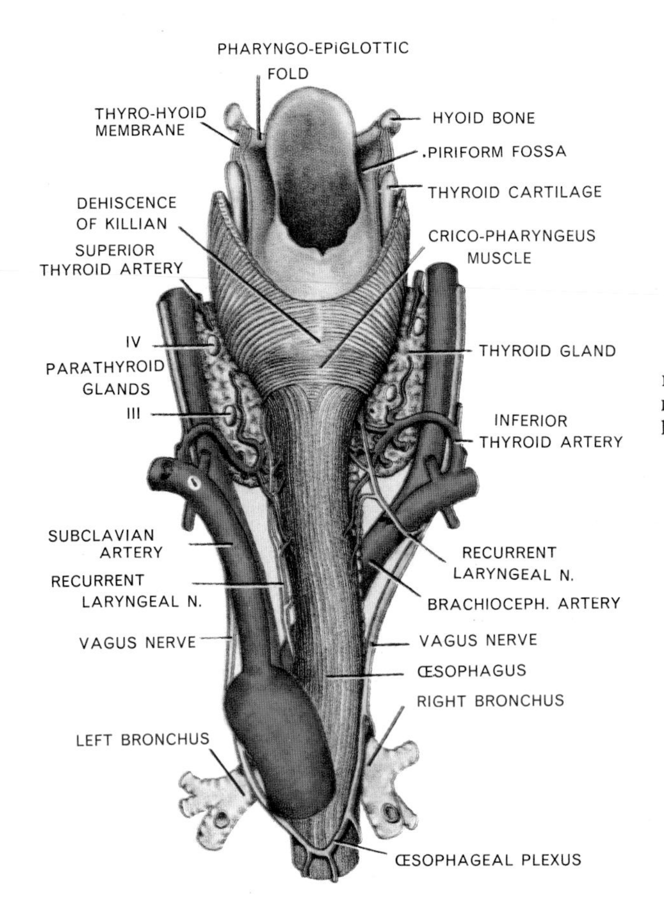

PLATE 42. THE LARYNGO-PHARYNX AND UPPER ŒSOPHAGUS FROM BEHIND. The carotid arteries are displaced somewhat laterally to expose the thyroid gland.

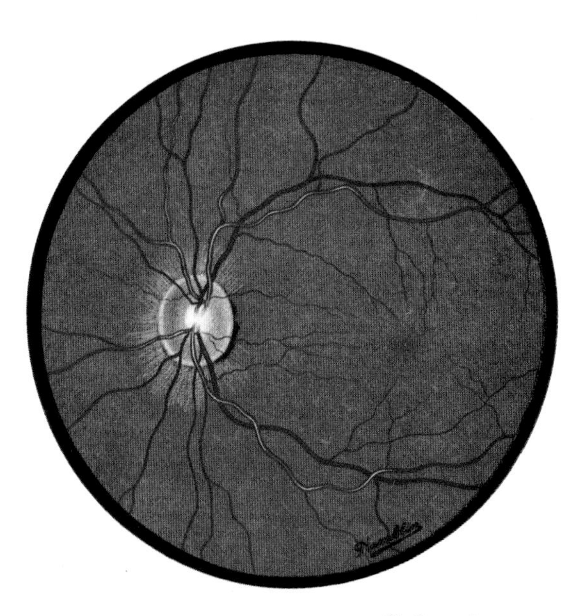

PLATE 43. THE FUNDUS OCULI. (Left eye.)

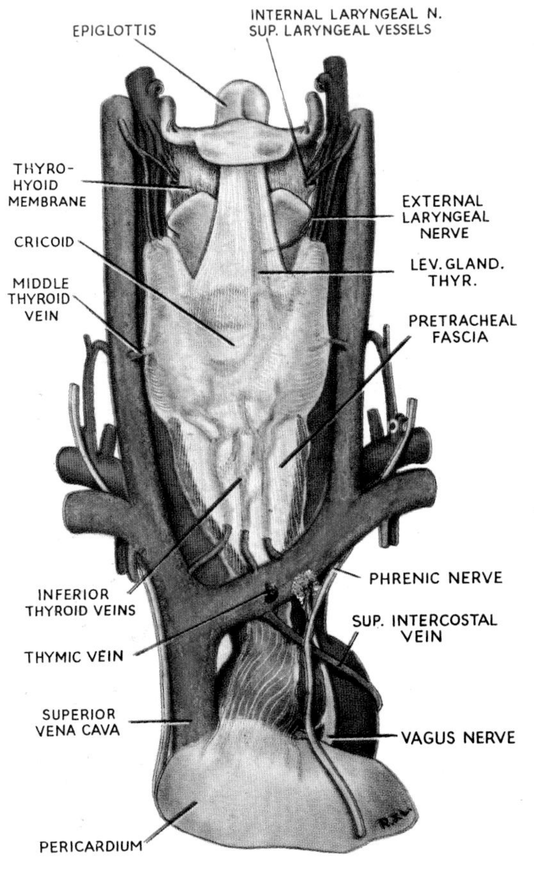

PLATE 44. THE PRETRACHEAL FASCIA, EXPOSED BY REMOVAL OF THE STERNUM AND THE INFRAHYOID STRAP MUSCLES.

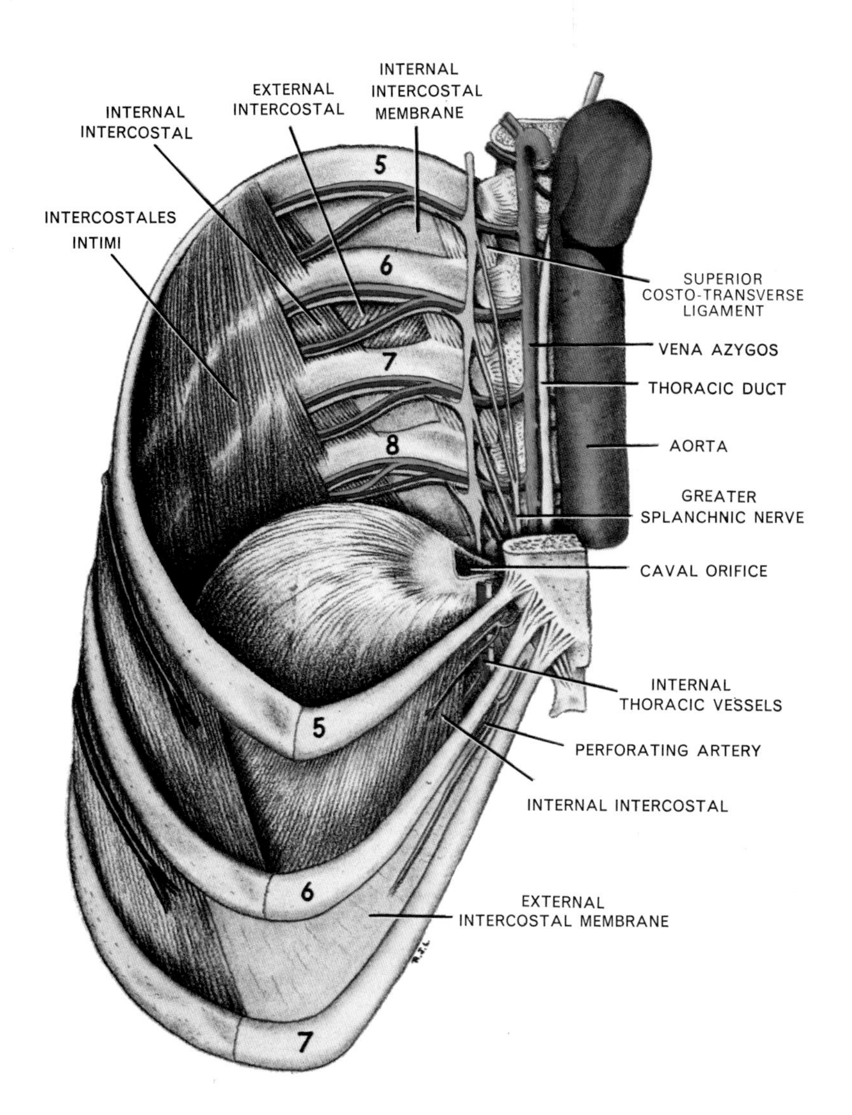

PLATE 45. THE THORACIC WALL. The paravertebral gutter is exposed by removal of the subcostal muscles. The internal intercostal membrane has been removed from the 6th space. The sympathetic trunk (unlabelled) lies near the rib heads, anterior to the intercostal vessels.

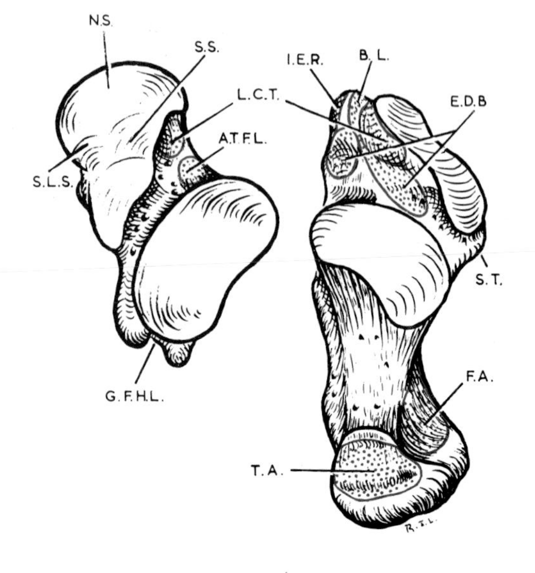

PLATE 46. THE UPPER SURFACE OF THE LEFT CALCANEUS, WITH THE TALUS TURNED LATERALLY TO EXPOSE ITS UNDER SURFACE. **Calcaneus: I.E.R.** Inferior extensor retinaculum. **B.L.** Bifurcate ligament. **E.D.B.** Extensor digitorum brevis. **S.T.** Sustenaculum tali. **F.A.** Flexor accessories. **T.A.** Tendo Achillis. **Talus: L.C.T.** Ligamentum cervicis tali. **A.T.F.L.** Anterior talo-fibular ligament. **G.F.H.L.** Groove for flexor hallucis longus. **S.L.S.** Spring-ligament surface. **N.S.** Navicular surface. **S.S.** Sustentacular surface.

Section 1.

TISSUES AND STRUCTURES

Some general remarks will, it is hoped, assist the student to a better understanding of the various tissues and structures that make up the body. It is very fruitful to study microscopic anatomy, to see how beautifully structure is adapted to function in all the tissues of the body, and in organs too. By 'microscopic anatomy' is meant the nature of *tissues* rather than cells. The minute structure of the cell is, by and large, beyond the scope of this book.

Identification of Histological Sections. This is called for in many examinations, and in clinical laboratory work too. Moreover it is satisfying, and helpful to the understanding of structure, to recognize a tissue by a single confident glance. While it may occasionally be necessary to study individual cells by a high power view (e.g., the epithelium in Figs. 1.18 and 1.19), in general it is advisable to use the *lowest* possible magnification consistent with seeing the tissue. If you cannot identify a histological section you will generally need a *lower* magnification. A higher magnification of a smaller field will seldom help. Much can be learned from a strong hand lens, and from a naked eye survey of the section. Do not forget that microscopy gives only a two dimensional view of three dimensional reality.

The Skin. Do not confuse the epidermis with the skin. The essential skin is the **dermis** (Fig. 1.3**D**) or corium, a strong and tough fibrous tissue rich in blood vessels, lymphatics and nerves. When dried it makes greenhide, when tanned it makes leather. It cannot function without its surface layer of cells, the **epidermis**. If the epidermis is lost the moist dermis loses lymph or blood and becomes invaded by bacteria; an ulcer is formed. The epidermis consists of several layers of cells, which become flatter towards the surface, i.e., squamous stratified epithelium. The surface cells change into a dead horny layer that is rubbed off in scales (dandruff). Figure 1.2 shows the changes in the cells as they are extruded to the surface. Squamous stratified epithelium covers surfaces that are subject to wear and tear.

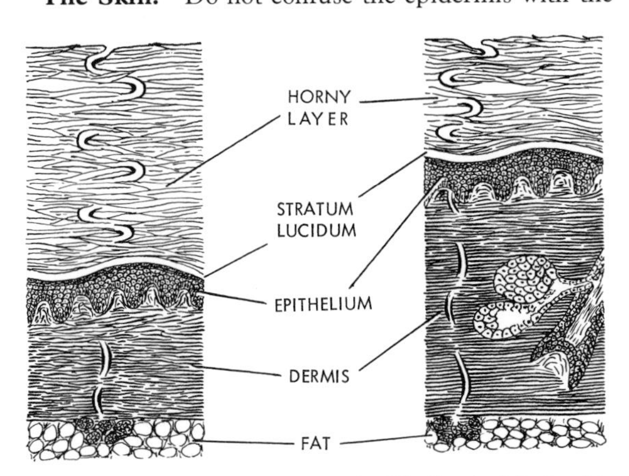

FIG. 1.1. SKINS OF EQUAL THICKNESS (× 30). On the left the skin of the sole; the thickness is due mainly to the dead stratum corneum (horny layer). On the right the skin of the back; the thickness is due mainly to the living dermis. The epithelium is of comparable thickness in each.

The horny layer of the skin is softened by the greasy secretion of sebaceous glands and, in most climates, it is moistened somewhat by the watery secretion of sweat glands. Fat solvents or emulsifiers remove the grease and leave the horny layer stiff and harsh. Undue contact with water macerates the keratin, which by imbibition becomes soft, thick and white ('washerwoman's fingers').

The thickness of the whole skin depends on two factors, namely the thickness of the horny layer and the thickness of the dermis itself (Fig. 1.1). On palms and soles the horny layer is responsible for the great thickness, the dermis being here rather thin. These areas are subject to great wear, hence the thickness of the protective horny layer and the absence of hairs. The thickness of the dermis usually differs between flexor and extensor surfaces. It tends to be thicker on extensor surfaces. The thick skin of the back makes sole leather, the thin belly skin is more suitable for upper leather, gloves, etc. The character of flexor and extensor skin differs in other respects than mere thickness. Flexor skin of the limbs tends to be less hairy than extensor skin and usually it is far more sensitive, having a richer nerve supply.

The colour of the skin in blondes depends on the amount of blood in the capillaries of the dermis, and on how much the keratin obscures this. Where keratin is thin the skin is red (e.g., the red margin of the lips); where keratin is thick the skin is white (e.g., palms). *Pigmentation* is produced by mobile cells containing

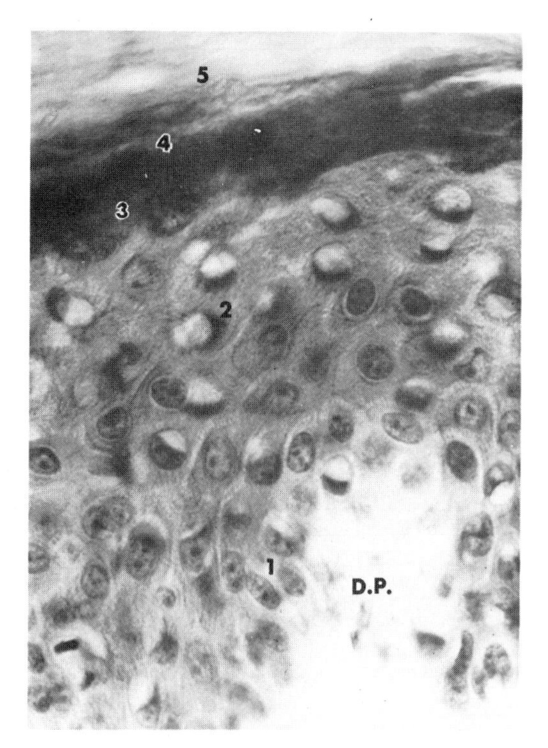

FIG. 1.2. THE LAYERS OF THE EPIDERMIS. (High power photomicrograph of **E** in Fig. 1.3.) The deepest layer (**1**) lies on the skin itself, here projected into a dermal papilla (**D.P.**). This layer is of columnar cells that undergo mitosis; it is the germinal layer—*stratum germinativum*. **2.** The daughter cells so produced lie in many layers (strata), the cells gradually flattening towards the surface—the *stratum mucosum, prickle-cell layer,* or *Malpighian layer.* **3.** Granules of a prekeratin called eleidin accumulate, and these cells form a granular layer—*stratum granulosum.* **4.** On the surface of this is a thin clear layer—*stratum lucidum.* **5.** Finally a layer of keratin, *stratum corneum,* of varying thickness forms the horny layer. This is rubbed off in scales.

pigment; these are the melanocytes. They lie between the cells of the epidermis, usually in the deepest layers. Some occur also in the dermis. Some of these cells are colourless, but they show a positive 'dopa' reaction (i.e., they convert **D**ihydr**O**xy**P**henyl**A**lanine into melanin).

The skin is bound down to underlying structures to a variable extent. On the dorsum of the hand or foot it can be pinched up and moved readily. On the palm and sole this is impossible, for here the dermis is bound firmly to the underlying aponeurosis, a necessary functional requirement to improve the grip of the hand and foot.

The *creases* in the skin are flexure lines over joints. The skin always folds in the same place. Along these flexure lines the skin is thinner and is bound more firmly to the underlying structures (usually deep fascia). The site of the flexure lines does not always correspond exactly to the topography of the underlying structures. For example, the anterior flexure line for the hip lies below the inguinal ligament; the posterior flexure line is not influenced by the oblique lower border of gluteus maximus and lies horizontally, to make the fold of the buttock.

Lines of cleavage in the skin have been known to exist since Langer described them in 1861. Their existence is due to the fact that the collagen fibres of the human dermis lie mostly in parallel bundles (this is not so in many animals). A conical object plunged through human skin splits the dermis and leaves a linear wound when it is withdrawn; the parallel collagen fibres have been forced apart without being ruptured. Surgical incisions made along Langer's lines heal with a minimum of scar tissue, incisions across the lines heal with a heaped up or broad scar. The directions of Langer's lines all over the body are of obvious importance to every surgeon, yet few surgeons and fewer anatomists know of them. Where crease lines exist (i.e., near joints) the cleavage lines usually coincide. Elsewhere in the body the cleavage lines tend to be longitudinal in the limbs and circumferential in the neck and trunk (Fig. 1.4).

Yellow elastic fibres occur in the dermis and impart to it an elasticity that gradually diminishes with advancing years, as the elastic fibres progressively atrophy.

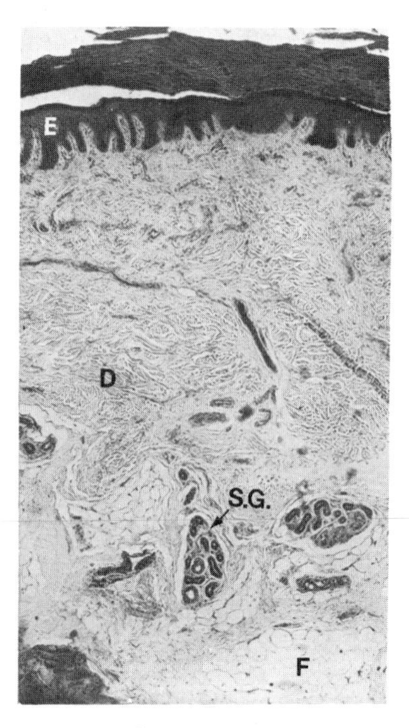

FIG. 1.3. HAIRLESS SKIN (human). The parallel collagen fibres (Langer's lines) are here cut transversely. **F.** Subcutaneous fat. **S.G.** Sweat glands. **D.** Dermis (i.e., *skin*). **E.** Epidermis. (Photomicrograph, low power.)

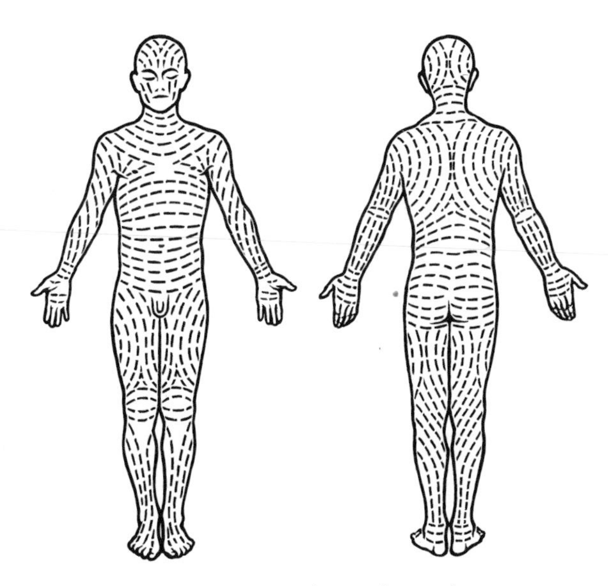

FIG. 1.4. THE CLEAVAGE LINES (LANGER'S LINES) OF THE SKIN.

Senile facial wrinkles lie at right angles to the line of pull of the underlying muscles (horizontal wrinkles on the brow, 'crow's foot' wrinkles at the lateral canthus, vertical wrinkles on both lips). Incisions along these wrinkles heal with minimal scarring.

Glands of the Skin. In man *sweat glands* are found in hairy and hairless skin alike. They are long tubular glands that lie coiled against the under surface of the dermis, in contact with the subcutaneous fat. The ducts pass through the dermis in a gentle spiral, but in the epidermis they dilate and become more closely coiled; this feature is particularly seen in the stratum corneum of palms and soles. The whole thickness of the dermis may be lost, but if the underlying sweat glands are intact they can regenerate a new epithelium; if the sweat glands are lost, skin grafting is required. *Sebaceous glands* occur only in hairy skin, and they differ from sweat glands in position and appearance. Each gland is a cluster of large pale cells, like a bunch of grapes. The cells are pale because they secrete an oily material (sebum), and in man this is delivered by the duct directly into the side of the adjacent hair follicle. Some few sebaceous ducts, even in man, open directly on the surface. These may become choked ('blackheads') or infected (acne). Sebaceous glands lie within the thickness of the dermis (Fig. 1.1).

Hairs grow from epithelial follicles, which are downgrowths of the epidermis. The follicles lie obliquely, and penetrate to the depths of the dermis or even into the subcutaneous tissue. The extremity of the follicle is bulbous, and it is indented by a vascular papilla of fibrous tissue. The follicle has a rich sensory nerve supply. A smooth muscle, the *arrector pili*, is attached to the under surface of each follicle, and slopes beneath it up towards the surface of the dermis. The muscle is supplied by sympathetic fibres, and is stimulated to contract by cold. Its contraction makes the hair lie more nearly vertical. In mammals this thickens the fur coat they wear, but in man the only result is minute pimpling of the skin, to produce 'goose flesh'. In birds (which are warm-blooded) a similar muscle causes erection of the down feathers.

Nails are keratin structures which grow from a transverse infolding of the epidermis into the dermis. They are like a much hardened stratum corneum.

Embryology of the Skin. The appendages of the skin are the sweat glands, the nails, the hair follicles and the sebaceous glands. All these are formed as downgrowths from the surface epithelium. The epidermis and the downgrowths are all of ectodermal origin. The fibrous tissue (collagen) and the yellow elastic fibres of the dermis are derived from mesoderm.

Subcutaneous Tissue. The skin is connected to the underlying bones or deep fascia by a layer of areolar tissue that varies widely in character in different species. In some animals it is loose and tenuous with a minimum of fat, so that it is a simple matter to skin the animal. In others, including man, fat is plentiful and fibrous bands in the fat tether the skin to the deep fascia. Such an animal is more difficult to skin; it has a blanket of fat beneath the skin, called the panniculus adiposus. The panniculus adiposus is well developed in man, and in it nerves, blood vessels and lymphatics pass to the skin.

The term superficial fascia is so ingrained in nomenclature that there is no hope of discarding it. Yet the tissue bears no possible resemblance to the other so-called 'deep' fasciæ, and the names panniculus adiposus, subcutaneous tissue or subcutaneous fat are greatly to be preferred.

In the panniculus adiposus are flat sheets of muscle called the *panniculus carnosus.* The degree of their development varies widely in different animals. In domestic quadrupeds such as sheep and horses the sheet is present over most of the body wall. It can be seen on the carcass in a butcher's shop, lying on the surface of the fat, generally incised in parallel slits to make an attractive pattern. It can be seen in action when a horse twitches the skin over its withers. The essential point about the panniculus carnosus is that one end of each muscle fibre is attached to the skin, the other end being usually attached to deep fascia or bone.

In man the sheet is well developed and highly differentiated to form the muscles of the scalp and face including the platysma, and remnants persist in such subcutaneous muscles as the palmaris brevis and as unstriped muscle in the corrugator cutis ani, in the dartos sheet of the scrotum and in the subareolar muscle of the nipple.

Deep Fascia. The limbs and body wall are wrapped in a membrane of fibrous tissue called the deep fascia. It varies widely in thickness. In the ilio-tibial tract of the fascia lata, for example, it is very well developed, while over the rectus sheath and external oblique aponeurosis of the abdominal wall it is so thin as to be scarcely demonstrable. In other parts, such as the face and the ischio-rectal fossa, it is entirely absent. A feature of the deep fascia of the body and limbs is that it never passes freely over bone but is always anchored firmly to the periosteum. A pin thrust into a muscle will pass through skin, panniculus adiposus and deep fascia, one thrust into a subcutaneous bone will pass through skin, panniculus adiposus and periosteum only (Fig. 1.5).

The deep fascia serves for attachment of the skin by way of fibrous strands in the subcutaneous tissue. In a few places it gives attachment to underlying muscles, but almost everywhere in the body the muscles are free to glide beneath it as they lengthen and shorten.

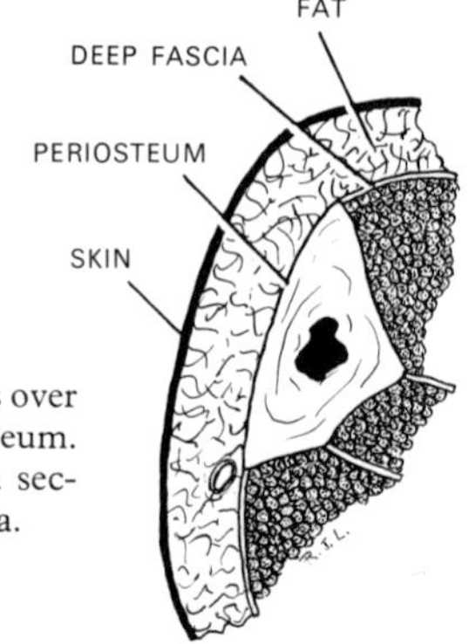

FIG 1.5. Deep fascia does not cross over bone; it blends with the periosteum. This principle is illustrated by a section through the shaft of the tibia.

Deep fascia is very sensitive. Its *nerve supply*, and that of subcutaneous periosteum where no deep fascia exists, is that of the overlying skin. The nerves to muscles do not supply the investing layer of deep fascia, but only the fibrous tissue of deep intermuscular spaces, and deep periosteum.

Fascia in General. As well as the investing layer of deep fascia on the surface of the body there are many other fascial layers in deeper parts. It is one of the greatest misfortunes of descriptive nomenclature that the term fascia has come to be applied to structures of widely differing character. On the one hand a named fascia may be a well developed membrane (e.g., fascia lata) while, on the other hand, it may be nothing more than a loose and indefinite collection of areolar tissue quite impossible to demonstrate as a membrane (e.g., fascia transversalis). In general it may be said that where fasciæ lie over non-expansile parts (e.g., muscles of the pelvic wall, prevertebral muscles) they are well developed membranes readily demonstrable, able to be sutured after incision; but where they lie over expansile parts (e.g., muscles of the pelvic floor, cheek, pharynx) they do not exist as demonstrable membranes, being indefinite and thin collections of loose areolar tissue, often too tenuous to retain a suture.

Descriptive anatomy has been further complicated by the naming as fascia of a third type of structure, the epimysium (*v.i.*). This areolar tissue clothing muscles is very important, because it is impervious to fluid collections and directs pus along the tissue spaces between individual muscles; but it does not clarify the understanding of anatomy that special names have been given to certain of these muscle envelopes. For example, the so-called 'bucco-pharyngeal fascia' has no existence as a separate structure; it is merely the epimysium on the surface of the buccinator and pharyngeal constrictor muscles.

Ligaments. Ligaments are made of two kinds of tissue. White fibrous tissue (collagen) comprises most of the ligaments of the body. It has the physical property of being non-elastic and unstretchable. Only if subjected to prolonged strain will white fibrous tissue elongate, and undue mobility is then possible in the joints (e.g., in flat foot, contortionists). White fibrous tissue ligaments are so arranged that they are never subjected to prolonged strain, with the curious exception of the sacro-iliac ligaments and the intervertebral discs, which are never free from the strain of the whole weight of the body except in recumbency.

The second type of ligament is composed of elastic tissue, which regains its former length after stretching. It is yellow in colour, hence the name of the ligamenta flava between the laminæ of the vertebræ. The capsular ligaments of the joints of the auditory ossicles are made of yellow elastic tissue.

Much has been written on the phylogenetic aspect of ligaments. Interest in the subject dates from the post-Darwinian enthusiasm for seeking 'vestiges' of a more primitive ancestor. Indeed most ligaments in the body have by one author or another been counted as 'degenerated' tendons of muscles possessed by a supposed ancestor (fish, reptile or even bird!) and if one is to believe all the speculations it is almost to suppose that the primitive ancestor of the mammals had no ligaments at all! The subject of ligaments is touched upon under the heading of joints (p. 15).

Raphés. A raphé is an interdigitation of the short tendinous ends of fibres of flat muscle sheets. It can be elongated passively by separation of its attached ends. There is, for example, no such structure as a pterygo-mandibular *ligament*; if there were, the mandible would be fixed, since ligaments do not stretch. The buccinator and superior constrictor interdigitate in the pterygo-mandibular *raphé*, the length of which varies with the position of the mandible. The mylo-

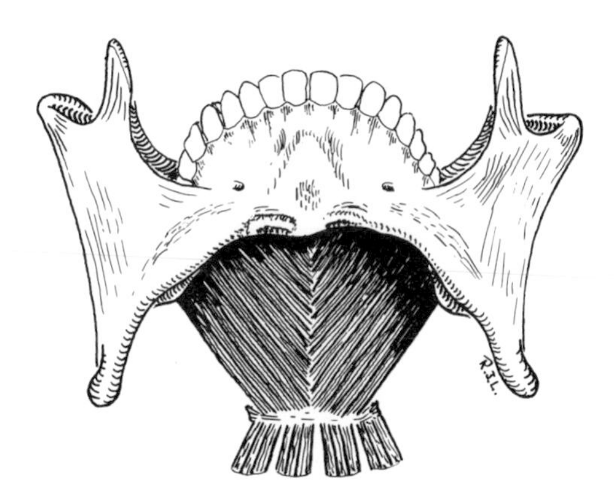

FIG. 1.6. The right and left halves of the mylo-hyoid muscle interdigitate in a midline raphé which extends from the mandible to the hyoid bone.

hyoid raphé (Fig. 1.6), pharyngeal raphé and anococcygeal raphé are further examples.

Cartilage. Most of the gristle in the body is composed of **hyaline cartilage**. The living cells of this tissue lie in a ground substance whose main feature is its complete avascularity. The ground substance is not truly structureless; the presence of fibres can be demonstrated by special methods. The substance is capable of a certain amount of deformation without fracture, it is not rigid like bone; it has great resistance to wear. It clothes the articular surfaces of almost all the synovial joints, certainly of all the weight-bearing ones. It is incapable of repair when fractured.

Fibro-cartilage is badly named, being all white fibrous tissue and no cartilage. To the naked eye it is white and glistening, like hyaline cartilage, but microscopically it consists of densely packed collagen fibres lying in parallel lamellæ, between which are many oval cells with well developed nuclei; but there is generally no cartilage around them. Fibro-cartilage is found in discs in joints and on the articular surfaces of the clavicle and mandible. Both hyaline cartilage and fibro-cartilage tend to calcify and even ossify in later life.

Yellow elastic cartilage has a ground substance that contains elastic fibres. Its cells are large and plentiful. It can be easily distorted, and just as easily springs back to its original shape when at rest. It is functionally ideal for the skeletal framework of the human pinna, the auditory tube and the epiglottis. It never calcifies or ossifies.

Fibro-cartilage has an ordinary blood supply (rather sparse because its metabolic rate is low) but hyaline and elastic cartilage have no capillaries, their cells being nourished by diffusion of lymph.

Muscles

The ordinary skeletal musculature, the red meat of the butcher, consists of non-branching striated muscle fibres, bound together by a loose areolar tissue rich in nucleated connective tissue cells. This connective tissue is condensed like the skin of a sausage on the surface of all muscles, forming a membrane of varying thickness and density well known to every dissector; it is the material dissected away and discarded in the process of 'cleaning' a muscle for demonstration purposes. The membranous envelope, or *epimysium*, is impervious to the spread of fluid such as pus. It is seldom of such a nature as to warrant special description as a named fascia. Nevertheless, many such epimysia have been named, especially in the neck, to the complication of descriptive anatomy and to the perplexity of the student. It is no over-simplification to say that no epimysium should ever be named, and that fasciæ should be labelled only where they exist in their own right as membranes *per se*, distinct from muscle envelopes.

The Form of Muscles. The disposition of the individual fibres in a muscle can be in one of only two ways, namely, parallel or oblique to the line of pull of the whole muscle. In the former maximum range of mobility is assured, in the latter the range of mobility is less but increased force of pull of the muscle is correspondingly gained. A good example of a muscle with parallel fibres is provided by sartorius. In flexing the knee and hip and laterally rotating the hip the muscle is contracted to its shortest length; reversing the movements elongates the muscle by 30 per cent. Other examples of muscles with parallel fibres are rectus abdominis, the infrahyoid group, the extrinsic eye muscles, the anterior and posterior fibres of the deltoid, the flank muscles of the abdomen, and the intercostals.

Muscles whose fibres lie oblique to the line of pull of the whole muscle fall into four patterns:

1. **Unipennate Muscles.** The tendon forms along one margin of the muscle and all the fibres slope into one side of the tendon, giving a pattern like a feather split longitudinally. A good example is flexor pollicis longus.

2. **Bipennate Muscles.** The tendon forms centrally, usually as a fibrous septum which enlarges distally to form the tendon proper. Muscle fibres slope into the two sides of the central tendon, like an ordinary feather. An example is rectus femoris (in which muscle the fibres slope *upwards* towards the central septum).

3. **Multipennate Muscles.** These are of two varieties: (*a*) a series of bipennate masses lying side by side, as in the acromial fibres of the deltoid, the subscapularis, etc.; (*b*) a cylindrical muscle within which a central tendon forms. Into the central tendon the sloping fibres of the muscle converge from all sides. An example is the tibialis anterior.

Mechanics of Muscle Form. In the case of a muscle whose fibres run parallel with its line of pull a given shortening of muscle fibres results in equal shortening of the whole muscle. In the case of unipennate and multipennate muscles a given shortening of muscle fibres results in less shortening of the whole muscle. The loss of shortening is compensated by a corresponding gain in force of pull. Obliquity of pull of a contracting fibre involves a loss of mechanical efficiency. But the *number* of oblique fibres is much greater than the number of longitudinal fibres required to fill the volume of a long muscle belly. The greater number of oblique fibres, though each fibre loses some efficiency, results in an overall gain of power in the muscle as a whole. Such muscles are found where great power and less range of movement are needed.

The Surface Appearance of Muscles. Whether or not the clinician should have a minutely detailed knowledge of osteology is arguable. Certainly it is of more practical value to know the surface appearance of those muscles that are distinctive. There are many such muscles. Beneath the transparent epimysium there is a great difference, not only in the general shape and contour but also in the amounts of red flesh and white fibres that produce their surface appearance. Some muscles are wholly fleshy, some largely aponeurotic, while many have a quite characteristic mixture of the two. Such variations provide an illustration of the relation of form to function. If the surface of a muscle bears heavily on an adjacent structure it will be covered by a glistening aponeurosis; where there is no pressure there is usually flesh. Examples are manifold, and in this book the surface appearances of many muscles are described. Here one may use the rectus femoris as a good example. The anterior surface of this bipenniform muscle is fleshy where it lies beneath the fascia lata, being aponeurotic only at its upper end, where it plays against a fibro-fatty pad that separates it from sartorius. Its deep surface is exactly the reverse. At the upper end is flesh, but the remainder of the posterior surface is wholly aponeurotic, where the muscle plays heavily on a corresponding aponeurosis of the anterior surface of vastus intermedius. The advantage of knowing the surface characteristics of muscles should be obvious to both physician and surgeon. In the case of the physician for example, the diagnosis of 'rheumatic' pain or tenderness will often hinge on whether the site is over aponeurosis or flesh, for where there is an aponeurosis there is a bursa. Such bursæ are often very extensive and are usually open at one end, so that effused fluid never distends them, but a 'dry' inflammation comparable to 'dry' pleurisy will produce pain on movement and tenderness on pressure over these aponeuroses. The surgeon sees muscles far more often than bones, and instant recognition of a muscle by its surface appearance gives great confidence and accuracy at operation.

Origins and Insertions of Muscles. There is no reality in these terms, though the sanctity of long usage and failure to find satisfactory substitutes force their continued use. Which end of a muscle remains fixed and which end moves depend on circumstances, and vary with most muscles.

The insertion of a tendon when, as usually, it is near a joint, is almost always into the epiphysis. An exception is the tendon of adductor magnus, the insertion of which into the adductor tubercle is bisected by the epiphyseal line of the femur.

Bone Markings. Fleshy origins generally leave no mark on the bone, though often the area is flattened or depressed and thus visible on the dried bone (e.g., pectoralis major on the clavicle). Contrary to usual teaching, insertions of pure tendon, like the attachments of ligaments, almost always leave a *smooth* mark on the bone, though the area may be raised into a plateau or depressed into a fossa (spinati, tibialis anterior, ligamentum patellæ, cruciate ligaments on femur, psoas, obturator tendons on femur, etc.). Rough marks are made where there is an admixture of flesh and tendon, or where there is a lengthy insertion of aponeurosis (e.g., ulnar tuberosity, gluteal crest, linea aspera).

A characteristic of flat muscles that arise from flat bones and play over their surfaces is that the muscle origin does not extend to the edge of the flat bone. The origin of the muscle is set back from the edge of the bone in a curved line. Between the edge of the bone and the curved line is a bare area, over which the contracting muscle slides. This allows a greater range of movement of the contracting muscle fibres. The bare area is invariably occupied by a bursa, and such bursæ are always of large size. The bursa may communicate with the nearby joint (e.g., subscapularis, iliacus) in which case infection of one cavity necessarily involves the other. Some of these bursæ remain separate from the nearby joint (e.g., supraspinatus, usually infraspinatus, obturator internus). The temporalis muscle is an exception to this rule, for its fibres arise from the whole of the temporal fossa down to the infratemporal crest, and there is no bursa beneath it.

Phylogenetic Degeneration of Muscles. In the period immediately following the publication of Darwin's *Origin of Species* the new concept of evolution produced many naïve ideas. The enthusiastic search for 'vestiges' and the comparison of similar structures found in different species gave rise to a great deal of speculation. Much of that speculation persists into the present day. The speculation may coincide with the truth or it may not; but it should be remembered that it is no more than speculation. With that proviso in mind, it may be permitted to describe what is generally accepted to be the series of changes that take place in a muscle that is being lost from the species. The most common change is that the muscle belly shortens and the tendon correspondingly lengthens (e.g., plantaris). A further change is that the distal end of the tendon becomes attached to both bones of the joint across which it passes. The muscle thus comes to be inserted into a more proximal bone, and the doubly attached distal part of the tendon 'degenerates' into a ligament (e.g., adductor magnus and medial ligament of knee, long head of biceps and sacro-tuberous ligament). An interesting example is supplied by the coccygeus muscle and the sacro-spinous ligament, which are merely the pelvic muscular and gluteal ligamentous surfaces of the same structure.

Actions of Muscles. A muscle may act as a prime mover, and it is the movement so produced that is conventionally described as the 'action' of the muscle. But a single muscle rarely contracts alone, and its action is influenced accordingly by its companions in contraction.

The origin of a muscle acting on a joint is very often so proximal that the muscle crosses, in fact, two joints (e.g., biceps and triceps both cross the shoulder joint). In such cases, and there are many, the action of these muscles on the proximal joint is merely to steady it while acting as prime movers on the more distal joint. Movement of the proximal joint (produced by other prime movers) changes the length of the muscle and so affects its action on the distal joint. When a muscle is passively elongated its subsequent contraction is much stronger. For example, in the fully flexed knee gastrocnemius is short and acts only weakly on the ankle joint, but in propulsion the extending knee lengthens gastrocnemius which then plantar-flexes the ankle joint much more strongly. Similarly, the long flexors of the fingers are affected by the position of the wrist (p. 104).

To appreciate muscle actions it should be remembered that it is excessively rare for any muscle, or group of muscles with similar actions, not to have an opponent. It makes for greater simplification if muscles are therefore studied in opposing groups, and no group of muscles should ever be thought of without thinking of the opposing group in the same context.

Moreover, muscles acting as prime movers on a certain joint have a different action when a more distal segment of the limb is in motion. In such a case they act usually as synergists, to brace and steady proximal joints while distal joints are in action. For instance, the short scapular muscles are in almost constant contraction to stabilize the shoulder joint during movements of elbow, wrist and fingers. Acting as prime movers they rotate the humerus, medially or laterally as the case may be, but this is a much less frequent event in the everyday use of the upper limb.

Muscles act synergically in yet another way. This is to cancel out the unwanted secondary effects produced by contraction of the prime movers. For example, triceps contracts during supination of the flexed forearm; this is synergic, to cancel out the unwanted flexion of elbow and shoulder that would otherwise be produced by the biceps, while in no way opposing the supinating effect of biceps on the radius.

Assessment of Muscle Action. It is impossible to have objective proof of the primary action of most muscles; 'common sense' is the final arbiter. For whatever the means of studying muscle action, common sense is still needed to assess the results. Instruments cannot distinguish between contraction as prime mover and contraction as synergist. There are several available methods of study:

(1) 'Common sense' study of the mechanics involved. This is usually simple across hinge joints (e.g., brachialis flexes the elbow joint), but it is difficult when the muscle pulls obliquely across universal joints (e.g., iliopsoas across the hip joint). Common-sense reasoning may be fallacious because relevant mechanical factors have been overlooked.

(2) The muscle may be seen or felt in contraction in the living model. This method covers only a limited field, for many muscles are not palpable.

(3) The muscle or its tendon may be pulled on in the dead. This is often impossible (e.g., in flat sheets of muscle like the levator ani), and is usually unconvincing because the normal opponents and synergists are out of action.

(4) Stimulation of a muscle or its motor nerve. But here the movement produced may be abnormal because the usual opponents and synergists are not stimulated into action.

(5) The effects of paralysis. In many cases failure results from the appearance of 'trick movements' produced by other muscles not formerly used in the movement being studied.

(6) Electromyography. This modern method is more popular than its reliability warrants. From surface electrodes and even from intramuscular electrodes spread of current from nearby muscles cannot always be ruled out. Nor can the machine distinguish between primary and synergic contraction.

(7) Comparative anatomy. An assessment of similar muscles in other animals will often throw light on the action of the muscle in man.

(8) After excessive use contraction of the muscle may be painful for a few days. It is sometimes instructive to notice what movements are painful in these circumstances (e.g., tibialis anterior after excessive walking, latissimus dorsi after excessive coughing, abdominal wall after excessive tennis, etc.).

The most fruitful approach to the complicated study of the actions of muscles on joints is, after the mechanical anatomy of the muscles has been studied, to analyse the basic movements that take place in the joints themselves. These movements are determined by the shape of the articulating surfaces and the direction of the ligaments of the joints, and most of them are easy to observe in the living body. Having established the direction and range of the basic movement, he who knows the muscles of the body can often say with confidence which muscles can and which cannot produce the movement. And if the muscles known to produce the movement have other secondary actions, these must be neutralized by the synergic contraction of *muscles which do not oppose the primary movement.*

Action of Paradox. A multiplicity of common movements are aided by gravity. The opposing muscles of such movements are then in contraction, and 'pay out rope' against the force of gravity. Thus, in sitting from standing, the hip and knee are flexed by the controlled elongation of the contracted extensors of those joints; here the quadriceps is the *flexor* of the knee. Such reversal of muscular action when 'paying out rope' against gravity is known as the *action of paradox*. This must never be overlooked in clinical investigations of muscle function.

Blood Supply of Muscles. Muscles have a rich blood supply. The arteries and veins usually pierce the surface in company with the motor nerve. From the muscle belly vessels pass to supply the adjoining tendon. Lymphatics run back with the arteries to regional lymph nodes.

Nerve Supply of Muscles. The flat muscles of the body wall are perforated by cutaneous nerves on their way to the skin. Such nerves do not necessarily supply the muscles. In the case of the limbs, however, it is a fact that whenever a nerve pierces a muscle it supplies that muscle, and the motor branch leaves the nerve proximal to the muscle. As a matter of *morphological* fact, limb nerves do not pierce muscles at all, but pass actually in planes between distinct morphological masses that have fused together. Whenever a nerve pierces a muscle, suspect a morphological pitfall (e.g., coraco-brachialis, supinator, sterno-mastoid).

Note that all muscles in a limb are supplied by branches of the limb plexus, and that flexor muscles derive their nerve supply from anterior divisions and extensor muscles from posterior divisions of the nerves of the plexus. This matter is considered fully on p. 24.

Sensory Supply of Muscles. The nerve to a muscle in the body wall or in a limb contains some 40 per cent of sensory fibres. These innervate muscle spindles and mediate proprioceptive impulses. They are indispensable to properly co-ordinated muscle contraction.

The nerves supplying the ocular and facial muscles (III, IV, VI and VII cranial nerves) contain no sensory fibres. Proprioceptive impulses are conveyed from the muscles by local branches of the trigeminal nerve. The spinal part of the accessory nerve likewise contains no sensory fibres, and proprioceptive impulses are conveyed from the sterno-mastoid by C2 and 3 and from the trapezius by C3 and 4. The hypoglossal nerve is also purely motor. It is a curiosity of anatomy that until recently it was held that the tongue muscles have no proprioceptive innervation; but muscle spindles have now been demonstrated in the tongue. Their innervation has not been conclusively worked out, but it is probably the lingual nerve.

Lower Motor Neurone. A muscle fibre is made to contract in only one way, namely, by impulses descending from the cell body of the lower motor neurone through its axis cylinder. Flaccid paralysis of the muscle results from loss of the cell body (e.g., poliomyelitis) or by division or disease of the nerve fibre.

The lower motor neurone may be fired off by three main pathways, *viz.*:

1. Upper motor neurone (prime movement)
2. Extrapyramidal system (synergic contraction)
3. Sensory pathways (reflex muscle tone).

The *upper motor neurone*, with its cell body in the opposite motor cortex, initiates voluntary movement. Here the muscle in question contracts as a prime mover. Interruption of this pathway results in hemiplegia (or monoplegia), with its typical spastic paralysis and increased muscle jerks.

The *extrapyramidal system* consists of a complicated system of 'feed back' circuits including the basal ganglia and cerebellum. Impulses from the premotor area of the opposite cortex fire off the lower motor neurone, causing the muscle to contract as a synergic partner, modifying the action of the prime movers. Interruption of this pathway results in cerebellar ataxia, Parkinsonism (basal ganglia), etc.

Sensory pathways form an essential background to co-ordinated voluntary movements. They mediate proprioceptive information from the muscle, its tendon, and the capsule and ligaments of the joint being acted upon. These pathways are stimulated in the clinical investigation of muscle jerks. Their interruption results in locomotor ataxia (tabes), with loss of muscle jerks.

Microscopic Anatomy. Striated muscle consists of fibres that never branch. They are banded by striations that pass completely through the whole thickness of the fibre. The nuclei are small (of the order of 1/20th of the diameter of the fibre) and placed peripherally, on the surface of the fibre. They are dark from concentrated chromatin. Fibre size varies greatly within a single muscle. Hypertrophy of the muscle after physical culture consists of increase in diameter of all the fibres up to their maximum size; no new fibres are called into existence (Fig. 1.7).

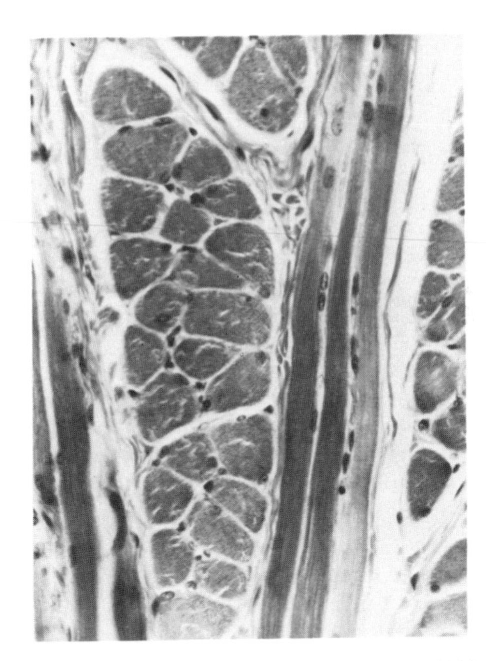

FIG 1.7. STRIATED MUSCLE. (Photomicrograph ×480.) The muscle nuclei are small and lie on the surface of their fibre. Other surface nuclei are those of fibroblasts. Only in the tongue (shown here) may longitudinal and transverse fibres be seen in the same field.

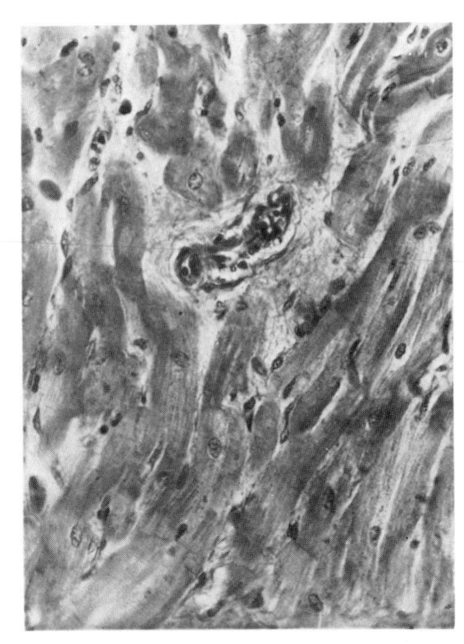

FIG 1.8. CARDIAC MUSCLE. (Photomicrograph ×480.) The whorls and spirals of heart muscle show on section a mixture of longitudinal, oblique and transverse fibres. The nuclei are larger than those of striated muscle and lie *central* in the fibre. Do not be deceived by small nuclei on the surface of fibres; these are the nuclei of fibroblasts.

Cardiac muscle, too, is striated. The fibres branch and rejoin, and appear to form a syncytium (no cell boundaries). The nuclei are large (about one quarter the diameter of the fibre) and placed centrally in the fibre. They are relatively pale, their chromatin being dispersed. Nuclei of fibroblasts between the muscle fibres show as dark chromatin-filled nuclei near the surface of the muscle fibres; they must not be mistaken for the peripheral nuclei of striated muscle (Fig. 1.8).

Smooth muscle consists of long spindle-shaped cells lying parallel. A rod-shaped nucleus lies central in each cell. The parallel cells join to make muscle fibres that are usually large enough to see with the naked eye. In tubes that undergo peristalsis these fibres are arranged in circular and longitudinal fashion (e.g., the alimentary canal, the ureter, the bicornuate uterus of multiparous animals). The circular fibres milk along a wave of peristalsis while the longitudinal fibres pull the whole wall proximally over the contents of the tube. In viscera that undergo a mass contraction, without peristalsis (bladder, human uterus) the fibres are arranged in whorls and spirals and do not lie in demonstrable layers.

Cardiac muscle is similarly arranged in whorls and spirals; each chamber of the heart empties by a mass contraction, not by peristalsis.

Nerve supply. Cardiac muscle and all smooth muscles are supplied by autonomic nerves. Unlike the lower motor neurone which goes direct to striated muscle, these autonomic fibres relay peripherally on their way to the muscle.

Physiology of Muscle. All striped muscle in the body (except the heart) is supplied by somatic nerves, and is often called voluntary muscle. The heart and all unstriped muscle in the body are supplied by autonomic nerves and the latter is often called involuntary muscle. The terms voluntary and involuntary are not strictly accurate in this context, since there is no voluntary control over much of the striated muscle (e.g., the upper third of the œsophagus). An essential difference lies in the *rapidity of contraction* in response to a stimulus. Striped muscle (including the heart) responds with an 'immediate' and rapid contraction, smooth muscle responds more slowly with a 'delayed' and more leisurely contraction. A further distinction is that striated muscle contracts, weight for weight, many times more forcibly than smooth muscle.

Furthermore, smooth muscle has much greater ability to elongate without paralysis (e.g., the bladder or stomach from emptiness to full distension). Striped muscles cannot elongate beyond one-third of their resting length without damage, and they are so attached to the skeleton that elongation does not exceed this amount. Note that the hamstrings are exceptional; with the knee extended their elongation limits hip flexion. (Is this why they tear more commonly than most other muscles?)

Tendons

Tendons consist of longitudinally arranged collagen fibres. They may be cylindrical in shape or flat. The aponeuroses of the abdominal muscles are wide sheets of tendon. Tendons are supplied with blood from two main sources, reinforced from a third source in the case of long tendons. Descending vessels from the muscle belly anastomose longitudinally with ascending vessels from the periosteum at the tendon insertion. In long tendons intermediate vessels from a neighbouring artery reinforce the longitudinal anastomosis.

Synovial Sheaths. Where tendons bear heavily on neighbouring structures, and especially where they pass around loops or pulleys of fibrous tissue or bone which change the direction of their pull, they are lubricated by being provided with a synovial sheath. The parietal layer of the sheath is firmly attached to the surrounding structures, the visceral layer firmly fixed to the tendon, and the visceral and parietal layers glide on each other, lubricated by a thin film of synovial fluid secreted by the synovial sheath. The visceral and parietal layers join each other. Usually they do not enclose the tendon cylindrically. It is as though the tendon were pushed into the double layers of the closed sheath from one side (Fig. 1.9). In this way blood vessels are able to enter the tendon to reinforce the longitudinal anastomosis. In other cases blood vessels perforate the sheath and raise up a synovial fold like a little mesentery, called a mesotendon or vinculum (e.g., flexor tendons of digits).

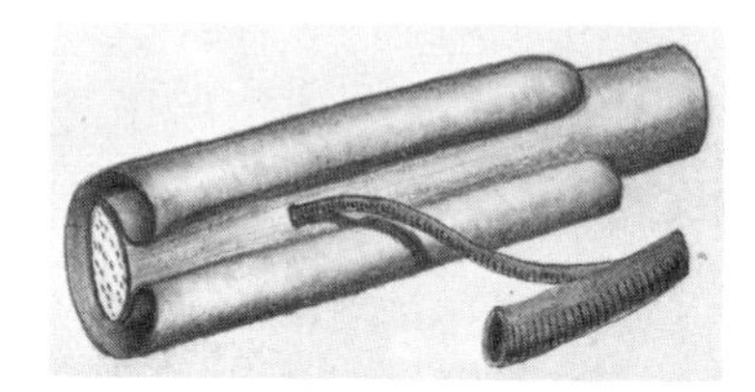

FIG. 1.9. The 'booster' supply of arterial blood to a long tendon passing through a gap in the synovial sheath.

Sesamoid fibrocartilages. These are common, and fairly constant, in tendons or in joint capsules where they bear upon bony prominences. Why some of them ossify is a mystery. The resultant **sesamoid bones** are very inconstant and variable.

Bone

Bone, for all its rigidity and high calcium content, is a *living tissue* and, unlike avascular hyaline cartilage, it is a bloody tissue. It has greater reparative power than any other tissue in the body except blood.

Bone contains one-third of organic matter. This consists of collagen fibres, which lie in a groundwork of calcific inorganic matter (calcium phosphate and carbonate, with a little fluoride and magnesium). To the stony-hard brittleness of the inorganic matter the fibres contribute a rugged elasticity, so that bone is immensely strong. It can withstand a crushing pressure of more than 2 tons to the square inch. If the organic matter be destroyed by burning, the inorganic residue ('animal charcoal'), though retaining the original shape of the bone, is very brittle.

Macroscopically bone exists in two forms, compact and cancellous. **Compact bone** is hard and dense, and resembles ivory, for which it is often substituted in the arts. True ivory is dentine. **Cancellous bone** consists of a spongework of trabeculæ, arranged not haphazardly but in a very real pattern best adapted to resist the local strains and stresses. If for any reason there is an alteration in the strain to which cancellous bone is subjected there is a re-arrangement of the trabeculæ. The moulding of bone results from the resorption of existing bone by phagocytic osteoclasts and the deposition of new bone by osteoblasts; but it is not known how these activities are controlled and co-ordinated.

The marrow cavity in long bones and the interstices of cancellous bone are filled with marrow, red or yellow as the case may be. This marrow apparently has nothing whatever to do with the bone itself, being merely stored there for convenience. At birth all the marrow of all the bones is red, active hæmopoiesis going on everywhere. As age advances the red marrow atrophies and is replaced by yellow, fatty marrow, with no power of hæmopoiesis. This change begins in the distal parts of the limbs and gradually progresses proximally. By young adult life there is little red marrow remaining in the limb bones, and that only in their cancellous ends; ribs, sternum, vertebræ and skull bones contain red marrow throughout life.

Microscopically all bone is made up of lamellæ. The lamellæ consist of collagen fibres lying in a calcified material; adjacent lamellæ are held together by interchange of fibres. Bone cells lie scattered between the lamellæ.

(a) **Cancellous Bone.** There is no *microscopic* difference between this bone and compact bone. There is only one tissue that is bone; 'cancellous bone' is a purely naked-eye description of the spongy architecture. Each spicule of cancellous bone consists of undulating lamellæ which are bound surface to surface with each other.

Between lamellæ bone cells occupy small holes (*lacunæ*) which communicate freely with each other and with the bone marrow by minute tunnels (*canaliculi*). The canaliculi are occupied by protoplasmic processes from the bone cells, which thus communicate with each other. It is usual to see two types of cells on opposite surfaces of the trabeculæ of cancellous bone (Fig. 1.10).

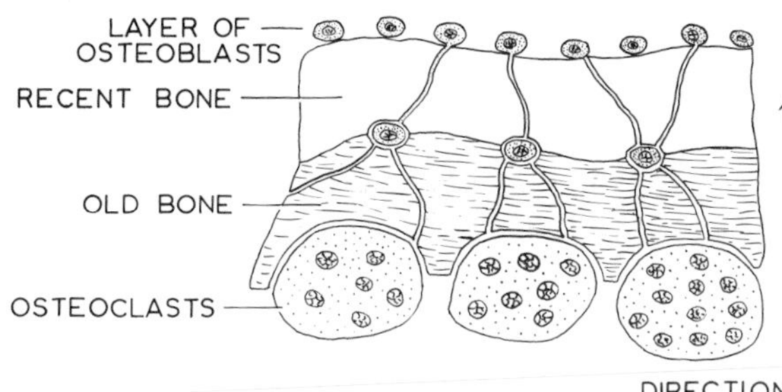

FIG. 1.10. A SECTION ACROSS A TRABECULA OF CANCELLOUS BONE. Moulding of the bone is brought about by osteoblasts and osteoclasts. Osteoblasts become imprisoned between the lamellæ, where they persist as bone cells, communicating with each other by fine protoplasmic processes which occupy the canaliculi.

One is the *osteoblast*. The osteoblasts form a single layer of cells which respond to functional needs by laying down new bone. On the opposite side of the trabeculæ the *osteoclasts*, large multinucleated phagocytes, excavate the bone and give its surface a scalloped appearance. Thus alteration in size and shape (moulding) of the bone and directional change in the trabeculæ are brought about.

(*b*) **Compact Bone.** Here the bone is dense and contains no marrow spaces. There are two varieties of compact bone. In one the lamellæ, bound surface to surface, are concentric with the periosteum. All subperiosteal bone is like this, including the thin layer of compact bone enclosing the cancellous structure of the ends of long bones.

Haversian Bone. In the shafts of long bones the lamellæ are arranged in concentric cylinders around small blood vessels. The cylinders lie parallel with the long axis of the bone. Bone corpuscles lie in lacunæ between the lamellæ; they communicate with each other and with the central canal along canaliculi that contain their long branching processes. The cylinders are called Haversian systems. The spaces between adjacent Haversian systems are occupied by lamellæ that for the most part curve concentrically with the periosteum on the surface of the bone. Lamellæ are held together by interchange of fibres. Many of these fibres pass transversely through the lamellæ, like tiny nails driven in—they are known as **Sharpey's fibres.** Blood vessels from the periosteum supply the bone—their branches pierce the lamellæ to enter the Haversian canals. The tunnels made by these vessels are known as *Volkmann's canals* (Fig. 1.11).

Periosteum. Bone surfaces are covered with a thick layer of fibrous tissue, in the deeper parts of which the blood vessels run. This layer is the periosteum and the nutrition of the underlying bone substance depends on the integrity of its blood vessels. The periosteum is itself osteogenic, its cells differentiating into osteoblasts when required. In the growing individual new bone is laid down under the periosteum, and even after growth has ceased the periosteum retains the power to produce new bone when it is needed, e.g., in the repair of fractures. The periosteum is united to the underlying bone by Sharpey's fibres, particularly strongly over the attachments of tendons and ligaments. Periosteum does not, of course, cover the articulating surfaces of the bones in synovial joints; it is reflected from the articular margins, to join the capsule of the joint.

Nerve Supply of Periosteum. Subcutaneous periosteum is supplied by the nerves of the overlying skin. In deeper parts the local nerves, usually the motor branches to nearby muscles, provide the supply. Periosteum in all parts of the body is very sensitive.

Blood Supply of Bones. In the adult the nutrient artery of the shaft of a long bone usually supplies little more than the bone marrow. The compact bone of the shaft and the cancellous bone of the ends are supplied by branches from the periosteum, especially numerous beneath muscular and ligamentous attachments. Before union with the shaft an epiphysis is supplied from the circulus vasculosus of the joint (p. 15). Veins are numerous and large in the cancellous red marrow bones (e.g., the basi-vertebral veins) and run with the arteries in Volkmann's canals in compact bone. Lymphatics are present, but scanty; they drain to the regional lymph nodes of the part.

Development of Bone. Bone develops by three different processes namely, ossification in cartilage,

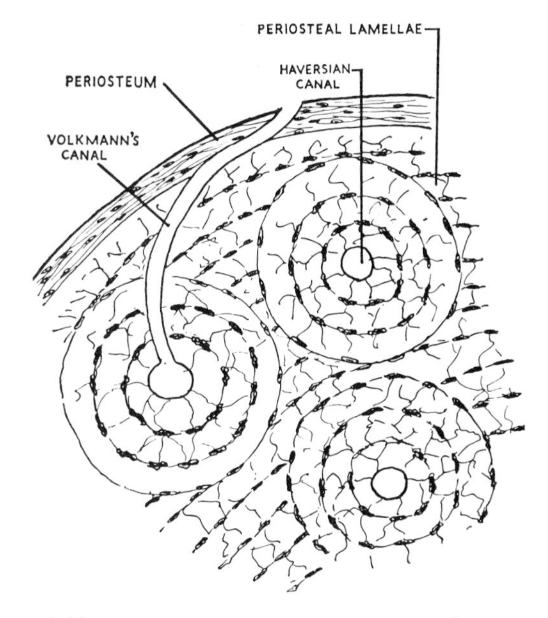

FIG. 1.11. A CROSS SECTION THROUGH COMPACT (HAVERSIAN) BONE.

ossification in membrane and ossification in secondary cartilage. There is no essential difference between them, since the same kind of osteoblast produces the same kind of bone in all three cases. It seems probable that membranous ossification occurs in the embryo in bones required urgently for support or protection of essential organs, for the characteristic feature of membranous ossification is the extreme rapidity with which the process takes place. Ossification in cartilage is an altogether more leisurely affair. In general the long bones of the skeleton ossify in cartilage, while the bones of the vault of the skull, the face and the clavicle ossify in membrane.

Membranous Ossification. This is a process whereby osteoblasts simply lay down bone in fibrous tissue. There is no cartilage precursor. As well as the bones of the vault of the skull, face, etc., it should be noted that growth in thickness of cartilage bones (i.e. subperiosteal ossification) is a process of membranous ossification.

Cartilaginous Ossification. The bones of the skeleton first show as condensations of mesenchyme. At the sixth week most of these condensations become converted into hyaline cartilage, which forms a little model of the future bone. Hyaline cartilage is bloodless, its cells nourished by diffusion. **Centres of ossification** appear in the cartilage; bone is a bloody tissue and a branch from a neighbouring artery supplies the centre. This becomes the nutrient artery. In most of the long bones the shaft is ossified at the eighth week. The ends of the bone remain cartilaginous (the **epiphyses**) and enter into the formation of joints of the limbs. They ossify at various periods (p. 15). There must be some causal factor for the date of ossification of an epiphysis, but it is quite unknown. It is no measure of a man's understanding of anatomy that he can memorize dates of epiphyseal ossifications and fusions. They seem as haphazard as telephone numbers, and equally can be looked up in the book when required.

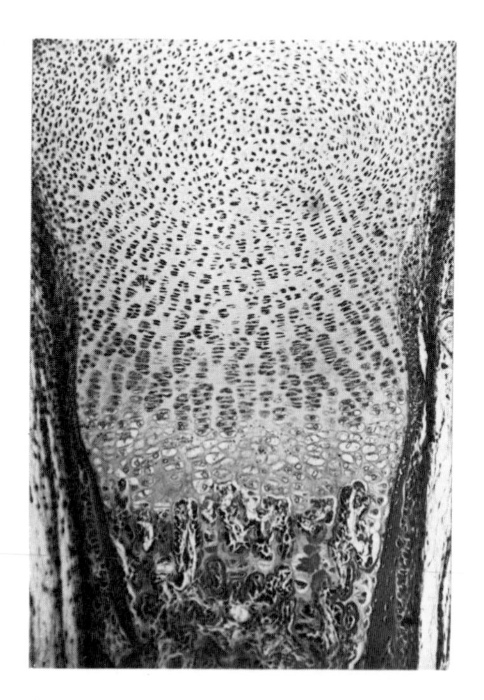

FIG. 1.12. OSSIFICATION OF HYALINE CARTILAGE. In the cartilage columns of cells radiate away from the growing centre. In each column the cells enlarge as they grow older. The cells then die and they and the calcified cartilage around them are replaced by newly growing bone (Photomicrograph ×140). Compare the diagram of Fig. 1.13.

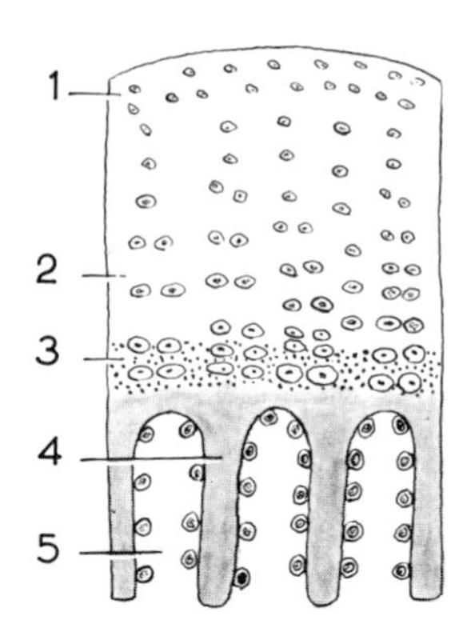

FIG. 1.13. OSSIFICATION OF HYALINE CARTILAGE IN THE EPIPHYSIS AT THE END OF A LONG BONE (cf. Fig. 1.12). 1. Articular surface (hyaline cartilage). 2. Columns of growing cartilage cells. The ground substance is hyaline. 3. Zone of calcified cartilage and dead cells. 4. Trabeculæ of new bone, with osteoblasts on the surface. 5. Marrow space.

Microscopic Anatomy. Hyaline cartilage undergoes certain well-marked changes before ossification (Fig. 1.12). At the site of the future ossific centre the cartilage cells begin to divide. The daughter cells so produced are laid down into parallel columns of gradually enlarging cells. The regularity of these columns depends on an adequacy of vitamin D. At the distal end of each column the cells die, and at the same time there is a precipitation of calcium salt in the surrounding ground substance. The zone of calcified cartilage is essential to normal ossification in hyaline cartilage. All these changes are avascular—now the process becomes vascularized, for loops of capillary blood vessels grow to the edge of the calcified cartilage. Phagocytes eat away the calcified cartilage and dead cells, osteoblasts on the capillary loops lay down trabeculæ of lamellated bone, many osteoblasts becoming imprisoned between the laminæ they have produced and persisting as the bone cells (osteocytes)

Secondary Cartilage. This is the name given to a special type of cartilage that develops in certain growing membrane bones. It is best seen in the neck of the mandible and the sternal end of the clavicle. Both these

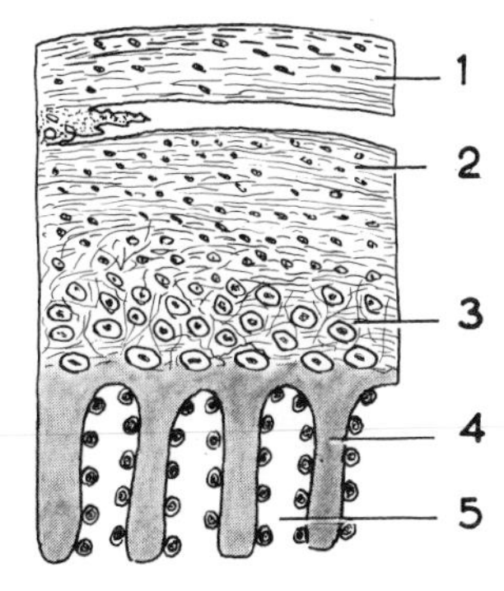

FIG. 1.14. OSSIFICATION IN SECONDARY CARTILAGE (occurs only in head of mandible and both ends of clavicle). 1. Intra-articular disc of fibro-cartilage. 2. Articular surface (fibro-cartilage). 3. Secondary cartilage. The ground substance is fibrous. 4. Trabeculæ of new bone, with osteoblasts on the surface. 5. Marrow space.

bones are primarily ossified in membrane. Their articular surfaces are covered with a dense mass of glistening fibrous tissue (fibro-cartilage). It looks like cartilage to the naked eye but there is no cartilage matrix. It is identical with the intra-articular disc in structure. Between the bundles of fibrous tissue are many cells. At some distance beneath the articular surface the cells divide, enlarge, and come to lie close together in a groundwork of cartilage that is not hyaline, for it contains many fibres. This 'secondary cartilage' differs in appearance from hyaline cartilage in that its cells are larger and more tightly packed and that the matrix is much more fibrous (Fig. 1.14). The secondary cartilage in the neck of the mandible persists until growth of the mandible is complete. It is rather like the epiphysis at the end of a long bone, but unlike the clavicle it has no secondary centre of ossification.

Note that secondary cartilage has nothing to do with the production of secondary cartilaginous joints (*v. inf.*). It is found only in the growing mandibular joint and at both ends of the growing clavicle. It is significant that these are the only places in the body where membrane bones form synovial joints. It may well be that a cap of cartilage beneath the articulating surface of the growing bone allows of adjustment between joint surfaces in harmony with the ever-changing stresses produced by alterations in size and direction of muscle pull. Perhaps the epiphyses at the ends of cartilage bones serve a like purpose, for other bones grow in size and alter in shape without epiphyses (e.g., skull and face bones, upper end of ulna).

Joints

The learning of a number of ill-understood Greek names is no help to the understanding of joints. Union between bones can be in one of three ways: by fibrous tissue, by cartilage or by synovial joints (Fig. 1.15).

Fibrous joints exist between bones or cartilages. The surfaces are simply joined by fibrous tissue. Degree of movement depends on the area of the joint surface and the length of the uniting fibres. Fibrous joints unite the bones of the vault of the skull at the sutures; these gradually ossify (from within outwards) as the years pass by. A fibrous joint unites the lower ends of tibia and fibula; this does not ossify.

Cartilaginous joints are of two varieties, primary and secondary. A **primary cartilaginous joint** is one where bone and hyaline cartilage meet. The junction of bone and cartilage in ossifying hyaline cartilage provides an example. Thus all epiphyses are primary cartilaginous joints, as are the junctions of ribs with their own costal cartilages. All primary cartilaginous joints are quite immobile. They are very strong. The adjacent bone may fracture, but the bone-cartilage interface will not separate.

A secondary cartilaginous joint (symphysis) is a union between bones whose articular surfaces are covered with a thin lamina of hyaline cartilage. The hyaline laminæ are united by fibro-cartilage, histologically a dense fibrous tissue with no recognizable cartilage in it. There is frequently a cavity in the

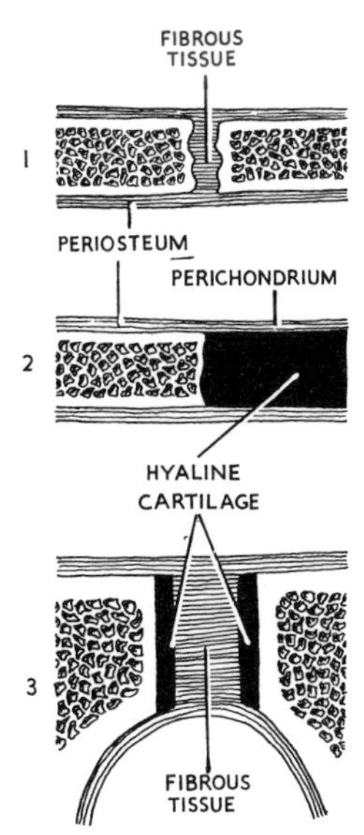

FIG. 1.15. VARIETIES OF JOINTS. (Sectional views, diagrammatic.) 1. Fibrous joint. 2. Primary cartilaginous joint. 3. Secondary cartilaginous joint (symphysis).

fibro-cartilage, but it is never lined with synovial membrane and it contains only tissue fluid. Examples are the symphysis pubis and the joint of the sternal angle (between the manubrium and the body of the sternum). An *intervertebral disc* is a secondary cartilaginous joint, but here the cavity in the fibro-cartilage contains a gel (p. 460).

A limited amount of movement is possible in secondary cartilaginous joints, depending on the amount of fibrous tissue within them. In spite of the name 'cartilaginous' they have nothing in common with the 'primary' cartilaginous joint.

Classification of Joints. Since the understanding of function is the chief reason for the study of structure, it is very tempting to classify joints on a functional basis. Yet such a classification is strangely sterile. Certainly at one end of the scale lie all primary cartilaginous joints, for in none of them does any movement take place. But for the rest, the amount of movement in a joint bears no relationship to its structural classification. Some fibrous joints are quite mobile (e.g., those between the tubercles of the 11th and 12th ribs and the transverse processes of their vertebræ, or those between the vault bones of the fœtal skull), while some synovial joints (e.g., sacro-iliac) possess no significant movement at all. Synovial joints are in general the most mobile, but the direction and range of their movements are very different one from another, and depend on the shape of the articulating surfaces and the nature of the ligaments and muscles of the particular joint.

Furthermore, the nature of pathological change in a joint depends on the anatomical structure of the joint, not on the degree of movement therein.

For once a purely structural classification is the best. This is, as already stated:

1. Fibrous joints
2. Primary cartilaginous joints
3. Secondary cartilaginous joints
4. Synovial joints.

Structure of Synovial Joints

To appreciate the anatomy of synovial joints it is advisable to consider their embryological development. One purpose of this study is to show that every synovial joint possesses a *capsule* which in the embryo was attached at the site of the epiphyseal line but which in many adult joints, has wandered from this attachment. Thickenings of the capsule develop as named *ligaments*. Other thickened bands develop independently of the capsule as ligaments in their own right, sometimes separated by some distance from the true capsule, at other times fused with it in the adult joint. A third type of ligament is said to represent a phylogenetically degenerated tendon that formerly passed freely across the joint but now has become attached to the proximal bone and bound to it as a ligament across the joint. A further event in joint development is the persistence of intra-articular mesenchyme in the form of synovial membrane, fibrocartilages and other intra-articular joint structures.

Embryology of Synovial Joints. The early limb bud (at four weeks, 3 mm length) consists of a swelling on the lateral part of the trunk, covered with ectoderm and filled with undifferentiated mesenchyme rather like the Wharton's jelly of the umbilical cord. Condensations of the cells of the mesenchyme rapidly appear in shapes recognizable as those of the adult bones, in the fœtal position of flexion. In other parts of the body similar condensations of mesenchyme outline the future bones. The cells on the surface of these bone rudiments are condensed as a membrane continued from one bone to the next across the more loosely packed mesenchyme of the developing joint space. At the sixth week the mesenchyme of the bone rudiments rapidly turns into hyaline cartilage. The surface membrane of cells, now called *perichondrium*, stretches from one bone rudiment to the next, as the future *joint capsule*. The future joint cavity remains filled with mesenchyme. At the eighth week (25 mm) ossification of the hyaline cartilage of the shafts of long bones occurs. Each long bone thus comes to possess a shaft of bone covered with the original membrane of mesenchymal cells, which now changes its name from perichondrium to *periosteum*. In the two ends of each long bone the original hyaline cartilage persists as the epiphyseal cartilage. At the junction of bony shaft and cartilaginous epiphysis the periosteum loses contact to enclose the future joint cavity as a tubular sleeve of fibrous tissue called the capsule (Fig. 1.16). The mesenchyme in the joint cavity does not liquefy until about the fourth month, when free movements become possible in the joint. By this time the limb muscles are adequately developed, and 'quickening of the womb' occurs. This joint mesenchyme develops into the synovial membrane that lines the joint cavity and any intra-articular structures such as joint fibro-cartilages. The cartilaginous epiphysis at the end of the bone ossifies by one or more centres, separately from the shaft. The date of commencement of epiphyseal ossification varies widely from bone to bone. The joint surface of the epiphysis does not ossify, persisting as the hyaline articular cartilage on the end of the bone. This is continuous with the cartilaginous plate that separates the bony epiphysis from the bony shaft. The plate between epiphysis and shaft appears on the surface; it is called the epiphyseal line. This **epiphyseal line** gives attachment to the capsule in the fœtus; the capsular attachment may later wander on to the epiphysis itself or on to the shaft, causing the adult epiphyseal line to be extracapsular or intracapsular (PLATE 7).

Synovial Membrane. This lines the capsule and invests all non-articulating surfaces within the joint. That is to say, it is *attached around the articular margin* of each bone of the joint. This is true of every synovial joint.

Synovial Fluid. The characteristic property of synovial fluid is its viscosity, conferred by mucopolysaccharide (hyaluronic acid). However, synovial joints should not be compared with machine bearings, which require hydro-dynamic lubrication. Under slow-moving weight-bearing, hyaline cartilage on joint surfaces possesses an inherent slipperiness greater than that of a skate on ice.

Intra-articular Fibro-cartilages. Discs or menisci of fibro-cartilage are found in certain joints, usually but not always in contact with bones that have developed in membrane. They may be complete or incomplete. They occur characteristically in joints in which two separate movements take place, each movement occurring in its own compartment. Intra-articular fibro-cartilages are not weight-bearing and they do not act as shock absorbers. When incomplete, as in the knee, they act as swabs to spread synovial fluid.

Fatty pads are found in some synovial joints, occupying spaces where bony surfaces are incongruous. Covered in synovial membrane, they probably function as swabs to spread synovial fluid. Examples occur in the hip (the Haversian fat pad) and the talo-calcaneo-navicular joint (p. 184) and in the infrapatellar fold and alar folds of the knee joint (p. 162).

Blood Supply of Joints. Hyaline cartilage is bloodless, bone is a bloody structure. As a centre of ossification appears, blood vessels grow in from any convenient artery. The nutrient vessel to the shaft of a long bone branches generally from the main artery of the particular limb segment concerned and, as elongation of the bone proceeds, the artery gets carried obliquely away from the growing end of the bone. The end of the shaft in contact with the epiphyseal plate of cartilage is known as the *metaphysis.* In the metaphysis the terminal branches of the nutrient artery of the shaft are end arteries, subject to the pathological phenomena of embolism and infarction; osteomyelitis in the child most commonly involves the metaphysis for this reason (Fig. 1.16).

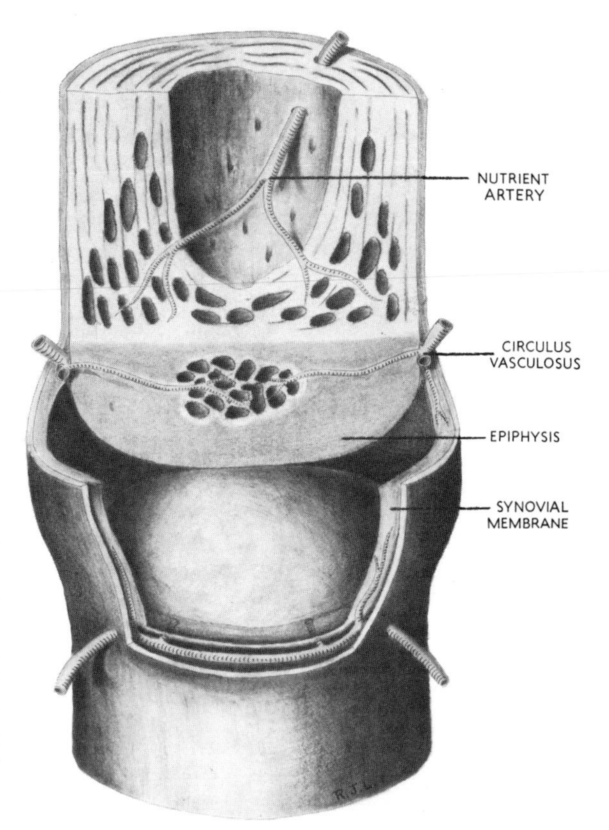

FIG. 1.16. THE BLOOD SUPPLY OF A SYNOVIAL JOINT. Part of the capsule is removed and the upper bone is shown in longitudinal section. The nutrient artery to the shaft terminates by end arteries in the metaphysis. The ossific centre of the epiphysis and the synovial membrane are supplied from the circulus vasculosus of William Hunter. These two arterial systems do not anastomose with each other until bony union takes place between the epiphysis and the shaft.

The cartilaginous epiphysis has, like all hyaline cartilage, no blood supply. The synovial membrane, joint mesenchyme and its derivatives are supplied from a vascular plexus that surrounds the epiphysis and sends branches to the joint structures. It lies between the capsule and the synovial membrane at their attachment to the epiphyseal line. It was described by William Hunter (John's brother) as the **circulus vasculosus.** As ossification of the cartilaginous epiphysis begins, branches from the circulus vasculosus penetrate to the ossific centre. They have no communication across the epiphyseal plate with the vessels of the shaft. Not until the epiphyseal plate ossifies, at cessation of growth, are vascular communications established. Now the metaphysis contains no end arteries and is no longer subject to infarction from embolism; osteomyelitis, common in the child in this region, no longer has any particular site of election in the bone.

Nerve Supply of Joints. The capsule and ligaments possess a rich nerve supply. Synovial membrane has no nerve supply. *Hilton's Law:* The motor nerve to a muscle tends to give a branch of supply to the joint which the muscle moves and another branch to the skin over the joint.

Stability of Joints

Factors contributing to the stability of joints may be conveniently analysed under the headings bony, ligamentous and muscular. Atmospheric pressure is a negligible factor in most joints. The importance of the three factors is nearly always in the ascending order of bone, ligament and muscle.

Bony Contours. In such a firm ball and socket joint as the hip, or a mortise joint like the ankle, bony contours play an important part; but in most joints (e.g., sacro-iliac, shoulder, knee, mandibular, arches of foot) they contribute nothing at all to stability.

Ligaments. Ligaments are an important factor in

most joints, acting for the most part in preventing over-movement and in guarding against sudden accidental stresses. They are, however, usually of no avail in guarding against continuous stress (e.g., in supporting the arches of the foot), since they are composed of white fibrous tissue (collagen) which, once stretched, tends to remain elongated. This is well seen in 'double-jointed' contortionists, in whom the usually restricting ligaments are kept permanently elongated by practice. Elastic tissue in ligaments gives a different picture, since such ligaments shorten after elongation. The ligamenta flava and the ligaments of the joints of the auditory ossicles are composed of yellow elastic tissue, an important factor in maintaining stability of the joints concerned. Yellow elastic fibres compose the ligamentum nuchæ of long-necked animals, but not of man, whose skull is balanced on the vertical cervical spine and whose ligamentum nuchæ consists merely of a thin sheet of fibrous tissue, a median intermuscular septum.

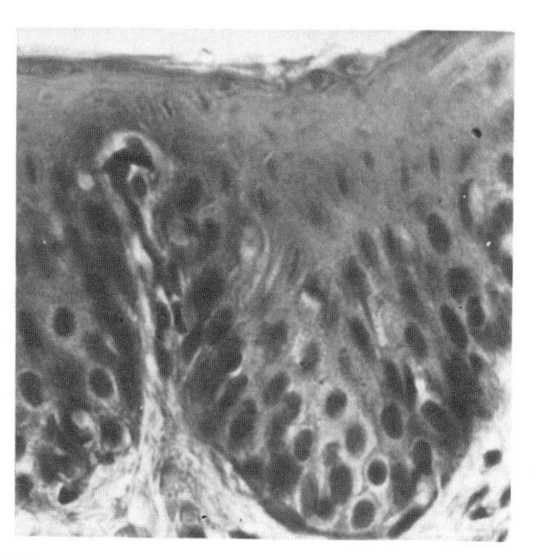

FIG. 1.18. SQUAMOUS STRATIFIED EPITHELIUM (human œsophagus). The flat surface cells are dying and are constantly being shed. Since there is no cornification there is no granular layer (contrast epidermis, Fig. 1.2). (Photomicrograph, high power.)

Muscles. In almost all joints muscles are the most important, and in many an indispensable factor in maintaining stability. Such joints as the knee and shoulder are very unstable without their muscles; wasting of vastus medialis after synovial effusion, for example, is followed by a feeling of insecurity in the knee in a matter of hours. The short muscles of the scapula are indispensable as fixators of the shoulder joint in all movements of the upper limb. Muscles are essential to the maintenance of the arches of the foot. There are many other examples. The matter of stability is discussed under individual joints in the course of this book.

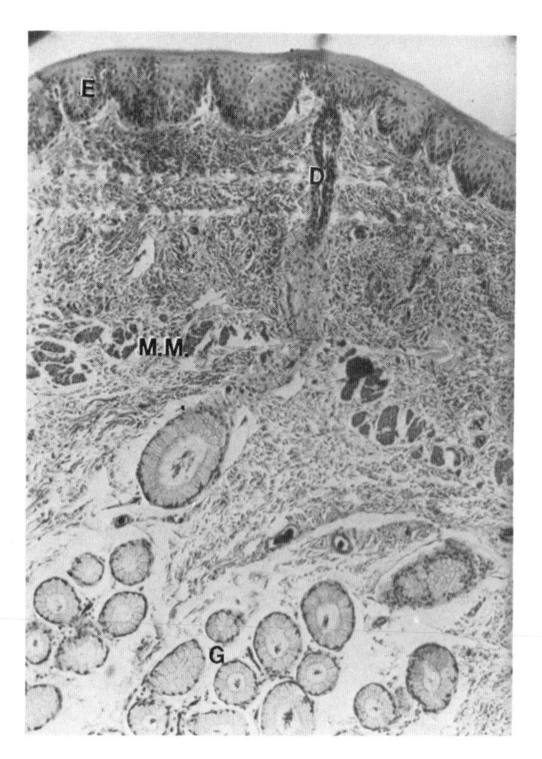

FIG. 1.17. ŒSOPHAGEAL MUCOUS MEMBRANE (Photomicrograph, low power). The loose-woven expansile fibrous tissue is intersected by thick, discrete bundles of muscle, the muscularis mucosæ (**M.M**). **G.** Mucous glands in the submucosa. **D.** A duct traversing the mucous membrane. **E.** The lining of squamous stratified epithelium.

Mucous Membranes

Like the skin on the external surface, mucous membranes line the internal surfaces of the body. A mucous membrane, like the skin, consists of a sheet of fibrous tissue surfaced by epithelium. The sheet of fibrous tissue is known as the corium. Many *but not all* secrete mucus on the surface. The word 'mucosa' is a diminutive of 'mucous membrane'.

The *corium* varies greatly in thickness and density. In general it is thicker and looser in tubes that dilate greatly (e.g., gut), thinner and denser in tubes that do not dilate (e.g., ductus deferens). Throughout the whole of the alimentary canal, from hypopharynx to anal canal, there is a sheet of smooth muscle lying in the corium, called the *muscularis mucosæ*. The depths of the corium, between the muscularis mucosæ and the muscle wall of the gut, constitute the *submucosa*. When there is no muscularis mucosæ (e.g., bladder) it is inaccurate to speak of a submucous layer. Glands of various types lie in the mucous membrane, or in the submucosa, and their ducts perforate the epithelium to carry the secretions to the surface.

The *epithelium* of a mucous membrane varies according to functional needs. In the mouth, pharynx and œsophagus it is stratified squamous (wear and tear), in

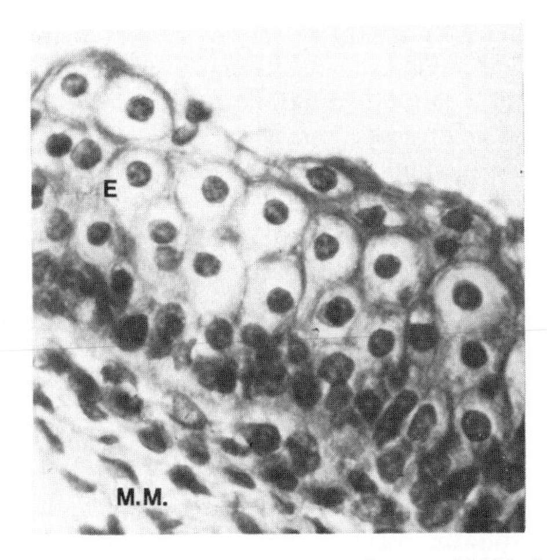

FIG. 1.19. TRANSITIONAL EPITHELIUM (human bladder). The surface cells are (were) alive; there is much cytoplasm around each nucleus. The nuclei in the mucous membrane beneath the epithelium are of fibroblasts. **E.** Transitional epithelium. **M.M.** Mucous membrane. (Photomicrograph, high power.)

the rest of the alimentary canal it is columnar (absorbent) with an increasing number of goblet cells towards the lower end (lubricant and anti-bacterial). The large intestine epithelium is all goblet cells.

Respiratory epithelium covers the mucous membrane of the nose, nasopharynx, larynx (below the vocal cords), trachea, bronchi and bronchioles. It is ciliated columnar epithelium with many goblet cells. The cilia waft away the surface film of mucus with its entrapped particles (dust, bacteria, smoke). Cilia on the epithelium and mucous glands in the mucous membrane go hand in hand, neither being of use without the other (Fig. 6.20, p. 399). They cease together; in man this is at bronchioles with a calibre of 1 mm.

Urinary epithelium is transitional in type. It is the only urine-proof epithelium in the body. It lines the renal pelvis, the ureter, bladder and most of the urethra. Its surface cells are living, to withstand the baneful osmotic and other effects of the urine. They are not normally shed. The surface cells of stratified squamous epithelium are dead (perhaps cornified) and are shed as a result of wear and tear.

Serous Membranes

Mobile viscera in the abdominal and thoracic cavities are covered with a smooth membrane that flows from their walls to line the walls of the containing cavity. The peritoneal, pericardial and pleural cavities are thus potential slit-like spaces between the *visceral layer* that clothes the viscus and the *parietal layer* that lines the walls of the cavity. The two layers slide readily on each other, lubricated by a film of tissue fluid (lymph). No glands exist to produce a lubricating secretion, and the surfaces are covered with a mosaic of very flat cells, like tiles, usually called mesothelium because it is derived from the mesoderm of the cœlomic cavity. The membranes themselves consist of fibrous tissue, like skin. These serous membranes are usually very adherent to the viscera, from which they can be stripped only with some difficulty. The parietal layer is attached to the wall of the containing cavity by loose areolar tissue and in most places can be stripped away easily. The word 'serosa' is a diminutive of 'serous membrane'.

Nerve Supply. The parietal layer of all serous membranes is derived from the somatopleure, and is supplied segmentally by spinal nerves. The visceral layer, developed from the splanchnopleure, possesses no sensory supply (p. 39).

Synovial membrane lining joint capsules and tendon sheaths is in many respects similar, but the fluid produced is not lymph; it is much more viscous (contains mucopolysaccharide). The surface cells are incomplete, so that in some places the bare fibrous tissue of synovial membrane constitutes the working surface. The mechanism of production of synovial fluid is still uncertain. Synovial membrane itself has no nerve supply. Articular nerves supply the capsule and ligaments only.

Blood Vessels

Perhaps more than any other arrangement in the body vascular patterns vary within extremely wide limits. This is not surprising when one thinks for a moment of the way in which the adult pattern is reached. The embryo is little more than a bloody sponge, possessing a vast and intricate network of anastomosing veins. Arteries are sprouting, new vessels replacing the old, while the venous pattern is reached by the disappearance of most of the network of veins in the embryo. The wonder is that such uniformity of pattern is ever attained. The site of origin of a branch from the parent trunk is very variable; the field of supply is much less variable (this applies to peripheral nerves, too).

An interesting feature of embryology is found in the fate of the vessels supplying an organ that migrates far from the site of its original appearance. In one case the vessels may persist, merely elongating to reach the organ, as in the gonads; in another case a series of new vessels appear, replacing the old as the organ migrates, as in the kidney. The kidney is developed at the pelvic brim, where it is supplied by that prolongation of the dorsal aorta known as the median sacral artery. As the kidney 'climbs' up the posterior abdominal wall it is supplied successively from the internal iliac and common iliac arteries and then by a series of branches from the abdominal aorta until the definitive adult position is reached. One or more aortic branches may persist as 'abnormal' renal arteries.

It is true of all parts of the body that veins are bigger than their corresponding arteries, for the simple hydrodynamic reason that the rate of flow is so much slower in them. For the same reason they are often double, as the venæ comitantes of peripheral limb arteries. Note that in both the hand and the foot most of the venous return is by way of the dorsum, to escape the pressure of the palm and sole which would constrict the veins. In the proximal parts of limbs the venæ comitantes unite into a single large vein (axillary, popliteal).

Large veins have 'dead space' around them, to allow for the great dilatation that takes place during increased blood flow. The axillary vein lies in front of the prevertebral and axillary fasciæ where it has adequate room to expand, the femoral vein lies alongside the femoral canal, into which it can readily expand. The internal jugular vein has no carotid sheath around it, unlike the carotid artery. The pulmonary veins lie in the lower part of the lung root, where they can expand between the two lax layers of the pulmonary ligament. The dead space alongside large veins commonly contains the regional lymph nodes (e.g., deep inguinal nodes alongside femoral vein, lateral axillary nodes medial to the axillary vein, deep cervical nodes along the internal jugular vein).

Anastomosis of Arteries. The anastomoses between various arteries of the body are of great importance in medicine and surgery alike. They are of two types.

1. **Actual,** where arteries meet end to end. The test for such an anastomosis is that the cut vessel spurts from both ends. Examples are found in the labial branches of the facial arteries, the intercostal arteries, the uterine and ovarian arteries, and the arteries of the greater and lesser curvatures of the stomach.

2. **Potential.** Here the anastomosis is by terminal arterioles. Given sufficient time the arterioles can dilate to take sufficient blood, but with sudden occlusion of a main vessel the immediate anastomosis may be inadequate to nourish the part. Examples are seen in the coronary arteries, most of the limb anastomoses in the region of joints, and the cortical arteries of the cerebral hemispheres.

End Arteries. In many cases there is no precapillary anastomosis between adjacent arteries, and here interruption of arterial flow necessarily results in gangrene or infarction. Examples are found in the liver, spleen, kidney, medullary branches of central nervous system, and the vasa recta of the mesenteric arteries.

Tortuosity of Arteries. Many arteries are very tortuous (facial, splenic, posterior inferior cerebellar, internal carotid, etc.). This may be to permit elongation, an explanation not tenable for the cerebellum nor for the bony course of the internal carotid; and the facial and splenic *veins* are each as straight as an arrow. The most convincing explanation is that the extra length of a tortuous artery delays its blood flow into terminal anastomotic branches, in order not to overshoot the anastomosis but to collide with neighbouring blood at the correct place.

Arterio-Venous Shunts. Short-circuiting channels between terminal arterioles and primary venules have been demonstrated in many parts of the body. In such places as the kidney and stomach they are simple vessels, in fingertips they are complex structures called glomus bodies. In either case they can short-circuit arteriolar blood directly into venules, by-passing the capillary bed. The reason for their existence and the mechanism of their control are both unknown.

Lymphatic Vessels. In considering the blood supply to any organ or part it makes for a more complete picture if arterial input is always linked with venous *and lymphatic* return. Not all the blood entering a part returns by way of veins; much of it becomes tissue fluid and returns by way of lymphatics.

Course of Lymphatics. There is a simple principle that clarifies the whole picture of the lymphatics of the body. It is that superficial lymphatics (i.e., in subcutaneous tissues) follow veins while deep lymphatics follow arteries. This is true of almost any part of the body except the tongue, which is quite exceptional. The fact that lymphatics accompany arteries seems to be beneficial, for it is probable that the pulsations of arteries compress the nearby lymphatics and so 'milk' the lymph along. If this is so it would certainly account for the multitude of valves with which lymphatics are provided.

It must be realized that tissue fluid first enters an intricate plexus of minute lymphatic vessels (capillaries) from which there are usually alternative routes to regional lymph nodes. A given drop of lymph may go by one of several paths, depending on local and even distant pressures, and it is only when it reaches a definitive lymphatic vessel with its valves that its subsequent route becomes unalterable. Thus accounts and illustrations of lymphatic drainage vary in different textbooks and atlases. The question, 'Which description is correct?' has no answer, since all may be correct under different circumstances.

Clinical spread of disease (e.g., infection, neoplasm) by lymphatics does not necessarily follow strictly anatomical pathways. Lymph nodes may be by-passed by the disease process. If lymphatics become dilated by obstruction their valves may be separated and reversal of lymph flow can then occur. Lymphatics communicate with veins freely in many parts of the body; the termination of the thoracic duct may be ligated with impunity, for the lymph finds its way satisfactorily into more peripheral venous channels.

Lymphatic Tissue

Aggregates of lymphocytes occur in many places in the body, and everywhere conform to a basic pattern, the *lymphatic follicle*. The follicle, just visible to the naked eye, is a spherical collection of lymphocytes with a pale centre known as the *germinal centre*. It is paler than the periphery of the follicle because the lymphocytes are here more loosely packed and because their nuclei are less dense.

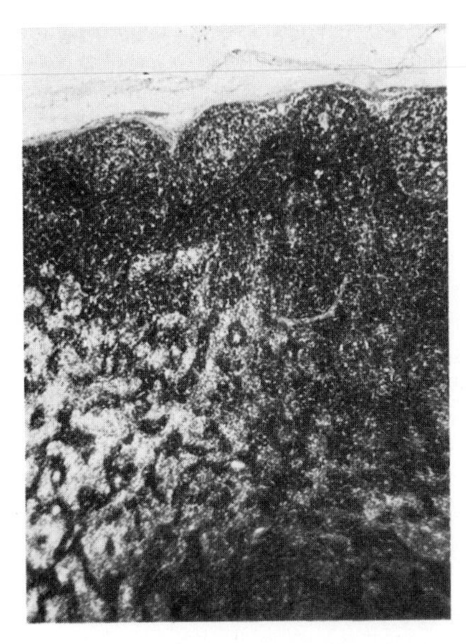

FIG. 1.20. A LYMPH NODE ($\times$ 50). Under the thin capsule lies a clear subcapsular space. The cortex consists of massed lymphatic follicles. The medulla is a network.

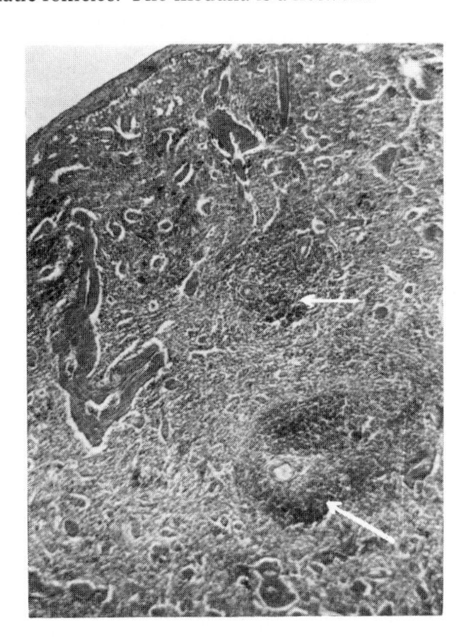

FIG. 1.21. THE SPLEEN ($\times$ 50). The thick capsule is in contact with the underlying pulp, which lies in a framework of trabeculæ projected in from the capsule. Each arrow points to a Malpighian corpuscle (i.e., a lymphatic follicle perforated by an arteriole). The Malpighian corpuscles are scattered through the pulp; there is no division into cortex and medulla.

Isolated lymphatic follicles occur in the mucous membrane and submucosa of the alimentary canal in all parts from the pharynx to the anal margin. They are aggregated into visible collections in the tonsils, Peyer's patches of the ileum, and in the appendix. Isolated follicles occur also in the mucous membrane of the airway and in the lungs (p. 246). Lymphatic follicles are characteristic parts of lymph nodes, spleen and tonsils. The unaccustomed eye can mistake sections of *thymus* and *parathyroid* for lymphatic tissue (the differences are pointed out on pp. 227 and 369).

Lymph Nodes. The microscopic anatomy of a lymph node can be observed best through a magnifying glass (e.g., the reversed eyepiece of a microscope) and the more highly magnified it is the less obvious become the differences between its parts. The ideal section will pass through the hilum, and the typical lymph node will then be seen to be kidney-shaped. The node is loosely enclosed in a capsule which is attached to the surface only at the hilum. Elsewhere the capsule is separated from the surface of the node by a subcapsular lymph space into which the afferent lymphatics empty. The substance of the node consists of a cortex and a medulla (Fig. 1.20).

The *cortex* occupies the whole of the convexity of the surface, and consists of several layers of lymphatic follicles, each with its pale germinal centre. The cortex is penetrated by fibrous trabeculæ from the capsule and each of these is surrounded by a space continued in

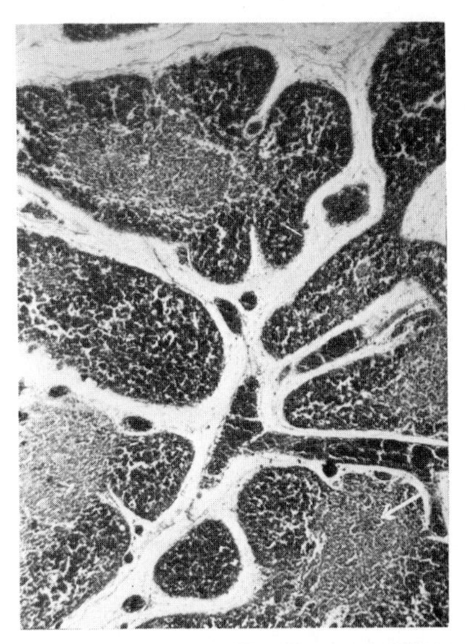

FIG. 1.22. THE THYMUS ($\times$ 50). The thymus is lobulated; each lobule contains a pale medulla (thymocytes) enclosed in a dark cortex (lymphocytes). The white arrow points to a Hassall's corpuscle in the pale medulla.

from the subcapsular lymph space. Fibres from these trabeculæ interlace in a network to form the *medulla*, which reaches the surface at the hilum. The threads of the network in the medulla are encrusted with large, pale, phagocytic cells (histiocytes) and similar large cells lie free in the lymph in the interstices of the network. Occasional multi-nucleated giant cells occur. All this constitutes part of the reticulo-endothelial system. From the multiple afferent lymphatics the lymph passes into the subcapsular space, penetrates the follicles of the cortex and the interstices of the medulla to emerge at the hilum by fewer but larger efferent lymphatics. The artery and vein perforate the hilum and small vessels ramify along the trabeculæ. The lymphatic follicles in the cortex are relatively avascular, only a few capillaries are seen among them. The interstices of the medulla are likewise avascular, except in *hæmo-lymph nodes*. Here the medulla contains erythrocytes mixed with the reticulo-endothelial cells, resembling the pulp of the spleen, but the lymphatic follicles of the cortex remain relatively avascular. Hæmo-lymph nodes, characteristic of the sheep, are few in man, being found only in the retroperitoneal tissues high up on the posterior abdominal wall.

Functionally lymph nodes are doubly protective. Lymphocytes elaborate antibodies, while the phagocytes of the medulla engulf particles (smoke, bacteria, etc.).

Spleen (see p. 305). The spleen resembles a hæmo-lymph node. The chief differences are the scattered arrangements of the lymphatic follicles (there is no cortex in the spleen) and the fact that each follicle in the spleen is perforated by a central arteriole (Fig. 1.21). There is no subcapsular space.

Tonsil. The tonsil is surfaced by the mucous membrane of the pharynx in which it lies. Squamous stratified epithelium on the surface of lymphatic follicles signifies tonsil. Apart from this difference of situation (no *lymph node* lies beneath mucous membrane) there is a noteworthy difference in structure, for a tonsil consists of lymphatic follicles only. It is like a lymph node without a medulla, and it lacks a subcapsular lymph space, too. The mucous membrane is invaginated into the substance of the tonsil in the form of crypts. Neighbouring mucous glands commonly open into the depths of the crypts, possibly to flush them free of debris. Inspissated mucus and macerated squames often plug the crypts. Tonsils are protective against dissolved antigens.

THE NERVOUS SYSTEM

There is only one nervous system. It supplies the body wall and limbs (somatic) and the viscera (autonomic). Its plan is simple. It consists of afferent (sensory) and efferent (motor) pathways, with association and commissural pathways to connect and coordinate the two. There is no more than this, in spite of the many pages here devoted to its study.

The functional unit of the nervous system is a nerve cell-body with its processes, the whole being known as a **neurone.**

The processes of nerve cells conduct impulses in only one direction. Those conveying impulses to the nerve cell are called *dendrites*, the process conveying the impulse away from the nerve cell is called an *axon*. A nerve cell usually gives rise to only one axon, but the latter may branch to reach many different destinations.

Pathways are established in the nervous system by connexions between neurones. The axon of one neurone either forms a synapse with dendrites of the recipient neurone or it arborizes directly around the cell body of the recipient neurone.

Cell bodies with similar functions show a great tendency to group themselves together into ganglia or nuclei or nerve centres. Similarly the processes from such aggregations of cell bodies tend to run together in bundles, forming tracts within the brain and spinal cord, and nerves outside the brain and cord. A fundamental difference between tracts inside the central nervous system and peripheral nerves outside the central nervous system lies in the absence of Schwann cells (and reparative power) in the central nervous system.

The influence which causes like cell bodies to group together, and like fibres to do the same, is completely unknown—so it is called *neurobiotaxis*.

Peripheral Nerves. These are myelinated (medullated) or nonmyelinated. The peripheral process of the cell body is known as the axis cylinder. Each myelinated axis cylinder lies in a delicate tube (neurilemma) which is constricted at intervals (nodes of Ranvier). The neurilemma is easily seen by light microscopy (Fig. 1.23) but it does not seem to be there in electron micrographs (actually it is the surface layer of Schwann cytoplasm plus the basement membrane that surrounds this). Between adjacent nodes each segment contains a nucleated cell (Schwann cell).

Myelinated nerves have a sheath of myelin between axis cylinder and neurilemma. The myelin is interrupted at the nodes, and ceases at the terminal part of the axis cylinder near the end organ. In each segment a Schwann cell is wrapped in fine layers around the axis cylinder, with myelin sandwiched between the spiral layers of Schwann cytoplasm. The Schwann nucleus lies in the surface layer of Schwann cytoplasm (i.e., the neurilemma). Nonmyelinated nerves have no (or very little) myelin; several nonmyelinated fibres may be enveloped by a single Schwann cell.

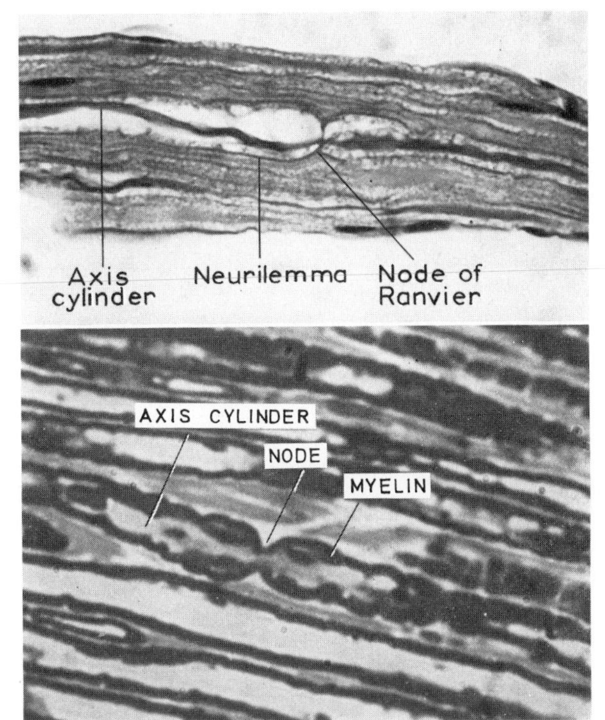

FIG. 1.23. A MYELINATED NERVE AND THE 'NEURILEMMA' AS SEEN BY THE CAMERA IN HIGH-POWER MICROGRAPHS. In the upper figure, a paraffin section, the shrunken axis cylinder is densely stained (haematoxylin and eosin). The 'neurilemma', pinched in at the node of Ranvier, is in reality the outer layer of Schwann cytoplasm, an artefact. In the lower photograph, an epoxy resin section stained with methylene blue, the pale axis cylinder is not shrunken, and the myelin sheath is seen turned in at the node. There is no 'neurilemma'. *Section by courtesy of Dr. Douglas Silva.*

Schwann cells are indispensable to regeneration of divided axis cylinders. The *tracts* inside the central nervous system are myelinated (making the white matter white) but there are no Schwann cells and no regeneration occurs clinically after division of fibres.

Peripheral nerves consist of masses of axis cylinders, each in its neurilemmal tube. Longitudinal fibres of collagen lie between the tubes, giving considerable tensile strength to the whole nerve; this collagen is called *endoneurium*. Fibres and endoneurium are bound together by fibrous tissue into bundles; the binding fibrous tissue is known as *perineurium*. The small bundles are in turn bound together into larger bundles, the whole surrounded by a fibrous-tissue sheath known as *epineurium*.

The spaces between the bundles generally contain fat. In many cases the peripheral part of a nerve is just as big as its proximal part, due to the increase in fat content as the nerve fibre content lessens; this is particularly noticeable in many of the cutaneous nerves of the limbs.

Blood Supply of Nerves. Peripheral nerve trunks in the limbs are supplied by branches from local arteries. The sciatic nerve in the buttock and the median nerve at the elbow have each a large branch from the inferior gluteal and common interosseous arteries respectively, each the former axial artery of the limb. Elsewhere, however, regional arteries supply nerves by a series of longitudinal branches which anastomose freely within the epineurium, so that nerves can be displaced widely from their beds without risk to their blood supply. The phrenic nerve, finding no regional artery in its course through the mediastinum, takes its own artery with it (the pericardiaco-phrenic artery, p. 216).

General Principles of Nerve Supply. Once the nerve supply to a part is established in the embryo it never alters thereafter, unlike the vascular supply. However far a structure may migrate in the developing fœtus it always drags its nerve with it. Conversely, the nerve supply to an adult structure affords visible evidence of its embryonic origin.

Nerve Supply to the Body Wall

Posterior Primary Rami (Divisions). The body wall is supplied segmentally by spinal nerves. Each spinal nerve, on emerging from its intervertebral foramen, divides immediately into an anterior and a posterior primary ramus (Fig. 1.24). The posterior primary ramus passes backwards and supplies the extensor muscles of the vertebral column and skull, and to a varying extent the skin that lies over them.

In the trunk *all* the muscles of the sacro-spinalis group that lie deep to the thoraco-lumbar fascia, *and no others*, are supplied segmentally by the posterior primary rami of the spinal nerves. In the neck, splenius and all the muscles deep to it are similarly supplied (Fig. 1.25).

The cutaneous distribution of the posterior primary rami extends further out than the extensor muscles—almost to the posterior axillary lines. It is to be remembered that:

1. Each posterior primary ramus divides into a medial and a lateral branch (Fig. 1.24). Both branches of *all* the posterior primary rami supply muscle, but only one branch, either medial or lateral, reaches the skin.
2. C1 has no cutaneous branch, and the posterior primary rami of the lower two nerves in the cervical and lumbar regions of the cord likewise fail to reach the skin. All twelve thoracic and five sacral nerves reach the skin.
3. In the upper half of the body the medial branches, and in the lower half the lateral branches, of the posterior primary rami provide the cutaneous branches (Fig. 1.25).
4. No posterior primary ramus ever supplies skin or muscle of a limb in any vertebrate.

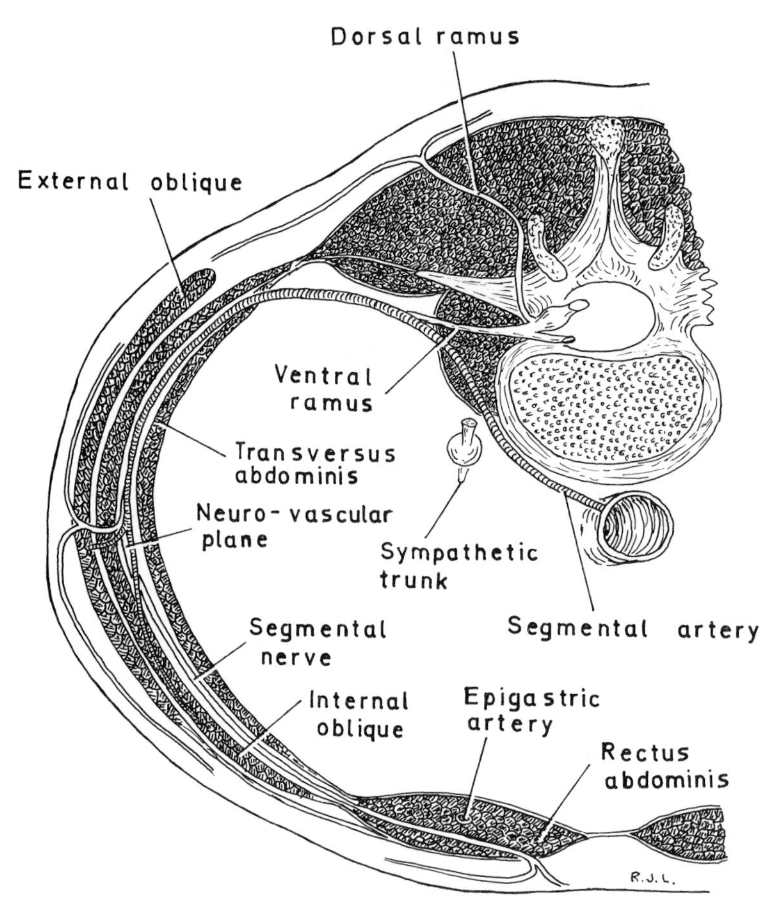

FIG. 1.24. DIAGRAM OF THE NEURO-VASCULAR PLANE OF THE BODY WALL, LYING BETWEEN THE INNERMOST AND INTERMEDIATE OF THE THREE MUSCLE LAYERS. Where the spinal nerve and segmental artery cross, the nerve lies nearer the skin. The segmental nerve goes to the rectus and its overlying skin, but the artery ends in the flank muscles. The sympathetic trunk lies vertically within the circle of the segmental artery. The attachments of the lumbar fascia (p. 307) and the abdominal muscles (p. 256) and the formation of the rectus sheath (p. 257) are shown.

Anterior Primary Rami (Divisions). The anterior primary rami supply the skin at the sides and front of the neck and body.

In the neck only C2, 3 and 4 take part, by branches from the cervical plexus. The skin of C5, 6, 7, and 8 and Th. 1 clothes the upper limb, innervated via the brachial plexus from these segments.

In the trunk the skin is supplied in strips or zones in regular sequence from Th. 2 to L1 inclusive. The intercostal nerves have each a lateral branch to supply the sides and an anterior terminal branch to supply the front of the body wall (Fig. 1.26). The lower six thoracic nerves pass beyond the costal margin obliquely downwards to supply the skin of the abdominal wall. Each nerve throughout the whole of its course supplies a strip of skin that overlies it. The area of skin supplied by a single segment of the spinal cord is called a *dermatome*. On the body wall adjacent dermatomes overlap considerably, so that interruption of a single segment produces no anæsthesia (Fig. 1.26).

Each spinal nerve contains a mixture of sensory and motor fibres. The lateral branch and anterior terminal branch of the *intercostal* nerve contain sensory fibres only. They perforate muscles, without supplying them, to reach the deep fascia and skin. Most of the motor fibres of the intercostal nerves run in the collateral branches (p. 215).

Muscular Distribution. The anterior primary rami supply the prevertebral flexor muscles segmentally by separate branches from each nerve (e.g., longus capitis, scalene muscles, psoas, quadratus lumborum, piriformis). The anterior rami of the twelve thoracic nerves and L1 supply the muscles of the body wall segmentally. Each intercostal nerve supplies the muscles of its intercostal space, and the lower six nerves pass beyond the costal margin to supply the muscles of the anterior abdominal wall. The first lumbar nerve (ilio-hypogastric and ilio-inguinal nerves) is the lowest spinal nerve to supply the anterior abdominal wall. In the thoracic wall only the collateral branch is motor, but in the abdominal wall both the main nerve and its collateral branch are motor.

The muscle below L1 is no longer in the body wall; it has migrated into the thigh to become specialized into the quadriceps and adductor group of muscles (femoral and obturator nerves).

Neuro-vascular Plane. The nerves of the body wall, accompanied by their segmental arteries and veins,

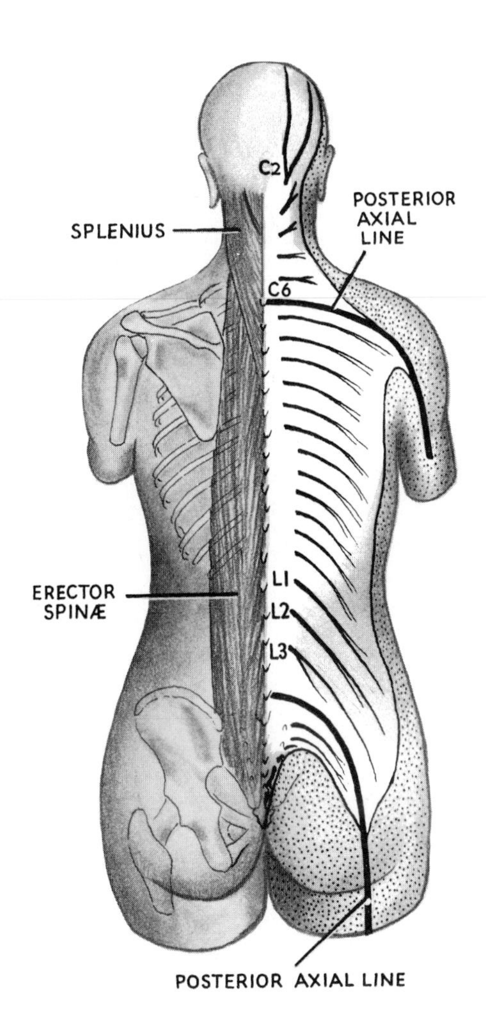

FIG. 1.25. THE DISTRIBUTION OF THE POSTERIOR PRIMARY RAMI. **On the right** the cutaneous distribution is shown (medial branches, to clear the scapula, down to Th. 6, lateral branches below this). The stippled areas of skin are supplied by anterior primary rami. **On the left** the muscular distribution is shown. It is to erector spinæ, and to splenius and the muscles deep to it.

spiral around the walls of the thorax and abdomen in a plane between the middle and deepest of the three muscle layers (p. 211). In this neuro-vascular plane the nerves lie below the arteries as they run around the body wall. But the nerves cross the arteries posteriorly alongside the vertebral column and again anteriorly near the ventral midline, and at these points of crossing a definite relationship is always maintained between the two. *The nerve always lies nearer the skin.* The spinal cord lies nearer the surface of the body than the aorta, and as a result the spinal nerve makes a circle that surrounds the smaller arterial circle. The arterial circle is made of the aorta with its intercostal and lumbar arteries, completed in front by the internal thoracic and the superior and inferior epigastric arteries. As a part of the same arterial pattern the vertebral arteries pass up to the cranial cavity. The spinal nerves, as they emerge from the intervertebral foramina, pass laterally behind the vertebral artery in the neck, behind the posterior intercostal arteries in the thorax, behind the lumbar arteries in the abdomen and behind the lateral sacral artery in the pelvis. The anterior terminal branches of the spinal nerves similarly pass in front of the internal thoracic and the superior and inferior epigastric arteries. (Fig. 1.24.)

The sympathetic trunk does not lie in the neuro-vascular plane but it is useful to remember that it runs vertically within the arterial circle. From the base of the skull to the coccyx the sympathetic trunk lies anterior to the segmental vessels (vertebral, posterior intercostal, lumbar and lateral sacral arteries) which, as already seen, lie anterior to the spinal nerves (Fig. 1.24).

Sympathetic Fibres. Every spinal nerve without exception, from C1 to the coccygeal nerves, carries postganglionic (non-medullated, grey) sympathetic fibres which 'hitch-hike' along the nerves and accompany all their branches. They leave the spinal nerve only at the site of their peripheral destination. They are in the main vaso-constrictor in function, though some go to sweat glands in the skin (sudomotor) and to the arrectores pilorum muscles of the hair roots

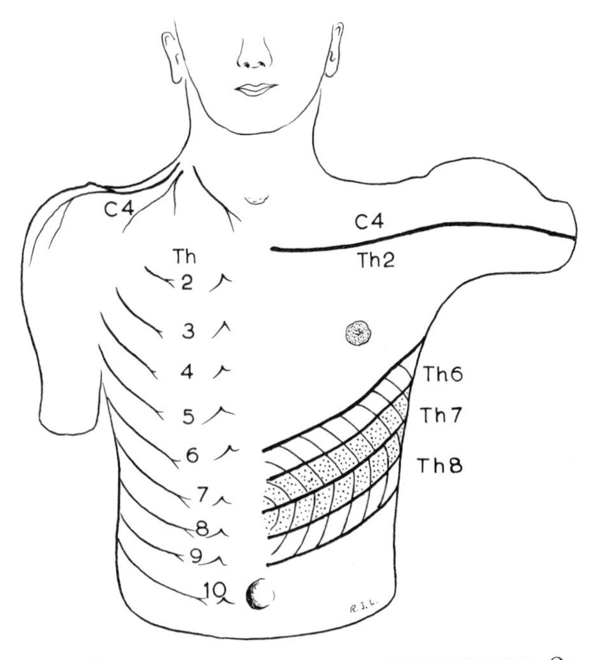

FIG. 1.26. OVERLAP OF DERMATOMES ON THE BODY WALL. On the right the supraclavicular and thoracic nerves are shown, on the left the anterior axial line is indicated. The dermatomes of Th.6 and Th.8 meet each other, completely covering Th.7. Division of an intercostal nerve therefore gives rise to no anæsthesia on the trunk.

(pilomotor). In this way the sympathetic system innervates the whole body wall and all four limbs. This is chiefly for the function of temperature regulation. The visceral branches of the sympathetic system have a different manner of distribution (p. 33).

The body wall has been seen to be supplied segmentally by spinal nerves (Fig. 1.26). A longitudinal strip posteriorly is supplied by posterior primary rami, a lateral strip by the lateral branches of the anterior primary rami, and a ventral strip by the anterior terminal branches of the anterior primary rami. In all vertebrates the limb buds grow out from the lateral strip supplied by the lateral branches of the anterior primary rami (Fig. 1.28) and these lateral branches, by their anterior and posterior divisions, form the plexuses for supply of the muscles and skin of the limbs (Fig. 1.27). The posterior divisions supply extensor muscles and skin of the limbs. The anterior divisions generally represent fusion of the anterior division of the lateral branch with the anterior terminal branch. They supply flexor muscles and skin of the limbs.

Nerve Supply of Limbs

Limb Plexuses. Each limb consists of a flexor and an extensor compartment, which meet at the pre-axial and postaxial borders of the limb. These borders are marked out conveniently by veins. In the upper limb the cephalic vein lies at the pre-axial and the basilic vein at the postaxial border. In the lower limb extension and medial rotation, which replace the early fœtal position of flexion, have complicated the picture. The great saphenous vein marks out the pre-axial and the small saphenous vein the postaxial borders of the limb.

Nerve Supply of the Limbs. In all vertebrates the skin and muscles of the limbs are supplied by plexuses. The plexuses are formed from the anterior primary rami, never from the posterior primary rami, of spinal nerves. The spinal nerves entering into a limb plexus come from enlarged parts of the cord, the cervical enlargement for the brachial plexus and the lumbar enlargement for the lumbo-sacral plexus. The enlargements are produced by the greatly increased mass of grey matter in the anterior horns at these levels (Fig. 1.35).

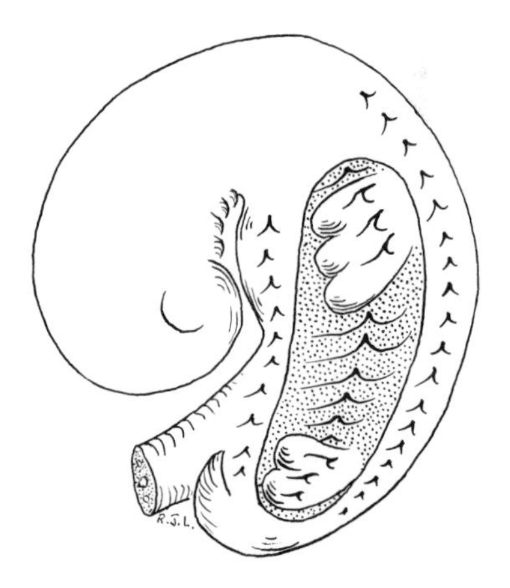

FIG. 1.28. THE ORIGIN OF THE LIMB BUDS. The body wall is innervated in three longitudinal strips. The embryonic limb buds grow out from the lateral strip, supplied by the lateral branches of the anterior primary rami (see Fig. 1.27).

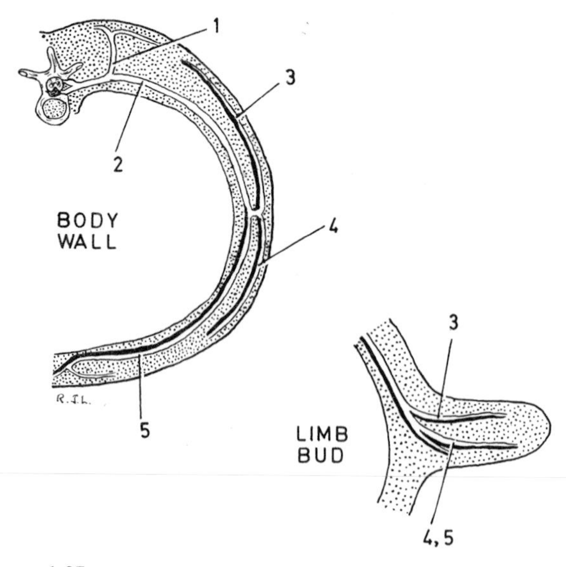

FIG. 1.27. A TYPICAL BODY-WALL NERVE AND ITS INCORPORATION INTO A LIMB PLEXUS. 1. Posterior primary ramus. 2. Anterior primary ramus. 3. In body wall, posterior branch of lateral branch of spinal nerve; in limb bud, posterior division of limb plexus. 4. In body wall, anterior branch of lateral branch of spinal nerve; in limb bud, anterior division of limb plexus. 5. Anterior terminal branch of spinal nerve: in limb bud this is incorporated into the anterior division of the limb plexus.

The constituents of every limb plexus divide into anterior and posterior divisions. The anterior divisions of the limb plexus supply the flexor compartment and the posterior divisions supply the extensor compartment of the limb (Fig. 1.29). The flexor compartment has a richer nerve supply than the extensor compartment. Flexor skin is more sensitive than extensor skin; it has a richer sensory innervation, especially in the distal parts of a limb. Flexor muscles are quicker-acting and under more precise voluntary control; they are finer-fibred (and more tender to eat) than the coarse extensor musculature. Flexor muscles have a richer innervation than extensor muscles. The richer innervation, both sensory and motor, of the flexor compartment shows in the fact that the *most caudal root of a limb plexus is distributed entirely to the flexor compartment of the limb.* This is Th. 1 in the brachial plexus (apart from a small area of skin innervated by the posterior cutaneous nerve of the arm none of this segment supplies the extensor compartment) and S3 in the lumbo-sacral plexus—no S3 enters the extensor compartment of the lower limb (peroneal nerve). Note the composition of the two parts of the sciatic nerve:

1. tibial (medial popliteal) division L4, 5—S1, 2, 3
2. peroneal (lateral popliteal) division L4, 5—S1, 2

Double Innervation of Muscles. In a few cases muscles near the pre-axial or postaxial border of a limb receive a double nerve supply. Generally they are flexor muscles that receive a supply from the nerve of the extensor compartment. This apparent contradiction is due to embryological causes, namely that the muscle concerned was developed in the extensor compartment of the fœtal limb but lies, for functional reasons, in the flexor compartment of the adult limb, bringing its nerve with it. The lateral portion of brachialis (supplied by the radial nerve) and the short head of biceps femoris (supplied by the peroneal part of the sciatic nerve) are examples of flexor muscles supplied by extensor compartment nerves, and in each case the remainder of the muscle is, in fact, supplied by a flexor compartment nerve.

In the arm the intermuscular septum between the flexor and extensor compartments does not represent the fœtal septum. In the fœtus the ulnar nerve was in the flexor compartment, while brachio-radialis, the lateral part of brachialis, and the radial nerve were in the extensor compartment. The adult septum imprisons the ulnar nerve in the extensor compartment above the elbow and brachio-radialis, the lateral part of brachialis and the radial nerve in the flexor compartment (Fig. 1.30).

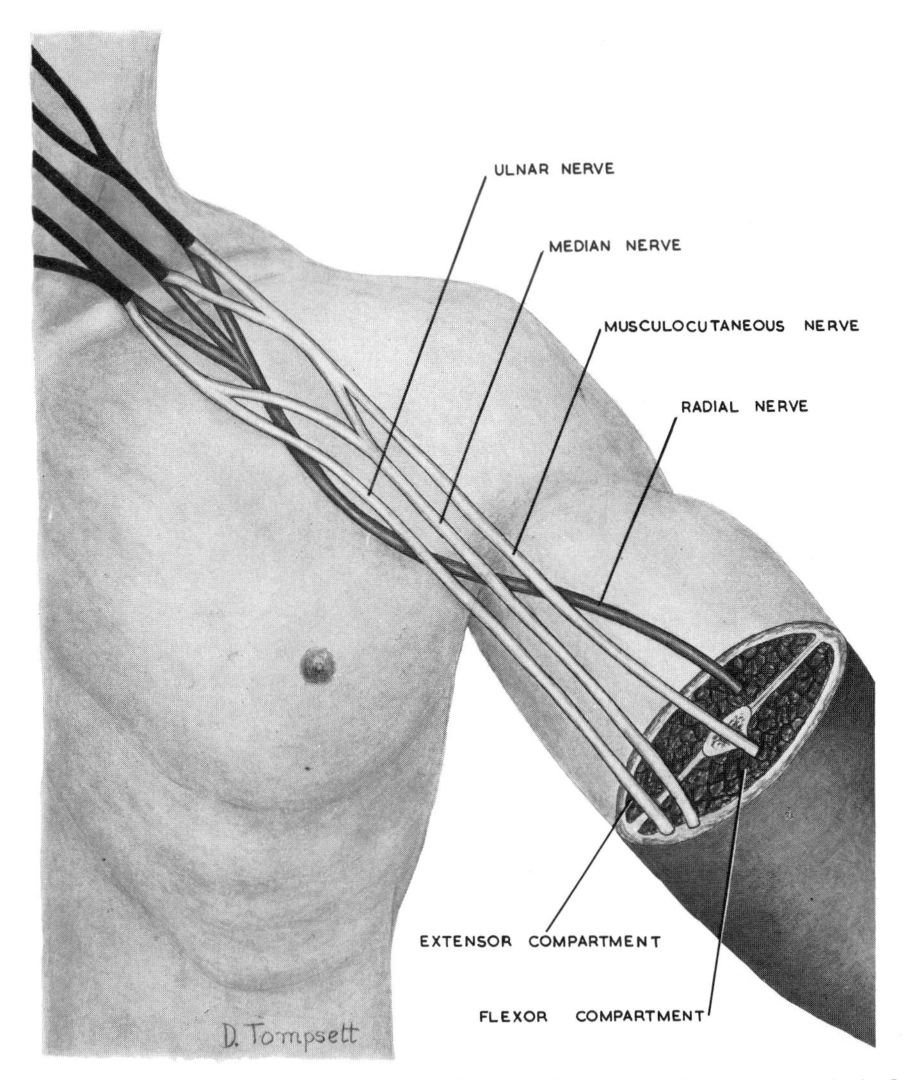

FIG 1.29. THE INNERVATION OF THE UPPER LIMB. Nerves derived from anterior divisions of the trunks supply the flexor compartment, nerves from the posterior divisions supply the extensor compartment.

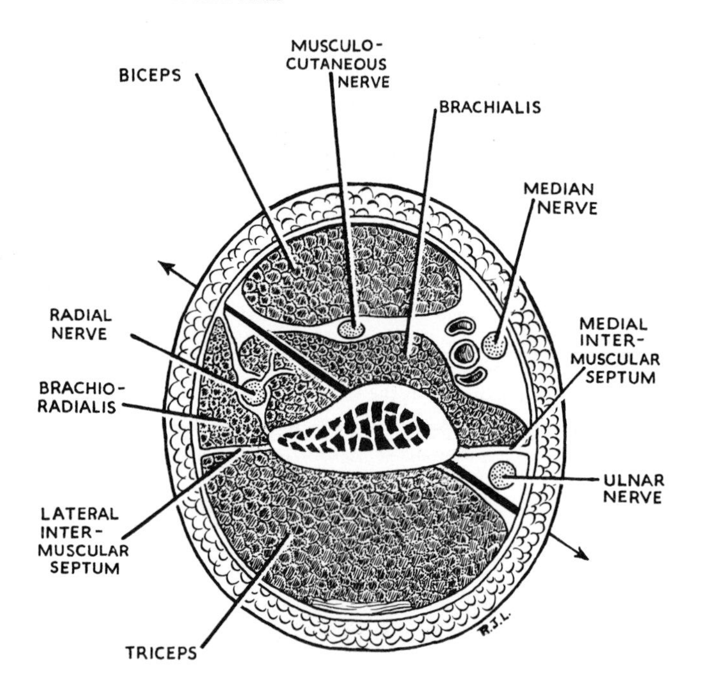

FIG. 1.30. CROSS SECTION OF THE ARM ABOVE THE ELBOW. The arrows indicate the site of the fœtal intermuscular septum between the flexor and extensor compartments. In the fœtus the radial nerve and its adjoining muscles lay in the extensor compartment, the ulnar nerve in the flexor compartment.

Segmental Innervation of the Skin. Dermatomes

The area of skin supplied by a single spinal nerve is called a dermatome. On the trunk adjacent dermatomes overlap considerably, and so it is on the limbs except at the axial lines. The skin that envelops a limb is drawn out over the developing limb from the trunk, and study of the trunk dermatomes near the root of a limb shows that here many of the dermatomes are missing; they lie out along the limb. The line of junction of two dermatomes supplied from discontinuous spinal levels is known as an **axial line,** and such axial lines extend on the limbs.

In the **upper limb** the anterior axial line runs from the angle of Louis across the second costal cartilage and down the front of the limb almost to the wrist (Fig. 1.31). Similarly a posterior axial line is said to run from the vertebra prominens across the back and down the back of the arm only as far as the insertion of the deltoid muscle.

The central dermatome of the limb plexus clothes the peripheral extremity of the limb. More cranially disposed segments are distributed to dermatomes along the pre-axial border of the limb and more caudally disposed dermatomes lie along the postaxial border. There is much overlap between adjacent dermatomes but *no overlap across axial lines*, a point of great clinical value to remember when investigating cutaneous paræsthesia of segmental origin. Always test across axial lines. Note that the dermatomes lie in orderly numerical sequence when traced distally down the front and proximally up the back of the anterior axial line (C5, 6, 7, 8, and Th. 1) and that these dermatomes belong to the nerves of the brachial plexus. In addition, skin has been 'borrowed' from the neck and trunk to clothe the proximal part of the limb (C4 over the deltoid muscle, Th. 2 for the axilla) (Fig. 1.31). The 'borrowing' of skin resembles the borrowing of a book by a friend ('borrowed, never to be returned').

Considerable distortion occurs to the dermatome pattern of the **lower limb** for two reasons. Firstly

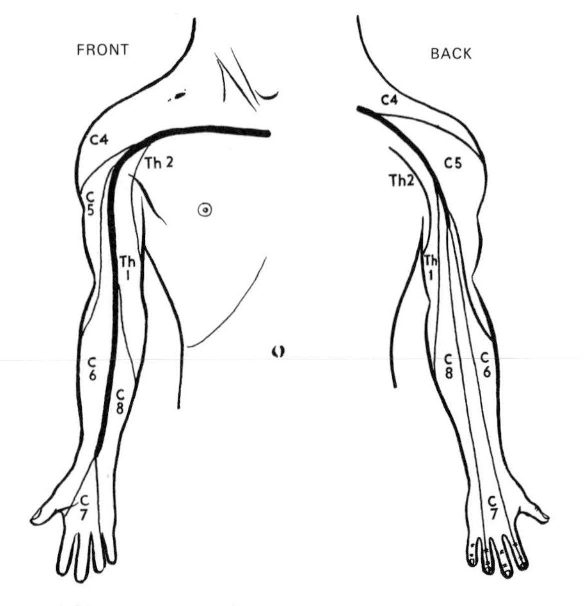

FIG. 1.31. THE AXIAL LINES AND DERMATOMES OF THE RIGHT UPPER LIMB (approximate).

the limb, from the fœtal position of flexion, is medially rotated and extended, so that the anterior axial line is caused to spiral from the root of the penis (clitoris) across the front of the scrotum (labium majus) around to the back of the thigh and calf in the midline almost to the heel (Fig. 1.32). Secondly, a good deal of skin is 'borrowed' from the trunk, and it is all borrowed on the cranial side (there is no skin caudal to the limb except a small area of peri-anal skin). In the upper limb there has been symmetrical cranial and caudal borrowing of skin (a little of C4 and a little of Th. 2) but in

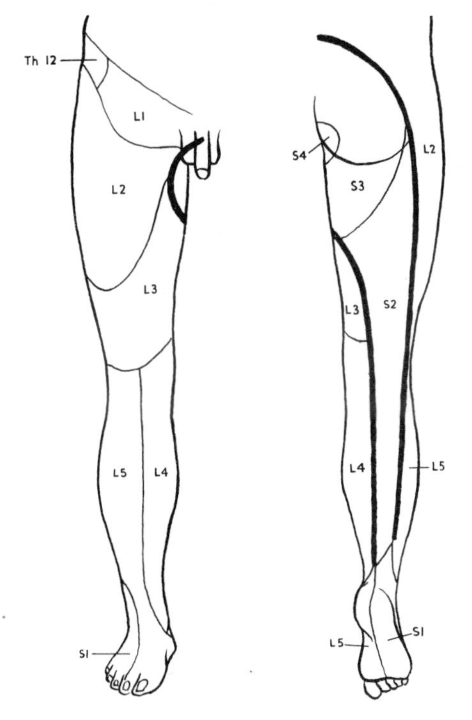

FIG. 1.32. THE AXIAL LINES AND DERMATOMES OF THE RIGHT LOWER LIMB (approximate).

the lower limb four dermatomes, all on the cranial side, have been borrowed from beyond the limits of the limb plexus (these are Th. 12, L1, 2 and 3). The anterior axial line has been described. The posterior axial line is said to run from the fourth lumbar interspace (the site of a lumbar puncture) in a bold convexity to the back of the thigh down to the head of the fibula and thence half-way down the calf.

As in the upper limb the central root of the limb plexus is most peripheral. This is S1 and it lies along the outer side of the foot, dorsum and sole, excluding the big toe. One walks on S1. One sits on S3, for it lies around the anus in a wide semicircle between the axial lines (S4 supplies the immediate peri-anal skin). S2 is the narrow strip up the middle of the calf and hamstrings joining S1 and S3 and limited by the axial lines. L4 is over the tibia and L5 over the fibula but extends also to include the big toe. The other dermatomes are as shown in Fig. 1.32; they are borrowed skin from the trunk. Note that, as in the upper limb, the dermatomes can be traced in numerical sequence down in front and up behind the anterior axial line (L1, 2, 3, 4, 5, and S1, 2, 3).

Thus each anterior primary ramus, passing through the ramifications of its limb plexus, is paid off into the appropriate cutaneous nerves and finally reaches its allotted area of skin (i.e. the dermatome). The **sympathetic grey ramus** that joins each spinal nerve accompanies it to its dermatome (where it is temperature-regulating). Thus knowledge of the segmental supply of an area of skin tells also the site of the cell body of the sympathetic nerve in that area (see p. 33).

A practical application of the anterior axial line arises in spinal analgesia. A 'low spinal' (caudal) anæsthesia anæsthetizes skin of the posterior two-thirds of the scrotum or labium majus (S3), but to anæsthetize the anterior one-third of the scrotum or labium L1 must be involved, an additional seven spinal segments higher up.

It must be emphasized that considerable ignorance still exists concerning dermatomes, and dermatome charts of the limbs are probably today about as accurate as maps of the world were in the sixteenth century. Original dermatome charts made by Sherrington, Head and Foerster are in process of modification in the light of investigations of segmental lesions resulting from intervertebral disc protrusions. It is probable that posterior axial lines (as mentioned and illustrated in this book) do not, in fact, exist. The evidence for the existence of anterior axial lines is much more convincing. Difficulty of investigation arises in the main from the blurring of pattern due to overlap of adjacent dermatomes, and is enhanced by the rarity of a lesion that is complete in one single nerve root while yet involving no others.

Segmental Innervation of Muscles. Myotomes

The regular sequence of myotomes in the trunk (p. 22) appears at first sight to be broken up into a bewildering complexity of myotomes in the limbs (a myotome being the amount of muscle supplied by one segment of the spinal cord). Knowledge is required on two points: (1) what is the segmental supply of this particular limb muscle? and (2) what is the total muscular distribution of this particular spinal segment? It is an enormous task to attempt the memorizing of muscle and nerve lists, and what is more, it is unnecessary, since the underlying plan of segmental innervation is very simple. It is based on the following four facts:

(1) Most muscles are supplied equally from two

adjacent segments (but some, especially in the upper limb, are uni-segmental).

(2) Muscles sharing a common primary action on a joint irrespective of their anatomical situation are all supplied by the same (usually two) segments.

(3) Their opponents, sharing the opposite action, are likewise all supplied by the same (usually two) segments and these segments usually run in numerical sequence with the former.

Put in other words this is to say that there are **spinal centres for joint movements,** and that these spinal centres tend to occupy four continuous segments in the cord. The upper two segments innervate one movement, the lower two segments innervate the opposite movement. For example, the spinal centre for the elbow is in C5, 6, 7, 8 segments and C5, 6 supply the flexors and C7, 8 the extensors of the joint. If the elbow be flexed while the numbers 5, 6 are recited, and extended to the numbers 7, 8, one only needs to perform this simple movement to the numbers 5–6 and 7–8 a few times to learn that biceps, brachialis and brachio-radialis (the prime flexors of the elbow) are supplied by C5, 6 and that triceps (the prime extensor of the elbow) is supplied by C7, 8.

(4) This is obviously a great simplification, but the situation is even simpler. It is based on the principle that for joints more distal in the limbs the spinal centre lies lower in the cord. For a joint one segment more distal in the limb the centre lies, *en bloc*, one segment lower in the cord.

Summarizing these principles (four-segmented spinal centres lying one segment lower in the spinal cord for each joint more distal in the limb) let us contemplate the **lower limb.**

The hip centre is 2, 3, 4, 5 (lumbar).

The knee, one joint distal, has its centre one segment lower, namely 3, 4, 5, 1 (lumbar and sacral).

The ankle, one joint more distal, has its centre one segment still lower, namely 4, 5, 1, 2 (lumbar and sacral).

This is perfectly regular according to the rule. It only remains to state what the movements are. If the numbers are recited (in pairs) contracting the anterior muscles first, we get:

Hip		**Knee**		**Ankle**	
2, 3	Flex	3, 4	Extend	4, 5	Dorsi-flex
4, 5	Extend	5, 1	Flex	1, 2	Plantar-flex

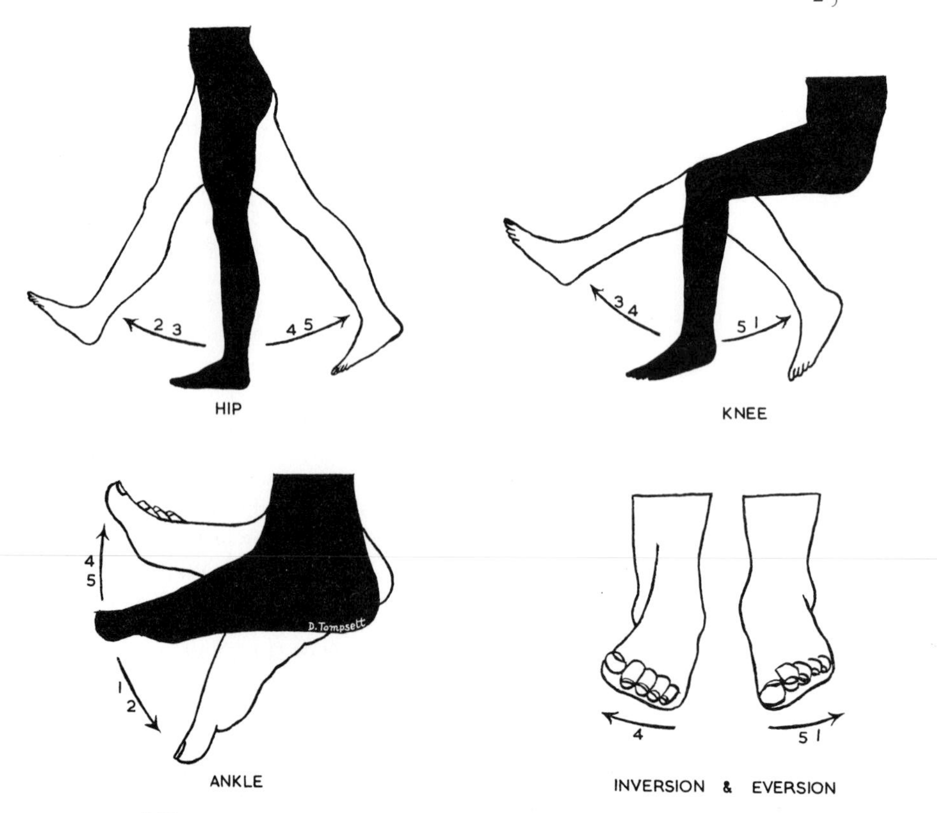

FIG. 1.33. THE SEGMENTAL INNERVATION OF THE MOVEMENTS OF THE LOWER LIMB.

It is cumbersome to print all this in black and white, but very easy to demonstrate it (similar, for instance, to teaching the steps of a dance or the movements of an exercise). If the reader will perform the above simple movements (Fig. 1.33) while reciting the numbers that go with each, in a few moments he will know the essentials of the segmental innervation of the muscles of the lower limb.

To illustrate by a few examples:

Iliacus (flexes hip) L2, 3
Biceps (flexes knee) L5, S1
Soleus (plantar-flexes ankle) S1, 2.

The above are simple flexion-extension movements and, indeed, cover all knee- and ankle-moving muscles. At the hip, however, other movements than flexion-extension are possible, but *all are innervated by the same four segments*. Thus:

Adduction or medial rotation (same as flexion) L2, 3.
Abduction or lateral rotation (same as extension) L4, 5.

There remains one other movement of the lower limb, namely **inversion** and **eversion** of the foot, the formula being:

Invert foot L4.
Evert foot L5.
S1.

Here is a movement innervated by a single segment. It is important to note this. Tibialis anterior and tibialis posterior invert the foot and both are innervated by L4 segment. Tibialis anterior is also a dorsi-flexor and L4, 5 (from the formula already given for dorsi-flexion) is approximately correct. Tibialis posterior, however, lies deep among the plantar flexors of the ankle (S1, 2) and it is illuminating to find this solitary muscle innervated from quite a different level from its neighbours —but a level that coincides with its inverting partner, the anterior tibial.

Only one more point need be explained about the lower limb; it concerns the muscles of the buttock. The simple formula given above shows extension, abduction and lateral rotation of the hip as L4, 5 and this covers all the muscles of the buttock. It is, indeed, accurate enough for most purposes in clinical investigation, but for those readers who require more detailed information the following analysis has greater precision:

Muscles of the Buttock

Nerve to quadratus femoris and inferior gemellus L4, 5, S1—Superior gluteal nerve (glutei medius, minimus and tensor fasciæ latæ).
Nerve to obturator internus and superior gemellus L5, S1, 2—Inferior gluteal nerve (gluteus maximus).

In short, obturator internus and gluteus maximus are supplied by 5, 1, 2, all the other buttock muscles by 4, 5, 1.

The **upper limb** has three of its joint movements controlled uni-segmentally, but otherwise conforms to the general plan of segmental innervation. The movements, with the segments involved (Fig. 1.34), are as follows:

Shoulder	Abduct and laterally rotate C5.
	Adduct and medially rotate C6, 7, 8.
Elbow	Flex C5, 6.
	Extend C7, 8.
Forearm	Pronate C6.
	Supinate C6.
Wrist only	Flex C6, 7.
	Extend C6, 7.
Fingers and thumb (long tendons)	Flex C7, 8.
	Extend C7, 8.
Hand (intrinsic muscles)	Th. 1.

If these movements are performed to the numbers, it will be seen that the segmental innervation of all the upper limb muscles can be learned in a few minutes. The formula is not quite so simple as that of the lower limb, probably because in the upper limb much more precise movements are constantly being employed, and the spinal centres have broken up into separate nuclei to control these (neurobiotaxis). Thus below the elbow the plan does not conform to the basic pattern of four spinal segments for each joint. Flexion and extension share the same two segments; these are C6, 7 for the wrist and C7, 8 for the digits. But the rule holds that the more distal joints are innervated from lower centres in the cord.

Note that flexion and extension of the shoulder are also included in the above formulæ, for the muscles involved (deltoid, pectorals, latissimus dorsi), are already accounted for. Deltoid is C5 and the others are C6, 7, 8. Note also that flexor carpi ulnaris, C6, 7 by the formula, is supplied via the ulnar nerve from just these segments. The fibres from C6 and 7 leave the lateral cord of the brachial plexus and join the ulnar nerve in the axilla.

The motor spinal centres for the joints of the limbs lie in cell aggregates in the lateral parts of the anterior grey horns in the cervical and lumbar enlargements of the cord. The medial parts of the anterior grey horns in the limb enlargements are in line with the slender grey horns of the thoracic region and, like them, supply segmentally the extensor and flexor muscles of the

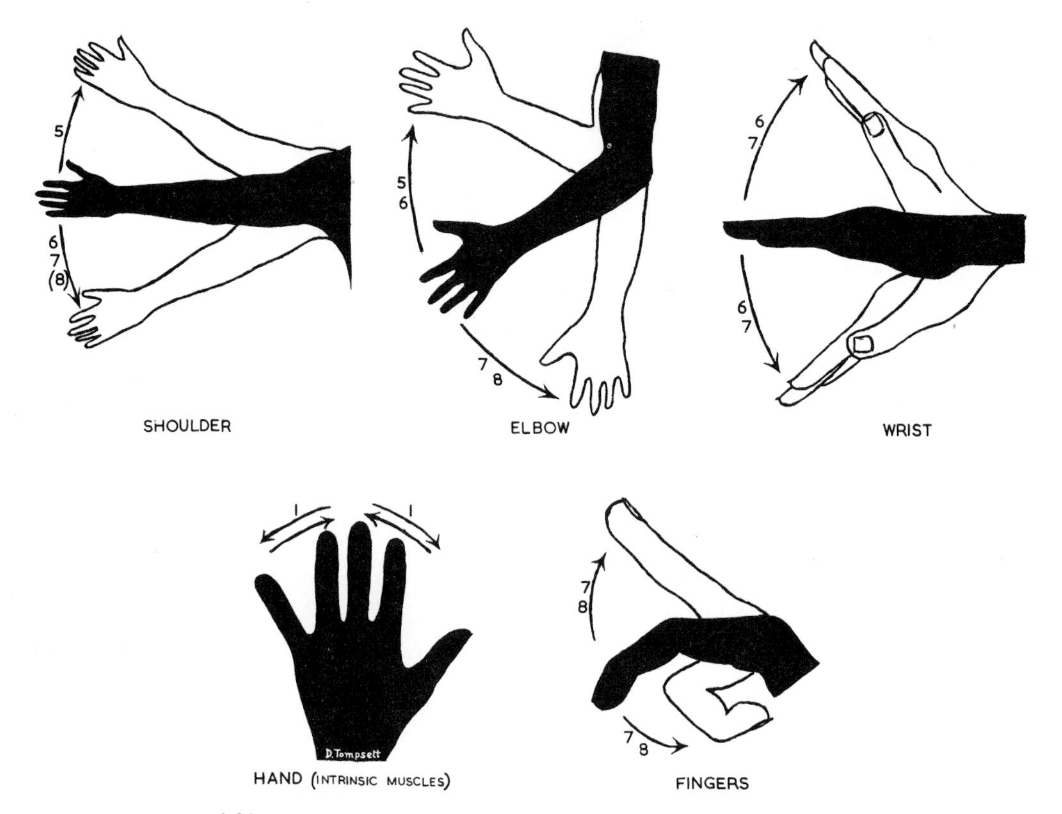

FIG. 1.34. THE SEGMENTAL INNERVATION OF THE MOVEMENTS OF THE UPPER LIMB.

vertebral column and have nothing to do with the muscles of the limbs (Fig. 1.35).

Nerve Supply of the Head and Neck

Cutaneous Supply. The cutaneous supply of the head and neck is very simple (Fig. 1.36). Posterior primary rami of cervical nerves supply a broad band along the extensor surface of the neck, which extends up to the vertex (Fig. 1.25). C1 has no cutaneous branch; the highest nerve in this area of skin is C2. The face is supplied by the trigeminal (V) nerve and the sides and front of the neck by C2, 3, and 4. C3 supplies the cylindrical part of the neck, C2 the expanded part above this, with considerable overlap. C4 extends down over the clavicle, acromion and spine of the scapula to meet the dermatome of Th. 2. The missing segments (C5, 6, 7, 8, and Th. 1) enter the brachial plexus to supply the skin of the upper limb. The face itself is supplied in three strips by the three divisions of the trigeminal nerve. They meet at the angles of the eye and mouth. The fifth nerve has three divisions and the number of their cutaneous branches is 5, 3, 3 (Fig. 1.37). They are:

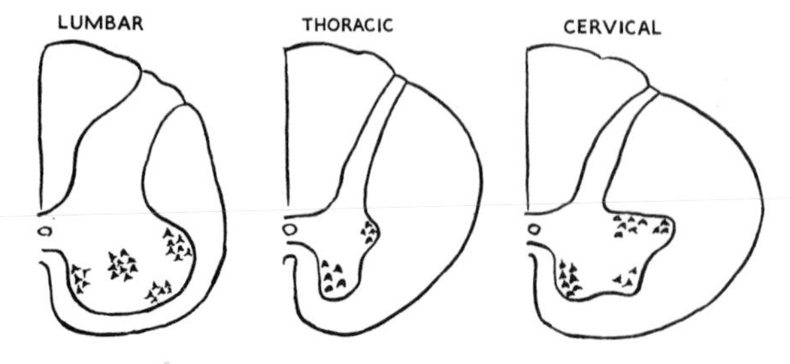

FIG. 1.35. CROSS SECTIONS OF THE SPINAL CORD TO SHOW THE GROUPS OF MOTOR CELLS IN THE ANTERIOR HORNS. In the lumbar and cervical enlargements the lateral groups supply the muscles of the limbs. In the thoracic region the lateral horn contains sympathetic connector cells. In all regions of the cord the medial group of anterior horn cells supplies the longitudinal flexor and extensor muscles of the trunk.

1st division (ophthalmic)
1. Lacrimal
2. Supra-orbital
3. Supratrochlear
4. Infratrochlear
5. External nasal (anterior ethmoidal)

2nd division (maxillary)
1. Infra-orbital
2. Zygomatico-facial
3. Zygomatico-temporal

3rd division (mandibular)
1. Mental
2. Buccal
3. Auriculo-temporal.

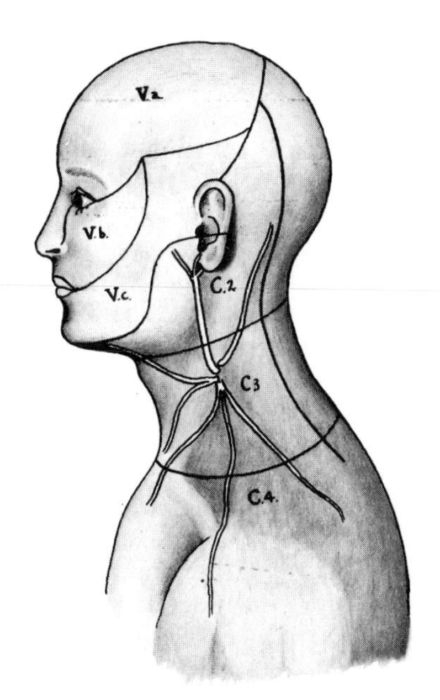

FIG. 1.36. THE SEGMENTAL INNERVATION OF THE SKIN OF THE FACE AND NECK. Note the extension of V.c. skin to the temple and the consequent encroachment of C2 (great auricular nerve) over the angle of the jaw.

It is to be particularly noted that the lines of junction between the areas supplied by the three parts of the trigeminal nerve curve steeply upwards from the angles of the eye and mouth, and that a considerable area of cervical skin encroaches on the face over the parotid gland.

This arrangement is developmental. The areas of skin supplied by the three divisions of the trigeminal nerve originally met each other along lines that extended horizontally back from the angles of the eye and the mouth. The skin supplied by the mandibular nerve (V.c.) lay over the mandible. The area now supplied by the auriculo-temporal nerve (V.c.) was at first over the angle of the mandible. As the cranial cavity expands over the growing brain it draws face skin over it, so that beard skin is pulled up over the temple, and neck skin is drawn up to replace it. The temple is so named because by its greying hairs it is the first part to show the passage of time (*Tempus fugit*); it was originally beard skin.

Nerve Supply of the Dura Mater. The supratentorial dura mater is supplied by the trigeminal nerve, the subtentorial dura by IX and X. It is interesting to note that the dura surrounding the foramen magnum in the posterior cranial fossa is supplied by C1, 2 and 3. This is spinal dura that has been drawn up to cover the expanding brain as the outside skin has been drawn up to cover the expanding cranium.

Deep Supply of the Head and Neck. Most of this is clarified when considered in conjunction with the development of the pharyngeal arches (p. 40).

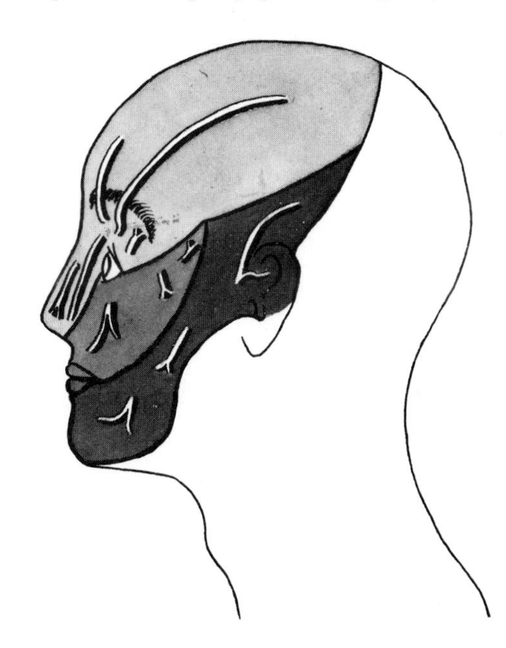

FIG. 1.37. THE CUTANEOUS BRANCHES OF THE THREE DIVISIONS OF THE TRIGEMINAL NERVE. They are 5, 3 and 3 in number.

THE AUTONOMIC NERVOUS SYSTEM

The topographical anatomy of the various parts of the autonomic nervous system will be considered in the appropriate regions of the body, but the system itself can best be understood by reviewing it as a whole. To this end it is desirable to commence with a brief survey of the somatic nervous system (the nerve supply of the body generally) to serve as a standard for comparison.

Somatic Nervous System. The *central nervous system* comprises the brain and spinal cord; the *peripheral nervous system* consists of cranial and spinal nerves that supply sensory and motor fibres to the body. The division into central and peripheral nervous systems is highly artificial, involving as it does the arbitrary division of many neurones between the two systems. To take an example, consider the skin of the big toe. Supplied by L5, the sensory axis cylinder runs in the sciatic nerve to the posterior root ganglion of the fifth lumbar nerve. The cell body lies here, and its central process runs up in the cauda equina and spinal cord to the gracile nucleus in the medulla. Here is a single neurone stretching from big toe to skull, with its cell body half-way between the two. The peripheral process of the neurone lies in the peripheral nervous system, the central process enters the central nervous system;

yet it is only one neurone. The division into central and peripheral nervous systems, however, is convenient in topographical description and serves, too, as a reminder that regeneration of nerve fibres can occur in the peripheral nervous system but never in the central nervous system (p. 21). The great disadvantage is that such a division tends to blur in one's mind the essential continuity of the two systems. *Never think of a nerve fibre without knowing where its cell body lies; think in whole neurones, never in mere axis cylinders.*

Arrangement of Neurones. The *somatic nervous system* is concerned with reception of special sensations (smell, sight, taste, hearing and equilibration) and tactile sensations from the body wall and limbs, the integration of the same, and the control of the striated skeletal muscle of the body wall and limbs. It is arranged in a plan of sensory, connector and motor pathways which are built up by connexions (synapses) between neurones. The **sensory neurone** has its cell body in a ganglion. The ganglion lies in the peripheral nervous system; in spinal nerves it is on the posterior root, in cranial nerves on the nerve itself. The sensory cell in the ganglion is unipolar (i.e., its process divides T-fashion).

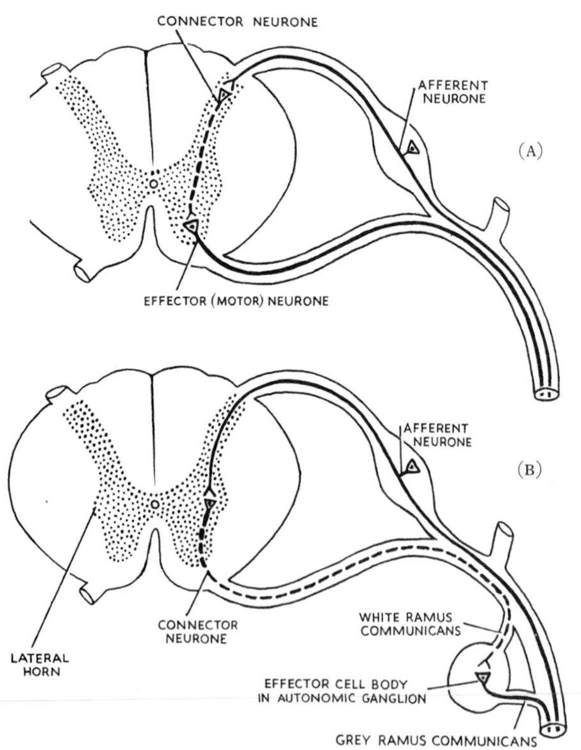

FIG. 1.38. A. THE THREE NEURONES OF A SPINAL REFLEX. B. THE SAME THREE NEURONES COMPRISE A SYMPATHETIC REFLEX BUT THE SITE OF THE CELL BODIES IS DIFFERENT. The connector cell body lies in the lateral grey horn and the effector (motor) cell body lies in an autonomic ganglion. The connector axis cylinder (white ramus) leaves the spinal nerve distal to the posterior primary division. The grey ramus usually joins the spinal nerve *proximal* to the white ramus, not distally as simplified in diagram B.

The central process of the sensory neurone may direct the impulse to one of several chief destinations, namely, the sensory cortex, the cerebellum, the brain stem or the spinal cord. The last-named is the simplest example, that of a **spinal reflex** (Fig. 1.38).

The **connector (internuncial) neurone** has its cell body in the central nervous system. The central process of a sensory neurone arborizes around it.

The integration and co-ordination of afferent and efferent sides of the system is carred out at different levels. Sensory and connector neurones at the same level constitute, with the motor neurone, a simple reflex arc (Fig. 1.38). At higher levels the basal ganglia (thalamus and corpus striatum) join the circuit and, finally, at the highest level, cortical circuits are added.

The **motor neurone** is activated by the connector neurone or from higher levels. Its cell body lies in the central nervous system, ventrally near the midline, in both brain stem and spinal cord. The cell bodies of the motor neurones are, in fact, aggregated into the motor nuclei of the cranial nerves (III, IV, V, VI, VII, IX, X, XI and XII) and the groups of cells in the anterior grey columns of the spinal cord.

The **autonomic nervous system** is built upon the same functional plan; the only topographical difference from the somatic system is in the situation of the cell bodies of the connector and motor neurones. The autonomic nervous system consists of two parts, the sympathetic and parasympathetic systems. The sympathetic nervous system is connected to the thoracic part of the spinal cord and the parasympathetic system is connected to certain cranial and sacral nerves.

In both parts of the autonomic system local reflexes occur. At higher levels the *hypothalamic centres* are the counterpart of basal ganglia. Cortical levels are stimulated on the sensory side (one is conscious of pain in a blood vessel in a limb or of pain in a hollow viscus) but in most individuals there is no cortical control of the efferent side, apart from the inhibition of emptying reflexes of bladder and rectum. The autonomic system is, in fact, the involuntary nervous system.

The autonomic nervous system is concerned in innervating the non-skeletal muscles and many of the glands of the body; that is to say, it is visceral in its distribution. All non-striated muscle is supplied by the autonomic system (blood vessels, walls of the alimentary, respiratory, uro-genital systems, etc.) as well as the highly striated muscle of the heart. The glands supplied include the salivary, alimentary, respiratory, sweat glands, etc. In most of the viscera, both parts of the autonomic system give branches of supply and their effects are, in such cases, antagonistic.

Parasympathetic control is exemplified by a man sitting at stool reading the newspaper. His lens is accommodated, his pupils constricted. His face is flushed, mouth moist, pulse slow, blood pressure low. Bladder and gut are contracting, the perineal sphincters relaxed. Sympathetic control is exemplified by the same man storming the citadel in the face of the enemy. Pupils are dilated, the face pale, mouth dry, the pulse and blood pressure perhaps 180 each. The hollow viscera are inhibited and the perineal sphincters tightly closed (usually). As well as these functional differences the two systems differ in their anatomical arrangements. In both systems the motor cells are outside the central nervous system. In the sympathetic system they are in ganglia associated with the sympathetic trunks; in the parasympathetic system the motor cell-body is usually more peripheral, namely in the wall of the viscus itself. In both systems the fibre of the connector cell passes from the central nervous system to a motor cell body in a ganglion; this is the preganglionic fibre and it is medullated (white). The fibre of the motor cell leaves the ganglion (postganglionic fibre) as a non-medullated or grey fibre.

The efferent side of both sympathetic and parasympathetic systems thus consists of preganglionic (white) fibres of connector cells which lie in the brain stem or spinal cord, of ganglia, and of the postganglionic (grey) fibres of the motor cell bodies that lie in the ganglia.

The Sympathetic Nervous System

As already explained, this consists of afferent, connector, and motor pathways. Since the connector cell bodies are restricted to the thoracic region of the cord, it is well to start here.

The lateral horn of grey matter in the spinal cord extends from the lower part of C8 to the upper part of L2 segments. It contains the connector cell bodies for the whole of the sympathetic system (Fig. 1.38). The peripheral processes of these connector cell bodies pass out in the anterior (i.e., motor) roots of the spinal nerves Th. 1 to L2 inclusive. They pass into the spinal nerves and, beyond the junction of the posterior primary rami, leave the spinal nerves to enter the sympathetic trunk. These connector cell fibres are medullated; thus each spinal nerve from Th. 1 to L2 inclusive, gives a *white ramus communicans* to the sympathetic trunk.

The sympathetic trunk extends alongside the vertebral column from the base of the skull to the coccyx (Fig. 1.42). It is a series of ganglia joined by nerve fibres. The ganglia contain the motor cells, around which the connector fibres (white rami) synapse. The ganglia consist of 3 cervical, 11 thoracic, 4 lumbar and 4 sacral (except in the neck, one less than the spinal nerves) but this is irregular and adjacent ganglia often fuse. Only the ganglia from Th. 1 to L2 receive white rami communicantes (Fig. 1.39); the ganglionated trunk beyond these levels is formed by the continuation upwards and downwards of the fibres of the white rami.

The sympathetic ganglia give rise to two sets of branches with two destinations, somatic and visceral. In this way *all parts of the body* are innervated. No living part of the body lacks a sympathetic supply (remember, for example, that all blood vessels are so innervated).

The **somatic branches** of the ganglia are non-medullated; they are distributed to every spinal nerve from C1 to the coccygeal nerve (Fig. 1.40). *Every spinal nerve thus receives a grey ramus communicans*, which 'hitch-hikes' along the nerve to supply vaso-constrictor fibres to the arterioles, and sudomotor and pilomotor fibres to the skin, in the somatic area of distribution of the nerve. The cell body of each fibre lies in the ganglion from which the grey ramus arises. That is to say no grey fibres run in the sympathetic trunk itself.

In reality there are the same number of ganglia as of spinal nerves, but fusion of adjacent ganglia has reduced the number. Eight cervical ganglia have fused into three. Thus the superior cervical ganglion gives four grey rami—to the first four cervical nerves (cervical plexus). The middle cervical ganglion gives two grey rami—to C5 and 6, and the inferior cervical ganglion gives two grey rami—to C7 and 8 (brachial plexus, see p. 107). Below this each ganglion gives a grey ramus to its corresponding spinal nerve; where two ganglia have fused into one this ganglion gives off two grey rami—to the corresponding spinal nerves.

Every ganglion of the sympathetic chain has a visceral branch.

The **visceral branches** all arise high from the ganglionated sympathetic trunks and descend steeply to form plexuses for the viscera (Fig. 1.41). Thus cardiac branches arise from the three cervical ganglia to descend into the mediastinum to the *cardiac plexuses*. From the upper four thoracic ganglia branches descend to the *œsophageal* and *pulmonary plexuses*. From the lower thoracic ganglia the three splanchnic nerves pierce the diaphragm to reach the *cœliac plexus*, from the upper lumbar ganglia the hypogastric (presacral) nerves descend to the *superior hypogastric plexus* and this divides to enter the *inferior hypogastric (pelvic) plexus* on the lateral wall of the pelvis. The inferior hypogastric (pelvic) plexus is joined by visceral branches from all the sacral ganglia.

The sympathetic visceral plexuses thus formed are joined by parasympathetic nerves (vagus down to the cœliac plexus, pelvic parasympathetics (from S2 and 3) to the inferior hypogastric plexuses; see Figs. 5.46, p. 315 and 5.68, p. 343).

The three cervical ganglia give off their somatic

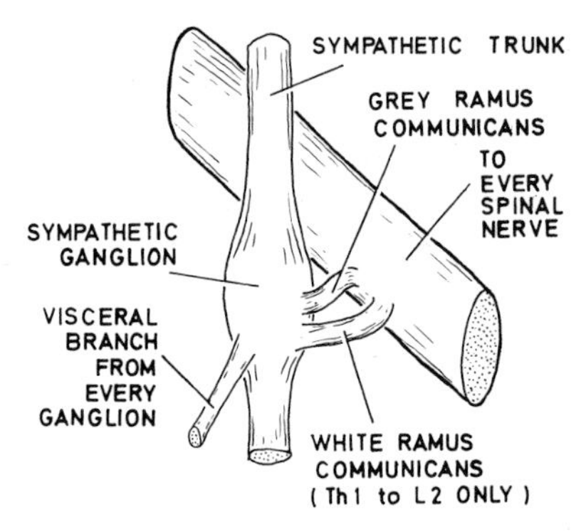

FIG. 1.39. THE CONNEXIONS OF A SYMPATHETIC GANGLION. EVERY GANGLION HAS A GREY RAMUS AND A VISCERAL BRANCH. ONLY THOSE FROM Th. 1 to L. 2 HAVE A WHITE RAMUS COMMUNICANS.

branches to the cervical nerves and their visceral branches to the cardiac plexus according to the general plan. But in addition to these branches the three cervical ganglia, and *only these three*, give off **vascular branches.** The superior cervical ganglion gives branches to both internal and external carotid arteries. These are somatic (to the skin area of the trigeminal nerve) and visceral (dilator pupillæ, smooth muscle of levator palpebræ superioris, nasal and salivary glands, etc.). The middle cervical ganglion gives branches that hitch-hike along the inferior thyroid artery to supply the lower larynx and trachea and the hypopharynx and upper œsophagus; these are visceral only. Fibres from the inferior cervical ganglion run on the vertebral artery. These are vascular only, to the branches of the artery itself.

The branches to the head along the arteries are, of course, all postganglionic (grey); so, too, are those to the cardiac, pulmonary, and œsophageal plexuses. On the contrary, the splanchnic nerves are all preganglionic (white), destined to relay in the cœliac ganglia. The

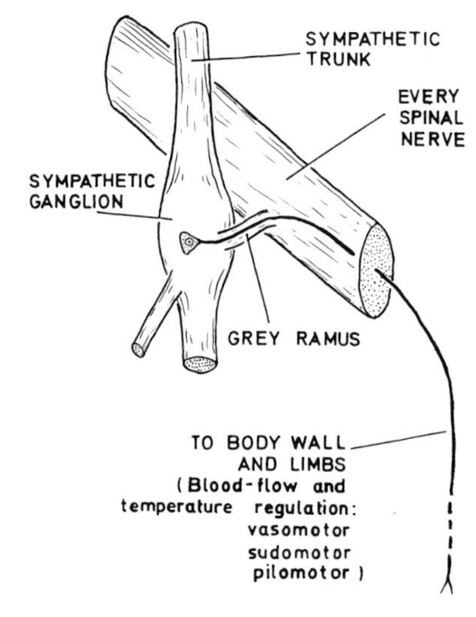

FIG. 1.40. THE SOMATIC DISTRIBUTION OF THE SYMPATHETIC CHAIN (UNIFORM PATTERN).

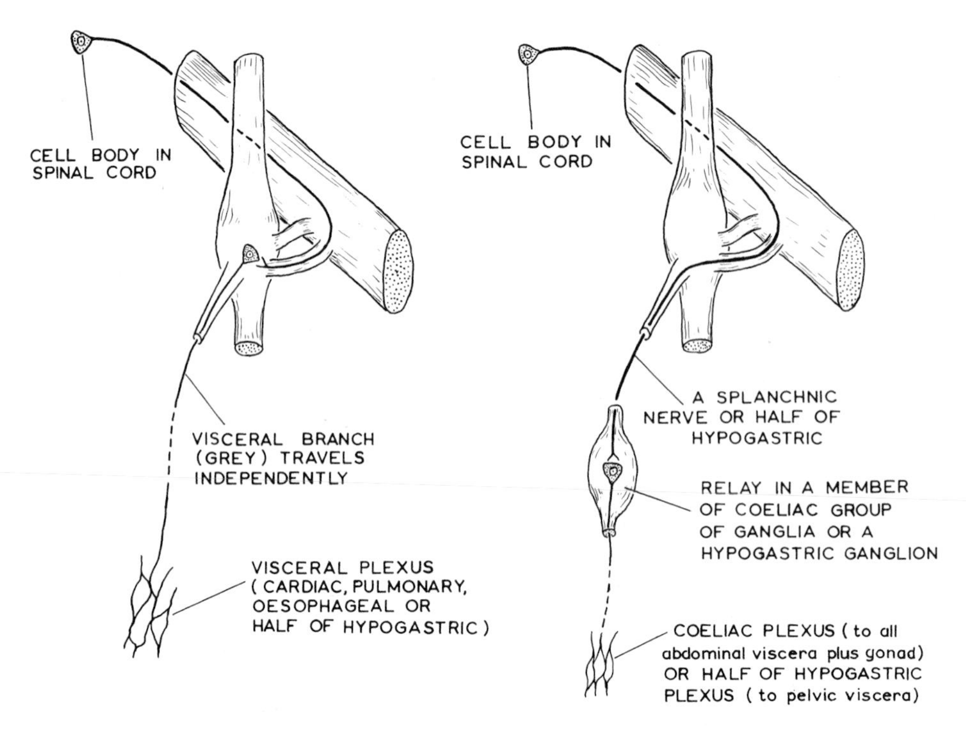

FIG. 1.41. THE VISCERAL DISTRIBUTION OF THE SYMPATHETIC CHAIN (TWO PATTERNS).

hypogastric (presacral) nerves are a half and half mixture of grey and white fibres, the latter relaying in the inferior hypogastric plexuses. The visceral branches of the ganglionated trunks reach their plexuses independently, but the branches from the plexuses travel along arteries to reach the viscera.

The **spinal levels of the connector cells** are indicated in Fig. 1.43). Knowledge is by no means exact, but it will be seen that in general the body is represented in upright order from head to perineum between the limits of Th. 1 and L.2.

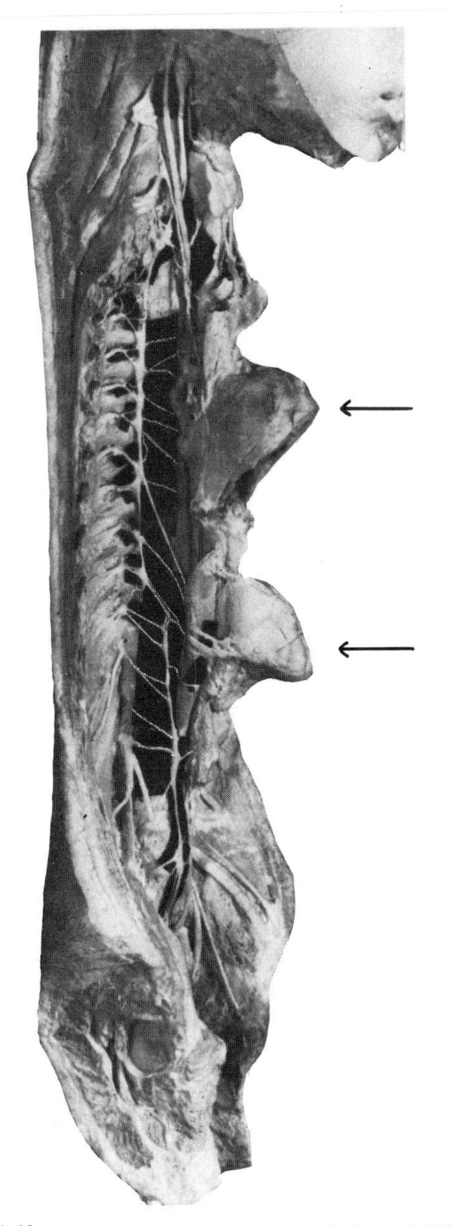

FIG. 1.42. THE SYMPATHETIC TRUNK OF A NEWBORN CHILD. Note the thoracic trunk lying back on the necks of the ribs, while the lumbar trunk lies forward on the bodies of the vertebræ. The upper arrow points to the base of the heart, the lower to the suprarenal gland. Photograph of specimen D.873.1, R.C.S. Museum.

Now that the efferent side has been dealt with it is easier to explain the afferent side of the sympathetic system. Although many details are unknown, in general it may be said that **afferent sympathetic fibres** occur in all areas, somatic and visceral, and that their cell bodies lie in posterior root ganglia at approximately the same segmental levels as the connector cells (Fig. 1.43).

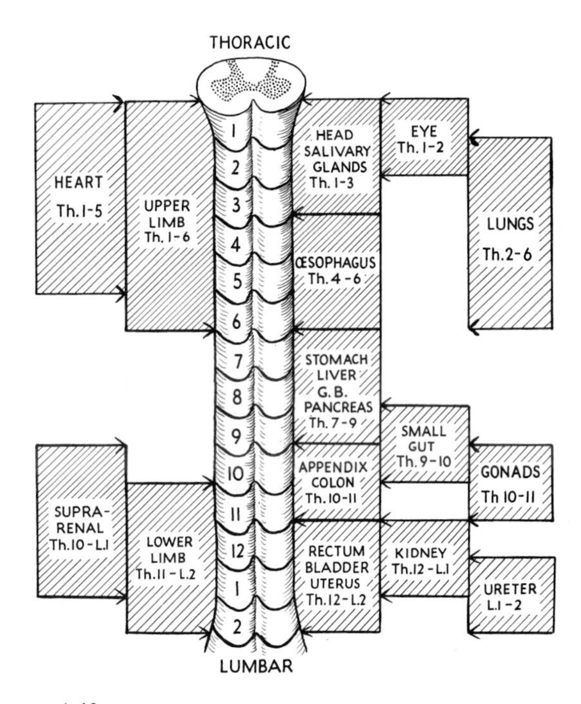

FIG. 1.43. THE SPINAL LEVELS OF THE SYMPATHETIC CONNECTOR CELLS.

The peripheral processes of these sensory sympathetic neurones travel in two different ways: (1) most from trunk and body wall travel by blood vessels and ordinary somatic spinal nerves; (2) most from viscera travel by sympathetic nerves and plexuses and ganglionated trunks and white rami. That is to say, in both somatic and visceral regions of the body the afferent fibres run mainly the same course as the efferent sympathetic fibres.

The central processes of the sympathetic afferent cells are distributed to connector cells in the lateral grey horn (for sympathetic spinal reflexes) or ascend to the hypothalamus and sensory cortex, by spinal pathways that are not yet well known.

The Parasympathetic Nervous System

All parts of the body, both somatic and visceral, receive a sympathetic supply (e.g., all blood vessels are so innervated). By contrast the parasympathetic system has *no somatic distribution*, and while its distribution is *wholly visceral*, not all viscera are so innervated (e.g., suprarenal and gonad, which have only a sympathetic supply).

Parasympathetic efferent fibres accompany the visceral branches of the sympathetic plexuses enumerated above. Their connector cells lie in the brain stem in the parasympathetic (Edinger-Westphal) part of III and in the superior and inferior salivatory nuclei of the nervus intermedius, IX and X. In the sacral cord they lie at S2 and 3 or S3 and 4 levels. The motor cells from III, and the salivatory nuclei lie in the four parasympathetic ganglia of the head and are discussed below. The connector fibres of X synapse with motor cells that lie in the walls of the viscera supplied (the cardiac plexus, pulmonary plexus, and cœliac plexus provide pathways for the passage of vagal fibres to their synapses with motor cells in the walls of the heart, lung, alimentary tract, etc.).

The nervi erigentes are parasympathetic white rami that leave the sacral nerves 2 and 3 or 3 and 4 in the pelvis and pass forwards to enter into the formation of the inferior hypogastric (pelvic) plexus. They pass with the branches of that plexus to all derivatives of the cloaca (bladder, uterus, rectum, etc.). They synapse around motor cell bodies in the walls of these viscera. A nerve passes upwards on each side from the inferior to the superior hypogastric plexus; the two nerves join and pass up from the hypogastric plexus to the inferior mesenteric artery; so the parasympathetic supply to the hindgut is distributed along the branches of the artery.

There is no evidence that parasympathetic efferent fibres accompany the sympathetic efferents to the body wall and limbs.

Parasympathetic Afferents. Cranial parasympathetic afferent fibres run in the nervus intermedius and in IX and X. Their cell bodies are in the sensory ganglia of these nerves. The central processes connect with the cranial nuclei and the hypothalamus.

Sacral parasympathetic afferent fibres run in the nervi erigentes. Their cell bodies are in the posterior root ganglia of the sacral nerves 2, 3 and 4. They bring impulses from bladder, prostate, cervix, vagina, rectum. It is of interest to note the skin area (dermatomes) of these segments. Pain from the derivatives of the cloaca may be referred down the back of the thigh (posterior cutaneous nerve, S2) and be incorrectly diagnosed as sciatica.

The Cranial Autonomic Ganglia

These are not 'cranial'; they lie in the face.

They are four in number. The ganglia themselves belong to the cranial part of the parasympathetic system; that is to say, *no cell bodies and synapses other than parasympathetic are to be found in them.* Each ganglion possesses a parasympathetic 'motor' root which carries the preganglionic connector fibres. This is the essential functional part of the ganglion. But sympathetic and sensory fibres, bound for similar destinations, enter each ganglion and pass through it. The fibres in the sympathetic and sensory roots have no functional connexion with the ganglia; in their peripatetic course they usually, however, share the peripheral branches with the issuing postganglionic parasympathetic fibres.

Plan of the Ganglia. All four ganglia are very similar in plan (Fig. 1.44).

The **motor (parasympathetic) root** arises from a connector cell in the brain stem. Three brain stem nuclei contain the connector cells. The Edinger-Westphal part of III is projected to the ciliary ganglion.

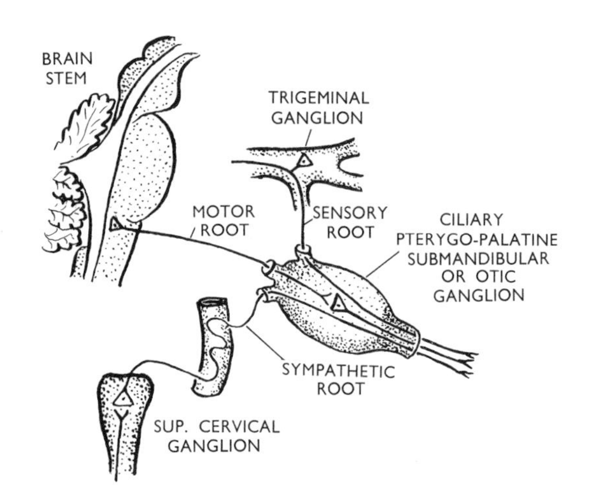

FIG. 1.44. PLAN OF CONNEXIONS OF THE CRANIAL PARASYMPATHETIC GANGLIA. Only the motor root relays in the ganglion (see PLATE 29).

The superior salivatory nucleus (nervus intermedius) is projected to both pterygo-palatine and submandibular ganglia while the inferior salivatory nucleus (IX) is projected to the otic ganglion (PLATE 29).

The **sensory root** is the peripheral process of a cell body in the trigeminal ganglion. For the ciliary ganglion it runs in the ophthalmic division of V, for the pterygo-palatine in the maxillary division of V and for the submandibular and otic ganglia in the mandibular division of V.

The **sympathetic root** is the postganglionic (grey, non-medullated) fibre passing from the superior cervical ganglion on the internal carotid artery to the ciliary and pterygo-palatine ganglia, and on the external carotid artery to the submandibular and otic ganglia. The connector cell bodies lie in the upper three thoracic segments.

Connexions of the Ganglia. The topographical arrangements of the four ganglia are dealt with separately under the regions concerned. A summary of their connexions is included here for reference, not for memorizing at this stage (PLATE 29).

Ciliary Ganglion (p. 437). See PLATE 29.

Motor Root (Parasympathetic). This, as in all the ganglia the essential root, leaves the inferior division of the third cranial nerve in the branch to the inferior oblique muscle. Its fibres come from cell bodies in the cranial part of the third nerve nucleus (Edinger-Westphal nucleus).

Sympathetic Root. This comes from the cavernous plexus of postganglionic fibres whose cell bodies lie in the superior cervical ganglion.

Sensory Root. This is a branch from the naso-ciliary nerve; the cell bodies are in the trigeminal ganglion.

Distribution of Branches. Via the short ciliary nerves, all three are distributed to the globe of the eye.

Parasympathetic fibres from the motor root relay in the ganglion and are distributed to the sphincter pupillæ and ciliary muscles; sympathetic fibres pass without relay to the intra-ocular blood vessels; sensory fibres go to the whole globe, including the cornea. (*N.B.* Sympathetic fibres to the dilator pupillæ muscle travel mostly in the long ciliary nerves, and few pass through the ganglion. For all four ganglia the sympathetic fibres are vaso-constrictor).

Pterygo-palatine (Spheno-palatine) Ganglion (p. 397). See PLATE 29.

Motor Root (Parasympathetic Root). This enters the ganglion in common with the sympathetic root in the nerve of the pterygoid canal. The connector cells of these preganglionic fibres lie in the superior salivatory nucleus; the fibres leave the brain stem in the nervus intermedius, join the seventh nerve and issue from it as the greater (superficial) petrosal nerve.

Sympathetic Root. This enters the ganglion in common with the parasympathetic root in the nerve of the pterygoid canal. The cell bodies are in the superior cervical ganglion, from which fibres pass to the internal carotid plexus, leaving this plexus in the deep petrosal nerve.

Sensory Root. This comes from the maxillary nerve, as two short, thick branches; their cell bodies lie in the trigeminal ganglion.

Distribution of Branches. The parasympathetic postganglionic fibres supply the lacrimal gland by a peripatetic course, wandering first along the zygomatic nerve then leaving it in the middle of the orbit to run with the lacrimal nerve into the lacrimal gland. The remainder of the parasympathetic excitor fibres run with the other branches of the ganglion to supply the mucous glands of the nose, naso-pharynx, paranasal sinuses and palate. The sympathetic fibres go, as vaso-constrictor nerves, to the mucous membrane of the nose, naso-pharynx, paranasal sinuses and palate; the sensory fibres supply sensation to the same area. (*N.B.* A few fibres of the greater (superficial) petrosal nerve are gustatory—from the palate—with their cell bodies in the geniculate ganglion).

Submandibular Ganglion (p. 413). See PLATE 29.

Motor Root (Parasympathetic Root). This enters the ganglion as a branch from the lingual nerve. The preganglionic connector cells lie in the superior salivatory nucleus; their fibres leave the brain stem in the nervus intermedius. They join the facial nerve and leave it in the chorda tympani, which unites with the lingual nerve.

Sympathetic Root. This comes from the plexus around the facial artery (grey rami from the superior cervical ganglion).

Sensory Root. This is a branch of the lingual nerve; the cell bodies are in the trigeminal ganglion.

Distribution of Branches. Postganglionic parasympathetic fibres go to the sublingual gland; many preganglionic fibres pass through the ganglion to synapse with cells lying in the hilum of the submandibular gland. Sympathetic branches supply both sublingual and submandibular glands; the sensory branches of the mandibular nerve supply both glands.

Otic Ganglion (p. 392). See PLATE 29.

Motor Root (Parasympathetic root, Secreto-motor root). The preganglionic connector cells lie in the inferior salivatory nucleus. They leave the brain stem in the ninth nerve, and pass via its tympanic branch (Jacobson's nerve) to the tympanic plexus. Here they pick up some fibres from the nervus intermedius part of the facial nerve. The two groups unite and emerge from the tympanic plexus, without relay, in the lesser (superficial) petrosal nerve. (*N.B.* The otic ganglion, unlike the other three ganglia, has an additional motor root. This is a branch from the nerve to the medial pterygoid. It is an ordinary muscular motor nerve, which passes through the ganglion without synapse, to supply the two tensor muscles—tensor palati and tensor tympani).

Sympathetic Root. This comes from the plexus on the middle meningeal artery, the cell bodies of these fibres lying in the superior cervical sympathetic ganglion.

Sensory Root. Fine twigs enter the ganglion from the auriculo-temporal nerve; their cell bodies are in the trigeminal ganglion.

Distribution of Branches. With the obvious exception of the two motor nerves to the two tensor muscles, the branches of the ganglion are distributed chiefly to the parotid gland, travelling via the auriculo-temporal nerve.

Summary of Distribution (PLATE 29)

The *ciliary ganglion* supplies the eyeball. The *pterygo-palatine ganglion* supplies the palate and all above this (nose and naso-pharynx, paranasal sinuses, the lacrimal gland in the orbit and the labial glands of the upper lip). The *submandibular ganglion* is for all the glands in the floor of the mouth. The *otic ganglion* is for glands that open into the vestibule of the mouth (parotid, buccal, molar, and the labial glands of the lower lip).

Caudal to these four parasympathetic ganglia the oro-pharynx (which includes the posterior one-third of the tongue) is supplied via IX by secreto-motor fibres that relay in tiny ganglia scattered in the mucous membrane. Caudal to the oro-pharynx neurobiotaxis has not worked, and X and the nervi erigentes relay in scattered cells close to the smooth muscle or glands.

EARLY EMBRYOLOGY

The development of most of the organs and systems is touched upon in the text descriptions of the regions concerned. A brief account of the early division of the embryo and the differentiation of the face and mouth parts is included here.

Differentiation of the Ovum

The human being comes into existence as a single cell, the fertilized ovum. Nine months later he is born as a fully developed normal baby. These nine months may conveniently be divided into two phases. The first, or embryonic, is that period (about two months) during which differentiation of tissues and formation of structures are occurring; the second, or fœtal, is that period during which growth and maturation of these structures take place, to prepare the fœtus for its emergence into a separate existence. There is much variation between the times of appearance of the different organs and structures in the developing homunculus; indeed, some structures do not differentiate until long after birth. Embryonic and fœtal stages merge and overlap to some extent, but by the end of the second month the embryo has taken on a recognizably human form, and it is substantially true to say that after this date development consists in the maturing of already formed organs and structures. That is to say, practically all the changes of differentiation which take place do so in the first two months.

The fertilized ovum grows by cell division; the mass of cells so formed is known as the **morula,** from its resemblance to a mulberry. After a few days it is so large that the central cells can no longer receive adequate nourishment from the surface. Accordingly, a cystic space develops within the mass of dividing cells; this facilitates interchange by diffusion of nourishment and waste products between the embryonic cells and the maternal tissues. The morula having developed this cavity (the *extra-embryonic cœlom,* as it is called) becomes known as the **blastocyst.** At this stage implantation into the uterine mucosa takes place, probably about nine days after fertilization. The outer layer of cells in the blastocyst is destined to become placental; it is called **trophoblast,** and the cells merge into a syncytium. The remainder of the cells are concentrated at one end of the blastocyst, where they are attached to the inner layer of the trophoblast. They are known as the *inner cell mass.* Within this inner cell mass a further two cavities appear, **amnion** and **yolk sac** respectively (Fig. 1.45). These two bubbles within the original bubble are separated by a mass of cells called the **embryonic plate,** from which the tissues and organs of the embryo will differentiate. On the amniotic aspect of the embryonic plate is ectoderm, on the yolk sac aspect is endoderm. Between them is primary mesoderm, later to be added to by the appearance of so-called secondary mesoderm, derived from cells of the ectoderm. An axial rod of cells grows in from the ectoderm towards the endoderm; this is the **notochord,** and alongside it the embryo now possesses bilateral symmetry. On the dorsal (ectodermal, amniotic) surface of the embryonic plate a groove develops. Its edges unite to form a tube, the **neural tube,** from which the nervous system develops. Alongside the notochord and neural tube the mesoderm lies in three longitudinal strips. That nearest the midline is the paraxial mesoderm; it becomes segmented into masses of cells called **mesodermal somites** (Fig. 1.46). The somites produce (1) the *sclerotome,* medially, which surrounds the neural tube and notochord, producing the vertebræ and dura mater, and (2) the *myotome* or muscle plate, laterally, which produces the muscles of the body wall.

The intermediate strip of mesoderm is segmented. It is known as the **intermediate cell mass;** it projects ventrally between the other two strips (Fig. 1.47). It

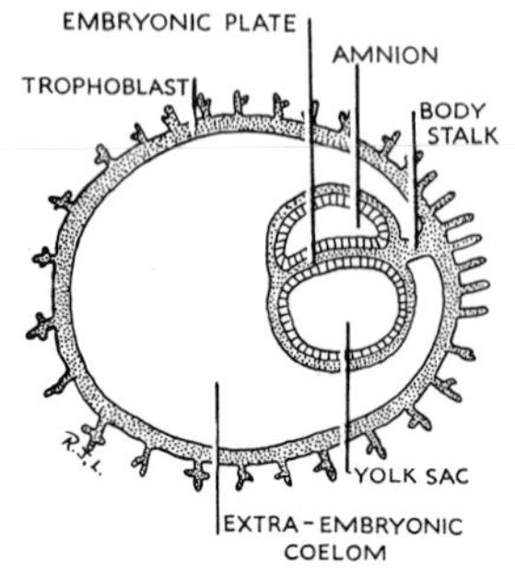

FIG. 1.45. DIAGRAM OF A LONGITUDINAL SECTION THROUGH AN EARLY EMBRYO.

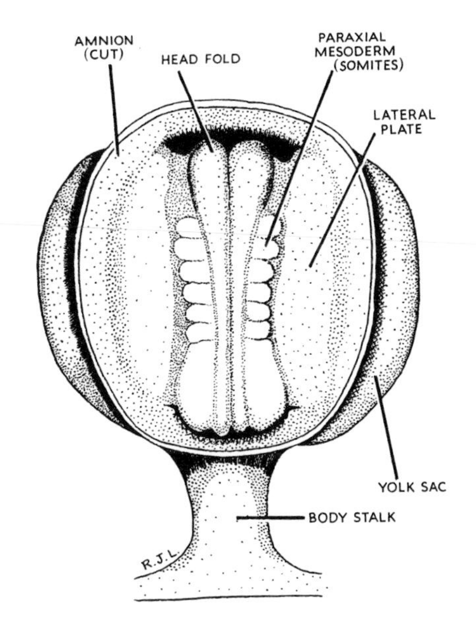

FIG. 1.46. DIAGRAM OF DORSAL VIEW OF EMBRYONIC PLATE AFTER REMOVAL OF AMNION (after Keith). At a slightly later stage a ventral view of this embryo would resemble Fig. 1.56A.

produces the uro-genital system (p. 319). The most lateral strip of mesoderm is unsegmented; it is called the **lateral plate** (Fig. 1.46).

Very early the embryo begins to curl up, a result of the more rapid growth of the dorsal (ectodermal) surface. The embryo becomes markedly convex towards the amnion and correspondingly concave towards the yolk sac. As the lateral plate curls around to enclose the yolk sac its mesoderm becomes split into two layers by a space that appears within it. The space is the beginning of the intra-embryonic **cœlom** or body cavity. The inner layer is called the **splanchnopleure;** it is innervated by the autonomic nervous system. It encloses the yolk sac in an hour-glass constriction; the part of the yolk sac outside persists in the umbilical cord as the vitelline duct, the part inside the embryo becomes the alimentary canal. The outer layer of the lateral plate is called the **somatopleure;** it is innervated by the somatic (spinal) nervous system. Into it the paraxial myotomes migrate in segments to produce the flexor and extensor muscle layers of the body wall. The cœlomic cavity at first includes pleura and peritoneum in one continuum. The lining of the cavity is mesodermal; note that the parietal layer of the serous membranes is supplied by somatic nerves, while the visceral layer is innervated, if at all, by autonomic nerves.

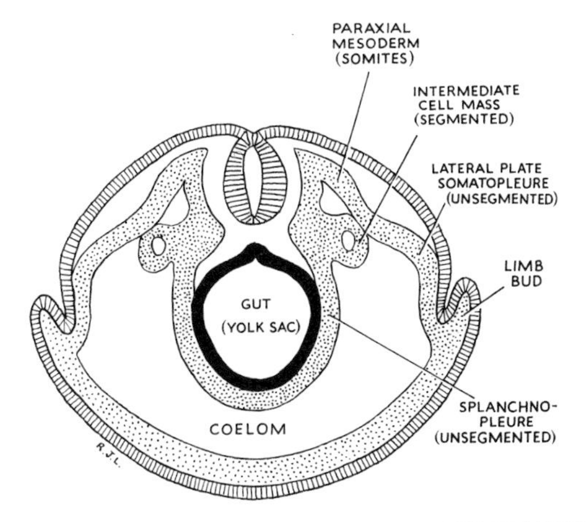

FIG. 1.47. DIAGRAM OF CROSS SECTION THROUGH THE EMBRYO ILLUSTRATED IN FIG. 1.46 AT A SLIGHTLY LATER STAGE.

The **limb buds** grow from the lateral plate mesoderm and their muscles develop *in situ.* Although the lateral plate mesoderm is unsegmented the motor fibres that grow into it from the spinal cord limb plexuses arrange their distribution in a very definite segmental pattern (p. 27).

The **septum transversum** consists of the mass of mesoderm lying on the cranial aspect of the cœlomic cavity. Its cranial part contains the pericardial cavity, the walls of which develop into the pericardial membranes and the anterior part of the diaphragm. It is invaded by muscles from the fourth cervical myotome; they produce the muscle sheet of the dome of the diaphragm. The caudal part of the septum transversum is invaded by the developing liver, which it surrounds as the ventral mesogastrium. The septum transversum later descends, taking the heart with it, to the final position of the diaphragm.

The folding of the embryo is impeded to some extent at the tail end by the presence of the body stalk, which later becomes the umbilical cord. The greatest amount of folding occurs at the head end of the embryo (Fig. 1.28, p. 24). By the end of the first fortnight the forebrain capsule is folded down over the pericardium and a mouth pit, the **stomodœum,** shows as a dimple between the two. Within the body of the embryo the gut cavity extends headwards dorsal to the pericardium, as far forwards as the *bucco-pharyngeal membrane*, which closes the bottom of the mouth pit. The bucco-pharyngeal membrane breaks down and disappears so early (third week) that its former site cannot be made out with certainty in the later embryo or adult. Cranial to the site of the membrane the mouth pit is lined with ectoderm; this includes the region of all the mandibular and maxillary teeth, and probably the submandibular and sublingual glands, and perhaps even the anterior two-thirds of the tongue. Rathke's pouch arises from this ectoderm and forms the anterior lobe of the pituitary gland. Caudal to the bucco-pharyngeal membrane is the pharynx, lined with endoderm and lying dorsal to the pericardium.

Mesodermal condensations (**pharyngeal arches**) form in the side walls of the primitive pharynx and grow around it towards each other ventrally, where they fuse in the midline. In this way a series of horseshoe-shaped arches come to support the pharynx. These are indicated in Fig. 1.28, p. 24. The most cranial pharyngeal arch, the mandibular, separates the mouth pit from the pericardium; as the latter moves caudally by a process of differential growth in the neck the floor of the pharynx elongates and comes successively to possess six of these pharyngeal arches, separated from each other at the side of the pharynx by pouches. The fourth and fifth pouches open into the pharynx by a common groove on each side; thus there are but four pouches separating the six arches (Fig. 1.48). On the

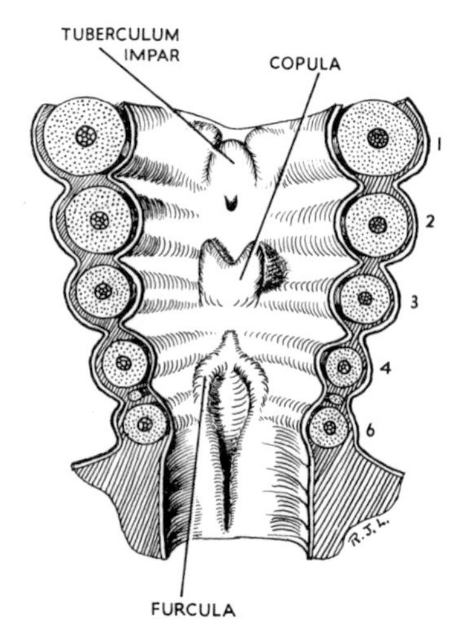

FIG. 1.48. THE THE FLOOR OF THE DEVELOPING PHARYNX, VIEWED FROM THE DORSAL ASPECT. The pharyngeal arches are numbered. The foramen cæcum lies in the midline between the first and second arches.

external aspect of the pharynx four ectodermal depressions correspond in position with the four endodermal depressions within. The four thin parts, between the arches, break down in fishes to form the gill slits or gill clefts; in mammals, however, the ectoderm and endoderm not only remain intact but are actually separated by mesoderm. The mammalian pharyngeal arches and pouches are taken to be homologous with the gill apparatus of fish. It clarifies the understanding of much of the anatomy of the neck to consider the derivatives of the pharyngeal arches and of the pharyngeal pouches, and the midline structures derived from the floor of the pharynx.

Derivatives of the Pharyngeal Arches

In man the fate of the arches is known in greater detail at the cranial end of the pharynx and as we proceed caudally our knowledge becomes less precise. In general it may be stated that the first arch develops into the mandible and maxilla, the second into part of the hyoid bone, the third into the remainder of the hyoid bone, and the fourth, fifth and sixth into the cartilages of the larynx. In each arch a central bar of *cartilage* forms and *muscle* differentiates from the mesoderm around it. An artery and cranial nerve are allocated to the supply of each arch and its derivatives. Vascular patterns are very changeable during development, but a nerve supply once established remains constant. Outgrowths from the intervening pouches, mostly glandular, are supplied by the artery and nerve of the arch nearest to their place of origin.

Mandibular Arch. The right and left halves of the first arch fuse ventrally in the midline. Chondrification in the mesoderm produces Meckel's cartilage. Cranially a bump appears on each side of the arch; this is the maxillary process. The two grow towards the midline, meet each other and the medial and lateral nasal processes, and so produce the upper jaw and palate (p. 45). The dorsal end of Meckel's cartilage produces the incus and malleus, the anterior ligament of the malleus and the spheno-mandibular ligament. The latter is the fibrous perichondrium of Meckel's cartilage, remaining after the cartilage has disappeared. The lingula at the mandibular foramen is a small persistent part of the cartilage; so too are the chin bones (ossa mentalia) occasionally seen persisting at the midline. Around the intervening part of Meckel's cartilage the mandible, ossifying in membrane at the sixth week, extends in two symmetrical halves (Fig. 1.49). Some time after birth the cartilage disappears. Ectodermal and endodermal derivatives of this arch are the mucous membrane and glands (but not the muscle) of the anterior two-thirds of the tongue, mesodermal derivatives are the muscles of mastication (masseter, temporal and pterygoids), the mylo-hyoid and anterior belly of digastric, the two tensor muscles, tensor palati and tensor tympani—in a word, all the muscles supplied by the **mandibular nerve.** The artery is the first aortic arch, part of which persists as the maxillary artery (Fig. 1.54).

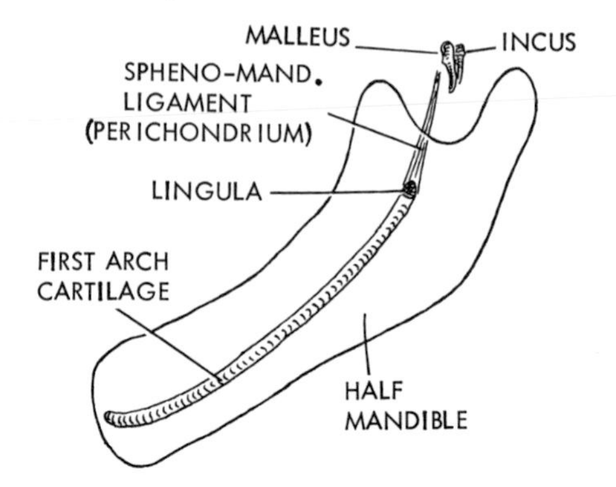

FIG. 1.49. THE DERIVATIVES OF THE FIRST ARCH CARTILAGE.

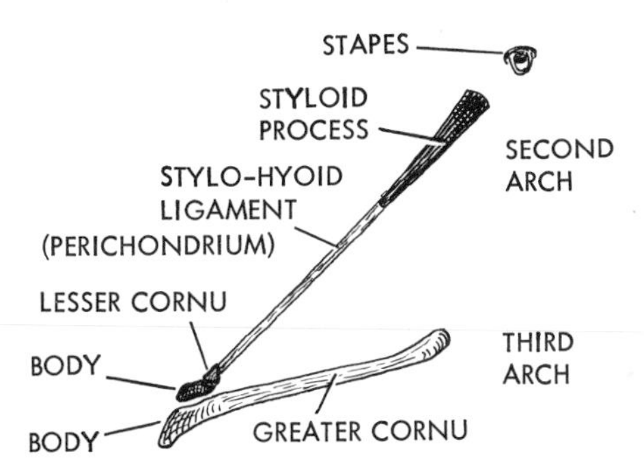

FIG. 1.50. THE DERIVATIVES OF THE SECOND AND THIRD ARCH CARTILAGES.

Hyoid Arch. The stapes, styloid process, lesser cornu and upper part of the body of the hyoid bone are all derived from the cartilage of the second arch. The stylo-hyoid ligament is the persistent perichondrium of the intervening part of the second arch cartilage. Derived from the remaining mesoderm are the stapedius, stylo-hyoid and posterior belly of the digastric muscles, and all the muscles of 'facial expression' including the buccinator and platysma; that is, all the muscles supplied by the **facial nerve,** the nerve of this arch. Part of the second arch artery persists as the stapedial artery.

Third Pharyngeal Arch. The cartilage of this arch, sometimes known as the thyro-hyoid arch, forms the greater cornu and inferior part of the body of the hyoid bone. Its nerve is the **glosso-pharyngeal,** its artery persists in part as the internal carotid, whose carotid sinus is supplied by a branch of IX. Derived from the arch are the stylo-pharyngeus muscle and the mucous membrane and glands of the posterior third of the tongue, the valleculæ and the anterior surface of the epiglottis.

Fourth, Fifth and Sixth Arches. The fifth arch is small, soon disappears, and may be ignored. The fourth and sixth make the muscles and cartilages of the larynx. The nerves are the **superior** and **recurrent laryngeal nerves** respectively. The fourth arch artery persists as the right subclavian artery on the right and the arch of the aorta on the left. The fifth arch artery disappears, the sixth persists as the pulmonary artery ventrally and the ductus arteriosus dorsally on the left side; on the right side the dorsal part of the sixth arch and the whole of the fifth arch arteries disappear, thus allowing the recurrent laryngeal nerve to pass cranially as far as the fourth arch, i.e., the right subclavian artery (Fig. 1.54), under which it loops. On the left side the recurrent laryngeal nerve loops around the ductus arteriosus (ligamentum arteriosum) in the concavity of the aortic arch.

Pretrematic Nerves. The nerve of an arch has an anterior (pretrematic) branch which supplies certain structures cranial to the arch. Thus the glosso-pharyngeal has its tympanic branch supplying mucous membrane of the tympanic cavity produced from the endoderm of the first and second pouches, and the vagus has its auricular branch supplying ectoderm of part of the tympanic membrane and external acoustic meatus.

Lateral Derivatives of the Pharyngeal Pouches

Except for the first, each pouch grows laterally into a dorsal and a ventral diverticulum.

First Pouch. This is the only pouch in which ectoderm and endoderm remain in any sort of contact, viz., at the tympanic membrane, where the mesoderm separating them is minimal. In the other pouches the ectoderm and endoderm are finally widely separated. The endoderm of the first pouch is prolonged laterally, via the auditory tube, to form the middle ear and mastoid antrum (the second pouch gives a contribution to the middle ear). The external pharyngeal groove becomes deepened to form the external acoustic meatus.

Second Pouch. The *dorsal diverticulum* assists the first pouch in the formation of the tympanic cavity, taking its nerve (facial branch to the tympanic plexus) and a pretrematic nerve (the tympanic branch of the glosso-pharyngeal) with it. The *ventral diverticulum* of the pouch develops the tonsillar crypts and the supratonsillar fossa from its endoderm, the surrounding mesoderm contributing the lymphatic tissue of the palatine tonsil. The nerve supply of these derivatives is the glosso-pharyngeal, overlapped in part by the facial.

Third Pouch. *Dorsally* the inferior parathyroid (parathyroid III) and *ventrally* the thymic rudiment grow from this pouch. The latter progresses caudally and joins with that of the other side to produce the bi-lobed thymus gland. In its descent the thymic bud draws parathyroid III in a caudal direction, so that ultimately the latter lies inferior to parathyroid IV, which is derived from the fourth pouch. Externally the cervical sinus overlies the pouch and it is probable that some of the thymic rudiment is derived from the ectoderm thereof. From this thymic bud the medulla of the thymic lobule, including Hassall's corpuscles, is derived; the lymphocytes of the cortex of the lobule differentiate from the surrounding mesoderm.

Fourth Pouch. *Dorsally* parathyroid IV is derived from the endodermal lining of this pouch. *Ventrally* the pouch is attached to the thyroid gland, thus preventing parathyroid IV (the superior parathyroid of the adult) from descending. Whether any of the lateral lobe of the thyroid develops from this pouch is still unknown. Some embryologists believe that part of the thymus develops from the ventral diverticulum of this pouch.

Fifth Pouch. This usually undergoes regression but may form the ultimo-branchial body, whose fate is in any case very uncertain. It is possible that the interstitial cells of the thyroid gland (producing thyrocalcitonin) may originate from this source.

The Cervical Sinus. Concurrently with the growth of the above derivatives from the endoderm of the pouches a change takes place externally in the overlying ectoderm. The second arch increases in thickness and grows caudally, covering in the third, fourth and sixth arches and meeting skin caudally to these. Thus a deep groove is formed, which becomes a deep pit, the cervical sinus. The lips of the pit meet and fuse and the imprisoned ectoderm disappears. Persistence of this ectoderm gives rise to a branchial cyst. A branchial fistula sometimes results from breaking down of the endoderm in, usually, the second pouch; the track then runs from the region of the tonsil, between the external and internal carotid arteries, and reaches the skin anterior to the sterno-mastoid muscle (Fig. 1.55).

It is this formation of the cervical sinus and subsequent obliteration of ectoderm that joins skin over the mandible (V c) to neck skin (C 2). The ectoderm supplied by VII, IX, X and C 1 lines the cervical sinus.

Ventral Derivatives of the Floor of the Pharynx

From the floor of the mouth and pharynx are derived the tongue, thyroid gland and larynx (Fig. 1.48).

The trachea separates from the œsophagus, which lies caudal to the pharynx. The caudal end of the trachea sprouts into the bronchi and lungs.

Development of the Tongue. Three buds from the first arch and one from the third contribute to the formation of the skin, mucous glands and lymphatic tissue of the tongue. Occipital myotomes contribute the musculature and bring their nerve, the hypoglossal, with them as they migrate ventrally around the carotid arteries.

Mandibular Arch Component. From the pharyngeal surface of the first arch the centrally placed tuberculum impar and a pair of lateral rudiments develop. All three fuse to form the anterior two-thirds of the tongue, supplied by the nerve of this arch, the mandibular, and the chorda tympani branch of the nervus intermedius. It is not certain whether the anterior two-thirds of the tongue, thus derived, is ectodermal or endodermal.

Third Arch Component. From the endodermal surface of the third arch a swelling (the copula of His) grows forward, pushes the second arch ventrally, and unites directly with the mandibular arch components of the tongue. This forms the posterior third of the tongue, including the vallate papillæ, which are supplied by the nerve of the third arch, the glossopharyngeal. Note that the sulcus terminalis of the adult tongue is *not* the embryological junction of the two parts of the tongue. Between the two components of the tongue is the foramen cæcum of the adult. In the embryo the foramen is by no means blind; it leads into the thyro-glossal duct. The meeting of first and third arch components of the tongue obliterates the second arch endoderm and with it the sensory area of the facial nerve. A few sensory fibres of VII persist at the supratonsillar recess.

Musculature of the Tongue. The first and third arches contribute the mucous membrane, mucous glands, lymphatic and connective tissue to the tongue, but no muscle. The muscle has an extraordinary origin, viz., from three or four occipital myotomes immediately cranial to the first cervical segment. These migrate ventrally around both carotid arteries (but medial to the internal jugular vein), dragging behind them a trail of motor axons which make up the hypoglossal nerve.

Development of the Thyroid Gland. A simple downgrowth, by a process of evagination, leads from the foramen cæcum ventrally between the first and second arches then caudally in front of the remaining arches as far back as the commencement of the trachea. This is the **thyro-glossal duct.** From its distal extremity the bi-lobed thyroid gland grows out; a portion of the distal extremity often remains as the pyramidal lobe. The lateral lobes of the thyroid are firmly attached to the fourth pharyngeal pouch. It is probable that some of the thyroid gland develops from this source, especially the interstitial cells (see p. 369). Remnants of the thyroglossal duct may persist, and give rise to cysts. The course of the duct is shown in Fig. 1.51; note the

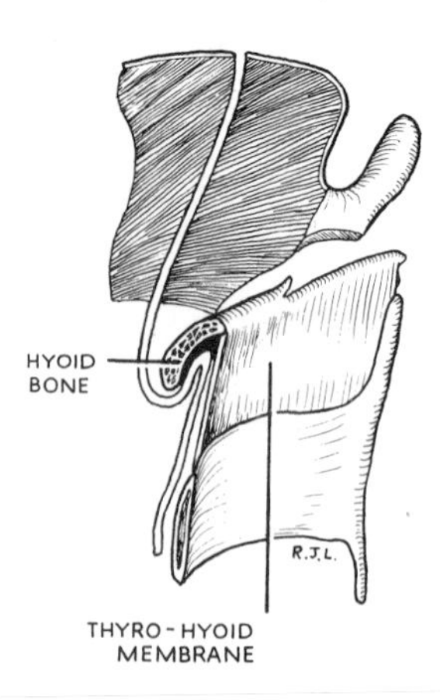

FIG. 1.51. THE COURSE OF THE THYRO-GLOSSAL DUCT. If a remnant of the duct persists it usually does so at the kink behind the hyoid bone.

apparent passage posterior to the hyoid which is so frequently described at operation. This is a common site of thyro-glossal cyst. The duct may, however, occasionally pass behind or, more rarely, through the hyoid bone.

Development of the Respiratory System. In the ventral wall of the œsophagus a groove appears. The cephalic end of this gutter is limited by the **furcula,** a ridge in the shape of a wishbone which lies in the caudal part of the floor of the pharynx (Fig. 1.48). The ridges which limit the gutter grow towards each other and, by their fusion, convert the gutter into a tube. This tube, the trachea, then separates from the œsophagus and buds out into the bronchi and lungs at its caudal end. The process of fusion does not, of course, affect the furcula, which persists as the aperture of the larynx whose cartilages, including that of the epiglottis, derive from the underlying fourth and sixth pharyngeal arches.

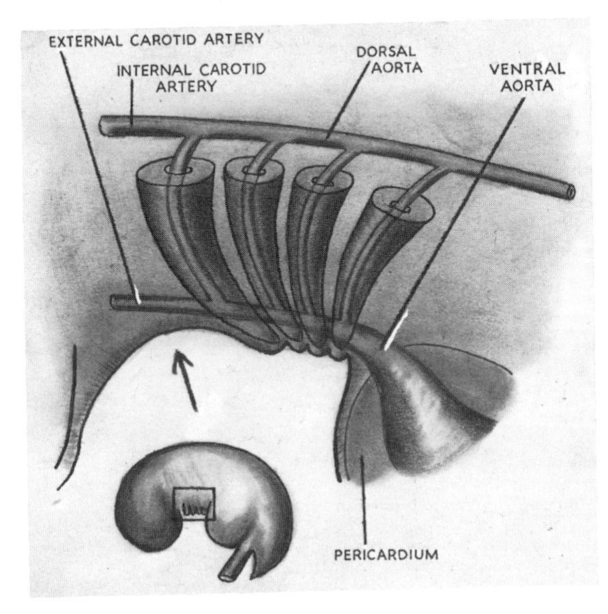

FIG. 1.52. THE FIRST FOUR PHARYNGEAL ARCH ARTERIES VIEWED FROM THE LEFT SIDE. The diagram is an enlargement of the rectangular area shown in the small figure; the arrow points to the stomodæum. Each pharyngeal arch contains an artery that passes from the ventral aorta to the dorsal aorta of its own side. The carotid arteries pass on to the head and neck.

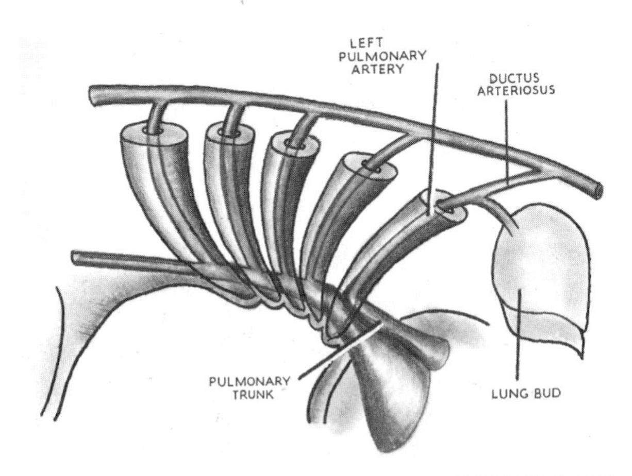

FIG. 1.53. THE ORIGIN OF THE SIXTH ARCH ARTERIES FROM THE PULMONARY TRUNK. The artery gives a branch to the lung bud, but most of its blood is short-circuited to the dorsal aorta by the ductus arteriosus.

The Pharyngeal Arch Arteries

From the primitive heart tube a ventral aorta divides right and left into two branches which curve back caudally as the two dorsal aortæ which are essentially continued into the two umbilical arteries. As the pharyngeal arches develop a vessel in each arch joins the ventral to the dorsal aortæ. Thus six aortic arches are to be accounted for (Fig. 1.52).

The first and second arch arteries disappear early,

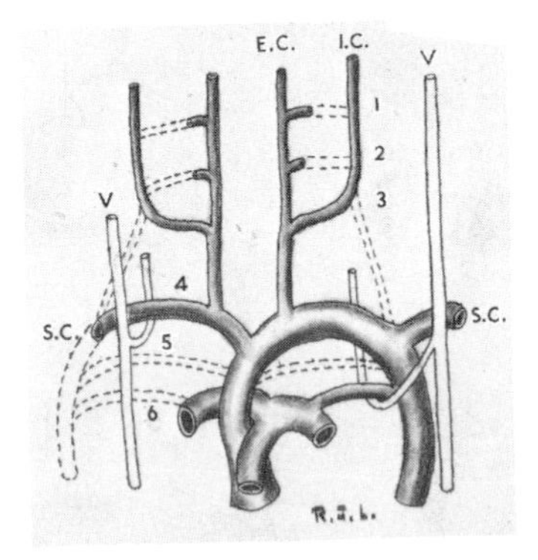

FIG. 1.54. THE FATE OF THE ARCH ARTERIES. Ventral view. E.C. External carotid artery. I.C. Internal carotid artery. S.C. Subclavian artery. V. Vagus nerve. The dotted lines indicate arteries that disappear. The numbers indicate the arch arteries. The embryological symmetry of the recurrent laryngeal nerves is replaced by final asymmetry; the left nerve hooks around the ductus arteriosus, the right nerve hooks around the subclavian artery.

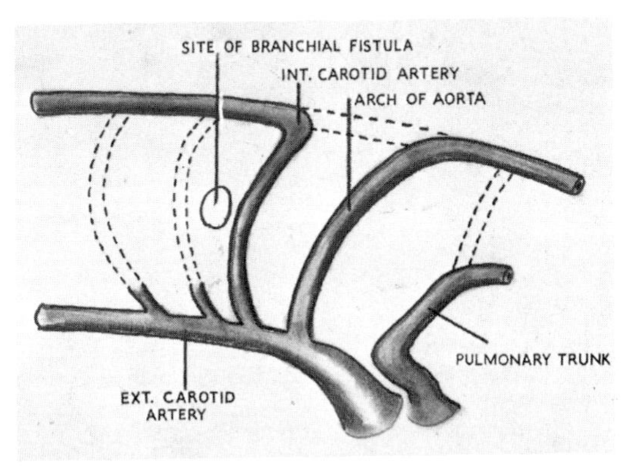

FIG. 1.55. A branchial fistula results from breakdown of a pharyngeal pouch. It is usually the second pouch, and then the fistula opens in the pharynx above the tonsil. The fistula always lies between the external and internal carotid arteries, and the sterno-mastoid muscle is always behind it, for the muscle migrates from a dorsal site. The arteries are viewed from the left side as in Fig. 1.53; the dotted lines indicate arteries that fail to persist.

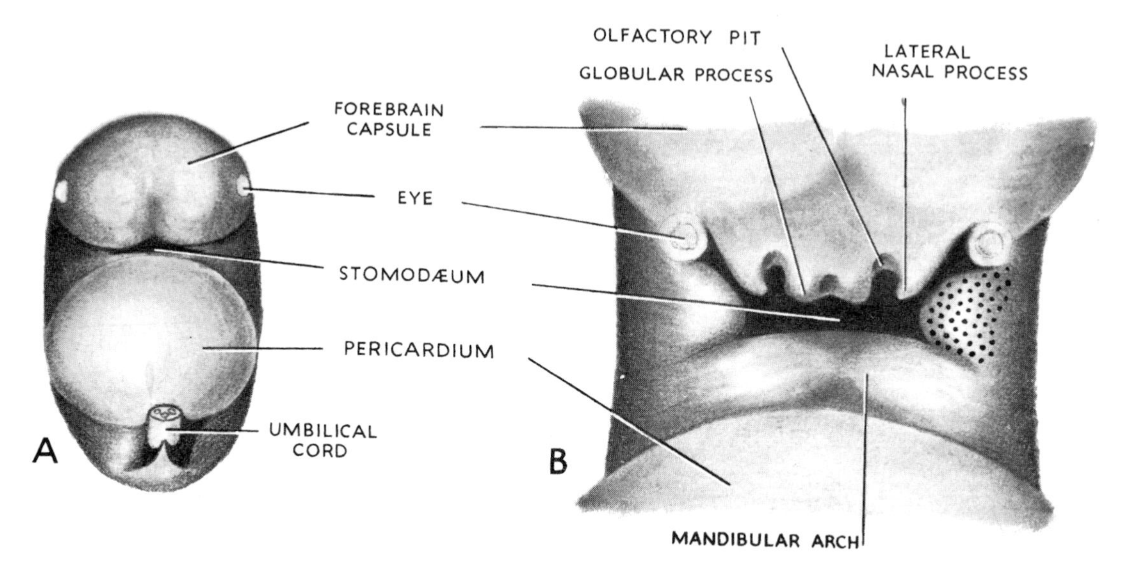

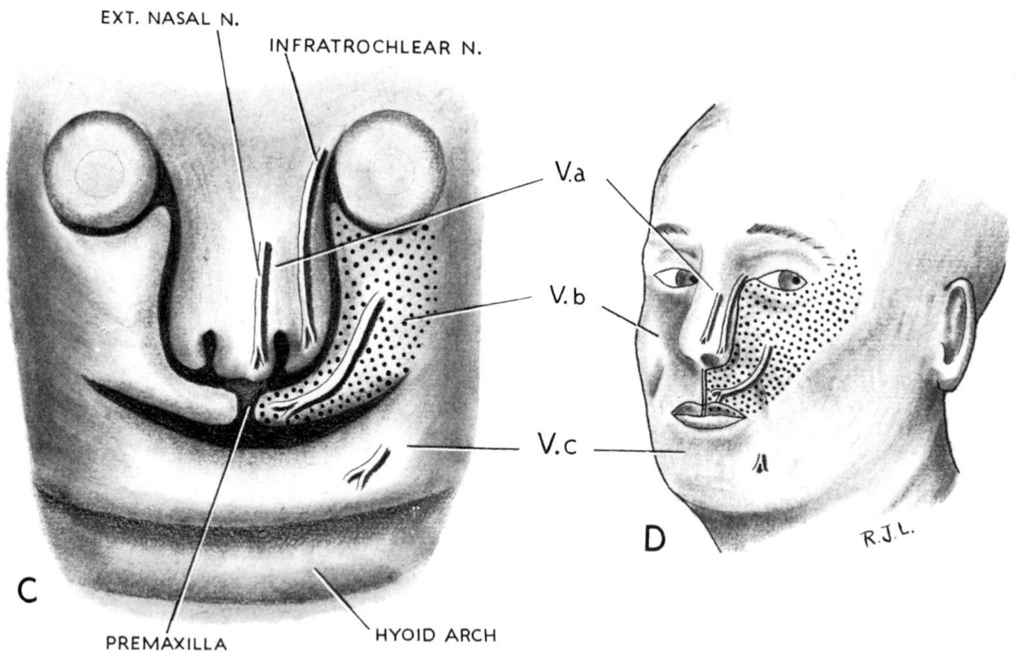

FIG. 1.56. STAGES IN THE DEVELOPMENT OF THE HUMAN FACE.

A. Ventral view of an early embryo, after the formation of the head fold. The slit-like stomodæum appears between the forebrain capsule and the pericardium.

B. Ventral view of the stomodæum at the next stage. From the forebrain capsule the fronto-nasal process grows down; it is divided by a pair of olfactory pits into a median and two lateral nasal processes. The median nasal process is divided by a midline pit into two globular processes. The stomodæum is separated from the pericardium by the mandibular arch, from each side of which a maxillary process grows (the left maxillary process is stippled in Figs. B, C and D).

C. The median nasal process produces the tip of the nose (supplied by the external nasal nerve). The lateral nasal process produces the ala of the nose (supplied by the infratrochlear nerve). The maxillary processes (supplied by the maxillary nerve) grow together across the premaxilla. DIAGRAMMATIC. No actual clefts exist in the normal embryo.

D. Final result. The whole of the upper lip, *including the philtrum*, is derived in man from the maxillary processes. The derivatives of the fronto-nasal process are supplied by V.a (ophthalmic nerve), those of the maxillary process by V.b (maxillary nerve) and those of the mandibular arch by V.c (mandibular nerve).

their only remnants in the adult being the maxillary and stapedial arteries respectively. The fifth disappears entirely. The third remains as part of the internal carotid artery. The fourth on the right persists as the subclavian artery, on the left as the arch of the aorta. By the time the sixth artery appears the primitive heart tube has been divided by the spiral septum into aorta and pulmonary trunk. It is to the pulmonary trunk that the sixth arch arteries are connected ventrally (Fig. 1.53). Dorsally they communicate with the dorsal aortæ. The pulmonary arteries grow out from the sixth arch. The dorsal part of the sixth arch artery disappears on the right side but persists on the left as the ductus arteriosus which thus connects the left pulmonary artery to the arch of the aorta (Fig. 1.54).

Formation of Mouth Parts

The stomodæum is present at the third week, when somites appear and the bucco-pharyngeal membrane breaks down. The fronto-nasal process and the beginning of mandibular and hyoid arches appear in the fourth week, when the embryo is 3 mm ($\frac{1}{8}$ in) long. The teeth begin as the dental lamina at the sixth week, when the embryo is 12 mm ($\frac{1}{2}$ in) long. Calcification of the first dentine, soon followed by the enamel, begins in the fifth month.

Development of the Face

The stomodæum is bounded by the mandibular arch, which produces the floor of the mouth, lower jaw and lower lip, and the two maxillary processes. Its cranial boundary is the forebrain capsule, from which the fronto-nasal process grows. The **fronto-nasal process** is indented by two **nasal pits,** which divide it into a **median** and two **lateral nasal processes.** The median process is characterized by a pair of convex **globular processes** (Fig. 1.56B). The lateral nasal processes unite with them to encircle the nostril. The maxillary and lateral nasal processes encircle the eye and meet together along the line of the naso-lacrimal duct. The maxillary processes in man unite in the midline below the nostril to produce the *whole of the upper lip* and the maxillæ. The fronto-nasal process produces the premaxilla, which emerges on the facial skeleton in lower animals but is covered over by medial extensions of the maxilla in man (Fig. 1.57).

The upper part of the face is completed by fusion of the fronto-nasal and maxillary processes in the above manner, but at first the face is only a mask, for behind it the nasal septum and palate have yet to be completed. The developing tongue lies against the floor of the cranium. A midline flange (the nasal septum) grows down from the base of the forebrain capsule (which is the mesenchymal precursor of the skull). From each maxillary process a flange known as the **palatal process** grows medially across the dorsum of the tongue. The two palatal processes and the nasal septum meet and unite from before backwards, thus separating the nasal cavities from each other and from the mouth. So is formed the **nasal capsule,** in mesoderm. Chondrification of the nasal capsule occurs, and by the sixth week the nasal walls and hard palate are outlined by a thin layer of hyaline cartilage. Ossification of the cartilaginous nasal capsule begins almost at once, spreading from several centres. In the upper part of the wall of the nose the hyaline cartilage is itself replaced by bone, but in the lower part of the septum and in the hard palate there is a deposition of membrane bone on each surface of the hyaline cartilage. The cartilage, thus sandwiched between two layers of membrane bone, is not absorbed until some time after birth.

The nerve supply of all these structures, *inside and outside,* is derived from the fifth cranial (trigeminal) nerve. The fronto-nasal process and its derivatives are supplied by V.a. (the ophthalmic division), the maxillary process and its derivatives by V.b. (the maxillary division) and the lower jaw by V.c. (the mandibular division).

Defects of Development. The commonest abnormalities are hare-lip and cleft palate, which may or may not co-exist. Hare-lip is almost always lateral; the cleft runs down from the nostril. The median part of the lip is derived in these cases from the opposite maxillary process or, perhaps, from the fronto-nasal process which latter normally does not produce any part of the upper lip. Hare-lip may be bilateral, in which case the central part of the lip, between the two clefts, is obviously derived, abnormally, from the fronto-nasal process.

Cleft Palate. This may be partial or complete. The two palatal processes unite with each other (and with the nasal septum) from before backwards. Arrest of union thus results in a posterior defect that varies from the mildest form of bifid uvula to a complete cleft from uvula to gum margin. In the latter case the cleft almost always runs between premaxilla and maxilla and involves the jaw between the lateral incisor and canine teeth. Irregular formations of incisor and canine teeth, however, often accompany these defects of

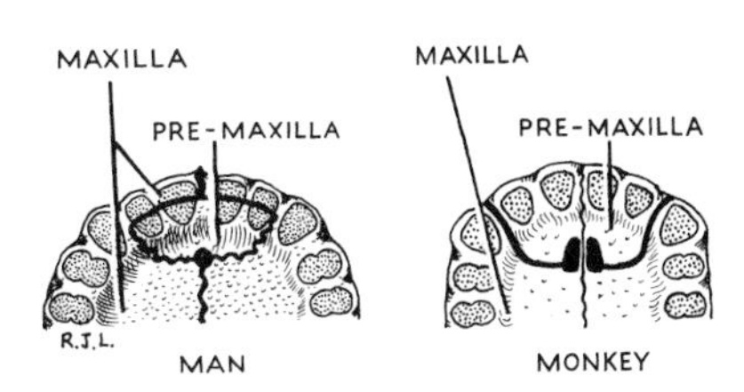

FIG. 1.57. THE HARD PALATE OF MAN COMPARED WITH THAT OF A MONKEY. In man the maxillæ meet across the premaxilla. In the monkey the premaxilla lies on the face. Drawn from specimens in the Hunterian Museum R.C.S.

palatal development. Very rarely a midline cleft may separate the two halves of the premaxilla.

A less common defect arises from the failure of fusion of the lateral nasal process with the maxillary process, producing a groove on the face along the line of the naso-lacrimal duct. Reduplication of various forms produces monsters, completely two-headed or less grotesque, while fusion results in such conditions as cyclopia; most museums display examples of these gross defects of development.

THE ANATOMY OF THE CHILD

The proportions of the newborn child differ markedly from the form of the adult. Some of its organs and structures are well developed and even of full adult size (e.g., the internal ear) while others have yet to develop (e.g., cerebro-spinal tracts to become myelinated, teeth to erupt, secondary sex characters to appear).

It is proposed to consider several features of a general nature, followed by accounts of certain special parts of the body, but before doing so it seems convenient to discuss first the circulatory changes which take place in the neonatus at the time of birth.

The Fœtal Circulation

The fœtal blood is oxygenated in the placenta, not in the lungs. The economy of the fœtal circulation is improved by three short-circuiting arrangements, all of which cease to function at the time of birth. The three short-circuiting structures are the ductus venosus, the foramen ovale and the ductus arteriosus.

The Ductus Venosus. Oxygenated blood returns from the placenta by the (left) umbilical vein, which joins the left branch of the portal vein in the porta hepatis. This oxygenated blood short-circuits the capillaries of the liver; it is conveyed directly to the inferior vena cava by a channel called the *ductus venosus*. The ductus venosus lies along the inferior surface of the liver, between the attached layers of the lesser omentum. After birth, when blood no longer flows along the thrombosed umbilical vein, the blood in the ductus venosus clots and the ductus venosus becomes converted into a fibrous cord, the *ligamentum venosum*, lying deep in the cleft bounding the caudate lobe of the liver. The intra-abdominal part of the umbilical vein persists as a fibrous cord, the *ligamentum teres*. The two are continuous.

The Foramen Ovale. The interatrial septum of the fœtal heart is patent, being perforated by the *foramen ovale*. Blood brought to the right atrium by the inferior vena cava is directed by its 'valve' through the aperture in the interatrial septum and so enters the left atrium. The oxygenated placental blood is thus made to by-pass the right ventricle and the airless lungs, and is directed into the left ventricle and aorta and so to the carotid arteries.

After birth this aperture, the foramen ovale, is closed by approximation and overlap of flanges of cardiac muscle which act from below and above like shutters. The two flanges are the septum primum and the septum secundum (see p. 238). The two flanges overlap and adhere together, so closing the interatrial septum. The site of union is marked by a shallow depression, the *fossa ovalis*, in the right side of the interatrial septum. After closure of the foramen ovale all the blood in the right atrium perforce passes into the right ventricle and so to the lungs.

The continuing patency of the interatrial septum until birth and its prompt closure after the first breath enters the lungs provide an admirable example of correlation between form and function.

The Ductus Arteriosus. It has already been noted that oxygenated blood in the umbilical vein passes via the ductus venosus, inferior vena cava and right atrium through the foramen ovale to the left side of the heart and so to the head. Venous blood from the head is returned by way of the brachio-cephalic veins to the superior vena cava. In the right atrium this venous blood stream crosses the stream of oxygenated blood brought there via the inferior vena cava. The two streams of blood scarcely mix with each other. The de-oxygenated blood from the superior vena cava passes through the right atrium into the right ventricle and so into the pulmonary trunk. It now short-circuits the airless lungs by the *ductus arteriosus*. This is a thick artery joining the left branch of the pulmonary trunk to the aorta, distal to the origin of the three branches of the aortic arch. The de-oxygenated blood thus passes distally along the aorta and via the umbilical arteries to the placenta to be re-oxygenated.

After birth the ductus arteriosus is occluded by contraction of its muscular walls. It persists as a fibrous band, the **ligamentum arteriosum,** which connects the commencement of the left pulmonary artery to the concavity of the arch of the aorta. After the closure of the ductus arteriosus blood from the right ventricle perforce circulates through the lungs. The ductus is caused to contract into closure by the stimulus, acting locally, of a raised oxygen tension.

General Features of the Newborn

Of all the mammalian newborn the human is perhaps the least graceful. Kittens and puppies, lambs and foals, and most other newborn mammals possess a beauty and grace in their playful movements that are lacking in the human young. Man is born in a more immature form and requires some months to reach the kitten or puppy stage.

In comparison with the adult the neonatus is much more fully developed at its head end than at its tail end. The large head and massive shoulders stand out in marked contrast to the smallish abdomen and poorly developed buttocks and lower limbs. The edentulous jaws and shallow maxillæ combine to produce a face short in its vertical extent, with the cheeks bulging forwards to accommodate their tissues. The bulging cheeks of the baby face are not entirely lost until eruption of the permanent teeth and the rapid increase in size of the maxillary sinuses between the ages of six and seven years.

The newborn baby has no visible neck; its lower jaw and chin touch its shoulders and thorax. Gradually the neck elongates and the chin loses contact with the chest. The head thus becomes more mobile, both in flexion-extension and in rotation.

The abdomen is not prominent at birth but becomes gradually more and more so. The 'pot-belly' of the young child is due mainly to the large liver and the small pelvis; the pelvic organs lie in the abdominal cavity. In later childhood the pelvic organs and much of the intestinal tract sink into the developing pelvic cavity and the rate of growth of the abdominal walls outpaces that of the liver. In this way the disposition of the viscera and the contour of the abdominal wall reach the adult proportions, and the bulging belly flattens.

The limbs are disproportionately developed. The upper limb is well-developed at birth, but its movements are ill-controlled and ataxic. Fingers can be flexed and hyperextended and there is a very powerful grasping reflex, so that the neonatus can hold itself suspended by the grip of its hand. The hand takes many months to become the chief tactile organ—until this time the lips are used for feeling, the hand functioning merely as the prehensile organ to convey objects to the mouth for examination. The pathetically developed buttocks and short legs give little indication of the form and function they will later attain.

Some Special Features of the Newborn

The Skull as a Whole. The most striking feature of the neonatal skull is the disproportion between cranium and facial skeleton; the cranium is very large in proportion to the face.

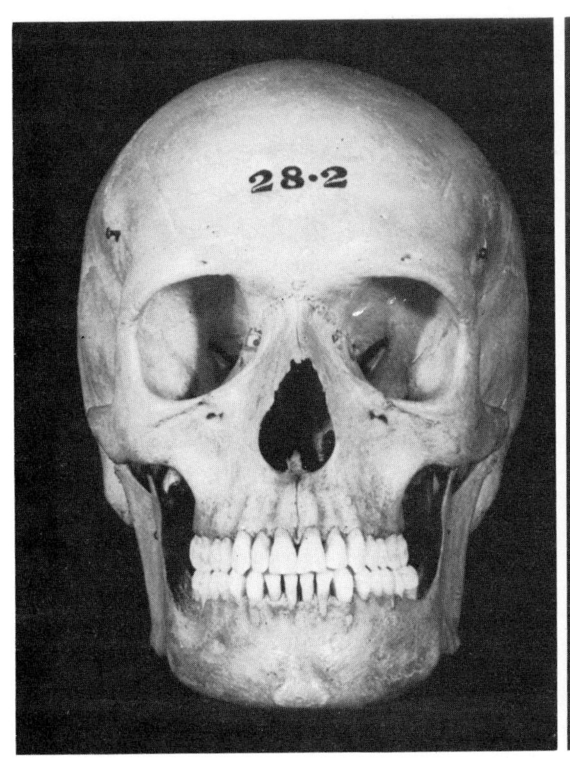

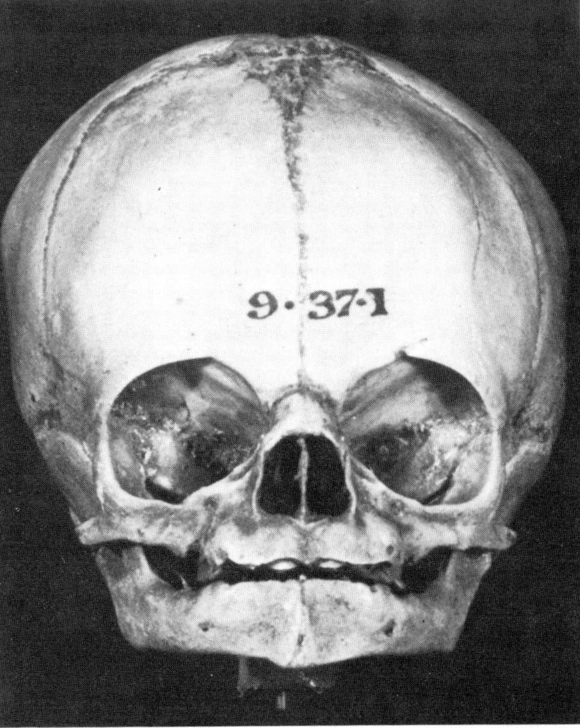

FIG. 1.58. NORMAL ADULT AND FŒTAL SKULLS (photograph). The fœtal skull is projected to the same vertical height as that of the adult.

Note the disproportion in the vertical extent of the face. In the adult the distance from the lower margin of the orbit to the lower border of the mandible is thrice the vertical diameter of the orbit. In the fœtal skull the distance from the lower margin of the orbit to the lower border of the mandible is equal to the vertical diameter of the orbit.

In Fig. 1.58 the photograph of a full-term fœtal skull has been enlarged to the same vertical projection as a normal adult skull and this procedure shows in striking manner the disproportion between the two. The rotundity of the fœtal skull in comparison with the adult is mainly the result of under-development of the face. The cranial skeletons in this frontal view are of much the same proportions. There is a very great difference in the facial skeletons. In the fœtal skull the vertical diameter of the orbit equals the vertical height of maxilla and mandible combined. In the adult skull the growth of the maxillary sinuses and the grow of alveolar bone around the permanent teeth has so elongated the face that the vertical diameter of the orbit is only one-third of the vertical height of maxilla and mandible combined.

The bones of the vault of the skull and face are developed by ossification in membrane, the base of the skull by ossification in cartilage. Most of the separate skull and face bones are ossified by the time of birth but they are mobile on each other and are fairly readily disarticulated in the macerated skull. Their mobility on each other is most marked in the vault and the ability to overlap provides that moulding of the cranium which so generally occurs during parturition.

The bones of the vault (frontal, parietal, occipital, squamous, etc.) do not interdigitate in sutures, as in the adult, but are separated by linear attachments of fibrous tissue and, at their corners, by larger areas known as fontanelles. The anterior and posterior fontanelles are the most readily examined, for they lie in the midline of the vault.

The **anterior fontanelle** lies between four bones. The two parietal bones bound it behind, the two halves of the frontal bone lie in front. The two halves of the frontal bone unite by the second year; the anterior fontanelle is closed by this time—clinically it is usually not palpable after the age of eighteen months. The extent of the anterior fontanelle is indicated in the adult skull by relatively straight parts of the coronal and sagittal sutures (p. 553).

The **posterior fontanelle** lies between the apex of the squamous part of the occipital bone and the posterior edges of the two parietal bones. It is closed at the end of the first year.

Growth of the vault of the skull takes place by deposition of bone around the edges of the separate bones, which come to interdigitate with each other along the various sutures. But in addition to this appositional growth there is interstitial growth in thickness, *accompanied regularly throughout by a moulding of each bone.* That is to say, absorption of inner surfaces of bone by osteoclasts accompanies the laying down of new external surfaces of bone by osteoblasts (Fig. 1.10, p. 11). An adult parietal bone, for example, is the surface of a larger sphere than that of the neonatal parietal bone, though some slight persistence of the infantile curvature is indicated at the parietal eminence. In the vault only compact bone is present at birth—with subsequent growth the interior of the bones becomes excavated into cancellous bone (the *diploë*), and red bone marrow fills the interstices therein.

The Temporal Bone and Ear

The temporal bone develops in four separate pieces, two in membrane and two in cartilage. The squamous and tympanic bones develop in membrane, the petro-mastoid and styloid process in cartilage.

The **squamous part** resembles the adult and increases in size by appositional growth combined with a continuous process of moulding.

The **tympanic part** is present at birth as the C-shaped *tympanic ring*, applied to the under surface of the petrous and squamous parts and enclosing the tympanic membrane, which is slotted into it. The external acoustic meatus of the newborn is wholly cartilaginous. The tympanic membrane is almost as big as in the adult, but faces more downwards and less outwards than the adult ear drum; it therefore seems somewhat smaller and lies more obliquely when viewed through the otoscope. The tympanic ring elongates by growth from the lateral rim of its whole circumference, the tympanic plate so produced forming the bony part of the external acoustic meatus and pushing the cartilaginous part of the meatus laterally, further from the ear drum. The adult bony meatus is thus twice as long as the cartilaginous part.

As the tympanic plate grows laterally from the tympanic ring the tympanic membrane tilts and comes to face rather more laterally and less downwards than in the neonatus. Growth of bone from the C-shaped tympanic ring is at first more rapid anteriorly and posteriorly, and less rapid inferiorly. The growing anterior and posterior bony flanges usually join and enclose an irregular foramen (of Huschke) between them, but subsequent growth of bone obliterates this gap.

The **petro-mastoid part** encloses the internal ear and tympanic antrum, all parts of which are *full adult size at birth.* The mastoid process is not developed and the stylo-mastoid foramen is near the lateral surface of the skull, covered by the thin fibres of sterno-mastoid —the issuing facial nerve is thus unprotected and vulnerable at birth. The mastoid process develops with growth of the sterno-mastoid muscle as the child commences to move its head, and air cells grow into it from the mastoid (tympanic) antrum. The mastoid process is aërated in this way at the beginning of the second year (the same time as the frontal sinus appears).

As already stated the cochlea, vestibule and semicircular canals are full adult size at birth. The middle ear is roofed in by a plate of bone, the tegmen tympani, which projects laterally from the petrous bone. At birth the tegmen tympani is not fully grown and it does not cover the geniculate ganglion of the facial nerve; the ganglion is in contact with the dura mater of the middle cranial fossa. The chorda tympani leaves the middle ear and emerges from the base of the skull between the tympanic ring and the under surface of the squamous bone. As the tympanic ring grows laterally and becomes the scroll-shaped tympanic plate the chorda tympani comes to emerge between it and the squamous part of the temporal bone—i.e., through the squamo-tympanic fissure. By this time the tegmen tympani of the petrous part of the temporal bone has grown across the geniculate ganglion and curved downwards to form the lateral wall of the canal for the tensor tympani muscle. Its growing edge peeps out from the medial part of the squamo-tympanic fissure, dividing that fissure into petro-squamous in front and petro-tympanic behind (Fig. 8.4, p. 559). The chorda tympani is caught behind the down-growing tegmen tympani and thus emerges from the base of the skull through the petro-tympanic fissure.

The mastoid antrum is full adult size at birth and is generally covered by some 3 mm of petrous bone deep to the floor of the suprameatal triangle. After ossification of the squamo-petrous suture new bone from the squamous 'flows down' over the developing mastoid process, and at the rate of 1 mm a year buries the mastoid antrum more deeply. This growth stops at twelve years; thereafter the antrum lies 15 mm from the surface. The squamo-mastoid 'suture' becomes an irregular line above the margins of the mastoid process (Fig. 6.19, p. 393).

The Face

The **maxilla,** between the floor of the orbit and the gum margin, is very shallow at birth and is full of developing teeth. The antrum (maxillary sinus) is a narrow slit excavated into its medial wall. Eruption of the deciduous teeth allows room for excavation of the antrum beneath the orbital plate but the maxilla grows slowly until the permanent teeth begin to erupt at six years. At this time it 'puts on a spurt' of growth. The rapid increase in size of the antrum and the growth of the alveolar bone occur simultaneously with increased depth of the mandible. These factors combine to produce a rapid elongation of the face.

The hard palate grows backwards to accommodate the extra teeth, but the forward growth of the base of the skull at the occipital sutures outstrips it and prevents the hard palate from approaching the cervical vertebræ. So the naso-pharyngeal isthmus is kept open, and in fact increases in size. After eruption of the first permanent molar (six years) the suture between basi-occiput and ex-occiput closes but forward growth of the base of the skull continues at the spheno-occipital cartilage and this latter does not ossify until well after the third molar is either erupted or fully formed (twenty-five years of age).

The paranasal sinuses are rudimentary at birth. Their development and growth are considered on p. 403. Growth of the face occurs by overall increase in size of all the face bones; most of the sutures slope downwards and backwards, so that growth at the sutures forces the face downwards and forwards away from the base of the skull.

The **mandible** is in two halves at birth; the fibrous joint (miscalled symphysis menti) between them ossifies at the end of the first year. The body of the mandible is practically 'full of teeth', so the mental foramen lies near its lower border. After eruption of the permanent teeth the foramen lies halfway between the upper and lower borders of the bone. In the edentulous jaw absorption of the alveolar margin leaves the mental foramen nearer the upper border of the mandible, so exposing the mental nerve to the pressure of an artificial denture (Fig. 1.59). Elongation of the body of the mandible is required to accommodate the erupting teeth and keep them in occlusion with the increasing number of maxillary teeth. Elongation of the ramus of the mandible is also required, for the jaws must separate to allow room for the eruption of the teeth. With growth in width of the face the mandible widens correspondingly. This overall growth of the mandible takes place, as in all bones of the body, by a process of moulding which results from the harmonious deposition of new bone and resorption of old bone where required. An 'epiphysis' at the neck of the mandible (composed of secondary cartilage) allows ready moulding of the condyle of the bone for accommodation with the changing size and direction of the articular cartilage and temporal bone.

These growth changes produce alterations in the contour of the mandible that are very characteristic (Fig. 1.59). At birth the angle is obtuse and the coronoid process lies at a higher level than the condyle; the latter lies in line with the upper border of the body. The resting mandible is depressed and the tongue lies between the edentulous gums, and the adult rest position of the mandible is not reached until the age of about 2 years. After eruption of the permanent teeth the angle is more nearly a right angle, and the well-developed condyle lies higher than the coronoid process. In the edentulous mouth the angle of the mandible does not alter very significantly, but gradual moulding of the neck ultimately depresses the condyle to a lower level than the coronoid process.

These age variations in the relative upward protrusion of coronoid process and condyle are much more apparent than real, for they are measured in relation

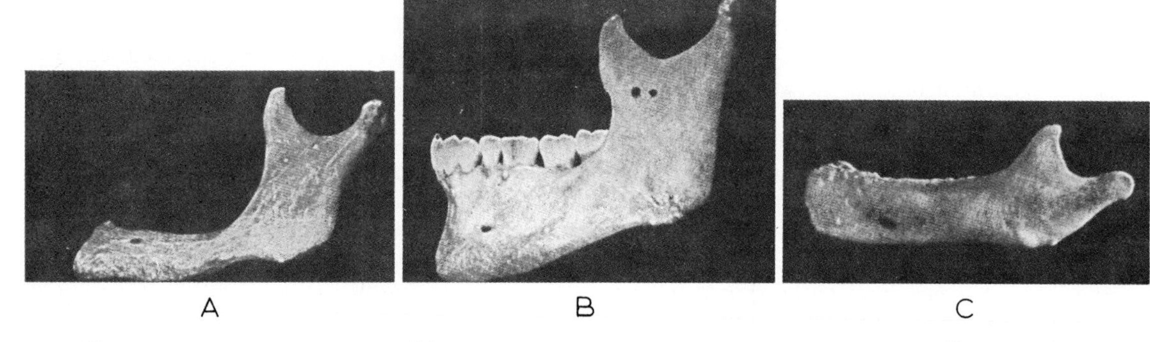

FIG. 1.59. THE NORMAL ADULT MANDIBLE (**B**) COMPARED WITH THE SENILE EDENTULOUS MANDIBLE (**A**) AND THE NEONATAL MANDIBLE (**C**). Note the relative levels of coronoid process and condyle. These are as seen on a museum shelf; that is, in the *occluded* position. Compare these levels in the *resting* position of Fig. 1.60. Note also the position of the mental foramen between upper and lower borders of the mandible.

to the working position of the mandible, with the occlusal surface horizontal (Fig. 1.59). In the resting position, however, the infantile and edentulous senile mandible are in each case more open than in the normal adult (Fig. 1.60), and this accounts for the apparent disproportion of the two bony processes.

The tongue is relatively large at birth and has a blunt tip that cannot be extruded. The newborn is 'tongue-tied' and only slowly does the tip of the tongue elongate. The hard palate is high and the orifice of the auditory tube lies at the same level; as the nasal septum grows in height the palate descends and leaves the tubal orifice above it in the naso-pharynx.

In common with lymphatic tissue elsewhere in the body the palatine and naso-pharyngeal tonsils (adenoids) tend to be exuberant in the child. Their relatively large size should never in itself be taken as evidence of infection.

The Neck

The newborn baby has no visible neck. The subsequent elongation of the neck is accompanied by growth changes in the covering skin; an incision over the lower neck in an infant usually results in the scar lying over the upper sternum by later childhood.

The left brachio-cephalic (innominate) vein crosses the trachea so high in the superior mediastinum that it may encroach above the sternal notch into the neck, especially if it is engorged and the head extended; this should be remembered by the surgeon performing tracheotomy upon the young child.

The shortness of the neck of the newborn involves a higher position of its viscera. The epiglottis and larynx lie relatively nearer the base of the tongue and their descent is slow; they reach their adult levels only after the seventh year. The larynx and trachea are of small bore at birth. The vocal cords are about 5 mm (0.2 in) long by the end of the first year. Laryngitis and tracheitis in infancy thus carry far more risk of respiratory obstruction than they do in later years. Up to the age of puberty there is no difference between the male and female larynx. At puberty the male larynx increases rapidly in size and the vocal cords elongate from 8 to 16 mm within the year, resulting in the characteristic 'breaking' of the voice. Castration or failure of testicular hormone prevents this change taking place.

The Thorax

The thoracic cage of the child differs from that of the adult in being more barrel-shaped. A cross section of the infant thorax is nearly circular; that of the adult is oval, the transverse being thrice the length of the antero-posterior diameter. The large thymus extends from the lower part of the neck through the superior into the anterior mediastinum. The thymus atrophies at puberty. The ribs lie more nearly horizontal, so the cage is set at a higher level than in the adult. This results in the shortness of the neck already noticed. The high thorax involves a higher level of the diaphragm, with consequent increase of abdominal volume. Descent of the thoracic cage as the ribs take up their adult obliquity is the chief cause of the elongation of the neck.

The Abdomen

At birth the liver is relatively twice as big as in the adult and its inferior border is palpable below the costal margin. The kidneys are always highly lobulated at birth and grooves on the surface of the adult organ frequently persist as visible signs of the original fœtal lobulation. The suprarenal is enormous at birth, nearly as large as the kidney itself (Fig. 1.42, p. 35). The cæcum is conical and the appendix arises from its apex in the fœtus, and this arrangement is usually still present at birth. During infancy and early childhood the lateral wall of the cæcum balloons out and the base of the appendix comes to lie posteriorly on the medial wall.

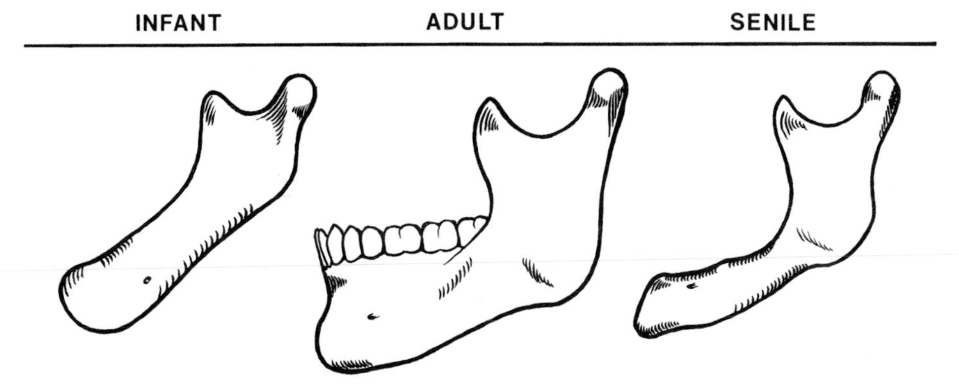

FIG. 1.60. THE AGE CHANGES IN THE RESTING POSITION OF THE MANDIBLE. The position of the mandibular joint and coronoid process does not change. (Tracing from Fig. 1.59; the infant mandible has been enlarged to adult proportions.)

The appendiceal mucous membrane is packed with massed lymphatic follicles in the child. These become much more sparse in later life. The pelvic cavity is very small at birth and the fundus of the bladder lies above the symphysis pubis (Fig. 5.12, p. 264).

The Upper Limb

As already noticed, the upper limb is more fully developed than the lower limb at birth. The grasping reflex of the hand is very pronounced. Growth in length occurs more at the shoulder and wrist than at the elbow. Amputation through the humerus in a young child requires a very generous flap of soft tissue lest the growing bone should later protrude through the stump.

The Lower Limb

At birth the lower limb is not only poorly developed, but occupies the fœtal position of flexion, a position which is maintained for six months or more. In preparation for standing and walking the limb not only becomes more robust, but undergoes an extension and medial rotation that carry the flexor compartment around to the posterior aspect of the limb. The inverted foot of the newborn gradually becomes everted harmoniously with these changes in position of the knee and hip joints. Growth of the limb proceeds more rapidly at the knee than at the hip or ankle. It is not symmetrical across the lower epiphysis of the femur, and 'knock-knee' (genu valgum) is normal in the child.

The Vertebral Column

Until birth the column is C-shaped. This is imposed by constriction *in utero*. After birth the column is so flexible that it readily takes on any curvature imposed by gravity. The cervical curve opens up into a ventral convexity when the infant holds up its head and the lumbar curve opens up into a ventral convexity when the infants walks. The extension of the hip that accompanies walking tilts the pelvis forwards, so that the axis of the pelvic cavity is no longer in line with that of the abdominal cavity. This forward tilt of the pelvis necessitates a high degree of forward curvature (lordosis) of the lumbar spine in order to keep the body vertical in the standing position.

The spinal cord extends to L3 at birth and does not 'rise' to L2 until adult years.

Section 2. The Upper Limb

General Plan. The upper limb of man is built for prehension. The hand is a grasping mechanism, with four fingers flexing against an opposed thumb. The hand is, furthermore, the main tactile organ. It is provided with a rich nerve supply, particularly of the flexor compartment. This is discussed on p. 24, and a comparison between upper and lower limbs is made on p. 126.

In grasping, the thumb is equal in functional value to the other four fingers; loss of the thumb is as disabling as loss of all four fingers. In order to be able to grasp in any position the forearm is provided with a range of 180 degrees of pronation and supination, and at the elbow has a range of flexion and extension of like amount. In addition very free mobility is provided at the shoulder joint, and this mobility is further increased by the mobility of the pectoral girdle through which the upper limb articulates with the axial skeleton.

On second (or subsequent) revision of upper limb anatomy there is much to be said for reversing the conventional order and studying the hand first and then proceeding proximally along the limb.

THE PECTORAL GIRDLE

Morphology of the Pectoral Girdle. It is a birthright of all vertebrates to possess four limbs. The limbs are connected to the axial skeleton by means of bones known as the pectoral and pelvic girdles. The pelvic girdle is formed by the two hip bones. It is firmly fixed to the vertebral column; this is a necessity, since the hind limb serves for propulsion of the animal. The **pectoral girdle,** on the other hand, does not articulate with the vertebral column. In many vertebrates it does not articulate with the axial skeleton at all; in these it consists only of a shoulder blade slung in muscle, the clavicle being absent or so rudimentary as to be functionless. The muscles act as shock absorbers, the weight of the bounding body being received on a fore limb that articulates with a very mobile shoulder blade. This arrangement is typical of the fleet-footed quadrupeds (horse, dog, cat, etc.); such creatures generally have no power of abduction of the fore limb. In other vertebrates the pectoral girdle articulates with the thoracic cage; these are generally animals that require a free abduction of the fore limb (e.g. for burrowing, swimming, climbing, flying). In these creatures the pectoral girdle is steadied and braced on the thorax during adduction by a bony strut, the clavicle. The rest of the girdle remains free to move about this strut.

The hip-bone, or os innominatum, consists of three parts, the ilium, ischium and pubis, which meet and fuse together at the acetabulum. There are likewise three parts of the shoulder blade. The shoulder blade is called the scapula in descriptive anatomy, but in morphological language the name scapula is restricted to the dorsal part of the bone (Fig. 2.9). The ventral part of the shoulder blade is called the coracoid bone; the two meet at an epiphyseal line that crosses the upper part of the glenoid fossa. Scapula proper and coracoid are the counterparts of ilium and ischium. The counterpart of the pubis is represented by a tiny piece of bone (the pre-coracoid) that ossifies separately at the tip of the coracoid process; it takes no part in the formation of the shoulder joint.

The clavicle, ossifying in membrane, is an added bone and has no counterpart in the os innominatum. The shoulder blade is attached to the clavicle by strong ligaments (coraco-clavicular) and the clavicle is attached to the first costal cartilage by a strong ligament (costo-clavicular). Forces from the upper limb are transmitted, by the clavicle, to the axial skeleton through these ligaments, and neither end of the clavicle normally transmits much force.

Embryology. Muscles developed in the upper limb are supplied by branches from the brachial plexus. During their development some of these muscles migrate for considerable distances to gain attachment to the trunk; but they continue to be innervated from the brachial plexus. Motor supply, once established in the embryo, never changes thereafter. An example is latissimus dorsi which, notwithstanding its very wide attachment to the trunk, still retains its supply from the posterior cord of the brachial plexus. Other muscles, not developed in the limb, migrate from the trunk and gain attachment to the girdle; e.g., sterno-mastoid and trapezius, supplied by the accessory nerve, attach themselves to clavicle and scapula. The nerve supply tells the story of the development of any muscle.

Movements of the Pectoral Girdle. The essential functional requirement of the pectoral girdle is mobility on the thorax, to enhance the mobility of the shoulder joint. And it is a fact that all movement between humerus and glenoid cavity, except very slight movement, is accompanied by an appropriate movement of the scapula itself. Nor can the scapula move without making its supporting strut, the clavicle, move also. Generally speaking the shoulder joint, the acromio-clavicular and sterno-clavicular joints all move together in harmony, providing a kind of 'thoraco-humeral articulation.' It is important to appreciate this, for functional defects in any part of the 'thoraco-humeral articulation' must impair the whole. But to understand the mechanisms involved, it is essential to analyse the composite movement into its constituent basic parts, and this is done with the gleno-humeral joint on p. 72, and with the clavicular joints on pp. 59 and 60.

The bones of the pectoral girdle are described on p. 111.

Muscle Attachments. The muscular attachments between pectoral girdle and trunk are direct and indirect.

Direct attachment of the pectoral girdle to the trunk is provided by muscles that are inserted into the clavicle or scapula from the axial skeleton. These muscles are pectoralis minor, subclavius, trapezius, the rhomboids, levator scapulæ and serratus anterior. Indirect attachment to the axial skeleton is secured by the great muscles of the axillary folds (pectoralis major and latissimus dorsi); these muscles, by way of the upper end of the humerus, move the pectoral girdle on the trunk. Their descriptions are included with the muscles of the pectoral girdle.

The muscular attachments between upper limb and pectoral girdle include the deltoid and short scapular muscles, which are inserted about the upper end of the humerus, and the biceps and long head of triceps which, running over the humerus, are inserted beyond the elbow joint into the bones of the forearm. All these muscles are important factors in giving stability to the very mobile shoulder joint across which they lie.

Muscles of the Pectoral Girdle

Pectoralis Major. From a wide sterno-clavicular origin the muscle converges on the upper humerus, folding on itself where it forms the anterior axillary wall. It is attached to the humerus by means of a trilaminar tendon. A **clavicular head** arises from the medial half of the clavicle over a smooth flattened area which is easily seen on most clavicles. Running almost horizontally laterally the fibres of this head lie in a groove on the manubrial part of the muscle, from which they are quite separate. They are inserted by the anterior lamina of the tendon into the lateral lip of the bicipital groove of the humerus, into the anterior lip of the deltoid

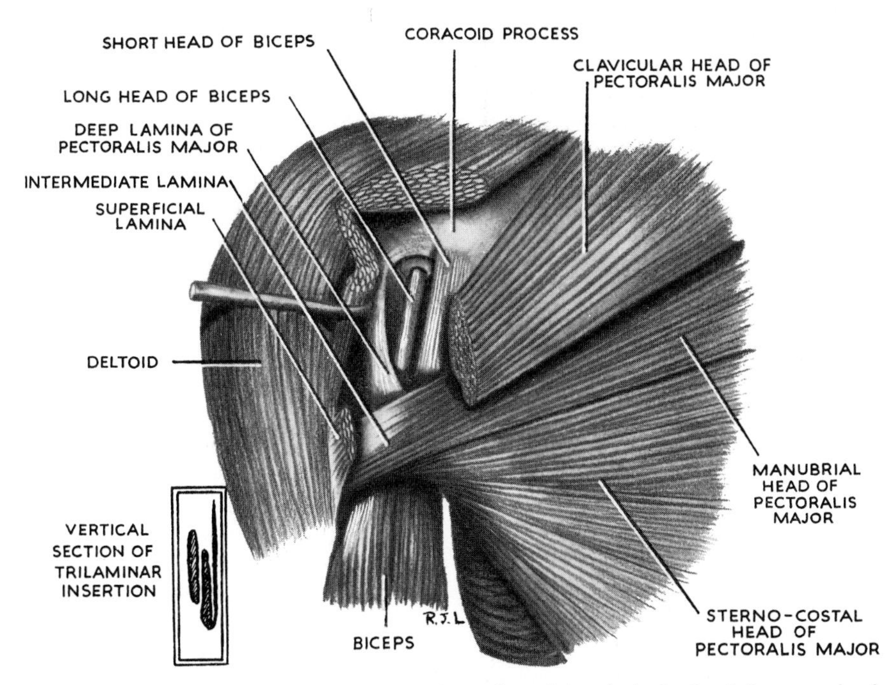

FIG. 2.1. THE TRILAMINAR INSERTION OF PECTORALIS MAJOR. A portion of the clavicular head (i.e., anterior lamina) has been removed. The deltoid muscle is incised and retracted to expose the underlying bilaminar tendon of the pectoralis major. In radical mastectomy the manubrial and sterno-costal heads and their bilaminar tendon can be excised while the clavicular head is left intact.

tuberosity and, below that, into the deep fascia of the arm.

The **sterno-costal head** arises from the lateral part of the anterior surface of the manubrium and body of sternum, and from the aponeurosis of the external oblique muscle over the upper attachment of rectus abdominis. On the deep surface of this sheet of muscle, and in continuity with it, fibres arise by a series of slips from the upper six costal cartilages. The manubrial fibres are inserted by tendon into the lateral lip of the bicipital groove behind (i.e., deep to) the clavicular fibres as far down as the upper part of the deltoid tuberosity. They form the intermediate lamina of the insertion. The sterno-costal fibres, arising from below the sternal angle, course upwards and laterally to be inserted progressively higher into the posterior lamina of the tendon, producing the rounded appearance of the anterior axillary fold. The fibres which arise lowest of all are thus inserted highest of all in the posterior leaf of the tendon. The uppermost limit of the insertion of the posterior leaf is, by a crescentic fold, into the capsule of the shoulder joint (Fig. 2.1). The anterior leaf is not inserted as high as this, its uppermost limit being on a level with the surgical neck of the humerus.

Morphology. The muscle is derived embryologically from upper limb myotomes, but morphologically it belongs to the outer layer of the three primitive layers of the body wall, represented in the abdomen by the external oblique (p. 211). One body in twenty shows on dissection the presence of vertical musculo-aponeurotic fibres on the surface of the pectoralis major alongside the sternum (Fig. 2.2). This is the **rectus sternalis muscle,** a derivative of the superficial layer of the rectus abdominis (see p. 211); it is supplied segmentally by intercostal nerves. Its upper fibres usually fuse with the sternal tendon of the sterno-mastoid muscle.

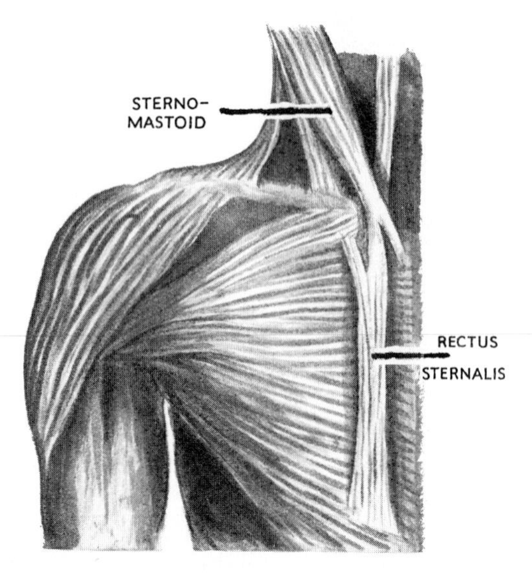

FIG. 2.2. THE RECTUS STERNALIS MUSCLE. Sketched from a dissection.

Nerve Supply. The muscle is supplied from the brachial plexus via the medial and lateral pectoral nerves, so named because of their origins from the medial and lateral cords of the plexus. The medial pectoral pierces pectoralis minor (and therefore supplies it) and the lateral pectoral pierces the clavi-pectoral fascia medial to the pectoralis minor. Both nerves enter the deep surface of pectoralis major. The relations of the two pectoral nerves are the reverse of their names as they enter the deep surface of the muscle. The segmental innervation of the muscle is from all five roots of the brachial plexus, but the essential segments are C6, 7 and 8 (p. 29). A separate branch of the lateral pectoral nerve (C6) supplies the clavicular fibres.

Action. The muscle is a medial rotator of the arm and, in combination with the muscles of the posterior axillary fold, particularly the latissimus dorsi, a powerful *adductor of the arm.* It is especially well developed in climbing and flying animals. The sterno-costal fibres are the chief adductors; the clavicular head of the muscle draws the arm forward into flexion at the shoulder joint. With the upper limb fixed in abduction the muscle is a useful accessory muscle of inspiration, drawing the ribs upwards towards the humerus.

Pectoralis Minor. Like pectoralis major this is a muscle derived from the outermost of the three sheets of the primitive body wall. In contradistinction to pectoralis major, whose costal fibres arise from cartilage, this muscle arises from bone, usually from the third, fourth and fifth ribs, by fleshy slips beneath pectoralis major. The muscle converges in triangular fashion towards the coracoid process. It is inserted by a short thick tendon into the medial border and upper surface of the coracoid process of the scapula (PLATE 1). Variations in its costal origin are common, the muscle being often prefixed and occasionally post-fixed (PLATE 26).

Nerve Supply. The pectoralis minor is supplied on its deep surface from the medial pectoral nerve, which pierces it to reach the overlying pectoralis major (PLATE 1).

Action. The function of the muscle is to assist serratus anterior in protraction of the scapula, keeping the anterior (i.e., glenoid) angle in apposition with the chest wall as the vertebral border is drawn forwards by serratus anterior. The muscle is elongated when the scapula rotates in full abduction of the arm; its subsequent contraction assists gravity in restoring the scapula to the rest position.

Subclavius. The muscle arises from the costo-chondral junction of the first rib and is inserted into the subclavian groove on the inferior surface of the clavicle, filling the groove completely. The muscle thus lies almost horizontally. It is enclosed by the upper attachment of the clavi-pectoral fascia (Fig. 2.3).

Nerve Supply. The nerve to subclavius from the brachial plexus, an anterior branch from the roots (C5 and 6), runs down from the point of formation of the upper trunk to enter the posterior surface of the muscle.

Action. To stabilize the clavicle in movements of the peripheral parts of the upper limb.

The **clavi-pectoral fascia** is a sheet of membrane filling in the space between clavicle and pectoralis minor, limited laterally by the coracoid process and passing medially to fuse with the external intercostal membrane of the upper two spaces. It splits above to enclose subclavius, being attached to the anterior and posterior ridges which limit the subclavian groove on the under surface of the clavicle. These two layers reappear above the clavicle as the lower attachment of the investing layer of deep cervical fascia (p. 362), but there is no physical continuity between them, each fascia being firmly attached to the clavicle above and below, not passing over the bone (Fig. 2.3).

At the lower border of subclavius the two layers fuse into a well-developed band, the **costo-coracoid ligament,** stretching from the knuckle of the coracoid to the first costo-chondral junction. From this ligament the fascia stretches as a loosely felted membrane to the upper border of pectoralis minor, which muscle is enclosed by the splitting of the membrane into anterior and posterior leaves. These leaves are very thin and amount to little more than the epimysium of the muscle. They rejoin below pectoralis minor to extend downwards as the **suspensory ligament of the axilla.** The latter is attached to the deep fascia over the floor of the axilla, and by its tension maintains the axillary hollow.

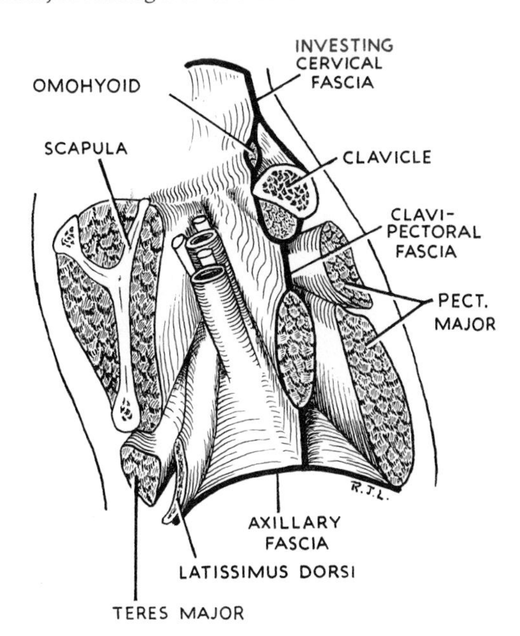

FIG. 2.3. VERTICAL SECTION THROUGH THE LEFT AXILLA LOOKING LATERALLY TOWARDS THE ARM, TO SHOW THE ATTACHMENTS OF THE CLAVI-PECTORAL AND CERVICAL FASCIÆ. The clavi-pectoral fascia encloses subclavius and is attached to the under surface of the clavicle; its two layers reappear above the clavicle in the investing layer of the deep cervical fascia. The clavi-pectoral fascia splits to enclose pectoralis minor, below which it reappears as the suspensory ligament of the axilla, attached to the axillary fascia. Between anterior and posterior axillary walls is the neuro-vascular bundle of the upper limb.

The clavi-pectoral fascia is almost covered by pectoralis major and the anterior fibres of the deltoid, but a small extent appears in the **infraclavicular fossa** in the interval between these two muscles. In this situation it is pierced by four structures—two passing inwards, two passing outwards. Passing inwards are lymphatics from the infraclavicular nodes to the apical nodes of the axilla, and the cephalic vein; passing outwards are the acromio-thoracic axis and the lateral pectoral nerve. The acromio-thoracic axis has four main branches of distribution: clavicular, humeral, acromial and pectoral. These frequently pierce the fascia separately; their corresponding veins, however, join the cephalic vein anterior to the fascia. This is a characteristic arrangement in anatomy (cf. branches of femoral artery below the groin, piercing fascia lata separately, veins joining great saphenous vein before latter pierces cribriform fascia, p. 132).

Trapezius. Like sterno-mastoid, omo-hyoid, and levator scapulæ, this muscle, though attached to the pectoral girdle, is not derived from myotomes of the limb bud. In other words, it is not supplied from the brachial plexus.

The muscle arises in the midline from skull to lower thorax and converges on the outer part of the pectoral girdle, which it rotates upwards. More precisely, its origin extends from the medial third of the superior nuchal line to the spine of the vertebra prominens (C7 or Th. 1), finding attachment to the ligamentum nuchæ between the external occipital protuberance and the vertebra prominens. Below this the origin extends along the spinous processes and supraspinous ligaments of all twelve thoracic vertebræ. Opposite the upper thoracic spines the muscle shows a triangular aponeurotic area, which makes a diamond with that of the opposite side (Fig. 2.6).

The occipital fibres are inserted into the lateral third of the clavicle at its posterior border; from above downwards the fibres can be traced into their insertion along the medial border of the acromion and the superior lip of the crest of the scapular spine. The part of the muscle which arises from the lowest half-dozen thoracic spinous processes is inserted by a narrow recurved tendon into the medial end of the spine at the 'deltoid' tubercle; this tendon slides over the bare area at the base of the spine of the scapula, a bursa intervening. The recurved insertion of trapezius, not the deltoid muscle, produces the 'deltoid' tubercle (Figs. 2.6, 2.16).

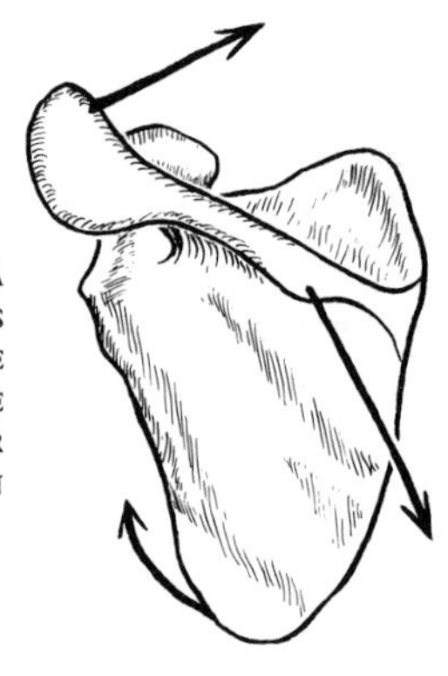

FIG. 2.4. ROTATION OF THE SCAPULA BY BOTH UPPER AND LOWER PARTS OF TRAPEZIUS ACTING ON THE 'BUTTERFLY NUT', ASSISTED BY THE FORWARD PULL OF THE LOWER FIBRES OF SERRATUS ANTERIOR ON THE INFERIOR ANGLE.

Nerve Supply. From the accessory nerve (its spinal roots only) and branches from the cervical plexus (C3 and 4), the latter being only proprioceptive (cf. sternomastoid, p. 365). These nerves cross the posterior triangle to enter the deep surface of trapezius.

Action. Rotation of the scapula, so that the glenoid fossa faces upwards, is produced by both upper and lower fibres. The upper fibres elevate the acromion, the lower depress the medial end of the scapular spine; each factor is complementary to the other in producing rotation of the scapula, like turning a butterfly nut (Fig. 2.4). This action is strongly assisted by the lowest four digitations of serratus anterior, which are inserted into the inferior angle of the scapula. The antagonists of the trapezius are the rhomboids and levator scapulæ, much weaker because gravity assists them. The muscles of the anterior and posterior axillary folds (pectoralis major and latissimus dorsi) should also be counted as opponents, since by adducting the arm they indirectly rotate the scapula back to its position of rest.

Latissimus Dorsi. This muscle, covering such a large area of the back, is characterized by its very wide origin and its very narrow insertion. It is a derivative of the upper limb myotomes, being supplied by a branch of the brachial plexus.

The origin commences above, at the spine of the seventh thoracic vertebra, and extends downwards along the spinous processes and supraspinous ligaments of all the lumbar and sacral vertebræ. Fleshy in the thoracic portion, the origin becomes aponeurotic in the lumbar and sacral region, and fuses with the posterior lamella of the lumbar fascia, by which lamella it also arises from the central ridge on the posterior part of the crest of the ilium (Fig. 2.5). Lateral to this it arises by flesh from the posterior part of the outer lip of the iliac crest. The upper border of the flat sheet of muscle runs horizontally, and is covered by the lower triangular part of trapezius, and flows over the inferior angle of the scapula, from which a few fibres arise to join the muscle (Fig. 2.6). The lateral border of the muscle, thicker and more rounded than its thin upper border, runs vertically upwards, being reinforced by four slips from the lowest four ribs, whose fibres of origin interdigitate with those of the external oblique in this situation. This lateral border of latissimus dorsi forms a boundary of the lumbar triangle (p. 256). The muscle converges towards the posterior axillary fold, of which it forms the lower border. The fibres, sweeping spirally around the lower border of teres major, are replaced by a narrow tendon of paper thinness which is inserted into an inch of the floor of the bicipital groove (Fig. 2.10). The surfaces of the muscle, anterior and posterior, are reversed at the insertion of the tendon as a result of the spiral turn around the teres major.

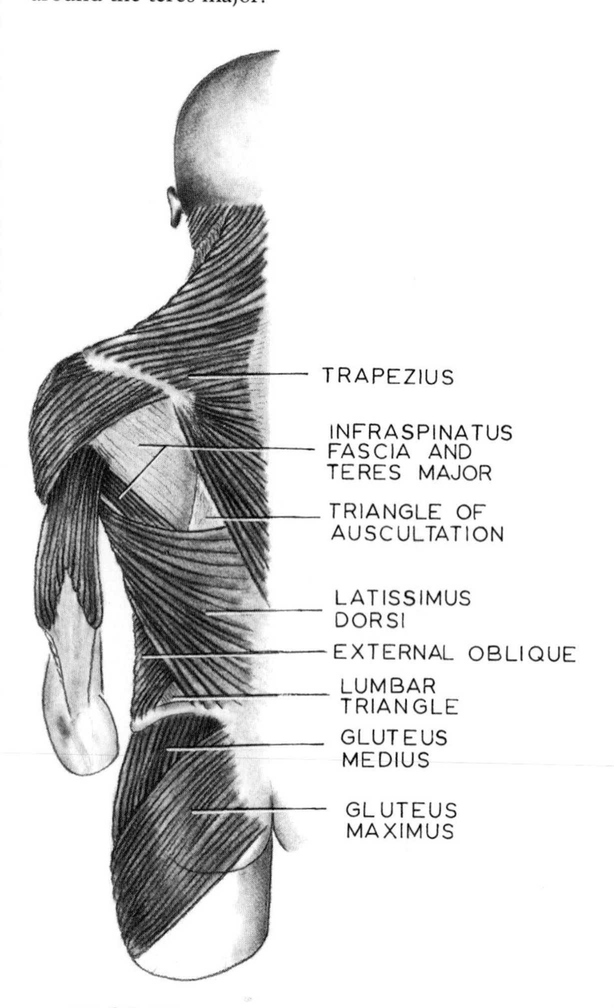

FIG. 2.5. THE MUSCLES ON THE BACK OF THE TRUNK.

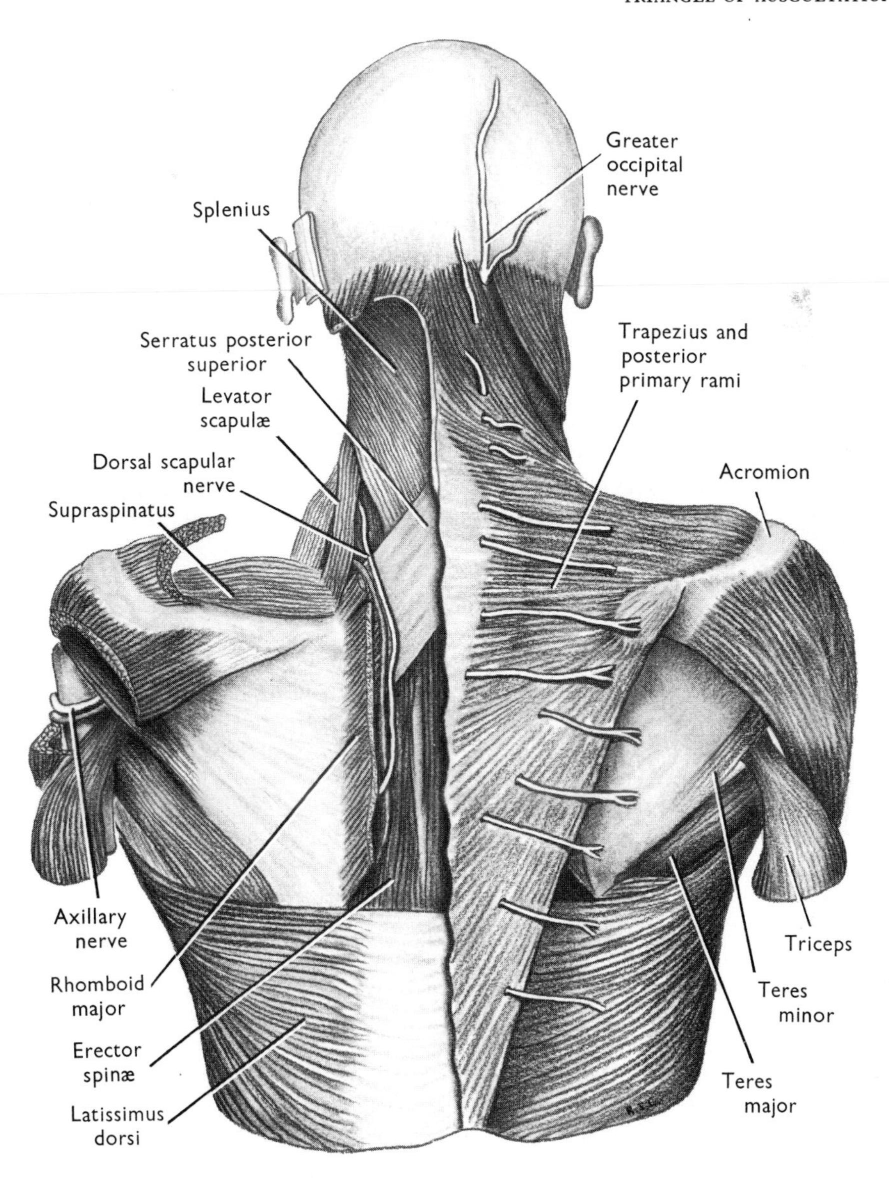

FIG. 2.6. THE PECTORAL GIRDLE MUSCLES (POSTERIOR VIEW). Drawn from Specimen S.350 in R.C.S. Museum.

Nerve Supply. The thoraco-dorsal (nerve to latissimus dorsi) is a branch of the posterior cord of the brachial plexus. It is vulnerable in operations on the axilla, for in its course down the posterior wall it slopes forwards to enter the medial surface of the muscle just behind its anterior border (PLATE 22). The latissimus dorsi is developed in the extensor compartment of the limb and migrates to its wide attachment on the trunk, taking its nerve with it; the segments are C6, 7 and 8.

Action. It extends the shoulder joint and medially rotates the humerus (e.g., folding the arms behind the back, or scratching the opposite scapula), but in combination with pectoralis major it is a powerful adductor. Especially used in restoring the upper limb from abduction above the shoulder, it is essentially the climbing muscle.

Its costal fibres of origin can act as accessory muscles of inspiration, elevating the lower four ribs towards the fixed humerus. But the remainder of the muscle, sweeping from the vertebral column around the convexity of the postero-lateral chest wall, compresses the lower thorax and is an accessory muscle of *expiration.* These fibres sometimes become sore after prolonged attacks of severe coughing.

Triangle of Auscultation. The upper horizontal border of the muscle forms with the vertebral border of the scapula and the lateral border of trapezius the triangle of auscultation (Fig. 2.5). A fascial sheet which encloses latissimus dorsi and the rhomboids floors in the triangle superficial to the seventh rib and sixth and seventh intercostal spaces. The triangle is so named because deep to it, on the left side, is the cardiac orifice of the stomach, where the splash of swallowed liquids was timed in cases of œsophageal obstruction in pre-Roentgen days.

Rhomboid Major. The muscle arises from four vertebral spines (Th. 2, 3, 4, 5), and the intervening supraspinous ligaments. Its insertion into the scapula (Fig. 2.6) extends from the inferior angle to the *upper* part of the triangular area at the base of the scapular spine; a fibrous arch receives the fibres of the muscle between these two points. The fibrous arch is often only loosely attached to the vertebral border of the scapula except at its ends (cf. white line of levator ani muscle).

Rhomboid Minor. This is a narrow ribbon of muscle parallel with the above, arising from two vertebral spines (C7 and Th. 1) and inserted into the medial border of the scapula above the triangular area and below the attachment of levator scapulæ (Fig. 2.16).

The Dorsal Scapular Nerve (nerve to the rhomboids). From the brachial plexus a posterior branch from C5 passes through scalenus medius, runs down deep to levator scapulæ (which it supplies) and lies on the serratus posterior superior muscle to the medial side of the descending branch of the transverse cervical artery (Fig. 2.6). It supplies each rhomboid on the deep surface.

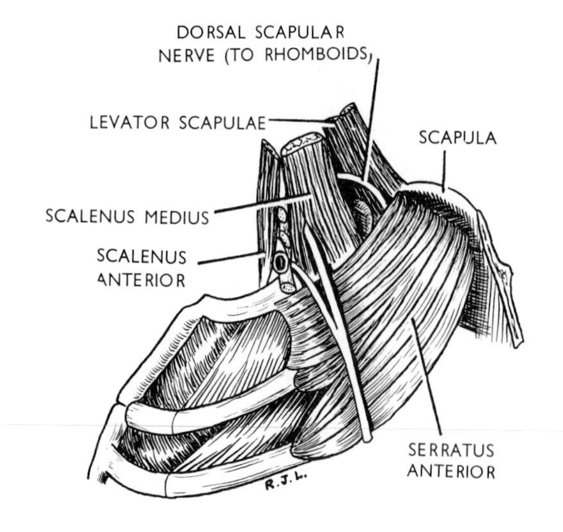

FIG. 2.7. THE LONG THORACIC NERVE (TO SERRATUS ANTERIOR). The branches from C5 and 6 fuse within scalenus medius and emerge as a single trunk which is joined in the axilla by the branch from C7; the junction takes place over the first digitation of serratus anterior. The second rib gives origin to half the first digitation and all the second digitation of serratus anterior. Drawn from a dissection.

Action of Rhomboids. They draw the vertebral border of the scapula medially and upwards. They are antagonists to the rotatory action of trapezius; they contract with trapezius in squaring the shoulders., i.e. retracting the scapula.

Levator Scapulæ. A strap-like muscle which appears in the floor of the posterior triangle, it arises from the posterior tubercles (i.e. from true transverse processes) of the upper four cervical vertebræ by slender slips of tendon. The four slips fuse into the long belly, which usually is longitudinally split. It courses downwards to be inserted into the upper angle of the scapula. It is not derived from the upper limb myotomes, but from the external of the three layers of the body wall (p. 211).

Nerve Supply. From the cervical plexus (C3 and 4, anterior primary rami), reinforced by the dorsal scapular nerve (nerve to the rhomboids) (C5).

Action. Same as rhomboid muscles.

The **serratus anterior** is a broad sheet of thick muscle which clothes the side wall of the thorax. It forms the medial wall of the axilla. It arises by a series of digitations from the upper eight ribs. The first digitation, *which appears in the posterior triangle*, arises from the outer border of the first rib and also from the rough impression which characterizes the second rib. Over its upper border passes the neuro-vascular bundle from the posterior triangle to the axilla. The second digitation arises from the second rib and is inserted, together with the first, into the upper angle of the scapula. This part of the muscle is innervated from C5. The third and fourth digitations arise from the third and fourth ribs and form a thin sheet of muscle which spreads out to be inserted into the length of the costal surface of the scapula along a narrow strip at its vertebral border. This part of the muscle is innervated from C6. The lower four digitations arise from the fifth, sixth, seventh and eighth ribs, interdigitating with the slips of origin of external oblique at the anterior angles of the ribs. Thick, fleshy muscles, they converge strongly on the inferior angle of the scapula. They are supplied by C7 (Fig. 2.11). The muscle is covered by a strong well-developed fascia.

Nerve Supply. The long thoracic nerve (nerve to serratus anterior) arises from the roots of the brachial plexus (C5, 6, 7). The branches from C5 and 6 join in the scalenus medius muscle and emerge from its lateral border as a single trunk which enters the axilla by passing over the first digitation of serratus anterior. The contribution from C7 also passes over the first digitation of serratus anterior and joins the former nerve on the medial wall of the axilla (i.e., on the surface of serratus anterior) to form the nerve to serratus anterior (Fig. 2.7). The nerve lies behind the midaxillary line (i.e., behind the lateral branches of the intercostal arteries) on the surface of the muscle, deep to the fascia, and is thus protected in operations on the axilla. The muscle is supplied segmentally; C5 into the upper two digitations, C6 into the next two, and C7 into the lower four digitations (PLATE 1).

Action. The whole muscle contracting *en masse* protracts the scapula (punching and pushing), thus effectively elongating the upper limb. A further important action is of the lower four digitations, which powerfully assist the trapezius in rotating the scapula laterally and upwards in raising the arm above the level of the shoulder. In all positions the muscle keeps the vertebral border of the scapula in firm apposition with the chest wall. Paralysis results in 'winged scapula.'

Joints of the Pectoral Girdle

The **sterno-clavicular joint** (Fig. 2.8) is a synovial joint, separated into two cavities by an intervening fibro-cartilage. The articular surface on the manubrium sterni is set at an angle of 45 degrees with the perpendicular, and is markedly concave from above downwards. It is covered with hyaline cartilage. The articular surface on the sternal end of the clavicle, flattened or slightly concave, is continued over the inferior surface of the shaft in a high percentage of cases, for articulation with the first costal cartilage. It is covered with dense fibrous tissue (known as fibro-cartilage) identical in structure with the disc. A capsule invests the articular surfaces like a sleeve. To this capsule the fibro-cartilage is attached, thus dividing the joint into two separate cavities. Rarely the fibro-cartilage is perforated. The disc is attached also above to the medial end of the clavicle and below to the manubrium, as though to restrain the sternal end of the clavicle from tilting upwards as the weight of the arm depresses the acromial end.

Ligaments. The *capsule* is thickened above and behind. Its weakest part is anteriorly, where infection tends to point. Anterior dislocation is more common than posterior, though both are rare.

Accessory ligaments strengthen the joint. The *interclavicular ligament* joins the upper borders of the sternal ends of the two clavicles and is itself firmly attached to the suprasternal notch. The *costo-clavicular (rhomboid) ligament* binds the clavicle to the first costal cartilage just lateral to the joint; it is an accessory ligament thereof. It is in two laminæ, usually separated by a bursa, which are attached to the anterior and posterior lips of the rhomboid impression on the clavicle. The fibres of the anterior lamina run upwards and laterally, of the posterior lamina upwards and medially (these are the same directions as those of the external

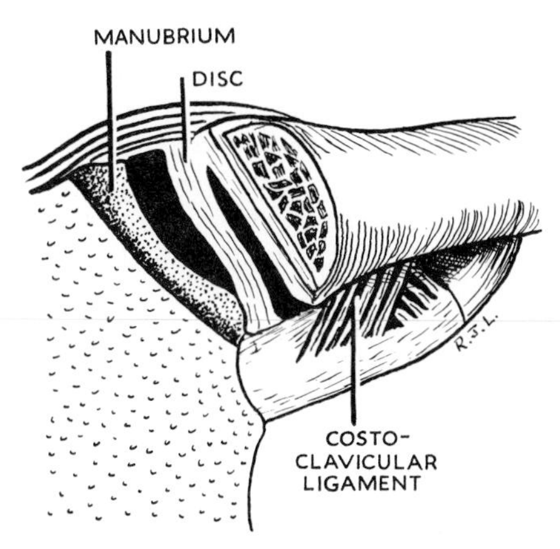

FIG. 2.8. THE LEFT STERNO-CLAVICULAR JOINT OPENED FROM THE FRONT.

and internal intercostal muscles). The ligament is very strong (Fig. 2.8). *It is the major stabilizing factor of the sterno-clavicular joint.*

Nerve Supply. The medial supraclavicular nerves (C3 and 4) from the cervical plexus give articular branches to the capsule and ligaments.

Movements. The fulcrum of movements at this joint is *not the sternal end of the clavicle*, but the rhomboid ligament. As the lateral end of the clavicle moves, its medial end moves in the opposite direction. This can be readily demonstrated by simple palpation (feel your own!). Elevate the acromial end by shrugging the shoulder; the sternal end moves down. Only in complete elevation of the acromial end, as in full abduction of the arm, when the *medial* end of the clavicle can be depressed no further, is the rhomboid ligament put on full stretch. Depress the acromial end of the clavicle by drooping the shoulder; the sternal end moves up. Upward movement of the sternal end is halted by the interclavicular ligament, and especially by the intra-articular disc (Fig. 2.8). Protrude the acromial end of the clavicle by hunching the shoulders forward; the sternal end moves back. Retract the acromial end by squaring the shoulders; the sternal end moves forward. The clavicle moves about the rhomboid ligament like a see-saw in both the horizontal and the coronal planes.

A further movement takes place at the sterno-clavicular joint, namely rotation. Rotation of the clavicle is passive; there are no rotator muscles. It is produced by rotation of the scapula and transmitted to the clavicle through the coraco-clavicular ligaments (see below). Palpate the forward convexity of the clavicle. Flex the arm in the sagittal plane and continue upwards in this plane to full abduction above the head. Restore the arm and carry it back in the sagittal plane into full extension. The rotation of the clavicle can be easily felt—it amounts, in fact, to some 40 degrees.

Elevation and depression of the acromial end of the clavicle, resulting in movements downwards and upwards respectively of the sternal end, cause movement between the clavicle and its fibro-cartilage. Forward and backward movements in the horizontal plane result in movements between the fibro-cartilage and manubrium, the former moving with the sternal end of the clavicle. Similarly, in rotary movements (abduction of the arm above the head) the fibro-cartilage moves with the clavicle.

The *stability* of the joint is maintained by the ligaments, most especially the rhomboid ligament. It takes all strain off the joint, transmitting stress from clavicle to first costal cartilage. The latter is itself immovably fixed to the manubrium by a primary cartilaginous joint (see p. 213).

The **acromio-clavicular joint** is a synovial joint between the overhanging lateral end of the clavicle and the underhanging medial border of the acromion. A fibro-cartilage, usually incomplete, crosses the joint cavity. A sleeve of *capsule* surrounds the articular surfaces; its fibres are not strong.

Accessory Ligaments. The **coraco-clavicular ligament,** extremely strong, is the principal factor in providing stability to the joint. It consists of two parts, conoid and trapezoid (Fig. 2.9). Examine the clavicle and scapula. The *conoid ligament*, an inverted cone, extends upwards from the knuckle of the coracoid to a wider attachment around the conoid tubercle, on the under surface of the clavicle. The *trapezoid ligament* is attached to the ridge of the same name on the upper surface of the coracoid process and extends laterally, in an almost *horizontal plane*, to the trapezoid ridge on the under surface of the clavicle (Figs. 2.9 and 2.58, p. 111).

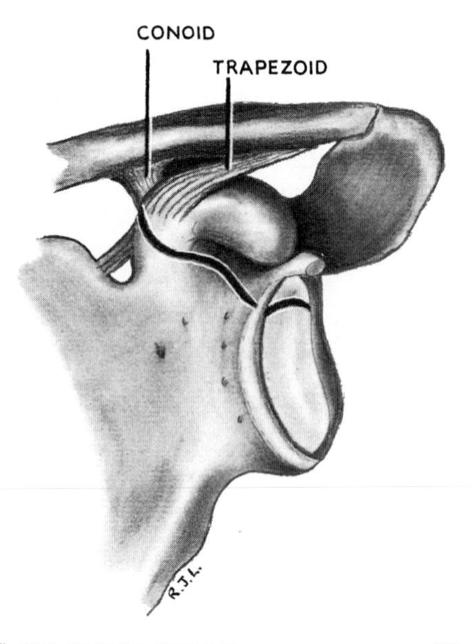

FIG. 2.9. THE GLENOID ANGLE OF THE LEFT SCAPULA. The two parts (conoid and trapezoid) of the coraco-clavicular ligament are shown. The black line marks the epiphysis between the dorsal scapula proper and the ventral coracoid bone (p. 52). The labrum glenoidale and long head of biceps are *in situ*. Sketched from a dissection.

Nerve Supply. Lateral supraclavicular nerves (C4) from the cervical plexus.

Movements. These are passive (cf. superior tibio-fibular joint, p. 168); no muscle connects the bones to move the joint. Muscles which move the scapula cause it to move on the clavicle. Scapular movements on the chest wall can be analysed into three basic ones. These are (1) protraction and retraction around the chest wall, (2) rotation, and (3) elevation or depression. These basic movements can be combined in varying proportions, and each of these transmits, through ligaments, *corresponding movements to the clavicle.* All movements of the scapula involve movements in the joint at either end of the clavicle.

Horizontally, in protraction and retraction of the tip of the shoulder, the scapula hugs the thoracic wall, held to it by serratus anterior and pectoralis minor. The scapula moves in a circle of a shorter radius (*viz.* the radius of the upper thorax) than the length of the clavicle. Hence movement takes place between the acromion and the fibro-cartilage. The axis of this movement is vertical and passes through the conoid ligament. The acromium glides to and fro on the tip of the clavicle.

In abduction of the arm the scapula does not retain its position relative to the clavicle but rotates around the conoid ligament as it swings forwards on the chest wall, and movement takes place between the fibro-cartilage and the clavicle. The axis of scapular rotation passes through the conoid ligament and the acromio-clavicular joint; the scapula swings to and fro like a pendulum below these two fixed points on the clavicle. The total range of scapular rotation *on the chest wall* is about 60 degrees, but only 20 degrees of this occurs between the scapula and the clavicle. The coraco-clavicular ligaments are then taut, and transmit the rotating force to the clavicle, whose rotation then accounts for the remainder of scapular rotation on the chest wall. In both movements at this joint the fulcrum around which the scapula swings is the coraco-clavicular ligament.

Elevation (shrugging the shoulders) is produced by the upper fibres of trapezius together with levator scapulæ and the rhomboids, mutually neutralizing their rotatory effects. Depression of the scapula is produced by the lower fibres of trapezius and the lateral fibres of latissimus dorsi. Elevation and depression move the medial end of the clavicle (*see above*) but they scarcely move the acromio-clavicular joint.

The *stability* of the joint is provided by the coraco-clavicular ligament. The scapula and upper limb hang suspended from the clavicle by the conoid ligament (assisted by the deltoid, biceps and triceps muscles). Forces transmitted medially from the upper limb to the glenoid fossa are transmitted from scapula to clavicle by the trapezoid ligament and from clavicle to first rib by the rhomboid ligament. Thus a fall on outstretched hand or elbow puts no strain on either end of the clavicle at the joints. If the clavicle fractures as a result, it always does so between these ligaments.

THE AXILLA

The axilla is the space between the upper arm and the side of the thorax, bounded in front and behind by the axillary folds, communicating above with the posterior triangle of the neck and containing neurovascular structures, including lymph nodes, for the upper limb and the side wall of the thorax. Its *floor* is the deep fascia extending from the fascia over the serratus anterior to the deep fascia of the arm, attached in front and behind to the margins of the axillary folds, and supported by the suspensory ligament (Fig. 2.3). Its *anterior wall* is completed by pectoralis major, pectoralis minor, subclavius and the clavi-pectoral fascia. The *posterior wall* extends lower; it is formed by subscapularis and teres major, with the tendon of latissimus dorsi winding around the latter muscle. The *medial wall* is formed by serratus anterior. The anterior and posterior walls converge laterally to the lips of the bicipital groove of the humerus (Fig. 2.3).

The apex is bounded by clavicle, scapula and the outer border of the first rib; it is the channel of communication between axilla and posterior triangle.

The axilla virtually does not exist when the arm is fully abducted. Its folds disappear as their muscles run tangentially along the humerus, and the axillary hollow is replaced by a bulge.

The **subscapularis muscle** arises from the medial four-fifths of the costal surface of the scapula, from the intermuscular septa which raise bony ridges on this surface and from the concave axillary border of the scapula up to the infraglenoid tubercle. The multipennate muscle converges over the bare area at the glenoid angle of the scapula, its musculo-tendinous fibres sliding over this bone and separated from it by a bursa. This *subscapular bursa* extends laterally and communicates with the cavity of the shoulder joint through a gap in the anterior part of its capsule. Lateral to this the tendon fuses with the capsule of the shoulder joint and is inserted into the lesser tuberosity of the humerus and the medial lip of its bicipital groove for half an inch below (Figs. 2.10, 2.11). The muscle is covered by a dense fascia which is attached to the scapula at the margins of its origin.

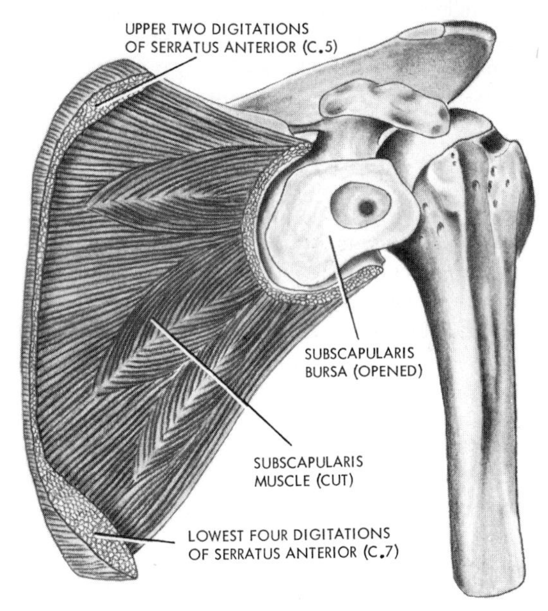

FIG. 2.11. THE LEFT SUBSCAPULARIS BURSA. The bursa is exposed by removal of the overlying subscapularis fibres. The anterior wall of the bursa is opened to expose the communication with the shoulder joint. The insertion of serratus anterior is in position, and the segmental innervation of the muscle fibres is shown in brackets. Compare Fig. 2.10.

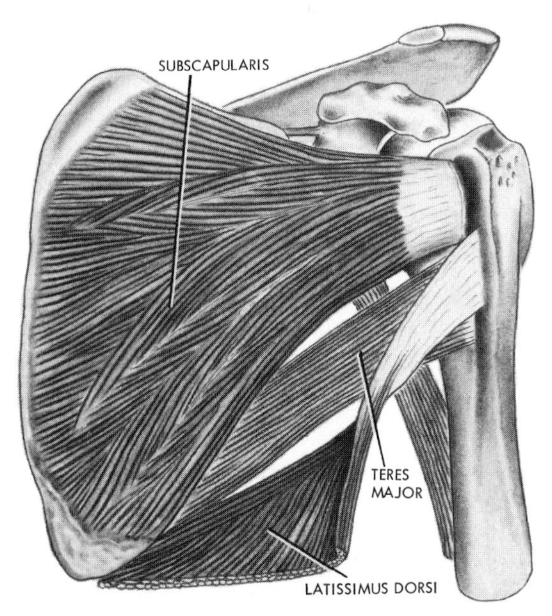

FIG. 2.10. THE MUSCLES OF THE POSTERIOR WALL OF THE LEFT AXILLA (ANTERIOR VIEW). The long head of triceps passes behind teres major and makes three spaces, one quadrilateral and the other two triangular. Compare Fig. 2.16.

Nerve Supply. The upper and lower subscapular nerves (C6 and 7) from the posterior cord of the brachial plexus.

Action. With the other short scapular muscles the subscapularis gives stability to the shoulder joint, assisting in fixation of the upper end of the humerus during movements of elbow, wrist, and hand. Acting as a prime mover, it is a medial rotator of the humerus.

Teres Major Muscle. This muscle is an offspring of subscapularis. It has migrated around to the dorsal surface of the scapula, where it arises from an oval area on the lateral side of the inferior angle. It runs edge to edge with the lower border of its parent subscapularis, and is inserted in continuity with it into the medial lip of the bicipital groove. The ribbon-like tendon of latissimus dorsi winds around its lower border and comes to lie in front of the upper part of the muscle at its insertion (PLATE 22).

Nerve Supply. The lower subscapular (C6 and 7), one of the nerves which supply its parent subscapularis, lies in the angle between subscapular and circumflex scapular arteries, and enters the anterior surface of the muscle.

Action. It assists the other short muscles in steadying the upper end of the humerus in movements at the periphery of the limb; acting alone it is an adductor and medial rotator of the humerus at the shoulder joint. With teres minor it holds down the upper end of the humerus as deltoid pulls up the bone into abduction (Fig. 2.21).

The **quadrilateral space** lies between subscapularis and teres major in the posterior wall of the axilla. It is bounded laterally by the humerus and medially by the long head of triceps (Fig. 2.10). It transmits the axillary nerve, with the posterior circumflex humeral artery and vein inferiorly. Viewed from behind, the quadrilateral space is bounded above by the teres minor muscle; its three other boundaries are the same (Fig. 2.16). Two *triangular intermuscular spaces* are seen. The one, medial to the quadrilateral space, transmits the circumflex scapular artery, the other, below the teres major, between humerus and triceps, transmits the radial nerve and profunda brachii vessels.

Contents of the Axilla

The axilla transmits the neuro-vascular bundle of the upper limb.

The three cords of the brachial plexus are formed behind the clavicle and enter the upper part of the axilla above the artery (PLATE 26). The cords approach the artery and embrace its second part, the part which lies under cover of the pectoralis minor, lying medial, lateral and posterior to it in the manner indicated by their names. The axillary vein lies throughout on the medial side of the artery and nerves. In the fibro-fatty areolar tissue of the axilla lie several groups of lymph nodes.

The Axillary Artery. This is the continuation of the third part of the subclavian artery. It enters the apex of the axilla by passing over the first digitation of serratus anterior, at the outer border of the first rib, behind the midpoint of the clavicle. It is invested in fascia, the axillary sheath, projected down from the prevertebral fascia (Fig. 6.11, p. 377). It continues at the lower border of teres major by becoming the brachial artery. It is conveniently divided into three parts by the pectoralis minor; the part above, the part behind, and the part below. The second part, that behind the pectoralis minor, is readily to be recognized in a dissected axilla, where pectoralis minor has been removed, as that part of the artery which is clasped by the three cords of the plexus laterally, medially and posteriorly. The shape of the artery depends upon the position of the arm. With the arm at the side the artery has a bold curve with its convexity lateral, and its third part is clasped by the two heads of the median nerve. With the arm laterally rotated and abducted, as in operations upon the axilla, the artery pursues a straight course and the two heads of the median nerve lie loosely upon its third part.

Branches. Apart from some inconstant superior thoracic twigs to the upper axillary walls, it may be said that the first part has one branch, the second part two, and the third part three branches. These are:

1. *Superior thoracic artery*, from the first part, runs forwards to supply both pectoral muscles.

2. Acromio-thoracic and lateral thoracic arteries from the second part. The *acromio-thoracic artery* skirts the upper border of pectoralis minor to pierce the clavi-pectoral fascia, often separately in its four terminal branches (clavicular, humeral, acromial and pectoral). These branches radiate away at right angles from each other.

The *lateral thoracic artery* follows the inferior border of pectoralis minor, close to the pectoral lymph nodes, on the fascia over serratus anterior, supplying branches to pectoralis minor and major and, in the female, being an important contributor of blood to the breast.

3. Subscapular and the two humeral circumflex arteries from the third part. The *subscapular artery* runs down the posterior axillary wall, supplies latissimus dorsi and serratus anterior and terminates at the inferior angle of the scapula. A dorsal branch, the *circumflex scapular artery*, passes through the posterior wall of the axilla between subscapularis and teres major, medial to the long head of the triceps (i.e., through a triangular space, Fig. 2.10). The *anterior circumflex humeral artery* runs deep to coraco-brachialis and both heads of biceps (giving here an ascending branch which runs up the bicipital groove to the long tendon of biceps and the capsule of the shoulder joint), and passes around the

surgical neck of the humerus to anastomose with the *posterior circumflex humeral artery*. This latter, a much larger branch of the axillary artery, passes through the posterior axillary wall between subscapularis and teres major lateral to the long head of triceps, between it and the humerus (i.e., through the quadrilateral space). It is accompanied above by the axillary nerve and, like it, supplies the deltoid. It also gives branches to the long and lateral heads of triceps and the shoulder joint, and anastomoses with the profunda brachii artery.

The Axillary Vein. The venæ comitantes of the brachial artery are joined by the basilic vein above the lower border of the posterior wall of the axilla. They form the axillary vein, which courses upwards on the medial side of the axillary artery and leaves the axilla by passing through its apex anterior to the third part of the subclavian artery. Over the upper surface of the first rib, in front of scalenus anterior, it continues as the subclavian vein. Tributaries of the third and second parts of the axillary vein are the same as the branches of the artery. Into the first part (i.e., above pectoralis minor) the cephalic vein enters after having pierced the clavi-pectoral fascia. There is no axillary sheath around the vein, which is free to expand during times of increased blood flow.

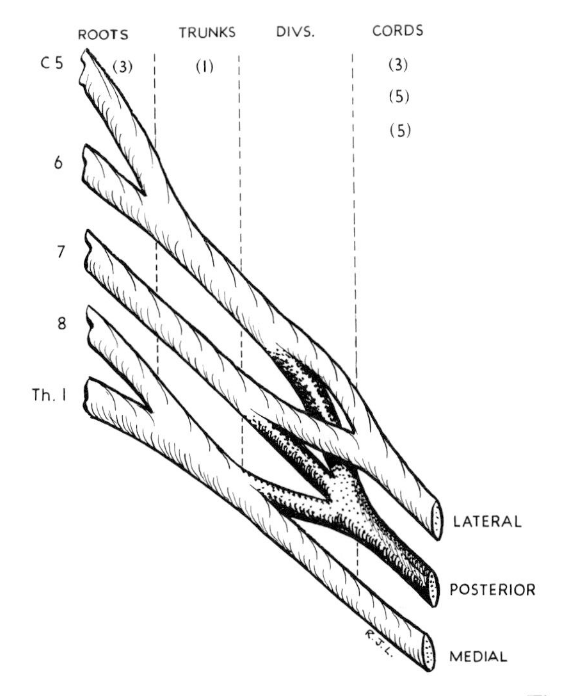

FIG. 2.12. THE FORMATION OF THE LEFT BRACHIAL PLEXUS. The figures in brackets denote the number of branches from its constituent parts.

The five nerves C5, 6, 7 and 8 and Th. 1 are the anterior primary rami of the spinal nerves. After giving off branches to the prevertebral muscles they constitute the **roots** of the brachial plexus.

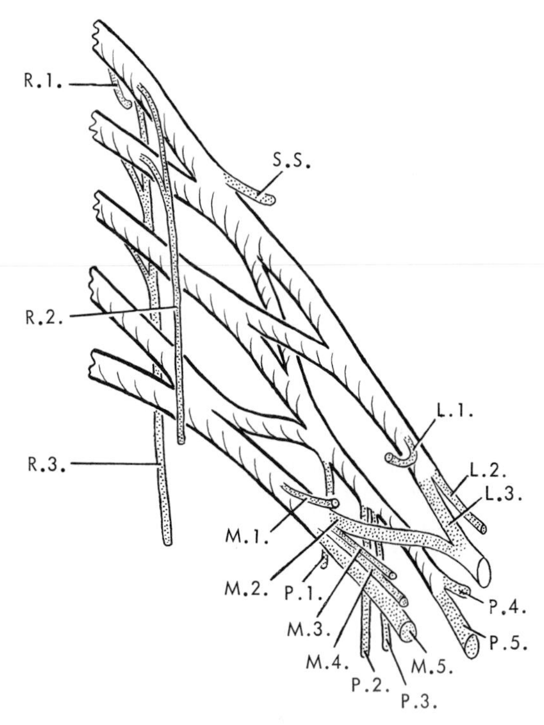

FIG. 2.13. PLAN OF BRANCHES OF THE BRACHIAL PLEXUS.

Branches of roots: R.1. Dorsal scapular (nerve to rhomboids). R.2. Nerve to subclavius. R.3. Long thoracic (nerve to serratus anterior).

Branch of upper trunk: S.S. Suprascapular nerve.

Branches of lateral cord: L.1. Lateral pectoral. L.2. Musculo-cutaneous. L.3. Lateral root of median nerve.

Branches of medial cord: M.1. Medial pectoral. M.2. Medial root of median nerve. M.3. Medial cutaneous nerve of arm. M.4. Medial cutaneous nerve of forearm. M.5. Ulnar nerve.

Branches of posterior cord: P.1. Upper subscapular. P.2. Thoraco-dorsal (nerve to latissimus dorsi). P.3. Lower subscapular. P.4. Axillary (circumflex). P.5. Radial nerve.

The Brachial Plexus

Five roots contribute to the formation of the plexus for the upper limb. They are the fibres that remain in the anterior primary rami of C5, 6, 7, and 8. and Th. 1 after these have given their segmental supply to the prevertebral and scalene muscles. They are to divide into anterior and posterior divisions to supply the flexor and extensor compartments respectively (p. 24), but before doing so they unite to form three trunks in the following manner. Of the five roots of the plexus the upper two unite to form the upper trunk, the lower two unite to form the lower trunk, and the central root runs on as the middle trunk (Fig. 2.12). The five roots lie behind scalenus anterior muscle, the trunks emerge from its lateral border and cross the posterior triangle of the neck. Each of the three trunks divides into an anterior and a posterior division behind the clavicle. Here, at the outer border of the first rib, the upper two anterior divisions unite to form the lateral cord, the anterior division of the lower trunk runs on as the medial cord, while all three posterior divisions unite to form the posterior cord. These three cords enter the axilla above the first part of the artery, approach and embrace its second part, and give off their branches around its third part.

It is important to appreciate the situation of the constituents of the plexus, *viz.*: roots between scalene muscles, trunks in the triangle, divisions behind the clavicle, and cords in the axilla. When the basic pattern has been learned it is a simple matter to put in the branches of the plexus. They consist of 3 branches from the roots and 3, 5 and 5 from the lateral, medial and posterior cords respectively. The only exception to the 3, 5, 5 rule is in the branch from the trunks. It is only one, the suprascapular nerve, from the upper trunk, in the posterior triangle. Everywhere else in the plexus the number of branches follows the 3, 5, 5 pattern (Fig. 2.13).

Branches from the Roots

Three in number, *viz.*: the dorsal scapular nerve, the nerve to subclavius, and the long thoracic nerve, they arise successively from C5, C5, 6 and C5, 6, 7, and pass downwards behind, in front of, and behind the roots in that order.

The **dorsal scapular nerve** (nerve to the rhomboids) arises from the posterior aspect of C5, enters scalenus medius, appears at its posterior border, and courses downwards beneath levator scapulæ, lying on serratus posterior superior. It is accompanied by the descending scapular vessels. It supplies both rhomboids on their deep surfaces. It usually gives a branch to levator scapulæ (Fig. 2.6).

The **nerve to subclavius** arises from the roots of C5 and C6 just as they combine to form the upper trunk. It passes down in front of the trunks and the subclavian vessels to enter the posterior surface of subclavius (PLATE 35). It may carry some aberrant phrenic nerve fibres (p. 364).

The **long thoracic nerve** (nerve to serratus anterior) arises from the posterior aspects of C5, 6 and 7. Branches of C5 and 6 enter scalenus medius, unite in the muscle, emerge as a single trunk from its lateral border and pass down into the axilla. On the surface of serratus anterior (the medial wall of the axilla) this is joined by the branch from C7 which has descended in front of scalenus medius. The nerve passes down posterior to the midaxillary line and supplies serratus anterior muscle segmentally (p. 58 and Fig. 2.7).

Branch from the Trunks

The solitary branch from the trunks is the **suprascapular nerve,** which arises from the upper trunk in the lower part of the posterior triangle. It can be seen above the clavicle as a large nerve leaving the upper trunk and passing back to disappear beneath the border of trapezius. It passes through the suprascapular foramen (beneath the transverse scapular ligament) and supplies supraspinatus, descends lateral to the scapular spine with the suprascapular vessels and supplies infraspinatus and gives a twig to the shoulder joint. It enlarges just before passing beneath the ligament (possibly due to increased vascularity) (PLATE 26).

Branches from the Lateral Cord

Three in number, they are the lateral pectoral, musculo-cutaneous and lateral head of the median nerve.

The **lateral pectoral nerve** pierces the clavi-pectoral fascia to supply pectoralis major (PLATE 1). It communicates across the axillary artery with the medial pectoral nerve. It has no cutaneous branch.

The **musculo-cutaneous nerve** leaves the lateral cord, runs obliquely downwards and sinks into the coraco-brachialis muscle, giving a twig of supply to it before entering the muscle (PLATE 1).

Lateral Head of the Median Nerve. The continuation of the lateral cord, it is joined by the medial head of the median nerve, which latter crosses the axillary artery; the two heads embrace the artery and, when the arm is pulled down to depress the shoulder, may in some cases compress the vessel.

Branches from the Medial Cord

Five in number, they are the medial pectoral, medial head of the median nerve, ulnar nerve, and the two cutaneous nerves, to the arm and forearm respectively.

The **medial pectoral nerve** is so named because it is a branch of the medial cord. It enters the deep surface of pectoralis minor, giving a branch of supply before doing so, perforates the muscle and enters the pectoralis major, in which it ends by supplying the lower costal fibres. It is joined by a communication from the lateral pectoral nerve which passes across the axillary artery. Like the lateral pectoral nerve it has no cutaneous branch.

Medial Head of the Median Nerve. The continuation of the medial cord, it crosses the axillary artery to join the lateral head (PLATE 1).

Medial Cutaneous Nerve of the Arm. The most medial of all the branches, it runs down on the medial side of the axillary vein and supplies skin over the front and medial side of the arm.

Medial Cutaneous Nerve of the Forearm. Runs down between artery and vein and supplies skin over the lower part of the arm and the medial side of the forearm (PLATE 1).

Ulnar Nerve. The largest branch of the medial cord. It runs down on a posterior plane between artery and vein, and is the most posterior of the structures which run down the medial side of the flexor compartment of the arm. It receives a branch from the lateral cord (C6, 7) in over 90 per cent of cases. These fibres are given off in the forearm as the motor branch to flexor carpi ulnaris (see p. 80).

Branches from the Posterior Cord

Five in number, they are: upper subscapular, thoraco-dorsal nerve (nerve to latissimus dorsi), lower subscapular, axillary (circumflex), and radial nerves (PLATE 22).

Upper Subscapular Nerve. A small nerve which enters the upper part of subscapularis.

The **thoraco-dorsal nerve** (nerve to latissimus dorsi) a large nerve which runs down the posterior axillary wall, crosses the lower border of teres major and enters the deep surface of latissimus dorsi, well forward near the border of the muscle. It comes from high up behind the subscapular artery, but as it descends to enter the muscle it lies in front of the artery (PLATE 22). It is thrown into prominence in the position of lateral rotation and abduction of the humerus and is thus in danger in operations on the lower axilla.

Lower Subscapular Nerve. A fairly large nerve which supplies the lower part of the subscapularis. It gives a separate twig to teres major, which runs in the angle between the subscapular and circumflex scapular arteries.

Axillary (circumflex) Nerve. A large branch. It supplies nothing in the axilla, and leaves the axilla immediately it branches from the plexus. It runs around the neck of the humerus. None the less, its name has been changed from 'circumflex' to 'axillary'. It leaves the posterior cord and passes between subscapularis and teres major, lateral to the long head of triceps, i.e., through the quadrilateral space (PLATE 22). Here it lies in contact with the surgical neck of the humerus, just below the capsule of the shoulder joint, with the posterior humeral circumflex vessels below it. Having given a branch to the shoulder joint, it divides into deep and superficial branches. The *deep branch* runs forward around the humerus in contact with bare bone and enters the deep surface of the deltoid to supply it; a few terminal twigs pierce the muscle and reach the skin. The *superficial branch* gives off the motor nerve to teres minor, then winds around the posterior border of deltoid and becomes cutaneous. It is here called the *upper lateral cutaneous nerve of the arm.* It supplies a few of the posterior fibres of the deltoid.

Radial Nerve. This is the continuation of the posterior cord, and is very large. It crosses the lower border of the posterior axillary wall, lying on the glistening tendon of latissimus dorsi (PLATE 22). It passes out of sight through a triangular space below the lower border of teres major, between the long head of triceps and the humerus. Before disappearing it gives nerves of supply to the long head of triceps, to the medial head by a nerve which accompanies the ulnar nerve along the medial side of the arm, and a cutaneous branch which supplies the skin along the posterior surface of the upper arm *(the posterior cutaneous nerve of the arm).*

Lymph Nodes of the Axilla

Contained in the fibro-fatty tissue of the axilla are many scattered lymph nodes. They are divisible into the following groups (see p. 129).

Pectoral group, lying along the medial wall of the axilla with the lateral thoracic artery, at the lower border of pectoralis minor. They receive from the upper half of the trunk anteriorly and from the *major part of the breast.*

Scapular group, lying along the medial wall of the axilla in the posterior part, i.e., along the subscapular artery. They receive from the upper half of the trunk posteriorly, and from the axillary tail of the breast.

Lateral group, lying along the medial side of the axillary vein. They receive from the upper limb.

Infraclavicular group, lying in the clavi-pectoral fascia in the interval between pectoralis major and deltoid. They receive from the upper part of the breast and, by lymphatics accompanying the cephalic vein, from the thumb and its web.

Apical group, lying in the apex of the axilla, receives from all the groups named above, and from the *central group* draining the floor of the axilla. The apical group drains by the subclavian lymph trunk through the apex of the axilla into supraclavicular nodes lying in the lower part of the posterior triangle of the neck. These drain into the thoracic duct or the right lymph trunk.

THE BREAST

The male breast throughout life and the immature female breast resemble each other. In both the nipple is small but the areola is fully formed. The breast tissue does not extend beyond the margin of the areola; it consists of a few ducts embedded in fibrous tissue. In a histological section alveoli are very sparse indeed. The lymph drainage of the immature breast is identical with that of the fully formed organ (*v. inf.*). At puberty the female nipple and breast both enlarge and thereafter retain the female form throughout life. The female form is very variable indeed, but the *size of the base of the breast is fairly constant.* It extends from the second to the sixth rib in the midclavicular line; it lies over the pectoralis major and extends beyond the border of that muscle to lie on the serratus anterior and external oblique. An axillary tail is sometimes present. From this circular base the breast protrudes or depends to a degree that varies within very wide limits. Asymmetry of the two breasts is not uncommon.

The resting (non-lactating) breast consists mainly of fibrous tissue. Glandular tissue is very sparse and consists almost entirely of ducts; alveoli are difficult to find in a histological section (Fig. 2.14). Prior to

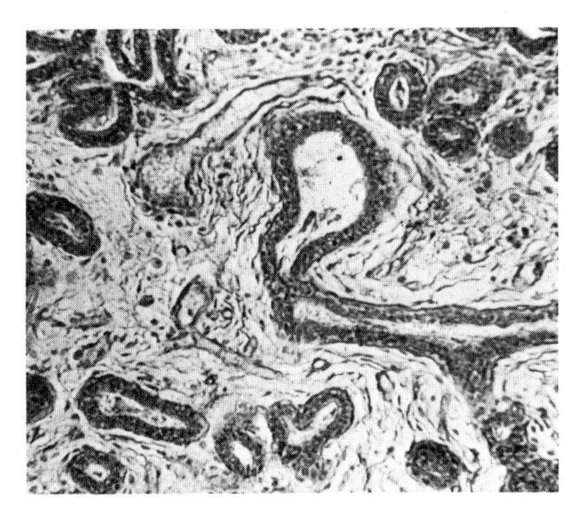

FIG. 2.14. THE NON-LACTATING BREAST. The substance of the breast consists of fibrous tissue, in which are embedded the ducts. Alveoli are very scarce.

lactation new alveoli bud off from the ducts into the fibrous tissue and the organ usually enlarges significantly. The main ducts number about 15; they open separately on the summit of the nipple. Each is dilated into an ampulla beneath the areola. Each main duct drains a lobe of the breast; the organ is divided by fibrous tissue septa that radiate out from the centre. Each lobe is irregularly lobulated. The whole breast is embedded in the subcutaneous fat, which usually obscures the lobules from sight and touch. There is no fat beneath the nipple and areola.

Beneath the breast is a condensation of *superficial fascia*, the continuation upwards of the fascia of Scarpa. Between this fascia and the deep fascia over pectoralis major is a *submammary space* in which the lymphatics run. Submammary infusions can readily be given into this space. The axillary tail when present lies in the medial wall of the axilla and may be a discrete mass very poorly connected with the duct system. Usually it lies in the subcutaneous fat, which is condensed around it—very rarely it may penetrate the deep fascia of the floor of the axilla.

Support of the Breast. The young breast is protuberant, the older breast pendulous. The former is supported by fibrous tissue strands connecting the deep fascia with the overlying skin (dermis); these are the ligaments of Astley Cooper. When atrophic they allow the organ to droop, when contracted from the fibrosis around a carcinoma they cause pitting of the skin.

The **blood supply** is derived from the *lateral thoracic artery* by branches that curl around the border of pectoralis major and by other branches that pierce the muscle. The *internal thoracic artery* sends branches through the intercostal spaces beside the sternum; those of the second and third spaces are particularly large. Similar perforating branches arise from the intercostal arteries. Pectoral branches of the *acromio-thoracic artery* supply the upper part of the breast. Venous return simply follows the above-mentioned arteries. Note, however, that the venous blood, like that from the thyroid gland, is received into the large veins that receive blood also from the vertebræ and thoracic cage—spread of malignancy by veins can thus involve these bones (p. 465).

Lymph Drainage. This is so important in connection with the spread of malignant disease that it should be studied carefully.

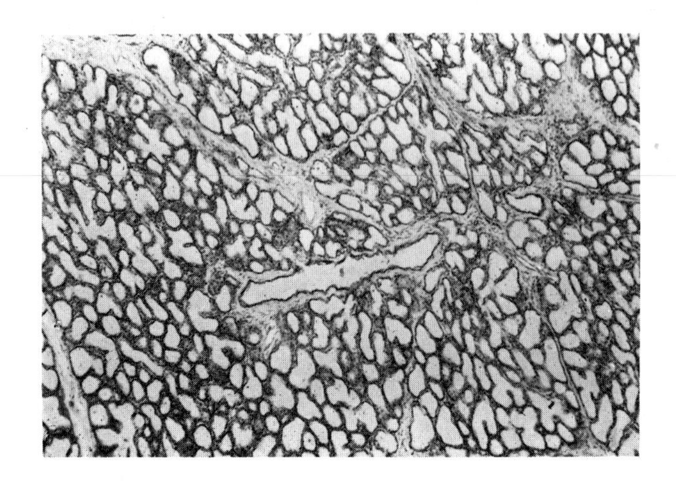

FIG. 2.15. THE LACTATING BREAST. Lobules of milk-containing alveoli distend the gland. Distinguish from the thyroid gland by the presence of ducts.

As elsewhere in the body, so in the breast, the lymph capillaries make a richly anastomosing network continuous with the lymph capillaries of neighbouring structures—in this case those of the opposite side and of the abdominal wall. Lymph from the breast may thus radiate away to any point of the compass according to local pressures of clothing or of the examining surgeon's hand. Only when the lymph leaves the lymphatic capillaries and enters valved lymphatic vessels is its subsequent course irrevocable. Most of the lymph of the breast drains, in fact, to the axilla.

The superficial parts of the breast drain to a subareolar plexus, the deep parts drain to a submammary plexus; both communicate freely through the breast. The submammary plexus lies in the deep fascia over pectoralis major and serratus anterior.

From these plexuses lymph from most of the breast drains to the pectoral group of axillary nodes (the axillary tail when present drains into the scapular group). The upper convexity drains to the infraclavicular group. The medial convexity, communicating across the midline with the submammary plexus of the opposite breast, drains through the intercostal spaces to nodes along the internal thoracic artery and thence to the mediastinal nodes. The inferior convexity communicates through the abdominal wall with lymphatic capillaries in the extraperitoneal areolar tissue and so through the diaphragm to mediastinal nodes. In tranquil states lymph does not drain along these communicating channels, but pressure of an examining clinician's hand may cause it to do so. In any case malignant disease may spread along these communicating vessels.

THE SCAPULAR REGION

From the posterior surface of the blade of the scapula three muscles converge to the greater tuberosity of the humerus, being inserted in turn into three impressions thereon (Fig. 2.16). They lie hidden, for the most part, under deltoid and trapezius (Fig. 2.6).

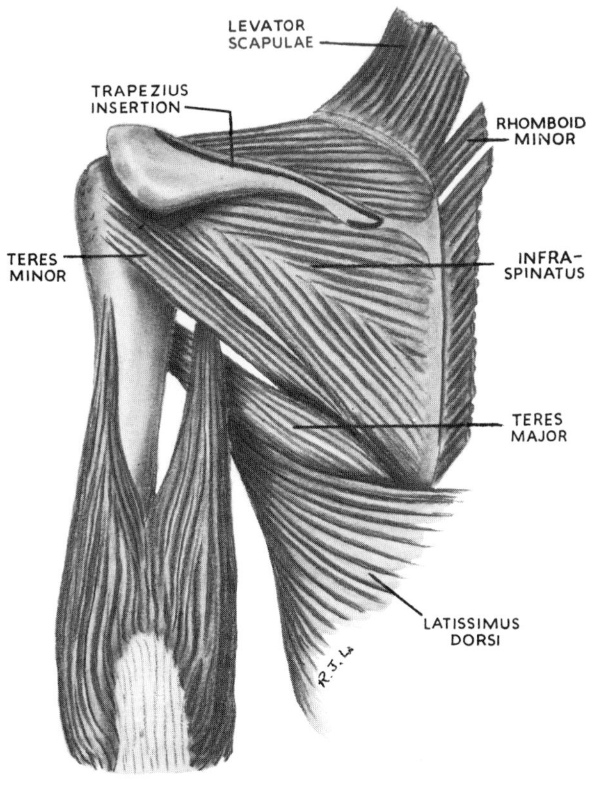

Supraspinatus. This muscle fills snugly the whole of the supraspinous fossa of the scapula. It arises from the medial three-fourths of the fossa and from the upper surface of the scapular spine. It is bipennate (Fig. 2.17), an arrangement that gives this bulky muscle great force of pull. A bursa separates it from the lateral fourth of the fossa. The tendon develops deep to the muscle as it crosses the superior part of the capsule of the shoulder joint; the two become blended together, to the great advantage of the capsule. The tendon passes on to be inserted into the smooth facet on the upper part of the greater tuberosity (Fig. 2.60, p. 114).

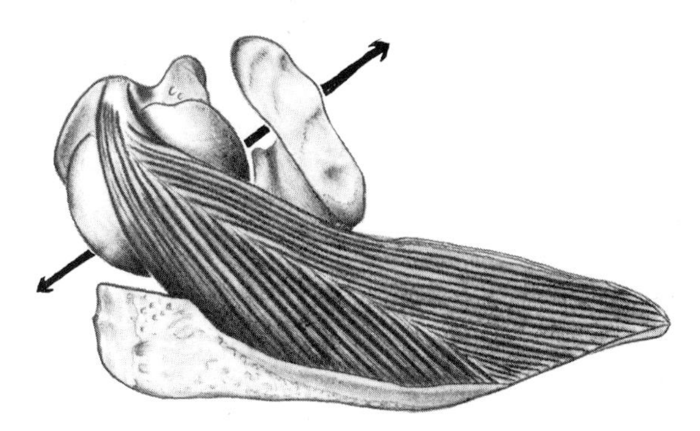

FIG. 2.17. SUPRASPINATUS. The very powerful bipennate muscle is here exposed by removal of the acromion process and the clavicle (left side, viewed from above). Supraspinatus is seen to be the prime abductor in or near the coronal plane of the resting humerus when the arm is by the side, acting around an axis indicated by arrows. Drawn from Specimen S.30 in R.C.S. Museum.

FIG. 2.16. POSTERIOR VIEW OF THE MUSCLES ATTACHED TO THE LEFT SCAPULA. The rhomboid major is inserted below rhomboid minor down to the inferior angle of the scapula. The long head of triceps makes with teres major a quadrilateral and two triangular spaces. Compare Fig. 2.10.

Nerve Supply. The suprascapular nerve, branching from the superior trunk of the brachial plexus (C5), passes beneath the transverse ligament of the scapula and supplies the muscle. It passes on to the infraspinous fossa, around the lateral border of the spine, in contact with the bone and in company with the suprascapular artery.

Action. The muscle is of primary importance in bracing the head of the humerus to the glenoid fossa, to give stability during the action of other muscles, especially the deltoid. Inspection of the mechanics of this powerful muscle makes it apparent that, with the arm at the side, it initiates abduction in the coronal plane, acting as a prime mover (Fig. 2.17). This was long ago believed, and it is borne out clinically.

Infraspinatus. The muscle arises from beneath a dense fascia from the medial three-fourths of the infraspinous fossa. Fibrous intramuscular septa give added attachment to the multipennate fibres, which converge to slide freely over the bare area of the scapula at its glenoid angle. A bursa lies between the two. Here tendon replaces the muscle fibres, and is blended with the capsule of the shoulder joint, greatly increasing its strength. The tendon is inserted into the smooth area on the central facet of the greater tuberosity of the humerus (Fig. 2.60, p. 114).

Nerve Supply. From the suprascapular (C5) branch of the upper trunk of the brachial plexus. Passing under the transverse scapular ligament and giving a branch to supraspinatus, the nerve passes around the lateral border of the scapular spine and enters the infraspinous fossa to be distributed to the muscle.

Action. Primarily acting to brace the head of the humerus against the glenoid fossa, giving stability to the joint in movements of the peripheral parts of the limb, the muscle is also a lateral rotator of the humerus.

Teres Minor. The muscle is completely hidden beneath the deltoid. It arises from an elongated oval area at the axillary border of the scapula. It passes upwards and laterally, edge to edge with the lower border of infraspinatus, to the lowest facet on the greater tuberosity of the humerus and to a thumb's breadth of bone below this. It passes behind the origin of the long head of triceps. At the lower border it lies edge to edge with teres major at its origin, but the latter muscle leaves it by passing forward in front of the long head of triceps (Fig. 2.16). Triceps long head thus makes a triangular space between these muscles on the scapular side and a quadrilateral space on the humeral side.

Nerve Supply. A branch from the superficial division of the axillary nerve (C5), from the posterior cord of the brachial plexus.

Action. It assists the other small muscles around the head of the humerus in steadying the shoulder joint in movements of peripheral parts of the limb. More specifically, it can act as a lateral rotator and weak adductor of the humerus. With the teres major it holds down the head of the humerus against the upward pull of the deltoid during abduction of the shoulder (Fig. 2.21).

Infraspinatus Fascia. The infraspinatus and teres minor muscles lie together beneath a very strong membrane which is firmly attached to bone at the margins of these muscles. It is attached above to the lower border of the scapular spine beneath the deltoid muscle, and medially along the vertebral border of the scapula. It is attached to the axillary border of the scapula along the sinuous ridge that can be seen running down from the infraglenoid tubercle. Examine a scapula and note that this ridge runs back to the dorsal surface of the bone above the origin of teres major. The latter muscle does not lie beneath the fascia (Fig. 2.5, p. 56).

The importance of the infraspinatus fascia is not confined to its value as a landmark in surgical exposures of this region, but lies in the fact that in fracture of the blade the resulting hæmatoma is confined beneath the fascia. A large rounded swelling that is *limited to the margins of the bone* is diagnostic of fracture of the blade of the scapula.

Deltoid Muscle. This powerful muscle arises from a long strip of bone and converges triangularly to its insertion on the deltoid tuberosity. It covers the shoulder joint like a cape and its convex shape is due to the underlying upper end of the humerus. The anterior and posterior fibres are long, and in parallel bundles. The intermediate, or acromial fibres are multipennate. The muscle arises from the anterior border and upper surface of the flattened lateral one-third of the clavicle, from the whole of the lateral border of the acromion and from the inferior lip of the crest of the scapular spine as far medially as the 'deltoid' tubercle. On the lateral border of the acromion four ridges may be seen; from them four fibrous septa pass down into the muscle. The deltoid tuberosity on the humerus is U-shaped, with a central vertical ridge. From the ridge and limbs of the U three fibrous septa pass upwards between the four septa from the acromion. The spaces between the septa are filled with a fleshy mass of oblique muscle fibres which are attached to contiguous septa. The multipennate centre of the deltoid so formed has a diminshed range of contraction, but a correspondingly increased force of pull. The anterior and posterior fibres, arising from the clavicle and the scapular spine, are not multipennate. They converge on the limbs of the U-shaped deltoid tuberosity of the humerus. Their range of movement is greater but the force of their pull is less than that of the multipennate central portion of the muscle.

Nerve Supply. The axillary (circumflex) nerve (C5), a branch of the posterior cord of the brachial plexus.

Action. Abducts the arm after supraspinatus has initiated the movement. Most of this work is done by the multipennate acromial fibres. The anterior and posterior fibres act like guy ropes to steady the arm in the abducted position. The anterior fibres acting alone assist pectoralis major in flexing the arm, the posterior fibres assist latissimus dorsi in extending the arm.

Scapular Anastomosis (Fig. 2.18). The transverse cervical artery from the subclavian, usually the first part, occasionally the third part, has a descending branch (the *descending scapular artery*) which accompanies the dorsal scapular nerve. It runs down the vertebral border of the scapula to its inferior angle. The *suprascapular artery* from the subclavian, usually the first part (thyro-cervical axis in common with the transverse cervical) but occasionally the third part, crosses over the transverse ligament of the scapular notch, passes through the supraspinous fossa, turns around the lateral border of the spine of the scapula and supplies the infraspinous fossa as far as the inferior angle. The *subscapular artery*, branching from the third part of the axillary, supplies the subscapularis muscle in the subscapular fossa as far as the inferior angle. Its *circumflex scapular branch* enters the infraspinous fossa on the dorsal surface of the bone. All the vessels anastomose, thus connecting the first part of the subclavian with the third part of the axillary artery. The companion veins form corresponding anastomoses.

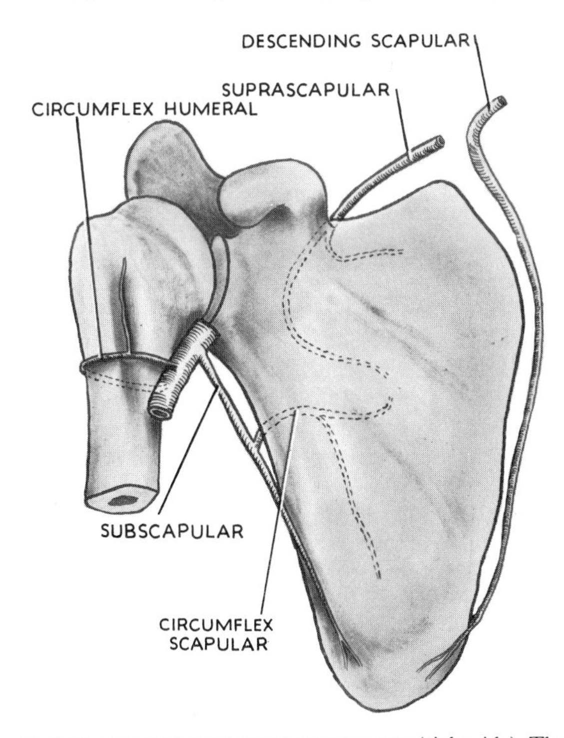

FIG. 2.18. THE CIRCUMSCAPULAR ANASTOMOSES (right side). The descending scapular and suprascapular arteries are from the first part of the subclavian. The subscapular artery is from the third part of the axillary. They and the circumflex scapular artery all anastomose on both ventral and dorsal surfaces of the scapula.

The Shoulder Joint

In all tetrapods there is a scapula which is dorsal and a coracoid which is ventral. The humerus always articulates at the junction of the two elements. In the human shoulder, scapula and coracoid articulate with each other across the joint line (Fig. 2.9, p. 60). The presence of an epiphyseal cartilage across a joint line is beneficial in that it facilitates adjustment of the joint surfaces during growth of the bone ends (shoulder, hip, elbow).

The shoulder joint is a synovial joint of the ball and socket variety. There is marked disproportion between the large round head of the humerus and the small shallow glenoid fossa. The *glenoidal labrum*, a ring of fibro-cartilage attached to the margins of the glenoid fossa, deepens slightly but effectively the depression of the glenoid fossa (Fig. 2.9).

The **capsule** of the joint is attached to the scapula beyond the supraglenoid tubercle and the margins of the labrum. It is attached to the humerus around the articular margins of the head except inferiorly, where its attachment is to the neck of the humerus a finger's breadth below the articular margin. At the upper end of the bicipital groove the capsule bridges the gap between the greater and lesser tuberosities, being here named the **transverse ligament.** A gap in the anterior part of the capsule allows communication between the synovial membrane and the subscapularis bursa (Fig. 2.11). A similar gap is often present posteriorly, allowing communication with the infraspinatus bursa. The fibres of the capsule all run horizontally between scapula and humerus. The capsule is thick and strong but it is very lax, a necessity in a joint so mobile as this. Near the humerus the capsule is greatly thickened by fusion of the tendons of the short scapular muscles. The long tendon of biceps is intracapsular.

The *synovial membrane* is attached around the glenoidal labrum and lines the capsule. It is attached to the articular margin of the head of the humerus and covers the bare area of bone that lies within the capsule at the upper end of the shaft. It 'herniates' through the hole in the front of the capsule to communicate with the subscapularis bursa and sometimes it communicates with the infraspinatus bursa. It invests the long head of biceps in a tubular sleeve that is reflected back along the tendon to the transverse ligament and adjoining floor of the bicipital groove (Fig. 2.19). The synovial sleeve glides to and fro with the long tendon of biceps during abduction–adduction of the shoulder, as shown in Fig. 2.19.

The *gleno-humeral ligaments* are scarcely worthy of mention, being slight thickenings above and below the hiatus in the anterior part of the capsule. They are visible only from within the joint cavity.

The **coraco-humeral ligament** is quite strong. It runs from the under surface of the coracoid process laterally across the capsule, to which it becomes attached at the margin of the greater tuberosity, and along the transverse ligament.

Coraco-acromial Ligament. From the medial border of the acromion, in front of the acromio-clavicular articulation, a strong flat triangular band fans out to the lateral border of the coracoid process. It lies above the head of the humerus and serves to increase the surface upon which the head of the humerus is supported. It is separated from the 'rotator cuff' by the subacromial bursa (Fig. 2.19).

Subacromial Bursa (Subdeltoid Bursa). A large bursa lies under the coraco-acromial ligament, to which its upper layer is attached. Its lower layer is attached to the tendon of supraspinatus and the capsule.

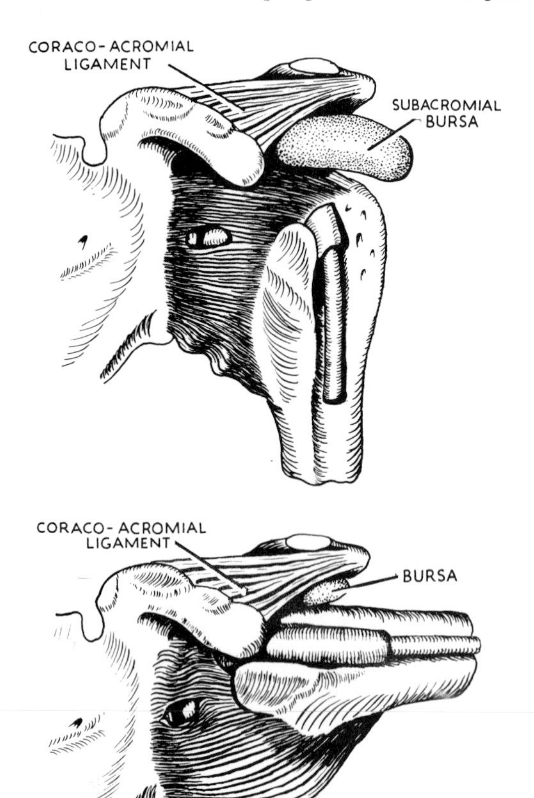

FIG. 2.19. THE LEFT SUBACROMIAL BURSA. With the arm abducted (lower figure) the bursa is withdrawn beneath the acromion, but the long tendon of biceps and its synovial sheath protrude further from the joint.

It extends beyond the lateral border of the acromion with the arm at the side, but is rolled inwards under the acromion when the arm is abducted. Tenderness over the greater tuberosity of the humerus beneath the deltoid muscle which disappears when the arm is abducted indicates subacromial bursitis (Fig. 2.19).

Stability of the Shoulder Joint. The shoulder joint, thus far described, is seen to be a very unstable structure. The head of the humerus is much larger than the glenoid cavity (Fig. 2.20), and the joint capsule, though strong, is very lax. True it is that the concavity of the glenoid fossa, deepened by the labrum, is a significant stabilizing factor. But this is only because the tonus of the short scapular muscles holds the head in close apposition. Fracture of the labrum results in dislocation. Stability is increased by the coraco-acromial arch, the fusion of tendons of scapular muscles with the capsule of the joint, and the tonus of the muscles attaching the humerus to the pectoral girdle.

The Coraco-acromial Arch. Upward displacement of the head of the humerus is prevented by the overhanging coracoid and acromion processes and the coraco-acromial ligament that bridges them. The whole constitutes the coraco-acromial arch and, lubricated by the subacromial bursa, functions mechanically as an 'articular surface' of the shoulder joint. The arch is very strong. Upward thrust on the humerus will never fracture the arch; the clavicle or the humerus itself will fracture first.

Tendons of Scapular Muscles. The tendons of subscapularis, supraspinatus, infraspinatus and teres minor are not only attached very near the joint but actually in part fuse with the lateral part of the capsule. This is an indispensable factor in adding stability to the joint. The fused mass of tendon and lateral part of capsule is known surgically as the **'rotator cuff'.** The rotator cuff prevents the lax capsule and its lining synovial membrane from being nipped. There is no cuff inferiorly, and here the capsule is attached well below the articular margin to prevent its being nipped.

Muscles attaching Humerus to Pectoral Girdle. All these muscles assist by their tonus in maintaining the stability of the joint. Especially active in this respect are the long heads of biceps and triceps muscles.

The *long head of biceps*, arising from the supraglenoid tubercle, sinks in through the capsule of the embryonic joint and is intracapsular in the mature joint. It leaves the capsule beneath the transverse ligament across the upper part of the bicipital groove. The tendon acts as a strong support over the head of the humerus.

The *long head of triceps* is of importance during abduction of the joint, for in this position it lies immediately beneath the head of the humerus at the lowest part of the joint. This is the weakest part of the joint; the long head of triceps is thus of very great importance in giving stability to the abducted humerus.

Atmospheric pressure has been suggested as a factor. The area of the glenoid is about 1 sq in. Thus a pressure of 15 lb is available to prevent *distraction* of the bones. But in very few positions is distraction likely to occur. Atmospheric pressure has no influence in preventing side to side sliding movements along the plane of the apposed surfaces. In the shoulder joint there is little tendency to separation of the joint surfaces; displacement of the head of the humerus upwards, downwards, forwards and backwards is the ever-present possibility and atmospheric pressure has no effect in preventing these movements.

Nerve Supply. The capsule is supplied by branches from the axillary, musculo-cutaneous and suprascapular nerves. Each illustrates Hilton's law (see p. 15).

Movements. The articular surface of the head of the humerus is four times the area of the glenoid; thus there is considerable freedom for a variety of movements. Movements at the joint are accompanied by movements of the scapula on the thoracic wall and by consequential movements of the clavicle.

These various movements of the shoulder joint itself, however, can be understood only if each is analysed into its constituent basic parts. These basic movements are only three, namely (i) flexion and extension, (ii) adduction and abduction, and (iii) rotation. Note that *circumduction* is merely a rhythmical combination in orderly sequence of flexion, abduction, extension and adduction (or the reverse), and for purposes of analysis it is not an elementary movement.

It is of clinical value to appreciate the fact that the medial epicondyle of the humerus faces in the same

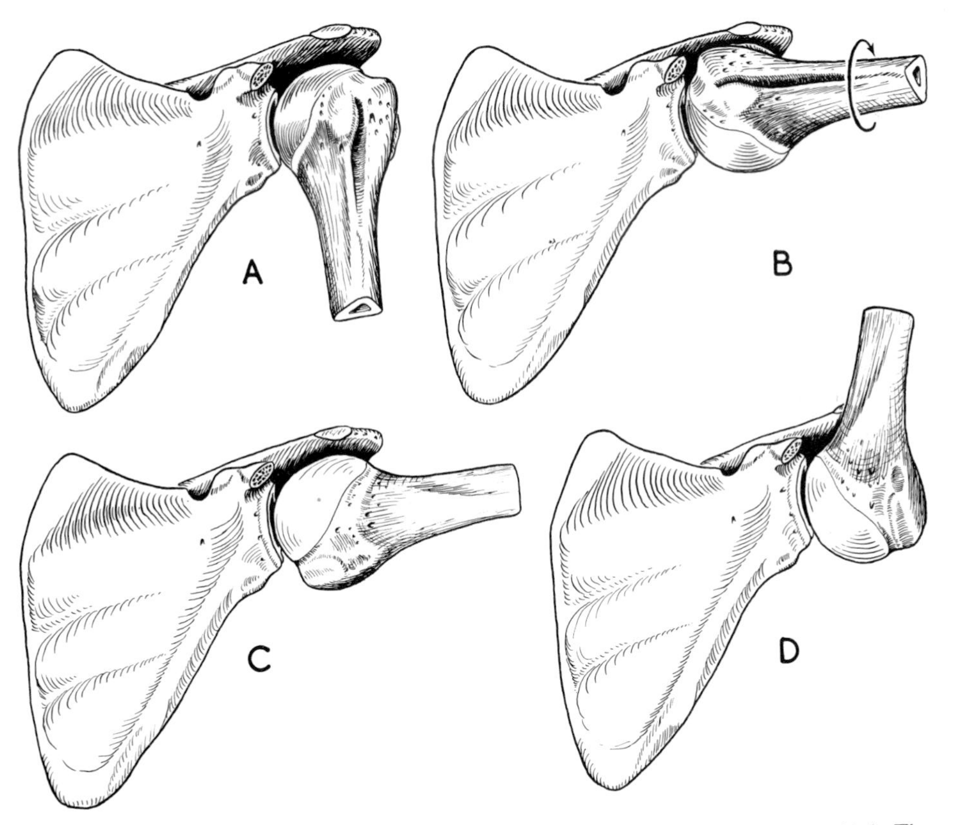

FIG. 2.20. THE BONY FACTORS IN ABDUCTION AT THE SHOULDER JOINT. (The coracoid process is amputated.) A. The rest position. Very little surface of the head of the humerus is available for adduction, but there is enough available above the glenoid for abduction to about 90 degrees. B. Abduction to 90 degrees. All available surface of the humerus is used up and further abduction would cause dislocation. C. Lateral rotation of the humerus from position B rolls the spare surface from below to above the glenoid (such free rotation is limited in the living by tension in the rotator cuff). Now adduction dislocates the head of the humerus. D. Full abduction is possible from the rotated position in C. The posterior surface of the humerus is now in view.

These movements take place in the plane of the paper (the coronal plane of the body), but the final position D can be reached directly from A by movement at right angles to the paper (flexion in the sagittal plane carried up to full abduction).

Full abduction of the shoulder joint illustrated in D cannot be reached in the living; it is limited to about 120 degrees. Scapular rotation accounts for the remaining 60 degrees.

direction as the articular surface of the head. Note that this is not exactly medial, but rather medial and somewhat backwards. The glenoid cavity does not face exactly laterally, but peeps forward a little around the convexity of the thoracic wall, and the articular surface on the head of the humerus looks back towards it. The purist may claim that flexion and extension should be described as movements in the plane of the joint cavity, which is oblique to the sagittal plane, and that abduction and adduction should be defined as movements in a plane at right angles to this, which is oblique to the coronal plane. But for purposes of analysis of these movements the difference is negligible.

Analysis of Shoulder Joint Movements. Examine the humerus and scapula. Any of the basic movements under review is limited in its range (apart from tension in ligaments or muscles) by the available articular surface on the head of the humerus. When glenoid cavity and humeral head lie margin to margin, further movement is impossible without dislocation of the joint.

Abduction of the joint requires study. With the arm at the side (Fig. 2.20A) *adduction* is very limited in range, not only by compression of soft parts but also because very little surface of the humerus is available. *Abduction* from this position is limited to about 90 degrees. Fixation here is not due to bony interlocking between greater tuberosity and acromion, but to the fact that *no further articular surface is available on the humerus* (Fig. 2.20B). Further abduction in this position can only result in dislocation of the head of the humerus. The available free surface of humeral head lies below the joint. If this free surface is now put above the joint by lateral rotation of the humerus (Fig. 2.20C), further abduction is made possible (Fig. 2.20D). Note that this final position of full abduction in lateral rotation (Fig. 2.20D), reached through the coronal plane, can be achieved in the sagittal plane by a simple pendulum movement of the humerus, carrying the normal forward swing of flexion straight up into full abduction. *Flexion* is free and direct from A to D (Fig. 2.20) with no rotation of the humerus. *Abduction* from A to D is possible only with lateral rotation of the humerus. In each case the final position of the humerus is the same.

Rotation of the humerus can be performed, about its long axis, in any position. It is limited by the extent of the articular surface on the humerus, and by simultaneous tension in the appropriate part of the capsule.

Abduction in the coronal plane is initiated by supraspinatus (Fig. 2.21). Deltoid tends to pull the shaft upwards, without abduction. When some abduction is produced by supraspinatus then, and only then, do the acromial fibres of deltoid pull with increasing mechanical advantage as the humerus abducts. Meanwhile the teres major and teres minor hold the head down against the upward pull of deltoid (Fig. 2.21). Here the downward pull on the head of the humerus of the subscapularis, teres major and minor, and the infraspinatus acts as a stabilizing couple against the upward pull of the deltoid.

Adduction is produced by the above short scapular muscles when deltoid relaxes, and the movement is much strengthened by contraction of the great muscles of the axillary walls, pectoralis major and latissimus dorsi.

Flexion is produced by the clavicular head of pectoralis major and the anterior fibres of the deltoid.

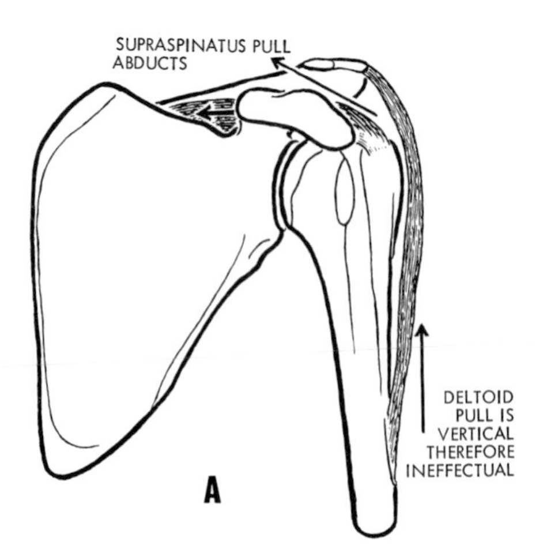

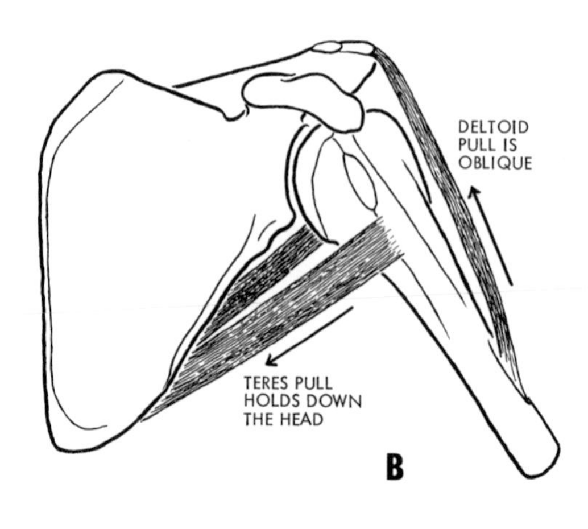

FIG. 2.21. ABDUCTION AT THE SHOULDER JOINT. **A.** With the arm at the side supraspinatus is the initiator of abduction. In this position its tendon, curving down somewhat over the head of the humerus, pulls actually *upwards* on the greater tuberosity (see Fig. 2.17). **B.** After abduction is achieved by supraspinatus deltoid acts as a powerful abductor; teres major and teres minor prevent upward movement of the head meanwhile.

Extension is done by the latissimus dorsi, posterior fibres of deltoid and the long head of the triceps.

Rotation is produced chiefly by the short scapular muscles (infraspinatus and teres minor, C5, producing lateral rotation, while subscapularis and teres major, C6, 7, produce medial rotation).

THE ARM

The Flexor Compartment

Like the thigh the arm consists of flexor and extensor compartments. Part of the flexor compartment in each limb is occupied by the adductor musculature. In the thigh this is differentiated into adductors longus, brevis, and that part of magnus innervated from the obturator nerve. In the arm the adductor musculature is represented by the rather insignificant coraco-brachialis muscle; the function of adduction has been taken over by the well-developed muscles in the anterior and posterior axillary walls. The nerve of the flexor compartment is the musculo-cutaneous.

Coraco-brachialis Muscle. Functionally unimportant the muscle nevertheless shows several interesting morphological and anatomical characteristics. It is the counterpart in the arm of the adductors (longus, brevis, magnus) of the thigh. It arises from the apex of the coracoid process, where it is fused with the medial side of the short head of biceps. The tendon is continued into a muscular belly of varying development which is inserted into the medial border of the humerus. The lower extent of the insertion is marked by the nutrient foramen of the humerus, for the nutrient branch of the brachial artery runs along the lower border of the muscle. The upward extent of the insertion cannot be seen on most bones, the muscle usually leaving no impression.

Morphology. In some animals this is a tricipital muscle. In man the lowest head is suppressed and the upper two have fused together, to imprison the musculo-cutaneous nerve between them. Persistence of the lower head in man is associated with the presence of the *ligament of Struthers.* Occasionally present in man is a *supratrochlear spur*, projecting down from the antero-medial aspect of the lower humerus; in an antero-posterior X-ray it can be seen only when thrown into

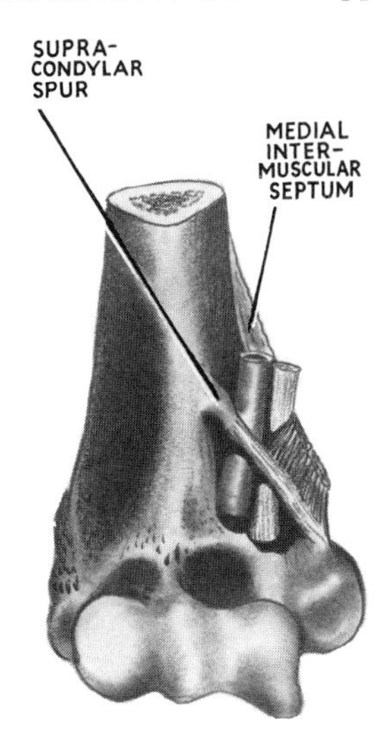

FIG. 2.22. THE SUPRATROCHLEAR SPUR AND THE LIGAMENT OF STRUTHERS. Drawn from a specimen in R.C.S. Museum.

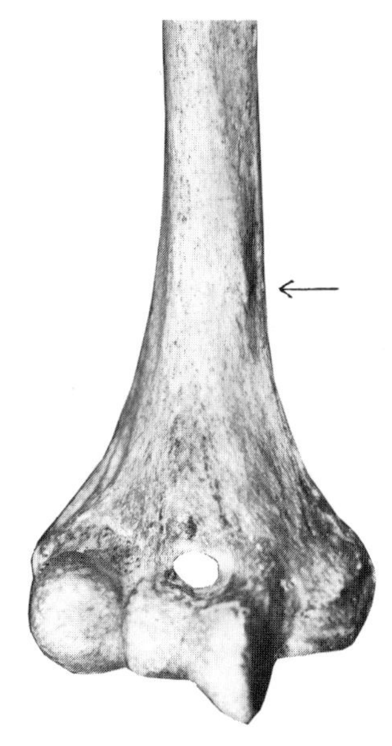

FIG. 2.23. LOWER END OF RIGHT HUMERUS. The arrow points to a supratrochlear spur. A small spur is as important as a large one; it is the ligament of Struthers which comes from it that matters. The foramen between supratrochlear and olecranon fossæ is of neither morphological nor practical importance. (Photograph.)

profile by medial rotation of the humerus. When a supratrochlear spur is present it is connected by a fibrous band (the ligament of Struthers) to the medial epicondyle. Attached to this band are the origin of part of pronator teres and the insertion of abnormally low fibres (the third head) of coraco-brachialis. Beneath the ligament pass the median nerve or brachial artery or both; this resembles the endepitrochlear foramen of carnivores and some other mammals. Pressure on artery or nerve by the ligament of Struthers may very rarely cause an irritative spasm of the vessel or a median nerve palsy (Fig. 2.22). Such a palsy can be readily distinguished from that seen in the 'carpal tunnel syndrome', for weakness of the relevant forearm flexor muscles and paræsthesia over the thenar eminence are added to the simpler picture of loss of median nerve conduction at the wrist (p. 99).

Nerve Supply. The muscle is supplied by the musculo-cutaneous nerve (C7). The main trunk of the nerve passes through the muscle, first giving off its branch of supply (PLATE 1, p. *i*).

Action. Compared to the morphological interest of this muscle its action is negligible. It is a weak adductor of the shoulder joint, the main adductors of which are pectoralis major and latissimus dorsi.

Biceps Muscle. The **long head** of this muscle arises from the supraglenoid tubercle and adjoining part of the labrum glenoidale of the scapula (Fig. 2.9, p. 60). The rounded tendon passes through the synovial cavity of the shoulder joint, surrounded by synovial membrane, and emerges beneath the *transverse ligament* at the upper end of the bicipital groove. The membrane pouts out below the ligament to an extent which varies with the position of the arm, being greatest in full abduction. Complete range of mobility of the tendon under the transverse ligament between full adduction and abduction of the shoulder joint is 6 cm. The long tendon develops outside the capsule, then sinks through it, then hangs on a mesotendon which subsequently breaks down. This embryological progression may become arrested at any stage. The **short head** arises from the apex of the coracoid process to the lateral side of the tendon of coraco-brachialis. Each tendon expands into a fleshy belly; the two bellies lie side by side, loosely connected by areolar tissue, but do not merge until just above the elbow joint, below the main convexity of the muscle belly. A tendon, flat from side to side, lies across the elbow joint, converges into a flattened cord which rotates (anterior surface turning laterally) as it passes through the cubital fossa to its insertion into the posterior border of the bicipital tuberosity of the radius (Fig. 2.31). A bursa separates the tendon from the anterior part of the tuberosity. From the medial border of the tendon, at the level of the elbow joint, the **bicipital aponeurosis** is inserted by way of the deep fascia of the forearm into the subcutaneous border of the upper end of the ulna. This aponeurosis has a sharp concave upper margin which can be felt tensed when the supinated forearm is flexed to a right angle. The main tendon, sliding in and out of the cubital fossa during pronation and supination, has a total range of movement of 6 cm.

Nerve Supply. The muscle is supplied by the musculo-cutaneous nerve (C5, 6).

Action. If the elbow is extended the muscle is a simple flexor thereof, but in any position of the elbow short of full extension the biceps is a powerful supinator of the forearm. In ordinary supination the triceps contracts to prevent flexion at the elbow joint. In full supination the bicipital aponeurosis is pulled taut and the biceps then pulls on *both* bones of the forearm to flex the elbow. Unopposed by triceps the biceps, while supinating the forearm, flexes the elbow and shoulder joints. It 'puts in the corkscrew then pulls out the cork'.

Brachialis Muscle. The muscle arises from the front of the lower two-thirds of the humerus and the medial intermuscular septum. Its upper fibres *clasp the deltoid insertion*; some fibres arise from the lower part of the spiral groove. The broad muscle flattens to cover the anterior part of the elbow joint and is inserted by mixed tendon and muscle fibres into the coronoid process and tuberosity of the ulna (Fig. 2.32, p. 82).

Nerve Supply. From the musculo-cutaneous (C5, 6). Some of the lateral part of the muscle is innervated by a branch of the radial nerve, an indication that it was developed in the extensor compartment of the fœtal limb (p. 25).

Action. A flexor of the elbow joint. The muscle, together with biceps, is used just as frequently as an *extensor* of the elbow by the action of paradox (p. 8), 'paying out rope' against gravity. 'It picks up the drink and puts down the empty glass.'

Medial Intermuscular Septum. This septum extends along the medial supracondylar line behind the coraco-brachialis insertion and fades out above, between that muscle and the long head of triceps. It gives origin to the most medial fibres of brachialis and the medial head of triceps, and is pierced by the ulnar nerve and ulnar collateral artery.

The **lateral intermuscular septum** extends along the lateral supracondylar line and fades out behind the insertion of deltoid. Both brachio-radialis and extensor carpi radialis longus extend out from the humerus to gain attachment to the septum in front, and posteriorly the *medial* head of triceps arises from it. It is pierced by the radial nerve and profunda brachii artery (anterior descending branch).

Nerves and Vessels of the Arm

The neuro-vascular bundle from the axilla passes into the flexor compartment of the arm. Some branches (circumflex vessels and nerve) are given to the extensor compartment in the axilla, which they leave through the quadrilateral space. In the upper part of the arm the radial nerve and profunda brachii artery leave to enter the extensor compartment through a triangular gap between the long head of triceps, the humerus and the lower margin of teres major. The remainder of the neuro-vascular structures pass through the flexor compartment.

Brachial Artery. This is the continuation of the axillary artery, which vessel has the axillary vein on its medial side. In the arm the vein is seen to be formed from venæ comitantes to the brachial artery, strongly reinforced above by the basilic vein, which perforates the deep fascia at a point opposite the deltoid insertion. The brachial artery has the median nerve lateral to it above, but the nerve crosses very obliquely in front of the artery and lies on its medial side below. The ulnar nerve, posterior to the artery above, leaves it in the lower part of the arm and slopes backwards through the medial intermuscular septum. The vessel is superficial throughout its course in the arm, lying immediately deep to the deep fascia. It passes into the cubital fossa (Fig. 2.34).

Branches. (*a*) The *profunda brachii* leaves through the lower triangular space, runs in the spiral groove with the radial nerve, supplies the triceps and ends as anterior and posterior anastomotic branches which join the cubital anastomosis.

(*b*) *Muscular* to flexor muscles and *nutrient* to the humerus.

(*c*) *Ulnar collateral*, accompanying the ulnar nerve.

(*d*) *Supratrochlear*, which divides into anterior and posterior branches to join the cubital anastomosis.

(*e*) *Terminal*, radial and ulnar. The axillary artery may bifurcate in the axilla or upper arm.

Veins of the Arm. Venæ comitantes accompany the brachial artery and all its branches. In addition, the **basilic** and **cephalic veins** course upwards through the subcutaneous tissue. The former perforates the deep fascia in the middle of the arm, the latter lies in the groove between deltoid and pectoralis major and ends by piercing the clavi-pectoral fascia to enter the axillary vein (Fig. 2.24 and PLATE 26, p. *xiv*).

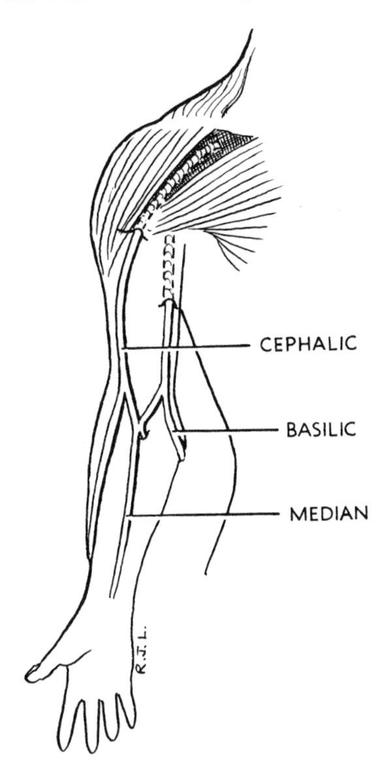

FIG. 2.24. THE SUPERFICIAL VEINS OF THE UPPER LIMB. The cephalic and basilic veins drain the dorsal venous arch of the hand (i.e., most of the blood from the palm).

Median Nerve. The nerve is formed at the lower border of the axilla by the union of its medial and lateral heads, from the corresponding cords of the brachial plexus. The axillary artery is clasped between the two heads, the medial head crossing in front of the vessel. The commencement of the nerve is thus lateral to the artery. Passing distally through the arm the nerve lies in front of the brachial artery and at the elbow is found on its medial side. The nerve has no branches in the arm (PLATE 1, p. *i*).

Musculo-cutaneous Nerve. Giving a twig to the shoulder joint and a branch to the coraco-brachialis muscle the nerve proceeds between the two conjoined parts of the coraco-brachialis (p. 73), and comes to lie between the biceps and brachialis muscles, accompanied by many branches of the brachial artery and tributaries of the brachial veins. While lying in this intermuscular plane it supplies both muscles. Its remaining fibres, purely cutaneous now, appear at the lateral margin of the biceps tendon as the lateral cutaneous nerve of the forearm (PLATE 1). The musculo-cutaneous is the nerve of the flexor compartment of the arm, supplying all three muscles therein.

Ulnar Nerve. Lying posterior to the vessels this nerve inclines backwards away from them and pierces the medial intermuscular septum in the lower third of the arm, accompanied by the ulnar collateral artery and a branch of the radial nerve to the medial head of the triceps muscle (the 'ulnar collateral' nerve). It gives no branch in the arm; its branch to the elbow joint comes off as it lies in the groove behind the medial epicondyle of the humerus.

Medial Cutaneous Nerve of the Arm. Lying anterior to the vessels this nerve pierces the deep fascia in the upper part of the arm and supplies the skin on the front and medial side of the upper part of the arm (Fig. 2.56, p. 106).

Medial Cutaneous Nerve of the Forearm. Commencing on the medial side of the vessels this nerve passes anterior to them and pierces the deep fascia on the medial side of the arm together with the basilic vein. It supplies skin over the lower part of the arm; the main part of the nerve passes into the medial side of the forearm (Fig. 2.56, p. 106).

Intercosto-brachial Nerve. The skin of the axilla is supplied by the lateral cutaneous branch of the second intercostal nerve. This nerve, known as the intercosto-brachial nerve, extends for a variable distance into the skin on the medial side of the upper arm (Fig. 2.56, p. 106). Not infrequently the lateral cutaneous branch of the third intercostal nerve also extends out to supply the skin of the axilla.

Delto-pectoral Lymph Nodes. These lie along the cephalic vein in the delto-pectoral groove, and receive lymphatics accompanying the vein along the pre-axial border of the limb. They are outlying members of the infraclavicular group (p. 65).

Supratrochlear Lymph Nodes. These lie in the subcutaneous fat just above the medial epicondyle. They drain the superficial tissues of the ulnar side of the forearm and hand, the afferent lymphatics running with the median basilic vein and its tributaries. Their efferent vessels pass to the lateral group of axillary nodes. They partake in the general lymphatic enlargement which accompanies secondary syphilis, a useful diagnostic point.

Extensor Compartment of the Arm

The extensor compartment is occupied by the triceps muscle, through which run the radial nerve and profunda artery. The ulnar nerve runs through the lower part of the extensor compartment.

Triceps Muscle. The three heads of this muscle lie lateral, medial and deep; but they are not so named. The medial head is called the long head, and the deep head is called the medial head, an unfortunate error in nomenclature which results in few students appreciating the position of the medial head (*v. inf.*). The **long head** arises from a low tubercle at the upper end of the axillary border of the scapula, the infraglenoid tubercle. The **lateral head** arises from the lateral lip of the spiral groove, above the deltoid tuberosity of the humerus. The long and lateral heads converge and fuse into a flattened tendon which lies superficially on the lower part of the muscle, and is inserted into the posterior part of the upper surface of the olecranon (Fig. 2.25). The **medial head** arises on the medial side of the spiral groove of the humerus. Thus in its upper part it lies medial to the lateral head, with radial nerve and profunda brachii vessels between them. Below the spiral groove, however, the origin widens to include the whole posterior surface of the humerus and of *each* intermuscular septum. The fibres of the medial head that arise from the lateral intermuscular septum actually occupy a *more lateral position*, during contraction, than those of the lateral head. The fibres of the medial head are inserted partly into the olecranon and partly into the deep part of the flattened tendon described above. A few fibres are inserted into the posterior part of the capsule of the elbow joint to prevent its being nipped in extension of the forearm (cf. articularis genu).

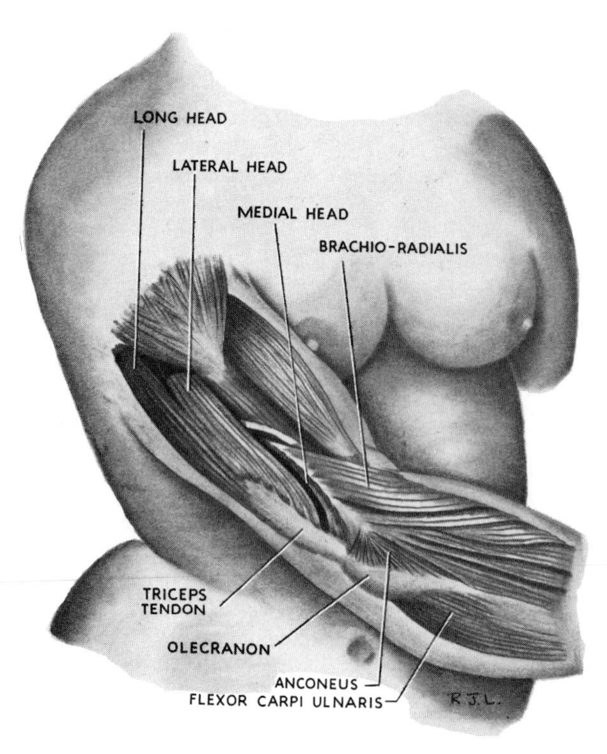

FIG. 2.25. THE CONSTITUTION OF THE TRICEPS MUSCLE. The so-called medial head extends to the lateral margin at the lower part of the muscle. Muscles drawn on a copy of Leonardo's *Leda*.

The nerve supply is from the radial (C7, 8).

Action. The muscle is the extensor of the elbow joint. The long head is an important factor in stabilizing the abducted shoulder joint, and it aids in extending the shoulder joint.

The Radial Nerve. Leaving the axilla as described at p. 65 the nerve accompanies the profunda brachii vessels into the spiral groove. Here the neuro-vascular bundle lies on the humerus between the lateral and medial heads of the triceps; the uppermost fibres of origin of the brachialis floor in the lower part of the spiral groove. Above, the nerve gives off branches to medial and lateral heads of triceps and a long branch which runs down through the medial head into the anconeus muscle. In the groove are given off the lower lateral cutaneous nerve of the arm and the posterior cutaneous nerve of the forearm. It is characteristic of the radial nerve to give off its branches at a level considerably proximal to the part to be innervated. The nerve leaves the extensor compartment by piercing the upper third of the lateral intermuscular septum and so gains the cubital fossa (p. 81). The nerve is easily palpated (feel your own!). It can be rolled on the humerus beneath the finger tip one-third of the way down from the deltoid tuberosity to the lateral epicondyle.

Profunda Brachii Artery. This is the vessel of supply to the triceps. At the lateral intermuscular septum it divides into anterior and posterior branches; only the former pierces the septum. Each runs down-

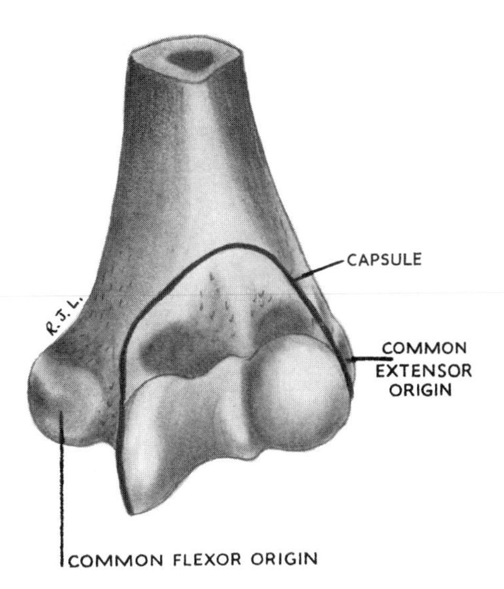

FIG. 2.26. THE LOWER END OF THE LEFT HUMERUS, to show the line of attachment of the capsule of the elbow joint. (See also Fig. 2.62, p. 115.)

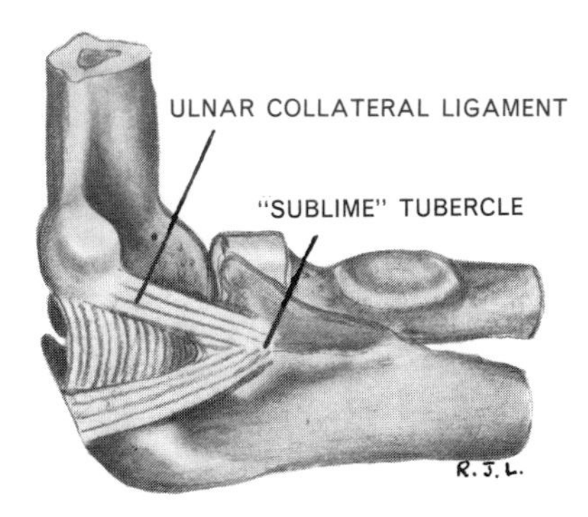

FIG. 2.27. THE ULNAR COLLATERAL (MEDIAL) LIGAMENT OF THE LEFT ELBOW JOINT.

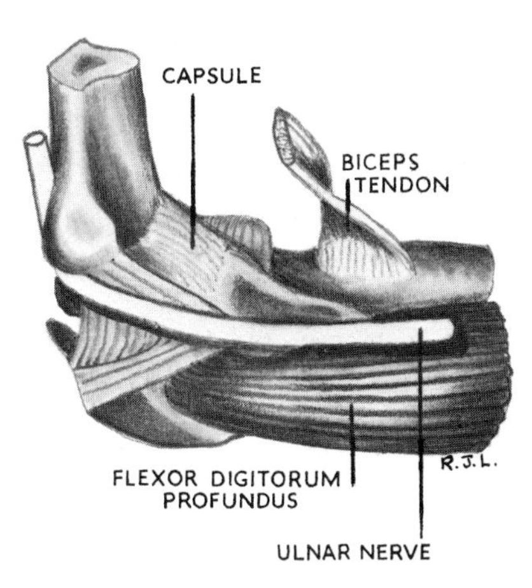

FIG. 2.28. THE VULNERABLE POSITION OF THE ULNAR NERVE ALONGSIDE THE 'SUBLIME' TUBERCLE.

wards, in front of and behind the elbow, to join the cubital anastomosis (Fig. 2.34).

Ulnar Nerve. The nerve courses through the lower part of the extensor compartment and disappears into the forearm by passing between the humeral and ulnar heads of origin of flexor carpi ulnaris. It lies in contact with the bone in the groove behind the medial epicondyle, then lies against the medial ligament of the elbow joint, to which it gives a twig of supply (Fig. 2.41).

The Elbow Joint

This is a synovial joint of the hinge variety between the lower end of the humerus and the upper ends of radius and ulna. It communicates with the superior radio-ulnar joint, in contrast to the wrist, which does not communicate with the inferior radio-ulnar joint.

The **articular surfaces** are coated with hyaline cartilage.

The lower end of the **humerus** shows the prominent conjunction of capitulum and trochlea. The *capitulum*, for the head of the radius, is a portion of a sphere. It projects forward, and also downwards, where its lower border lies at the distal extremity of the humerus. In contrast the *trochlea*, which lies medial, is a grooved surface that extends around the lower end of the humerus to the posterior surface of the bone. The groove of the trochlea is limited medially by a sharp and prominent ridge and laterally by a lower and blunter ridge that blends with the articular surface of the capitulum (Fig. 2.26). Thus is produced a tilt on the lower end of the humerus that accounts in part for the carrying angle of the elbow. Fossæ immediately above the capitulum and trochlea receive the head of the radius and coronoid process of the ulna in full flexion; posteriorly a deep fossa receives the olecranon in full extension (Fig. 2.62, p. 115).

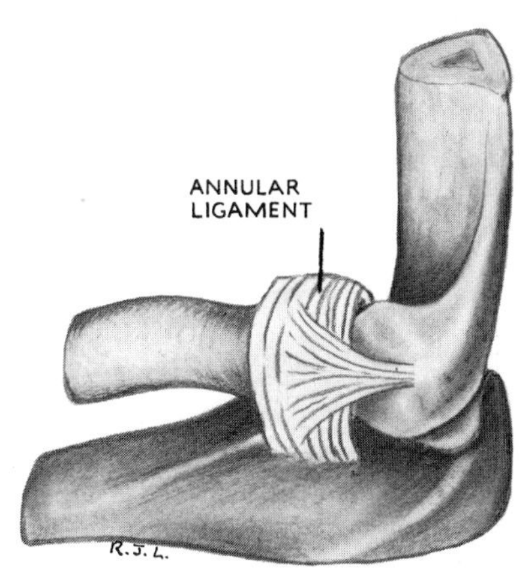

FIG. 2.29. THE RADIAL COLLATERAL (LATERAL) LIGAMENT OF THE LEFT ELBOW JOINT, JOINING THE LATERAL EPICONDYLE TO THE ANNULAR LIGAMENT.

The upper surface of the cylindrical head of the **radius** is spherically concave to fit the capitulum.

The upper end of the **ulna** shows the deep trochlear notch. A curved ridge joins the prominences of coronoid process and olecranon; the ridge fits the groove in the trochlea of the humerus. The obliquity of the shaft of the ulna to this ridge accounts for most of the carrying angle at the elbow (Fig. 2.63, p. 118). There are commonly two separate articular surfaces in the trochlear notch, one on the olecranon and the other on the coronoid process.

The **capsule** is attached to the humerus at the margins of the lower rounded ends of the articular surfaces of capitulum and trochlea, but in front and behind it is carried up over the bone above the coronoid and olecranon fossæ (Figs. 2.26 and 2.62, p. 115). Distally, the capsule is attached to the trochlear notch of the ulna at the edge of the articular cartilage, and to the annular ligament of the superior radio-ulnar joint (Fig. 2.28). It is not attached to the radius.

Synovial Membrane. The capsule and annular ligament are lined with synovial membrane, which is attached to the articular margins of all three bones. The synovial membrane thus floors in the coronoid and olecranon fossæ on the lower end of the humerus, and bridges the gap between the radial notch of the ulna and the neck of the radius. The quadrate ligament (p. 85) prevents herniation of the synovial membrane between the anterior and posterior free edges of the annular ligament.

Ligaments. The **ulnar collateral (medial) ligament** of the elbow joint is triangular. Of its three constituents the *anterior band* is the strongest. It passes from the medial epicondyle of the humerus to the sublime tubercle on the medial border of the coronoid process. The *posterior band* joins the coronoid process and the medial border of the olecranon. A *middle band* connects these two and lies more deeply; it lodges the ulnar nerve on its way from the arm to the forearm (Fig. 2.28). The **radial collateral (lateral) ligament** is a flattened band attached to the humerus below the common extensor origin; it fuses with the annular ligament of the head of the radius. The anterior and posterior ligaments are merely thickened parts of the capsule. The **annular ligament** is attached to the margins of the radial notch of the ulna, and clasps the head *and neck* of the radius in the superior radio-ulnar joint. It has no attachment to the radius, which remains free to rotate in the annular ligament.

Nerve Supply. The joint is supplied by the musculo-cutaneous, median, ulnar and radial nerves (Hilton's Law).

Movements. The only appreciable movement possible at the elbow joint is the simple hinge movement of flexion and extension. This movement does not take place in the line of the humerus, for the axis of the hinge lies obliquely. The extended ulna makes with the humerus an angle of about 170 degrees. This so-called '*carrying-angle*' fits the elbow into the waist when the arm is at the side, and it is significant that the obliquity of the ulna is more pronounced in women than in men. The line of upper arm and forearm, however, becomes straightened out when the forearm is in the usual working position of almost full pronation (Fig. 2.35, p. 85). During pronation-supination of the forearm there is some rocking movement of the ulna on the trochlea (see p. 87).

THE FLEXOR COMPARTMENT OF THE FOREARM

The flexor muscles in the forearm are arranged in two groups, superficial and deep. The five muscles of the superficial group cross the elbow joint; the three muscles of the deep group do not. The flexor compartment is much more bulky than the extensor compartment, for the necessary power of the grip. The extensor muscles merely release the grip.

Superficial Muscles

These five muscles are distinguished by the fact that they possess a common origin from the medial epicondyle of the humerus, at its anterior surface. Three of the group have additional areas of origin. The **common origin** attaches itself to a smooth area on the anterior surface of the medial epicondyle (Fig. 2.26); there is but little flesh attached to the humerus, for at this point the intermuscular septa and the muscles themselves are fused into a tendinous mass. With the heel of the hand placed over the opposite medial epicondyle, palm lying on the forearm, the fingers point down along the *five superficial muscles*—thumb for pronator teres, index for flexor carpi radialis, middle finger for flexor digitorum superficialis, ring finger for palmaris longus, little finger for flexor carpi ulnaris.

Pronator Teres. Arising from the common origin and from the lower part of the medial supracondylar ridge the superficial belly is joined by the deep head, which latter arises from the medial border of the coronoid process of the ulna just distal to the sublime tubercle. The median nerve lies between the two heads. The muscle, forming the medial border of the cubital fossa, runs distally across the front of the forearm to be inserted by a flat tendon into the most prominent part of the lateral convexity of the radius, between the lower ends of the anterior and posterior oblique lines. The attachment lies just *behind* the lateral profile of the radius (Figs. 2.41, 2.64, p. 119). Supplied by the median nerve (C6) the muscle pronates the forearm and flexes the elbow. In simple pronation, the flexing action of the muscle requires opposition by triceps. Its oblique origin from the humerus is functional; it exerts an adducting pull on the radius (and thus on the ulna) to oppose the abduction of anconeus (see p. 88).

Flexor Carpi Radialis. Arising from the common origin the fleshy belly lies distal to pronator teres. In the middle of the forearm the flesh gives way to a flattened tendon which becomes rounded at the wrist, where it first perforates the flexor retinaculum, and then lies in the groove of the trapezium. It passes deep to the tendon of flexor pollicis longus and is inserted into the bases of metacarpals II and III (symmetrically with extensors longus and brevis). Quite frequently it gives a slip of insertion into the scaphoid bone. The absence of fleshy fibres distinguishes the tendon above the wrist from that of flexor pollicis longus, which is 'feathered' by fleshy fibres joining it from the radial side (Figs. 2.30, 2.32). Supplied by the median nerve (C6, 7), the muscle is a flexor and radial abductor of the wrist.

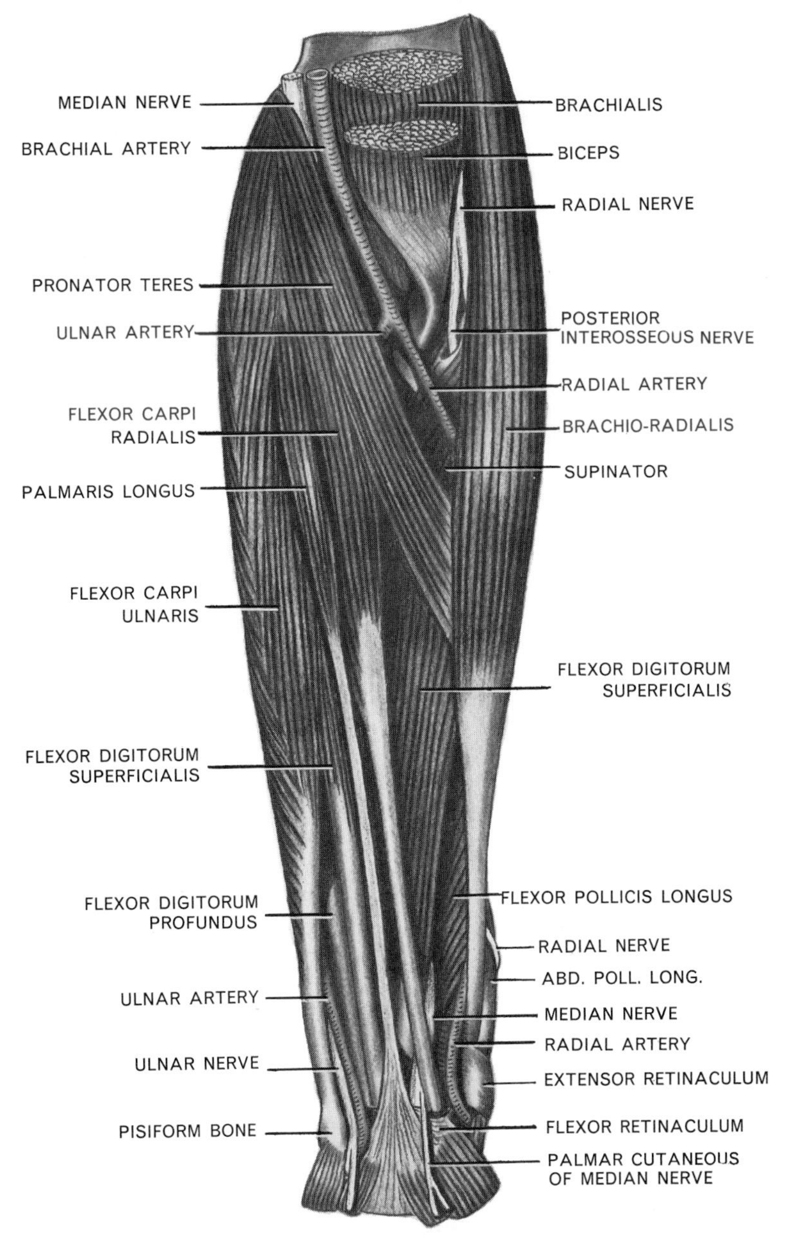

FIG. 2.30. THE FLEXOR COMPARTMENT OF THE LEFT FOREARM. SUPERFICIAL MUSCLES.

Flexor Digitorum Superficialis (sublimis). It arises from the common origin, the medial ligament of the elbow joint, and the sublime tubercle on the medial border of the coronoid process of the ulna. From this point a fibrous arch continues the origin across to the radius, where the muscle arises from the whole length of the anterior oblique line. The fleshy belly is partly hidden above by the other superficial flexors, and is therefore frequently described as being in an intermediate layer. Its oblique origin, in continuity from the medial epicondyle to the insertion of pronator teres, forms the upper limit of the space of Parona. Above the wrist the tendons appear on each side of palmaris longus tendon. The flesh which gives rise to the tendons for the middle and ring fingers is superficial across the wide origin of the muscle. That giving rise to the tendons of the index and little fingers arises more deeply from the common flexor origin and ulnar collateral (medial) ligament of the elbow. It lies beneath the belly for the middle two fingers, and is divided by a transverse tendinous intersection into an upper and two lower bellies. The tendons pass in the above order beneath the flexor retinaculum, i.e., middle and ring fingers superficial to index and little finger. Their course in the palm and insertion into the middle phalanges is considered on p. 103. In the forearm the muscle has the median nerve plastered to its deep surface by areolar tissue; the commencing profundus tendon for the index finger can readily be mistaken for the nerve at operation or in a dissection. Supplied by the median nerve (C7, 8) the muscle is a flexor of the proximal interphalangeal joints, and secondarily of the metacarpo-phalangeal and wrist joints.

Palmaris Longus. Functionally negligible, this muscle is of morphological interest. It is absent in one-tenth of individuals. It is phylogenetically degenerating and shows the characteristics of this, viz., short belly and long tendon (cf. plantaris). Replacement of its distal tendon by ligament (palmar aponeurosis) is a further characteristic of degeneration (cf. coccygeus, sacrospinous ligament). It is supposed that the muscle once existed as a flexor of the proximal phalanges, with its tendons lying in the palm superficial to those of flexor digitorum superficialis, and splitting around them to be attached to the proximal phalanges. The muscle arises from the common origin and its tendon is inserted, adherent across the front of the flexor retinaculum, into the palmar aponeurosis. It is supplied by the median nerve (C6, 7) and flexes the wrist.

Flexor Carpi Ulnaris. Arising from the common origin the muscle receives a further contribution from a wide aponeurosis which arises from the medial border of the olecranon and the upper three-fourths of the subcutaneous border of the ulna. This aponeurosis lies superficial to the belly of flexor digitorum profundus—the latter gives the bulk to the medial side of the upper forearm. It lies edge to edge, at the subcutaneous border of the ulna, with the aponeurosis of extensor carpi ulnaris (Fig. 2.39). The ulnar nerve enters the flexor compartment of the forearm by passing between the humeral and ulnar heads of this muscle. The tendon of insertion of flexor carpi ulnaris passes to the pisiform. The pisiform is morphologically a sesamoid bone in the tendon; by way of the piso-hamate and piso-metacarpal ligaments the muscle acts on a wider insertion. Its symmetry with the insertion of extensor carpi ulnaris into the base of the fifth metacarpal is maintained by way of the piso-metacarpal ligament. It is supplied by the ulnar nerve; but in 95 per cent of cases the fibres have come from the lateral cord or direct from the middle trunk by a communication in the axilla. The segmental value of this communication is C6, 7, not the ordinary C8, Th. 1 of the remainder of the ulnar nerve. It is a flexor of the wrist, an ulnar adductor when acting with extensor carpi ulnaris, and a fixator of the pisiform during contraction of the hypothenar muscles.

The Cubital Fossa. This is by definition the triangular area between pronator teres, brachio-radialis and a line joining the humeral epicondyles. It possesses a roof and floor, and certain structures pass through it.

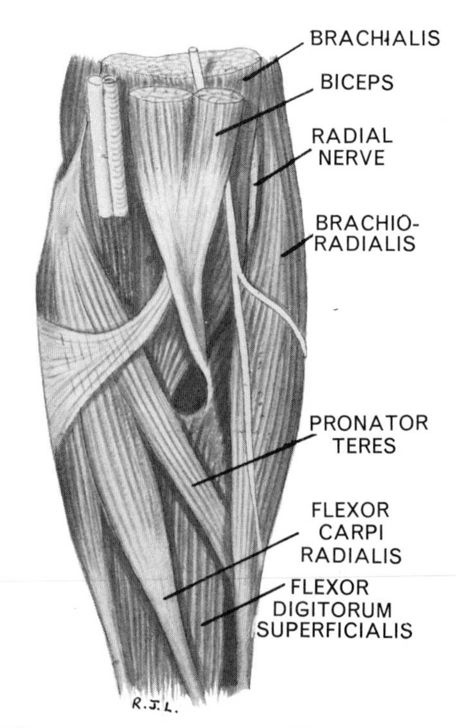

FIG. 2.31. THE LEFT CUBITAL FOSSA. The musculo-cutaneous nerve emerges lateral to the biceps tendon and descends as the lateral cutaneous nerve of the forearm. The bicipital aponeurosis fans away from the medial border of the biceps tendon. The median nerve and brachial artery have been divided alongside the biceps tendon (see PLATE 2).

The roof is formed by the deep fascia of the forearm, reinforced on the medial side by the bicipital aponeurosis. On the bicipital aponeurosis lies the median basilic vein with the medial cutaneous nerve of the forearm on its medial side; the aponeurosis separates these structures from the underlying median nerve and brachial artery. Laterally on the roof lie the lateral cutaneous nerve of the forearm and the median cephalic vein. The floor is formed in the main by the brachialis muscle; below and laterally the supinator clasps the neck of the radius. If the bordering muscles are separated there can be seen in the fossa, from medial to lateral side, the median nerve, brachial artery, tendon of biceps, posterior interosseous and radial nerves. The median nerve and brachial artery are discussed below (pp. 83, 84).

The *tendon of the biceps*, rotating so that its anterior surface faces laterally, passes deeply through the fossa lateral to the midline and disappears between the two bones of the forearm (Fig. 2.28). Its range of movement from full supination to full pronation is 6 cm. A bursa separates it from the anterior part of the bicipital tuberosity of the radius. The *radial nerve* appears high in the fossa between the brachialis and the brachioradialis muscles. It gives a branch to each and, lower down, another branch to extensor carpi radialis longus. It then divides into its two terminal branches, viz., radial and posterior interosseous nerves. The latter gives branches to extensor carpi radialis brevis and supinator and disappears from the fossa by passing between the two heads of origin of the supinator muscle. The radial nerve passes down the forearm under cover of the brachio-radialis (p. 84).

Deep Flexor Muscles

The group consists of three named muscles, viz., flexor digitorum profundus, flexor pollicis longus and pronator quadratus. The two former are subdivisions of a common deep flexor from which a third, flexor indicis profundus, appears to be phylogenetically emerging (Fig. 2.32).

Flexor Digitorum Profundus. The most powerful and the bulkiest of the forearm muscles, it arises by fleshy fibres *from the medial surface of the olecranon* (Fig. 2.28), from the upper three-fourths of the medial and anterior surfaces of the ulna as far distally as pronator quadratus, and from a narrow strip of interosseous membrane. The tendon for the index separates in the forearm; the three other tendons are still partly attached to each other as they pass across the carpal bones in the flexor tunnel and do not become detached from each other until they reach the palm (Fig. 2.32). At this point of separation the four lumbricals take origin. They are described in the section on the hand (p. 99). The muscle is innervated by the anterior interosseous branch of the median nerve and by the ulnar nerve. Characteristically these nerves share equally; the bellies which merge into the tendons for index and middle fingers being supplied from the median, the ring and little fingers from the ulnar nerves. The corresponding lumbricals are similarly supplied.

This distribution of 2 : 2 between median and ulnar nerves occurs in only two-thirds of individuals (60 per cent). In the remaining 40 per cent the median and ulnar distribution is 3 : 1 or 1 : 3 equally (20 per cent each). Whatever the variation, however, the rule is that each lumbrical is supplied by the same nerve which innervates the belly of its parent tendon. The segmental value for the flexor digitorum profundus is C7, 8; the action of the muscle is to flex the terminal interphalangeal joints and, still acting, to roll up the fingers and wrist. It is the great gripping muscle. Extension of the wrist is indispensable to the full power of contraction of the muscle (Fig. 2.54, p. 105).

Flexor Pollicis Longus. This muscle arises from the anterior surface of the radius below the anterior oblique line and above the insertion of pronator quadratus, and from an adjoining strip of interosseous membrane. Some fibres, arising in common with flexor digitorum superficialis, join the muscle from either the medial or the lateral border (or both) of the coronoid process of the ulna. The oblique cord represents phylogenetically degenerate fibres of the upper part of the muscle. The tendon forms on the ulnar side of this unipennate muscle; it should be noted that the tendon receives fleshy fibres into its radial side to a point just above the wrist, a useful point in distinguishing it from the flexor carpi radialis in window dissections or in wounds above the wrist (Fig. 2.32). The tendon passes in the carpal tunnel deep to that of the flexor carpi radialis, then spirals around its ulnar side to become superficial. It extends into the thumb to be inserted into the base of the distal phalanx. It is supplied by the anterior interosseous branch of the median nerve (C7, 8) and its action is indicated by its name.

Pronator Quadratus. Arising from the lower fourth of the ulna, especially from the sinuous ridge on its antero-medial aspect, the fibres are inserted, superficially into the ridge on the antero-lateral border of the radius above the styloid process, more deeply into the anterior surface of the lower end of the radius, and most deeply into the triangular interosseous area just above the ulnar notch. The interosseous membrane clings to the posterior border of this triangular area, the membrana sacciformis lies just in front of it, and the muscle is inserted into the remainder of the area. It is supplied by the anterior interosseous branch of the median nerve (C6). The superficial fibres pronate the forearm (note the extent of separation of the ridges of attachment in pronation and supination of the dried bones). The deepest fibres possess almost the same length in pronation and supination, and their action would appear to be chiefly concerned with maintaining apposition of the lower ends of radius and ulna.

The Space of Parona. In front of pronator quadratus and deep to the long flexor tendons of the fingers is a space into which the proximal parts of the flexor synovial sheaths protrude. The space is limited proximally by the oblique origin of flexor digitorum superficialis. The space becomes involved in proximal extensions of synovial sheath infections; it can easily be drained through radial and ulnar incisions to the side of the flexor tendons.

The Neuro-Vascular Pattern in the Forearm

It is desirable to have an outline of the general arrangement of the deep arteries and nerves of the forearm. A nerve runs down each border of the forearm (radial and ulnar nerves). The brachial artery divides into branches (radial and ulnar arteries) that run down to approach these nerves but do not cross them. The

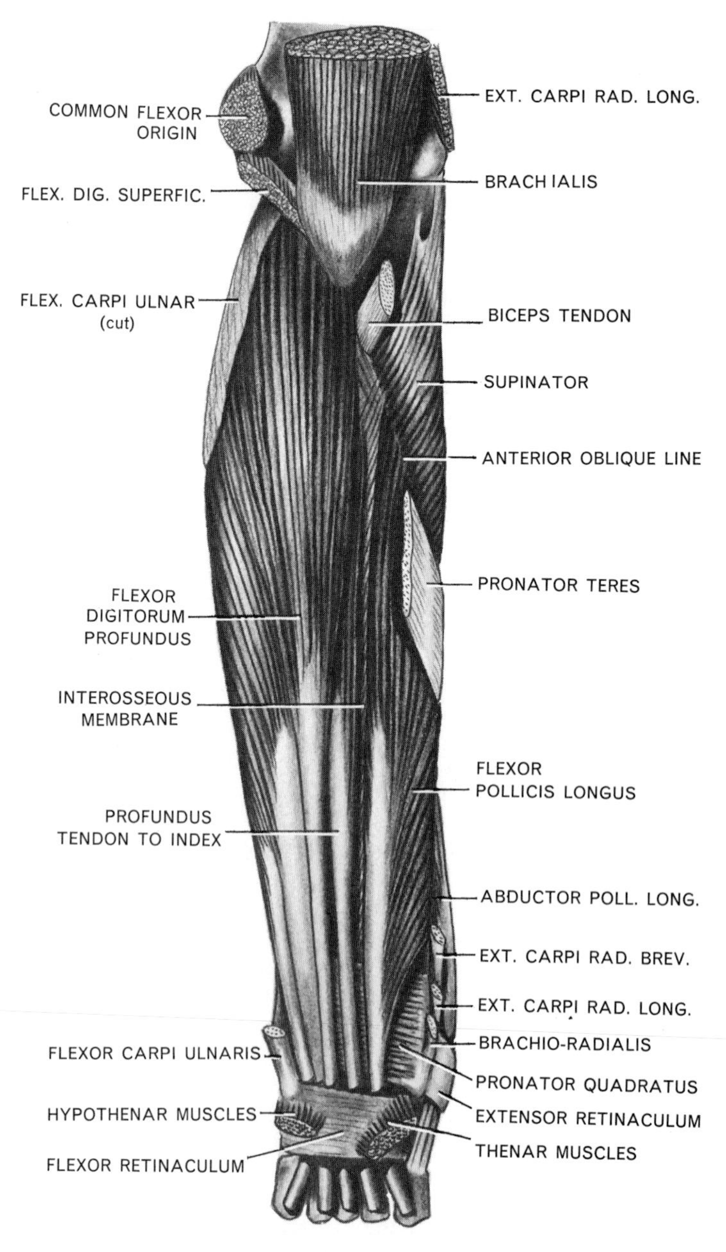

FIG. 2.32. THE FLEXOR COMPARTMENT OF THE LEFT FOREARM. DEEP MUSCLES.

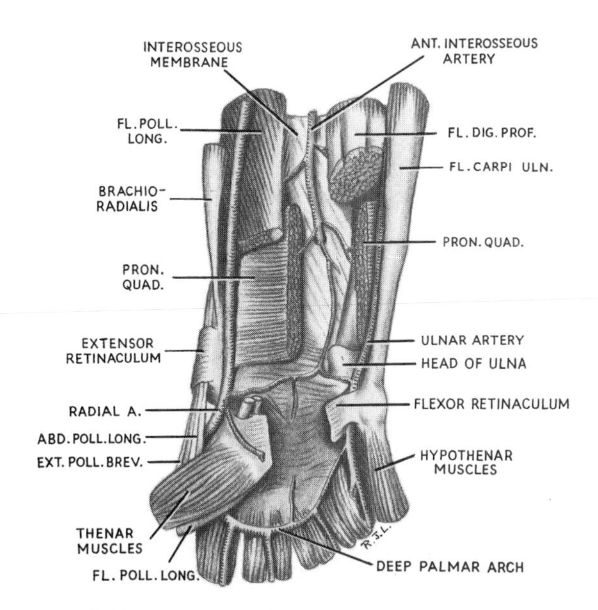

FIG. 2.33. DEEP STRUCTURES IN FRONT OF THE RIGHT WRIST.

median nerve, on the deep surface of flexor superficialis, crosses the ulnar artery to lie between the two. Radial and ulnar arteries supply the hand; they run down directly into deep and superficial palmar arches. The arterial supply for the forearm comes from the common interosseous branch of the ulnar, which divides into posterior and anterior interosseous arteries. The posterior interosseous artery is rather a failure. Assisted at first by branches of the anterior interosseous that pierce the interosseous membrane, it later fails and is replaced by the anterior interosseous artery, which pierces the membrane to enter the extensor compartment (Fig. 2.33). Anterior (from median) and posterior (from radial) interosseous nerves, on the other hand, remain in their own compartments right down to the wrist, supplying muscles, periosteum and carpal ligaments. Neither nerve reaches the skin.

Three nerves share in the supply of the muscles of the forearm and each nerve passes between the two heads of a muscle. The median nerve passes between the two heads of pronator teres and the ulnar nerve passes between the two heads of flexor carpi ulnaris. These two nerves share in the supply of the muscles of the flexor compartment. The muscles of the extensor compartment are supplied by the posterior interosseous nerve, which enters the compartment by passing between the two layers of the supinator muscle.

The Vessels of the Flexor Compartment

The brachial artery enters the forearm by passing into the cubital fossa in the midline; halfway down the fossa it divides into radial and ulnar arteries.

The **radial artery** passes distally medial to the biceps tendon, across the supinator, over the tendon of insertion of the pronator teres, the origin of the flexor pollicis longus, and the lower part of the radius. It disappears beneath the tendons of abductor pollicis longus and extensor pollicis brevis to cross the anatomical snuff box (Fig. 2.42). In the upper part of the forearm it is covered anteriorly by the brachio-radialis muscle. In the middle third of its course it has the radial nerve lateral to it.

The **ulnar artery** disappears from the cubital fossa by passing deep to the deep head of pronator teres (PLATE 2) and beneath the fibrous arch of the flexor digitorum superficialis near the median nerve. It then leaves the median nerve and lies on flexor digitorum profundus with the ulnar nerve to its ulnar side and passes down over the front of the wrist into the palm. Its chief branch is the **common interosseous,** which divides into anterior and posterior interosseous branches. The **anterior interosseous artery** lies deeply on the interosseous membrane between flexor digitorum profundus and flexor pollicis longus, supplying each. Perforating branches pierce the interosseous membrane to supply the deep extensor muscles. Nutrient vessels are given to both radius and ulna. The artery passes posteriorly through the interosseous membrane at the level of the upper border of pronator quadratus (cf. peroneal artery in the leg) (Fig. 2.33).

The **posterior interosseous artery** disappears by passing backwards through the interosseous space above the upper end of the interosseous membrane but distal to the oblique cord (cf. anterior tibial artery in the leg).

Anastomosis Around the Elbow Joint. Recurrent branches, in some cases double, arise from radial, ulnar and interosseous arteries and run upwards both anterior and posterior to the elbow joint, to anastomose with descending articular branches of the profunda brachii, ulnar collateral and supratrochlear arteries (Fig. 2.34).

Anastomosis Around the Wrist Joint. Both radial and ulnar arteries give off anterior and posterior carpal branches. These anastomose with each other deep to the long tendons, forming the anterior and posterior carpal arches. The **anterior carpal arch** lies transversely across the wrist joint (Fig. 2.33); it supplies the carpal bones and sends branches distally into the hand to anastomose with the deep palmar arch. The **posterior carpal arch** lies transversely across the distal row of carpal bones (p. 94). It sends dorsal metacarpal arteries distally into each metacarpal space; these divide to supply the fingers and they anastomose through the interosseous spaces with the palmar digital and metacarpal branches of the palmar arches. Thus a free anastomosis is established between radial and ulnar arteries through the carpal and palmar arches (Fig. 2.33).

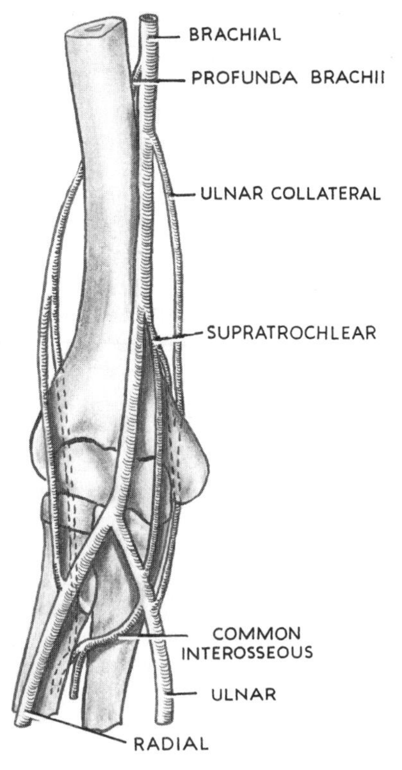

FIG. 2.34. THE ANASTOMOSES AROUND THE ELBOW JOINT (right side). The brachial artery gives off the profunda, ulnar collateral and supratrochlear arteries. These three branches anastomose with recurrent branches from the radial, ulnar and posterior interosseous arteries.

Veins of the Forearm. The **deep veins** are plentiful and accompany the arteries, usually by double venæ comitantes which anastomose freely with each other. They drain the forearm but bring relatively little blood from the hand.

The **Superficial Veins** (Fig. 2.24). Most of the blood from the palm of the hand passes through to a superficial venous arch on the dorsum. From the radial side of this arch the **cephalic vein** runs up along the pre-axial border of the limb. It runs in the upper arm lateral to biceps, in the delto-pectoral groove, and perforates the clavi-pectoral fascia to drain into the axillary vein. From the ulnar side of the dorsal venous arch the **basilic vein** runs up the postaxial border of the limb. It pierces the deep fascia half-way between elbow and axilla and joins the brachial veins to form the axillary vein.

The **median vein** drains subcutaneous tissue of the front of the wrist and forearm. It divides at the elbow into *median cephalic* and *median basilic*; the latter receives a deep vein. These two veins open into the cephalic and basilic veins respectively. The deep communicating vein joins the median basilic vein and makes it a larger vessel than the median cephalic. The median cephalic is much less movable in the subcutaneous tissues than the median basilic, and therefore is often more convenient to use in intravenous therapy in spite of its smaller size.

Lymphatics of the Forearm. As elsewhere in the body the superficial lymphatics follow veins, the deep ones follow arteries. From the ulnar side of the hand and forearm the subcutaneous lymphatics run alongside the basilic vein to the supratrochlear nodes. From the radial side the lymphatics run alongside the cephalic vein to the delto-pectoral and infraclavicular nodes. From the deep parts of the hand and forearm and from the supratrochlear nodes lymphatics pass to the lateral group of axillary nodes.

The Nerves of the Flexor Compartment

The **lateral cutaneous nerve of the forearm** pierces the deep fascia at the level of the elbow joint and supplies the pre-axial border of the forearm, by anterior and posterior branches, as far distally as the ball of the thumb. The **medial cutaneous nerve of the forearm** is roughly symmetrical with the lateral cutaneous nerve, supplying the postaxial border of the forearm by anterior and posterior branches as far distally as the wrist. Their areas of cutaneous distribution meet at the anterior axial line (p. 26), where there is minimal overlap (Fig. 2.56, p. 106).

The **radial nerve,** the cutaneous continuation of the main radial nerve, runs from the cubital fossa on the surface of supinator, pronator teres tendon and flexor digitorum superficialis on the lateral side of the forearm under cover of the brachio-radialis muscle. While under cover of the latter muscle it lies on the radial side of the radial artery. It leaves the flexor compartment of the forearm by passing backwards under the tendon of brachio-radialis a couple of inches (a few cm) above the radial styloid, and breaks up into two or three branches which can be rolled on the surface of the tautened tendon of extensor pollicis longus. They are distributed to the radial two-thirds of the dorsum of the hand and the proximal parts of the dorsal surfaces of thumb, index, middle and half of the ring fingers (Fig. 2.57, p. 107).

The **median nerve** while still in the cubital fossa gives a branch to pronator teres then disappears between the two heads of that muscle. Beyond the muscle it supplies flexor carpi radialis and palmaris longus, and then joins the ulnar artery at the fibrous arch of the superficial flexor. Deep to flexor digitorum superficialis it passes distally in the midline of the forearm closely attached by areolar tissue to the deep surface of the muscle, which it supplies. It emerges from the lateral border of the muscle (Fig. 2.30), between palmaris longus and flexor carpi radialis tendons, then passes beneath the flexor retinaculum and enters the hand (p. 98). It is nourished by the **median artery,** a branch of the anterior interosseous artery, which was the original axial artery of the fœtal limb. It may persist as a large vessel.

The median nerve gives off an **anterior interosseous** branch which runs down with the artery of the same name and supplies flexor digitorum profundus (usually the bellies which move index and middle fingers), flexor pollicis longus and pronator quadratus, also the interosseous membrane and the periosteum of the radius and ulna. It ends in the anterior part of the capsule of the wrist joint and the carpal joints, which it supplies. It has no cutaneous branch.

Just above the flexor retinaculum the median nerve gives off a **palmar branch** to the skin over the thenar muscles.

The **ulnar nerve** enters the forearm from the extensor compartment by passing between the humeral and ulnar heads of origin of flexor carpi ulnaris. It is more easily compressed against the medial surface of the coronoid process (Fig. 2.28) than against the humerus where it lies behind the medial epicondyle. In the forearm the nerve lies under cover of the flattened aponeurosis of flexor carpi ulnaris with the ulnar artery to its radial side. This neuro-vascular bundle lies on flexor digitorum profundus. Branches of supply are given to flexor carpi ulnaris and the ulnar half (usually) of flexor digitorum profundus. The branch to flexor carpi ulnaris contains C6, 7 fibres brought to the ulnar nerve in the axilla; the branch to flexor digitorum profundus contains C7, 8 fibres.

The ulnar nerve and ulnar artery emerge from beneath the tendon of flexor carpi ulnaris just proximal to the wrist and pass across the flexor retinaculum into the hand. Before emerging each gives off a dorsal branch which passes medially between the tendon of flexor carpi ulnaris and the lower end of the ulna. The dorsal branch of the ulnar nerve supplies the dorsum of the hand (Fig. 2.57, p. 107) and of the ulnar one and a half fingers proximal to their nail beds. The dorsal branch of the ulnar artery enters the posterior carpal arch.

Radio-Ulnar Articulations

The **superior joint** has been mentioned in connection with the elbow joint (p. 78). The essential structure here is the *annular (orbicular) ligament* which imprisons the head of the radius. The annular ligament is attached to the radial notch of the ulna and its fibres encircle the head and *neck* of the radius. The ligament has no attachment to the radius, which is free to rotate within it. Superiorly it blends with the capsule of the elbow joint. The superior radio-ulnar joint and the elbow joint form one continuous synovial cavity. The synovial membrane is attached to the radius at the lower margin of the cylindrical articular surface; it is supported between the ulna and the radius by the *quadrate ligament*. This stretches between the neck of the radius,

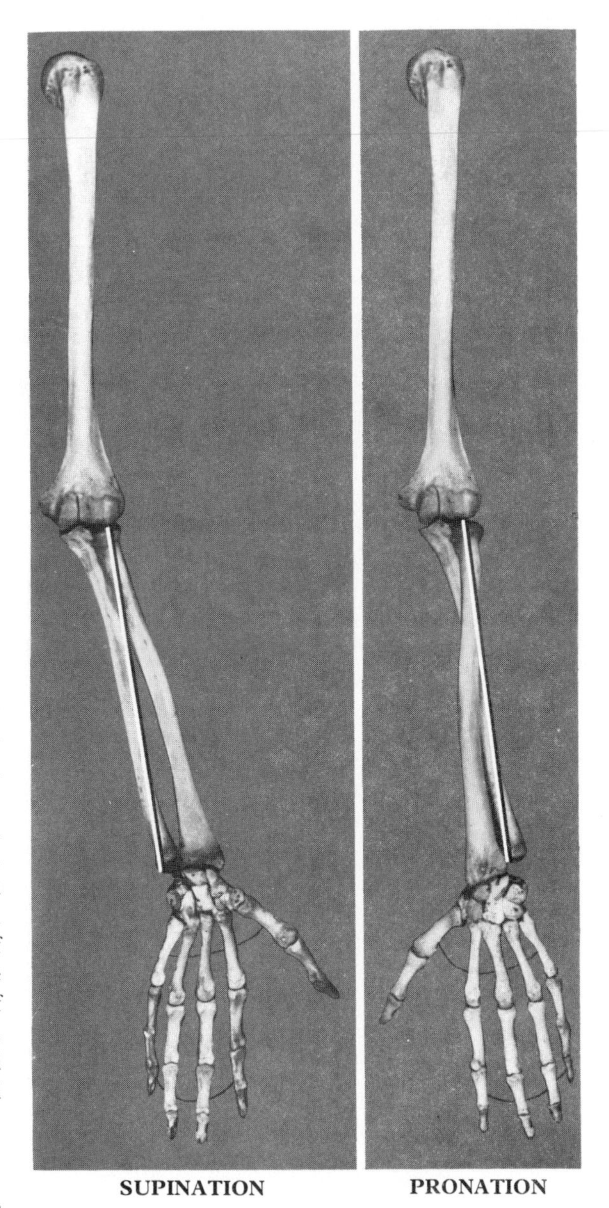

FIG. 2.35. THE CARRYING ANGLES OF THE ELBOW AND THE WRIST (photograph). The bones of the left upper limb viewed from in front. The axis of pronation-supination is shown; it is identical in both supination and pronation if the humerus does not move. In supination the third metacarpal lies parallel with the shaft of the humerus; the forearm itself lies oblique to these, making the carrying angles at the elbow and the wrist. In pronation the axis of the forearm comes into line with the shaft of the humerus and with the axis of pronation-supination; the carrying angle at the elbow apparently disappears, though in fact the ulna has not moved.

proximal to the bicipital tuberosity, and the upper part of the supinator fossa of the ulna just distal to the radial notch. Its fibres run criss-cross, so that some are tense while others relax; its overall tension remains constant in all positions of supination and pronation.

The **inferior radio-ulnar joint** is closed distally by a triangular fibro-cartilage, which is attached by its base to the ulnar notch of the radius and by its apex to a small fossa at the base of the ulnar styloid. The capsule is loose and pouches upwards between the two bones behind the surface of the deep part of pronator quadratus, forming the *membrana sacciformis.* It is unusual for the fibro-cartilage to be incomplete; in such rare cases, of course, the joint communicates with the wrist joint (Fig. 2.44).

The **interosseous membrane** joins the interosseous borders of the two bones. Its fibres run from the radius down to the ulna at an oblique angle, and are supposed to have an effect in transmitting thrust from the wrist to the elbow via lower end of radius to upper end of ulna and so to the humerus. It tends to be relaxed when the weight is suspended from the hands; but in this case, of course, the weight is suspended by muscles, not by bones or ligaments. It is *taut in pronation* and lax in supination.

The *oblique cord* represents a degenerated portion of the flexor pollicis longus muscle. Its fibres run in opposite obliquity to those of the interosseous membrane. The cord slopes upwards from just below the radial tuberosity to the side of the ulnar tuberosity, distal to the quadrate ligament. The posterior interosseous vessels pass through the gap between the oblique

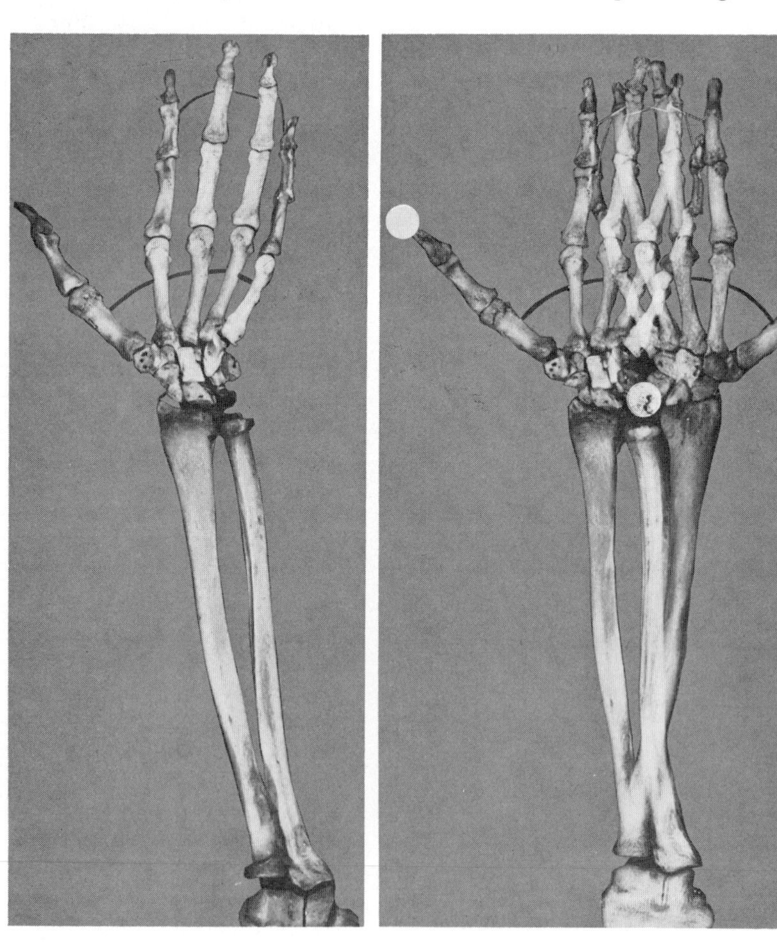

FIG. 2.36. THE MECHANISM OF PRONATION.

A. The articulated bones of the left forearm photographed in supination.

B. Double exposure photograph superimposing on **A** the pronated radius and hand. The ulna has been kept stationary against the marker disc at its lower end. The axis of pronation passes through the tip of the middle finger in this photograph. Note the wide excursion of the tip of the thumb away from its marker disc. The metacarpals and phalanges can be identified where superimposed by identifying the bases of the metacarpals from the thumb side and then tracing distally to the digits. Note that with a fixed ulna the centre of the palm must move, and the grip on a rotating fixture is lost. Compare Fig. 2.37.

cord and the upper end of the interosseous membrane.

Nerve Supply. In accordance with Hilton's law (p. 15) both joints and the interosseous membrane are innervated from the median nerve and its anterior interosseous branch (the pronator muscles being the operative structures).

Movements. Examine the articulated bones of the upper limb. The *basic* movements of pronation and supination are produced by movements of radius (and hand) around an immobile ulna. The axis around which the radius rotates obviously passes through the centre of curvature of each radio-ulnar joint. The *axis of pronation-supination* is oblique along the supinated forearm—it joins the centre of the head of the radius and the base of the styloid process of the ulna, and when prolonged passes near the little finger. It lies in the line of the shaft of the humerus, hence the carrying angle of the elbow (Fig. 2.35). For the ulna, as a result of the opposite curvatures of its upper and lower articular surfaces, must lie oblique to the axis of pronation-supination and therefore oblique to the shaft of the humerus, and this obliquity constitutes the carrying angle.

The axis of pronation-supination is fixed in relation to the ulna, but the *axis of the forearm itself* is not fixed in relation to the ulna. The axis of the forearm runs from the mid-point between the epicondyles of the humerus down to the mid-point between the styloid processes of the radius and ulna. The axis of the forearm is therefore constantly changing with the position of the bones throughout the range of pronation and supination. This can easily be seen by inspection of the dry bones and can be confirmed in the living forearm. It is illustrated in Fig. 2.35. In supination the axis of the forearm is parallel with the ulna; it is oblique to the shaft of the humerus (carrying angle). In pronation the axis of the forearm crosses the ulna obliquely and lies parallel with the axis of pronation-supination; it is in line with the shaft of the humerus though the ulna has not moved. The carrying angle exists so that the forearm may be in line with the humerus in the working position, which is that of almost full pronation. Few acts except carrying are performed in the fully supinated position.

A simple experiment is worth performing. Flex the elbow to a right angle and lay the forearm on a table. Supination and pronation in this position are seen to cause a wide movement of the thumb from right to left; note also that the *anterior surface* of the distal extremity of the ulna remains facing towards the ceiling in all positions of the hand, even when the latter lies palm downwards. During this experiment the ulna remains fixed and the hand moves around the axis of the little finger. Now repeat the movement with the flexed forearm free in mid air. The axis of rotation of the forearm now no longer passes through the little finger, but has moved to the middle finger. The lower extremity of the ulna still faces forward as before; but it moves into slight adduction and abduction, during supination and pronation, in a manner beautifully co-ordinated. To the basic rotation of the radius around an immobile ulna is superadded a movement of the ulna itself; the actual motion performed by wrist and hand is the resultant of these two movements.

Ulnar movement is produced in two ways. In the usual working position *with elbow flexed* (e.g., turning a screwdriver, a doorknob) abduction and adduction

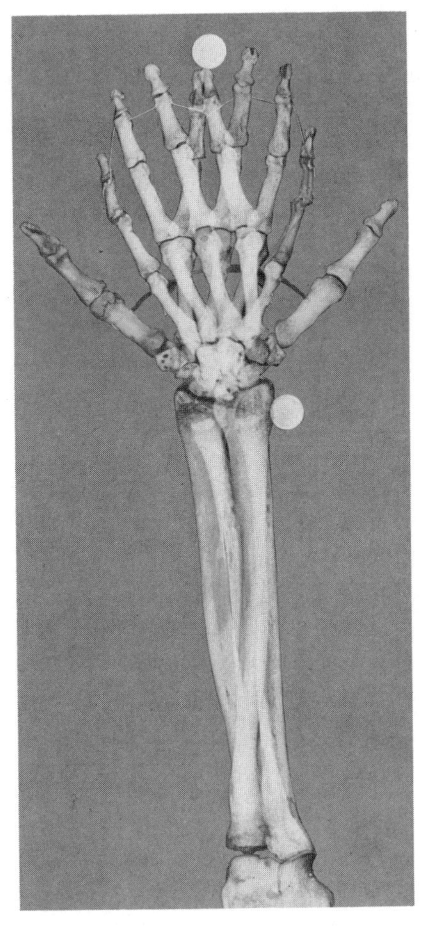

FIG. 2.37. TO ILLUSTRATE ULNAR ABDUCTION DURING PRONATION. The hand articulated as in Fig. 2.36. Double exposure photograph of hand pronated around tip of index finger, marked by a disc. A second marker disc shows the edge of the ulna in supination; in pronation the ulna has moved lateral to this and can be seen overlying the shadow of the radius. Thumb excursion is less than in Fig. 2.36. To achieve pronation around the tip of the index finger the necessary ulnar abduction is brought about by rotation of the shaft of the humerus, not shown in the photograph. Note that the centre of the palm has not moved away. It is only by this ulnar movement that a rotating grip can be maintained on fixed objects such as door-knobs, screwdrivers, etc.

occur between ulna and trochlea. Contraction of the anconeus causes slight abduction and pronation of the ulna, and these movements accompany pronation of the forearm. Lecomte (1874) named the anconeus the pronator of the ulna. Ulnar abduction is opposed by the humeral head of pronator teres. This oblique pull adducts the radius (and hence the ulna). Ulnar movement is illustrated in Fig. 2.37.

In the rarer movements around a more laterally placed centre (lateral to the centre of the palm) ulnar movement *at the trochlea* is insufficient. Ulnar abduction and adduction with the elbow flexed are then produced by rotation of the shaft of the humerus, under the action of the short scapular muscles.

The forearm can be pronated and supinated around any finger tip placed on a fixed point. Ulnar movement is least when the little finger-tip is fixed, greatest when index finger or thumb is fixed. The axis of rotation at the level of the wrist can thus be made to pass through any point between the styloid processes of radius and ulna; and each styloid process describes a semicircle around that point (Fig. 2.38). Ulnar abduction and adduction are each accompanied by slight extension and then flexion at the elbow in order to bring this about. Thus during pronation the first half of the movement is accompanied by abduction and slight extension of the ulna, and the second half by further abduction and slight flexion of the ulna. Likewise during supination the first half of the movement is accompanied by adduction and slight extension of the ulna, and the second half by further adduction and slight flexion of the ulna.

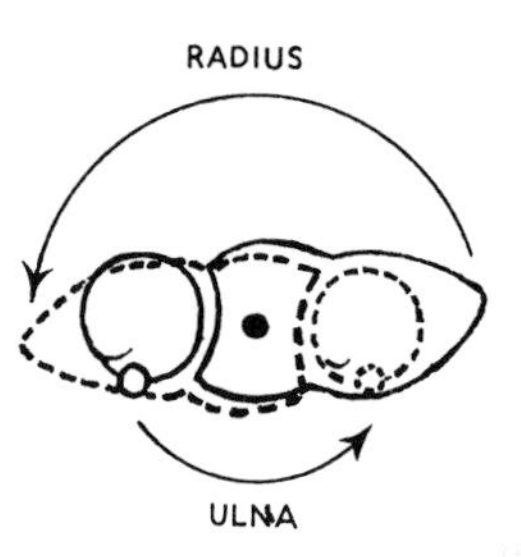

FIG. 2.38. LOWER ENDS OF THE LEFT RADIUS AND ULNA OUTLINED IN SUPINATION. PRONATION AROUND THE FIXED BLACK DOT AT THE LOWER END OF THE RADIUS RESULTS IN THE BONES TAKING UP THE DOTTED OUTLINES. THE CURVED LINES INDICATE THE PATHS OF RADIAL AND ULNAR STYLOIDS AS THE LOWER END OF THE ULNA MOVES INTO ABDUCTION.

Summarizing, simple pronation and supination about an immobile ulna are rare and unnatural movements. Almost always they are accompanied by synergic movements at shoulder and elbow to produce simultaneous movement of the ulna.

THE EXTENSOR COMPARTMENT OF THE FOREARM

A dozen muscles occupy the extensor compartment. At the upper part are the anconeus muscle which is superficial, and the supinator muscle, which is deep. From the lateral part of the humerus arise *three muscles* that pass along the radial border and *three* that pass along the posterior surface of the forearm. At the lower end of the forearm these two groups are separated by *three* muscles that emerge from deeply in between them and go to the thumb. Finally, *one* muscle for the forefinger passes deeply to reach the back of the hand. The nerve of the extensor compartment is the posterior interosseous nerve, which reaches it by passing around the radius (compare the peroneal nerve in the leg); the artery is the posterior interosseous, which gains the extensor compartment by passing between the two bones (cf. anterior tibial artery). The artery is small and the blood supply of the posterior compartment is reinforced by the anterior interosseous artery.

The six long muscles that come from the lateral side of the humerus have not enough area available at the lateral epicondyle. Two of them arise high above this, from the lateral supracondylar ridge and the lateral intermuscular septum.

Brachio-Radialis Muscle. Arising from the upper two-thirds of the lateral supracondylar ridge, the muscle passes along the pre-axial border of the forearm to a tendon that is inserted at the base of the styloid process of the radius. The muscle forms the lateral border of the cubital fossa; passing down the forearm it overlies the radial nerve and the radial artery as they lie together on supinator, pronator teres tendon, flexor digitorum superficialis and flexor pollicis longus. The lower part of the tendon is covered by abductor pollicis longus and extensor pollicis brevis as they spiral down to the thumb (Fig. 2.39). It is supplied by the radial nerve (C5, 6) above the elbow joint (before the posterior interosseous nerve is given off). Its action is to flex the elbow joint. It pulls on the radial styloid process, which is thus made the leading point as the forearm flexes; the ulnar side of the forearm passively follows, falling into the position of mid-pronation. This is purely passive. The muscle has no power as a prime mover to cause either pronation or supination.

Extensor Carpi Radialis Longus. Arising from the lower third of the lateral supracondylar ridge of the humerus the muscle passes down the forearm, behind the brachio-radialis and beneath the thumb muscles, to be inserted as a flattened tendon into the base of the second metacarpal. It is supplied by the radial nerve (C6, 7) above the elbow, before the posterior interosseous branch is given off. Its action is indicated by its name. It is indispensable in the action of 'making a fist', and its tendon can be felt and often seen when this is done (p. 104).

Common Extensor Origin. Examine a humerus. The smooth area on the *front* of the lateral epicondyle is for the attachment of the common extensor origin (Fig. 2.26, p. 77). From it arise the fused tendons of extensor carpi radialis brevis, extensor digitorum communis, extensor minimi digiti and extensor carpi ulnaris. All four muscles pass to the posterior surface of the forearm. When the forearm is extended and supinated they spiral around the upper end of the radius; behind this rounded mass of muscle is an elongated pit in which lies the head of the radius. In the *working position* of the forearm (flexed and half pronated) however, these muscles pass straight from the front of the lateral epicondyle into the forearm, as can be seen by an examination of the bones. It is for this reason that the common origin is on the *anterior* surface of the lateral epicondyle. It is to be noted that all four tendons are fused to each other and to the deep fascia near their origin from the humerus.

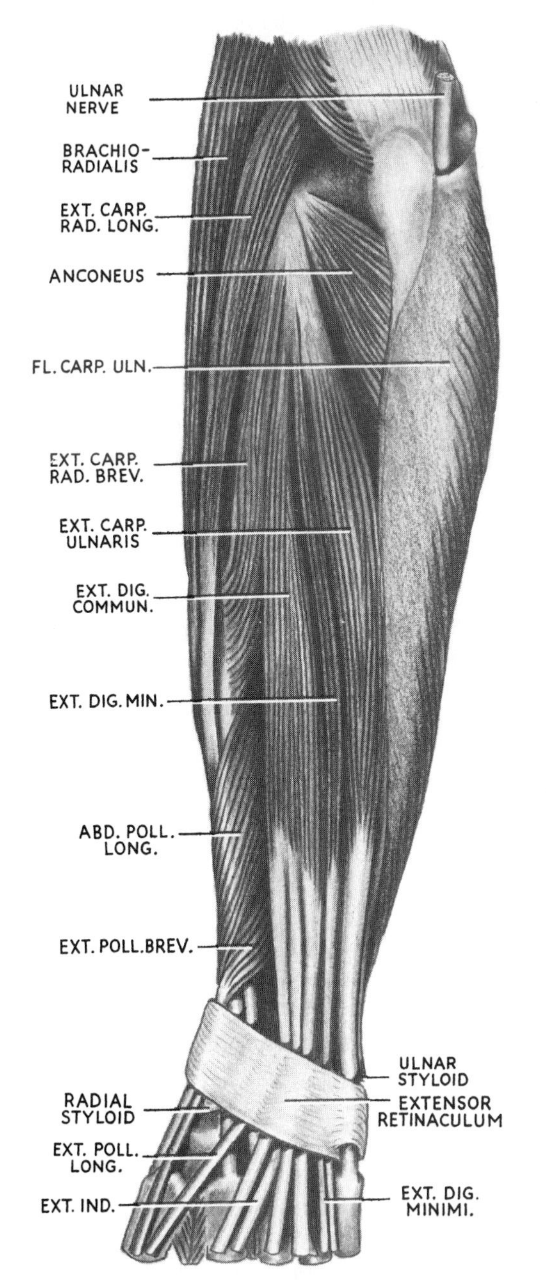

FIG. 2.39. THE EXTENSOR MUSCLES OF THE LEFT FOREARM.

Extensor Carpi Radialis Brevis. This muscle arises from the common extensor origin on the front of the lateral epicondyle of the humerus, passes down behind and deep to its fellow longus, and by a flattened tendon is inserted into the base of the third metacarpal. The lower part of the tendon is crossed by the three extensor muscles of the thumb. The muscle is supplied in the cubital fossa by the posterior interosseous (C6, 7) before the nerve pierces the supinator muscle. Its name indicates its action. It is indispensable in 'making a fist' (p. 104). Note that it and the longus are inserted into the same metacarpals as flexor carpi radialis and compare the symmetry of insertion of flexor and extensor carpi ulnaris muscles.

Extensor Digitorum Communis. Arising from the common extensor origin the muscle expands into a rounded belly in the middle of the forearm, diverging from the three muscles on the radial side and separated from them by the emergence of the thumb extensors (Fig. 2.39). The four tendons that develop pass under the extensor retinaculum crowded together, and overlying the tendon of extensor indicis. On the back of the hand the tendons spread out towards the fingers. Commonly the fourth tendon is fused with that to the ring finger, and reaches the little finger only by a vinculum that passes across near the metacarpo-phalangeal joint. Other vincula join adjacent tendons in a variable manner. The extensor expansions and their insertions into the phalanges are considered with the hand (p. 104). The muscle is supplied by the posterior interosseous nerve on the back of the forearm (C7, 8). It is an extensor of the digits; its action is discussed in detail on p. 104.

Extensor Minimi Digiti. Arising in common with the extensor digitorum communis the belly of the muscle separates after some distance and then becomes tendinous. Passing beneath the extensor retinaculum on the dorsal aspect of the radio-ulnar joint the tendon usually splits, the two halves lying side by side on the fifth metacarpal bone as they pass to the little finger (Fig. 2.43). The tendon of extensor digitorum communis to the little finger commonly joins it by a vinculum near the metacarpo-phalangeal joint. It becomes an expansion on the dorsum of the little finger, behaving as the other extensor expansions (p. 104). The muscle is supplied by the posterior interosseous nerve (C7, 8) in the forearm and its name indicates its action.

Extensor Carpi Ulnaris. There is an origin from the common extensor tendon; as the muscle slopes downwards it is completed by an aponeurotic sheet of origin from the subcutaneous border of the ulna. This aponeurosis arises in common with that of flexor carpi ulnaris, the two passing in opposite directions into the extensor and flexor compartments. The tendon of the muscle lies in the groove beside the ulnar styloid as it passes on to be inserted into the base of the fifth metacarpal. Note the symmetry of insertion with flexor carpi ulnaris, which acts on the fifth metacarpal by way of the piso-metacarpal ligament. The muscle is supplied by the posterior interosseous nerve (C6, 7) in the forearm and its action is indicated by its name. It is indispensable in 'making a fist' (p. 104). Acting with flexor carpi ulnaris it produces ulnar adduction at the wrist joint.

Anconeus. This muscle arises from a smooth facet on the lower extremity of the lateral epicondyle. It fans out to its insertion on the lateral side of the olecranon (Fig. 2.41). Its lateral fibres are vertical, in the axis of the forearm, but its upper fibres are horizontal, passing transversely across the forearm. The vertical fibres may act as a weak extensor of the elbow but the horizontal upper fibres plainly can have no extensor action. Their contraction could abduct and rotate the ulna were such a movement possible at the elbow joint. A slight amount of such movement is in fact possible, and the muscle contracts during pronation of the forearm (p. 88). It is supplied by the radial nerve (C7, 8) by a branch that leaves the trunk in the spiral groove and passes through the triceps muscle with the nerve to its medial head.

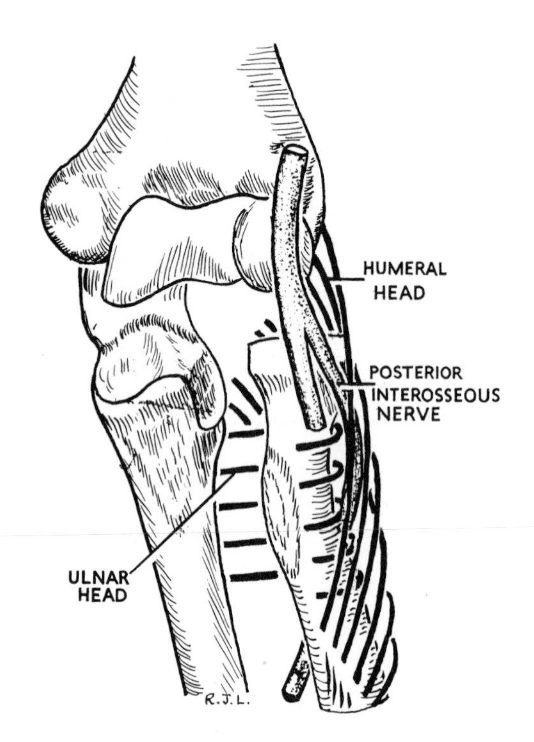

FIG. 2.40. DIAGRAM OF THE TWO LAYERS OF THE LEFT SUPINATOR MUSCLE. The posterior interosseous nerve passes between the two layers. (After Testut.)

Supinator. It is important to observe that this muscle consists of two parts with different origins and passing in different directions. The posterior interosseous nerve passes between these two parts as it leaves the cubital fossa to enter the back of the forearm.

The deep part of the supinator consists of fibres that arise from the supinator crest and fossa of the ulna and wrap horizontally around the radius to be inserted into the area between the anterior and posterior oblique lines. The superficial part of the muscle arises from the distal border of the lateral epicondyle (just in front of anconeus), from the radial collateral (lateral) ligament of the elbow joint and from behind the supinator crest above abductor pollicis longus (Fig. 2.41). The fibres slope downwards to overlie the horizontal deep fibres and reach the radius just above the anterior oblique line (Fig. 2.40).

The posterior interosseous nerve pierces the muscle by passing between its two parts, and therefore supplies it (p. 8). The branches to both parts of the supinator leave the posterior interosseous nerve in the cubital fossa before the nerve enters the muscle. Note that the biceps is the powerful supinator of the forearm; the supinator is to be regarded rather as a muscle that fixes the forearm in supination. Only when the elbow is completely extended is the supinator the prime mover for the action of supination, which is much weaker in this position.

Abductor Pollicis Longus. This muscle arises from an oblique area on both bones of the forearm and the intervening interosseous membrane. Its radial origin lies immediately distal to the posterior oblique line. The ulnar origin, in line with it, is more proximal in the forearm (Fig. 2.41). The muscle belly emerges from the depths and spirals around the radial extensors of the wrist and brachio-radialis to reach the base of the metacarpal bone of the thumb, into which it is inserted by a round tendon. This tendon is split longitudinally into three or four separate bands, one of which commonly passes beyond the metacarpal bone and expands into the belly of the abductor pollicis brevis. The muscle is innervated by the posterior interosseous nerve (C6, 7) in the forearm. Its name indicates that it abducts the thumb; its old name was *extensor* ossis metacarpi pollicis. Former anatomists considered that it extended the thumb. Its unaided action lies somewhere between abduction and extension. It can assist in flexing and abducting the wrist.

Extensor Pollicis Brevis. This muscle arises below abductor pollicis longus from the radius and the adjacent interosseous membrane. It spirals from the depths of the forearm around the radial extensors and brachio-radialis, in contact with abductor pollicis longus, whose tendon it somewhat overlies on the radial border of the snuff box. Its slender tendon lies with the tendon of abductor pollicis longus at the radial border of the snuff box and then passes along the dorsal surface of the metacarpal bone of the thumb, and is inserted into the base of the proximal phalanx. It is supplied by the posterior interosseous nerve (C7, 8) in the forearm and its action is to extend the proximal phalanx of the thumb (Fig. 2.42).

Extensor Pollicis Longus. This muscle arises from the ulna just distal to abductor pollicis longus. Thus it extends higher into the forearm than extensor pollicis brevis. It extends more distally also into the thumb, being inserted into the base of the distal phalanx. Its long tendon changes direction as it hooks around the radial tubercle (of Lister), whence it forms the ulnar boundary of the snuff box. In this situation the tendon is supplied with blood by local branches of the *anterior* interosseous artery. Their occlusion after Colles' fracture may lead to necrosis and spontaneous rupture of the tendon—hammer thumb. Such a rupture is not due to wearing through of the tendon as it grates over the fragments.

There is no extensor expansion on the thumb; the tendon of extensor pollicis longus is stabilized on the dorsum of the thumb by receiving expansions from abductor pollicis brevis and adductor pollicis. The muscle is supplied by the posterior interosseous nerve (C7, 8) and its action is to extend the terminal phalanx of the thumb, and to draw the thumb back from the opposed position (Figs. 2.41, 2.42).

Extensor Indicis. This small muscle arises from the ulna just distal to the former muscle (Fig. 2.41). Its tendon remains deep and passes across the lower end of the radius covered by the bunched-up tendons of extensor digitorum communis with which it shares a common synovial sheath. From here it passes over the dorsal surface of the metacarpal bone of the index finger lying to the ulnar side of the communis tendon (Fig. 2.43). It joins the dorsal expansion of the index finger. The muscle is innervated by the posterior interosseous nerve (C7, 8) and its action is to extend the index finger, as in pointing.

The Anatomical Snuff Box. If the thumb is fully extended the extensor tendons are drawn up, and a concavity appears between them on the radial side of the wrist. The 'snuff box' lies between the extensor pollicis longus tendon and the tendons of extensor pollicis brevis and abductor pollicis longus (Fig. 2.42). The cutaneous branches of the radial nerve cross these tendons, and they can be rolled on the tight tendon of extensor pollicis longus. The radial artery, deep to all three tendons, lies in the floor. Bony points readily palpable in the snuff box are the radial styloid proximally, and the base of the thumb metacarpal distally; between the two the scaphoid and trapezium can be felt.

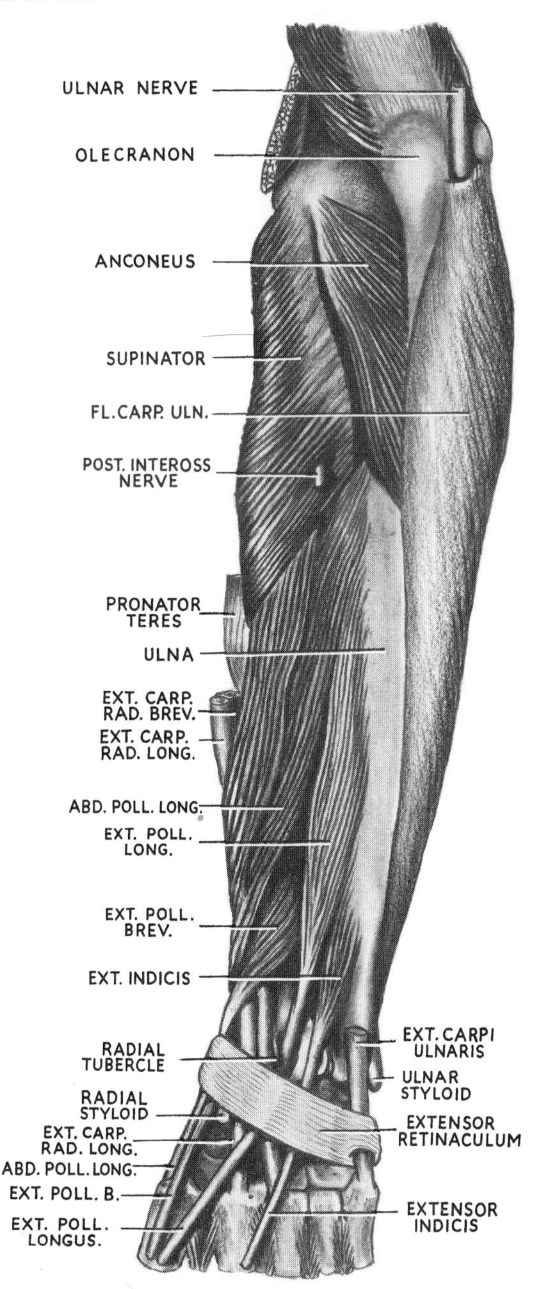

FIG. 2.41. THE DEEP MUSCLES OF THE EXTENSOR COMPARTMENT (LEFT FOREARM) (cf. Fig. 2.64, p. 119).

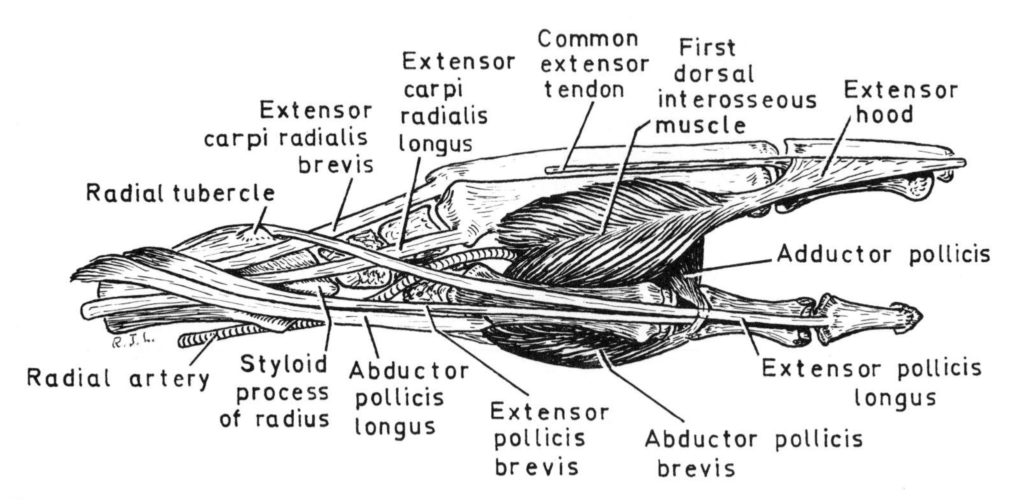

FIG. 2.42. THE LEFT ANATOMICAL SNUFF BOX. It lies between the extensor tendons of the thumb. Its bony floor is the radial styloid, the scaphoid, the trapezium and the base of the first metacarpal. The floor is crossed by the radial artery (cf. Fig. 2.67, p. 121).

Posterior Interosseous Nerve. After passing through the supinator muscle between its two laminæ the nerve appears in the extensor compartment of the forearm and passes downwards over the abductor pollicis longus origin. It now dips deeply to reach the interosseous membrane on which it passes between the muscles as far as the wrist joint. Here it ends in a small nodule from which branches supply the wrist joint and intercarpal ligaments. The nerve supplies the muscles which arise from the common extensor origin and the deep muscles of the extensor compartment. It is sensory to the interosseous membrane and periosteum of the radius and ulna, and to the dorsal periosteum and ligaments of the carpal bones. It has no cutaneous branch.

Posterior Interosseous Artery. This vessel gains the extensor compartment by passing between the bones of the forearm above the interosseous membrane and below the oblique cord. It accompanies the posterior interosseous nerve and supplies the deep muscles of the extensor compartment. At the lower border of extensor indicis this small vessel is generally exhausted. Distal to this level the arterial supply of the extensor compartment is furnished by the *anterior* interosseous artery, which pierces the interosseous membrane just above the upper border of pronator quadratus. The anterior interosseous artery then passes distally to end on the back of the wrist in the dorsal carpal anastomosis.

Extensor Retinaculum. It is important to appreciate the exact attachments of this ligament. A ribbon-like band, less than an inch wide, it lies *obliquely* across the extensor surface of the wrist joint (Fig. 2.43). Its proximal attachment is to the radius at the antero-lateral border above the styloid process, at the lateral border of pronator quadratus. It is *not* attached to the ulna; its upper border passes by the styloid process of the ulna. It is attached to the pisiform and triquetral bones. It is a thickening in the deep fascia of the forearm. Immediately proximal to it the deep fascia is attached to the styloid process of the ulna but the oblique fibres which pass from here to the radial side are not attached to the radius, being held away from the bone by the extensor muscles of the thumb.

If the extensor retinaculum were attached to both bones of the forearm it would be over 30 per cent longer in pronation than in supination, as is shown by measurement on the bones. Such elongation is obviously not possible, and in fact the retinaculum, attached to the radius and carpus and free from the ulna, maintains a constant tension throughout the whole range of pronation and supination, while the radius and carpus move together around the lower end of the ulna. Thus the retinaculum holds down the extensor tendons, like a wrist strap, in all positions of the forearm bones.

From the deep surface of the extensor retinaculum fibrous septa pass to the bones of the forearm, dividing the extensor tunnel into six compartments. The most lateral compartment lies over the lateral surface of the radius at its distal extremity. Through this compartment pass the tendons of abductor pollicis longus and extensor pollicis brevis, each usually lying in a *separate synovial sheath*. The next compartment extends as far as the radial tubercle (of Lister), and conveys the tendons of the radial extensors of the wrist (longus and brevis) each lying in a separate synovial sheath. The groove on the ulnar side of the radial tubercle lodges the tendon of extensor pollicis longus, which lies within its own compartment invested with a synovial sheath. Between this groove and the ulnar border of the radius is a shallow depression in which all four tendons of extensor digitorum communis lie, crowded together over the tendon of extensor indicis. All five tendons in this compartment are invested with a common synovial sheath. The next compartment lies over the radio-ulnar joint and transmits the tendon, often double, of extensor minimi digiti in its synovial sheath. Lastly, the groove near the base of the ulnar styloid transmits the tendon of extensor carpi ulnaris (in its synovial sheath) which then passes into its own compartment on the dorsal surface of the carpus to reach the fifth metacarpal bone (Fig. 2.64, p. 119).

The Dorsum of the Hand

Beyond the extensor retinaculum the extensors of the wrist (two radial and one ulnar) are inserted into the proximal part of the hand, at the bases of their respective metacarpal bones. Lying more superficially, the extensor tendons of the fingers fan out over the dorsum of the hand, attached to the deep fascia of this region and interconnected near the metacarpal heads by a variable arrangement of oblique fibrous bands known as *vincula*. The deep fascia and the subjacent extensor tendons roof in a subfascial space that extends across the whole width of the hand. Passing to the index finger are the tendons of the extensor communis and the extensor indicis lying together, while the tendon of extensor digiti minimi to the little finger is usually split longitudinally into two (Fig. 2.43). The two tendons over the second metacarpal bone belong to two separate muscles, while the two tendons usually seen over the fifth metacarpal belong to only one muscle. The tendon of extensor communis to the little finger lies over the metacarpal bone of the *ring* finger.

The **cutaneous innervation** of the dorsum of the hand is by the terminal branches of the radial and of the dorsal cutaneous branch of the ulnar nerve. These share the hand $3\frac{1}{2}:1\frac{1}{2}$ (Fig. 2.57, p. 107), though a $2\frac{1}{2}:2\frac{1}{2}$ distribution is not uncommon. The nerves

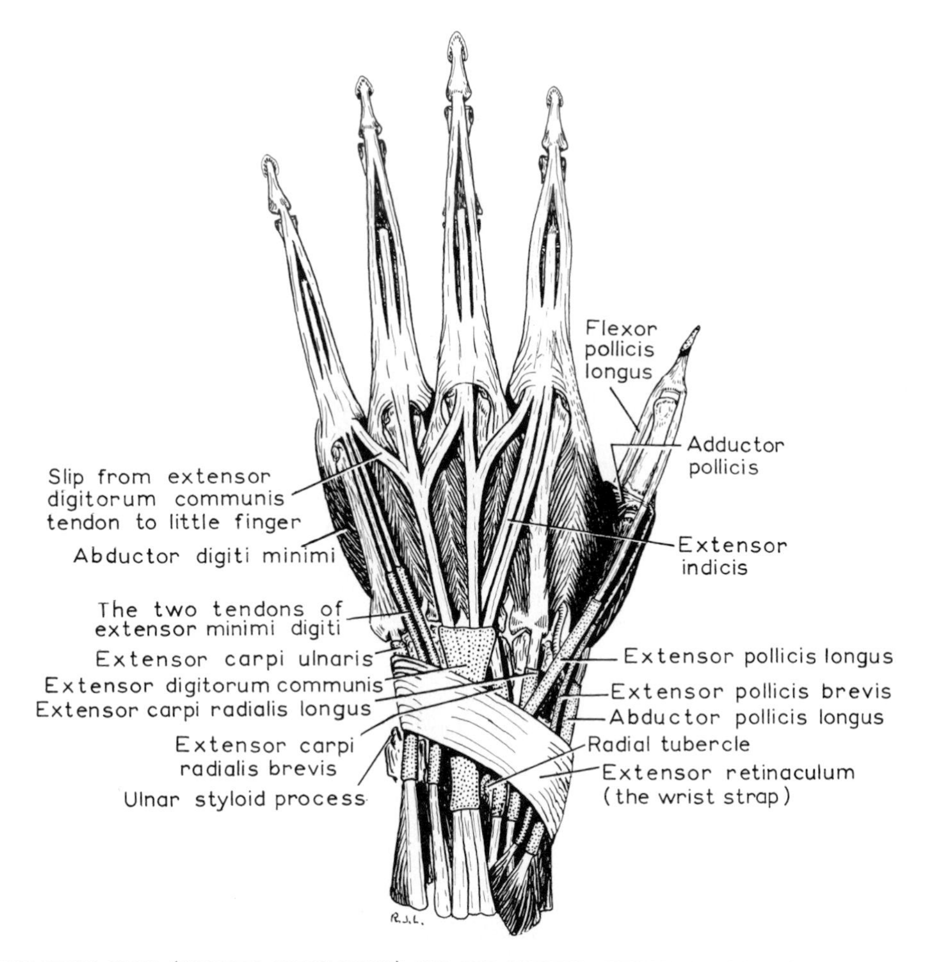

FIG. 2.43. THE WRIST STRAP (EXTENSOR RETINACULUM) AND THE SYNOVIAL SHEATHS OF THE EXTENSOR TENDONS. Note the obliquity of the extensor retinaculum.

pass to the dorsum of each digit, where they stop short of the nail beds (the nail beds are supplied $3\frac{1}{2} : 1\frac{1}{2}$ by the nerves of the flexor skin, the median and superficial ulnar nerves).

The **posterior carpal arch** is an arterial anastomosis between the radial, ulnar and *anterior* interosseous arteries. It lies on the back of the carpus and sends dorsal metacarpal arteries distally in the intermetacarpal spaces, deep to the long tendons. These split at the webs to supply the dorsal aspects of adjacent fingers. They communicate through the interosseous spaces with the palmar metacarpal branches of the deep palmar arch. Companion veins bring blood from the palm into the **dorsal venous arch.** The latter lies in the subcutaneous tissue (i.e. superficial to the extensor tendons) proximal to the metacarpal heads and drains on the radial side into the cephalic vein and on the ulnar side into the basilic vein.

The Wrist Joint

This is a simple synovial joint whose bony surfaces are formed by the proximal row of carpal bones and the distal surface of the radius. The fibro-cartilage between the radius and ulna is within the joint but it does not transmit thrust from the hand. Examine the lower end of the radius. The triangular facet whose apex is the styloid process is for articulation with the scaphoid. The rectangular area next to it is for the lunate. Except in extreme degrees of ulnar adduction the triquetral is non-articular, bearing against the ulnar collateral ligament. A simple capsular ligament surrounds the joint and is thickened on the radial and ulnar sides by collateral ligaments, and in general is much thicker in front than behind. Strong ligaments run transversely from the lower end of the radius to the lunate (Fig. 2.44).

Movements at the joint are flexion and extension, ulnar adduction and radial abduction. These four movements occurring in sequence produce circumduction. Only passive rotation is possible in the long axis of the forearm, no active rotation takes place. The four movements are carried out by combinations of muscle groups. Thus flexion is done by flexor carpi radialis and flexor carpi ulnaris as prime movers, assisted on occasion by the flexors of fingers and thumb and abductor pollicis longus. Extension is done by the radial extensors (longus and brevis) and the ulnar extensor as prime movers assisted on occasion by the extensors of fingers and thumb. Radial abduction is carried out by abductor pollicis longus and, when the wrist is displaced from the midline, by flexor carpi radialis and the two radial extensors acting together. Similarly ulnar adduction is brought about by simultaneous contraction of flexor and

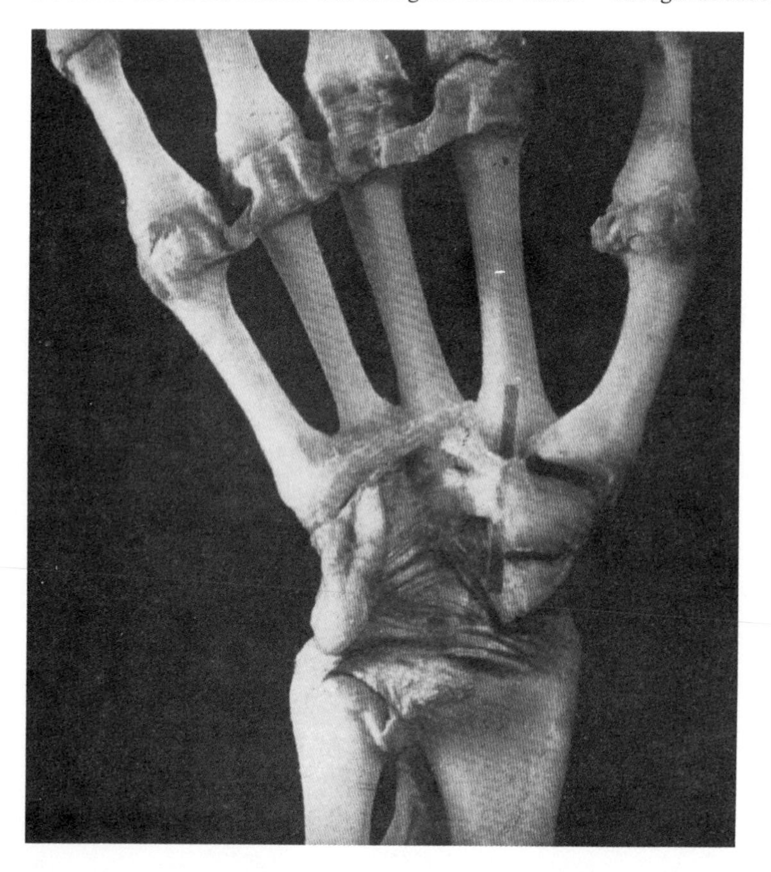

FIG. 2.44. THE INTEROSSEOUS LIGAMENTS OF THE WRIST AND CARPUS. The inferior radio-ulnar joint has been opened; the membrana sacciformis projects proximally between radius and ulna. The anterior margin of the fibro-cartilage is exposed. The piso-hamate ligament passes distally in two bands to the ulnar border and to the hook of the hamate (the piso-metacarpal ligament has been removed). Across the wrist joint strong fibres pass *transversely* from the radius to the lunate. A glass rod occupies the tunnel for the tendon of flexor carpi radialis; the tunnel is roofed in by a strong ligament joining the ridge on the trapezium to the capitate. The deep transverse ligament of the palm is seen joining the metacarpal heads. Photograph of Specimen S 15 A in R.C.S. Museum.

extensor carpi ulnaris. In the rest position the axis of the metacarpus (third metacarpal) is not in line with the axis of the forearm. With the supinated forearm by the side the axis of the hand is parallel with the humerus, thus compensating for the carrying angle at the elbow. Lateral movements of the wrist are symmetrical about this axis, but they are *ipso facto* not symmetrical about the axis of the forearm, from which the metacarpus can be brought further into ulnar adduction than into radial abduction. In the former movement there is some gliding of the carpal bones across the lower end of the radius but in the latter the carpal bones, rather, rotate each about its own axis. In flexion-extension movements the *hand* can be flexed to 90 degrees but extended not much beyond 45 degrees. The range of flexion is enhanced by movement at the midcarpal joint. At the wrist joint itself (radio-carpal joint) the range of extension is actually greater than that of flexion (see p. 120).

It is worth noting that the most usual movement of the wrist is one of extension combined with radial abduction, and of flexion combined with ulnar adduction. Hammering in a nail illustrates this movement, and exactly similar movements of the wrist, in the working position of the forearm, occur in a host of everyday acts like eating and drinking, washing and dressing, writing, etc. Indeed, pure flexion-extension and abduction-adduction are unusual movements. Since extension-abduction is an antigravity movement in the normal working position, this may explain the presence of two radial extensors where one serves on the ulnar side, and there is only one flexor each for radial and ulnar sides.

THE HAND

The hand of man differs from the manus of other creatures in being a grasping mechanism combining great strength with finely controlled accuracy and at the same time serving as the chief tactile organ. Apart from the free mobility of the thumb its component parts differ little from those of the foot; indeed their similarity is very striking (p. 128). The great difference lies in the richness of the cortical connexions, both sensory and motor, in the case of the hand. The anatomical structures to be studied in the hand are simple enough and differ not at all between the unskilled labourer, the skilled craftsman and the artist. They comprise the same four layers as in the foot. Beneath the skin of the palm lies the strong palmar aponeurosis. Next lie short muscles of thumb and little finger, then the long flexor tendons. More deeply is the adductor muscle of the thumb, then the metacarpal bones with their interosseous muscles. The long flexor tendons provide the power of the grip, the short intrinsic muscles of the hand are responsible for adjusting the position and for carrying out the finer skilled movements of the digits. The back of the hand has little but skin and sinew covering the bones.

The **skin** of the palm is characterized by flexure creases (the 'lines' of the palm) and the papillary ridges, or 'fingerprints', which occupy the whole of the flexor surface. Perhaps the latter serve to improve the grip; certainly they increase the surface area. Sweat glands abound. The little **palmaris brevis** muscle is attached to the dermis; it is part of the panniculus carnosus (p. 3). It lies across the base of the hypothenar eminence and is the only muscle supplied by the superficial branch of the ulnar nerve. It may improve the grip by steadying the skin on the ulnar side of the palm.

Elsewhere the skin is steadied by its firm attachment to the palmar aponeurosis. Fibrous bands connect the two and divide the subcutaneous fat into myriads of small loculi, forming a 'water-cushion' capable of withstanding considerable pressure. When cut the tension causes some bulging of these fatty loculi.

The skin of the dorsum of the hand is quite different; it can be picked up from the underlying tendons and moved freely over them—except in a very podgy hand. Though there is little but bone and sinew on the dorsum, large veins lie beneath the skin—they drain from the palm, so that the pressure of gripping does not impede venous return.

The Palmar Aponeurosis. This strong unyielding ligament is phylogenetically the degenerated tendon of palmaris longus (p. 6). It extends, in continuity with the tendon, from the distal border of the flexor retinaculum, whence it fans out in a thick sheet towards the bases of the fingers. It divides into four slips, one for each finger. Each slip divides into two bands over the proximal end of the fibrous flexor sheath; they are inserted into the deep transverse ligament of the palm, and into the bases of the proximal phalanges and the fibrous flexor sheaths (Fig. 2.45). Some strands from the aponeurosis pass up on each side of the finger. When the fingers are forcibly extended the soft tissues of the palm can be seen bulging in the three intervals between the four slips, just proximal to the interdigital webs.

Over the hypothenar muscles the deep fascia is much thinner than the palmar aponeurosis and it is thinnest of all over the thenar muscles. This is in keeping with the increased mobility of the metacarpal bone of the thumb (note that the plantar aponeurosis over the big toe muscles is thick; the first metatarsal has very little freedom of movement). The function of the palmar aponeurosis is purely mechanical. It gives firm attachment to the skin of the palm to improve the grip, and

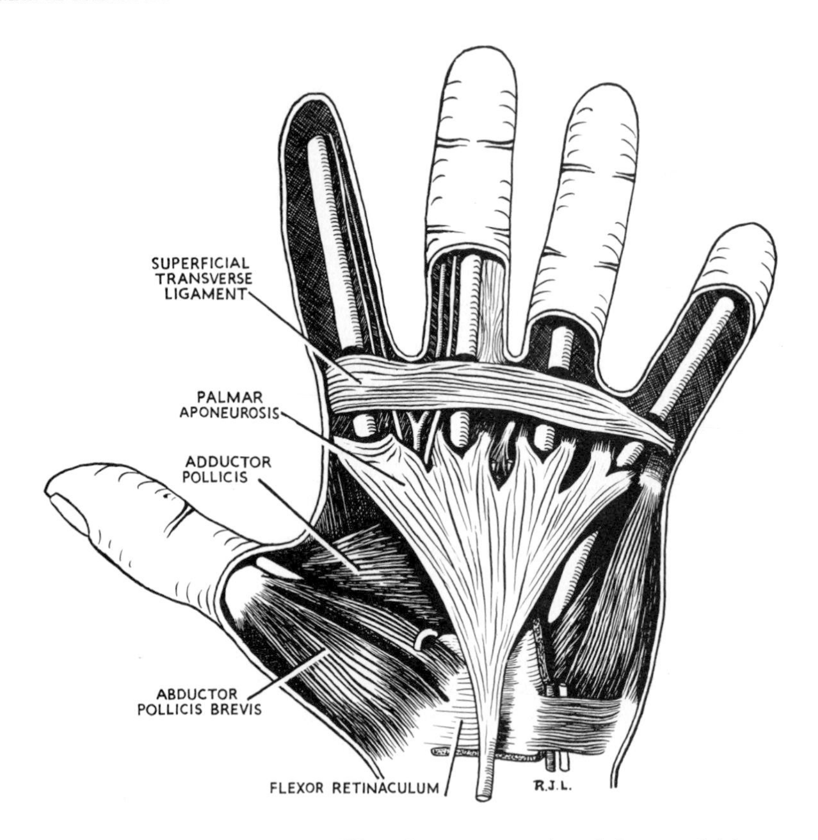

FIG. 2.45. A SUPERFICIAL DISSECTION OF THE LEFT PALM. The palmar aponeurosis and the superficial transverse ligament of the palm are shown. The bifurcation of the metacarpal arteries in the web is distal to that of the digital nerves.

it protects the underlying tendons. Contraction of the aponeurosis and its digital slips in Dupuytren's contracture results in a fixed flexion of the fingers concerned (usually the ring and little fingers).

The **flexor retinaculum** is a strong band that is attached on the radial side to the tubercle of the scaphoid and ridge of the trapezium and on the ulnar side to the pisiform and hook of the hamate. Note that these four bony points are all palpable in the living hand. There is a further attachment on the radial side, where a deep partition bridges the groove on the trapezium (Fig. 2.44). The fibro-osseous tunnel so formed transmits the tendon of flexor carpi radialis and its synovial sheath (Fig. 2.46).

The muscles of the thenar and hypothenar eminences arise from the retinaculum and several structures pass across it. The tendon of palmaris longus is fused to the midline of the retinaculum as it passes distally to expand into the palmar aponeurosis. The ulnar nerve lies on the retinaculum alongside the pisiform bone, with the ulnar artery on its lateral side (PLATE 3). The nerve divides into a superficial (cutaneous) and a deep (muscular) branch at the distal border of the retinaculum; the latter is accompanied by a small deep branch of the artery. On the radial side of the palmaris longus tendon the palmar branch of the median nerve and the superficial palmar branch of the radial artery cross the retinaculum. The radial side of the retinaculum is pierced proximally by the tendon of flexor carpi radialis over the scaphoid; further distally the tendon lies deeply in its fibro-osseous tunnel on the trapezium.

The Carpal Flexor Tunnel. The carpus is deeply concave on its flexor surface. This bony gutter is converted into a tunnel by the flexor retinaculum. The long flexors of the fingers and thumb run through the tunnel. The four tendons of the superficial flexor are separate from each other and pass through in two rows, middle and ring finger tendons lying in front of index and little finger tendons. The tendons of flexor digitorum profundus, on the other hand, lie all on the same plane deep on the carpal bones; but here only the tendon to the index finger is yet separated, the other three being adherent to each other as they run in the carpal tunnel, and not splitting free until they gain the palm (Fig. 2.32, p. 82). All eight tendons of the superficial and deep flexors share a common synovial sheath. It does not invest them completely, but is reflected from their radial sides, where arteries of supply gain access to the tendons. It is as though the tendons had been invaginated into the sheath from the radial side (Fig. 2.46). The tendon of flexor pollicis longus lies in its own synovial sheath as it passes through the radial side of the flexor tunnel. The median nerve passes beneath the flexor retinaculum on the lateral side of flexor digitorum superficialis (middle finger tendon), between it and flexor carpi radialis.

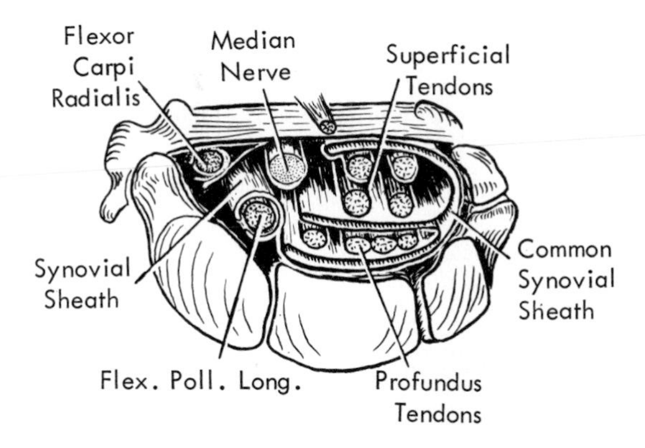

FIG. 2.46. THE LEFT CARPAL TUNNEL, LOOKING DISTALLY TOWARDS THE HAND. The synovial sheaths are open on the radial side, for access of blood supply to the flexor tendons. (Diagrammatic: the tendons in their sheaths and the median nerve are in reality tight-packed in the tunnel.)

The Thenar Eminence. The thenar eminence is made up of three short muscles whose origin is essentially from the flexor retinaculum. The most radial of these is the **abductor pollicis brevis.** It arises from the flexor retinaculum and the tubercle of the scaphoid and is inserted into the radial side of the base of the proximal phalanx and the tendon of extensor pollicis longus (Fig. 2.42). It is supplied by the median nerve (Th. 1) and its action is to abduct the thumb (moving it in a plane at right angles to the palm), to abduct the proximal phalanx at the metacarpo-phalangeal joint and slightly to rotate the proximal phalanx; it is thus an indispensable aid in *opposition of the thumb (v. inf.)*.

The **flexor pollicis brevis** lies to the ulnar side of the former muscle. It arises from the flexor retinaculum and is inserted into the radial sesamoid of the thumb and so to the radial border of the proximal phalanx. Irregularities of origin are fairly common and have been given many names. The least confusing descriptive nomenclature is to call any fibres that are inserted into the radial sesamoid flexor pollicis brevis, no matter what their origin; fibres inserted into the ulnar sesamoid should not be named flexor pollicis brevis, but considered part of either adductor pollicis or the first palmar interosseous. The flexor pollicis brevis is innervated by the median nerve (Th. 1) except in rare cases where a twig from the ulnar nerve may supply it. Its action is to flex the proximal phalanx and draw the thumb across the palm.

The **opponens pollicis** lies deep to the former two muscles. It arises from the flexor retinaculum and is inserted into the whole of the radial border of the metacarpal bone of the thumb (Fig. 2.49). It is supplied by the median nerve (Th. 1) and its action is to oppose the metacarpal bone of the thumb.

Note that the branch from the median nerve to the three thenar muscles hooks around the distal border of the flexor retinaculum and ascends, superficial to the tendon of flexor pollicis longus, to the thenar eminence (Fig. 2.47). Incision of the sheath of flexor pollicis longus over the metacarpal bone will divide this nerve if the incision is not kept sufficiently distal.

Flexion and Opposition of the Thumb. Hold the hand in the neutral position of complete relaxation. Note that the plane of the four finger nails is at right angles to the plane of the thumbnail. Flexion of the thumb across the palm maintains the plane of the thumbnail; opposition, on the other hand, rotates the metacarpal and phalanges in such manner that the thumbnail lies parallel with the nail of the opposed finger, as thumb and finger lie in contact pad to pad. Rotation of the metacarpal at its saddle-shaped joint with the trapezium is produced by the skew pull of the opponens muscle. Opposition is assisted by the abductor brevis, which abducts and slightly rotates the proximal phalanx at the metacarpo-phalangeal joint.

Confirm this by observing the movement in your own thumb. At rest the thumb phalanges are in line with their own metacarpal bone, but in full opposition there is notable angulation at the metacarpo-phalangeal joint. This abduction carries the pad of the thumb still further towards the ulnar side of the hand. The slight rotation of the proximal phalanx produced at the same time by abductor pollicis brevis aids in bringing the pad of the thumb into contact with that of the opposed finger.

The Hypothenar Muscles. In name these are the same as the three thenar muscles. The **abductor digiti minimi** is the most ulnar of the group; it arises from the pisiform bone and flexor retinaculum and is inserted into the ulnar side of the base of the proximal phalanx and into the extensor expansion. A sesamoid bone is common in its tendon. The abductor muscle abducts the little finger towards the ulnar side, away from the midline of the palm. The **flexor digiti minimi** arises from the flexor retinaculum and is inserted into the base of the proximal phalanx. The **opponens digiti minimi** arises from the flexor retinaculum and the hook of the hamate and is inserted into the ulnar border of the fifth metacarpal bone. It is badly named, for rotation of the fifth metacarpal (i.e., opposition) is scarcely possible.

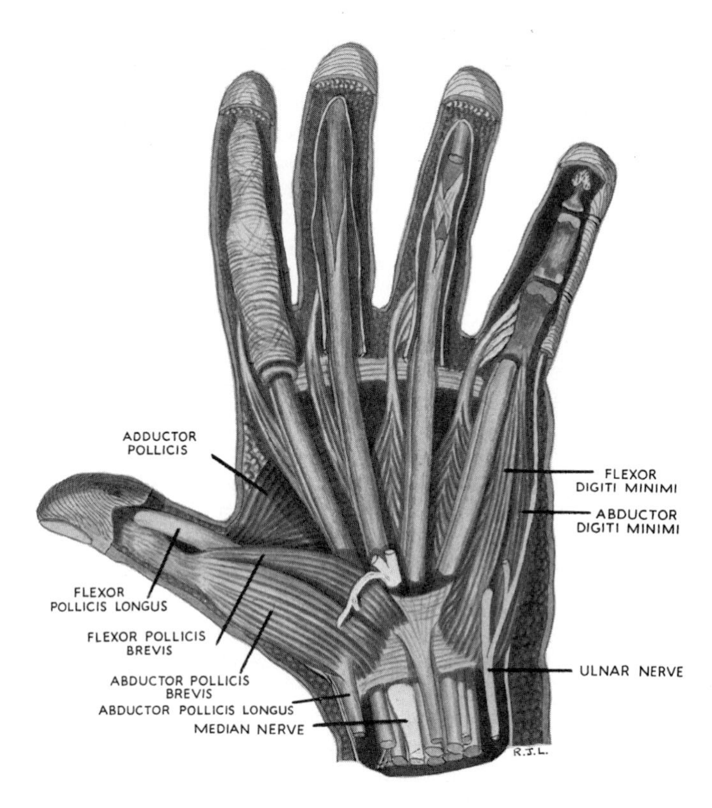

FIG. 2.47. A DISSECTION OF THE LEFT PALM. Note the median nerve passing beneath the flexor retinaculum distal to which a branch recurves to the thenar muscles. The first and second lumbricals (median nerve) are unicipital, the third and fourth (ulnar nerve) are bicipital. The index finger shows the fibrous flexor sheath, with the synovial sheath bulging proximally. The middle finger shows the long tendons exposed by incision of the flexor sheath. In the ring finger the profundus tendon has been removed; in the little finger all is removed down to the phalanges.

Yet the muscle is symmetrical with opponens pollicis; it helps in cupping the palm (Fig. 2.49).

All three hypothenar muscles help to cup the palm and assist in the grip on a large object. They are supplied by the deep branch of the ulnar nerve (Th. 1).

Superficial Palmar Arch. This is an arterial arcade that lies superficial to everything in the palmar compartment, i.e., in contact with the deep surface of the palmar aponeurosis. It is formed by the direct continuation of the ulnar artery beyond the flexor retinaculum. On the radial side it is usually completed by the superficial palmar branch of the radial artery. It lies across the centre of the palm, level with the distal border of the outstretched thumb. From its convexity four *palmar digital arteries* pass distally. The most medial of these passes to the ulnar side of the little finger. The remaining three pass to the webs between the fingers, where each divides into two vessels that supply adjacent fingers. The thumb side of the index finger and the thumb itself are not supplied from the superficial arch (PLATE 3).

Digital Nerves. Lying immediately deep to the superficial palmar arch are the digital nerves. They pass distally to the webs, between the slips of the palmar aponeurosis. Here the nerves lie superficially; they are destined essentially for the sensitive pads at the finger-tips. They end dorsally by supplying all five nail beds (Fig. 2.57). The digital arteries, lying dorsal to the digital nerves along the fingers, are directed towards the nail bed, where they are free from the pressure of gripping.

The superficial branch of the **ulnar nerve** divides into two branches; the medial one supplies the ulnar side of the little finger, the lateral the cleft and adjacent sides of little and ring fingers (PLATE 3).

The **median nerve** enters the palm beneath the flexor retinaculum and divides into two branches (Fig. 2.48). The *medial branch* divides into two and supplies the cleft and adjacent sides of ring and middle fingers and the cleft and adjacent sides of middle and index fingers. The latter branch supplies the second lumbrical muscle. The *lateral branch* supplies the radial side of the index, the whole of the thumb and its web on the palmar surface and distal part of the dorsal surface. The branch to the index supplies the first lumbrical. A *muscular branch* curls upwards from around the distal border of the flexor retinaculum to supply the thenar muscles.

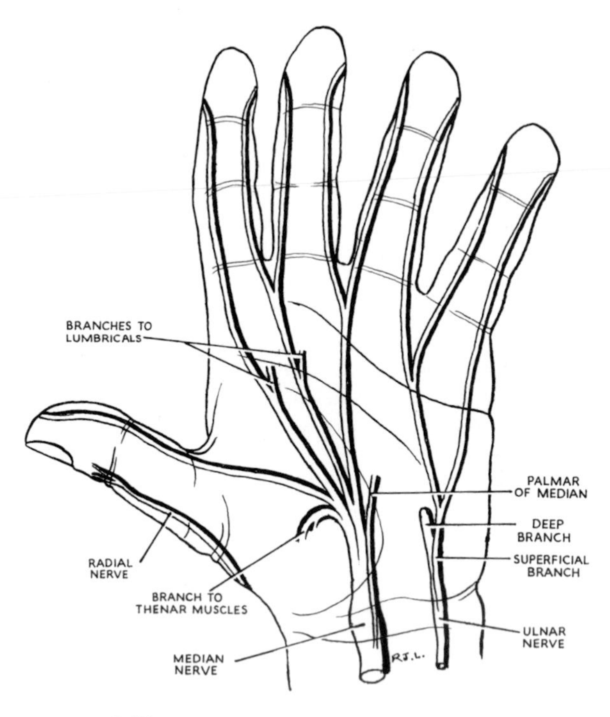

FIG. 2.48. THE CUTANEOUS INNERVATION OF THE PALM.

Carpal Tunnel Syndrome. In the tightly crowded flexor tunnel the median nerve can be compressed, especially by long-continued swelling in the synovial sheaths. Division of the flexor retinaculum relieves the symptoms, which are of wasting and weakness of the thenar muscles (with loss of power of opposition of the thumb) and anæsthesia over three and a half digits on the thumb side of the hand. There is no anæsthesia over the thenar eminence itself, for this area of skin is supplied by the palmar cutaneous branch of the median nerve, or occasionally by the lateral cutaneous nerve of the forearm. Either of these nerves enters the palm superficial to the retinaculum, and so escapes compression.

The carpal tunnel syndrome must be distinguished from median nerve palsy due to the compression of a ligament of Struthers (p. 73). In the latter case the palmar cutaneous branch will be affected, and in addition weakness of the relevant flexor muscles in the forearm (e.g., flexor pollicis longus) is a notable feature. In the carpal tunnel syndrome the terminal phalanx of the thumb can be flexed with normal power; in compression from the ligament of Struthers this power is lost.

The Long Flexor Tendons. In the palm the tendons of flexor superficialis immediately overlie those of flexor profundus digitorum. From the latter tendons the four lumbrical muscles arise. The superficial tendons overlie the profundus tendons as they pass, in pairs, into the fibrous flexor sheaths of the fingers. Their synovial sheaths are considered on p. 101.

The Lumbrical Muscles. From each of the four profundus tendons a lumbrical muscle passes along the radial side of the metacarpo-phalangeal joint. On the palmar surface of the deep transverse ligament of the palm, each develops a tendon which runs in a fibrous canal to reach the extensor expansion on the dorsum of the first phalanx. Characteristically, the two ulnar lumbricals are innervated by the ulnar nerve and the two radial lumbricals by the median nerve. The proportion of ulnar and median distribution to the lumbricals follows that of the parent bellies of the tendons in the forearm. Variations of 1 : 3 and 3 : 1 are not uncommon (each occurs in 20 per cent of cases). It is interesting to note, though impossible to explain, that lumbricals supplied by the ulnar nerve are bicipital, each arising by two heads from adjacent profundus tendons, while those supplied by the median nerve are unicipital and arise from one tendon only (Fig. 2.47).

The Deep Palmar Arch. This is an arterial arcade that is produced by the terminal branch of the radial artery, which gains the palm by passing between the oblique and transverse heads of adductor pollicis (Fig. 2.49). It runs across the palm at a level with the proximal border of the outstretched thumb. The deep branch of the ulnar nerve lies within its concavity. The arch is completed by the deep branch of the ulnar artery. From its convexity three palmar metacarpal arteries pass distally and in the region of the metacarpal heads anastomose with the digital branches from the superficial arch. Branches perforate the interosseous spaces to anastomose with the dorsal metacarpal arteries. Their accompanying veins drain most of the blood of the palm into the dorsal venous arch (p. 94). Branches from the anterior carpal arch anastomose with the deep palmar arch (Fig. 2.33, p. 83).

Adductor Pollicis Muscle. This muscle lies deeply in the palm in contact with the metacarpals and interossei. The *transverse head* arises from the whole length of the palmar border of the third metacarpal (Fig. 2.49) whence the muscle converges, fan-shaped, to the ulnar sesamoid of the thumb, and so to the ulnar side of the base of the proximal phalanx and the tendon of extensor pollicis longus (PLATE 4). There is an *oblique head* that arises from the bases of the second and third metacarpals and their adjoining carpal bones (trapezoid and capitate) by a crescentic origin that embraces the insertion of flexor carpi radialis. The fibres of this head run edge to edge with the transverse head and converge with it on the ulnar sesamoid. The muscle is supplied by the deep branch of the ulnar nerve (Th. 1) and its action is to approximate the thumb to the index, whatever the original position of the thumb.

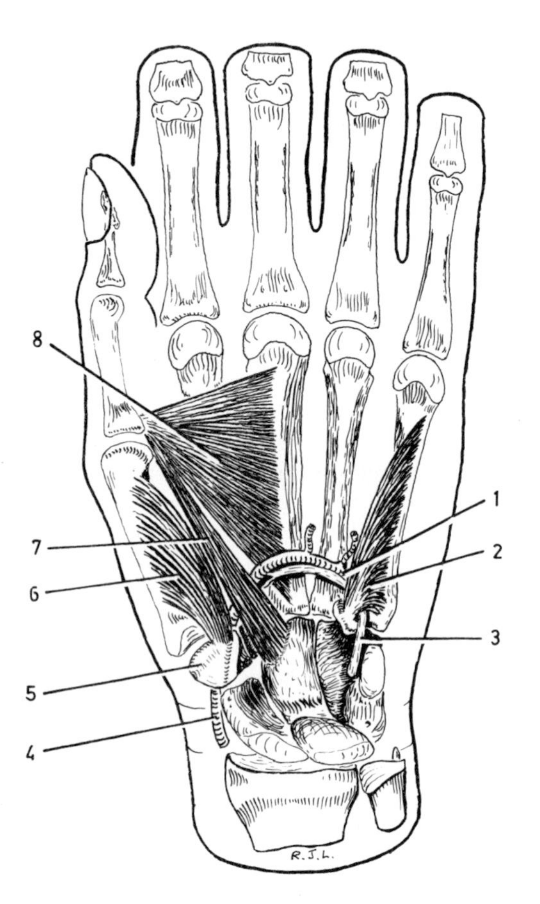

FIG. 2.49. INTRINSIC MUSCLES. The two opponens muscles and adductor pollicis are shown. 1. Deep palmar arch (radial artery). 2. Opponens minimi digiti. 3. Deep branch of ulnar nerve (passing deep to opponens minimi digiti and lying on the hook of the hamate). 4. Radial artery. 5. Trapezium. 6. Opponens pollicis. 7. Oblique head and 8. Transverse head of adductor pollicis.

Radial Artery in the Hand. The radial artery leaves the lower end of the radius and slopes across the snuff box over the trapezium and so passes into the hand between the two heads of the first dorsal interosseous muscle (Fig. 2.42). Lying now between this muscle and adductor pollicis it gives off two large branches. The *arteria radialis indicis* passes distally between the two muscles to emerge on the radial side of the index finger, which it supplies. The *arteria princeps pollicis* passes distally along the metacarpal bone of the thumb and divides into its two palmar digital branches at the metacarpal head. The main trunk of the radial artery now passes into the palm between the oblique and transverse heads of adductor pollicis to form the deep palmar arch (Fig. 2.33, p. 83).

The Ulnar Nerve in the Hand. The ulnar nerve leaves the forearm by emerging from beneath the tendon of flexor carpi ulnaris and passes distally on the flexor retinaculum alongside the radial border of the pisiform bone. Here it divides into superficial and deep branches (PLATE 3). The **superficial branch** supplies the little palmaris brevis muscle (portion of the panniculus carnosus) then is carried on as the digital nerves to the ulnar one and a half fingers. The **deep branch** passes deeply into the palm between the heads of origin of flexor and abductor minimi digiti. It passes between the origin of opponens minimi digiti and the fifth metacarpal bone (Fig. 2.49) lying on the hook of the hamate, whose distal border it indents with a shallow groove. Passing down to the interossei, it arches deeply in the palm within the concavity of the deep palmar arch. The deep branch of the ulnar nerve (Th. 1) gives motor branches to the three hypothenar muscles, the two lumbricals on the ulnar side, all the interossei and both heads of adductor pollicis. Compare the ulnar nerve in the hand with the lateral plantar in the foot. The cutaneous distribution is identical, but in the foot the superficial branch of the lateral plantar nerve, unlike that of the ulnar, supplies three muscles, and the deep branch supplies three instead of only two lumbricals.

Interosseous Muscles. The interossei are in two groups, palmar and dorsal. The former are small and arise from only one (their own) metacarpal bone; the latter are larger and arise from both metacarpal bones of the space in which they lie. The dorsal interossei are visible from the *palmar* aspect of the interosseous spaces. It is easy to recall the attachments of the interossei by appreciating their functional requirements. The formula 'PAD and DAB' indicates that palmar adduct and dorsal abduct the fingers *relative to the axis of the palm*, which is the third metacarpal bone and middle finger (in the foot the axis passes through the second digit).

The **palmar interossei** adduct the fingers. Thus the thumb requires no palmar interosseous, already possessing its own powerful adductor pollicis muscle. Nevertheless a few fibres are sometimes found passing from the base of the metacarpal of the thumb to the base of its proximal phalanx; when present these fibres represent the first palmar interosseous muscle. Similarly the middle finger has no palmar interosseous; it cannot be adducted towards itself. The second, third and fourth palmar interossei arise from the *middle finger side* of the metacarpal bone of the index, ring and little fingers and are inserted into the same side of the extensor expansion and proximal phalanx of each respective finger.

The **dorsal interossei,** more powerful than the palmar, abduct their own fingers away from the midline of the palm. The thumb and little finger already possess their proper abducting muscles in the thenar and hypothenar eminences. Thus there are dorsal interossei attached only to index, middle and ring fingers. In the case of index and ring fingers they are inserted into the side of the finger away from the middle finger. In the case of the middle finger itself a dorsal interosseous is present on each side. All four dorsal interossei arise by two heads, one from each bone bounding the interosseous space (Fig. 2.50).

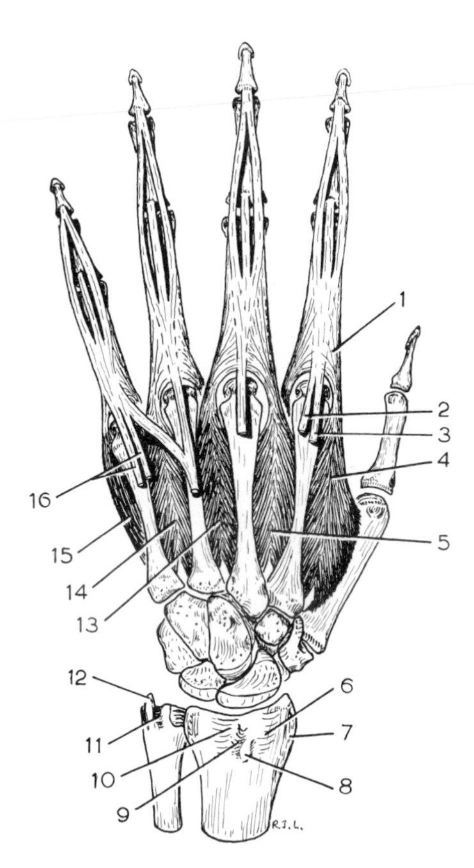

FIG. 2.50. THE FOUR DORSAL INTEROSSEI. 1. Extensor hood. 2. Tendon of extensor indicis. 3. Tendon of extensor digitorum communis. 4. First dorsal interosseous muscle. 5. Second dorsal interosseous muscle. 6. Groove for tendons of extensors carpi radialis longus and brevis. 7. Groove for tendons of abductor pollicis longus and extensor pollicis brevis. 8. Radial tubercle. 9. Groove for tendon of extensor pollicis longus. 10. Groove for common extensor tendons and extensor indicis. 11. Groove for tendon of extensor carpi ulnaris. 12. Ulnar styloid process. 13. Third dorsal interosseous muscle. 14. Fourth dorsal interosseous muscle. 15. Abductor digiti minimi. 16. Double tendon of extensor digiti minimi (cf. Fig. 2.43).

The tendons of palmar and dorsal interossei all pass on the posterior side of the deep transverse ligament of the palm to reach their distal attachments. They are inserted chiefly into the appropriate side of the extensor expansion (p. 104) but partly also into the base of the proximal phalanx. Characteristically, all the interossei are supplied by the deep branch of the ulnar nerve (Th. 1), but an occasional variant is for the first dorsal interosseous to be supplied by the median nerve (Th. 1).

The actions of the interossei are discussed on p. 104.

Fibrous Flexor Sheaths. From the metacarpal heads to the distal phalanges all five digits are provided with a strong unyielding fibrous sheath in which the flexor tendons lie. In the case of the thumb the fibrous sheath is occupied by the tendon of flexor pollicis longus alone. In the case of the four fingers the sheaths are occupied by the tendons of the superficial and deep flexors, the former splitting to spiral around the latter within the sheath. The proximal ends of the fibrous sheaths of the fingers receive the insertions of the four slips of the palmar aponeurosis. The sheaths are strong and dense over the phalanges, weak and lax over the joints (PLATE 4).

Synovial Flexor Sheaths. In the carpal tunnel the flexor tendons are invested with synovial sheaths that extend proximally into the lower part of the forearm and proceed distally to a varying extent (PLATE 4). In the case of flexor pollicis longus the sheath extends from above the flexor retinaculum to the insertion of the tendon into the terminal phalanx of the thumb. The tendons of the superficial and deep flexors are together invested with a common synovial sheath that is incomplete on the radial side. This common sheath commences a short distance above the wrist and extends down into the palm. In the case of the little finger, it is continued along the whole extent of the flexor tendons to the terminal phalanx. Next to the first, the fifth metacarpal is the most mobile bone of the palm. The common flexor sheath ends over the remaining three sets of tendons just distal to the flexor retinaculum. In the case of a very mobile fourth metacarpal (e.g., in violin players) the sheath may extend to the terminal phalanx of the ring finger, like that of the little finger in the average hand. The common flexor sheath communicates at the level of the wrist with the sheath of flexor pollicis longus in 50 per cent of individuals. In the case of index, middle and ring fingers, where the common sheath ends beyond the flexor retinaculum, a separate synovial sheath lines the fibrous flexor sheath over the phalanges. There is thus a short distance of bare tendon for index, middle and ring fingers in the middle of the palm. It is from this situation that the lumbrical muscles arise (PLATE 4). The fourth lumbrical obliterates the synovial sheath along its origin from the tendon to the little finger.

The Palmar Spaces. The palmar aponeurosis, fanning out from the distal border of the flexor retinaculum, is triangular in shape. From each of its two sides a fibrous septum dips downwards into the palm. That from the ulnar border is attached to the palmar border of the fifth metacarpal bone. Medial to it is the *hypothenar space*, not important surgically since it contains no long flexor tendons, but encloses only the hypothenar muscles. The remaining part of the palm is divided into two spaces by the septum that dips in from the radial border of the palmar aponeurosis to the palmar surface of the middle metacarpal bone. This septum lies obliquely and separates the thenar space on its radial side from the midpalmar space beneath the palmar aponeurosis. Normally the septum passes deeply between the flexor tendons of index and middle fingers; that is to say, the flexor tendons of the index finger overlie the thenar space. In some cases, however, the septum passes deeply between the index tendons and the first lumbrical; in these cases the first lumbrical overlies the thenar space whilst the flexor tendons to the index finger overlie the midpalmar space.

The Midpalmar Space (Fig. 2.51). The space is limited superficially by a thin fascia that lies deep to the common synovial sheath and the flexor tendons. The space is floored in by the interossei and metacarpals of the third and fourth spaces. Its sides are formed by the septa dipping in from the borders of the palmar aponeurosis. Distally the space is continued along the canals in which the tendons of the lumbrical muscles lie, on the palmar side of the deep transverse ligament of the palm, to the webs of the fingers. Proximally the space is closed by the firm attachment of the parietal layer of the common flexor synovial sheath to the walls of the carpal flexor tunnel.

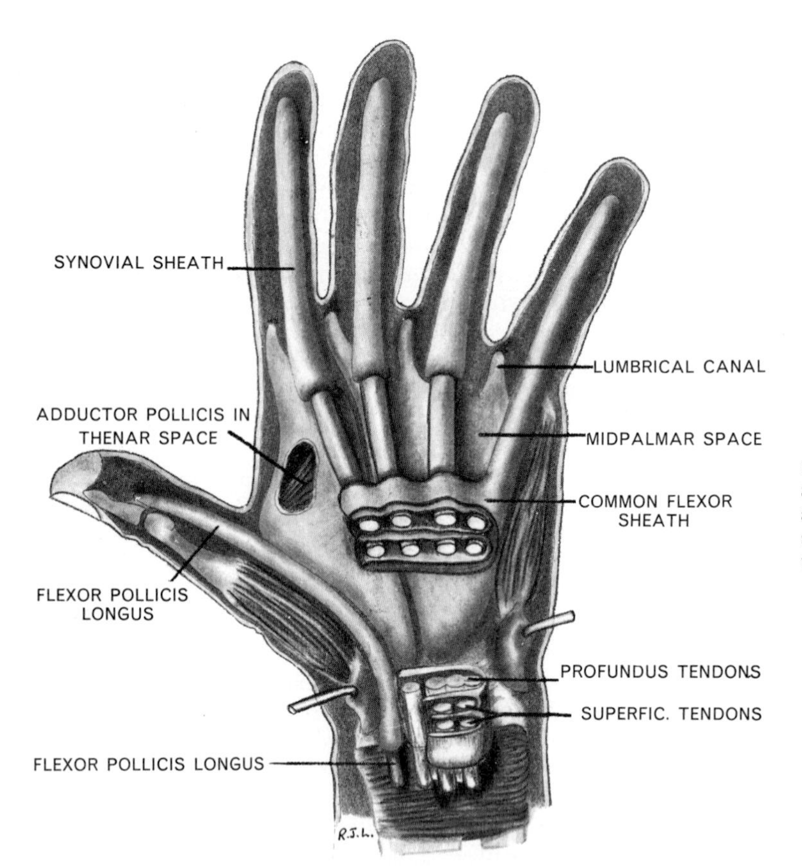

FIG. 2.51. THE PALMAR SPACES AND SYNOVIAL SHEATHS. The flexor retinaculum is split and retracted to expose the divided synovial sheaths in the carpal flexor tunnel.

The Thenar Space (Fig. 2.51). This space lies on the thumb side of the septum that joins the border of the palmar aponeurosis to the palmar border of the third metacarpal bone. It always contains the first lumbrical muscle, and generally also it is overlaid by the flexor tendons to the index finger. In the palm, it is bounded posteriorly by the adductor pollicis muscle, but around the distal border of this muscle, in the web of the thumb, it is continuous with the slit-like space that lies between adductor pollicis and the first dorsal interosseous muscle. The thenar space is closed at the wrist, in a manner similar to the midpalmar space, by fusion of the parietal layer of the synovial sheaths with the walls of the carpal tunnel.

The Anatomy of a Web. The three webs of the palm lie between the four slips of attachment of the palmar aponeurosis. From the skin edge they may be said to extend proximally as far as the metacarpo-phalangeal joints, a distance of about an inch and a half (4 cm). Between the palmar and dorsal layers of the skin lie the superficial and deep transverse ligaments of the palm, the digital vessels and nerves, and the tendons of the interossei and lumbricals on their way to the extensor expansions. The web is filled in with a packing of loose fibro-fatty tissue.

The **superficial transverse ligament of the palm (interdigital ligament)** lies beneath the palmar skin across the free margins of the webs (Fig. 2.45). Its fibres lie transversely in a flat band, best developed on the radial side and thinning out over the more mobile little finger. Its two ends, and the parts that lie across the fibrous flexor sheaths, are continued along the sides of the fingers, tethering the skin to the phalanges. The ligament supports the fold of skin at the web; it is well developed in quadrupeds whose forefeet are webbed.

The **digital vessels and nerves** lie immediately deep to the interdigital ligament, a point to be remembered in making web incisions for palmar space infections. Here the nerves lie on the palmar side of the arteries.

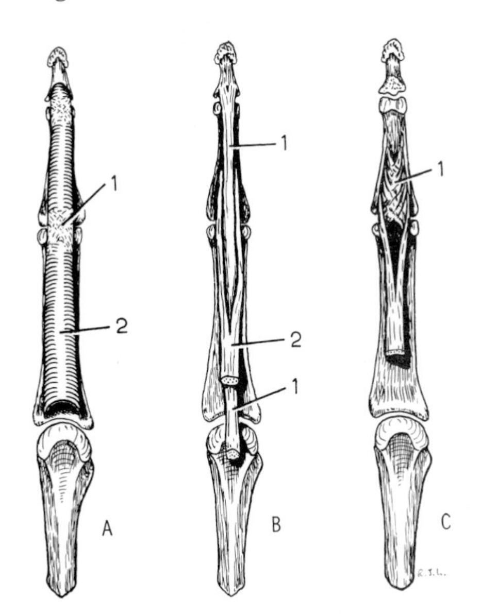

FIG. 2.52. FLEXOR TENDON INSERTIONS. **A.** THE FIBROUS FLEXOR SHEATH. Its fibres are oblique and slender (1) across the interphalangeal joints but transverse and strong (2) across the phalanges. **B.** The two flexor tendons exposed by removal of the fibrous flexor sheath. 1. Flexor digitorum profundus perforates flexor superficialis (2) and passes to the base of the distal phalanx. **C.** Flexor profundus tendon removed to expose the gutter-shaped chiasma (1) in the tendon of flexor digitorum superficialis.

The **lumbrical tendon** lies beneath the vessels, passing distally along the lumbrical canal into the radial side of the dorsal expansion of its own finger. The **lumbrical canal** is a condensation of fibro-fatty tissue that surrounds the tendon. The canal is a distal diverticulum of the palmar space.

The *deep transverse ligament of the palm* joins the palmar ligaments of the metacarpo-phalangeal joints (p. 106). It lies an inch (3 cm) proximal to the superficial transverse ligament (Fig. 2.44).

The *interosseous tendons* lie on the dorsal side of the deep transverse ligament.

The **web of the thumb** lacks both superficial and deep transverse ligaments, a factor contributing to the mobility of the thumb. The deep fascia passes across from palmar to dorsal surfaces of the web and beneath it lie the transverse head of adductor pollicis and the first dorsal interosseous muscle. From the slit-like space between them emerge the arteries radialis indicis and princeps pollicis. Each hugs its own digit and the central part of the web can be incised without risk to either vessel. The thenar space lies beneath the deep fascia on the palmar surface of adductor pollicis; the first lumbrical passes through the space to the radial side of the index finger.

Digital Attachments of the Long Tendons

The Flexor Tendons. The tendon of flexor digitorum superficialis enters the fibrous flexor sheath on the palmar surface of the tendon of flexor digitorum profundus. It divides into two halves, which flatten a little and spiral around the profundus tendon and meet on its deep surface in a chiasma (a partial decussation). This forms a tendinous bed in which lies the profundus tendon. Distal to the chiasma the superficialis tendon is attached to the sides of the whole shaft of the middle phalanx (Fig. 2.52).

The profundus tendon enters the fibrous sheath deep to the superficialis tendon, then lies superficial to the partial decussation of the latter, before passing distally to reach the base of the terminal phalanx. In the flexor sheath both tendons are invested by a common synovial sheath that possesses parietal and visceral layers. Each tendon receives blood vessels from the palmar surface of the phalanges. The vessels are invested in synovial membrane. These vascular synovial folds are known as the **vincula,** and each tendon possesses two, the short and long (Fig. 2.53). The profundus tendon has its short vinculum in the angle close to its insertion. Its long vinculum passes from the tendon between the two halves of the superficialis tendon (proximal to the chiasma) to the palmar surface of the proximal phalanx. The superficialis tendon has a short vinculum near its attachment into the sides of the middle phalanx. The long vinculum of the superficialis tendon is double, each half of the tendon possessing a vinculum just distal to its first division, passing down to the palmar surface of the proximal phalanx. The short vincula, attached to the capsules of the interphalangeal joints, serve to pull these out of harm's way during flexion. Only the long vinculum is vascular.

The Extensor Tendons. The extensor tendons to the four fingers have a characteristic insertion. Passing across the metacarpo-phalangeal joint, the tendon is partly adherent to the articular margins, and indeed in this situation its deepest fibres form the posterior capsule of the joint. The bulk of the tendon, however, passes freely across the joint, and broadens out on the dorsal surface of the proximal phalanx. This flat part now divides into three slips; the central slip passes on to the base of the middle phalanx. The two marginal slips diverge around the central slip. Each receives a strong attachment from the tendons of the interosseous and lumbrical muscles, forming a broad **extensor expansion.** The palmar border of the extensor expansion is free and lies virtually in the long axis of the finger.

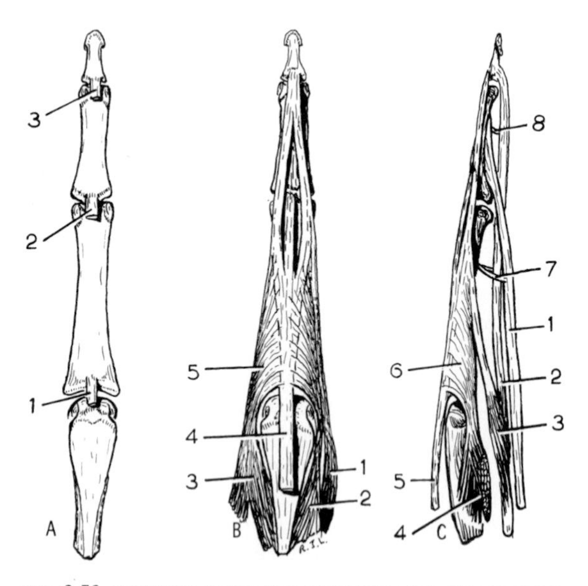

FIG. 2.53. EXTENSOR DIGITORUM COMMUNIS AND THE EXTENSOR HOOD. **A.** The bony insertions of extensor digitorum communis (1, 2, 3) into the base of each phalanx. **B.** Dorsal view of the insertion of extensor communis tendon and the extensor hood. 1. Lumbrical muscle. 2. Interosseous muscle. 3. Interosseous muscle. 4. Extensor digitorum communis tendon lying free on metacarpal bone. 5. Extensor hood. **C.** Lateral view. 1. Tendon of flexor digitorum superficialis. 2. Tendon of flexor digitorum profundus. 3. Lumbrical muscle. 4. Interosseous muscle. 5. Extensor digitorum communis tendon. 6. Extensor hood. 7. Vincula longia to tendon of flexor superficialis. 8. Vinculum longium to tendon of flexor profundus.

More proximally the fibres from the interossei and lumbricals radiate across the dorsum of the proximal phalanx to complete the extensor expansion, whose most proximal fibres pass transversely across the base of the phalanx. The fused interosseous tendons and marginal slips of the communis expansion, passing distally across the middle phalanx, converge to be inserted together into the base of the distal phalanx (Fig. 2.53).

Certain fibrous bands are attached to the extensor apparatus; they are tightened by appropriate positions of the phalanges. The oblique *retinacular ligaments* are bilateral strong narrow bands attached near the head of the proximal phalanx, straddling the fibrous flexor sheath. Each passes palmar to the proximal interphalangeal joint and joins the marginal slip of the extensor tendon. Extension of the proximal joint draws them tight and limits flexion of the distal joint. Flexion of the proximal joint slackens them and permits full flexion of the distal joint. The two joints thus passively tend to assume similar angulations.

Long Tendons of the Thumb. On the flexor aspect there is only one tendon, that of flexor pollicis longus invested by its synovial sheath as it passes to the distal phalanx. On the extensor surface the tendons of extensor pollicis brevis and longus are each inserted separately (into the proximal and distal phalanx). There is no extensor hood as in the four fingers, but the extensor pollicis longus tendon receives a fibrous expansion from both abductor pollicis brevis and adductor pollicis (Fig. 2.42, p. 92). These expansions serve to hold the long extensor tendon in place on the dorsum of the thumb.

Mechanism of Gripping. The action of 'making a fist' or of gripping an object is primarily one of contracting the long flexor tendons of the fingers and thumb; but synergic contraction of the extensors of the wrist is indispensable to an efficient grip.

As the long flexor tendons contract the interphalangeal joints are flexed. This movement is overlapped and followed by the flexion of the metacarpophalangeal joints. The thumb is opposed and then flexed against the fingers. Simultaneously with the closing of the digits the *wrist is moved into extension* by contraction of its radial and ulnar extensors. The length of the flexor muscles is thereby increased and their power of contraction greatly enhanced. A powerful grip cannot be exerted with the wrist in full flexion, neither can the fingers be closed into a fist (Fig. 2.54).

Most of the long flexor musculature arises from radius and ulna. The humeral head of flexor superficialis, however, passes across the elbow joint. Its flexing action on the elbow during gripping is counteracted by synergic contraction of the triceps.

Actions of Interossei and Lumbricals. The **interossei** are inserted into the proximal phalanges

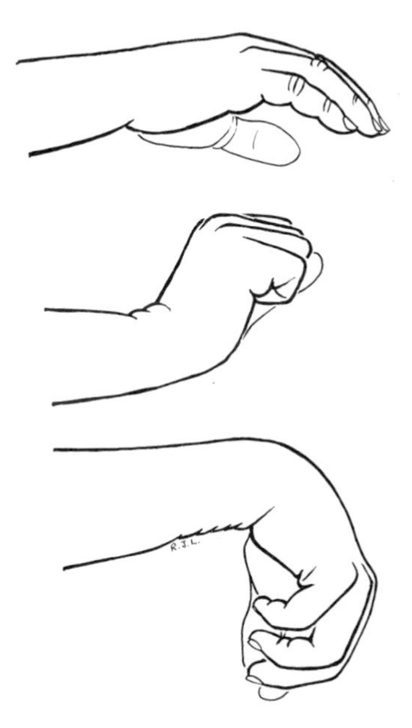

FIG. 2.54. THE MECHANISM OF GRIPPING. UPPER FIGURE—the right hand relaxed. MIDDLE FIGURE—Synergic extension of the wrist accompanies the act of 'making a fist'. LOWER FIGURE—in the fully flexed wrist the grip is weak and it is not possible to make a fist.

and into the extensor expansions. Contracting as palmar or dorsal groups, respectively, they adduct or abduct the fingers away from the midline of the palm. When palmar and dorsal interossei contract together the adducting and abducting effects cancel out. Flexion of the metacarpo-phalangeal joints results. The interossei are indispensable for the combined movement of flexion of the metacarpo-phalangeal joint and simultaneous extension of the interphalangeal joints; but the manner of their action is not perfectly clear. Opposing theories exist. The earliest idea, a very simple one, was that the muscles act as prime movers, and this seems reasonable when one contemplates the anatomy of their insertions. The proximal fibres of the extensor hood run transversely across the first phalanx. Tension of these fibres will bend the phalanx towards the palm. The fibres on the edge of the dorsal expansion run almost along the axis of the finger, and are traceable distally to their insertion into the dorsal aspect of the base of the distal phalanx (Fig. 2.53). Tension on these fibres would naturally be expected to extend the distal phalanx. The difficulty in accepting this rather simple explanation is emphasized by the observation that the extensor digitorum communis is indispensable to the action of extending the terminal phalanx with full force. In radial nerve palsy, or if the communis tendon is cut on the dorsum of the hand, the distal phalanx cannot be extended with full force even though the interossei are normal.

The probable explanation of the effect of the interossei *on the terminal phalanges* is that they are synergists to the prime mover the extensor communis. If the interossei are paralysed (claw hand) the pull of the communis tendon is wholly expended on the metacarpo-phalangeal joint which is hyperextended and there is too much slack in the distal part of the tendon for it to act on the terminal phalanx. If, however, the metacarpo-phalangeal joint is held flexed, the communis tendon, passing around its dorsal curve as around a pulley, is lengthened, and can now extend the distal phalanx. Clinical evidence from peripheral nerve injuries of radial, median and ulnar nerves, indicates that the essential action of the interossei is to flex the metacarpo-phalangeal joints to allow the extensor digitorum communis to extend the interphalangeal joints.

The action of the **lumbricals** is more obscure. In ulnar palsy all the interossei are paralysed but the lumbricals of only the little and ring fingers are out of action. The claw hand of ulnar paralysis affects these fingers most, index and middle fingers being less affected because their lumbrical muscles are intact. From this fact and from their anatomical attachment into the extensor expansion it seems certain that the lumbricals have the same action as the interossei, namely to flex the metacarpo-phalangeal joints. In closing the fist the profundus tendon is withdrawn proximally into the palm as the terminal phalanx is flexed. This draws the lumbrical attachment proximally also; it seems absurd to imagine that in this simple movement the lumbrical actually *extends* the terminal phalanx, opposing the pull of its parent tendon. Here it surely acts simply through the extensor expansion as a flexor of the proximal phalanx.

A further combined effect of the interossei and lumbricals may be to displace the marginal slips of the common extensor tendon a fraction more widely—enough at least to relax the central slip, so that the whole extensor pull on the phalanges is transmitted by the marginal slips to the distal phalanx. Much work has been done on the actions of these little muscles, but it seems certain that the last word has by no means yet been said.

Joints of the Carpus. An S-shaped **midcarpal joint** forms a continuous synovial space between the two rows of carpal bones, and this extends proximally and distally between adjacent carpal bones as a continuous intercarpal joint. A similar synovial joint lies between the distal row of carpal bones and the metacarpal bones of the four fingers. This **carpo-metacarpal joint** commonly communicates with the intercarpal joint.

The joint between hamate and fifth metacarpal is the most mobile of the four and the slight flexion possible here aids in 'cupping' the palm. The carpo-metacarpal joint of the *thumb* is a separate synovial cavity between the trapezium and first metacarpal bone. The joint surfaces are reciprocally saddle-shaped. The capsule is loose and permits independent movement of the whole thumb.

The **metacarpo-phalangeal joints** are synovial joints. They allow of flexion and extension, abduction and adduction. The *palmar ligaments* are strong pads of fibro-cartilage, which limit extension at the joint. They are joined together by transverse bands that together constitute the *deep transverse ligament of the palm*—or of the *metacarpal heads* (Fig. 2.44, p. 94). These joints (the 'knuckle joints') lie on the arc of a circle; hence the extended fingers diverge from each other, the flexed fingers crowd together into the palm. A single finger flexed points in towards the tubercle of the scaphoid.

The **interphalangeal joints** are pure hinge joints, no abduction being possible. Extension is limited by palmar ligaments and by the obliquity of the collateral ligaments; the latter are attached eccentrically, and tighten in extension and in flexion. The joint acts like a toggle switch (Fig. 2.55).

FIG. 2.55. THE COLLATERAL LIGAMENT OF AN INTERPHALANGEAL JOINT.

SUMMARY OF INNERVATION

The Cutaneous Innervation of the Upper Limb

The segmental supply (dermatomes) of the upper limb has been considered on p. 26. Most of the cutaneous nerves have been described in the preceding pages but their distribution may with advantage be summarized here (Fig. 2.56).

The skin over the shoulder is supplied by the supra-clavicular nerves (C4) from the cervical plexus. This is neck skin that during development has moved on to the pre-axial border of the root of the limb. Similarly the postaxial border is clad with skin developed on the trunk. The floor of the axilla and a variable area of the medial surface of the arm is supplied by the lateral

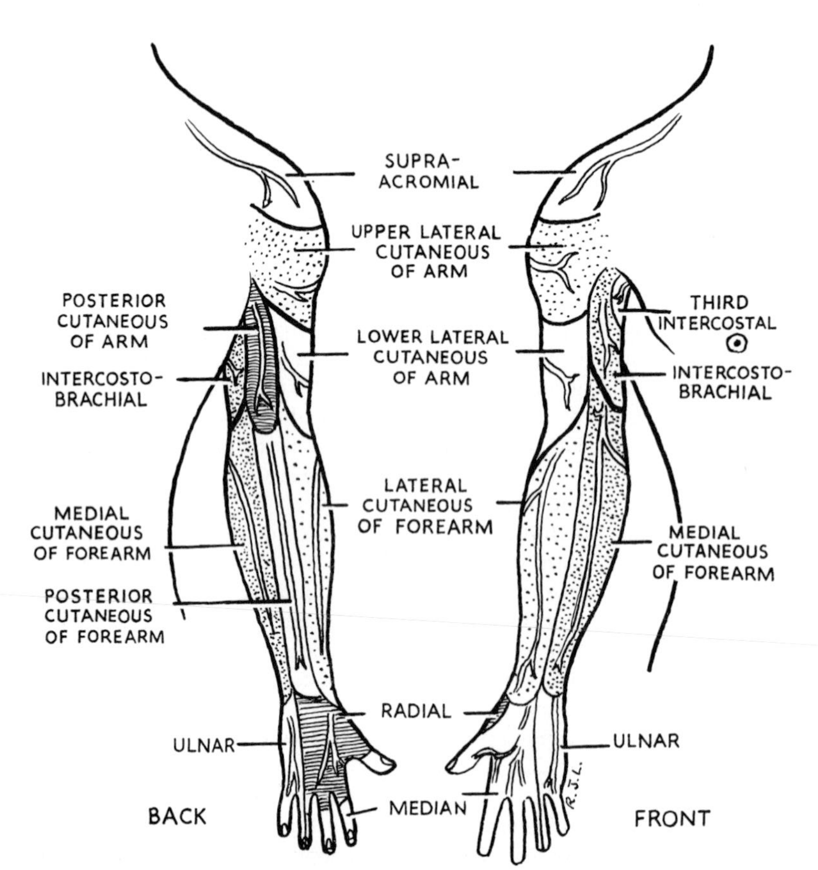

FIG. 2.56. THE CUTANEOUS NERVES OF THE RIGHT UPPER LIMB. In the figure the intercosto-brachial replaces the medial cutaneous nerve of the arm. This is not uncommon. Compare the dermatomes, Fig. 1.31, p. 26.

branch of the second intercostal nerve, the intercostobrachial nerve. Occasionally the lateral branch of the third intercostal nerve extends to supply skin on the floor of the axilla (Fig. 2.56).

The lateral surface of the arm is supplied by the upper lateral cutaneous and the lower lateral cutaneous nerves, branches of the axillary and radial nerves respectively. Posteriorly over the triceps a strip of skin is supplied by the posterior cutaneous nerve of the arm, a branch of the radial nerve. The intercostobrachial nerve and the medial cutaneous nerve of the arm supply the medial and anterior surfaces of the arm. It should be noted that the medial cutaneous nerve of the forearm supplies some skin of the arm just above the cubital fossa.

In the forearm a strip of skin posteriorly is supplied by the posterior cutaneous nerve of the forearm, a branch of the radial nerve. Medial and lateral sides are supplied by the medial and lateral cutaneous nerves of the forearm (from brachial plexus and musculocutaneous nerves respectively). Each of these divides symmetrically into anterior and posterior branches, which extend down to the wrist.

The palm of the hand is supplied on the ulnar side by the superficial branch of the ulnar nerve. Towards the centre of the palm the skin is innervated by the palmar branch of the median nerve, which perforates the deep fascia proximal to the flexor retinaculum. The palmar surfaces of the fingers and thumb are supplied by digital branches of the ulnar and median nerves. The ulnar digital branches supply the ulnar one and a half fingers on their palmar surfaces, their tips, and their dorsal surfaces over the distal one and a half phalanges. The median nerve by its digital branches supplies the radial three and a half digits on their palmar surfaces, tips, and dorsal surfaces over the distal one and a half phalanges. The dorsal surface of the hand is supplied on the ulnar side (one and a half fingers) by the dorsal branch of the ulnar nerve. The radial three and a half digits and the web of the thumb are supplied by the terminal cutaneous branches of the radial nerve.

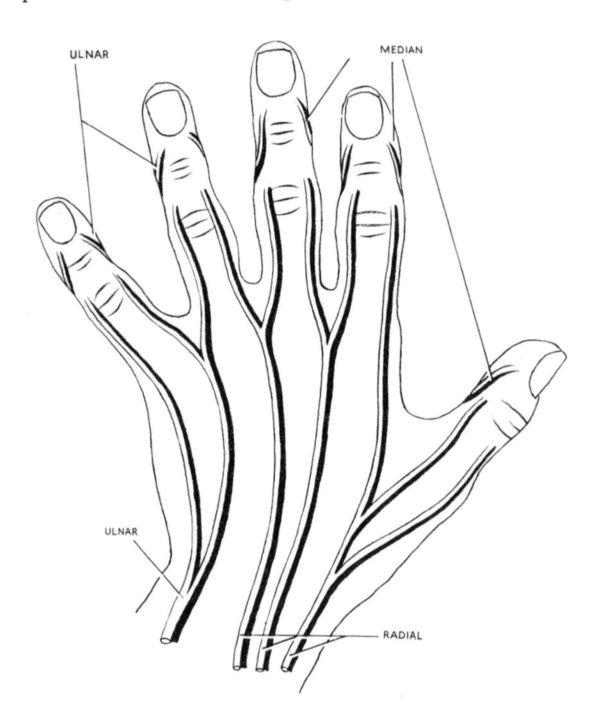

FIG. 2.57. THE CUTANEOUS INNERVATION OF THE DORSUM OF THE LEFT HAND.

The Muscular Innervation of the Upper Limb

In the upper arm the flexor compartment is supplied by the musculo-cutaneous nerve, the extensor compartment by the radial. The flexor compartment of the forearm is supplied chiefly by the median nerve, replaced by the ulnar for one and a half muscles (flexor carpi ulnaris and half of flexor digitorum profundus). The extensor compartment of the forearm is supplied by the radial nerve or its terminal branch, the posterior interosseous. The intrinsic muscles of the hand are supplied chiefly by the ulnar nerve, the median nerve supplying only the muscles of the thenar eminence and the radial two lumbricals.

The segmental innervation of the muscles of the upper limb is considered on p. 29.

Sympathetic Innervation of the Upper Limb

A grey ramus (Fig. 1.40, p. 34) joins each root of the brachial plexus, and these sympathetic fibres hitch-hike through the plexus and its branches and *remain in the nerves until very near their area of supply*. For example, the grey fibres to an arteriole in a finger-tip run in the digital nerves and not along the digital arteries. The grey sympathetic fibres remain invariably with the *same segmental fibres* which they originally joined.

Thus *each dermatome* of the upper limb is supplied via the cutaneous nerves by hitch-hiking grey fibres from cell bodies in the *appropriate sympathetic ganglion* (dermatomes C5 and 6 from cell bodies in the middle cervical ganglion, C7 and 8 from the inferior cervical ganglion and Th. 1 from the first thoracic, or stellate, ganglion). In the skin the grey rami innervate arterioles, sweat glands and the arrectores pilorum muscles (for temperature regulation).

The deep sympathetic supply to the limb is by a *series of grey fibres* that join the main artery from adjacent nerves. These form a peri-arterial plexus, but it is for only the local branches of the artery; the fibres in the peri-arterial plexus do not descend along the artery for any considerable distance. Thus the operation of peri-arterial sympathectomy denervates only local branches of the artery, and has been long since abandoned for this reason.

The preganglionic (white) fibres for the upper limb come from cell bodies in the upper six thoracic segments of the spinal cord. They ascend in the sympathetic trunk to synapse in the ganglia mentioned above.

The Nerves of the Upper Limb

The branches of the brachial plexus have been studied region by region in the previous pages, but it is convenient also to summarize each nerve as a whole. Details of relationships are to be found in the accounts of the regions concerned; the following is a review of only the major facts concerning each nerve.

In dealing with small nerves the branches can often be described with the nerve itself, but in the case of longer nerves with several branches it is much better to consider first the course of the nerve itself, and afterwards to recount the branches one by one in proper order.

Branches of the Roots (three in number)

(1) The **dorsal scapular nerve** (C5) runs down deep to levator scapulæ and the two rhomboids, supplying all three muscles. Lying on serratus posterior superior, it forms a neuro-vascular bundle with the descending scapular vessels alongside the vertebral border of the scapula (Fig. 2.6, p. 57).

(2) The **nerve to subclavius** (C5, 6) passes down over the trunks of the plexus and *in front* of the subclavian vein. If it contains accessory phrenic fibres these join the phrenic nerve in the superior mediastinum.

(3) The **long thoracic nerve** (C5, 6, 7) forms on the first digitation of the serratus anterior muscle and runs vertically downwards just behind the midaxillary line, i.e., just behind the lateral cutaneous branches of the intercostal nerves and vessels, beneath the deep fascia over the muscle. Its fibres are paid off segmentally. C5 supplies the upper two digitations (which pull on the upper angle of the scapula) C6 the next two (vertebral border of scapula) and C7 the lowest four digitations (pulling on inferior angle of scapula, i.e., rotation of scapula for abduction of arm).

Branch of the Upper Trunk (one only)

The **suprascapular nerve** (C5), prominent beneath the fascial floor of the posterior triangle, passes beneath the transverse scapular ligament (the suprascapular vessels lie above the ligament). The neuro-vascular bundle supplies supraspinatus, infraspinatus, and the posterior part of the capsule of the shoulder joint.

Branches of the Lateral Cord (three in number)

(1) The **lateral pectoral nerve** (C6, 7) passes through the clavi-pectoral fascia and supplies the upper fibres of pectoralis major.

(2) The **musculo-cutaneous nerve** (C5, 6, 7) is muscular to the flexors in the arm and cutaneous in the forearm. It supplies coraco-brachialis, then pierces that muscle to slope down in the plane between biceps and brachialis, supplying both muscles. Emerging at the lateral border of the biceps tendon, it pierces the deep fascia at the flexure crease of the elbow. It is now called the *lateral cutaneous nerve of the forearm*, and supplies skin from elbow to wrist by an anterior and a posterior branch along the radial border of the forearm.

(3) The lateral head of the median nerve (C6, 7) is joined by the medial head to form the complete nerve (*v. inf.*).

Branches of the Medial Cord (five in number)

(1) The **medial pectoral nerve** (C8) gives a branch to pectoralis minor, then pierces that muscle to supply the lower (i.e., sternocostal) fibres of pectoralis major.

(2) The medial head of the median nerve (C8, Th. 1) crosses the axillary artery to join the lateral head. The **median nerve** (C6, 7, 8, Th. 1) thus formed supplies most of the flexor muscles of the forearm, but only the three thenar muscles and two lumbricals in the hand. It is cutaneous to the flexor surfaces and nails of the three and a half radial digits and a corresponding area of palm.

Course of the Median Nerve. It leaves the axilla and slopes in front of the brachial artery beneath the deep fascia of the arm. At the elbow it lies medial to the artery beneath the bicipital aponeurosis. It descends between the two heads of pronator teres and beneath the fibrous arch of flexor digitorum superficialis. Adherent to the deep surface of the muscle, it emerges on the radial side, lying between flexor carpi radialis and palmaris longus tendons before passing through the carpal tunnel into the hand.

Branches of the Median Nerve. It has no branches in the arm, but gives a twig to the elbow joint. Here, in the cubital fossa, it supplies *muscular branches* to pronator teres, palmaris longus, flexor carpi radialis and flexor digitorum superficialis, and also gives off the *anterior interosseous nerve*, which descends on the interosseous membrane to the wrist. The anterior interosseous is the nerve of the deep flexor compartment, for it supplies the radial half (usually) of flexor digitorum profundus, all of flexor pollicis longus and pronator quadratus and is sensory to the interosseous membrane, periosteum of radius and ulna, and the wrist and intercarpal joints. The *palmar cutaneous* branch of the median nerve pierces the deep fascia just above the flexor retinaculum and supplies more than half of the thenar side of the palm.

The median nerve divides in the carpal tunnel and

enters the hand as a lateral and a medial branch. The lateral branch gives a muscular branch (Th. 1) which recurves around the distal border of the flexor retinaculum to supply the three thenar muscles (abductor brevis and flexor brevis, and opponens pollicis). Then it breaks up into three palmar digital branches (C6), two for the thumb and one for the index, this last one supplying also the first lumbrical (Th. 1). The medial branch divides into two palmar digital nerves for the adjacent sides of the second and third clefts (C7). The branch for the second cleft supplies also the second lumbrical (Th. 1). The five *palmar digital* branches supply the nail beds as well as the flexor skin of the radial three and a half digits.

(3) The **medial cutaneous nerve of the arm** (Th. 1) is rather a failure. It is sometimes replaced entirely by the intercosto-brachial nerve. It runs down with the axillary vein to pierce the deep fascia and supply skin medial to the anterior axial line, but fails to reach the elbow.

(4) The **medial cutaneous nerve of the forearm** (C8, Th. 1) is a much bigger nerve than the last. It runs down between axillary artery and vein and pierces the deep fascia half way to the elbow (often in common with the basilic vein). It supplies the lower part of the *arm* above the elbow and then divides into anterior and posterior branches to supply the skin along the ulnar border of the forearm down to the wrist. In the forearm it is symmetrical with the lateral cutaneous nerve (musculo-cutaneous) and the two meet without overlap along the anterior axial line. Their territories are separated posteriorly by the posterior cutaneous branch of the radial nerve.

(5) The **ulnar nerve** (C7, 8, Th. 1) is the direct continuation of the medial cord (C8, Th. 1). In 95 per cent of cases it picks up C7 fibres in the axilla from the lateral cord (these fibres are for flexor carpi ulnaris). The nerve is destined for some flexor muscles on the ulnar side of the forearm and the skin of the ulnar one and a half digits, but it is to be remembered chiefly as the nerve for most of the intrinsic muscles of the hand (i.e., those concerned in fine finger movements).

Course of the Ulnar Nerve. Running down behind the brachial artery it inclines backwards and pierces the medial intermuscular septum. Lying in front of the nerve to the medial head of triceps, it now descends along the triceps to lie on the shaft of the humerus between the medial epicondyle and the olecranon. It passes between the two heads of flexor carpi ulnaris and thus enters the flexor compartment of the forearm. Crossing the sublime tubercle, it descends on flexor digitorum profundus, under cover of flexor carpi ulnaris. Here it is joined by the ulnar artery. The two emerge from beneath the tendon of flexor carpi ulnaris just above the wrist and cross the flexor retinaculum alongside the pisiform bone. On the retinaculum the nerve divides into its terminal superficial and deep branches.

Branches of the Ulnar Nerve. Several twigs to the elbow joint bind the nerve close against the lower end of the humerus. In the forearm the nerve supplies flexor carpi ulnaris (C7) and the ulnar half (usually) of flexor digitorum profundus (C8). It has a *palmar cutaneous* branch (C8) which pierces the deep fascia above the flexor retinaculum to supply skin over the hypothenar muscles. A *dorsal cutaneous* branch (C8) winds around the lower end of the ulna deep to the tendon of flexor carpi ulnaris, crosses the triquetral bone, and is distributed to the dorsal skin of one and a half fingers and a corresponding area of the back of the hand. On the fingers it falls short of the nail beds. Not uncommonly it supplies two and a half instead of one and a half fingers.

The ulnar nerve divides on the flexor retinaculum alongside the pisiform bone. The *superficial branch* (C8) runs forward beneath palmaris brevis (which it supplies) and is distributed to the ulnar one and a half fingers, including their nail beds, by two digital branches. One of them supplies the ulnar side of the little finger, the other goes to the fourth cleft, and bifurcates to supply the adjacent sides of ring and little fingers.

The *deep branch* (Th. 1) passes deeply between abductor and flexor minimi digiti, then between opponens and the fifth metacarpal. It supplies all three hypothenar muscles. Crossing the palm in the concavity of the deep palmar arch it supplies the two ulnar lumbricals and all the interossei, both palmar and dorsal. It ends by supplying adductor pollicis.

Branches of the Posterior Cord (five in number)

(1) The **upper subscapular nerve** (C6, 7) passes straight back into the upper part of the muscle.

(2) The **thoraco-dorsal nerve** (nerve to latissimus dorsi) (C6, 7, 8) inclines forwards and enters the deep surface of latissimus dorsi just behind the anterior border; it is vulnerable in operations on the scapular lymph nodes, for its terminal part lies anterior to them and to the subscapular artery (PLATE 22, p. *xi*).

(3) The **lower subscapular nerve** (C6, 7) supplies the lower part of subscapularis and the teres major.

(4) The **axillary** (circumflex) **nerve** (C5) exemplifies Hilton's Law by supplying deltoid, the shoulder joint and skin over the joint. It passes backwards through the quadrilateral space, lying above its artery and vein, in contact with the neck of the humerus just below the capsule of the shoulder joint, which it supplies. Through the quadrilateral space it divides. The *posterior branch (superficial)* supplies teres minor and winds around the posterior border of deltoid, supplying a small part of the muscle. As the *upper lateral cutaneous nerve of the arm* it supplies skin over the lower half

of deltoid, below the supra-acromial nerves. The *anterior branch (deep)* lies on the humerus between teres minor and the lateral head of triceps and supplies the deltoid muscle. A few fine twigs pierce the deltoid to reach a small area of skin over the centre of the muscle.

(5) The **radial nerve** (C5, 6, 7, 8, Th. 1) is the nerve of the extensor compartments of the arm and forearm, supplying skin over them and on the dorsum of the hand. Its branches tend to arise high above their destination.

Course of the Radial Nerve. A direct continuation of the posterior cord, it passes beyond the posterior wall of the axilla and enters a triangular space below teres major. It spirals across the medial head of triceps to lie in the spiral groove on bare bone, deep to the lateral head of triceps. It pierces the lateral intermuscular septum and can be rolled on the humerus here, one-third of the way down from deltoid tuberosity to lateral epicondyle. In the flexor compartment of the lower arm it is held away from bare bone by brachialis, and it descends in the intermuscular slit between brachialis and brachio-radialis. After giving off the posterior interosseous branch, the rather slender remnant, purely cutaneous now, retains the name of radial nerve. It runs down the flexor compartment of the forearm on supinator, the tendon of pronator teres, flexor digitorum superficialis, and flexor pollicis longus under cover of brachio-radialis. It winds around the lower end of the radius deep to the tendon of brachio-radialis, crosses abductor pollicis longus, extensor pollicis brevis and extensor pollicis longus to reach the back of the hand.

Branches of the Radial Nerve. These are best considered region by region.

(*a*) *Branches in the Axilla* (3). Here the nerve lies just under the deep fascia. The *posterior cutaneous nerve of the arm* (C8 and all of Th. 1 which the radial nerve carries) passes back medial to the long head of triceps and pierces the deep fascia to supply a strip of skin along the extensor surface of the arm down to the elbow. Two *muscular branches* (C7, 8) reach the triceps. One sinks into the long head, the other runs down under the deep fascia of the arm, behind the ulnar nerve, to enter the lower part of the medial head. (This branch to the medial head was formerly misnamed the ulnar collateral nerve.)

(*b*) *Branches in the Spiral Groove* (4). While in the extensor compartment of the arm two muscular and two cutaneous branches are given off. High up in the groove a branch passes to the lateral head and another to the medial head of the triceps (C7, 8); the latter has fibres which pass distally, deep to triceps, to supply the anconeus. The *lower lateral cutaneous nerve of the arm* (C5) is given off just before the radial nerve pierces the lateral intermuscular septum. It passes up to the skin between the septum and the lateral head of triceps to supply skin over the lateral surface of the arm down to the elbow. In common with it arises the *posterior cutaneous nerve of the forearm* (C6, 7), which runs straight down behind the elbow to supply a strip of skin over the extensor surface of the forearm as far as the wrist. (In the working position of pronation this skin faces towards the *front* of the body.)

(*c*) *Branches in the Flexor Compartment* (4). While lying in the slit between brachialis and brachio-radialis, the main trunk gives a small branch to the lateral part of brachialis (not the main supply to this muscle) and supplies brachio-radialis (C5, 6) and extensor carpi radialis longus (C7, 8). At the bend of the elbow it gives off the posterior interosseous branch, and then continues on as the terminal cutaneous branch already described.

(*d*) *The posterior interosseous nerve* (C6, 7, 8) supplies extensor carpi radialis brevis and supinator in the cubital fossa, and then spirals down around the upper end of the radius between the two layers of supinator to enter the extensor compartment of the forearm. It crosses abductor pollicis longus, dips down to the interosseous membrane and runs to the back of the wrist. In the extensor compartment it supplies seven more muscles; three extensors from the common extensor origin (extensor digitorum communis, extensor digiti minimi, and extensor carpi ulnaris), the three thumb muscles (abductor pollicis longus, extensor pollicis brevis and extensor pollicis longus) and the extensor indicis. It is sensory to the interosseous membrane, periosteum of radius and ulna and the wrist and carpal joints on their extensor surfaces.

(*e*) The terminal *cutaneous branches* are two or three nerves which cross the roof of the snuff box and reach the radial three and a half digits (falling short of the nail beds) and a corresponding area of the back of the hand. These nerves and the dorsal branch of the ulnar nerve commonly share $2\frac{1}{2} : 2\frac{1}{2}$ instead of the more usually described $3\frac{1}{2} : 1\frac{1}{2}$ digits.

THE OSTEOLOGY OF THE PECTORAL GIRDLE AND UPPER LIMB

Only after the anatomy of the upper limb has been mastered can the bones be studied with proper appreciation.

The Clavicle

The bone is longer (5 inches or 14 cm plus) in the broad-shouldered male, and its curvatures are usually more pronounced than in the female. The medial two-thirds or more is rounded, rather quadrilateral, in section and convex forwards to clear the neuro-vascular bundle of the upper limb, at the apex of the axilla (PLATE 1). The lateral one-third or less is flat, and curves back to meet the scapula. The upper surface is smoother than the lower, especially in the lateral flat part. The bone lies horizontal and is subcutaneous; it is crossed by the supraclavicular nerves.

The *sternal end* (p. 59) is occupied by a facet for the disc of the sterno-clavicular joint. The disc is attached along the upper and posterior margin of the articular surface in a curved groove. The articular area below this extends to the under surface, for articulation with the first costal cartilage. The articular surface is covered in life by very dense fibrous tissue (called 'fibro-cartilage') identical with that of the disc. The capsule and synovial membrane are attached around the margin of the articular surface.

On the *posterior surface* of the shaft the origin of sterno-hyoid extends in a line beyond the capsule attachment, but leaves no mark on the bone. Lateral to this the subclavian vessels lie nearby, and further laterally the newly-formed lateral cord of the brachial plexus lies behind bare bone (PLATE 1). Here the nutrient artery, a branch of the suprascapular, runs laterally into the bone. The posterior border of the flat lateral one-third gives attachment to the occipital fibres of trapezius; this muscle encroaches somewhat on the upper surface, leaving a polished mark. The lateral extremity shows a small overhanging oval facet for the acromio-clavicular joint (p. 60).

Anteriorly the flat lateral part gives origin to deltoid, which muscle encroaches like trapezius somewhat on the upper surface, leaving a concave polished area limited by a low ridge. A 'deltoid' tubercle sometimes projects into the muscle near the medial limit of this origin. The anterior surface of the medial half of the clavicle is flattened slightly by the origin of pectoralis major, faint ridges marking the limits of attachment of the muscle.

The *upper surface* receives the interclavicular ligament alongside the capsule attachment. The clavicular head of sterno-mastoid arises from a wide area on the medial one-third or more of this surface (Fig. 6.54, p. 443) but usually leaves no very apparent mark on the bone. Two layers of cervical fascia (p. 362) surround sterno-mastoid and trapezius and are attached separately to the bone between these muscles, at the base of the posterior triangle (Fig. 2.3, p. 55).

The *under surface* shows some very characteristic features (Fig. 2.58). Lateral to the articular surface, which usually encroaches from the sternal end, there is an oval plateau (or a pit) for attachment of the costo-clavicular (rhomboid) ligament. From here a groove for the subclavius muscle extends as far as the conoid tubercle. A nutrient foramen commonly extends laterally in this groove. The lips of the groove give attachment to the clavi-pectoral fascia (Fig. 2.3, p. 55). The posterior lip extends to a prominent rounded boss near the posterior border of the flat part. This boss is the conoid tubercle and it lies buried in the base of the conoid ligament (Fig. 2.9, p. 60). The conoid tubercle lies close to the knuckle of the coracoid process. A synovial joint sometimes exists between the two bones, within the conoid ligament. From the conoid tubercle a rough irregular ridge, the trapezoid ridge, extends obliquely across the bone to end near the articular facet. It provides attachment for the strong trapezoid ligament (p. 60).

The *movements* of the clavicle are discussed on p. 59. Remember that the axis of movement is the costo-clavicular ligament, so the sternal end moves in the opposite direction to the acromial end.

Clinical considerations. The bone is easily seen and felt. Fracture is common; when due to indirect violence

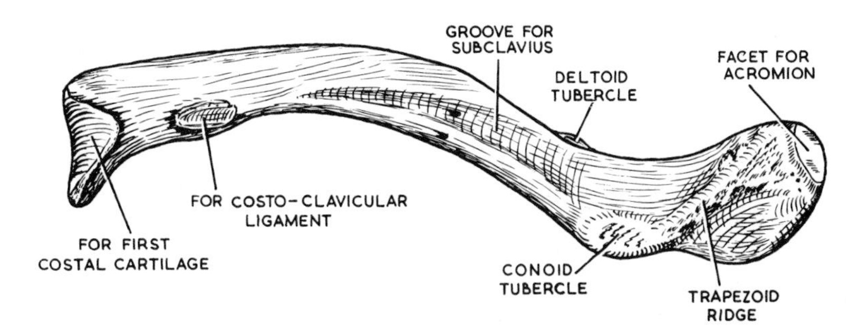

FIG. 2.58. THE UNDER SURFACE OF THE LEFT CLAVICLE.

the break is always between the costo-clavicular and coraco-clavicular ligaments, each of which is stronger than the clavicle itself.

Ossification. The bone appears in membrane before there is yet any cartilage in the embryo. It is the first bone of the skeleton. Two centres, which ossify at the fifth week, rapidly fuse; elongation, which occurs at the sternal end, is in secondary cartilage (p. 12). An epiphysis at the sternal end appears radiologically during the late 'teens and fuses in little more than a year (the story still told of appearance at 20 years and fusion at 25 is wrong).

The Scapula

The shoulder blade is triangular. The glenoid angle is thick, and projected upwards into the rugged coracoid process (Fig. 2.9, p. 60). The lateral border is thick down to the inferior angle. The rest of the blade is composed of thin, translucent bone. From the upper part of the dorsal surface a triangular spine projects back and extends laterally as a curved plate of bone, called the acromion, over the shoulder joint (Fig. 2.20, p. 71).

The **costal surface** is concave, and marked by 3 (or 4) ridges that converge from the medial border towards the glenoid angle. These give attachment to fibrous septa from which the multipennate fibres of subscapularis arise. This muscle is attached to the concavity of the costal surface and to the *lateral border* of the scapula back to a sinuous ridge that runs down from the infraglenoid tubercle (*v. inf.*). The thick bone of the glenoid angle is bare, separated from the overlying muscle by the subscapularis bursa; the ridges have faded out here. Vascular foramina perforate the bone over the bare area. The medial margin of the costal surface receives the insertion of serratus anterior (Figs. 2.10 and 2.11, p. 61). The first two digitations are attached to a smooth, flat, elongated triangle from the superior angle down to the base of the spine. The next two digitations are thinned out from this level down to the inferior angle, while the last four digitations converge to a roughened area on the costal surface of the inferior angle. A ridge separates the attachment of serratus anterior from the concavity of the subscapularis origin, and this ridge gives attachment to the strong fascia on the surface of subscapularis.

The *upper border* of the blade slants down to the root of the coracoid process, where it dips into the little scapular notch, which lodges the suprascapular nerve. The notch is bridged by the transverse scapular ligament, and the inferior belly of omo-hyoid arises from this ligament and the nearby upper border of the blade. The fascia over subscapularis is attached along the upper border of the bone. The *medial border*, from superior to inferior angle, gives edge to edge attachment to levator scapulæ, rhomboid minor and rhomboid major (p. 58).

The *lateral (axillary) border* extends from the glenoid fossa to the inferior angle. Just below the glenoid fossa it shows a rough triangular plateau, the infraglenoid tubercle (this may be depressed into a fossa), and this gives origin to the long head of triceps. From this area a sharp ridge runs down and turns over the dorsal surface of the inferior angle above the origin of teres major. Subscapularis arises from the groove on the costal surface of this ridge and its fascia is attached to the ridge (p. 61).

The **dorsal surface** of the blade is divided by the backwardly projecting spine into a small supraspinous and a large infraspinous fossa. The supraspinous fossa lodges supraspinatus, which muscle is covered in a strong fascia attached to the upper margin of the spine and the upper border of the blade. The muscle arises from the whole area of the spine and adjoining blade, but not from the glenoid angle. The large infraspinous fossa is concave on the under surface of the spine, and there is a groove alongside the lateral border, but between these two hollows the thin blade is convex in conformity with the concavity of the costal surface. Infraspinatus arises from the whole of this fossa (but not from the glenoid angle). Teres major arises from a large oval area at the inferior angle, and teres minor from an elongated narrower area dorsal to the ridge on the lateral border (Fig. 2.16, p. 67). This origin of teres minor is commonly bisected by a groove made by the circumflex scapular vessels. Infraspinatus and teres minor are covered by a thick fascia that is attached to bone at the margins of these two muscles. Along the ridge on the lateral border this attachment is shared with the fascia over subscapularis.

The **spine** is thick as it projects back from a horizontal attachment on the dorsal surface of the blade. It is twisted a little, so its posterior border slopes upwards towards the acromion. Its free lateral border curves widely out as a buttress to the inferior surface of the acromion, arching behind the upper part of the glenoid angle. The suprascapular vessels and nerve run here to reach the infraspinous fossa. The **acromion** itself is rectangular, and carries the facet for the clavicle at the anterior end of its medial border (Fig. 2.59). Along the medial border of the acromion and the upper margin of the spine is a bevelled surface, 5 mm wide, for the insertion of trapezius (Fig. 2.16, p. 67). Some 3 cm short of the medial border of the scapula this curves down and laterally to form a lip on the inferior margin of the spine. This lip is called the 'deltoid' tubercle; it receives the lowermost fibres of trapezius, which converge here. The medial end of the spine, alongside the border of the blade, is triangular and smooth. The tendon of trapezius plays over it, lubricated by a bursa. The 'deltoid' tubercle, though made by

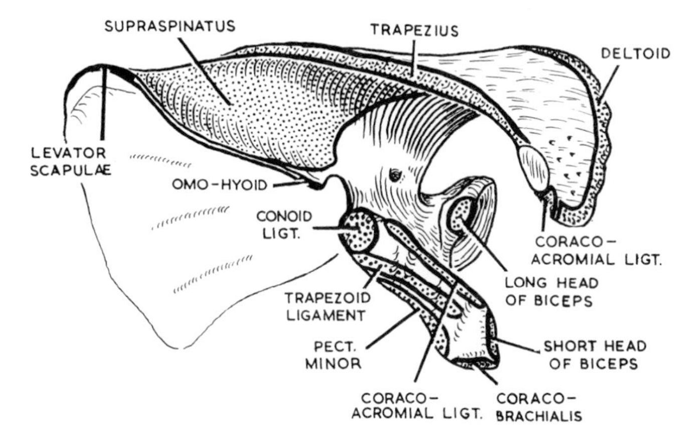

FIG. 2.59. THE LEFT SCAPULA VIEWED FROM ABOVE.

the trapezius attachment, marks the beginning of the attachment of deltoid, for from this place the muscle arises along the inferior margin of the spine and, in continuity, from the posterior, lateral and anterior borders of the acromion up to the clavicular facet. Along its lateral border the acromion shows four or more vertical ridges (Fig. 2.59) for attachment of septa in the multipennate central mass of the deltoid muscle. The junction of lateral and anterior borders of the acromion, 2 cm lateral to the acromio-clavicular joint, forms the prominent bony 'tip of the shoulder', and from here the length of the upper limb may be measured. In front of the facet for the clavicle the fibres of the coraco-acromial ligament converge (Fig. 2.19, p. 70). Beneath this ligament and the bare bone of the acromion the subacromial (subdeltoid) bursa lubricates the tendon of supraspinatus and the rotator cuff. Above the coraco-acromial ligament the deltoid origin continues across the acromion (Fig. 2.59) to the lateral part of the clavicle.

The **glenoid angle** of the scapula is wedge-shaped, broadening from the narrow neck out to the prominent margins of the glenoid fossa. The upper part of this wedge-shaped bone is projected upwards as the base of the coracoid process (Fig. 2.18, p. 69). Two morphological elements are fused here (see p. 52), and their epiphyseal junction crosses the upper part of the pear-shaped glenoid fossa (Fig. 2.9, p. 60). The glenoid fossa does not face directly lateral, but peeps forwards a little around the convexity of the chest wall. The prominent lips of the fossa are raised higher by the attachment of the glenoidal labrum. At the upper limit (the 'stem' of the pear), just on the base of the coracoid, lies the supraglenoid tubercle, for attachment of the long head of biceps within the shoulder-joint capsule. The capsule is attached to the labrum and to the surrounding bone. The infraglenoid tubercle lies on the glenoid angle, outside the capsule. There are no other attachments to the glenoid angle, whose bare bone is perforated by nutrient vessels, and separated from the tendons of subscapularis, supraspinatus and infraspinatus by a bursa beneath each.

The **coracoid process** rises from its broad base and curves forward like a pointing finger. Its tip is palpable in the delto-pectoral groove below the clavicle. The process is for muscle and ligament attachments, the ligaments being indispensable to the stability of the pectoral girdle. The conoid ligament is attached by its apex to the knuckle of the coracoid process (Fig. 2.59). From here a smooth narrow line runs towards the tip for attachment of the trapezoid ligament. Lateral to this the margin of the process gives attachment to the base of the coraco-acromial ligament, which is a functional 'socket' for the head of the humerus. The weaker coraco-humeral ligament sweeps from the under surface of the coracoid process to the anatomical neck of the humerus. The pectoralis minor is attached along the medial border of the coracoid process for about 2 cm behind its tip. The tip itself is bevelled to an arrow shape. From the lateral facet arises the tendon of the short head of biceps and, conjoined with it, from the medial facet comes the origin of coracobrachialis.

The *movements* of the scapula can be analysed into protraction-retraction, rotation and elevation-depression (p. 60).

Clinical considerations. Most of the scapula is easily palpable from behind. The tip of the coracoid process can be felt deep in the infraclavicular fossa.

Paralysis of serratus anterior causes 'winging' of the scapula; the weight of the flexed upper limb draws the vertebral border away from the chest wall. At the same time rotation is impaired, for trapezius alone is not quite adequate; full abduction of the upper limb is impossible.

Ossification. Mesenchyme chondrifies at the sixth week, when the whole scapula becomes cartilaginous. A bony centre appears in the eighth week at the thick part of the glenoid angle and gradually enlarges. At birth the blade and spine are ossified, but the acromion, coracoid process, medial border and inferior angle are still composed of hyaline cartilage. Secondary centres appear in these places, and around the lower margin of the glenoid fossa, at about puberty, and all are fused by 25 years. The centre at the base of the coracoid process ossifies at 10 years and fuses, across the glenoid fossa, soon after puberty.

The Humerus

The shaft expands above into an upper end whose articular surface looks up and back. The lower part of the shaft curves gently forwards to a flat lower end projected into medial and lateral epicondyles, between which lies the articular surface of the elbow joint. The medial epicondyle projects in the same direction as the articular surface of the head. The humerus at rest lies with its articular head facing backwards as well as medially. *The bone on a table lies in some degree of lateral rotation.*

The *upper end*, expanded above the shaft, consists of the convex articular surface and, anterior to this, the tuberosities. The articular surface is commonly called the *head*, in which case the articular margin must be called the 'neck', and this is usually referred to as the 'anatomical neck'. It is the thickest, not the narrowest, part of the upper end. The commonsense 'neck', a common site for fracture, somewhat below the epiphyseal line, is at the upper end of the shaft and is referred to as the 'surgical neck'. Thus have we become enslaved by our own words, for the humerus remains the same whatever names are chosen for its several parts.

The **head** (i.e., the articular surface) forms about one-third of a sphere and is about four times the area of the glenoid fossa of the scapula. It is coated with hyaline cartilage. At rest its lower and anterior quadrant articulates with the glenoid, giving a good range of lateral rotation and abduction from this position (confirm this point on the dry bones). In Fig. 2.20A, p. 71, the humerus was deliberately drawn in slight lateral rotation, to expose more articular surface to view.

The capsule of the shoulder joint, bridging the bicipital groove at its attachment to the transverse ligament, is attached to the articular margin except medially, where it extends down along the shaft for 2 cm, here enclosing the epiphyseal line. The synovial membrane lines the capsule and is attached to the articular margin. The long tendon of biceps is enclosed in a sheath of synovial membrane, one end of which is attached to the transverse ligament and underlying articular margin (the other end of the sheath is attached around the supraglenoid tubercle of the scapula).

The **lesser tuberosity** projects prominently forwards, and is continued downwards as the medial lip of the bicipital groove. An undulating area of smooth bone indicates the insertion of the tendon of subscapularis, while edge to edge below this teres major is received into the medial lip for a distance of nearly two inches (Fig. 2.10, p. 61). The *bicipital groove* lies on the anterior surface of the upper end (Fig. 2.35, p. 85). It is bridged above by a transverse ligament beneath which the long tendon of biceps leaves the

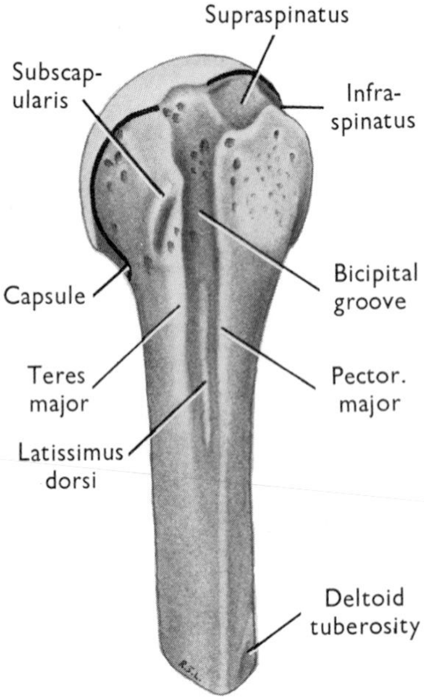

Supraspinatus
Subscapularis
Infraspinatus
Capsule
Bicipital groove
Teres major
Pector. major
Latissimus dorsi
Deltoid tuberosity

FIG. 2.60. THE UPPER END OF THE LEFT HUMERUS (ANTERIOR VIEW).

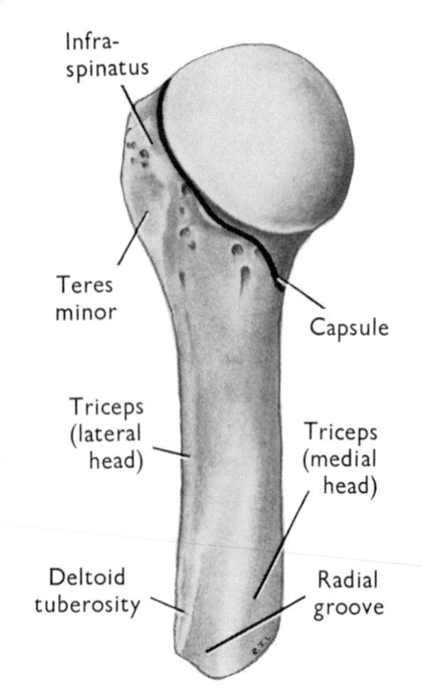

FIG. 2.61. THE UPPER END OF THE LEFT HUMERUS (POSTERIOR VIEW).

joint (Fig. 2.19, p. 70). The floor of the groove receives the ribbon-like tendon of latissimus dorsi.

The **greater tuberosity** is bare bone, perforated by vessels, except at its projecting junction with the head. Here three smooth facets receive the tendons of scapular muscles. Superiorly is the facet for supraspinatus. Behind this lies a smooth facet for infraspinatus, while posteriorly the lowest facet receives teres minor, whose tendon is inserted also into the shaft below the facet. Below this tendon the bare bone lies in contact with the axillary (circumflex) nerve and its vessels. The lateral lip of the bicipital groove extends down from the anterior margin of the greater tuberosity to run into the anterior margin of the deltoid tuberosity. It receives the folded tendon of pectoralis major, whose posterior lamina is attached as high up as the joint capsule (i.e., the 'anatomical neck'). The anterior lamina extends only as high as the 'surgical neck'.

The **shaft** is triangular in section. Viewed from in front the lateral lip of the bicipital groove runs into the anterior margin of the deltoid tuberosity and is continued down in the midline of the bone. Lateral to this, the bare bone of the greater tuberosity continues down to the deltoid tuberosity, nearly halfway down the shaft. The tuberosity is a V-shaped prominent ridge, with a smaller ridge between, the three giving attachment to fibrous septa in the multipennate acromial fibres of the deltoid. Below the deltoid tuberosity the lower end of the radial groove spirals down. The posterior margin of the groove runs down as the lateral supracondylar ridge and curves forwards into the lateral epicondyle (Fig. 2.29, p. 77). The ridge gives attachment to the lateral intermuscular septum. The medial lip of the bicipital groove continues down into the medial supracondylar ridge, which at its lower end curves into the prominent medial epicondyle. The ridge gives attachment to the medial intermuscular septum. Level with the lower part of the deltoid tuberosity the nutrient foramen, directed down towards the elbow, lies just in front of this medial border of the humerus. Above this foramen, opposite the deltoid tuberosity, coraco-brachialis is inserted and sometimes leaves a linear roughness. The flexor surface of the humerus, between the supracondylar ridges, gives origin to brachialis and this extends upwards to embrace the deltoid tuberosity both in the lower part of the spiral groove and between the tuberosity and the insertion of coraco-brachialis. The occasional supratrochlear spur on this surface is discussed on p. 73.

Viewed from behind the shaft appears twisted, but this is due only to the spiral groove behind and below the deltoid tuberosity. The lateral lip of this groove extends upwards from the posterior margin of the deltoid tuberosity; the lateral head of triceps arises here, extending up almost to the axillary nerve below the teres minor insertion. The lateral supracondylar ridge, as already noted, spirals up behind the radial groove to fade out on the posterior surface of the shaft level with the upper limit of the deltoid tuberosity. The medial head (i.e., the *deep* head) of triceps arises high up here and from the whole shaft between the supracondylar ridges down almost to the olecranon fossa. This head arises too from the medial and lateral intermuscular septa (p. 74). Between the origins of medial and lateral heads the spiral groove lies bare, and here the radial nerve lies in contact with the bone. The profunda brachii vessels run with it.

The **lower end** of the humerus carries the articular surface for the elbow joint and is projected into medial and lateral epicondyles for attachment of muscles for the flexor and extensor compartments of the forearm. Anterior and posterior appearances are quite different. The articular surface, coated with hyaline cartilage, shows the conjoined capitulum and trochlea (Fig. 2.23, p. 73). The *capitulum*, for articulation with the head of the radius, is a section of a sphere. It projects forwards and inferiorly from the lateral part and is bounded by a prominent ridge from the nonarticular bone. The *trochlea*, unlike the capitulum, extends also to the posterior surface. Its medial margin is a high ridge curving prominently from front to back around the lower end of the humerus, and this carries the side-to-side concavity of the trochlea with it into rather more than a semicircle. Laterally a lower ridge runs

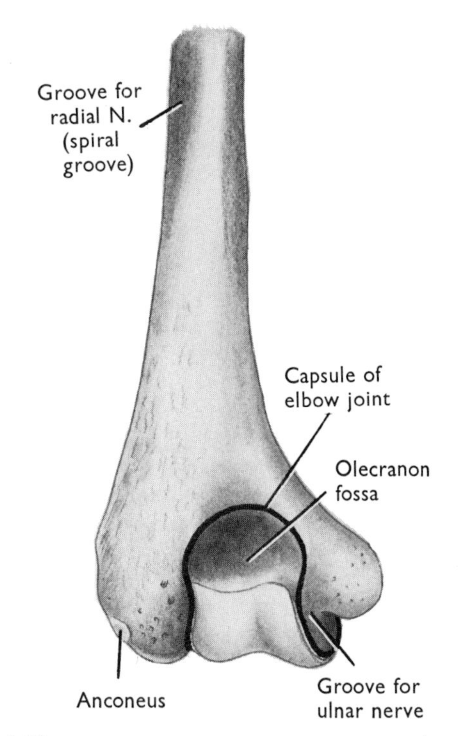

FIG. 2.62. THE LOWER END OF THE LEFT HUMERUS (POSTERIOR VIEW). See also Fig. 2.26, p. 77.

on into the capitulum in front and below. Behind this, beyond the capitulum, the ridge bounds the trochlea and arches over the olecranon fossa towards the medial epicondyle on the back of the bone. The difference in prominence between the two ridges gives a tilt to the trochlea which accounts for the carrying angle at the elbow.

On the anterior surface the shaft shows a shallow coronoid fossa separated by translucent bone from the depths of the olecranon fossa; this is above the concave part of the trochlea. The anterior border of the humerus runs down to the lateral ridge on the trochlea (Fig. 2.26, p. 77) and here, above the capitulum, a shallow radial fossa accommodates the head of the radius in full flexion. The capsule of the elbow joint is attached to the ridges that form the margins of capitulum and trochlea and to the shaft of the humerus above coronoid and radial fossæ in front and the olecranon fossa behind. The synovial membrane, as in all synovial joints, is attached to the articular margins. It thus clothes the floors of these fossæ before being reflected to the deep surface of the capsule.

The **medial epicondyle,** more prominent than the lateral, shows an undulating smooth facet on its anterior surface for the common flexor origin of forearm muscles. Pronator teres arises from the medial supracondylar ridge just above this. Posteriorly, between the smooth epicondyle and the curving ridge of the trochlea, is a groove which lodges the ulnar nerve, here in contact with the *shaft* of the humerus. Deeply on the distal border of this epicondyle is a small facet for attachment of the ulnar collateral (medial) ligament of the elbow joint.

The **lateral epicondyle** shows an undulating smooth facet on its *anterior* surface for the common extensor origin of forearm muscles, *viz.* extensor carpi radialis brevis, extensor digitorum communis, extensor digiti minimi and extensor carpi ulnaris (p. 89). Above this the lateral supracondylar ridge gives attachment to the lateral intermuscular septum. Here brachio-radialis arises from the upper two-thirds and extensor carpi radialis longus from the lower one-third of the ridge. At the distal border of the epicondyle the radial collateral (lateral) ligament of the elbow is attached, and around this the superficial fibres of supinator arise. Behind supinator is a shallow pit for anconeus, whose fibres arise also above this, in continuity with the lowest fibres of the medial (deep) head of triceps.

Clinical considerations. By appropriate movements and positioning, most of the bone can be palpated in the living. It is useful to note that the medial epicondyle points in the same direction as the head. Three main nerves are vulnerable as they lie in contact with bare bone: the axillary at the surgical neck, the radial in the spiral groove and the ulnar behind the medial epicondyle.

Movements. At the *shoulder* flexion-extension, abduction-adduction and rotation occur (p. 71). At the *elbow* there is a hinge movement of flexion-extension (p. 78). During pronation-supination the head of the radius rotates on the capitulum, and there is some rocking of the ulna in the trochlea for ulnar abduction and pronation (p. 87).

Ossification. The whole is cartilaginous at the sixth week. A primary centre appears in the centre of the shaft at the eighth week. Upper and lower ends are cartilaginous at birth. *Secondary centres* appear at both ends. For the *upper end* they are in the head during the first year, greater tuberosity at the third and lesser tuberosity at the fifth year. These three fuse by the seventh year into a single bony epiphysis, hollowed to fit a conical projection of the bony shaft. The epiphyseal line skirts the bone across the lowest margin of the articular surface, cutting across the tuberosities. This is the growing end of the bone; fusion occurs about twenty years. At the *lower end* four centres appear. There is one for the capitulum and lateral ridge of the trochlea at the second year, one for the medial epicondyle at the fifth year, the remainder of the trochlea at the twelfth and the lateral epicondyle at the thirteenth year. The medial epicondyle remains a separate centre, separated by a downward projection of the shaft (in contact with the ulnar nerve) from the other three, which fuse together. Union with the shaft occurs at about eighteen years.

The Radius

This bone *carries the hand*, and is stabilized against the ulna for pronation-supination and against the humerus for flexion-extension of the forearm. From a cylindrical head the bone tapers into a narrow neck. The shaft becomes increasingly thick as it curves down to the massive lower extremity.

The **head** is cylindrical and is covered with hyaline cartilage. A spherical hollow forms the upper surface, to fit the capitulum. The cylindrical circumference is continuous with this hollow, and is deepest on the medial side of the curvature; it articulates with the radial notch of the ulna and with the annular ligament. The synovial membrane of the elbow joint is attached to the articular margin, but head and neck are free of any capsule attachment, to rotate in the clasp of the annular ligament. The narrow *neck* is enclosed by the tapered lower margin of the annular ligament, below which the loose fibres of the quadrate ligament are attached.

The **shaft** is characterized at its upper end by an oval prominence, the bicipital tuberosity, projecting towards the ulna. The biceps tendon is attached along the posterior lip, and a bursa lies against the anterior surface of the tuberosity. In full pronation the tuberosity rotates back and looks laterally (Fig. 2.35, p. 85), winding the biceps tendon around the radius. From the anterior margin of the bicipital tuberosity the anterior oblique line forms a ridge that runs down to the point of greatest convexity of the shaft. It is mirrored by a similar, but less prominent, posterior oblique line on the extensor surface of the shaft. Between the two oblique lines the shaft is cylindrical and receives the insertion of supinator, with the posterior interosseous nerve off the bone, sandwiched between the two layers of the muscle. At the apex of the two lines is a longitudinal ridge or pit, 2 cm long, for the tendon of pronator teres. This lies at the point of greatest convexity of the radius, just *behind* the lateral profile of the supinated bone; invisible in a strictly anterior view, it shows in a posterior view of the radius (Fig. 2.64).

Below the bicipital tuberosity the shaft is pinched into a ridge for the interosseous membrane of the forearm. The oblique cord is attached at the upper end and passes up to the ulna; the fibres of the interosseous membrane pass *down* to the ulna. On the flexor surface the flexor digitorum superficialis (sublimis) is attached to the anterior oblique line, and flexor pollicis longus arises from a length of bone below this (and from the interosseous membrane) down to pronator quadratus. The lower part of the flexor surface is hollowed out above the expanded lower extremity; here pronator quadratus is inserted into the lower one-fifth of the shaft. On the extensor surface below the posterior oblique line is the oblique origin of abductor pollicis longus (this origin extends across the interosseous membrane up to the ulna). Below this extensor pollicis brevis arises obliquely from a slightly hollowed-out area and from the membrane. The lower part of the shaft is bare on the extensor surface (Fig. 2.41, p. 91).

The **lower extremity** expands from the lower end of the shaft, and is best studied surface by surface. The ulnar surface shows a notch for articulation with the head of the ulna. Above this is a triangular area enclosed by anterior and posterior ridges into which the interosseous border divides. The interosseous membrane is attached to the posterior ridge. In front of it lies the membrana sacciformis (p. 86) and then the deepest fibres of pronator quadratus are attached to the front of the triangle. Inferiorly is the articular surface for the wrist joint, two concave areas covered with hyaline cartilage. The ulnar (medial) surface is square, and articulates with the lunate. It continues into the hyaline cartilage of the ulnar notch, but in the intact wrist the triangular fibro-cartilage is attached to the right-angled border between the two and divides the inferior radio-ulnar and wrist joints from each other. The lateral concave area is triangular, with its apex on the styloid process. It articulates with the scaphoid. The capsule and synovial membrane are attached to the articular margins and to the fibro-cartilage. The anterior (flexor) surface of the lower extremity shows a prominent ridge below the pronator quadratus hollow. Between this ridge and the articular margin is a smooth area to which is attached the very strong anterior ligament of the wrist joint. The fibres of this ligament pass *transversely* to the medial side of the carpus (Fig. 2.44, p. 94).

Laterally the lower extremity is projected into the pyramidal styloid process, into whose base the tendon of brachio-radialis is inserted. Between this and the ridge of the pronator hollow the tendons of abductor pollicis longus and extensor pollicis brevis lie, in their synovial sheaths, in a single compartment beneath the extensor retinaculum. On the posterior surface beyond the styloid process a broad groove leads to the prominent radial tubercle (Lister's tubercle). It lodges, each in a separate sheath, the flat tendons of extensors carpi radialis longus and brevis, sharing a single compartment beneath the extensor retinaculum. On the ulnar side of the radial tubercle is a narrow groove that lodges the tendon of extensor pollicis longus in its synovial sheath, in a separate compartment. Between this and the ulnar notch is another broad groove. In the one compartment lie the four tendons of extensor digitorum communis, with the tendon of extensor indicis beneath them on the bare bone. All five tendons share a common synovial sheath. The tendon of extensor minimi digiti, in its sheath, crosses the radio-ulnar joint here.

The extensor retinaculum, a thickening of deep fascia, is attached to the prominent ridge on the lateral side of the pronator hollow (i.e., on the anterior surface of the radius, Fig. 2.33, p. 83) and sweeps obliquely across to the ulnar side of the carpus. Septa pass from it to the radius, making compartments for the extensor tendons as noted above. Four compartments lie on the lower end of the radius, and a fifth over the radio-ulnar joint. The sixth and last compartment lies over the head of the ulna.

The radial collateral (lateral) ligament of the wrist is attached to the radial styloid process.

Clinical considerations. The head of the radius is palpable in the hollow below the elbow on the extensor surface of the forearm; it can be felt rotating in pronation-supination movements. It is stabilized by the snug fit (around the neck) of the tapering annular ligament; a sudden powerful jerk on the hand of a child may avulse the head from the ligament, like pulling a cork out of a bottle. The expanded lower end takes the full thrust of the hand; Colles' fracture here is common. The lower end of the bone is easily palpable between the tendons, and the styloid process can be felt in the upper part of the snuff-box.

Ossification. The whole appears in cartilage at the sixth week, and a centre appears in the middle of the shaft at the eighth week. At birth both ends are cartilaginous; the lower is the growing end. A centre appears in the lower end early in the second year, and fuses at twenty years. This epiphyseal line is extracapsular; it runs transversely through the base of the styloid process and lies above the ulnar notch. The centre for the head appears at four years and fuses at eighteen; the epiphyseal line is at the junction of head and neck.

The Ulna

This is the stabilizing bone of the forearm. It obtains a good grip of the humerus, and on this foundation the radius and hand move in pronation-supination to secure the appropriate working position for the hand. The ulna tapers in the reverse way to the radius; it is massive above and small at its distal extremity.

The **upper extremity** has two projections, with a saddle-shaped articular surface between them. They are the olecranon and coronoid process, and they grip the trochlear surface of the humerus.

The **olecranon** is the proximal extension of the shaft, and in extension of the elbow it is lodged in the olecranon fossa of the humerus. Its upper surface is square, and receives the tendon of triceps over a wide area, which is smooth. Its anterior border forms a sharp undulating lip at the articular margin; the capsule of the elbow joint is attached just behind this lip. The posterior surface, triangular in shape, is subcutaneous; the olecranon bursa lies on it. The sides of the triangle are continued below the apex into the sinuous subcutaneous border of the shaft; they give attachment to the deep fascia of the forearm. The medial surface of the olecranon, gently concave, is continued down to the flexor surface of the shaft. The upper fibres of flexor digitorum profundus arise here, covered by the thin sheet of flexor carpi ulnaris, whose ulnar origin is by an aponeurosis that extends up to the medial side of the subcutaneous triangle (Fig. 2.39, p. 89). At the upper angle of this area is a small smooth facet for the posterior band of the ulnar collateral (medial) ligament of the elbow (Fig. 2.27, p. 77). The lateral surface of the olecranon continues down to the extensor surface of the shaft; anconeus is inserted here (Fig. 2.41, p. 91).

The **coronoid process** projects forwards from the upper end of the shaft. Its anterior lip, like that of the olecranon, is thin and abuts on the articular margin. Medially this lip shows a prominent smooth elevation, the 'sublime' tubercle. Flexor digitorum superficialis is attached here, and beneath this the ulnar collateral (medial) ligament of the elbow. The ulnar nerve lies in contact with the sublime tubercle (Fig. 2.28, p. 77). The anterior surface of the coronoid process is concave and receives the insertion of brachialis. The medial border of the process gives a linear origin to the deep head of pronator teres (PLATE 2, p. *i*). An occasional origin is given to flexor pollicis longus from either border of the coronoid process.

The lateral surface of the coronoid process carries a concave cylindrical facet (called radial notch) for the head of the radius and this surface, covered with hyaline cartilage, continues into that of the trochlear surface. The anterior and posterior margins of the notch give attachment to the annular ligament, while the quadrate ligament is attached to the shaft just below the notch. Just below this the oblique cord is attached.

Between the projections of the olecranon and coronoid process is a deeply saddle-shaped surface, the **trochlear notch.** Convex from side to side, it is concave from top to bottom and fits the trochlea of the humerus. It is covered with hyaline cartilage, which is sometimes

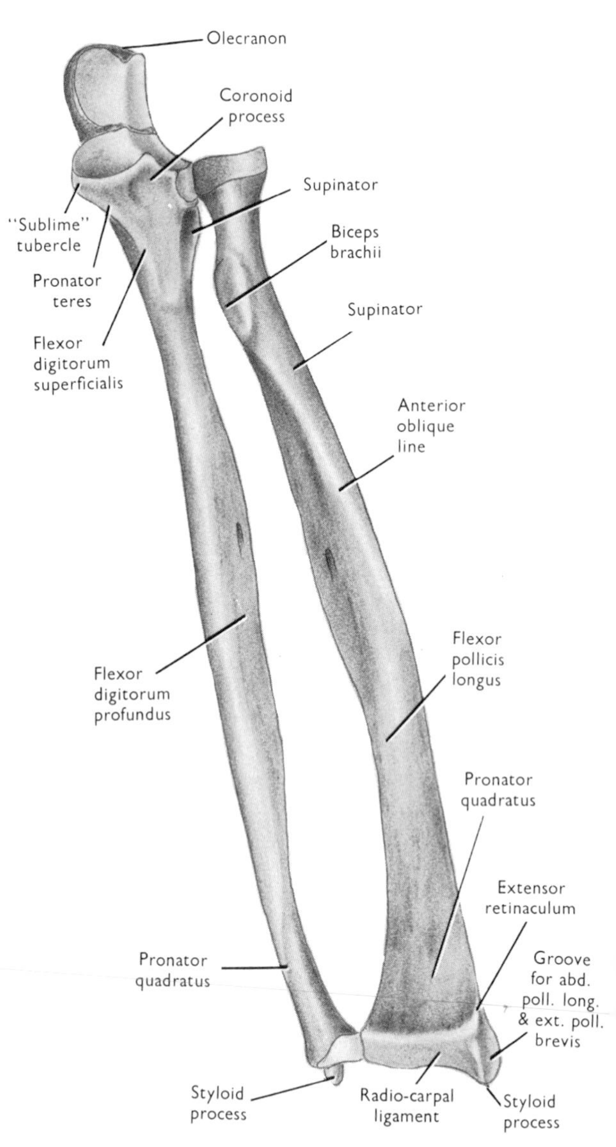

FIG. 2.63. THE BONES OF THE LEFT FOREARM IN FULL SUPINATION (ANTERIOR VIEW).

constricted in hour-glass shape or even separated into two surfaces, one on the olecranon and one on the coronoid process. The capsule and synovial membrane of the elbow joint are attached around the margins of the trochlear notch and extend to the radial notch. Thus the elbow and superior radio-ulnar joints form one cavity.

The **shaft** is angled somewhat laterally from the line of the trochlear notch (Fig. 2.35, p. 85) to form the carrying angle. It is projected at its middle into a prominent ridge, the interosseous border. The ridge fades out at the lower one-fifth of the shaft, but is continued up to the posterior lip of the radial notch.

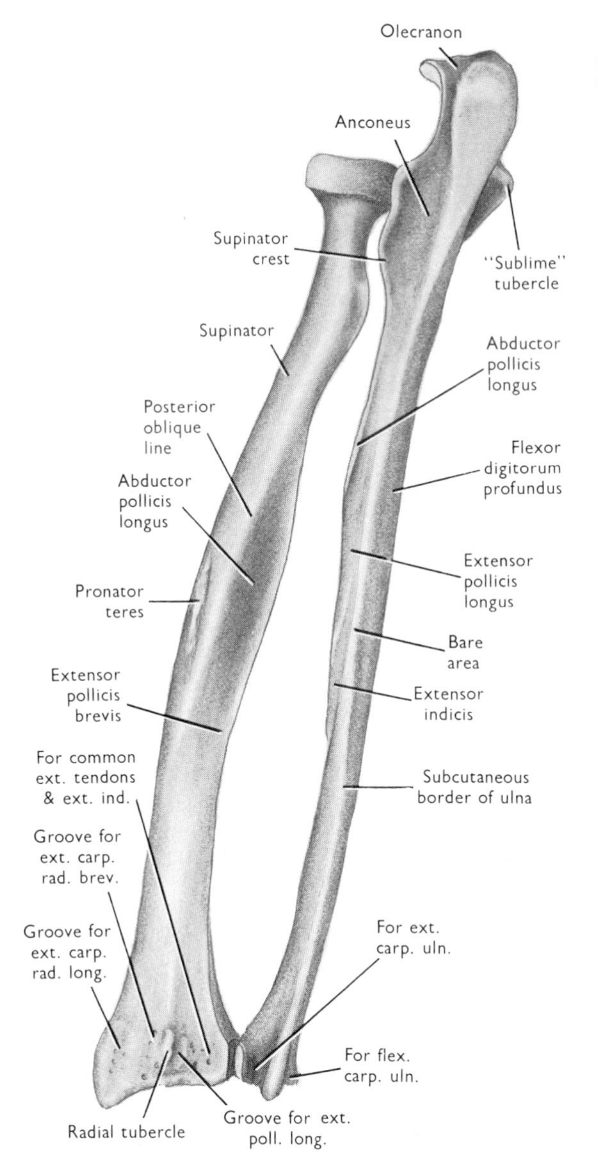

FIG. 2.64. THE BONES OF THE LEFT FOREARM IN FULL SUPINATION (POSTERIOR VIEW). Cf. Fig. 2.41, p. 91.

Here, just below the notch, the ridge is prominent and is named the supinator crest. Between it and the lower part of the coronoid process is a concavity called the supinator fossa. The ulnar fibres of the supinator muscle arise from the crest and the fossa. The interosseous membrane is attached to the sharp interosseous border and extends down to the inferior radio-ulnar joint. This divides the shaft into a narrow flat extensor surface passing to the nearby subcutaneous border, and an extensive convex flexor surface passing around to the same border. Only two muscles are attached to the flexor surface. Flexor digitorum profundus, the most massive muscle of the forearm, arises high on the medial concavity of the olecranon and from the upper three-quarters of the shaft between subcutaneous and interosseous borders, and from the interosseous membrane. The bulky upper part of the belly curves the subcutaneous border into a convexity towards the radius. The subcutaneous border recovers from this curvature and extends vertically down to the styloid process. This border gives attachment to the aponeurotic head of flexor carpi ulnaris in its upper three-quarters and to the deep fascia of the forearm in its whole length. Running up from the anterior part of the styloid process is a ridge that dies out in the lower one-fifth of the shaft. Pronator quadratus arises from the radial side of this ridge and passes transversely beneath the long flexor tendons (Fig. 2.33, p. 83). The narrow flat extensor surface lies between the interosseous and subcutaneous borders. It extends down from the lateral surface of the olecranon. Anconeus is received into its upper part. Below this a vertical ridge divides the surface. Between the ridge and the interosseous border three muscles leave shallow impressions on the bone. They are abductor pollicis longus, extensor pollicis longus and extensor indicis (see Fig. 2.41, p. 91). Between the ridge and the subcutaneous border the bone is bare; extensor carpi ulnaris plays over it. This muscle has an aponeurotic origin from the subcutaneous border below the insertion of anconeus.

The **lower extremity** expands into a small rounded prominence; it is named the **head** of the ulna. Its distal surface and more than half of its radial circumference, covered with hyaline cartilage, form the inferior radio-ulnar joint. A small pit at the base of the styloid process gives attachment to the apex of a triangular fibro-cartilage whose base is attached to the radius. This divides off the wrist joint, for the capsule is attached to the fibro-cartilage. The capsule and synovial membrane of the inferior radio-ulnar joint are attached around the articular margins but project proximally a little, deep to pronator quadratus, as the membrana sacciformis. The styloid process projects distally as a continuation of the subcutaneous border. Judged from the axis of the forearm it does not project distally as far as the radial styloid, but with the *supinated* forearm hanging free the carrying angle at the elbow gives an obliquity to radius and ulna that results in the two styloid processes lying at the same horizontal level (Fig. 2.63). The ulnar collateral (medial) ligament of the wrist joint is attached to the tip of the styloid process. A groove alongside the process, on the extensor surface, lodges the tendon of extensor carpi ulnaris, in its sheath, beneath the extensor retinaculum (Fig. 2.41, p. 91).

Clinical considerations. The olecranon is subcutaneous and is easily palpable. Its position in relation to the epicondyles of the humerus depends on the angle of flexion-extension of the elbow joint. In extension the tip lies in line with the epicondyles, in full flexion the three bony points make an equilateral triangle. The sinuous subcutaneous border is palpable down to the styloid process. It is visible in the supinated forearm, because the encircling deep fascia is attached to it and makes a groove between the bulging flexor and extensor muscles. The muscles themselves are held down to this border of the bone by the diverging aponeurotic origins of flexor and extensor carpi ulnaris.

Ossification. Cartilage at the sixth week, a centre for the shaft appears at the eighth week. The head is not ossified until the sixth year. This is the growing end and does not fuse with the shaft until twenty years. The upper end is curious in that it shows only a small epiphysis at the proximal surface of the olecranon; this appears at about the eighth year and fuses at eighteen years or earlier. The massive upper end of the ulna grows in size and adapts its shape to the trochlea as one mass of bone, unaided by growth and adjustment at an epiphyseal line.

The Hand

The wrist and hand bones should be studied in an accurately articulated specimen, to see the pattern of the skeleton as a whole. The purpose of such study is twofold. First to become acquainted with the places of attachment of ligaments, tendons and muscles, and second to analyse the movements that occur between the bones. The old-fashioned necessity to tell an examiner to which side an isolated carpal bone belongs, memorized for the occasion and rightly forgotten after the event, has happily disappeared from most modern curricula. Such detail is of little or no clinical use, and adds nothing to the understanding of function.

It is better first to study only the bony features, then afterwards note the places of attachment of soft parts. The articulated bones are seen to form a carpus (eight bones), a metacarpus of five bones articulated with it, and the phalanges of the five digits articulated with the metacarpal heads. Their morphology is discussed on p. 127.

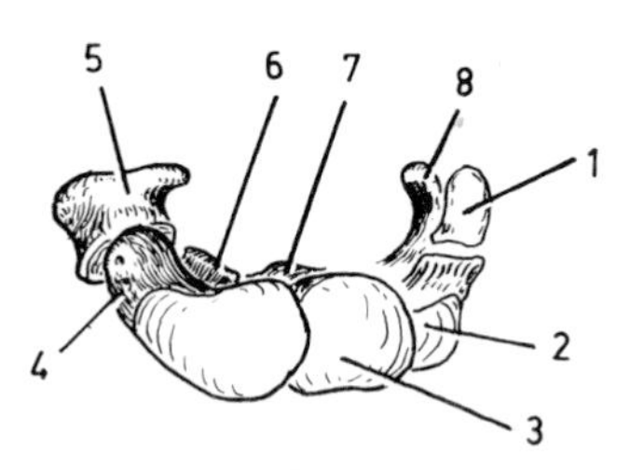

FIG. 2.65. THE LEFT CARPUS. View of the wrist surface, looking distal towards the palm. 1. Pisiform. 2. Triquetral. 3. Lunate. 4. Scaphoid. 5. Trapezium. 6. Trapezoid. 7. Capitate. 8. Hamate. The flexor surface is deeply concave and this arch is maintained by the flexor retinaculum.

The Carpus

The eight wrist bones articulated together form a semicircle, the convexity of which is proximal and articulates with the forearm. The diameter of the semicircle is distal, and articulates with the metacarpal bases (Fig. 2.66). The flexor surface of the carpus is deeply concave to accommodate the flexor tendons (Fig. 2.65). The extensor surface is gently convex and the extensor tendons pass across it.

The eight carpal bones lie in two rows, a proximal and a distal. The proximal row consists of the scaphoid, lunate and triquetral, which together form the convexity of the semicircle. The fourth bone, the pisiform, completes the proximal row by articulating with the front of the triquetral, and so builds up the flexor concavity of the carpus. The palmar concavity of the carpus is maintained by the 'tie-beam' effect of the flexor retinaculum (p. 96 and Fig. 2.46, p. 97).

The scaphoid and lunate together articulate with the radius, to whose distal extremity they are bound by strong ligaments (Fig. 2.44, p. 94). The triquetral, carrying the pisiform, itself articulates with the medial ligament of the wrist joint. Thus in pronation-supination the carpus moves with the radius, and upward thrust from the hand is carried wholly to the lower end of the radius, not at all to the head of the ulna. The cavity of the wrist joint, formed by radius and fibro-cartilage, is completed by the three proximal bones. The pisiform articulates with the triquetral by a separate synovial joint. The resting position of the wrist joint is in ulnar adduction, the carrying angle of the wrist (p. 94). But at rest there is also some extension; flexion and extension at the wrist joint are equal in range *from this rest position.* Thus *along the line of the radius* extension at the wrist joint has a wider range than flexion; consequently the articular surfaces of the three proximal carpal bones at the wrist joint extend further on their dorsal than on their ventral surfaces (Figs. 2.66 and 2.68).

The four bones of the distal row are the trapezium, trapezoid, capitate and hamate. They articulate with the bones of the proximal row by an S-shaped **midcarpal**

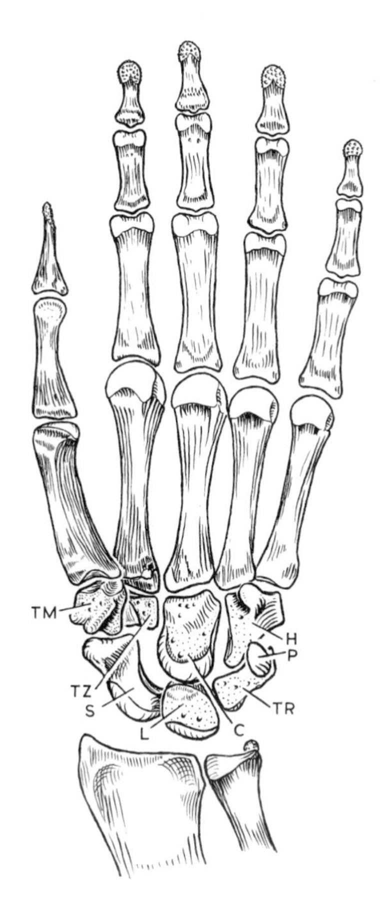

FIG. 2.66. PALMAR VIEW OF THE BONES OF THE LEFT HAND. The carpal bones are separated to obtain a fuller view of each, thereby displacing distally the thumb and index metacarpals and phalanges. **H.** Hamate. **P.** Pisiform. **TR.** Triquetral. **C.** Capitate. **L.** Lunate. **S.** Scaphoid. **TZ.** Trapezoid. **TM.** Trapezium.

joint. At the midcarpal joint flexion is of greater range than extension, so that in the average western individual the metacarpus can be flexed to a right angle with the forearm but extended to only 45 degrees with it. East of Suez the range of extension is much greater (observe the wrist and finger movements of both male and female Eastern dancers). The bones of each row articulate with each other by **intercarpal joints** that extend proximally and distally from the S-shaped midcarpal joint. These all have one continuous synovial space, normally separate from the wrist joint but commonly extending distally into the carpo-metacarpal space, especially around the trapezoid. The four bones of the distal row make a transverse joint line, with which the five metacarpal bones articulate. The joint between the trapezium and thumb metacarpal is a separate synovial space, but the remaining four metacarpals have a continuous space along their articulations with the carpus and between their own bases.

The Carpal Bones. Before observing the site of attachment of ligaments, tendons and muscles, a brief survey of the individual bones can be made, *as they lie together in the articulated skeleton.*

The **scaphoid** (boat-shaped) bone resembles no boat ever built by human hands. Its radial surface is a convex articular area that extends well on the dorsal surface. There is a narrow convexity for the lunate, from which the articular surface sweeps distally into a concavity ('boat-shaped') for the capitate, and is continued more distally into a triangular convexity for trapezium and trapezoid. The tubercle is a blunt prominence to the thumb side of the distal surface, and is palpable as a clinical landmark. The narrow non-articular waist of the bone is palpable in the snuff-box distal to the radial styloid (Fig. 2.42, p. 92). It is perforated, especially on its dorsal surface, by vascular foramina. These are more numerous distally, so that fracture across the waist sometimes results in avascular necrosis of the proximal fragment. Flexor carpi radialis sometimes give a slip of attachment to the scaphoid, and this tendon then carries some blood vessels to the bone.

The **lunate** (semilunar) bone shows a cylindrical proximal facet for the radius, extended more on the dorsal than the ventral surface. To each side there is a

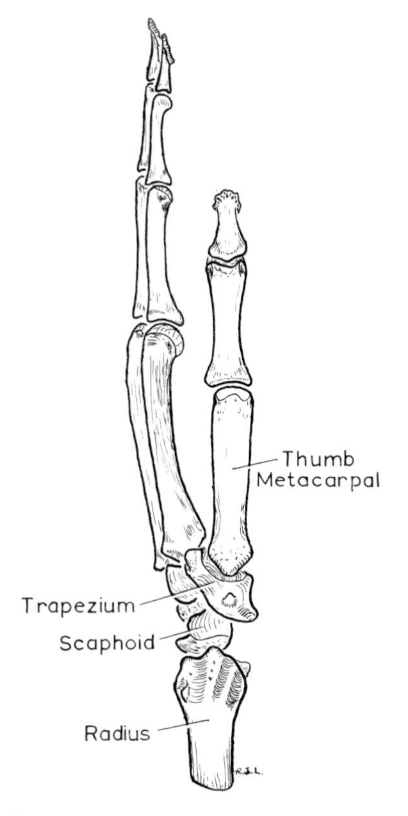

FIG. 2.67. THE BONES OF THE LEFT HAND IN LATERAL VIEW.

facet for the adjoining bones (scaphoid and triquetral) of the proximal row, and these are continued distally into an antero-posterior concavity for the capitate. Only the broad anterior and narrow posterior surfaces are non-articular, and these are perforated, especially posteriorly, by vascular foramina (Fig. 2.68).

The **triquetral** (cuneiform) bone shows a slight concavity for the lunate, projected into a proximal convexity for the ligamentous surface of the wrist joint, and a distal concavity for the hamate. The ventral surface has an oval facet distally for the pisiform. The rest of the ventral surface, as well as the medial and dorsal surfaces, are non-articular. A smooth-surfaced tubercle on the distal part of the medial surface is produced by the attachment of the ulnar collateral (medial) ligament of the wrist joint (Fig. 2.68).

The **pisiform** has a flat surface for articulation with the triquetral, and the convexity of the remainder of the bone leans somewhat towards the radial side, over the concavity of the carpus.

The **trapezium** articulates with the adjacent trapezoid, and these together by concave facets fit the distal convexity of the scaphoid. A distal articular surface, saddle-shaped, is for the thumb metacarpal, and this is a separate synovial joint. The trapezium articulates narrowly with the tubercle on the base of the index-finger metacarpal. There is a prominent ridge (the crest) lying obliquely on its flexor surface. The extensor surface, on the lateral convexity of the non-articular area, carries a prominent tubercle for attachment of a carpo-metacarpal ligament (Fig. 2.69).

The **trapezoid** lies wedged between trapezium and capitate, articulating proximally with the scaphoid and distally with the index-finger metacarpal. Its dorsal surface is four times the area of its ventral; both these non-articular surfaces are perforated by vessels.

The **capitate** (os magnum) is the largest of the carpal bones. It lies between the hamate medially and the trapezoid and scaphoid laterally. Proximally its convex surface fits the distal concavity of the lunate. Distally it articulates with the base of the middle and a tiny part of the ring-finger metacarpals. Anterior and posterior surfaces, non-articular, show many vascular foramina.

The **hamate,** alongside the capitate, makes with it a proximal convexity for the S-shaped midcarpal joint. The triquetral articulates with this, while distally the fourth and fifth metacarpals have their own surfaces, conjoined. Broad dorsal and narrow medial surfaces are non-articular. The anterior surface, likewise non-articular, is projected as a hook-like flange that overhangs the ventral concavity of the carpus distal to the pisiform (Fig. 2.66).

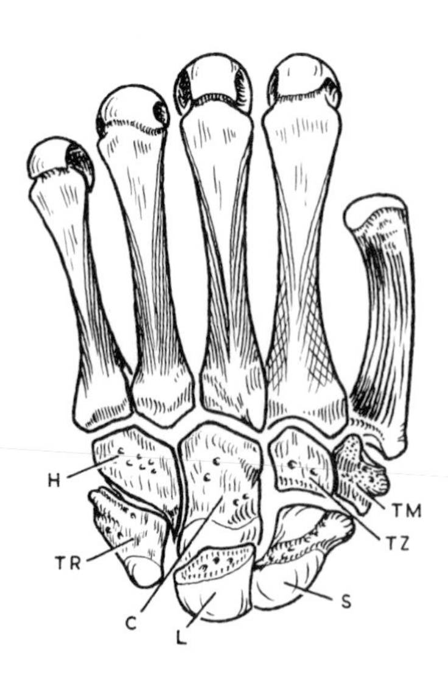

FIG. 2.68. THE LEFT CARPUS AND METACARPUS (Dorsal view). **TM.** Trapezium. **TZ.** Trapezoid. **S.** Scaphoid. **L.** Lunate. **C.** Capitate. **TR.** Triquetral. **H.** Hamate.

The Metacarpus

The thumb metacarpal is shorter and thicker than the others. Its base has a saddle-shaped facet for the trapezium. The convex facet on its head is not so boldly rounded as those of the other four metacarpals (Fig. 2.66); the flexor margin of the facet is grooved for the sesamoids of the thenar muscles. The shaft is set at right angles to the plane of the other four (Fig. 2.44, p. 94), so that its flexor surface faces across the palm. This surface has a rounded longitudinal ridge that divides it into two longitudinal concavities. The borders between flexor and extensor surfaces are sharp. The extensor surface is rather flat.

The remaining four metacarpals show expanded bases by which they articulate with the distal row of carpal bones and with each other. The middle metacarpal shows a prominent styloid process that projects dorsally into the angle between capitate and trapezoid. (Fig. 2.68). The shaft of each shows a ridge along its flexor surface. On the extensor surface is a long flat triangle with its base against the head and its apex prolonged proximally as a ridge which marks the shaft. The head carries a boldly rounded articular facet which extends further on the flexor than the extensor surface. Adjacent surfaces of the heads are pitted by deep smooth fossæ; behind these fossæ lie dorsal tubercles, at the base of each triangular flat area on the shaft (Fig. 2.68). The four metacarpal bones together form a gentle concavity for the palm. Their heads form the knuckles of the fist; the four heads make a gentle convexity distally and also dorsally.

The Phalanges

Two phalanges form the thumb, three form each finger. Each of the five proximal phalanges has a concave facet on the base, for the head of its own metacarpal. Intermediate and distal phalanges carry a facet on each base that is divided by a central ridge into two concavities (Fig. 2.66). The heads of the proximal and intermediate phalanges are correspondingly trochlea-shaped, with their facets on the distal and flexor surfaces, not on the extensor surface. Each terminal phalanx expands distally into the ungual tuberosity, roughened on the flexor surface for attachment of the digital fibro-fatty pad.

Sesamoid Bones

The pisiform is morphologically a sesamoid bone in the tendon of flexor carpi ulnaris, but it is rightly included in the carpus as a functional constituent of its bony skeleton. In the thumb a pair of sesamoid bones articulate with the flexor surface of the metacarpal head. That on the radial side lies in the tendon of flexor pollicis brevis, that on the ulnar side in the tendon of adductor pollicis. Sesamoid bones are commonly found also at the other metacarpal heads, especially the fifth and second, lodged in the palmar capsule of the metacarpo-phalangeal joints, and occasionally at the interphalangeal joints. Cartilaginous at first, they ossify generally soon after puberty (see p. 10).

Ossification. Unlike the tarsus the carpus is all cartilaginous at birth. Each carpal bone ossifies from one centre. The largest bone, the capitate, ossifies first (first year) and the smallest, the pisiform, ossifies last (tenth year). The others ossify in sequence, according to their size, at approximately yearly intervals (hamate, triquetral, lunate, trapezium, scaphoid, trapezoid), so the whole carpus, except the sesamoid pisiform bone, is ossified by the seventh year. The shafts of all the metacarpals and phalanges ossify *in utero*, so that at birth there is a cartilaginous epiphysis at the *base* of every bone except the metacarpals of the palm (second, third, fourth and fifth), where the epiphysis is at the head. Note that the thumb metacarpal ossifies like a phalanx. The cartilaginous epiphyses ossify at the second to third year and fuse at twenty years. The ungual tuberosity of each terminal phalanx ossifies in membrane.

Attachments. Returning now to the articulated hand, the attachments of ligaments, tendons and muscles can be studied.

Ligaments. The capsule of the wrist joint is attached to the articular margins of scaphoid, lunate and triquetral. At the radio-carpal joint very strong ligaments unite the lower end of the radius to the flexor surfaces of scaphoid and lunate distal to the capsule (Fig. 2.44, p. 94). On the extensor side strong bands connect the distal end of the radius to a smooth ridge on the triquetral; these are continuous with the bands of the radio-ulnar fibro-cartilage. The carpal bones are themselves united by interosseous ligaments, and the deep concavity of the carpus is further maintained by the transverse fibres of the flexor retinaculum, which acts as a tie-beam. The retinaculum is attached to the pisiform and hook of the hamate, and to the tubercle of the scaphoid and ridge of the trapezium. The extensor retinaculum passes obliquely down from the radius to the pisiform and triquetral. The capsule of the carpo-metacarpal joint of the thumb is greatly strengthened by a ligament that passes obliquely from the tubercle on the trapezium to the dorsal prolongation of the base of the metacarpal bone (Fig. 2.69). Strong piso-hamate and piso-metacarpal ligaments connect the pisiform to the hook of the hamate and to the base of the fifth metacarpal (Fig. 2.44, p. 94). The other carpo-metacarpal joints are reinforced by strong anterior and posterior and interosseous ligaments, so that the synovial joints are almost immobile, especially in the case of the index- and middle-finger metacarpals.

The metacarpal heads of the palm (second to fifth) are united by the deep transverse ligament of the palm (p. 106). This is attached across the front of the palmar ligaments, which are thick bands of dense fibrous tissue attached to pits at the metacarpal heads and to the bases of the proximal phalanges. These ligaments show grooves for the flexor tendons, and may hold sesamoid bones (Fig. 2.44, p. 94). Collateral ligaments stabilize the metacarpo-phalangeal joints; they are attached to the pits and dorsal tubercles at the sides of the metacarpal heads and to the smooth facets at the sides of the bases of the proximal phalanges. Interphalangeal ligaments form similar collateral bands (Fig. 2.55, p. 106).

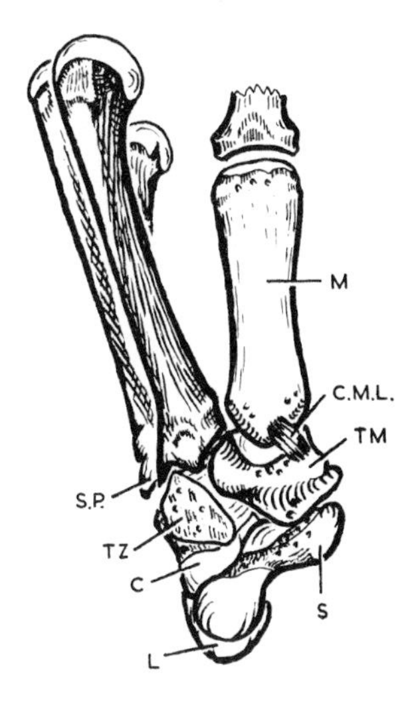

FIG. 2.69. THE LEFT CARPUS AND METACARPUS (Lateral view). **M.** Thumb metacarpal bone. **C.M.L.** Carpo-metacarpal ligament. **TM.** Trapezium. **S.** Scaphoid. **L.** Lunate. **C.** Capitate. **TZ.** Trapezoid. **S.P.** Styloid process of middle metacarpal bone. Cf. Fig. 2.42, p. 92.

The fibrous flexor sheaths bridge the grooves in the palmar ligament; each is then attached along the sharp margin of proximal and intermediate phalanges and surrounds the insertion of the profundus tendon on the terminal phalanx. Distal slips of the palmar aponeurosis are attached not only to the fibrous flexor sheaths but also to the bases and along the sides of the proximal phalanges.

No palmar aponeurosis extension reaches the thumb, but the two phalanges give attachment to a fibrous flexor sheath for the tendon of flexor pollicis longus.

Tendons. The tendons of forearm muscles are those of the wrist and digits. The flexor carpi radialis lies in the groove beside the ridge of the trapezium, being held there by a band uniting the ridge with the capitate (Fig. 2.44, p. 94). It then passes on to tubercles at the bases of the second and third metacarpals. It sometimes gives a slip to the tubercle of the scaphoid. Flexor carpi ulnaris is received into the pisiform, whence piso-hamate and piso-metacarpal ligaments prolong the insertion of the muscle. Flexor digitorum profundus is received into the proximal part of the base of the terminal phalanx of each finger; flexor pollicis longus is attached similarly to the terminal phalanx of the thumb. Flexor superficialis (sublimis) beyond its chiasma is inserted by two slips alongside the sharp borders of each intermediate phalanx.

Extensor carpi ulnaris is attached to a smooth-surfaced tubercle at the base of the fifth metacarpal. Extensor carpi radialis brevis is received into the styloid process of the middle metacarpal and the surface of bone distal to this. Extensor carpi radialis longus has an attachment to the base of the index-finger metacarpal. The long extensors of the fingers spread into the extensor expansions (p. 104). All four receive also the interossei and lumbricals. The index finger expansion receives extensor indicis and the little finger extensor minimi digiti as well. The extensor communis tendon is inserted partly into the base of the proximal phalanx, completing here the dorsal capsule of the metacarpo-phalangeal joint, and then runs on to divide into a central slip for the base of the intermediate phalanx and two marginal slips that unite to be inserted into the base of the terminal phalanx.

The thumb receives an extensor tendon into the base of each of its three bones (Fig. 2.42, p. 92). That of the metacarpal is nowadays named the abductor pollicis longus; it is inserted to the front of the tubercle, at the junction of lateral and anterior margins of the base. The extensor pollicis brevis tendon runs with it, but passes on across the flat dorsal surface of the metacarpal to reach the base of the proximal phalanx. The extensor pollicis longus tendon passes separately across the wrist to reach the base of the terminal phalanx; this tendon is stabilized across the metacarpo-phalangeal joint by expansions from the abductor pollicis brevis and adductor pollicis. There is no extensor expansion on the thumb.

The tendons of the intrinsic muscles are best studied with the muscles themselves (*v. inf.*).

Muscles. No muscles, only tendons, cross the wrist. All muscle fibres in the hand belong to intrinsic muscles.

The three muscles of the thenar eminence arise from the radial side of the flexor retinaculum. Abductor pollicis brevis has a few fibres coming from the tubercle of the scaphoid and the crest of the trapezium. Flexor brevis and opponens pollicis arise also partly from the crest of the trapezium. The deeply lying opponens is inserted into the ridge along the radial border of the thumb metacarpal and into the longitudinal groove alongside this. Abductor brevis and flexor brevis are inserted together into the base of the proximal phalanx, on its radial side; a sesamoid bone usually lies in the tendon of the short flexor. Adductor pollicis arises deep in the palm. The large transverse head arises from the ridge along the palmar surface of the shaft of the middle metacarpal. The oblique head arises from the bases of the second and third metacarpals distal to the insertion of flexor carpi radialis and also from the adjoining capitate. The two bellies converge across the web of the thumb to the base of the proximal phalanx on its index-finger side; a sesamoid bone commonly lies in the tendon of the oblique head.

The hypothenar muscles arise from the ulnar side of the flexor retinaculum and the adjoining bones. The deeply placed opponens digiti minimi arises also from the hook of the hamate. The deep branch of the ulnar nerve passes beneath the opponens before crossing the hook of the hamate. Flexor digiti minimi arises also from the hook of the hamate and abductor digiti minimi from the pisiform; thus the deep branch of the ulnar nerve passes between the two bony attachments before dipping under opponens. Opponens digiti minimi is inserted into the palmar surface of the fifth metacarpal alongside the ulnar border of the shaft. The other two muscles are inserted into the base of the proximal phalanx distal to the attachment of the collateral ligament, on the ulnar margin.

The interossei arise from the shafts of the metacarpals. The palmar interossei are relatively small, and the first is sometimes absent. They arise from the flexor surface of their own metacarpal, from the groove that faces towards the middle metacarpal (they are adductors of the fingers towards the middle finger). The dorsal interossei arise from both bones that enclose their inter-osseous space. They are attached to the longitudinal grooves on the *flexor* surfaces of the shafts and extend around to the dorsal surface up to the margin of the flat triangle and proximal ridge already noted. The flat triangles proximal to the metacarpal heads are bare bone, and the extensor tendons play across these areas. The interossei are inserted chiefly into the extensor expansions, but some of the fibres of both palmar and dorsal tendons are received into the base of the proximal phalanx.

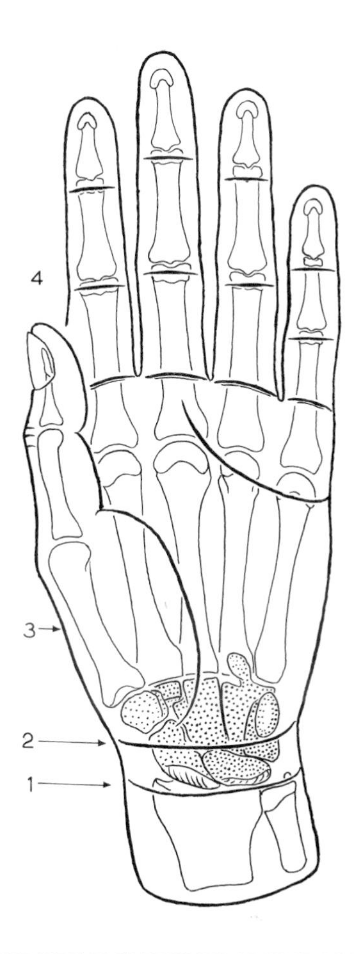

FIG. 2.70. THE SURFACE PROJECTIONS OF SOME BONY FEATURES OF THE WRIST AND HAND. **1.** Proximal skin crease: wrist joint. **2.** Distal skin crease: intercarpal joint, but pisiform bone and tubercle of scaphoid lie at this crease. The carpal bones are stippled. **3.** Metacarpus. Note the metacarpal heads so far proximal to the interdigital webs. **4.** Phalanges.

Clinical Considerations. Many of the bony features are visible or palpable in the living hand, and it is of far more use to be familiar with these landmarks than to study detailed osteology of the separate bones. On the flexor surface the tubercle of the scaphoid and the pisiform are visible in the fully extended wrist, and are palpable in the distal skin crease. The proximal skin crease marks the radio-carpal joint (Fig. 2.70). The ridge of the trapezium and the hook of the hamate are each palpable deep to their overlying muscles. The prominence of the metacarpal heads can be felt in the palm; they lie along the distal skin crease of the palm, an inch proximal to the webs between the fingers. In the snuff-box, distal to the radial styloid, the scaphoid and trapezium can be felt, and the tubercle on the base of the thumb metacarpal is very prominent. The lunate, too deep to be readily felt on the flexor surface, is easily palpable midway between radial and ulnar styloids on the extensor surface; it rolls prominently under the palpating finger-tip during full flexion of the wrist. The styloid process of the middle metacarpal is very prominent in a line distal to the radial (Lister's) tubercle. It can often be seen.

Further clinical considerations will arise in a study of movements.

Movements. These have been dealt with already (wrist, p. 94; intercarpal, pp. 95 and 120; thumb, p. 97; fingers, p. 105).

It is worth noting the line of thrust from the hand to the radius. While leaning on the fully extended hand the weight is taken principally by the distal row of carpal bones, but these are braced by the palmar ligaments and the flexor radio-carpal ligaments, which act as a spring-like shock absorber, preventing hyperextension of the wrist joint. If the lower end of the radius fractures, the distal fragment is momentarily free to be pulled into 'hyperextension' by these ligaments, and this dorsal rotation of the distal fragment which precedes impaction of the fragments is the characteristic displacement in Colles' fracture.

Section 3. The Lower Limb

The Limb Girdle

The pelvic girdle consists of the two hip-bones. They articulate with each other in front (the symphysis pubis) and each is firmly fixed to the lateral mass of the sacrum. The rigid bony pelvis thus produced transmits the body weight through the acetabulum to the lower limb and likewise transmits the propulsive thrust of the lower limb to the body. In sitting, body weight is transmitted to the ischial tuberosities and the legs are free to rest.

GENERAL PLAN OF THE LOWER LIMB

The lower limb of man is built upon the same plan as the upper limb, and this is true of the hind limb and fore limb of all vertebrates. Modifications for functional needs produce very great differences in form and proportions (e.g., in the leg and wing of a bird or bat), but the basic pattern is the same. The student of human anatomy is well advised to study the upper and lower limbs together, since their similarities are so striking. Thigh and upper arm, leg and forearm, tarsus and carpus, foot and hand are all very similar in their bones, their muscles, their vessels and their nerves. They are developed from identical patterns in the embryonic limb buds.

It is difficult to compare the upper and lower limbs before the anatomy of each is well understood. The following summary will be intelligible to the reader who has studied the anatomy of the upper limb, but it will be better understood after the anatomy of the lower limb has also been mastered.

Thigh and Arm. The femur and humerus are articulated to the axial skeleton via the limb girdles. The femur articulates with three bones (ilium, ischium and pubis) at the acetabulum, the humerus with only two bones (scapula proper and coracoid) at the glenoid fossa. The acetabulum looks *back* and down; the femoral neck slopes up and *forward* to it, for the forward thrust of propulsion. The glenoid fossa looks forward around the chest wall; the head of the humerus faces back to it, to take the *backward* thrust of the upper limb. The hip joint is more stable than the shoulder joint at the expense of a certain loss of mobility in circumduction. The hind limb has become more extended and medially rotated from the fœtal position of flexion than the upper limb, so that the extensor muscles of the thigh (quadriceps) lie in front while the extensor muscle of the arm (triceps) lies behind. Flexor and extensor compartments exist in both thigh and arm, with an abductor mass of muscle covering the hip and shoulder joints. The 'deltoid' of the hip consists of gluteus maximus and tensor fasciæ latæ, connected together by an intervening fascia. Beneath both deltoid muscles lie short muscles connecting the upper end of the bone to the limb girdle, namely the short scapular muscles in the upper limb and the glutei medius and minimus, obturator muscles, etc., in the lower limb. In each case these short muscles are rotators, but their chief and most usual function is that of adjusting and stabilizing hip and shoulder while more distal parts of the limb are in motion.

In each limb certain of the flexor compartment muscles have become separated into an adductor compartment. In the thigh the adductors have been specialized into adductors longus, brevis and magnus, gracilis and part of pectineus, while in the arm the adductor mass of muscle has degenerated phylogenetically into the small coraco-brachialis muscle condensed around the musculo-cutaneous nerve. This is not to say that adduction of the arm has been lost; indeed, the opposite is the case, for adduction is a very powerful movement in man. It is served by a new mass of muscle, the great sheets of the anterior and posterior walls of the axilla, pectoralis major and latissimus dorsi. In each limb the true adductor muscles are derived from the flexor compartment, and are thus supplied by anterior divisions of the anterior primary rami of the limb plexuses (obturator nerve and musculo-cutaneous nerve). The walls of the axilla, on the other hand, are supplied from both anterior and posterior divisions of the anterior primary rami; the anterior wall from the medial and lateral pectoral nerves and latissimus dorsi from the posterior divisions (posterior cord of brachial plexus). The latissimus dorsi is an extensor as well as an adductor (see p. 24).

The arterial supply of the thigh is essentially from the profunda femoris artery, the femoral artery itself passing on in Hunter's adductor canal to supply the leg and foot. A similar vessel, the profunda brachii, accompanies the radial nerve in the spiral groove though the brachial artery itself, unlike the femoral, takes a considerable share, by other branches, in the supply of the arm. The venous and lymphatic returns are very similar in the two limbs and are dealt with together on p. 129.

The pattern of nerve supply appears very different, for the reason that the embryological derivation of the muscles is somewhat different. Muscles of the upper limb are supplied from the brachial plexus, but the thigh extensors, functionally the counterpart of the triceps brachii, have a different embryological origin and thus a different nerve supply. The five nerves of the lower limb plexus are L4, 5, S1, 2, 3 in the sacral plexus, and they do not supply the quadriceps and adductors, which are derived from muscle 'borrowed from the trunk'. This 'borrowed' muscle separates into extensor and flexor groups, whose nerves, the femoral and obturator, arise from the lumbar plexus and have no counterpart in the upper limb. Quadriceps femoris and triceps brachii are functional counterparts but they are not morphologically homologous.

Leg and Forearm. Two bones support the structure of these limb segments. The pre-axial bones, on the thumb and hallux side, are radius and tibia, the postaxial bones are ulna and fibula. Both pre-axial and postaxial bones take part in forming the elbow joint and supination and pronation are free movements in the forearm. Only the pre-axial bone takes part in forming the knee joint and pronation and supination are not possible in the leg. The two bones grip the talus at the ankle joint, which is thus greatly stabilized. The elbow is more stable than the knee from the purely bony point of view, but the latter joint is adequately strengthened by ligaments and tendons.

Despite the great functional modifications and differences between hand and foot, there is a most striking parallelism between the muscles of the forearm and leg. In the extensor compartment it is to be noted that the three tendons to the thumb (terminal phalanx, proximal phalanx and base of metacarpal) are inserted in a homologous way with those to the great toe (terminal phalanx, proximal phalanx and base of metatarsal) though their fleshy origins are very different in forearm and leg. The homologous muscles are extensor pollicis longus and extensor hallucis longus, extensor pollicis brevis and extensor hallucis brevis (of extensor digitorum brevis) and abductor pollicis longus and tibialis anterior.

The extensor tendons to the digits are similar. The extensor digitorum communis muscle arises from the humerus, the extensor digitorum longus arises in many animals from the femur though in man its origin has descended below the knee to the fibula. There are no real counterparts in the leg of the radial extensors of the wrist (extensors carpi radialis longus and brevis and extensor carpi ulnaris), though *functionally* there are similar extensors (i.e., dorsiflexors) of the ankle in tibialis anterior and peroneus tertius.

In the flexor compartments of these limb segments the muscle homologies are striking. Flexor digitorum superficialis has a wide origin from both forearm bones and from a fibrous arch between them. Soleus has a similar origin in the leg. Their insertions into the intermediate phalanges are similar, though in the case of soleus the muscle mass has been interrupted by the backward projection of the calcaneus upon which soleus acts as a flexor of the ankle joint, and a further muscle belly persists in the sole of the foot (flexor digitorum brevis) as the flexor of the intermediate phalanges. Superficial to these muscles the phylogenetically degenerating palmaris longus and plantaris muscles, each with a small belly and a long tendon, are traceable into the palmar and plantar aponeuroses. The gastrocnemius may be regarded as homologous with the carpal flexors, though it is unrealistic to pursue such similarities too far.

The vessels and nerves of forearm and leg are strikingly similar and especially is this so in the extensor compartments. The posterior interosseous nerve reaches the extensor compartment of the forearm by winding around the radius (pre-axial bone) while the peroneal nerve reaches the extensor compartment of the leg by winding around the fibula (postaxial bone). The vessels in each case pass between the two bones of the forelimb. The posterior tibial vessels and nerve pass beneath the fibrous arch of soleus while the ulnar artery and median nerve pass beneath the fibrous arch of flexor digitorum superficialis.

Tarsus and Carpus. These consist of similar bones, though variations in size and arrangement make their form very dissimilar. The carpus is a mechanism for articulating hand to radius and allowing circumduction to occur, the tarsus is adjusted as a mobile weight-bearing and propulsive mechanism.

Morphology. The primitive mammalian arrangement of the bones is:

1. A proximal row of three bones (radiale or tibiale, intermediale, ulnare or fibulare).

2. A distal row of five bones, one for each metacarpal or metatarsal.

3. A central bone (os centrale) between the two rows.

In all tetrapods it is normal to find the two postaxial bones of the distal row fused into one bone, the hamate for fourth and fifth metacarpals and cuboid for fourth and fifth metatarsals. The human carpal scaphoid is the fused os radiale and os centrale, and fracture of the scaphoid usually separates these two elements. In the tarsus the os centrale becomes the navicular. In the carpus the os intermediale becomes the lunate, in the tarsus it becomes the lateral tubercle of the posterior process of the talus (Figs. 3.1 to 3.4).

The functional differences between wrist and ankle should be noted. In addition to flexion and extension the wrist has abduction and adduction, while at the ankle joint movement is limited to flexion and extension. Movements between the talus and the other tarsal bones provide the important side-to-side movements of eversion and inversion, and these replace the movements of pronation and supination which occur in the forearm. The movements of inversion and eversion are essential to an efficient 'grip' of the foot when walking across uneven or laterally sloping smooth surfaces, or when leaning over while turning at speed on horizontal surfaces.

Foot and Hand. While the hand consists of metacarpals and phalanges the foot includes also the tarsal bones. In the hand the first metacarpal with its two phalanges has 'broken free' and can be opposed to the other four, thus enabling objects to be gripped. The thumb is thus, functionally, as important as all four other digits combined. In contrast, the foot is relatively fixed, and the first metatarsal bone has little power of free movement. All five metatarsals are weight bearing and in propulsion it is the big toe side from which the 'take off' is made. This has resulted in a shifting of the axis of the human foot towards the tibial side and the longitudinal axis of the foot passes through the *second metatarsal*, while in the hand the axis remains the *middle* (i.e., *third*) *metacarpal*. In spite of these bony and functional differences the arrangements of muscles, vessels and nerves in the sole and palm are strikingly similar. In each a powerful aponeurosis gives concavity to palm and sole and serves for attachment of skin to improve the grip. The hand and foot each contain four muscle layers. Beneath the aponeurosis lies a group of small (intrinsic) muscles, beneath these the long flexor tendons pass to the digits. Deeper still are the adductor

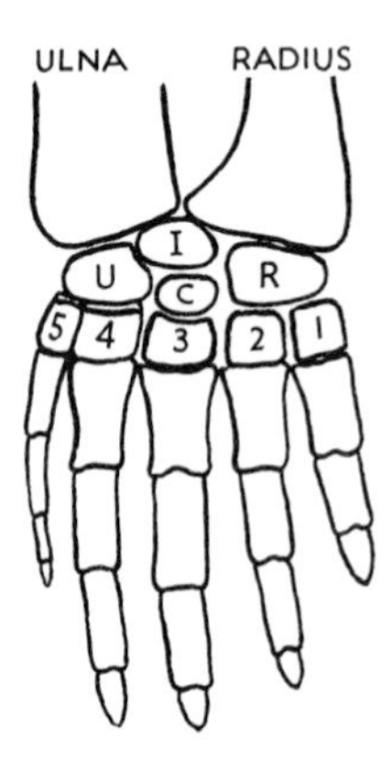

FIG. 3.1. THE MANUS OF THE WATER TORTOISE, SHOWING THE PRIMITIVE ARRANGEMENT OF THE CARPAL BONES IN VERTEBRATES. R. Os radiale. I. Os intermediale. C. Os centrale. U. Os ulnare. 1–5. The carpal bones of the distal row.

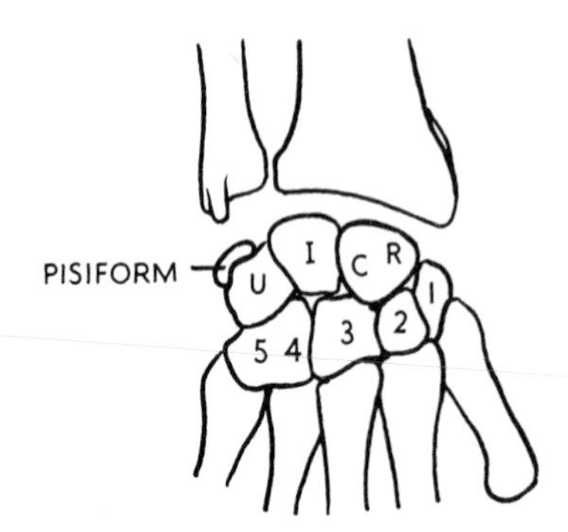

FIG. 3.2. THE HUMAN CARPUS. The ossa radiale and centrale have fused to form the scaphoid. The fourth and fifth bones of the distal row have fused to form the hamate. The pisiform appears as an extra bone, a sesamoid bone in the tendon of flexor carpi ulnaris; it has no counterpart in the foot.

FIG. 3.3. THE PRIMITIVE ARRANGEMENT OF THE TARSAL BONES (cf. Fig. 3.1). T. Os tibiale. I. Os intermediale. C. Os centrale. F. Os fibulare. 1–5. The tarsal bones of the distal row.

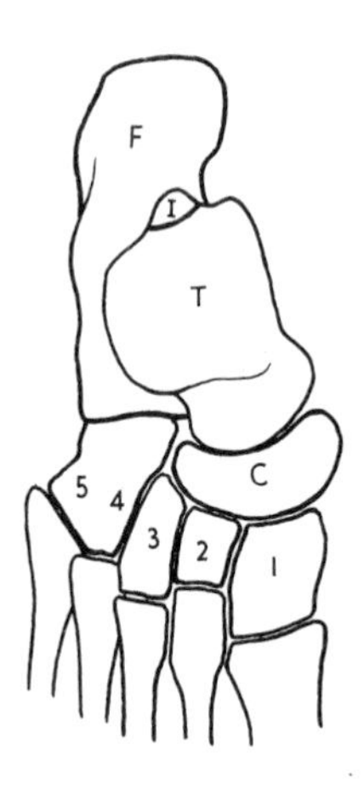

FIG. 3.4. THE HUMAN TARSUS. The os intermediale has fused with the os tibiale to form the talus. The os fibulare has become the calcaneus, the os centrale the navicular. The fourth and fifth bones of the distal row have fused to form the cuboid.

muscles of thumb or big toe, while deep to this third layer lie the metacarpals or metatarsals with their intervening interosseous muscles.

The *nerves* of the sole and palm are almost identical in their sensory and motor distribution. The lateral plantar nerve corresponds with the ulnar nerve in the palm and the medial plantar nerve corresponds with the median nerve in the hand.

Cutaneous Branches. The skin of the postaxial one and a half digits is supplied in the sole by the lateral plantar nerve, and in the palm by the ulnar nerve. The skin of the pre-axial three and a half digits is supplied in the sole by the medial plantar nerve and in the palm by the median nerve. In each digit of the hand and foot the nerves of the flexor skin extend around the tip to the dorsal surface, where they supply the extensor skin up to a level proximal to the nail bed.

Motor Branches. The distribution of the medial plantar and median nerves is similar, for each supplies intrinsic flexor and abductor muscles of the first digit (big toe or thumb). The medial plantar nerve supplies the first lumbrical muscle, the median nerve usually supplies the first two lumbrical muscles. There is less resemblance between the motor distribution of the lateral plantar nerve and the ulnar nerve, though each divides into a superficial and a deep branch. The superficial branch of the ulnar supplies no muscle except the little palmaris brevis, while that of the lateral plantar supplies flexor minimi digiti and the interossei of the fourth space; but in each case the deep branch supplies all the remaining muscles of sole and palm.

The dorsum of the foot contains the extensor tendons lying bound together by the deep fascia and this is so in the hand also. The extensor digitorum brevis of the foot has its counterpart in the hand, but the latter serves only the index finger. It is the extensor indicis, arising from the ulna.

There is much similarity in the vascular patterns. The dorsal venous arches are similar in the hand and foot, and each collects most of the blood from palm and sole.

Venous Return in the Lower and Upper Limbs

In each limb the venous return is by way of large veins that lie in the subcutaneous tissue and by deep veins that accompany the arteries.

The **superficial veins** commence in a dorsal venous arch that lies over the heads of the metatarsals or metacarpals. The arch receives much of the blood from the sole or palm, not only around the margins of foot or hand, but also *by veins that perforate the interosseous spaces.* Thus pressure on the sole in standing or on the palm in gripping fails to impede venous return. A vein leaves each side of the dorsal arch and runs up in the subcutaneous tissue. These veins lie along the pre-axial and postaxial borders of the limbs. The pre-axial vein of the lower limb is the great saphenous, that of the upper limb the cephalic. Each runs proximally to the root of the limb, where it pierces the deep fascia to open into the main vein of the limb. The postaxial vein of the lower limb is the small saphenous, that of the upper limb the basilic. Each runs up in the subcutaneous tissue of the forelimb and pierces the deep fascia to join, in the case of the former the popliteal vein, and in the case of the latter the brachial vein (Figs. 2.24, p. 75 and 3.5).

The **deep veins** accompany the arteries. In foot and hand, in leg and forearm, they consist of venæ comitantes, a vein lying to each side of the artery. They communicate by frequent transverse channels. Towards knee and elbow the venæ comitantes flow into a single vein, which remains single throughout the rest of its course. In the lower limb the popliteal vein is a single one, in the upper limb the veins do not unite into a single trunk until about the middle of the arm.

Lymphatic Drainage of Lower and Upper Limbs

In each limb the rule applies that superficial lymphatics follow veins, while deep lymphatics follow arteries. The lymph is taken to lymph nodes lying in groin and axilla. The superficial lymphatics of the *lower limb* accompany the great and small saphenous veins (Fig. 3.6). Infections of the superficial tissues of the foot are common; lymphadenitis, when it occurs, is limited almost exclusively to the inguinal nodes and popliteal lymphadenitis is quite rare. This is because practically all the superficial tissues and skin of the lower limb drain to the groin along the route of the great saphenous vein; only a very small area of skin and subcutaneous tissue over the heel drains alongside the small saphenous vein to the lymph nodes in the popliteal fossa.

The superficial lymphatics along the course of the great saphenous vein drain into nodes lying in a vertical chain along the termination of the vein. The remaining nodes of the groin, lying lateral and medial to the saphenous opening, receive lymph from the lateral side of the trunk and from the back below the waist and from the anterior abdominal wall and perineum respectively (cf. the three groups in the axilla).

All the lymph from the superficial inguinal nodes passes by efferent vessels through the cribriform fascia to deep inguinal nodes, lying beneath the fascia lata to the medial side of the femoral vein. These receive also the deep lymphatics of the lower limb that have ascended alongside the arteries, some having passed through the popliteal nodes. In turn they send their efferents through the femoral canal to the nodes along the external iliac artery.

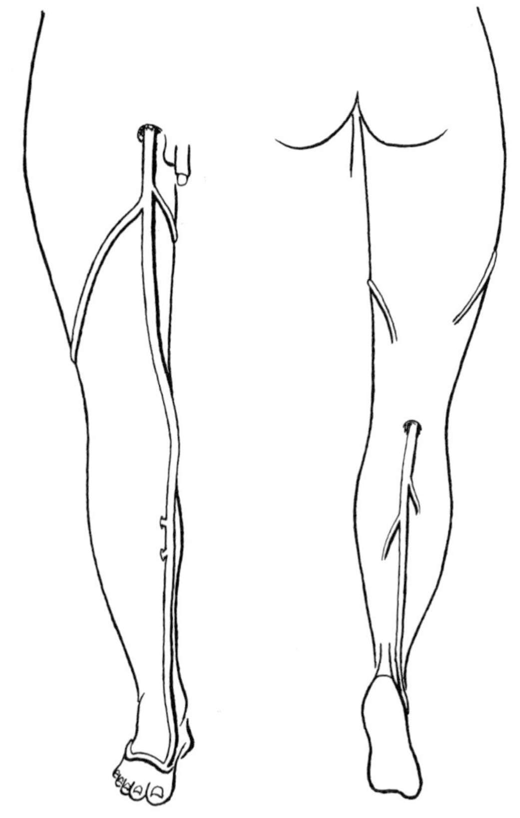

FIG. 3.5. THE SUPERFICIAL VEINS OF THE LOWER LIMB. The great saphenous vein drains from the medial side of the foot to the femoral vein. The small saphenous vein drains from the lateral side of the foot to the popliteal vein.

In the *upper limb* the grouping of the lymph nodes in the axilla and the areas they drain show great similarity (p. 65). The lateral group of nodes receives practically all the lymph from the upper limb. The nodes lie along the medial side of the axillary vein and correspond to the vertical group of superficial inguinal nodes. The anterior or pectoral group of axillary nodes receive lymph from the breast and anterior body wall as far down as the waist (umbilicus) and they correspond with the medial group of superficial inguinal nodes. The posterior or scapular group of axillary nodes receive lymph from the back as far down as the waist and they correspond with the lateral group of superficial inguinal nodes. Just as all three groups of superficial inguinal nodes send their lymph to the deep inguinal group, so all three groups of axillary nodes send theirs to the apical group.

Nerve Supply of the Lower and Upper Limbs

The plan of innervation of the limbs and the segmental supply of the skin and muscles are considered on p. 24.

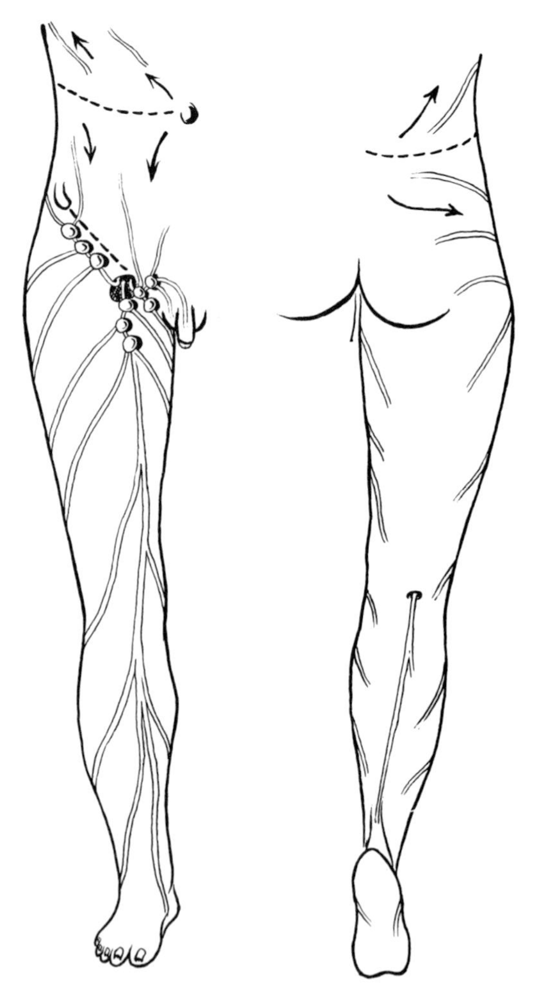

FIG. 3.6. THE SUPERFICIAL LYMPHATIC DRAINAGE OF THE LOWER LIMB. The superficial inguinal nodes receive all the lymph from the subcutaneous tissues below the waist.

THE FRONT OF THE THIGH

Subcutaneous Tissue. The fat of the front of the thigh contains cutaneous nerves, lymphatic vessels and nodes, the termination and tributaries of the great saphenous vein and the cutaneous branches of the femoral artery. Scarpa's fascia (the membranous layer of the 'superficial fascia' of the abdominal wall) extends into the upper part of the thigh.

Cutaneous Nerves (Fig. 3.7). The cutaneous branches of the lumbar plexus that supply the thigh are derived from the first three lumbar nerves. They are the ilio-inguinal, femoral branch of the genito-femoral nerve, and the medial, intermediate and lateral cutaneous nerves of the thigh. In addition the superficial branch of the obturator nerve supplies an area of skin above the medial side of the knee. All these nerves supply fascia lata as well as skin.

The **ilio-inguinal nerve** is the collateral branch of the ilio-hypogastric, both being derived from the first lumbar nerve. Like the collateral branch of an intercostal nerve, it has no lateral but only a terminal cutaneous distribution. In the anterior abdominal wall it lies in the neuro-vascular plane between the internal oblique and transversus abdominis muscles, pierces internal oblique and supplies its lower fibres, and passes down beneath the external oblique (Fig. 5.10, p. 262) to emerge on the front of the cord through the superficial inguinal ring. Piercing the external spermatic fascia its chief distribution is to the skin of the root of the penis and the anterior one-third of the scrotum, but it supplies also a small area of thigh below the medial end of the inguinal ligament.

The **genito-femoral nerve** is derived from the first and second lumbar nerves, but fibres from only L1 pass into the *femoral branch*. This branch is given off from the nerve as it lies on psoas major (or minor if this is present). It runs down on the external iliac artery and passes beneath the inguinal ligament into the femoral sheath. It pierces the anterior wall of the sheath and the overlying fascia lata below the middle of the inguinal ligament (Fig. 3.17). It supplies most of the skin over the femoral triangle. The femoral branch thus described is wholly cutaneous in its distribution. The genital branch enters the spermatic cord (p. 311) and does not reach the skin.

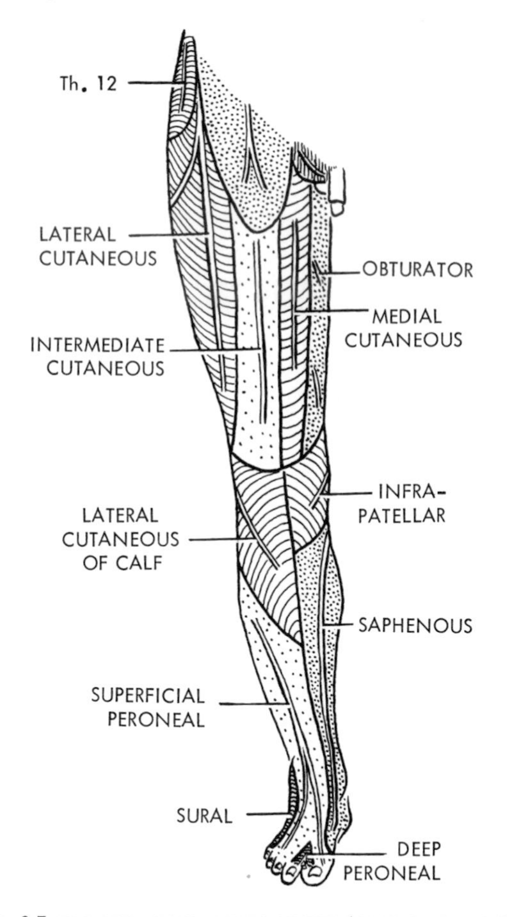

FIG. 3.7. THE CUTANEOUS NERVES OF THE FRONT OF THE RIGHT LOWER LIMB. The ilio-inguinal nerve supplies scrotal skin and the adjacent thigh, the genito-femoral nerve supplies skin over the upper part of the femoral triangle. The obturator nerve may be restricted to an area just above the knee, being replaced above this by the medial cutaneous nerve of the thigh, as in Fig. 3.25.

The **medial cutaneous nerve of the thigh** is a branch of the anterior division of the femoral nerve. Inclining medially across the femoral vessels it pierces the deep fascia (fascia lata) above the upper border of sartorius which, on the medial side of the thigh, is well below the inguinal ligament. It supplies the medial side of the thigh, and its terminal twigs join the patellar plexus (p. 132).

The **intermediate cutaneous nerve of the thigh** has a common origin with the medial cutaneous nerve from the anterior division of the femoral nerve. It passes vertically downwards beneath the fascia lata, usually pierces the upper border of sartorius, and then pierces the fascia lata at a higher level than the medial cutaneous nerve to supply the front of the thigh as far down as the knee, where its terminal twigs join the patellar plexus.

The **lateral cutaneous nerve of the thigh** is a branch of the lumbar plexus derived from L2 and 3. Passing from the lateral border of psoas major across the iliac fossa it lies at first behind the fascia iliaca; but approaching the inguinal ligament it inclines forwards and is incorporated within the substance of the iliac fascia, which is here a thick tough membrane. The nerve now pierces the inguinal ligament, where it lies free in a fibrous tunnel a centimetre to the medial side of the anterior superior iliac spine (Fig. 3.8). It is possible that meralgia paræsthetica may be associated with irritation of the nerve in this fibrous canal. In those individuals in whom fibres of transversus abdominis arise from the fascia iliaca, freeing the nerve from the tunnel in the inguinal ligament is not enough to cure the pain. The nerve must be freed also from the fascia iliaca, to prevent irritation from the pull of the contracting fibres of transversus.

The nerve enters the thigh deep to the fascia lata, and divides into anterior and posterior branches which pierce the fascia lata separately an inch or two below the lateral end of the inguinal ligament. The anterior branch contains all the L3 that is in the nerve, and is distributed along the antero-lateral surface of the thigh, and its terminal twigs enter the patellar plexus. The posterior branches, containing only L2 fibres, pass down

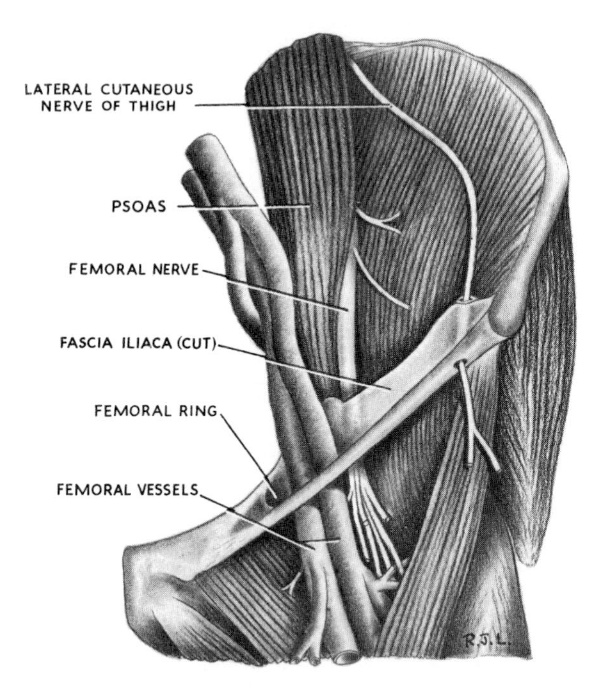

FIG. 3.8. THE LATERAL CUTANEOUS NERVE OF THE THIGH. It perforates the inguinal ligament near the anterior superior iliac spine. In the last inch of its course across the iliac fossa the nerve runs in the fascia iliaca; some fibres of transversus abdominis arise from this part of the fascia. Note the branches to iliacus from the femoral nerve. Sketched from a dissection.

the thigh along the postero-lateral aspect (along the ilio-tibial tract).

The **obturator nerve** sends a twig from its superficial division into the subsartorial plexus, whence cutaneous branches pass to skin and fascia lata over the adductor magnus tendon above the medial condyle of the femur (Fig. 3.25, p. 145). Direct branches from the superficial division of the obturator nerve frequently supply the skin on the medial side of the thigh above this level (Fig. 3.7).

The Patellar Plexus (Prepatellar Plexus). This is a fine network of communicating twigs in the subcutaneous tissue over and around the patella and ligamentum patellæ. It is formed by the terminal branches of the medial and intermediate cutaneous nerves of the thigh, of the anterior branch of the lateral cutaneous nerve of the thigh, and of the infrapatellar branch of the saphenous nerve (p. 142).

Cutaneous Arteries. Four cutaneous branches of the femoral artery are found in the subcutaneous tissue below the inguinal ligament.

The **superficial circumflex iliac artery** pierces the fascia lata lateral to the saphenous opening and passes up below the inguinal ligament to the anastomosis at the anterior superior iliac spine.

The **superficial epigastric artery** may pierce the fascia lata or emerge through the saphenous opening. It crosses the inguinal ligament and is distributed towards the umbilicus to the skin and fat of the lower abdominal wall.

The **superficial external pudendal artery** emerges from the saphenous opening and passes medially, in front of the spermatic cord (round ligament) to supply skin and superficial tissue of the scrotum (labium majus).

The **deep external pudendal artery** emerges more medially and passes behind the spermatic cord (round ligament) to supply the skin of the scrotum (labium majus).

Great (Long) Saphenous Vein. It commences at the medial end of the dorsal venous arch (p. 168), courses upwards in front of the medial malleolus, passes behind the knee, and spirals forwards around the medial convexity of the thigh and ends by passing through the cribriform fascia to open into the medial side of the femoral vein (Fig. 3.10). It contains over a dozen valves. It does not drain the medial side of the leg between tibia and tendo calcaneus; here three veins pierce the deep fascia to enter the deep veins of the calf (Fig. 3.9).

Its tributaries in the femoral triangle will need to be tied off in particular operations for varicose veins. Like venous patterns elsewhere they are variable; the usual arrangement is shown in Fig. 3.10. The lateral and medial superficial femoral veins drain wide areas of the thigh and join the saphenous vein a little distance below the saphenous opening. At this opening the saphenous vein receives four tributaries that correspond with the four branches of the femoral artery and, in addition, a

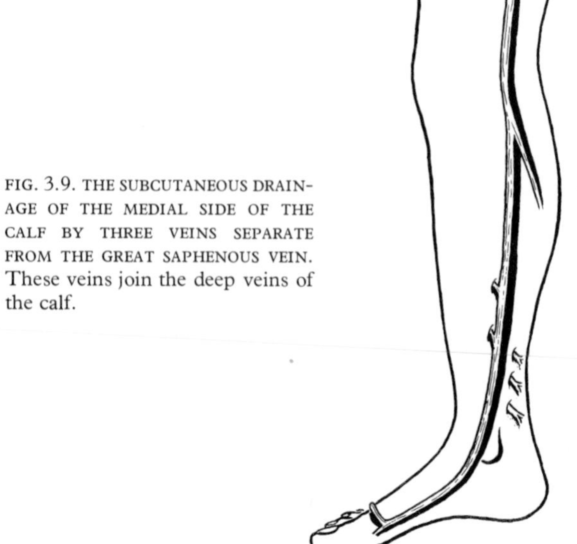

FIG. 3.9. THE SUBCUTANEOUS DRAINAGE OF THE MEDIAL SIDE OF THE CALF BY THREE VEINS SEPARATE FROM THE GREAT SAPHENOUS VEIN. These veins join the deep veins of the calf.

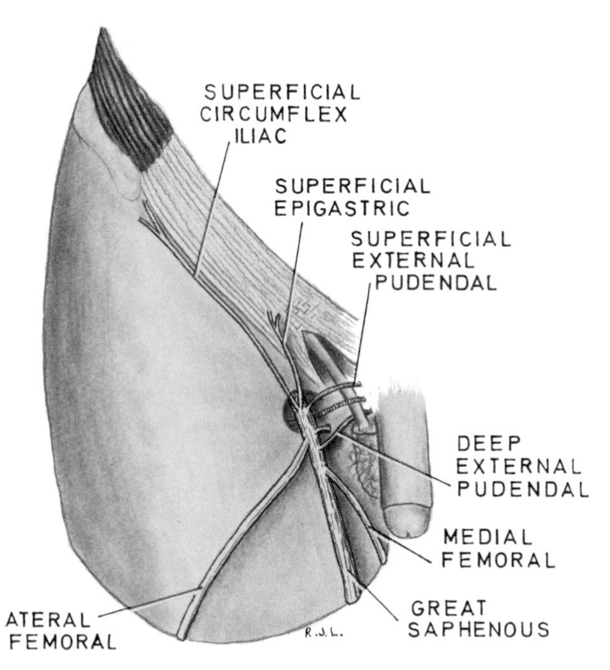

FIG. 3.10. THE TRIBUTARIES OF THE GREAT (LONG) SAPHENOUS VEIN IN THE FEMORAL TRIANGLE.

deep branch which pierces the fascia lata over the adductor longus muscle. The superficial external pudendal artery lies between the superficial and deep external pudendal veins; though usually behind (Fig. 3.10), it may lie in front of the terminal part of the saphenous vein.

Deep Connexions. Along the medial side of the calf some four or five perforating veins connect the great saphenous vein with the deep veins of the calf. Two or three of these pierce the medial edge of soleus to join the plexus within that muscle (PLATE 6, p. *iii*). The valves in these perforating veins are directed inwards. Venous blood in the foot flows from deep veins (in the sole) to superficial veins on the dorsum of the foot. In the leg there is no such flow from deep veins to the saphenous veins; indeed, much of the saphenous blood passes through the perforating veins to be pumped up the deep veins by the contractions of soleus and the other muscles of the calf.

Lymph Vessels and Lymph Nodes (Fig. 3.6). Large lymphatic vessels accompany the great saphenous vein from the foot, leg and thigh. Numerous large vessels also spiral around the outer side of the thigh to converge on the **superficial inguinal nodes.** These consist of a number of nodes arranged irregularly in the subcutaneous fat of the femoral triangle. Inspection of the dissected nodes shows no obvious arrangement into groups yet they are rightly described as belonging to three groups, for they drain three distinct areas. The *vertical group*, lying lateral to the termination of the great saphenous vein, receives lymphatics from the deep fascia (and everything superficial to it) of the lower limb. The *lateral group*, lying below the lateral part of the inguinal ligament, receives lymph from the buttock, flank and back below the level of the waist. The *medial group* lies below the medial end of the inguinal ligament, and a node or two may encroach on the anterior abdominal wall. They receive from below the umbilicus and medial to a line drawn vertically upwards from the anterior superior iliac spine. More important, the medial group receives lymph also from the perineum; it is to be noted that the perineum includes the anal canal below Hilton's white line, the urethra and the external genitalia of both sexes (*but excluding the testes*). The efferent lymphatics from all three groups converge towards the saphenous opening and pass through the cribriform fascia to enter the deep inguinal nodes, lying medial to the femoral vein. Their passage through the fascia covering the saphenous opening produces a number of holes which give rise to the sieve-like appearance denoted by the name 'cribriform'.

Scarpa's Fascia. The membranous layer of the superficial fascia of the abdominal wall extends into the thigh and becomes fixed to the fascia lata at the flexure skin crease of the hip joint, where attachment of skin to fascia lata limits its further descent. The attachment extends laterally from the pubic tubercle below the inguinal ligament. It should be noted that the saphenous opening lies below this line, so that a femoral hernia, emerging from the saphenous opening, can never come to lie in the space beneath Scarpa's fascia. The hernia emerges into ordinary subcutaneous fat and can therefore never become very large (Fig. 3.11).

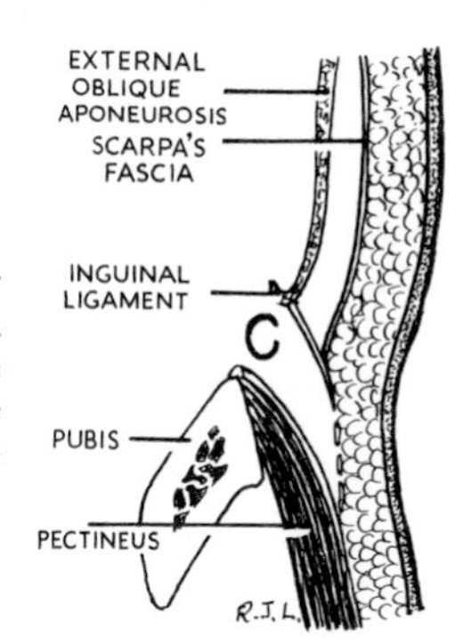

FIG. 3.11. VERTICAL SECTION THROUGH THE FEMORAL CANAL. Scarpa's fascia is attached below the inguinal ligament to the fascia lata. The femoral canal (C in the figure) is connected through the cribriform fascia with only the subcutaneous fat.

The Fascia Lata

The fascia lata encloses the thigh, like a stocking whose top is too large. A vertical slit has been cut at the top of the stocking, so to speak, and the cut edges overlapped to fit the bony pelvis. The fibres of the fascia lata lie vertically in the long axis of the thigh. Examine the os innominatum or, better still, an articulated bony pelvis. Trace the attachments of the fascia lata to the pelvic girdle. The line encloses all the thigh and hamstring muscles.

Commence with the cut edge A (Fig. 3.12) and attach it to the pubic tubercle. The top of the 'stocking' of fascia lata traced laterally is seen to be attached to the inguinal ligament, which it draws into a downward convexity when the thigh is extended. From the anterior superior iliac spine the top of the 'stocking' is attached to the external lip of the iliac crest, splitting to enclose the tensor fasciæ latæ muscle, as far back as the tubercle of the iliac crest. From the tubercle the single layer of fascia is attached to the iliac crest as far back as the posterior gluteal line, where it again splits to enclose gluteus maximus. The two layers that enclose the gluteus maximus are very thin indeed, but are nevertheless in continuity with the rest of the fascia lata. Attached to bone at the edges of the gluteus maximus origin on the ilium and sacrum, and to the sacrotuberous ligament, the fascia lata may now be traced along the convexity of the ischial tuberosity, the ischiopubic ramus and the body of the pubic bone, between the attachments of the adductors and gracilis on the one hand and the perineal muscles on the other. Passing over the origin of adductor longus in the angle between pubic crest and symphysis the top of the 'stocking' is now attached to bone just below the pubic tubercle, and, further laterally, it passes beneath the pubic tubercle and inguinal ligament, lying on the surface of pectineus to be attached to the pectineal line as far laterally as the

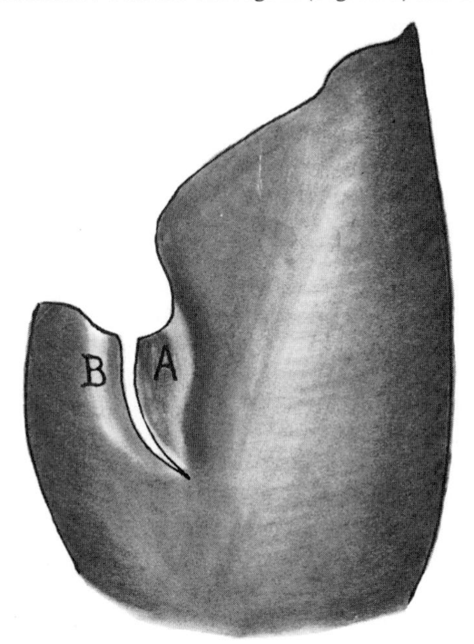

FIG. 3.12. THE FASCIA LATA OVER THE LEFT FEMORAL TRIANGLE, TO BE FITTED TO THE INGUINAL LIGAMENT AND THE BONY PELVIS OF FIG. 3.14.

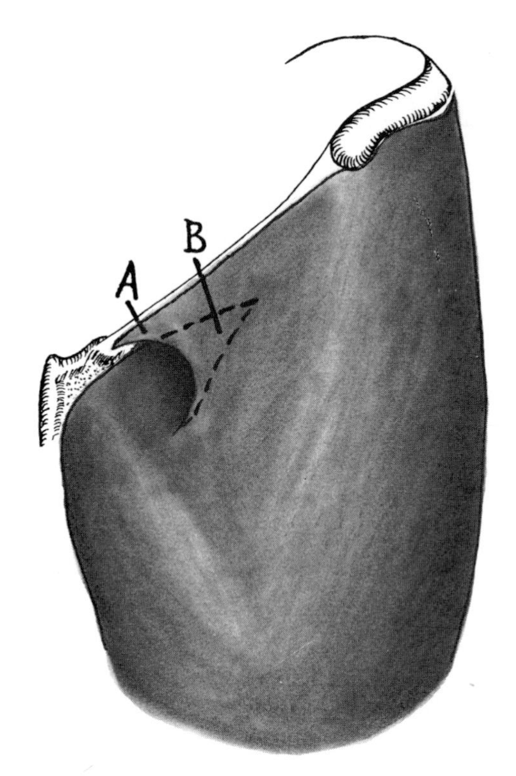

FIG. 3.13. 'A' IS ATTACHED ALONG THE INGUINAL LIGAMENT TO THE PUBIC TUBERCLE. 'B' IS TUCKED BENEATH 'A' ALONG THE PECTINEAL LINE, THUS FORMING THE SAPHENOUS OPENING.

cut edge B (Fig. 3.13). The cut edge A thus appears as a crescentic margin to the **saphenous opening,** which is merely the oblique space lying between the cut edges of the top of the 'stocking'. The deep part of B (Fig. 3.13) lies on pectineus beneath the femoral sheath, and is prolonged downwards in front of the adductor muscles

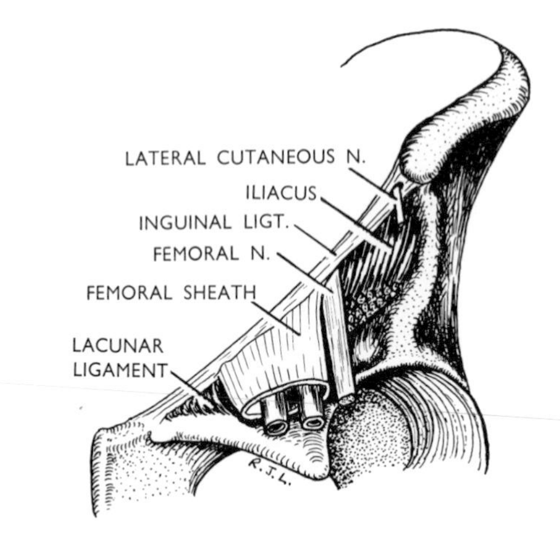

FIG. 3.14. THE FEMORAL SHEATH PROJECTS INTO THE THIGH BENEATH THE INGUINAL LIGAMENT; THE FEMORAL NERVE LIES TO ITS LATERAL SIDE. The iliac crest and inguinal ligament receive the fascia lata (cf. Fig. 3.13).

as the medial intermuscular septum. To change the simile the edges of A and B of the slit in Fig. 3.13, are overlapped like lapels of a double-breasted jacket and the great saphenous vein passes through the gap like a hand reaching into the inside pocket. The saphenous opening is covered in by a loose fascia attached laterally to the falciform edge and medially to the fascia lata where it lies over adductor longus. This fascia is pierced by the great saphenous vein and the efferent lymphatics from the superficial inguinal lymph nodes, giving it a sieve-like appearance whence it derives the name of *cribriform fascia.*

The fascia lata is attached below to the patella and to the inferior margins of the tibial condyles and to the head of the fibula. Over the popliteal fossa it is strengthened by transverse fibres and continues below into the deep fascia of the calf. The fascia lata has been seen to enclose two muscles, the tensor fasciæ latæ and gluteus maximus. These two muscles converge towards each other below the greater trochanter of the femur. The *deep half of the lower half* of gluteus maximus is inserted into the gluteal crest of the femur. The remaining three-fourths of gluteus maximus is inserted with tensor fasciæ latæ into the ilio-tibial tract. The triangular interval between them is filled by the continuation of the fascia lata described above, the whole constituting the 'deltoid of the hip joint' reminiscent of the deltoid muscle covering the shoulder joint.

Tensor Fasciæ Latæ. The muscle arises from the external lip of the iliac crest between the anterior superior iliac spine and the tubercle of the iliac crest. It is a thin sheet of muscle at its origin; it is triangular in shape and becomes thicker at its insertion into the ilio-tibial tract. Its anterior border lies edge to edge with sartorius at the anterior superior iliac spine, below which the two muscles diverge and allow rectus femoris to emerge between them (PLATE 5). It is supplied by the superior gluteal nerve (L4, 5, S1), which crosses the buttock and ends in the muscle. Its action is to pull upon the ilio-tibial tract (*v. inf.*).

The Ilio-Tibial Tract. This is a thickening of the fascia lata that commences at the level of the greater trochanter, where three-fourth of gluteus maximus and the tensor fasciæ latæ are inserted into it. It passes vertically down the postero-lateral aspect of the thigh, crosses the lateral condyle of the femur and is inserted into a smooth circular facet one centimetre in diameter on the anterior surface of the lateral condyle of the tibia (Fig. 3.46, p. 166). When the knee is straight the tract passes in front of the axis of flexion; thus it maintains the knee in the extended position. It is not an extensor of the flexed knee, however; in the right-angled knee it passes behind the axis of flexion. The lateral intermuscular septum in the lower part of the thigh is attached to the ilio-tibial tract and the mass of vastus lateralis can be seen bulging in front of it. The tract

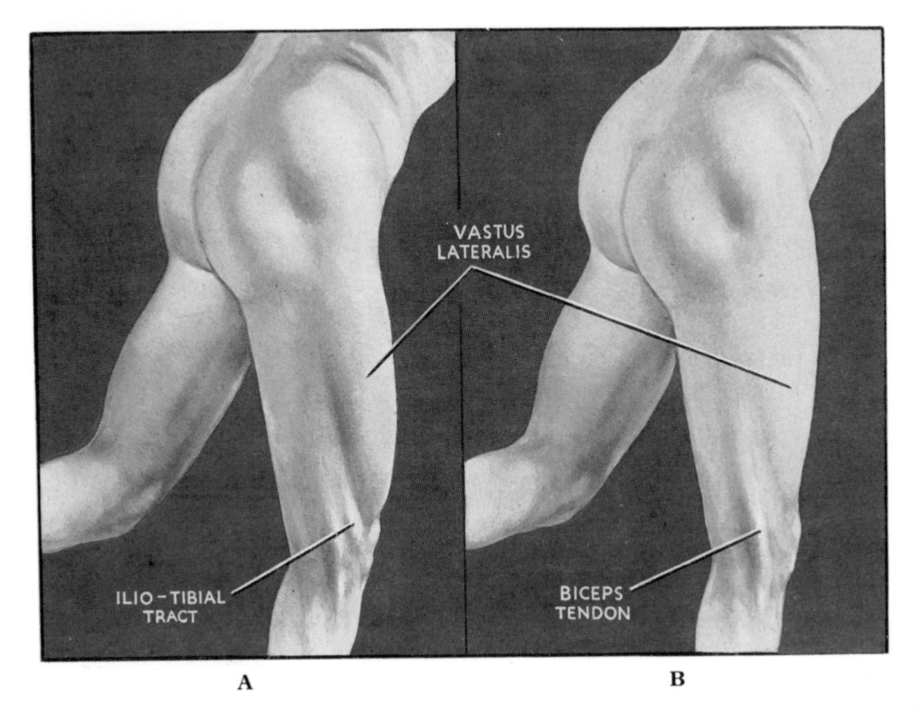

FIG. 3.15. THE ACTION OF THE ILIO-TIBIAL TRACT. A. The tract is acting as the only antigravity support at the knee; the flaccid vastus lateralis bulges over the tract. B. Contraction of the quadriceps abolishes the groove made by the ilio-tibial tract, because vastus lateralis is now in firm contraction. Drawn from a model.

is an inch wide and its chief value lies in the stabilizing influence it has on the knee joint. It is particularly in action when the slightly flexed knee is bearing the weight of the body, and is thus constantly in use in the appropriate phases of walking and running. In rising from the sitting position gluteus maximus extends the hip and then, as the knee is extended by quadriceps, the ilio-tibial tract operates to assist quadriceps (feel it on yourself or a friend).

In leaning forward with the knee slightly flexed the ilio-tibial tract may provide the only antigravity force supporting the knee joint. The antigravity pull on the ilio-tibial tract increases as the body leans further and further forwards at the hip. This elongates the powerful gluteus maximus, which contracts more and more strongly as the movement proceeds. In this posture the quadriceps can be quite relaxed, and if the patella is palpated it can be moved freely from side to side, while in the thigh the relaxed vastus lateralis bulges prominently beyond the groove indented by the taut ilio-tibial tract (Fig. 3.15).

The **lateral intermuscular septum** is a strong layer extending into the thigh from the deep surface of the ilio-tibial tract. It is attached to the lateral lip of the linea aspera. From its upper end it passes backwards to the fascia lata, forming a *midline* septum between vastus lateralis and the origin of the hamstrings (Fig. 3.18). Its lower part lies more obliquely and is attached to the ilio-tibial tract. It prevents lateral movement of the ilio-tibial tract, which forms therefore a gutter behind the bulging vastus lateralis (Fig. 3.15).

The Front of the Thigh. When the fascia lata is removed from the front of the thigh the underlying muscles are exposed (Fig. 3.16). The most superficial of all is sartorius, a parallel-sided ribbon of muscle that swings obliquely across the thigh. The adductor longus is prominent on the medial side, and the two muscles enclose, with the inguinal ligament, Scarpa's femoral triangle.

The Femoral Triangle. This is defined as the triangle that lies between the inguinal ligament, sartorius, and the *medial* border of adductor longus. The floor of the triangle is not flat, but gutter-shaped, and the hollow can be seen when the thigh is flexed. The femoral nerve and vessels lie in the gutter. The muscles lying in the floor of the femoral triangle are the iliacus, psoas, pectineus and, in the narrow interval between pectineus and adductor longus, a glimpse is had of adductor brevis with the anterior division of the obturator nerve lying on it. All these muscles pass to the posterior aspect of the femur, hence the gutter shape of the floor of the triangle (Fig. 3.16).

Sartorius arises from the upper centimetre of the anterior border of the ilium, immediately beneath the anterior superior iliac spine. Its parallel fibres extend for the whole length of the muscle. It spirals obliquely across the thigh, passes downwards on the fascial roof of the adductor canal, lies near the posterior aspect of the medial condyle of the femur, whence its tendon proceeds to be inserted into the upper end of the subcutaneous surface of the tibia, in front of gracilis and semitendinosus tendons and separated from them by a bursa (PLATE 6, p. *iii*). It is pierced usually by two nerves, the intermediate cutaneous nerve of the thigh below its origin and the infrapatellar branch of the saphenous nerve above its insertion. A branch from the anterior division of the femoral nerve (L2, 3 and 4) supplies it and its action is to draw the lower limb into the sitting tailor's position (thigh flexed and laterally rotated, knee flexed). It is difficult to understand the need for this muscle in man, since much stronger thigh and knee flexors exist. Perhaps its purpose is to act as a guy rope on the pelvis (see p. 166).

The **adductor longus** arises from a circular area in the angle between pubic crest and symphysis by a strong round tendon, sometimes ossified ('rider's bone'). The muscle rapidly becomes fleshy and flattens out to be inserted by an aponeurotic flat tendon into the lower

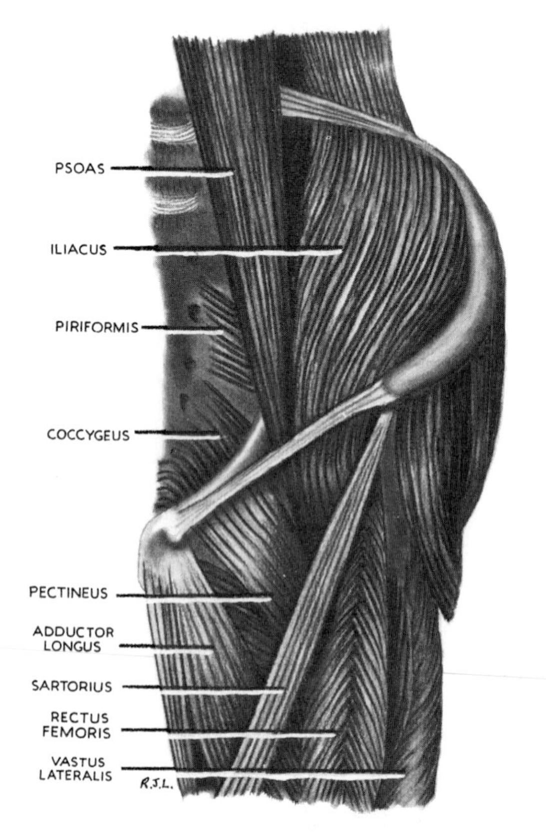

FIG. 3.16. THE MUSCLES OF THE LEFT FEMORAL TRIANGLE. Drawn from Specimen S 105 in R.C.S. Museum.

two-thirds of the linea aspera (middle third of shaft of femur). It is supplied by the anterior division of the obturator nerve (L2, 3). It is a powerful adductor and *medial* rotator of the extended femur (p. 151).

Iliacus. The muscle arises from the hollow of the iliac fossa up to the inner lip of the iliac crest and encroaches across the sacro-iliac joint to arise also from the anterior sacro-iliac ligament, where it overlies the ala of the sacrum. It is triangular in shape; its fibres converge medially towards the lateral margin of the psoas and pass out of the iliac fossa bulging beneath the lateral part of the inguinal ligament, where the femoral nerve lies on it, and curves backwards behind the femoral vessels to be inserted into the front of the psoas tendon and a small area of femoral shaft just below the lesser trochanter. The lower part of the iliac fossa is bare of muscle, the concave gap being occupied by a bursa the size of the hollow of the palm. This is called the *iliac bursa*, and it may communicate through a gap in the capsule with the cavity of the hip joint. It was formerly misnamed the psoas bursa. The muscle is supplied in the iliac fossa (Fig. 3.8) from the femoral nerve (L2, 3) and, with psoas, it is a powerful flexor of the hip joint and *medial* rotator of the extended femur (p. 151).

Psoas. The muscle arises from the lumbar spine (see p. 306). It passes into the thigh below the middle of the inguinal ligament, at which level the femoral artery lies on it. Its rounded tendon is inserted into the lesser trochanter. The *iliac bursa* extends deep to psoas and passes forwards somewhat between the two muscles, whose unequal lengths cause intermuscular gliding here. Both iliacus and psoas pass across the front of the capsule of the hip joint, with the lower part of the bursa lying between. The bursa may communicate with the hip joint through a gap in the capsule that lies between the ilio-femoral ligament of Bigelow and the pubo-femoral ligament (Fig. 3.30, p. 150). The psoas is supplied segmentally from lumbar nerves (chiefly L2 and 3) and with iliacus it is a powerful flexor and *medial* rotator of the femur (p. 151).

'Ilio-psoas' Muscle. The common insertion of the two muscles and the similarities of their actions result in the combined mass being often referred to as the 'ilio-psoas', but the term is inaccurate and has nothing to recommend it except economy of expression. Psoas is supplied segmentally by lumbar nerves (2, 3). It is a body-wall muscle, part of the prevertebral rectus, that has migrated down, for functional reasons, to become attached to the femur. Iliacus is supplied by the femoral nerve (L2, 3). It is a thigh muscle that has migrated up to the iliac fossa for functional reasons. Both psoas and iliacus share a common action on the hip joint, but psoas flexes also the lumbar vertebræ, while iliacus does not. The term 'ilio-psoas' is permissible only when referring to the movements produced at the hip joint (flexion and medial rotation).

Iliac and Psoas Fasciæ. In the abdomen each muscle is covered with strong fascia—this is for attachment of the peritoneum. Apart from the prolongation at the posterior wall of the femoral sheath, these fasciæ are attached to the inguinal ligament. There is no peritoneum below this level, and thus no need for a fascia over the muscles in their course through the femoral triangle.

Pectineus. This quadrilateral muscle arises from the pectineal line of the pubis and from a narrow area of bone below (Fig. 3.24). It slopes backwards down to the upper end of the femoral shaft, where it is inserted into a vertical line between the spiral line and gluteal crest, below the lesser trochanter (Fig. 3.27). The muscle is covered anteriorly by an infolding of fascia lata that passes beneath the falciform margin of the saphenous opening (B in Fig. 3.13) and is continued below as a thin sheet of areolar tissue, the *medial intermuscular septum*, which is attached to the linea aspera and separates the vastus medialis from the adductors. The femoral vein and femoral canal lie on the surface of the muscle, while adductor brevis and the anterior division of the obturator nerve lie behind it (PLATE 5). The muscle and its relations are worthy of note; it is the key to the femoral triangle. It is supplied from the anterior division of the femoral nerve (L2, 3) by a branch that passes behind the femoral sheath. Occasionally it receives a twig from the obturator nerve (L2, 3), indicating its double origin in such cases from flexor and extensor compartments (p. 25). Its action is to flex, adduct and *medially* rotate the femur (p. 151).

The Femoral Sheath. The extraperitoneal areolar tissue of the abdominal and pelvic cavities varies very much in density. Over expansile parts it is loose and cellular (e.g., transversalis fascia on the anterior abdominal wall) while over non-expansile parts it is often condensed into a thick tough membrane (e.g., fascia iliaca). The aorta and its branches and the inferior vena cava and its tributaries lie within this fascial envelope, while the spinal nerves emerge from the intervertebral foramina behind it. Thus the vessels that emerge from the abdominal cavity into the thigh must pierce the fascial envelope, while the femoral nerve does not do so.

The femoral vessels, passing beneath the inguinal ligament, draw around themselves a funnel-shaped prolongation of the extraperitoneal fascia derived from the transversalis fascia in front and the psoas fascia behind (Fig. 3.14). This prolongation of fascia fuses with the adventitia of the artery and vein about an inch below the inguinal ligament; it is called the femoral sheath. The presence of the sheath allows freedom for the femoral vessels to glide in and out beneath the inguinal ligament during movements of the hip joint. The sheath

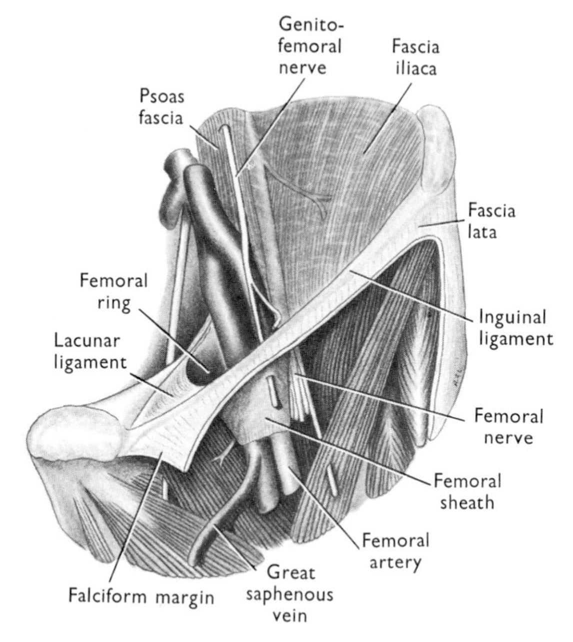

FIG. 3.17. THE FEMORAL SHEATH AND THE FEMORAL RING (LEFT SIDE). Sketched from a dissection.

does not protrude into the thigh in the fœtal position of full flexion, but is drawn down below the inguinal ligament in the adult position of extension of the thigh (Fig. 3.17).

The femoral nerve crosses the iliac fossa lying in the gutter between psoas and iliacus, behind the fascia iliaca. Lateral to the femoral sheath the transversalis fascia and fascia iliaca fuse with the inguinal ligament, and the femoral nerve thus enters the thigh outside the femoral sheath. The sheath itself is separated from the medial side of the femoral vein by a space, the **femoral canal,** that serves two purposes. Firstly, it is the route by which the efferent lymph vessels from the deep inguinal nodes pass to the abdomen and secondly it provides a loose 'dead space' into which the femoral vein can expand during times of increased venous return from the lower limb. The femoral canal is widest at its abdominal end, where its opening is known as the **femoral ring,** which has four boundaries. Anteriorly lies the medial part of the inguinal ligament, medially the crescentic edge of the lacunar ligament, posteriorly the pectineal line of the horizontal ramus of the pubic bone, covered by an expansion from the conjoint tendon that fuses with the periosteum and is called the pectineal ligament (or ligament of Astley Cooper), while laterally lies the femoral vein. The canal contains a lymph node (of Cloquet) which in the female is said to drain directly from the clitoris, and in the male from the glans penis.

The anterior wall of the femoral sheath is pierced by the femoral branch (L1) of the genito-femoral nerve, which runs on the anterior surface of the external iliac artery into the sheath (Fig. 3.17).

The Femoral Artery. The artery enters the thigh at a point midway between the anterior superior iliac spine and the symphysis pubis, just medial to the deep inguinal ring, which lies above the midpoint of the inguinal ligament. Here it lies on the psoas major tendon. It emerges from the femoral sheath and courses downwards to disappear beneath sartorius where the muscle crosses the lateral border of adductor longus. It has four small branches below the inguinal ligament (p. 132) and just below the termination of the femoral sheath gives off a large branch, the profunda femoris, the chief artery of the thigh.

Profunda Femoris. The artery arises from the lateral side of the femoral artery and then spirals deep to it and passes downwards deep to adductor longus, whose upper border thus separates the femoral and profunda femoris arteries in a similar way to that in which the upper border of adductor brevis separates the anterior and posterior divisions of the obturator nerve.

Branches. The **medial circumflex artery** arises from the medial side of the profunda (occasionally from the femoral artery higher up). It encircles the medial side of the femur, keeping as *close to the bone* as neighbouring muscles will allow. It will be seen passing posteriorly between pectineus and psoas, and its further course will be considered later (p. 144).

The **lateral circumflex artery** arises from the lateral side of the profunda artery or separately from the femoral artery, and passes laterally, keeping as *near the femur* as neighbouring muscles will allow. It passes between the many branches of the femoral nerve, dividing them for descriptive purposes into anterior and posterior divisions, and disappears from the femoral triangle beneath sartorius to lie deep to rectus femoris. Here it breaks up into three branches.

The *ascending branch* runs up on the vastus lateralis, gives a branch to the trochanteric anastomosis and passes on towards the anterior superior iliac spine, where it ends by anastomosing with the superficial and deep circumflex iliac, an iliac branch of the ilio-lumbar and the superior branch of the superior gluteal artery.

The *transverse branch* passes across the vastus lateralis and winds around the femur to form one limb of the cruciate anastomosis (p. 149).

The *descending branch* slopes steeply downwards, with the nerve to vastus lateralis, in a groove between the anterior edge of the vastus lateralis and the vastus intermedius. It supplies both muscles and ends by sending twigs to the anastomosis around the lower end of the femur.

The profunda femoris artery gives off four perforating branches, the fourth being the termination of the vessel. The upper two arise in the femoral triangle and pass immediately posteriorly, the first above adductor brevis, and the second through it (p. 144).

The Femoral Vein. The vein enters the lower angle of the femoral triangle, where it lies posterior to the artery. In its course through the femoral triangle it ascends posteriorly and comes to lie on the medial side of the artery at the lower limit of the femoral sheath. It receives a tributary corresponding to the profunda femoris artery and just below the femoral sheath the great saphenous vein joins its medial side (Fig. 3.17). Within the sheath it passes under the inguinal ligament and runs along the brim of the pelvis as the external iliac vein. It has valves just above the junctions with the profunda and great saphenous veins.

The Femoral Nerve. This is the nerve of the extensor compartment of the thigh, and is formed from the *posterior* divisions of the anterior primary rami of the lumbar nerves 2, 3 and 4, the same segments as the obturator nerve, but the adductor muscles are derived from the flexor muscles of the thigh, so their nerve is derived from *anterior* divisions of the nerves (p. 24). Lying in the iliac fossa between psoas and iliacus muscles the femoral nerve enters the thigh by passing deep to the inguinal ligament at the lateral edge of the femoral sheath, which separates it from the femoral artery. After an inch or so it breaks up into a number of branches, through which passes the lateral circumflex artery, separating the branches into superficial and deep (Fig. 3.8).

The **superficial division** gives off two cutaneous and two muscular branches. The cutaneous branches are the intermediate and medial cutaneous nerves of the thigh. The muscular branches are given off to sartorius and pectineus. The former enters the upper part of sartorius, the latter passes medially behind the femoral sheath to enter the anterior surface of the pectineus (the pectineus is often supplied also on its deep surface by a branch from the obturator nerve).

The **deep division** supplies muscular branches to the quadriceps muscle and one cutaneous branch, the saphenous nerve, to the skin of the medial side of the leg and foot (PLATE 5).

1. The *nerve to rectus femoris* is usually double; the upper nerve gives a proprioceptive branch to the hip joint (Hilton's law).

2. The *nerve to vastus medialis* is the largest of the muscular branches, equalling in size the saphenous nerve itself. It passes down on the lateral side of the femoral artery and at a point just below the apex of the femoral triangle, i.e., in the upper part of the adductor canal, it sinks into the vastus medialis, which it supplies, while much of the nerve continues downwards to supply the capsule of the knee joint. Of all the muscular branches of the femoral nerve this branch to the vastus medialis contains most proprioceptive fibres to the knee joint, a fact which accounts for its large size. The nerve is often double.

3. The *nerve to vastus lateralis* slopes steeply downwards with the descending branch of the lateral femoral circumflex artery and enters the anterior border of vastus lateralis.

4. The *nerve to vastus intermedius* passes deeply to enter the upper fleshy part of its anterior surface.

5. The **saphenous nerve** leaves the femoral triangle at its lower angle and in the subsartorial canal passes across the front of the femoral artery to reach its medial side. It gives twigs to the subsartorial plexus and leaves the canal by passing beneath the posterior border of sartorius (PLATE 5).

Quadriceps Femoris. This muscle is the extensor of the knee joint. It arises by four heads that are inserted into the tuberosity of the tibia. A sesamoid bone in the tendon is called the patella (or knee-cap) and the tendon above the bone is named the quadriceps tendon while that part below the bone is known as the ligamentum patellæ (PLATE 5).

The quadriceps is the biggest muscle in the body. Its four parts are the rectus femoris and the three vasti—lateralis, intermedius and medius.

The **rectus femoris** arises from the ilium by two heads (Fig. 3.29). The *reflected head* arises from a shallow concavity above the acetabulum; it is the primary head, present in quadrupeds. The *straight head* arises from the upper half of the anterior inferior spine, just above the ilio-femoral ligament; it is a secondary attachment of the muscle associated with the erect posture of man, and is not present in quadrupeds. The rectus femoris is a spindle-shaped muscle, bipenniform. Above the patella it flattens to form the anterior lamina of the quadriceps tendon. The posterior surface of the muscle is clad in a thick glistening aponeurosis that glides on the anterior surface of vastus intermedius. The anterior surface of the muscle is fleshy except above, where it is covered with a similar aponeurosis—here the muscle bears on a fibro-fatty pad beneath the upper end of sartorius. The edges of the muscle belly itself are free but the medial border of the tendon receives an insertion of vastus medialis.

The **vastus lateralis** has an extensive linear origin from the upper half of the intertrochanteric line, the lateral lip of the linea aspera and the upper two-thirds of the lateral supracondylar line of the femur. It also arises from the lateral intermuscular septum, a strong membrane that lies in the *posterior midline* of the upper thigh (Fig. 3.18). The muscle is largest here and tapers off as it spirals down to be received into the quadriceps tendon (Fig. 3.32, p. 153). Its deep surface is clad in a glistening sheet of aponeurosis equally as strong as the fascia lata itself; the sheet glides on the underlying vastus intermedius. The anterior edge of the muscle lies free

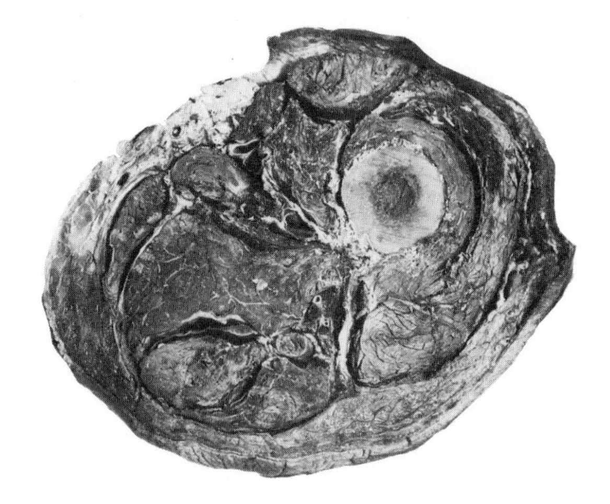

FIG. 3.18. CROSS SECTION THROUGH THE UPPER THIRD OF THE THIGH (Photograph).

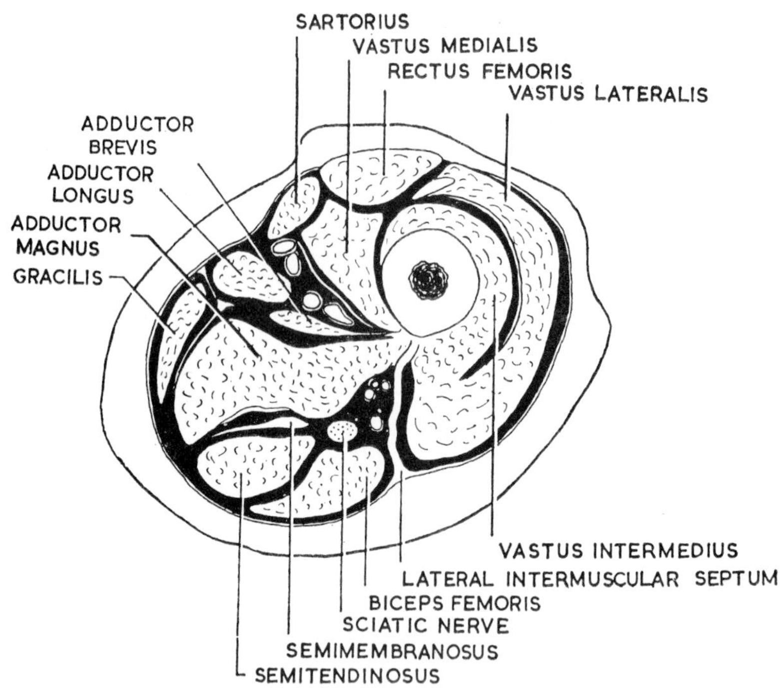

FIG. 3.19. KEY TO FIG. 3.18.

on vastus intermedius; the descending branch of the lateral circumflex artery and the nerve to vastus lateralis lie in the shallow gutter between the two.

The **vastus intermedius** arises from the greater trochanter and from the anterior and lateral surfaces of the upper two-thirds of the shaft of the femur—the medial surface of the femoral shaft is bare bone. The anterior surface of the muscle is covered by an aponeurosis which is continued down into the quadriceps tendon.

The **articularis genu** is a flat ribbon of muscle which arises beneath the vastus intermedius and is inserted into the upper convexity of the suprapatellar bursa (Fig. 3.39, p. 160).

The **vastus medialis** arises from the spiral line and the medial lip of the linea aspera and from the tendon of adductor magnus below the hiatus for the femoral vessels. The muscle slopes around the medial surface of the femur. Its anterior border is free above, lying on vastus intermedius; below this it is inserted into the aponeurosis of vastus intermedius, and lower still into the tendon of rectus femoris (PLATE 5), while the lowest fibres of all, lying nearly horizontal, are inserted directly into the medial border of the patella. These fibres are indispensable to the stability of the patella (Fig. 3.23).

The **quadriceps tendon** is tri-laminar. The aponeurosis of the insertion of vastus lateralis lies sandwiched between that of vastus intermedius and rectus femoris;

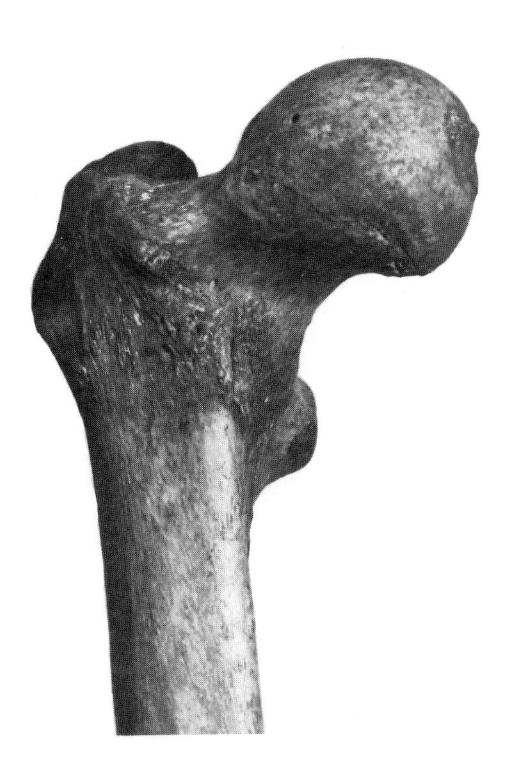

FIG. 3.20. THE UPPER END OF THE RIGHT FEMUR (ANTERIOR VIEW).

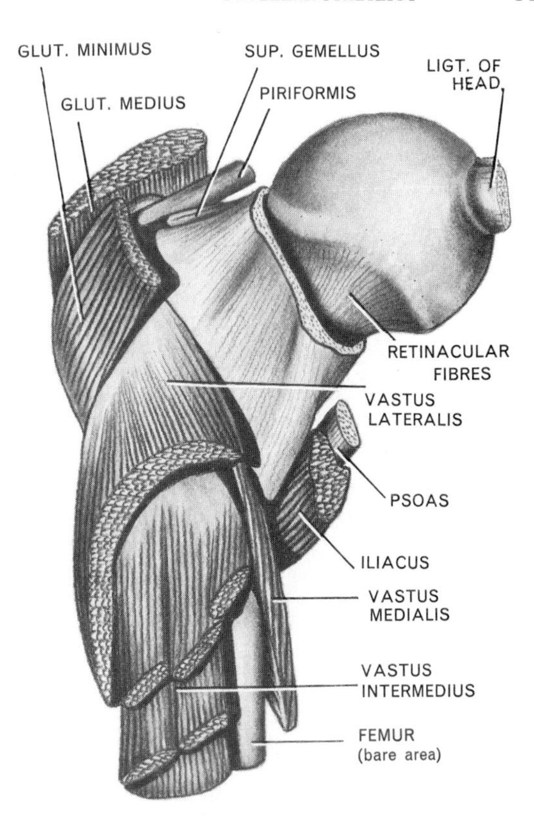

FIG. 3.21. THE ATTACHMENTS AT THE UPPER END OF THE RIGHT FEMUR.

the three laminæ are fused together by mutual interchange of fibres. A thin sheet passes across the front of the patella into the ligamentum patellæ and the retinacula.

The **ligamentum patellæ** connects the lower border of the patella with the smooth convexity on the tuberosity of the tibia (Fig. 3.46, p. 166). The *patellar retinacula* are fibrous expansions from the quadriceps which connect the sides of the patella with the lower margins of the condyles of the tibia. They are attached also to the sides of the ligamentum patellæ.

Nerve Supply. All four heads are supplied by the posterior division of the femoral nerve.

Action. The muscle is the extensor of the knee joint, making with the buttock and calf muscles the great propulsive mass for walking, running, jumping, etc. The attachment of rectus femoris to the pelvic girdle has a stabilizing effect on the hip joint and can assist ilio-psoas to flex the hip. In propulsive movements hip extension elongates the rectus femoris, which thus contracts more powerfully in the simultaneous extension of the knee.

Stability of the Patella. The patella is a sesamoid bone. It is mobile from side to side. The ligamentum patellæ is vertical. The pull of the quadriceps is oblique, in the line of the shaft of the femur; when the muscle contracts it tends to draw the patella laterally. Three factors discourage this lateral dislocation; they are the usual bony, ligamentous and muscular factors that control the stability of any bone. The bony factor consists in the forward prominence of the lateral condyle of the femur (Fig. 3.22), the ligamentous factor is the tension of the medial patellar retinaculum, but they are in themselves incapable of preventing lateral displacement of the patella. The lowest fibres of vastus medialis, inserted into the border of the bone, hold the patella

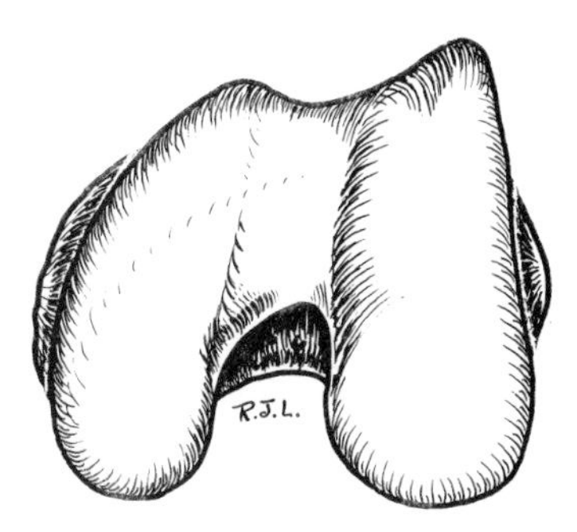

FIG. 3.22. THE LOWER END OF THE LEFT FEMUR, TO SHOW THE ASYMMETRY OF THE CONDYLES. The prominence of the lateral condyle discourages lateral displacement of the patella.

medially when the quadriceps contracts (Fig. 3.23). These fibres of vastus medialis are indispensable to the stability of the patella. They waste very rapidly after effusion into the knee joint; the patella is then carried

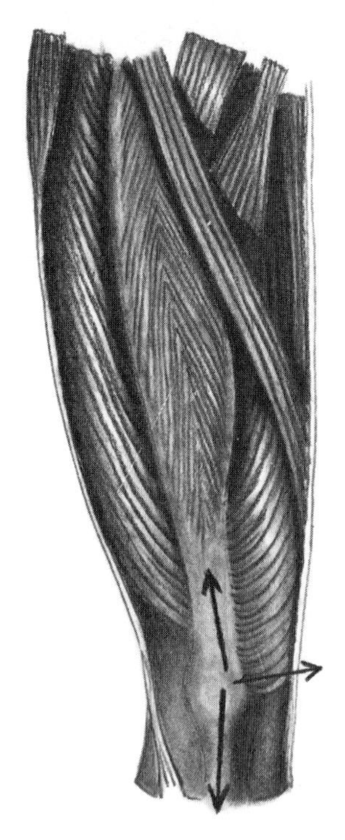

FIG. 3.23. THE PATELLA STABILIZED BY MUSCLE BALANCE.

laterally by contraction of the quadriceps, and proprioceptive impulses from the joint are interpreted as an unstable knee.

The Adductor Canal. The canal is known also as Hunter's canal or the subsartorial canal. There is a gutter-shaped groove between vastus medialis and the front of the adductor muscles, below the apex of the femoral triangle. The gutter is roofed in by a fascia which contains in its meshes the subsartorial plexus. The canal so formed contains the femoral artery and vein, the saphenous nerve and, in the upper part, the nerve to vastus medialis. Sartorius lies on the fascial roof. The adductors in the floor of the canal are the adductor longus above and the adductor magnus below. The **subsartorial plexus** receives small branches from the intermediate cutaneous nerve of the thigh, the saphenous nerve and the anterior division of the obturator nerve. The plexus supplies the overlying fascia lata and an area of skin above the medial side of the knee. The femoral vessels leave the canal by passing into the popliteal fossa through the hiatus between the hamstring and adductor parts of the adductor magnus muscle. The vein spirals gradually around the artery. At the adductor magnus hiatus the vein is lateral (i.e., next to the femur) but ascends posteriorly until in the femoral triangle it lies medial to the artery. This is in keeping with the rotation medially of the lower limb from the fœtal position, and is further reflected in the spiral manner in which the saphenous nerve passes across the femoral artery. *At all levels in the thigh the artery lies between saphenous nerve and femoral vein* (PLATE 5). Below the hiatus in the adductor magnus the canal is occupied by the saphenous nerve and the *descending genicular artery* (anastomotica magna). This branch leaves the femoral artery just above the hiatus and, passing downwards, divides into a superficial saphenous artery that accompanies the saphenous nerve and a deep muscular branch that enters the vastus medialis and joins the arterial anastomosis around the knee (Fig. 3.35, p. 157).

The saphenous nerve passes out of the canal by escaping from beneath the posterior border of the sartorius (in front of the gracilis) whence it passes downwards with the great saphenous vein. Just before leaving the canal the **infrapatellar branch** is given off; this nerve pierces the sartorius muscle and joins the patellar plexus (PLATE 5).

THE ADDUCTOR COMPARTMENT

The contents of this compartment of the thigh are separated from the extensor compartment by the medial intermuscular septum, but there is no septum dividing them from the flexor (hamstring) compartment. The muscles consist of the gracilis and the three adductors, longus, brevis, and magnus; while deeply lies the obturator externus. The nerve of the compartment is the obturator, and the artery is the profunda femoris.

The **medial intermuscular septum** lies deep to the femoral vessels and is a downward continuation of the fascia on the pectineus, i.e., of the fascia lata in the floor of the saphenous opening. It is quite thin. It is attached to the linea aspera, and passes over the anterior surfaces of pectineus, adductor longus and the lower part of adductor magnus in the floor of the adductor (Hunter's) canal. Its medial edge is attached to the fascia lata. It is a thin, cellular layer, amounting to little more than the areolar tissue found on the surface of any muscle; this is in marked contrast with the dense lateral septum (Fig. 3.18).

Gracilis Muscle. This muscle arises as a flat sheet from the edge of the inferior ramus of the pubis, just beneath the fascia lata. Its origin extends inferiorly to encroach a little on the ischial ramus, but it is supplied by the obturator, the nerve of the pubis. The sheet of muscle narrows in triangular fashion and in the lower third of the thigh is replaced by a cylindrical tendon which is inserted into the subcutaneous surface of the shaft of the tibia just behind sartorius, from whose tendon it is separated by a bursa. With the hamstring muscles it is a flexor and a medial rotator of the flexed knee joint and with the other adductors it is an adductor of the hip joint. It does not appear to be an important muscle and the reason for its existence is not very clear (see p. 166). Older anatomists gave the pair of muscles the whimsical name of 'custodes virginitatis'.

Adductor Longus (see p. 136).

Adductor Brevis. The muscle arises from the body and inferior ramus of the pubic bone, deep to pectineus and adductor longus. It widens in triangular fashion to be inserted into the upper part of the linea aspera immediately lateral to the insertion of pectineus and above that of adductor longus. The anterior division of the obturator nerve supplies it and passes vertically downwards on its anterior surface; the posterior division of the obturator nerve passes down behind it. The oblique upper border of adductor brevis thus lies between the two divisions of the obturator nerve in the same way as the upper border of adductor longus lies between the femoral and the profunda femoris vessels.

Adductor Magnus. This is really a composite muscle formed by fusion of two muscle masses, a true adductor muscle and a hamstring muscle; hence the absence of a septum between the adductor and hamstring compartments of the thigh. The adductor mass is supplied by the obturator nerve and the hamstring mass by the sciatic nerve (the tibial component, the nerve of the flexor compartment) and the hiatus for the passage of the femoral artery and vein lies between the two. Trace the origin on the os innominatum from behind forwards. Observe that the ischial tuberosity is divided by a transverse ridge into an upper smooth and a lower rough part. The upper smooth part gives origin to the hamstring muscles (Fig. 3.31). The lower rough part is subdivided by a longitudinal crest into two surfaces that slope upwards away from each other. The medial of these receives the sacro-tuberous ligament and is bounded above by a crescentic ridge. The lateral surface of the tuberosity gives origin to the fibres of adductor magnus that pass vertically down the thigh to the *adductor tubercle.* These are the true hamstring components of the muscle and are supplied by the sciatic nerve. In many vertebrates they are inserted into the tibia, the muscle then being known as the 'presemimembranosus'. In man the distal part of the tendon has 'degenerated' into the tibial collateral ligament. Returning to the os innominatum the origin of the adductor magnus can be traced in continuity along the

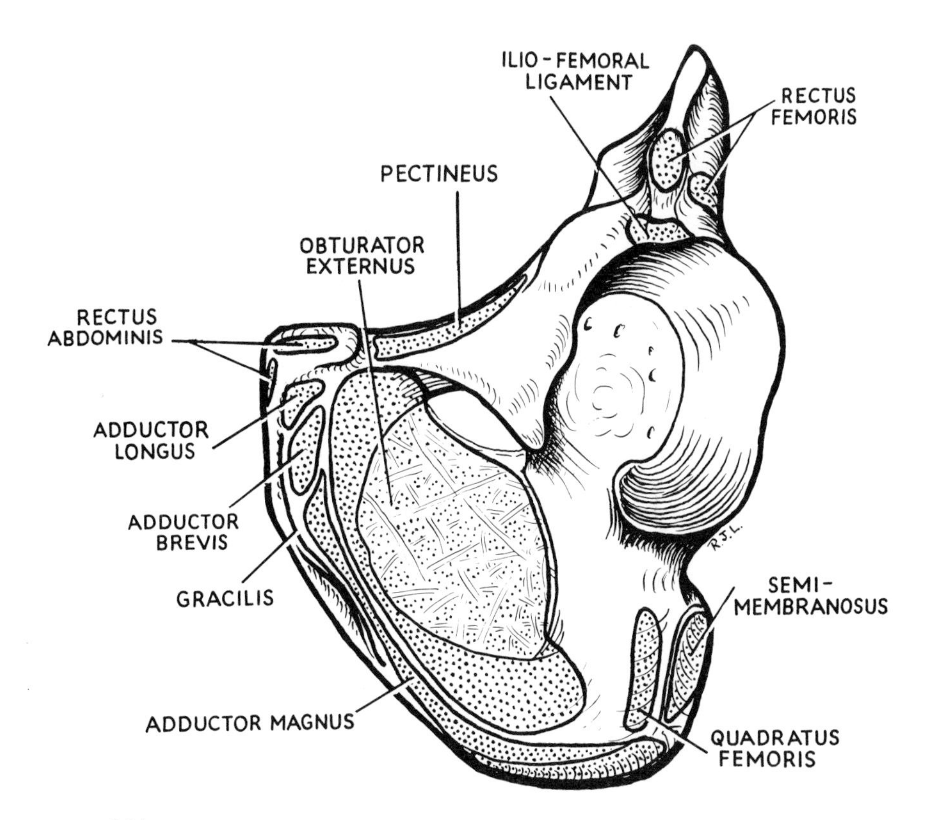

FIG. 3.24. THE ADDUCTOR SURFACE OF THE LEFT HIP BONE, TO SHOW THE MUSCLE ATTACHMENTS.

ischio-pubic ramus between the obturator foramen (whose edge gives origin to obturator externus) and the origin of adductor brevis (Fig. 3.24). As one passes progressively forwards along the bony origin, one should note that the fibres therefrom are inserted progressively higher along the *linea aspera* and *gluteal crest* of the femur. The upper border of the muscle lies horizontal, edge to edge with the lower border of quadratus femoris, and the horizontal branch of the medial circumflex artery passes between them to reach the cruciate anastomosis. This is on a level which bisects the lesser trochanter (Fig. 3.33, p. 155).

Actions of the Adductor Muscles. All muscles of the adductor group adduct the thigh towards the midline of the trunk. The more anteriorly placed fibres (e.g., all of adductor longus and part of adductor brevis) are also medial rotators of the extended thigh (see p. 151).

Nerve Supply of Adductors. Except for the 'pre-semimembranosus' part of adductor magnus, all the adductor group is supplied by the obturator nerve; obturator externus and adductor magnus by the posterior division, the remainder by the anterior division. The segments concerned are chiefly L2 and 3.

Blood Supply of the Adductors. The muscles of the adductor compartment, like those of the rest of the thigh, are supplied by the profunda femoris artery and its branches. The obturator artery assists higher up.

Profunda Femoris Artery. The artery enters the adductor compartment by spiralling down from the lateral side of the femoral artery to pass behind the upper border of adductor longus. It has four perforating branches, the fourth being the terminal branch. It gives rise, near its origin, to the lateral and medial circumflex arteries.

Medial Circumflex Artery. Like the lateral circumflex, this artery circles around the femoral shaft keeping as close to the bone as neighbouring muscles will permit. It arises in the femoral triangle and passes posteriorly, near the femur, between pectineus and the psoas tendon. Running above the upper border of adductor brevis it then meets the contiguous borders of adductor magnus and quadratus femoris. It supplies branches to all these muscles and ends by dividing on the quadratus femoris into an ascending and a horizontal branch, which follow the upper and lower borders of quadratus femoris. The *ascending branch*, passing along the upper border of quadratus femoris, follows the obturator externus to the trochanteric anastomosis, while the *horizontal branch* passes back between quadratus femoris and adductor magnus to form the medial limb of the cruciate anastomosis.

The four **perforating arteries** pass backwards through the adductor muscles, the first above and the fourth below the adductor brevis, and the second and third through it. They supply the adductor muscles and the hamstrings, and *end in the vastus lateralis muscle*. The terminal twigs make a series of anastomoses between each other, with the cruciate anastomosis above, and the popliteal artery below. These communications along the back of the thigh are by terminal arterioles (see p. 18).

Obturator Externus. The muscle arises from the whole of the obturator membrane and from the anterior bony margin of the obturator foramen. Both membrane and muscle fall short of the obturator notch above, thereby forming a short canal for the passage of the obturator nerve and vessels (Fig. 3.24). The muscle passes laterally and posteriorly beneath the neck of the femur where it narrows into a tendon that spirals in contact with the back of the femoral neck to be inserted on the *medial* surface of the great trochanter into the deep pit called the trochanteric fossa. The capsule of the hip joint extends along the neck of the femur only as far as the place where obturator externus tendon is in contact with periosteum, namely half the neck of the femur, whereas in front the capsule of the hip joint includes the whole of the neck of the femur (Figs. 3.27, 3.33).

The muscle is supplied by the obturator artery and the posterior division of the obturator nerve. Its action is, with the other short muscles around the hip joint, to stabilize and support the proximal part of the limb. Its line of pull passes behind the hip joint; acting as a prime mover it is a lateral rotator of the femur.

The **obturator artery,** on emerging from the obturator foramen, divides into medial and lateral branches that encircle the origin of obturator externus and anastomose with each other and with the medial circumflex artery. From the lateral branch the *articular* twig to the hip joint arises; it enters the transverse notch and runs in the ligament of the head of the femur to supply a small scale of bone in the region of the fovea capitis femoris (see p. 150).

The **obturator nerve** divides in the obturator notch into anterior and posterior divisions. The former passes above the obturator externus. The latter passes through the upper border of the muscle, giving off a branch to supply it before doing so.

The **anterior division,** giving an articular branch to the hip joint, descends in the thigh behind the adductor longus, which it supplies. Passing over the anterior surface of adductor brevis it supplies that muscle and gracilis and ends in the subsartorial plexus, whence branches supply the skin over the medial side of the thigh. Direct branches to the skin are often given off at a level above the subsartorial plexus (Fig. 3.7, p. 131).

The **posterior division** emerges through obturator externus (having already supplied that muscle), and passes vertically downwards on adductor magnus deep to the other adductor muscles. It supplies adductor magnus and gives off a fine terminal branch which runs with the femoral artery through the hiatus in the muscle to the popliteal fossa and supplies the capsule of the knee joint by passing in with the middle genicular artery (Hilton's law, p. 15). Note that the anterior and posterior divisions straddle the adductor brevis muscle in a manner similar to that in which the femoral and profunda femoris vessels straddle adductor longus.

THE GLUTEAL REGION

The buttock lies behind the pelvis, from the walls of which muscles, nerves and vessels emerge into the lower limb. These emerging structures are covered by gluteus maximus.

Subcutaneous Tissue. The panniculus adiposus is well developed in the gluteal region and gives to the buttock its characteristic convexity. The fold of the buttock is the transverse skin crease for the hip joint and is not caused by the oblique lower border of the gluteus maximus. The *blood supply* of the skin and fat is derived from perforating branches of the superior and inferior gluteal arteries, and the lymphatic drainage is into the lateral group of the superficial inguinal lymph nodes. The *cutaneous nerves* of the buttock are derived from posterior and anterior primary rami. The posterior primary rami of the upper three lumbar nerves slope downwards over the iliac crest to supply the upper skin of the buttock. The posterior primary rami of all five sacral nerves are cutaneous. The upper three supply the skin of the natal cleft, the lower two, with the coccygeal nerve, supply skin over the coccyx (Fig. 1.25, p. 23).

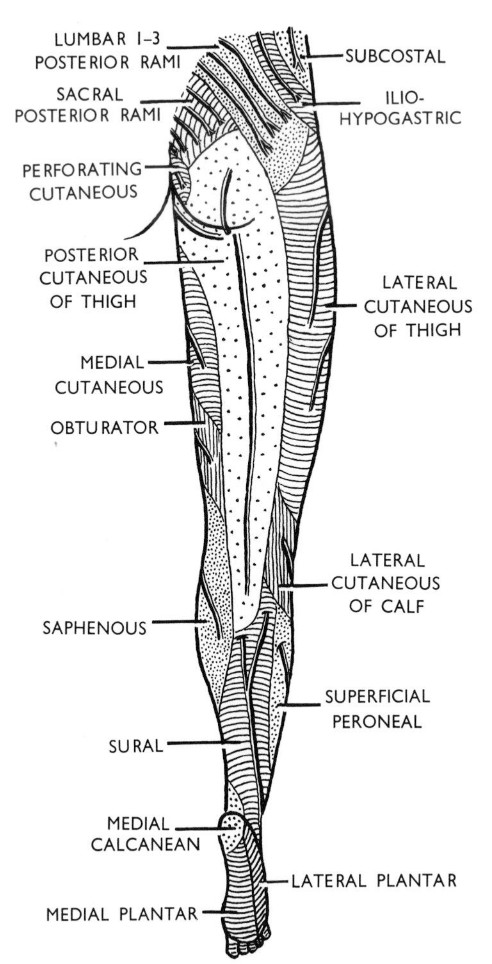

FIG. 3.25. THE CUTANEOUS NERVES OF THE BACK OF THE RIGHT LOWER LIMB. The medial cutaneous nerve of the thigh may be replaced here by the obturator nerve as in Fig. 3.7.

The anterior primary rami are derived from widely separated segments. The upper part of the lateral skin is supplied by lateral cutaneous branches of the subcostal and ilio-hypogastric nerves (Th. 12 and L1), the lower part by branches of the lateral cutaneous nerve of the thigh (L2). The perforating cutaneous nerve and branches of the posterior cutaneous nerve of the thigh supply the lower central part of the buttock. The former is derived from two branches of the sacral plexus (S2 and 3) and the nerve passes backwards to pierce the sacro-tuberous ligament and the lower part of gluteus maximus to reach the deep fascia and skin. Thus the segments between L2 and S2 are not represented in the skin of the buttock; their dermatomes lie peripherally, providing the skin of the limb (Fig. 1.32, p. 27). The fault so produced is responsible for the posterior axial line. There is a considerable mass of clinical evidence that L5 and S1 in fact supply buttock skin up to the midline of the body; if this really is so, it means that the named cutaneous nerves of the buttock carry fibres from more segments than has hitherto been suspected (see p. 27).

Gluteus Maximus. This is the largest and the most superficial of the gluteal muscles, and it is characterized by the large size of its fibres. A thick flat sheet of muscle, it slopes from the pelvis down across the buttock at 45 degrees. Its oblique lower border takes no part in forming the gluteal fold, which is the horizontal skin crease of the hip joint with bulging fat above it. It arises from the gluteal surface of the ilium behind the posterior gluteal line, from the lumbar fascia, from the lateral mass of the sacrum below the auricular surface, and from the sacro-tuberous ligament. The deep half of its lower half lies behind the greater trochanter and is inserted into the gluteal crest of the femur (Fig. 3.27). The remaining three-fourths of the muscle, enclosed by a splitting of the fascia lata, is inserted into the upper end of the ilio-tibial tract in common with the tensor fasciæ latæ. Its nerve supply is derived from the inferior gluteal nerve alone (L5, S1, 2). The nerve enters its deep surface on the medial side, nearer its origin than its insertion. Its blood supply comes from both superior and inferior gluteal arteries and its veins form a well marked plexus beneath the muscle.

Its action on the femur is a combination of lateral rotation and extension at the hip joint, while through the ilio-tibial tract its contraction supports the extended knee. It can be felt in contraction in standing with hip and knee each slightly flexed, in which case it is a powerful antigravity muscle. In this position the muscle can be felt in ever-increasing contraction as the body bends forwards more and more at the hip. It comes into play as an extensor of the hip joint chiefly at the extremes of hip movement, as in running, climbing stairs, etc., and it is called into play little if at all in the mid-position of the hip joint, as in quiet walking, when the main extensors of the hip are, in fact, the hamstrings. It is the chief antigravity muscle of the hip during the act of sitting down from standing, in which case, by the action of paradox, it is the *flexor* of the hip joint (p. 8).

Gluteus Medius. This muscle can be seen in its entirety only when the gluteus maximus is removed. It arises from the gluteal surface of the ilium between the middle and posterior gluteal lines. Note that the middle gluteal line meets the iliac crest halfway between the tubercle and the anterior superior spine; thus with the os innominatum held in the correct anatomical position it can be seen that the free anterior borders of gluteus medius and gluteus minimus lie edge to edge. The gluteus medius converges in triangular fashion towards the lateral surface of the greater trochanter, into which it is inserted along an oblique line that slopes downwards and forwards from the apex of the trochanter (Fig. 3.27). Its tendon of insertion gives a strong expansion that crosses the capsule of the hip joint to blend with the upper end of the ilio-femoral ligament (Fig. 3.30). A bursa separates the upper part of the lateral surface of the greater trochanter from the tendon.

Gluteus Minimus. The muscle arises under cover of gluteus medius from the gluteal surface of the ilium between the middle and inferior lines, whence its fibres converge to a J-shaped area on the anterior surface of the greater trochanter (Fig. 3.21). Its anterior border lies edge to edge with that of gluteus medius from origin to insertion.

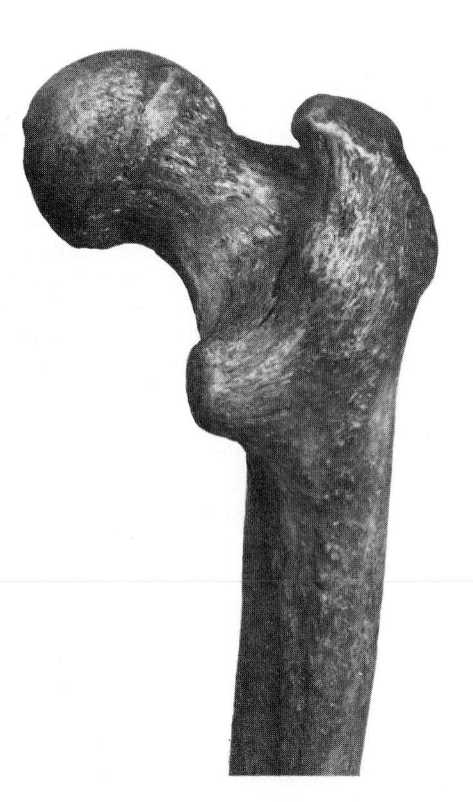

FIG. 3.26. THE UPPER END OF THE RIGHT FEMUR (POSTERIOR VIEW).

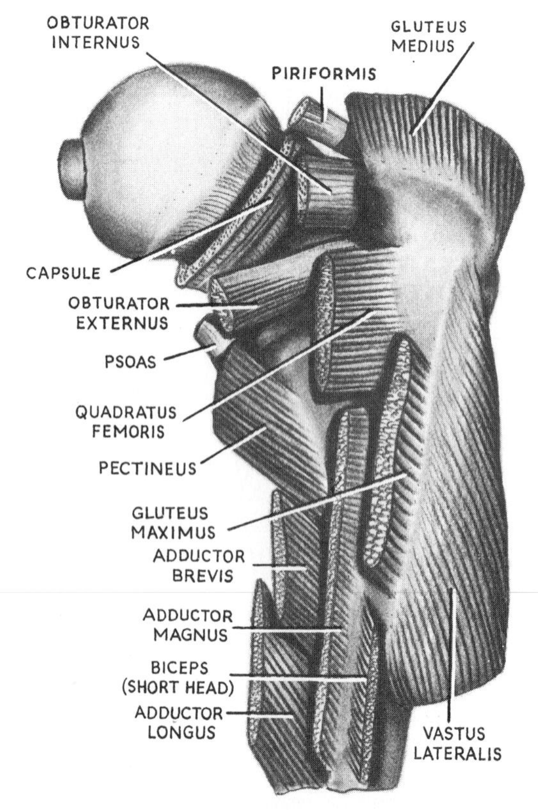

FIG. 3.27. THE ATTACHMENTS AT THE UPPER END OF THE RIGHT FEMUR.

The glutei medius and minimus are supplied by the superior gluteal nerve and artery. Their *action* is to abduct the hip joint. This is not a very common event; the two muscles are constantly called into play, however, to *prevent adduction* at the hip when the weight of the body is supported on one leg. Thus in walking the muscles of the two sides are alternately contracting. They can be palpated above the upper border of gluteus maximus—feel them in walking and note their contraction when their own leg bears weight. If they are paralysed the gait is markedly affected, the trunk swaying from side to side towards the weight-bearing limb to prevent downward tilting of the pelvis on the unsupported side.

The **superior gluteal nerve** (L4, 5, S1) emerges from the great sciatic notch above the upper border of piriformis and immediately disappears beneath the posterior border of gluteus medius (Fig. 3.28). It runs forwards below the middle gluteal line in the space between gluteus medius and gluteus minimus, supplies both, and ends by sinking into the substance of tensor fasciæ latæ. It has no cutaneous distribution.

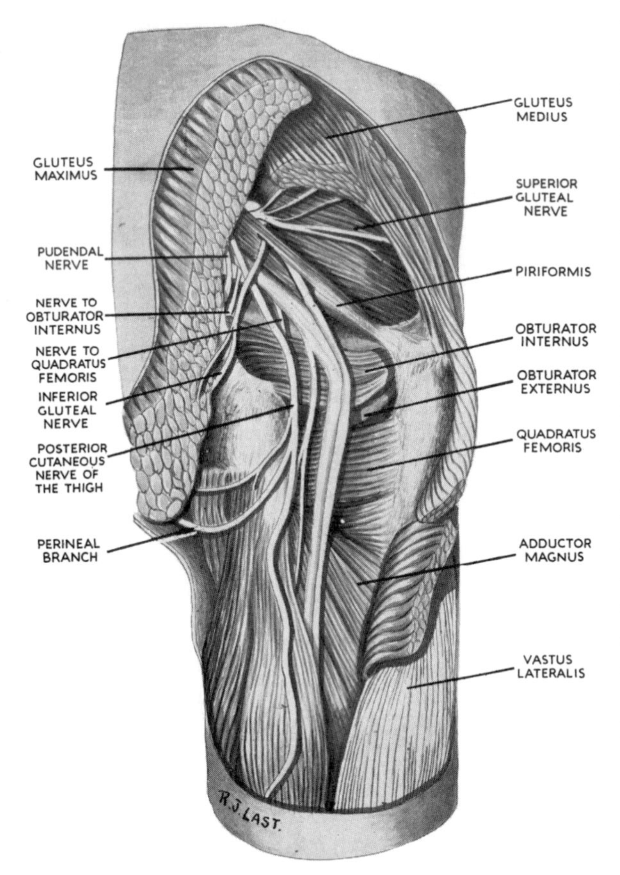

FIG. 3.28. THE NERVES OF THE RIGHT BUTTOCK. In this case there is a high division (i.e., a failure of fusion) of the tibial and common peroneal components of the sciatic nerve, and the inferior gluteal nerve accompanies the common peroneal nerve through the substance of piriformis. Drawn from Specimen S 131 in R.C.S. Museum.

The **superior gluteal artery** emerges from the pelvis above the upper border of piriformis and, unlike the nerve, gives off a *superficial branch* which sinks into the deep surface of the overlying gluteus maximus, which it penetrates to supply the overlying skin. The *deep branch* passes laterally in the space between gluteus medius and gluteus minimus and divides into an upper and a lower branch. The upper reaches the anastomosis at the anterior superior iliac spine, the lower joins the trochanteric anastomosis.

The **piriformis muscle** is important in that its relations provide the key to the understanding of the arrangement of the structures in the gluteal region. The muscle arises from the middle three pieces of the sacrum and passes laterally behind the sacral plexus to emerge through the great sciatic notch, which it almost completely fills. Some additional fibres arise from the upper margin of the notch (Fig. 3.63, p. 190). In the buttock its upper border lies alongside gluteus medius, its lower border alongside the superior gemellus. It converges into a rounded tendon which is inserted into the apex of the greater trochanter (that is, the posterior inturned end of its upper border). Some of its fibres run forwards along the whole length of the upper border of the greater trochanter, outlining with the tendons of gluteus medius and gluteus minimus a triangle of insertion on the greater trochanter (Fig. 3.28).

The *nerve supply* of piriformis is segmental from the anterior primary rami of S1 and 2. The *action* of the muscle is to aid the other short muscles in adjusting and stabilizing the hip joint, especially in abduction.

A number of structures emerge from the pelvis into the gluteal region; to do so they pass above or below the piriformis muscle. Above the upper border emerge the superior gluteal nerve and vessels. Below the lower border emerge the inferior gluteal nerve and vessels, the pudendal nerve and vessels, the nerve to obturator internus, and the sciatic nerve with the posterior cutaneous nerve of the thigh on its surface and the nerve to quadratus femoris deep to it.

The **inferior gluteal nerve** (L5, S1, 2) leaves the pelvis beneath the lower border of piriformis and sinks into the deep surface of gluteus maximus (Fig. 3.28). Its course in the buttock is short. It enters the muscle towards its origin (medial end of muscle). It has no cutaneous distribution.

The **inferior gluteal artery** appears in the buttock between piriformis and the superior gemellus. It breaks up into muscular branches which supply piriformis, obturator internus, and gluteus maximus. It sends anastomotic branches to the trochanteric and cruciate anastomoses. One branch, *the companion artery to the* *sciatic nerve*, is of interest as being the remnant of the original axial artery of the limb in the embryo, resembling in this respect the artery to the median nerve from the anterior interosseus artery. The artery supplies the sciatic nerve and rarely may persist as a very large vessel.

The **pudendal nerve** (S2, 3, 4) makes but a brief appearance in the buttock (Fig. 3.28). On emerging from beneath piriformis the nerve turns forward around the sacro-spinous ligament, on which it lies just medial to the spine of the ischium. It leaves the buttock by passing forward through the space between the sacro-tuberous and sacro-spinous ligaments (the lesser sciatic foramen) (Fig. 5.73, p. 349).

The **internal pudendal artery** follows a similar course to that of the nerve, and lies on its lateral side. It crosses the tip of the ischial spine, against which it can be compressed to control arterial hæmorrhage in the perineum. A companion vein lies on each side of the artery.

The **nerve to obturator internus** (L5, S1, 2) lies still more laterally and loops around the base of the ischial spine and then passes forward to sink into the muscle, deep to its fascia, in the side wall of the ischio-rectal fossa (Figs. 3.28, 5.73, p. 349). It supplies also the superior gemellus.

The **lesser sciatic foramen** lies between the sacro-tuberous and sacro-spinous ligaments, and is bounded laterally by the concave part of the ischium that lies between the ischial spine and ischial tuberosity. The obturator internus muscle, as it emerges from the lateral wall of the pelvis, plugs the lateral part of the foramen and the strong fascia over the muscle (parietal pelvic fascia) is attached to both ligaments and does not project into the buttock. The medial part of the lesser sciatic foramen forms a funnel-shaped orifice which leads forwards into the pudendal canal (of Alcock); the internal pudendal vessels and nerve entering the foramen from the buttock are thus directed into the canal (PLATE 13, p. *vi*).

The **sciatic nerve** (L4, 5, S1, 2, 3) emerges from below the piriformis muscle more laterally than the inferior gluteal and pudendal nerves and vessels (Fig. 3.28). It lies upon the ischium over the posterior part of the acetabulum. It is in contact with bone at a point one-third of the way up from the ischial tuberosity to the posterior superior iliac spine. It is the nerve of the ischium or 'ischiadic' nerve. It passes vertically down over obturator internus and quadratus femoris to the hamstring compartment of the thigh, where it disappears under cover of the biceps femoris. In the buttock it lies under cover of gluteus maximus midway between the greater trochanter and the ischial tuberosity (Fig. 3.33).

Its tibial and common peroneal components usually separate in the upper part of the popliteal fossa but occasionally there is a high division and the two components may leave the pelvis separately, in which case the common peroneal nerve (L4, 5, S1, 2) will be found piercing the lower part of piriformis while the tibial nerve (L4, 5, S1, 2, 3) emerges from beneath the muscle in the ordinary way (Fig. 3.28).

The **posterior cutaneous nerve of the thigh** (S2, 3) emerges from beneath piriformis and in its course in the buttock it lies on the sciatic nerve under cover of gluteus maximus. Below the buttock the posterior cutaneous nerve passes vertically down the midline of the back of the thigh to beyond the level of the knee (Fig. 3.25). It lies *beneath the fascia lata*, superficial to the hamstrings, which separate it from the sciatic nerve. Its **branches** are (*a*) *recurrent*, a branch or two curling around the lower border of gluteus maximus to supply fascia and skin over the prominent convexity of the buttock (Fig. 3.25); (*b*) *perineal*, a long branch which winds medially and forward between gracilis and the fascia lata at the root of the limb to supply the posterior part of the scrotum or labium majus, and (*c*) *perforating*, a series of twigs which emerge to supply a strip of fascia lata and skin down the midline of the limb to the level of the bellies of the gastrocnemius, midway between knee and ankle.

It is significant that the segments (S2, 3) of this nerve are also those of the pelvic parasympathetic nerves which supply the derivatives of the cloaca. Pain from pelvic disease is often referred over the distribution of the posterior cutaneous nerve of the thigh, and such pain along the back of the thigh and calf must be distinguished from sciatica.

The **nerve to quadratus femoris** (L4, 5, S1) lies on the ischium beneath the sciatic nerve (Fig. 3.28). It passes over the back of the hip joint, to which it gives an *articular branch* (Hilton's law). In this situation it lies deep to obturator internus and the gemelli, which thus separate it from the overlying sciatic nerve. It runs downwards to sink into the deep (anterior) surface of quadratus femoris. It supplies also the inferior gemellus.

The **obturator internus** muscle plugs the lateral part of the lesser sciatic foramen and makes a right-angled bend around the lesser sciatic notch of the ischium. Its deep surface is strongly tendinous and its pressure leaves grooves on the fresh bone, which is here covered with fibro-cartilage. The bursa beneath the muscle on the side wall of the pelvis extends backwards to lubricate the muscle as it plays over the lesser sciatic notch. As the muscle emerges into the buttock it is reinforced by additional muscle fibres arising from the margins of the lesser sciatic notch. These are the superior and inferior gemelli and they form, with the obturator internus, a *tricipital tendon* which is inserted into the *medial surface* of the greater trochanter where the bone rises above the neck of the femur and the trochanteric fossa (Fig. 3.27). The somewhat faceted *smooth* area of this insertion can be seen on inspection of the femur. The **superior gemellus** arises from the spine of the ischium and is supplied by the nerve to obturator internus, while the **inferior gemellus** arises from the ischial tuberosity at the margin of the lesser sciatic notch and is supplied by the nerve to quadratus femoris (Fig. 3.63, p. 190).

The tricipital tendon of obturator internus and the gemelli lies horizontal in the buttock, its upper border alongside piriformis, its lower border edge to edge with quadratus femoris. The sciatic nerve passes down on its surface, the nerve to quadratus femoris deep to it. The actions of piriformis, obturator internus and quadratus femoris are primarily synergic, to act with the other short muscles of the hip to adjust and stabilize the joint. Acting as prime movers they are lateral rotators of the femur.

The **quadratus femoris,** true to its name, is a rectangular muscle. It arises from the ischial tuberosity immediately below the inferior gemellus, and passes laterally. It is inserted into the quadrate tubercle of the femur and into a line (not visible on most bones) that passes vertically downwards therefrom to a level that bisects the lesser trochanter (Fig. 3.27). It should be noted that the 'quadrate tubercle' is a heaping up of bone at the site of fusion of the epiphysis of the greater trochanter (cf. the ilio-pectineal eminence of the os innominatum) and is not itself produced by the pull of the quadratus femoris. The upper and lower borders of the muscle are horizontal and parallel and they lie edge to edge with the inferior gemellus above and the free upper border of adductor magnus below.

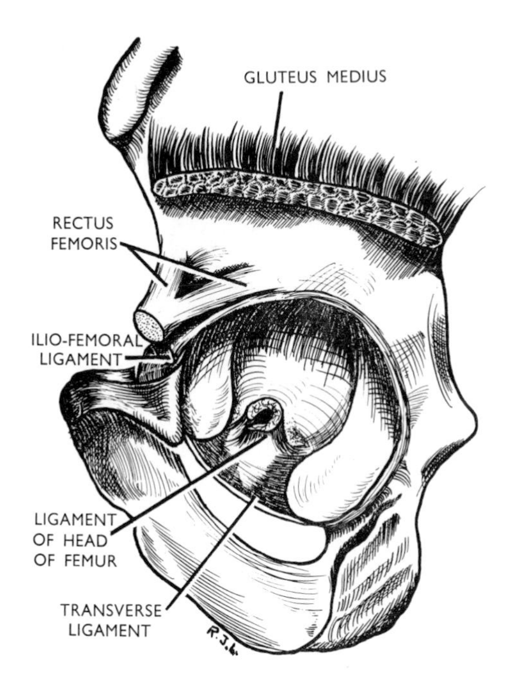

FIG. 3.29. THE LEFT ACETABULUM AND ADJACENT STRUCTURES. Sketched from a dissection.

The Trochanteric Anastomosis. This provides the main source of blood for the supply of the head of the femur. The anastomosis lies near the trochanteric fossa. It is formed by anastomosis of the descending branch of the superior gluteal artery with the ascending branches of both lateral and medial femoral circumflex arteries. The inferior gluteal artery usually joins the anastomosis. Branches from the anastomosis pass along the femoral neck beneath the retinacular fibres of the capsule (p. 150).

The Cruciate Anastomosis. At the level of the middle of the lesser trochanter the transverse branch of the medial circumflex artery meets the transverse branch of the lateral circumflex artery at the lower border of the insertion of quadratus femoris. They are joined by an ascending branch of the first perforating artery and the cross is completed above by a descending branch of the inferior gluteal artery.

The Hip Joint

In general it can be said that in all joints stability and range of movement are in inverse proportion to each other; the hip joint provides a remarkable example of a high degree of both. It is a synovial joint of the ball and socket variety. Its stability is largely the result of the adaptation of the articulating surfaces of acetabulum and femoral head to each other, and its great range of mobility *results from the femur having a neck* that is much narrower than the equatorial diameter of the head.

The **acetabulum** is formed by fusion of the three component bones of the os innominatum (Fig. 3.67, p. 196). Ilium, ischium, and pubis meet at a Y-shaped cartilage which forms their epiphyseal junction. This epiphysis closes after puberty (cf. the glenoid fossa of the scapula). The site of union can be seen on the os innominatum by a heaping up of bone at the ilio-pectineal eminence and at the meeting place of ilium and ischium. These sites indicate the position of the upper limbs of the Y-shaped epiphyseal cartilage; the stem of the Y is vertical and passes through the acetabular notch to the obturator foramen. The acetabular articular surface, covered with hyaline cartilage, is a C-shaped concavity. Its peripheral edge is deepened by a rim of dense fibrous tissue which encloses the femoral head beyond its equator, thus increasing the stability of the joint. This rim is named the **labrum acetabulare,** and it is continued across the acetabular notch, which it plugs to produce the **transverse ligament.** The transverse ligament gives attachment to the ligament of the head of the femur (Fig. 3.29).

The central non-articular part of the acetabulum is occupied by a pad of fat known as the Haversian pad.

The spherical head of the femur is adapted to the concavity of the articular surface of the acetabulum. The neck of the femur is narrower than the equatorial diameter of the head and considerable movement in all directions is possible before the femoral neck impinges upon the labrum acetabulare. The presence of a relatively narrow neck is the mechanical factor responsible for the wide range of movement in such a stable joint. The shoulder joint, mechanically, is in marked contrast with the hip joint in this respect. The head of the femur is covered with hyaline cartilage which, in many cases, encroaches a little on the anterior surface of the neck for articulation with the acetabulum when the hip is flexed (Fig. 3.20, p. 141). The non-articulating convexity of the head is excavated into a pit (the *fovea*) for attachment of the **ligament of the head of the femur** (ligamentum teres), whose other end is attached to the transverse ligament (Fig. 3.29).

The **capsule** of the joint is attached circumferentially around the labrum acetabulare and transverse ligament, whence it passes laterally, like a sleeve, to be attached to the neck of the femur. In front it is attached to the intertrochanteric line, but behind it extends for only half this distance, being attached halfway along the femoral neck (Fig. 3.27). The capsule is loose but extremely strong.

From these attachments the fibres of the capsule are reflected back along the neck of the femur, intimately blended with the periosteum, to the articular margin of the femoral head. This reflected part constitutes the **retinacular fibres,** which bind down the nutrient arteries that pass, chiefly from the trochanteric anastomosis, along the neck of the femur to supply the major part of the head. Fracture of the femoral neck within the capsular attachment necessarily ruptures the retinacular fibres and the vessels, causing avascular necrosis of the head.

Synovial Membrane. As in all joints, this is attached to the articular margins. From its attachment around the labrum acetabulare and transverse ligament it lines all the capsule and is reflected back along the neck of the femur, where it invests the retinacular fibres up to the articular margin of the head of the femur. The Haversian fat pad and the ligament of the head are likewise invested in a sleeve of synovial membrane that is attached to the articular margins of the concavity of the acetabulum and of the fovea on the femoral head.

Occasionally a perforation in the anterior part of the capsule, between the ilio-femoral and pubo-femoral ligaments, permits communication between the synovial cavity and the iliac bursa (Fig. 3.30).

Blood Supply. The capsule and synovial membrane are supplied from nearby vessels. The head and intracapsular part of the neck receive their blood from two sources. The ligament of the head contains an arterial twig from the obturator artery; this vessel supplies the ligament and the major part of the head in the young bone. As age advances this artery shrivels, and in later years it supplies only the thin flake of bone into which the ligament is inserted. The major part of the head is then supplied by the arteries in the retinacula mentioned above.

Nerve Supply. The three nerves of the pelvic girdle and lower limb supply the hip joint (Hilton's law). The femoral nerve via the nerve to rectus femoris, the sciatic nerve via the nerve to quadratus femoris and the obturator nerve directly from its anterior division all innervate the capsule and retinacular fibres.

The fibrous capsule is strengthened by three **ligaments** which spiral around the long axis of the femoral neck. The ligaments arise one from each constituent bone of the os innominatum. If the femur is flexed and laterally rotated in such a manner as to restore the fœtal position the ligaments are 'unwound' and lie parallel with the femoral neck, and are thus relaxed. The opposite movement (extension and medial rotation) draws them tight.

The **ilio-femoral ligament (Y-shaped ligament of Bigelow)** is the strongest of the three. The stem of the 'Y' arises from the lower half of the anterior inferior iliac spine and from the acetabular rim. The diverging limbs are attached to the upper and lower ends of the intertrochanteric line, each to a low tubercle visible on most bones. It is difficult to display the ligament as a separate entity, and it appears, as indicated in Fig. 3.30, to be merely a single band adherent to the

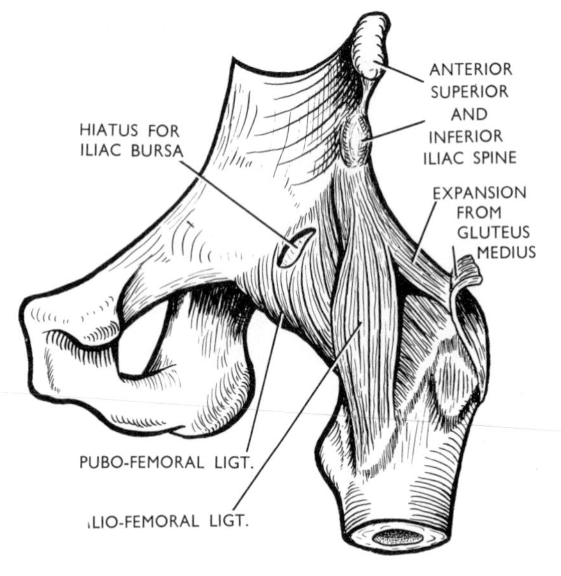

FIG. 3.30. THE LEFT HIP JOINT. Drawn from Specimen S 91.2 in R.C.S. Museum.

capsule (as illustrated by Bigelow himself, who never described it as being Y-shaped!). The ligament limits extension at the hip joint and is of clinical interest in that it forms the fulcrum or axis around which the neck of the femur rotates in dislocation of the hip joint, as originally noted by Bigelow.

The **pubo-femoral ligament** passes from the ilio-pubic eminence and obturator crest to the capsule on the inferior part of the neck of the femur. Abduction combined with extension of the hip pulls it tight.

The **ischio-femoral ligament** is the weakest of the three. It arises from the postero-inferior margin of the acetabulum, and its fibres, passing laterally to the capsule, spiral upwards and are continued into a band of fibres that run in the capsule transversely around the neck of the femur. They form the *zona orbicularis* (Fig. 3.27). Very few fibres reach the femur, and the name *ischio-capsular ligament* is more accurate.

Movements of the Hip Joint. Examine an os innominatum with its own femur. As in any ball and socket joint, movement is possible in any direction. Flexion, extension, adduction and abduction are free; a combination of all four produces circumduction. In addition, medial and lateral rotation of the femur occur. It is essential to the understanding of movements at the hip that the presence of the neck of the femur should be constantly borne in mind. For the purpose of analysing the mechanics of hip movement it is permissible and desirable to picture the neck of the femur as lying horizontal and the shaft of the femur as lying vertical. Any movement of the shaft of the femur is accompanied by a quite different movement of the neck and head, and in this respect the mechanics of hip and shoulder are dissimilar.

Flexion of the hip is a term used to denote flexion of the thigh; when the thigh is flexed upon the trunk the head of the femur rotates about a transverse axis that passes through both acetabula. The muscles responsible for flexion are psoas major and iliacus, assisted by rectus femoris, tensor fasciæ latæ, sartorius and pectineus. Flexion is limited by the thigh touching the abdomen, or by tension of the hamstrings if the knee is extended.

Extension of the thigh, the reverse of the above movement, is performed by gluteus maximus at the extremes of the movement and by the hamstrings in the intermediate stage. The movement is limited by tension in the ilio-femoral ligament.

Adduction and **abduction** of the thigh produce similar movements in femoral shaft and neck; the femoral head rotates in the acetabulum about an antero-posterior axis. Adduction is produced by contraction of the pectineus, adductors longus, brevis and magnus and the gracilis. It is limited by contact with the other leg or, if the latter is abducted out of the way, by the tension of glutei medius and minimus (as in bending sideways on one leg). Abduction is produced by contraction of the glutei medius and minimus, assisted by piriformis. It is limited by tension in the adductors and in the pubo-femoral ligament.

Rotation of the Thigh. In considering rotation of the thigh it is absolutely essential to study the *movement occurring in the neck of the femur*. The femoral shaft does not rotate about its own axis (unless the femoral neck is fractured), and in this respect it differs completely from the humerus. During rotation of the thigh the femoral neck swings backwards and forwards like a gate on its hinges, and the femoral head rotates in the acetabulum about a vertical axis as does a hinge. As the femoral neck swings to and fro like a gate on its hinges 'rotation' of the femoral shaft occurs.

The direction of rotation of the shaft depends on whether the thigh is extended or flexed. Examine the dry bones with the femoral head in the acetabulum. With femur extended, forward movement of the neck produces medial rotation of the shaft; with femur fully flexed along the abdomen forward movement of the neck produces lateral rotation of the shaft. Thus the muscles that produce rotation of the extended femur *reverse their actions* on the fully flexed thigh. But rotary movements of the fully flexed thigh are not usually performed and are in any case rather limited in range; in the following account the muscle actions are therefore dealt with *in the extended position of the femur*.

Lateral Rotation. Any muscle passing obliquely or transversely across the back of the hip joint, and therefore behind the vertical axis around which the femoral head rotates, must act as a lateral rotator of the extended thigh. Gluteus maximus is the most powerful prime mover among the lateral rotators. The other lateral rotators serve rather as stabilizers of the hip joint. They are gluteus medius and gluteus minimus (posterior fibres), piriformis, obturator internus and the gemelli, and the quadratus femoris.

The power of lateral rotation is considerable. The movement is limited by tension of the pubo-femoral ligament and the anterior fibres of the capsule.

Medial Rotation. Any muscle whose line of pull passes across the front of the hip joint, and therefore anterior to the vertical axis around which the femoral head rotates, must act as a medial rotator of the *extended* thigh, and this is quite irrespective of the site of its insertion.

That the anterior fibres of the glutei medius and minimus act as medial rotators is agreed by all observers. That psoas major, iliacus, pectineus and adductor longus are medial rotators is not so easily seen. But their lines of pull all pass in front of the hip joint, and the fact that they are attached towards the back of the femur is irrelevant. They pull the lesser trochanter and linea aspera forwards, hence the greater trochanter moves forwards, the neck of the femur moves forwards like a gate and medial rotation of the femur occurs. If the neck of the femur is fractured, the mechanism is altered. The femur is now free to rotate about its own axis, and since these muscles pass backwards along its medial side their pull causes lateral rotation of the femur, as every orthopædic student knows.

A simple experiment illustrates this. Stand the femur upright on a table, with finger and thumb loosely around the shaft. Push the lesser trochanter forwards. The shaft rotates laterally and the head moves forwards (such a movement can occur in the body only if the neck is fractured). Now place the shaft in its correct obliquity, balanced by the tip of the forefinger on top of the head. Push the lesser trochanter forward; the shaft rotates medially, as it does in life from any forward pull on the lesser trochanter.

Medial rotation of the femur is too powerful a movement to be accounted for by the rather weak anterior fibres of the glutei medius and minimus. The powerful pull of the ilio-psoas, aided by pectineus and adductor longus, is mainly responsible for the movement. Flexion of the thigh is prevented meanwhile by contraction of the hamstrings (which have no rotatory action on the hip) and not by gluteus maximus, whose lateral rotating effect would oppose the medial rotation desired. Any undesired adducting effect of pectineus and adductor longus is likewise countered by the abducting effect of the anterior fibres of the middle and smallest glutei. These muscles (except ilio-psoas) can be palpated, and felt to be contracting if medial rotation is attempted against resistance.

Stability of the Hip Joint. The hip joint is very stable. Bony factors are responsible, the femoral head fitting snugly into the cup-shaped acetabulum, which is deepened by the labrum acetabulare. Ligaments play an important part; the ilio-femoral ligament is the strongest. The short muscles, especially the glutei medius and minimus, stabilize the joint while distal parts of the limb are in motion.

Relations of the Hip Joint. Anteriorly the iliac bursa lies over the capsule and extends upwards into the iliac fossa beneath the iliacus muscle. The psoas major tendon and iliacus muscle separate the capsule from the femoral artery and femoral nerve. Superiorly there is a cellular space between the capsule and the overhanging gluteus minimus and piriformis muscles. Inferiorly the obturator externus muscle spirals back around the femoral neck. Posteriorly the tricipital obturator internus and gemelli separate the sciatic nerve from the capsule. Medially the acetabular fossa is thin and translucent; this bone forms part of the lateral wall of the pelvis and in the female the *ovary* lies hard by, separated only by obturator internus and the obturator nerve and vessels and the peritoneum.

THE FLEXOR (HAMSTRING) COMPARTMENT OF THE THIGH

The flexor or hamstring compartment of the thigh extends from the buttock to the back of the knee. It is separated from the extensor compartment by the lateral intermuscular septum, but there is no septum dividing it from the adductor compartment, for the adductor magnus is a muscle consisting of fused flexor and adductor components.

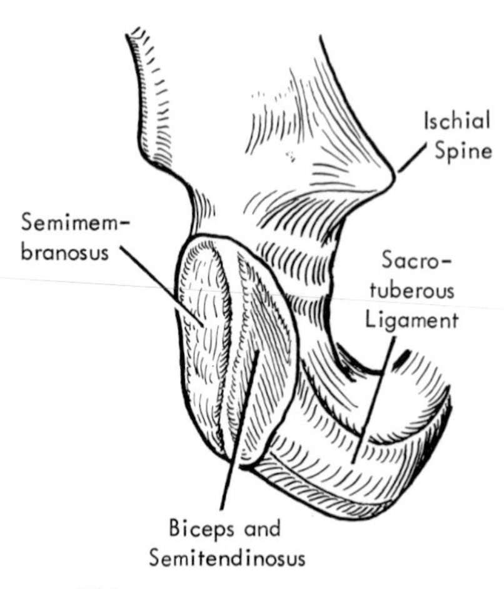

FIG. 3.31. THE LEFT ISCHIAL TUBEROSITY.

The *cutaneous nerve supply* is by the posterior cutaneous nerve of the thigh, which runs vertically downwards just beneath the fascia lata to end halfway down the calf. It sends a series of perforating branches through the fascia lata to supply it and the overlying skin. The spinal segment is S2.

The *fascia lata* is relatively thin at the upper part, but is thickened and strengthened over the popliteal fossa by transverse fibres. The hamstring muscles all arise from the ischial tuberosity. They are the semimembranosus, semitendinosus and biceps. The sciatic nerve, which supplies them, descends deep to them. The ischial fibres of adductor magnus which descend to the adductor tubercle are a part of the hamstrings and, like them, are supplied by the sciatic nerve.

The Hamstring Muscles

The **semimembranosus muscle** extends from the ischial tuberosity to the medial condyle of the tibia. The muscle has some important features that are worth studying. It arises from a smooth facet on the lateral part of the ischial tuberosity (Fig. 3.31), above the part that bears weight in sitting. It derives its name from the fact that this origin is a long flat tendon, or 'membrane', that extends down from the ischial tuberosity for a full six inches (15 cm). The tendon is rounded on its lateral margin, sharp at its medial margin, so that it resembles a hollow-ground razor. In cross section through the amputated thigh its appearance is very characteristic (Fig. 3.18, p. 140). In wounds below the buttock the tendon provides a very useful landmark, for a finger's breadth from its lateral margin the sciatic nerve lies on the adductor magnus, deep to the long head of the biceps (Fig. 3.32). The flat tendon passes deep to semitendinosus and the long head of biceps, and muscle fibres commence to arise from its sharp medial edge some 15 cm below the ischial tuberosity. Below this level the whole tendon is soon replaced by a bulky belly on the surface of which the cord-like tendon of semitendinosus lies. The muscle belly is replaced just behind the medial condyle of the femur by a strong cylindrical tendon which is inserted into the posterior surface of the medial condyle of the tibia (p. 156).

Three expansions diverge from its insertion (*a*) along the medial surface of the tibial condyle; (*b*) obliquely across the capsule of the knee joint (the oblique popliteal ligament); and (*c*) downwards over the popliteus to the soleal line (Fig. 3.36).

The **semitendinosus muscle** arises, in common with the long head of biceps, from the medial facet on the ischial tuberosity (Fig. 3.31). The fleshy belly diminishes in size from above downwards; it lies on the flat tendon of origin of semimembranosus. The semitendinosus is well named, for halfway down the thigh its belly is replaced by a cord-like tendon that lies in a reciprocal gutter on the surface of the muscular belly of semimembranosus.

The fleshy belly of each muscle lies against the tendon of the other, so that together they form a cylindrical mass. The tendon of semitendinosus passes behind the medial condyle of the femur and then curves forwards as it descends to the tibia, being inserted behind gracilis into the upper part of the subcutaneous surface of the tibia (Fig. 3.35).

The **biceps femoris** is so named because it possesses two heads of origin. The long head arises, in common with semitendinosus, from the medial facet on the ischial tuberosity. It passes down to be joined by the short head. This has a long origin, from the whole length of the linea aspera and from the upper part of the lateral supracondylar line of the femur above the superior lateral genicular artery. The origin of the short head is by an aponeurosis which is rapidly replaced by muscle fibres that slope steeply downwards to fuse with the long head. The single tendon so formed is inserted into the head of the fibula in front of the styloid process and across the tibio-fibular joint to encroach a little on the condyle of the tibia. The tendon, at its insertion, is folded around the lateral ligament of the knee joint (Fig. 3.42).

Nerve Supply of the Hamstrings. These three muscles of the flexor compartment of the thigh are supplied, with the ischial part of adductor magnus, by the tibial component of the sciatic nerve (L5, S1), the nerve of the flexor compartment. There is an exception in the case of the short head of biceps, which is supplied by the common peroneal part of the sciatic nerve, the nerve of the extensor compartment of the leg, but the segments are the same (L5, S1). The explanation is that the short head of biceps was developed in the extensor compartment but migrated to the flexor compartment for functional reasons, retaining its nerve supply.

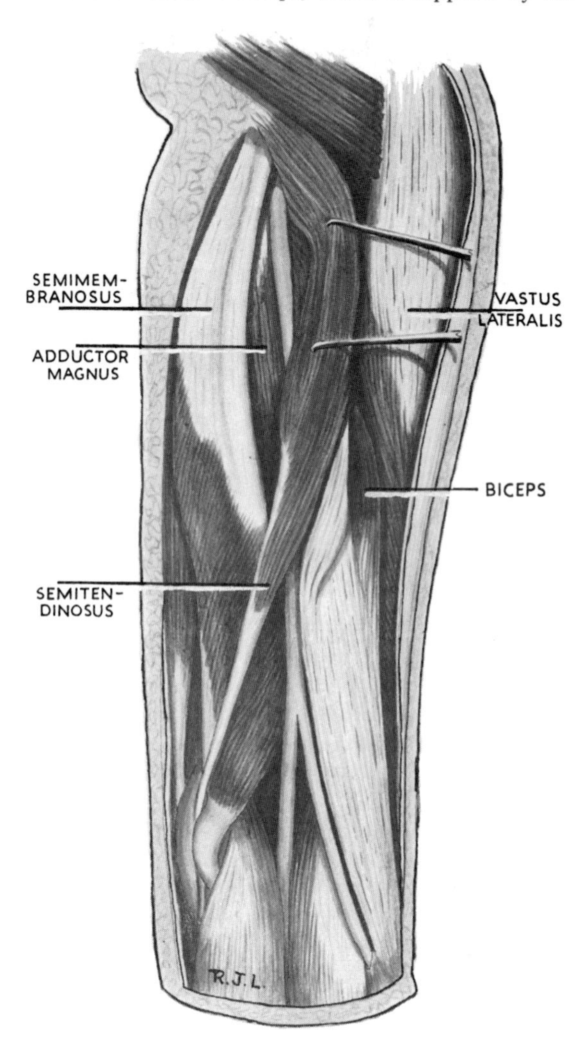

FIG. 3.32. THE RIGHT HAMSTRING MUSCLES. The semitendinosus is retracted laterally. The sciatic nerve lies on adductor magnus lateral to semimembranosus tendon. Drawn from a dissection.

Blood Supply of the Hamstrings. The hamstring

compartment receives its blood supply mainly from the profunda femoris artery by way of the four perforating branches. These pierce the adductor magnus and give off large branches to the hamstrings and the overlying fat and skin, then pass laterally through the lateral intermuscular septum to end in the substance of vastus lateralis. The blood supply of the upper part of the hamstrings is derived from the inferior gluteal artery, that of the lower part from the popliteal artery. These and the perforating branches of the profunda femoris form a series of anastomoses along the back of the femur by means of their terminal arterioles. The highest is the cruciate anastomosis, in the buttock.

Actions of the Hamstrings. The hamstrings are essentially the flexor muscles of the knee joint, but they also have an extensor action on the hip joint, which is especially important in walking. When the knee is held extended by the contracting quadriceps the hamstrings extend the hip, particularly when the position of the hip joint is intermediate between full flexion and full extension. At the extreme ranges of hip movement, as in running, the extensor action of the hamstrings is enhanced by gluteus maximus. Note that the two 'semi' muscles are inserted medially and the two heads of the biceps laterally into the upper part of the leg. When the knee is flexed the alternate contraction of these muscles produces rotation at the knee joint. The 'semi' muscles are medial rotators, the biceps is a lateral rotator of the tibia on the femur. On the weight-bearing tibia the 'semi' muscles produce lateral rotation of the flexed femur, while the two heads of the biceps rotate the femur medially.

The **sciatic nerve** runs vertically through the hamstring compartment, lying deep to the long head of biceps, between it and the underlying adductor magnus (Fig. 3.32). At the apex of the popliteal fossa, a hand's breadth or more above the knee joint, it divides into its tibial and common peroneal components. A high division is not at all uncommon and, indeed, the two parts of the nerve may leave the pelvis separately, in which case the common peroneal nerve pierces the lower margin of the piriformis (Fig. 3.28).

The nerve supplies the semimembranosus, semitendinosus, part of adductor magnus and the long head of the biceps from its tibial portion, the short head of biceps from its common peroneal portion. The sciatic nerve is supplied with blood from the inferior gluteal artery; the *arteria comitans nervi ischiadici* represents the remnant of the original axial artery of the limb and it may persist as a large vessel.

THE POPLITEAL FOSSA

In the living, the back of the flexed knee is hollow between the ridges made by the tensed hamstring tendons. In the extended knee the hamstring tendons lie against the femoral condyles and the fat of the popliteal space bulges the roof of the fossa.

The popliteal fossa is a diamond-shaped space behind the knee. It is limited above by the diverging semimembranosus and the semitendinosus on the medial side and the tendon of biceps on the lateral side.

The lower part of the 'diamond' is occupied by the heads of the gastrocnemius and is opened up only when they are artificially separated. The roof of the fossa is formed by the fascia lata, which is here strongly reinforced by transverse fibres. It is pierced by the small saphenous vein and the posterior cutaneous nerve of the thigh. The floor is provided, from above downwards, by the popliteal surface of the femur (the area between the medial and lateral supracondylar lines), the capsule of the knee joint and the popliteus muscle covered by its fascia. The popliteal artery and vein and the tibial and common peroneal nerves pass through the fossa. A small group of popliteal lymph nodes lie alongside the popliteal vein.

Examine a dissected popliteal fossa in the museum (PLATE 19). The surest way to distinguish right from left is to observe the common peroneal nerve emerging superficially across its lower *lateral* boundary. Both nerves enter the apex of the fossa near the surface.

The **common peroneal** (lateral popliteal) **nerve** slopes downwards medial to the biceps tendon (where it is easily palpable in the living) and disappears into the substance of peroneus longus to lie on the neck of the fibula, against which it can be rolled in the living (confirm this on yourself). It lies successively upon the plantaris, the lateral head of the gastrocnemius, the capsule of the knee joint (with popliteus tendon within) and the fibular origin of soleus. In its course it gives off the following branches: (*a*) the *lateral sural nerve* which pierces the roof of the fossa and runs downwards in the subcutaneous fat to join the sural nerve below the bellies of gastrocnemius, (*b*) the *lateral cutaneous nerve* of the calf, which pierces the roof of the fossa over the lateral head of gastrocnemius and supplies skin over the upper part of the peroneal and extensor compartments of the leg, (*c*) superior and inferior *genicular nerves* which travel with the arteries of the same name and supply the capsule of the knee joint on its lateral aspect and, finally, (*d*) the *recurrent genicular nerve* which arises in the substance of peroneus longus, perforates tibialis anterior (supplying its upper lateral fibres) and supplies the capsule of the superior tibio-fibular joint and the knee joint. The common peroneal nerve ends by dividing, in the substance of peroneus longus, into the deep peroneal (anterior tibial) and superficial peroneal (musculo-cutaneous) nerves.

The **tibial** (medial popliteal) **nerve** runs vertically down *along the middle of the fossa* and disappears by passing deeply between the heads of the gastrocnemius. If the latter are separated the nerve is seen passing, with the popliteal vessels, beneath the fibrous arch in the origin of soleus (cf. the median nerve in the cubital fossa passing beneath the fibrous arch in flexor digitorum superficialis). Below this fibrous arch it enters the calf. The nerve gives *motor branches* to all the muscles that arise in the popliteal fossa, namely, to plantaris, both heads of the gastrocnemius, soleus, and popliteus. The last branch hooks around the lower border of popliteus to enter its deep (tibial) surface.

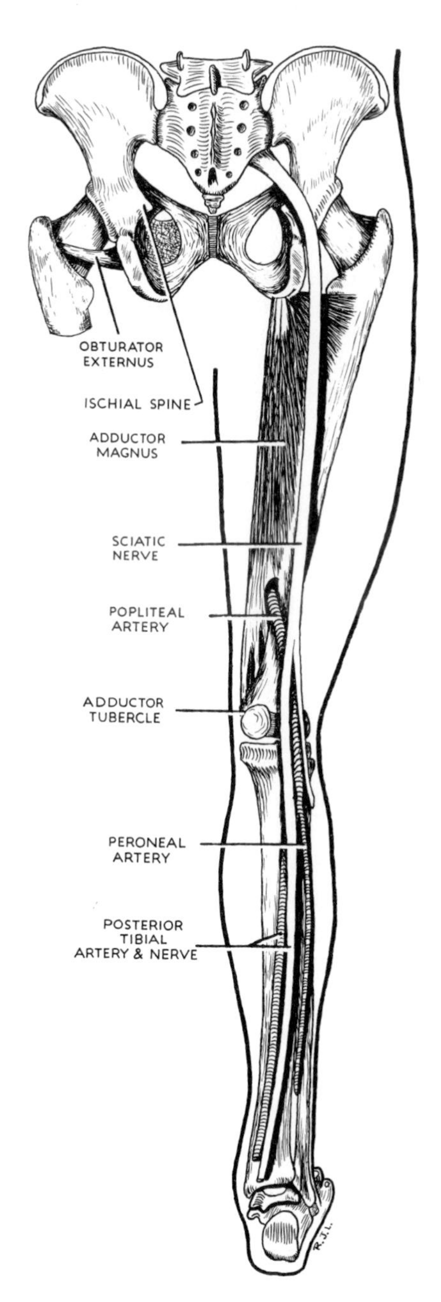

FIG. 3.33. THE RIGHT SCIATIC NERVE. The relations of the popliteal and posterior tibial arteries to the tibial nerve are shown. The popliteal vein is not shown; at all levels it lies between the artery and the nerve.

The tibial nerve here has only one cutaneous branch, the *sural nerve*. It runs vertically down in the narrow chink between the two heads of gastrocnemius and pierces the deep fascia halfway down the calf, where it replaces the posterior cutaneous nerve of the thigh. Passing down in the subcutaneous fat it joins the lateral sural nerve below the bellies of gastrocnemius.

Articular branches, the *genicular nerves*, are three in number. They accompany the superior and inferior medial genicular arteries to supply the medial ligament and medial part of the capsule of the knee joint, and the middle genicular artery to pierce the oblique popliteal ligament and supply the cruciate ligaments.

The **Popliteal Artery** (Fig. 3.33). Throughout the whole of its course the popliteal artery lies deep in the fossa. It extends from the hiatus in adductor magnus (a hand's breadth above the knee) to the fibrous arch in soleus (a hand's breadth below the knee) and is about 8 in (20 cm) long. It enters the fossa on the medial side of the femur; there it lies not only deep but medial to the sciatic nerve and its vertical continuation the tibial nerve. As it passes downwards it is convex laterally, coming to lie lateral to the tibial nerve. Below the fibrous arch in soleus, as the posterior tibial artery, it returns to the medial side of the nerve (Fig. 3.33). At all levels the popliteal vein *lies between the artery and the nerve*. The artery lies at first on the popliteal surface of the femur, separated from it by a little cellular fat, then on the posterior ligament of the knee joint to which it is firmly fixed by some overlying fibrous tissue and by the short middle genicular artery and, below this, it lies free on the fascia over the popliteus muscle. It passes under the fibrous arch in soleus and immediately divides into anterior and posterior tibial arteries. In the fœtus the vessel passes between the popliteus muscle and the tibia and this condition may persist in the adult (Fig. 3.35).

Muscular branches are given to the muscles in the popliteal fossa; that to the gastrocnemius is a very constant vessel, a single trunk which passes backwards to branch in Y-shaped fashion to the two heads of the muscle. The **genicular arteries** are five in number, upper and lower, lateral and medial and a middle (Figs. 3.34, 3.35).

The medial and lateral upper genicular arteries encircle the lower end of the femur, the medial and lateral lower arteries encircle the tibia, while the *middle genicular artery* pierces the posterior ligament of the knee joint to supply the cruciate ligaments. The genicular branch of the posterior division of the obturator nerve (p. 145), having travelled with the popliteal artery, enters the capsule alongside the middle genicular artery.

The *upper medial genicular artery* passes obliquely upwards across the medial head of gastrocnemius and crosses the medial supracondylar line deep to the tendon of adductor magnus (PLATE 19).

The *upper lateral genicular artery* crosses the lateral head of gastrocnemius and passes upwards to cross the lateral supracondylar line just beneath a small tubercle that marks the lower limit of attachment of the short head of biceps femoris and the lateral intermuscular septum. The two upper genicular arteries anastomose over the front of the femur and patella with the descending branch of the lateral circumflex artery and the deep branch of the descending genicular (anastomotica magna) artery, and, over the front of the tibia, with the two lower genicular arteries.

The *lower medial genicular* artery courses obliquely downwards under the medial head of gastrocnemius and passes forwards, beneath the medial ligament of the knee. The *lower lateral genicular artery* runs horizontally outwards under the lateral head of gastrocnemius, crosses the popliteus tendon and passes deep to the lateral ligament of the knee. Thence it passes forwards *lying upon the lateral meniscus* of the knee joint (Fig. 3.34).

The **popliteal vein** lies, at all levels, between the artery and the tibial nerve. It is formed by union of the two venæ comitantes of the anterior and of the posterior tibial arteries, and it receives tributaries that accompany the branches of the popliteal artery. In addition, it receives the *small saphenous vein*, which pierces the roof of the fossa.

The **popliteal lymph nodes** consist of a few scattered nodes lying about the termination of the small saphenous vein. They lie beneath the deep fascia. They receive from a small area of skin just above the heel by a few superficial afferents which run with the small saphenous vein and pierce the roof of the fossa, and from the deep structures of the calf by deep afferents that accompany the posterior tibial vessels. They send their efferents alongside the popliteal and femoral vessels to the deep inguinal nodes.

The **semimembranosus tendon** is received mainly into the concavity on the back of the medial condyle of the tibia, but an expansion passes forwards along

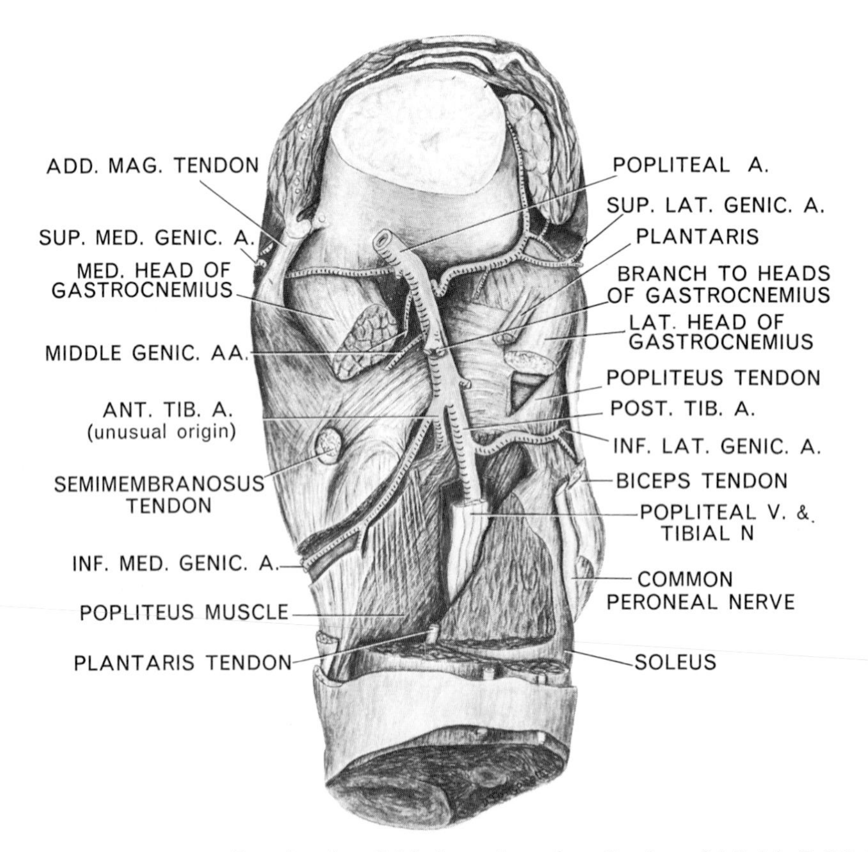

FIG. 3.34. THE GENICULAR ARTERIES. Posterior view of right knee, drawn from Specimen S 142 A in R.C.S. Museum.

the medial surface of the condyle beneath the free posterior border of the tibial collateral (medial) ligament of the knee and separated from it by a bursa. A second expansion passes obliquely upwards from the tendon to be attached to the popliteal surface of the femur immediately above the lateral condyle. It is firmly attached to the capsule of the knee joint and is fenestrated for the passage of the middle genicular artery and its companion veins. It is named the **oblique popliteal ligament.** A third expansion passes vertically downwards to form a strong fascia over the popliteus muscle; it is attached to the soleal line of the tibia. Its lower part gives origin to soleus. A bursa lies between the semimembranosus tendon and the capsule of the knee joint, and often communicates with the bursa under the medial head of gastrocnemius; in such cases the semimembranosus bursa thereby communicates with the knee joint (Fig. 3.43).

The Popliteus Muscle. The fleshy fibres of this muscle arise from the popliteal surface of the tibia above the soleal line and below the tibial condyles. The muscle slopes upwards and laterally towards the cord-like tendon, which is attached to a pit just below the epicondyle on the lateral surface of the lateral condyle of the femur. The tendon lies within the capsule of the knee joint, entering it beneath a falciform free edge, the arcuate ligament, to which the superficial fibres of the muscle are attached (Fig. 3.36). Only half of the popliteus muscle continues into this tendon; the upper half of the muscle ends in a short flat tendon which is *inserted into the posterior convexity of the lateral meniscus* (Fig. 3.37).

The popliteus bursa lies deep to the tendon, where the tibia is grooved. The muscle is supplied by a branch of the tibial nerve which winds around its lower border and sinks into its deep (anterior) surface. This nerve also supplies the tibio-fibular joint and, by a long slender branch, the interosseous membrane of the leg. This conforms with Hilton's Law, on the supposition that the popliteus muscle was once the pronator of the bones of the leg (cf. pronator teres), and its nerve thus sent proprioceptive fibres to the articulations on which the muscle acted. If you believe this supposition you would believe anything (which human ancestor, immediate or remote, could pronate the bones of its leg?).

The action of popliteus is to rotate the knee and simultaneously to draw the lateral meniscus posteriorly. In the fully extended knee the femur has rotated medially on the tibia to 'lock' the joint. The femur, from the extended position, is rotated laterally by popliteus to 'unlock' the joint. Further lateral rotation of the femur involves excursion of its lateral condyle backwards across the tibial plateau. The lateral meniscus is pulled backwards in advance of the femoral condyle

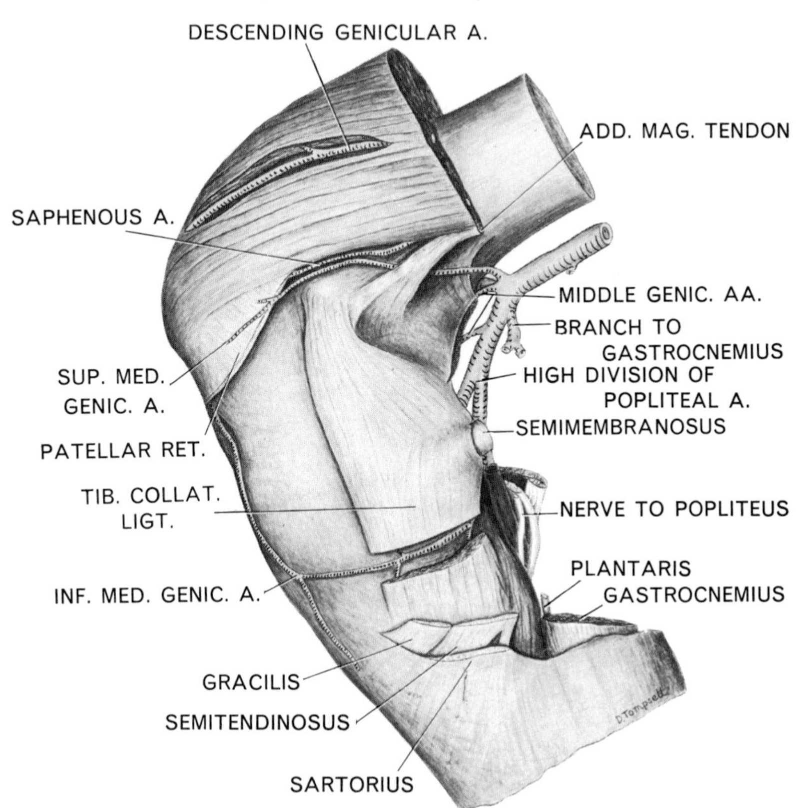

FIG. 3.35. THE GENICULAR ARTERIES. Medial view of right knee, drawn from Specimen S 142 A in R.C.S. Museum.

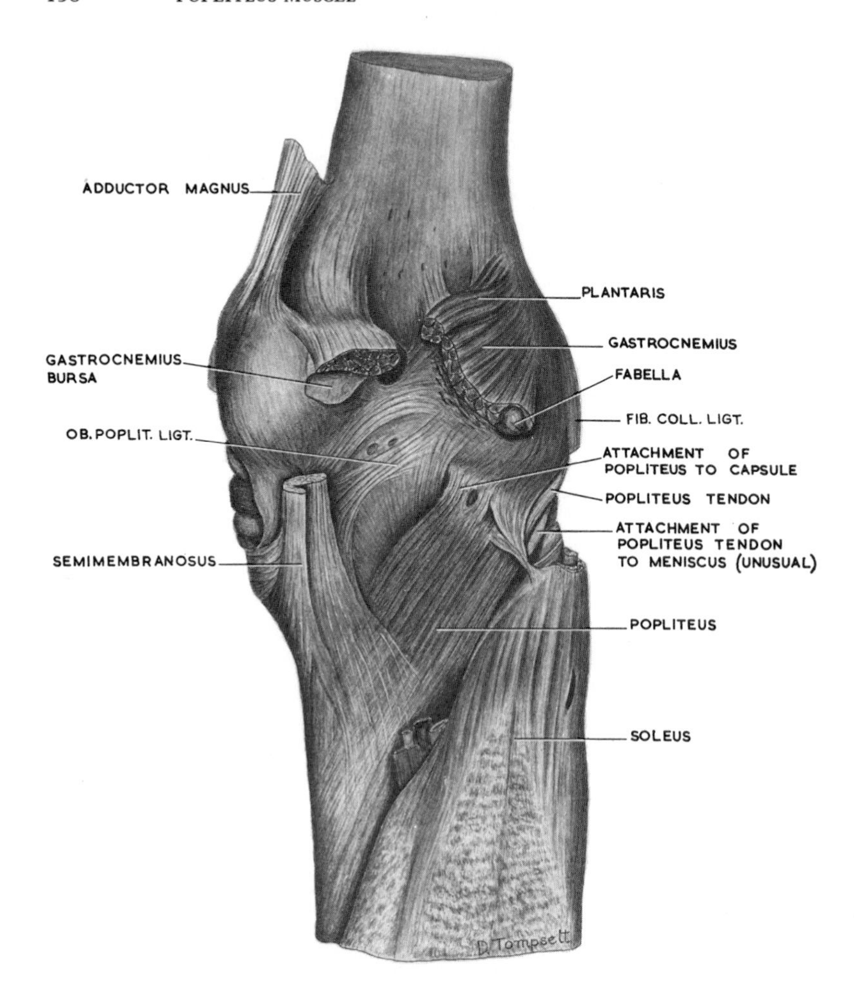

FIG. 3.36. DISSECTION OF RIGHT KNEE JOINT FROM BEHIND. The *fabella* is a sesamoid bone not infrequently present in the origin of the lateral head of gastrocnemius. Illustration of Specimen S 94 C in R.C.S. Museum.

by contraction of the popliteus. The posterior margin of the lateral condyle of the tibia is rounded off, so that the posterior convexity of the lateral meniscus can be pulled down over it out of harm's way. The mobile lateral meniscus, its position thus controlled by the popliteus, is therefore relatively immune to impaction between femur and tibia. The medial meniscus lacks such muscular control and is more liable to rupture by being caught between the two bones.

To summarize, the popliteus is a lateral rotator of the femur on the tibia and simultaneously a retractor of the lateral meniscus. Its femoral tendon is attached at the axis of the hinge joint of the knee; consequently it is not even a weak flexor of the knee joint (p. 162).

Thus it is more realistic to consider popliteus as *arising* from the fixed weight-bearing tibia, and moving the femur and the lateral meniscus by its *insertion* into these two. The terms 'origin' and 'insertion' have no objective reality; they are terms of convenience (see p. 6). If the tibia is not weight-bearing there is no need for a popliteus muscle (e.g., the fruit bat, see p. 165).

The Knee Joint

The knee is a synovial joint between femur and tibia. The joint can flex and extend like a hinge. Extension is for propulsion, and flexion is used prior to this and also to absorb the shock (by quadriceps) in landing. In addition the *flexed* knee can rotate, as in change of direction at speed. This active rotation is a matter of choice, and is not to be confused with the passive and inevitable rotation that occurs in straightening the knee in the 'screw-home' mechanism (p. 164). During all these movements the knee is adapted to be weight-bearing and stable in any position.

Bony Contours. The plateau of the *tibia* possesses two separate articular facets, each slightly concave. The medial facet lies wholly on the upper surface of the condyle, but the lateral facet curves back over the posterior margin of the tibial condyle (Fig. 3.73, p. 202). This bevelled margin allows withdrawal of the lateral meniscus by the popliteus muscle (p. 165). The *femur* has two condyles, separated posteriorly by a deep notch, but fusing anteriorly into a trochlear groove for articula-

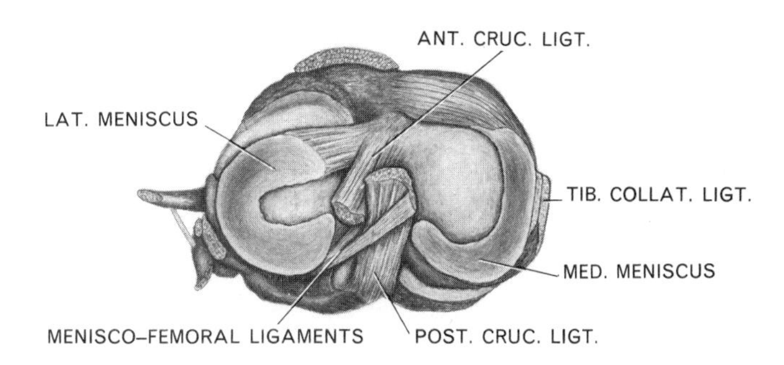

FIG. 3.37. DISSECTION OF LEFT KNEE, VIEWED FROM ABOVE AND FROM BEHIND. Illustration of Specimen S 108 B in R.C.S. Museum.

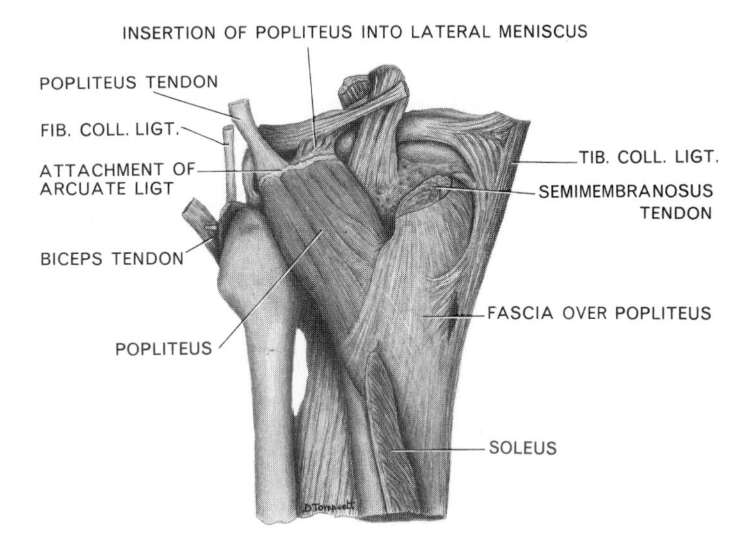

tion with the patella. The lateral ridge of the trochlear groove is very prominent (Fig. 3.22, p. 141). The curve of the femoral condyles is cam-shaped (in lateral profile); it is flatter on the end of the femur and more highly curved at the free posterior margin of each condyle (Fig. 3.44). The distal surface of the medial condyle is narrower, longer and more curved than the lateral condyle (Fig. 3.22). This is for the screw-home movement (p. 164). The articular surface of the *patella* is divided by a vertical ridge into a large lateral and a small medial surface; this latter is further divided by a vertical ridge into two smaller areas. The large lateral surface glides around *in contact with* the lateral condyle of the femur in all ranges of flexion. In extension the area next to it lies on the trochlea, and the most medial of the three surfaces is not in articulation with the femur (Fig. 3.38). In flexion this surface glides into articulation with the medial condyle, and the middle of the three surfaces lies free in the intercondylar notch of the femur. This can be readily demonstrated on the dry bones.

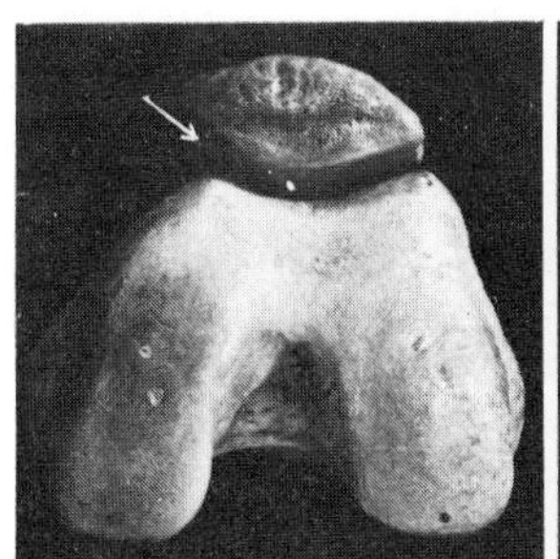
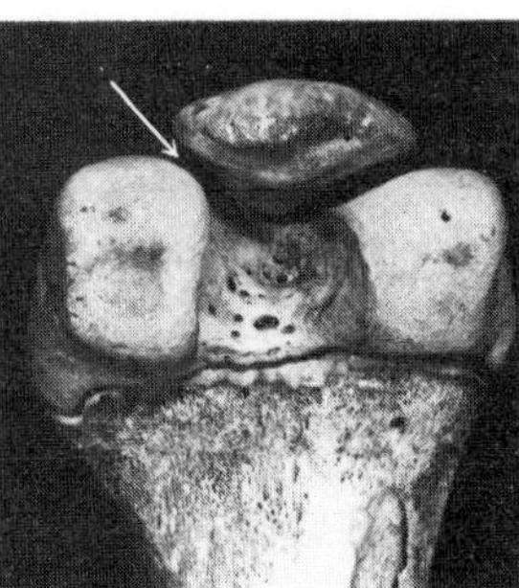

EXTENSION FLEXION

FIG. 3.38. THE INFERIOR SURFACES OF THE LEFT FEMUR AND PATELLA, VIEWED LOOKING UPWARDS AS FROM THE TIBIAL PLATEAU (Photograph). The medial facet on the patella (arrowed) is free in extension but articulates with the medial condyle of the femur in flexion.

Capsular Attachments. On the *femur* the capsule adheres below the epiphyseal line down to the articular margin (PLATE 7) except in two places. At the back it is attached to the intercondylar ridge at the lower limit of the popliteal surface, and on the lateral condyle it encloses the pit for the popliteus tendon (Fig. 3.44).

On the *tibia* the capsule is attached around the margins of the plateau (Fig. 3.39) except in two places. Posteriorly it is attached to the ridge between the condyles at the lower end of the groove for the posterior cruciate ligament. Laterally the capsule is not attached to the tibia but is prolonged down over the popliteus tendon to the styloid process on the head of the fibula (Fig. 3.40).

The edges of this prolongation are the arcuate ligament posteriorly and the short external lateral ligament anteriorly. The *arcuate ligament* is the edge of the capsule that arches down from the lateral meniscus to the styloid process of the fibula. The superficial fibres of the popliteus muscle are attached to it (Fig. 3.36).

The thickness of the adult capsule varies very much. From the lower margin of the patella to the anterior margin of the plateau of the tibia it is excessively thin. It is invaginated together with the synovial membrane by a pad of fat whose herniation into the joint raises up a median fold called the infrapatellar fold. The original capsule above the patella, between it and femur, perforates when the infant walks and so communicates with a large suprapatellar bursa which lies deep to the quadriceps tendon, extending in the adult a hand's breadth above the joint. Thus the capsule has two *main* gaps in it, the one allowing popliteus tendon to enter, the other communicating with the suprapatellar bursa. Anteriorly the capsule is interrupted by a circular gap whose margins are attached to the patella.

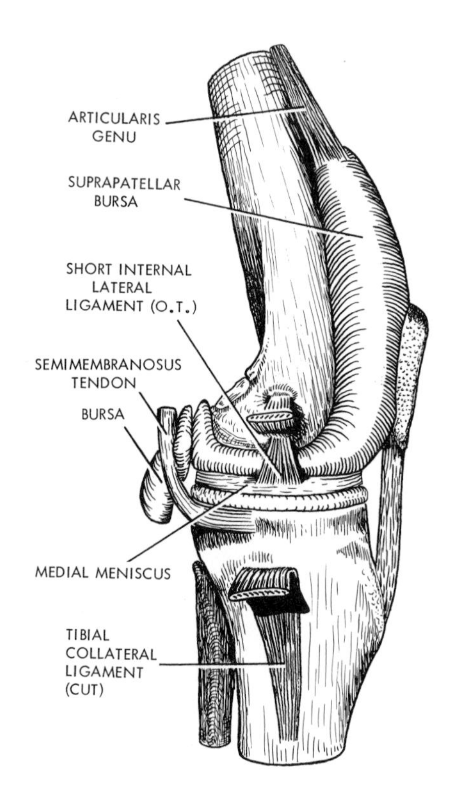

FIG. 3.39. MEDIAL VIEW OF THE LEFT KNEE JOINT (cf. Fig. 3.43).

Capsular Thickenings. Medially and laterally the capsule is greatly thickened, forming two ligaments that are unrecognized in current nomenclature. In Old Terminology they were named the short lateral ligaments (internal and external). The **short internal lateral ligament** lies deep to the tibial collateral (medial) ligament of the knee joint. It is a thickening of capsule which extends from the medial epicondyle of the femur to the medial meniscus; it holds the medial convexity of the meniscus firmly to the femur (Fig. 3.39). It blends with the medial ligament at the femoral attachment, but below this the two are separated by a bursa. Elsewhere around the convexity of the meniscus the capsule, attached above to femur and below to tibia, is thin and lax; it is known as the **coronary ligament.** The **short external lateral ligament** is a cord-like thickening of the capsule on the lateral side of the joint, just anterior to the arcuate ligament. Its femoral attachment is on the epicondyle, where it blends with the tendon of popliteus (Fig. 3.40). Its lower attachment is on the medial border and styloid process of the upper end of the fibula. It is not infrequently absent or poorly developed. Elsewhere on the lateral aspect of the joint the capsule, attached to the lateral meniscus, is thin and lax both above and below. This lax capsule is known as the **coronary ligament,** and it allows great mobility of the lateral meniscus.

Ligaments. The capsule is reinforced by four main ligaments, namely the patellar retinacula, the tibial and fibular collateral ligaments and the oblique popliteal ligament. The *patellar retinacula* extend from the patella to the lower margins of the condyles of the tibia; they are fibrous expansions from the quadriceps tendon and from the lower margins of the vasti medialis and lateralis. In front of the collateral ligaments they blend with the capsule; further anteriorly they are attached to the margins of the ligamentum patellæ, below the patellar attachment of the capsule. They are, of course, not attached to the femur and *must not be confused with the capsule.* Deep to them lie the (extracapsular) fat pad and the deep infrapatellar bursa (Fig. 3.46, p. 166).

The **tibial collateral** (medial) **ligament** is attached to the epicondyle of the femur below the adductor tubercle and to the subcutaneous surface of the tibia a hand's breadth below the knee. It is a broad, *triangular* band of great strength (Fig. 3.35). Its anterior margin lies free except at its attached extremities and is not attached to the medial meniscus, being separated from it and the condyle of the tibia by a bursa. Its posterior margins converge to be inserted into the medial meniscus; *only these marginal fibres and no others* are attached to the meniscus (Fig. 3.42). Over the condyle of the tibia the ligament is separated from bone by the forward extension from the semimembranosus tendon and the intervening bursa. Below this, over the upper surface of the shaft of the tibia, the ligament is separated from bone by the passage of the inferior medial genicular vessels and nerve (Fig. 3.35). Hence the extension of the medial ligament so far below the knee joint to find a firm attachment; the condyle of the tibia is free to rotate beneath the upper part of the ligament. From its tibial attachment the ligament slopes back a little as it passes up to be inserted *behind* the axis of flexion of the femoral condyle (Fig. 3.39). It is thus drawn taut by (and limits) extension of the knee and its terminal 'screw-home' rotation.

The **fibular collateral** (lateral) **ligament** is attached to the lateral epicondyle of the femur in continuity with the short external lateral ligament. It slopes down and back to the head of the fibula. It lies free from the capsule and lateral meniscus, being separated from the meniscus by the tendon of popliteus inside the joint and the inferior lateral genicular vessels outside the joint. It is a strong cord-like ligament. It is attached just behind the axis of flexion of the femoral condyle and is drawn taut by (and limits) extension and the terminal 'screw-home' movement of the knee.

The **oblique popliteal ligament** is a thick rounded band of great strength, perforated by the middle genicular vessels. It is an expansion from the insertion of semimembranosus which slopes up to the popliteal surface of the femur. It blends with the capsule above the lateral condyle of the femur, and in the intercondylar notch rather above its margin, so that a prolongation upwards of synovial membrane extends a little on the popliteal surface of the femur. Loose bodies may lodge here, behind the upper ends of the cruciate ligaments, and elude discovery at operation. The obliquity of this ligament limits rotation-extension in the 'screw-home' or locked position (Fig. 3.36).

The **intra-articular structures** include the cruciate ligaments, the menisci, and the femoral tendon of popliteus. Of these only the menisci are intrasynovial.

The **cruciate ligaments** consist of a pair of very strong ligaments connecting tibia to femur and they lie within the capsule of the knee joint, but not within the synovial membrane. It is as though they had been herniated into the synovial membrane from behind, carrying forward over themselves a fold which invests their anterior and lateral surfaces but leaves their posterior surfaces uncovered. They are named from their tibial origins. The *anterior cruciate ligament* is attached to the anterior part of the tibial plateau in front of the tibial spine and extends upwards and backwards to a smooth impression on the lateral condyle of the femur well back in the intercondylar notch. The *posterior cruciate ligament* is attached to the posterior part of the head of the tibia between the condyles and passes forwards medial to the anterior cruciate ligament. It is attached to a smooth impression on the medial condyle of the femur well forward in the intercondylar notch (Figs. 3.37 and 3.68, p. 198). The two cruciate ligaments cross like the limbs of the letter X. They are

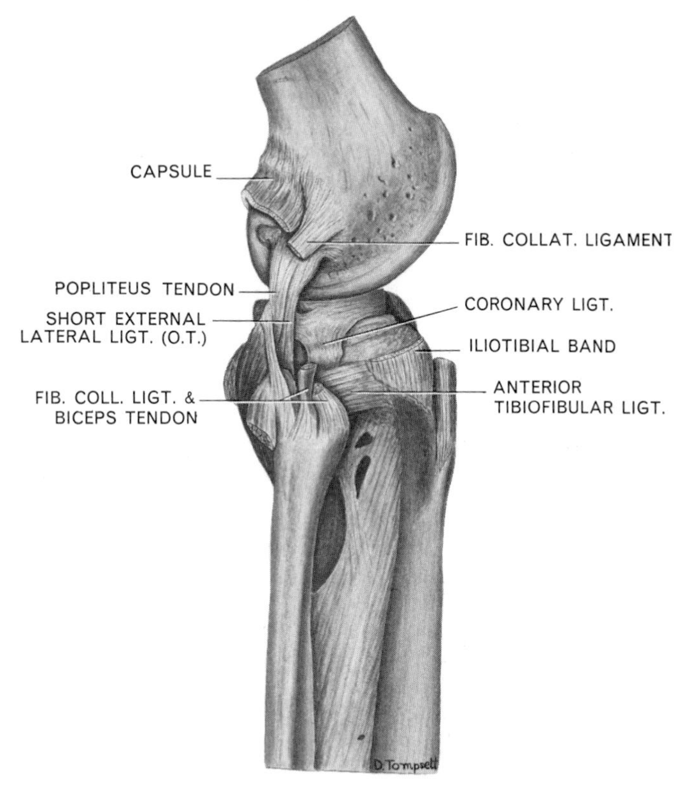

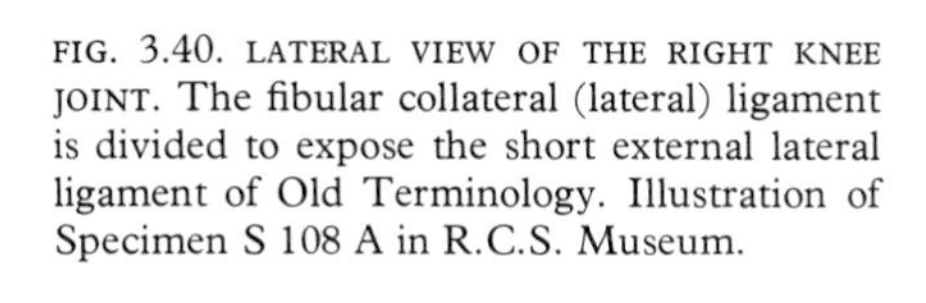

FIG. 3.40. LATERAL VIEW OF THE RIGHT KNEE JOINT. The fibular collateral (lateral) ligament is divided to expose the short external lateral ligament of Old Terminology. Illustration of Specimen S 108 A in R.C.S. Museum.

essential to the antero-posterior stability of the knee joint, especially in the flexed position.

The rôle of each cruciate ligament must be studied separately.

The *posterior cruciate ligament* prevents the femur from sliding forwards off the tibial plateau (Fig. 3.41). In the weight-bearing flexed knee it is the *only* stabilizing factor for the femur and its attached quadriceps (popliteus is too weak to be of significant help). In walking downhill or downstairs the upper knee is flexed and weight-bearing while the lower knee is straight as its foot reaches down to find support. Thus with a ruptured posterior cruciate ligament the patient leads with the damaged leg at each downward step.

The *anterior cruciate ligament* prevents backward displacement of the femur on the tibial plateau, but this is unlikely to happen. The anterior cruciate ligament has a much more important role, that of limiting extension of the lateral condyle of the femur and of then causing medial rotation of the femur in the 'screw-home' position of full extension. This is discussed on p. 164.

The **menisci** were once named the *semilunar cartilages*. Histologically they are composed of dense fibrous tissue with no cartilage and are more accurately called menisci. They are C-shaped, and triangular in cross section. They are avascular except at their attachments. The medial meniscus is the larger, with an open curve whose horns enclose the curve of the lateral meniscus. Each lies on the upper surface of its respective tibial condyle. The **medial meniscus** is fixed at its anterior and posterior horns by fibrous tissue to the tibia. Its circumference is connected by the capsule of the joint to the femur and tibia. The capsular attachment to the tibia is lax, that to the femur is strong on the medial side (the short internal lateral ligament). The **lateral meniscus** is likewise fixed to the tibia at both its horns; in addition its posterior convexity is slung by fibrous tissue ligaments to the femur. These slings are attached to the medial condyle of the femur in front of and behind the attachment of the posterior cruciate ligament, forming the anterior and posterior menisco-femoral ligaments (Fig. 3.42). The circumference of the meniscus is attached by very lax capsule (the coronary ligaments) to the articular margins of femur and tibia except beneath the tendon of popliteus. Here there is a gap in the coronary ligament; through it the popliteus tendon and bursa pass. The posterior convexity of the lateral meniscus receives the insertion of a flat tendon derived from the *upper half of the popliteus muscle* (Fig. 3.37).

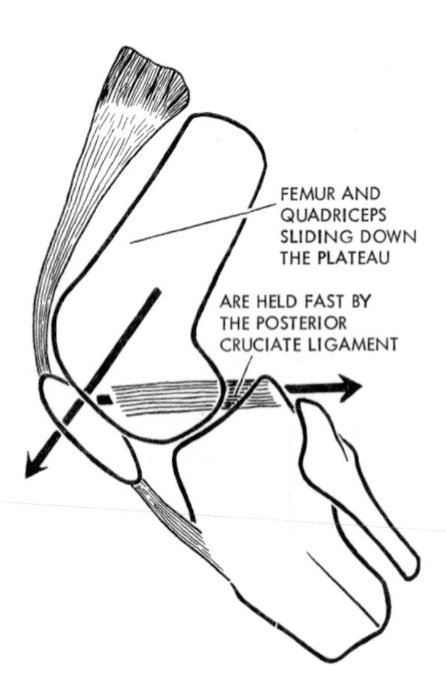

FIG. 3.41. WEIGHT-BEARING ON THE BENT KNEE. The whole body-weight is suspended on the posterior cruciate ligament. This is the position of the upper knee during progression downhill.

The *transverse ligament* is a variable band that, when present, passes across between the anterior horns of the medial and lateral menisci. The delicate capsule is attached to it.

The functions of the menisci are discussed below.

The femoral **tendon of popliteus** lies between the capsule and the synovial membrane. The tendon does not lie free within the cavity of the knee joint, but is adherent to the capsule. The adherent tendon makes a prominent ridge on the inner surface of the capsule. The ridge is invested with the synovial membrane of the joint cavity both above and below the lateral meniscus. Between the upper and lower synovial reflexions the bare lateral meniscus is in contact with the bare tendon of popliteus, and the meniscus is often grooved by the tendon. Occasionally the tendon is partly attached to the meniscus in the manner shown in Fig. 3.36.

Attachment of Synovial Membrane. This does not coincide with the capsular attachments, because of the intra-articular structures. On the *femur* it lines the intercondylar notch and on the lateral condyle is separated from the capsule by the attachment of popliteus tendon, which lies between the two. That is to say, it is attached all around the articular margin of the femur. On the *tibia* it is attached to the articular margins of medial and lateral condyles, and is reflected forwards over the anterior cruciate ligament from these margins. A fold extending from here to the inferior margin of the patella is known as the *infrapatellar fold*; an *alar fold* extends both medially and laterally from it. The infrapatellar fold and alar folds are produced by an extra-synovial fat pad and they adapt their shape to the contours of the bones in different positions of the knee. By keeping the synovial membrane in contact with the articular surfaces of the femoral condyles they act as Haversian fat pads (p. 15).

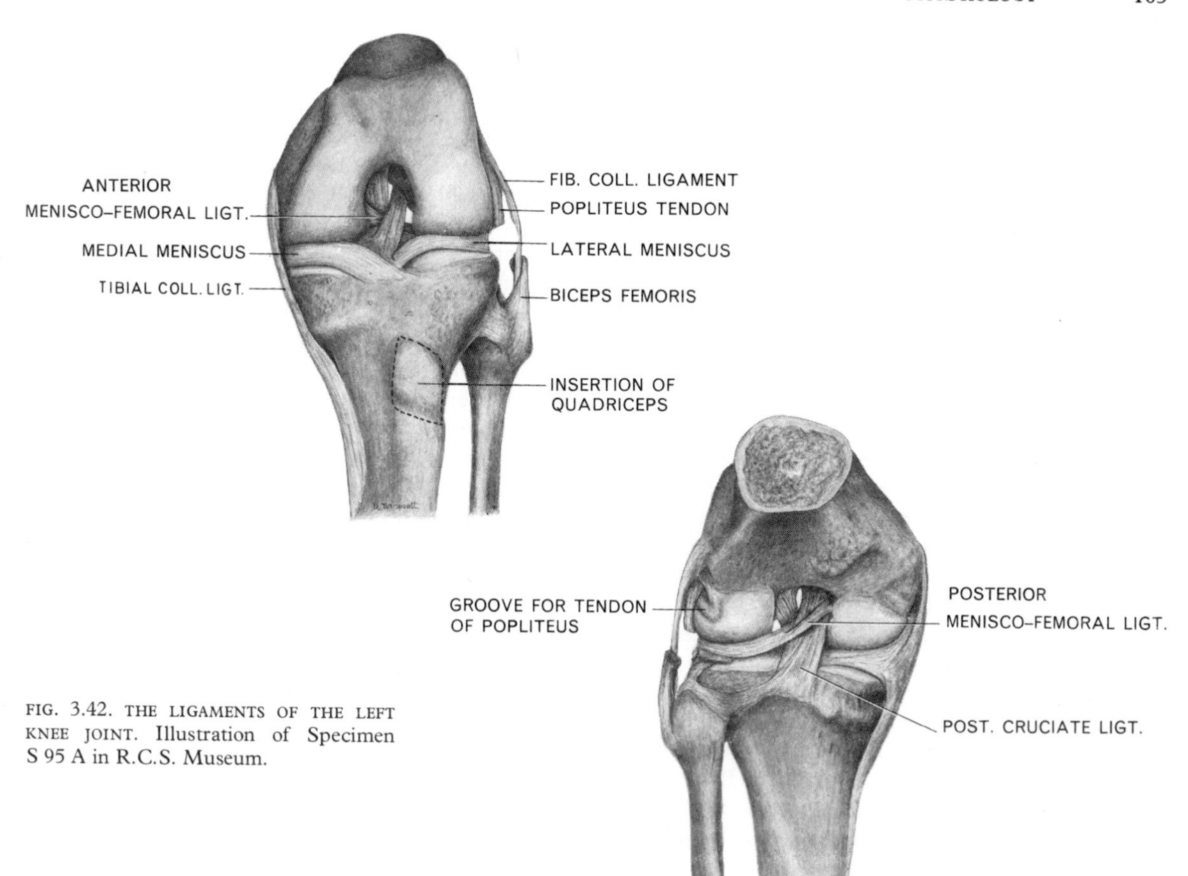

FIG. 3.42. THE LIGAMENTS OF THE LEFT KNEE JOINT. Illustration of Specimen S 95 A in R.C.S. Museum.

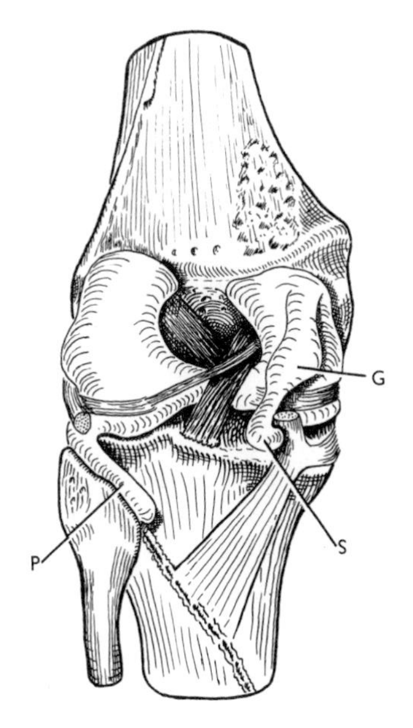

FIG. 3.43. THE SYNOVIAL MEMBRANE OF THE KNEE (Posterior view of left knee after removal of the capsule). **P.** Bursa deep to popliteus tendon. **G.** Bursa deep to medial head of gastrocnemius, communicating with **S,** the semimembranosus bursa. Drawn from Specimen S 110 A in R.C.S. Museum.

Communications with bursæ characterize the synovial cavity of the knee joint. The communication with the suprapatellar bursa has already been noted. A herniation of synovial membrane beneath the rounded tendon of popliteus produces the banana-shaped *popliteus bursa* lying in the gutter between the tibia and the head of the fibula (Fig. 3.43). The bursa beneath the medial head of gastrocnemius always and that beneath the lateral head usually, communicate with the joint. The bursa under the medial head of gastrocnemius usually communicates also with the semimembranosus bursa (p. 172), thereby connecting the latter bursa with the cavity of the knee joint (Fig. 3.43).

Morphology of the Knee Joint. The medial ligament is said to be the phylogenetically degenerated tendon of adductor magnus and the lateral ligament that of peroneus longus. The two menisco-femoral ligaments form the sole posterior attachment of the lateral meniscus in almost all mammals; the attachment of the posterior horn to the tibia is a secondary event in man, associated with the upright posture.

Blood Supply. The capsule and joint structures are supplied from the anastomoses around the knee. The chief contributors are the five genicular branches of the popliteal artery, of which the middle genicular supplies the cruciate ligaments.

Nerve Supply. In accordance with Hilton's Law the joint is supplied from the femoral, especially its branch to the vastus medialis, from the sciatic by the genicular branches of the tibial and common peroneal, and from the obturator nerve by the twig from its posterior division, which accompanies the femoral artery through the gap in the adductor magnus into the popliteal fossa.

Movements. The movements of the knee joint are flexion, extension and rotation. *Active* rotation is possible only in a flexed knee. The passive rotation that occurs in the 'screw-home' movement of full extension is something quite different. There is no active rotation of the extended knee.

Flexion is performed by the hamstrings and this is limited by compression of the soft parts behind the knee. *Extension* is performed by the quadriceps and is limited by the tension of the anterior cruciate ligament, the oblique popliteal ligament and the collateral ligaments, but these four ligaments do not tighten simultaneously. As the knee moves into full extension the anterior cruciate ligament is the first to become taut. *Extension of the lateral condyle of the femur is thus terminated.* Further extension of the medial condyle is made possible by passive rotation forwards of the lateral condyle around the radius of the taut anterior cruciate ligament. This forces the medial condyle to glide backwards into its own full extension. The medial condyle has a longer and more curved articular surface than the lateral condyle for this very reason (Fig. 3.22, p. 141). This medial rotation of the femur on the tibial plateau tightens the oblique popliteal ligament; the medial and lateral ligaments of the knee joint are set slightly obliquely and are tightened simultaneously. All three become taut and limit further rotation. This 'screw-home' movement is said to *lock the joint*, and the description is just. In this position the joint is slightly hyperextended and all four ligaments are taut—the anterior cruciate preventing further extension and the other three also preventing further rotation. The knee is completely rigid. It must be emphasized that these rotary movements are purely passive and result from the skew pull of the obliquely set ligaments. They occur whether the extending force on the knee is active (quadriceps contraction) or passive.

It should be noted in passing that full extension of the knee is impossible with a flexed hip, but this is merely because of the pull of the hamstring muscles.

From the 'screw-home' or 'locked' position lateral rotation of the femur must precede flexion; this *lateral rotation is produced by the popliteus.* The 'untwisted' knee can now be flexed by the hamstrings.

In the flexed position all the above four ligaments are relaxed, and a smaller femoral surface articulates with the tibial plateau; thus active rotation is possible and is produced by the hamstrings contracting alternately.

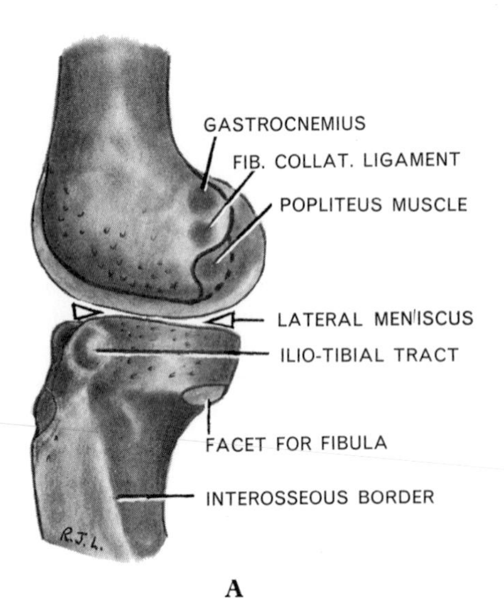

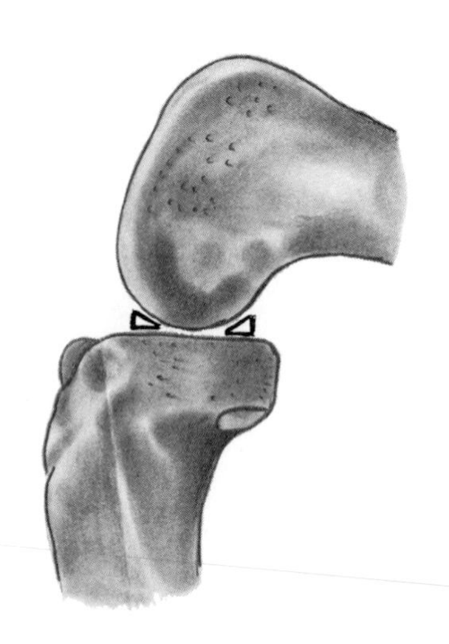

FIG. 3.44. THE ARTICULAR CONTOURS OF THE KNEE JOINT. Lateral view of left knee (semi-diagrammatic), A in extension and B in flexion. In the left figure the heavy line on the femur shows the attachment of the capsule of the knee joint. The interrupted line shows the attachment of the synovial membrane, separated from the capsule by the femoral tendon of popliteus.

The Rôle of the Menisci. The menisci play no significant part in simple flexion and extension; they are passively opened up in extension because a broader femoral condyle separates their anterior and posterior convexities (Fig. 3.44). Flexion and extension take place above the menisci, in the upper compartment of the joint.

The chief rôle of the menisci is in rotation; the only mammal unable to rotate the knee (the fruit bat) is the only mammal to lack menisci and popliteus muscle! In rotation the menisci move with the femur; that is to say, rotation takes place in the lower compartment of the joint.

The surfaces of the menisci are bare, with no synovial membrane. These surfaces keep gentle contact in all knee positions with the articular surfaces of femur and tibia. Thus the menisci spread synovial fluid, and this seems to be the reason for their existence. They are never subjected to weight-bearing thrust, except by accident, in which case they rupture.

The Medial Meniscus. It is true that in rotation the extremities (horns) of the meniscus are immobile on the tibia, but the medial convexity of the meniscus moves with the femur (held to it by the short internal lateral ligament and the posterior fibres of the short upper part of the tibial collateral (medial) ligament; Fig. 3.39). A spiral distortion of the meniscus thus accompanies rotation; this makes the meniscus prone to rupture. If the medial meniscus were fixed around its whole circumference to the tibia, the rotating femur would have to ride up and down the slope of its anterior and posterior curvatures. Such impaction of the meniscus between the two bones obviously does not, in fact, occur. A simple experiment confirms the mobility of the meniscus *on the tibia.* Sit with the knee flexed to a right angle. Rotate the foot and tibia medially; the medial meniscus can be felt level with the articular margins of femur and tibia, filling the gap between them. Now turn the toes out and rotate the tibia laterally; palpate the joint line. The edge of the tibial condyle *and the tibial plateau* can be felt; the convexity of the meniscus has not come forward with the tibia.

The Lateral Meniscus. The extremities are attached to the tibia but the posterior convexity of the meniscus is attached to the femur by the two menisco-femoral ligaments. Lateral rotation of the femur pulls by these ligaments on the posterior convexity of the meniscus, which thus tends to be drawn medially and forwards against the posteriorly advancing lateral femoral condyle. The meniscus would be crushed were it not for the strong attachment of popliteus, which draws the posterior convexity of the meniscus back and down over the slope of the tibial surface, out of harm's way. Medial rotation of the femur is accompanied by relaxation of the popliteus, and elastic recoil of the meniscus. This muscle control on the position of the meniscus renders the structure relatively immune to injury. The lateral meniscus is vastly more mobile than its medial colleague. Mobility *per se* is no insurance against impaction; the most mobile of all, a loose body in the joint, readily becomes impacted between femur and tibia. The lateral meniscus is relatively immune to injury because its mobility is *controlled* by the two menisco-femoral ligaments and by the insertion of the popliteus muscle.

Stability of the Knee Joint. Bony contours contribute nothing to the antero-posterior stability of this joint, but the spine of the tibia prevents sideways gliding of femur on tibia (try it on the dry bones!). The cruciate ligaments are indispensable to antero-posterior stability in flexion. Lateral stability and stability in extension are provided by the collateral and oblique popliteal ligaments. Muscle function is an indispensable factor; the vasti by their expansions (the retinacula) contribute greatly. Vastus medialis is indispensable to the stability of the patella (p. 141). The ilio-tibial tract (gluteus maximus and tensor fasciæ latæ) stabilizes the slightly flexed knee (p. 135).

THE FRONT OF THE LEG

The front of the leg includes the subcutaneous surface of the tibia on the medial side and the extensor muscular compartment on the lateral side.

The **cutaneous nerves** are derived from the femoral nerve over the tibia and from the common peroneal nerve over the extensor compartment (Fig. 3.7, p. 131). The *saphenous nerve* gives off its infrapatellar branch to supply the subcutaneous periosteum of the upper end of the tibia and the overlying skin and then descends with the great saphenous vein, with which it passes in front of the medial malleolus. It ends on the medial side of the foot at the bunion region—the metatarso-phalangeal joint. Halfway down the shin the nerve usually divides into equal halves which lie on either side of the saphenous vein for the rest of their course (PLATE 6). The *lateral cutaneous nerve* of the calf, a branch of the common peroneal, supplies deep fascia and skin over the upper parts of the extensor and peroneal compartments and the *superficial peroneal nerve* replaces it over the rest of these surfaces.

The subcutaneous surface of the tibia has subcutaneous fat attached to its periosteum; there is no deep fascia covering it. The great saphenous vein and the saphenous nerve lie in the fat, accompanied by numerous lymphatic vessels which pass up from the foot to the vertical group of superficial inguinal nodes. In this part of its course the great saphenous vein is connected with the deep veins by communicating channels that perforate the deep fascia on the medial side of the calf and enter the soleus muscle (see p. 132).

The upper end of the subcutaneous surface of the shaft of the tibia receives the tendons of three muscles that converge from the three constituent bones of the os innominatum. They are sartorius (supplied by the femoral, the nerve of the ilium), gracilis (supplied by the obturator, the nerve of the pubis) and semitendinosus (supplied by the sciatic, or ischiadic, the nerve of the ischium) in that order from before backwards. The three tendons are separated by a bursa which lies deep to the flattened sartorius tendon (PLATE 6).

It is noteworthy that all three muscles, running up from the tibia, are as widely separated above as the bony pelvis will allow. Are they three 'guy ropes', helping to stabilize the bony pelvis? If so, it would explain the existence of sartorius and gracilis. Of sartorius we say it is a weak flexor and rotator of both hip and knee—but why have such a weak muscle when each joint already possesses its own powerful flexors and rotators? And gracilis is counted as a weak adductor of the thigh; why have a weak adductor when three other powerful adductors are already there? If indeed guy ropes were to be installed between the shin bone on which we stand and the mobile pelvis high above, they could not be situated to better advantage than are these three muscles (Fig. 3.45).

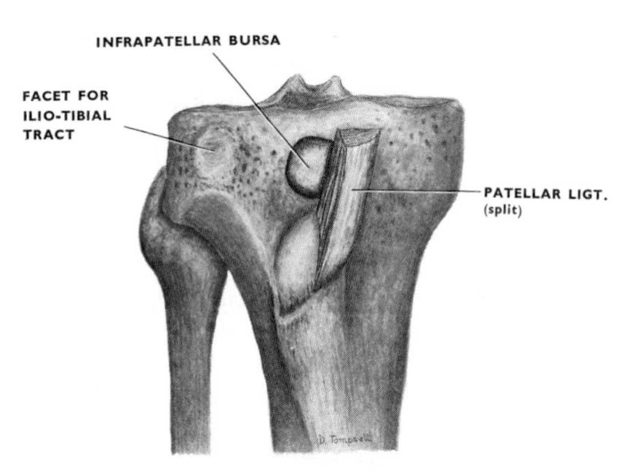

FIG. 3.46. THE HEAD OF THE RIGHT TIBIA.

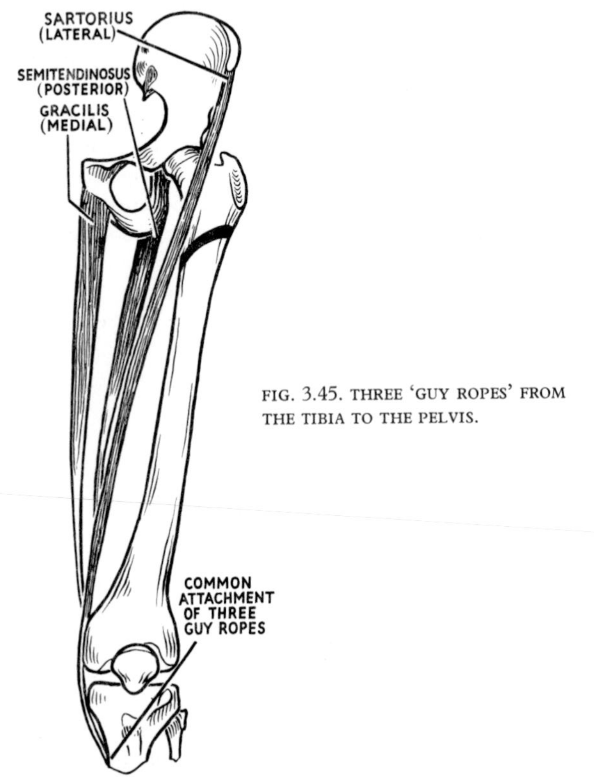

FIG. 3.45. THREE 'GUY ROPES' FROM THE TIBIA TO THE PELVIS.

The **ligamentum patellæ** is received into a *smooth* area that lies obliquely on the tibial tuberosity. This insertion is situated entirely upon the epiphysis of the upper end of the tibia. The patellar retinacula, fibrous expansions from the tendon of the quadriceps femoris, are inserted into the edges of the patella and ligamentum patellæ and into the inferior borders of the tibial condyles. The prepatellar bursa (swollen in housemaid's knee) lies in front of the patella, but there are two infrapatellar bursæ. The superficial infrapatellar bursa (swollen in clergyman's knee) lies in front of the ligamentum patellæ, the deep infrapatellar bursa lies between the ligament and the upper part of the head of the tibia (Fig. 3.46).

The **deep fascia** of the leg covers only muscles, being attached to periosteum at all places where bone is subcutaneous. Below the tibial condyles it encloses the muscles of the leg, being attached to anterior and posterior borders of the tibia. Above the ankle it is attached also to the lateral malleolus and the triangular subcutaneous area of the fibula. Two intermuscular septa pass from its deep surface to become attached to the fibula. They enclose the peroneal compartment. Between the anterior intermuscular septum and the tibia lies the extensor compartment, while between the posterior intermuscular septum and the tibia posteriorly lies the much more bulky flexor compartment or calf of the leg.

The Extensor Compartment of the Leg

The compartment comprises the space between the deep fascia and the interosseous membrane, bounded medially by the extensor surface of the tibia and laterally by the extensor surface of the fibula and the anterior intermuscular septum. In its lower extent the deep fascia does not encircle the leg but is attached to the subcutaneous border of the fibula above the lateral malleolus. Here the fascia is thickened to form the **superior extensor retinaculum,** which is characterized by the fact that the tibialis anterior tendon perforates it, splitting off a lamina, so that there is a double attachment to the tibia. As it lies in the channel the tendon of tibialis anterior possesses a synovial sheath which continues down to its insertion. The other extensor tendons pass deep to the superior extensor retinaculum and at this level possess no synovial sheath. The muscle bellies lie lateral to the tibia but their tendons pass across its lower end beneath the superior retinaculum (Fig. 3.47).

The **tibialis anterior** muscle has a spindle-shaped belly that moulds the anterior subcutaneous border of the tibia into a reciprocal concavity. It arises from the upper two-thirds of the extensor surface of the tibia, from the interosseous membrane and, especially strongly, from the upper part of the deep fascia overlying it. The multipennate fibres converge conically downwards into a central tendon that pierces the superior extensor retinaculum, from which level the tendon is invested with a continuous synovial sheath. The pressure of the tendon rounds off the anterior border of the lower end of the tibia. Tendon and sheath are slung by the inferior extensor retinaculum (p. 169) and the tendon is inserted into a *smooth* facet at the antero-inferior angle of the medial cuneiform and into the adjacent part of the first metatarsal bone. A small bursa separates the tendon from the upper part of the facet.

Note that the tibialis anterior and peroneus longus are symmetrical in their insertions into opposite sides of the adjoining medial cuneiform and first metatarsal base, but tibialis anterior is inserted mainly into the cuneiform while peroneus longus tendon is inserted mainly into the metatarsal.

The muscle is supplied by the deep peroneal and recurrent genicular nerves (L4) and its action is a combined dorsiflexion of the ankle joint and inversion of the foot (p. 185).

The **extensor hallucis longus** muscle arises from the middle two-fourths of the fibula and the adjacent interosseous membrane. (Flexor hallucis longus, too, arises from the fibula.) The muscle lies deep at its origin, but emerges between tibialis anterior and extensor digitorum longus in the lower part of the leg. It passes beneath the superior extensor retinaculum, and is slung by the inferior extensor retinaculum, where it receives a separate synovial sheath. It passes along the medial side of the dorsum of the foot and is inserted into the base of the terminal phalanx of the great toe.

It is supplied by the deep peroneal nerve (L4, 5) and its action is to dorsiflex (anatomically this is to extend) the great toe. Secondarily it is a dorsiflexor of the ankle. Note that dorsiflexion of the great toe 'winds up' the plantar aponeurosis around the 'pulley' of the head of the metatarsal bone and so increases the concavity of the medial longitudinal arch of the foot (p. 187).

The **extensor digitorum longus** arises from the upper three-fourths of the extensor surface of the fibula and from a small area of the tibia across the superior tibio-fibular joint. Much of the muscle gains origin from the anterior intermuscular septum and, to a less extent, from the deep fascia overlying it. It forms its four tendons over the lower part of the tibia, beneath the superior extensor retinaculum. They are slung together by the inferior extensor retinaculum, and are here enclosed with the tendon of peroneus tertius in a common synovial sheath. The four tendons diverge slightly, superficial to extensor digitorum brevis, just beneath the deep fascia on the dorsum of the foot. They are inserted into the lateral four toes (Fig. 3.47).

Their mode of insertion is precisely the same as that of the extensor communis tendons in the hand. The tendon divides into three slips over the proximal phalanx, the central slip being inserted into the base of the middle phalanx. The two side slips reunite after being joined by the tendons of the interossei and lumbricals and are inserted into the base of the distal phalanx. The muscle is supplied by the deep peroneal nerve (L4, 5) and its action is to extend (i.e., dorsiflex) the lateral four toes.

The **peroneus tertius** muscle arises from the lower third of the fibula below the extensor digitorum longus. It is a unipennate muscle whose tendon forms anteriorly, in contact with the superior extensor retinaculum. It passes through the stem of the inferior retinaculum, where it shares the synovial sheath of extensor digitorum longus and is inserted into the dorsum of the base of the fifth metatarsal bone and, by a falciform extension, into the superior surface of that bone as far forwards, in many cases, as its neck (Fig. 3.48).

It is supplied by the deep peroneal nerve (L5). Its action is to dorsiflex and evert the foot. In spite of its small size it possesses fairly good mechanical advantage in dorsiflexion, since its tendon passes some distance in front of the axis of movement of the ankle and its insertion is so far forward on the foot.

The **deep peroneal** (anterior tibial) **nerve** arises within peroneus longus, over the neck of the fibula, at the bifurcation of the common peroneal nerve. It spirals around the neck of the fibula deep to the fibres of extensor digitorum longus, and so reaches the interosseous membrane, on the lateral side of the anterior tibial vessels. With them it lies between extensor digitorum longus and tibialis anterior, the only two muscles in the upper part of the extensor compartment. In the middle of the leg the neuro-vascular bundle lies on the interosseous membrane between the tibial and fibular muscles, i.e., tibialis anterior and extensor hallucis longus. The latter muscle crosses the bundle, so that over the lower end of the tibia the tendons of two muscles lie on each side of the bundle (tibialis anterior and extensor hallucis longus medially, and extensor digitorum longus and peroneus tertius laterally). At this level the deep peroneal nerve has regained its position lateral to the vessels.

The deep peroneal (anterior tibial) nerve supplies the four muscles of the extensor compartment of the leg, and is sensory to the periosteum of the extensor surfaces of tibia and fibula.

The **anterior tibial artery,** formed at the bifurcation of the popliteal artery in the calf, passes forwards above the upper border of the interosseous membrane to reach the extensor compartment. In doing so it lies nearer the fibula than the tibia, with a companion vein on each side. The fibular companion vein may leave a notch in the fibula, visible in a radiograph. The artery with its companion veins runs vertically downwards on the interosseous membrane and crosses the lower end of the tibia midway between the malleoli. It gives off a recurrent branch to the arterial anastomosis around the upper end of the tibia, supplies the muscles of the extensor compartment and gives malleolar branches to both malleolar regions.

Relations of the Anterior Tibial Artery. The tibialis anterior lies to its medial side throughout. Extensor digitorum longus and peroneus tertius lie to its lateral side throughout. Extensor hallucis longus crosses it from fibular to tibial sides. The deep peroneal nerve reaches it from the lateral side, runs in front of it in the crowded space of the middle of the leg and returns to its lateral side below. The *anterior tibial veins* run, one on each side of the artery, in close contact with it and anastomose by cross channels at frequent intervals.

The Tibio-fibular Joints

The fibula articulates with the femur in the embryo, but differential growth of tibia and fibula results in the latter sinking below the level of the plateau of the tibia and making a separate tibio-fibular articulation. This is a synovial joint, with capsule and synovial membrane. The lower ends of the two bones are strongly bound together by ligamentous fibres to form a fibrous joint and the two bones are further held together by an interosseous membrane which connects their interosseous borders. The nature of the tibio-fibular articulations varies widely from animal to animal, and supression of upper or lower end of the fibula is quite common.

The presence of a synovial joint at the upper ends of the two bones in man indicates movement, but this movement is entirely passive and depends upon the variable shape of the talus. The fibular malleolus maintains intimate contact with the lateral surface of the talus in all positions of the ankle joint. The lateral surface of the talus is usually convex in an antero-posterior direction, so that dorsiflexion and plantar flexion at the ankle joint produce rotation of the fibula around its own axis. In such cases the superior tibio-fibular joint surfaces are more nearly horizontal, to allow rotation to occur. If the lateral surface of the talus is plane, the superior tibial and fibular facets are more nearly vertical.

The **superior tibio-fibular joint,** with these variations in the direction of its surface, is surrounded by a capsule whose fibres are thickened anteriorly and posteriorly and from the tibia slope downwards to the fibula. The joint cavity may communicate posteriorly with the bursa under the popliteus tendon and thence with the knee joint.

The **interosseous membrane** consists of strong fibres that slope steeply from the tibia down to the fibula. They do more than bind the bones together, for they resist downward movement of the fibula when the powerful fibular muscles pull on it (flexor hallucis longus, the peronei, extensors hallucis and digitorum longus). The only muscle exerting an upward pull on the fibula is the biceps femoris.

The **inferior tibio-fibular joint** is a fibrous joint, the two bones being strongly bound together by fibres that occupy the triangular area on each bone at the lower end of the interosseous border (Fig. 3.74, p. 203).

THE DORSUM OF THE FOOT

The skin of the dorsum of the foot is supplied by the superficial peroneal nerve, assisted slightly by the deep peroneal, saphenous, and sural nerves (Fig. 3.7, p. 131). The large veins form a **dorsal venous arch** which receives most of its blood by marginal and interosseous tributaries from the sole of the foot. The dorsal venous arch, lying over the heads of the metatarsals, drains from its ends into the great and small saphenous veins.

The **superficial peroneal** (musculo-cutaneous) **nerve** surfaces at the middle of the leg. It passes downwards over the peronei and divides above the ankle into medial and lateral branches which supply the skin of the dorsum of the foot. The medial branch further divides to supply the medial side of the dorsum of the great toe and the sides of the second cleft. The lateral branch divides to supply the third and fourth clefts. The nerve and its divisions are visible in a thin leg

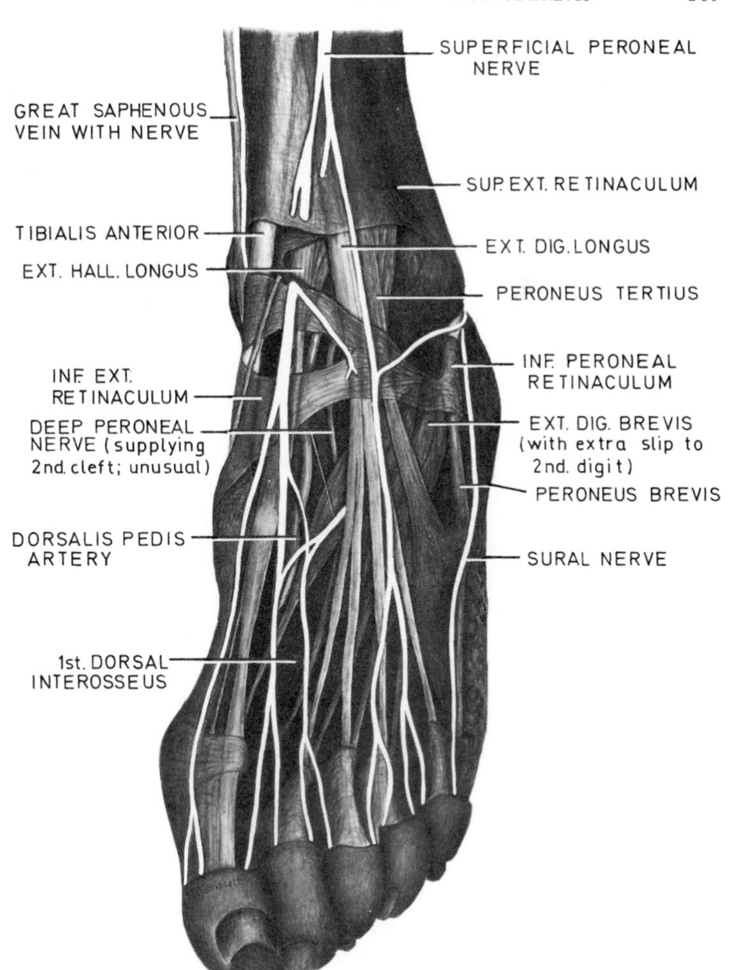

FIG. 3.47. THE DORSUM OF THE LEFT FOOT. Illustration of Specimen S 135 in R.C.S. Museum.

when stretched by plantar-flexion of the foot. The lateral side of the foot and lateral side of the little toe are supplied by the sural nerve, while the first cleft is supplied by the deep peroneal nerve. The terminal phalanges and toe-nails are supplied by the medial and lateral plantar nerves. The medial side of the foot as far forward as the bunion area (metatarso-phalangeal joint) is supplied by the termination of the saphenous nerve.

The subcutaneous layer of the dorsum of the foot, as of the hand, contains but little fat in most people, and the veins are consequently easily seen when distended.

The **deep fascia** on the dorsum of the foot binds down the underlying tendons, and a Y-shaped thickening beneath it (the inferior extensor retinaculum) prevents bow-stringing of the extensor tendons as they pass across the front of the ankle joint, acting like an ankle strap.

The **inferior extensor retinaculum** arises by a stem from the anterior part of the upper surface of the calcaneus, on the lateral border of the dorsum of the foot. From the stem two limbs diverge. The upper limb is attached to the medial malleolus, the lower limb arches across the tendons on the dorsum and blends with the plantar aponeurosis under the medial longitudinal arch of the foot. The tendons are usually said to pierce the extensor retinaculum, but this is not strictly true. Most of the inferior retinacular fibres sling around the tendons and return to be attached to the calcaneus, and only the superficial fibres pass across to the insertions mentioned at the medial malleolus and plantar aponeurosis (Fig. 3.51). Former names of the inferior extensor retinaculum were the 'sling ligament' or the 'frondiform ligament' of Retzius. All the extensor tendons are enclosed in synovial sheaths where they are slung in the inferior extensor retinaculum.

The **extensor digitorum brevis** is a muscle whose fleshy belly can be seen in most feet and felt in all. It arises from the upper surface of the calcaneus and from the deep surface of the stem of the Y-shaped inferior extensor retinaculum. It passes obliquely across the dorsum of the foot and gives off four tendons to the *medial* four toes.

The tendon to the great toe is different from the others and is deservedly given a special name, the **extensor hallucis brevis.** Its belly usually separates early from the main muscle mass and the tendon is inserted separately into the base of the proximal phalanx of the great toe. As in the thumb, so in the great toe there is no dorsal extensor expansion. The remaining three tendons are inserted into the dorsal extensor expansions of the second, third and fourth toes. All four tendons of the muscle pass deep to the tendons of the extensor digitorum longus. The muscle is supplied on its deep surface by the deep peroneal nerve and its action is to extend the medial four toes. It is particularly of value when the long extensor is out of action, in the fully dorsiflexed ankle, as in the moment just before the take off from the hind foot in walking, running, and jumping. The toes thus extended lengthen the grip of the foot on the ground, and this provides a faster propulsive take-off.

The anterior tibial artery, lying over the lower end of the tibia midway between the malleoli, extends forwards as the **dorsalis pedis artery.** This runs to the base of the first intermetatarsal space and passes down into the sole, where it joins the lateral plantar artery to complete the plantar arch. It lies between the tendon of extensor hallucis longus medially and the digital branch of the deep peroneal nerve laterally, and it is crossed by the tendon of extensor hallucis brevis. Its pulsation can easily be felt. It has three named branches. The **lateral tarsal artery** runs laterally beneath extensor digitorum brevis, to supply that muscle and the underlying tarsal bones. The **arcuate artery** runs laterally beneath the tendons of extensor digitorum brevis over the bases of the metatarsal bones. It gives off dorsal metatarsal arteries to supply the lateral three clefts. Each metatarsal artery gives off a perforating branch at the posterior and anterior end of its intermetatarsal space to communicate with the plantar arch and its digital branches. It is the accompanying perforating veins that are responsible for bringing much of the blood from the sole of the foot through the intermetatarsal spaces to the dorsal venous arch. The *first dorsal metatarsal artery*, a direct continuation of the dorsalis pedis, supplies the first cleft and the medial side of the dorsum of the great toe.

The **deep peroneal** (anterior tibial) **nerve** crosses the tibia lateral to the artery, midway between the malleoli. It passes forward, deep to the tendons, on the lateral side of the dorsalis pedis artery, to pierce the deep fascia and supply the first cleft. It gives off a branch which curves laterally beneath the muscle belly of extensor digitorum brevis and supplies this muscle and the underlying periosteum and joint capsules.

THE PERONEAL COMPARTMENT

This muscular compartment lies between the peroneal surface of the fibula and deep fascia of the leg and is bounded in front and behind by the anterior and posterior intermuscular septa. It contains the peronei longus and brevis and the superficial peroneal nerve. Its blood supply is derived from branches of the peroneal artery which pierce flexor hallucis longus and the posterior intermuscular septum. Its veins drain, for the most part, into the small saphenous vein.

The **peroneus longus** muscle arises from the upper two-thirds of the peroneal surface of the fibula, from the head of that bone and, across the superior tibiofibular joint, from a small area of the lateral tibial condyle. Its fibres take origin also from the intermuscular septa. The **peroneus brevis** muscle arises from the lower two-thirds of the fibula; in the middle third of the bone its origin lies in front of that of peroneus longus and the two muscles and their tendons maintain this relationship.

The broad tendon of peroneus brevis lies behind (and grooves) the lateral malleolus. The narrower tendon of peroneus longus lies on that of brevis and does not come into contact with the malleolus (Fig. 3.49). The two tendons pass forwards to the peroneal trochlea on the lateral surface of the calcaneus, which separates them. The tendon of brevis passes above the peroneal trochlea to be inserted into the tip of the tubercle (styloid process) at the base of the fifth metatarsal bone. The tendon of peroneus longus passes below the peroneal trochlea and enters the sole of the foot, lying against the posterior ridge of the groove on the cuboid bone. Here the tendon possesses a sesamoid fibro-cartilage which often ossifies. The tendon crosses the sole obliquely to be inserted into the base of the first metatarsal and the adjoining part of the medial cuneiform (Fig. 3.55).

The tendons are bound down at the lateral malleolus by the **superior peroneal retinaculum,** a band of deep fascia that extends from the tip of the malleolus to the calcaneus, and at the peroneal trochlea by the **inferior peroneal retinaculum** (Fig. 3.48). This is a band of fascia attached to the peroneal trochlea and to the calcaneus above and below the peroneal tendons. Its upper part is continuous with the stem of the Y-shaped inferior extensor retinaculum. The two tendons are enclosed in a common *synovial sheath* from above the lateral malleolus to the peroneal trochlea, where the sheath divides to accompany each tendon separately to its insertion.

The peronei longus and brevis are supplied by the superficial peroneal nerve and their action is to evert, and weakly plantar-flex, the foot. In addition, peroneus longus is a factor in maintaining the lateral longitudinal and transverse arches of the foot (p. 188).

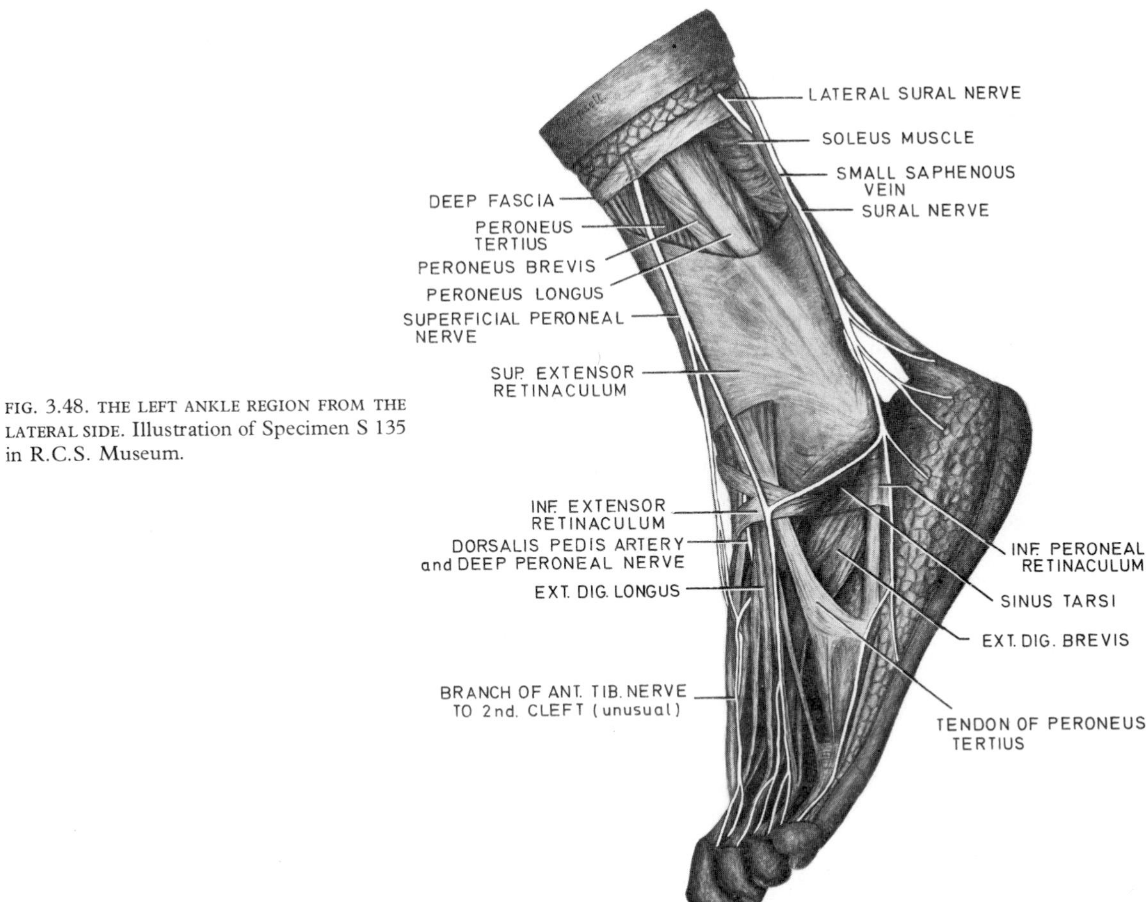

FIG. 3.48. THE LEFT ANKLE REGION FROM THE LATERAL SIDE. Illustration of Specimen S 135 in R.C.S. Museum.

The **superficial peroneal** (musculo-cutaneous) **nerve** is formed in the substance of peroneus longus at the division of the common peroneal nerve. It passes downwards in the muscle and emerges at its anterior border behind the anterior intermuscular septum. It supplies both peronei and pierces the deep fascia halfway down the leg and soon divides into medial and lateral branches. In thin individuals these can be seen or felt as ridges under the skin if they are put on the stretch by full plantar flexion of the foot (Fig. 3.48).

THE CALF

The skin of the calf is supplied by the termination of the posterior cutaneous nerve of the thigh halfway to the ankle. Below this level the sural and lateral sural nerves, from tibial and common peroneal nerves, supply the back and lateral side of the calf, and the saphenous nerve supplies the medial side (Fig. 3.25, p. 145).

The **small** (short) **saphenous vein,** draining the lateral side of the dorsal venous arch and the lateral margin of the foot, lies with the sural nerve behind the lateral malleolus. It passes upwards in the subcutaneous fat along the midline of the calf and pierces the deep fascia on the roof of the popliteal fossa to enter the popliteal vein (Fig. 3.5, p. 130). It communicates by several channels with the great saphenous vein.

The deep fascia is a continuation downwards from the popliteal fascia and is attached to the posterior border of the tibia. It surrounds the calf and the peroneal and extensor compartments to become attached to the anterior border of the tibia. Where the lower end of the fibula becomes subcutaneous the deep fascia is attached to the periosteum. The posterior intermuscular septum divides the calf from the peroneal compartment.

The deep fascia is thickened above the heel, where it is attached to the tibia and fibula across the tendo calcaneus (Achillis), forming a 'pulley' for the tendon and separated from it by a bursa. A further thickening of fascia bridges the deep flexor tendons and neurovascular bundle; it stretches from the medial malleolus to the back of the calcaneus, and is called the *flexor retinaculum* (PLATE 6).

The Gastrocnemius Muscle. Examine the lower end of a femur. The epiphyseal junction passes transversely through the adductor tubercle. The lateral head of gastrocnemius arises from the epiphysis, the medial head from both epiphysis and shaft. The plantaris arises from the shaft. The lateral head of gastrocnemius arises on the lateral surface of the lateral condyle, by a tendon that leaves a smooth pit above that of popliteus; the pits are separated by the epicondyle (Fig. 3.44). A few fleshy fibres arise from the condyle above the pit, towards the lower end of the lateral supracondylar line. The medial head of gastrocnemius arises by a tendon from a smooth shallow pit on the medial condyle at the lower end of the medial supracondylar line and by muscular fibres from an area of roughened bone on the popliteal surface of the shaft of the femur (Fig. 3.68, p. 198). In the fœtus the medial head, like the lateral, arises wholly from the epiphysis; the upward extension of this origin to encroach on the popliteal surface of the shaft of the femur is a postnatal event.

The two heads converge to lie side by side, where the broad bellies of the muscle have a dense aponeurosis beneath them, bearing on the soleus muscle. The medial head is longer at each end; it extends below the lateral head, as inspection of the living calf will show. The flat aponeurosis blends with that of soleus at the lower border of the lateral head. In the midline it blends with soleus aponeurosis by a criss-cross exchange of fibres. The medial half of the aponeurosis is separate from soleus down to the heel, and the slender tendon of plantaris lies between. A bursa lies between the medial head and the capsule over the medial condyle of the femur. It communicates with the knee joint and it may communicate also with the semimembranosus bursa (Fig. 3.43).

The aponeurosis forms, with that of soleus, the tendo calcaneus (Achillis), which is inserted into a smooth transverse area on the middle third of the posterior surface of the calcaneus. A bursa lies between it and the upper part of the calcaneus. A second bursa lies between it and the thickened deep fascia two inches above its insertion; inflammation here is frequently miscalled 'tenosynovitis'. There is no synovial sheath around the tendo calcaneus (PLATE 6).

The **plantaris** is a vestigial muscle showing the short belly and long tendon characteristic of phylogenetic degeneration. It arises from the shaft of the femur at the lower part of the lateral supracondylar line, lying edge to edge with the lateral head of gastrocnemius. Its slender tendon runs deep to the medial head of gastrocnemius and down the midline of the calf, between the aponeuroses of gastrocnemius and soleus, to the calcaneus at the medial side of the tendo calcaneus (Fig. 3.51). The tendon is flat as it lies sandwiched between the two aponeuroses, but it can be unravelled and proves to be a wide ribbon twisted spirally upon itself.

The gastrocnemius and plantaris are each innervated by the tibial nerve (S1, 2) and their action is to plantarflex the ankle joint. In conjunction with contraction of the knee extensors, they are drawn proximally as the knee extends and so elongated to increase their pull on the heel (see below).

The Soleus Muscle. There is great morphological resemblance between soleus in the leg and flexor digitorum superficialis in the forearm. Structurally they are different. Flexor digitorum superficialis has parallel fibres for range of movement of the fingers, soleus is multipennate for power in propulsion. The flexor digitorum brevis muscle in the sole of the foot can be regarded as the divorced distal part of soleus, cut off from it by the posterior projection of the calcaneus.

The muscle arises from the fibula and tibia, mostly from the latter bone. The upper fourth of the fibula, including the head of the bone, gives origin to the muscle, whence a fibrous arch carries it in continuity to the soleal line of the tibia and the fascia upon popliteus above this line, and so along the posterior border of the middle third of the tibia, that is, a hand's breadth below the lower end of the soleal line (Fig. 3.49). The muscle has a curious and characteristic structure. It is flat, and there is a dense aponeurosis upon either surface.

Between the two aponeurotic lamellæ lies the great bulk of the soleus, made up of muscle fibres that slope downwards from the anterior to the posterior lamella; these fleshy fibres are visible at the medial and lateral borders of the muscle. The posterior (superficial) lamella is continued at its lower end into the tendo calcaneus, and the muscle fibres of soleus are received into its deep surface down to within a short distance of the calcaneus (PLATE 6). The tendo calcaneus is received into a smooth transverse area across the middle third of the posterior surface of the calcaneus (PLATE 46, p. *xxiv*).

If the soleus is detached from its fibular origin and turned aside the anterior surface is brought into view. Here a slender bipennate muscle belly is attached to the centre of the aponeurosis. This belly lies on the neuro-vascular bundle, in the groove between flexor digitorum longus and flexor hallucis longus.

Perforating veins from the great saphenous vein enter the substance of soleus. The muscle contains a rich plexus of small veins, and these are pumped empty by contraction of the muscle, which in thus aiding venous return acts as a 'peripheral heart'.

The soleus is supplied by two separate branches of the tibial nerve (S1, 2), one from above the muscle in

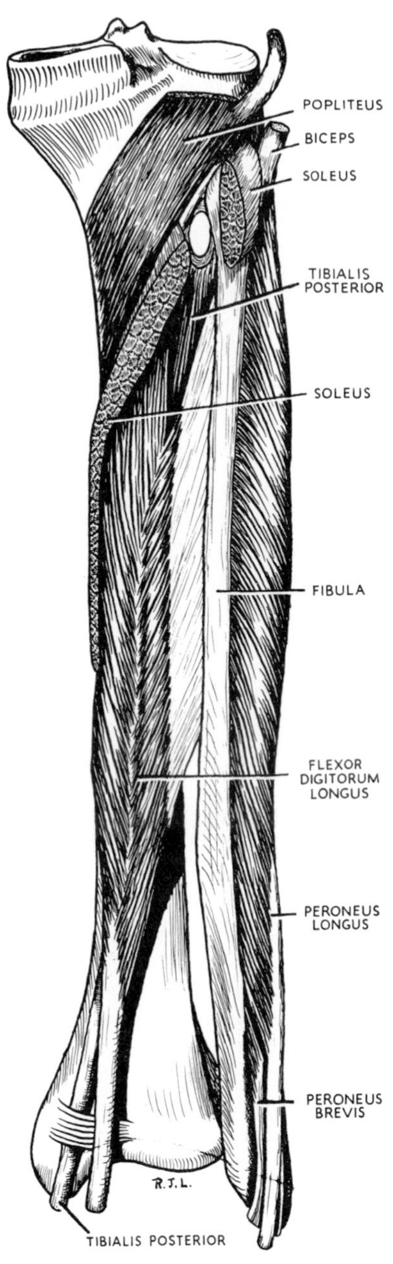

FIG. 3.49. THE FLEXOR DIGITORUM LONGUS MUSCLE. It is a bipennate muscle, arising by flesh from the tibia and by a broad aponeurosis from the fibula. Drawn from Specimen S 113 in R.C.S. Museum.

the popliteal fossa and one on its deep surface in the calf. In cases of intractable intermittent claudication both branches must be cut if soleus is to be completely denervated. Soleus is supplied with blood by branches of the posterior tibial artery that enter perforations in the aponeurosis on its deep surface. Passing under the fibrous arch between fibula and tibia is the neuro-vascular bundle (popliteal artery and vein and tibial nerve).

Soleus and gastrocnemius act together on the tendo calcaneus; the action of the weak vestigial plantaris is negligible. The powerful multipennate soleus is an anti-gravity muscle. In standing it contracts alternately with the extensor muscles of the leg to maintain balance. It is a very strong but relatively slow plantar flexor of the ankle joint, a necessary mechanical result of the obliquity of its multipennate fibres. The gastrocnemius bellies are not multipennate; their fibres all lie roughly vertical in the long axis of the leg. They provide the necessary 'whip-like' contraction that aids rapid propulsion. This action is greatly increased if the flexed knee is simultaneously extended, for the origin of the contracting muscle is thereby pulled upwards and the tendo calcaneus with it. In propulsion the powerful, multipennate soleus overcomes the inertia of the body weight (bottom gear), and when movement is under way the quicker-acting gastrocnemius greatly increases the speed of movement (top gear). One strolls along quietly mainly with soleus; one wins the long jump mainly with gastrocnemius (see p. 188).

Deep Muscles of the Calf

The deep muscles of the calf consist of flexor digitorum longus, flexor hallucis longus and tibialis posterior. Their three tendons pass under the flexor retinaculum into the sole of the foot. Those of tibialis posterior and flexor hallucis longus are parallel throughout their course, the former medial to the latter. The tendon of flexor digitorum longus takes an oblique course *superficial to both*. In the calf it lies medially, crosses tibialis posterior to lie under the flexor retinaculum between the two, and in the sole it crosses flexor hallucis longus to pass to the lateral four toes (Fig. 3.50).

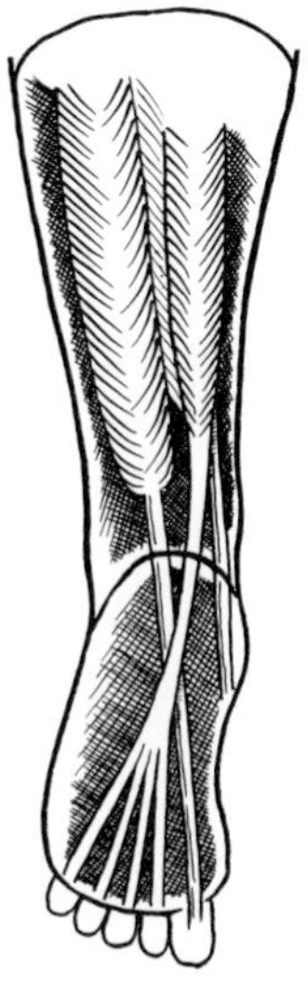

FIG. 3.50. BLACKBOARD DIAGRAM SHOWING THAT FLEXOR DIGITORUM LONGUS CROSSES TIBIALIS POSTERIOR IN THE LEG AND FLEXOR HALLUCIS LONGUS IN THE SOLE. TIBIALIS POSTERIOR AND FLEXOR HALLUCIS LONGUS ARE PARALLEL.

The flexor digitorum longus muscle is often described inaccurately. It is a bipennate muscle that arises from *both bones* of the leg.

Its tibial origin is by flesh from the posterior surface of the bone below the soleal line, but *it arises also from the fibula by a broad aponeurosis* whose oblique fibres are replaced by flesh to form the lateral half of the bipennate muscle (Fig. 3.49). The tendon forms centrally in the bipennate mass and slopes downwards across the tendon of tibialis posterior in the lower part of the leg. Passing beneath the flexor retinaculum it enters the sole of the foot, and crosses the tendon of flexor hallucis longus. At this point it divides into four tendons, the medial two of which *receive a strong slip from the tendon of flexor hallucis longus.* The four tendons at their commencement receive the insertion of the flexor accessorius muscle. More distally each gives origin to a lumbrical muscle. The tendons pass into the fibrous flexor sheaths of the lateral four toes, perforate the tendons of flexor digitorum brevis, and are inserted into the bases of the distal phalanges (for lumbricals, see p. 178).

The muscle is supplied by the tibial nerve (S1, 2). Its principal action is to plantar-flex the lateral four toes and, secondarily, to plantar-flex the ankle joint. Its tonus is, with that of the other deep calf muscles, indispensable in maintaining the longitudinal arch of the foot.

The **flexor hallucis longus** muscle is the bulkiest and most powerful of the three deep muscles of the calf and its importance is usually not given due credit. It is a multipennate muscle whose fibres arise from the flexor surface of the fibula *and from the adjoining aponeurosis of the flexor digitorum longus.* Below this the muscle arises in continuity from the whole flexor surface of the fibula and the lower part of the interosseous membrane (Fig. 3.73, p. 202).

The fibres spiral down to be inserted into a central tendon which escapes from the muscle just at the lower end of the tibia. 'Beef to the heel' describes the flexor hallucis longus (PLATE 6). The tendon grooves the posterior process of the talus and the under surface of the sustentaculum tali, from which it passes directly forwards like a bow-string beneath the arched medial border of the foot to be inserted into the base of the distal phalanx of the great toe (Fig. 3.61, p. 186).

It is crossed in the sole by the tendons of flexor digitorum longus and gives a strong slip to the medial two of these (those for the second and third toes). The peroneal artery runs down deep to it on the fibular aponeurosis of flexor digitorum longus. The artery is covered by a fibrous roof on the deep surface of flexor hallucis longus.

The muscle is supplied by the tibial nerve (S1, 2). Its principal action is to flex the great toe, and it is significant that this is the 'take-off' point, the last part to leave the ground in propulsion. It plantar-flexes the ankle joint simultaneously.

In addition, the pull of this powerful muscle is the most important single factor in maintaining the medial longitudinal arch of the foot (p. 187).

The **tibialis posterior muscle** arises from the interosseous membrane and the adjoining surfaces of both bones of the leg below the origin of soleus and from the fibular aponeurosis of flexor digitorum longus. It is a unipennate muscle, the tendon emerging from its medial side and lying in close contact with the edge of the tibia. The tendon grooves the back of the medial malleolus; the groove in life is lined with fibro-cartilage which is lost from the dried bone. It passes forward above the medial side of the sustentaculum tali, and is inserted into the tuberosity of the navicular. The glib statement of the average student that it is inserted into 'every bone of the tarsus except the talus' is misleading (see p. 180). Most of the 'expansions' that pass from its insertion on the navicular to other tarsal bones are ligaments in their own right, and not extensions of the tendon. Its blood supply is derived from the *peroneal artery* by branches that pierce the aponeurosis of flexor digitorum longus. Very few branches of the posterior tibial artery reach the muscle.

It is supplied by the tibial nerve (L4, see p. 29). Its action is to invert and adduct the forefoot and, since it passes behind the medial malleolus, to plantar-flex the ankle joint.

The **posterior tibial artery** arises at the lower border of popliteus, where the popliteal artery divides into anterior and posterior tibial branches. It passes under the fibrous arch in the origin of soleus and runs down on the fibular aponeurosis of flexor digitorum longus, between the muscle bellies of that muscle and flexor hallucis longus. It ends under the flexor retinaculum by dividing into medial and lateral plantar arteries at a slightly higher level than the point of bifurcation of the tibial nerve. It is accompanied throughout its course by a pair of venæ comitantes which communicate with each other at frequent intervals around the artery (Fig. 3.33, p. 155).

Its branches are (*a*) *nutrient* to the tibia, the largest single nutrient artery in either limb. The vessel pierces the fibular aponeurosis of flexor digitorum longus, passes medially beneath the muscle and passes downwards into the bone between the tibial origins of flexor digitorum longus and tibialis posterior.

(*b*) *Muscular* to the overlying soleus, tendo calcaneus and skin and

(*c*) *Peroneal.* This artery arises an inch below the commencement of the posterior tibial and runs into a tunnel whose fibrous roof gives attachment to muscle fibres of flexor hallucis longus. It gives muscular

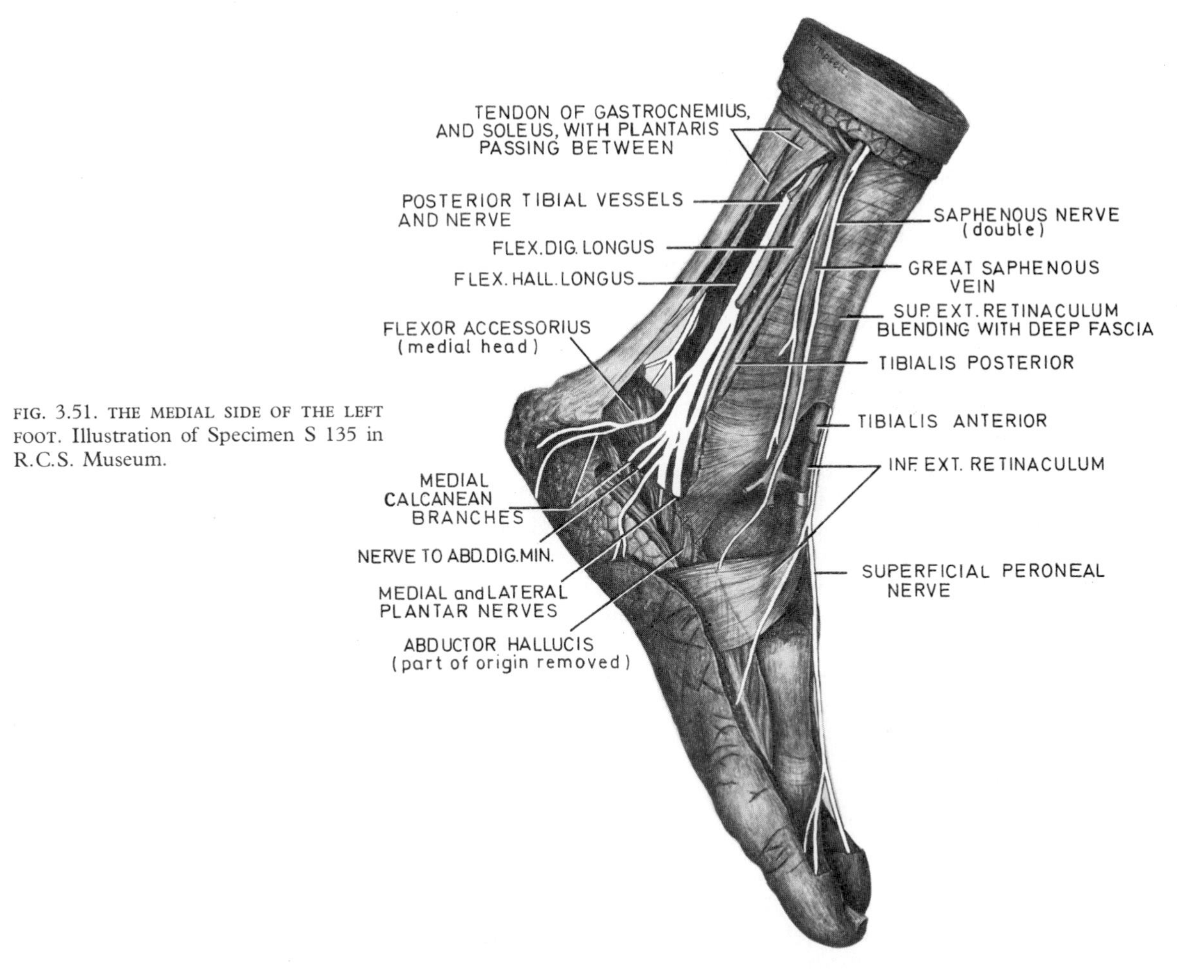

FIG. 3.51. THE MEDIAL SIDE OF THE LEFT FOOT. Illustration of Specimen S 135 in R.C.S. Museum.

branches that perforate the fibular aponeurosis of flexor digitorum longus to supply tibialis posterior and others that wind around the fibula to supply the peronei longus and brevis. It gives a nutrient artery to the fibula. It ends by dividing into a perforating branch which pierces the interosseous membrane to enter the extensor compartment and a lateral calcanean branch which ramifies behind the lateral malleolus and over the lateral side of the heel.

The (posterior) **tibial nerve** runs straight down the midline of the calf, deep to soleus, lying on the fibular aponeurosis of flexor digitorum longus. The posterior tibial artery is at first lateral to it, but passes deep to it at the origin of its peroneal branch and continues downwards on its medial side. The nerve ends under the middle of the flexor retinaculum by dividing into the medial and lateral plantar nerves, at a slightly lower level than the bifurcation of the artery.

It is the nerve of the flexor compartment, giving muscular branches to the deep surface of soleus, to flexors digitorum longus and hallucis longus, and tibialis posterior. It gives off several cutaneous twigs, the *medial calcanean nerves*, to the skin of the heel, including the weight-bearing surface (Fig. 3.51).

THE SOLE OF THE FOOT

The skin is supplied posteriorly, over the weight-bearing part of the heel, by the medial calcanean nerves. The sole is supplied from the medial and lateral plantar nerves by branches that perforate the plantar aponeurosis along each edge of the strong central portion. Blood vessels accompany all the cutaneous nerves of the sole. The plantar surfaces of the toes are supplied by digital branches of the medial (three and a half toes) and lateral (one and a half toes) plantar nerves. Note that the plantar digital nerves supply skin on the dorsum of each toe proximal to the nail bed. Compare the median and ulnar nerve distributions to the fingers; they are identical.

The **subcutaneous tissue** in the sole, as in the

palm, differs from that of the rest of the body in being more fibrous. Fibrous septa divide the tissue into small loculi which are filled with a rather fluid fat under tension, so that the cut tissue bulges. This makes a shock-absorbing pad, especially over the heel. The septa anchor the skin to the underlying plantar aponeurosis, to improve the 'grip' of the sole.

The **plantar aponeurosis** covers the whole length of the sole. It arises posteriorly from the medial and lateral tubercles of the calcaneus and from the back of that bone below the insertion of the tendo calcaneus. It fans out over the sole and is inserted by five slips, into each of the five toes. The digital slips bifurcate for the passage of the flexor tendons and are inserted around the edges of the fibrous flexor sheaths and into the transverse ligaments that bind together the metatarsal heads. From each edge of the plantar aponeurosis a septum penetrates the sole, separating the flexor digitorum brevis from the abductors of big and little toes. The septa are attached to the first and fifth metatarsal bones. The abductors of the big and little toes, lying along the margins of the sole, are covered by deep fascia that is much thinner than the dense plantar aponeurosis (Fig. 3.52).

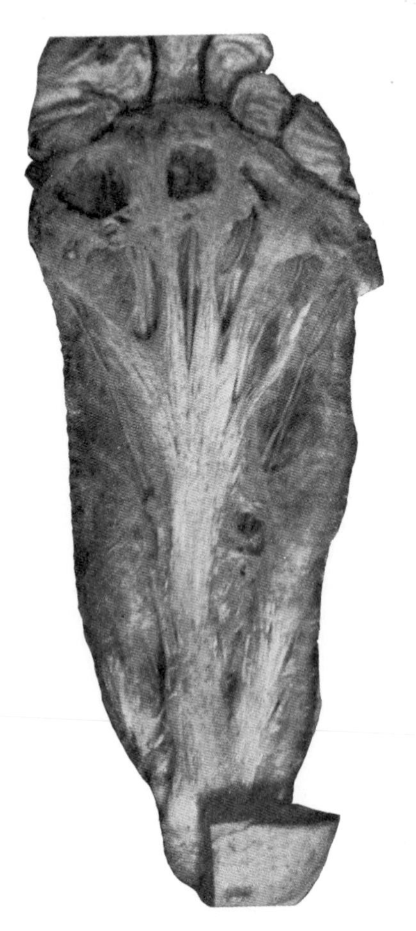

FIG. 3.52. THE PLANTAR APONEUROSIS OF THE LEFT FOOT. Photograph of Specimen S 117 A in R.C.S. Museum.

The **muscular layers** of the sole are much the same as those of the palm (see p. 95). Four layers are conventionally described. The superficial consists of three short muscles that cover the sole, beneath the plantar aponeurosis. Next lies the second layer, consisting of long tendons to the digits, and their connexions. The third layer consists of the short muscles of the great and little toes; it is confined to the metatarsal region of the foot. The fourth layer consists of both plantar and dorsal interossei and it includes also the tendons of peroneus longus and tibialis posterior. Apart from these last two tendons all pass to the toes, so naturally the shortest muscles lie deepest.

The plantar arteries and nerves lie between the first and second layers.

The First Layer

Three short muscles lie side by side along the sole of the foot. The central of these, flexor digitorum brevis, is the counterpart of flexor superficialis in the upper limb, and is represented in the palm by the four tendons of that muscle.

Flexor digitorum brevis arises from the medial tubercle of the calcaneus and its fleshy belly lies in contact with the strong central part of the plantar aponeurosis. It divides into four tendons which pass to the lateral four toes. Each tendon enters the fibrous flexor sheath on the plantar aspect of its digit, divides and spirals around the long flexor tendon, and partially reunites in a chiasma before dividing again to be inserted into the sides of the middle phalanx. This is an identical arrangement with that of the flexor digitorum superficialis. The muscle is supplied by the *medial plantar nerve* and its action is to flex the toes with equal effect in any position of the ankle joint.

The **abductor hallucis** arises from the medial tubercle of the calcaneus and high up from the deep fascia that overlies it (Fig. 3.60). It is inserted into the medial side of the proximal phalanx of the great toe. It is supplied by the medial plantar nerve. Its contraction abducts the great toe, an action largely lost in boot-wearing races, but retained in the bare-footed peoples.

The **abductor digiti minimi** has an unexpectedly wide origin, from both medial and lateral tubercles of the calcaneus, deep to the origin of flexor digitorum brevis. It lies side by side with the latter muscle, along the lateral margin of the foot. Its tendon is inserted into the lateral side of the proximal phalanx of the fifth toe. Some of its medial fibres are usually inserted into the tubercle of the fifth metatarsal bone. It is supplied by a branch from the main trunk of the lateral plantar nerve. Its contraction abducts the little toe, an action, however, which is retained but little in modern man.

The Second Layer

This consists of the long flexor tendons and their connexions in the sole.

The **tendon of flexor hallucis longus** passes forward like a bow-string beneath the medial longitudinal arch of the foot (Fig. 3.61). Posteriorly it lies in a groove beneath the sustentaculum tali. Further forward it is crossed by the tendon of flexor digitorum longus, to the medial two of whose divisions it gives off a strong slip. It next lies in a groove between the two sesamoids beneath the head of the first metatarsal bone, and finally is inserted into the base of the distal phalanx of the big toe. It is invested by a synovial sheath throughout its whole course in the foot.

The **tendon of flexor digitorum longus** enters the sole on the medial side of the tendon of flexor hallucis longus. At its point of division into its four tendons of insertion it crosses the tendon of flexor hallucis longus, which gives a strong slip to the tendons for the second and third toes. At this point it also receives the insertion of flexor accessories (Fig. 3.53). The four tendons pass forwards in the sole deep to those of flexor digitorum brevis and after giving off the lumbricals they enter the fibrous sheaths of the lateral four toes. Each tendon perforates the tendon of flexor digitorum brevis and passes on to be inserted into the base of the distal phalanx.

The **flexor accessorius** muscle arises by a large medial head which is fleshy and by a small lateral head which is tendinous. The medial head arises from the medial surface of the calcaneus almost as far back as its posterior border and from the medial tubercle (Fig. 3.51). The lateral head is a flat tendon which arises from the lateral tubercle of the calcaneus and converges to the muscle belly. The posterior part of the long plantar ligament is visible in the triangular interval between the two heads. The muscle belly is inserted into the tendon of flexor digitorum longus at the point where it is bound by a fibrous slip to the tendon of flexor hallucis longus and where it breaks up into its four tendons of insertion.

The flexor accessorius is supplied by the main trunk of the lateral plantar nerve. Its action, in pulling on

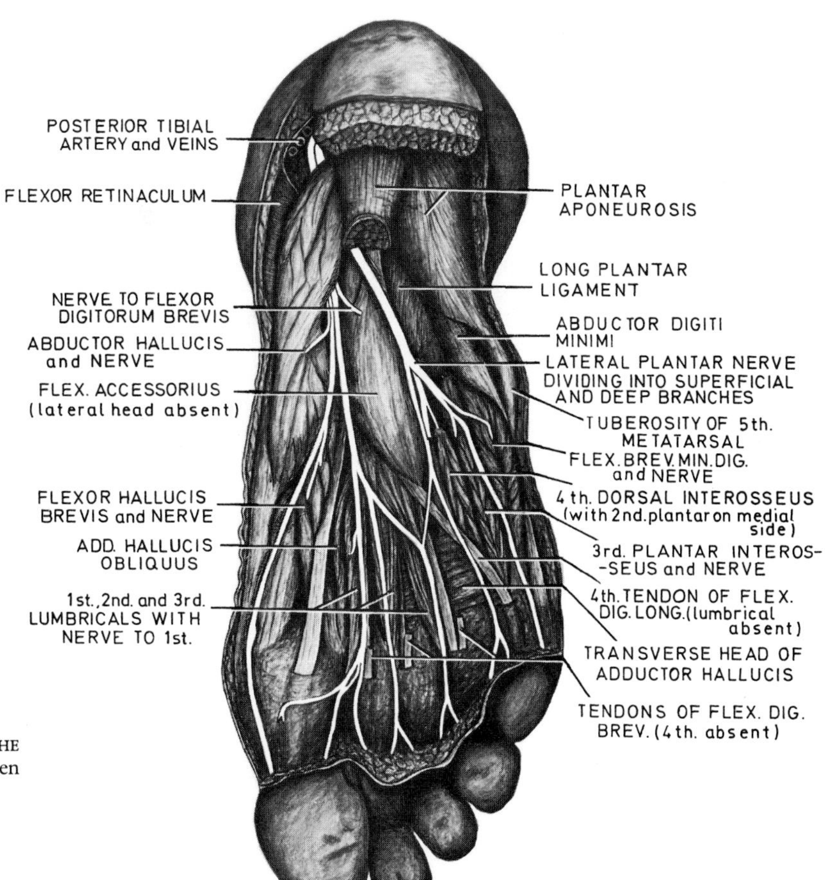

FIG. 3.53. THE PLANTAR NERVES OF THE RIGHT FOOT. Illustration of Specimen S 135.2 in R.C.S. Museum.

the tendons of flexor digitorum longus, provides a means of flexing the lateral four toes in any position of the ankle joint, particularly in full plantar-flexion (here the flexor digitorum longus is so shortened as to be out of action). It also straightens the pull of the long flexor tendons on the toes.

The muscle is known as *quadratus plantæ* in North America, but the name 'accessory flexor' is better, for it indicates the rôle of the muscle.

The **lumbrical muscles** arise from the tendons of flexor digitorum longus, pass forward on the medial (tibial, or pre-axial, as in the hand) sides of the metatarsophalangeal joints of the lateral four toes. Their tendons lie on the plantar surfaces of the deep transverse ligament of the metatarsal heads and pass dorsally to be inserted into the extensor expansions. As in the hand, a lumbrical supplied by the medial plantar (cf. median) nerve is unicipital, one supplied by the lateral plantar (cf. ulnar) nerve is bicipital. In the foot only the first lumbrical is supplied by the medial plantar nerve; it arises by a single head from its own tendon. The lateral three lumbricals are supplied by the lateral plantar nerve (deep branch) and each arises by two heads from the adjoining sides of the tendons.

The action of the lumbricals is to maintain extension of the digits at the interphalangeal joints while the flexor digitorum longus tendons are flexing the toes, so that in walking and running the toes do not buckle under.

The Third Layer

This consists, like the first layer, of three muscles but they are confined to the metatarsal region of the foot. Two act on the big toe, one on the little toe.

The **flexor hallucis brevis** muscle lies against the under surface of the metatarsal bone of the first toe (Fig. 3.54). It arises from the under surface of the cuboid by a slender slip and from the under surfaces of all three cuneiforms, blending here with the expansions from the insertion of tibialis posterior. The belly of the muscle splits into two parts whose edges are in

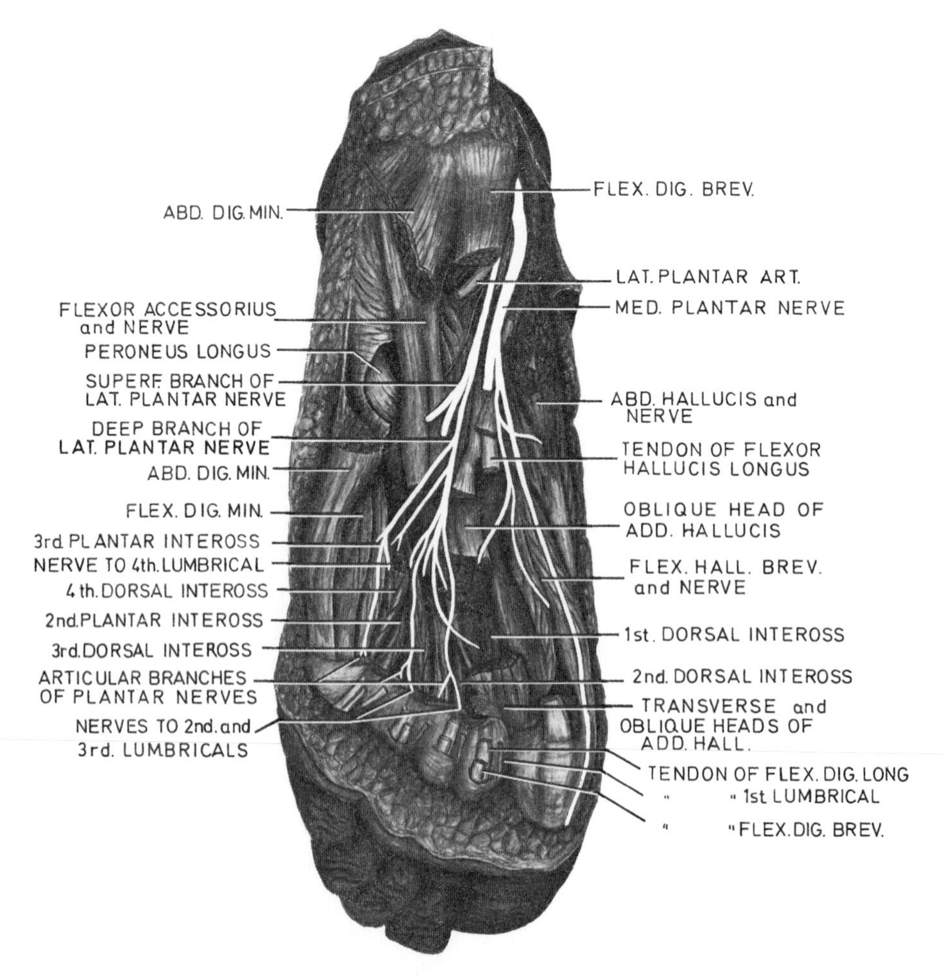

FIG. 3.54. THE DEEP BRANCH OF THE LEFT LATERAL PLANTAR NERVE. Illustration of Specimen S 135.1 in R.C.S. Museum.

contact. They are inserted, each by way of a sesamoid bone, into the medial and lateral sides of the base of the plantar surface of the proximal phalanx of the great toe. The medial insertion blends with that of abductor hallucis, the lateral with that of adductor hallucis. It is supplied by the medial plantar nerve and its action is to flex the proximal phalanx of the big toe.

The **adductor hallucis** muscle has two heads, a large oblique and a small transverse. The oblique head arises anterior to flexor hallucis brevis, from the long plantar ligament where it roofs over the peroneus longus tendon, and from the bases of the second, third and fourth metatarsal bones (Fig. 3.53). The slender transverse head has no bony origin; it arises from the deep transverse ligament and from the under surfaces of the lateral four metatarso-phalangeal joints. The two heads unite in a short tendon which is inserted, with the lateral insertion of flexor hallucis brevis, into the lateral side of the plantar surface of the base of the proximal phalanx of the big toe (Fig. 3.54).

The deep branch of the lateral plantar nerve sinks into the muscle and supplies it. The muscle draws the big toe towards the axis of the metatarsus and thus assists in maintaining the transverse arch.

The **flexor digiti minimi brevis** arises from the base of the fifth metatarsal bone and the adjoining fibrous sheath of peroneus longus. The muscle belly lies along the under surface of the fifth metatarsal bone and its tendon is inserted into the base of the proximal phalanx medial to the insertion of abductor minimi digiti (Fig. 3.53). It is supplied by the superficial branch of the lateral plantar nerve.

The Fourth Layer

The fourth layer of muscles consists of the interossei in the intermetatarsal spaces. The tendons of tibialis posterior and peroneus longus, lying deeply against the under surface of the tarsus, are conveniently included in this layer.

The Interosseous Muscles. The actions of the

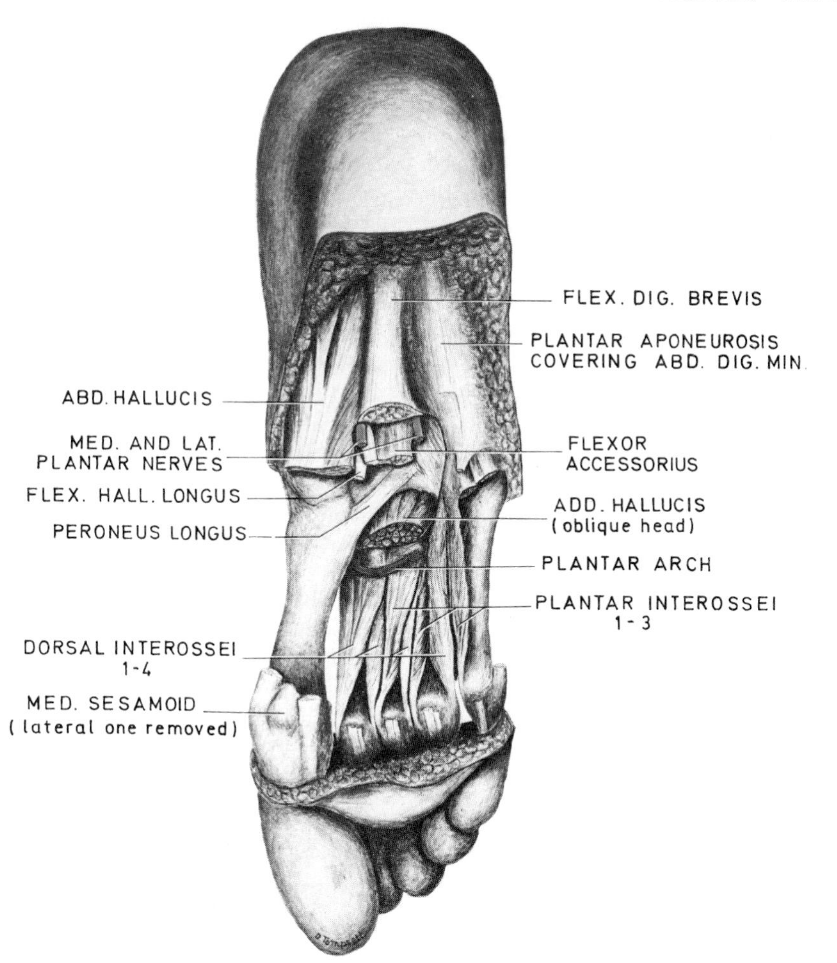

FIG. 3.55. THE INTEROSSEOUS MUSCLES OF THE RIGHT FOOT EXPOSED IN A DEEP DISSECTION OF THE SOLE. Illustration of Specimen S 122 in R.C.S. Museum.

interossei of the foot, as of the hand, are indicated by the formula 'PAD and DAB'. That is to say, the plantar adduct and the dorsal abduct, but with the important difference that the longitudinal axis of the foot has shifted pre-axially, and lies along the second metatarsal bone and the phalanges of the second toe. As in the hand, a plantar interosseous arises from the metatarsal bone of its own toe, a dorsal interosseous arises by two heads from the two metatarsals between which it lies. By use of the formula 'PAD and DAB' the attachments of the interossei may be inferred if memory fails.

The plantar adducting muscles adduct towards the second toe. The first toe has its own adductor mechanism (oblique and transverse heads of adductor hallucis) so that only the lateral three toes require adducting interossei. The first, second and third plantar interossei, each arising from only the metatarsal bone of its own digit, are inserted by tendons into the medial (tibial or pre-axial) sides of the third, fourth and fifth digits. The three tendons pass dorsal to the deep transverse ligament of the metatarsal heads and are inserted chiefly into the bases of the proximal phalanges, though each possesses an additional attachment into the dorsal extensor expansion.

The dorsal abducting muscles abduct away from the line of the second toe. The big and little toes each possess an abductor muscle. The second requires an abducting muscle on each side while the third and fourth require a single muscle each to abduct the digit laterally, away from the second toe. Each arises from both bones of its own intermetatarsal space. The first is inserted on the big toe side of the second toe, the second, third and fourth on the little toe side of the second, third and fourth toes. The tendons are inserted chiefly into the bones of the proximal phalanges, though each gives an extension also to the dorsal extensor expansion.

The interossei, both plantar and dorsal, are crowded into the intermetatarsal spaces. The first space has the first dorsal interosseous muscle only. The other three spaces contain both dorsal and plantar interossei but in each of them the dorsal is the bulkier muscle, arising from both bones of the space, and a view from the plantar aspect of the sole displays as much dorsal as plantar interosseous muscle (Fig. 3.55).

All interossei are supplied by the lateral plantar nerve. Those of the fourth space are supplied by the superficial branch, all the remainder by the deep branch.

The **tendon of peroneus longus** crosses the sole obliquely. It lies against the posterior ridge, not deeply in the groove, of the cuboid and its bearing surface is protected at the lateral margin of the foot by a sesamoid fibro-cartilage which may be ossified. As it crosses the groove of the cuboid it is held in position by the long plantar ligament which is firmly attached to the anterior and posterior ridges to bridge the groove. Emerging from this tunnel the tendon proceeds to its insertion at the base of the first metatarsal and adjoining area on the lateral surface of the medial cuneiform (Fig. 3.55). A synovial sheath accompanies it throughout its course.

The **tendon of tibialis posterior** is inserted into the tuberosity of the navicular. It lies above the sustentaculum tali and spring ligament. From its insertion many bands of fibres are traceable to other parts of the foot; usually described as insertions of the muscle they are rather in the nature of ligaments. From the navicular a few pass to the sustentaculum tali; all three cuneiforms, the floor of the groove in the cuboid and the bases of the second, third and fourth metatarsals receive strong bands of fibres.

The Vessels and Nerves

These are derived from the posterior tibial neuro-vascular bundle in the calf. Posterior tibial artery and nerve divide, each into medial and lateral plantar branches, under cover of the flexor retinaculum, the artery higher than the nerve, so that on medial and lateral borders of the sole the *artery is more marginal than the nerve*. Where they cross, the nerve is nearer the skin; and each artery is accompanied by a pair of venæ comitantes. The medial plantar nerve is larger than the lateral; it supplies fewer muscles (though their bulk is considerable) than the lateral plantar nerve but it supplies much more skin. The medial plantar artery, on the other hand, is smaller than its fellow; it gives rise to no plantar arch. In the hand there are two palmar arches, a superficial from the ulnar artery, and a deep from the radial artery; but in the foot there is only one plantar arch, derived from the lateral plantar artery.

The neuro-vascular plane of the sole lies between the first and second layers, upon the long tendons.

The **medial plantar artery** runs forward on the marginal (medial) side of the **medial plantar nerve** under cover of the muscles of the first layer. Both give off many branches to the sole, which perforate the plantar aponeurosis in the interval between the abductor hallucis and flexor digitorum brevis. The artery supplies these two muscles and the structures on the medial side of the foot and its digital supply is restricted practically to the big toe.

The nerve supplies these two muscles, and also the flexor hallucis brevis and the first lumbrical; in addition it gives off digital cutaneous branches that supply the medial three and a half toes on their plantar surfaces and on their dorsal surfaces proximal to the nail beds. Its most lateral cutaneous branch communicates with the neighbouring lateral plantar digital branch across the plantar surface of the fourth metatarso-phalangeal joint, where pressure on the nerve may give rise to the painful condition known as metatarsalgia (Fig. 3.53).

The **lateral plantar artery** crosses the sole obliquely, on the marginal (lateral) side of the nerve, just deep to the first layer of the sole, towards the base of the fifth metatarsal bone.

Both artery and nerve give off cutaneous branches to the sole that perforate the plantar aponeurosis in the interval between flexor digitorum brevis and abductor digiti minimi. The artery gives off a branch that accompanies the superficial branch of the nerve, but its main trunk accompanies the deep branch of the nerve to form the plantar arch. The **plantar arch** curves convexly forwards, across the bases of the fourth, third and second metatarsals and is joined in the proximal part of the first intermetatarsal space by the dorsalis pedis artery (Fig. 3.55).

Perforating branches from the arcuate artery (anterior tibial) join the arch in the proximal ends of the other three interosseous spaces. From the convexity of the plantar arch plantar metatarsal arteries run forwards and bifurcate to supply the four webs and digits. Anterior perforating arteries from the plantar metatarsal arteries reinforce the dorsal metatarsal arteries. The **veins** accompanying the perforating arteries take most of the blood from the sole and from the interosseous muscles to the dorsal venous arch.

The **lateral plantar nerve** crosses the sole obliquely just deep to the first layer of muscles. It supplies the flexor accessorius and abductor digiti minimi and sends perforating branches through the plantar aponeurosis to supply skin on the lateral side of the sole. Near the base of the fifth metatarsal bone it divides into superficial and deep branches. The *superficial branch* supplies the fourth cleft and communicates with the medial plantar nerve and, by a lateral branch, supplies the skin of the lateral side and distal dorsum of the little toe. Unlike the superficial branch of the ulnar nerve, this branch supplies three muscles, namely the flexor digiti minimi and the two interossei of the fourth space (third plantar and fourth dorsal). The *deep branch* lies within the concavity of the plantar arch and ends by sinking into the deep surface of the oblique head of adductor hallucis. It gives off branches to the remaining interossei, to the transverse head of adductor hallucis and to the three lateral (bicipital) lumbricals. The branch to the second lumbrical passes dorsal to the transverse head of adductor hallucis and recurves ventrally to enter the lumbrical (Figs. 3.53, and 3.54).

THE ANKLE JOINT

This synovial joint is usually described as being of the hinge variety; but its movements are not quite those of a simple hinge, for the axis of rotation is not fixed but changes between the extremes of plantar flexion and dorsiflexion. The articulating surfaces are covered with hyaline cartilage. The weight-bearing surfaces are the upper facet of the talus and the inferior facet of the tibia. Stabilizing surfaces are those of the medial and lateral malleoli, which grip the sides of the talus. The joint is enclosed in a capsule lined with synovial membrane. The *capsule* is attached to the articular margins of all three bones except the anterior part of the talus, where it is fixed some distance in front of the articular margin, on the neck of the bone. Posteriorly the capsule, on its way up to the tibia, is attached also to the posterior tibio-fibular ligament.

The *synovial membrane* is attached to the articular margin of the talus and clothes the intracapsular part of the neck. Elsewhere it is attached to all articular margins, as in any synovial joint. Occasionally the joint cavity extends up a little between tibia and fibula, into the inferior tibio-fibular ligament.

Strong medial and lateral ligaments strengthen the joint. The **deltoid ligament,** on the medial side, is in two layers. The *deep lamina* is narrow, extending from the tibial malleolus to the side of the talus, inserted there into the concavity below the comma-shaped articular surface (Fig. 3.57). Its shape is rectangular. Only the *superficial lamina* is triangular, like a delta. It fans downwards from the borders of the tibial malleolus and its lower margin has a continuous attachment from the medial tubercle of the talus (a weak band)

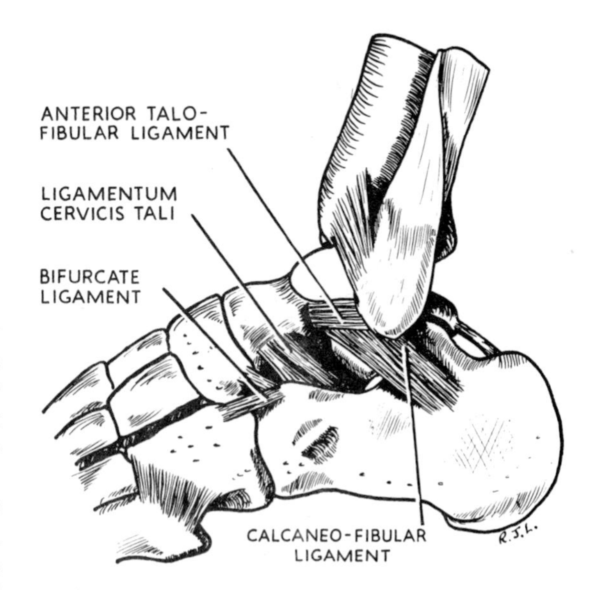

FIG. 3.56. LATERAL VIEW OF THE LEFT ANKLE JOINT. Drawn from Specimen S 97 B in R.C.S. Museum.

along the edge of the sustentaculum tali and spring ligament (very strong) to the tuberosity of the navicular (a weak band) (Figs. 3.59, 3.60 and 3.61).

The **lateral ligament** consists of three separate bands, radiating from the lateral malleolus. Anterior and posterior bands pass to the talus, the intermediate band to the calcaneus. The *anterior talo-fibular ligament* joins the anterior border of the lateral malleolus to the neck of the talus. It is a flat band. The *calcaneo-fibular ligament* is a rounded cord extending from the front of the tip of the malleolus down and back to the lateral surface of the calcaneus. The *posterior talo-fibular ligament* lies horizontally between the malleolar fossa and the lateral tubercle of the talus. It is strong. Above it lies the **posterior tibio-fibular ligament,** which articulates with the talus. In plantar flexion these two ligaments lie edge to edge, but in dorsiflexion they diverge like the blades of an opening pair of scissors (Fig. 3.57).

The blood supply of the capsule and ligaments is derived from anterior and posterior tibial arteries and the peroneal artery, and the nerve supply is by the deep peroneal and tibial nerves.

Movements. Examine a talus. Its upper facet, slightly concave from side to side, is convex anteroposteriorly. It is broad in front and narrow behind. The lateral facet, while gently concave from above downwards, is convex from front to back in most bones; it articulates with the lateral malleolus. The upper and medial surfaces of the talus are in contact with the tibia and its medial malleolus in all positions of the joint. The lateral malleolus adapts itself to the lateral surface of the talus; this involves rotary and lateral movements of the lower end of the fibula, which cause reciprocal movements at the upper end of the bone, in the superior tibio-fibular synovial joint.

In full plantar flexion the smallest area of the talus is in contact with the tibia, but even in this position the amount of inversion and eversion possible at the ankle joint is very small indeed. For all practical purposes the ankle may be regarded as a true hinge joint. The axis of rotation is not horizontal, but slopes downwards and laterally. It passes through the lateral surface of the talus just below the apex of the articular triangle and through the medial surface at a higher level, just below the concavity of the comma-shaped articular area. It passes through the malleoli just above their apices. In truth, though the fact is of no practical application, the axis changes during movement, for the upper convexity of the talus is not the arc of a circle, but rather of an ellipse. The obliquity of the axis involves a slight movement resembling inversion in full plantar flexion, and the reverse, resembling slight eversion, in full dorsiflexion, but these apparent movements are not of true inversion and eversion (*v. inf.*).

Plantar flexion is produced most powerfully and essentially by the gastrocnemius and soleus. The long flexor tendons and the long and short peronei all have a secondary action of flexion on the ankle joint. Dorsiflexion (extension) is produced by the tibialis anterior and peroneus tertius, aided on occasion by the toe extensors.

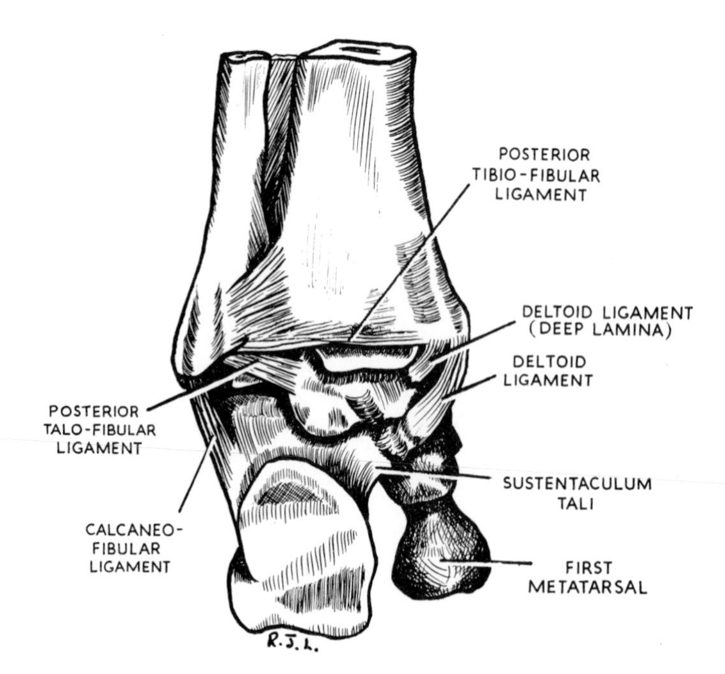

FIG. 3.57. THE LEFT ANKLE AND HEEL FROM BEHIND. Drawn from Specimen S 97 B in R.C.S. Museum.

THE JOINTS OF THE TARSAL BONES

By far the most important joints in the tarsus are those between the talus, calcaneus and navicular and between the calcaneus and cuboid.

Examine the articulated bones of the foot. The synovial joint between the anterior surface of the calcaneus and the cuboid is seen to lie transversely across the foot in line with the joint between the head of the talus and the navicular. These two joints together are usually referred to as the **mid-tarsal joint.** This is a functional description, for a small amount of inversion and eversion of the fore part of the foot takes place across these joints; but it is essential to appreciate that while the lateral joint is separate and possesses its own capsule, the medial part is not a separate joint but is only the anterior part of a continuous synovial cavity that extends backwards below the head of the talus, and includes the spring ligament, the anterior part of the upper surface of the calcaneus, and the sustentaculum tali. The head of the talus, in other words, is a ball which articulates with a socket composed of the posterior concavity of the navicular (part of the 'mid-tarsal joint') and the spring ligament and sustentaculum tali (part of the 'subtalar joint').

There is a third joint, separate in its own capsule, between the body of the talus and the calcaneus, behind the ball and socket joint and completing, with it, the 'subtalar joint' (Fig. 3.60).

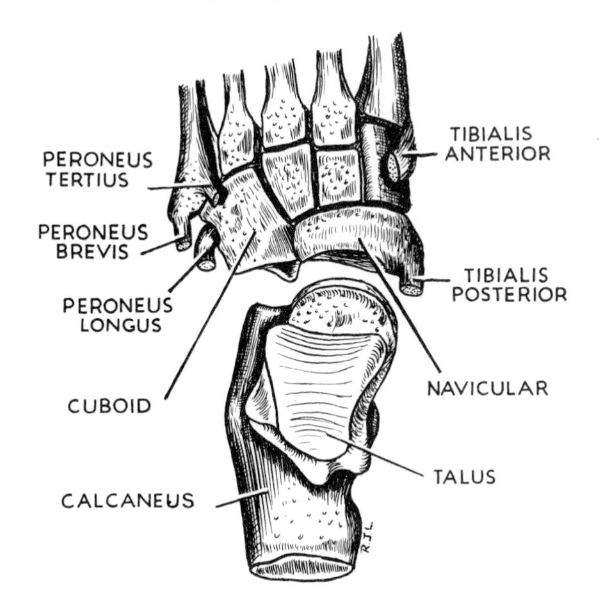

FIG. 3.58. THE MID-TARSAL JOINT. The joint surfaces are separated. All five tendons of inversion and eversion act on the foot in front of the mid-tarsal joint. Inversion and eversion of the calcaneus are produced by ligaments that cross the mid-tarsal joint (spring ligament, long and short plantar ligaments, etc.).

The **calcaneo-cuboid joint** is a synovial joint whose surfaces are gently undulating. It is surrounded by a simple capsule, thickened above and below. Accessory ligaments on its plantar surfaces are the long and short plantar ligaments, and the lateral limb of the bifurcate ligament strengthens it medially. Simple gliding movement takes place during inversion and eversion of the foot.

The **short plantar ligament** is a thick bundle which fills in the adjacent hollows in front of the anterior tubercle of the calcaneus and behind the posterior ridge of the cuboid. It is covered over by the long plantar ligament.

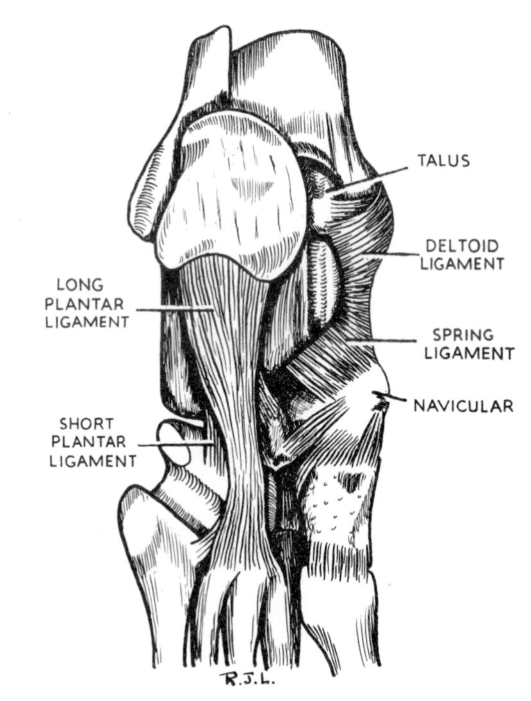

FIG. 3.59. THE LEFT TARSUS FROM BELOW. Drawn from Specimen S 97 B in R.C.S. Museum.

The **long plantar ligament** covers the plantar surface of the calcaneus. It extends from the posterior tubercles (i.e., medial and lateral tubercles) on the calcaneus to the anterior tubercle of that bone. Thence it covers the short plantar ligament, and its deeper fibres are attached to the posterior ridge of the cuboid. Its superficial fibres bridge the groove of the cuboid, making a fibrous roof over the peroneus longus tendon, and are attached to the anterior ridge of the cuboid and extend forwards to the bases of the central three metatarsal bones. It is covered by the flexor accessorius, and its posterior part is visible in the gap between the medial fleshy and lateral tendinous heads of that muscle.

The **talo-calcaneo-navicular joint** is a synovial joint of the ball and socket variety. The ball is the head of the talus and the socket comprises two bones and two ligaments (Fig. 3.60). The bones are the navicular and calcaneus. The posterior surface of the navicular has an articular surface which is concave reciprocally with the anterior convexity of the head of the talus.

The anterior end of the upper surface of the calcaneus has a concave facet, and the sustentaculum tali a similar one (the two are often fused into a single concavity) for articulation with the inferior convexity of the head of the talus. On the head of the talus, between its navicular and calcaneal surfaces, lies cartilage that articulates with neither bone. Here the talus articulates with the spring ligament medially and the calcaneo-navicular limb of the bifurcate ligament laterally. All these structures are enclosed in a single capsule. One end of the capsular sleeve is attached to the neck of the talus around the articular margin on the head. The other end of the sleeve is attached along the upper surface of the navicular, the medial edge of the spring ligament, the posterior limits of the articular facets on sustentaculum tali and body of calcaneus, the medial limb of the bifurcate ligament and so back to the upper surface of the navicular. The socket of the joint is closed below by the spring and bifurcate ligaments. Between the two lies a pad of fat, covered with synovial membrane, that acts as a swab to spread synovial fluid on the moving head of the talus (cf., the Haversian fat pad in the acetabulum).

The **bifurcate ligament** arises from the upper surface of the calcaneus just behind the anterior margin of that bone, under cover of the extensor digitorum brevis muscle at the front of the sinus tarsi. From the single origin two limbs diverge slightly from each other. The medial limb is attached to the navicular, near the infero-lateral part of its articular margin. It forms part of the socket in which the head of the talus lies and gives attachment to the capsule of the talo-calcaneo-navicular joint. It is always present. The lateral limb of the bifurcate ligament is often absent. It is extra-capsular, and is attached to the upper surface of the cuboid, being no more than an accessory ligament of the calcaneo-cuboid joint.

The **inferior calcaneo-navicular (spring) ligament** is a very strong band that joins the whole thickness of the edge of the sustentaculum tali to the navicular between its tuberosity and the articular margin. Its upper surface, which extends to the articular margins of both bones, articulates with the antero-inferior part of the head of the talus. Its lower fibres extend well under the sustentaculum tali and lie almost transversely across the foot (Fig. 3.59).

The **talo-calcaneal joint** lies behind the talo-calcaneo-navicular joint. It is a synovial joint, oval in shape, with its long axis lying obliquely from postero-

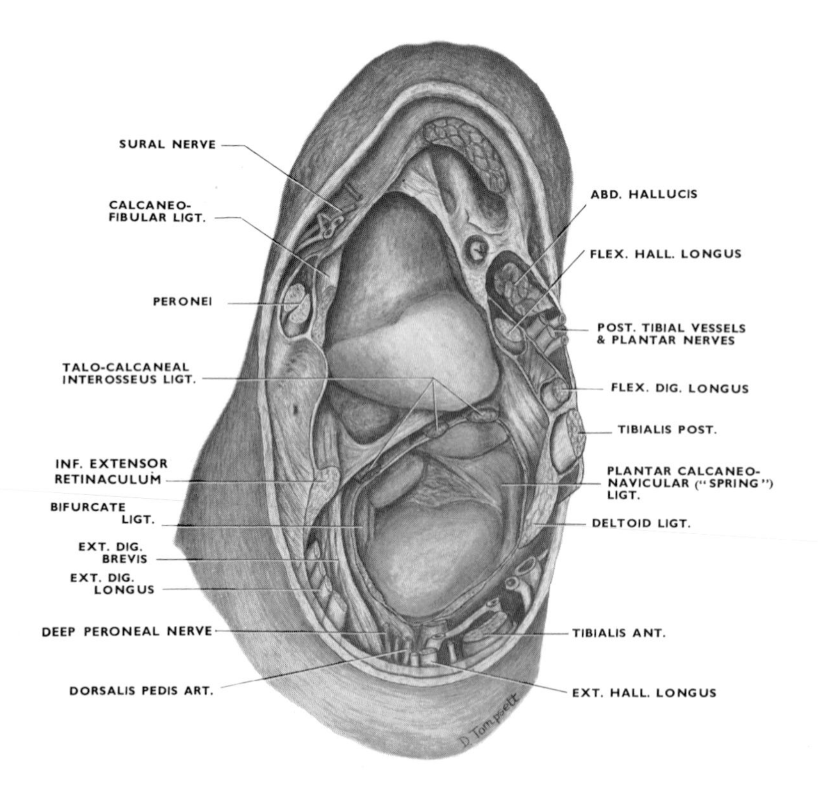

FIG. 3.60. THE RIGHT SUBTALAR JOINT EXPOSED BY REMOVAL OF THE TALUS. Illustration of Specimen S 100.1 in R.C.S. Museum.

medial to antero-lateral. The under surface of the talus is concave from side to side, reciprocally with the convex articular facet on the calcaneus (PLATE 46, p. *xxiv*). The capsule and synovial membrane are attached to the articular margin.

The **tarsal canal,** often misnamed the sinus tarsi, lies obliquely between the talo-calcaneo-navicular joint and the talo-calcaneal joint. It is a cylindrical canal formed by the semi-cylindrical grooves between the articular surfaces on the two bones. Anteriorly and posteriorly the canal is occupied by strong, flat bands, the anterior and posterior interosseous talo-calcaneal ligaments, whose fibres diverge upwards from each other in V-shaped manner. They are in reality thickenings of the joint capsules along the tarsal canal. They leave *smooth* marks on each bone alongside the articular facets. The central portion of each bony gutter is perforated by vascular foramina.

The **sinus tarsi,** properly so-called, consists of the concavity between talus and calcaneus at the lateral end of the tarsal canal. It is occupied by a very strong cord-like band, the *ligamentum cervicis tali*. The ligament of the neck of the talus leaves *smooth* facets on calcaneus and talus. Further laterally the sinus tarsi is occupied by the flesh of extensor digitorum brevis where it arises from the calcaneus, and the sinus is closed in finally by the stem of the Y-shaped inferior extensor retinaculum. The sinus tarsi forms the bowl and the tarsal canal forms the stem of a funnel-shaped space between the talus and the calcaneus (Fig. 3.56 and PLATE 46, p. *xxiv*).

Inversion and Eversion of the Foot

After the mid-tarsal and subtalar joints have been studied, the movements of inversion and eversion which they allow can be appreciated. The ability to invert and evert the foot confers on man a corresponding ability to walk across uneven surfaces that slope sideways, and loss of this power produces an appreciable disability when progressing across rough ground. A similar disability is noted in walking on smooth surfaces if these slope sideways, for such progression requires the upper foot to be everted and the lower inverted. Moreover, in turning at speed the movements are essential, in order to lean sideways on a foot whose sole is flat on the ground.

Mostly these inversion-eversion movements are performed on a foot anchored to the ground, with the leg bones and thus the whole body inverting and everting above it. It is difficult to analyse the movements from a fixed foot: so the analysis is best made with the foot free of the ground. The malleoli lock the talus, and the suspended foot inverts and everts around it.

The movement of inversion (raising the medial border of the foot) is produced, self-evidently, by any muscle that is attached to the medial side of the foot. Tibialis anterior and tibialis posterior are responsible, assisted by extensor and flexor hallucis longus on occasion. Tibialis anterior dorsi-flexes and tibialis posterior plantar-flexes the foot at the ankle joint and these opposite effects cancel each other out when the two muscles combine to produce an uncomplicated inversion of the foot. Inversion is a movement of supination (cf., supination of the forearm), and it is accompanied by adduction of the fore part of the foot.

The movement of eversion (raising the lateral border of the foot) is produced, self-evidently, by any muscle that is attached to, or pulls upwards upon, the lateral side of the foot. The peronei longus, brevis, and tertius are responsible. The former two, whose tendons pass behind the lateral malleolus, are plantar-flexors, the last is a dorsi-flexor, of the ankle joint. These opposite effects cancel each other out when the three muscles combine to produce a simple eversion of the foot. Eversion is a movement of pronation (cf., pronation of the forearm), and it is accompanied by abduction of the fore part of the foot.

All the muscles producing inversion and eversion are attached to the fore part of the foot, anterior to the **mid-tarsal joint** (Fig. 3.58). It is therefore evident that in both inversion and eversion the beginning of the movement must occur wholly at the mid-tarsal joint, and that only when the bones of this joint are fully 'wound up' by tension of their ligaments is the rotatory force transmitted, passively by ligaments, to the subtalar joints. In fact, relatively little movement is possible at the mid-tarsal joint, and *most of the full range of inversion and eversion occurs at the subtalar joint.*

The calcaneus and cuboid are firmly connected by the long and short plantar ligaments, which limit the range of mobility at the mid-tarsal joint; when they and the spring ligament of the mid-tarsal joint are taut they transmit the rotatory force to the calcaneus. This bone then rotates (i.e., inverts or everts) under the talus, which is firmly wedged against the tibia between the malleoli and cannot therefore be inverted or everted.

The **subtalar joint** is much more mobile than the mid-tarsal joint. The head and body of the talus articulate with the front and body of the calcaneus each by a separate synovial joint. These two joints are concavo-convex in opposite directions, as are the joints at the upper and lower ends of radius and ulna. Similar movements of supination and pronation occur in each. Between radius and ulna the axis of rotation passes of necessity through the centres of the circular facets around which rotation takes place. The axis thus passes through the centre of curvature of the upper end of the radius and the centre of curvature of the lower end of the ulna (p. 87). In an exactly similar way the

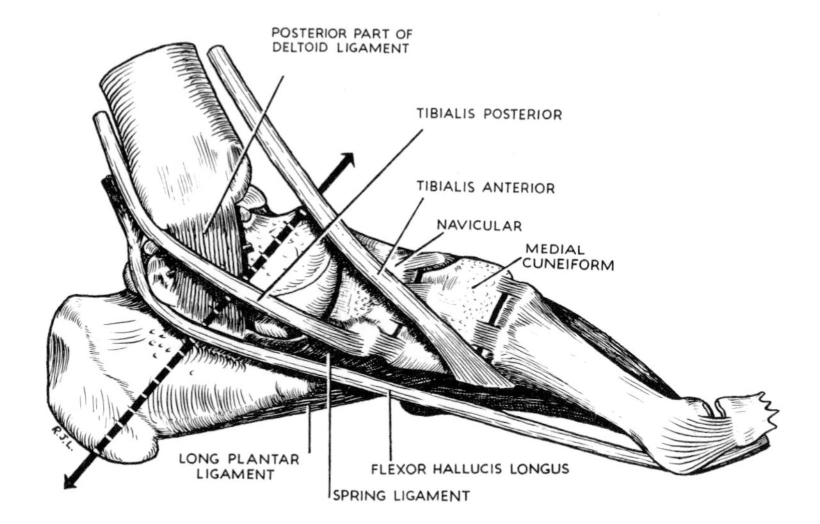

FIG. 3.61. THE AXIS OF INVERSION AND EVERSION OF THE FOOT AT THE SUBTALAR JOINT INDICATED BY THE INTERRUPTED ARROW. The inverting tendons pull at right angles around the axis (cf. Fig. 3.48 for the everting peroneal tendons). Drawn from Specimens S 97 B and S 112 in R.C.S. Museum.

axis of subtalar movement passes through the centres of the oppositely curved subtalar joints; the only dissimilarity is that these two joints are close together, the radio-ulnar joints far apart. The centre of the convex posterior facet on the calcaneus lies beneath it, inside the calcaneus. The centre of the convex head of the talus lies in the neck of the talus (Fig. 3.61).

If these points be joined, the axis of rotation in the subtalar joint is found to be an oblique line passing from the lateral tubercle of the calcaneus forwards, upwards and slightly medially through the neck of the talus, in its passage bisecting the tarsal canal.

Inversion is a movement of supination about this axis, and due to the obliquity of the axis it is necessarily accompanied by adduction of the fore foot while eversion, the movement of pronation, is likewise accompanied by abduction of the fore foot.

In ordinary use of the foot in walking over rough ground, inversion and eversion take place while the long axis of the foot continues to point in the desired direction. The adduction and abduction movements are masked by reciprocal rotation of the leg at the knee (if the knee is flexed) or of the whole lower limb at the hip (if the knee is extended). These synergic movements, under extrapyramidal control, are precisely similar in their nature to those of elbow and shoulder which accompany pronation and supination of the forearm.

When the direction of the axis of rotation of inversion and eversion is appreciated it can be seen at once that the lines of pull of the muscles lie at right angles to its obliquity, so that the muscles act to the best mechanical advantage. The balanced actions of these muscles combine in different patterns to produce ordered movements of the ankle and tarsal joints.

Mechanically there are four lines of pull.

1. Tibialis anterior, which inverts (and therefore adducts) the foot at the tarsal joints and dorsi-flexes the foot at the ankle joint.
2. Peroneus tertius, which everts (and therefore abducts) the foot at the tarsal joints and dorsi-flexes the foot at the ankle joint.
3. Tibialis posterior, which inverts (and therefore adducts) the foot at the tarsal joints and plantar-flexes the foot at the ankle joint.
4. Peronei longus and brevis, which evert (and therefore abduct) the foot at the tarsal joints and plantar-flex the foot at the ankle joint.

These four combine in groups of two to produce their common movement and cancel out their opposing movements. Thus result dorsiflexion (1 and 2), inversion (1 and 3), eversion (2 and 4) and plantar flexion (3 and 4) of the foot as a whole, though it must be remembered that there is an added powerful mechanism (antigravity and propulsive) producing plantar flexion of the foot as a whole, namely the pull on the heel of the soleus and gastrocnemius through the tendo calcaneus.

The ligamentum cervicis tali and the calcaneo-fibular band of the lateral ligament of the ankle joint both lie parallel with the axis of inversion-eversion (Fig. 3.56) and thus exert no restraining influence on these movements.

Tarso-metatarsal Joints. The osteology of the tarsal bones and the anatomy of the tarso-metatarsal joints are considered in the account of the osteology of the foot (pp. 204 and 208).

The **metatarsus** is much more rigid than the metacarpus. The first tarso-metatarsal joint possesses its own capsule and synovial membrane and is capable of some movement in a vertical plane to conform with movements in the medial longitudinal arch of the foot, and it becomes hyperextended in flat foot; but the joint movements in no way compare with those of the carpometacarpal joint of the thumb and no opposition of the big toe is possible. The second tarso-metatarsal joint is immobile, the base of the metatarsal being firmly fixed between the anterior ends of the medial and lateral cuneiforms. This is a result of the shifting of the axis of the foot towards its medial side, and the second toe forms the line for adduction and abduction of the digits. The immobility of the second metatarsal and the slenderness of its shaft are contributory factors in 'spontaneous' fracture ('march fracture') of this bone.

The first metatarso-phalangeal joint is the site of hallux valgus. The big toe has no dorsal extensor expansion nor fibrous flexor sheath; its long tendons are held in position by strands of deep fascia. If the phalanges become displaced laterally and the fibrous bands give way, the pull of extensor hallucis longus, like that of extensor hallucis brevis, becomes oblique to the long axis of the toe and tends to increase the deformity.

The Foot as a Supporting Mechanism

Examine the bones of an accurately articulated foot, a lateral X-ray of the foot and the imprint of a wet foot on the bathroom floor. The heel, lateral margin of the foot, the ball of the foot (the part underneath the metatarsal heads) and the pads of the distal phalanges touch the ground. The medial margin of the foot arches up between the heel and the ball of the big toe, forming a visible and obvious **medial longitudinal arch.** The lateral margin of the foot is in contact with the ground, but its constituent bones do not bear with equal pressure on the ground. As on the medial side, so on the lateral side, there is a bony longitudinal arch extending from the heel to the heads of the metatarsal bones, but the **lateral longitudinal arch** is much flatter than the medial.

The constituent bones of the arches can be seen in the articulated foot. The *medial longitudinal arch* consists of calcaneus, talus, navicular, the three cuneiform bones and their three metatarsal bones. The pillars of the arch are the back of the calcaneus posteriorly and the heads of the medial three metatarsal bones anteriorly. The *lateral longitudinal arch* consists of calcaneus, cuboid and the lateral two metatarsal bones.

The *transverse arch* is, in reality, only half an arch, being completed as an arch by that of the other foot. It consists of the bases of the five metatarsal bones and the adjacent cuboid and cuneiforms. The heads of the five metatarsal bones lie flat upon the ground and can scarcely be described as taking part in the transverse arch, though the first and fifth heads bear more weight than the others.

The factors maintaining the integrity of the arches of the foot are identical with those responsible in any joint of the body, namely, bony, ligamentous and muscular, but their relative importance is different in the three arches.

The Medial Longitudinal Arch. No bony factor is responsible for maintaining the stability of this arch. The head of the talus is supported on the sustentaculum tali, but this is a negligible factor. Ligaments are important, but *are inadequate alone* to maintain the arch. The most important ligament is the plantar aponeurosis, stretching like a tie beam or bowstring between the supporting pillars of the arch. If it is shortened by extension of the toes, especially the big toe, it draws the pillars together and so heightens the arch. Next in importance is the spring ligament, for it supports the head of the talus. If it stretches it allows the navicular and calcaneus to separate and so the head of the talus, the highest point of the arch, sinks lower between them. All the interosseous ligaments, especially those which hold the talus and calcaneus together at their articulating surfaces, are of assistance in maintaining the arch.

Muscles are indispensable to the maintenance of the medial longitudinal arch, for if they are paralysed, or weakened beyond a certain point, the ligaments alone are unable to maintain the arch.

Examine a living foot. The arch cannot be pressed flat; the ligaments are too strong. Static body weight will not flatten the arch; indeed the muscles are not in action when standing still. Consider the inertia of the body-weight to be overcome at the take-off and the momentum to be absorbed in landing. In each of these the forces are vastly greater than the mere weight of the body. The arch now acts like a semi-elliptic spring. If the muscles are too weak the ligaments eventually stretch and flat foot results.

The most efficient means of maintaining an arch such as this is to tie its two pillars together. The most important muscular supporting structures are therefore those which run longitudinally beneath the arch. *Overwhelmingly efficient in this respect is the tendon of flexor hallucis longus.* The muscle is the bulkiest of those beneath the soleus in the calf and from its powerful multipennate mass the tendon runs beneath the sustentaculum tali and thence passes straight forward, below the arch, to lie between the sesamoid bones on the metatarsal head before it reaches its insertion at the distal phalanx of the big toe. This is not only the bulkiest of the deep calf muscles, but it can spare a slip to assist the pull of its weaker sister the flexor digitorum longus, and it is highly significant that this slip *acts only on the tendons to the second and third toes*, the remaining members of the medial longitudinal arch.

Thus the flexor hallucis longus acts as a bowstring along the medial longitudinal arch from the sustentaculum tali to the medial three digits. It is not called into play during short periods of standing, for generally the weight is borne well back on the heel and the pads of the toes are not pressed on the ground. Meanwhile the ligaments support the arch. But during prolonged standing the ligaments 'tire', and relief is obtained by pressing the pads of the toes to the ground. During movements of propulsion and during landing on the feet the inertia and momentum of the body weight throw a vastly greater strain on the arch, and this strain is taken up by the contraction of the flexor hallucis longus muscle and the tension of its tendon.

The short muscles in the first layer that are inserted into the medial three toes (abductor hallucis, and the medial half of flexor digitorum brevis) likewise assist in maintaining the arch.

The Sling of Tibialis Anterior and Peroneus Longus. The tendons of these muscles are inserted into the same two bones (medial cuneiform and first metatarsal bone) and they are often compared to a sling that supports the longitudinal arches of the foot, like a skipping rope pulled up beneath the instep of a shoe. But there is a significant difference—the tendons are not free like a skipping rope to slide under the foot, but are fixed to bone and they exert opposite effects upon the medial longitudinal arch. Tibialis anterior, by the upward pull of its tendon, may have some slight influence in maintaining the medial longitudinal arch, but the peroneus longus tendon has the reverse effect, as is so well seen in the everted *flat* foot of peroneal spasm in children. The lowest point in the tendon is at its sesamoid fibrocartilage, where it bears on the cuboid; from this point it rises further from the ground as it slopes across to its insertion. Its pull tends to evert and abduct the foot and *lower the medial side of the foot.*

Tibialis anterior and tibialis posterior have a significant influence on the arch, but mechanically different from the bowstring effect of the longitudinally directed muscles and tendons in the sole. They act by their tendency to invert and adduct the foot, in other words to raise the medial border from the ground, and have no direct pull tending to approximate the pillars of the arch, and are consequently less important factors in maintaining its integrity.

The Lateral Longitudinal Arch. No bony factor contributes to the stability of this arch, but ligaments play a relatively more important part than in the case of the medial arch.

The plantar aponeurosis in its lateral part acts as a bowstring beneath the arch and the plantar ligaments are also significant. The short plantar ligament is very thick, filling the adjacent concavities of calcaneus and cuboid. The long plantar ligament, though a thinner layer, extends from the heel to both ridges of the cuboid and the metatarsal bases and helps to maintain the slight concavity of the lateral arch.

The peroneus longus tendon, tending to depress the medial longitudinal arch, at the same time pulls upwards on the lateral longitudinal arch and is the most important single factor in maintaining its integrity. The tendons of flexor digitorum longus to the fourth and fifth toes (assisted by flexor accessorius) and the muscles of the first layer of the sole (lateral half of flexor digitorum brevis and abductor digiti minimi) also assist strongly, by preventing separation of the pillars of the arch.

The Transverse Arch. The intermediate and lateral cuneiforms are wedge-shaped, and in this single respect the bones are adapted to the maintenance of the transverse arch of the foot. The lateral cuneiform overhangs the cuboid and thus rests on it to some extent. But the medial cuneiform is wedge-shaped the wrong way for an arch, and it is evident that bony factors in fact play but little part in maintaining the transverse arch. The ligaments that bind together the cuneiforms and bases of the five metatarsal bones are much more important, and the most important factor of all is the tendon of peroneus longus, the pull of which tends to approximate medial and lateral borders of the foot across the sole. At the heads of the metatarsal bones a shallow arch is maintained by the deep transverse ligament of the metatarsal heads, by the transverse fibres that bind together the digital slips of the plantar aponeurosis and, perhaps, by the transverse head of the adductor hallucis muscle, but the last is a very slender band.

The Foot as a Propulsive Mechanism

If the foot were entirely rigid it would serve as a propulsive member by plantar flexion at the ankle joint and, indeed the contraction of soleus and gastrocnemius is the chief factor responsible for propulsion in walking, running and jumping. But the propulsive action of these great muscles of the calf is enhanced by arching of the foot and flexion of the toes. In walking the weight of the foot is taken successively on the heel, lateral border, and the ball of the foot and the last part to leave the ground is the anterior pillar of the medial longitudinal arch, and the medial three digits. In running the heel remains off the ground, but the take-off point is still the anterior pillar of the medial longitudinal arch.

While the heel is rising from the ground the medial toes are gradually extended. The extended big toe pulls the plantar aponeurosis around the first metatarsal head and so heightens the arch; at the same time it elongates the flexor hallucis longus and flexor digitorum longus muscles, and this increases the force of their subsequent contraction.

Contraction of the toe flexors, long and short, increases the force of the take-off by the pressure of the toes on the ground. Contraction of the long flexors also aids plantar flexion at the ankle joint.

The most powerful muscle acting in this way is the flexor hallucis longus, which acts on the big toe through

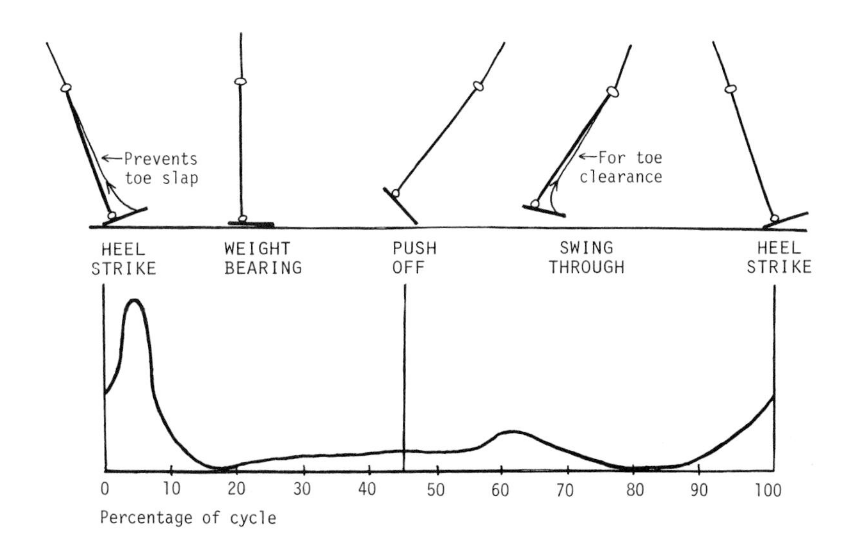

FIG. 3.62. PHASIC ACTION OF THE PRETIBIAL MUSCLE GROUP DURING LEVEL WALKING. The contraction during heel strike (shock-absorbing) is much greater than that needed for toe clearance during the swing-through phase. Adapted from *U.C.L.A. Manual of Above-knee Prosthetics*, by courtesy of Dr. Frances S. Grover.

its tendon of insertion and on the second and third toes by the slip it gives to the flexor digitorum longus. The short flexor muscles of the sole strongly assist the long flexors in heightening the arch and in flexing the toes—the lumbricals prevent the toes from buckling under when pulled upon by flexor digitorum longus. The propulsive force so produced adds greatly to that of the soleus and gastrocnemius. The tendo calcaneus plantar-flexes a very mobile, not a rigid, foot. If the tendo calcaneus is activated by a 'bottom gear' and 'top gear' mechanism (soleus and gastrocnemius), the flexors of the tarsus and toes provide an 'overdrive'.

The Foot as a Shock-absorbing Mechanism

When landing at speed, as for example from a height, the toes and then the forefoot take the first thrust of the weight before the heel strikes the ground. It is, in fact, the reverse time sequence of taking off at maximum speed. But even in quiet walking, with heel and toe sequence, muscles play an essential part in absorbing the shock of landing. At the moment of heel strike the pretibial group, notably tibialis anterior, contract as they elongate (action of paradox, p. 8) to lower the forefoot more gently to the ground (Fig. 3.62).

The Cutaneous Innervation of the Lower Limb

The dermatomes of the lower limb have been considered on p. 26. The cutaneous nerves themselves are described with the various regions of the lower limb, and are illustrated in Figs. 3.7, p. 131 and 3.25, p. 145.

The Muscular Innervation of the Lower Limb

The borrowed muscle of the thigh is supplied by the femoral nerve (extensor compartment) and the obturator nerve (adductor compartment). The glutei are supplied by the gluteal nerves, superior and inferior. The hamstrings (except short head of biceps), calf muscles and sole of the foot are supplied by the tibial component of the sciatic nerve. In the leg the peronei are supplied by the superficial peroneal nerve, while the anterior tibial group are supplied by the deep peroneal nerve, which also supplies extensor digitorum brevis.

The segmental innervation of the muscles of the lower limb is considered on p. 28.

The Nerves of the Lower Limb

The lower limb is innervated basically from the sacral plexus, but a good deal of skin and muscle ('borrowed from the trunk') is supplied from the lumbar plexus. The nerves concerned are dealt with in their various regions, but it is convenient also to summarize the whole course of each nerve. This is done for the branches of the lumbar plexus on p. 356, and for those of the sacral plexus on p. 357.

OSTEOLOGY OF THE LOWER LIMB

The Pelvic Girdle

A summary of the bony pelvis is on p. 322 as an introduction to the study of the pelvic walls and cavity. The osteology of the sacrum is studied with the vertebral column (p. 473).

The Hip Bone (Os Innominatum)

This is formed of three bones, which fuse in a Y-shaped epiphysis at the hip-joint socket. The pubis and ischium together form an incomplete bony wall for the pelvic cavity; their outer surfaces give attachment to thigh muscles. The ilium forms the pelvic brim between the hip joint and the joint with the sacrum; above the pelvic brim it is prolonged, broad and wing-like, for the attachment of ligaments and large muscles. The anterior two-thirds of the projecting ilium, thin bone, forms the iliac fossa, part of the posterior abdominal wall. The posterior one-third, thick bone, carries the auricular surface for the sacrum and, behind this, is prolonged for strong sacro-iliac ligaments which bear the body weight. The outer surface of the ilium gives attachment to buttock muscles. The ischium and pubis together lie in approximately the same plane; the plane of the ilium is at nearly a right angle with this.

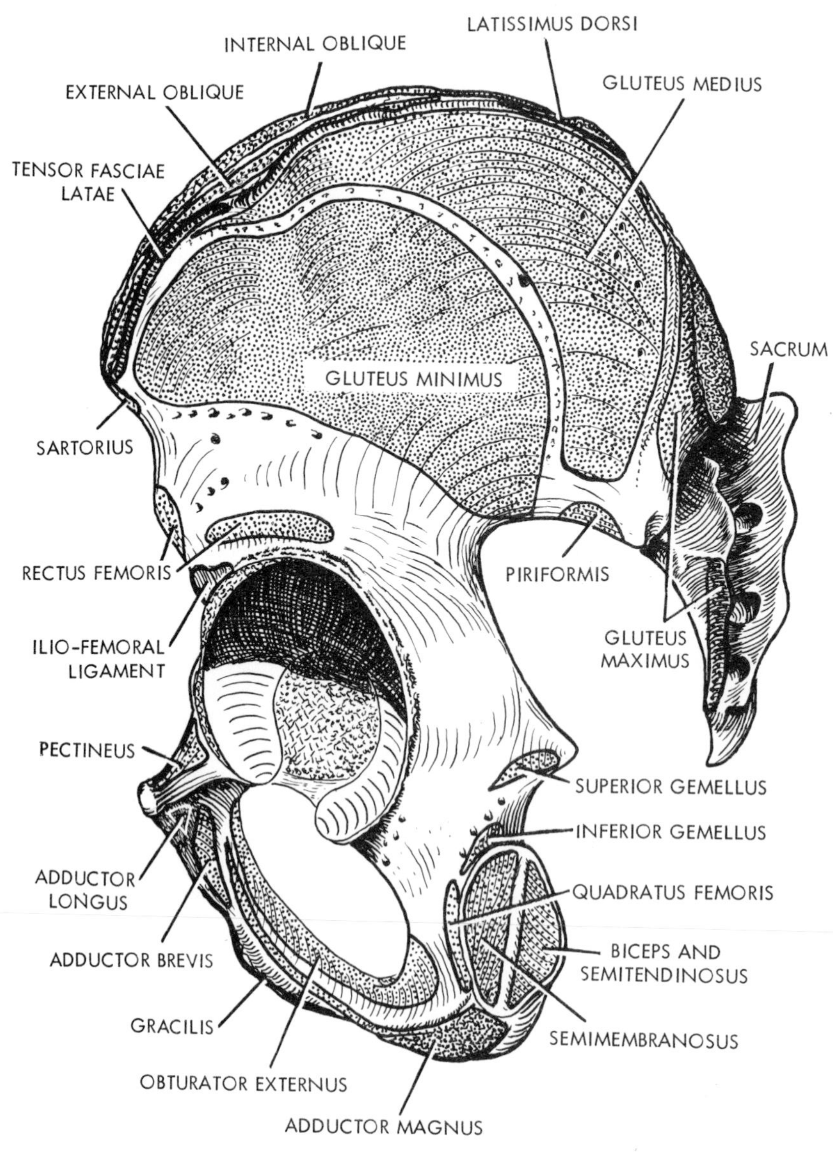

FIG. 3.63. THE LEFT HIP-BONE FROM THE LATERAL SIDE AND SOMEWHAT FROM BEHIND.

Note the correct anatomical position of the bone. The pubic tubercle and anterior superior iliac spine lie in the same vertical plane, the upper border of the symphysis pubis and the ischial spine lie in the same horizontal plane, and the symphysis pubis is vertical (Fig. 5.54, p. 323).

Outer Surface of Os Innominatum. This should not be studied until the anatomy of the thigh and buttock is familiar.

The hip-joint socket is called the **acetabulum** ('vinegar-cup'). It is a concave hemisphere whose axis is not strictly horizontal but is directed also downwards and slightly backwards along the axis of the femoral neck. Its inferior margin is lacking (Fig. 3.63), so the vinegar would run out of the cup. The margins of this *acetabular notch*, curved in outline, are lined with hyaline cartilage, which continues as a wide band inside the concave surface of the acetabulum. It is widest above, opposite the notch. This wide upper part of the articular surface is the weight-bearing area, lying like a cap over the femoral head (*remember the anatomical position*, Fig. 5.54, p. 323). The articular surface does not quite reach the rim of the acetabulum, nor does it floor in the whole concavity. The non-articular bone in the acetabular fossa is paper-thin and translucent. This bone is prolonged down below the acetabular notch as a narrow rim skirting the obturator foramen (Fig. 3.63). Here pubis and ischium meet. Pubis and ilium meet at the anterior pole of the acetabulum (the place is indicated by the ilio-pectineal eminence). Ilium and ischium meet just above the posterior pole of the acetabulum. Thus the broad weight-bearing part of the articular surface is carried *entirely by the ilium* (see Fig. 3.67).

Attachments of the Acetabulum. The acetabular notch is filled by the *transverse ligament*, thus completing the concavity of the articular surface (Fig. 3.29, p. 149). Beyond the articular margin the rim of the acetabulum gives attachment to a dense fibrous-tissue ring called the *labrum acetabulare*. The capsule of the hip joint is attached to the acetabular labrum and to the surrounding bone; inferiorly it is attached to the transverse ligament. The *ligament of the femoral head* (ligamentum teres) is attached to the transverse ligament, not to the bone of the acetabulum. Synovial membrane is attached to the concave inner margin of the articular surface, whence it covers the Haversian fat pad in the acetabular fossa and is continued along the ligamentum teres to the fovea on the femoral head. Synovial membrane is likewise attached around the convexity of the articular margin and to the outer margin of the transverse ligament, whence it lines the hip-joint capsule down to the femur.

Outer Surface of Ilium. Above the acetabulum the ilium rises wedge-shaped along an anterior border to the **anterior superior spine.** Behind the acetabulum it passes up as a thick bar of weight-bearing bone, curves back to form the **great sciatic notch,** and continues to a posterior margin that lies between **posterior inferior** and **posterior superior spines.** The upper border of the ilium, between anterior and posterior superior spines, has a bold upward convexity and is at the same time curved from front to back in a sinuous bend. This border is known as the iliac crest. The anterior part of the crest is curved outwards, and the external lip is built up into a more prominent convexity 2 inches (5 cm) behind the anterior superior spine; this is the **tubercle** of the iliac crest.

The **gluteal surface** of the ilium is undulating, convex in front and concave behind, in conformity with the curvature of the iliac crest. It shows three curved gluteal lines. The most prominent is the *posterior gluteal line*, a low crest passing down from the iliac crest to the front of the posterior inferior spine. The area behind this line gives attachment to part of gluteus maximus. The *middle gluteal line* is a series of low tubercles coming from the iliac crest an inch (2 cm) behind the anterior superior spine; it is convex upwards, curving back just below the tubercle of the iliac crest and then downwards towards the great sciatic notch. Gluteus medius arises from the ilium between this and the posterior gluteal line, up to the outer lip of the iliac crest. The *inferior gluteal line* is less prominent. It curves from below the anterior superior spine towards the apex of the great sciatic notch. Tensor fasciæ latæ arises from the gluteal surface just below the iliac crest, between anterior superior spine and tubercle. Below this the area between the middle and inferior gluteal lines is occupied by the attachment of gluteus minimus. Below the inferior gluteal line is an area of multiple vascular foramina, then a narrow non-perforated strip along the upper margin of the acetabulum for the reflected head of rectus femoris (Fig. 3.29, p. 149).

The fusion of ilium and ischium is marked by a rounded elevation between the acetabulum and the great sciatic notch. Above this the ilium forms the major part of the notch. Piriformis emerges here, almost filling the notch and arising in part slightly from its upper margin (Fig. 3.63).

The **anterior border** of the ilium shows a gentle S-bend. Its upper narrow part gives attachment to the sartorius, at the anterior superior spine and for a finger's breadth below this. The lower half of this border is projected into the **anterior inferior iliac spine.** This shows a prominent oval facet for the straight head of rectus femoris. Below the facet is a smoother area for the ilio-femoral ligament, which is attached also somewhat behind this, above the acetabular margin (Fig. 3.64).

The **posterior border** of the ilium is a rounded bar of bone between the posterior superior and posterior inferior spines (Fig. 6.81, p. 474). It gives attachment to the sacro-tuberous ligament.

The Iliac Crest. The posterior part of the crest is thicker than the rest. Running forward from the posterior superior spine is a ridge which is traceable into the external lip at the upper end of the posterior gluteal line. External to the ridge the bevelled surface of the iliac crest continues down to the gluteal surface behind the posterior gluteal line. Gluteus maximus arises from the whole of this area.

In the articulated pelvis the ridge along the posterior part of the iliac crest is seen to be in line, below the posterior superior iliac spine, with the transverse tubercles of the lowest three sacral vertebræ (Fig. 6.81, p. 474). Along the ridge and these transverse tubercles the posterior lamella of the lumbar fascia is attached. The aponeurotic origin of latissimus dorsi is fused with the posterior lamella along the ridge, but the muscle itself extends forward of this along the external lip of the iliac crest, nearly halfway to the anterior superior spine (Fig. 2.5, p. 56). Internal to the ridge and its attached posterior lamella of lumbar fascia the ilio-costalis arises along a visible strip of non-perforated bone. Between this strip and the auricular surface the convex face of the bone shows lines of massed tubercles for the attachment of the strong sacro-iliac ligaments (Fig. 3.66).

At the lateral margin of ilio-costalis the strong ilio-lumbar ligament is attached to the iliac crest. Quadratus lumborum arises from the ilio-lumbar ligament and extends lateral to this to arise in continuity from a visible smooth strip that extends to a point one-third of the way forward from the posterior superior spine. This point is commonly marked by a low tubercle, for the attachment of the conjoined lamellæ of the lumbar fascia that enclose quadratus lumborum (p. 307). In the anatomical position of the bone this is the *highest part of the iliac crest.* From here the internal oblique and transversus abdominis muscles are attached side by side. The internal oblique is attached along the centre of the crest, the transversus to the inner lip of the crest, both extending to the anterior superior spine. External oblique is attached to the outer lip of the iliac crest in its anterior half; the gap between it and latissimus dorsi forms the base of the lumbar triangle (p. 256). The inguinal ligament, attached to the anterior superior spine, extends across to the pubic tubercle. The fascia lata of the thigh is attached along the whole length of the external lip of the iliac crest, splitting to enclose the narrow origin of tensor fasciæ latæ, and splitting also around gluteus maximus, where its deep layer is attached to the posterior gluteal line.

Outer Surface of Pubis. (Fig. 3.64). The body of the pubis, quadrilateral in shape, is projected laterally as a **superior ramus** which joins ilium and ischium at the acetabulum and an **inferior ramus** which fuses with the ischium below the obturator foramen. The symphyseal surface of the body is oval in shape; it is coated with a layer of hyaline cartilage for the secondary cartilaginous joint that constitutes the symphysis pubis (p. 355). The upper border of the body, gently convex, is called the **pubic crest.** It is marked laterally by a forward-projecting prominence, the **pubic tubercle.** From the pubic tubercle two ridges diverge laterally into the superior ramus. The upper ridge, sharp, is called the **pectineal line;** it forms part of the pelvic brim, and joins the arcuate line of the ilium. The lower ridge, more rounded, is called the **obturator crest.** It passes downwards into the anterior margin of the acetabular notch, where it becomes very prominent. Between these ridges the surface of the superior ramus can be traced to its junction with the ilium. Here lies a rounded prominence, the **ilio-pectineal eminence,** the site of the cartilaginous epiphysis between these parts of the hip bone. Below the obturator crest the pubic ramus is grooved obliquely by the obturator sulcus, which lodges the obturator nerve in contact with the bone. The obturator vessels lie below the nerve as the neuro-vascular bundle passes over the obturator membrane. Behind this sulcus a sharp border runs down to form the anterior margin of the obturator foramen. The inferior ramus of the pubis is marked by an everted crest. The line of junction with the ischial ramus sometimes shows as a low ridge; it is halfway between the ischial tuberosity and the pubic crest.

Attachments to the Pubis. The pubic crest gives origin to the rectus abdominis and, in front of this, the pyramidalis. The narrow linea alba is attached to the fibrous tissue of the symphysis pubis, not to bone, but lateral to this the conjoint tendon is attached over these muscles to the front of the pubic crest, across to the pubic tubercle and, in continuity, along the sharp pectineal line for about 2 cm. This same extent of the pectineal line receives also the upper margin of the lacunar ligament. The periosteum of the pectineal line lateral to this is thickened by the pectineal (Astley Cooper's) ligament (Fig. 5.8, p. 261). Pectineus arises deep to the lacunar ligament, from just below the pectineal line, between pubic tubercle and ilio-pectineal eminence. The lateral prominence of the obturator crest, near the acetabular margin, gives attachment to the pubo-femoral ligament. The pubic tubercle receives the attachment of the inguinal ligament and its so-called reflected part, which is really the posterior crus from the opposite side. The lateral crus of the superficial ring is attached with the inguinal ligament to the pubic tubercle. The medial crus, in front of the conjoined tendon, is inserted into the front of the pubic crest alongside the symphysis (Fig. 5.1, p. 255).

The rounded tendon of adductor longus arises from

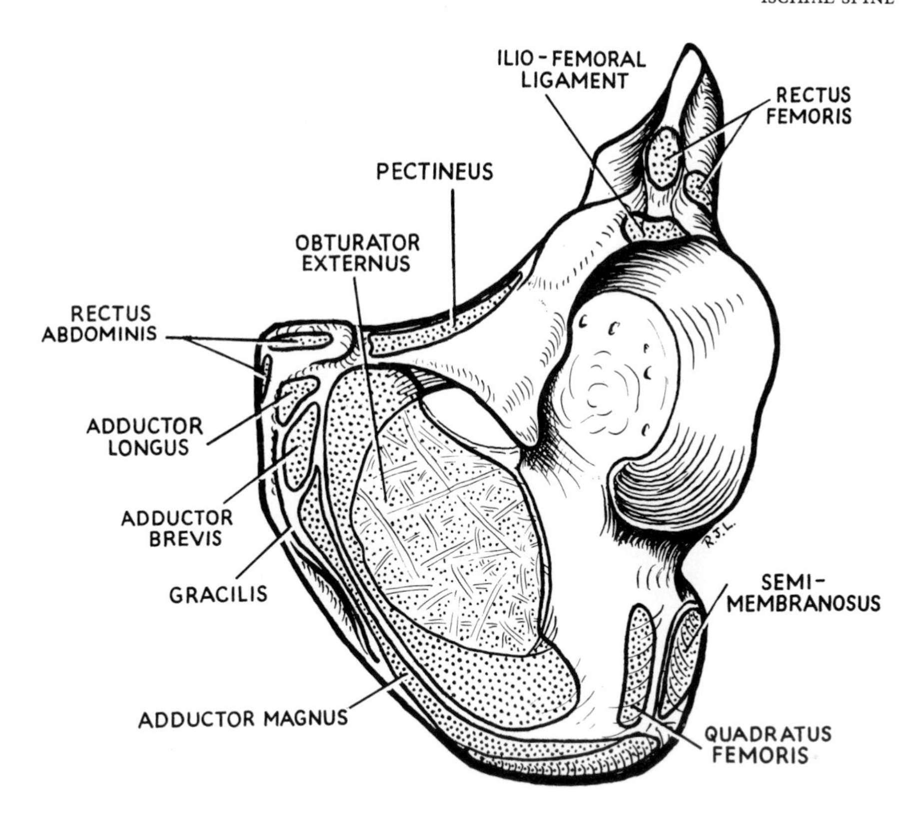

FIG. 3.64. THE ADDUCTOR SURFACE OF THE LEFT HIP BONE, TO SHOW THE MUSCLE ATTACHMENTS.

the front of the body of the pubis, in the angle between the pubic crest and symphysis; bone spurs may be found at this attachment (rider's bone). Below it the linear origin of gracilis extends down along the margin of the everted crest of the inferior ramus to reach the ischial ramus. Deep to adductor longus and gracilis the adductor brevis arises from the body of the pubis. Extending up along the inferior ramus the pubic fibres of adductor magnus arise deep to gracilis, and deeper still is the obturator externus. These muscles leave no distinguishing marks on the bone; their extents of origin have been studied, and are indicated in Fig. 3.64. The fascia lata is attached by its deep lamina to the pectineal line over the surface of pectineus (p. 134), and below the pubic tubercle along the front of the body of the pubis to the everted crest. It encloses adductor longus and gracilis, and separates them from the external genitalia in the superficial perineal pouch. Colles' fascia is attached to the pubis alongside the fascia lata.

Outer Surface of Ischium. Like the pubis the ischium is an L-shaped bone, and the two lie in the same plane. An upper thick portion of the ischium joins with pubis and ilium at the acetabulum and extends down to the ischial tuberosity; it supports the sitting weight. It is called the **body** of the ischium. A lower, thinner bar, the **inferior ramus,** joins the inferior ramus of the pubis to enclose the obturator foramen.

Behind the acetabulum observe again the low elevation at the line of fusion of ischium and ilium. At the margin of the acetabulum, where the capsule of the hip joint is attached, the fibres of the ischio-femoral ligament sweep from this upper part of the ischium upwards to encircle the capsule around the neck of the femur as the zona orbicularis. More medially the upper part of the body of the ischium completes the lower part of the great sciatic notch. Here the sciatic nerve, with the nerve to quadratus femoris deep to it, lies on the ischium. This site of emergence of the nerve into the buttock lies one-third of the way up from the ischial tuberosity to the posterior superior iliac spine.

The **spine** of the ischium projects medially to divide the greater from the lesser sciatic notch. The sacrospinous ligament is attached to its margins and tip, converting the great sciatic notch into the great sciatic foramen. The pudendal nerve lies on the ligament just medial to the spine. The internal pudendal vessels cross the tip, while the nerve to obturator internus lies on the base of the spine. The superior gemellus takes origin from the spine. The lesser sciatic notch lies between the spine and the ischial tuberosity (Figs. 3.65, 5.73, p. 349).

It is bridged by the sacro-tuberous ligament, which with the sacro-spinous ligament converts the notch into the lesser sciatic foramen (Fig. 5.55, p. 323). Obturator internus emerges through this foramen into the buttock, and the internal pudendal vessels and nerve pass forward into the perineum. The lesser sciatic notch is floored in life by thickened periosteum, grooved by tendinous fibres on the deep surface of the muscle (Fig. 3.66), and lubricated by a bursa. The inferior gemellus arises from the upper margin of the ischial tuberosity, above the hamstrings (Fig. 3.63).

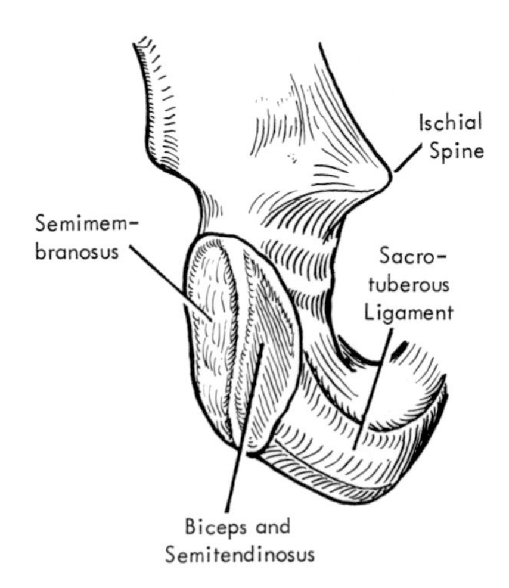

FIG. 3.65. THE LEFT ISCHIAL TUBEROSITY (POSTERIOR VIEW).

The **ischial tuberosity** is a rugged prominence whose convex posterior surface is divided transversely by a low ridge. An oval smooth area above this is divided by a vertical ridge into two areas, a lateral and a medial (Fig. 3.65). Semimembranosus tendon is attached to the lateral area, semitendinosus and the long head of biceps to the medial area.

Between semimembranosus area and obturator foramen the lateral surface gives a vertical linear origin, 3 cm long, to quadratus femoris, which leaves no mark on the bone (Fig. 3.64). Below the transverse ridge the ischial tuberosity shows a longitudinal crest; this supports the sitting body. The fascia lata of the thigh is attached to this ridge. The lateral bevelled surface here gives origin to the fibres of the hamstring part of adductor magnus (p. 143). The medial bevelled surface receives the sacro-tuberous ligament. The ischial tuberosity curves forward from the rugged weight-bearing part into the slender inferior ramus. Adductor magnus has a continuous origin along this ramus up to the pubic part of the muscle.

The **obturator foramen** is ringed by the sharp margins of pubis and ischium, those of the pubis overlapping each other in a spiral to form the obturator sulcus. The obturator membrane is attached to the margin of the foramen, but not to the obturator sulcus. Obturator externus arises from the outer surface of the membrane, a little of the ramus of the ischium and a good deal from the pubis (Fig. 3.64).

Inner Surface of Os Innominatum (Fig. 3.66)

This should not be studied until the pelvic walls (p. 323) and perineum (p. 344) are understood.

The pelvic brim is made by the top of the pubic crest and the pectineal line, continued up along the rounded border of the ilium (called the arcuate line) to the top of the auricular surface. This curved pelvic brim slopes up at sixty degrees. Below the brim lies the pelvic cavity, above it is the iliac fossa in the abdominal cavity.

Inner Surface of Ilium. The ilio-costalis area and the lines of tubercles on the ligamentous area have already been examined during the study of the outer surface. The **auricular area** really is ear-shaped; it extends from the pelvic brim to the posterior inferior iliac spine. Its surface undulates gently from convex above to concave below and is roughened by numerous tubercles and depressions. It is covered with hyaline cartilage and forms a synovial joint (immobile) with the ala of the sacrum. The capsule and synovial membrane are attached to the articular margin. In later years fibrous bands often join the articular surfaces within the joint space.

The **iliac fossa** is a gentle concavity in the ala of the ilium in front of the sacro-iliac joint and its ligaments. Its deepest part, high in the fossa, is composed of paper-thin translucent bone. The iliacus arises, up to the inner lip of the iliac crest, over the whole area down to the level of the anterior inferior iliac spine. The lower one-third of the fossa is bare bone, separated by a large bursa from the overlying iliacus. The fibres of iliacus converge to pass over a broad groove between the ilio-pectineal eminence and the anterior inferior spine. The iliacus fascia is attached around the margins of the muscle: to the iliac crest, the arcuate line and ilio-pectineal eminence (and to the inguinal ligament). Psoas major passes freely along the pelvic brim and crosses the ilio-pectineal eminence. The 'psoas' bursa deep to it is the iliacus bursa already mentioned. The psoas fascia is attached to the arcuate line and the ilio-pectineal eminence. Psoas minor, when present, has a tendon which flattens out to be inserted into the pelvic brim at the ilio-pectineal eminence.

Inner Surface of Pubis. The body and superior ramus are bare bone, with no attachments. The abnormal obturator artery and/or vein cross behind the ramus and the obturator nerve lies in contact with the ramus in the obturator sulcus. Levator ani (pubo-coccygeus) is attached to the junction of body and inferior ramus

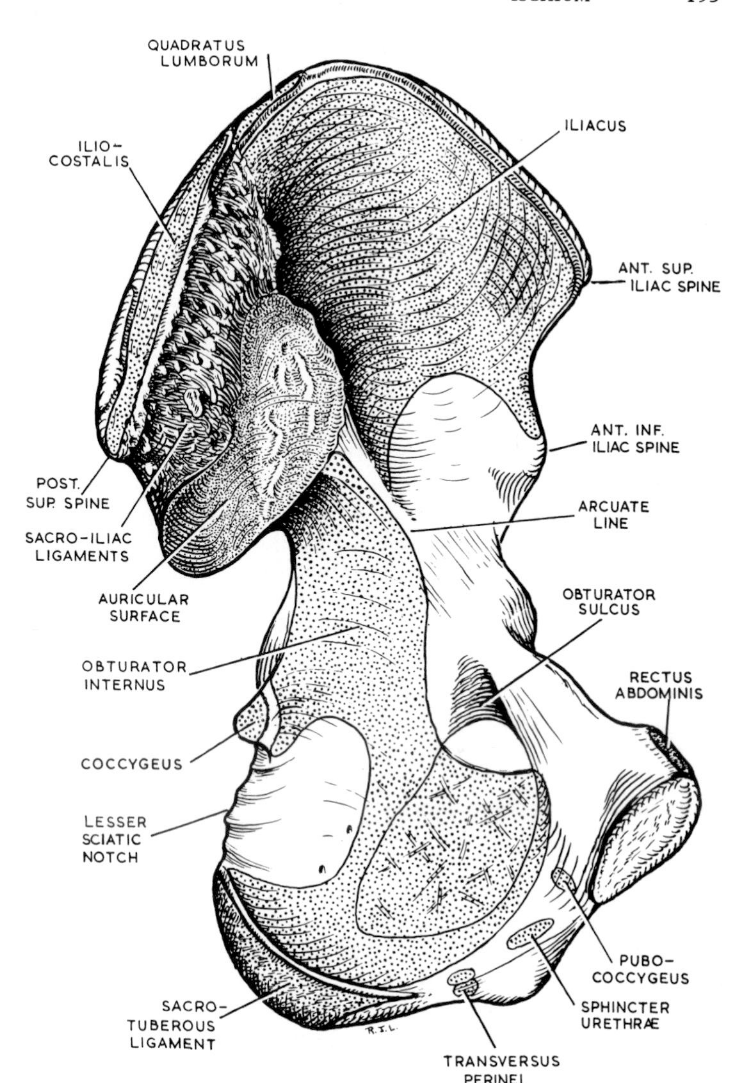

FIG. 3.66. THE LEFT HIP-BONE FROM THE MEDIAL SIDE.

level with the lower border of the symphysis. More medially the pubo-prostatic (pubo-vesical in the female) ligaments are attached at this level. The inferior ramus thus lies in the perineum. Immediately below the symphyseal surface the subpubic ligament is attached; the deep dorsal vein of the penis (clitoris) lies in the midline below it. Extending down from the symphyseal surface is a ridge of bone to which the perineal membrane is attached. External to this is the everted margin already noted. Here the crus of the corpus cavernosum is attached, with the ischio-cavernosus muscle. Between the perineal membrane and the obturator foramen the inferior ramus forms the wall of a forward prolongation of the ischio-rectal fossa. Here the dorsal nerve and artery of the penis (clitoris) pass forwards below levator ani, and here the bone gives attachment to fibres of sphincter urethræ. At the margin of the obturator foramen the obturator internus muscle is attached (see below).

Inner Surface of Ischium. The body of the ischium is very gently concave. From its posterior border projects the prominent ischial spine. This divides the greater from the lesser sciatic notch. To it are attached the coccygeus muscle and the white line that gives attachment to the pelvic floor (levator ani, Fig. 5.56, p. 323). Between the ischial spine and tuberosity the lesser sciatic notch is grooved by the tendinous fibres on the deep surface of obturator internus (Fig. 3.66). The grooves and ridges are made more prominent in the fresh bone by thickening of the periosteum. As the ischial tuberosity curves forwards into the ramus it is characterized by a falciform ridge. This ridge and the rough area below it receive the sacrotuberous ligament. Obturator internus arises from the body of the ischium and above it up to the arcuate line on the ilium (pelvic brim) to the margin of the great sciatic notch, from the obturator membrane below the obturator sulcus, from the obturator margin of the inferior ramus of the pubis and from the ramus of the ischium down to the falciform ridge (Fig. 5.67, p. 340). A gap in the origin extends up under the muscle from the lesser sciatic notch (Fig. 3.66) and here a bursa lies on the bare bone beneath the muscle. The fascia over obturator internus is attached to bone at the margins of the muscle. The pudendal canal on the obturator internus fascia lies just above the falciform ridge on the ischial tuberosity (Fig. 5.73, p. 349). The transverse muscles of the perineum are attached at the anterior extremity of the ramus.

Clinical Considerations. Two big nerves touch the hip bone. The sciatic nerve lies at the inferior margin of the great sciatic notch (ischium) and the obturator nerve in the obturator sulcus (pubis). The pudendal nerve lies close to the tip of the ischial spine and then runs into the ischio-rectal fossa just above the falciform margin of the ischial tuberosity. The nerve to quadratus femoris lies on the ischium from the great sciatic notch to the muscle, and the nerve to obturator internus crosses the base of the ischial spine.

The whole length of the iliac crest is palpable. The posterior superior iliac spine makes a characteristic dimple in the skin of the buttock. The widest part of the bony pelvis is between the tubercles of the iliac crests. This is not the highest part of the iliac crest, whose vertex lies 3 inches (7·5 cm) behind the tubercle. A horizontal plane through the highest points cuts the vertebral column between the 4th and 5th lumbar spinous processes, the site of election for lumbar puncture, and a starting point for counting the spinous processes with accuracy. The pubic tubercle is easily palpable (behind the spermatic cord for the comfort of the male examinee!) and the ischial tuberosity is a useful landmark, too. The ischial spine is palpable deep in the buttock, and pressure on it may control perineal bleeding by compressing the internal pudendal artery.

Sex Differences. These can be surveyed in the articulated pelvis (p. 322). The female pelvis is made roomier by changes in the os innominatum, and the isolated bone should be surveyed from behind forwards. The ala of the ilium has been drawn out to widen the female pelvis; the great sciatic notch is near a right angle in the female, much less in the male, and the female bone may show a pre-auricular sulcus below the arcuate line. The female ischial spine lies in the plane of the body of the ischium; the male spine is inverted towards the pelvic cavity. The female obturator foramen is triangular, the male foramen is oval in outline. The *surest single feature* is, however, that the distance from the pubic tubercle to the acetabular margin is greater than the diameter of the acetabulum in the female, equal or less in the male bone.

Ossification. The bone develops in cartilage. Three primary centres appear, one for each bone, near the acetabulum. The centre for the weight-bearing ilium appears first, at the 2nd month, followed by the ischium at the 3rd and pubis at the 4th month of fœtal life. At birth the acetabulum is wholly cartilage (Fig. 3.67), and while the ilium is a broad blade of bone the ischium and pubis are no more than tiny bars of bone buried in the cartilage. Growth of these three bones causes them later to approximate each other in a Y-shaped cartilage in the acetabulum. The ischial and pubic rami fuse with each other at about 7 years. Around the whole bone there remain strips of hyaline cartilage as follows: the whole length of the iliac crest, anterior inferior iliac spine, body of pubis at the symphysis, ischial tuberosity and sometimes ischial spine. The Y-shaped cartilage ossifies to close the acetabulum soon after puberty (say 15 years) and at the same time bony centres appear in the peripheral strips of cartilage. These fuse with the main bone when growth of the whole body ceases (say 25 years).

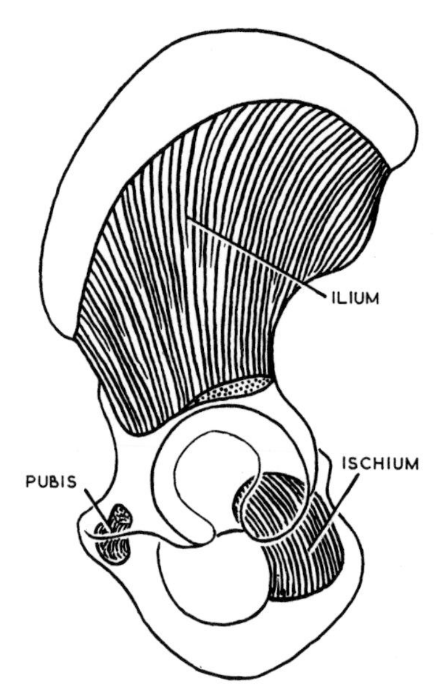

FIG. 3.67. THE LEFT HIP-BONE AT BIRTH. Drawn from Specimen A 161 in R.C.S. Museum.

The Femur

Note first the general features and the *anatomical position* of the bone. Lay it back on the table. This shows the forward convexity of the shaft, whose posterior concavity is buttressed by a strong ridge, the linea aspera, in its middle third. The head is above the table, because the neck has an angle of torsion with the shaft of 30 degrees. The neck of the femur passes from the head *backwards* as it slopes down to the shaft. This slope of the neck of the femur is in line with the forward and upward propulsive thrust of normal progression (walking, leaping, etc.). Hold the bone vertically, then place both condyles on the table. This produces an inclination of the shaft which is the true standing position; in life the femoral condyles rest horizontally on the plateau of the tibia. The inclination of the shaft is greater in the female, with her shorter legs and wider pelvis. In the long-legged male the shaft is nearer vertical, and the neck lies more nearly in line with the shaft (say 140 degrees). In the female the neck makes a less obtuse angle with the shaft (say 120 degrees). In the child the position is similar to that of the adult female; it is a matter of the length of the femur and the width of the pelvis.

The **head** of the femur, capped with hyaline cartilage, is more than half a sphere. Its medial convexity has a pit, the fovea, for the ligament of the head. Anteriorly the articular cartilage extends on the neck (Fig. 3.20, p. 141). This extension does not articulate with the acetabulum in standing because it is for weight-bearing in the *flexed* hip (try it on the dry bones). The synovial membrane of the hip joint, lining the capsule and covering the retinacular fibres on the neck, is attached to the articular margin of the head.

The **neck** of the femur is an upward extension of the shaft, angled as already noted. The angle is strengthened by the *calcar femorale*, a flange of compact bone projecting like a spur into the cancellous bone of the neck and adjoining shaft from the concavity of their junction, well in front of the lesser trochanter. The neck joins the shaft at the greater trochanter above and the lesser trochanter below. Many vascular foramina, directed towards the head, perforate the upper and anterior surfaces of the neck. Grooves and ridges on the surface indicate the attachment of retinacular fibres, reflected from the attachment of the hip-joint capsule to the articular margin of the head. These fibres hold down the arteries to the head (mostly from the trochanteric anastomosis) and their rupture may result in avascular necrosis of the head of the femur in *intracapsular* fracture of the neck. The neck joins the greater trochanter in front along a rough line called the trochanteric line. The capsule of the hip joint is attached to the line; the anterior surface of the neck with its adherent retinacular fibres is wholly intracapsular (Fig. 3.21, p. 141). The back of the neck joins the greater trochanter at a prominent rounded ridge, the trochanteric crest. Here the capsule of the hip joint is attached only halfway to the crest; the lower part of the neck alongside the crest is bare bone, over which the tendon of obturator externus plays (Fig. 3.27, p. 146). The trochanteric line, as well as receiving the anterior part of the capsule of the hip joint, gives attachment to the ilio-femoral ligament (of Bigelow), whose thickest part is received into a low tubercle at the lower end of the line. The pubo-femoral ligament is received into the lower surface of the neck alongside the capsular attachment. The ischio-femoral ligament reaches only the zona orbicularis; it is misnamed ('ischio-capsular' ligament is a more accurate name).

The Trochanters. These are for muscle attachments; each is an epiphysis that ossifies separately from the shaft.

The **greater trochanter** projects up and back from the convexity of the junction of neck and shaft. Its upper border is projected into an inturned apex posteriorly (Fig. 3.26, p. 146); this carries the upper part of the attachment of gluteus medius. Piriformis is attached here and spreads forward along the upper border deep to gluteus medius. More anteriorly the medial surface of the upper border shows smooth facets for the tricipital tendon of obturator internus and the gemelli. The apex of the trochanter overlies a deep pit, the trochanteric fossa. The bottom of the pit is smooth for the attachment of obturator externus tendon. The anterior surface of the greater trochanter shows a J-shaped ridge for gluteus minimus tendon (Fig. 3.20, p. 141). The lateral surface shows a smooth oblique strip, 1 cm wide, sloping down from the apex of the greater trochanter to the middle of the J-shaped ridge. This is for the tendon of gluteus medius. An expansion from this tendon passes from the apex of the trochanter across to the ilium to blend with the ilio-femoral ligament (Fig. 3.30, p. 150). The prominent convexity of the trochanter below gluteus medius forms the widest part of the hips. It is covered by the beginning of the ilio-tibial tract, where gluteus maximus is received. This plays freely over a bursa on the bare bone. Posteriorly the apex of the trochanter is continued down as the prominent trochanteric crest to the lesser trochanter. Nearly halfway down the crest is an oval eminence called the quadrate tubercle. Quadratus femoris is attached here, but it does not make the tubercle, which is a heaping up of bone at the epiphyseal junction. Quadratus femoris is attached to the quadrate tubercle and vertically below it down to a level that bisects the lesser trochanter.

The **lesser trochanter** lies back on the lowest part of the neck. Its rounded surface, facing medially, is smooth for the reception of the psoas major tendon. Iliacus is inserted into the front of the tendon and into the bone below the lesser trochanter (Fig. 3.21, p. 141).

The **shaft** of the femur is characterized by the linea aspera along its middle third posteriorly. This narrow ridge has medial and lateral lips. Note particularly their manner of formation. The trochanteric line slopes across the front of the neck and shaft at their junction, and continues down below the lesser trochanter as a spiral line that runs into the medial lip of the linea aspera (Fig. 3.26, p. 146). The medial lip continues on as a medial supracondylar ridge to the adductor tubercle on the medial condyle. On the back of the shaft below the greater trochanter is a vertical ridge (sometimes a groove) for gluteus maximus. This is called the gluteal crest and it runs down into the lateral lip of the linea aspera, and this lip is continued on as the lateral supracondylar line to the lateral epicondyle.

Attachments to the shaft. The trochanteric line in its lower half gives an aponeurotic origin to vastus medialis, which arises in continuity below this along the spiral line, medial lip of the linea aspera and upper one-third of the medial supracondylar line (the lowest part of this muscle comes from adductor magnus tendon, not from bone). The medial surface of the femoral shaft is bare bone, over which vastus medialis plays. Vastus lateralis arises in continuity from the upper half of the trochanteric line, the lower part of the J-shaped ridge of gluteus minimus, the gluteal crest, the lateral lip of the linea aspera and the upper two-thirds of the lateral supracondylar line. Vastus intermedius, arising from the front and lateral surfaces of the upper two-thirds of the shaft, extends up to the angle between the trochanteric line and gluteus minimus. Articularis genu arises a quarter of the way up the shaft below the lower limit of vastus intermedius. Thus the three vasti of the extensor compartment enclose almost the whole circumference of the shaft (Fig. 3.18, p. 140).

Note the attachments in the angle between the gluteal crest and the spiral line, above the linea aspera (Fig. 3.27, p. 146). Psoas major and iliacus have already been seen. Pectineus is received into this area, behind and below iliacus. Adductor magnus is inserted alongside the gluteal crest with its upper limit, edge to edge with quadratus femoris, lying at a level that bisects the lesser trochanter. Below this adductor magnus is attached to the middle part of the linea aspera and then to the medial supracondylar line down to the adductor tubercle. There is a gap in the supracondylar attachment, a hand's breadth above the knee, through which the femoral vessels pass into the popliteal fossa, and here the supracondylar line is obliterated. The gluteal crest (or groove) receives the deep half of the lower half of gluteus maximus.

Crowded into the narrow ridge between adductor magnus and vastus lateralis, the short head of biceps femoris arises by a flat sheet of aponeurotic fibres from the whole length of the linea aspera (i.e., middle one-third of shaft of femur), and passes down into the hamstring compartment. Between adductor magnus and vastus medialis are inserted adductor brevis and adductor longus. Adductor brevis is received into the upper one-third of the linea aspera and extends above this behind pectineus. Adductor longus is received as an aponeurotic sheet into the lower two-thirds of the linea aspera. The popliteal surface of the femur between the supracondylar lines is bare. The anterior surface of the lower shaft is likewise bare, with the suprapatellar pouch in contact with periosteum deep to quadriceps tendon for a hand's breadth above the knee joint.

The **lower extremity** of the femur carries the two condyles, separated behind by an intercondylar notch but joined in front by a trochlear surface for the patella. The lateral condyle projects further forward than the medial, thus helping to stabilize the patella (Fig. 3.22, p. 141). The articular surface of the trochlea, covered with hyaline cartilage, extends higher above the knee on the lateral than on the medial condyle. At the distal surface the lateral condyle is broad and straight, the medial narrow and curved. Both are almost flat antero-posteriorly, but boldly curved on their posterior convexities (Fig. 3.44, p. 164). They are joined, below the popliteal surface of the shaft, by an intercondylar ridge that encloses the intercondylar fossa. Just above this ridge the capsule and oblique popliteal ligament of the knee are attached. In the fossa the cruciate ligaments are attached to smooth areas: the anterior cruciate ligament far back on the lateral condyle alongside the articular margin, the posterior far forward on the medial condyle alongside the articular margin (Fig. 3.68). The medial condyle shows on its convex non-articular medial surface a shallow pit the bottom of which is smooth for the tibial collateral ligament. This is the epicondyle. Above it lies the adductor tubercle at the lower end of the medial supracondylar line. On the posterior surface, between the adductor tubercle and the articular margin, is a smooth area for the tendinous fibres of the medial gastrocnemius. Above this a rough area is raised on the popliteal surface of the shaft by the muscular fibres of gastrocnemius. The lateral condyle

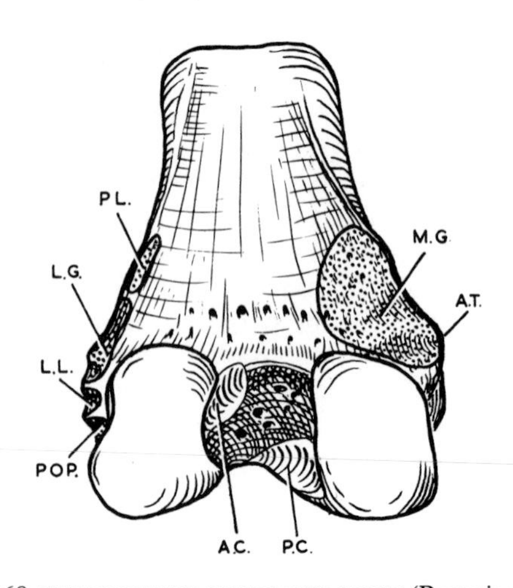

FIG. 3.68. THE LOWER END OF THE LEFT FEMUR (Posterior View). **M.G.** Medial head of gastrocnemius. **A.T.** Adductor tubercle. **P.C.** Posterior cruciate ligament. **A.C.** Anterior cruciate ligament. **POP.** Popliteus. **L.L.** Fibular collateral (lateral) ligament of knee. **L.G.** Lateral head of gastrocnemius. **PL.** Plantaris.

shows, towards the back of its non-articular lateral surface, a vertical arrangement of three smooth-floored pits (Fig. 3.44, p. 164). The upper pit is for the tendinous fibres of the lateral head of gastrocnemius. The muscular fibres of the lateral gastrocnemius arise in continuity from the lower half-inch of the lateral supracondylar line, which ends at the pit (Fig. 3.68). Above this plantaris arises from the line (Fig. 3.36, p. 158). The central pit is at the prominence of the convexity of this surface, called the lateral epicondyle; the fibular collateral ligament is attached to the pit. The lowermost pit receives the popliteus tendon; a groove behind the pit runs up to the articular margin for lodging the popliteus tendon when the knee is flexed.

The synovial membrane of the knee joint is attached to the articular margin. The capsule is attached to the articular margin except at two places. It is attached posteriorly to the intercondylar ridge, to imprison the cruciate ligaments and it is attached laterally above the pit and groove for the popliteus, to imprison the tendon within the knee joint. Its attachment across the trochlea is a very narrow flange, because here the capsule is widely perforated for communication of the synovial membrane with the suprapatellar pouch.

Clinical considerations. Only the greater trochanter is palpable at the upper end of the bone. The condyles are easily felt medially and laterally, and the adductor tubercle is the guide to the epiphyseal line at the lower end of the bone. The articular margin of the trochlea and the distal surfaces of the condyles are palpable if the knee is flexed.

Movements. Those at the hip are discussed on p. 151, at the knee on p. 164 and the patellar movements on p. 159.

Ossification. The whole femur ossifies in cartilage. A centre in the shaft appears at the eighth week of fœtal life. A centre for the lower end appears at the end of the ninth month (the day of birth) and its presence is acceptable medico-legal evidence of maturity. This is the growing end of the bone, and the epiphysis, which bisects the adductor tubercle (Fig. 3.38, p. 159), unites with the shaft after 20 years. A centre appears in the head at one year of age, greater trochanter at 3 and lesser trochanter at 12 years. These upper epiphyses fuse with the shaft at about 18 years of age. Note that the neck of the femur is the upper end of the shaft, not epiphysis.

The Patella

This sesamoid bone in the quadriceps tendon plays on the articular surface of the femur. Its edges form a rounded triangle, its anterior surface is gently convex. The lower border is projected down as an apex to the triangle. The articular surface, covered with hyaline cartilage, extends to the convex upper border but falls short of the lower apical part. Thus upper and lower borders can be identified. The articular surface has a vertical ridge dividing it into narrow medial and broader lateral areas; the bone laid down on a table lies on the broad lateral surface of the facet. So the right patella can easily be distinguished from the left. The narrow medial surface is further divided into two vertical strips (see p. 159). The upper border of the bone receives the quadriceps tendon, the medial border receives a flat tendon from the lowest fibres of vastus medialis. The quadriceps tendon sends a few superficial fibres across the front of the bone, grooving its anterior surface, and from each side extensions (patellar retinacula) from the vasti pass down to the ligamentum patellæ and to ridges on the tibia. The patella is attached distally to the tuberosity of the tibia by the ligamentum patellæ. This is attached to the patella across the whole non-articular lower border of the bone. The patellar retinacula extend down from the bone to the side of the ligamentum patellæ before sweeping away to the tibia. The patella is set like a plug in a gap in the anterior capsule of the knee joint. The capsule and synovial membrane are attached to the articular margin. The capsule is very short at its attachment to the upper border, being here widely perforated for the synovial membrane to communicate with the suprapatellar pouch. From the lower articular margin capsule and synovial membrane pass directly across to the margin of the tibial plateau (see below).

Movements are discussed on p. 159 and *stability* on p. 141.

Ossification. The bone forms in hyaline cartilage by a centre that appears at 3 years. Ossification is complete soon after puberty.

Clinical considerations. The patella remains in constant position relative to the tibia, held there by the ligamentum patellæ. The *lower* border of the patella is level with the knee joint (tibial plateau). Stellate fractures may show no displacement of fragments if the overlying quadriceps expansions and retinacula remain intact. The bone can be removed with surprisingly little disability.

The Tibia

This bone has a large upper end (the head) and a smaller lower end. The shaft is vertical in the standing position; the femur *inclines* up from the head of the tibia outwards to the acetabulum.

The **upper extremity** is widely expanded, and there is a prominent tuberosity projecting anteriorly from its lower part. The head is surfaced by very thin compact bone which is fragile particularly around the margins of the plateau. Try to secure a perfect specimen for study. Before studying the features of the upper end it is well to note the line of the epiphysis, to see whether

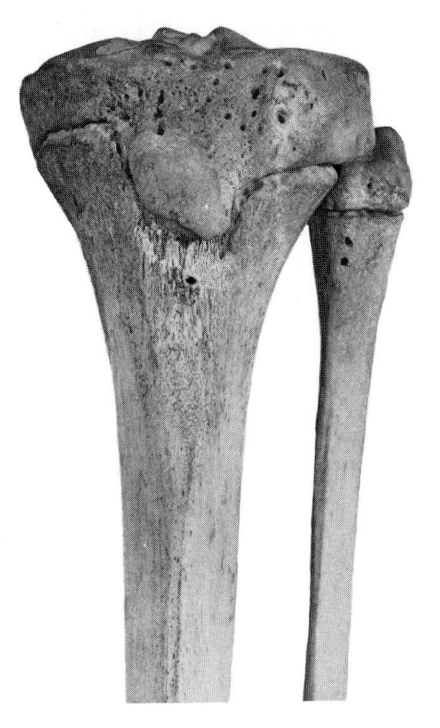

FIG. 3.69. THE EPIPHYSEAL LINES AT THE UPPER ENDS OF TIBIA AND FIBULA, LEFT SIDE, ANTERIOR VIEW. (Youth aged 18 years; photograph.)

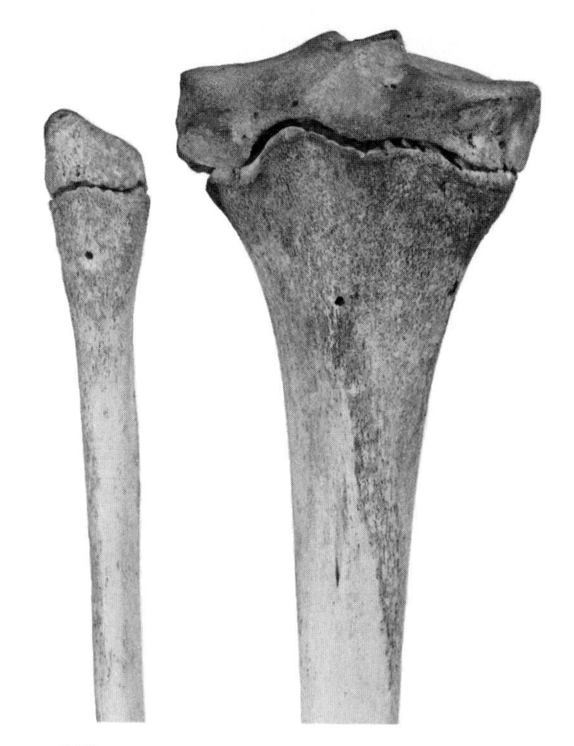

FIG. 3.70. THE UPPER ENDS OF THE LEFT TIBIA AND FIBULA, POSTERIOR VIEW. THE BONES ARE SEPARATED TO SHOW THE EPIPHYSEAL LINES. (Youth aged 18 years; photograph.)

structures are attached to it or to the shaft. The **epiphyseal line** cuts across the lower margin of each condyle at the back, and in front dips down to include the upper smooth part of the tuberosity (Figs. 3.69, 3.70). The chief points that will be noted are that the fibular facet and the attachments of semimembranosus and the ligamentum patellæ are on the epiphysis.

The knee-joint surface, or **plateau** of the tibia, shows a pair of gently concave condylar articular surfaces, for articulation with the menisci and the condyles of the femur (Fig. 3.37, p. 159). The surface on the medial condyle is oval (long axis antero-posterior) in conformity with the medial femoral condyle and meniscus. The lateral surface is more nearly circular, in conformity with the lateral femoral condyle and meniscus. The medial surface does not extend beyond the margin of the plateau. The lateral surface curves down over the margin to the posterior surface of the lateral condyle (this is for movement of the lateral meniscus, p. 165). Between the condylar surfaces the plateau is elevated into the **spine** (or intercondylar eminence), which is grooved antero-posteriorly between ridges on to which the articular surfaces rise. These little ridges on the spine are called the medial and lateral *intercondylar tubercles.* Nothing is attached to them. The non-articular areas in front of and behind the spine show well-marked facets for attachment of the horns of the menisci and the cruciate ligaments. In front of the spine is a large smooth area for attachment of the anterior cruciate ligament. The lateral margin of this area receives the anterior horn of the lateral meniscus, just in front of the lateral intercondylar tubercle. Further forward, at the margin of the tibial plateau, is a round smooth facet for the anterior horn of the medial meniscus. Behind the spine is a smooth area sloping down to an oblique ridge between the posterior convexities of the condyles. The posterior cruciate ligament is attached to the ridge and to the smooth slope above it. The posterior horn of the medial meniscus is attached to a deep slit behind

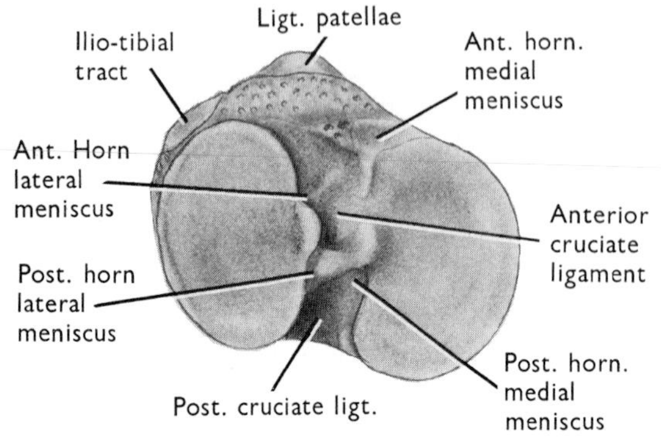

FIG. 3.71. THE PLATEAU OF THE LEFT TIBIA.

the medial intercondylar tubercle. The posterior horn of the lateral meniscus is attached, in front of the posterior cruciate ligament, just behind the lateral intercondylar tubercle (Fig. 3.71). The capsule of the knee joint is attached to the circumference of the tibial plateau except in two places. Where the tendon of popliteus crosses the margin of the tibia the capsule extends down to the head of the fibula. Between the condyles posteriorly the capsule is attached not to the margin of the plateau but to the ridge below the groove for the posterior cruciate ligament (Figs. 3.36, p. 158, 3.72). Between the menisci and the tibial plateau the capsule is often referred to as the 'coronary ligament'. The synovial membrane is attached to each articular

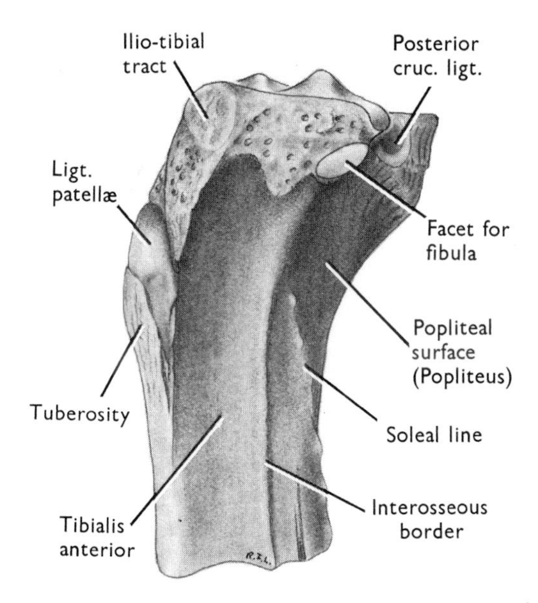

FIG. 3.72. THE UPPER END OF THE LEFT TIBIA (LATERAL VIEW).

surface, following the condylar margins alongside the spine to be draped over the attachment of the anterior cruciate ligament.

The **head** of the tibia is wide, and the **condyles** lean back over the popliteal surface of the shaft. The back of the medial condyle is deeply grooved for the semimembranosus insertion; this groove extends around the medial contour of the bone, to receive the expansion of the tendon beneath the tibial collateral ligament (Fig. 3.39, p. 160). The ligament is not attached to the condyle, but to the shaft lower down. The lateral condyle carries the facet for the head of the fibula (see p. 168). The capsule and synovial membrane of this tibio-fibular joint are attached to the articular margin. Just above this facet the lateral condyle may show a groove for the popliteus tendon, which plays across the bone here, lubricated by the popliteus bursa. Small upward extensions of peroneus longus and extensor digitorum longus usually encroach from the fibula on the lateral convexity of the condyle in front of the fibular facet, and an expansion from biceps femoris overlies them. The anterior surface of the lateral condyle shows a smooth facet, 1 cm in diameter, for the ilio-tibial tract (Fig. 3.72). Below it a ridge slopes down to the tuberosity. A similar ridge extends down from the medial condyle. Expansions from the vasti lateralis and medialis, known as patellar retinacula, are attached to these ridges. The **tuberosity** shows a smooth oval prominence set obliquely; it receives the quadriceps insertion via the ligamentum patellæ. The rough triangular area on the lower part of the tuberosity is subcutaneous. A bursa covers it, swollen in clergyman's knee (rare). The space between the capsular attachment and the patellar retinacula contains soft extracapsular fat that adapts its shape to the changing contours of the femoral condyles in flexion-extension of the knee. Deep to the ligamentum patellæ lies an infrapatellar bursa (Fig. 3.46, p. 166).

The **shaft** of the tibia is triangular in section. Its anterior and posterior borders, with the surface between them, are subcutaneous. The subcutaneous surface receives the tendons of sartorius, gracilis and semitendinosus at its upper end (see p. 166), and behind these the tibial collateral ligament is attached. The surface is continued at its lower end into the medial malleolus. The anterior border is sharp above, where it shows a medial convexity imprinted by tibialis anterior. This border becomes blunt below, where it continues into the anterior border of the medial malleolus. The blunt posterior border runs down into the posterior border of the medial malleolus. On its fibular side the tibia shows a sharp interosseous border which, over an inch (4 cm) from the lower end, splits into two. The interosseous border gives attachment to the interosseous membrane, whose oblique fibres slope down to the fibula. In front of the interosseous border is the extensor surface of the shaft. This surface runs from the upper end below the lateral condyle to spiral over the front of the lower end of the tibia. Tibialis anterior arises from the upper two-thirds or less of this surface. Below the muscle the extensor surface of the tibia is bare. It is crossed by tibialis anterior and extensor hallucis longus tendons. The flexor surface of the shaft lies behind the interosseous border. Its upper part is distinguished by the soleal line, which runs down obliquely from just below the tibio-fibular joint across this surface to meet the posterior border one-third of the way down. Popliteus arises from the popliteal surface of the tibia above the soleal line; the popliteus fascia, a downward extension from semimembranosus tendon, is attached to the line. The upper end of the soleal line often shows a tubercle for attachment of a fibrous band that passes to the head of the fibula, bridging the posterior tibial

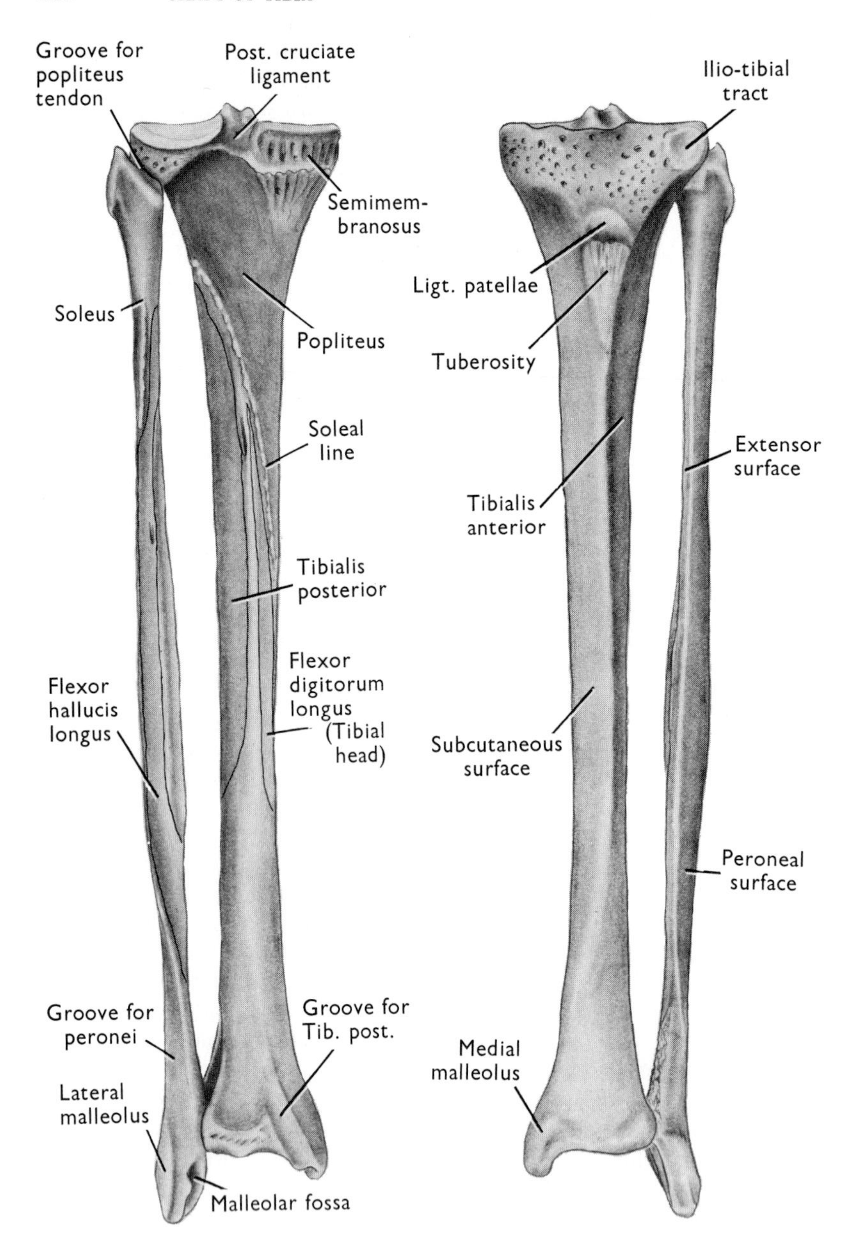

FIG. 3.73. THE BONES OF THE LEFT LEG (POSTERIOR AND ANTERIOR VIEWS).

neuro-vascular bundle. From this band, the soleal line, and the *middle third of the posterior border* the soleus arises in continuity. Below the soleal line the flexor surface shows a vertical ridge. Tibialis posterior arises between this ridge and the interosseous border (it crosses the interosseous membrane to arise also from the fibula). Medial to the ridge, between it and the posterior border, flexor digitorum longus arises. The upper part of this surface is perforated by a large nutrient foramen directed downwards. The lower part of this surface is bare bone, crossed by tibialis posterior and the overlying flexor digitorum longus.

The **lower end** of the tibia is rectangular in section. Medially the surface is subcutaneous, with the great saphenous vein and nerve crossing above the medial malleolus (PLATE 6, p. *iii*). Anteriorly the bare bone is crossed by the tendons of tibialis anterior and extensor hallucis longus and the anterior tibial neuro-vascular bundle and extensor digitorum longus (Fig. 3.47, p. 169). There may be a facet (the 'squatting' facet) just above the articular margin, for articulation with the neck of the talus in full dorsiflexion of the ankle joint. Laterally the surface is triangular between the ridges that diverge from the lower end of the interosseous border; this triangular area gives attachment to the strong inferior tibio-fibular ligament. The lower part of this surface may articulate with the fibula as a synovial upward continuation of the ankle joint. Posteriorly there is a groove behind the medial malleolus, made deeper in life by thickening of periosteum at its margins. The tendon of tibialis posterior, in its synovial sheath, is lodged here, bridged by a band of fibrous tissue (Fig. 3.49, p. 173). Alongside it the tendon of flexor digitorum longus crosses the bone, while on the fibular side the posterior surface has the lower end of the fleshy belly of flexor hallucis longus in contact with bare bone. The distal surface shows a saddle-shaped facet for the talus, the articular surface extending to line the medial malleolus. The capsule and synovial membrane of the ankle joint are attached to these articular margins (and to the posterior tibio-fibular ligament). The distal surface of the medial malleolus shows a smooth area for the deep lamina of the deltoid ligament, and the superficial part of the ligament is attached more superficially (the deltoid ligament is as thick as the bone of the malleolus). The upper limb of the inferior extensor retinaculum is attached superficially, to the front of the malleolus.

The fascia lata of the thigh is attached to the ridges running down to the tuberosity, along with the patellar retinacula. The deep fascia of the calf is attached to the anterior and posterior borders of the shaft, down to the medial malleolus. It is thickened at the lower end of the extensor compartment as the superior extensor retinaculum (p. 167), and from the back of the medial malleolus a band passes to the calcaneus as the flexor retinaculum (PLATE 6, p. *iii*).

Movements at the knee joint are discussed on p. 164 and at the ankle on p. 182. Tibio-fibular movements are dealt with on p. 168.

Clinical considerations. The tibia is easy to examine, for so much of the bone is subcutaneous. The lower third of the shaft is bare of any muscle or tendon attachments, has a correspondingly low blood supply, and may show delayed or imperfect union after fracture.

Ossification. The shaft ossifies in cartilage from a primary centre that appears in the 8th week of fœtal life. The upper epiphysis (the growing end) shows a centre immediately after birth. This joins the shaft at 20 years along the epiphyseal line already noted. A secondary centre for the tuberosity may appear about puberty. The lower epiphysis ossifies at the second year and joins the shaft at 18 years. The epiphyseal line passes a centimetre above the distal end of the shaft and includes the medial malleolus; it is extracapsular.

The Fibula

The slender shaft expands above into a quadrilateral head and below into a flattened malleolus. The head carries a small *oval* facet with a styloid process behind it and the malleolus carries a larger *triangular* facet with a deep malleolar fossa behind it. Thus can upper and lower ends be distinguished. Note that the malleolar facet faces medially towards the tibia and that the malleolar fossa is *behind* it. Short practice will enable right and left fibulæ to be recognized by *looking at* the malleolar facet without thumbing the malleolar fossa. Glance at the malleolar facet; the fossa is to the left in the left fibula, to the right in the right fibula (Fig. 3.74).

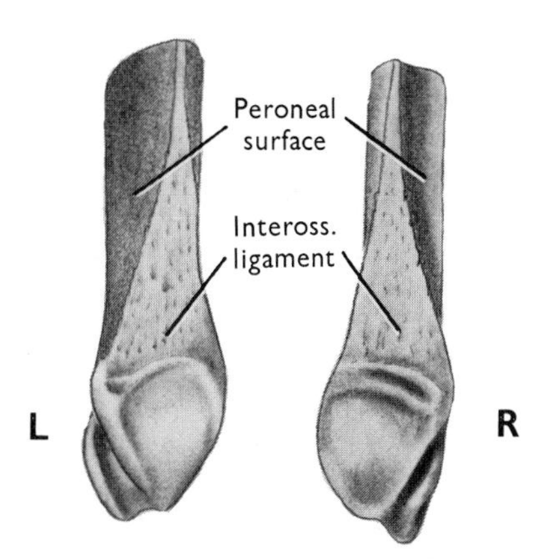

FIG. 3.74. THE LEFT AND RIGHT FIBULAS DISTINGUISHED BY A GLANCE AT THE MALLEOLAR FOSSA.

The bone is difficult to study, and should be placed against its companion tibia to appreciate properly its surfaces and attachments.

The **head** of the fibula carries an oval or round facet set obliquely on its upper surface; this articulates with a reciprocal facet on the upper epiphysis of the tibia. The capsule and synovial membrane of the superior tibio-fibular joint are attached to the articular margin. The back of the head is projected upwards as the styloid process. The arcuate ligament is attached to it (p. 160). The fibular collateral ligament is attached in front of the styloid process, with biceps tendon still further forward, on the upper surface (Fig. 3.42, p. 163). From the styloid process a groove for the popliteus tendon slopes down behind the articular surface.

The **shaft** has three surfaces, extensor, peroneal and flexor, in conformity with the compartments of the leg. The *peroneal surface* is smooth and easy to identify. The groove behind the lateral malleolus spirals up behind the triangular subcutaneous surface of the lower end to the lateral surface of the shaft and so up to the head. Peroneus brevis arises from the lower two-thirds of this surface and its tendon passes down to groove the back of the malleolus. Peroneus longus arises from the upper two-thirds (behind brevis where they overlap) and its tendon passes down behind that of peroneus brevis. The common peroneal nerve enters peroneus longus at the upper end of the shaft (the 'neck', below the head) and there divides. The superficial peroneal nerve, supplying the two muscles, passes down to emerge between them. The deep peroneal nerve pierces the anterior peroneal septum to reach the extensor compartment of the leg. These nerves, very close to the bone, are not in actual contact with it, being cushioned by a few deep fibres of peroneus longus. The ridges that border the peroneal surface give attachment to the anterior and posterior septa that enclose the peroneal muscles.

Identify the interosseous border as follows. Above the malleolar facet is a somewhat rough triangular surface for the interosseous tibio-fibular ligament (Fig. 3.74). From the apex of this triangular area the interosseous border passes up alongside the *anterior* border of the shaft, which is, in fact, the anterior peroneal ridge. The *extensor surface* is now seen to be extremely narrow, especially at its upper end, where in some bones the anterior and interosseous borders fuse. From the upper three-fourths of this narrow strip extensor digitorum longus takes origin, and here the deep peroneal nerve touches the bone beneath the muscle. In continuity with this muscle peroneus tertius arises from the lower third, where the extensor surface broadens somewhat. Deep to this (i.e., towards the interosseous border) extensor hallucis longus arises from the middle two-fourths of the fibula and the adjacent interosseous membrane.

The *flexor surface*, between the interosseous and posterior borders, is much wider. The middle third of this surface shows a vertical ridge which runs down into the interosseous border. It is called the medial crest; do not mistake it for the interosseous border. The medial crest is traceable up to the neck of the fibula. Between the medial crest and the interosseous border tibialis posterior is attached (it crosses the interosseous membrane to a similar area on the tibia). The medial crest gives attachment to what is often described as a deep intermuscular septum but which is in reality the fibular aponeurotic origin of flexor digitorum longus (Fig. 3.49, p. 173). The flexor surface between the medial crest and the posterior border of the fibula is for flexor hallucis longus. Below the medial crest (i.e., below the origin of tibialis posterior) flexor hallucis longus arises from the whole flexor surface and from the adjoining interosseous membrane as far as the inferior tibio-fibular fibrous joint. Here a spiral twist of the flexor surface matches the spiral twist of the peroneal surface. The upper part of the flexor surface, varying from a quarter to one-third, gives origin to soleus; a roughened 'soleal line' shows on many bones.

The **lower extremity,** or **malleolus,** projects further distally than the tibial malleolus. Its medial surface has a triangular facet for the talus. Capsule and synovial membrane of the ankle joint are attached to the articular margin except posteriorly, where they are carried on the posterior tibio-fibular ligament. The malleolar fossa behind this surface is perforated by foramina. It gives attachment to two ligaments that diverge to the tibia and the talus. They are the posterior tibio-fibular ligament, which articulates with the talus and carries the ankle-joint capsule, and the posterior band of the lateral ligament of the ankle, called the posterior talo-fibular ligament. The triangular subcutaneous area has a rounded lower margin with a smooth area in front for the anterior talo-fibular ligament and a similar area at the apex for the calcaneo-fibular ligament (Fig. 3.56, p. 181). Between the subcutaneous area and the malleolar fossa is a smooth groove for the tendon of peroneus brevis (the long tendon lies on it and does not touch the fibula). The two tendons are in a common synovial sheath, and bridged by the superior peroneal retinaculum, a band of fibres passing from the tip of the malleolus to the calcaneus.

Clinical considerations. The head and lower end are palpable, and the deep peroneal nerve can be rolled against the neck. It is vulnerable here, especially in tight bandaging. *Movements* of the fibula are discussed on p. 168. Distracting forces at the ankle joint commonly cause indirect fracture at the lower end (Pott's fracture). Fracture of the lower tibia may be accompanied by fracture at the *upper* end of the shaft of the fibula.

Ossification. The fibula ossifies in cartilage by a centre in the shaft which appears in the 8th week. There is an epiphysis at each extremity. The head, the growing end, is exceptional in ossifying later (4th year) than the lower end (2nd year). The upper epiphysis fuses with the shaft at 20 years, the lower before this (say 18 years).

The Foot

The principal joints of the tarsal bones have already been studied (pp. 181–189), and the movements of inversion and eversion should be well understood before osteological details of individual bones are studied. The tarsus is built from seven bones. It articulates by the talus at the ankle joint, and it also provides the mobility of inversion-eversion by its midtarsal and subtalar joints. Only one of the seven bones, the calcaneus, rests on the ground. The metatarsus articulates with the tarsus, and the metatarsal heads, especially the 1st and 5th, rest on the ground. The toes lie free to move in front of the weight-bearing metatarsal heads.

The Calcaneus

The heel bone is the largest of the tarsal bones and it is the first to ossify (p. 209). It articulates with the talus above and the cuboid in front. It is a rectangular block of bone, characterized by the sustentaculum tali, a shelf that projects from the upper border of its medial surface (PLATE 46).

The **upper surface** of the calcaneus carries articular surfaces on its anterior half. The sustentaculum tali and the anterior part of the body show a common articular surface, elongated and concave; this may be separated into two halves, but even so they occupy the one joint cavity with the head of the talus. Study of the talo-calcaneo-navicular joint (p. 184) will make clear the attachments to the margins of this articular surface. The capsule and synovial membrane of the talo-calcaneo-navicular joint are attached to only the postero-lateral margin (Fig. 3.60, p. 184). The antero-medial margin is in the joint space, with the spring ligament attached to the margin of the sustentacular surface and the synovial membrane over the fat pad attached to the articular margin on the body of the calcaneus. Behind the sustentacular facet an oblique groove (the floor of the tarsal canal, p. 185) passes laterally and forwards to a wide non-articular area which floors in the open part of the sinus tarsi. The groove gives attachment to the talo-calcaneal interosseous ligaments (Fig. 3.60, p. 184). In the sinus tarsi a smooth round facet alongside the lateral margin is for the ligamentum cervicis tali (Fig. 3.56, p. 181). The bifurcate ligament is attached to the anterior margin of the calcaneus. Behind this is the origin of extensor digitorum brevis and, at the lateral margin of the sinus tarsi, the inferior extensor retinaculum (frondiform ligament) is attached (Fig. 3.60, p. 184). Behind the tarsal groove and sinus is a convex articular surface, oval-shaped, its long axis parallel with the groove. Here the talus articulates by a separate joint; the capsule and synovial membrane are attached to the articular margin. These two joint surfaces on the anterior half of the calcaneus together constitute the inferior surface of the 'subtalar' joint, where inversion and eversion take place. The non-articular posterior half of the upper surface is saddle-shaped (with the rider facing forwards).

The **posterior surface** of the calcaneus has a smooth upper part for the tendo calcaneus. Its lower part, convex, is grooved longitudinally for attachment of the posterior fibres of the plantar aponeurosis, which sweep back from the under surface. The extreme upper part, below an upturned lip, may be bare for a bursa that lies here deep to the tendo calcaneus.

The **inferior surface** of the calcaneus shows two tubercles, a large medial and a smaller lateral, at its posterior end (Fig. 3.76). These form the weight-bearing part of the bone. The medial tubercle gives origin to abductor hallucis, flexor digitorum brevis and part of abductor digiti minimi, and the lateral tubercle to the rest of abductor digiti minimi. The lateral head of flexor accessorius arises from the lateral tubercle deep to these muscles. Superficial to these muscles the plantar aponeurosis is attached, some fibres sweeping up to grooves on the back of the bone as noted above. In front of the tubercles the narrow inferior surface is grooved by the attachment of the long plantar ligament. This surface tapers to a smooth anterior tubercle, with a fossa in front of it. The fossa fits a similar fossa behind the ridge on the cuboid; a finger-tip lies comfortably here when the two bones are articulated. The fossa is filled by the short plantar ligament, which is attached to the smooth area on the anterior tubercle.

The **anterior surface** of the calcaneus carries the articular surface for the cuboid, an undulating surface that is slightly lipped above and bevelled below.

The **lateral surface** of the calcaneus carries an oblique ridge below and behind the sinus tarsi (Fig. 3.56, p. 181). This is the *peroneal trochlea*, from which the inferior peroneal retinaculum bridges the sheathed tendons of peroneus brevis in the groove above and peroneus longus in the groove below the trochlea (Fig. 3.48, p. 171). Much further back the calcaneo-fibular band of the lateral ligament of the ankle joint slants back to its attachment; this generally leaves no distinguishing mark on the bone. The rest of the lateral surface is bare bone, perforated by numerous blood vessels.

The **medial surface** of the calcaneus is concave (PLATE 46). The fleshy medial head of flexor accessorius occupies the whole of this area, down to the medial tubercle. Above the concavity the **sustantaculum tali** projects; its under surface is deeply grooved by the tendon of flexor hallucis longus in its sheath. The rounded medial border of the sustentaculum tali gives attachment across its whole thickness to the spring ligament in front and the superficial lamina of the deltoid ligament behind (Fig. 3.59, p. 183). The tendon of flexor digitorum longus lies superficial to the sustentaculum.

Clinical considerations. The peroneal trochlea is palpable, and the weight-bearing medial and lateral tubercles can be felt through the thickness of the heel-pad. The sustentaculum, lying inferior to the medial malleolus, is more difficult to feel because its border is overlaid by the tendon of flexor digitorum longus.

The Talus

The talus carries the whole body weight. It lies on the weight-bearing calcaneus, below the tibia, and communicates thrust from the one to the other. The bone possesses a body which is prolonged forward into a neck and a rounded head. The upper surface of the body carries an articular area known as the trochlea. The **trochlea** is convex from front to back but with a shallow central groove (i.e., concave from side to side). The trochlea is usually broad in front and narrow behind, its lateral border curving backwards and medially (Fig. 3.58, p. 183). The posterior end of this lateral border is blunt, or bevelled, by the posterior tibio-fibular ligament, which here comes into contact

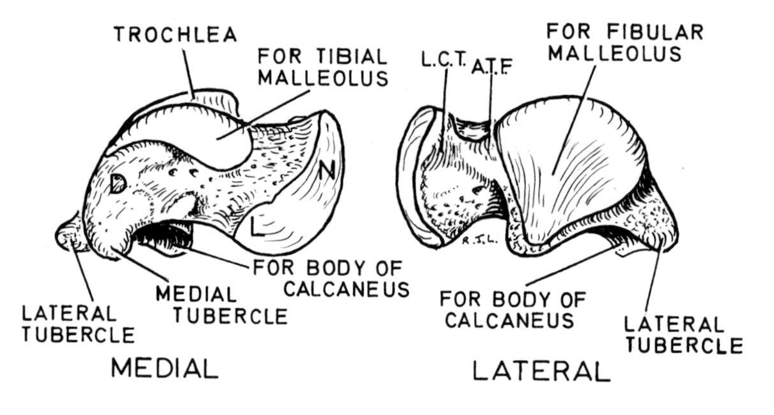

FIG. 3.75. MEDIAL AND LATERAL VIEWS OF THE LEFT TALUS.

Medial view. D. Area of attachment of deep lamina of deltoid ligament. **L.** Surface for articulation with anterior part of deltoid ligament (the part cut in Fig. 3.60). **N.** Area for articulation with navicular.

Lateral view. L.C.T. Area of attachment of ligamentum cervicis tali. **A.T.F.** Area of attachment of anterior talo-fibular ligament.

with the margin of the trochlea (Fig. 3.57, p. 182), particularly in dorsi-flexion of the ankle joint. The trochlear surface is continued down over each side of the body for articulation with the stabilizing malleoli. On the medial surface the articular area is comma-shaped, broad end anterior (Fig. 3.75). This area articulates with the malleolus of the tibia. In the concavity of the comma curve there are many vascular foramina. Behind these the deep lamina of the deltoid ligament is attached to a smooth area. On the lateral surface the articular area is much bigger, covering almost the whole surface. Triangular in outline, it is concave from above down and usually convex from front to back. This surface articulates with the fibular malleolus (Fig. 3.75).

Behind the trochlea the talus is projected into a **posterior process** which is deeply grooved by the tendon of flexor hallucis longus (the groove lies in line with the groove on the under surface of the sustentaculum tali). The posterior process projects as a pair of tubercles, one on either side of the flexor hallucis longus groove. The **lateral tubercle** is the most posterior part of the talus. This may form a separate ossicle, called the **os trigonum** (morphologically the os intermediale, the counterpart of the lunate bone of the wrist, p. 127). The lateral tubercle gives attachment to the posterior talo-fibular ligament, and this ligament lies in a groove below the articular margin almost to the apex of the fibular surface. The less prominent **medial tubercle** is blunt and rounded; it gives attachment to the posterior fibres of the deltoid ligament.

The capsule of the ankle joint is attached to the articular margin except in front, where its attachment encroaches forward on the neck of the talus. The synovial membrane is attached to the articular margin.

The under surface of the body of the talus (PLATE 46) has a large oblique facet, concave for articulation with the calcaneus. In front of this the **neck** is grooved to fit over the corresponding groove on the calcaneus to make the tarsal canal. The groove gives attachment to the interosseous talo-calcaneal ligament. Laterally, in the sinus tarsi, the neck carries a smooth round facet for attachment of the ligamentum cervicis tali, and behind this the anterior talo-fibular ligament is attached. The neck of the talus is very short. It is directed forwards and medially.

The rounded **head** of the talus is capped by a large articular surface facing forwards and downwards. Anteriorly the surface is convex for articulation with the navicular. Inferiorly it is flattened for articulation with the sustentaculum tali and the body of the calcaneus; a low ridge commonly separates this calcaneal area into two flat facets. The navicular convexity and the calcaneal flattening are separated from each other by a triangular convexity (base of the triangle at the neck of the talus) for the spring ligament and the deltoid ligament, according to the position of inversion or eversion of the foot. The convex navicular surface articulates laterally with the bifurcate ligament (Fig. 3.60, p. 184). The head of the talus is the ball of the ball and socket talo-calcaneo-navicular joint, whose capsule and synovial membrane are attached around the neck of the talus to the articular margin of the head.

Movements of the talus. The possible movements are three: (1) *above* the talus dorsi-flexion and plantar-flexion of the ankle joint, (2) *below* the talus inversion and eversion of the foot, (3) *in front* of the talus rotation of the navicular at the mid-tarsal joint. At the trochlea the hinge movements of dorsi-flexion and plantar-flexion impart passive rotary movements to the fibula (see p. 182). The talo-calcaneo-navicular joint is anatomically one cavity, but it functions as two separate joints (like the elbow and the superior radio-ulnar joint cavities). The talus articulates with the calcaneus and the navicular, and it can move on one of these bones *without necessarily moving on the other*. Movement on only the navicular is accompanied by movement at the calcaneo-cuboid joint (these two separate joints together form the mid-tarsal joint). Movement on only the calcaneus occurs below the head and body (these two separate joints together constitute the subtalar joint). These movements of inversion and eversion have been discussed on p. 185.

Stability. At the ankle joint bony factors are important, for the talus is wedged between the malleoli, and the shaft of the fibula is mobile for adjustment of its malleolus to the side of the talus. In dorsi-flexion of the ankle a wider area of the trochlea engages with the tibia, and the bones are in their most stable position. The subtalar joints depend entirely on ligaments for their stability.

Clinical considerations. The head of the talus is palpable, and becomes a visible prominence in the fully plantar-flexed foot. Many vascular foramina perforate the non-articular surfaces; in the tarsal groove and on the medial surface the arteries are from the posterior tibial, on the upper surface of the neck from the anterior tibial. Dislocation of the talus ruptures these vessels, and since *no muscle or tendon is attached to the talus* to carry blood to the dislocated bone, avascular necrosis is almost the rule after reduction of the dislocation.

The Cuboid

The bone is rather wedge-shaped, narrowest at the lateral margin and broadest medially where it articulates with the lateral cuneiform. Set in front of the calcaneus, its anterior surface articulates with the 4th and 5th metatarsal bones, thus forming the lateral longitudinal arch of the foot (Fig. 3.76). Medially it articulates with the lateral cuneiform, and sometimes with the navicular. Its calcaneal surface is undulating, slightly bevelled above and laterally and slightly lipped below and medially. Its anterior surface, concave from above down, is slightly convex from side to side. A low vertical ridge separates two facets, rectangular for the base of the 4th metatarsal and triangular for the base of the 5th. The medial surface shows an oval facet, high against its upper border, for the lateral cuneiform; the posterior end of this facet is occasionally prolonged for 2 or 3 mm for a tiny synovial joint with the navicular. The dorsal surface of the bone is bare, the margins here giving attachment to the joint capsules and interosseous ligaments. The short lateral border is projected into a small anterior and a large faceted posterior tubercle, with a deep groove between them for the tendon of peroneus longus. The facet on the posterior tubercle, smooth and round, is for the sesamoid cartilage or bone in the tendon. Inspection of the under surface shows that the tubercles on the lateral border are actually the extremities of a pair of oblique ridges on the inferior surface. The anterior of these is inconspicuous, being merely the prominent border of the metatarsal articular surface. The posterior is a prominent ridge in its own right, bridged by the long plantar ligament to make a fibro-osseous tunnel that lodges the peroneus longus tendon and its synovial sheath. Behind the ridge is a deep hollow to which is attached the short plantar ligament.

Movements are simple gliding of its synovial joints with the rise and fall of the arches of the foot.

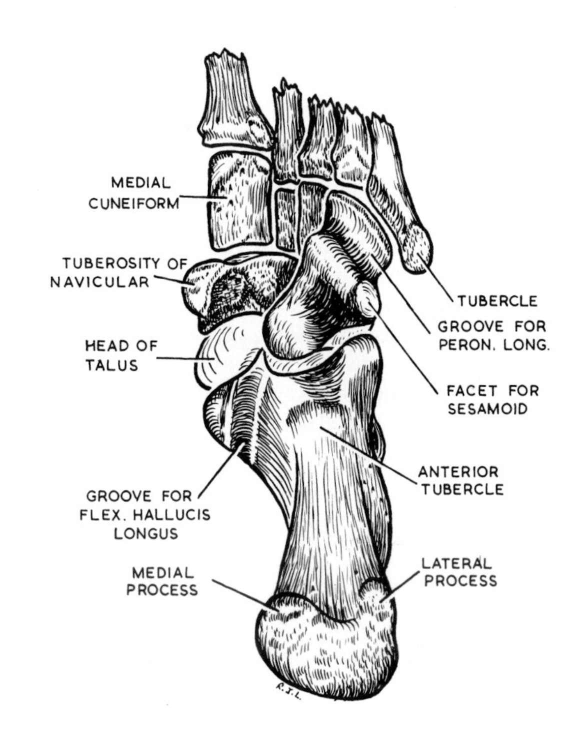

FIG. 3.76. INFERIOR VIEW OF LEFT TARSUS. The prominent (weight-bearing) posterior part of the calcaneus is called the tuberosity. Its two bulges, now officially named the medial and lateral processes, were formerly called 'tubercles', and this latter name, more accurate, is deliberately retained in the text.

The Navicular

In so far as it is concavo-convex this bone may be said to be boat-shaped, with a prominent medial tuberosity representing the prow of the boat (Fig. 3.58, p. 183). A proximal concave articular surface fits the convexity of the head of the talus. Here the upper and lateral margins give attachment to the capsule and synovial membrane of the talo-calcaneo-navicular joint. The inferior margin gives attachment to the spring ligament medially, to the bifurcate ligament laterally, and to the synovial membrane over the fat pad between these two (Fig. 3.60, p. 184). The distal convex articular surface is divided by a V-shaped ridge into three triangles, the medial one base down, the other two apex down. These articulate with the three cuneiform bones. The lateral facet may extend from the distal to the lateral surface of the navicular for articulation with the cuboid. The joint between the navicular and the three cuneiforms is a single synovial cavity that is continuous along either side of the intermediate cuneiform with the cavity along the bases of the 2nd to the 5th metatarsals.

The **tuberosity** is palpable in all feet and visible in most. Its smooth convexity gives insertion to tibialis posterior and from this bone 'expansions' diverge widely, to the cuboid ridge, to the cuneiforms and the metatarsal bases. Many of these bands are ligaments, and not really fibres from the tendon itself. Below the tuberosity is a smooth area, as wide as the thickness of the sustentaculum tali, for attachment of the spring ligament, which is of corresponding thickness.

The Cuneiform Bones

The cuneiform bones, true to their name, are all three wedge-shaped. The medial is the largest and lies edge upwards. The intermediate is the smallest and, with the lateral, lies edge downwards in the transverse arch (p. 188). All three articulate posteriorly with the navicular, and anteriorly each articulates with its metatarsal bone, thus completing the medial longitudinal arch (p. 187).

The Medial Cuneiform. The *medial surface* of the bone shows a smooth round facet at its anterior inferior angle. Tibialis anterior is inserted into the facet (and into a smaller facet on the adjoining base of the first metatarsal). A small area at the upper limit of the *lateral surface* articulates with the intermediate cuneiform and the base of the 2nd metatarsal. Below this articular area, in the sole of the foot, the lateral surface is largely occupied by the attachment of strong interosseous ligaments. At the anterior inferior angle is a low tubercle opposite a large smooth fossa on the adjoining base of the first metatarsal. Peroneus longus is inserted into the tubercle and into the metatarsal base. The *anterior surface*, wholly articular, is kidney-shaped. It forms a separate synovial joint with the base of the first metatarsal bone. Slight gliding occurs here with up and down movements of the medial longitudinal arch. Capsule and synovial membrane are attached to the articular margins. The *posterior surface* is occupied by a smaller pear-shaped facet for the navicular.

The **intermediate cuneiform** articulates by small synovial joints with the medial and lateral cuneiforms, and is bound to them by strong interosseous ligaments. It lies base upwards, and is the shortest of the three, so the base of the 2nd metatarsal is mortised between the medial and lateral cuneiforms.

The **lateral cuneiform,** base up, articulates laterally with the cuboid, to which it is bound by strong ligaments. It projects further distally than the intermediate cuneiform, thus articulating with the base of the 2nd metatarsal (Fig. 3.58, p. 183).

Note that the synovial joints between the three cuneiforms and between the lateral cuneiform and cuboid all lie at the upper limits of the adjacent surfaces. The broad lower parts of these surfaces give strong fixation to plantar interosseous ligaments that bind the bones firmly together in the transverse arch of the foot.

Each cuneiform articulates with the base of its own metatarsal bone. The first forms a *separate synovial cavity*. The other four metatarsal bases, articulating with intermediate and lateral cuneiforms and the cuboid, and with each other, share a common joint space. This space extends back alongside the intermediate cuneiform into the joint between the cuneiforms and the navicular.

The Metatarsal Bones

The **1st metatarsal** is a thick bone which transmits thrust in propulsion of the body. Proximally a kidney-shaped facet articulates with the medial cuneiform by a synovial joint. On the lateral side of the base a small facet against the upper pole of the kidney articulates with the 2nd metatarsal base, and a smooth fossa opposite the lower pole of the kidney receives peroneus longus tendon. The shaft is flattened dorsally, but inferiorly it is braced by a ridge that arches up between base and head. The head has a distal articular surface that extends convexly on the dorsal surface and extends still more on the plantar surface, where it shows a pair of parallel deep grooves for the sesamoid bones. These bones receive short muscles (abductor, flexor, adductor) whose tendons pass to the proximal phalanx; the sesamoid bones are themselves attached to the capsule of the metatarso-phalangeal joint.

The **2nd to 5th metatarsals** have slender shafts, and there is no flat dorsal triangular area as on the metacarpals. The mortise for the base of the 2nd has already been noted. The metatarsal shafts give attachment to the interosseous muscles. Plantar interossei arise from the shaft of their own bone (3rd, 4th and 5th metatarsals) and adduct their toes towards the 2nd toe. Dorsal interossei arise from both bones of their space, and abduct the 2nd, 3rd and 4th toes away from the line of the 2nd metatarsal bone (Fig. 3.55, p. 179). The metatarsal heads are united by a series of deep transverse ligaments that bind them together.

The base of the 5th metatarsal is prominent, lateral to the joint with the cuboid, as a proximally directed tubercle (styloid process) which receives the tendon of peroneus brevis (Fig. 3.76). The dorsal surface of the base, next to the base of the 4th, receives peroneus tertius, whose tendon extends along the top of the shaft for a variable distance, even up to the neck of the bone (Fig. 3.48, p. 171).

Movements. See p. 186.

The Phalanges

As in the hand, there are two for the pre-axial digit and three each for the others. Each metatarso-phalangeal joint is greatly strengthened by a thick pad of fibro-cartilage on its plantar surface. The pad serves as a connecting ligament and at the same time is the plantar part of the capsule of the synovial joint. Sesamoid bones occur irregularly in these pads. Collateral ligaments, as in the hand, reinforce the sides of the joints. A similar arrangement exists in the interphalangeal joints.

Each phalanx of the big toe receives a separate extensor tendon into its base. The distal phalanx receives extensor hallucis longus, the proximal extensor hallucis brevis (part of extensor digitorum brevis). There are similar flexor attachments on the plantar surface; flexor hallucis longus is inserted into the base of the distal phalanx, and flexor hallucis brevis, via the two sesamoids, into the sides of the base of the proximal phalanx. The other four toes receive tendons as do the fingers. On the plantar surface flexor digitorum brevis splits into a chiasma for insertion into the intermediate phalanx, and flexor digitorum longus passes through this to reach the distal phalanx. The little toe muscles abductor and flexor digiti minimi brevis are inserted into the base of the proximal phalanx. On the dorsal surface an extensor expansion exists as in the hand, formed from extensor digitorum longus, the lumbricals and part of the interossei. The tendons of extensor digitorum brevis join the extensor expansions on the 2nd, 3rd and 4th toes (and sometimes the 5th). The tendons of the interossei are inserted directly into the bases of the proximal phalanges as well as into the extensor expansions.

The **attachments** to the bones of the foot have been noted during the study of the tarsal bones and the phalanges. There remain a few points to note in the tarso-metatarsal region. Strong plantar interosseous ligaments unite adjacent bones and help to support the arches of the foot. The three short muscles of the third layer of the sole arise here, chiefly from the cuboid and the bases of the 2nd, 3rd and 4th metatarsals (p. 178).

Ossification of the Foot. All the foot bones ossify in cartilage. Three bones of the tarsus are ossified at birth. The calcaneus begins to ossify at the 6th month and the talus at the 7th month of fœtal life. The cuboid ossifies in the 9th month, and the presence of this centre is acceptable medico-legal evidence of maturity (it is a convenience to look for this by a simple incision into the side of the foot of a stillborn fœtus). The navicular ossifies in the 4th year. The cuneiforms do not ossify in the order of their size; the lateral one ossifies in the 1st year, the medial in the 3rd and the small intermediate in the 4th year. Metatarsals and phalanges ossify by shaft centres *in utero*, and their epiphyses are as in the hand (the epiphysis of the first metatarsal is at the base, that of each of the other four is in the head). These epiphyses ossify 2 or 3 years later than those in the hand, their centres appearing about the 5th year; but they join earlier (say 18 years).

There is a secondary centre on the posterior surface of the calcaneus; it is a thin plate of bone that appears about the 10th year and joins at 18 years. The lateral tubercle of the talus, the tubercle at the base of the 5th metatarsal and the tuberosity of the navicular sometimes ossify as separate centres.

Section 4. The Thorax

THE BODY WALL

Though very different functionally, the wall of the thorax and the wall of the abdomen are one, topographically and developmentally. They are best considered together.

The Skin of the Body Wall. The skin varies in texture, tending to be thin in front and thick behind. Distribution of hair varies with sex, age and race. Natural lines of cleavage of the skin are very constant, and are of tremendous importance to the cosmetic appearance of healed incisions. An incision along a cleavage line will heal as a hair-line scar, virtually invisible; an incision across the lines will tend to heal with either a wide or a heaped-up scar. The cleavage lines run almost horizontally around the body wall (Fig. 1.4, p. 3).

Subcutaneous Tissue. This is the same as elsewhere in the body (p. 3). Fat is contained in loculi, whose fibrous walls connect the overlying dermis to the underlying deep fascia. There is an exception over the dilatable part of the body wall, namely the anterior abdominal wall and lower part of the thoracic wall in front of the midaxillary lines. Here the fibrous septa of the subcutaneous tissue are condensed beneath the fat into a thin but strong membrane, the fascia of Scarpa.

The **fascia of Scarpa** allows the subcutaneous fat (unnecessarily called the 'fascia of Camper') to slide freely over the underlying thoracic wall, rectus sheath, and external oblique aponeurosis. It fades out over the upper thoracic wall and along the midaxillary lines. Below, over the thighs, it is attached by fibrous strands that connect the dermis to the fascia lata along the flexure skin crease of the hip, extending from the pubic tubercle obliquely outwards below the inguinal ligament. The fascia of Scarpa is attached to the sides of the body of the pubic bone and is continued over the penis and scrotum, where it receives a different name, the fascia of Colles (p. 352).

Blood Supply of the Subcutaneous Tissue. The intercostal and the lumbar arteries pass forward in the neuro-vascular plane (p. 22) to supply the flanks; the internal thoracic and the superior and inferior epigastric arteries supply the ventral midline tissues. From all these arteries cutaneous branches pass to the superficial fat and skin. The *venous return* from the subcutaneous tissue does not follow the arteries. The blood is collected by an anastomosing network of veins that radiate away from the umbilicus. Below this level they pass to the great saphenous vein in the groin, above the umbilicus they run up to the lateral thoracic vein and so to the axillary vein. A few at the umbilicus drain the lower part of the ligamentum teres. The upper part of this ligament drains to the left portal vein; these *para-umbilical veins* may distend in portal obstruction, giving rise, if the distension spreads to the subcutaneous veins, to the *caput medusæ*. There is a channel of anastomosis between the supra-umbilical and the infra-umbilical veins that opens up in some cases of portal obstruction. In these cases a tortuous mass of subcutaneous varicose veins may extend in the midclavicular line from the upper thorax to the lower abdomen.

Lymphatic return from the subcutaneous tissue and skin follows the veins, to axillary and superficial inguinal nodes. From above the level of the umbilicus, lymph from the front of the body goes to the pectoral lymph nodes, from the back of the body to the scapular nodes. (For lymph drainage of breast, see p. 66.) From below the umbilicus lymph from the anterior aspect of the abdominal wall and perineum goes to the medial group of superficial inguinal nodes, and from the lateral and posterior aspects of the abdominal wall to the lateral group of superficial inguinal nodes.

Cutaneous Innervation of the Body Wall (p. 21). Above the second rib, the skin is supplied by supraclavicular branches of the cervical plexus (C4). Below this level a midline strip of ventral skin is supplied by the *anterior terminal branches* of the spinal nerves from Th. 2 to L1, which latter supplies suprapubic skin. A broad lateral strip is supplied by the *lateral cutaneous branches* of the spinal nerves from Th. 2 or 3 to L1; these branches emerge in the midaxillary line. Their anterior and posterior branches each innervate an oblique zone of skin in regular sequence. Note that the lateral cutaneous branches of Th. 12 and the ilio-hypogastric nerve descend over the iliac crest to supply the skin of the buttock. The ilio-inguinal nerve has no lateral cutaneous branch; it is the collateral branch of the ilio-hypogastric, both coming from the first lumbar nerve. A posterior strip of skin is innervated by the *posterior primary rami* of spinal nerves, by their medial branches in the upper part and their lateral branches in the lower part (Fig. 1.25, p. 23).

Morphology of the Body Wall Muscles

The body wall of all vertebrates consists essentially of three layers of muscles. The muscles encircle the body cavity, their fibres lying in different slopes of obliquity. The intermediate layer is reinforced with bony condensations (the ribs) to a cranio-caudal extent that varies with the species. Ribs never intersect any but the intermediate layer of muscles. The external layer passes *outside the ribs*, while the innermost layer passes *inside the ribs*. The three layers are all innervated segmentally by anterior primary rami. They may be identified in the human abdominal wall as the external oblique, internal oblique and transversus muscles.

Over the thoracic wall the **external layer** passes outside the ribs and the sheet of muscle becomes broken up for functional needs into separate muscle masses for attachment to the pectoral girdle and upper limb. The flat muscles on the chest wall supplied segmentally by spinal nerves belong to this layer (e.g., serratus anterior, levator scapulæ and the rhomboids) and the other great muscles of the axillary folds (pectoralis major and latissimus dorsi), though derived embryologically from the upper limb belong, morphologically, to this same muscle group.

The **intermediate layer** in the thorax, confined to the rib layer and passing neither outside nor inside the ribs, is broken up for functional needs into two muscle sheets, with the fibres lying at right angles to each other. That they slope in similar directions to the fibres of the external oblique and internal oblique muscles of the abdominal wall is coincidental and gives no clue to their morphology. Both external and internal intercostal muscles belong, with the ribs, to the intermediate layer of the body wall. The external oblique muscle itself is attached, outside the ribs, external to the external intercostal muscle.

In submammalian species the **innermost layer** encircles the body cavity inside the bony cage of the thorax and, as a continuous sheet, surrounds the abdominal cavity and floor of the pelvis (levator ani). It is the compressor of the cœlom. The lungs lie within it and air has to be pumped into them at a pressure higher than that of the atmosphere, thus discouraging at the same time venous return to the lung.

In the mammals the cranial part of the innermost layer has descended caudally to produce the muscular septum known as the diaphragm. The lungs lie cranial to it, in the thorax, and air is sucked into them at a pressure less than that of the outside air. This 'negative' pressure simultaneously sucks venous blood into the thorax, a more efficient mechanism than that of the creatures which have no diaphragm. In the mammals not the whole of the transversus sheet descends to the diaphragm; some of its fibres remain as thin layers of muscle within the ribs, forming the transversus thoracis group of muscles (p. 214). The mammals, with their diaphragms and temperature-regulating mechanisms (and of course their breasts) are a great improvement on their non-mammalian precursors. 'This year's model excels last year's.'

Towards the midline, ventrally and dorsally, the three layers change direction to make muscles that run longitudinally. **Ventrally** the longitudinal muscle is formed by fusion together of all three layers. It begins behind the symphysis menti as the genio-hyoid muscle, and is traceable downwards as the thyro-hyoid and the other infra-hyoid strap muscles (sterno-hyoid, omohyoid and sterno-thyroid). There is a gap where the sternum and costal cartilages intervene, then rectus abdominis begins and extends to the symphysis pubis. Traces of the ventral rectus muscle appear in this gap in one body in twenty; longitudinal fibres lie in front of the sternal fibres of pectoralis major, forming the rectus sternalis muscle (p. 54). The rectus complex is supplied by spinal nerves; note that from the inferior genial tubercles of the mandible down to the symphysis pubis the nerve supply is segmental. In the neck it is from C1, 2, 3 (C1 fibres 'hitch-hiking' via the hypoglossal nerve), and in the abdomen from the lower six thoracic nerves.

Dorsally the three layers turn longitudinally, but remain quite separate from each other. The *innermost layer* (transversus layer) persists as the 'prevertebral rectus', that is to say, the prevertebral muscles of the neck, the crura of the diaphragm, psoas and the piriformis; they are all supplied segmentally by anterior primary rami of spinal nerves. The *intermediate layer* is represented by muscles that lie entirely between ribs and costal elements. Quadratus lumborum and the scalene muscles belong to this group.

The erector spinæ is divided into separate muscles in layers, for functional reasons, but the whole mass is derived entirely from the *external muscle layer*.

All the muscles of the body wall are supplied segmentally by spinal nerves. The erector spinæ group, derived from the external layer, is supplied exclusively by posterior rami and it is the only muscle which these rami innervate. All the remaining muscles are supplied segmentally by anterior primary rami. These spinal nerves, with the segmental arteries, run in the *neurovascular plane*. This is the space between the intermediate and the innermost muscle layers and it is continuous throughout the whole extent of the body wall, from thoracic inlet to pelvic brim. In the thorax it lies between the transversus thoracis group and the internal intercostal muscles, in the abdomen between the transversus abdominis and internal oblique muscles (Fig. 1.24, p. 22).

THE THORACIC WALL

The skin and subcutaneous tissue of the thoracic wall having been studied (p. 210) it now remains to consider the muscles and skeletal structures. The skeleton of the thoracic wall consists of the twelve thoracic vertebræ, the twelve pairs of ribs and costal cartilages and the sternum. The thoracic cavity is roofed in above the lung apices by the suprapleural membrane and is floored by the diaphragm. The floor is highly convex (domes of the diaphragm), so that the volume of the thoracic cavity is much less than inspection of the bony cage would suggest. The liver and spleen and the upper parts of the stomach and both kidneys lie in the abdominal cavity covered by ribs.

The Ribs

Ribs are to breathe with. The purpose of studying their joints and their muscles is to understand how this is brought about.

Ribs are not primarily protective (heart and lungs are no more sensitive to external violence than 'unprotected' abdominal viscera—*vide* modern thoracic surgery). Perhaps ribs are protective in fish—against increased hydrostatic pressure. In snakes they are organs of locomotion—the rib-ends act as feet inside the skin, instead of outside as in the invertebrate centipede. But in air-breathing mammals the primary function of ribs is respiratory.

The Articulations of the Ribs

The ribs articulate posteriorly with the vertebral column and anteriorly with their own costal cartilages. The upper seven costal cartilages articulate with the sternum, the next three articulate each with the costal cartilage above it, and the last two lie free.

The Costo-vertebral Joints. The posterior end of each rib articulates with the vertebral column in two places, the head and the tubercle. Examine a typical rib. The **head** possesses two articular facets that slope away from each other, separated by a ridge. Each facet articulates by a synovial joint; the lower facet with the costal facet on the body of its own vertebra and the upper facet with that on the vertebra above (Fig. 4.1). The ridge between the two is attached to the intervertebral disc by a fibrous ligament. In addition to the capsular ligaments of the two synovial joins three ligaments are attached to the head of the rib. Together they may be called the **tri-radiate ligament.** The upper fasciculus passes across the joint to the body of the vertebra above, the lower fasciculus passes to the body of the vertebra below. The central fasciculus runs horizontally, deep to the anterior longitudinal ligament, across the intervertebral disc, with which it blends, to join with fibres from the other side (Fig. 4.1). This arrangement, known as the **hypochordal bow,** is of morphological interest, since it explains the anterior arch of the atlas vertebra (p. 463). In this vertebra the body (or more accurately speaking, the centrum) is absent, forming the dens by fusion with the centrum of the axis vertebra. The anterior arch of the atlas consists of the ossified hypochordal bow joining the two costal elements of this vertebra.

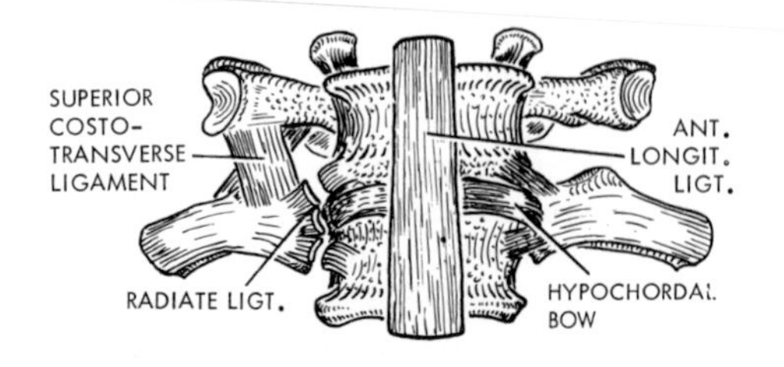

FIG. 4.1. ANTERIOR VIEW OF THE COSTO-VERTEBRAL JOINTS.

The **tubercle** of a typical rib has two facets (p. 251). The medial facet, covered with hyaline cartilage, articulates with a reciprocal facet near the tip of the transverse process of its own vertebra. The lateral facet gives attachment to a ligament. The capsule of the costo-transverse synovial joint is strengthened by three accessory ligaments.

The **superior costo-transverse ligament** (Fig. 4.1) passes in two laminæ from the crest on the neck of the rib to the transverse process of the vertebra above, the **inferior costo-transverse ligament** occupies the space between the neck of the rib and the transverse process of its own vertebra, while the **lateral costo-transverse ligament** attaches the tip of the transverse process to the lateral or ligamentous facet of the tubercle.

The tri-radiate and the costo-transverse ligaments are stronger than the rib, which will fracture before ever the head or tubercle could dislocate.

Exceptional Ribs. The first rib articulates with the upper border of the first thoracic vertebra only, never coming into contact with the vertebra above. Of the two 'floating' ribs, the eleventh and twelfth, each articulates with only its own vertebra, and makes no synovial joint with the transverse process, to which it is attached by ligament only (see p. 252).

Movements (see also p. 218). Rotation of the necks of the ribs occurs at all twelve costo-vertebral joints, and this results in elevation and depression of the anterior ends of the ribs. The articular surfaces of the upper six costo-transverse joints are curved. The articular facet on the tubercle of the rib is convex, and the facet at the tip of the transverse process of the vertebra is reciprocally concave. While this shape of the joint permits rotation of the neck of the rib, it prevents up and down excursion of the tubercle; thus the true bucket-handle movement does not occur in the upper six ribs (*v. inf.*).

The articular surfaces of the 7th to the 10th costo-transverse joints are flat. Thus the tubercles of the ribs of the costal margin can move up and down and so permit the bucket-handle movement in addition to rotation around the neck of each rib.

Costo-chondral Joints. Every rib makes with its costal cartilage a primary cartilaginous joint. The costal cartilage represents no more than the unossified anterior part of a rib. The anterior end of the bony rib is deeply concave to receive the reciprocally convex end of the costal cartilage. No movement takes place at these joints.

Chondro-sternal Joints. The first costal cartilage is exceptional in that it articulates with the manubrium by a primary cartilaginous joint. The cartilage is short and thick and capable of very little distortion. Thus the manubrium and the first ribs are fixed to each other

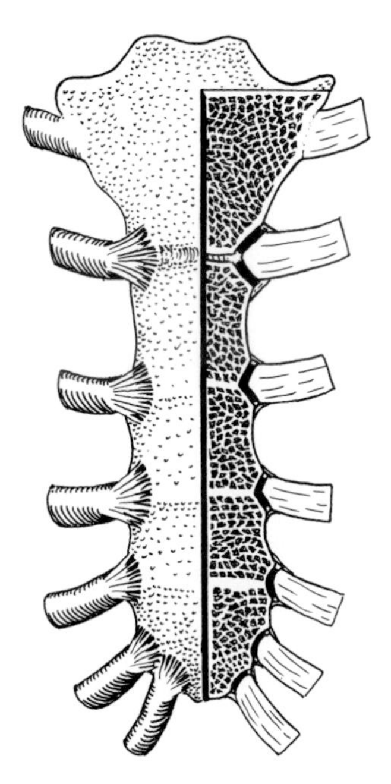

FIG. 4.2. THE ANTERIOR HALF OF THE STERNUM REMOVED ON THE LEFT SIDE TO EXPOSE THE CHONDRO-STERNAL JOINTS. THE FIRST IS A PRIMARY CARTILAGINOUS JOINT. THE SECOND IS A SYNOVIAL JOINT WITH TWO SEPARATE CAVITIES. THE REMAINDER ARE SINGLE SYNOVIAL JOINTS IN THE ADULT.

and move together as one. This fixation is necessary to give stability to the clavicle and so to the pectoral girdle and upper limb (p. 60).

The next six costal cartilages each articulate with the sternum by a synovial joint—a single cavity except in the case of the second, where the cavity between the cartilage and manubrium is separate from that between the cartilage and body (Fig. 4.2). The eighth, ninth and tenth costal cartilages articulate, each with the cartilage above, by a single synovial joint.

Development of Ribs and Sternum. The ribs appear as mesodermal condensations in the somatopleure (p. 39). Chondrification ensues and almost immediately ossification begins towards the back and spreads rapidly forwards. A segment remains unossified —it persists as the costal cartilage. The ribs grow ventrally from the vertebral column and their anterior ends are united on each side by a bar of cartilage. The protruding pericardium at first prevents these bars from fusing; later they meet in front of the pericardium and form the cartilaginous body of the sternum. The *manubrium* is probably developed *in situ* from the mesoderm of the pectoral girdle. Knowledge on this point is incomplete. However this may be, a continuous cartilaginous plate is formed. Five bony centres (often double) appear in this plate of cartilage from above downwards during the fifth, sixth, seventh, eighth and ninth fœtal months. The upper centre forms the manubrium. The others form separate bones known as *sternebræ*. These fuse with each other from below upwards during adolescence. The body of the sternum, thus fused into one plate of bone, normally never fuses with the manubrium, even in advanced old age. The two are united by a secondary cartilaginous joint. In childhood each costal cartilage articulates with two sternebræ, just as the posterior end of the rib articulates with two vertebræ. There is a separate synovial joint for each sternebra. When the sternebræ fuse with each other the two synovial cavities coalesce. Since fusion does not occur at the sternal angle (between manubrium and body) the two separate cavities persist here.

The Muscles

The **muscles of the thoracic wall** lie in the same three morphological layers as those of the abdominal wall. The outer or external oblique layer is broken up and specialized into separate muscles for attachment to the upper limb, and consists of such muscles as the pectorals, rhomboids and serratus anterior, muscles that are innervated segmentally in zones. The intermediate or internal oblique layer is split into segments by the ribs, and the muscle is specialized into two layers, the external and internal intercostal muscles which slope in different obliquities. The fibres of the external intercostal muscles slope in the same direction as those of the external oblique muscles, but the two are not homologous. *Both sets of intercostal fibres* belong, with the ribs, to the intermediate layer. The innermost or transversus layer is broken up into three groups of muscles, the subcostales, intercostales intimi and the sterno-costalis. This incomplete layer may be referred to, as a whole, as the transversus thoracis group. Between it and the middle layer is the neuro-vascular plane, continuous with that of the abdominal wall; in it run intercostal vessels and nerves, with their collateral branches (PLATE 45).

External Layer of Muscles. Two small muscles of the external layer do not attach themselves to the pectoral girdle, but remain confined to the thoracic wall. They are the posterior serratus muscles. Supplied like the rest of the layer by anterior primary rami, they have migrated posteriorly, and lie on the surface of the erector spinæ mass. Each arises from four spinous processes, two in the thorax and two beyond it, and each is inserted into four ribs just lateral to the erector spinæ, i.e., just lateral to the posterior angles of the ribs.

The **serratus posterior superior** arises from the spinous processes of the lowest two cervical and the upper two thoracic vertebræ and from the intervening supraspinous ligaments. The flat sheet of muscle slopes downwards on the surface of splenius and is inserted just lateral to the angles of the 2nd, 3rd, 4th and 5th ribs. Many tendinous fibres in the sheet of muscle give it a characteristic glistening appearance which provides a useful landmark in exposures of this region. The dorsal scapular nerve and the descending scapular vessels run down on the muscle, which is covered by levator scapulæ and the rhomboids (Fig. 2.6, p. 57).

The **serratus posterior inferior** arises from the lower two thoracic and the upper two lumbar spinous processes and from the intervening supraspinous ligaments. The origin is an aponeurosis which fuses with the posterior lamella of the lumbar fascia deep to latissimus dorsi. The flat sheet of muscle lateral to the aponeurosis slopes upwards in contact with the thoraco-lumbar fascia and is inserted just lateral to the angles of the lowest four ribs.

The serrati posterior are weak muscles of respiration. The superior muscle elevates the upper ribs (inspiration) while the inferior muscle depresses the lower ribs (expiration).

Intercostal Muscles

Examine a rib (Fig. 4.34, p. 251). Note that the subcostal groove is bounded externally by a lip that projects downwards; internally the groove possesses a rounded border. The upper border of the rib is smoothly rounded. The external intercostal muscle is attached between the sharp lower border of a rib and the upper border of the rib below; the internal intercostal muscle arises from the subcostal groove and is inserted with the external intercostal into the upper border of the rib below (Fig. 4.3).

External Intercostal Muscle. The fibres of this muscle pass obliquely downwards and forwards from the lower border of the rib above to the upper border of the rib below. The muscle extends from the superior costo-transverse ligament at the back of the intercostal space as far forwards as the costo-chondral junction; here it is replaced by the *external intercostal membrane*. This extends to the side of the sternum. Between the bony ribs is muscle, between the costal cartilages is membrane.

Internal Intercostal Muscle. The fibres of the muscle run downwards and backwards, from the subcostal groove to the upper border of the rib below. The muscle, unlike the external intercostal muscle, extends as far forwards as the side of the sternum; it is replaced posteriorly by the *internal intercostal membrane*, which extends from the angle of the rib to the superior costo-transverse ligament at the posterior limit of the space (PLATE 45).

The *actions* of the external and internal intercostal muscles, which are respiratory, are discussed on p. 220.

Transversus Thoracis Group of Muscles. In the lower vertebrates the innermost layer of muscle of the body wall lines the ribs, which develop in the intermediate layer. With the advent of a diaphragm in the phylogenetic scale, i.e., in placentalia, most of the innermost layer recedes caudally to form the diaphragm; some thin fibres remain, however, adhering to the ribs, in three sheets, together known as the transversus thoracis group.

Subcostales. Lying in the paravertebral gutter are fibres, each crossing more than one intercostal space, that are better developed below than above. The thin sheet of muscle so formed is separated from the posterior border of the intercostales intimi by a space across which the intercostal nerves and vessels are in contact with the parietal pleura.

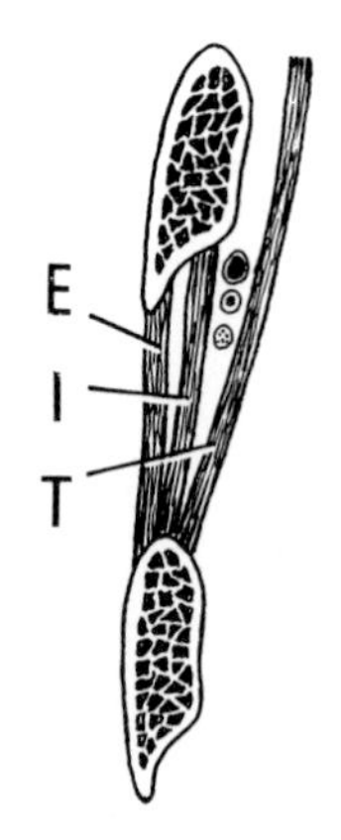

FIG. 4.3. BLACKBOARD SKETCH OF VERTICAL SECTION THROUGH AN INTERCOSTAL SPACE. E. External intercostal muscle. I. Internal intercostal muscle. T. Transversus thoracis sheet. The neuro-vascular bundle lies in the subcostal groove between the internal intercostal muscle and the transversus thoracis, i.e., in the neuro-vascular plane. The collateral branches lie in the same plane at a lower level.

Intercostales Intimi. Similar to the subcostales, the fibres of this sheet cross more than one intercostal space. The sheet lies within the lateral wall of the thorax, and is wider below than above.

Sterno-costalis. The muscle arises from the lower end of the sternum, whence digitations diverge on each side, one to each costal cartilage from the second to the sixth inclusive. This muscle is now officially named 'transversus thoracis' but its former name of 'sterno-costalis' is retained here because it is much more exact. The transversus thoracis *group* is the best name for including all three, because it conforms with the transversus abdominis muscle, with which it is morphologically identical.

Action. All three members of the transversus thoracis group are said to depress the ribs. They retain their submammalian function of compressing the body cavity around which they lie (p. 211). Their expiratory role is relatively unimportant, since they comprise but thin sheets of musculo-tendinous fibres. It is interesting to note that if sterno-costalis 'pulls the costal cartilages down towards the xiphoid process', it really pulls the xiphoid process up towards the costal cartilages—i.e., it is *inspiratory*, like the interchondral fibres of the internal intercostals.

Anatomy of a Typical Intercostal Space

Intercostal Nerve. The mixed spinal nerve, having emerged from the intervertebral foramen and given off its posterior primary ramus, passes around in the neuro-vascular plane, between the internal intercostal and the transversus thoracis group of muscles. It gives off a **collateral branch,** which is wholly concerned with the supply of the muscles of the space, the parietal pleura and the periosteum of the ribs and has no cutaneous branch. The main nerve itself has a **lateral cutaneous branch** and an anterior terminal branch. The lateral branch pierces the intercostal muscles and the overlying muscles of the body wall along the midaxillary line, and divides into an anterior and posterior branch to supply the skin over the space. The **anterior terminal branch** in the upper six spaces passes anterior to the internal thoracic artery and pierces the intercostal muscles to reach the skin. In its course around the space the intercostal nerve lies below the vein and artery, but all three are protected by the downward projection of the lower border of the rib. The collateral branch lies at a lower level and is not under protection of the rib. In its course around the body wall the nerve lies in a wider circle that embraces the narrower circle of the intercostal vessels (Fig. 1.24, p. 22). That is to say, where the nerve crosses the artery it will always be nearer the skin. At the back of the intercostal space the nerve crosses behind the artery; at the front of the space the nerve crosses in front of the internal thoracic artery. Between these points the nerve is confined in the neuro-vascular plane below the artery. The lower five intercostal nerves and their collateral branches slope downwards behind the costal margin into the neuro-vascular plane of the abdominal wall, which they supply (Fig. 5.5, p. 258).

The First Intercostal Nerve. This is a very small nerve. It courses around beneath the flat inferior surface of the first rib, some distance inside the external border of the rib, in contact with the endothoracic fascia and pleura (Fig. 4.32, p. 248). It supplies the intercostal muscles of the first space with motor and proprioceptive fibres, and the adjacent pleura and rib periosteum with sensory fibres. It is to be noted that the first intercostal nerve supplies no skin, lacking both lateral and anterior terminal cutaneous branches.

The Subcostal Nerve. Although arising in the thorax, the twelfth thoracic nerve quickly leaves by passing behind the lateral arcuate ligament into the abdomen, below the subcostal artery and vein (p. 310).

Intercostal Arteries. The upper two spaces are supplied by the **superior intercostal artery.** This is the descending branch of the costo-cervical trunk, which comes off from the second part of the subclavian artery (PLATE 11) behind the scalenus anterior muscle. It enters the thorax by passing across the front of the neck of the first rib; here it has the sympathetic trunk on its medial side, while the first thoracic nerve passes laterally across the first rib to join the brachial plexus. The supreme intercostal vein lies between the artery and the sympathetic trunk. At this point the sympathetic trunk frequently presents the stellate ganglion, a fusion of the first thoracic ganglion with the inferior cervical ganglion.

The remaining nine intercostal spaces are supplied each with a separate branch of the descending thoracic aorta. All eleven arteries constitute the posterior intercostal arteries. Each gives off a collateral branch, which passes around in the neuro-vascular plane at a lower level than the main trunk.

At the front of the intercostal space the internal thoracic artery in the upper six spaces and the musculophrenic artery in the lower five spaces give off two anterior intercostal arteries that pass backwards and make an end to end anastomosis with the posterior vessels.

Intercostal Veins. In each space there are one posterior and two anterior intercostal veins, accompanying the arteries of the same names. The anterior veins drain into the musculo-phrenic and internal thoracic veins. The posterior veins are not regular. In the lower eight spaces they drain into the azygos system; the azygos vein on the right and the hemiazygos and accessory hemiazygos on the left. The first space is drained by the highest, or **supreme intercostal vein,** which opens normally into the vertebral vein, or sometimes the brachio-cephalic vein, of its own side. The blood from the second and third intercostal spaces is collected into a single trunk on each side, the **superior intercostal vein.** That on the right drains simply into the azygos vein. That on the left runs forward over the arch of the aorta, lateral to the vagus nerve and medial to the phrenic nerve, to empty into the brachio-cephalic vein (PLATES 8 and 10).

Lymphatics of an Intercostal Space. The lymphatic vessels of the intercostal space conform to the general rule that deep lymphatics follow arteries. From the front of the space vessels pass to the anterior intercostal nodes that lie along the internal thoracic artery; from the back of the space they drain to posterior intercostal nodes (p. 240).

The Internal Thoracic (Mammary) **Artery.** From the first part of the subclavian artery, the internal thoracic artery passes vertically downwards a finger's breadth from the border of the sternum. It gives off two anterior intercostal arteries in each intercostal space. At the costal margin it divides into the *superior epigastric* and *musculo-phrenic arteries.* The former passes lateral to the xiphisternal fibres of the diaphragm to enter the rectus sheath behind the muscle. The latter passes along the costo-diaphragmatic gutter and gives off two anterior intercostal arteries in each space; it ends by piercing the diaphragm to ramify on its abdominal surface. The internal thoracic artery is accompanied by two venæ comitantes that empty into the brachio-cephalic (innominate) vein.

The artery gives off a *pericardiaco-phrenic branch* that runs with the phrenic nerve and supplies branches to the nerve itself, pleura and fibrous and parietal pericardium.

Perforating branches emerge towards the surface from each intercostal space. They are especially large in the second and third spaces of the female for supply of the breast. Thus the internal thoracic artery supplies the anterior body wall from the clavicle to the umbilicus.

The Suprapleural Membrane. This is a rather dense fascial layer attached to the inner border of the first rib and costal cartilage. It is not attached to the neck of the first rib, across which the first thoracic nerve passes to join the brachial plexus (PLATE 11). It has the cervical dome of the pleura attached to its under surface, and when traced medially it is found to thin out and disappear into the mediastinal pleura. It is flat and lies in the oblique plane of the thoracic inlet, projecting scarcely at all into the root of the neck. Lying on it are the subclavian vessels and other structures seen in the root of the neck (p. 376). Its posterior attachment is to the transverse process of the seventh cervical vertebra, where muscular fibres may be found in it. This is the *scalenus minimus* (or *pleuralis*) muscle, and the suprapleural membrane is to be regarded as the flattened out tendon of the muscle. The function of the membrane is to give a rigidity to the thoracic inlet that prevents distortion during respiratory changes of intrathoracic pressure. If it were not a flat and unyielding structure, it and the neck structures above it would be 'puffed' up and down during respiration. Actually it is moved only in forced respiration.

The Diaphragm

The diaphragm is a thin sheet of muscle caudal to the lungs. Its purpose is essentially for inspiration. It is present only in placentalia.

Morphologically the diaphragm is a derivative of the innermost (transversus) layer of the muscles of the body wall. In the mammals this innermost sheet descended from the thoracic inlet to the lower margin of the thoracic cage. Embryologically a similar descent occurs. Being derived from the innermost layer, its fibres arise in continuity with those of transversus abdominis from *within* the costal margin. It is completed behind the costal origin by fibres that arise from the arcuate ligaments and the crura. From the circumference of this oval origin the fibres arch upwards into a pair of domes and then descend to a central tendon.

Form of the Diaphragm. Viewed from in front the diaphragm curves up into right and left domes. The right is higher than the left, ascending in full expiration as high as the nipple (fourth space), while the left dome reaches the fifth rib (Fig. 4.6).

The central tendon is level with the lower end of the sternum (sixth costal cartilage). Viewed from the side the profile of the diaphragm resembles an inverted J, the long limb extending up from the crura (upper lumbar vertebræ) and the short limb attached to the xiphisternum (Th. 8 vertebra). Viewed from above the outline is kidney-shaped, in conformity with the oval outline of the body wall which is indented posteriorly by the vertebral column (Fig. 1.24, p. 22).

Origin. Trace the origin from behind. The *crura* are strong tendons attached to the bodies of the upper lumbar vertebræ at the lateral margins of their anterior convexities, alongside the psoas muscle. The large right crus is fixed to the upper three lumbar vertebræ and the discs between them, the smaller left crus to the upper two lumbar vertebræ and the intervening disc. Muscle fibres radiate from each crus, overlap, and pass vertically upwards before curving forwards into the central tendon. Some of the fibres on the abdominal surface of the right crus slope up to the left and surround the œsophageal orifice in a sling-like loop (PLATE 17). The *medial arcuate ligament* is a thickening in the psoas fascia. It extends from the upper part of the body of the second lumbar vertebra to a ridge on the anterior surface of the transverse process of the first lumbar vertebra, at the lateral margin of psoas. From this ridge the *lateral arcuate ligament* extends across to the twelfth rib at the lateral border of quadratus lumborum; it is a thickening in the anterior lamella of the lumbar fascia. The muscle of the diaphragm arises alongside the crus from the medial and lateral arcuate ligaments. Further laterally a digitation comes from the tip of the 12th rib, thence around the costal margin a digitation arises from each costal cartilage up to the 7th. These muscular slips all arise from *within* the costal cartilages, interdigitating with the slips of origin of transversus abdominis. Finally, in front, the muscle sheet is completed by fibres that pass backwards from the xiphisternum to the central tendon. These are the shortest muscle fibres of the diaphragm; the longest fibres are those which arise from the 9th costal cartilage.

The **central tendon** is shaped like the club on a playing card. A rounded leaf, placed nearer the front than the back, is fused on each side with a similar leaf that extends back towards the paravertebral gutter. The tendon, consisting of interlaced fibres, is inseparable from the fibrous pericardium, with which it is embryologically identical. Near the junction of central and right leaves it is pierced by a foramen, the caval opening, to which the adventitial wall of the inferior vena cava is very strongly attached (PLATE 17).

The Openings of the Diaphragm. Three main orifices exist.

The *aortic opening* is opposite Th. 12 vertebra. It is a midline arch between the overlapping right and left crural fibres. The epimysium at its margin is somewhat thickened and has been named the *median arcuate ligament*. No muscle fibres arise from it. The aortic opening transmits the aorta; the azygos vein is to the right and the thoracic duct leading up from the cisterna chyli is between them.

The *œsophageal opening* is opposite Th. 10 vertebra an inch to the left of the midline, behind the 7th left costal cartilage. It lies in the fibres of the left crus, but a sling of fibres from the *right* crus passes on the abdominal surface of these to loop around it. It transmits the œsophagus, which is firmly attached by fibrous tissue to the sling. This 'phrenico-œsophageal ligament' may stretch or weaken and allow part of the cardiac orifice of the stomach to slide up into the mediastinum. This is an acquired hiatus hernia, not to be confused with the congenital variety (p. 218). The vagal trunks and the œsophageal branches of the left gastric artery, veins and lymphatics perforate the fibrous tissue to accompany the œsophagus.

The *caval opening* is opposite Th. 8 vertebra just to the right of the midline, behind the 6th right costal cartilage. It lies in the central tendon as stated above. The right phrenic nerve pierces the central tendon alongside the inferior vena cava at this opening (the two are separated in the mediastinum by the fibrous pericardium).

Structures Passing Through the Diaphragm. The greater, lesser, and least splanchnic nerves pierce each crus, the sympathetic trunk passes behind the medial arcuate ligament, the subcostal nerve and vessels pass behind the lateral arcuate ligament, while the left phrenic nerve pierces the left dome by several branches to supply it on the abdominal surface. The neurovascular bundles of the 7th to the 11th intercostal spaces pass between the digitations of the diaphragm and transversus abdominis into the neuro-vascular plane of the abdominal wall. Finally the superior epigastric vessels pass between the xiphisternal and costal (seventh) fibres of the diaphragm. Extra-peritoneal lymphatic vessels on the abdominal surface pass through the diaphragm to lymph nodes lying on its thoracic surface, mainly in the posterior mediastinum.

Nerve Supply of the Diaphragm. Each phrenic nerve, overwhelmingly C4, supplies the muscle of the dome on its abdominal surface. Peripherally a few proprioceptive fibres pass in from the intercostal nerves; the crura, derivatives of the prevertebral rectus (*q.v.*), are supplied by the lower intercostal nerves, not by the phrenic.

Blood Supply of the Diaphragm. The costal margin of the diaphragm is supplied by the lower five intercostal and the subcostal arteries. The main mass of fibres rising up from the crura are supplied on their *abdominal* surface by right and left phrenic arteries from the abdominal aorta (PLATE 17). The tiny pericardiaco-phrenic artery, supplying the phrenic nerve and the pleura and the fibrous pericardium, is almost spent as it reaches the diaphragm.

Actions. The major role of the diaphragm is inspiratory, but it is used also in abdominal straining.

Inspiration. When the fibres contract in tranquil inspiration only the domes descend; this sucks down the lung bases and does not disturb the mediastinum. In a deeper breath further descent of the domes, below the level of the central tendon, can depress the central tendon from Th. 8 to Th. 9 level. This stretches the mediastinum (traction on pericardium and great vessels) and no further descent of the tendon is possible. Further contraction of the muscle (maximum inspiration) now everts the ribs of the costal margin in the 'bucket-handle' movement (Fig. 4.7).

Note the beneficial effect on the three main openings. As the diaphragm contracts intra-abdominal pressure tends to rise, and the *caval orifice* (in the central tendon) is pulled widely open to assist venous return via the inferior vena cava. The *œsophageal orifice* is held closed by the pinch-cock contraction of the muscle sling from the right crus, to discourage regurgitation of stomach contents. The *aortic opening* is unaffected because it is not within the diaphragm.

Expiration. Whether expiration is tranquil or forced (coughing, sneezing, blowing, etc.) the diaphragm is *wholly* passive, its relaxed fibres being elongated by pressure from below (see p. 221).

Abdominal Straining. For evacuation of a pelvic effluent (defæcation, micturition, parturition) its contraction aids that of the abdominal wall in raising intra-abdominal pressure. It is much weaker than the powerful obliques, transversus and recti, so for maximum pressure a deep breath is held by a closed glottis, and the diaphragm is prevented from undue elevation by being pressed up against a cushion of compressed air. Forcible escape of some of this air causes the characteristic grunt.

During heavy lifting in the stooping position abdominal straining is beneficial. With the breath held and intracœlomic pressure raised as above, the vertebral column cannot easily flex; it is as though an inflated football filled the body from pelvic brim to thoracic inlet. The weight of the stooping trunk is supported on the football, freeing erector spinæ to use all its power to lift the weight. Such acts are similarly accompanied on occasion by the characteristic grunt.

Development of the Diaphragm. The septum transversum contains the heart and liver, across the *ventral* part of the body cavity. Between heart and liver, myotomes from C4 migrate to form the muscle fibres of the domes. They bring their nerve (phrenic) with them. The septum transversum descends caudally, bringing down the muscles and trailing the phrenic nerves behind these. Dorsally the peritoneal cavity communicates with a recess, the pleural cavity, on either side of the mediastinum. The lung buds grow into the pleural cavities. The canals between peritoneal and pleural cavities are closed by a pair of *pleuro-peritoneal folds*, which grow in from the dorso-lateral body wall. Their failure to fuse on one or other side with the transverse septum results in a diaphragmatic hiatus. On the left side this may contain the stomach as a congenital hernia, not to be confused with an acquired œsophageal-hiatus hernia in later life. The muscle fibres in these folds and in the crura develop *in situ*; morphologically they belong to the transversus layer of the body wall musculature (p. 211).

MOVEMENTS OF THE THORACIC CAGE

All three diameters of the thorax are increased during inspiration.

The **antero-posterior diameter** is increased by elevation of the ribs. The 11th and 12th ribs are free to move independently, but the other ten, all joined behind *and in front*, necessarily all rise and fall together. The thoracic operculum is raised during inspiration. The manubrium moves upwards with the first rib; its lower border projects anteriorly during elevation. If the body of the sternum were fixed to the manubrium, its lower end would project forward prodigiously, a manifest impossibility, for it is held back by the ribs. Hence the persistence into advanced old age of the hinge joint between manubrium and body. The body of the sternum rises with a hinge movement at this joint, and its upper and lower ends project forwards equally during inspiration (Fig. 4.4). The manubrio-sternal joint at the extremes of respiratory excursion angulates through some 7 degrees. It is a secondary cartilaginous union (p. 13). If the joint becomes ankylosed thoracic expansion is virtually lost (as in emphysema), and only diaphragmatic respiration is possible. Due to the obliquity of the ribs elevation of the sternum carries it forward and thus the antero-posterior diameter of the thorax is increased. The up and down movements of the body of the sternum and all the ribs attached to it is aptly termed the **'pump-handle' movement** (Fig. 4.4).

The **transverse diameter** of the thorax is increased in two ways, one passive the other active. The **passive increase** in transverse diameter is due to the shape of the ribs themselves and the axis around which they hinge during inspiration. The axis is not transverse across the body but, passing through the head and tubercle of the rib, lies obliquely backwards from the midline (Fig. 4.5). Thus the downward-sloping rib is elevated not only antero-posteriorly but also laterally. The upward-sloping costal cartilages are thereby made to lie less obliquely. This 'lateral spread' of the ribs (Fig. 4.5) increases in range from the 5th rib downwards because the costal cartilages become progressively more oblique. It does not occur in the upper four pairs, since their costal cartilages are too short to permit such separation from the midline. These upper four pairs, then, rotate around a transverse axis passing through their tubercles, and as the anterior end of the shaft rises and falls each rib-head glides up and down slightly. Thus the transverse diameter of the upper part of the thorax does not alter. The **active increase** in transverse diameter is brought about by the **'bucket-handle' movement** of the lower ribs. In this movement the lower ribs rotate about an axis that passes through the anterior and posterior extremity of each, like lifting up the fallen handle from the side of a bucket. As these ribs are everted in the bucket-handle movement their tubercles rise upwards a little; this movement is made possible by the flatness of the costo-transverse joints of the ribs of the costal margin (see p. 213). The bucket-handle movement appears to be produced mainly by the diaphragm, for when the central tendon cannot descend further, contraction of the muscle fibres pulls upwards on the costal margin.

The bucket-handle movement takes place only in the ribs of the costal margin. It begins *towards the later stages*

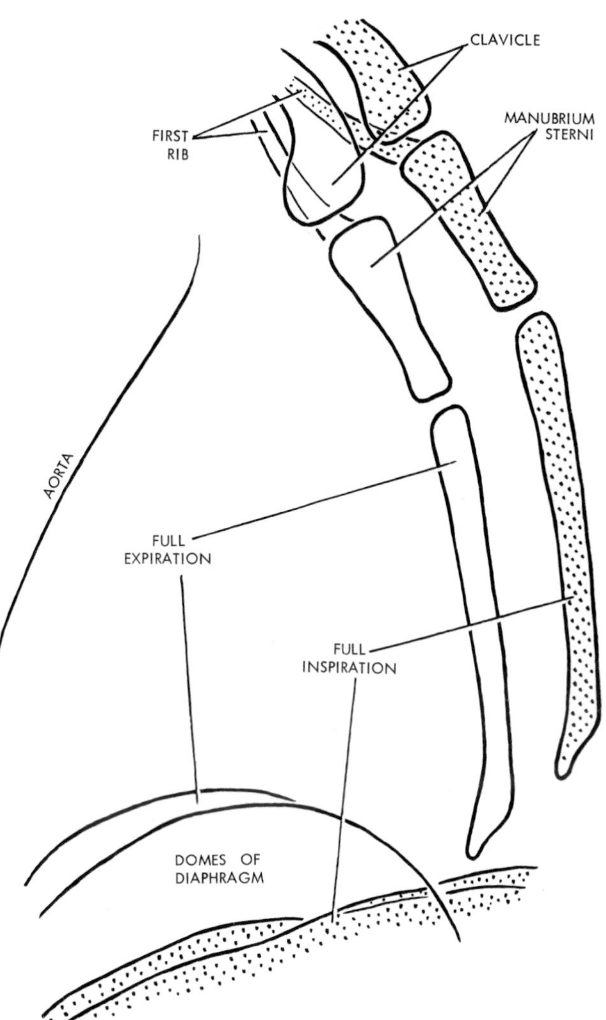

FIG. 4.4. LATERAL X-RAY VIEW OF THE THORACIC CAVITY OF A HEALTHY YOUNG MALE, SHOWING MAXIMUM EXCURSION SIMULTANEOUSLY OF THE CHEST WALL AND OF THE DIAPHRAGM (tracing, reduced to $\frac{1}{2}$). In full expiration the manubrio-sternal angle measures 161°; this changes to 154° in full inspiration. The anterior fibres of the domes of the diaphragm are attached to the cartilages of the costal margin (not shown by X-rays).

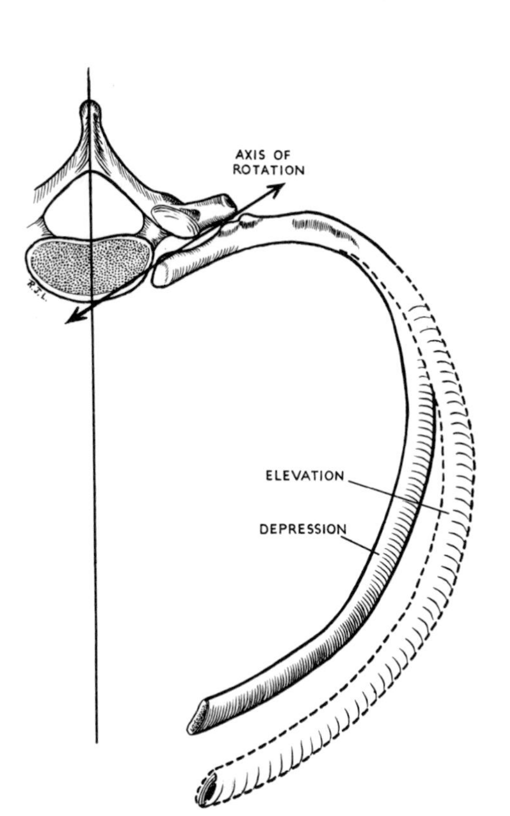

FIG. 4.5. THE AXIS OF ROTATION OF A RIB IS OBLIQUE TO THE MIDLINE OF THE BODY; HENCE ELEVATION INCREASES NOT ONLY THE ANTERO-POSTERIOR BUT ALSO THE TRANSVERSE DIAMETER OF THE THORAX. Remember that the shaft of the rib slopes down at 45 degrees; this may not be apparent in the diagram.

of full inspiration and continues to the end. It does not occur in tranquil inspiration. Neither does it occur in the dyspnœa of exercise, for here inspiration is never deep enough.

The 12th rib, held down by quadratus lumborum, does not take part in the elevation of the other ribs. This results in fixing the posterior fibres of the diaphragm and increasing the vertical diameter of the thorax. In expiration the quadratus lumborum relaxes and the 12th rib rises a little towards the 11th rib.

'Chest expansion' as measured with a tape *under the armpits* during forced inspiration is deceptive in that most of it, especially in trained athletes, is produced by contraction of the muscles of the axillary walls.

The **vertical diameter** of the thorax is increased by descent of the diaphragm (see above).

Movements of the Abdominal Wall. Since the volume of the abdominal cavity remains constant, the abdominal wall moves in accordance with changes in the thoracic cavity. Diaphragmatic inspiration and rib inspiration occur simultaneously, but each in itself produces opposite movements in the abdominal wall. In purely diaphragmatic breathing, with the ribs motionless, descent of the diaphragm is accompanied by passive protrusion of the relaxed abdominal wall; indeed no descent is possible without such protrusion. Ascent of the diaphragm is accompanied by retraction of the abdominal wall; indeed, it is the active contraction of the abdominal wall muscles that forces the relaxed diaphragm up. This to-and-fro movement of the abdominal wall is usually called 'abdominal respiration'. It consists of alternate contraction of the diaphragm and the anterior abdominal wall.

In 'thoracic respiration' the movements of the abdominal wall are purely passive. If the ribs are elevated the diaphragm is elevated with the up-going costal margin and the abdominal wall is sucked in. With descent of the costal margin the abdominal wall moves forwards again.

'Thoracic' inspiration makes the upper abdominal wall hollow, 'abdominal' inspiration makes the wall protrude. The ordinary simultaneous rib and diaphragm movements can be so balanced that the abdominal wall does not move at all. Thus may respiration function quite well in tight corsets, plaster casts, etc. In children and many women thoracic movement is greater than diaphragmatic movement. In men diaphragmatic movement is greater, especially as the years go by.

Action of the Intercostal Muscles. Each external intercostal muscle is a thin sheet; but if removed and rolled up it makes a mass equal in thickness to the sterno-mastoid muscle. It is much more powerful than is commonly supposed. Since the fibres pass obliquely downwards and forwards the muscle lifts the ribs in a powerful inspiratory movement.

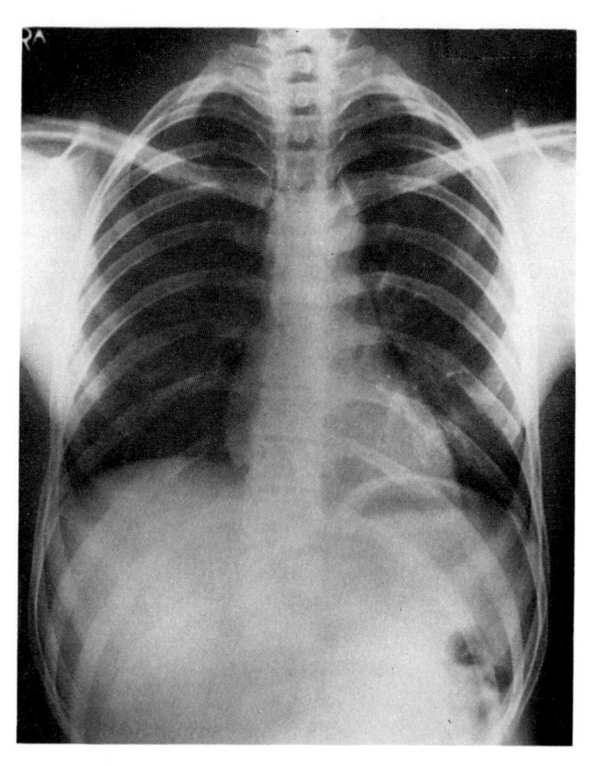

FIG. 4.6. ANTERO-POSTERIOR X-RAY PHOTOGRAPH OF THE THORAX IN EXTREME EXPIRATION. HEALTHY MALE OF TWENTY-ONE YEARS.

The eleven external intercostal muscles by their contraction elevate the twelve ribs; there is no necessity for an upward pull on the first rib. The mechanics of the movement involve the upward excursion of any two ribs if the external intercostal muscle that lies between them shortens its fibres (Fig. 4.8). The movement is a function of the downward slope of the bony ribs and the obliquity of the muscle fibres. The costal cartilages have an obliquity in the opposite direction, and it is significant that no external intercostal muscle lies between them. The internal intercostal muscle that intervenes between the costal cartilages has the same mechanical effect of elevation when its fibres shorten—cartilage and muscle both slope in the opposite obliquity to that of the bony rib and external intercostal fibres. The interchondral fibres of the internal intercostal muscles are, then, also inspiratory, but the lateral fibres, passing obliquely downwards and backwards, are expiratory. They slope in opposite obliquity to the external intercostal muscle and their action is likewise opposite.

The contraction of the intercostal muscles prevents both indrawing of the intercostal spaces during inspiration and bulging of the spaces during expiration; the occurrence of these movements in any space is diagnostic of paralysis of the intercostal muscles.

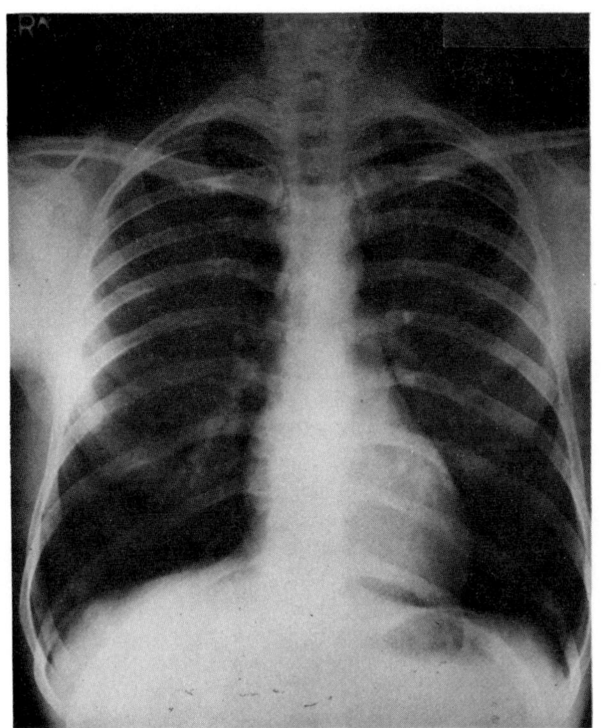

FIG. 4.7. X-RAY PHOTOGRAPH OF THE THORAX IN FIG. 4.6 TAKEN IN FULL INSPIRATION. THE ANTERIOR ENDS OF THE RIBS ARE ELEVATED AND THE LOWER RIBS ARE EVERTED. THE DESCENT OF THE DOMES OF THE DIAPHRAGM IS GREATER THAN THE DESCENT OF THE CENTRAL TENDON.

Accessory Muscles of Inspiration. Distinction must be made between full inspiration in a tranquil state (as in testing for vital capacity) and increased pulmonary ventilation (i.e., more litres per minute). *Full inspiration* can be achieved by intercostals and diaphragm alone. The added use of accessory muscles sucks in no more air, for the ribs are already fully elevated (Fig. 4.7). But the process of elevation by intercostals alone is slow. *Increased pulmonary ventilation* is a matter not only of breathing more deeply but of breathing more rapidly. The chest is not expanded to its utmost limit, but the accessory muscles are required to overcome the resistance of the airway, to suck in and blow out the air *more quickly* at each breath. So the mouth is opened to remove the resistance of the nasal passages, discarding in emergency the beneficial warming, cleaning and moistening effects of the nasal conchæ.

Remember that the muscles of respiration function without respite, on working days and holidays, 24 hours a day for seven days a week, for three-score years and ten, *without fatigue*. Only the heart can match this performance. The reason is that in a tranquil state only one muscle fibre in 10 is contracting. This fibre *does* become fatigued; then fibres 2, 3, 4 and the others up to 10 take over the function in rhythmical sequence. More than one fibre in ten can be stimulated into contraction, and the respiratory force is thereby correspondingly increased.

If *all fibres* of the intercostals and diaphragm are used, increased ventilation is produced in both depth and rate, up to 50 litres per minute. For further increase of ventilation, the accessory muscles are needed. It is better to classify the accessory muscles of inspiration under the headings of (1) arms free and (2) arms fixed.

It is evident that any muscles that bend the head and neck forward, down towards the thoracic inlet, will pull the thoracic inlet up if the head is fixed. With the strong extensor muscles fixing the head, sternomastoid is the strongest elevator of the inlet, for it is the biggest muscle and is placed furthest forward. It is aided by the scalenes. Together they haul the thoracic cage upwards. In addition to these, the muscles attaching the upper limb to the trunk are inspiratory if the arm is fixed; these are pectorales major and minor, serratus anterior and the costal fibres (ribs nine to twelve) of latissimus dorsi (p. 56). In orthopnœa, therefore, adequate support should be provided for the fixation *in abduction* of the upper limbs.

Muscles of Expiration. Expiration is produced by elevation of the diaphragm and/or depression of the ribs and sternum. **Elevation of the diaphragm** occurs when its relaxed fibres are pressed on from below. In the erect position this pressure comes from contraction of the muscles of the anterior abdominal wall. In the supine position elevation of the diaphragm is wholly passive, for the weight of the abdominal viscera

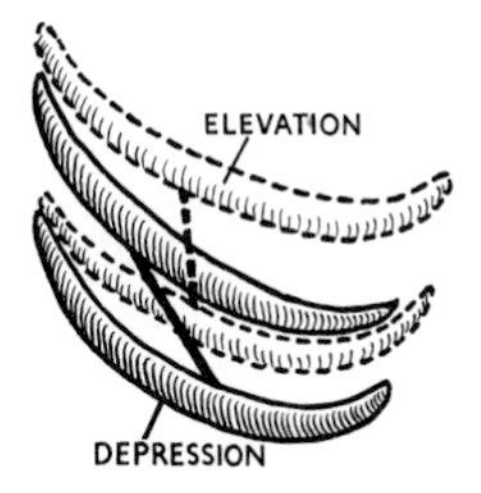

FIG. 4.8. AN EXERCISE IN GEOMETRY, TO ILLUSTRATE THE ELEVATING ACTION OF THE EXTERNAL INTERCOSTAL MUSCLES. The heavy outline (depression) represents two ribs connected by a fibre of the external intercostal muscle. The interrupted outline represents the two ribs in elevation. In spite of the fact that the two ribs are themselves further apart, the muscle fibre is shorter, because it now lies more nearly at a right angle between the ribs.

pushes the relaxed diaphragm into the thorax. In *forced expiration* the upward excursion of the diaphragm occurs with an energy directly proportional to the increased contraction of the abdominal wall. Contraction of the external oblique and transversus abdominis prevents eversion of the lower ribs and simultaneous contraction of the rectus abdominis holds the sternum down. The bony thoracic cage is thus prevented from passive expansion as intrathoracic pressure rises.

Depression of the ribs and sternum in tranquil expiration is wholly passive. It is due to the elastic recoil of the chest wall and of the lungs. The expulsive force on the thoracic wall can be increased by the lateral and posterior fibres of the internal intercostal muscles. In *forced expiration* depression of the thoracic cage results from the downward pull of the rectus abdominis and abdominal oblique muscles on the ribs. The abdominal muscles at the same time, by raising intra-abdominal pressure, elevate the diaphragm.

Forced expiration may be prolonged (as in blowing a trumpet, asthma, speech-making) or momentary (as in coughing, sneezing). In coughing, it immediately precedes the sudden opening of a closed glottis, and in sneezing a similar blast of compressed air explodes through the nose (p. 401).

THE CAVITY OF THE THORAX

The cavity of the thorax is completely filled laterally by the lungs, each lying in its pleural cavity. The space between the pleural cavities occupying the centre of the thoracic cavity is known as the mediastinum. It contains the heart and great blood vessels, the œsophagus, the trachea and its bifurcation, the thoracic duct, the phrenic and vagus nerves. It is a very mobile area, the lungs and heart being in rhythmic pulsation, and the œsophagus dilating with each bolus that passes down it. Hence there is but a minimum of loose connective tissue between the mobile structures. They lie in mutual contact, but the spaces between them can be readily distended by inflammatory fluid, neoplasm, etc. These loose spaces of the mediastinum, moreover, connect freely with those of the neck. Mediastinitis is an ever-present danger in infective collections in the neck.

The prevertebral and pretracheal fasciæ extend from the neck into the superior mediastinum. The former is attached over the fourth thoracic vertebra, the latter blends with the arch of the aorta. Thus neck infection in front of the pretracheal fascia is directed into the anterior mediastinum while behind the prevertebral fascia infection is imprisoned in the superior mediastinum in front of the vertebral bodies (Fig. 4.9). Elsewhere in the neck infection is directed through the superior into the posterior mediastinum.

Divisions of the Mediastinum. There is a plane of division to which the whole topography of the mediastinum can be related, namely a plane passing horizontally through the angle of Louis, i.e., the joint between the manubrium and the body of the sternum (Fig. 4.9). From the second costal cartilages, this plane passes backwards to the lower border of the 4th thoracic vertebra. Above, between it and the thoracic inlet, lies the superior mediastinum. Below the plane, the inferior mediastinum is conveniently divided into three compartments by the fibrous pericardium—a part in front, the anterior mediastinum, a part behind, the posterior mediastinum, and the pericardium itself, containing the heart and roots of the great arteries and veins. Note that the anterior and posterior mediastina are in direct continuity with the superior mediastinum; their separation from it is purely descriptive, not anatomical. The plane passes through the bifurcation of the trachea, the concavity of the arch of the aorta, and just above the bifurcation of the pulmonary trunk.

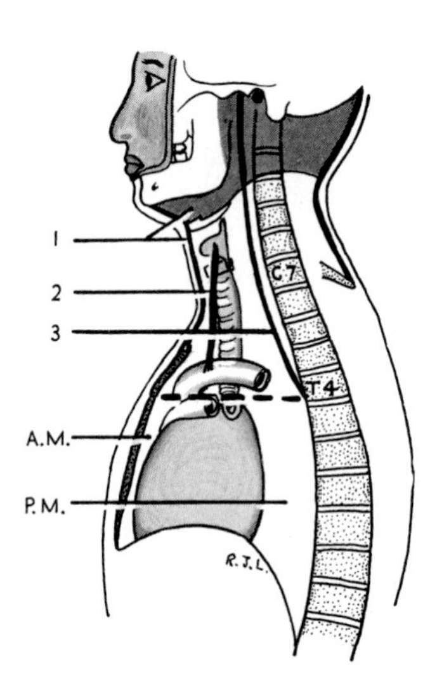

FIG. 4.9. BLACKBOARD SKETCH OF THE DIVISIONS OF THE MEDIASTINUM. The diagram shows the continuity between the tissue spaces of the neck and the mediastinum. 1. Investing layer of deep cervical fascia. 2. Pretracheal fascia. 3. Prevertebral fascia. The superior mediastinum lies above the interrupted line that passes from the angle of Louis to Th. 4 vertebra. **A.M.** Anterior mediastinum. **P.M.** Posterior mediastinum. The anterior mediastinum is continuous through the superior mediastinum with the pretracheal space of the neck, up to the larynx. The posterior mediastinum is continuous through the superior mediastinum with the retropharyngeal and paratracheal space of the neck, up to the skull base.

On the plane the azygos vein enters the superior vena cava, and the thoracic duct reaches the left side of the œsophagus in its passage upwards from the abdomen. Also lying in the plane are the ligamentum arteriosum and both superficial and deep cardiac plexuses.

SUPERIOR MEDIASTINUM

General Topography. The essential point about the disposition of the great veins and arteries of the superior mediastinum is their **asymmetry.** The veins are on the right, the arteries on the left. Structures themselves symmetrical, be they midline like the trachea, or bilateral like the apices of the lungs or the phrenic and vagus nerves, thus have asymmetrical relationships on the right and left sides. On the right side they are in contact with veins, on the left side with arteries. Veins expand enormously, large arteries not at all, during increased blood flow. Thus there is much 'dead space' on the right, none on the left, and it is into this space on the right side that tumours of the mediastinum or liquid collections tend to project.

The superior mediastinum is wedge-shaped (Fig. 4.9). The anterior boundary is the manubrium. The posterior boundary is much longer, due to the obliquity of the thoracic inlet. It consists of the bodies of the first four thoracic vertebræ; this wall is concave towards the mediastinum.

At the thoracic inlet the œsophagus lies against the body of the first thoracic vertebra. The trachea lies on the œsophagus and itself touches the jugular (sternal) notch. The midline of the inlet is thus wholly occupied by these two tubes. At the inlet the apices of the lungs lie laterally, separated by vessels and nerves passing between the superior mediastinum and the neck. The concavity of the arch of the aorta lies in the plane of the angle of Louis; that is to say, the arch of the aorta lies *wholly in the superior mediastinum, behind the manubrium.* Examine an antero-posterior radiograph of the thorax (Fig. 4.6). The arch of the aorta is seen in profile. It does not pass from right to left, but from in front at the manubrium *backwards* to the body of the 4th thoracic vertebra. It arches over the beginning of the left bronchus and the bifurcation of the pulmonary trunk. Thus the three great branches from the arch, the brachio-cephalic (innominate), left common carotid, and left subclavian arteries, pass upwards on the left side of the trachea, the brachio-cephalic (innominate) sloping gradually upwards to the right, so that the thoracic part of the trachea is clasped by an asymmetrical V (Fig. 4.10). These great arteries keep the left vagus nerve and apex of the left lung away from contact with the trachea. On the right side there is no structure to separate the trachea from the right vagus and apex of the right lung. Tracheal breath sounds are thus conducted, by direct contact, to the right apex. On the left side the great arteries intervene, and breath sounds are here more subdued, being conducted only along the bronchial tree.

The veins entering the superior mediastinum are the right and left brachio-cephalic (innominate), each formed by confluence of the internal jugular with the subclavian vein. They lie in front of the arteries and are asymmetrical. The right brachio-cephalic (innominate) vein runs vertically downwards; the left vein passes, on the other hand, almost horizontally across the superior mediastinum to join the right (Fig. 4.10). The confluence of the brachio-cephalic veins produces the superior vena cava, which passes vertically downwards behind the right edge of the sternum, anterior to the right bronchus. Note that on the right side of the mediastinum the venous channels open into the right atrium, which forms the right border of the heart (PLATE 8). On the left side there are arterial channels, which are continuous with the left side of the heart, i.e., the right and left ventricles (PLATE 10). The phrenic nerves

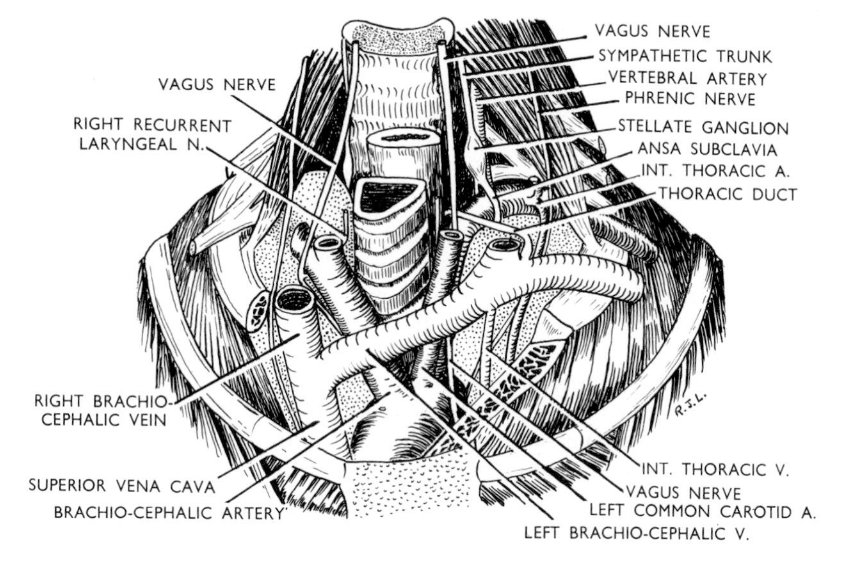

FIG. 4.10. THE SUPERIOR MEDIASTINUM, EXPOSED BY REMOVAL OF THE MANUBRIUM STERNI AND THE FIRST RIGHT COSTAL CARTILAGE. Drawn from Specimen S 233 B in R.C.S. Museum.

have, therefore, different relationships, the right to venous, and the left to arterial structures.

The Arch of the Aorta. Emerging from the pericardium the ascending aorta approaches the manubrium and then arches *backwards* over the left bronchus to reach the body of the 4th thoracic vertebra just to the left of the midline. From its upper convexity arise the three great arteries for the head and upper limbs, namely the brachio-cephalic, left common carotid and subclavian arteries (PLATE 10). The arch is crossed on its left side by the phrenic and vagus nerves as they pass downwards in front of and behind the lung root respectively. Between them lie the sympathetic and vagus branches to the superficial cardiac plexus. The left superior intercostal vein passes forwards across the arch superficial to the vagus, deep to the phrenic, to empty into the left innominate vein. The left recurrent laryngeal nerve hooks around the ligamentum arteriosum to pass upwards on the right side of the arch of the aorta, in the groove between the trachea and œsophagus. The pulmonary trunk bifurcates into right and left pulmonary arteries in the concavity of the arch. On the right side of the arch lie the trachea and œsophagus.

The **brachio-cephalic trunk** (innominate artery) arises a little to the left of the midline of the body. It slopes upwards across the trachea to the back of the right sterno-clavicular joint, where it divides into the right common carotid and right subclavian arteries. It has no branches apart from the rare thyroidea ima artery, which may arise from it or directly from the arch of the aorta. The termination of the left brachio-cephalic vein lies in front of the artery (Fig. 4.10). The trachea is in contact with the sternal notch at the thoracic inlet, but it slopes back and the manubrium slopes forward below this, and the brachio-cephalic artery and left brachio-cephalic vein occupy the space thus provided (PLATE 10).

The **left common carotid artery** arises just behind the brachio-cephalic trunk from the upper convexity of the aortic arch. It passes straight up alongside the trachea into the neck (PLATE 10). It has no branches in the mediastinum.

The **left subclavian artery** arises just behind the left common carotid artery. The two run upwards together. The subclavian artery arches to the left over the apex of the lung, which it deeply grooves. It leaves the left common carotid at a point directly behind the left sterno-clavicular joint. It has no branches in the mediastinum.

The **brachio-cephalic** (innominate) **veins** are formed behind the sterno-clavicular joints by confluence of the internal jugular and subclavian veins. It should be noted that each internal jugular vein lies lateral to the common carotid artery, in front of the scalenus anterior muscle. The subclavian vein, running medially into the brachio-cephalic vein, lies lateral to and then below the scalenus anterior. Medial to the scalenus anterior, in front of the first part of the subclavian artery, the vein has been joined by the internal jugular vein and is thence called the brachio-cephalic vein. This part of each brachio-cephalic vein receives tributaries corresponding to the branches of the first part of the subclavian artery (vertebral, inferior thyroid, internal thoracic and, on the left side only, superior intercostal).

The **right brachio-cephalic** (innominate) **vein** commences behind the right sterno-clavicular joint and runs vertically downwards; it is a persistent part of the right anterior cardinal vein of the embryo. It receives the right lymph duct and right jugular and subclavian lymph trunks, which normally enter independently of each other.

The **left brachio-cephalic** (innominate) **vein** passes almost horizontally across the superior mediastinum to join the right vein at the lower border of the first right costal cartilage (Fig. 4.10). It is a persistent anastomotic channel between the two anterior cardinal veins of the embryo; the left anterior cardinal vein has almost entirely disappeared. In the infant the left brachio-cephalic vein projects slightly above the jugular notch, and may do so in the adult if the vein is distended, especially if the head and neck are thrown back. The vein is then vulnerable to suprasternal incisions (e.g., tracheotomy). The commencement of the vein receives the thoracic duct, which often divides into two or three branches that join the vein separately. In addition to the vertebral and internal thoracic veins the left brachio-cephalic vein receives most of the blood from the inferior thyroid plexus of veins. The left superior intercostal vein joins it near the midline of the body, and a large thymic vein commonly enters nearby (PLATE 44).

The thymus or its remnants lie in front of the vein and the pretracheal fascia passes down behind it; the fascia directs a retrosternal goitre into the space between the vein and the brachio-cephalic artery and trachea.

The **superior vena cava** commences at the lower border of the first right costal cartilage by confluence of the two brachio-cephalic veins. It passes vertically downwards behind the right border of the sternum and, piercing the pericardium, enters the upper border of the right atrium at the lower border of the third right costal cartilage. Behind the angle of Louis, opposite the second right costal cartilage, it receives the azygos vein, which has arched forwards over the root of the right lung. The superior vena cava is formed as the persisting right anterior cardinal vein, but below the entrance of the azygos vein it represents the persisting right duct of Cuvier of the embryo (p. 308).

Ligamentum Arteriosum. This is the shrivelled fibrous remnant of the ductus arteriosus of the fœtus, a channel that short-circuited the lungs. It passes from the very commencement of the left pulmonary artery to the concavity of the aortic arch, beyond the point where the left subclavian artery branches off. It lies almost horizontally. The left recurrent laryngeal nerve hooks around it. The superficial cardiac plexus lies anterior to it, the deep cardiac plexus on its right, in front of the left main bronchus.

The Cardiac Plexuses. There are two plexuses, superficial and deep, but this is merely a descriptive division and *functionally they are one*. Their branches enter the pericardium to accompany the coronary arteries (vasomotor) and also to reach the sinu-atrial node (cardio-inhibitor and cardio-accelerator), and the atrio-ventricular node and bundle.

Superficial Cardiac Plexus. This is a sympathetico-parasympathetic plexus formed by union of the inferior of the two cervical cardiac branches of the left vagus nerve with the cardiac branch of the left superior cervical sympathetic ganglion. It lies in front of the ligamentum arteriosum.

Deep Cardiac Plexus. This larger plexus receives contributions from the right vagus nerve by its cervical cardiac branches, from the left vagus by its superior cervical branch, and one from each recurrent laryngeal nerve, and sympathetic fibres from the remaining five cervical ganglia, namely the middle and inferior left ganglia, and all three right ganglia. It lies to the right of the ligamentum arteriosum, in front of the left bronchus, at the bifurcation of the pulmonary trunk.

The vagus branches are preganglionic (they relay at the sinu-atrial node) while the sympathetic fibres are postganglionic from cell bodies in the cervical ganglia.

The *sensory supply* to the heart (e.g., pain of angina) is sympathetic. The thoracic viscera (œsophagus, trachea, lungs) have their sensory supply from the vagus. The heart is not a 'viscus' in this sense but merely an enlarged blood vessel with special muscle in its walls. Like all blood vessels its sensory supply is sympathetic.

The **trachea** commences in the neck below the cricoid cartilage at the level of the lower border of C6 vertebra (p. 369). Entering the thoracic inlet in the midline, the trachea lies in contact with the manubrium and is normally said to bifurcate at its lower border. This is true of the recumbent cadaver. In the living erect the bifurcation descends 2 inches (5 cm) lower than this in full inspiration (Fig. 4.11). Its total length is thus about 6 inches (15 cm). Its wall is a fibro-*elastic* membrane, whose patency is maintained by C-shaped rings of hyaline cartilage. The gaps lie posteriorly and are closed by a sheet of unstriped muscle (the *trachealis muscle*) whose contraction diminishes the calibre of the tube. The trachealis muscle likewise prevents overdistension of the trachea when pressure is raised (in abdominal straining, prior to a cough, etc.). The trachea is lined with respiratory

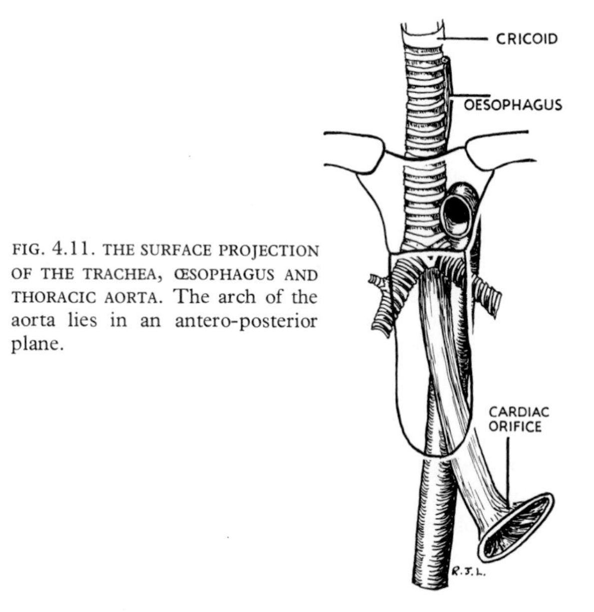

FIG. 4.11. THE SURFACE PROJECTION OF THE TRACHEA, ŒSOPHAGUS AND THORACIC AORTA. The arch of the aorta lies in an antero-posterior plane.

mucous membrane; the corium contains plentiful mucous and serous glands. It is supplied by the inferior thyroid artery, its veins drain to the brachio-cephalic veins, and its lymphatics run to nodes along the trachea and to the postero-inferior group of deep cervical nodes.

Nerve supply. This is by the vagus (recurrent laryngeal nerves) and the sympathetic. The vagus is sensory (cell bodies in inferior ganglion) and secreto-motor (cell bodies in inferior salivatory nucleus, with local relay) to the mucous membrane, and motor to the trachealis muscle (cell bodies in dorsal nucleus, with local relay). Sympathetic fibres, vaso-constrictor, reach the trachea on the inferior thyroid artery from cell bodies in the middle cervical ganglion.

Functions. The structure of the trachea conforms to functional needs.

1. The wall of the trachea is *elastic* because it must stretch. The trachea is stretched into elongation during swallowing. Elevation of the larynx elevates its upper end; the bifurcation does not move. Elastic recoil of the trachea restores its original length, pulling the larynx down to its rest position. Normally there is no call on sterno-thyroid to depress the larynx, and swallowing is unimpaired by loss of this muscle. Per contra, pulling down on the bifurcation by sudden descent of the diaphragm, pericardium and aortic arch produces the clinical sign of 'tracheal tug'.

2. The softness of the elastic wall must be strutted open by bars of *hyaline cartilage* to prevent collapse during inspiration.

3. The curved bars ('rings') of hyaline cartilage are incomplete so that the diameter may be controlled by the *trachealis muscle*. Cine-radiography of a cough shows 30 per cent increase in transverse diameter produced by

compressed air in the trachea while the vocal cords are shut, but 10 per cent narrowing of the resting diameter at the instant the cords open. Like the choke barrel of a shot-gun this greatly increases the explosive force of the blast of compressed air.

4. The *mucous membrane* shares with the other respiratory mucous membranes the property of trapping particulate matter in a surface film of mucus. The soiled mucus is beaten upwards to the larynx by the cilia of the surface epithelium. From the larynx it is expelled by coughing. Serous glands in the mucous membrane humidify the air.

Relations (Fig. 4.10). In the *neck* the relationships are symmetrical (p. 369). Here in the *superior mediastinum* great asymmetry results from the lop-sided arrangement of the great arteries and veins.

Posterior. In the superior mediastinum the trachea lies in contact with the front of the œsophagus. The left recurrent laryngeal nerve runs up in the gutter between the two, and supplies each (the right recurrent laryngeal nerve exists only in the neck).

Anterior. The trachea is covered by the pretracheal fascia. The brachio-cephalic artery crosses it obliquely, and in front of this and the pretracheal fascia, just below the jugular notch, the left brachio-cephalic vein lies almost horizontal (Fig. 4.10). The lung edges in their pleural sacs come to overlap these vessels behind the lower part of the manubrium sterni (Fig. 4.26, p. 243).

On the left. The arch of the aorta below, and the left common carotid and subclavian arteries above this, lie in contact. The vagus and phrenic nerves are held away from the trachea by these arteries (PLATE 10), and in contact with them is the mediastinal pleura and upper lobe of the left lung.

On the right. The vagus lies on the trachea, crossed at the lower end by the arch of the azygos vein (PLATE 8). These are covered by the mediastinal pleura and upper lobe of the right lung, which lies in direct contact with the trachea. The superior vena cava, with the phrenic nerve hugging its right side, lies further forward.

At the bifurcation. The pulmonary trunk branches into right and left pulmonary arteries rather to the left of the tracheal bifurcation, in front of the left bronchus. Here lies the deep cardiac plexus. The right pulmonary artery crosses just below the tracheal bifurcation, and between the two is the tracheo-bronchial group of lymph nodes (see Fig. 4.12).

The **bronchi** are asymmetrical, the right being one-third wider and a little shorter than the left. At the bifurcation an antero-posterior ridge, the *carina*, lies to the left of the midline. The right bronchus slopes more steeply (25 degrees off vertical) than the left (45 degrees off vertical), so that foreign bodies of whatever shape are statistically more likely to enter the right bronchus. The upper lobe bronchus leaves the right main bronchus outside the hilum of the lung; the left bronchus divides only after entering the hilum (Fig. 4.11). In structure the bronchi are identical with the trachea. Their distribution in the lung is considered on p. 244. They are supplied by the bronchial arteries. The right main bronchus drains to the azygos vein and the left to the accessory (superior) hemiazygos vein.

The Phrenic Nerve. Arising principally from C4 in the neck, the nerve passes down over the anterior scalene muscle across the dome of the pleura behind the subclavian vein. It runs through the mediastinum in front of the lung root. Each nerve lies in the thorax *as far lateral as possible*, being in contact laterally with the mediastinal pleura throughout the whole of its course. Their medial relations, however, are asymmetrical. The **right phrenic nerve** is in contact with venous structures throughout the whole of its course. The right brachio-cephalic vein, the superior vena cava, then the right atrium, and the inferior vena cava, lie to its medial side. It passes beside the caval opening in the central tendon of the diaphragm to supply that structure on its abdominal surface (PLATE 8). The **left phrenic nerve** has the great arteries springing from the arch of the aorta to its medial side. It crosses the arch of the aorta lateral to the superior intercostal vein and passes laterally across the pericardium over the left ventricle towards the apex of the heart (PLATE 10). It passes through the diaphragm to supply the muscle on its abdominal surface.

The phrenic nerve is motor to the dome of the diaphragm and sensory to the parietal pleura of the mediastinum, the fibrous pericardium, the parietal layer of the serous pericardium and the central parts of the diaphragmatic pleura and peritoneum. It supplies branches to the areolar tissue of the mediastinum, but not to the viscera. In a word, it is the *nerve of the septum transversum.* The nerve is supplied with blood by the pericardiaco-phrenic artery, a branch of the internal thoracic which accompanies it (with companion veins) to the diaphragm.

Pain referred from the diaphragmatic peritoneum (C4) is classically felt in the shoulder tip (C4), but pain from thoracic surfaces supplied by the phrenic nerve (pleura, pericardium) is usually located there, albeit vaguely.

The Vagus Nerve. In their descent through the thorax the vagus nerves are, so to speak, *attempting to reach the midline* at all levels. Thus the right vagus is in contact with the trachea, while the left is held away from that structure by the great arteries that spring from the arch of the aorta. The left nerve crosses the arch deep to the left superior intercostal vein, and the right nerve lies on the trachea deep to the arch of the azygos vein. Each vagus passes down behind the lung root, here giving off a large contribution to the pulmonary plexuses. The nerves now pass onwards to achieve their object of reaching the midline by entering into the œsophageal plexus (*q.v.*). In the plexus they become mixed, and the right and left vagal trunks, as they leave the plexus, contain fibres from each vagus. On the arch of the aorta the left vagus nerve *flattens out* and gives off its **recurrent laryngeal branch** (PLATE 10). This nerve hooks around the ligamentum arteriosum, and, passing up on the right side of the aortic arch, ascends in the groove between trachea and œsophagus. The right recurrent laryngeal nerve is given off in the root of the neck and hooks around the right subclavian artery. From the right vagus nerve thoracic cardiac branches enter the deep cardiac plexus. Cardiac branches are given off from each recurrent laryngeal nerve; they join the deep cardiac plexus. The recurrent laryngeal nerves supply the whole trachea and the adjacent œsophagus (i.e., above the lung roots).

The *œsophagus* lies against the vertebræ at the back of the superior mediastinum. The *thoracic duct* lies on its left side. Both structures pass through the posterior mediastinum; they are described on pp. 238 and 240.

THE ANTERIOR MEDIASTINUM

This space, little more than a potential one, lies between the pericardium and sternum. It is overlapped by the anterior edges of both lungs. It is continuous through the superior mediastinum with the pretracheal space of the neck. Its principal content is the thymus gland, which may lie, however, only in the anterior part of the superior mediastinum.

The Thymus Gland. The gland is developed from the ventral diverticulum of the third pharyngeal pouch (p. 41). The bi-lobed structure descends anterior to the left brachio-cephalic vein to occupy a variable extent of the anterior mediastinum. Relatively large in childhood, the gland atrophies during adolescence. The upper pole of each lobe remains in the neck deep to sterno-thyroid, encapsulated by a condensation of fibrous tissue superficial to the pretracheal fascia (transcervical thymectomy is thereby facilitated). Its arterial supply is by multiple small branches from the internal thoracic artery that give no trouble at operation. The venous drainage, by multiple small venules into the internal thoracic veins, corresponds to the arterial input, but in addition there is frequently found a wide short vein, easily avulsed, that passes directly backwards from the gland into the left brachio-cephalic vein (PLATE 44).

Microscopic Anatomy. To the unaccustomed eye the thymus resembles lymphatic tissue, but in fact a section of the gland is ridiculously easy to recognize, and with the naked eye alone. It is a lobulated structure, and *each lobule possesses a dark cortex and a pale medulla*. The lobules are several millimetres in diameter. The medulla of the thymic lobule is made up of small round cells called thymocytes; these are endodermal. The cortex is a thick layer of densely packed lymphocytes not themselves arranged in any pattern. They are derived from the mesoderm into which the developing thymus grows. Hassall's corpuscles, consisting of lamellated keratinous bodies, are found only in the medulla, never in the lymphocytes of the cortex. Their significance is unknown. A low-power section is illustrated in Fig. 1.22, p. 19.

THE MIDDLE MEDIASTINUM

This division of the mediastinum strictly includes the pericardium and its contents only, but it is convenient here to describe also the lung roots.

The Lung Roots. When the lungs are removed, the cut surface that remains of the root is different on the two sides.

The Left Lung Root (PLATE 10). The upper part is occupied by the left pulmonary artery lying within the concavity of the arch of the aorta. Below and behind it is seen the left bronchus as it slopes downwards from the bifurcation of the trachea. There are two pulmonary veins, one in front of and the other below the left bronchus, and both in contact with the upper end of the pulmonary ligament. These structures, which constitute the lung root, are enclosed in a sleeve of pleura; a sleeve that is too big for the lung root, as a coat cuff is too big for the wrist, so that it hangs down in an

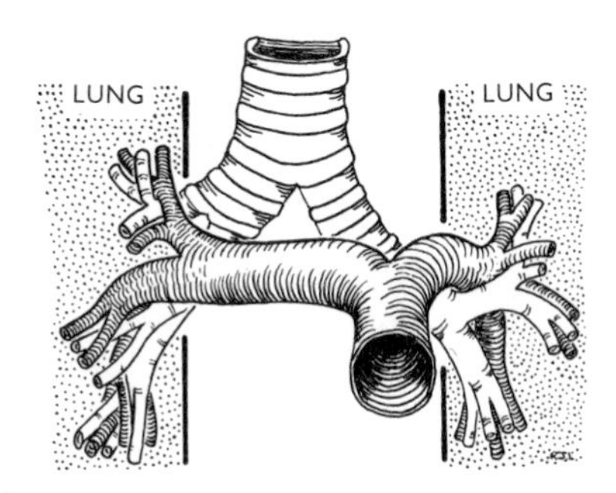

FIG. 4.12. THE ASYMMETRY OF THE PULMONARY ARTERIES. The pulmonary trunk bifurcates (in the concavity of the aortic arch) in front of the left main bronchus. The left pulmonary artery spirals over the left main bronchus to descend behind the lobar bronchi. The right pulmonary artery crosses below the tracheal bifurcation. Its descent behind the lower lobe bronchus is delayed because the artery is held anterior at the lung root by the upper lobe bronchus.

empty fold. This fold is called the *pulmonary ligament*. It is, of course, not a ligament, neither is it concerned with the lung. It provides the necessary freedom of 'dead space' for the structures of the lung root. Its two layers separate on descent of the lung root during inspiration, and on enlargement of the pulmonary veins during increased pulmonary blood flow.

The Right Lung Root (PLATE 8). Here in general the arrangement of the structures is similar to that on the left side, but the bronchus to the upper lobe (the eparterial bronchus) leaves the main stem bronchus outside the lung. Thus there are found the upper lobe bronchus and its accompanying pulmonary artery above the level of the main stem bronchus. Only two pulmonary veins leave the hilum of the right lung. They are disposed as on the left, one in front of and one below the main bronchus, at the upper end of the 'pulmonary ligament'. The root of the right lung lies within the curve of the azygos vein. It is surrounded by a sleeve of pleura that is too big for it, and depends below as a reflected fold of empty pleura, called the right *pulmonary ligament*. Its function is the same as that of the left pulmonary ligament.

At the root of each lung bronchial arteries enter and bronchial veins leave. Hilar lymph nodes lie around the main bronchus. The small anterior and the large posterior pulmonary plexuses surround the lung root.

The Pulmonary Arteries. These are asymmetrical in both length and position as they approach the lung roots (Fig. 4.12). The pulmonary trunk no sooner shakes itself free of its attachment to the fibrous pericardium than it bifurcates, in the concavity of the aortic arch, in front of the left main bronchus. The left pulmonary artery, attached to the concavity of the aortic arch by the ligamentum arteriosum, quickly spirals over the top to reach the back of the left main bronchus. The right pulmonary artery, longer than the left, passes across below the carina and at the lung root is held anterior to the right main bronchus by the upper lobe bronchus (PLATES 8 and 10).

THE PERICARDIUM

Do not confuse the *fibrous* pericardium with the *serous* pericardium. They are separate entities, with separate functions (p. 229).

The Fibrous Pericardium. This fibrous sac enclosing the heart is conical in shape. Its apex is fused with the roots of the great vessels at the base of the heart. Its broad base overlies the central tendon of the diaphragm, with which it is inseparably blended. Both it and the central tendon are derived from the septum transversum, and can be regarded as one structure. The fibrous sac is connected to the upper and lower ends of the sternum by weak *sterno-pericardial ligaments*. It is supplied with blood by the pericardiaco-phrenic and internal thoracic arteries.

The Serous Pericardium. A layer of serosa lines the fibrous pericardium, whence it is reflected around the roots of the great vessels to cover the entire surface of the heart. Between these parietal and visceral layers there are two sinuses, namely the transverse sinus and the oblique sinus of the pericardium. In order to understand the presence of these two spaces, it is desirable to recall the **development of the heart.** The heart first appears as a vascular tube lying free in a space, the pericardial cavity, within the septum transversum. The heart tube is suspended from the dorsal wall of this cavity by a *dorsal mesocardium*, a simple 'mesentery' consisting of two layers reflected around the tube and lining the cavity (Fig. 4.13A). It is to be noticed that the arterial and venous ends of the tube are each surrounded by a simple sleeve of visceral pericardium. No matter how many vessels may be derived from either end of this tube, they must all continue to be enclosed together in this single tube of serous pericardium. In fact, the arterial end splits into two, the aorta and the pulmonary trunk, while the venous end splits into six veins, the four pulmonary veins and the superior and inferior venæ cavæ. With elongation, bending and rotation of the heart tube (Fig. 4.13C), the two arteries and the six veins pierce the adult pericardium in a pattern that can best be appreciated by study of the interior of the pericardium after the heart has been removed (Fig. 4.13D). It will be seen that the **oblique sinus** of the pericardium is no more than a cul-de-sac between the two left pulmonary veins and the right pulmonary veins and inferior vena cava. Its anterior wall is formed by the posterior wall of the left atrium, between the four pulmonary veins. The presence of the **transverse sinus** is due to the fact that the dorsal mesocardium breaks down, thus producing a communication from right side to left dorsal to the heart tube, between its arterial and venous ends (Fig. 4.13B). The transverse and oblique sinuses are separated from each other by a double fold of serous pericardium. A finger in the transverse sinus will pass across the pericardial cavity behind both aorta and pulmonary trunk, but in front of the superior vena cava on the right side, and the left auricular appendage on the left. It is through the transverse sinus that a temporary ligature is passed to occlude pulmonary trunk and aorta during pulmonary embolectomy and other cardiac operations.

Nerve Supply of the Pericardium. The fibrous pericardium is supplied by the phrenic nerve. The parietal layer of serous pericardium that lines it is

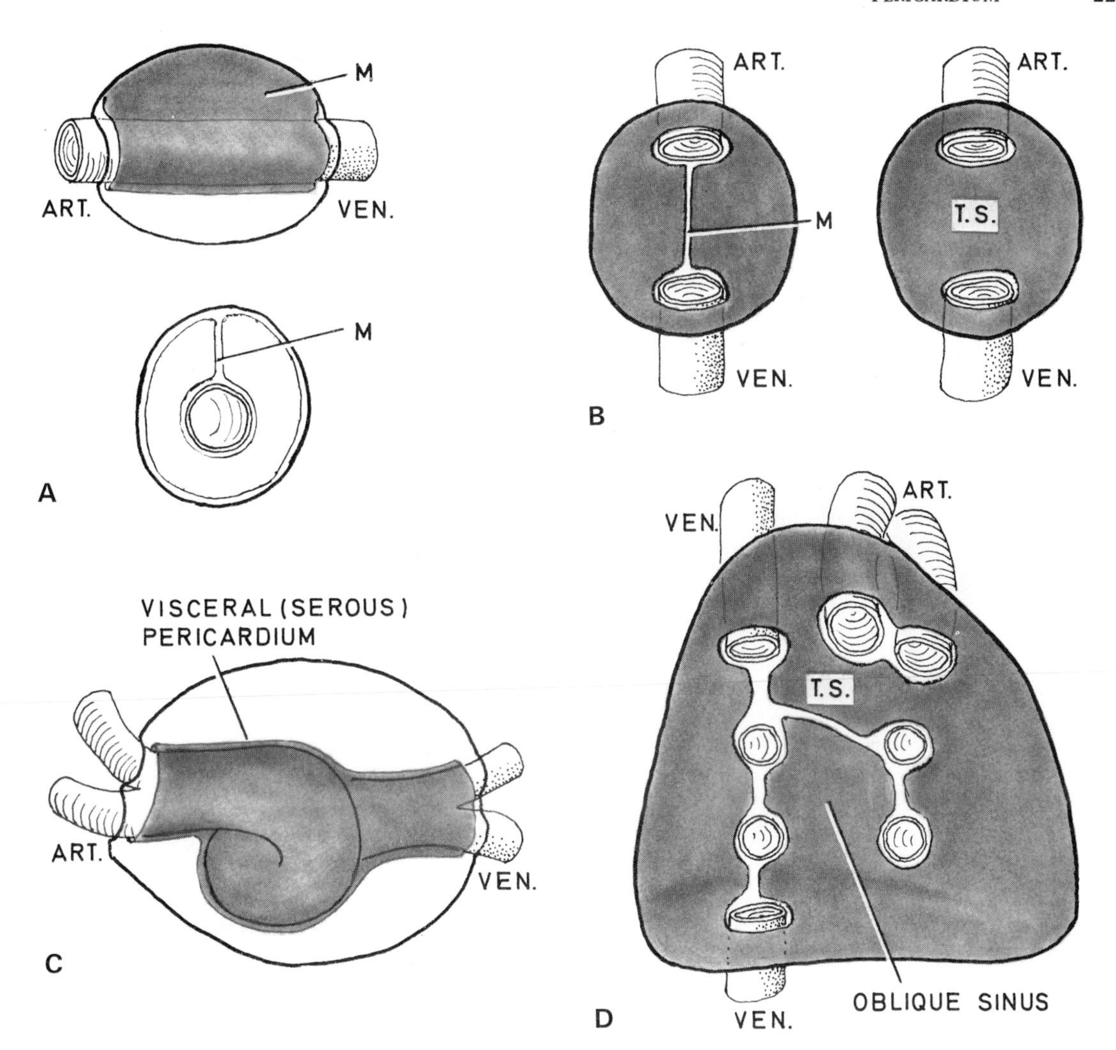

FIG. 4.13. THE DEVELOPMENT OF THE SINUSES OF THE SEROUS PERICARDIUM.

A. THE SIMPLE HEART TUBE SUSPENDED IN THE FIBROUS PERICARDIUM BY THE DORSAL MESOCARDIUM (M) OF SEROUS MEMBRANE. ART. Arterial end, VEN. Venous end of heart tube.

B. VENTRAL VIEW AFTER REMOVAL OF THE HEART TUBE. LEFT FIGURE SHOWS DORSAL MESOCARDIUM STILL PRESENT. RIGHT FIGURE SHOWS APPEARANCE OF TRANSVERSE SINUS AFTER ABSORPTION OF DORSAL MESOCARDIUM (T.S.): EACH END OF THE HEART TUBE REMAINS SURROUNDED BY A SINGLE SLEEVE OF SEROUS PERICARDIUM.

C. THE ELONGATION AND ROTATION OF THE HEART TUBE AFTER ABSORPTION OF THE DORSAL MESOCARDIUM. AT THE ARTERIAL AND VENOUS ENDS THE SEPARATE VESSELS ARE SURROUNDED BY THE SAME SINGLE SLEEVE AS IN FIG. B. (Lateral view.)

D. VENTRAL VIEW OF ADULT PERICARDIUM AFTER REMOVAL OF HEART (cf. Fig. 4.16). VENOUS AND ARTERIAL TRUNKS ARE STILL EACH SURROUNDED BY A SINGLE SLEEVE OF SEROUS PERICARDIUM. BETWEEN THE TWO LIES THE TRANSVERSE SINUS (T.S.). THE OBLIQUE SINUS IS A CUL-DE-SAC RESULTING FROM THE WIDE SEPARATION OF THE SIX VEINS. THUS ALL INTERNAL SURFACES OF THE FIBROUS PERICARDIUM ARE MADE SLIPPERY BY THE LINING OF PARIETAL SEROUS PERICARDIUM (THERE ARE NO 'BARE AREAS').

similarly innervated, but the visceral layer on the heart surface is insensitive. Pain from the heart (angina) originates in the muscle or the vessels and is transmitted by the sympathetic. The pain of pericarditis originates in the parietal layer only, and is transmitted by the phrenic nerve.

Function of the Pericardium. The fibrous sac provides slippery surfaces for the heart to beat inside and the lungs to move outside. Apart from the passage of the great vessels, all *inner surfaces* are slippery; the oblique sinus permits pulsation of the left atrium (a kind of 'cardiac bursa' like the omental bursa behind the stomach). Apart from the bare area in front (from the midline halfway to the apex beat) and a bare strip posteriorly for the œsophagus, the whole of the *outer surface* of the fibrous sac is clothed with densely adherent parietal pleura, on which the breathing lungs glide.

THE HEART

Obtain a heart; post-mortem room, dissecting room, or butcher's shop in descending order of preference.

Position of the Heart. Hold the heart in the position it occupies in the body. The *right border* consists entirely of right atrium; the *inferior border* passing to the left is continued by the right ventricle. The *left border* as seen from in front is a narrow strip of left ventricle (Fig. 4.14). The *anterior surface* of the heart consists of the right atrium, the vertical atrio-ventricular groove, and the right ventricle, with a narrow

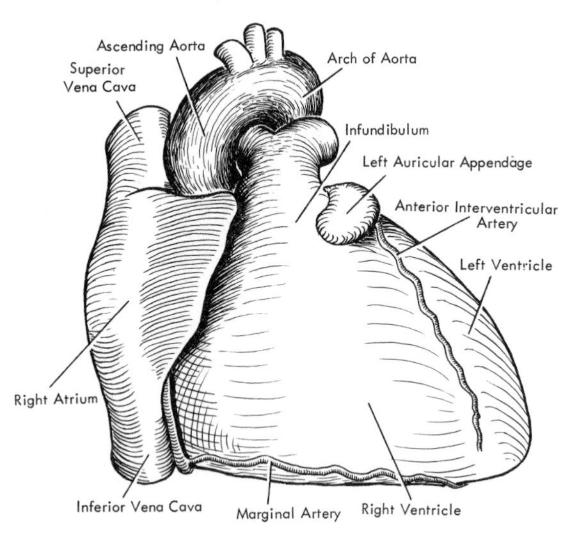

FIG. 4.14. THE ANTERIOR (STERNO-COSTAL) SURFACE OF THE HEART.

strip of left ventricle appearing on the horizon of the left border. The tip of the left auricular appendage peeps over this border. The inferior, or *diaphragmatic surface* of the heart (Fig. 4.15) consists of the right atrium receiving the inferior vena cava, the antero-posterior atrio-ventricular groove and to the left of this the ventricular surface is made up of one-third right ventricle and two-thirds left ventricle, separated by the inferior interventricular branch of the right coronary artery. The left atrium is attached behind the right atrium, but at a higher level; it does not extend down to the diaphragmatic surface. The *posterior surface* (or base) of the heart (Fig. 4.16) consists almost entirely of the left atrium, receiving the four pulmonary veins. From it the left ventricle converges to the left towards the apex. To the right a narrow strip of right atrium forms the horizon, while below a small part of the posterior wall of the right atrium receives the coronary sinus (Fig. 4.16).

This position of the heart as described and illustrated differs somewhat from the usual accounts, because the heart is here described in diastole in the position of full expiration, which is the cadaveric position, where it can be studied at leisure. The position of the heart varies a little between systole and diastole. The roots of the great vessels fix the base of the heart, but the ventricles are free to move within the pericardium. The ventricles are narrower and slightly rotated during systole.

During descent of the central tendon of the diaphragm in full inspiration the apex of the heart descends more than the relatively fixed base, and the heart occupies a somewhat more vertical position (Fig. 4.7). In full expiration the ascent of the diaphragm forces the heart into the more horizontal position characteristic of the cadaver (Fig. 4.6), but these differences in position are not very great.

Surface Projection of the Heart. Clinically the surface projection of the heart on the thoracic wall is studied under two heads. The **superficial projection** is determined by mapping out the area of superficial cardiac dullness. This is, in reality, the area bounded by the resonance on percussion of the sharp anterior edges of the lungs. The area indicates the amount of pericardium exposed beneath the chest wall between the lung edges. Normally it lies within the right border of the sternum and extends from the 4th costal cartilage in a convexity to the left down to the '*apex beat*'. The latter lies in the 5th space just medial to the mid-clavicular line (i.e., just medial to the line of the nipple, some $3\frac{1}{2}$ inches (9 cm) from the midline of the sternum). The apex beat is the palpable or audible impulse on the chest wall made by that part of the heart which is not covered by the left lung. It is medial to the actual apex of the heart. The **deep projection** of the heart on the chest wall is larger than the superficial area of pericardium exposed between the lung edges. It

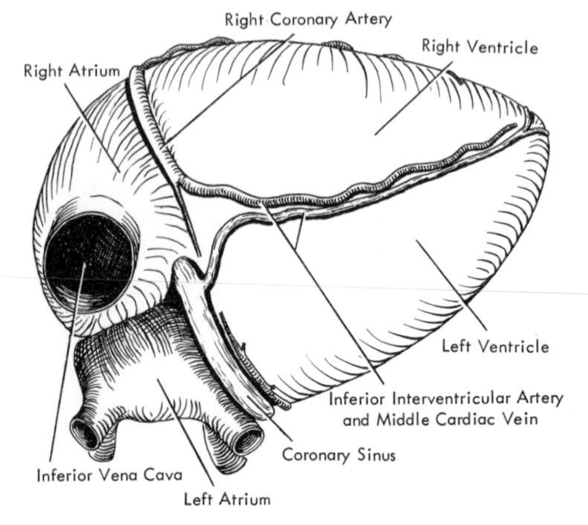

FIG. 4.15. THE INFERIOR (DIAPHRAGMATIC) SURFACE OF THE HEART.

can be determined by deep or heavy percussion and more accurately by radiography. It varies with body build and habitus, but an average projection lies over the middle four thoracic vertebræ. The gently convex right border extends from the 3rd to the 6th right costal cartilages half an inch or more to the right of the sternum. The gently convex inferior border extends from the 6th right cartilage to just beyond the apex beat—the line cuts the xiphisternal junction. The convex left border extends from the 2nd left intercostal space an inch from the sternum to just beyond the apex beat (Figs. 4.6, 4.7).

Skeleton of the Heart. The two atria and the two ventricles are attached to a pair of conjoined fibrous rings which, in the form of a figure 8, bound the atrio-ventricular orifices. The 'figure 8' lies on its side, very nearly in the sagittal plane. A small bone, the *os cordis*, lies at the junction of the two rings in some animals. To the fibrous skeleton the muscle of the heart is attached; the muscle fibres encircle the chambers of the heart in a series of whorls and spirals. The atria lie to the right and the ventricles to the left of the fibrous skeleton and there is no muscular continuity between the two. The atrio-ventricular conducting bundle is the only physiological connection between atria and ventricles across the fibrous ring. The interventricular septum is attached to the fibrous skeleton, which here projects a little to form the small *membranous part* of the septum. The bases of the cusps of the tricuspid and mitral valves are attached to the fibrous skeleton (Fig. 4.18).

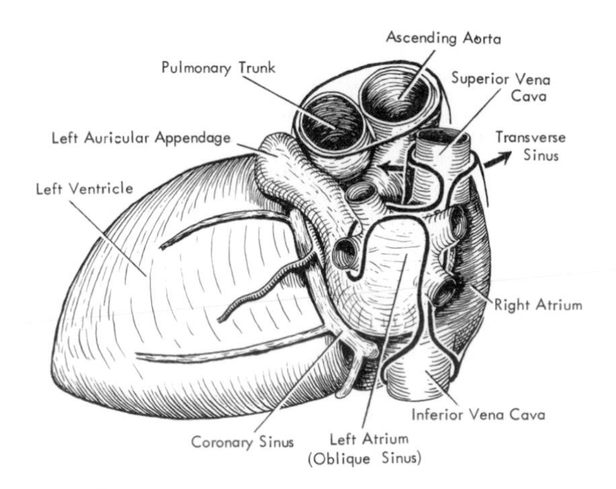

FIG. 4.16. THE HEART VIEWED FROM BEHIND. THE CUT EDGES OF THE SEROUS PERICARDIUM ARE SHOWN IN HEAVY LINE. THE POINTER TO THE LEFT ATRIUM INDICATES ALSO THE OBLIQUE SINUS OF THE PERICARDIUM, A CUL-DE-SAC BETWEEN THE PULMONARY VEINS. THE FOUR PULMONARY VEINS AND THE TWO VENÆ CAVÆ ARE ALL SIX ENCLOSED IN A SINGLE SLEEVE OF SEROUS PERICARDIUM. THE PULMONARY TRUNK AND AORTA ARE LIKEWISE ENCLOSED IN A SINGLE SLEEVE; BETWEEN THE TWO SLEEVES LIES THE TRANSVERSE SINUS OF THE PERICARDIUM. COMPARE THE CUT EDGES OF THE PERICARDIUM IN FIG. 4.13D. Sketched from a dissection.

The Chambers of the Heart. Clinically the heart is described as consisting of right and left sides. The right side propels blood to the lungs and the left side propels blood around the systemic circulation; separate failure of either side produces a chain of symptoms peculiar to itself. Anatomically, however, it should be noted that the 'right side' lies much more *in front of* than to the right of the 'left side', for the interatrial and interventricular septa lie almost in the coronal plane (Fig. 4.17).

The four chambers of the heart, the two atria and the two ventricles, possess anatomical features that will now be considered, in the order in which the circulating blood flows through them.

The Right Atrium (Fig. 4.14). This elongated chamber lies, between superior and inferior venæ cavæ, along the right border of the heart. Its lower end is almost completely occupied by the orifice for the inferior vena cava but its upper end is prolonged to the left of the superior vena cava as the auricular appendage. This overlies the commencement of the aorta and the upper part of the right atrio-ventricular groove and, with the left auricular appendage, it clasps the infundibulum of the right ventricle. The left atrium lies behind the right atrium. From the angle between the superior vena cava and the right auricular appendage a shallow groove sometimes descends. This is the *sulcus terminalis*. It is produced, when present, by the projection into the **cavity** of the right atrium of a vertical ridge of heart muscle called the *crista terminalis*. The

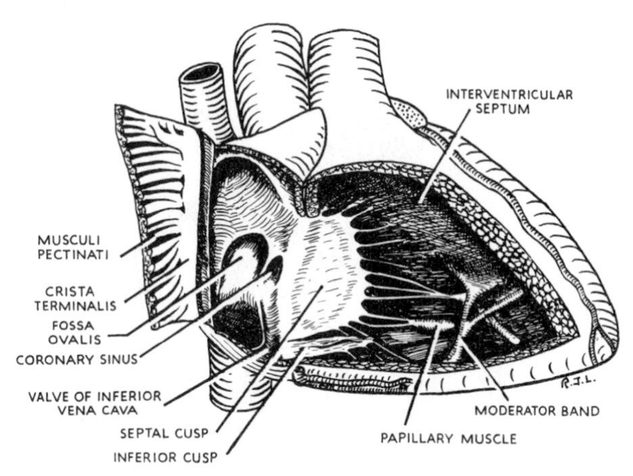

FIG. 4.17. THE INTERIOR OF THE HEART FROM IN FRONT. THE ANTERIOR WALL OF THE RIGHT VENTRICLE IS REMOVED, WITH THE ANTERIOR CUSP OF THE TRICUSPID VALVE. THE ANTERIOR WALL OF THE RIGHT ATRIUM IS INCISED AND HINGED TOWARDS THE RIGHT TO EXPOSE THE MUSCULI PECTINATI AND THE CRISTA TERMINALIS. NOTE THAT THE INTERVENTRICULAR AND INTERATRIAL SEPTA LIE ACROSS THE BODY (CORONAL PLANE). Sketched from a dissection.

interior of the right atrium is smooth to the right of the crista terminalis but between the crista and the blind extremity of the auricular appendage the myocardium is projected into a series of horizontal ridges like the teeth of a comb. These are the *musculi pectinati* (Fig. 4.17). This rough area represents the true auricular chamber of the embryonic heart. The remainder of the atrial cavity, smooth walled, is produced by incorporation of the right horn of the sinus venosus.

The opening of the inferior vena cava is guarded by a ridge, the remains of the valve of the inferior vena cava, that is continued upwards towards the opening of the coronary sinus. The *opening of the coronary sinus* lies near the septal cusp of the tricuspid valve. It is big enough to admit the tip of the little finger of the owner of the heart (big man, big heart, big little finger). The **interatrial septum** forms the posterior wall of the right atrium above the opening of the coronary sinus. Towards its lower part is a shallow saucer-shaped depression, the *fossa ovalis*. This is the closed foramen ovale of the fœtal heart, an opening that allowed blood from the inferior vena cava to flow directly through into the left atrium. The crescentic upper margin of the fossa ovalis is called the *anulus ovalis*.

The **right ventricle** (Fig. 4.17) projects to the left of the right atrium. The *atrio-ventricular groove* between the two is vertical over the front of the heart and antero-posterior on the inferior surface. It lodges the right coronary artery and is usually filled with fat. The right ventricle narrows as it passes upwards towards the commencement of the pulmonary trunk.

The interior of the cavity, whose walls are much thicker than those of the atrium, is thrown into a

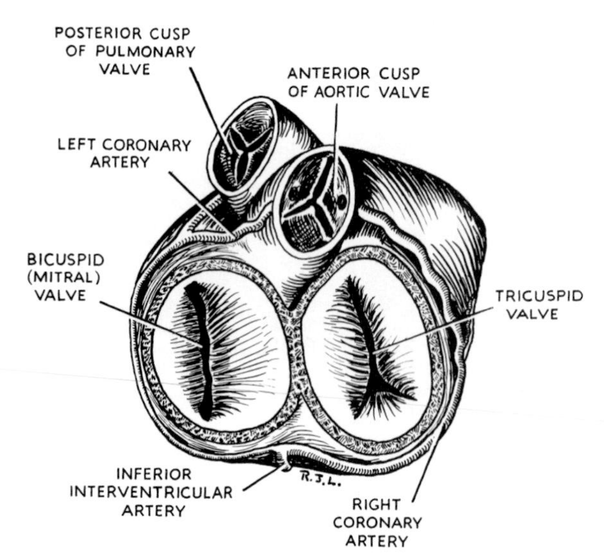

FIG. 4.18. THE ATRIO-VENTRICULAR ORIFICES VIEWED AFTER REMOVAL OF BOTH ATRIA. Sketched from the right side, looking left towards the apex of the heart.

series of muscular ridges, the *trabeculæ carneæ*. One of these ridges has broken free and lies in the cavity attached by its two ends to the interventricular septum and the anterior wall of the ventricle. This is the *trabecula septomarginalis (moderator band)*; it transmits part of the right branch of the conducting bundle.

The *tricuspid valve* guards the atrio-ventricular orifice. It has three cusps and admits the tips of three fingers (the bicuspid valve has two cusps and admits the tips of two fingers). The three cusps are attached by their bases to the fibrous atrio-ventricular ring and are arranged to lie against the three walls of the ventricle, septal, inferior and anterior. Their edges and ventricular surfaces receive the attachment of inelastic cords (*chordæ tendineæ*) that cross each other by diverging from small conical elevations of muscle (*papillary muscles*) that project from the trabeculæ carneæ.

The cavity of the right ventricle is flattened by the forward bulge of the interventricular septum. Thus the anterior wall and septum are of equal area, while the inferior wall is much narrower. The inferior cusp of the tricuspid valve is correspondingly smaller than the other two (Fig. 4.18).

The cavity of the ventricle continues upwards into a narrowing funnel-shaped approach to the pulmonary orifice. The walls of this part, the *infundibulum*, are thin and smooth, lacking trabeculæ carneæ. Here fibrous and elastic tissue progressively replace the heart muscle.

The Pulmonary Orifice. This fibrous ring at the commencement of the pulmonary trunk is guarded by three cusps, each semilunar in shape. They are arranged one posterior and two anterior. The pulmonary orifice lies at a higher level than the aortic orifice (Fig. 4.18).

The Pulmonary Trunk. Commencing at the pulmonary orifice, which is superior to the aortic orifice (Fig. 4.18), the wide pulmonary trunk arches backwards to the left of the ascending aorta. The two lie in a gentle spiral enclosed together in a common sleeve of serous pericardium in front of the transverse sinus. The pulmonary trunk passes backwards and emerges from the fibrous pericardium to bifurcate in the concavity of the arch of the aorta into right and left pulmonary arteries, which enter the lung roots with their corresponding bronchi. The trunk is some 2 inches (5 cm) long; its first $1\frac{1}{2}$ inches (4 cm) lie in the pericardial cavity.

A bulging of the wall of the pulmonary trunk lies above each cusp of the pulmonary valve. The three sinuses serve to lodge the cusps away from the blood stream during ventricular systole, and perhaps by creating backflow eddies help to close the cusps more quickly at ventricular diastole. Similar sinuses lie above the aortic cusps; two of them give origin to the coronary arteries.

The **left atrium** is applied to the interatrial septum

behind the right atrium. Its inferior margin lies a little above that of the right atrium, whose posterior wall here receives the coronary sinus. It lies on the posterior surface (base) of the heart and from it the left ventricle slopes away to the apex. It has a small *auricular appendage* that projects from its upper border and passes to the left over the atrio-ventricular groove and peeps around the convexity of the left side of the infundibulum (Fig. 4.14). The four pulmonary veins enter it symmetrically, one above the other on each side. They are enclosed, with the superior and inferior venæ cavæ, in a common sleeve of serous pericardium. The posterior surface of the left atrium between the pulmonary veins forms the anterior wall of the oblique sinus of the pericardium (Fig. 4.16). The fibrous pericardium separates this surface from the œsophagus.

The cavity of the left atrium is smooth-walled except in the tiny auricular appendage; here muscular ridges indicate that the appendage was the original auricular chamber of the embryonic heart. All the smooth-walled portion is derived by incorporation of the embryonic pulmonary veins into the atrial cavity. In the embryo two veins from each lung unite to form the right and left pulmonary veins which open by a common pulmonary vein into the left auricle. The common pulmonary vein and the right and left pulmonary veins dilate and are incorporated into the wall of the atrium, which thus receives four separate veins. Should this process be arrested during embryonic development, the atrium may receive three, two, or only one pulmonary vein. Three veins in place of the usual four are fairly commonly found, but two or only one are almost unknown except in abortions, because they are accompanied by other defects of cardiac development incompatible with life.

The Bicuspid (Mitral) Valve. This two-cusped valve admits the tips of two fingers (the tricuspid valve, with three cusps, admits the tips of three fingers). The cusps are a large anterior (or septal) and a small posterior. The former lies between the atrio-ventricular (i.e., mitral) and the aortic orifices. The bases of the cusps are attached to the margins of the atrio-ventricular orifice (Fig. 4.18). Usually the attachments of the two cusps are continuous around the orifice, but sometimes they fail to meet and a small accessory cusp fills the gap between them. The mitral cusps are smaller in area and thicker than those of the tricuspid valve and consequently are not ballooned back so much into the atrium during ventricular systole. The septal cusp of the mitral valve is thicker and more rigid than the posterior cusp.

The Left Ventricle. The walls of this cavity are three times as thick as those of the right ventricle. The *interventricular septum* bulges into the cavity of the right ventricle, so that in cross section the left ventricle is circular, the right crescentic (Fig. 4.19). Trabeculæ carneæ are well developed but are all attached through their whole length to the ventricular walls; there is no moderator band. From their extremities the papillary muscles project into the cavity of the ventricle, sending from their apices the chordæ tendineæ to the mitral cusps. The posterior cusp receives the chordæ tendineæ on both its margin and its ventricular surface, but since blood is squirted across both surfaces of the septal cusp the chordæ tendineæ are attached to it only along its margins. The chordæ tendineæ diverge from two large papillary muscles, one on the inferior wall and the other on the superior wall of the left ventricle (Fig. 4.20). The upper and right end of the septal wall is smooth; between the smooth part and the anterior cusp of the mitral valve is the *aortic vestibule*, which leads up to the aortic orifice.

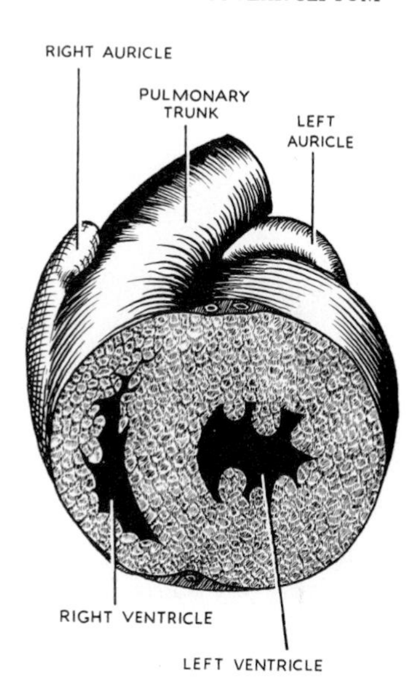

FIG. 4.19. SAGITTAL SECTION OF THE HEART, VIEWED FROM THE LEFT SIDE, LOOKING RIGHT. IN SECTION THE LEFT VENTRICLE IS CIRCULAR, THE RIGHT CRESCENTIC. THE WALLS OF THE LEFT VENTRICLE ARE THREE TIMES AS THICK AS THOSE OF THE RIGHT VENTRICLE. Sketched from a dissection.

Open a heart and place the left index finger in the aortic vestibule, its tip passing through the aortic valve. Anterior to the finger is the interventricular septum tapering down abruptly to the membranous part. Posterior is the septal cusp of the mitral valve. Now pass the right index finger from the left atrium into the left ventricle. The two index fingers lie touching, pointing in opposite directions; only the septal cusp of the mitral valve separates them. This is the pathway of the blood stream.

The Interventricular Septum. The interventricular septum lies vertically from side to side across the body; that is, the cavity of the right ventricle lies

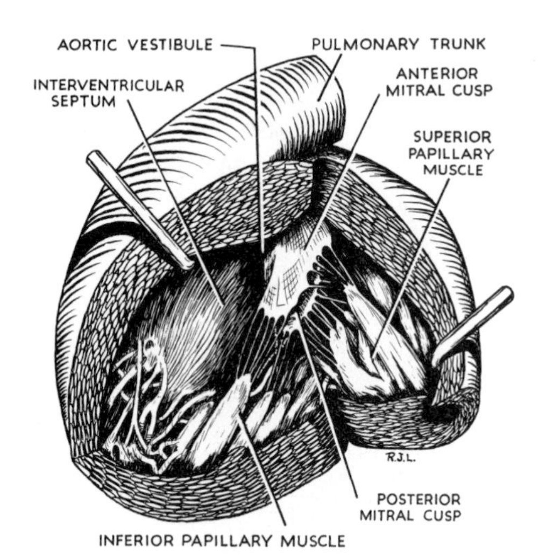

FIG. 4.20. THE CAVITY OF THE LEFT VENTRICLE. The superior and inferior ventricular walls have been incised just behind the interventricular septum and the cut edges pulled apart. The cavity of the ventricle is viewed from the apex of the heart, looking towards the right side. Drawn from a dissection.

in front of it and that of the left ventricle behind it. It is marked on the surface of the heart by the interventricular branches of right and left coronary arteries. Its muscle wall, equal in thickness to that of the left ventricle, bulges forward into the cavity of the right ventricle. At its attachment to the fibrous skeleton (conjoined atrio-ventricular rings) it is thinner and more fibrous. This is the *membranous* part of the septum, and the aortic vestibule lies between it and the anterior cusp of the mitral valve (Figs. 4.20 and 4.18).

The Aortic Orifice. This lies at a lower level than the pulmonary orifice and rather to its right side. It is guarded by three semilunar cusps, one anterior and two posterior (Fig. 4.18).

Structure of Heart Valves. The cusps of the tricuspid and mitral valves are all similar. They are flat and their free edges are serrated in outline by the attachment of the chordæ tendineæ. On closure of the valves during ventricular systole the cusps do not merely meet edge to edge, but come into mutual contact along strips of their auricular surfaces near the serrated margins. This contact and the pull of the marginal chordæ tendineæ prevent eversion of the free edges into the cavity of the atrium, and the centrally attached chordæ tendineæ limit the amount of ballooning of the cusps towards the atrium. The tricuspid and mitral valves are kept competent by active contraction of the papillary muscles, which pull on the chordæ tendineæ during ventricular systole.

The cusps of the pulmonary and aortic valves are similar to each other. The free edge of each cusp contains a central fibrous nodule from each side of which straight edges slope at 120 degrees from each other to the attached base of the cusp. Three cusps lying edge to edge thus close the circular orifice. The cusps are cup-shaped, and from their free and attached margins they bulge down in a globular convexity towards the cavity of the ventricle. During ventricular diastole pressure of blood above the valves distends the cusps, so that their free edges are forced together. Competence of the pulmonary and aortic valves is thus a passive phenomenon, the result of mutual pressure between the distended cusps, and depending on the integrity of their straight edges.

Microscopically the valves of the heart are composed of fibrous tissue in which many elastic fibres are found. They are covered on each surface by vascular endothelium.

The Ascending Aorta. Immediately above the aortic orifice the wall bulges to make the aortic sinuses, one above each cusp. From the anterior sinus the right coronary artery emerges; from the left posterior sinus the left coronary artery. Above the sinuses the aorta emerges to the right of the pulmonary trunk and as it passes upwards slants a little forward towards the manubrium before curving backwards at the commencement of the arch. Here the fibrous pericardium is blended with its wall. The ascending aorta, over 2 inches (5 cm) long, makes a gentle spiral with the pulmonary trunk; the two vessels lie within the fibrous pericardium and are enclosed in a common sleeve of serous pericardium in front of the transverse sinus (Fig. 4.16 and PLATE 9).

Great Vessels. In their course through the pericardial cavity the great vessels are invested with a reflexion of serous pericardium between the parietal and visceral layers. The two arteries share a common sleeve of serous pericardium in which they lie completely free in front of the transverse sinus. The pulmonary trunk inside the fibrous pericardium measures 1½ inches (4 cm) while the ascending aorta, on the convexity of their spiral curve, is slightly longer.

The six veins share a common sleeve of serous pericardium. The four pulmonary veins and the inferior vena cava are all ½ inch (1 cm) or less in length, opening into their atria immediately they are free from the fibrous pericardium. The superior vena cava, on the other hand, courses for over an inch (3 cm) through the pericardial cavity before opening into the right atrium.

Surface Projection of the Valves. The surface projection of the heart itself has been considered on p. 230. Now that the valves of the heart have been studied their surface projections should be noted. They all lie behind the sternum, making a line with each other that is nearly vertical. The bases of tricuspid and mitral valves, attached to the atrio-ventricular ring, are indicated by vertical lines over the lower part of the sternum. The tricuspid valve lies behind the midline of the lower sternum, the mitral valve, overlapping it, lies higher and somewhat to the left. The aortic and pulmonary orifices lie behind the left border of the sternum at the 3rd costal cartilage; the pulmonary is the higher of the two (Fig. 4.21).

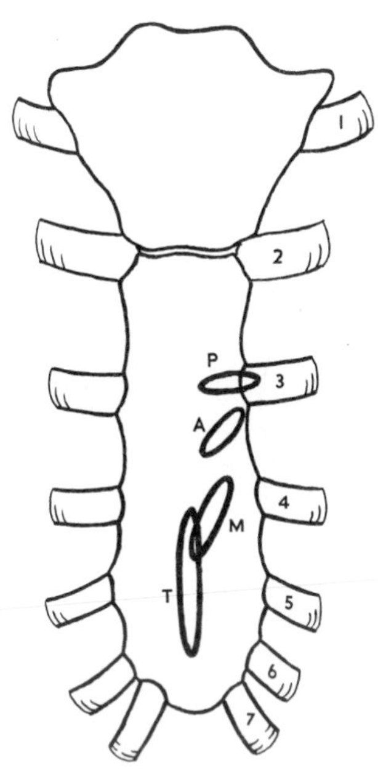

FIG. 4.21. THE SURFACE PROJECTION OF THE HEART VALVES. P. Pulmonary valve. A. Aortic valve. M. Mitral valve. T. Tricuspid valve.

It should be pointed out that knowledge of the surface markings of the heart valves has no application in auscultation of the heart. The normal heart sounds are produced by closure of heart valves, and the opening snap of the mitral valve may also be audible, but these sounds are by no means heard best directly over the valves concerned. They are heard best where the blood-containing chambers lie nearest to the chest wall. For the tricuspid valve this is over its surface (Fig. 4.21) but for the mitral valve it is at the apex beat, where the cavity of the left ventricle lies nearest the surface. For the aortic valve it is where the ascending aorta lies nearest the surface, at the right sternal margin in the second intercostal space. For the pulmonary valve it is at the left sternal margin at the 3rd costal cartilage (over the infundibulum of the right ventricle).

Blood Supply of the Heart

The Coronary Arteries. These emerge, one on each side, from behind the pulmonary trunk. They run in the atrio-ventricular grooves. The right coronary artery runs down over the front of the heart, the left coronary runs down over the back of the heart.

Right Coronary Artery (Fig. 4.22). Arising from the anterior aortic sinus the artery passes between the right auricular appendage and the infundibulum of the right ventricle. Passing now *vertically downwards* in the atrio-ventricular groove the artery turns backwards at the inferior border of the heart and runs posteriorly. It gives off branches to both atrium and ventricle as it passes vertically downwards. At the inferior border the *marginal branch* passes to the left along the right ventricle. On the diaphragmatic surface of the heart the *inferior interventricular* branch is given off. This large vessel passes along the interventricular groove to the apex of the heart. Its name of 'posterior' interventricular artery is a misnomer. The part of the right coronary artery now remaining is much smaller than its interventricular branch. It passes backwards to anastomose by terminal arterioles with the termination of the left coronary artery at the lower part of the left atrium (PLATE 9).

Left Coronary Artery (PLATE 9). Arising from the left posterior aortic sinus the vessel emerges between the left auricular appendage and the infundibulum of the right ventricle to pass backwards around the atrio-ventricular groove. It gives off the *anterior interventricular* branch near the upper border of the heart. This large vessel passes downwards in the interventricular groove to anastomose at the apex with the terminal branches of the inferior interventricular artery. The parent trunk, not much narrowed, passes down in the atrio-ventricular groove. It is named the *circumflex branch*. It gives off branches to the posterior wall of the left ventricle and runs on to anastomose with the

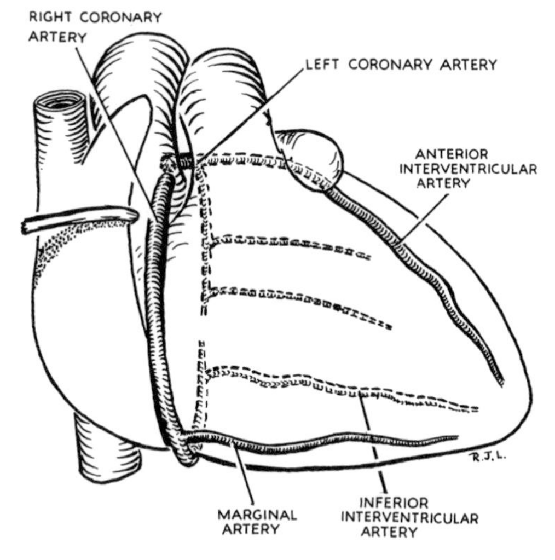

FIG. 4.22. THE CORONARY ARTERIES. The right atrium is displaced to the right to open up the atrio-ventricular groove. Anterior view.

termination of the right coronary artery, below the coronary sinus. Here in 40 per cent of individuals it gives off a sizable branch which, running up over the posterior surface of the left atrium between the upper left pulmonary vein and the auricular appendage, ends in the auricular appendage of the *right* atrium at the sinu-atrial node (PLATE 9).

Anastomoses of the Coronary Arteries. Arteriolar anastomoses exist between the terminations of the right and left coronary arteries in the atrio-ventricular groove and between their interventricular branches at the apex. These anastomoses on the surface of the heart are insignificant. In the interventricular septum and in the posterior wall of the left ventricle there are *very free anastomoses* between the interventricular arteries; these, however, are *by arterioles only*. Thus the time factor in occlusion is all important; in slow occlusion there is time for healthy arterioles to open up, in abrupt occlusion there is not. If the interventricular arteries (from right and left coronary arteries) meet at the apex, this provides maximum anastomosis between the two. If the meeting place of the interventricular arteries falls short of the apex, above or below, this diminishes the potential anastomotic area. In 10 per cent of individuals the inferior as well as the anterior interventricular artery is a branch of the left coronary; in these cases there is *no anastomosis* between right and left coronary arteries.

Potential anastomoses exist between the coronary arteries and pericardial arteries around the roots of the great vessels. These pericardial arteries are derived from the pericardiaco-phrenic, the bronchial and the internal thoracic arteries. In *very* rare instances one of these may open up to replace a coronary artery.

Distribution of the Coronary Arteries. The right and left coronary arteries vary a good deal in the manner in which they share the supply of blood to the heart. The right ventricle is supplied by the right coronary artery except at the upper margin of its anterior surface, where it is supplied by branches of the anterior interventricular artery (left coronary, PLATE 9). The left ventricle is supplied by the left coronary artery except for a narrow strip of its diaphragmatic surface behind the interventricular groove, where it is supplied by the inferior interventricular artery (right coronary). The two interventricular arteries share the supply of the interventricular septum, usually about equally.

The source of blood for the supply of the atria is more variable. The anterior surface of the right atrium is supplied from the right coronary artery. The posterior surface and auricular appendage of the left atrium are supplied from the left coronary artery. The *right* auricular appendage receives in 40 per cent of cases from the *left* coronary artery a branch that ascends over the back of the left atrium (PLATE 9). The *lower* part of the left atrium and the interatrial septum are supplied by branches of the *right* coronary artery that spring from near the origin of the inferior interventricular artery. Terminal branches of the left coronary artery share this area in a variable manner.

The sinu-atrial node is supplied by a branch of the right coronary artery in 60 per cent of cases, and from the left coronary artery in 40 per cent. The atrio-ventricular node and the conducting bundle of His are supplied by the inferior interventricular artery, which arises in 90 per cent of cases from the right coronary and in only 10 per cent from the left coronary artery.

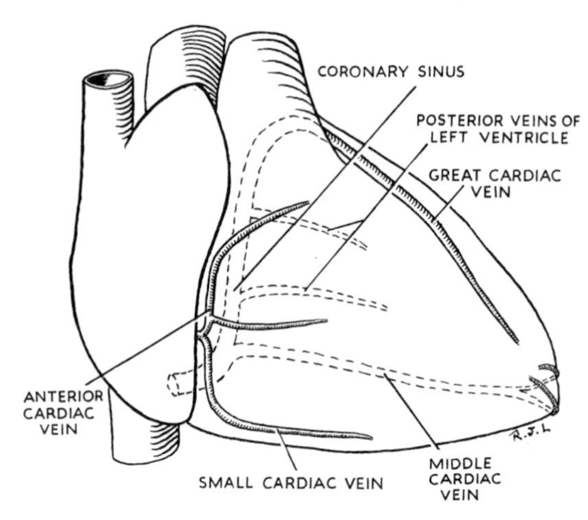

FIG. 4.23. THE VEINS OF THE HEART. The small cardiac vein, which accompanies the marginal branch of the right coronary artery, normally opens with other anterior cardiac veins directly into the anterior wall of the right atrium.

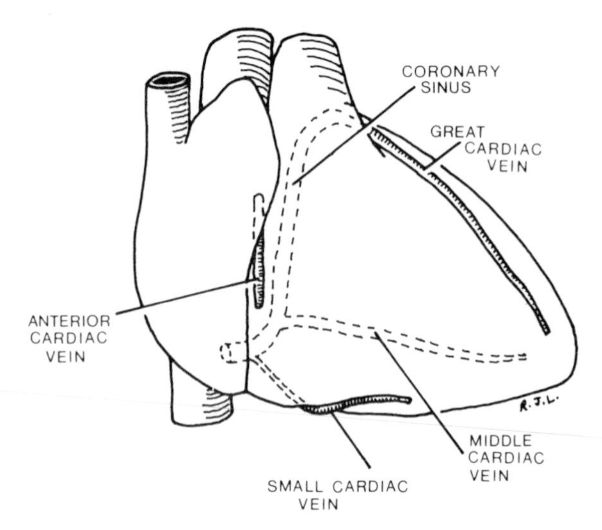

FIG. 4.24. THE VEINS OF THE HEART. The small cardiac vein here runs back to join the coronary sinus. This is an occasional variation that was formerly described, incorrectly, as the normal pattern (cf. Fig. 4.23).

The Veins of the Heart. The **coronary sinus** (PLATE 9) lying in the posterior part of the atrio-ventricular groove, opens at its right (i.e., its *lower*) end into the posterior wall of the right atrium (Fig. 4.17). Embryologically it may be said to commence at a valve alongside the entrance of the oblique vein of Marshall (PLATE 9). It is a wide-bored vessel about 3 cm long. It receives 60 per cent of the heart's blood. Three main veins open into it. The **great cardiac vein,** which accompanies the anterior interventricular artery, is joined by the upper of two large posterior veins of the left ventricle and flows into its commencement. The lower posterior vein of the left ventricle joins the coronary sinus independently. The **middle cardiac vein** accompanies the inferior interventricular branch of the right coronary artery and opens near the termination of the coronary sinus. The **small cardiac vein** was formerly said to open at the termination of the sinus; it accompanies the marginal branch of the right coronary artery (Fig. 4.23). It opens into the anterior wall of the right atrium and is, in fact, merely the lowest and largest of the anterior cardiac veins.

The Anterior Cardiac Veins. The anterior surface of the heart is drained by a series of parallel veins that run across the surface of the right ventricle to open into the right atrium. The lowest and largest of these runs with the marginal artery (right coronary), and is called the small cardiac vein. The anterior cardiac veins drain independently into the right atrium by dipping into the anterior atrio-ventricular groove. In this vertical groove it is common to find a vein (lying at right angles to the anterior cardiac veins) which receives two or more of the veins and itself opens into the right atrium (Fig. 4.24). The anterior cardiac veins drain most of the remaining 40 per cent of the heart's blood. The rest is received by the venæ cordis minimæ.

Venæ Cordis Minimæ. In all four chambers of the heart there are many small veins that open directly into the cavity. There are numerous arterio-venous shunts in the walls of all four chambers. Most of these open via the venæ cordis minimæ.

Lymph Drainage. The lymphatics of the heart drain back along the coronary arteries, emerge from the fibrous pericardium along with the aorta and pulmonary trunk, and empty into the tracheo-bronchial lymph nodes and mediastinal lymph trunks.

The Conducting System of the Heart

The Sinu-atrial Node (The Pacemaker). The histology of the node differs from that of the rest of the myocardium. Not only does it possess a rich supply of nerve fibres (sympathetic) and nerve-cells (vagus), but the actual heart muscle fibres are finer and more densely striated than elsewhere. Anatomically it occupies a large saddle-shaped area to the left of the sulcus terminalis. It extends for a centimetre along the sharp upper border of the right auricular appendage and tapers off down the crista terminalis for two centimetres below this and occupies a corresponding area on the posterior wall of the appendage. It is an isolated area, timed by the opposing influences of the sympathetic (accelerator) and vagus (depressor) impulses. From this area conduction of the impulse is carried by heart muscle fibres only.

The Atrio-ventricular Node. This small nodule of conducting tissue lies in the interatrial septum just above the attachment of the septal cusp of the tricuspid valve, near the opening of the coronary sinus. From it the **conducting bundle (of His)** crosses the atrio-ventricular ring and descends along the inferior border of the membranous part of the interventricular septum. Passing along the upper border of the muscular part of the septum it divides into right and left branches.

The Right Branch. This bundle of conducting tissue passes down the right side of the interventricular septum. A large part of it is carried across the moderator band to the anterior wall of the right ventricle. The fibres of the branch are continued into the *Purkinje fibres*. These lie immediately beneath the endocardium. They are large, pale fibres, striated only at their margins. They usually possess double nuclei. In some animals, notably the sheep, they are very prominent but in man they are less numerous because they quickly merge into the ordinary cardiac muscle fibres. They are part of the conducting system of the heart.

The Left Branch. This descends on the left side of the interventricular septum and spreads out over the wall of the left ventricle to merge into the Purkinje tissue.

Development of the Heart. Primitive blood vessels are laid down by angioblasts on the wall of the yolk sac. Two such vessels fuse together to make a single heart tube which develops muscle fibres in its wall and becomes pulsatile. It elongates, bends and rotates (p. 228). Ingrowing septa divide the tube into four chambers, which enlarge by incorporation of parts of the vessels to which they are joined.

Partition into chambers. A pair of *endocardial cushions* grow across the tube and meet each other, between the atrial and ventricular cavities. Their union divides the atrio-ventricular canal into two, the tricuspid and mitral orifices, from which the cusps of the valves grow.

The ventricular cavity is partitioned by a flange that grows in and meets the endocardial cushions; it becomes the *interventricular septum.* Meanwhile the ventricles have incorporated the bulbus cordis up to the ventral aorta (Fig. 1.52, p. 43), which itself becomes separated by the ingrowth of a spiral septum into pulmonary trunk and ascending aorta (Fig. 1.53, p. 43).

The atrial cavity is partitioned, but the septum is at all times patent, to pass oxygenated placental blood directly into the left atrium. The *septum primum* grows down to the endocardial cushions, but perforates dorsally before meeting them. Then the *septum secundum* grows down on its right but fails to meet the endocardial cushions. The overlap of the edges of the two septa is the *foramen ovale*. The septa adhere after the lungs function at birth. The edge of the septum secundum persists as the annulus (anulus) ovalis, the adherent septum primum forming the thin floor of the fossa ovalis.

Enlargement of chambers. The ventricles incorporate the bulbus cordis and the left atrium incorporates the pulmonary veins (p. 233). The right atrium enlarges by incorporation of the right horn of the sinus venosus. Thus the right vitelline vein (inferior vena cava) and right duct of Cuvier (superior vena cava) are drawn in. The opening of the duct of Cuvier has a valve with right and left cusps. They fuse cranially to make a ridge called the *septum spurium*. The left cusp fuses with the interatrial septum (septum secundum), while the right cusp and septum spurium persist as the ridge called crista terminalis. Thus the auricular appendage (roughened by musculi pectinati) is the original embryonic chamber, and the smooth-walled part of the atrium is the incorporated right horn of the sinus venosus. The rest of the sinus venosus persists as the coronary sinus. On the left side the horn and duct of Cuvier atrophy and may persist as a vestige dear to embryologists, the oblique vein of Marshall, which enters the commencement of the coronary sinus from the wall of the left atrium of the adult heart (PLATE 9).

Congenital Defects of the Heart. Cardiac malformations occur in 6 per 1000 live births. Of these the commonest is a ventricular septal defect. There follow patent ductus arteriosus, atrial septal defect, and the *tetrad of Fallot*, which causes the effort dyspnœa and the cyanosis of the 'blue baby'. The tetrad consists of (1) a defect in the membranous part of the interventricular septum, (2) stenosis of the pulmonary trunk, (3) the aortic orifice lying astride the incomplete interventricular septum, (4) the right ventricle being much hypertrophied. Early recognition followed by operation has greatly reduced mortality rates in young infants.

THE POSTERIOR MEDIASTINUM

The posterior mediastinum is the space posterior to the pericardium and to the diaphragm. It is continuous directly, via the posterior part of the superior mediastinum, with the tissue spaces behind the pretracheal fascia and in front of the prevertebral fascia, that is, with the retropharyngeal space and the spaces lateral to trachea and œsophagus, between these tubes and the carotid sheaths. It is bounded posteriorly by the thoracic vertebræ (fourth to twelfth) and anteriorly above by the pericardium (left atrium) and below by the posterior sloping fibres of the diaphragm (Fig. 4.9, p. 222). It contains some structures proper to itself and others that pass between the superior mediastinum and the abdomen. These latter structures will be described as a whole during their intrathoracic course.

The Descending Aorta. This great arterial trunk commences at the lower border of the 4th thoracic vertebra, where the arch of the aorta ends. At first to the left of the midline, the vessel slants gradually to leave the posterior mediastinum in the midline at the level of the 12th thoracic vertebra by passing behind the diaphragm between the crura. It gives off from each side nine *intercostal arteries* that arch upwards to enter the posterior ends of the lower nine intercostal spaces. Visceral branches are given off as the bronchial and œsophageal arteries. The *bronchial arteries* are very variable. Usually there are two on the left and one on the right. A second artery on the right is common; when present it arises from the first aortic intercostal artery (i.e., the artery of the 3rd intercostal space). An origin from the subclavian or the internal thoracic artery is not uncommon. The bronchial arteries supply the tracheo-bronchial lymph nodes before entering the hilum of the lung (p. 246). The *œsophageal* arteries are four or five in number; they are small vessels.

The Œsophagus. This muscular tube extends from the cricoid cartilage (at the level of the 6th cervical vertebra) to the cardiac orifice of the stomach (at the level of the 10th thoracic vertebra and the left 7th costal cartilage). It is 10 inches (25 cm) long.

The cervical portion of the œsophagus, lying in front of the prevertebral fascia, inclines slightly to the left of the midline, but the œsophagus enters the thoracic inlet in the midline in front of the body of the 1st thoracic vertebra. Passing downwards now through the superior mediastinum the tube is slightly to the left of the midline behind the left bronchus, which may indent it slightly in a radiograph of a barium swallow. The œsophagus, all this time in contact with the vertebral bodies, now inclines forward with a concavity more marked than that of the vertebral column, passes in front of the descending thoracic aorta, in contact with the pericardium, and pierces the diaphragm one inch (2·5 cm) to the left of the midline, opposite the body of the 10th thoracic vertebra. This level can be indicated on the anterior body wall on the 7th left costal cartilage a thumb's breadth from the side of the sternum. Fibres from the *right* crus of the diaphragm sweep around the œsophageal opening in a sling-like loop. The intra-abdominal part of the œsophagus varies in length according to the tone of its muscle and the degree of distension of the stomach. It averages less than 1 cm. In the superior mediastinum the œsophagus is crossed by the arch of the aorta on its left side, and the vena azygos on its right side. The thoracic duct lying at first behind it to the right, ascends with an inclination to the left, and in the superior mediastinum lies on the prevertebral fascia to the left of the œsophagus. The mediastinal pleura touches the œsophagus in places, particularly on the right side, where low down there is a pocket of pleura between the œsophagus and the aorta (PLATE 8, p. *iv*) but nowhere is the pleura attached to the œsophagus.

Structure. The muscular wall consists of an inner circular and an outer longitudinal layer which are of striped muscle in the upper third and unstriped muscle in the lower two-thirds of the tube. There is no sharp line of demarcation between these two areas; there is considerable overlap of the two types of muscle. The striped muscle provides rapid contraction so that the bolus is quickly passed well into the œsophagus, and the larynx may safely open to resume breathing. In the lower reaches of the œsophagus the smooth muscle provides a more leisurely peristaltic wave. Except for the short intra-abdominal segment there is no serous covering of the tube. The mucous membrane is thick and in the collapsed state thrown into longitudinal folds. There is a thick muscularis mucosæ. The surface epithelium of the mucous membrane is squamous stratified. The mucosa contains scattered lymphatic follicles. In the submucosa are mucous glands that are rather sparse and found for the most part at the upper and lower ends of the tube. The outer longitudinal layer of muscle is attached by fibrous strands to the midline ridge on the back of the lamina of the cricoid cartilage, and to the arytenoid cartilages. The inner circular layer is in direct continuity with the crico-pharyngeus muscle (p. 416).

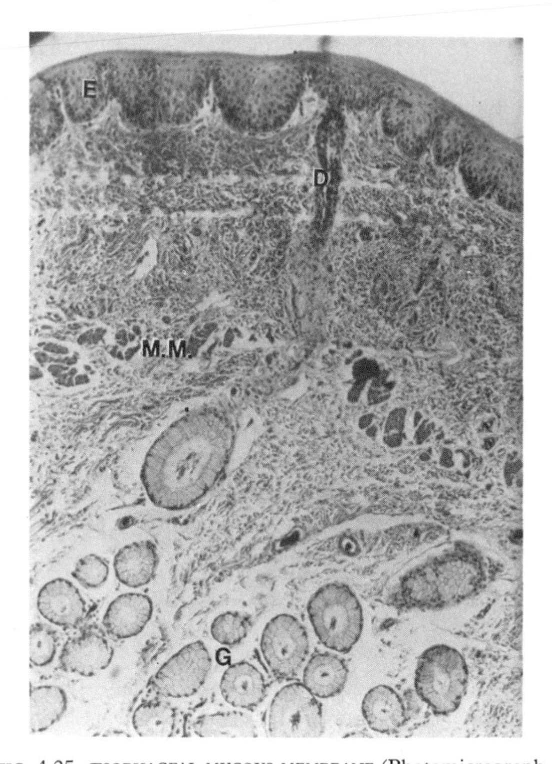

FIG. 4.25. ŒSOPHAGEAL MUCOUS MEMBRANE (Photomicrograph, low power). The loose-woven expansile fibrous tissue is intersected by thick, discrete bundles of muscle, the muscularis mucosæ (**M.M.**). **G.** Mucous glands in the submucosa. **D.** A duct traversing the mucous membrane. **E.** The lining of squamous stratified epithelium.

Blood Supply of the Œsophagus. The upper œsophagus, from the cricoid cartilage down to the level of the arch of the aorta in the superior mediastinum, is supplied by the inferior thyroid arteries, the middle portion by œsophageal branches from the aorta, and the lower part by the œsophageal branches of the left gastric artery. Venous return from the upper part is to the brachio-cephalic veins, from the middle part to the azygos veins, and from the lower reaches by œsophageal tributaries of the left gastric vein, which empties into the portal vein. Thus there exists, in the lower reaches of the œsophagus, an anastomosis between portal and systemic venous systems. This anastomosis lies level with the central tendon (Th. 8 vertebra) well above the œsophageal hiatus in the diaphragm. In cases of portal obstruction varicosities of these veins occur, and their rupture may give rise to serious or fatal hæmorrhage. Lymph drainage of the œsophagus is by way of lymphatics that follow the arteries. Thus from the lymphatic follicles in the mucous membrane the cervical portion drains into the postero-inferior group of deep cervical lymph nodes near the origin of the inferior thyroid artery, behind the carotid sheath. The middle portion drains into pre-aortic nodes of the posterior mediastinum along the œsophageal arteries, and thence to the tracheo-bronchial group and mediastinal lymph trunks. The lower part drains along the left gastric artery to the pre-aortic nodes of the cœliac group.

Nerve Supply of the Œsophagus. The upper part of the œsophagus is supplied by the recurrent laryngeal nerve and by sympathetic fibres from cell-bodies in the middle cervical ganglion running in on the inferior thyroid arteries. The lower half of the tube is supplied from the *œsophageal plexus*, a wide-meshed network of fibres that encircles the tube below the level of the lung roots (PLATE 10, p. *v*). The parasympathetic constituents of this plexus are the two vagi. The sympathetic contribution is by way of grey rami that come from the upper four thoracic ganglia of the sympathetic trunk.

The *motor* supply is from the vagus (cell bodies in the nucleus ambiguus for the upper striped-muscle part, cell bodies in the dorsal nucleus with local relay in the lower smooth-muscle part). The *secreto-motor* supply is likewise from the vagus (cell bodies in the inferior salivatory nucleus with relay in the mucous membrane). The vagus is also *sensory* to the whole œsophagus (cell bodies in the inferior ganglion). *Vaso-motor* fibres are provided by the sympathetic. Note that in the neck and mediastinum the alimentary and respiratory tubes share an identical innervation.

Both vagi and the sympathetic fibres become intermixed in the plexus. The sympathetic fibres are used up to supply the œsophagus, but most of the vagal fibres pass onwards as the right and left vagal trunks. These, generally two on the right and one on the left, pass through the œsophageal opening of the diaphragm. The two right nerves join the cœliac plexus (p. 314) and the left passes on the anterior surface of the stomach to the gall-bladder and liver. Each vagal trunk contains fibres derived from both vagus nerves.

Lymph Nodes of the Thorax. In the posterior mediastinum the nodes are arranged in series with those of the abdomen. The visceral *pre-aortic nodes* lie in front of the aorta and drain the middle reaches of the œsophagus. The somatic *para-aortic nodes* lie alongside the aorta and extend laterally into the posterior intercostal spaces, where they are known as the *posterior intercostal nodes*. They drain the parietes (intercostal spaces). The anterior ends of the intercostal spaces drain into nodes along the *internal thoracic* artery; these are known as the *anterior intercostal nodes* and they may become involved in cancer of the breast. They drain directly into the brachio-cephalic veins.

The lower members of the internal thoracic group lie on the diaphragm between pericardium and xiphisternum. The lower members of the para-aortic group in the posterior mediastinum likewise lie on the diaphragm. In addition a *middle diaphragmatic group* lies on each dome of the diaphragm; they drain from the extraperitoneal areolar tissue beneath the diaphragm and, on the right side, from the superficial part of the bare area of the liver.

The upper members of the posterior intercostal group drain into the thoracic duct or into the right lymph trunk; the lower (intercostal and diaphragmatic) members drain into a *descending intercostal trunk* that passes downwards to join the cisterna chyli in the aortic orifice of the diaphragm.

Heart and lungs drain into the *tracheo-bronchial lymph nodes*, which lie around the trachea and its bifurcation. These send their efferents to a right and left *mediastinal lymph trunk*, which may join the thoracic duct but usually open directly into the brachio-cephalic vein of their own side.

There are thus three upgoing lymph channels on each side, namely the internal thoracic trunk (anterior intercostal) alongside the sternum, the thoracic duct and right lymph trunk alongside the vertebræ (these drain the thoracic wall) and between them the mediastinal lymph trunks alongside the trachea (these drain lungs, midpart of œsophagus, and heart). All six trunks communicate freely with each other.

The Thoracic Duct. This lymphatic vessel commences in the cisterna chyli, on a level with the body of the 12th thoracic vertebra between the aorta and the vena azygos. It passes upwards to the right of the aorta between the crura of the diaphragm and comes to lie against the right side of the œsophagus. Inclining up to the left, alongside the aorta, it passes behind the œsophagus to reach its left side at the superior mediastinum. It lies anterior to the intercostal branches of the aorta. Passing now vertically upwards it finally arches forwards across the dome of the left pleura to enter the point of confluence of the left internal jugular and subclavian veins (Fig. 4.10, p. 223). It normally divides into two or three separate branches, all of which open at the angle between these two veins. There are no valves at the termination, and blood flows to and fro in the duct as pressures alter with respiratory movements. At post-mortem it is common to find the terminal inch or two of the thoracic duct full of blood. Commencing with the accumulated lymph from the lower half of the body, it receives in its course through the thorax lymph from the left posterior intercostal nodes (posterior half of the left thoracic wall). In the neck it receives the left jugular and subclavian lymph trunks and thus finally comes to drain all the lymph of the body except that from the right arm and the right halves of thorax and the head and neck.

Lymph from the posterior right thoracic wall enters the **right lymph trunk.** The right upper limb drains into the right subclavian trunk and the right side of the head and neck drains into the right jugular lymph trunk. These three trunks may join together and open into the commencement of the right brachio-cephalic vein or they may remain separate and open independently into the great veins (jugular and subclavian).

The Azygos System of Veins

The thoracic wall and upper lumbar region are drained by the posterior intercostal and lumbar veins into the azygos veins. These consist of longitudinal trunks on right and left sides, similar to the posterior cardinal veins of the embryo. There is a single trunk on the right. On the left side the posterior cardinal vein disappears (p. 309) and is replaced by longitudinal veins that persist from the embryonic prevertebral venous plexuses. These are the hemiazygos veins.

The Azygos Vein. This vessel represents the persistent right posterior cardinal vein of the embryo. Thus it begins embryologically at the inferior vena cava just above the renal veins, but this part is usually an avascular fibrous cord. The union of the ascending lumbar vein with the subcostal vein of the right side is the functional (blood-containing) commencement of the azygos vein. The vessel goes through the aortic opening of the diaphragm under shelter of the right crus and passes upwards lying on the sides of the vertebral bodies, on a plane posterior to that of the œsophagus (PLATE 8, p. *iv*). Descent of the heart and pericardium in the embryo caused the right posterior cardinal vein to arch over the right bronchus. This the azygos vein does at the level of the 4th thoracic vertebra and passing forward of the œsophagus the vessel enters the superior vena cava. It receives the lower eight intercostal veins and at its convexity the superior intercostal vein joins it. It receives the bronchial veins from the right lung and some veins from the middle third of the œsophagus. The two hemiazygos veins join it.

The Hemiazygos Veins. These are two, superior and inferior, lying longitudinally on the left side of the bodies of the thoracic vertebræ. They may communicate with each other, but characteristically drain separately from their adjoining ends behind the œsophagus into the azygos vein. They receive the lower eight intercostal veins, four each. The inferior is now named simply the *hemiazygos vein*, and the superior one is called the *accessory hemiazygos vein*. The accessory hemiazygos vein receives the bronchial veins from the left lung and some veins from the middle third of the œsophagus.

The Thoracic Sympathetic Trunk

(Fig. 1.42, p. 35)

Lying on the necks of the ribs, just lateral to their heads, and anterior to the intercostal vessels and nerves, is the ganglionated trunk of the thoracic part of the sympathetic outflow (PLATE 45, p. *xxiv*). It is described as possessing twelve ganglia, one for each intercostal nerve, but characteristically there are fewer, the result of fusion of adjacent ganglia. Thus the first ganglion is commonly fused with the inferior cervical ganglion to form the stellate ganglion. It is simplest, however, to describe them as though they consisted of twelve discrete ganglia. Each receives a white ramus from its corresponding spinal nerve, and this emerges from the anterior primary ramus of the nerve. After relay in the ganglion a postganglionic grey ramus is given to each thoracic nerve, and this usually lies medial to the white ramus (Fig. 1.38, p. 32).

The heart is supplied with sympathetic fibres from the cervical ganglia through the superficial and deep cardiac plexuses (*q.v.*). The remaining thoracic viscera (lungs and œsophagus) are supplied from the upper four thoracic ganglia. The **œsophageal plexus** is described on p. 239. The **pulmonary plexuses** are two in number for each lung. The small anterior pulmonary plexus lies in front of the lung root, the larger posterior pulmonary plexus lies behind the lung root. The vagal component is given off as the vagus nerve passes downwards behind the root of the lung.

The splanchnic nerves, three in number, come from the lower eight ganglia. The *lowest*, or **least splanchnic nerve** leaves the 12th ganglion. The **lesser splanchnic nerve** comes from the next two ganglia namely the 10th and 11th. The **greater splanchnic nerve** is formed by branches from the remainder of the lower eight ganglia, namely the 5th to the 9th inclusive. Each pierces the crus of its own side to relay in the cœliac ganglia (p. 314).

The thoracic trunk is continued upwards over the neck of the 1st rib as the cervical sympathetic trunk. It crosses the neck of the first rib well on the medial side, near the head. The main part of the 1st thoracic nerve, passing out of the thorax into the neck to join the brachial plexus, crosses the neck of the 1st rib more laterally. Between the two at this level lie the supreme intercostal vein medially, and the superior intercostal artery laterally. Classical description of the ganglia consists of describing the 1st thoracic ganglion as lying below the neck of the 1st rib, and the inferior cervical ganglion lying in the root of the neck just above the neck of the rib, postero-medial to the vertebral artery. More often than not, however, these two ganglia are fused into a single mass, the *stellate ganglion*, overlying the neck of the 1st rib. It is very variable in size, and may measure as much as 2 cm in length and 0·5 cm in diameter (PLATE 11).

The thoracic trunk is continued downwards into the abdomen by passing behind the medial arcuate ligament of the diaphragm, lying on the front of the fascia of the upper part of the psoas muscle. It makes an abrupt step forwards from the neck of the 11th (or even the 12th) rib to the front of the body of the 1st lumbar vertebra, and this part of the trunk is very slender (Fig. 1.42, p. 35).

THE PLEURA

Like peritoneum the pleura is a membrane of fibrous tissue surfaced by a single layer of very flat cells to make it slippery. It clothes each lung and lines the containing cavity.

The **parietal layer** of the pleura lines the thoracic wall (rib cage), to which it is attached by areolar tissue named the endothoracic fascia. It covers the thoracic surface of the diaphragm, from which it ascends over the pericardium to cover the mediastinum and to lie under the suprapleural membrane at the thoracic inlet. This is one continuous sheet. From its mediastinal layer a cuff of membrane is projected around the lung root and passes on to invest the surface of the lung. This is the **visceral layer** of the pleura; it extends into the depths of the interlobar clefts. The pleural cavity is a completely closed space. The visceral pleura on the lung surface is in contact with parietal pleura, but the anterior and inferior reflexions of the parietal pleura extend further than the lung edge, to allow space for lung expansion. In these situations the costal parietal pleura is in contact anteriorly with mediastinal and inferiorly with diaphragmatic parietal pleura. The function of the pleura, like that of pericardium and peritoneum, is to provide two frictionless surfaces between a mobile structure and the containing walls of its cavity; the surfaces are lubricated by a thin film of tissue fluid. The cuff of pleura projected around the lung root is too big for it, as a coat cuff is too big for the wrist. It hangs down below as an empty fold. This fold is called the *pulmonary ligament*; an ill-chosen name for it has nothing to do with the lung and is not a ligament. It provides 'dead space' into which the lung root descends with descent of the diaphragm, and, more important, into which the pulmonary veins can expand during increased venous return from the lungs, as in exercise. It is interesting to note that the two pulmonary veins lie at the lower part of the lung root, just above the empty fold of the so-called pulmonary ligament. Large veins always have 'dead space' near them (e.g., to the right of the superior vena cava, and in the femoral canal alongside the femoral vein).

The Endothoracic Fascia. Outside the pleura, and lining the thoracic wall, is a layer of loose areolar tissue similar to the transversalis fascia of the abdomen, which binds the parietal pleura to the inner side of the chest wall and transversus thoracis group of muscles where these exist. This loose areolar tissue has been named the endothoracic fascia; it is almost unworthy of anatomical description, being nowhere membranous, and consisting merely of the fibrous tissue that attaches the pleura to the chest wall. A similar layer binds the pleura to the diaphragm. No such arrangement exists on the fibrous pericardium. Here the slippery pleura is thin, and so adherent that the two cannot be separated.

Nerve Supply of the Pleura. The parietal pleura is supplied by somatic nerves, *viz.*, the collateral intercostal nerves, segmentally in zones on its costal extent; the diaphragmatic pleura is supplied by the phrenic nerve over the domes, and by intercostal nerves around its periphery. The mediastinal pleura is supplied by the phrenic nerve. The visceral pleura has only an autonomic (i.e., vaso-motor) supply—it is insensitive to ordinary stimuli.

Surface Marking of the Pleura (Fig. 4.26). The parietal pleura lines the costal walls of the thorax; seen from in front its lateral surface marking is the horizon of the thoracic cage. It projects above the medial third of the clavicle for a distance of over an inch. This convexity, well seen in an X-ray view of the chest, is due to the obliquity of the thoracic inlet (Fig. 4.6, p. 220). The neck of the 1st rib lies well above the clavicle, and the surface marking of the dome of the pleura (and the apex of the lung) is merely the projection of the inner border of the 1st rib. There is normally no encroachment of lung or pleura above the oblique plane of the thoracic inlet except for a minor amount of bulging above the subclavian groove, beneath the suprapleural membrane just in front of the neck of the 1st rib (PLATE 11, p. *v*).

Tracing the pleura now from behind the sterno-clavicular joint, downwards behind the sternum and around the costo-diaphragmatic gutter, there is a point to be noted at each of the *even-numbered ribs* (2, 4, 6, 8, 10, 12) as follows. The line of pleural reflexion slopes downwards from the sterno-clavicular joint to meet its fellow at the *second* rib level, that is, at the sternal angle of Louis. Lying together or even overlapping, they pass vertically behind the sternum down to the *fourth* costal cartilage. Here the right pleura continues vertically, but the left arches out and descends lateral to the border of the sternum, half-way to the apex of the heart. Each turns laterally at the *sixth* costal cartilage, and passing around the chest wall crosses the midclavicular line at the *eighth* rib, and the midaxillary at the *tenth* rib. This lower border is the costo-diaphragmatic recess; it falls somewhat short of the costal margin between sternum and midaxillary line. It crosses the *twelfth* rib at the lateral border of sacrospinalis muscle and passes in horizontally to the lower border of the *twelfth* thoracic vertebra. There is thus a triangle of pleura in the costo-vertebral angle below the 12th rib, behind the upper pole of the kidney, a fact to be noted in wounds in this region (Fig. 5.47, p. 316).

Surface Markings of the Lungs (Fig. 4.26). On the costal walls and supraclavicular region, the lung coincides with the pleura (*v. sup.*). The anterior border of the right lung falls very little short of the pleura, lying within the lateral margin of the sternum; that of the left lung, on the contrary, curves laterally to uncover the area of superficial cardiac dullness, namely from the 4th costal cartilage out to the apex beat in the 5th space just medial to the midclavicular line. From these points the lower border of the lung lies, nearly horizontal around the chest wall, two ribs

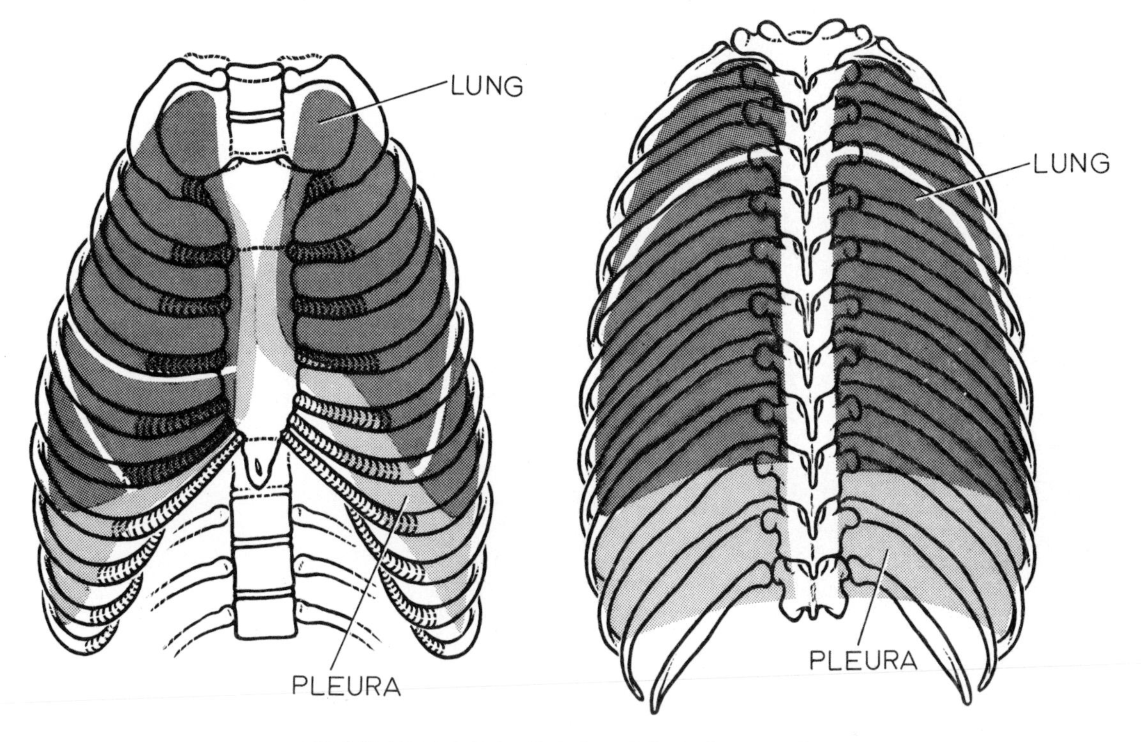

FIG. 4.26. THE SURFACE MARKINGS OF THE LUNGS AND PLEURÆ.

higher than the pleural reflexion; that is to say, at the midclavicular line the 6th rib, midaxillary line the 8th rib, and posteriorly the 10th rib. The fissures of the lung are indicated on the body surface by a line joining the spine of Th. 3 vertebra (i.e., the posterior end of the *fifth* rib) to the 6th rib in the midclavicular line. More simply, this fissure between upper and lower lobes can be taken as approximately the 5th rib (just slightly higher on the left side) and this line is marked out by the vertebral border of the scapula when the arm is fully abducted above the head. On the right side the 4th costal cartilage overlies the fissure between upper and middle lobes; continued horizontally this line meets the oblique fissure in the midaxillary line.

THE LUNGS

The lung surface is mottled, and in colour is pink or grey according to the atmosphere in which it has lived. The lung is crepitant to the touch.

The lung conforms to the shape of the cavity which contains it (Fig. 4.27). It has a convex costal surface and a concave diaphragmatic surface, separated from each other by a sharply-angled inferior border. The posterior border of each lung is generously rounded to fit the paravertebral gutter, and is continued up to the convex apex. The anterior border is thin and sharp; on the left side the lower part of this border is deeply concave—the *cardiac notch.* The mediastinal surfaces differ somewhat. On the left side the cardiac notch is seen to be the anterior margin of a deep concavity produced by the pericardium in front of the hilum; the arch and descending aorta make a deep groove on the lung surface around the hilum. On the right the cardiac impression is much shallower; a groove for the azygos vein curves over the hilum. The apices are grooved by the subclavian arteries (Figs. 4.30, 4.31).

The *lung roots* are considered on p. 227.

Fissures. The *oblique fissure* extends from the surface of the lung to the hilum and divides the organ into separate upper and lower lobes which are connected only by the lobar bronchi and vessels. On the right lung a *horizontal fissure* passes from the anterior margin into the oblique fissure to separate a wedge-shaped middle lobe from the upper lobe. The visceral pleura, clothing the surface of the lung, extends inwards to line the depths of the fissures. The middle lobe of the right lung is completely separate from the upper lobe in only about one-third of individuals; in the remainder the fissure separating it from the upper lobe is incomplete (Fig. 4.27) or even absent (10 per cent). In the

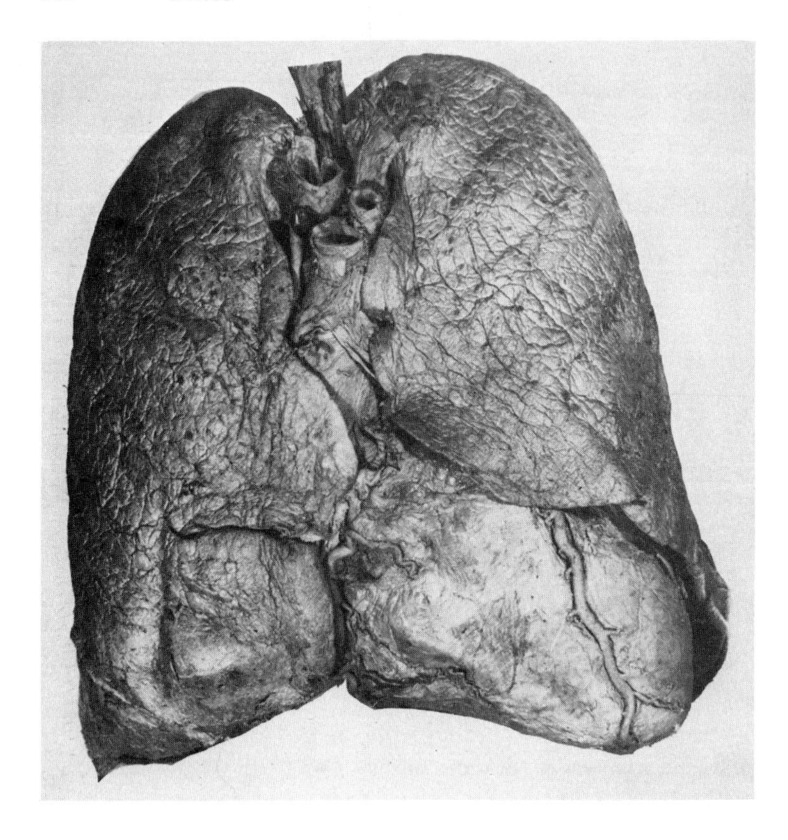

FIG. 4.27. THE HEART AND LUNGS REMOVED AFTER FIXATION *in situ*. THE RIGHT CORONARY AND LEFT INTERVENTRICULAR ARTERIES ARE DISSECTED. NOTE THE INCOMPLETE FISSURE BETWEEN RIGHT UPPER AND MIDDLE LOBES. (Photograph.)

left lung the boundary of the cardiac notch, extending down to the *lingula*, represents the middle lobe, fused always to the upper lobe (Fig. 4.31).

The oblique fissure of each lung makes a slippery surface between the two halves. This promotes easier and *more uniform* expansion of the whole lung. Movements of the chest wall and diaphragm are of greater range towards the base of the lung. The great elasticity of the lung encourages equal expansion throughout, but the apex is greatly helped to expand by the lower part of the upper lobe being so near the diaphragm and lower chest wall. Descent of segments 4 and 5 improves expansion of segments 1, 2 and 3 (Figs. 4.30, 4.31). Thus the *obliquity* of the great fissure is functional, for if the lung were divided into two halves by a horizontal fissure this would have less effect on apical expansion.

Internal Structure

Because the left lung grows into a smaller cavity some of its bronchi are delayed in their separation from each other.

Each lobe of the lung is made up of segments, and these are similar on the two sides. However, the pattern of division of the bronchial tree to aërate the segments is different on the two sides. That is to say, the asymmetrical bronchi ultimately reach, in the segments, a fairly symmetrical destination. One should distinguish clearly between the study of the plan of branching of the bronchial tree on the one hand, and the position of the segments on the other hand.

The **main bronchus** on each side supplies the three lobes of the lung. It extends from the carina at the bifurcation of the trachea down to the origin of the *middle* lobe bronchus. Beyond this level only the lower lobe bronchus continues on. The measurements of the cast in Fig. 4.28 are:

Right main bronchus: $1\frac{3}{4}$ in (4·4 cm) long.
Left main bronchus: $2\frac{1}{4}$ in (5·7 cm) long.

In the majority of casts, however, the two main bronchi are more nearly equal in length than this. Note that right and left main bronchi are also known as **principal bronchi.**

The Bronchial Tree. The right main bronchus gives off its three lobar bronchi separately; indeed the upper lobe bronchus leaves it outside the hilum of the lung (Fig. 4.28). The left main bronchus gives off three lobar bronchi, but the upper two are fused for a short distance before separating into the upper lobe bronchus and the lingular lobe bronchus (Fig. 4.29).

Segmental Bronchi. Each *lobar bronchus* in turn gives off branches, the *segmental bronchi*; each of

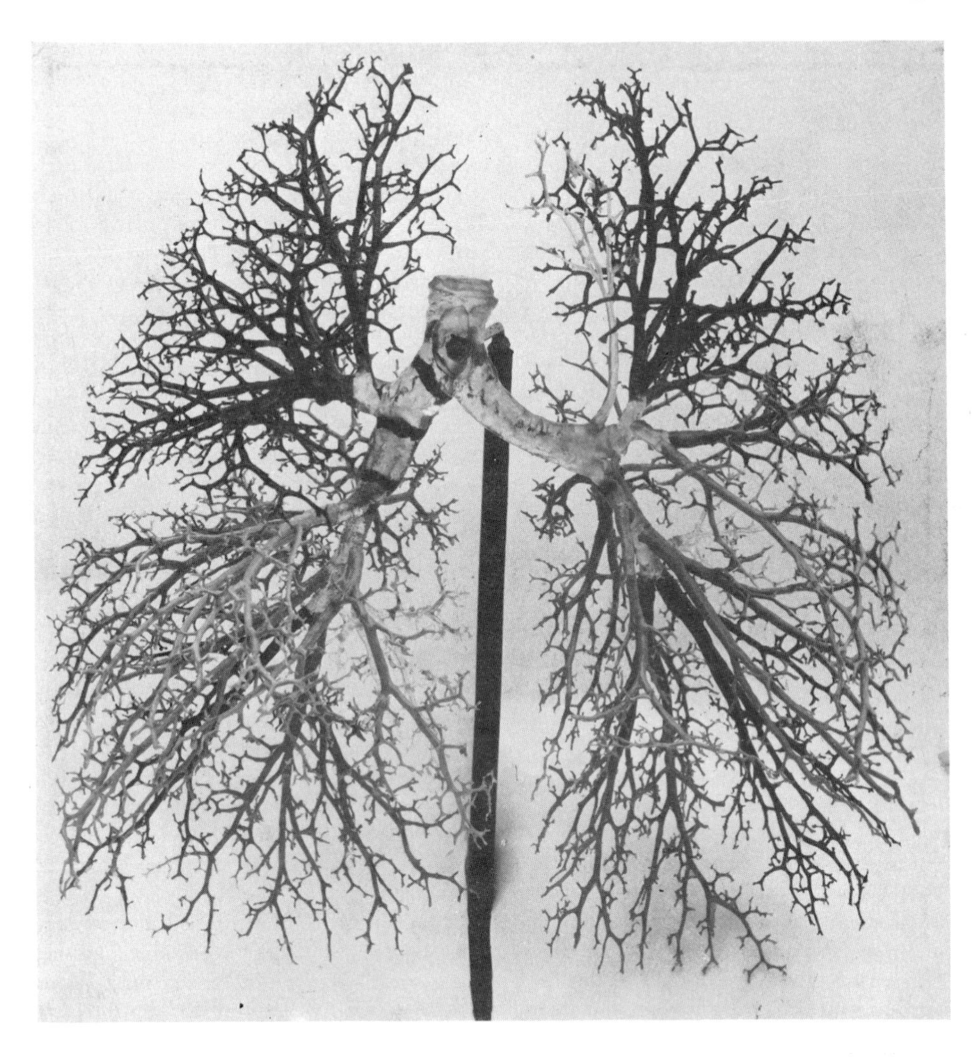

FIG. 4.28. PHOTOGRAPH OF A CAST OF THE BRONCHIAL TREE. THE BRONCHO-PULMONARY SEGMENTS ARE PAINTED IN DIFFERENT TONES OF GREY. NOTE THE ABNORMAL (UNPAINTED) BRANCH TO THE LEFT UPPER LOBE. DARK BANDS INDICATE THE DESCRIPTIVE DIVISIONS OF THE RIGHT MAIN BRONCHUS INTO UPPER AND LOWER SEGMENTS.

these supplies a segment of the lung (Fig. 4.29). These **broncho-pulmonary segments** are roughly pyramidal in shape, their apices towards the hilum, their bases lying on the surface of the lung (Figs. 4.30, 4.31). They are given the same names and numbers as the segmental bronchi. The segments differ on the two sides only in the different positions of the two segments of the middle lobe.

Each upper lobe bronchus gives rise to three segmental bronchi, the middle lobe bronchi each give off two segmental bronchi, and each lower lobe bronchus gives off five. These aërate the lung in two halves, five above and five below the oblique fissure. Ten in number, the **segmental bronchi** have been named *and numbered* as follows:

Right Lung	Left Lung
Upper lobe	**Upper lobe**
1. Apical	1.} Apico-posterior
2. Posterior	2.}
3. Anterior	3. Anterior
Middle lobe	**Lingular lobe**
4. Lateral	4. Superior
5. Medial	5. Inferior
Lower lobe	**Lower lobe**
6. Apical (superior)	6. Apical (superior)
7. Medial basal	7.} Medial basal
8. Anterior basal	8.} Anterior basal
9. Lateral basal	9. Lateral basal
10. Posterior basal	10. Posterior basal.

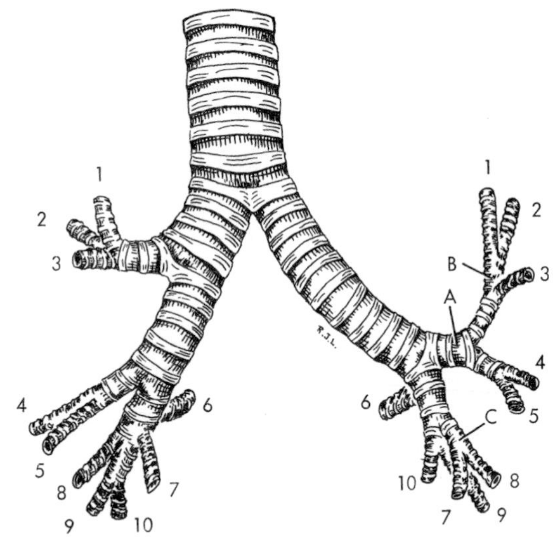

FIG. 4.29. THE BRANCHES OF THE BRONCHI. The two main (principal) bronchi give off lobar bronchi. There are three on each side; on the right they are separate, but on the left the upper lobe bronchus is fused with the middle (lingular) lobe bronchus (**A**) until they have room to separate. On the left there are two further examples of 'two bananas in one skin'. Segmental bronchi 1 and 2 come from a common stem (**B**, called apico-posterior). Segmental bronchi 7 and 8 do likewise (**C**) and for this reason 7 was formerly not recognized and was erroneously thought to be absent on the left side. Compare Fig. 4.12 for the course of the pulmonary arteries.

The Upper Lobe. On the **right** this is supplied by the upper lobe bronchus, which separates from the right principal bronchus outside the hilum of the lung. It divides into three (apical, posterior and anterior) for the three segments of the upper lobe. On the **left** the upper lobe bronchus branches from the left main bronchus inside the hilum of the lung and *in common with the bronchus to the lingula.* The two having separated, the upper lobe bronchus divides into two, a combined apico-posterior and an anterior. The apico-posterior bronchus soon separates into apical and posterior branches, one for each segment. Thus the left upper lobe has the same three segments as on the right side.

The Middle Lobe. On the **right** the bronchus leaves the front of the right main bronchus about an inch (3 cm) below the upper lobe bronchus. This is only slightly above the opening of the apical bronchus to the lower lobe. The bronchus divides to supply lateral (small) and medial (large) segments of the middle lobe. On the **left** the lingular lobe is joined with the lower division of the upper lobe. Its bronchus arises in common with the upper lobe bronchus and divides to supply superior and inferior segments. The cardiac excavation into the left lung has not only obliterated the fissure between the middle (i.e., the lingular) lobe and the upper lobe, but has rotated the lingula so that the lateral and medial segments of the right side have become superior and inferior here on the left.

The Lower Lobe. The apical segment of the lower lobe is supplied by a bronchus which is the *first posterior branch of the bronchial tree*—in the supine position inhaled liquids enter it. On each side the apical segmental bronchus comes off close below the middle lobe bronchus. It is shown in PLATE 42, p. *xxiii*. Note its projection on the chest wall (Figs. 4.30, 4.31). A stethoscope must be placed alongside the *base of the scapular spine* to listen to this segment (at the inferior angle of the scapula only segment number 10 is heard). The basal part of the lower lobe is supplied by four branches, which are medial, anterior, lateral and posterior. The medial basal (or cardiac) segment of the left lung is better understood nowadays as a result of wider experience in interpreting bronchial casts and bronchograms. It is always present but it is small because the pericardial excavation suppresses this part of the lung. Furthermore it comes off in common with the anterior basal segmental bronchus (Fig. 4.29C). This is the third example of 'two bananas in one skin', the others being shown in Fig. 4.29 (A and B).

Thus the patterns of right and left bronchial trees are mirror images, but the heart on the left side gives less room. So the left upper and middle lobe bronchi, and the left apical and posterior segmental bronchi, and the medial and anterior basal segmental bronchi are joined together for a while, before they find room to branch apart.

Blood Supply. The bronchial tree receives its own arterial supply by the *bronchial arteries*, direct branches of the aorta (usually one on the right and two on the left). They supply the bronchi from the carina to their ultimate ramifications (respiratory bronchioles) and nourish also the connective tissue and the visceral pleura. The *bronchial veins* drain into the azygos vein on the right and the accessory (superior) hemiazygos on the left. The *alveoli* contain within their walls a rich capillary plexus which is fed with deoxygenated blood by the *pulmonary artery*. The pulmonary artery divides with the bronchi; every bronchus is accompanied by a branch of the artery. The artery supplies no bronchus but it does supply the alveoli, giving them all they need except oxygen, of which they have more than enough. There is no anastomosis between the bronchial and pulmonary arteries, though peripherally there is some overspill of bronchial capillaries into the alveolar capillaries. The *pulmonary veins* are formed from tributaries which do not closely follow the bronchi but tend to run in the intersegmental septa. Two pulmonary veins leave each hilum, one from above and one from below the oblique fissure.

Lymph drainage is towards the hilum, from the pleura along the bronchi and pulmonary artery. Lym-

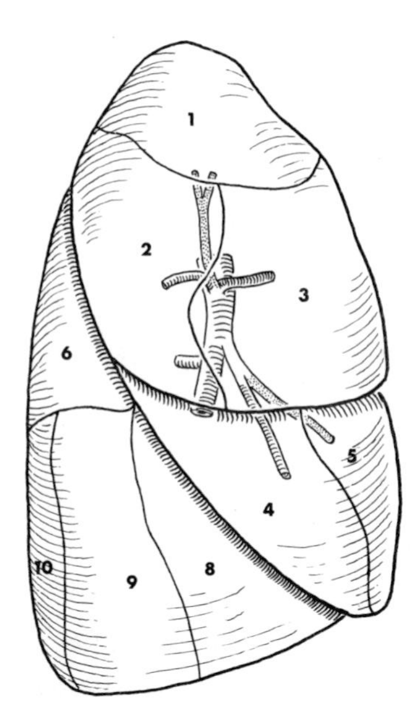

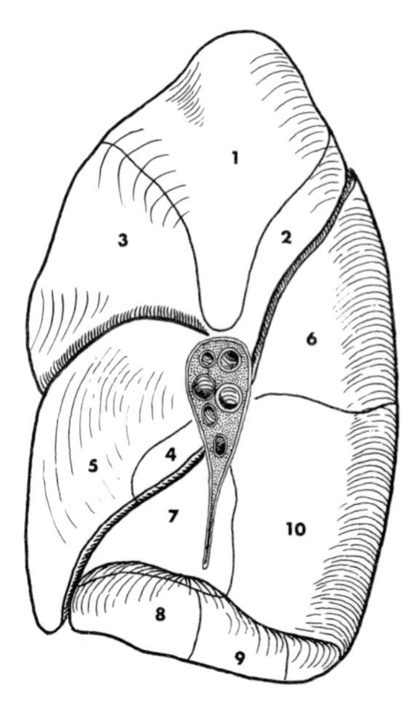

FIG. 4.30. THE LATERAL AND MEDIAL SURFACES OF THE RIGHT LUNG TO SHOW THE BRONCHO-PULMONARY SEGMENTS AND THE DISTRIBUTION OF THE UPPER AND MIDDLE LOBE BRONCHI AND THE POSTERIOR ORIGIN OF NO. 6 SEGMENTAL BRONCHUS.

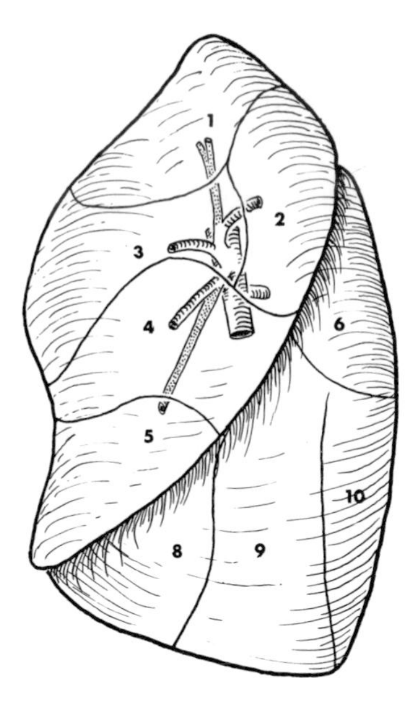

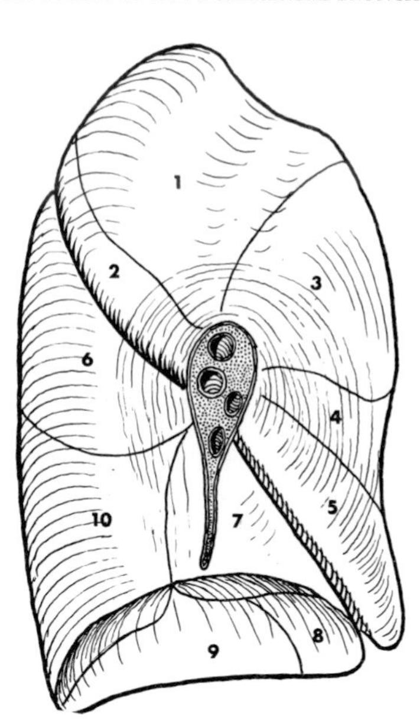

FIG. 4.31. THE LATERAL AND MEDIAL SURFACES OF THE LEFT LUNG TO SHOW THE BRONCHO-PULMONARY SEGMENTS AND THE DISTRIBUTION OF THE UPPER LOBE AND LINGULAR BRONCHI AND THE POSTERIOR ORIGIN OF NO. 6 SEGMENTAL BRONCHUS.

phatic tissue occurs as isolated follicles beneath the pleura and in the mucous membrane of the bronchi, usually straddling the bifurcations. Lymph nodes in the hilum lie just within and just without the lung substance. These *hilar nodes* drain to the tracheo-bronchial group, thence upwards by mediastinal lymph trunks to the brachio-cephalic veins.

Nerve Supply. The pulmonary plexus at the hilum of each lung sends its autonomic fibres along the bronchial tree. The parasympathetic fibres (vagus nerve) are afferent (cell bodies in the inferior ganglion) and efferent (cell bodies in dorsal nucleus, with relay in the bronchial mucosa). The vagal efferents are motor to the smooth muscle of the bronchi and the pulmonary arterioles, shutting both down for economy of effort. The sympathetic fibres relay in the upper four thoracic ganglia; their connector cells lie in Th. 2 to 6 segments of the cord. The sympathetic efferents are dilator to the pulmonary arterioles and to the bronchi, to accommodate the higher cardiac output and increased pulmonary ventilation that accompany 'storming the citadel' (p. 33).

Structure of the Lung. The lung itself is made up of air-conducting tubes (air ducts) and respiratory tissue. The segmental bronchi divide and redivide within the lung. Down to a calibre of 1 mm in the human lung they are lined with respiratory mucous membrane (with mucous and serous glands to clean and humidify the air) and surfaced by ciliated columnar epithelium (Fig. 6.20, p. 399). Beyond this level there are no glands in the mucous membrane and no cilia on the columnar epithelium. There are usually two generations of these *conducting bronchioles.* The terminal air-conducting bronchioles divide into *respiratory bronchioles,* so named because in places their walls are distended by respiratory alveoli and because their lining epithelium is a mixture of columnar air-conducting cells and flat, squamous respiratory cells. Each respiratory bronchiole divides into two or three generations. The terminal respiratory bronchioles divide into paired air sacs, usually four pairs to each bronchiole. Each air sac is long, narrow and has sinuous curves that intertwine with adjacent sacs in a three-dimensional tangle. The walls of the air sacs are studded with the bulgings of adjacent alveoli. Each alveolus, whether in the wall of a respiratory bronchiole or in a terminal air sac, is a distended thin-walled cavity in which gaseous interchange takes place. The wall of the alveolus is richly supplied with capillaries from the pulmonary (not the bronchial) artery.

Microscopic Anatomy. Whatever the mode of death of the subject, it is very difficult to secure a section of lung that is histologically normal *in all areas.* Areas of collapse often lie alongside areas of over-distension; in many cases an exudate may have coagulated within the alveoli. The section can be mistaken for a secreting gland, with its (air) ducts and alveoli. The key to recognition is that *every duct is accompanied by an artery* (pulmonary) of equal size, an arrangement that occurs in no gland in the body. Down to a calibre of 1 mm the ducts are lined with well-staining ciliated columnar epithelium, the walls contain occasional pieces of cartilage, and the mucous membrane has mucous and serous glands. The respiratory bronchioles are lined with plain columnar epithelium and their mucosa contains no glands. The alveoli are thin-walled and in outline polygonal from mutual pressure. Smooth muscle fibres are to be seen in the ducts and elastic fibres abound in ducts and alveoli. The epithelium of the normally distended alveoli is too flattened to be visible in section; in collapsed alveoli or in the fœtal lung it is of low columnar type.

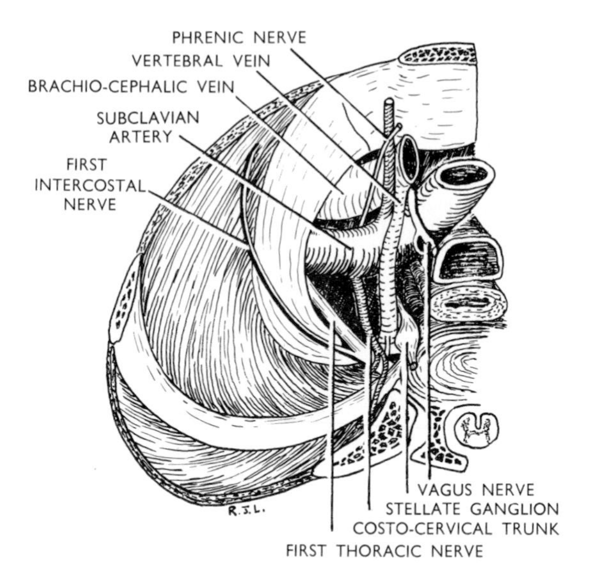

FIG. 4.32. THE RIGHT SIDE OF THE THORACIC INLET VIEWED FROM BELOW, SHOWING THE RELATIONS OF THE APEX OF THE RIGHT LUNG. Sketched from Specimen S240. 1 in R.C.S. Museum.

The **relations of the apices** are best studied with the root of the neck (p. 376). The relations of the right apex are illustrated in Fig. 4.32.

OSTEOLOGY OF THE THORACIC CAGE

The Articulated Thoracic Skeleton

Study the general features in an articulated skeleton (Fig. 4.26), remembering that cartilage cannot be preserved dry. Make allowances for 'costal cartilages' that may be distorted. The cage is convex, narrow above and below like a barrel, but flattened from front to back. In conformity with the general convexity of the thoracic wall the 1st rib has an upper surface and a sharp external border; its pleural surface looks downwards, instead of inwards like the others. The vertebral column projects into the cage, leaving a deep paravertebral gutter on either side. A cross section is shaped like a kidney, with the vertebral body making the hilum. The ribs slope down from the vertebral column towards the sternum at an angle of 45 degrees; *note this*, and remember it later when handling individual ribs.

The angles show as a line of ridges, one at the posterior convexity of each rib, lateral to the transverse processes of the thoracic vertebræ. A posterior angle scarcely exists on the 1st rib, where it coincides with the tubercle, but from here down the posterior angle lies progressively farther lateral. The ridge at the angle lies vertical across the obliquity of the rib (Fig. 4.34). It is made by the attachment of the thoraco-lumbar fascia that encloses erector spinæ; this muscle is widest behind the 12th rib and narrows as it passes upwards (Fig. 6.74, p. 467).

The term 'angle' always refers to the *posterior* angle of a rib. There is an anterior 'angle', but this is much less pronounced, and cannot be made out on every rib. It lies between 1 and 2 inches from the anterior extremity of the rib and consists of a low plateau, triangular in shape (apex up). It is for the origin of external oblique (hence absent from the upper four ribs) and is usually best seen on the 6th and 7th ribs (Fig. 4.33).

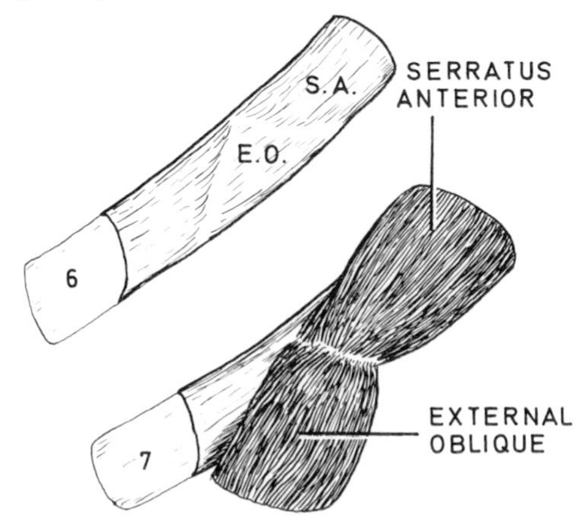

FIG. 4.33. THE ANTERIOR ANGLE OF RIB 6 AND THE ATTACHMENTS ON RIB 7. S.A. Serratus anterior. E.O. External oblique.

The anterior ends of the ribs fall progressively short of the midline from above down. The upper seven articulate with the sternum, which is to say that these seven costal cartilages become progressively longer from the 1st to the 7th. The upper three cartilages lie nearly horizontal, but below this the cartilages, themselves bent, pass progressively more obliquely upwards as they converge towards the lower end of the sternum (Fig. 4.26). The costal margin is made by the cartilages of the 7th, 8th, 9th and 10th ribs. Below this the 11th and 12th ribs, each capped by an apex of cartilage, float free in muscles.

While studying the articulated chest wall it is useful to study also the sternum, the better to understand its attachments.

The Sternum

Like the ribs the sternum is made of cancellous bone which throughout life is filled with hæmopoietic marrow. The thin layer of compact bone on the surface is scored and perforated by multiple minute vessels. The sternum is in two parts, manubrium and body, connected by a secondary cartilaginous joint that normally never ossifies. Movement at this joint is essential to movement of the ribs. The xiphoid process projects down from the lower end of the body.

The *manubrium* is a flat four-sided bone broader above than below (PLATE 11). Its upper margin is concave, the jugular notch, and this receives the interclavicular ligament. It is spanned by the investing layer of deep cervical fascia, attached to the anterior and posterior borders of the notch (p. 362). Each upper angle of the manubrium is scooped out into a concavity for the sterno-clavicular joint. The articular surface, covered in hyaline cartilage, lies in contact with the disc, while the capsule and synovial membrane are attached to the articular margin. Below the clavicular surface the lateral border is excavated for about 2 cm to receive the first costal cartilage (a primary cartilaginous joint). At the inferior angle is a 'demifacet', a small cartilage-covered area for articulation (synovial) with the upper part of the second costal cartilage. The inferior border of the manubrium is irregular and undulating; like the reciprocal upper border of the body it is coated with hyaline cartilage for the manubrio-sternal secondary cartilaginous joint. The anterior surface gives origin to pectoralis major alongside the lateral border, and more medially the tendon of sterno-mastoid is attached; neither of these marks the bone. Posteriorly sterno-hyoid arises from a strip of bone alongside the clavicular margin while below this the origin of sterno-thyroid curves alongside the border of the bone below the facet for the first costal cartilage.

Most of the posterior surface is bare. It lies against the lung margins in their pleural sacs, between which the left brachio-cephalic vein touches the bone unless thymic remnants lie between.

The *body* of the sternum (gladiolus) carries articular facets along its lateral border. The lower part of the second costal cartilage articulates with a demifacet at the upper angle, making a synovial joint separate from that with the manubrium. Below this the border of the sternum is indented by five more cartilage-covered facets. These make single synovial joints with the costal cartilages (Fig. 4.2, p. 213). The facets for the 6th and 7th costal cartilages may coalesce, especially in the female. Between the costal facets the lateral border of the sternum gives attachment to the external intercostal membrane and the internal intercostal muscle. The bone of this lateral border receives nutrient branches from the internal thoracic artery. Three faint ridges cross the anterior surface between the facets for the 3rd, 4th and 5th cartilages; they mark the lines of fusion of adjacent sternebræ (p. 213). Pectoralis major arises widely from the anterior surface almost to the midline, and sterno-costalis arises from the posterior surface low down. From the posterior surface the weak sterno-pericardial ligaments pass, above and below, into the fibrous pericardium. From the posterior margin of the gladiolus the *xiphisternum (xiphoid cartilage)* projects downwards for attachment of the linea alba. It represents the unossified lower end of the body, with which it forms a primary cartilaginous joint. The xiphisternum may ossify late in life.

The ossification of the sternum is dealt with on p. 213, and its movements have already been studied (p. 218).

Returning now to the thoracic cage note the multiple attachments of large sheets of muscle. Lateral to the sternum, pectoralis major has slips of origin from the upper six costal cartilages. Serratus anterior is attached to the upper eight ribs; the first rib has only a thin external border available, so the first digitation spans the intercostal space along a fibrous band and is attached, with the second digitation, to a rough plateau which characterizes the second rib. The next digitations arise behind pectoralis minor, and the last four interdigitate with external oblique. So the eight digitations at the border of serratus anterior form a bold forward convexity. Pectoralis minor is attached to the 3rd, 4th, and 5th ribs (often also to the 2nd) just in front of serratus anterior. External oblique arises from the anterior angles of the lower eight ribs; below serratus anterior it interdigitates with latissimus dorsi on the last four ribs. Rectus abdominis is attached to the anterior surfaces of the 5th, 6th and 7th costal cartilages and the lower border of the 7th. Lateral to this the lower borders of the last six costal cartilages receive the internal oblique. Transversus abdominis and the diaphragm interdigitate on the internal surfaces of the lower six costal cartilages. The serratus posterior muscles, lying on the thoraco-lumbar fascia, are inserted just lateral to the ridges at the posterior angles of their respective ribs—the 2nd, 3rd, 4th and 5th ribs for the superior and the last four ribs for the inferior muscle. Between the posterior angles and the transverse processes (i.e., the tubercles of the ribs) the ilio-costalis and longissimus parts of erector spinæ are attached, and a levator costæ is inserted into the upper border of each rib just lateral to its tubercle. The scalene muscles and quadratus lumborum are attached to the thoracic cage, above and below; they belong with the intercostal muscles to the intermediate layer of the body wall (p. 211). Their attachments will be noted in the study of the individual ribs. On the internal surfaces of the ribs the subcostalis sheet lies in the paravertebral gutter, and the intercostales intimi lie in the lateral curve of the thoracic wall; each sheet widens from above down, but neither leaves any mark on the ribs. Between these sheets of muscle parietal pleura clothes the ribs and costal cartilages, except at the cardiac notch alongside the sternum behind the 5th and 6th left costal cartilages. Sterno-costalis is attached behind the medial ends of the 2nd to 6th costal cartilages.

Clinical considerations. In the living the angle of Louis is visible and palpable. The *second* costal cartilage can be easily felt at either side of the angle; the ribs must be counted down from the angle as the only certain guide to their numerical identification. The downward inclination of the ribs is indicated by the vertebral levels of the sternum. The sternal notch lies at the upper border of the 3rd, the angle of Louis at the lower border of the 4th, and the lower end of the sternum at the lower border of the 8th thoracic vertebræ. The surface markings of the heart (p. 230), pleura and lungs (p. 242) have already been studied. Movements of the ribs are considered on p. 212.

Ossification of ribs (see also p. 213). Cartilage is laid down in mesoderm at the 6th week and a bony centre appears at the angle at the 8th week of fœtal life. The head and tubercle remain cartilaginous, and secondary centres appear at both places at about 15 years and fuse late (say 25 years).

Individual Ribs

The 1st, 11th and 12th ribs are atypical. A typical rib from the middle of the series has a head, a neck and a shaft. A tubercle projects posteriorly from the end of the neck. The shaft slopes down and laterally to an angle and then curves forward. The upper border of the shaft is blunt. Lateral to the angle the lower border projects down as a sharp ridge sheltering a subcostal groove. These features identify right from left ribs. They will be studied soon, but first lay the separate ribs in order on a table. The upper six lie reasonably flat, but the 7th, 8th and 9th show a degree of torsion at their angles that lifts the posterior end of each in a slope of 30 degrees from the table. The 10th, 11th and 12th lie flat. The upper six ribs are bent into a tight curve, so that the end of the shaft has turned parallel with the neck of the rib (as in Fig. 4.5, p. 219) to reach the front of the chest at a relatively short costal cartilage. The lower six ribs show an opening out of the curve, which in their case is completed by the long costal cartilages at the front of the chest.

The necks of the ribs are of equal length (except the 11th and 12th), but the distance from tubercle to angle increases from above down as already noted in the articulated thoracic cage. This distance becomes equal to the neck length at the 5th and 6th rib. The tubercles of the typical ribs carry two oval facets. The medial one (for a synovial joint) of the upper six ribs is convex and projected backwards but in the next four ribs is flat and faces backwards and slightly downwards.

From all the above features it is easy to identify a rib of the upper, middle or lower part of the series.

A Typical Rib (Fig. 4.34). The rib lies on the table in an unnatural position. *Pick it up* and hold it in its anatomical position; the shaft slopes down at 45 degrees and the ridge at the angle is almost vertical. *The angle is the most posterior part of the rib. Study the rib in this position.*

The *head* is bevelled by two articular facets that slope away from a dividing ridge. The lower facet is vertical; it articulates with the upper border of its own vertebra. The upper facet faces up to an overhanging facet on the lower border of the vertebra above. Each makes a synovial joint, and the cavities are separated by a ligament attached to the ridge on the head. The capsule and synovial membrane of each separate joint are attached to the articular margin. The head projects

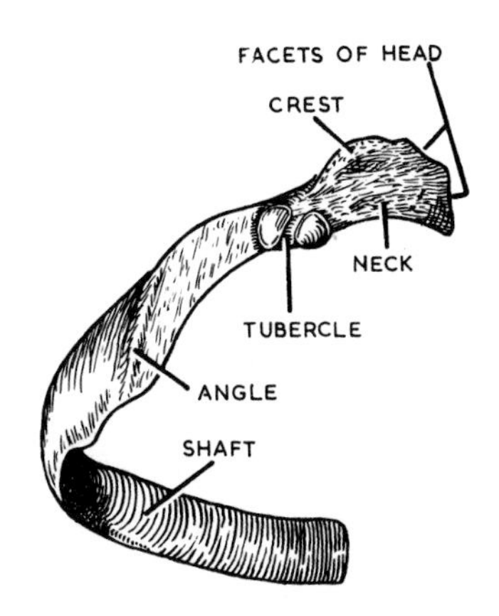

FIG. 4.34. A TYPICAL RIB (LEFT SIDE, POSTERIOR VIEW).

forward a little as it expands from the neck. The triradiate ligament is attached here, to the front of the head, and the sympathetic trunk lies in contact (PLATE 45, p. *xxiv*), plastered in position by the adherent parietal pleura. Ribs *atypical as to the head* are the 1st, 11th and 12th—each makes a single synovial joint, with only its own vertebra.

The *neck* is flattened, with the upper border curving up into a thin, prominent ridge known as the crest (*crista colli costæ*). The two laminæ of the superior costo-transverse ligament are attached to the crista colli costæ, whence they pass up to the transverse process of the vertebra above (Fig. 4.1, p. 212). The inferior costo-transverse ligament is attached between the back of the neck and the transverse process of its own vertebra. The front of the neck is clad in pleura. The neck of the rib inclines backwards when traced from head to tubercle.

The *tubercle* shows two smooth facets (Fig. 4.34). In the anatomical position they lie medial and lateral. The medial facet is covered with hyaline cartilage and makes a synovial joint with the transverse process of its own vertebra. Its shape and direction vary in upper and lower levels as already noted. The joint capsule and synovial membrane are attached to the articular margin. The lateral facet, smooth-surfaced, receives the lateral costo-transverse ligament from the tip of its own transverse process. Ribs *atypical as to the tubercle* are the 11th and 12th (sometimes the 10th); no synovial joint exists here, only the ligament, and usually there is no tubercle on the rib.

The *shaft* slopes down and back to the angle and there twists forward in its characteristic curvature. The upper border is blunt. It gives attachment in its whole length to the external intercostal muscle from the rib above. Deep to this the internal intercostal muscle arises from the upper border between the anterior extremity and the angle, while between angle and tubercle the internal intercostal membrane is attached. Deep to this layer the subcostales and intercostales intimi sweep across the internal surface of the shaft, some of the fibres attaching themselves to the upper border. Elsewhere the internal surface of the shaft is in contact with parietal pleura. The lower border of the shaft is sharp, especially lateral to the angle, where it hangs down to produce a well-marked subcostal groove. Further forward the subcostal groove dies away, and the lower border of the rib becomes blunt like the upper border. The external intercostal muscle arises from the sharp lower border and the internal intercostal is attached in front of the angle to the subcostal groove; anteriorly they are together attached to the blunt lower border. Between angle and tubercle the internal intercostal muscle is replaced by the internal intercostal membrane. The neuro-vascular bundle lies just below its own rib, deep to the internal intercostal membrane and muscle, high up in the space under shelter of the projecting sharp lower border of the rib. Between the anterior and posterior angles the external surface of the shaft is bare, and serratus anterior plays over it on the upper eight ribs. The anterior end of the rib is excavated into a concave fossa; cancellous bone comes to the surface here. The fossa is plugged by the costal cartilage in an immovable primary cartilaginous joint.

The Costal Cartilages

These form primary cartilaginous joints at the extremities of all twelve ribs. The *first* is short and thick, and forms also a primary cartilaginous joint with the manubrium. It articulates with the clavicle and gives attachment to the sterno-clavicular disc and joint capsule and, lateral to this, the costo-clavicular (rhomboid) ligament. The suprapleural membrane is attached to its upper border.

Below this the cartilages increase in length down to the 7th, which is the longest. From the 5th to the 10th they are bent from a downward slope in line with the rib to an upward slope towards the sternum (PLATE 45, p. *xxiv*). Their medial ends form synovial joints (p. 213). On the 11th and 12th ribs there is a pointed tip of cartilage. All cartilages, especially the first, tend to calcify and even ossify in patches after middle age. The costal cartilages give attachment to the external intercostal membranes and the internal intercostal muscles of their spaces, and the lower six give origin to the diaphragm and transversus abdominis from their inner surfaces. The attachments of sterno-costalis behind the 2nd to 6th, and of pectoralis major in front of the upper six costal cartilages have already been noted.

Particular Ribs

The Twelfth Rib. The gently curved 12th rib varies a good deal in length. In conformity with the barrel-shaped contour of the thoracic cage the concavity of the shaft looks up a little instead of directly inwards like the others. The head has a single facet for the 12th thoracic vertebra; the capsule and synovial membrane are attached to the articular margins. A short constriction forms a neck which passes imperceptibly into the shaft. There is usually no tubercle, and since the 12th transverse process is very short the ligament from its tip is attached nearer the head than in a typical rib (the 11th rib is similar in this respect). A wide area of ridges on the convex surface of the shaft shows the attachment of sacro-spinalis; lateral to this external oblique and latissimus dorsi arise and diverge from each other. *There is no subcostal groove.* The thin upper part of quadratus lumborum is attached to the medial part of the lower border; its markings may be difficult to identify. It does not extend as far out as sacro-spinalis. The upper ends of the anterior and middle lamellæ of the lumbar fascia are attached to the rib around the muscle. At the lateral limit of quadratus (and one may have to guess where this is) the lateral arcuate ligament is attached. From here to the tip of the costal cartilage the diaphragm and transversus abdominis arise from the inner surface. From the diaphragm origin to the rib head the inner surface is clad by pleura which dips below the shaft in the costo-diaphragmatic recess (p. 242). Lateral to quadratus the internal oblique is attached to the lower border. The external and internal intercostal muscles are attached along the upper border; there is commonly no posterior intercostal membrane in the 11th intercostal space.

The **eleventh rib** resembles the 12th in having a single facet for the head and a short neck with no tubercle, but its shaft is longer and thicker than the 12th, and it shows a well-marked subcostal groove.

Movements. Unlike the upper ten ribs, which must all move up or down together, the 12th and the 11th are free to move independently. When maximum effort is required by the diaphragm these two ribs can be depressed to draw down that part of the diaphragm which is attached to them and to the lateral arcuate ligament.

Lumbar ribs and *cervical ribs* are mentioned on p. 459.

The First Rib. The borders and pleural surface of this rib form part of the study of the thoracic cage, but its upper surface can be understood only when the anatomy of the root of the neck is known. The shaft is curved on the flat and, in conformity with the general convexity of the thoracic cage, the neck slopes down a little towards the head from the plane of the shaft. This can be *seen*, but if in doubt lay the rib on the table; head and anterior extremity both touch the

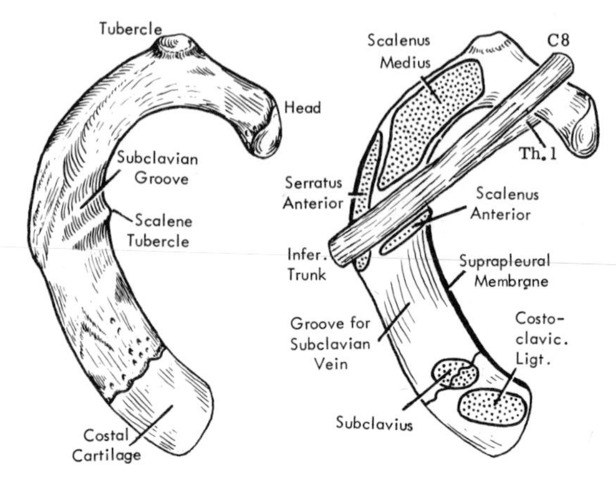

FIG. 4.35. THE FIRST RIB (RIGHT SIDE, VIEWED FROM ABOVE).

surface. Laid upside down the neck slopes up to the head, which lies above the table. Thus can right and left sides be identified. Lift the rib from the table and study it *in its anatomical position*. The plane of the shaft is at 45 degrees, and the tubercle is the most posterior and the highest part of the rib (Fig. 4.35).

The *head* is small and carries a single facet for the synovial joint it makes with the upper part of the body of the first thoracic vertebra. Capsule and synovial membrane are attached to the articular margin. A radiate ligament attached to the head reinforces the anterior part of the capsule. A slender *neck* slopes backwards *and upwards* to join the shaft. The anterior primary rami of C8 (above) and Th. 1 (below) lie in contact with the medial part of the neck. The sympathetic trunk (or stellate ganglion) lies in contact with the anterior border of the neck alongside the head. Lateral to it the supreme intercostal vein and then the superior intercostal artery lie in contact, and more lateral still the first thoracic nerve lies in front of the neck and inner border of the shaft (PLATE 11, p. *v*). The cervical dome of the pleura and the apex of the lung hold these vessels and nerves against the front of the neck (Fig. 4.32).

The rib broadens at the junction of neck and shaft, and here a prominent *tubercle* projects back to form the most posterior convexity of the rib. It is a fusion of tubercle and angle. Medially it has a cylindrical facet for a corresponding concavity on the first transverse process. The usual synovial joint exists here, with capsule and synovial membrane attached to the articular margins. The lateral prominent part of the tubercle receives the lateral costo-transverse ligament and the costalis and longissimus parts of erector spinæ (Fig. 6.73, p. 466).

The under surface of the *shaft*, crossed obliquely by the small first intercostal nerve and vessels (Fig. 4.32) is covered by adherent parietal pleura. The external and internal intercostal muscles are attached together to the outer rim of this surface. Some undulations in the shaft often produce broad grooves that can be mistaken for the subclavian grooves on the upper surface; if the surfaces are reversed the rib will be allotted to the opposite side. This mistake can be avoided by remembering the slope of the neck already studied. The outer border is blunt between the tubercle and subclavian groove. The anterior end of this blunt part is at the most lateral part of the convexity of the rib; this gives origin to part of the first digitation of serratus anterior. The posterior end of the blunt border is bare bone, and here scalenus posterior slides across the rib deep to serratus anterior. Forward of the subclavian groove the sharp external border is encroached on by the external intercostal muscle. The concave internal border of the shaft gives attachment to the suprapleural membrane in front of the subclavian groove (PLATE 11, p. *v*).

The *upper surface* of the shaft, sloping down at 45 degrees, is at the root of the neck. It is grooved obliquely at its greatest lateral convexity (PLATE 11). The groove is called the subclavian groove. It lodges the lower trunk of the brachial plexus. The fibres in contact with the rib are all Th. 1, and the C8 fibres lie above them, not yet intermingled. Note the direction of the groove; prolonged backwards it leads up to the neck of the rib, along the line of Th. 1. The subclavian artery has its upward convexity lying more transversely. The artery does not lie in the groove, and it touches only the outer border of the rib. Between the groove and the tubercle the large quadrangular area of the upper surface gives attachment to scalenus medius. At the front of the groove the inner border

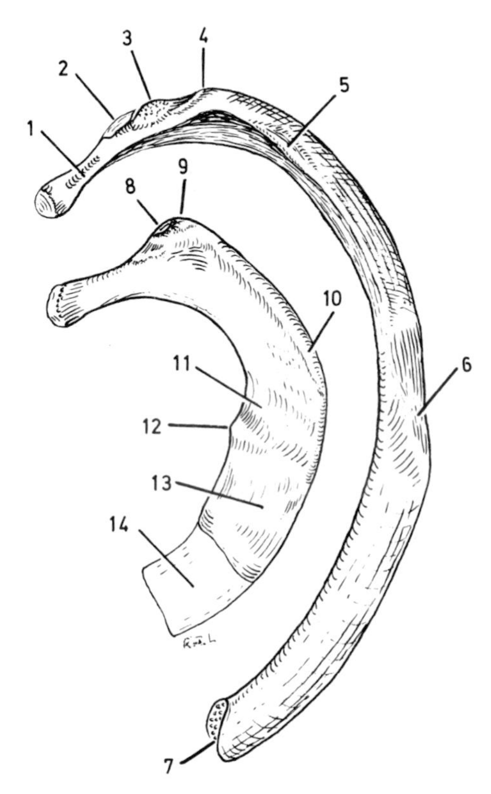

FIG. 4.36. THE FIRST AND SECOND RIBS (LEFT SIDE, SUPERIOR SURFACE).

SECOND RIB. 1. Crista colli costæ. 2. Articular facet of tubercle. 3. Ligamentous facet of tubercle. 4. Angle. 5. Ridge for scalenus posterior. 6. Serratus anterior ($1\frac{1}{2}$ digitations). 7. Fossa for costal cartilage.

FIRST RIB. 8. Articular facet. 9. Ligamentous facet (at one with the 'angle'). 10. Serratus anterior ($\frac{1}{2}$ digitation). 11. 'Subclavian' groove. 12. Scalene tubercle. 13. Groove for subclavian vein. 14. First costal cartilage.

is projected into a spur called the scalene tubercle. The ribbon-like tendon of scalenus anterior is attached to the tubercle and extends across the upper surface halfway to the external border. In front of the subclavian groove, down the slope of the rib, is a second groove in which the subclavian vein lies in contact with the bone. The anterior end of the shaft expands into a concavity for the first costal cartilage. From the upper surface of this junction subclavius arises from both bone and cartilage.

The Second Rib. (Fig. 4.36). This has a more slender shaft, and it makes a much larger curvature than the first. It lies with its head slightly above the table. In the anatomical position its concave surface faces inwards as well as downwards. It possesses an angle about half an inch (1·5 cm) from its tubercle. The angle, not the tubercle, is the most posterior part of its curvature. The most lateral part of the curve shows a characteristic 'bump'—a rough tuberosity on the outer surface of the shaft, which in some bones resembles the callus of an old fracture. This is for the second and part of the first digitation of serratus anterior. Further back, an inch or more (3 cm) lateral to the angle, the upper border is drawn up into a smaller 'bump' for the attachment of scalenus posterior. The other features of the 2nd rib are as those of a typical rib; the subcostal groove is present but poorly developed.

Section 5. The Abdomen

THE ANTERIOR ABDOMINAL WALL

The skin and subcutaneous tissues of the anterior abdominal wall have been dealt with as part of the body wall (p. 210).

The Muscles of the Anterior Abdominal Wall

The three muscle layers of the body wall (p. 211) are separate in the flanks, where they are known as the external oblique, internal oblique and transversus abdominis muscles. The layers fuse together ventrally to form the rectus abdominis muscle.

External Oblique Muscle. The muscle arises by eight digitations, one from each of the lower eight ribs at their anterior angles. The anterior angles of the ribs lie just lateral to their anterior extremities and are usually not very well marked (Fig. 4.33). The lower four slips interdigitate with the costal fibres of latissimus dorsi, the upper four with a corresponding number of the digitations of serratus anterior (PLATE 1). From its fleshy origin the muscle fans out to a very wide insertion, much of which is aponeurotic. The muscle has three borders lying free, a posterior muscular and a superior and inferior aponeurotic.

Starting from behind at the 12th rib the muscle may be traced to its insertion as fleshy fibres into the anterior half of the outer lip of the iliac crest. At the anterior superior iliac spine the muscular fibres give way to an aponeurosis; indeed no flesh is to be found below a line joining this point to the umbilicus (Fig. 5.5). The limit of the fleshy fibres is visible in an athlete as a graceful curve. The aponeurotic fibres, directed obliquely downwards and forwards, are attached to the pubic tubercle and, interdigitating with each other above the symphysis pubis, cross the front of the rectus abdominis to the whole length of the linea alba up to the xiphisternum. The free upper border of this aponeurosis, extending from the 5th rib to the xiphisternum, runs horizontally. It is the only structure in the anterior sheath of the rectus muscle above the costal margin. From it arise the lowermost fibres of pectoralis major. The free lower border, lying between the anterior superior iliac spine and the pubic tubercle, forms the inguinal ligament (Fig. 5.5).

The **inguinal ligament** (Poupart's ligament) extends from the anterior superior iliac spine to the pubic tubercle. Its edge is rolled inwards to form a gutter; the lateral part of this gutter gives origin to part of the internal oblique and transversus abdominis muscles. To the inguinal ligament is attached the fascia lata of the thigh. When the thigh is extended the fascia lata pulls the inguinal ligament downwards into a gentle convexity.

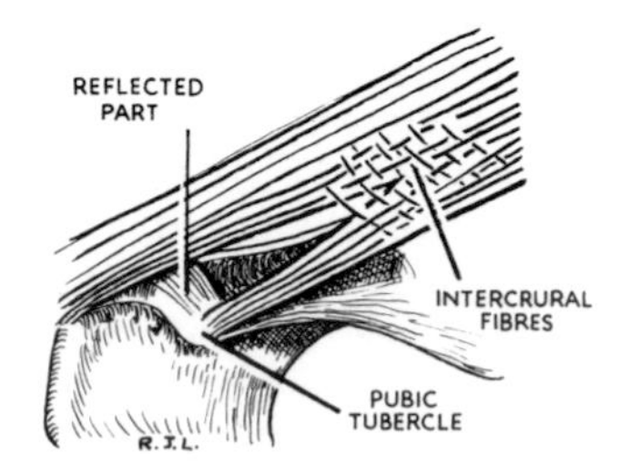

FIG. 5.1. THE SUPERFICIAL INGUINAL RING (LEFT SIDE).

Just above and lateral to the pubic tubercle is an oblique V-shaped gap, the superficial inguinal ring, in the aponeurosis (Fig. 5.1). This gap extends down to the pubic crest, medial to the tubercle; the aponeurosis is attached to the pubic crest only in its medial part, alongside the symphysis pubis. From the medial end of the inguinal (Poupart's) ligament the **lacunar ligament** (Gimbernat's ligament) extends backwards and *upwards* to the pectineal line. Its crescentic free edge is

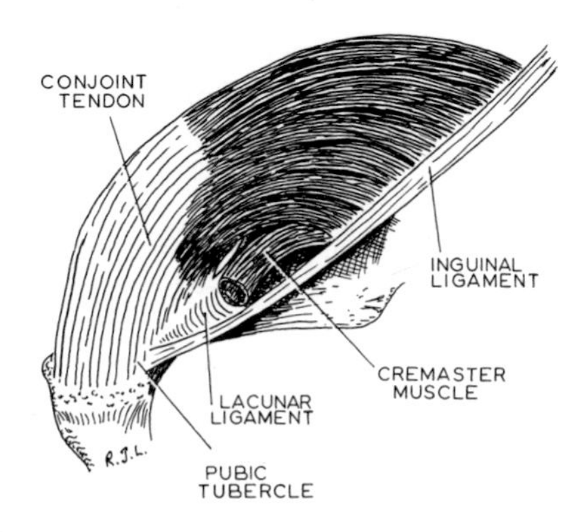

FIG. 5.2. THE INSERTION OF THE INTERNAL OBLIQUE MUSCLE INTO THE CONJOINT TENDON. Note that the nearly vertical conjoint tendon lies at right angles with the nearly horizontal lacunar ligament.

the medial margin of the femoral ring. From the pubic tubercle, fibres may be traced upwards and medially, behind the spermatic cord, to interdigitate in the linea alba with those of the opposite side. This is called the **'reflected'** part of the ligament; it is not important surgically, but it is, in fact, a reinforcement of the attachment of the aponeurosis of the *opposite* side. The French rightly call it the 'posterior crus'. Lastly, near the apex of the superficial inguinal ring are fibres running at right angles to those of the aponeurosis, the **intercrural fibres,** that blend and prevent the crura from separating (Fig. 5.7).

The Lumbar Triangle. Behind the free posterior border of the external oblique muscle is a triangle floored in by the internal oblique muscle and bounded by the anterior border of the latissimus dorsi and the iliac crest, the *lumbar triangle of Petit*, the site of the rare lumbar hernia (Fig. 2.5, p. 56).

Internal Oblique Muscle. The fleshy fibres arise from the whole length of the conjoined lamellæ of the lumbar fascia, from the intermediate area of the anterior two-thirds of the iliac crest and from the lateral *two-thirds* of the inguinal ligament (the rolled-in lower border of the external oblique aponeurosis). From the lumbar fascia the muscle fibres run upwards along the costal margin, to which they are attached, becoming aponeurotic at the tip of the 9th costal cartilage. Below the costal margin, the aponeurosis splits around the rectus muscle, the two lamellæ rejoining at the linea alba. The tendinous intersections of the rectus muscle are adherent to the anterior lamella. At a point an inch below the umbilicus the posterior lamella ends in a curved free margin, concave downwards, the semicircular fold (of Douglas). Below this point, the aponeurosis passes wholly in front of the rectus muscle, to the linea alba (Fig. 5.5).

The muscle fibres that arise from the inguinal ligament are continued into an aponeurosis that is attached to the crest of the pubic bone and, more laterally, to the pectineal line. The internal oblique has thus a free lower border, which arches over the spermatic cord—laterally the margin consists of muscle fibres in front of the cord; medially the margin consists of tendinous fibres behind the cord. The flat tendon, attached to the pectineal line, is fused with a similar arrangement of the transversus aponeurosis to form the conjoint tendon (Fig. 5.2).

Transversus Abdominis Muscle. The muscle has a very long origin, in continuity from the whole costal margin, lumbar fascia, iliac crest and inguinal ligament. From the costal margin a fleshy slip arises *inside* each costal cartilage, interdigitating with the costal origin of the diaphragm; in continuity with the lowest costal fibres the muscle arises from the conjoined lamellæ of the lumbar fascia lateral to the quadratus

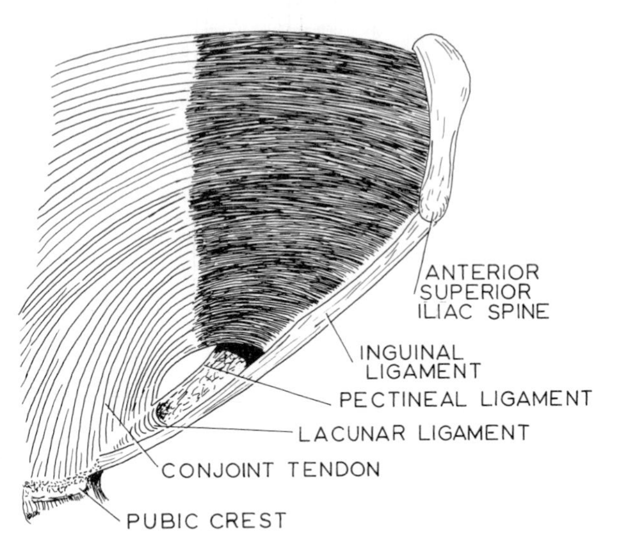

FIG. 5.3. THE LOWER EXTENT OF TRANSVERSUS ABDOMINIS.

lumborum, then from the internal lip of the iliac crest in front of this (the anterior two-thirds of the crest) *from the fascia over iliacus* and from the lateral *half* of the inguinal ligament, deep to the internal oblique. The muscle fibres become aponeurotic and pass behind the rectus to fuse with the internal oblique aponeurosis into the linea alba. Below the semicircular fold of Douglas (see under internal oblique muscle) the aponeurosis passes wholly in front of the rectus muscle, behind the aponeurosis of the internal oblique, which it accompanies laterally as the conjoint tendon on the pubic crest and along the pectineal line behind the spermatic cord (Fig. 5.3).

The Neuro-vascular Plane. The abdominal wall is supplied segmentally by all the lower six thoracic nerves and the first lumbar nerve. The intercostal nerves pass behind the costal margin, between the interdigitations of the diaphragm and transversus abdominis muscle and pass around to pierce the posterior rectus sheath (Fig. 1.24, p. 22). Each nerve is accompanied by its collateral branch in the neuro-vascular plane (Fig. 5.5, left side of specimen), but only the main nerve itself gives off a lateral cutaneous branch (Fig. 5.5, right side of specimen).

The external oblique is muscular in only its upper part and is supplied segmentally by the 7th to 11th intercostal nerves. The other two muscles are supplied by all six lower thoracic nerves (7th to 12th inclusive) and also by L1 via both ilio-hypogastric and ilio-inguinal nerves, the last-named supplying the lower border of each muscle where it arches over the spermatic cord into the conjoint tendon. The rectus abdominis receives no L1 fibres, being supplied segmentally by the lowest six thoracic nerves (7th to 12th inclusive). Branches of the musculo-phrenic artery and, lower down, the lumbar arteries, accompany the nerves in the flanks. The nerves and vessels run in the plane between transversus and internal oblique muscles. This is the neuro-vascular plane, which is continuous with that of the thoracic wall. The vessels supply only the flank muscles—the rectus abdominis has its own vascular arrangement, the epigastric arteries (p. 258).

The **lymphatic drainage** of the abdominal wall is, superficially, in quadrants; to the pectoral group of axillary nodes above the umbilicus, to the medial and lateral groups of superficial inguinal nodes below that level. The deeper parts of the wall drain into vessels in the extraperitoneal (transversalis) fascia; above the umbilicus these pierce the diaphragm to reach mediastinal nodes, below the umbilicus they run to the external iliac and para-aortic nodes. There are, rather surprisingly, no lymph nodes in the abdominal wall to correspond with the intercostal nodes of the thoracic wall.

Rectus Abdominis Muscle. The muscle arises by two heads, a medial from in front of the symphysis pubis and a lateral from the upper border of the pubic crest by a relatively small tendon; but the belly rapidly thickens. The two muscles lie edge to edge in the lower part, but broaden out above, and are there separated

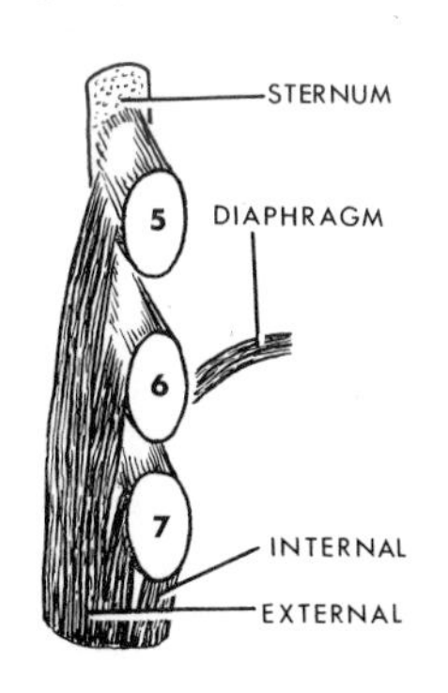

FIG. 5.4. THE UPPER ATTACHMENTS OF THE RECTUS ABDOMINIS, SEEN FROM THE LEFT SIDE, TO SHOW THE THREE MORPHOLOGICAL LAYERS OF THE MUSCLE. 'External' indicates the external oblique layer passing over the fifth, sixth and seventh costal cartilages, 'internal' is the intermediate layer attached to the costal margin, whilst the xiphisternal fibres of the diaphragm represent the innermost layer.

from each other by the linea alba. They are inserted into the thoracic cage. The bulk of the muscle passes in front of the costal cartilages and is attached to the 7th, 6th and 5th cartilages. This represents the external oblique layer. Some of the fibres are inserted into the costal margin (i.e., the lower border of the 7th costal cartilage) representing the internal oblique layer. The transversus layer is represented by the xiphisternal fibres of the diaphragm (Fig. 5.4). Typically three *tendinous intersections* are found in the muscle, one at the umbilicus, one at the xiphisternum, and one between these two. The muscle is formed by fusion of mesodermal somites as indicated by its regular segmental innervation. The tendinous intersections represent lines of fusion of myotomes; this explains the fact that such intersections are sometimes found below the umbilicus. The tendinous intersections interchange fibres to blend inseparably with the anterior rectus sheath. They occupy only the superficial part of the rectus and do not penetrate to the posterior surface of the muscle. There is no connexion posteriorly between the muscle and its sheath (Fig. 5.5). The contracting rectus abdominis can be seen as bulgings between the tendinous intersections in an individual who is not too fat.

Rectus Sheath (Fig. 5.6). The aponeurosis of the internal oblique splits to enclose the rectus muscle. Thus the external oblique aponeurosis is directed in front and the transversus aponeurosis behind the muscle. They fuse to form its sheath. An inch below the umbilicus this arrangement ceases, and from this level downwards all three aponeuroses pass in front of the muscle. There is thus a free lower margin to the posterior sheath; it is concave and named the **semicircular fold of Douglas.** Below the umbilicus the aponeuroses of internal oblique and transversus fuse completely in the sheath, but that of external oblique fuses only over the medial part of the sheath and lies free over the linea semilunaris. Above the costal margin there is no posterior sheath, the rectus adhering to the underlying costal cartilages, and in this area the anterior sheath consists of the external oblique aponeurosis only. Between the two recti all three aponeuroses fuse into the linea alba, which extends from symphysis pubis to xiphisternum.

The splitting of the internal oblique aponeurosis along the lateral border of the rectus muscle forms a relatively bloodless line known as the **linea semilunaris.** It curves up from the pubic tubercle to the costal margin at the tip of the 9th costal cartilage in the transpyloric plane (Fig. 5.5).

The sheath braces back the rectus muscle when the whole abdominal wall contracts. There is part of the sheath behind the muscle above the umbilicus to give a firm attachment for the parietal peritoneum, undisturbed by contraction of the rectus muscle (cf., psoas and iliacus fasciæ). It is significant that the muscle fibres are not attached to the posterior sheath, over which they slide freely during contraction. Below, where the peritoneum is more loosely attached above the bladder, the sheath passes wholly in front of the rectus, the better to brace and support the muscle.

The **linea alba** lies in the midline of the anterior abdominal wall. It is a strong fibrous structure formed by the fusion of the aponeuroses of all three muscles of the abdominal wall. Above the symphysis pubis it is very narrow, for here the two recti are in contact with each other behind it. From just below the umbilicus to the xiphisternum it broadens out between the recti. Here the fibres form a felted membrane that is very tough (Fig. 5.5). The linea alba is strongly attached below to the symphysis pubis and above to the xiphisternum (xiphoid process).

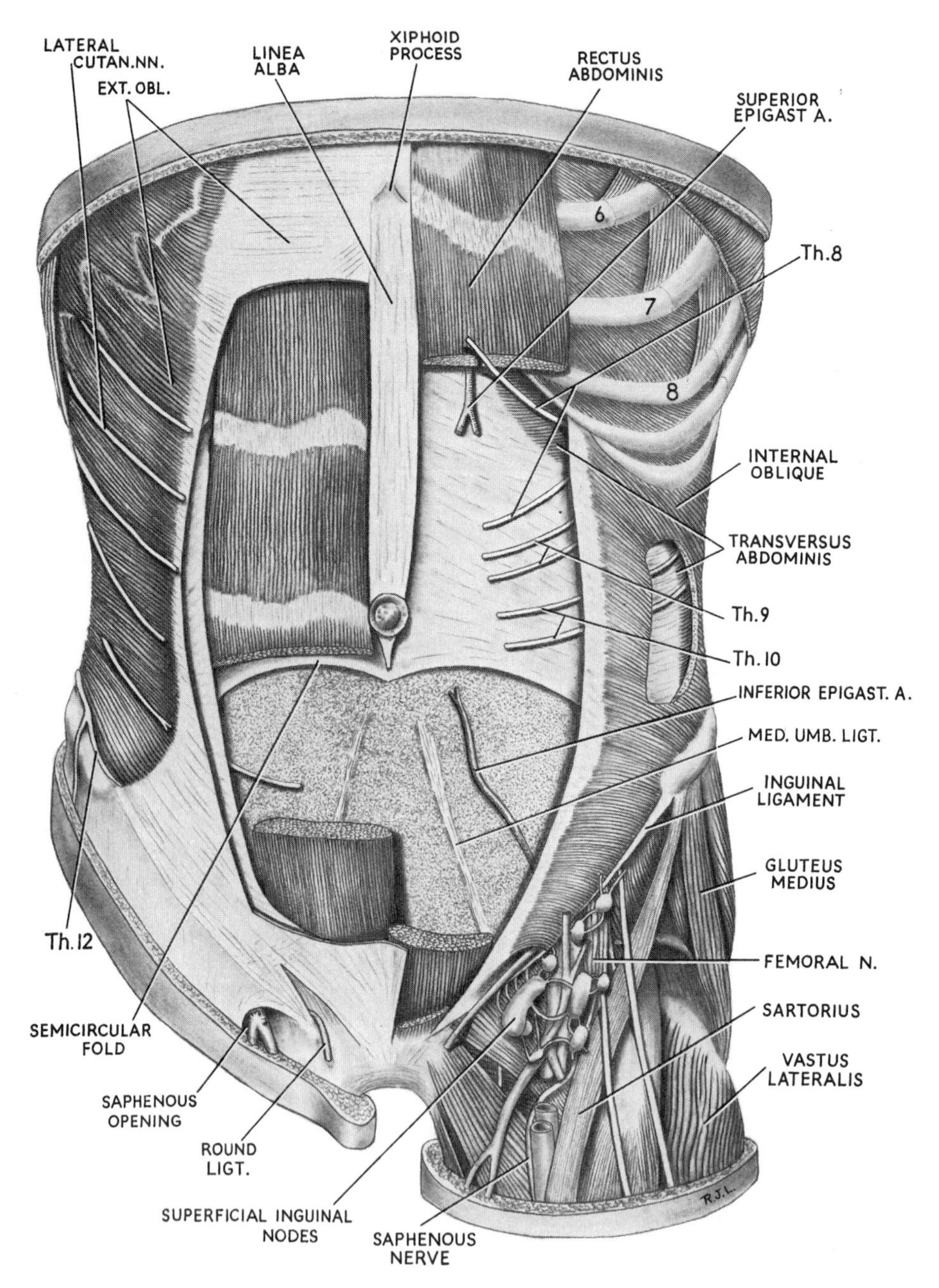

FIG. 5.5. THE ANTERIOR ABDOMINAL WALL OF A YOUNG FEMALE. On the left side the external oblique muscle is removed, and a window cut in the internal oblique exposes the neuro-vascular plane. The anterior layer of the rectus sheath and part of the rectus muscle have been removed on each side. Drawn from Specimen S.351 in R.C.S. Museum.

Blood Supply of the Rectus Muscle. The *superior epigastric artery*, one terminal branch of the internal thoracic, enters the rectus sheath by passing between the xiphisternal and highest costal fibres of the diaphragm. It supplies the muscle and anastomoses in it with the *inferior epigastric artery*. This latter vessel leaves the external iliac at the inguinal ligament, passes upwards behind the conjoint tendon, slips over the semicircular fold of Douglas and so enters the rectus sheath. Veins accompany these arteries, to the internal thoracic and external iliac veins respectively. Tissue fluid from the rectus abdominis is drained via lymphatics that accompany the arteries to the anterior intercostal nodes above and the external iliac nodes below.

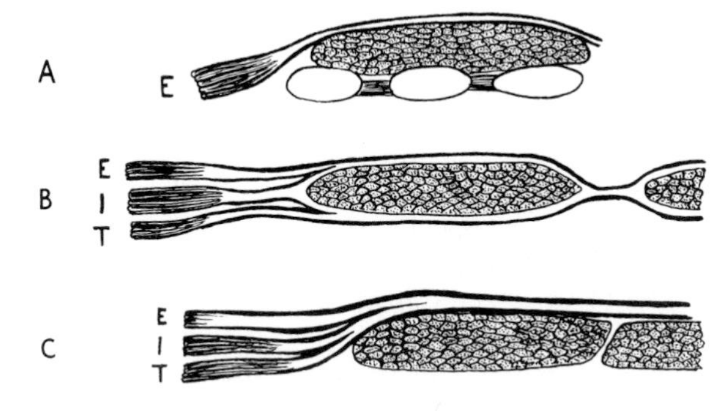

FIG. 5.6. THE FORMATION OF THE RECTUS SHEATH. A. Horizontal section above the costal margin; only the external oblique 'E' and its aponeurosis exist here. B. Horizontal section between umbilicus and costal margin. The aponeurosis of the internal oblique 'I' splits around the rectus, taking external oblique 'E' in front and transversus 'T' behind the muscle. All fuse between the recti in the linea alba. C. Horizontal section below the semicircular fold of Douglas. All three aponeuroses pass in front of the overlapping muscles.

The **pyramidalis muscle** arises from the pubic crest between the rectus abdominis and its sheath. It converges with its fellow into the linea alba an inch or more above its origin. It is innervated by Th. 12. It is often absent.

Actions of the Abdominal Muscles. The muscles of the anterior abdominal wall play four main roles: (1) to move the trunk, (2) to depress the ribs (expiration), (3) to compress the abdomen (evacuation, expiration, heavy lifting) and (4) to support the viscera (intestines only). The abdominal wall, moving to and fro with breathing, conforms to the volume of the abdominal contents. Its *shape* is determined by the tonus of its own muscles. The subumbilical pull of healthy flank muscles keeps its lower part flat by holding back the lower recti.

Moving the Trunk. It must be remembered that the muscles are attached to the thoracic cage and the bony pelvis; evidently their action is to approximate the two. They are the great *flexor muscles* of the vertebral column in its lumbar and lower thoracic parts. Rectus abdominis is the most powerful flexor, but all four oblique muscles, acting symmetrically, are important partners. The oblique muscles themselves are *abductors* and *rotators* of the trunk. The external oblique of one side acts with the internal oblique of the other side to do this, an indispensable postural adjunct to almost all one-armed movements (e.g., tennis, boxing, etc., etc.).

Depressing the Ribs. The recti and obliques (and *not* transversus) approximate the ribs to the pelvic girdle. If erector spinæ prevents thoraco-lumbar flexion this provides a powerful expiratory force (e.g., coughing, blowing the trumpet, speech-making). Added to this is the abdominal compression (aided by transversus) that elevates the diaphragm to increase the expiratory effort.

Compressing the Abdomen. Flexion of the vertebral column is prevented by the erector spinæ muscles. The recti and obliques compress the abdominal cavity; in this they are aided strongly by transversus abdominis, which has no flexing action on the spine. If the diaphragm is relaxed, it is forced up, and this is discussed with *expiration* on p. 221. At the same time the levator ani helps to hold the pelvic effluents closed. The reverse occurs in *evacuation* of the pelvic effluents. Here the diaphragm contracts to resist upward displacement, but it is a far weaker muscle than the abdominal wall, and in forceful compression it is prevented from rising by 'holding the breath' (i.e., by closure of the glottis, and of perhaps the mouth and naso-pharynx). This is discussed on p. 218.

Supporting the Viscera. If the abdominal wall is split or removed only the intestines spill out. Severe visceroptosis occurs with a normal abdominal wall. At most the wall supports the intestines, and the heavy upper abdominal viscera (liver, spleen, kidneys, even the stomach) are supported by other factors.

The Inguinal Canal

The canal is an oblique intermuscular slit lying above the medial half of the inguinal ligament. It commences at the deep inguinal ring, ends at the superficial inguinal ring, and transmits the spermatic cord in the male and the round ligament of the uterus in the female. Its anterior wall is formed by the external oblique aponeurosis, assisted laterally by a portion of the internal oblique muscle. Its floor is the inrolled lower edge of the inguinal ligament, reinforced medially by the lacunar ligament (Gimbernat's ligament) and fusing more laterally with the transversalis fascia. Its roof is formed by the lower edges of the internal oblique and transversus abdominis muscles, which arch over from in front of the cord laterally to behind the cord medially, where their conjoined aponeuroses, constituting the conjoint tendon, are inserted into the pectineal line of the pubic bone. The posterior wall of the canal is formed by the strong conjoint tendon medially and the weak transversalis fascia laterally.

Anterior Wall of the Inguinal Canal. The fibres of the external oblique aponeurosis run parallel with their lower border, the inguinal ligament. Above its medial end they diverge from each other to make a V-shaped opening, the superficial inguinal ring (Fig. 5.1). The lateral crus of this opening is attached to the pubic tubercle, the medial crus to the pubic crest near the symphysis. The intervening part of the pubic crest receives no attachment from the external oblique aponeurosis. At the point of junction of the crura are shining fibres, running at right angles across the external oblique aponeurosis. They bind the crura together, and

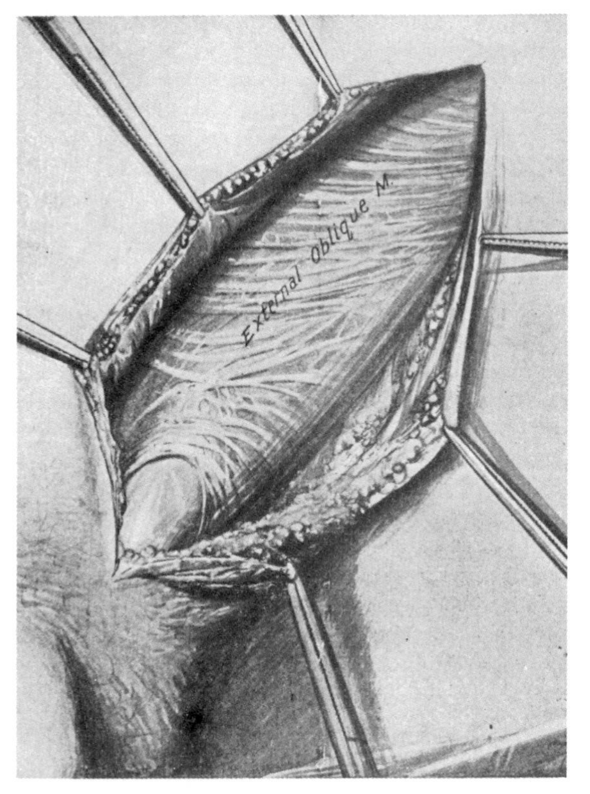

FIG. 5.7. THE ANTERIOR WALL OF THE INGUINAL CANAL AS EXPOSED IN THE OPERATION OF HERNIORRHAPHY. SCARPA'S FASCIA IS RETRACTED IN HAEMOSTATS. THE SPERMATIC CORD IS SEEN EMERGING FROM THE SUPERFICIAL RING. THE INTERCRURAL FIBRES ARE SEEN LACING ACROSS THE EXTERNAL OBLIQUE APONEUROSIS.

serve as a visible landmark to the superficial ring at operation. They are the intercrural fibres (Fig. 5.7).

Superficial Inguinal Ring. This is a triangular gap, with sides of aponeurosis (the crura) and a base of bone, the pubic crest. It transmits the spermatic cord, to which it contributes a fascial covering, the *external spermatic fascia.* The crura can be demonstrated only by sharp dissection, cutting the external spermatic fascia free. Due to the obliquity of the superficial ring, the cord, which passes vertically downwards after emerging therefrom, *overlies the pubic tubercle.* For this reason the pubic tubercle is not readily palpable from in front, and for the satisfaction of the examiner and the comfort of the subject, should be palpated in the living by a finger that invaginates the scrotum *behind the cord.* From the attachment of the lateral crus (i.e., the pubic tubercle) some fibres pass upwards behind the cord and behind the medial crus to blend in the rectus sheath with those from the opposite side. They are known as the reflected part of the inguinal ligament (p. 256). They are not reflected fibres, and have nothing to do with the inguinal ligament. They constitute an additional attachment, *a posterior crus,* of the aponeurosis of the *opposite* external oblique.

The anterior wall of the inguinal canal is reinforced laterally by the lowest muscle fibres of the internal oblique. The deep inguinal ring lies above the midpoint of the inguinal ligament; the internal oblique fibres extend medial to this, for they arise from the lateral two-thirds of the ligament (Fig. 5.10).

The Floor of the Canal. Hold a hip bone in the position it occupies during life, anterior superior iliac spine and symphysis pubis in the same vertical plane. Picture the inguinal ligament, joining anterior superior spine to pubic tubercle. Note that the pectineal line of the pubic bone lies *superior* to it. Thus the lacunar (Gimbernat's) ligament filling the angle between inguinal ligament and pectineal line, passes *upwards* from the ligament to the bone. Its abdominal surface faces forwards as well as upwards. Its femoral surface faces backwards as well as downwards. It lies in the floor of the inguinal canal. Lateral to its attachment the incurved edge of the inguinal ligament forms a gutter which floors in the inguinal canal. The transversalis fascia is fused with this part of the inguinal ligament.

The Roof of the Canal. This is formed by the arched lower borders of the internal oblique and transversus abdominis muscles. Each arises from the hollow of the inrolled lower edge of the inguinal ligament, but their precise attachments need to be appreciated separately. The internal oblique muscle arises by fleshy fibres from the lateral *two-thirds* of the inguinal ligament. The fibres arch medially and downwards, merging into a flat aponeurosis. The most lateral fibres, those arising from just below the anterior superior iliac spine, arch downwards to reach the symphysis pubis, in front of the rectus abdominis. The remaining fibres (trace them medially along their origin at the inguinal ligament) arch concentrically within the former, passing in front of rectus abdominis along the pubic crest as far as the pubic tubercle and then extending laterally along the pectineal line as far as the crescentic edge of Gimbernat's ligament (Fig. 5.2). These lateral fibres, joining the underlying transversus aponeurosis, constitute with them the **conjoint tendon.** The transversus abdominis lies more laterally at its origin, coming from only the lateral *half* of the inguinal ligament, by fleshy fibres deep to those of the internal oblique. They arch downwards in front of the rectus abdominis, where the fleshy fibres are replaced by tendinous ones, to make an aponeurosis that is attached along the pubic crest and extends out along the pectineal line, fusing with the aponeurosis of the internal oblique (Fig. 5.8). Note that the conjoint tendon and the lacunar (Gimbernat's) ligament, attached in common to the pectineal line, *lie in planes at right angles to each other.* The deep inguinal ring lies in the angle between the edge of trans-

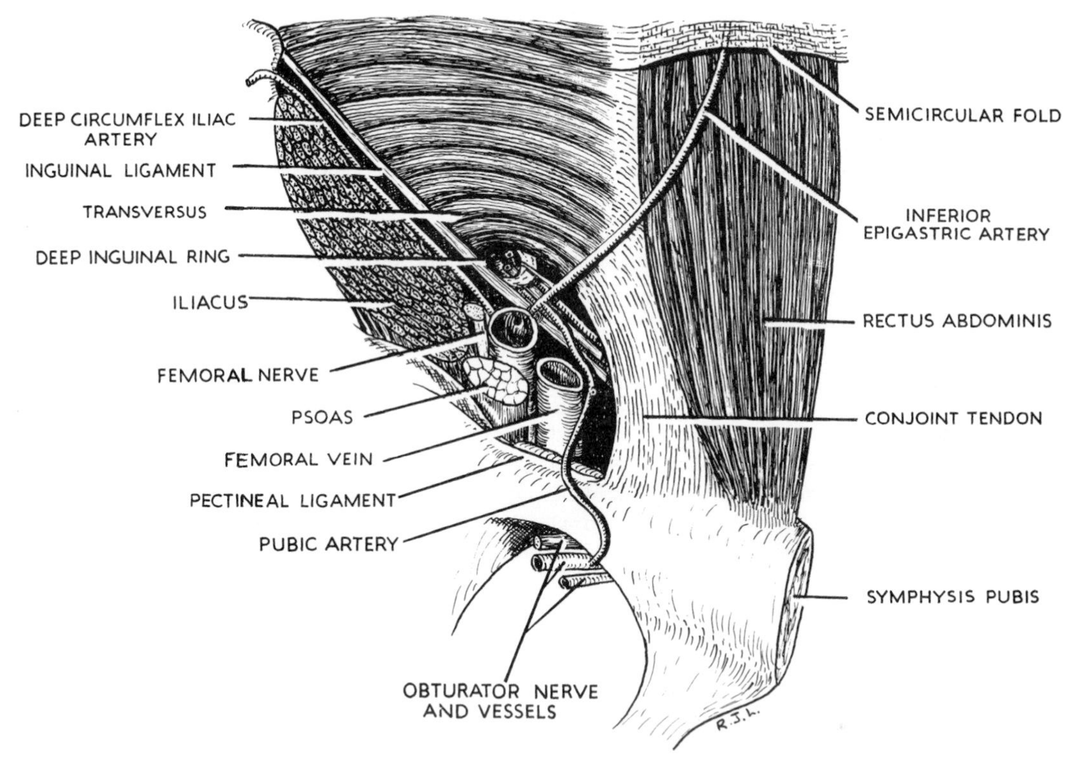

FIG. 5.8. THE ANTERIOR ABDOMINAL WALL VIEWED FROM WITHIN. Left side, peritoneum removed.

versus and the inguinal ligament. Since the internal oblique muscle arises a little more medially than this, it lies in front of the deep ring. The muscular arch of the roof, starting in the anterior wall of the canal, passes over the cord and, becoming tendinous, passes down behind the cord, in the posterior wall of the canal, to reach the pectineal line. Note the oblique course of the canal through the muscular layers of the abdominal wall.

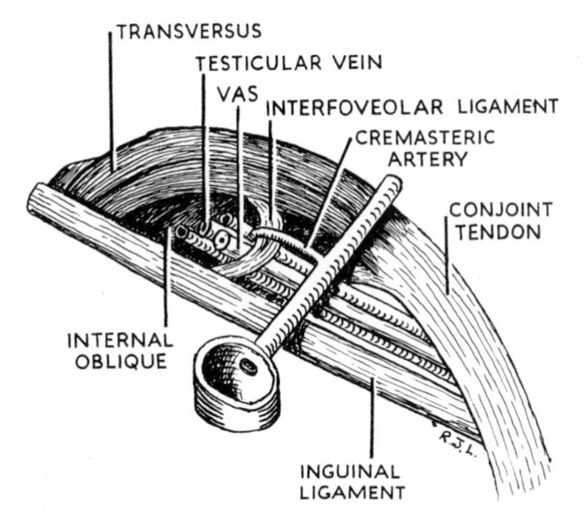

FIG. 5.9. THE LEFT INGUINAL CANAL FROM BEHIND. Enlarged from Fig. 5.8.

The lowermost fibres of internal oblique and transversus are supplied by the ilio-inguinal nerve. Their contraction tightens the conjoint tendon and lowers the roof of the canal, like pulling down a shutter. Thus division of the ilio-inguinal nerve above this level (as in a split-muscle incision for appendicectomy) leads to a direct inguinal hernia—the conjoint tendon bulges when intra-abdominal pressure rises.

Posterior Wall of the Canal. Medially this consists of the strong conjoint tendon, as already explained. Lateral to the conjoint tendon there is only a thin posterior wall, the gap between the arched roof (internal oblique and transversus muscles) and the floor (the inguinal ligament) being covered over by the weak areolar tissue of the transversalis fascia and peritoneum. The integrity of the inguinal canal depends upon the strength of the anterior wall in the lateral part and of the posterior wall in the medial part, and provided the abdominal muscles are of good tone and their aponeuroses unyielding no *direct* herniation of viscera can take place. The deep and superficial inguinal rings lie at opposite ends of the inguinal canal, $2\frac{1}{2}$ in (6 cm) from each other, and the intervening part of the canal is pressed flat when the aponeuroses are under tension and the intra-abdominal pressure raised. The fascia transversalis in the posterior wall is strengthened, moreover, by the presence in front of it of certain tendinous, and sometimes muscular, fibres derived from the transversus abdominis muscle. These fibres constitute the **interfoveolar ligament.** They arch down from the lower border of transversus around the vas to the inguinal ligament, and constitute the functional medial edge of the deep ring (Fig. 5.9). They extend for a variable distance medially across the posterior wall of the inguinal canal and contain a variable amount of muscle fibres. The muscle is well developed in animals that withdraw the testis from the scrotum during their winter of sexual inactivity (e.g., the Canadian moose).

The Deep Inguinal Ring. Lying above the midpoint of the inguinal ligament this opening in the transversalis fascia is bounded laterally by the angle between the transversus muscle fibres and the inguinal ligament. Its medial border is the transversalis fascia where this structure is projected along the canal, like a sleeve from the armhole of a coat, as the internal spermatic fascia. The transversalis fascia is here thickened a little to make the medial border of the ring, but is greatly strengthened by the interfoveolar ligament.

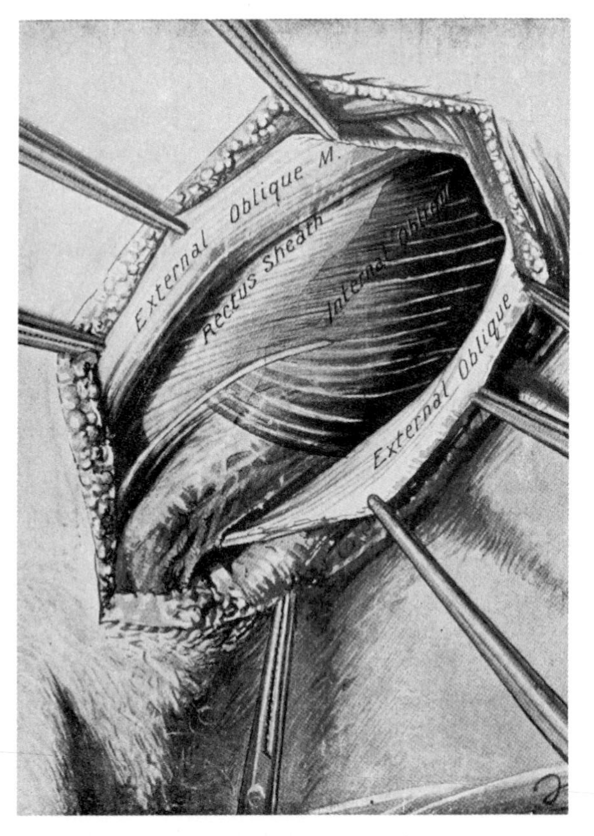

FIG. 5.10. THE EXTERNAL OBLIQUE APONEUROSIS DIVIDED, TO SHOW THE INTERNAL OBLIQUE MUSCLE FIBRES REINFORCING THE LATERAL PART OF THE ANTERIOR WALL OF THE CANAL. THE ILIO-INGUINAL NERVE PIERCES THE LOWER PART OF THE INTERNAL OBLIQUE AND RUNS ON THE CORD DOWN THROUGH THE SUPERFICIAL RING. THIS PART OF THE ILIO-INGUINAL NERVE IS SENSORY ONLY, TO THE ANTERIOR SCROTAL (OR LABIAL) SKIN AND NEARBY THIGH SKIN.

Structures Deep to the Posterior Wall of the Canal. Crossing the posterior wall at the medial edge of the deep inguinal ring is the *inferior epigastric artery.* Lateral to the artery the vas in the male and the uterine round ligament in the female enter the canal by hooking around the interfoveolar ligament (Fig. 5.9). The artery is a branch of the external iliac just proximal to the inguinal ligament. It arises deep to the fascia transversalis, which it perforates obliquely as it slants upwards and medially. Lying superficial to the transversalis fascia the artery now passes over the semicircular fold of Douglas to enter the rectus sheath behind the muscle. At the deep ring the inferior epigastric artery gives off the *cremasteric branch* to supply that muscle and the coverings of the cord. By definition a hernial sac passing lateral to the artery (i.e., through the deep ring) is an *indirect hernia,* one passing medial to the artery is a *direct hernia*; the latter stretches out the conjoint tendon over itself and is therefore seldom large. The inferior epigastric artery also gives off a pubic branch to the periosteum of the superior pubic ramus. This anastomoses with the pubic branch of the obturator artery. If the obturator artery is absent, this anastomosis is opened up to form an artery named the *abnormal obturator artery,* though it is present in over 30 per cent of cases (p. 341). The obliterated umbilical artery, known in the adult as the *medial umbilical ligament,* passes obliquely across the posterior wall of the inguinal canal, medial to the inferior epigastric artery (Fig. 5.5).

The Spermatic Cord (Fig. 5.11)

The vas (ductus deferens), accompanied by its artery and the testicular artery and vein (the vein usually double), enters the deep ring. As these structures pass along the inguinal canal and emerge from the superficial ring, they receive a covering from every layer through which they pass, three coverings in all. The spermatic cord is to be understood as the vas and vessels invested by all three layers, and it is thus not complete until it emerges from the superficial ring. The first layer entered is the transversalis fascia at the deep inguinal ring, and its investment of vas and vessels constitutes a fascial tube called the *internal spermatic fascia.* Passing under the lower edges of transversus and internal oblique it receives a muscular contribution from each, together constituting the *cremasteric layer.* The transversus fibres spiral down the cord and return behind it to become attached to the pubic tubercle. The internal oblique fibres, a much larger contribution, spiral around the cord to return partly to the pubic tubercle, but mostly to the internal oblique itself. Both together

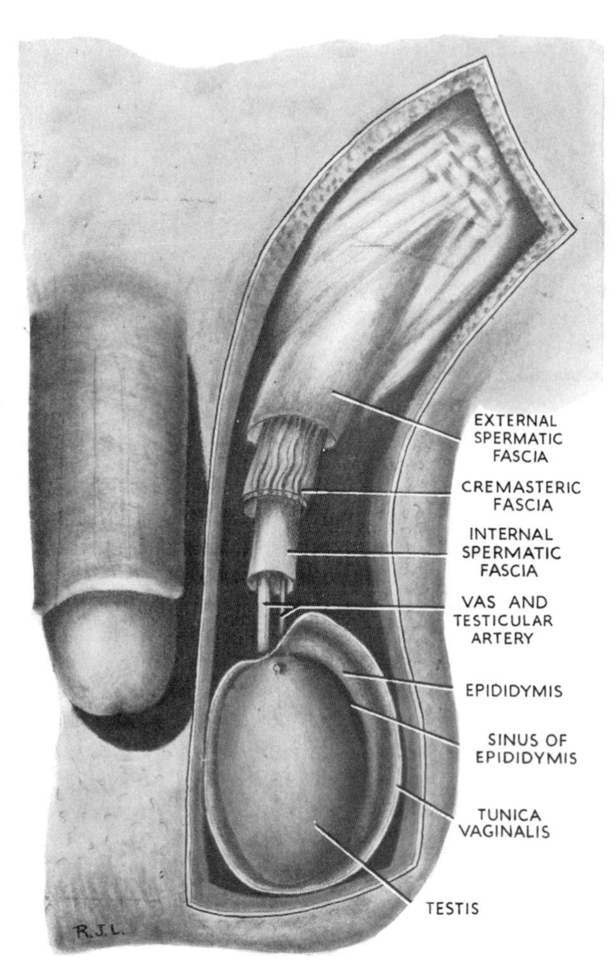

FIG. 5.11. THE LEFT TESTIS AND THE COVERINGS OF THE SPERMATIC CORD.

constitute the *cremaster muscle*. The areolar tissue between the muscle fibres is continued down the cord as the *cremasteric fascia*. The cremaster muscle is supplied by the genital branch of the genito-femoral nerve (L2), which is also sensory to the tunica vaginalis and the spermatic fasciæ. Emerging now through the superficial ring the cord receives its final covering, the *external spermatic fascia*, from the crura. The *ilio-inguinal nerve*, leaving the neuro-vascular plane, pierces the internal oblique muscle near its lower border (Fig. 5.10), runs between it and the external oblique aponeurosis *on the front of the cord*, leaves the superficial ring and supplies skin on the upper part of the thigh and the anterior third of the scrotum (labium majus).

The cremaster muscle and the coverings of the cord are supplied with blood by the cremasteric artery, a branch given off by the inferior epigastric at the deep ring. Venous return is by the cremasteric vein to the inferior epigastric vein, and lymphatics *of the coverings* (not of the testis) drain with the artery to the external iliac group of lymph nodes. The cord passes vertically downwards over the pubic tubercle and enters the scrotum. In the vertical part of its course in the scrotum the testicular veins form a rich longitudinal plexus, the **pampiniform plexus.** In the region of the epididymis there is an anastomosis between the testicular artery and the artery to the vas; it is inadequate in most cases, however, for the artery to the vas is too small to sustain the testis if the testicular artery be divided.

The Testis (Fig. 5.11)

The testis is an oval organ possessing a thick covering of fibrous tissue, the *tunica albuginea*. To its postero-lateral surface the epididymis is attached. The vas (ductus deferens) lies medial to the epididymis, connected to its inferior pole. The front and lateral surfaces of the testis lie free in a serous space formed by the over-lying *tunica vaginalis*, a remnant of the fœtal processus vaginalis. This serous membrane covers also the antero-lateral part of the epididymis. A slit-like space, which lies between testis and epididymis, called the *sinus of the epididymis*, thus lies within the tunica vaginalis. Testis, epididymis and tunica vaginalis lie in the scrotum surrounded by thin membranes, adherent to each other, that are downward prolongations of the coverings of the spermatic cord. Right and left sides are separated by the median septum scroti. The *appendix testis* is a sessile cyst 2 or 3 mm in diameter attached to the upper pole of the testis (Fig. 5.11), within the tunica vaginalis.

Internal Structure. The upper pole of the epididymis is attached high up on the postero-lateral surface of the testis. From this area fibrous septa radiate into the testis and reach the tunica albuginea. The rete testis is contained herein; the fibrous mass containing the rete is named the *mediastinum testis*. The septa radiating from the mediastinum divide the organ into some 400 spaces, each of which contains two (sometimes three or four) highly convoluted *seminiferous tubules*. Each tubule is 2 ft (60 cm) long. The tubules lie rather loosely between the tunica albuginea and fibrous septa so that the cut surface of the organ bulges with herniating tubules. The seminiferous tubules open into the *rete testis*, which is a network of intercommunicating channels lying in the mediastinum testis. From the rete the *vasa efferentia*, fifteen to twenty in number, enter the commencement of the canal of the epididymis, thus attaching the head of the epididymis to the testis. Each efferent tubule is coiled into a small *lobule* before connecting with the canal of the epididymis.

Microscopic Anatomy. Histological sections of the mature testis are easily recognized. The dense fibrous tissue of the tunica albuginea is thick, and fibrous septa divide the field into loculi. The seminiferous tubules, convoluted within the loculi, are cut in multiple sections. Each tubule shows several layers of cells, and the slender tails of spermatids project into the lumen. The named cells in the layers cannot always be distinguished with certainty. Between *supporting cells* the testicular cells are undergoing spermatogenesis. The basal cells are called *spermatogonia*—gonadal cells. These divide into *primary spermatocytes*—sperm cells. The next generation is by meiosis, to produce *secondary spermatocytes*—sperm cells with half the adult number of chromosomes. These mature into spermatozoa by migration of the nucleus to one end (the 'head') and thinning out of the cytoplasm into a tail that projects into the lumen. While still attached by the head, the cell is named a *spermatid.* Only when free in the lumen is it a separate creature—the *spermatozoon.* The spermatozoa may be seen free in the lumen of the tubules or they may have been washed away in the preparation of the slide. They vary greatly in appearance in different species.

Blood Supply. The testicular artery, from the aorta, runs in the spermatic cord, gives off a branch to the epididymis, and reaches the back of the testis, where it divides into medial and lateral branches. These do not penetrate the mediastinum testis, but sweep around horizontally within the tunica albuginea. Branches from these vessels penetrate the substance of the organ. Venules reach the mediastinum, from which several veins pass upwards in the spermatic cord and surround the testicular artery with a mass of intercommunicating veins, the *pampiniform plexus.* In the inguinal canal they join to form two *testicular veins.* From the deep inguinal ring one vein, usually, passes up near the testicular artery on the psoas muscle. The right joins the inferior vena cava, the left the renal vein. This is embryologically symmetrical, since in each case the segment of vein so joined represents a persistent part of the subcardinal vein. Lymphatics run back with the testicular artery to para-aortic nodes lying alongside the aorta at the level of origin of the testicular arteries (L2), i.e., just above the umbilicus.

Nerve Supply. The testis is supplied by sympathetic nerves. The connector cells lie in Th. 10 segment of the cord. Passing in the greater or lesser splanchnic nerve to the cœliac ganglion the efferent fibres synapse there. Postganglionic grey fibres reach the testis along the testicular artery. *Sensory* fibres ('testicular sensation') share the same sympathetic pathway. They run up along the testicular artery and through the cœliac plexus and lesser splanchnic nerve and its white ramus to cell bodies in the posterior root ganglion of Th. 10 spinal nerve. There is no parasympathetic supply to the testis.

Descent of the Testis. The testis develops in the peritoneum of the posterior abdominal wall (Fig. 5.12).

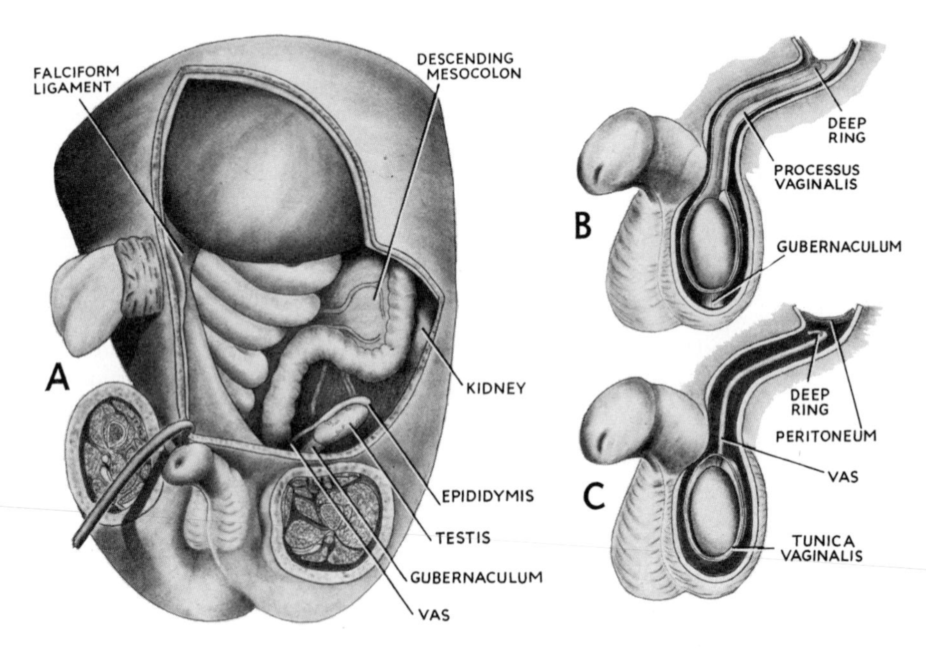

FIG. 5.12. THE DESCENT OF THE TESTIS. In A, drawn from a fœtus of six months, the testis lies on the posterior abdominal wall below the kidney. The vas descends from the lower pole (tail) of the epididymis. B and C are diagrams of subsequent events. In B, a neonatal specimen, the gubernaculum is under the testis in the scrotum, and the vas recurves from the lower pole (tail) of the epididymis behind the processus vaginalis. In C, the completed descent is figured; processus vaginalis and gubernaculum have disappeared.

Behind the peritoneum, at its lower pole, is attached a mass of mesoderm called the **gubernaculum.** During fœtal growth the testis descends, and by the 7th month it lies at the deep inguinal ring. By birth it is in the scrotum. It is preceded in its descent by the gubernaculum, which structure ultimately attaches itself to the scrotum before withering away. Descent may be delayed or arrested; in the latter case spermatogenesis is usually incomplete and the organ is more than usually liable to malignant change. Note that the anterior surface of the gonad is not covered by peritoneum; the bare organ projects into the cœlom. Everyone realizes that the ovary has no true peritoneal coat on the part that projects into the cœlom from the posterior leaf of the broad ligament. Merely the epithelium that surfaces the peritoneum is continued as a single layer over the ovary, forming the 'germinal' epithelium. Few appreciate the fact that the same is true of the testis; the tunica vaginalis has, strictly, only a parietal and no visceral layer. The surface of the tunica albuginea that lies in the tunica vaginalis is covered only by a single layer of flat cells, and there is no continuation of the fibrous peritoneum over it (Fig. 5.13). The arrangement resembles that of the eye, where the conjunctiva ends at the limbus and only the epithelium continues over the cornea. The tunica vaginalis is the cut-off part of the *processus vaginalis*, a herniation of peritoneum that extends to the bottom of the scrotum ready to receive the descending testis. The testis slides, so to speak, down the posterior abdominal wall, preceded by the retroperitoneal gubernaculum. It slides into a scrotal hernia which afterwards seals off spontaneously.

The Epididymis

This is a firm structure, attached behind the testis, with the ductus deferens to its medial side. It consists of a single tube, 6 m (20 ft) long, highly coiled and packed together by fibrous tissue. The mass so resulting has a large *head* (upper pole, globus major) and a small *tail* (lower pole, globus minor) connected by the intervening *body*, which is applied in crescentic manner to the back of the testis. A narrow slit between epididymis and testis is known as the *sinus of the epididymis*. The sinus, with the anterior half of the epididymis, lies within the tunica vaginalis. The head receives the vasa efferentia from the rete testis and is thus firmly attached to the testis. Elsewhere the epididymis has no functional connection with the testis, to which it is, however, fairly firmly bound by fibrous tissue. From the tail the *vas (ductus) deferens*, a direct continuation of the canal

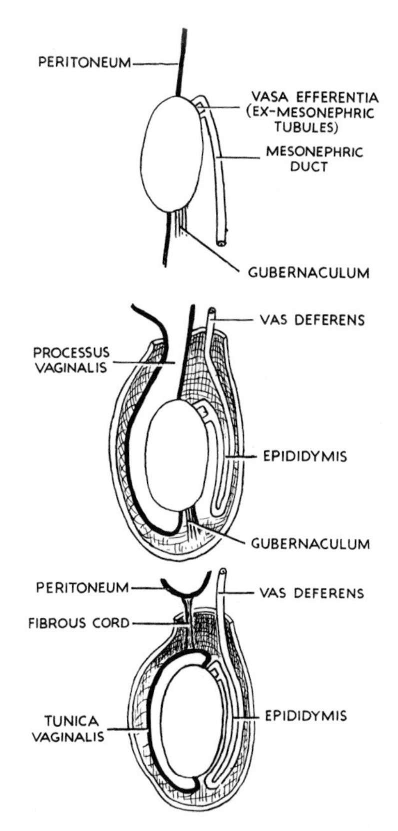

FŒTUS. The testis projects through the peritoneum into the cœlomic cavity. The vas (mesonephric duct) runs down.

NEONATUS. The testis has reached the scrotum via the processus vaginalis, and the vas runs up from the tail of the epididymis.

END RESULT. The testis lies in the tunica vaginalis; the processus vaginalis and gubernaculum have disappeared.

FIG. 5.13. THE STAGES OF TESTICULAR DESCENT SHOWN IN FIG. 5.12 REPRESENTED IN DIAGRAM.

of the epididymis, provided with a thick wall of smooth muscle, passes up medially. It enters the spermatic cord, passes through the inguinal canal, across the side wall of the pelvis just under the peritoneum, and crosses the pelvic cavity. It pierces the prostate and opens by the ejaculatory duct into the prostatic urethra (p. 331).

Microscopic Anatomy. Sections of the **epididymis** are easily recognized. One single coiled tube is cut many times; circles, ovals or comma-shaped outlines of a constant diameter are seen. The wall is of thin fibrous tissue, and the lining is of tall ciliated columnar epithelium. Fibrous tissue binds the coils together.

Ductus deferens. The ductus deferens, or vas, is characteristic. Its wall is very thick in contrast to the narrow lumen (Fig. 5.52). The thick wall is of smooth muscle, in longitudinal and circular arrangement; the fibres are not in separate layers but interwoven like a basket. The lining mucous membrane is a *thin* layer of *dense* fibrous tissue, surfaced with tall ciliated columnar epithelium like that of its parent epididymis. The tall 'cilia' which characterize the epithelium of the epididymis are, in fact, immotile ('large microvilli') and are generally known as 'stereocilia'. Similar stereocilia are found in the vas, ependymal epithelium and in the tympanic cavity.

Note on terminology. The 'vas' is now renamed the 'ductus deferens'. But sterilization of the male by division and ligation of the tube is still popularly known as 'vasectomy' or, more accurately, vasotomy, and has not yet been dubbed a 'deferential dochotomy'.

Blood Supply. The epididymis is supplied by a branch of the testicular artery. This enters the upper pole and runs down to the lower pole. It anastomoses with the tiny artery to the vas. By this branch to the epididymis an anastomosis exists between the testicular

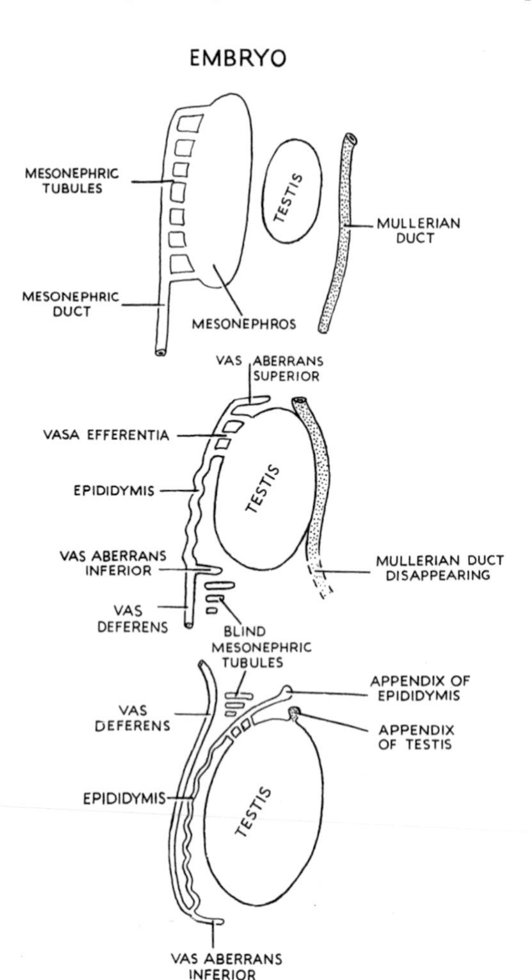

EARLIEST STAGE. The mesonephros drains urine into the mesonephric duct. The gonad develops between the mesonephros and the Mullerian duct.

LATER STAGE. The mesonephros vanishes but its duct persists. Some mesonephric tubules remain, to carry spermatozoa from testis to mesonephric duct (epididymis); they are the vasa efferentia. Other tubules persist, blind at one or both ends. The Mullerian duct is vanishing.

AFTER DESCENT. The adult relationships result from descent. The vas, formerly running downwards, now runs upwards from the epididymis; this carries the blind mesonephric tubules (paradidymis) from lower to upper pole.

FIG. 5.14. THE DEVELOPMENT OF THE VASA EFFERENTIA AND THE EPIDIDYMIS.

artery and the artery to the vas. But the artery to the vas is too small to carry an adequate supply to the testis, which thus atrophies if the testicular artery is divided.

Venous and lymphatic drainage are as for the testis (*q.v.*).

Nerve Supply. The epididymis is supplied, like the testis, by sympathetic fibres from the cœliac ganglion via the testicular artery.

Embryological Remnants. It is necessary to recall that the whole length of the single tube constituting the epididymis and vas is a persistent and much elongated part of the mesonephric (Wolffian) duct of the embryo. This duct receives the efferent tubules of the mesonephros (or middle kidney) (Fig. 5.14). When the mesonephros is replaced by the metanephros and disappears, some of its tubules persist and form a new attachment to the developing testis, forming the vasa efferentia and draining the products of the testis into the commencement of the Wolffian duct. Some mesonephric tubules persist without serving any function of drainage, since they are blind at one or both ends. Thus, above and below the epididymis blind tubules open into its canal. They are called *vasa aberrantia*. Their bulbous blind ends may form appendices of the epididymis. An upper one is relatively constant, the *appendix of the epididymis*. Above the epididymis, at the lower extremity of the spermatic cord, a mass of tubules, blind at each end, persists as the *paradidymis* (or organ of Giraldès). A cyst formed from an aberrant vessel will contain spermatozoa and thus be opalescent. A cyst formed from a tubule of the paradidymis cannot contain spermatozoa, and its fluid is thus crystal clear (Fig. 5.14). The paramesonephric (Mullerian) duct, developing into the oviduct (Fallopian tube) in the female, disappears in the male except at its two ends. The upper end persists as the *appendix testis*, the conjoined lower ends of the two ducts persist as the *utriculus masculinus* (p. 331).

THE ABDOMINAL CAVITY

The abdominal cavity is much more extensive than the impression gained from examination of the anterior abdominal wall. Much of it lies under cover of the lower ribs, for the domes of the diaphragm arch high above the costal margin. Hidden by the lower ribs are the whole of the liver and spleen, much of the stomach, and the upper poles of the kidneys and both suprarenals. The volume of the thoracic cavity is, correspondingly, much less than examination of the bony thorax would suggest. Furthermore, an appreciable amount of the abdominal cavity projects backwards into the pelvis, just in front of the buttocks. A perforating wound of the buttock can easily involve the pelvic cavity. The pelvic cavity accommodates not only its own pelvic organs (rectum, uterus, bladder, etc.), but also a goodly volume of intestine (sigmoid colon and ileum).

General Topography of the Abdomen

The alimentary canal and its three chief derivatives the liver, spleen and pancreas are developed in fœtal mesenteries which later alter their disposition as a result of fusion of adjacent leaves of peritoneum. The liver and spleen remain invested in peritoneum, but the pancreas becomes retroperitoneal.

The alimentary canal is invested unevenly. Parts of it are free to swing on peritoneal folds ('mesenteries'); other parts become plastered down to the posterior abdominal wall. The stomach is fixed at its two ends, but elsewhere swings free on 'mesenteries'. The duodenum is plastered down to the posterior abdominal wall, while the whole length of small intestine swings free on its own mesentery. Ascending and descending colons are both adherent to the posterior abdominal wall, but between the colic flexures the transverse colon is mobile on its own mesentery, the transverse mesocolon. The sigmoid (pelvic) colon swings free on a mesentery while, lastly, the rectum is plastered by peritoneum to the hollow of the sacrum.

The suprarenals, kidneys and ureters lie behind the peritoneum and possess no serous coat. The aorta and vena cava lie behind the peritoneum. The intestinal vessels run through the mesenteries to reach the gut.

The division of the abdominal cavity into regions by imaginary vertical and horizontal planes is archaic and gives no aid in understanding abdominal topography. By far the most useful plane is the transpyloric plane, and if the structures lying at this level are appreciated the topography of the rest of the abdominal organs becomes clear.

The **transpyloric plane** bisects the body between the jugular notch and the symphysis pubis. But the plane need not be defined with geometrical accuracy and it is sufficient to mark it as passing through a point midway between the xiphisternum and the umbilicus. It cuts each costal margin at the tip of the 9th costal cartilage, which is at the lateral border of the rectus abdominis (linea semilunaris); beneath this point on the right side the fundus of the gall bladder lies, on the left is the body of the stomach. The plane passes through the lower border of the first lumbar vertebra, where the spinal cord ends at the conus medullaris.

As its name implies, the plane passes through the pylorus. It must be noted that the pylorus is free on a mesentery, and therefore mobile. In the erect posture it hangs down over the front of the head of the pancreas, so the plane passes along the head, neck and body of that gland, just above the attachment of the transverse mesocolon. The supracolic compartment (liver, spleen, fundus of stomach) lies above the plane, the infracolic compartment (small intestine, colon) below it. The superior mesenteric artery leaves the aorta at this level, and the splenic vein runs transversely behind the pancreas in the plane. The hilum of each kidney lies at the plane, the right just below and the left just above it, level with the tips of the 9th costal cartilages.

THE PERITONEUM

Peritoneum is a fibrous membrane (i.e., a skin) whose surface is smoothed by a single layer of flat cells, a pavement epithelium.

Peritoneum lines the walls of the abdominal cavity. This is the **parietal peritoneum:** it clothes the anterior and posterior abdominal walls, the under surface of the diaphragm, and the cavity of the pelvis. It is attached to these walls by extraperitoneal areolar tissue which varies in both thickness and density in different places. Over expansile parts this areolar tissue is loose and cellular (e.g., transversalis fascia on the lower anterior abdominal wall) while over non-expansile parts it is often very thick (e.g., fascia iliaca, psoas fascia, parietal pelvic fascia); but loose or dense, thin or thick, these variously named fasciæ are part of the one continuous extraperitoneal connective tissue lying between the parietal peritoneum and the walls of the abdominal and pelvic cavities. On the posterior abdominal wall the dense psoas and iliac fasciæ and the anterior lamella of the lumbar fascia serve as firm bases upon which the extraperitoneal tissue can gain attachment. The posterior surfaces of retroperitoneal structures (pancreas, duodenum, ascending and descending colon) also gain a firm attachment to these fasciæ. Thus peritoneum and viscera have a firm anchorage undisturbed by the movements of contraction of the underlying muscles.

Nerve Supply. The *parietal peritoneum* is supplied segmentally by the spinal nerves that innervate the overlying muscles. Thus the diaphragmatic peritoneum is supplied centrally by the phrenic nerve (C4, hence referred pain and hyperæsthesia from this area at the tip of the shoulder). Peripherally the diaphragmatic peritoneum is supplied by intercostal nerves. The remainder of the parietal peritoneum is supplied segmentally by intercostal and lumbar nerves. In the pelvis the obturator nerve is the chief source of supply. The *visceral peritoneum* has no nerve supply, as far as is known. Pain from diseased viscera is due to muscle spasm, tension on mesenteric folds, or involvement of the parietal peritoneum.

The **visceral peritoneum** is the continuation of the parietal peritoneum, which leaves the posterior wall of the abdominal cavity to invest certain viscera therein. In some cases it merely passes over the front of the organ (e.g., duodenum, ascending and descending colon), in other cases it leaves the posterior wall as two leaves lying together to form a mesentery which completely invests the organ (e.g., small intestine). The liver and spleen are almost completely invested in visceral peritoneum (serous coat) and possess, therefore, double layers of peritoneum connecting them to the parietes. The upper parts of the pelvic organs (rectum, uterus, bladder) project into the pelvic cavity and are covered with visceral peritoneum, which hangs down in depressions or pouches between them.

The visceral peritoneum can stretch enormously in some places (e.g., over the distended stomach) while in other places it does not stretch but merely peels away from its nearby attachments (e.g., from the bladder fundus it peels away from the rectus abdominis as the bladder rises).

The serous-coated organs fill the abdominal cavity, so that visceral surfaces are in contact with each other or with the parietal peritoneum. The intraperitoneal space between them is only potential, not actual, and it contains in all merely a few millilitres of tissue fluid which moistens and lubricates the serous surfaces. This is the general *peritoneal cavity*, body cavity, or cœlom; it is opened up when incisions are made through the parietal peritoneum of the anterior abdominal wall. Another name for it is the *greater sac* of the peritoneal cavity, to distinguish it from the lesser sac. The *lesser sac*, now called *omental bursa*, is a smaller cavity which lies behind the stomach; it opens, as a diverticulum from the greater sac, through a narrow window called the opening of the lesser sac, epiploic foramen, or foramen of Winslow. The foramen lies between the first part of the duodenum and the under surface of the liver; it is bounded posteriorly by the parietal peritoneum of the posterior abdominal wall, which, in continuity, forms the posterior wall of the greater and lesser sacs. Anteriorly the foramen is bounded by the vessels and duct of the liver.

The lesser sac (now named *omental bursa*) exists behind the stomach, as a slippery surface for the changing volume of the stomach (PLATE 12B). The anterior wall of the stomach slides on the parietal peritoneum of the anterior abdominal wall, in the greater sac.

The disposition of the peritoneal folds and mesenteries that go to make up the lesser sac and other peritoneal pouches and reflexions is best appreciated by examination of the opened but undisturbed abdomen at an autopsy. In the dissecting room the inelastic and unyielding gut and peritoneum of the coagulated cadaver make such an examination everywhere difficult and in places impossible.

The Greater Sac

Examine, therefore, the abdomen at an autopsy, and check the peritoneal attachments with Fig. 5.17. Lying over the coils of the small intestine will be seen the greater omentum, the 'policeman of the abdomen', hanging like a vascular apron from the transverse colon. It is an apron of peritoneum, translucent or filled with fat according to the state of nutrition of the individual. It readily becomes adherent to any area of inflammation, moving to the site of infection and by its adhesion containing the same. It is difficult to believe that this is the reason for its existence. Perhaps its presence as an apron discourages coils of small intestine from crossing up in front of the transverse colon.

The Infracolic Compartment. Throw the greater omentum up over the costal margin (Fig. 5.16). This will displace the transverse colon upwards, and expose to view the lower layer of the *transverse mesocolon*, which divides the abdominal cavity into supracolic and infracolic compartments (Fig. 5.17). Note its attachment transversely across the posterior abdominal wall, towards the lower border of the retroperitoneal pancreas. On the right the transverse mesocolon is attached across the descending (second) part of the duodenum and it ends over the lower pole of the right kidney at the right colic, or hepatic, flexure. On the left the transverse mesocolon ends over the lower pole of the left kidney at the left colic, or splenic flexure. While here note the small transverse fold of peritoneum between the left (splenic) flexure and the diaphragm, the *phrenico-colic ligament*. It is in contact with the spleen, which lies in the supracolic compartment of the greater sac; below the ligament is the commencement of the left *paracolic gutter*, which lies wholly in the infracolic compartment. Trace the gutter down the lateral side of the contracted descending colon; it leads in front of the attachment of the pelvic mesocolon, across the pelvic brim, into the pelvis (Fig. 5.17).

Compare the *right paracolic gutter*. Unlike the left, its commencement lies in the supracolic compartment, between the upper pole of the right kidney and the under surface of the liver, where the peritoneum passes from one to the other. This is the *hepato-renal pouch* (of Rutherford Morison) and it leads directly downwards lateral to the ascending colon and cæcum to cross the pelvic brim into the cavity of the pelvis. There is no phrenico-colic ligament on the right side to interrupt the continuity of this gutter (Fig. 5.17).

Take the opportunity to examine the small intestine and the attachment of its mesentery. The jejunum has a larger lumen than the ileum, and its walls are thicker. Roll the wall of each between finger and thumb—the wall of the jejunum feels double, like palpating a shirt sleeve through a coat sleeve. (The inner sleeve is the thick mucous membrane of the jejunum.) The thinner wall of the ileum feels single, its mucous membrane is not palpable as a separate layer. Identify the direction of any loop of small intestine. Straighten the loop and trace each leaf of the mesentery down to its attachment on the posterior abdominal wall. The segment of intestine, lying parallel with the attachment of its own part of the mesentery, now leads downwards towards the right iliac fossa.

Examine the *mesentery*. Up to 20 feet long (6 m) at its intestinal border, it is but 6 in (15 cm) long at its attachment to the posterior abdominal wall. Its intestinal border is consequently greatly folded, like the hem of a very full skirt, and the attached small intestine is forced to lie in a series of adjacent coils. The attachment of the mesentery begins at the duodeno-jejunal flexure, crosses

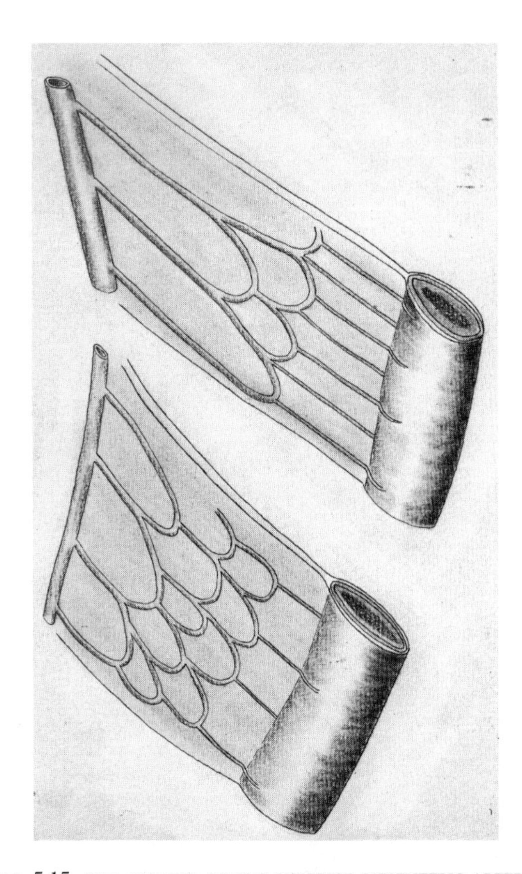

FIG. 5.15. THE ARCADES OF THE SUPERIOR MESENTERIC ARTERY. UPPER FIGURE: JEJUNUM. LOWER FIGURE: ILEUM. Contrast the high narrow windows between the vasa recta of the jejunum with the low broad windows between the vasa recta of the ileum.

the third part of the duodenum, where the superior mesenteric vessels lie between its folds, and slopes downwards across the aorta, inferior vena cava, right psoas muscle and ureter to the right iliac fossa over the iliacus fascia (Fig. 5.17). The flange of mesentery is short at each end, longest in the middle, where it measures some 6 in (15 cm) from its attachment to its intestinal border. It usually has fat between its layers, obscuring the mesenteric vessels and lymph nodes from sight. At the jejunal end the fat does not reach the intestine and the vasa recta make high narrow windows visible if held up to the light. At the ileal end the vasa recta are shorter and further apart, but the low broad windows in the mesentery are here usually obscured by encroachment of fat right up to the intestinal wall (Fig. 5.15).

Now examine the infracolic compartment between the ascending and descending parts of the colon. It is divided into right and left parts by the oblique attachment to the posterior abdominal wall of the mesentery of the small intestine. Press the coils of jejunum and ileum downwards and to the left. A triangular area of the peritoneum of the posterior abdominal wall is thus exposed. This is the **right infracolic compartment.** Its apex lies below, at the ileo-cæcal junction. Its right side is the ascending colon, its left side the attachment of the mesentery of the small intestine, and its base is the attachment of the transverse mesocolon. Examine the floor of the triangle (Fig. 5.17). At the right end the inferior pole of the right kidney can be seen and felt, crossed by the ascending branch of the right colic vessels. Just to the left of this the descending (second) part of the duodenum appears, passes downwards for a couple of inches and turns transversely across the posterior abdominal wall and ends by passing upwards to the duodeno-jejunal flexure, at which point the gut breaks free from under the peritoneum of the posterior abdominal wall and, as the jejunum, gains a mesentery. The duodeno-jejunal flexure, one of the angles of the left infracolic compartment, lies to the left of the midline, over the left psoas muscle, on a level with the second lumbar vertebra. Note that to

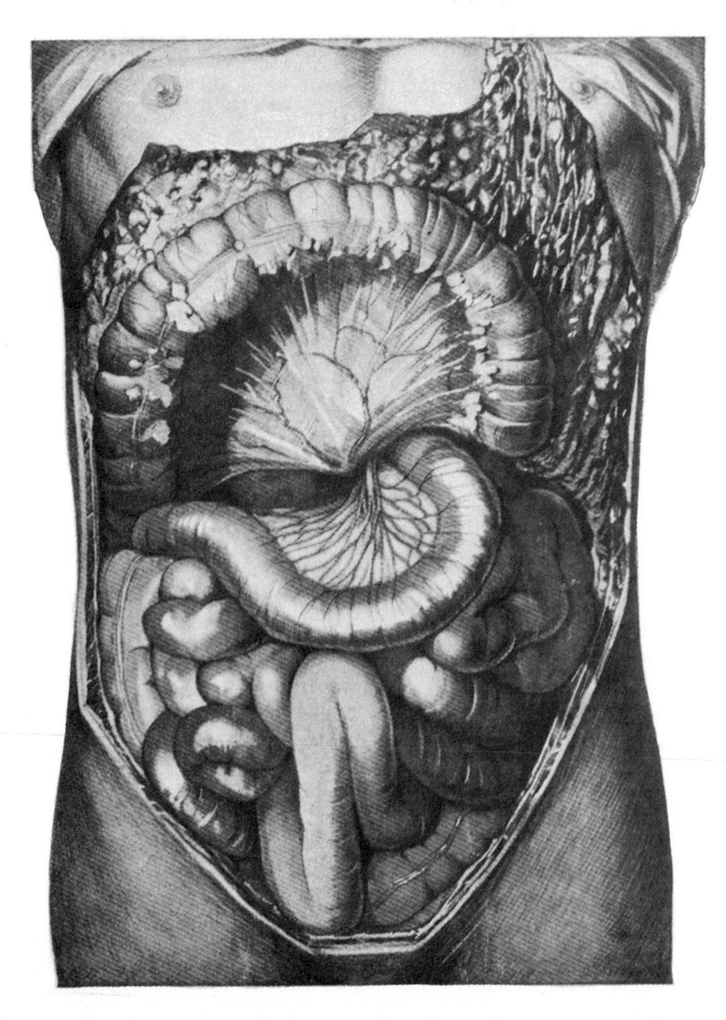

FIG. 5.16. THE GREATER OMENTUM THROWN UPWARDS OVER THE COSTAL MARGIN, TAKING TRANSVERSE COLON AND MESOCOLON WITH IT. THE INFRACOLIC COMPARTMENT IS EXPOSED READY FOR INSPECTION.

descend from the right infracolic compartment into the pelvis it is necessary to pass forwards over the lower ileum, and contrast this with the free passage into the pelvis on the left side.

The **left infracolic compartment** is exposed by displacing the coils of small intestine upwards and to the right. It is larger than the right infracolic compartment and is quadrilateral in shape. It widens below to pass in a smooth sweep across the pelvic brim into the cavity of the pelvis (Fig. 5.17). Its upper border is the attachment of the transverse mesocolon, between the duodeno-jejunal flexure (to the left of the midline) and the left colic (splenic) flexure.

The fourth or ascending part of the duodenum lies in this compartment, in the upper angle to the left of the mesentery of the small intestine and it runs up to the commencement of the jejunum. The paraduodenal fossæ (p. 293) lie here. At the lateral end of the upper border, the inferior pole of the left kidney can be seen and felt, in the angle between the transverse mesocolon and the splenic flexure. It is crossed by the ascending branch of the upper left colic vessels. The right border of the left infracolic compartment is provided by the attachment of the mesentery which slopes down to the right iliac fossa. The floor is shallowest in the midline, for the forward prominence of the lumbar vertebræ is here enhanced by the overlying aorta and inferior vena cava. Over the promontory of the sacrum the peritoneal floor dips down in a bold sweep to line the pelvic cavity.

The left border is provided by the descending colon which, as already noticed, is plastered down to the posterior abdominal wall and possesses no mesentery. Trace it into the left iliac fossa and examine the pelvic mesocolon.

The *pelvic mesocolon* has a Λ-shaped attachment. The limbs diverge from each other at the bifurcation of the common iliac vessels, which point lies on the pelvic brim over the left sacro-iliac joint. The upper limb passes forwards along the pelvic brim (over the external iliac vessels) halfway to the inguinal ligament (i.e., about 5 cm) while the lower limb slopes down into the hollow of the sacrum, where it reaches the midline in front of the third sacral vertebra, at the commencement of the rectum. At the apex of the junction of these two leaves, just beneath the peritoneum and lying over the bifurcation of the common iliac artery, lies the left ureter. Throw the pelvic colon upwards to expose this point.

Now replace the great omentum and transverse colon and proceed to examine the supracolic compartment, which is partly divided into right and left parts by the *falciform ligament* of the liver. This is a fold of peritoneum passing from the anterior abdominal wall and

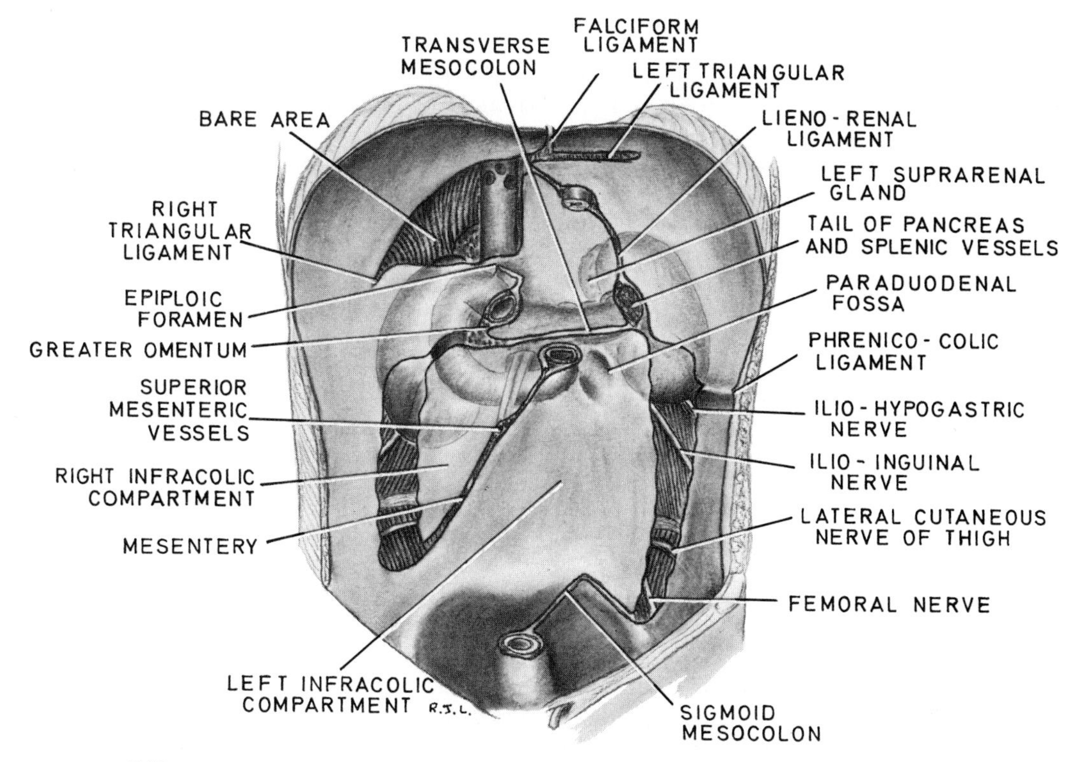

FIG. 5.17. THE POSTERIOR ABDOMINAL WALL, TO SHOW THE ATTACHMENTS OF THE PARIETAL PERITONEUM.

the abdominal surface of the diaphragm to the anterior surface of the liver. Its free lower edge contains a rounded cord, the *ligamentum teres,* which passes from the umbilicus to a notch in the inferior border of the liver to the *right* of the midline. The falciform ligament, a flat flange, thus passes from the ventral midline of the body to the right to reach the anterior surface of the liver, and it lies flat between these two adjacent surfaces. Examine its connexions and continuations at a later stage (p. 273).

The Supracolic Compartment. Most of the supracolic compartment lies under cover of the costal margin of the thoracic cage, below the diaphragm, and consequently cannot be exposed to view as readily as the infracolic compartment. Much of it has to be examined by palpation only. Examine first the attachments of the peritoneum of the greater sac. When the greater omentum and transverse colon were displaced upwards over the costal margin, the inferior leaf of the transverse mesocolon was exposed to view. Replacing them does not expose the superior leaf, however, for the *greater omentum passes between the transverse colon and the lower border, or greater curvature, of the stomach.* This part is the *gastro-colic omentum,* visible in the undisturbed abdomen. To the right it shortens and becomes plastered down over the head of the pancreas, between the first part of the duodenum and the transverse colon. To the right of this line of fusion the peritoneum on the posterior abdominal wall plasters down the descending (second) part of the duodenum and is traceable to the right over the right kidney into the hepato-renal pouch previously examined (Fig. 5.17). The greater omentum is attached to the greater curvature of the stomach, behind the fundus and up to the abdominal portion of the œsophagus. From here palpate its attachment across the diaphragm (behind the spleen) down to the upper pole of the left kidney, down the front of that organ to the splenic flexure and so back to the transverse mesocolon. Note that the right portion of the greater omentum stretches between the greater curvature of the stomach and the transverse colon; this is the gastro-colic omentum. The left portion of the greater omentum stretches between the greater curvature, the fundus and the abdominal œsophagus on the one hand and the diaphragm and left kidney on the other hand.

The greater omentum is two closely applied leaves of peritoneum enclosing blood-vessels and lymphatics. The anterior (outer) leaf is in the greater sac, the posterior (inner) leaf is in the lesser sac. The outer leaf surrounds the spleen, which lies between the fundus of the stomach and the left kidney, in the greater sac. The posterior or inner leaf passes smoothly across the hilum of the spleen and takes no part in investing that organ with its serous coat. The attachment of the spleen to the outer leaf divides the left part of the greater omentum into a part between stomach and spleen, the *gastro-splenic omentum* and a part between spleen and left kidney, the *lieno-renal ligament* (Fig. 5.49, p. 318). Above the spleen the greater omentum sweeps from the back of the fundus of the stomach directly to the diaphragm above the left kidney; this part is known as the *gastro-phrenic ligament.*

Summarizing, the greater omentum is attached to the convexity leading from the œsophagus around the greater curvature of the stomach to the first part of the duodenum. To the left it consists of the gastro-phrenic and of the continuous layers of gastro-splenic omentum and lieno-renal ligament and to the right, in continuity, of the gastro-colic omentum. The spleen projects from its left, or outer, layer into the greater sac.

The two layers of the greater omentum enclose the stomach and leave its upper border, the lesser curvature, to pass upwards. They form a peritoneal fold known as the **lesser omentum,** or *gastro-hepatic omentum.* Its attachment to the stomach extends from the right side of the abdominal œsophagus, along the lesser curvature and pylorus to the first inch of the duodenum. Its upper attachment passes from the œsophageal opening across the diaphragm to the right, behind the liver, to the left of the inferior vena cava, and so across to the under surface of the liver (Fig. 5.40, p. 301). It has a curved attachment to the lower surface of the liver, which extends around the margin of the caudate lobe and so to the porta hepatis where its two layers end by separating from each other to surround the structures at the hilum of the liver, namely the right and left branches of the hepatic artery and portal vein and the right and left hepatic ducts. From the right side of the porta hepatis down to the first inch of the duodenum the right edge of the lesser omentum lies free, its front and back layers joining at the free edge, where they enclose the hepatic artery, common bile duct and the portal vein. The inferior edge of the liver hangs down over the front of the lesser omentum and touches the stomach, so the lesser omentum can only be seen when the liver is lifted up away from the stomach. The anterior leaf of the lesser omentum lies in the greater sac, the posterior leaf lies in the lesser sac (Fig. 5.18).

The two peritoneal leaves of the lesser omentum, traced downwards from their attachment to the lesser curvature of the stomach, separate to clothe the anterior and posterior walls of that viscus and re-unite at the greater curvature to continue downwards as the greater omentum. In a precisely similar way these two leaves traced upwards from their attachment to the liver, separate to clothe that organ and re-unite above it to become attached to the diaphragm and the anterior abdominal wall, where they constitute the triangular, coronary and falciform ligaments of the liver.

The diaphragmatic attachment of the lesser omentum has already been traced to the left side of the inferior vena cava, where it passes forwards to the under surface of the liver. From this point the two layers that have enclosed the liver unite above that organ and pass forwards under the diaphragm, to which they gain their parietal attachment (Fig. 5.17). These two leaves, however, are not united to the diaphragm side by side, as in the case of the two leaves of the lesser omentum. They separate widely from each other before coming together again anteriorly, and the space between them forms the bare area (i.e., an area not covered with peritoneum) on the back of the liver and the corresponding surface of the diaphragm.

It will be necessary to trace the attachments of each leaf separately over the area where they are separated. It must be remembered that the convexity of the liver touches the concavity of the diaphragm, so that (except the falciform ligament) each leaf is short, and its attachments to the liver are mirrored by identical attachments to the adjacent diaphragm.

Start with the left (or anterior) leaf at the left of the vena cava (Fig. 5.36, p. 296). It joins the liver and diaphragm and passes upwards between the two. On the upper surface of the left lobe it makes a transverse excursion to the left and then doubles back on itself towards the midline. This double fold of peritoneum, produced by the left leaf, is called the *left triangular ligament*. Its anterior fold meets the right leaf and from here forwards the left and right leaves are together again. But it is now necessary to trace out the attachments of the right (or posterior) leaf. These are a little more difficult to understand. The attachments begin in the lesser sac and *cannot be confirmed until that cavity is examined.* Starting from the same point at the left of the inferior vena cava the right (or posterior) leaf is merely the continuation of the posterior leaf of the lesser omentum, and forms the upper limit of the lesser sac. Traced to the right it is seen to be attached to the posterior part of the diaphragmatic surface of the liver, vertically along the left border of the inferior vena cava (Fig. 5.37, p. 297 and Fig. 5.17). It sweeps immediately from the liver to the right crus of the diaphragm. This attachment forms the right limit of the upper recess of the lesser sac (omental bursa). To its left the caudate lobe touches the diaphragm in the midline, above the aortic orifice (both these surfaces are peritonealized). The attachment of this leaf now turns to the right in front of the inferior vena cava, at the opening into the omental bursa. Tracing its attachment to the right of the opening into the lesser sac, the narrow flange is now seen connecting the under surface of the liver far back to the upper pole of the right kidney; this is the hepato-renal pouch already examined. Continue to trace the attachments of the right leaf, which is here named the *inferior layer of the coronary ligament*.

The leaf, lateral to the upper pole of the right kidney, attaches the liver to the diaphragm and then doubles back on itself for a centimetre. This small fold is named the *right triangular ligament* (Fig. 5.37, p. 297). From this ligament the leaf now passes upwards and to the left, diverging widely in V-shaped manner from the inferior layer of the coronary ligament at the hepato-renal pouch, and leaving a wide area of liver and contiguous diaphragm bare. This leaf, known as the *superior layer of the coronary ligament*, passes in front of the caval orifice of the diaphragm and rejoins the left leaf below the level of the anterior end of the left triangular ligament, on the upper surface of the liver (Fig. 5.37). The bare areas of liver and diaphragm, enclosed between the folds that diverge from the right triangular ligament, are in contact with each other, and the loose areolar tissue between them provides continuity of lymphatic spaces from the surface of the bare area of the liver through the diaphragm to nodes in the posterior mediastinum. The bare area lies on the posterior surface of the right lobe of the liver, and is in contact with the diaphragm, the right suprarenal gland and, in a deep gutter, with the inferior vena cava.

The leaves that diverge from the right triangular ligament to upper and lower edges of the posterior surface of the liver are known together as the coronary ligament. The upper layer of the coronary ligament meets the left leaf, as already noted, on the upper surface of the liver. From this point forwards the two leaves lie side by side, as in the typical mesenteries. Together they form the **falciform ligament.** This is attached to the upper surface of the liver somewhat to the right of the midline, as far forward as the notch in the anterior border of the liver (Fig. 5.36, p. 296). Its diaphragmatic attachment is to the central tendon (base of fibrous pericardium) and is continued along the linea alba from the xiphisternum to the umbilicus. The falciform ligament is thus seen to slope from the midline of the anterior abdominal wall across to the right to the anterior surface of the liver. Its free lower edge is crescentic in shape and contains a rounded fibrous cord, the *ligamentum teres*. This is the remnant of the obliterated left umbilical vein of the fœtus, and it runs in the free edge of the falciform ligament from the umbilicus to the anterior surface of the liver, cuts through the inferior border of the liver and lies in a deep groove (the fissure for the ligamentum teres) on the under (visceral) surface of the liver as far as the left end of the porta hepatis.

The attachments of the falciform ligament, of the

coronary ligament and of the left triangular ligament can all be examined by palpation of the upper convexity of the liver.

Having inspected the dispositions of the peritoneal folds and fossæ of the greater sac in both supracolic and infracolic compartments, the student should now investigate the omental bursa (lesser sac).

The Omental Bursa (Lesser Sac)

This is a diverticulum from the greater sac; it extends down behind the stomach as far as the transverse mesocolon and is bounded below the stomach by the greater omentum. The **epiploic foramen (the foramen of Winslow)** lies behind the free edge of the gastro-hepatic omentum and it admits two fingers. Place two fingertips of the left hand, fingernails down, in the opening. Deep to the fingers the peritoneum of the posterior abdominal wall passes as a smooth layer from the hepato-renal pouch across to the left, above the duodenum, through the opening and into the posterior wall of the lesser sac. At the opening the inferior vena cava lies immediately behind the posterior peritoneum. Press the fingers up towards the diaphragm; they touch the inferior surface of the liver, at the caudate process. Press them downwards away from the liver; they touch the first inch of the duodenum. Lift the fingers and grasp the free edge of the opening; between fingers and thumb lie the portal vein behind, the common bile duct in front in the free edge, and the hepatic artery in front on the left of the bile duct (Fig. 5.40, p. 301).

Explore further into the omental bursa with one finger, sweeping it around. Behind the posterior wall to the left of the inferior vena cava is the aorta. Here it gives off the cœliac artery, two of whose branches may be felt. The hepatic artery curves down to the right behind the peritoneum and then curves up behind the first inch of the duodenum to enter the gastro-hepatic omentum. It raises a fold of peritoneum, the *pancreatico-duodenal fold*, which can be felt—to the left of the fold the finger tip passes steeply downwards behind the pylorus, as if passing over a step. The left gastric artery passes from the front of the aorta behind the posterior peritoneum up to the œsophageal opening to enter the gastro-hepatic omentum (Fig. 5.32, p. 292). It raises a palpable fold in the posterior wall of the omental bursa, called the *pancreatico-gastric fold*. These two folds together produce a slight hour-glass constriction of the omental bursa, beyond which the cavity becomes extensive, but the examining finger cannot reach its limits.

Examine the roof of the bursa. It is like a lean-to roof. The posterior wall of the bursa passes up over the aorta and diaphragm to the under surface of the liver. Traced from the right to the left this peritoneal attachment extends from the inferior layer of the coronary ligament (greater sac) through the foramen of Winslow into the upper limit of the omental bursa, and is traceable across the diaphragm to the diaphragmatic attachment of the lesser omentum as far as the œsophagus (Fig. 5.17). Anterior to this peritoneal attachment the caudate lobe of the liver slopes steeply downwards like a lean-to roof (Fig. 5.18). Trace the caudate lobe downwards to the attachment of the lesser omentum; this forms the *anterior wall* of the omental bursa below the lean-to roof. Above the lesser curvature the left boundary of the bursa is the attachment of the lesser (gastro-hepatic) omentum to the diaphragm between

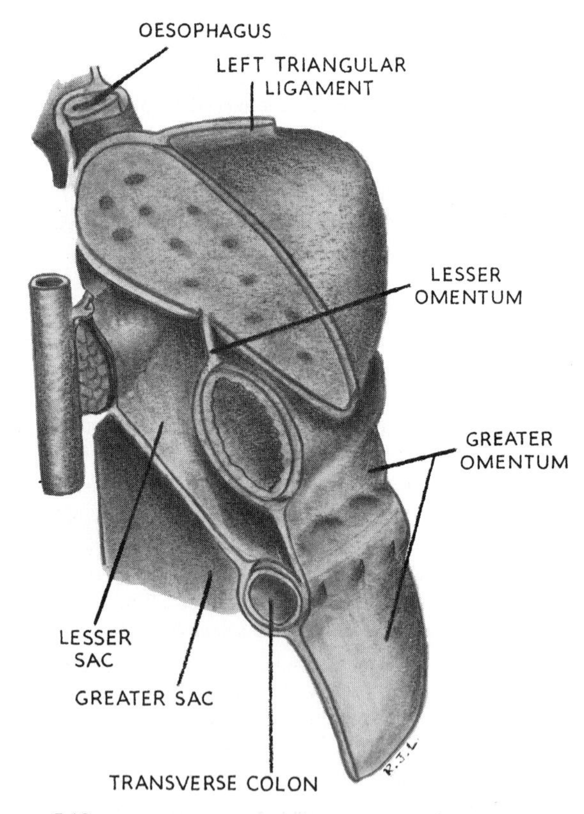

FIG. 5.18. A MIDLINE SAGITTAL SECTION THROUGH THE UPPER ABDOMEN, LOOKING LEFT INTO THE LESSER SAC. Blackboard sketch (see also PLATE 12).

the œsophageal opening and the inferior vena cava, already noted in the inspection of the greater sac (Fig. 5.17).

The body and fundus of the stomach continue the anterior wall below the gastro-hepatic omentum. The left boundary of the omental bursa below the œsophagus passes behind the cardiac orifice of the stomach, over the diaphragm to the lieno-renal ligament, which closes it off from the greater sac, down to the splenic flexure of the colon. The transverse mesocolon, a sloping shelf, forms the lower limit of the bursa, which is completed in front by the gastro-colic omentum (Fig. 5.18).

These last features cannot be reached by a finger passed through the epiploic foramen. Incise the gastro-colic omentum transversely and throw the stomach upwards (Fig. 5.26, p. 286). This manœuvre exposes the cavity of the omental bursa and now, for the first time, the upper leaf of the transverse mesocolon can be seen. It is attached towards the lower border of the pancreas. The structures to be seen or felt beneath the peritoneum of the *posterior wall* of the omental bursa can now be identified. Just above the attachment of the transverse mesocolon lies the body of the pancreas, extending up as high as the cœliac artery, whose left gastric branch runs from the upper border of the pancreas to the œsophagus, as already felt with the finger. The pancreatico-gastric fold can now be seen as well as felt. To the left of the fold the left suprarenal gland and upper pole of the left kidney can be felt and sometimes seen.

The peritoneal folds enclosing the omental bursa are thus identified as the lesser (gastro-hepatic) omentum, the greater omentum (gastro-colic omentum, lieno-renal ligament and gastro-splenic omentum) and the transverse mesocolon; and surgical access to the cavity of the omental bursa and posterior wall of the stomach may be gained by incision through any one of these folds (Fig. 5.18).

The lesser sac really is a 'bursa'. Its function is to provide a slippery surface for the necessary mobility of the *posterior surface* of the stomach (contrast the bladder, where increase of volume merely peels the peritoneum from behind the lower part of rectus abdominis, see p. 329).

After completing the examination of the peritoneum of the greater and lesser sacs, take the opportunity of examining the peritoneum of the pelvic cavity (see p. 339).

Development of the Gut and Formation of Peritoneal Folds

The disposition of the gut and its mesenteries in the early embryo is extremely simple. The more complex arrangement in the adult is due to elongation and consequent coiling of the alimentary canal and to fusion of certain adjacent peritoneal surfaces. Some slight knowledge of the embryology is not only of value in the satisfaction of knowing how the adult arrangement comes about, but has the considerable practical advantage of greatly clarifying understanding of the arterial supply and venous and lymphatic drainage of the whole alimentary system in the adult. The following embryological account, while true in substance, lacks accuracy in places for the purpose of simplification, and is not to be taken as a precise statement of the changes that actually occur in the growing embryo.

Picture the abdominal cavity just before the sixth week of embryonic life. The alimentary canal is a simple tube passing through to the hind end, its whole length supported by a *dorsal mesentery* attached in the midline in front of the aorta. *Three gut arteries* leave the aorta and pass ventrally to supply the tube. The most cranial passes in the dorsal mesogastrium to supply the foregut, the next passes through the dorsal mesentery to supply the midgut and the last passes through the dorsal mesocolon to supply the hindgut. They are the cœliac, the superior mesenteric and the inferior mesenteric arteries respectively, and they continue to supply the derivatives of these parts of the alimentary canal in the adult.

The foregut possesses, in addition, a *ventral mesogastrium* attached in the midline to the under surface of the diaphragm and the anterior abdominal wall down to the umbilicus. Its free caudal edge is crescentic and carries the left umbilical vein (Fig. 5.19).

The three derivatives of the foregut (liver, pancreas, spleen) are supplied with blood by the artery of the foregut, the cœliac artery. The liver develops as an outgrowth from the foregut at its junction with the midgut. A tube grows ventrally into the ventral mesogastrium, bifurcates, and cells proliferate from the blind end of the two divisions to form the two lobes of the liver, which is thus enclosed between the two layers of the ventral mesogastrium. The pancreas develops as two outgrowths, one into the ventral and one into the dorsal mesogastrium. These two parts subsequently fuse and exchange ducts by anastomosis, but the double origin of the pancreas is imprinted on the adult by the persistence of two pancreatic ducts, the main and the accessory. The spleen develops by proliferation of cells in the left leaf of the dorsal mesogastrium. It is not, properly speaking, a derivative of the foregut itself. The cœliac artery supplies the stomach and these three derivatives in the adult.

Lymphatic return from the alimentary canal is by lymph vessels that pass with the arteries and end ultimately in lymph nodes that lie in front of the aorta at the roots of the three gut arteries. Lymph from the mucous membrane of the alimentary canal passes through several filters. In the mucous membrane itself, from tonsils to anal margin, are lymphatic follicles. In the mesentery, at its gut margin, are lymph nodes (the 'epi-' group, e.g., epicolic). In the mesentery between its gut margin and the root of the artery are further nodes (the 'para-' group, e.g., paracolic nodes) while the pre-aortic nodes, inferior mesenteric, superior mesenteric, and cœliac are interconnected by lymph vessels. All the lymph thus ultimately reaches the cœliac nodes, whence it passes to the cisterna chyli. This simple lymphatic arrangement persists in the adult.

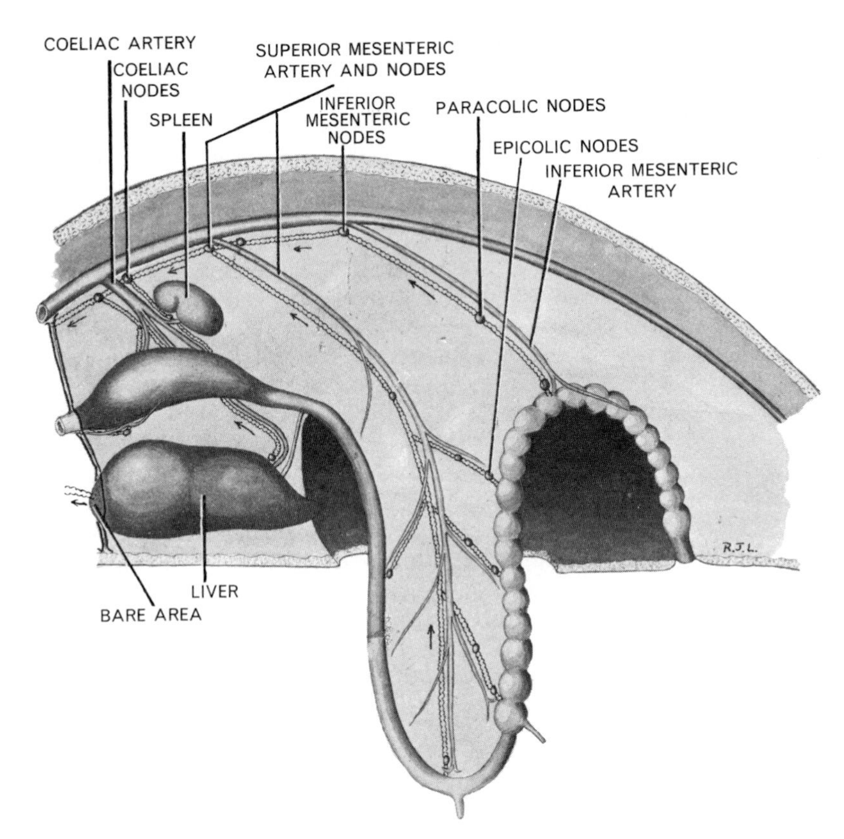

FIG. 5.19. A FANCIFUL REPRESENTATION OF THE DEVELOPING ALIMENTARY CANAL, VIEWED FROM THE LEFT SIDE. LYMPHATICS RUN BACK ALONG THE GUT ARTERIES TO PRE-AORTIC NODES.

By the end of the sixth week the liver has enlarged greatly and the gut has elongated, both to such an extent that the more leisurely growing abdominal walls cannot accommodate them. A loop of gut extrudes into the umbilical cord; it is called the 'physiological hernia' (Fig. 5.19). The loop remains in the umbilical cord for a full month. At the end of the tenth week the abdominal walls have grown enough to accommodate the abdominal contents and the hernia is reduced.

The herniated loop of gut is that supplied by the superior mesenteric artery and it is defined as the *midgut*. It is destined to produce all the small intestine and the proximal part of the colon, almost as far as the left colic flexure. The apex of the loop is at the attachment of the vitello-intestinal duct, the site of Meckel's diverticulum. The main trunk of the superior mesenteric artery is directed to the apex of the loop. Many branches run from it to the proximal limb of the loop, extending from the ventral pancreatic bud to Meckel's diverticulum. They persist as the jejunal and ileal branches. Only three branches run to the distal limb of the loop; all three persist in the adult as the ileo-colic, right colic and middle colic arteries. Their directions are altered considerably after the reduction of the physiological hernia and the rotation of the gut.

The Rotation of the Midgut. As the loop of midgut in the physiological hernia returns to the abdominal cavity it rotates so that the distal limb goes up on the left and the proximal limb goes down on the right, that is, *to the observer looking at the front of the abdomen, in an anti-clockwise direction* (Fig. 5.20). The distal loop, developing into colon, thus comes to lie anterior to the commencement of the proximal loop. The commencement of the proximal loop becomes, after some rotation, plastered to the posterior abdominal wall as the duodenum, and the mesentery of the transverse colon thus comes to lie across it (Fig. 5.21). The last part of the midgut to be re-included within the abdominal cavity is the cæcum, which lies first near the midline, high up. It grows then to the right, turns downwards at the right colic flexure and stops elongating at the right iliac fossa. It leaves a trail of large intestine to indicate its migration and drags the attached lower end of the ileum with it.

Rotation of the midgut loop occurs around the axis of the superior mesenteric artery, so in the adult the branches to the proximal loop (jejunal and ileal arteries) come off its left side while the three branches to the distal loop (colic arteries) leave its right side (Figs. 5.21, 5.25).

The simple dorsal mesentery of the midgut containing the superior mesenteric artery is, of course, much twisted and distorted during the return of the rotated loop of midgut and the subsequent migration of the

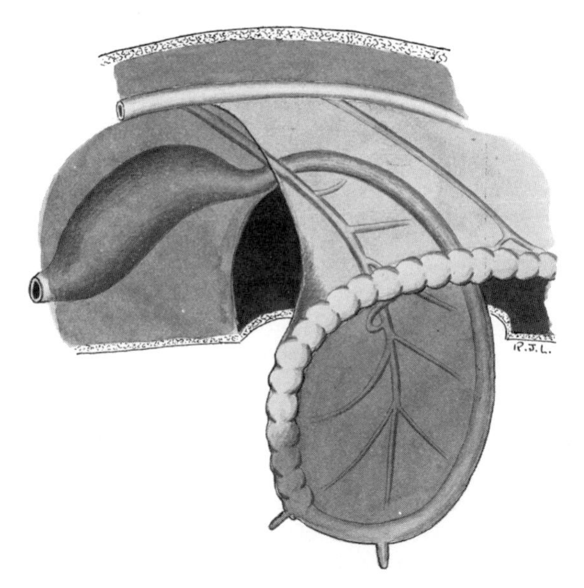

FIG. 5.20. THE ROTATION OF THE MIDGUT LOOP AROUND THE SUPERIOR MESENTERIC ARTERY (LEFT LATERAL VIEW).

cæcum. Its attachment to the *proximal loop* causes it to pass across the posterior abdominal wall from the commencement of the loop (duodenum) to the ileo-cæcal junction. The dorsal mesentery of the *distal loop* of the midgut hinges like a door across from the midline to the right. Its two layers come into contact with the parietal peritoneum in the right paravertebral gutter, so that three layers lie in the floor of the right infracolic compartment.

The deeper two fuse and are absorbed, the anterior (originally the right) layer remaining to floor in the right infracolic compartment, with the colic vessels lying immediately deep to it and *in front of everything else on the posterior abdominal wall.* The dorsal mesentery of the most distal part of the distal loop, pulled across transversely, does not fuse completely with the parietal peritoneum and persists, with the middle colic artery between its layers, as the transverse mesocolon (Fig. 5.21).

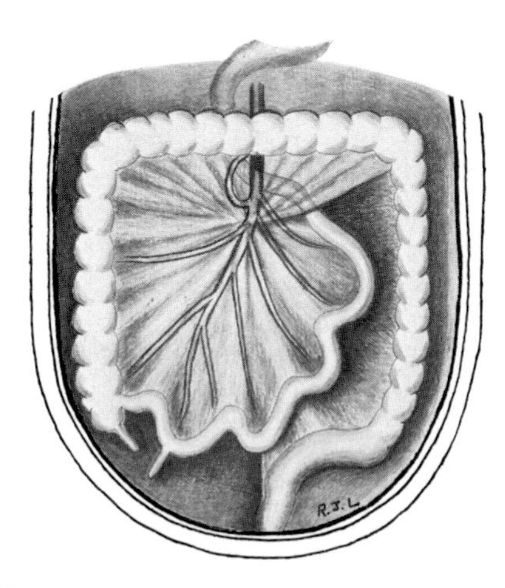

FIG. 5.21. THE RETURN OF THE PHYSIOLOGICAL HERNIA AND COMPLETION OF ROTATION OF THE MIDGUT (VENTRAL VIEW). THE MIDGUT LOOP ROTATES THROUGH THREE RIGHT ANGLES (270 DEGREES).

Movement of the Hindgut. As the midgut loop returns to the abdominal cavity, the hindgut swings on its dorsal mesocolon like a door across to the left (Fig. 5.22). The two layers of the mesocolon thus come to lie on the parietal peritoneum of the left paravertebral gutter. The left infracolic compartment is floored in by three layers of peritoneum. The two deeper layers fuse and are absorbed; the anterior (originally the right) layer persists, with the left colic vessels immediately beneath it and lying *in front of everything else on the posterior abdominal wall.* At the pelvic brim fusion of the layers is not complete and a small part of the intestinal edge of the dorsal mesocolon of the hindgut remains free as the sigmoid mesocolon of the adult.

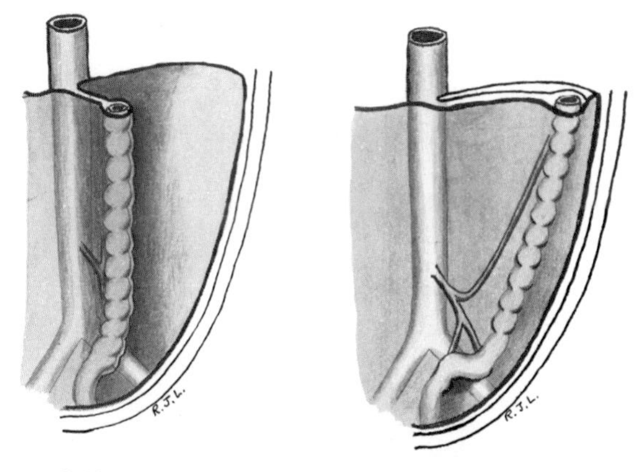

FIG. 5.22. THE HINDGUT BEFORE AND AFTER THE RETURN OF THE MIDGUT HERNIA (VENTRAL VIEW). THE MESOCOLON FUSES WITH THE PERITONEUM OF THE POSTERIOR ABDOMINAL WALL EXCEPT AT THE PELVIC BRIM, WHERE PART OF THE EMBRYONIC MESOCOLON PERSISTS AS THE SIGMOID MESOCOLON.

Growth of the Liver. The liver grows apace, and soon outstrips the ventral mesogastrium in which it lies. If it had remained between the layers of the ventral mesogastrium, the peritoneal attachments would have been simple; the liver would have been surrounded equatorially by a simple line of attachment of the two leaves of peritoneum that enclose it. The double leaf of peritoneum would have been attached to the ventral border of the foregut (i.e., from œsophagus to duodenum), to the under surface of the diaphragm, and to the middle of the ventral abdominal wall down to the umbilicus. The caudal border of this peritoneal flange would have swung free from umbilicus to duodenum; at this border the peritoneum encloses the left umbilical vein from umbilicus to liver and the common bile duct, hepatic artery, and portal vein from duodenum to liver (Fig. 5.19).

The growth of the liver beyond the bounds of this simple mesogastrium, and the simultaneous rotation of the stomach, tend to obscure the persisting simplicity of the peritoneal arrangements.

The liver grows caudally into the free edge of the ventral mesogastrium, which it pushes down until the umbilical vein (ligamentum teres in the adult) notches its inferior border and is enclosed in a deep groove on its under surface. This encroachment of the liver over the crescentic free margin of the ventral mesogastrium divides the latter into two separate parts, namely the falciform ligament between liver and anterior abdominal wall and the lesser omentum (gastro-hepatic omentum) between liver and stomach (Fig. 5.19). The lesser omentum is attached to the liver along a fissure that runs backwards behind the fissure for the ligamentum teres, a fissure that lodges the fibrous remnant of the ductus venosus (the ligamentum venosum), and this attachment marks the descriptive (but not the true) division of the under surface of the liver into right and left lobes. The hepatic attachment of the lesser omentum passes back to the left of the inferior vena cava, and so reaches the diaphragm. Its two leaves are attached to the diaphragm between this point and the œsophageal opening (Fig. 5.17).

Had the liver not outgrown the cranial part of the ventral mesogastrium, the latter would have been attached from this point along a simple line on the under surface of the diaphragm and anterior abdominal wall, and to the convexity of the upper surface of the liver. But this is not so. As the liver grew caudally to engulf the ligamentum teres it was contained within the layers of the ventral mesogastrium and merely distended them to incorporate them into its own serous coat. But the liver grows also towards the diaphragm and peels apart the two leaves of the ventral mesogastrium, and comes into contact with the diaphragm over the bare area. The right and left leaves of the ventral mesogastrium between the back of the liver and adjacent diaphragm are thus stripped away from each structure, and become separated widely from each other. The right leaf follows, along the posterior surface of the liver, the lower border of that organ (whence it is reflected over the right kidney at the hepato-renal pouch), then returns at the right triangular ligament along the upper border of the posterior surface of the liver to meet the left leaf. This divergent V-shaped fold is called the *coronary ligament* (p. 273).

The two leaves, right and left, touch on the upper surface of the liver. Further forwards the left leaf sweeps to the left and back again upon itself to make the left triangular ligament and thence forwards the two leaves

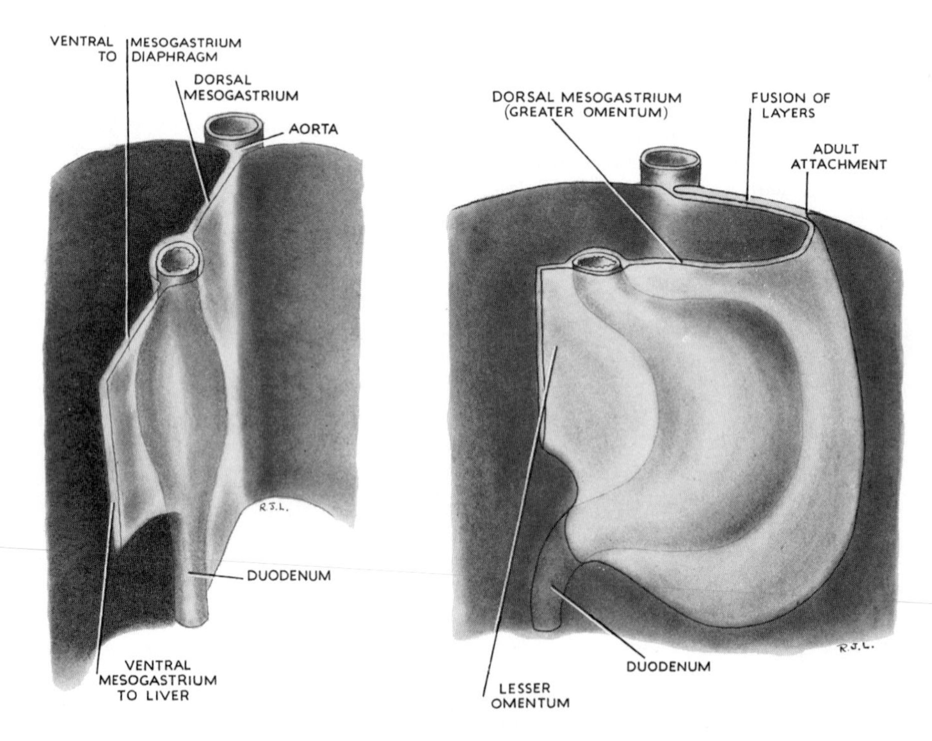

FIG. 5.23. THE ROTATION OF THE FOREGUT (VENTRAL VIEW AFTER REMOVAL OF THE LIVER). THE DORSAL MESOGASTRIUM BALLOONS DOWN AS THE GREATER OMENTUM. THE SPLEEN AND PANCREAS ARE NOT SHOWN. THE STOMACH ROTATES THROUGH 90 DEGREES, SO THAT ITS LEFT SURFACE BECOMES ANTERIOR.

lie together as the falciform ligament, which has already been studied.

The peritoneal flanges of the coronary ligament and left triangular ligament are short, and their attachments to liver and diaphragm are identical. The **falciform ligament** is wider, and its visceral and parietal attachments do not coincide. The liver attachment lies to the right of the midline, the parietal attachments lie in the midline; the ligament lies with its right surface against the anterior abdominal wall and its left surface against the liver.

Rotation of the Foregut. Coincident with the growth of the liver, the foregut rotates. The liver originally was ventral to the foregut, in the ventral mesogastrium, and both lay in the midline. As the liver grows it swings to the right, taking the ventral mesogastrium with it. The stomach swings across to the left and in doing so rotates (Fig. 5.23). It has already elongated and broadened, with its dorsal border becoming convex, and its ventral border concave. The distal end of the foregut, destined to become the duodenum, does not dilate in this manner, and its dorsal mesentery shortens. The duodenal part of the gut elongates into a loop which swings to the right and becomes plastered to the posterior abdominal wall (cf. the ascending and descending parts of the colon). At the same time its walls grow asymmetrically so that the ventral common bile duct and pancreatic duct are carried around to open on the medial wall (see Fig. 5.44, p. 309) in line with the duct of the dorsal diverticulum. The duodenum is now fixed in position; so, too, is the œsophagus at the diaphragm. Between these two fixed points as an axis, the dorsal convexity of the stomach rotates to the left. The dorsal convexity becomes the greater curvature, the original left side now faces anteriorly. The concave ventral border now becomes the lesser curvature, fixed by the lesser omentum (originally the ventral mesogastrium) to the under surface of the liver, and to the diaphragm between liver and œsophagus. The original right surface of the stomach now lies behind, against the peritoneum of the posterior abdominal wall, and the free edge of the gastro-hepatic omentum lies over the opening from the greater sac into this space behind the stomach, called the omental bursa (lesser sac).

Meanwhile, however, changes have occurred in the disposition of the dorsal mesogastrium. It will be remembered that it was attached to the midline of the posterior abdominal wall. As the dorsal border of the stomach swings to the left the dorsal mesogastrium hinges to the left from this attachment and adheres to the parietal peritoneum as far left as the front of the left kidney. Its left layer fuses with the peritoneum on the posterior abdominal wall and the two become absorbed (Fig. 5.23). The original right layer of the dorsal mesogastrium then lies on the posterior abdominal wall, with the left gastric and splenic branches of the cœliac axis (originally running in the dorsal mesogastrium) immediately deep to it. From the front of the left kidney, and from the diaphragm above it, the two layers of the dorsal mesogastrium pass to the œsophagus and the upper part of the greater curvature of the stomach. They form part of the greater omentum, and constitute the left boundary of the omental bursa (lesser sac) behind the stomach. The spleen projects from the left leaf into the greater sac, and it divides this part of the greater omentum into gastro-splenic and lieno-renal parts, as already seen.

The fate of the more caudal part of the dorsal mesogastrium remains to be stated. It is attached to the lower part of the greater curvature of the stomach, the inferior border of the pylorus and the first inch of the duodenum. Its dorsal attachment, in the midline, hinges to the left and becomes plastered to, and fused with, the peritoneum on the posterior abdominal wall over the pancreas. Below this it balloons down like an apron from the greater curvature (Fig. 5.23). It balloons over the transverse mesocolon and the transverse colon, and its posterior part returns to the posterior abdominal wall. Each part consists of the double layer of the original dorsal mesogastrium. The deeper part of the double layer fuses with the upper leaf of the transverse mesocolon and with the transverse colon itself, forming the inferior limit of the omental bursa. The superficial part of the double layer hangs down from the greater curvature directly to the transverse colon, to which it adheres, forming the anterior wall of the lower part of the omental bursa (PLATE 12).

From the transverse colon the two folds of the double layer hang down over the front of the coils of small intestine; adjacent leaves of peritoneum become practically fused to each other and can be separated only partially and with difficulty in the adult. They form the fat-containing apron generally known surgically as 'the omentum'.

The Blood Supply of the Alimentary Canal

When the disposition of the peritoneum in the adult is clear, the course of the three ventral branches of the aorta to the gut can be followed simply.

The Cœliac Artery. This is the artery of the foregut, and its three branches supply the alimentary canal down to the opening of the bile duct, and the derivatives, liver, spleen and pancreas. It arises from the front of the aorta high on the posterior abdominal wall, between the crura of the diaphragm, opposite the body of the twelfth thoracic vertebra. It is a short wide trunk, flanked by the cœliac group of the pre-aortic lymph nodes. The cœliac ganglia of the sympathetic system lie one on each side and they send nerves to the artery which are carried along all its branches.

The artery appears at the upper border of the pancreas and divides immediately into its three branches, behind the peritoneum of the posterior wall of the omental bursa. The splenic artery passes to the left. The left gastric and hepatic arteries pass in opposite directions to each other; the former upwards to the œsophageal opening, the latter down to the pylorus, and each raises a fold in the peritoneum. The two folds make a slight 'hour glass' constriction across the cavity of the omental bursa.

The **left gastric artery** runs behind the peritoneum from the upper border of the pancreas across the left crus of the diaphragm to the œsophageal hiatus. It raises the peritoneum into a ridge (the pancreatico-gastric fold), and gives off an *œsophageal branch* which runs up on the œsophagus to supply the lower reaches of that tube in the posterior mediastinum. It now enters between the two leaves of the lesser omentum (gastro-hepatic omentum) and turns to the right along the lesser curvature. It breaks into two parallel branches which anastomose end on with the two branches of the right gastric artery. The two arteries give off branches at right angles. These sink into the anterior and posterior walls of the stomach and anastomose very freely with similar branches from the arteries of the greater curvature. Such anastomoses between arteries in the gut wall cease beyond the stomach. In the small and large intestines the vasa recta that enter the gut wall are end arteries.

The **splenic artery** arises at the upper border of the pancreas and passes to the left. It is very tortuous. The crests of its waves appear above the pancreas, the troughs lie hidden behind its upper border. The artery runs behind the peritoneum and passes, with its vein and the tail of the pancreas, across the left crus and left psoas muscle to the hilum of the left kidney where it turns forward in the lieno-renal ligament to the hilum of the spleen. Here it breaks up into a 'handful of fingers'—four or five short branches that radiate as they sink into the splenic substance. The splenic artery is the main source of arterial supply to the pancreas. Several branches supply that gland—one large branch is named the *arteria pancreatica magna* (PLATE 16).

Just before the artery breaks up into its terminal splenic branches it gives off the left gastro-epiploic artery and the vasa brevia. All pass to the greater curvature of the stomach between the leaves of the gastro-splenic omentum. The **vasa brevia** consist of half a dozen short arteries that pass to the fundus of the stomach, which they supply. The **left gastro-epiploic artery** passes to the right along the greater curvature, between the two layers of the gastro-colic

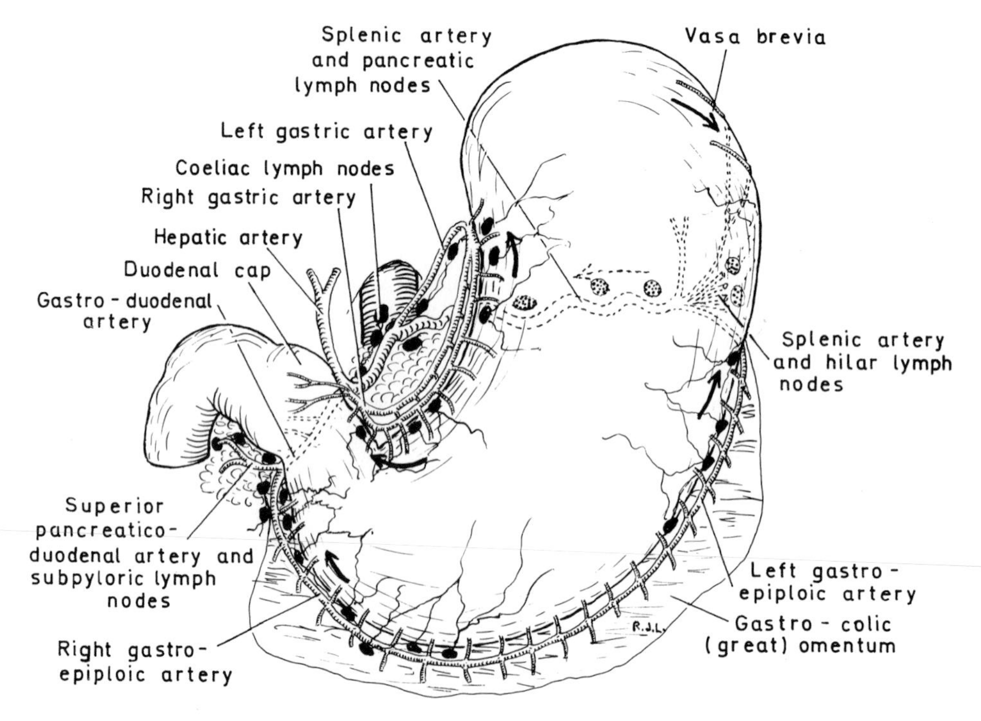

FIG. 5.24. THE ARTERIAL SUPPLY OF THE STOMACH, FROM THE THREE BRANCHES OF THE CŒLIAC ARTERY. The lymphatic drainage is shown. THE STOMACH WALL DRAINS IN FOUR QUADRANTS TO LYMPHATIC NODES ALONG THE GREATER AND LESSER CURVATURES. The arrows indicate the direction of lymph drainage, alongside the arteries against their direction of blood flow.

(greater) omentum. It usually anastomoses end on with the right gastro-epiploic artery in the gastro-colic omentum (Fig. 5.24), but occasionally it ends by sinking into the greater curvature. It lies about a centimetre from the stomach wall, and gives off branches at right angles. The upper branches pass in the gastro-colic omentum to the greater curvature and sink into the anterior and posterior walls of the stomach and anastomose freely with the corresponding branches of the gastric arteries on the lesser curvature. They are the *gastric branches*. The lower branches pass downwards between the leaves of the gastro-colic omentum, across the transverse colon and into the peritoneal apron of the greater omentum. Note that they take no part in supplying the transverse colon; their area of distribution is restricted to the greater omentum. They are the *epiploic branches* (Fig. 5.24).

The **hepatic artery** passes over the upper border of the pancreas, downwards and to the right behind the peritoneum of the posterior abdominal wall (in the lesser sac) as far as the first part of the duodenum. It turns forward at the opening into the lesser sac (epiploic foramen) and curves upwards into the space between the two layers of the gastro-hepatic (lesser) omentum. Here it meets the common bile duct and lies on its left side, both in front of the portal vein. All three lie in this position, between the duodenum and the porta hepatis, surrounded by the peritoneum at the free edge of the gastro-hepatic omentum. On reaching the porta hepatis, the hepatic artery divides into right and left branches to supply the right and left halves of the liver (Fig. 5.40, p. 301).

Quite commonly the hepatic artery ends at the porta hepatis by entering only the left half of the liver; in these cases the right half of the liver is supplied by a right hepatic artery which branches from the superior mesenteric artery (PLATE 16).

Branches. The **right gastric artery** leaves the hepatic as it turns into the lesser omentum. The vessel passes to the left between the layers of the lesser omentum, and divides into two branches which anastomose end on with the branches of the left gastric artery. They give off branches at right angles, which sink into the anterior and posterior walls of the stomach to supply it and anastomose freely with similar branches of the right gastro-epiploic artery.

The **gastro-duodenal artery** passes down behind the first part of the duodenum, to the left of the portal vein, and divides into two. The **right gastro-epiploic artery** passes forward between the first part of the duodenum and the pancreas, and turns to the left to enter between the two leaves of the greater (gastro-colic) omentum at their attachment over the front of the head of the pancreas. It runs to the left in the gastro-colic omentum a centimetre from the greater curvature and usually anastomoses end on with the left gastro-epiploic artery. It gives off branches at right angles to anterior and posterior walls of the stomach and to the greater omentum. Rarely the gastro-epiploic arteries may be double.

The other branch of the gastro-duodenal artery is called the **superior pancreatico-duodenal artery.** It, too, divides into two branches, which encircle the head of the pancreas in the concavity of the duodenum down to the entrance of the bile duct (PLATE 16, p. *viii*). They consist of a small anterior and a larger posterior branch, each of which anastomoses with similar branches of the inferior pancreatico-duodenal branch of the superior mesenteric artery. The entrance of the common bile duct marks the junction of foregut and midgut, and is the meeting place of the arterial distributions of the respective arteries, cœliac and superior mesenteric. The pancreatico-duodenal arteries supply the duodenum and the adjacent head of the pancreas.

Venous Drainage of the Foregut

It is convenient to study the venous return with the arterial supply, for the two are essentially similar. Right and left gastric, right and left gastro-epiploic veins run with the corresponding arteries. All this blood reaches the liver via the portal vein and, with the arterial blood of the hepatic artery, passes through the liver to be carried via the hepatic veins to the inferior vena cava.

The superior mesenteric vein passes up behind the pancreas and continues as the portal vein, a new name it receives after the splenic vein has joined it at a right angle behind the neck of the pancreas. The veins of the stomach empty into the nearest part of the vertical venous channel formed by the superior mesenteric and portal veins. They correspond with the arteries of the stomach with the single exception that *there is no gastro-duodenal vein.*

The lower third of the œsophagus in the posterior mediastinum drains downwards by **œsophageal veins,** through the œsophageal hiatus in the diaphragm, to the left gastric vein. The œsophagus above this level drains into the azygos system of veins. Here is a watershed from which venous blood is diverted into the systemic system and the portal system. In portal obstruction from any cause the œsophageal tributaries of the left gastric vein become distended beneath the mucous membrane of the lower œsophagus. Such 'œsophageal piles' may rupture and give rise to massive hæmorrhage.

The **left gastric vein** runs to the left along the lesser curvature up to the œsophagus, then passes around behind the peritoneum of the posterior wall of the lesser sac, with the left gastric artery. It passes down to the right above the hepatic artery and joins the portal vein at the upper border of the pancreas. The **right gastric vein** runs along the lesser curvature to the pylorus and, to the right, behind the first part of the duodenum, it empties into the portal vein. It receives the *prepyloric vein.* This small vein, lying vertically across the front of the gut, is a visible guide to the situation of the pylorus.

The **vasa brevia** and **left gastro-epiploic vein** run with the arteries through the gastro-splenic omentum to the hilum of the spleen, where they empty into the splenic vein.

The **splenic vein** begins in the hilum of the spleen by confluence of half a dozen tributaries from that organ. Having received the vasa brevia and the left gastro-epiploic vein it passes with the tail of the pancreas, below the splenic artery, through the lienorenal ligament to lie over the hilum of the left kidney. It is a large straight vein which passes to the right in contact with the posterior surface of the pancreas. In its course it lies on the hilum of the left kidney, the left psoas muscle and left sympathetic trunk, the left crus of the diaphragm, the aorta and superior mesenteric artery and the inferior vena cava. It lies in front of the left renal vein along the upper border of that vessel. In front of the inferior vena cava it joins the vertical trunk of superior mesenteric and portal veins at a right angle. It receives many tributaries from the tail, body, neck and head of the pancreas. As it lies in front of the left crus of the diaphragm it receives the inferior mesenteric vein from the hindgut.

The **right gastro-epiploic vein** runs to the right in the gastro-colic omentum until that peritoneal structure becomes fused to the peritoneum of the posterior abdominal wall over the head of the pancreas. It runs down over the front of the pancreas, behind the peritoneum, to join the superior mesenteric vein at the lower border of the neck of the pancreas.

The **superior pancreatico-duodenal veins** run up in the curve between the duodenum and the head of the pancreas, behind both, and join the portal vein at the upper border of the pancreas.

Lymphatic Drainage of the Foregut (see p. 285).

Blood Supply of the Midgut

The artery of the midgut is the superior mesenteric, which supplies the gut from the entrance of the bile duct to a level just short of the splenic flexure of the colon. In the embryo it is directed to the apex of the loop of midgut in the physiological hernia, the site of attachment of the vitello-intestinal duct. This is the site of Meckel's diverticulum—in the adult it is in the ileum 2 feet proximal to the cæcum and the artery terminates at this point. The rotation of the midgut loop occurs around this artery as its axis. The cranially directed branches (jejunal and ileal) to the proximal loop of the midgut thus face to the left in the adult, and the three caudally directed branches (ileo-colic, right colic and middle colic) face to the right (Fig. 5.25).

The **superior mesenteric artery** arises from the front of the aorta $\frac{1}{2}$ inch below the cœliac axis, at the level of the first lumbar vertebra. It is directed steeply downwards behind the splenic vein and the neck of the pancreas. With the superior mesenteric vein on its right side it lies on the left renal vein, then on the uncinate process, then on the third part of the duodenum. Here the two vessels enter the upper end of the mesentery of the small intestine. They pass down to the right along the root of the mesentery and end at the ileum 2 feet proximal to the cæcum (Fig. 5.25). Pressure of the superior mesenteric artery on the left renal vein (e.g., in visceroptosis) may produce left-sided varicocele in either sex, and pressure on the duodenum may give symptoms of chronic duodenal ileus.

The **inferior pancreatico-duodenal artery** is its first branch. It supplies the commencement of the midgut; that is to say, the duodenum below the entrance of the bile duct. It runs in the curve between the duodenum and the head of the pancreas, supplies both, and anastomoses with the terminal branches of the superior pancreatico-duodenal artery. The **right hepatic artery** occasionally arises with the inferior pancreatico-duodenal artery and runs up behind the portal vein and common bile duct (PLATE 16, p. *viii*) to supply the right half of the liver.

The **jejunal arteries** arise from the left of the main trunk and pass forward between the two layers of the mesentery (Fig. 5.25). They join each other in a series of anastomosing loops which form single arterial arcades for the upper part of the jejunum, double arcades further down. From the arcades straight arteries pass to the mesenteric border of the jejunum (Fig. 5.15). These arteries are long and close together, forming high narrow 'windows' in the intestinal border of the mesentery, visible because the mesenteric fat does not reach thus far into the mesentery. The straight vessels pass to one or other side of the jejunum and sink into its wall. They possess no pre-capillary anastomoses with each other (i.e., they are end arteries).

The **ileal arteries** enter the mesentery and form a series of arterial arcades. Four or five such arcades are so produced, the most distal one lying near the ileum. Straight vessels (vasa recta) pass to the mesenteric border of the ileum and sink into one or other wall, with no anastomoses between them (Fig. 5.15). They are shorter and farther apart than the vasa recta of the jejunum; the low broad 'windows' they produce in the intestinal edge of the mesentery are, however, usually hidden from view by mesenteric fat, which extends through the mesentery to its ileal attachment.

The arcades of the terminal ileal branch anastomose freely with those of the terminal part of the main trunk of the superior mesenteric artery.

The **ileo-colic artery** (Fig. 5.25) arises from the right side of the superior mesenteric trunk low down in the base of the mesentery. It runs therein to the ileo-colic junction, where it gives off an *ileal branch* which anastomoses with the terminal branch of the superior mesenteric artery and a *colic branch* which runs up along the left side of the ascending colon, behind the peritoneal floor of the right infracolic compartment, to anastomose with the right colic artery. The artery now divides to supply the cæcum. The *anterior cæcal artery* is the smaller of the two terminal branches and ramifies over the anterior surface of the cæcum. The *posterior cæcal artery* is the largest branch of the ileo-colic artery. It supplies the posterior wall of the cæcum, as well as the medial and lateral walls of that part of the gut. It gives off the *appendicular artery* (often double) which passes down behind the terminal part of the mesentery of the small intestine and draws up the inferior leaf thereof into a fold, the meso-appendix, as it passes towards the tip of the appendix (Fig. 5.34, p. 294).

The **right colic artery** (Fig. 5.25) arises in the root of the mesentery from the right side of the superior mesenteric artery, often in common with the ileo-colic artery. It runs to the right across the psoas muscle, right gonadal vessels and ureter and genito-femoral nerve, and the right quadratus lumborum muscle, just behind the peritoneal floor of the right infracolic compartment. It divides near the left side of the ascending colon into two branches. The descending branch runs down to anastomose with the colic branch of the ileo-colic artery. The ascending branch runs up across the inferior pole of the right kidney to the hepatic flexure where it anastomoses with a branch of the middle colic artery. From these two branches multiple vessels sink into the walls of the colon.

The **middle colic artery** (Fig. 5.25) is the highest branch from the right side of the superior mesenteric trunk. It arises as the artery emerges at the lower border of the neck of the pancreas and passes forwards between the two leaves of the transverse mesocolon. It lies to the right of the midline and at the intestinal border of the transverse mesocolon it divides into right and left branches which run along the transverse colon. The right branch anastomoses with the ascending branch of the right colic artery. The left branch supplies the transverse colon almost to the splenic flexure (the distal part of the midgut) where it anastomoses with a branch of the left colic artery. From these two branches multiple vessels sink into the walls of the colon. As the middle colic lies to the right of the midline it leaves a large avascular window to its left in the transverse

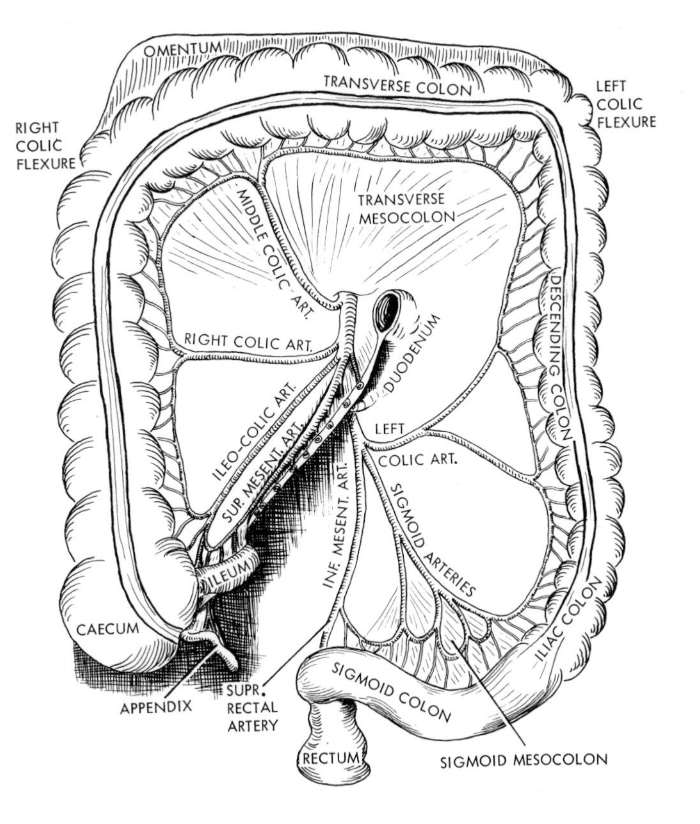

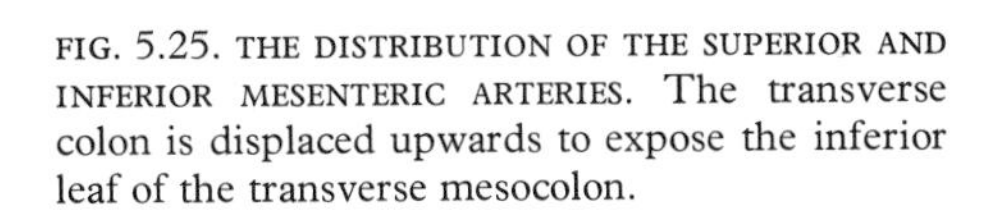
FIG. 5.25. THE DISTRIBUTION OF THE SUPERIOR AND INFERIOR MESENTERIC ARTERIES. The transverse colon is displaced upwards to expose the inferior leaf of the transverse mesocolon.

mesocolon. This window is the site of election for surgical access to the lesser sac (omental bursa) and the posterior wall of the stomach (Fig. 5.25).

Venous Drainage of the Midgut

This is quite regular. Each branch of the superior mesenteric artery is accompanied by a vein. All these veins flow into the **superior mesenteric vein,** a large trunk which lies to the right of the artery. It crosses the third part of the duodenum, runs between the uncinate process and the neck of the pancreas and passes straight up, as the portal vein, behind the first part of the duodenum. It is named portal vein above, and superior mesenteric vein below, the level of entry of the splenic vein, but the two represent *a single continuing venous trunk* (Fig. 5.44C, p. 309).

Lymphatic Drainage of the Midgut (see p. 286).

Blood Supply of the Hindgut

The artery of the hindgut is the inferior mesenteric and it supplies the whole extent of the hindgut. The muscle walls of the rectal ampulla and anal canal receive a reinforcement of arterial supply from the middle and inferior rectal arteries but the mucous membrane is supplied by the inferior mesenteric artery to the distal extremity of the hindgut, the muco-cutaneous junction at the lower end of the anal canal (Hilton's line).

The **inferior mesenteric artery** arises from the front of the aorta at the inferior border of the third part of the duodenum, opposite the third lumbar vertebra, at the level of the umbilicus. It is much smaller than the superior mesenteric artery. It runs obliquely down to the pelvic brim immediately beneath the peritoneal floor of the left infracolic compartment. It crosses the pelvic brim at the bifurcation of the left common iliac vessels over the sacro-iliac joint, at which point it converges on the ureter, at the apex of the Λ-shaped attachment of the pelvic mesocolon. In its course it lies on the aorta, left psoas muscle and sympathetic trunk and the left common iliac artery, and hypogastric nerves. Over the pelvic brim it continues along the pelvic wall in the root of the pelvic mesocolon as the superior rectal artery (p. 327). It does not cross the ureter where the two meet at the pelvic brim, but all its branches cross to the left in front of the ureter and the other structures in the floor of the left infra-colic compartment.

The *upper left colic artery* leaves the trunk and passes up to the left towards the splenic flexure, lying beneath the peritoneal floor of the left infracolic compartment. It crosses the left psoas muscle, left gonadal vessels, left ureter and genito-femoral nerve, and the quadratus lumborum muscle. It is crossed by the inferior mesenteric vein. It divides into two branches. The *upper branch* passes upwards across the inferior pole of the left kidney to the splenic flexure. The *lower branch* passes transversely to the descending colon. Each of the arteries divides into ascending and descending branches which anastomose with the left branch of the middle colic artery and with each other to continue the arterial 'circle' around the concavity of the large intestine.

The *lower left colic arteries* are two or three branches which leave the inferior mesenteric artery, close together just above the pelvic brim. They pass to the left behind the peritoneum of the iliac fossa and supply the lower part of the descending colon and the iliac colon, forming anastomosing loops with each other before doing so.

The *sigmoid arteries* are three or four branches which pass forwards between the layers of the sigmoid mesocolon, in which they form anastomosing loops from which vessels sink into the wall of the sigmoid colon.

The anastomoses around the concave border of the large intestine from ileo-colic junction to pelvi-rectal junction result in the formation of a single arterial trunk. It is sometimes called the *marginal artery.* This is a useful conception. The artery supplies the colon by vessels which sink into the walls. It is fed with blood from the colic branches of the superior and inferior mesenteric arteries. Its commencement anastomoses freely with the ileal branches of the superior mesenteric artery. Its lower part anastomoses less freely with the superior rectal artery. The anastomoses between the lowest sigmoid branch to the pelvic colon and the upper rectal branch of the superior rectal artery is sometimes by a small vessel. The site of this vessel is known as the '*critical point*'. It is called 'critical' because of the supposed inadequacy of blood supply coming up to the lower part of the pelvic colon from the superior rectal artery if all the sigmoid arteries are divided in excision of the pelvic colon. In such cases it was formerly recommended to leave the lowest sigmoid artery intact. Experience has shown that the 'critical point' is largely imaginary and that the anastomosis between superior rectal artery and lowest sigmoid artery is usually adequate to nourish the lower part of the pelvic colon even when all sigmoid arteries are divided (Fig. 5.25).

Blood Supply of the Rectum (see p. 327).

Venous Drainage of the Hindgut

The venous drainage of the rectum is considered in detail on p. 328. The superior rectal vein runs up in the root of the sigmoid mesocolon, on the left of the superior rectal artery, to the pelvic brim, above which it is named the **inferior mesenteric vein.** This receives tributaries identical with the branches of the inferior mesenteric artery. The vein itself runs vertically upwards well to the left of the artery, beneath the peritoneal floor of the left infracolic compartment. It lies on the left psoas muscle, in front of the left gonadal vessels, left ureter and genito-femoral nerve. At the upper limit of the left infracolic compartment, just below the attachment of the transverse mesocolon, it lies to the left of the duodeno-jejunal flexure. Here it curves towards the right and often raises up a ridge of peritoneum. This ridge may be excavated by a small recess of peritoneum which makes a shallow cave beneath it called the paraduodenal fossa (Fig. 5.33).

The inferior mesenteric vein now passes behind the lower border of the body of the pancreas, in front of the left renal vein, and joins the splenic vein. Occasionally it curves to the right still more sharply, and, passing behind the pancreas, below and parallel with the splenic vein, in front of the superior mesenteric artery, opens directly into the superior mesenteric vein.

Lymph Drainage of the Alimentary Canal

From the whole length of the alimentary canal the lymph vessels pass back along the arteries to lymph nodes that lie in front of the aorta at the origins of the gut arteries (Fig. 5.19). These comprise the cœliac, superior mesenteric and inferior mesenteric groups of lymph nodes. They drain into each other from below upwards, the cœliac group itself draining by two or three lymph channels into the cisterna chyli.

These pre-aortic lymph nodes are, however, the last in a series of lymph-node filters that lie between the mucous membrane of the gut and the cisterna chyli. The first filtering mechanism consists of isolated *lymphatic follicles* which lie in the mucous membrane of the alimentary canal from tonsils to anus. They are not very numerous in the œsophagus, are more numerous in the stomach, and become increasingly so along the small intestine. In the lower reaches of the ileum they become aggregated together into patches visible through the muscular wall. These lie on the antimesenteric border of the ileum and are oval in shape, with their long axes lying longitudinally along the ileum. They are called *Peyer's patches.* In the large intestine the lymph follicles in the mucous membrane are numerous, but isolated from each other. In the appendix they are aggregated; the appendix is a tonsil.

Lymph vessels pass from the follicles in the mucous membrane through the muscle wall of the gut to nodes that lie in the peritoneal attachments at the margin of the gut. The next groups of nodes lie along the arteries between the gut wall and the aorta, and finally the three pre-aortic groups receive lymph, from the hindgut, midgut and foregut respectively. Lymph from the lacteals of the villi of the small intestine (laden with emulsified fat) mostly by-passes nodes and passes directly to the cisterna chyli by way of two or three intestinal lymph trunks.

With this general plan of lymph drainage in mind it is simple to follow the precise pathways of lymph vessels and the situation of the lymph nodes.

Lymph Drainage of the Foregut. The lymph drainage of the *œsophagus* is considered on p. 239. Note that the lower portion of this tube drains with the œsophageal branch of the left gastric artery down through the œsophageal opening in the diaphragm to join the left gastric lymphatics.

The **stomach** drains to lesser and greater curvatures, in four quarters, which correspond in area to those supplied with blood from the right and left gastric and gastro-epiploic arteries (Fig. 5.24). Scattered lymph nodes lie with the arteries along the lesser and greater curvatures of the stomach, and they receive lymphatics which have perforated the stomach wall from the lymphatic follicles in the mucous membrane.

The lesser curvature drains to the left along the left gastric artery, through the pancreatico-gastric fold, where a few lymph nodes lie with the artery, into the cœliac lymph nodes. These lymphatics are joined by vessels coming down from the lower part of the œsophagus in the thorax. The lesser curvature drains to the right along the right gastric artery to its origin from the hepatic artery. Here, behind the first part of the duodenum, are some nodes which receive lymphatics from the liver and from the area of distribution of the gastro-duodenal artery. Their lymph vessels pass to the cœliac nodes, with the hepatic artery (Fig. 5.26).

The fundus and upper part of the greater curvature drain to lymph nodes lying in the attached margin of the gastro-splenic omentum. From these nodes vessels pass with the left gastro-epiploic artery and the vasa brevia to the group of lymph nodes in the hilum of the spleen. These also receive lymph from the spleen. Their efferent vessels pass to the right with the splenic artery to retro-pancreatic nodes, which receive also from the pancreas, and so with the splenic artery to the cœliac nodes. The lower part of the greater curvature drains into nodes lying in the attached margins of the gastro-colic omentum. From these nodes lymph vessels pass to the right along with the right gastro-epiploic artery to the gastro-duodenal artery behind the first part of the duodenum. A group of lymph nodes lie here, whence vessels pass with the hepatic artery to the cœliac group. This group of nodes behind the first part of the duodenum also receives lymph vessels that come up with the superior pancreatico-duodenal artery; they drain the foregut part of the *duodenum* (i.e., as far distally as the entrance of the bile duct) and the upper part of the head of the pancreas (Fig. 5.24).

The **cœliac group** of lymph nodes, surrounding the cœliac artery in front of the aorta, receives lymph thus from all the foregut and also from the superior and inferior mesenteric nodes, that is, from all the gut and

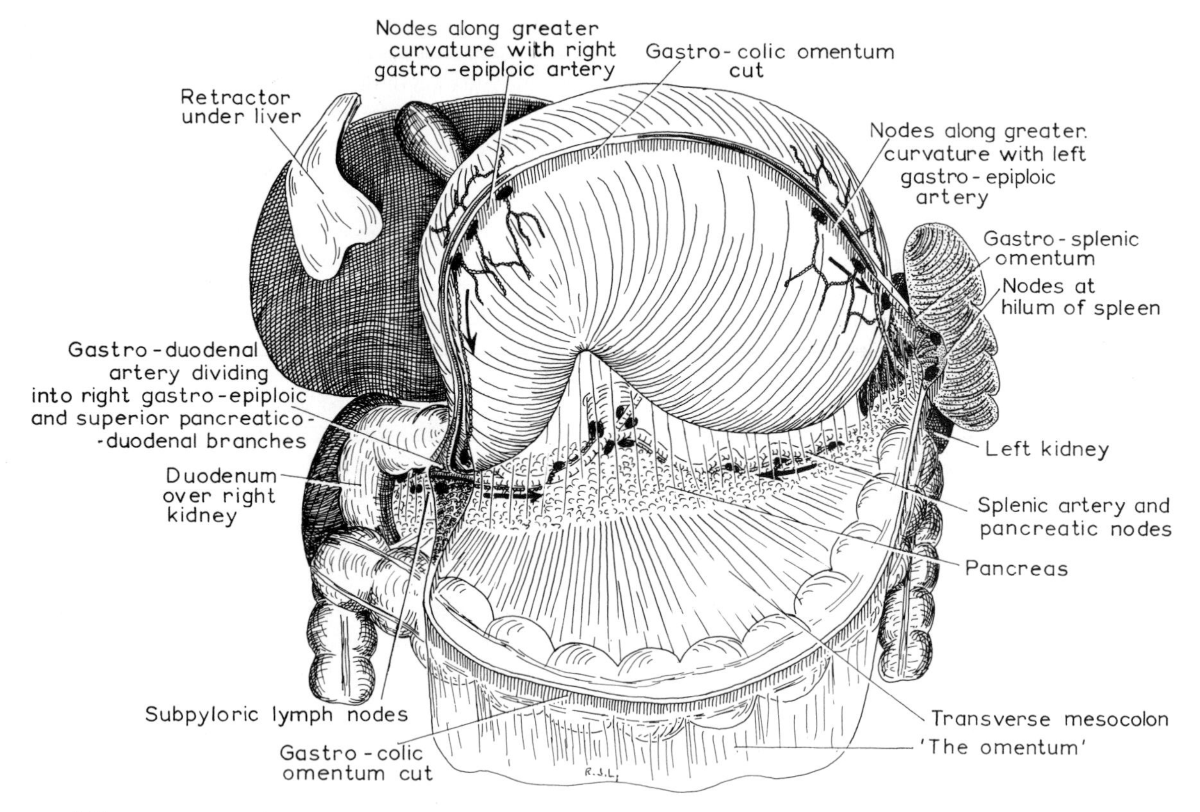

FIG. 5.26. THE STOMACH DISPLACED UPWARDS AFTER DIVISION OF THE GASTRO-COLIC OMENTUM. Lymph nodes are in solid black. Arrows indicate direction of lymph flow.

from liver, pancreas and spleen. The group drains directly into the cisterna chyli and so to the thoracic duct.

Lymph Drainage of the Midgut (Fig. 5.19). The same plan of lymph drainage applies here as in all the gut (see p. 285). The *duodenum* below the entrance of the bile duct, and the lower part of the head of the pancreas drain along the inferior pancreatico-duodenal artery to the superior mesenteric group of pre-aortic lymph nodes.

The *jejunum* and *ileum* drain from the lymphatic follicles of the mucous membrane through the muscle wall into the mesentery. Many nodes lie along the arterial arcades of the jejunal and ileal arteries and from these nodes lymphatics drain to the superior mesenteric group of pre-aortic lymph nodes.

The *appendix* drains from its lymphatic follicles through the muscle wall into nodes in the meso-appendix. These drain into paracolic nodes lying along the ileo-colic artery and so to the superior mesenteric group.

The *cæcum* and *ascending colon* drain from their lymphatic follicles into nodes lying along the left side of the gut (i.e., the attached border of the mesocolon in the embryo). These are the epicolic nodes. They drain into paracolic nodes lying along the ileo-colic and right colic arteries behind the peritoneal floor of the right infracolic compartment, and so to the superior mesenteric group of pre-aortic lymph nodes (Fig. 5.34).

The *transverse colon* (excluding the splenic flexure) drains from its lymphatic follicles through the muscle wall into the epicolic nodes lying in the intestinal border of the transverse mesocolon. These drain to paracolic nodes lying in the transverse mesocolon and thence to the superior mesenteric group.

The **superior mesenteric group** of pre-aortic lymph nodes surround the proximal part of the superior mesenteric artery behind the neck of the pancreas. They lie just below the cœliac group, into which they send their efferent lymph vessels. They receive all the lymph of the midgut and the efferent vessels from the inferior mesenteric group.

Lymph Drainage of the Hindgut (Fig. 5.19). The pattern of lymph drainage is the same here as in the remainder of the gut (p. 285).

From the *splenic flexure* down the *descending colon* and *pelvic colon* to the commencement of the *rectum* the lymph from the lymphatic follicles in the mucous membrane passes through the muscular wall of the gut by lymphatics which drain into the epicolic nodes, lying at the right border of the descending colon (i.e., at the attachment of the mesocolon in the embryo) and in the intestinal border of the sigmoid mesocolon. From the epicolic nodes vessels pass to paracolic nodes lying along the branches of the inferior mesenteric artery. The paracolic nodes of the splenic flexure and descending colon lie beneath the peritoneal floor of the left infracolic compartment, those of the pelvic colon lie in the root of the sigmoid mesocolon. From all the paracolic nodes the lymphatics converge towards the inferior mesenteric group.

The **inferior mesenteric group** of pre-aortic lymph nodes lie around the commencement of the inferior mesenteric artery beneath the peritoneal floor of the left infracolic compartment. They receive lymph from the hindgut (from splenic flexure to anal canal) and send their lymph by efferent vessels to the superior mesenteric group.

For lymph drainage of the *rectum* and *anal canal* see pp. 329 and 345.

Nerve Supply of the Alimentary Canal

All parts of the gut and its derivatives are innervated by parasympathetic and sympathetic nerves, which travel together along the gut arteries to reach their destination. Most come from the cœliac plexus, but the inferior hypogastric (pelvic) plexus contributes parasympathetic fibres to the hindgut.

The *parasympathetic fibres* accompanying the arteries are medullated (preganglionic) and they relay in the wall of the gut. They are motor to the gut and secreto-motor to the glands. Motor fibres for the muscle relay on cell bodies in a plexus between the circular and longitudinal layers (the myenteric plexus of Auerbach). Secreto-motor fibres for glands relay on cell bodies in a plexus in the submucosa (the plexus of Meissner).

The *sympathetic fibres* are vaso-motor, inhibitory to gut muscle, and sensory. They are motor to smooth-muscle sphincters (e.g., ileo-colic). All but the sensory fibres are non-medullated (postganglionic) from cell bodies in the cœliac ganglia. The sensory fibres are medullated. All sympathetic fibres are distributed (without relay) along the meshes of the plexuses of Auerbach and Meissner.

The cœliac plexus (p. 314) contributes parasympathetic fibres from the vagi to the foregut and its derivatives, and to the midgut. Likewise it contributes sympathetic fibres from the splanchnic nerves (after relay in the cœliac ganglia) to these same parts and to the hindgut as well. The parasympathetic supply of the hindgut comes up from the pelvic parasympathetics (nervi erigentes) to run out along the inferior mesenteric artery (p. 343).

Structure of the Alimentary Canal

General Features. The alimentary canal has its embryological origin in the yolk sac. It is lined by epithelium derived from the endoderm, and outgrowths from this epithelium (liver and pancreas) are thus endodermal in origin.

Macroscopically the alimentary canal is a tube of muscle lined with mucous membrane and covered, in most part, with peritoneum.

The **muscular wall** of the alimentary canal, from the upper end of the œsophagus to the lower end of the rectum is in two separate layers, an inner circular and an outer longitudinal. This arrangement is characteristic of all tubes that empty by orderly peristalsis and not by a mass contraction. In reality the muscle is in *two spiral layers*. The inner layer forms a close-wound spiral and is known as the *circular layer*, while the outer layer spirals in such gradual fashion that its fibres are virtually longitudinal, and it is called the *longitudinal layer*. The latter is separated into three discrete bundles in the colon. An innermost third layer reinforces the body of the stomach.

The two layers of the upper third of the œsophagus (cervical and superior mediastinal parts) consist of striated muscle. In the lower two-thirds of the œsophagus this is gradually replaced by smooth muscle, which constitutes the two layers of the remainder of the alimentary canal. Contraction of the striated muscle is brisk and rapid and a swallowed bolus is hurried through the pharynx past the laryngeal opening and well into the œsophagus, safe from regurgitation; only then does the slow, worm-like contraction of the smooth muscle propel it by a more leisurely peristalsis.

The structure of the **mucous membrane** of the alimentary canal is admirably adapted to its function; it is thick, but loosely woven, allowing a considerable degree of dilatation to occur. The mucous membrane can be readily stripped from the inner circular layer of the muscular wall; such separated mucous membrane, from the small intestine of the sheep, is manufactured into catgut. The mucous membrane is projected into folds which vary in their pattern. They are longitudinal in the pylorus, circular in the small intestine.

Microscopic Anatomy. The **mucous membrane** of the alimentary canal is a loosely woven network of fibrous tissue (collagen). It is lined with epithelium and divided into two layers by a thin sheet of smooth muscle called the *muscularis mucosæ*. That part of the mucous membrane between muscularis mucosæ and the inner circular layer of the wall of the alimentary canal is named the *submucosa*. Incidentally it should be noted that a mucous membrane possessing no muscularis mucosæ (e.g., bladder, vagina) can, by definition, have no submucosa.

The **œsophagus** is lined by squamous stratified

epithelium throughout its length, including the abdominal segment, right to the cardiac orifice of the stomach (Fig. 1.17, p. 16). Lymphatic follicles are scattered beneath the epithelium. Islands of columnar epithelium with gastric glands are occasionally found in the lower part of the œsophagus; these may give rise to peptic ulcer or columnar-celled carcinoma. The muscularis mucosæ is a thick layer broken up somewhat into separate bundles of muscle. Beneath it, in the submucosa, lie scattered mucous glands, restricted to the upper and lower ends; there are no such glands in the middle portion of the œsophagus. The only other parts of the alimentary canal containing mucous glands in the submucosa are the duodenum and the anal canal. Perhaps this deep position of the glands hinders free drainage of infection; certainly œsophagitis, duodenitis and proctitis are all likely to be very chronic.

General Histological Features. The alimentary canal from the cardiac orifice to the anal canal possesses many similar histological features. The mucous membrane is in all places thick but lax, and the muscularis mucosæ is a thin sheet of smooth muscle, its fibres disposed in circular fashion. The surface epithelium is a single layer of columnar cells. From the surface, simple test-tube shaped pits, or crypts, extend into the mucous membrane as deep as the muscularis mucosæ. In the stomach they form the gastric and pyloric glands, in the small intestine the crypts of Lieberkühn and in the whole of the large intestine the mucous crypts. They are lined with serous or mucus-secreting cells according to the functional needs of the various parts of the alimentary canal. Throughout the whole of the small intestine (duodenum, jejunum and ileum) the surface between the mouths of the crypts is projected into villi, long processes like the fingers of gloves, covered with columnar epithelium, and containing capillaries and a central straight lymph vessel (the lacteal) for absorption of nutriment. There are no villi in the stomach or the large intestine. In the mucous membrane, and occasionally in the submucosa, are scattered numerous lymphatic follicles.

Two plexuses of nerve fibres and cells lie in the wall. One lies in the submucosa (the plexus of Meissner) the other between the circular and longitudinal coats (the plexus of Auerbach). The nerve cells are parasympathetic, the nerve fibres are both sympathetic and parasympathetic (see p. 287).

The **stomach** is lined with columnar epithelium. The crypts, which are invaginated into the mucous membrane as far as the muscularis mucosæ, form glands of two types, producing acid and pepsin on the one hand and alkaline mucus on the other. Their relative distribution in the stomach wall varies with the functional needs (dietary and digestive) of the species. In man the acid-producing glands extend from the cardiac orifice over the fundus and body of the stomach. The pyloric antrum and pyloric canal contain glands of only the alkaline, mucous type. They are readily distinguished from each other. The **gastric gland** has a relatively short duct and a long acinus, and three or four acini commonly open into one duct. The duct is lined with similar epithelium to that on the surface, a single layer of columnar cells, many of which have a goblet-cell appearance from contained mucus. Their mucus content makes them take ordinary stains poorly. The acini are straight, and are lined with a single layer of spherical cells, which produce a serous secretion (pepsin-containing) and stain well with ordinary stains. In a low-power view the mucous membrane shows a pale edge (surface and mouths of the glands) and a broader deeply stained part (the gastric glands). The lumen of the gastric gland is not usually patent. Lying in the connective tissue between the adjacent peptic acini are scattered separate large cells with a clear acidophilic cytoplasm, called the parietal or *oxyntic cells.* They secrete hydrochloric acid. The **pyloric glands** possess long ducts and short acini. The ducts are patent and extend deeply into the mucous membrane. They are lined with mucus-producing, columnar epithelium similar to the surface epithelium. Beneath the ducts are short acini, lined with poorly staining mucus-secreting cells. The acini lie coiled beneath the ducts, and their lumen is usually patent (Fig. 5.28).

The **duodenum** possesses long slender villi, and the crypts extend from the bases of the villi to the muscularis mucosæ. The villi are covered with columnar epithelium containing many goblet cells, the crypts are lined with spherical cells which stain well. The whole of the submucosa is packed with mucous glands from muscularis mucosæ to the inner circular muscle of the bowel wall. These are *Brunner's glands,* and their ducts pierce the muscularis mucosæ to open in the depths of the crypts. Brunner's glands commence abruptly at the pyloro-duodenal junction and disappear gradually towards the duodeno-jejunal junction (Fig. 5.29).

The **jejunum** and **ileum** are similar. Each possesses villi, and in each the crypts penetrate through the mucous membrane to the muscularis mucosæ. Villi and crypts are both well developed in the jejunum and there are relatively few goblet cells in the columnar epithelium; in the ileum the villi and crypts become gradually shorter and goblet cells more numerous, but there is no abrupt change, and it is often impossible to identify a section as one or the other. At all levels the height of a villus equals the length of a crypt (Fig. 5.29).

Under the microscope it is difficult or impossible to distinguish jejunum from ileum, especially if the species is not known. In man the upper jejunum shows long slender villi and correspondingly long crypts of Lieber-

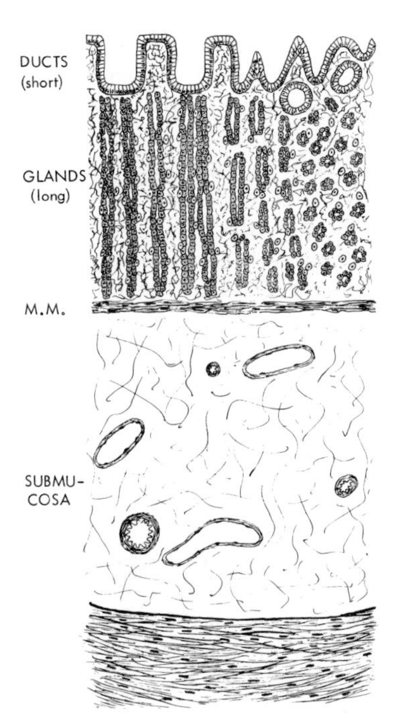

FIG. 5.27. The Body of the Stomach (Gastric-juice producing area).

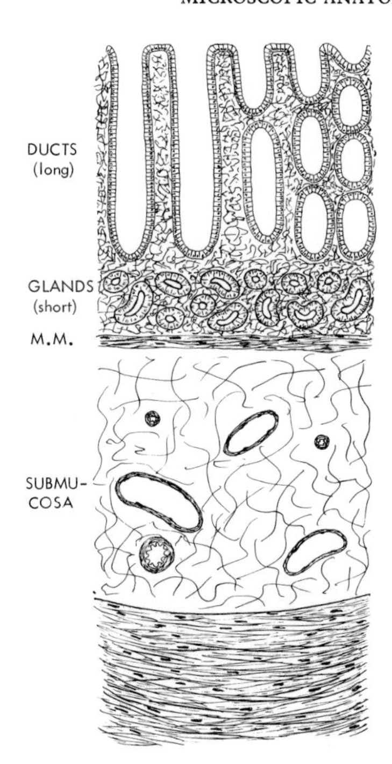

FIG. 5.28. The Pyloric End of the Stomach.

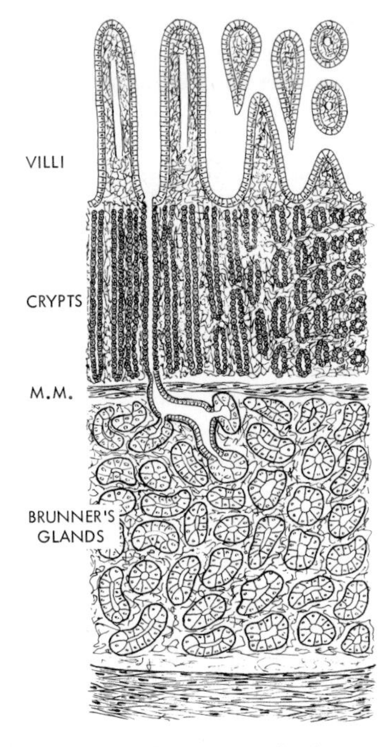

FIG. 5.29. The Duodenum (in the remainder of the small gut there are no Brünner's glands).

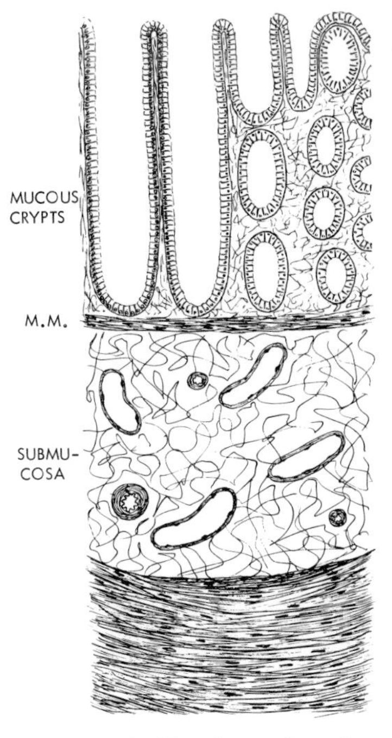

FIG. 5.30. The Large Intestine. (M.M. Muscularis mucosæ.)

FIGS. 5.27 to 5.30. ALIMENTARY MUCOUS MEMBRANE IS TYPICAL ENOUGH TO DISTINGUISH READILY THE DIFFERENT PARTS OF THE GUT. (Histology is a two-dimensional study of three-dimensional reality. Twisting of the mucous membrane in the paraffin block introduces obliquity of microtome section. In the diagrams a perpendicular section is indicated on the left, trailing off into increasing obliquity on the right.)

kühn (glandulæ intestinales). The terminal ileum shows much shorter villi and shorter crypts; the goblet-cell content of the crypts is very high, and the crypts resemble mucous crypts of the large gut. The transition is gradual along the intestine. Aggregated lymphatic follicles do not help to distinguish. They are not necessarily cut through (they are scattered only along the antimesenteric border) and they occur also in the jejunum (where they are smaller than in the ileum).

The **large intestine** is characteristic; the mucous membrane is histologically identical throughout its entire length. The one word *uniformity* describes it. The crypts are lined throughout by columnar epithelium, almost all of the cells being of the goblet variety, similar in every respect to the surface epithelium. There is no duct and acinus, as in the stomach, but one continuous epithelial lining. The crypt possesses a patent lumen. It penetrates to the muscularis mucosæ. The muscle wall of the large intestine shows the arrangement of the layers characteristic of the animal concerned; in man the longitudinal layer, except in the sigmoid colon and rectum, is in three parallel ribbon-like bands (tæniæ).

The **appendix** is lined with typical large bowel epithelium with mucous crypts projecting into the mucous membrane. They do not usually penetrate as far as the muscularis mucosæ, for the deeper part of the mucosa is occupied by an almost continuous ring of lymphatic follicles. The lymphatic tissue is profuse in the child, but becomes less in the adult and atrophies in old age. The muscle walls are thick in proportion to the lumen and the outer longitudinal layer forms a continuous band beneath the serosa.

TOPOGRAPHICAL ANATOMY OF THE ALIMENTARY CANAL

The **œsophagus** (p. 238) projects through the diaphragm at a level with the seventh costal cartilage a thumb's breadth to the left of the sternum. This is level with Th. 10 vertebra. The abdominal portion is very short, the length depending on muscle tonus and the weight of the stomach contents; the average is perhaps $\frac{1}{2}$ inch (one cm). It is bound firmly to the diaphragm by a 'ligament' of fibrous tissue. The right and left vagal trunks, usually double on the right, lie on its surface. It is invested by peritoneum which passes from it on the right to the diaphragm (the upper part of the lesser omentum) and on the left to the diaphragm (the upper part of the greater omentum). The posterior wall of the œsophagus is rather shorter than the anterior, for the orifice in the diaphragm lies very nearly vertical. This posterior part of the œsophagus can scarcely be said to possess a serous coat, for the fold of peritoneum that lies against it is not firmly attached, and is stripped up slightly when the abdominal œsophagus elongates (a miniature replica of the roof of the cave of Retzius).

The œsophagus enters the stomach at the cardiac orifice, which is protected by a functional though not very obvious anatomical sphincteric ring of muscle. The lower œsophagus often, though not always, shows a thickening of its circular coat. Closure of the lower end of the œsophagus by these circular fibres allows the thick mucosal ridges of the stomach lining to plug the orifice and so discourage reflux.

The looping fibres of the right crus which surround the œsophageal opening in the diaphragm help to close the lower œsophagus. This is an emergency mechanism which prevents regurgitation up the thoracic œsophagus should the cardiac sphincter give way under increased intra-abdominal pressure (cf. the 'external sphincter' of the bladder, i.e., the sphincter urethræ muscle. Here again is a quick-acting and stronger striated muscle to spring into instant contraction should the weaker smooth-muscle sphincter prove inadequate).

The **stomach** is a muscular bag, fixed at both ends, mobile elsewhere, and subject to great variations in size in conformity with the volume of its contents. Much of it lies under cover of the lower ribs. It consists of fundus, body, pyloric antrum and pylorus (Fig. 5.24, p. 280). The *fundus* is that part which projects upwards, in contact with the left dome of the diaphragm, *above the level of the cardiac orifice*. It is usually full of gas. The *body* extends from the fundus to the level of the *incisura angularis*, a constant notch in the lower part of the lesser curvature. The *pyloric antrum* extends from this level, narrowing gradually towards the pylorus. The *pylorus* is palpably thicker than the rest of the stomach wall and the pyloric canal is held closed by the tonus of the pyloric sphincter except when the latter relaxes to allow the stomach to expel a jet of its contents into the duodenum (Fig. 5.31). The mucous membrane of the stomach is smooth and very red—in the pyloric antrum it is thrown into longitudinal folds which flatten out when the organ is greatly distended.

The outer longitudinal and inner circular muscle coats completely invest the stomach; they are reinforced by an innermost oblique muscle coat, which is incomplete. Its fibres loop over the fundus, being thickest at the notch between the fundus and the œsophagus. They pass along the anterior and posterior walls of the organ in a direction oblique to its long axis, but they lie vertically when the body is erect, and thus obtain the best mechanical advantage in supporting the weight of the stomach contents. It is the contraction of the oblique coat which produces the *magenstrasse*. This is a pathway along the lesser curvature which allows liquids to pass along while the body and greater curvature are pinched off. Thus a meal of rice and curry remains

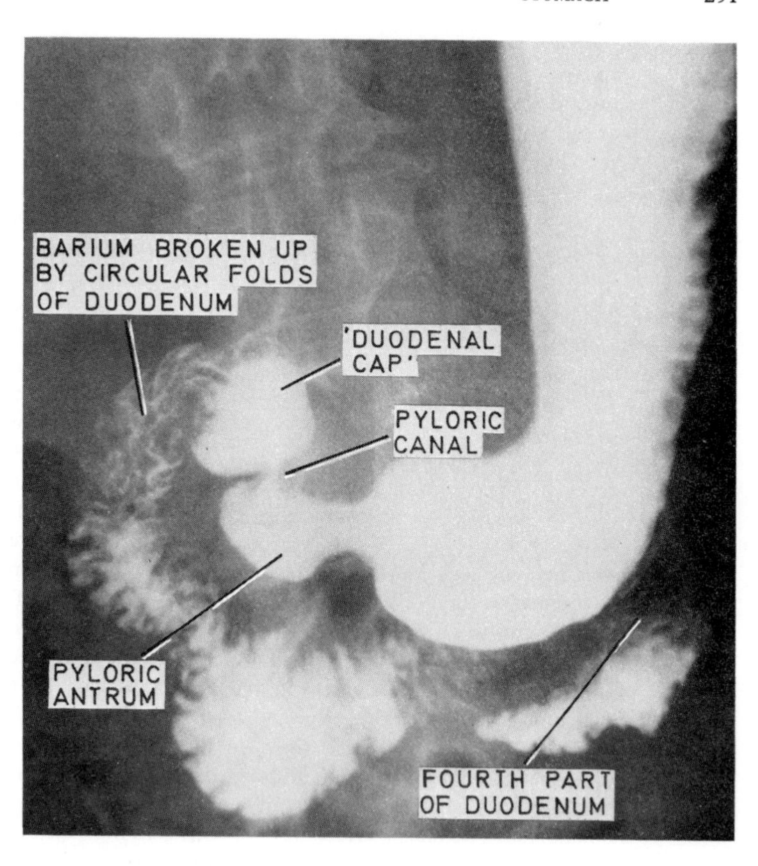

FIG. 5.31. RADIOGRAM OF STOMACH AND DUODENUM (by courtesy of Dr Seymour J. Reynolds).

undisturbed in the stomach while the following drink of iced water flows directly through to the duodenum.

The left vagal trunk passes on to the anterior surface of the stomach and sinks into its walls to supply the muscles and the glands. Further fibres pass in the lesser omentum to the porta hepatis. The right vagal trunk runs on the posterior wall of the stomach and many of its fibres supply the organ; the remainder pass from the cardiac orifice along the left gastric artery to the cœliac plexus (p. 314).

The stomach is completely invested in peritoneum, which passes in a double layer from its lesser curvature as the lesser omentum and from its fundus and greater curvature as the greater omentum.

The fundus and upper part of the body are fairly constant in shape and size, lying in contact with the diaphragm and the upper part of the stomach bed. That part of the stomach below the costal margin, namely the lower part of the body and the pyloric antrum, is much more variable in size and position, according to the build and habitus of the individual and the changing volume of his stomach contents. The stomach may project but little beyond the costal margin; on the other hand, the lower part of the greater curvature may lie on the pelvic floor. These are the extremes and are not common. Most usually the greater curvature lies only a little below the level of the umbilicus, being in any individual highest when the stomach is empty and the body supine, lowest when full and the body erect.

Relationships of the Stomach. The upper part of the lesser curvature is overlapped in front by the sharp inferior border of the left lobe of the liver; elsewhere the anterior surface is in contact with the diaphragm and the anterior abdominal wall. The fundus occupies the concavity of the left dome of the diaphragm.

The convexity of the greater curvature lies in contact with the transverse colon, the gastro-colic omentum being instrumental in making their curvatures conform to each other.

The posterior wall of the stomach lies with its serous coat in contact with the peritoneum of the floor of the lesser sac. If the stomach is removed, the *stomach bed* may be inspected (PLATE 14). It extends on the left of the œsophageal opening to the highest part of the dome of the diaphragm. The lesser sac is limited by the attachment of the greater omentum to the diaphragm and to the front of the left kidney (the lieno-renal ligament). The upper part of the greater curvature bulges to the left of this ligament and the stomach is here in contact with the spleen. Below this level the posterior surface of the stomach lies upon the downward-

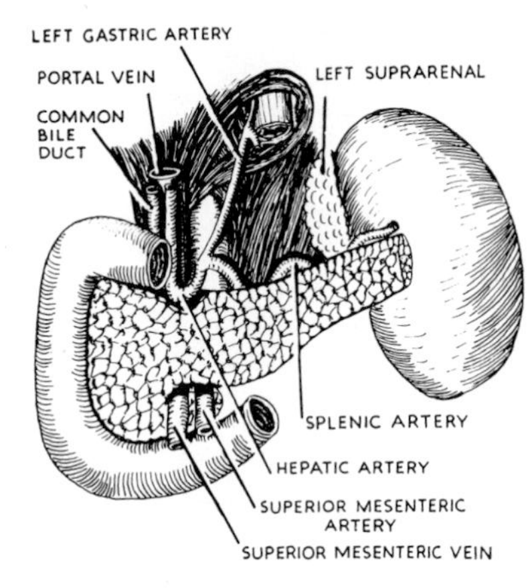

FIG. 5.32. THE BED OF THE STOMACH. (Rough diagram. For more detail see PLATE 14 and Fig. 5.48, p. 317.)

sloping transverse mesocolon. Above the attachment of the transverse mesocolon the body of the pancreas lies behind the peritoneum, the crests of the waves of the tortuous splenic artery appearing above its upper border. In front of the left crus of the diaphragm lies the crescentic left suprarenal gland, closely applied to the medial border of the left kidney. To its right, in the midline, lies the aorta with the short trunk of the cœliac artery dividing into its three divisions at the upper border of the pancreas. The cœliac artery lies between the cœliac ganglia and is surrounded by the cœliac lymph nodes and cœliac plexus.

The **pylorus** has some mobility, for it is enclosed between the peritoneum of the right extremities of both greater and lesser omenta. It hangs down over the head of the pancreas. The pylorus is a sphincter of smooth muscle. Its great thickness is due mainly to thickening of the circular muscle coat of the stomach, but is enhanced by the curling back here of many of the longitudinal fibres. Its canal is closed. The first part of the duodenum is widely patent; the duodenal aspect of the pylorus thus resembles the vaginal aspect of the cervix uteri, except that there are no fornices.

The arterial supply of the stomach is reviewed on p. 280, the venous drainage on p. 281 and the lymphatic drainage on p. 285.

The Duodenum (PLATE 14). The first inch of the duodenum is contained between the peritoneum of the lesser and greater omenta, but the remainder of this part of the gut is entirely retroperitoneal. The duodenum is a C-shaped tube curved over the convexity of the forwardly projecting aorta and inferior vena cava, while its descending limb lies more posteriorly in the right paravertebral gutter. Appreciation of the directions of the various parts of the duodenum is essential to the interpretation of radiograms of barium meals. The duodenum is divided into four parts, all of which run in different directions. The tube is 10 inches (25 cm) long and the lengths of its parts are 2, 3, 4 and 1 inches (say 5, 7, 10 and 3 cm).

The *first part* of the duodenum runs *backwards* and somewhat upwards from the pylorus; a fore-shortened view is consequently obtained in sagittal X-ray projections. The first 2 cm (i.e., the duodenal cap, *see below*) lies between the peritoneal folds of the greater and lesser omenta; it forms the lowermost boundary of the opening into the lesser sac. It lies upon the liver pedicle (common bile duct, hepatic artery and portal vein, PLATE 16). Behind this lies the inferior vena cava at the epiploic foramen. The neck of the gall-bladder touches the upper convexity of the duodenal cap (Fig. 5.40). The next 3 cm passes backwards and upwards on the right crus of the diaphragm and the right psoas muscle to the medial border of the right kidney. Its posterior surface is bare of peritoneum; this is to the right of the epiploic foramen (Fig. 5.17, p. 271). It touches the upper part of the head of the pancreas and is covered in front with peritoneum of the posterior abdominal wall. The inferior surface of the right lobe of the liver lies over this peritoneum.

The *second part* of the duodenum curves downwards over the hilum of the right kidney. It is covered in front with peritoneum, and crossed by the attachment of the transverse mesocolon, so that its upper half lies in the supracolic compartment to the left of the hepatorenal pouch (in contact with the liver) and its lower half lies in the right infracolic compartment medial to the inferior pole of the right kidney (in contact with coils of jejunum). It lies alongside the head of the pancreas (Fig. 5.48, p. 317).

Its postero-medial wall receives the common opening of the bile duct and main pancreatic duct at the *duodenal papilla* (papilla of Vater). The papilla lies about halfway along the second part, some 4 inches (10 cm) from the pylorus. It is guarded by the semilunar flap of mucous membrane which surmounts it. Two cm (less than one inch) proximal to the papilla is the small opening of the accessory pancreatic duct.

The *third part* of the duodenum curves *forwards* from the right paravertebral gutter over the slope of the right psoas muscle (gonadal vessels and ureter intervening) and passes over the forwardly projecting inferior vena cava and aorta to reach the left psoas muscle (PLATE 14). Its inferior border lies on the aorta at the commencement of the inferior mesenteric artery at the level of the umbilicus. Its upper border hugs the lower border of the pancreas. It is covered by the peritoneum of the posterior abdominal wall just below the transverse mesocolon. It is crossed by the superior mesenteric vessels and by the leaves of the commencement of the mesentery of the small intestine sloping down from the duodeno-jejunal flexure. It lies, therefore, in both right and left infracolic compartments (Fig. 5.17, p. 271). Its anterior surface is in contact with coils of jejunum.

The *fourth part* of the duodenum ascends to the left of the aorta, lying on the left psoas muscle and left lumbar sympathetic trunk, to reach the lower border of the pancreas, almost as high as the root of the transverse mesocolon (L2 vertebra). It is covered in front by the peritoneal floor of the left infracolic compartment and by coils of jejunum. It breaks free from the peritoneum that has plastered it down to the posterior abdominal wall and curves forwards and to the right as the *duodeno-jejunal flexure*. This pulls up a double sheet of peritoneum from the posterior abdominal wall, the mesentery of the small intestine, which slopes down to the right across the third part of the duodenum and posterior abdominal wall (Fig. 5.33).

There is no distinction between duodenum and jejunum save only the peritoneal arrangement. The duodenum is retroperitoneal, the jejunum has a mesentery (compare junction of rectum and sigmoid colon).

The duodeno-jejunal flexure is fixed to the left psoas fascia by fibrous tissue. It is said to be further supported by the *suspensory 'ligament' of the duodenum*. This is a thin band of smooth muscle (the *muscle of Treitz*); it descends from the right crus of the diaphragm in front of the aorta, behind the pancreas, and blends with the outer muscle coat of the flexure. It is usually impossible to find the muscle.

Macroscopic Appearance. The duodenum is a wide-bored tube, whose muscular walls are smooth and rather thin. The mucous membrane is thick and villous; it is thrown into numerous circular folds (plicæ circulares or valvulæ conniventes) in the retroperitoneal part of the duodenum (i.e., all parts except the first 2 cm). The smooth walls of the first 2 cm are thin, and they are held open by attachment to the pylorus, hence the *smooth* outline of the full shadow of barium in the 'duodenal cap' at X-ray examination (Fig. 5.31). From the duodenal cap onwards the plicæ break up the barium and its shadow (Fig. 5.31).

Microscopic Structure (see p. 288).

Blood Supply. The duodenum is supplied by the superior and inferior pancreatico-duodenal arteries, and in its first inch (2 cm) by multiple small branches from the hepatic and gastro-duodenal arteries. It is drained by corresponding veins into the portal vein and superior mesenteric vein. The veins from the multiple arteries to the first inch (2 cm) collect into the prepyloric vein (p. 282). The lymph drainage of the duodenum is by way of paraduodenal nodes and thence to the cœliac and superior mesenteric groups of preaortic lymph nodes (p. 285).

The Paraduodenal Fossæ. To the left of the duodeno-jejunal flexure certain peritoneal folds and evaginations are occasionally present. The most important surgically is the *paraduodenal fossa* proper (Fig. 5.33). This is a small evagination of peritoneum beneath the upper end of the inferior mesenteric vein; an incarcerated hernia in this fossa may obstruct and thrombose the vein, and there is danger of dividing the vein if the peritoneum has to be divided at operation to free the hernia. Folds of peritoneum may rise from the floor of the infracolic compartment to both upper and lower convexities of the flexure; the fossæ beneath such folds are known as the *superior duodenal* and

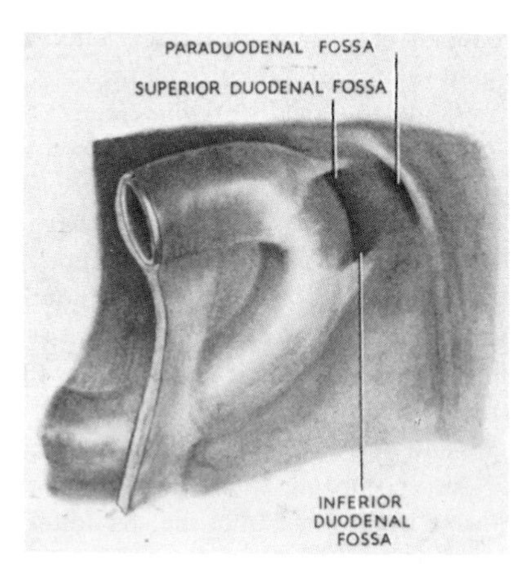

FIG. 5.33. THE PARADUODENAL FOSSÆ. They are only occasionally present.

inferior duodenal fossæ respectively. Finally a *retro-duodenal fossa* may be excavated behind the flexure, between the superior duodenal and inferior duodenal fossæ.

It should be noted that the mouths of these four fossæ all face inwards towards each other, and a hernia in any one may thus spread to involve others. This is not so with the paracæcal fossæ, whose mouths all face away from each other (p. 294).

The Small Intestine. The jejunum is wider-bored, thicker-walled, and redder than the ileum; but these differences are only relative, and a more useful method of distinguishing one from the other in the living is by rolling the wall of the intestine gently between finger and thumb. The wall of the jejunum is thick and double (the mucous membrane can be felt through the muscle wall, 'a shirt sleeve felt through a coat sleeve'), while the wall of the ileum is thin and single (the mucous membrane cannot be felt as a separate layer).

The lower reaches of the ileum are distinguished by the presence on the antimesenteric border of elongated whitish plaques in the mucous membrane, usually but not always visible through the muscle wall. These are the aggregated lymphatic follicles (Peyer's patches).

The jejunum lies coiled in the upper part of the infracolic compartment, the ileum in the lower part thereof and in the pelvis.

Meckel's diverticulum is present in 2 per cent of individuals, 2 feet (60 cm) from the cæcum and is 2 inches (5 cm) long. This useful mnemonic is two-thirds true; the length of the diverticulum is very variable. It may be a very small bulge projecting from the antimesenteric border of the ileum or it may be 6 inches long. Its blind end may contain gastric mucosa or liver or pancreatic tissue. Its apex may be adherent to the umbilicus or connected thereto by a fibrous cord, a remnant of the vitelline duct. Ulceration and perforation of the tip are not infrequent.

Blood Supply of the Small Intestine (see p. 282).

Lymph Drainage of the Small Intestine (see p. 286).

Microscopic Anatomy (see p. 288).

The Cæcum. This blind pouch of the large intestine projects downwards from the commencement of the ascending colon, below the ileo-cæcal junction (Fig. 5.34). Over the front and on both sides it is covered with peritoneum. The serous coat continues up behind it and is reflected downwards to the floor of the right iliac fossa. The retrocæcal peritoneal space may be shallow or deep according to the distance of the retrocæcal fold from the lower end of the cæcum. The space may be continuous across the iliac fossa or it may be interrupted by a peritoneal fold from one or other side of the posterior wall of the cæcum. Often there are two folds, forming between them a retrocæcal fossa in which the appendix may lie. As in the rest of the colon, the longitudinal muscle of the cæcum is restricted to three flat bands, between which the circular muscle layer constitutes the sacculated wall of the gut. The flat bands of longitudinal muscle (tæniæ) lie one anterior, one postero-medial and one postero-lateral. All three converge on the base of the appendix.

In the infant the cæcum is conical and the appendix depends from its apex. The lateral wall outgrows the medial wall and bulges down below the base of the appendix in the adult; the base of the appendix thus comes to lie in the postero-medial wall of the cæcum above its lower end, and the three tæniæ converge to this point (Fig. 5.35). The terminal inch or so (say 2 cm) of the ileum is commonly adherent to the left convexity of the cæcum, below the ileo-cæcal junction.

The cæcum lies on the peritoneal floor of the right iliac fossa, over the iliacus and psoas fasciæ and the femoral nerve. Its lower end lies at the pelvic brim. When distended its anterior surface touches the parietal peritoneum of the anterior abdominal wall, when collapsed coils of ileum lie between the two.

Blood Supply of the Cæcum (see p. 283).

Lymph Drainage of the Cæcum (see p. 286).

The Appendix. The appendix is attached by its base to the point of convergence of the three tæniæ coli on the postero-medial wall of the cæcum, below the ileo-cæcal opening. On the surface of the abdomen this point lies one-third of the way up the oblique line that joins the right anterior superior iliac spine to the umbilicus. The three tæniæ coli merge into a complete longitudinal muscle layer over the appendix.

The meso-appendix is a peritoneal fold enclosing the appendicular vessels. Its base is a prolongation of the left (inferior) layer of the mesentery of the terminal ileum, and its free crescentic edge contains the appendicular branch of the posterior cæcal artery. It is attached to the appendix from the base to the point of entry of the appendicular artery, a variable point which is usually near the tip but may be at any point along the appendix (Fig. 5.34).

The tip of the appendix commonly lies at or about the pelvic brim, the tube being about 2 inches (5 cm) long. Both length and position are very variable. The long appendix with a short meso-appendix is subject to wide variation in the position of its tip—the two commonest sites being the right pelvic wall and the retrocæcal fossa.

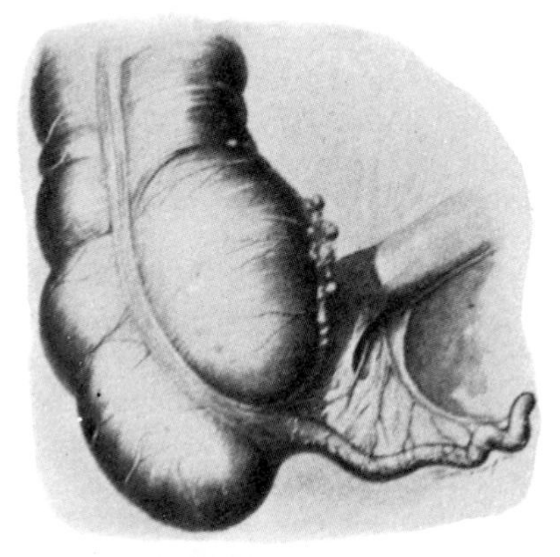

FIG. 5.34. THE ILEO-CÆCAL JUNCTION AND THE VERMIFORM APPENDIX. A SMALL BLOODLESS FOLD IS PRESENT. A GROUP OF EPICOLIC LYMPH NODES IS ON THE MEDIAL WALL OF THE CÆCUM.

Lymph Drainage of the Appendix (see p. 286).

Para-appendiceal (paracæcal) Fossæ. Peritoneal folds near the base of the appendix are sometimes found. The most anterior lies in front of the terminal ileum between the base of the mesentery and the anterior wall of the cæcum and is raised up by the contained anterior cæcal artery. This anterior cæcal fold is rather rare. Beneath this fold is often found a fold of peritoneum (the 'bloodless fold' of Treves) which runs from the terminal ileum towards the base of the appendix and becomes attached to the meso-appendix. Behind the bloodless fold is the meso-appendix (Fig. 5.34). Thus three parallel folds of peritoneum, the middle one bloodless, enclose two ileo-cæcal fossæ between them. Between the meso-appendix and the peritoneal floor of the right iliac fossa there is often a third recess and, finally, the retrocæcal fossa when present constitutes a fourth. The mouths of these four fossæ face away from each other, unlike the mouths of the paraduodenal fossæ (p. 293).

The Ileo-cæcal 'Valve'. The terminal ileum opens into the medial wall of the large intestine at the junction of the cæcum and ascending colon. An upper and a lower fold project within like a pair of pouting lips (Fig. 5.35). No valvular formation exists, and the opening is controlled by a certain thickening of the circular muscle of the terminal inch or two of the ileum which constitutes a sphincter.

The ascending colon, varying about 6 inches (15 cm) in length, extends upwards from the ileo-cæcal junction to the right colic (hepatic) flexure. The latter lies on the lateral surface of the inferior pole of the right kidney, in contact with the inferior surface of the liver. The ascending colon lies on the iliac fascia and the anterior lamella of the lumbar fascia, being connected and fixed to them by fibrous tissue of the extraperitoneal fascial envelope. Its front and both sides possess a serous coat, which runs laterally into the

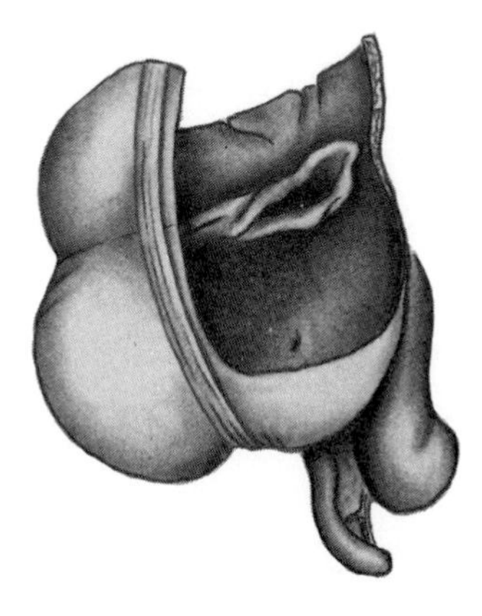

FIG. 5.35. THE POUTING OF THE ILEO-CÆCAL 'VALVE' SEEN FROM WITHIN THE CÆCUM. THE LUMEN OF THE APPENDIX OPENS INTO THE CÆCUM WELL BELOW THE 'VALVE'. Drawn from Specimen S.280 in R.C.S. Museum.

paracolic gutter and medially into the right infracolic compartment. Its original embryonic mesentery hinged across from the midline and became adherent to the peritoneum of the posterior abdominal wall, bringing the ileo-colic and right colic vessels within its folds. These vessels thus lie, in the adult, immediately beneath the peritoneum of the right infracolic compartment.

The tæniæ coli lie, in line with those of the cæcum, anteriorly, postero-laterally and postero-medially. These constitute the sole longitudinal muscle coat and the circular muscle coat is exposed between them. The ascending colon is sacculated, due to the three tæniæ coli being 'too short' for the bowel. If the tæniæ are divided between the sacculations the latter can be drawn apart and the bowel wall flattened.

Bulbous pouches of peritoneum, distended with fat, project in places from the serous coat. These are the *appendices epiploicæ*. The blood vessels supplying them from the mucosa perforate the muscle wall. Mucous membrane may herniate through these vascular perforations, a condition known as *diverticulosis*. Inflammation of these mucosal herniæ (diverticulitis) is not uncommon.

The transverse colon, normally over 18 inches (45 cm) long, extends from the hepatic to the splenic flexure in a loop which hangs down to a variable degree between these two fixed points. It is in contact with the anterior abdominal wall. The convexity of the greater curvature of the stomach lies in its concavity, the two being connected by the gastro-colic omentum. The *greater omentum* hangs down from its lower convexity, in front of the coils of small intestine; it is a continuation downwards of the gastro-colic omentum and it receives its blood supply from the epiploic branches of the gastro-epiploic arteries. The transverse colon is completely invested in peritoneum; it hangs free on the transverse mesocolon, which is attached from the inferior pole of the right kidney across the descending (second) part of the duodenum and the pancreas to the inferior pole of the left kidney. The splenic flexure lies, at a higher level than the hepatic flexure, well up under cover of the left costal margin.

The tæniæ coli continue from the ascending colon. Due to the looping downwards and forwards of the transverse colon from the flexures, which lie well back in the paravertebral gutters, some rotation of the gut wall occurs at the flexures, and the anterior tænia of ascending and descending colons lies posteriorly, while the other two lie anteriorly, above and below. The appendices epiploicæ are larger and more numerous than on the ascending colon.

For blood supply and lymph drainage see pp. 283 and 286.

The descending colon, less than 12 inches (30 cm) long, extends from the splenic flexure to the pelvic brim, and in the whole of its course is plastered to the posterior abdominal wall by peritoneum. It lies on the lumbar fascia and the iliac fascia, being connected to them by the fibrous tissue of the extraperitoneal fascial envelope of the abdomen. It ends at the pelvic brim about 2 inches (5 cm) above the inguinal ligament; the part lying in the left iliac fossa, from iliac crest to pelvic brim, is sometimes called the iliac colon.

The three tæniæ coli, in continuity with those of the transverse colon, lie one anterior and two posterior (medial and lateral). Appendices epiploicæ are numerous. Diverticulosis is common in this part of the colon.

In the embryo the descending colon possessed a midline dorsal mesocolon containing the left colic vessels between its layers. The mesocolon hinged to the left like a door and became fused with the parietal peritoneum of the posterior abdominal wall (Fig. 5.22, p. 277); thus in the adult the peritoneal floor of the left infracolic compartment is the right leaf of the original dorsal mesocolon, and the left colic vessels lie immediately beneath it.

For blood supply and lymph drainage see pp. 284 and 286.

The sigmoid colon (pelvic colon) extends from the iliac colon at the pelvic brim to the commencement of the rectum in front of the third piece of the sacrum. It is completely invested in the peritoneum and hangs free on a mesentery, the sigmoid (pelvic) mesocolon. It is usually less than 18 inches (45 cm) long, though great variations in length are common. There is no change in the gut wall between terminal sigmoid colon and upper rectum; the distinction is only of peritoneal attachment. Where there is a mesentery the gut is called sigmoid. Where the mesentery ceases the gut is called rectum. Compare the duodeno-jejunal junction (p. 293).

Like the rest of the large intestine, the commencement of the sigmoid colon is sacculated by three tæniæ coli, but these muscular bands are wider than elsewhere in the large gut, and meet to clothe the terminal part of the sigmoid in a complete longitudinal coat. The sigmoid colon possesses well-developed appendices epiploicæ. It lies, usually, in the pelvic cavity, coiled in front of the rectum, lying on the peritoneal surface of the bladder (and uterus).

The *sigmoid mesocolon* in the embryo had originally a midline dorsal attachment. It hinged to the left like that of the descending colon and became partially, but not completely, fused with the parietal peritoneum of the posterior abdominal and pelvic wall. The part remaining free is joined to the parietal peritoneum along a Λ-shaped base. The limbs of the Λ diverge from the bifurcation of the common iliac artery, over the sacro-iliac joint at the pelvic brim. The upper limb is attached to the external iliac artery along the pelvic brim from this point halfway to the inguinal ligament,

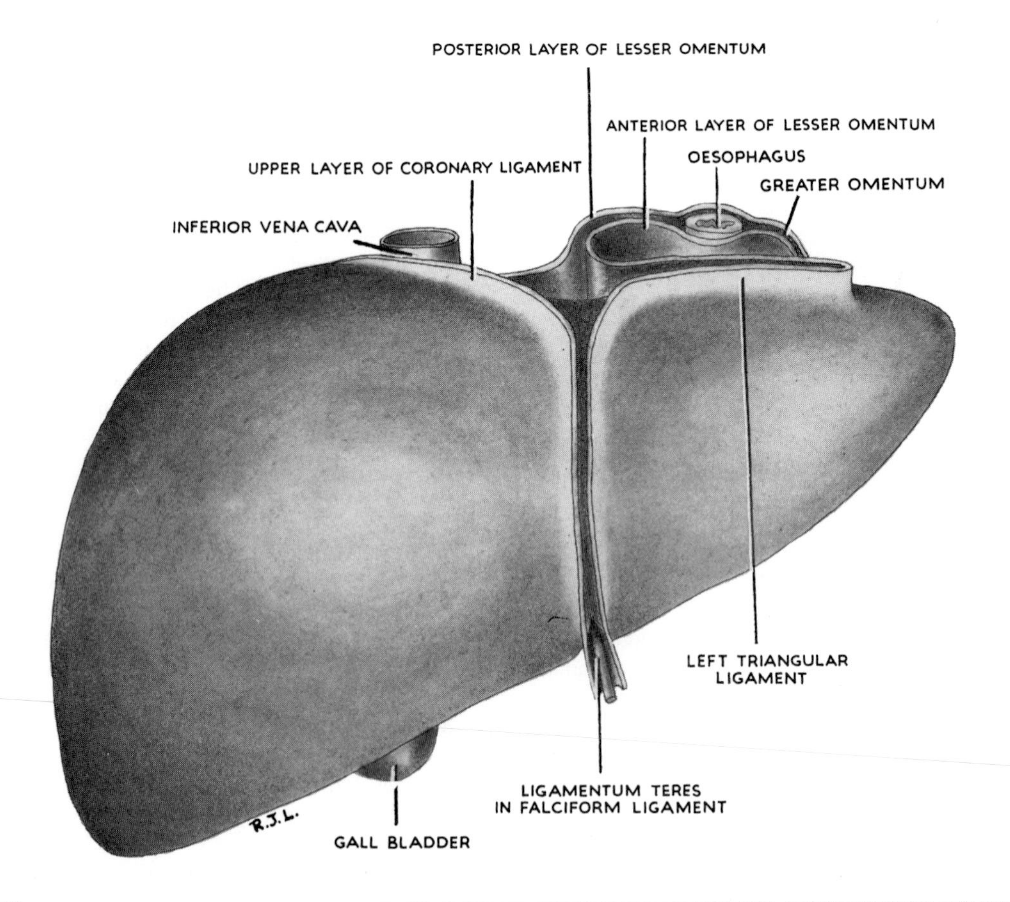

FIG. 5.36. ANTERIOR VIEW OF THE LIVER. THE ŒSOPHAGUS IS PULLED UPWARDS FROM ITS NORMAL POSITION BEHIND THE LEFT LOBE TO SHOW THE PERITONEAL ATTACHMENTS. All peritoneal edges seen here are attached to the diaphragm.

a distance of about 2 inches (5 cm). The base of the sigmoid mesocolon thus measures 4 or 5 inches (10 to 12 cm); its intestinal border, equal to the length of the sigmoid colon, is four times as long (say 16 inches (40 cm)). The sigmoid vessels lie between the layers of the sigmoid mesocolon.

For blood supply and lymph drainage see pp. 284 and 286.

THE LIVER

Inspect a liver. Its form has nothing to do with its function. The large wedge-shaped mass is merely a cast of the cavity in which it grows. Thus the liver has only two surfaces, diaphragmatic and visceral. The diaphragmatic surface is boldly convex, moulded to the diaphragm. The visceral surface, flat, slopes down to the right and forwards too (Fig. 5.37). Faint visceral impressions are moulded on this surface. From the diaphragmatic and visceral surfaces peritoneal folds pass respectively across to the diaphragm and down to the stomach; these persist from the ventral mesogastrium into which the developing liver grows (Fig. 5.19, p. 276).

Diaphragmatic Surface. This surface is for the most part covered in peritoneum, which peels off in places to join the adjacent diaphragm.

The *anterior view* shows along its horizon the bulging summit of the right side moulded by the right dome of the diaphragm (Fig. 5.36). Tapering to the left there is a bulge produced by the left dome. Between the two is a depression produced by the central tendon and the overlying heart. The sharp lower border slopes steeply up from right to left (along the costal margin and across the epigastrium). Over the anterior convexity the falciform ligament (p. 273) is attached from the centre of the depressed area down to the notch made by the ligamentum teres in the lower border. This notch is to the left of the fundus of the gall bladder, which peeps below the inferior border. The upper attachment of the falciform ligament sweeps to the left along the summit of the liver as the left triangular ligament (Fig. 5.36). It is a reduplication of the left leaf of the falciform ligament. The right leaf of the falciform ligament sweeps to the right, over the summit of the right dome, to pass

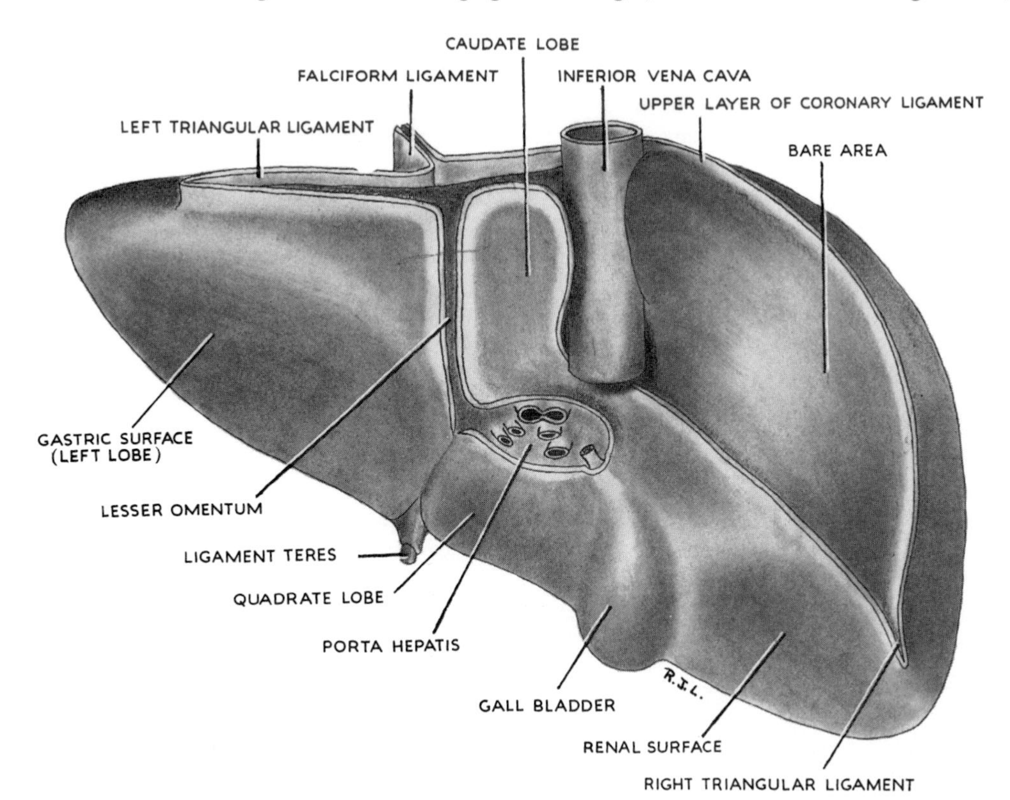

FIG. 5.37. POSTERIOR VIEW OF THE LIVER. THE UNDER SURFACE OF THE ORGAN, SLOPING DOWN TO THE ANTERIOR BORDER, IS VISIBLE FROM THIS ASPECT. THE PERITONEAL ATTACHMENTS ARE SHOWN. The cut edges around the porta hepatis and in the lower part of the lesser omentum are attached to the lesser curvature of the stomach; all other peritoneal edges seen here are attached to the diaphragm.

just in front of the inferior vena cava. It cannot be seen from in front.

Posterior view. Inspect the liver from behind (Fig. 5.37). On the posterior convexity of the diaphragmatic surface the inferior vena cava lies in a deep groove (sometimes a tunnel).

To the right is the *bare area*, which is triangular. Its base is the inferior vena cava, its upper and lower sides are the separated layers of the coronary ligament (p. 273) and its apex is the tiny right triangular ligament where these two layers meet. The lower layer of the coronary ligament is attached along the blunt rounded border between the diaphragmatic and visceral surfaces (from here the single sheet of peritoneum sweeps down over the right kidney into the paracolic gutter).

Trace the liver attachment of this inferior layer to the left, *in front of* the inferior vena cava, and thence up along its left side to the summit of the liver. Here it meets the right leaf that diverges from the falciform ligament (Fig. 5.37). Together they are now traceable along the deep groove cut into the diaphragmatic surface of the liver by the ligamentum venosum (Fig. 5.38). These two layers from the diaphragmatic surface of the liver are short and are attached to the diaphragm. They enclose the caudate lobe in the upper recess of the lesser sac (omental bursa). The caudate lobe is the only part of the diaphragmatic surface that is in the lesser sac. It touches the diaphragm in front of the thoracic aorta, just to the left of the inferior vena cava and to the right of the œsophagus (Fig. 5.17, p. 271). (The area can be seen stripped bare of peritoneum in PLATE 20.)

To the left of this the posterior surface of the liver tapers to its sharp left extremity.

The posterior view shows the visceral surface sloping down (Fig. 5.37) but this is foreshortened. It is better to view the visceral surface on the flat.

Visceral Surface (Fig. 5.38). Remember the slope, down and to the right. The inferior vena cava is vertical. This flat or slightly concave surface has a kidney-shaped outline in conformity with that of the body cavity. Centrally lies the hilum of the liver (porta hepatis). It is the cross-stroke of a capital H. The right limb (incomplete) of the H is made by the gall bladder and inferior vena cava, the left limb is made by the continuity of the grooves for the ligamentum teres and ligamentum venosum. The groove for the ligamentum venosum leaves the visceral surface and extends up to the diaphragmatic surface around the caudate lobe to meet the inferior vena cava, as already seen in the posterior view. The visceral surface is clothed in peritoneum which peels off as the lesser omentum.

The hilum is enclosed between the two layers of the lesser omentum; from its left end these two layers are attached to the ligamentum venosum lying deep in its groove (the lesser omentum passes down from this liver attachment and encloses the stomach and first inch of the duodenum between its layers).

The **porta hepatis** is the hilum of the liver. It is a

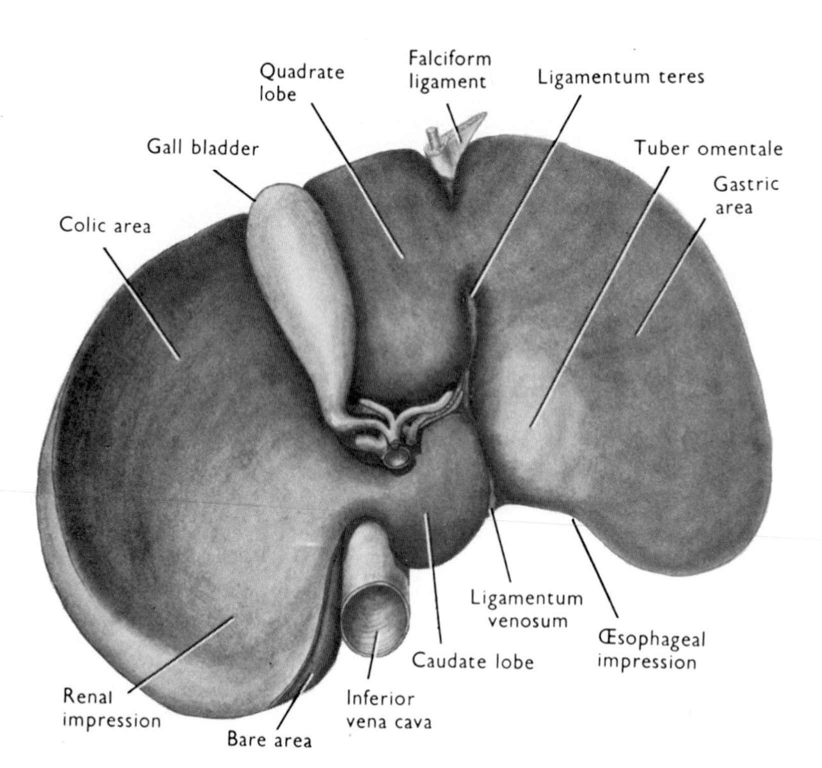

FIG. 5.38. THE VISCERAL SURFACE OF THE LIVER.

transverse slit, perforated by right and left hepatic ducts and vessels. The vessels are the ingoing hepatic artery and portal vein. They lie in the usual order, V.A.D., with the ducts in front (more accessible in surgery). The cystic duct lies in loose contact with the right end of the porta, and there are several lymph nodes here. These vessels, together with the nerves of the liver, lie enclosed between the layers of the free edge of the lesser omentum.

From the right end of the porta hepatis the gall bladder lies in a shallow fossa on the down-sloping visceral surface. Its neck is highest, its fundus lowest (see below).

The Lobes of the Liver. From early days the liver has been described in two lobes, but incorrectly. On the anterior surface the falciform ligament attachment divides the liver into named right and left 'lobes'; this is inaccurate. On the visceral surface the grooves for ligamentum teres and ligamentum venosum divide the liver into these inaccurately named right and left 'lobes'. The visceral surface to the left of these grooves was named the *left lobe*. The visceral surface from the sharp lower border to the porta hepatis, between ligamentum teres and gall bladder, is named the *quadrate lobe* (Fig. 5.38). The visceral surface behind the porta hepatis, enclosed by the ligamentum venosum and its attached lesser omentum, is named the *caudate lobe* (Fig. 5.38). This lies up on the diaphragmatic surface and it is joined by an isthmus of liver surface to the right lobe. The isthmus is called the *caudate process*; it lies at the upper limit of the epiploic foramen, between the porta hepatis and the groove for the inferior vena cava. The visceral surface to the right of the gall bladder and inferior vena cava is called the *right lobe*. The old description included the quadrate and caudate lobes in the right lobe.

The old anatomists knew but failed to understand why the right and left hepatic ducts are of equal diameter, as are the right and left branches of the hepatic artery and portal vein. They failed to appreciate that the liver is in two equal *halves*. The division runs along the plane of the gall bladder and inferior vena cava. The quadrate and caudate lobes are part of the *left half* of the liver. They are supplied by the *left* branches of the hepatic artery and portal vein and drain into the *left* hepatic duct (PLATE 15).

Relationships of the Visceral Surface. The *left lobe* shows anteriorly a concavity imprinted by the anterior wall of the stomach, while behind and above this is a convexity called the *tuber omentale* (Fig. 5.38). This, above the lesser curvature, touches through the lesser omentum a similar convexity on the neck of the pancreas. To the left of the tuber omentale above the gastric area the œsophagus sometimes leaves a shallow groove. The *quadrate lobe* slopes down in contact with the lesser omentum, the pyloric end of the stomach and, below this, the gastro-colic omentum. The *caudate lobe* lies at the upper limit of the lesser sac, not on the visceral surface but in contact with the peritoneum on the diaphragm immediately above the aortic orifice (in front of the thoracic aorta, to the left of the abdominal inferior vena cava). The *right lobe* is indented posteriorly by the upper pole of the right kidney, and in front of this lies in contact with the right colic (hepatic) flexure.

Blood Supply of the Liver. The liver receives blood from two sources. Arterial (oxygenated) blood is furnished by the hepatic artery, which divides into right and left branches in the porta hepatis. Venous blood is carried to the liver by the portal vein, which divides in the porta hepatis into right and left branches; this portal blood is laden with the products of digestion which have been absorbed from the alimentary canal, and which are metabolized by the liver cells.

Corrosion casts demonstrate the vascular architecture of the liver. The hepatic artery and portal vein have equal-sized right and left branches. They lie together as they ramify in each half of the liver, and they are everywhere accompanied by tributaries of the hepatic ducts; the three together lie in 'portal canals'. There is no communication between right and left halves of the liver; indeed, even within each half the arteries are end arteries (hence infarction of the liver).

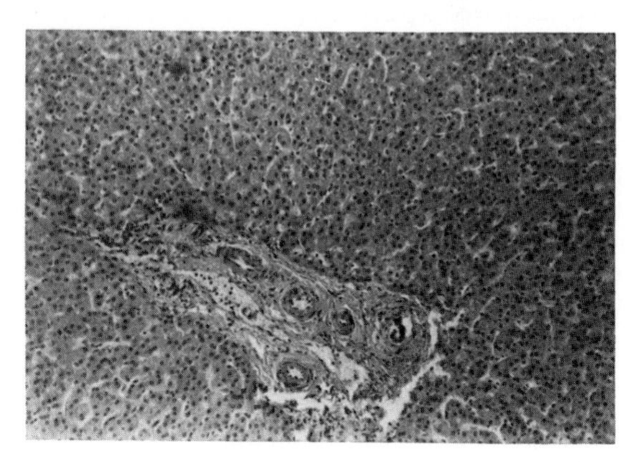

FIG. 5.39. A TYPICAL 'PORTAL CANAL' IN A HUMAN LIVER.

From the portal canals the blood passes between the liver cells, and in contact with them, to the centre of each lobule. Here the mixed hepatic and portal blood is received into a central vein. The central veins of all the lobules unite to form the hepatic veins.

The venous return differs in that it shows a mixing of right and left halves of the liver. Three main *hepatic veins*, high up near the diaphragmatic surface, drain into the inferior vena cava. A large central vein runs in the plane between right and left halves and receives from each. Further laterally lie a right and left vein (PLATE 15). All three open directly into the inferior vena cava (they have no extrahepatic course) near the upper surface of the liver, just below the central tendon of the diaphragm. They suspend the liver to the inferior vena cava and thus to the central tendon (*v. inf.*). Several accessory hepatic veins, very much smaller, commonly enter the vena cava below this.

The *lymphatics* of the liver drain into three or four nodes that lie in the porta hepatis (*hepatic nodes*). These nodes also receive the lymphatics of the gall bladder. They drain downwards alongside the hepatic artery to retropyloric nodes and so to the cœliac nodes. In some cases of carcinoma of the pylorus the retropyloric nodes become involved in spread of the disease, the afferent lymphatics become dilated, and *retrograde* spread of carcinoma may then involve the hepatic nodes. In such cases pressure of these nodes on a hepatic duct gives rise to obstructive jaundice. From the bare area the surface of the liver communicates with extraperitoneal lymphatics which perforate the diaphragm and drain to nodes in the posterior mediastinum. Similar communications exist along the left triangular and falciform ligaments from the adjacent liver surfaces.

The *nerve supply* of the liver is derived from both the sympathetic and vagus, the former by way of the cœliac ganglia, whence nerves run with the vessels in the free edge of the lesser omentum and enter the porta hepatis. Vagal fibres from the left vagal trunk reach the porta hepatis along the lesser curve of the stomach via the lesser omentum.

Development of the Liver (see also p. 277). The liver develops by proliferation of cells from the blind ends of a Y-shaped diverticulum which grows from the foregut into the septum transversum. The cranial part of the septum transversum becomes the pericardium and diaphragm. The caudal part becomes the ventral mesogastrium, and it is into this that the liver grows. At this stage the caudal part of the septum transversum transmits the vitelline veins which, by numerous anastomoses, form a rich venous plexus here. The proliferating liver cells break through the venous walls and grow freely in the blood-stream; thus in the adult liver the blood in the sinusoids is in direct contact with liver cells.

The original diverticulum from the endoderm of the foregut becomes the bile duct, its Y-shaped bifurcation produces the right and left hepatic ducts. A blind diverticulum from the common bile duct becomes the cystic duct and gall-bladder. The hepatic ducts divide and re-divide until finally liver cells grow, from the blind end of each, into the blood in the vitelline veins. The *embryological* centre of each liver lobule is a bile duct, but this is not the histological centre of the adult lobule. The lobules of the embryo fuse and are re-divided by the growth of fibrous septa along the bile ducts which thus lie at the periphery of the adult lobule.

Microscopic Anatomy of the Liver. The liver is divided into lobules which measure 1 or 2 mm in diameter. They are separated by an amount of areolar tissue which varies in thickness in different animals. There is very little connective tissue in the human liver. In the perilobular connective tissue run the branches of the hepatic artery and portal vein and the tributaries of the bile ducts; together they constitute the portal canals.

The lobules consist of columns or sheets of large cells which radiate out from the central vein to the periphery of the lobule. Between the sheets of liver cells are the blood sinusoids of the liver. Here the vascular endothelium and basement membrane are fenestrated to allow blood plasma to gain direct contact with liver cells. Kupffer cells (reticulo-endothelial cells) intervene between the liver cells and the blood in places. The bile is collected in tiny channels on the side of the liver cells away from the blood and taken to the tributaries of the bile ducts that lie in the periphery of the lobules; these tiny interlobular bile ducts are lined with columnar epithelium.

Stability of the Liver. The liver may enlarge, but it never falls down into the abdominal cavity. It is *supported by the hepatic veins.* The conventional claim that the liver is supported by the tone of the abdominal muscles and by atmospheric pressure is not tenable, as every surgeon knows. Incision of the abdominal wall equalizes atmospheric pressure within and without and at operation, especially under modern anæsthesia, the tone of the abdominal muscles is absent or negligible; yet even under these conditions the liver can at best be rotated only slightly. Post-mortem the liver cannot be displaced caudally until the inferior vena cava is divided.

The liver is suspended to the inferior vena cava by the hepatic veins, *which are entirely intra-hepatic in their course.* Thus the posterior (the heaviest) part of the liver could not descend without elongation of the vena cava. The thinner anterior edge of the liver is prevented from tilting downwards by the attachments of the left triangular ligament and the ligamentum teres and by resting on the underlying viscera (stomach and hepatic flexure of the colon).

The Biliary System. Bile is manufactured by the liver cells. It is collected in bile canaliculi in the lobules, flows along the portal canals in the bile-duct tributaries and so reaches the *right and left hepatic ducts,* which emerge at the porta hepatis. Here they join, and the *common hepatic duct* so produced passes down between the two peritoneal layers at the free edge of the lesser omentum. The common hepatic duct is soon joined by the cystic duct from the gall-bladder. The common bile duct is thus formed (Fig. 5.40). When retracted at operation the ducts descend below the liver (as in

Fig. 5.40) but at rest they lie in loose contact with the porta hepatis. The common bile duct begins here, at the porta.

The **gall-bladder** (Fig. 5.40) lies against the under surface of the right lobe. Its bulbous blind end, the *fundus*, projects a little beyond the sharp anterior margin of the liver and touches the parietal peritoneum of the anterior abdominal wall at the tip of the ninth costal cartilage, where the transpyloric plane crosses the right costal margin, at the lateral border of the right rectus abdominis muscle. The *body* of the gall-bladder, narrower than the fundus, passes backwards and *upwards* from this point towards the right end of the porta hepatis (Fig. 5.37). Here it narrows into a *neck*, from which the *cystic duct* lies against the porta hepatis to join the hepatic duct between the two layers of peritoneum that form the free edge of the lesser (gastro-hepatic) omentum. The cystic duct lies immediately in front of the right main branch of the hepatic artery (Fig. 5.38). The artery can be caught easily in a clamp placed on the cystic duct (a hazard during cholecystectomy).

The fundus and body of the gall-bladder are firmly bound to the under surface of the liver by connective tissue and many small cystic veins that pass from the gall-bladder into the liver substance. The peritoneum covering the liver passes smoothly over the gall-bladder. Occasionally the gall-bladder hangs free on a narrow 'mesentery' from the under surface of the liver, a condition that greatly facilitates the operation of cholecystectomy.

The fundus of the gall-bladder lies on the commencement of the transverse colon, just to the left of the hepatic flexure, while the body that lies behind it is in contact with the first part of the duodenum. The under surface of the liver is sloping, so *the neck of the gall-bladder lies at a higher level than the fundus.* It lies against the upper part of the free edge of the lesser (gastro-hepatic) omentum.

The gall-bladder (Figs. 5.38, 5.40) is a fibro-muscular sac which, histologically, shows a surprisingly small amount of smooth muscle in its wall. Its mucous membrane is a lax areolar tissue lined with a simple columnar epithelium. It is projected into folds which produce a honeycomb appearance in the body of the gall-bladder, but are arranged in a more or less spiral manner in the neck (the '*spiral valve*') just short of the cystic duct. There are no glands in the gall-bladder. In pathological conditions mucus is secreted by the columnar epithelium itself, the cells becoming goblet cells such as are found throughout the alimentary canal.

The *blood supply* of the gall-bladder is furnished by the *cystic artery*, a branch of the right hepatic artery; it passes behind the cystic duct and branches out over

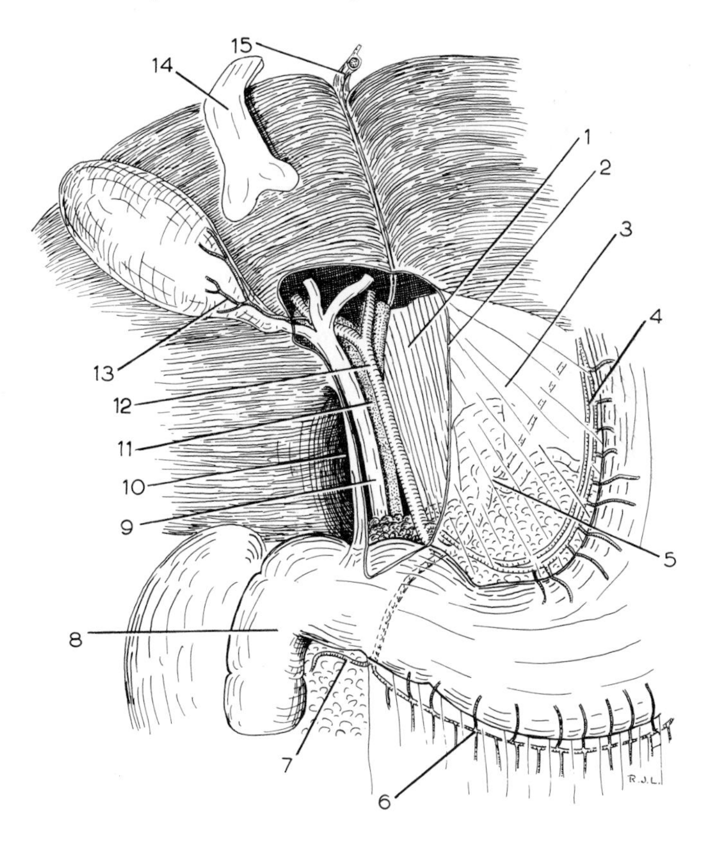

FIG. 5.40. THE FREE EDGE OF THE LESSER OMENTUM. 1. Posterior leaf of lesser (gastro-hepatic) omentum. 2. Window cut in anterior leaf of lesser omentum. 3. Lesser omentum. 4. Left gastric artery divided into two branches. 5. Cœliac axis seen through peritoneum, dividing at upper border of pancreas (tuber omentale) into left gastric, splenic and hepatic arteries. 6. Right gastro-epiploic artery. 7. Superior pancreatico-duodenal artery. 8. Duodenum on right kidney. 9. Common bile duct. 10. Epiploic foramen. 11. Portal vein. 12. Hepatic artery. 13. Cystic artery on gall-bladder. 14. Retractor under quadrate lobe of liver. 15. Ligamentum teres in falciform ligament.

the surface of the gall-bladder. Variations are common. The artery may arise from the main trunk of the hepatic artery or from the left branch of that vessel and in any case may pass in front of the cystic duct. Venous return is by multiple small veins into the substance of the liver and so to the hepatic veins. A *cystic vein* running from the neck of the gall-bladder into the portal vein is usually also present. The *lymphatics* of the gall-bladder drain to the lymph nodes in the porta hepatis and thence in the free edge of the lesser omentum to the cœliac group of pre-aortic lymph nodes.

The *sympathetic nerves* of the gall-bladder run from cell bodies in the cœliac ganglia along the hepatic artery. The *parasympathetic branches* reach the gall-bladder mainly in the left (i.e., anterior) vagal trunk along the lesser omentum.

The **common bile-duct** (Fig. 5.44) is 3 inches (7 to 8 cm) long and is best described in three parts. Its *upper third* lies in the free edge of the lesser (gastro-hepatic) omentum in the most accessible position for surgery—in front of the portal vein and to the right of the hepatic artery. Its *middle third* lies behind the first part of the duodenum, and slopes down to the right, away from the vertical portal vein. It leaves the hepatic artery here. It lies on the inferior vena cava. The *lower third* of the common bile-duct slopes down to the right behind the head of the pancreas. It lies in a deep groove, sometimes in a tunnel, on the posterior surface of the pancreas, in front of the right renal vein (PLATE 16). It opens, in common with the main pancreatic duct, into a spindle-shaped dilatation, called the ampulla (of Vater). Variations are very common, and often no dilatation (i.e., no ampulla) exists at the junction. The ampulla itself opens in the postero-medial wall of the second part of the duodenum at a small papilla four inches (10 cm) from the pylorus (see p. 292). The common opening of these two ducts is surrounded by a circular muscle (the *sphincter of Oddi*). Each duct, in addition, possesses its own sphincter, so that bile or pancreatic juice can be discharged independently into the duodenum. Some longitudinal fibres recurve into the papilla. Their contraction makes the papilla pout (a dilator mechanism). Some of the circular and longitudinal fibres blend with the muscle wall of the duodenum, to anchor the ampulla in place.

The Portal Vein. This vessel is merely the upward continuation of the superior mesenteric vein, which changes its name to portal vein after it has received the splenic vein behind the neck of the pancreas. It lies in front of the inferior vena cava, passes upwards behind the pancreas and the first part of the duodenum and loses contact with the inferior vena cava by entering between the two layers of the lesser (gastro-hepatic) omentum. It runs up in the free edge, a wide channel lying behind the bile-duct and the hepatic artery, and reaches the porta hepatis. Here it divides into a right and left branch which enter the respective halves of the liver. The portal vein is a vertical channel. In the free edge of the lesser (gastro-hepatic) omentum, the common bile-duct and hepatic artery run vertically in front of it (Fig. 5.40). Below the first part of the duodenum these structures are curved away from it. The common bile-duct curves to the right *behind the pancreas* and the hepatic artery curves to the left in *front of the pancreas* to the cœliac artery. The portal vein receives tributaries from the lower œsophagus and most of the stomach (p. 281).

The portal vein contains blood, obviously, from both splenic and superior mesenteric veins. In such a sluggish stream there is but little mixing of these two blood-streams, so that the right branch receives mostly superior mesenteric blood and the left branch mostly splenic (and inferior mesenteric) blood. The incidence of certain diseases on the right and left sides of the liver is thus explained. In the case of ingested liver poisons, their absorption from the small intestine into the tributaries of the superior mesenteric vein results in a greater concentration of poison reaching the right side of the liver, which may show toxic changes while the left side remains normal. On the other hand, in deficiencies of substances such as choline and methionine (containing labile methyl groups) the inferior mesenteric vein absorbs little or none, the small intestine already having absorbed most. The left half of the liver is thus deprived of an adequate supply and exhibits cirrhosis in the absence of such changes in the right side.

THE PANCREAS

(PLATES 14, 16, Fig. 5.48, p. 317)

The gland is of soft consistency, and its surface is finely lobulated. It is retort-shaped, tapering from a big head to a narrow tail, the whole being over 6 inches (15 cm) long.

The pancreas lies immediately behind the peritoneum of the posterior abdominal wall. The transverse mesocolon is attached to its anterior surface just above the inferior border; thus most of the gland lies in the supracolic compartment (in the lesser sac, forming part of the stomach bed), but a narrow strip along its inferior border lies in the infracolic compartment. It consists of head, neck, body and tail. The head and tail lie back in the paravertebral gutters, while the neck and body are curved boldly forward over the inferior vena cava and aorta in front of the first lumbar vertebra. The gland lies somewhat obliquely, sloping from the head upwards towards the tail.

The **head,** the broadest part of the pancreas, is moulded to the concavity of the duodenum, which it completely fills. It lies over the inferior vena cava and the right and left renal veins. Its posterior surface is deeply indented, and sometimes tunnelled, by the terminal part of the common bile-duct. The lower part of the posterior surface is prolonged, wedge-shaped to the left, behind the superior mesenteric vein and artery, in front of the aorta. This is the **uncinate process** (PLATE 16). The anterior surface of the head lies in both supracolic and infracolic compartments; some of this surface is bare, for the leaves of the greater omentum and of the transverse mesocolon are here wide apart at their attachments (Fig. 5.17, p. 271).

The **neck** of the pancreas is prolonged to the left from the upper part of the anterior portion of the head. It lies in front of the superior mesenteric vein and its direct upward continuation the portal vein, which are thus embraced between the neck and the uncinate process. The splenic vein joins the superior mesenteric vein here. The superior mesenteric artery touches the left side of the vein in front of the uncinate process (PLATE 14). The neck has a slight convexity to its left, the tuber omentale. Above the lesser curvature this touches the tuber omentale of the left lobe of the liver, the lesser (gastro-hepatic) omentum intervening. The transverse mesocolon is attached towards the lower border of the neck, which lies in the stomach bed of the lesser sac.

The **body** of the pancreas passes from the neck to the left, sloping gently upwards across the left renal vein and the aorta, the left crus of the diaphragm, the left psoas muscle and lower pole of the left suprarenal gland, to the hilum of the left kidney. Its upper border crosses the aorta at the origin of the short cœliac axis; the tortuous splenic artery passes to the left along the upper border of the body and tail, the crests of the waves showing above the pancreas, the troughs out of sight behind it (PLATE 14). Its lower border, alongside the neck, crosses the origin of the superior mesenteric artery. The splenic vein lies closely applied to its posterior surface; the inferior mesenteric vein joins the splenic vein behind the body of the pancreas in front of the left renal vein where it lies over the left psoas muscle. The transverse mesocolon is attached towards the lower part of the anterior surface; the body lies, therefore, behind the lesser sac, where it forms part of the stomach bed.

The **tail** of the pancreas passes forward from the anterior surface of the left kidney at the level of the hilum. Accompanied by the splenic artery, vein and lymphatics it lies within the two layers of the lieno-renal ligament and thus touches the hilum of the spleen.

The **blood supply** of the pancreas is derived chiefly from the splenic artery, which supplies the neck, body and tail. One large branch is named the *arteria pancreatica magna.* The head is supplied by the superior and the inferior pancreatico-duodenal arteries. Venous return is by numerous small veins into the splenic vein and, in the case of the head, by the superior pancreatico-duodenal vein into the portal vein and by the inferior pancreatico-duodenal vein into the superior mesenteric vein. The lymph drainage of the pancreas follows the course of the arteries. To the left of the neck the pancreas drains into the retropancreatic nodes. The head drains from its upper part into the cœliac group and from its lower part and uncinate process into the superior mesenteric group of pre-aortic lymph nodes.

The **pancreatic duct** is a continuous tube leading from the tail to the head, gradually increasing in diameter as it receives delicate tributaries on its way. It joins the common bile duct in a spindle-shaped dilatation, the ampulla of Vater. The ampulla opens on the surface of a tiny nipple, the duodenal papilla, that projects into the duodenum (PLATE 16). It drains the tail, body, neck and upper part of the head of the pancreas.

The **accessory pancreatic duct** drains the uncinate process and lower part of the head and crosses the main pancreatic duct to open in the duodenum at a small papilla situated 2 cm proximal to the duodenal papilla. The two ducts frequently communicate with each other (see development).

Microscopic Anatomy of the Pancreas. The pancreas is a lobulated gland composed of alveoli of serous cells with, in most sections, very few ducts. A histological section is distinguishable, in the absence of islets, by the characteristic staining reaction. In each alveolus the basal part of the cell is deeply stained and basophilic, while the central part of the alveolus, where the cells abut on the lumen, stains less heavily and is acidophilic. The nuclei, situated towards the basal part of each cell, are large and deeply stained. The ducts are lined with a simple columnar epithelium.

The *islets* appear in section as pale areas scattered in the deeply staining pancreatic tissue (Fig. 5.41). They vary in size, their diameter being from one to four times that of the pancreatic alveolus. They are composed of a tightly packed mixture of acidophilic and basophilic small round cells. They produce insulin. In the pancreas of many animals are found the large laminated pressure receptors of Pacini.

Pacinian corpuscles are rare in the human pancreas, but they are found in the retroperitoneal areolar tissue of the upper part of the abdomen. There are none in the lower part of the abdomen or in the pelvis. Surgical shock is produced much more readily in the upper

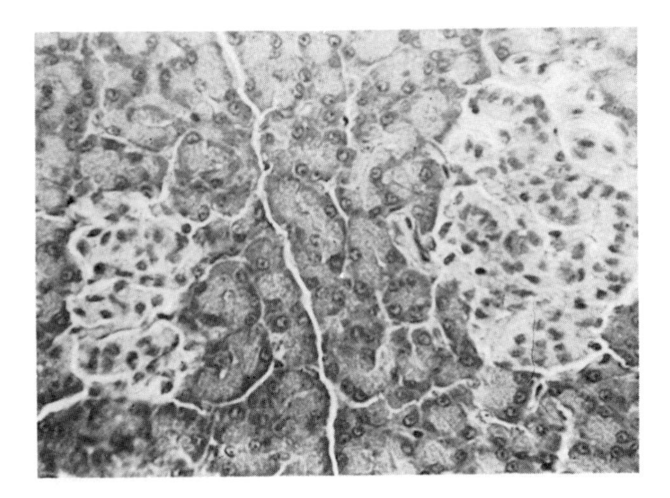

FIG. 5.41. ISLETS LYING IN PANCREATIC LOBULES.

abdomen than in the lower abdomen or pelvis, and it may be that traction exerted on peritoneal folds in the upper abdomen produces a fall of blood pressure by undue stimulation of these corpuscles. The reason for the existence of the corpuscles in the upper abdomen is difficult to understand. They register pressure, both positive and negative, and it is possible that they serve a protective function by causing reflex contraction of the abdominal wall to aid support of the heavy viscera of the upper abdomen when jarring movements of the body tend to displace these viscera and produce undue traction on their peritoneal attachments.

Development of the Pancreas. The pancreas develops as two separate buds, each an outgrowth of the endoderm at the junction of foregut and midgut. A *ventral bud* grows into the ventral mesogastrium in common with the outgrowth of the bile-duct and a *dorsal bud* grows independently from a separate duct into the dorsal mesogastrium. The duodenal portion of the gut subsequently rotates and becomes adherent to the posterior abdominal wall. It, with the pancreatic outgrowths, finally lies behind the peritoneum. The duodenal wall grows asymmetrically; the openings of the two ducts, originally diametrically opposite, are thus carried around into line with each other, and the two parts of the gland fuse into the single adult pancreas. The pancreatic alveoli develop by growth of cells from the terminal parts of the branching ducts. The islet cells appear to have an identical origin, but become separated from their parent ducts and undergo a complete change of secretory function.

It should be noted that the two parts of the pancreas have exchanged ducts. The ventral rudiment produced the lower part of the head and the uncinate process, which were connected at first with the bile-duct; these parts in the adult are drained separately by the accessory pancreatic duct. The remainder of the pancreas (upper part of head, neck, body and tail) developed in the dorsal mesogastrium from an independent duct; in the adult these parts drain by the main pancreatic duct in common with the bile-duct. The two duct systems anastomosed on fusion of the two parts of the gland, and persistence of anastomotic channels accounts for the interchange of drainage areas between the two ducts.

THE SPLEEN

The odd numbers 1, 3, 5, 7, 9, 11 summarize certain statistical features of the spleen. It measures 1 × 3 × 5 inches, weighs 7 oz and lies between the ninth and eleventh ribs (H. A. Harris). The spleen is a firm organ of a dull red colour, roughly (for those who dislike the above imperial measures) the size and shape of a clenched fist. The measurements quoted are average; the size of the spleen varies a good deal.

The spleen develops from the mesoderm of the left leaf of the dorsal mesogastrium and projects, in the adult, from the same part of the greater omentum. It is invested by the peritoneum of the left leaf, from which it projects into the greater sac. It lies below the diaphragm and its *diaphragmatic surface* is moulded into a reciprocal convexity. Its *hilum* lies in the angle between the stomach and left kidney, each of which impresses a concavity alongside the attached splenic vessels. Its long axis lies along the line of the tenth rib, and its lower pole does not normally project any further forward than the midaxillary line. A small *colic area* lies in contact with the splenic flexure and the phrenico-colic ligament. Its anterior border is notched, a relic of the fusion of the several 'splenules' from which the organ arises in the embryo.

Its visceral peritoneum, or serous coat, invests all surfaces (gastric, diaphragmatic, colic and renal) and at the hilum comes into contact with the right leaf of the greater omentum. The two leaves of the greater omentum, now in contact, pass from the hilum forwards to the greater curvature of the stomach (the gastro-splenic omentum) and backwards to the front of the left kidney (the lieno-renal ligament). The peritoneal attachment at the hilum of the spleen extends down towards the lower pole, and this attachment makes a ridge in the greater omentum. This attachment can easily be torn, accidentally, during splenectomy. The hilum of the spleen makes contact with the tail of the pancreas, which lies within the lieno-renal ligament (Fig. 5.49, p. 318).

If the spleen enlarges, its long axis extends down and forwards along the tenth rib, and its anterior border approaches the costal margin to the left of the greater curvature of the stomach. The spleen must at least double its normal size before its anterior border passes beyond the left costal margin. A palpable spleen is identified by the notch in its anterior border. In some diseases the spleen is grossly enlarged, and may extend across the upper abdomen to far beyond the umbilicus. Whatever the degree of enlargement the spleen glides in contact with the diaphragm and anterior abdominal wall in front of the splenic flexure, which remains anchored to the lower pole of the left kidney, and no colonic resonance is found on percussion over the organ. Retroperitoneal tumours (e.g., of the left kidney) do not displace the overlying colon and they are crossed by a band of colonic resonance, a further diagnostic point.

To identify the surfaces of the detached spleen hold the convexity of the organ (its diaphragmatic surface) in the hollow of the left hand. Rotate it until the notched anterior border lies to the front, near the thumb (Fig. 5.42). The concavity behind the notched anterior border is the gastric impression; it leads back to the low prominence of the hilum. Behind the hilum is the concave renal surface, while at the lower pole (at the tip of the little finger of the left hand) is the small colic impression.

FIG. 5.42. THE SPLEEN HELD IN THE LEFT HAND. THE THUMB LIES ALONG THE NOTCHED ANTERIOR BORDER, THE LITTLE FINGER SUPPORTS THE COLIC AREA AT THE LOWER POLE. THE VASCULAR PEDICLE AT THE HILUM SEPARATES THE CONCAVE GASTRIC AND RENAL AREAS. (Photograph.)

The *blood supply* of the spleen is provided by the splenic artery. It passes between the layers of the lieno-renal ligament and at the hilum of the spleen breaks up into a 'handful of fingers'—four or five branches which enter the hilum separately. Similar veins leave the hilum and unite to form the splenic vein. The lymph drains into several nodes lying at the hilum and thence, by way of the retropancreatic nodes, to the cœliac lymph nodes. The spleen is supplied from the cœliac plexus with sympathetic fibres only.

Microscopic Anatomy. The spleen is enclosed in a dense fibro-muscular capsule from which trabeculæ penetrate the substance of the organ. A bloody pulp fills the cavities between the trabeculæ; there is no subcapsular space (Fig. 1.21, p. 19). The pulp consists of all varieties of white cells, small and large, monocytes, histiocytes and multi-nucleated giant cells, lying on and between a network of reticulin fibres; this is an important part of the reticulo-endothelial system. Free erythrocytes abound, mixed with the conglomeration of white cells.

A most characteristic feature of the spleen is the presence of *Malpighian corpuscles* (Fig. 5.43). These consist of aggregates of lymphocytes just visible to the naked eye. Each is perforated by a small arteriole, though the arteriole is not seen in every Malpighian follicle cut in histological section, for the following reason. The branches of the splenic artery run in the trabeculæ and then branch into the splenic pulp. Each breaks up into a leash of straight vessels that enter blood sinusoids in the pulp. Each straight vessel is invested with a spherical mass of lymphocytes; the

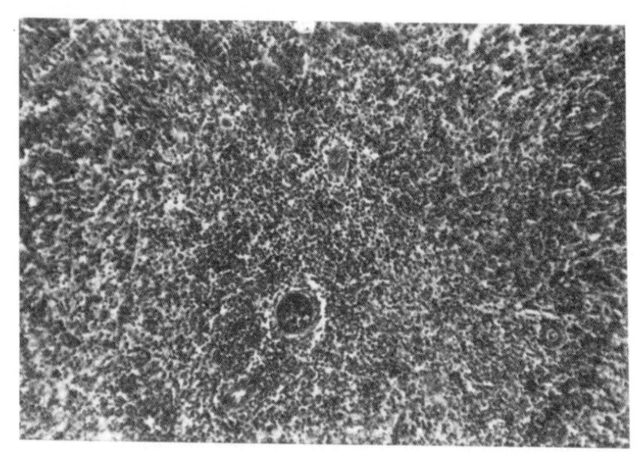

FIG. 5.43. A MALPIGHIAN CORPUSCLE, UNIQUE TO THE SPLEEN. A lymphatic follicle is perforated by an arteriole.

spleen is thus studded with masses of separate lymphatic follicles, each perforated by a straight arteriole. A section across the arteriole will show it lying within the lymphatic follicle; a section parallel with the arteriole will show the lymphatic follicle and no arteriole. The straight arterioles run in all directions, and on the law of averages it is necessary to inspect at most half a dozen follicles to find one showing the cut central arteriole. This arrangement is unique to the spleen; nowhere else in the body does an arteriole lie within a lymphatic follicle (Fig. 1.21, p. 19).

The straight arterioles open into thin-walled sinusoids which in turn drain into venules. There is no anastomosis between adjacent arterioles, i.e., the arteries themselves are end arteries and their obstruction leads to infarction of the spleen.

THE POSTERIOR ABDOMINAL WALL

The five lumbar vertebræ project forwards into the abdominal cavity; the lumbar spine has a normal lordosis (forward convexity). The midline forward projection is enhanced by the inferior vena cava and aorta, which lie in front of the bodies of the vertebræ. To each side of this convexity lie deep paravertebral gutters. They are floored in by the psoas and quadratus lumborum muscles and, below the iliac crest, by the iliacus muscle. The kidneys lie high up in the paravertebral gutters (PLATE 18).

The lumbar vertebræ are separated from each other by thick intervertebral discs, which unite them very strongly. A broad ribbon, the anterior longitudinal ligament, is attached anteriorly and crosses the lumbosacral prominence to become fused with the periosteum in the hollow of the sacrum.

The **psoas major muscle** lies in the gutter between the bodies and transverse processes of the lumbar vertebræ and passes downwards along the pelvic brim beneath the inguinal ligament into the thigh, where its tendon is attached to the lesser trochanter of the femur (PLATE 20). It surrounds and imprisons the lumbar plexus and is covered by a dense fascia. Its vertebral attachment is to the *discs above the five lumbar vertebræ*, the adjoining parts of the bodies of the vertebræ, and to fibrous arches that span the concavities of the sides of the vertebral bodies. Thus there is one continuous attachment from the lower border of Th. 12 to the upper border of L5 vertebræ. In addition, the muscle is attached to the medial ends of the transverse processes of the lumbar vertebræ. There are four fibrous arches; they span the concavities at the sides of the bodies of the upper four lumbar vertebræ. The four lumbar arteries and veins pass beneath the four arches and run laterally behind the psoas muscle; they are accompanied by sympathetic rami (p. 314).

A strong fascia, the *psoas fascia*, invests the surface of the muscle, attached to the vertebral bodies, the fibrous arches, and the transverse processes, and extends along the pelvic brim attached to the ilio-pectineal line at the margins of the muscle. It retains the pus of a psoas abscess, and spinal caries may present as a cold abscess in the groin.

There is a thickening in the psoas fascia curving obliquely from the body of the second lumbar vertebra to the transverse process of the first lumbar vertebra. This is the *medial arcuate ligament*, from which fibres of the diaphragm arise in continuity alongside the crus. The part of the psoas above this ligament is above the diaphragm, i.e., in the thorax. The ganglionated trunk of the sympathetic nervous system passes from thorax to abdomen beneath the medial arcuate ligament (PLATE 17).

Nerve Supply and Action (see p. 137).

The **psoas minor muscle,** present in only two out of every three individuals, is a slender muscle lying on the surface of psoas major. Its slender belly arises from Th. 12 and L1 vertebræ and its long tendon flattens out to blend with the psoas fascia behind the inguinal ligament and thus gains a bony attachment at the margin of psoas major to the arcuate line and ilio-pectineal eminence. Its action is to flex the lumbar spine, but so weakly as to be negligible in contrast with psoas major and rectus abdominis.

Quadratus Lumborum Muscle. The muscle is a flat sheet lying deep in the paravertebral gutter, edge to edge with psoas medially and transversus abdominis laterally. It lies in the anterior compartment of the lumbar fascia. It arises from the stout transverse process of the fifth lumbar vertebra, from the strong ilio-lumbar ligament and from a short length of the adjoining iliac crest. Its fibres pass upwards to the transverse processes (lateral to psoas) of the upper four lumbar vertebræ and, more laterally, to the inferior border of the twelfth rib. Its lateral border slopes upwards and medially and so crosses the lateral border of ilio-costalis, which slopes upwards and laterally. Its anterior surface is covered by the anterior lamella of the lumbar fascia. A thickening in front of this fascia passing from the first lumbar transverse process to the outer end of the twelfth rib constitutes the *lateral arcuate ligament.* The fibres of the diaphragm arise in continuity from the ligament. The subcostal neuro-vascular bundle (VAN, with vein above) emerges from the thorax beneath the ligament and slopes down across the lumbar fascia.

The muscle is supplied by Th. 12 and the upper three or four lumbar nerves. It represents the middle of the three muscular layers of the body wall; it is in series with the intercostal muscles (p. 211)

Its chief action is to prevent the diaphragm from elevating the twelfth rib and so wasting its contraction. By depressing the twelfth rib it aids descent of the contracting diaphragm. Additionally it is an abductor (lateral flexor) of the lumbar spine.

The **iliacus muscle** clothes the iliac fossa (p. 137).

Fascia of Posterior Abdominal Wall. It should be noted that each muscle of the posterior abdominal wall (quadratus lumborum, psoas and iliacus) is covered with a dense and unyielding fascia. These fasciæ provide a firm fixation for the peritoneum and retroperitoneal viscera of the posterior abdominal wall, undisturbed by the movements of contraction of the underlying muscles.

The Fascia Iliaca (Fig. 3.17, p. 138). The iliacus muscle is covered by the strong fascia iliaca; this is attached to bone at the margins of the muscle, and to the inguinal ligament. This fascia forms a floor to the abdominal cavity, and serves for the attachment of parietal peritoneum. Apart from its prolongation into the femoral sheath (p. 137) it does not extend into the thigh, for there is no peritoneum there.

The Lumbar Fascia (Fig. 1.24, p. 22). Three layers of tough fibrous tissue enclose two muscular compartments. The anterior and middle lamellæ occupy only the lumbar region, but the posterior lamella extends above this to the lower part of the neck. The quadratus lumborum occupies the anterior compartment, while the erector spinæ mass of muscle fills the posterior compartment. The *anterior lamella* extends from the front of the ilio-lumbar ligament and adjoining iliac crest up to the inferior border of the twelfth rib. Medially it is attached to the front of each lumbar transverse process near its root, adjoining the attachment of the psoas fascia. Laterally it blends with the middle lamella along the lateral border of quadratus lumborum; here the transversus abdominis and internal oblique muscles take origin. The *middle lamella* extends from the back of the ilio-lumbar ligament and adjoining iliac crest up to the twelfth rib. Medially it is attached to the tips of the lumbar transverse processes. Laterally it blends with both anterior and posterior lamellæ. The latter line of fusion is along the lateral border of the sacro-spinalis muscle. Note that quadratus lumborum and sacro-spinalis have lateral borders that slope in opposite obliquities and cross each other like the limbs of a very narrow X. The *posterior lamella* lies over the whole erector spinæ mass of muscle. It is attached medially to the spinous processes and supraspinous ligaments of all the sacral, lumbar and thoracic vertebræ. Its lateral margin traced from below upwards extends along the transverse tubercles of the sacrum to the ridge on the posterior part of the iliac crest. It slopes outwards to the twelfth rib, being attached across the lumbar region to the middle lamella along the lateral border of ilio-costalis. Above the twelfth rib its attachment is to the posterior angles of all the ribs; its lateral border over the thoracic cage thus slopes up medially. The thoracic part of the posterior lamella is known as the *thoraco-lumbar (lumbo-dorsal) fascia.* The posterior lamella is thick and strong over the lumbar region, being here reinforced by fusion of the aponeurotic origin of latissimus dorsi. Over the thorax it gradually becomes thinner and it fades out above the first rib over the extensor muscles of the neck, where it is replaced by the splenius muscle.

Vessels and Nerves of the Posterior Abdominal Wall. The posterior abdominal wall is supplied with blood by segmental arteries that lie in series with the intercostal arteries in the thorax. The spinal nerves, at their emergence from the intervertebral foramina necessarily pass into the substance of the psoas muscle. They give segmental branches of supply to the psoas and quadratus lumborum muscles and then unite to form the lumbar plexus within the substance of the psoas muscle.

The Vessels of the Posterior Abdominal Wall. The abdominal aorta has three groups of branches:

(*a*) Single ventral arteries to the gut and its derivatives (cœliac, superior and inferior mesenteric arteries),

(*b*) Paired branches to the other viscera (suprarenal, renal and gonadal arteries),

(*c*) Paired branches to the abdominal wall.

The **subcostal artery** leaves the thoracic aorta just above the diaphragm and passes below the twelfth rib, beneath the lateral arcuate ligament. It appears, between the subcostal nerve and vein, on the anterior surface of the lumbar fascia over the quadratus lumborum behind the kidney and passes laterally into the neuro-vascular plane of the anterior abdominal wall (between transversus abdominis and the internal oblique).

The **phrenic artery** leaves the aorta in the aortic orifice and slopes upwards across the crus of the diaphragm, which it supplies. It gives off a branch (the superior suprarenal artery) to the suprarenal gland. The phrenic veins join the azygos system of veins.

The **lumbar arteries,** four in number, leave the abdominal aorta opposite the bodies of the upper four vertebræ. Hugging the bone they pass beneath the lumbar sympathetic trunks and the fibrous arches in the psoas. On the right side the inferior vena cava overlies the lumbar arteries. Each artery gives off posterior and spinal branches and passes laterally through the psoas muscle. The three upper arteries pass laterally behind the quadratus lumborum muscle into the neuro-vascular plane between the transversus abdominis, and the internal oblique muscles. The fourth lumbar artery, like the subcostal, passes across to the neuro-vascular plane in front of the lower border of quadratus lumborum, along the upper margin of the ilio-lumbar ligament (PLATE 18).

The abdominal aorta bifurcates into two common iliac arteries in front of the body of the fourth lumbar vertebra. Consequently there is no fifth lumbar artery: its place is taken by the ilio-lumbar branch of the posterior division of the internal iliac artery.

The **ilio-lumbar artery** ascends from the pelvis in front of the lumbo-sacral trunk and passes laterally behind the obturator nerve and the psoas muscle. The *lumbar branch* supplies the psoas and quadratus lumborum and gives a spinal branch to the L5–S1 intervertebral foramen. The *iliac branch* runs into the iliac fossa, supplying the iliacus muscle and the iliac bone and ends in the anastomosis at the anterior superior iliac spine.

The **lumbar veins** accompany the lumbar arteries. They drain the lateral and posterior abdominal walls. The third and fourth empty into the inferior vena cava (those of the left passing behind the abdominal aorta), but the upper two join to form the *ascending lumbar vein*. This trunk passes vertically upwards, the right through the aortic orifice, the left through the left crus of the diaphragm, to join the subcostal vein and form the azygos and hemiazygos veins respectively.

Development of the Veins. The irregular course of the four lumbar veins is due to changes of the venous pattern in the embryo. It is better to study the picture of these venous changes as a whole: the opportunity is therefore taken at this stage to summarize the development of the whole venous system.

The early embryo is little more than a bloody sponge. From the vast intercommunicating network of veins, certain well-marked longitudinal channels return the blood to the heart. These longitudinal channels empty into the sinus venosus, which lies, with the heart, in the septum transversum. The septum transversum thus receives all the veins, which communicate freely with each other. Into this anastomosing network the liver cells proliferate.

The sinus venosus receives blood from *three sources*, and in the first instance there is bilateral symmetry in each of the three systems. The three sources of venous blood are the placenta, the yolk sac and the general body tissues of the embryo. The placenta returns venous (but oxygenated) blood by the right and left umbilical veins—after birth the umbilical system shrivels up, for it is no longer needed. The yolk sac returns blood by way of the right and left vitelline veins. The yolk sac becomes the alimentary canal and the vitelline veins become the portal system of veins. The general body tissues of the embryo are drained by the right and left cardinal veins, which ultimately become the caval system (superior vena cava and azygos veins).

In each of the three systems the longitudinal veins of one side disappear, wholly or in part, and the blood is returned by cross channels from that side to the persisting channel on the other side. Thus the *left* cardinal and *left* vitelline veins shrivel up and disappear, while the *right* umbilical vein does likewise.

Fate of the Umbilical Veins. The right umbilical vein shrivels up early. A fibrous remnant may survive in the umbilical cord, or the vein may disappear without trace. The left umbilical vein persists until birth. It passes from the umbilicus into the venous network of the septum transversum. After the liver takes shape, the vein joins the left branch of the portal vein, but its blood short-circuits the liver by passing along a venous shunt called the ductus venosus, which carries the stream to the termination of the right vitelline vein on the cranial side of the liver. This terminal segment of the right vitelline vein persists as the upper terminal portion of the inferior vena cava. After birth the left umbilical vein and its continuation the ductus venosus persist as fibrous cords, the ligamentum teres and the ligamentum venosum respectively.

Fate of the Vitelline Veins. Originally the vitelline veins, right and left, bring blood from the yolk sac to the venous plexus in the septum transversum and so to the sinus venosus. The liver cells grow into the venous plexus of the septum transversum, and the liver thus interrupts the vitelline veins. On the cranial side of the liver the right vitelline vein persist as the upper terminal segment of the inferior vena cava; the terminal part of the left vitelline vein shrivels up and disappears.

On the caudal side of the liver the vitelline veins, right and left, persist for a time. The splenic vein drains into the left vitelline vein. At this place there enters a further left vitelline vein tributary, destined to persist as the superior mesenteric vein (Fig. 5.44A). Right and left vitelline veins are connected, caudal to the liver, by cross channels. The cranial end of the left vitelline vein disappears and one cross channel persists dorsal to the future duodenum (Fig. 5.44B). It remains connected with the junction of splenic and superior mesenteric veins; that is, it becomes the portal vein, which vein includes also the terminal part of the right vitelline vein (Fig. 5.44B). The splenic vein now receives the inferior mesenteric vein. Not infrequently a more caudal cross channel persists also, in which case the inferior mesenteric vein is shunted to the right directly into the superior mesenteric vein, caudal to the entrance of the splenic vein.

The terminal part of the left branch of the portal vein is the terminal subhepatic part of the left vitelline vein (Fig. 5.44B).

Fate of the Cardinal Veins. The cardinal veins drain all the tissues of the embryo. There is an anterior and a posterior cardinal vein on each side. They join to form a common cardinal, or duct of Cuvier. The right and left ducts of Cuvier open into the sinus venosus. The anterior cardinal vein is formed by a vein from the head and neck (internal jugular) joined by a vein from the upper limb (subclavian).

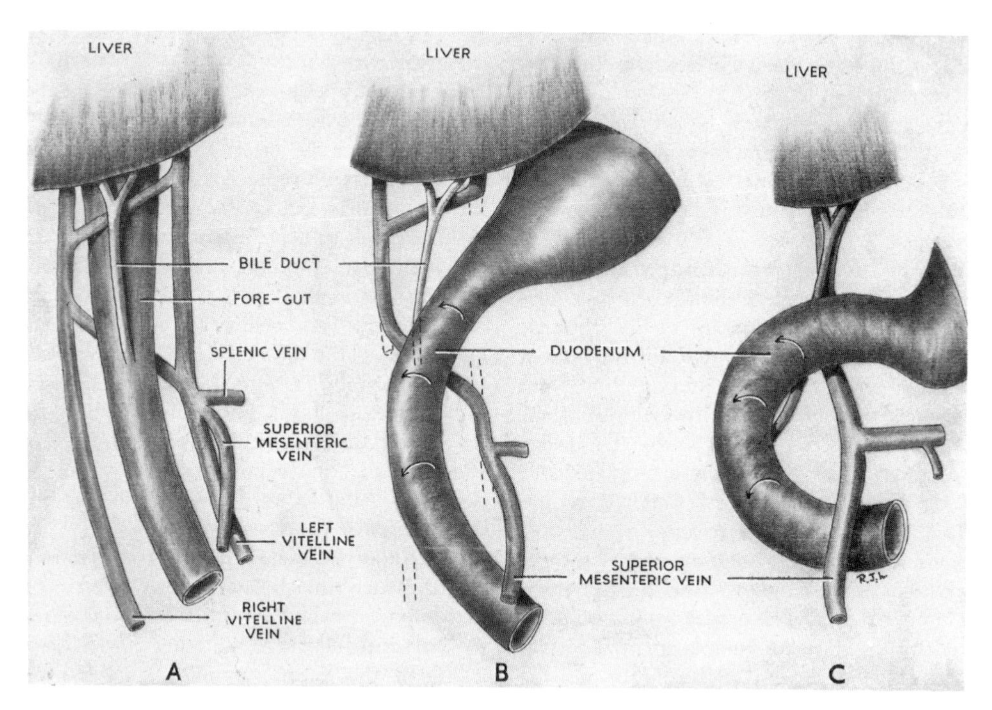

FIG. 5.44. THE DEVELOPMENT OF THE PORTAL VEIN AND THE ROTATION OF THE DUODENUM (ventral views). A. The right and left vitelline veins, running alongside the duodenum, communicate by anastomosing cross channels. B. The venous cross channels persist; disappearing parts of right and left vitelline veins are shown in dotted outline. The elongating duodenum is rotating, carrying the entrance of the bile duct around to the dorsal aspect. C. The final stage.

The posterior cardinal vein is formed by union of a vein from the lower limb (external iliac) and a vein from the pelvis (internal iliac). The sinus venosus is in the neck, and caudal to it the posterior cardinal veins receive the segmental veins of the thoracic and abdominal body wall (intercostal and lumbar veins).

In essence the subsequent changes consist in the obliteration of the left anterior and posterior cardinal veins, with the persistence of a cross channel at the cranial and caudal ends of the body. The cranial anastomotic channel is the left brachio-cephalic vein, the caudal channel is the left common iliac vein.

In the thorax the right anterior cardinal vein forms the right brachio-cephalic vein and the superior vena cava as far as the entry of the azygos vein (the posterior cardinal vein). From this point to the right atrium the superior vena cava is the persistent right duct of Cuvier. The proximal part of the right posterior cardinal vein becoms the azygos vein: descent of the septum transversum with its contained heart causes the azygos vein to curve over the root of the right lung. In the thorax the left anterior cardinal vein disappears; the persisting cross channel between it and the right cardinal vein becomes the left brachio-cephalic vein. The lower eight intercostal veins of the left side are received into the hemiazygos veins, which here replace the vanished left posterior cardinal vein.

In the abdomen the fate of the posterior cardinal veins is slightly more complicated. Several other longitudinal veins appear, the most important of which are the subcardinal veins. These appear, like loop lines on a single tram track, to the medial side of the posterior cardinal veins. They drain the paramesonephric ridge and its derivatives (suprarenals and gonads) and join the posterior cardinal veins above and below. At the level of the kidneys they communicate by a cross channel which persists in the adult as the left renal vein. Everything on the left disappears except this cross channel, into which the left suprarenal and left gonadal veins therefore empty. The disappearance of certain veins on the right results in the persistence of the **inferior vena cava** as a composite channel. Supracardinal veins appear as longitudinal channels in a plane dorsal to that of the aorta. The supracardinal or caudal segment of the inferior vena cava thus lies *dorsal*, and the subcardinal segment, the region of entrance of the renal veins, thus lies *ventral* to the abdominal aorta (PLATE 20). The caudal part of the inferior vena cava is the persisting right supracardinal vein and this receives the lower two pairs of lumbar veins. From this point to the renal veins the inferior vena cava represents the persisting right subcardinal vein: it is this part of the vessel which receives the right suprarenal and gonadal veins, in embryological symmetry with the left side, where a persisting segment of the left subcardinal vein, now incorporated in the left renal vein, receives the corresponding vessels.

Between the renal vessels and the entrance of the hepatic veins the inferior vena cava is a persisting embryonic anastomotic channel, while beyond the hepatic veins the terminal segment is the persisting right vitelline vein.

The Nerves of the Posterior Abdominal Wall. The segmental outflow from the spinal cord continues in series with the intercostal nerves of the thoracic wall. Th. 12 (the subcostal nerve) and all five lumbar nerves emerge in series from the intervertebral foramina, but they do not all share in the supply of the anterior abdominal wall. Only Th. 12 and L1 do so, in fact. L2, 3 and 4, after each giving a branch to the muscles of the posterior abdominal wall (psoas and quadratus lumborum) break up into anterior and posterior divisions which reunite to form nerves for the flexor and extensor compartments of the thigh. Thus is formed the lumbar plexus. The nerves to the thigh are the obturator (for the adductor compartment, a derivative of the flexor compartment of the thigh) and the femoral and lateral cutaneous nerves of the thigh (extensor compartment). The remainder of L4 and all L5 pass down to the sacral plexus for distribution to the lower limb (p. 342).

The **lumbar plexus,** formed from the anterior primary rami of the upper four lumbar nerves, lies within the substance of the psoas major, and its branches emerge therefrom. Most of the branches are for the supply of the lower limb, but in their passage across the posterior abdominal wall they give sensory branches to the parietal peritoneum.

The branches of the lumbar plexus are summarized on p. 356.

Nerves for the supply of the anterior abdominal wall cross the anterior surface of quadratus lumborum. They are the subcostal and the ilio-hypogastric and ilio-inguinal nerves (PLATE 20).

The **subcostal nerve** (Th. 12) passes from the thorax behind the lateral arcuate ligament, where it lies below the vein and artery. The whole neuro-vascular bundle slopes down (parallel with the twelfth rib) across the front of the anterior lamella of the lumbar fascia, bound to it by the cellular tissue of the extraperitoneal fascial envelope; here it lies behind the kidney. The subcostal is a large nerve, which disappears by passing through the transversus abdominis muscle to reach the neurovascular plane. It slopes down around the anterior abdominal wall, whose muscles it supplies, and ends by supplying the lower part of rectus abdominis and the pyramidalis muscle and the skin above them. Its lateral cutaneous branch pierces the oblique muscles and descends over the iliac crest to supply the skin of the anterior part of the buttock between the iliac crest and the greater trochanter.

The **ilio-hypogastric** and **ilio-inguinal nerves** lie in front of quadratus lumborum at a lower level. They both arise from the anterior primary ramus of L1. The ilio-inguinal nerve represents the collateral branch of the ilio-hypogastric and consequently has no lateral cutaneous branch (cf. the intercostal nerves). The nerves divide from a common stem. The point of division may be in the psoas muscle or lateral to it, in front of quadratus lumborum. In either case the nerves emerge from the lateral border of psoas major *behind the anterior lamella of the lumbar fascia.* As they slope across the quadratus lumborum, behind the kidney, they pierce the fascia and pass laterally in front of it to sink into the transversus abdominis muscle and run downwards and forwards, above the iliac crest, in the neuro-vascular plane.

The **ilio-hypogastric nerve** gives a lateral cutaneous branch which sinks below the iliac crest to supply skin of the upper part of the buttock behind the area supplied by the subcostal nerve. The nerve then slopes downwards in the neuro-vascular plane and pierces the internal oblique muscle above the anterior superior iliac spine. Sloping down between external and internal obliques it pierces the aponeurosis of the external oblique an inch or so above the superficial inguinal ring and ends by supplying the skin over the lower part of rectus abdominis (the skin of the mons).

The **ilio-inguinal nerve** represents the collateral branch of the ilio-hypogastric nerve. It runs parallel with the ilio-hypogastric at a lower level. Piercing the lower border of internal oblique it passes between the crura of the superficial ring in front of the cord, covered only by the external spermatic fascia. It supplies the anterior one-third of the scrotum, the root of the penis, and the upper and medial part of the groin, down to the anterior axial line. As it perforates the lower border of the internal oblique muscle it gives motor branches to those muscle fibres of internal oblique and transversus which are inserted into the free edge of the conjoint tendon. Division of the nerve, paralysing these muscle fibres and so relaxing the conjoint tendon, causes a direct inguinal hernia (Fig. 5.10, p. 262).

The **lateral cutaneous nerve of the thigh** is formed by union of fibres from the posterior divisions of the anterior primary rami of L2 and L3. It usually emerges from the lateral border of psoas major below the ilio-lumbar ligament, passing around the iliac fossa on the surface of the iliacus muscle *deep to the iliac fascia.* For a full inch or more (3 cm) above the inguinal ligament it slopes forward and lies imprisoned in the fibrous tissue of the iliac fascia (Fig. 3.8, p. 132). Fibres of transversus abdominis commonly arise from this fascia, and their contraction may pull on the nerve (meralgia paræsthetica). The nerve perforates the inguinal ligament a centimetre from the anterior superior iliac spine and so enters the thigh. The nerve may arise, not as stated above, but as a branch from the femoral nerve where the latter lies in the iliac fossa in the gutter between psoas and iliacus (PLATE 20). This is not surprising, since the femoral and lateral cutaneous nerves are derived from the same source (posterior divisions of the anterior primary rami of L2, 3, 4 and L2, 3 respectively). The nerve supplies the parietal peritoneum of the iliac fossa. Its peripheral distribution is considered on p. 131.

The **femoral nerve** is formed in the substance of the psoas major muscle by union of branches from the posterior divisions of the anterior primary rami of L2, 3, 4 (the obturator nerve is formed from the anterior divisions of the same nerves). It emerges from the lateral border of the psoas in the iliac fossa and runs down deep in the gutter between psoas and iliacus *behind the iliac fascia.* It gives two or three branches to the iliacus muscle (L2 and 3) and leaves the iliac fossa by passing beneath the inguinal ligament to the lateral side of the femoral sheath, on the iliacus muscle (PLATE 20). Its course in the thigh is considered on p. 139.

The **genito-femoral nerve** is formed in the substance of psoas major muscle by union of branches from L1 and 2. It emerges from the anterior surface of psoas major (and of psoas minor if it be present) and runs down on the muscle deep to the psoas fascia. In front of the fascia the left nerve is overlaid by the ureter, the gonadal vessels, the left inferior colic arteries and the inferior mesenteric vein and the peritoneum of the floor of the left infracolic compartment. The right nerve is overlaid by the ureter, gonadal vessels and the ilio-colic artery and the mesentery in the right infracolic compartment. Just above the inguinal ligament it perforates the psoas fascia and divides into genital and femoral branches (PLATE 18). The genital branch is composed of L2 and the femoral branch of L1 fibres.

The *genital branch* passes through the transversalis fascia and enters the spermatic cord. It supplies motor fibres to the cremaster muscle and sensory fibres to the spermatic fasciæ and to the tunica vaginalis of the testis. The *femoral branch* passes down in front of the femoral artery, pierces the femoral sheath and the fascia lata, and supplies the skin of the groin below the middle part of the inguinal ligament (Fig. 3.17, p. 138).

The **obturator nerve,** formed in the substance of psoas, appears in the pelvis and its course is considered on p. 341.

The Abdominal Aorta (PLATE 18). The descending thoracic aorta passes behind the diaphragm and appears in the abdomen between the crura of the diaphragm, on the front of Th. 12 vertebra. It passes downwards on the bodies of the lumbar vertebræ, inclining slightly to the left, and on the body of L4 it bifurcates into the two common iliac arteries; on the surface the point of bifurcation is 2 cm below and to the left of the umbilicus. From its posterior surface, at the bifurcation, the small **median sacral artery** runs, in the midline, over the promontory down into the hollow of the sacrum. This is morphologically the direct continuation of the aorta, vestigial in tail-less man, but large in big-tailed creatures (e.g., the crocodile).

The cœliac artery comes off opposite the body of Th. 12. At this level the cœliac ganglia lie on each side of the aorta, over the crura. Their branches make a rich plexus on the cœliac artery and on the aorta itself. Between the cœliac and the superior mesenteric arteries the aorta is overlaid by the splenic vein and the pancreas. Between the superior and inferior mesenteric arteries lie the left renal vein then the uncinate process of the pancreas and the third part of the duodenum, while below the inferior mesenteric artery the aorta is covered by the parietal peritoneum on the floor of the left infracolic compartment.

The **branches** of the abdominal aorta (PLATE 18) fall into three main groups. First the single ventral midline branches to the gut and its embryological derivatives; these are the cœliac (p. 279) and the superior (p. 282) and inferior mesenteric (p. 284) arteries. Second, the paired lateral branches to the parietes; these are the phrenic (p. 307) and four lumbar arteries (p. 307). Lastly, the paired visceral branches to the paired viscera; these are the suprarenal, renal and gonadal arteries.

The **suprarenal artery** arises from the aorta between its phrenic and renal branches. It runs laterally across the crus of the diaphragm to enter the gland. That on the left lies behind the posterior wall (parietal peritoneum) of the omental bursa, in the stomach bed, while the right suprarenal artery lies between the right crus and the inferior vena cava, behind the bare area of the liver.

The **renal artery,** a large vessel, arises at right angles from the aorta opposite the body of L2. The left artery is a little shorter than the right; it crosses the left crus and psoas muscle, behind and somewhat above the left renal vein, both of which are covered by the tail of the pancreas and the splenic vessels. The longer right artery crosses the right crus and psoas muscle behind the inferior vena cava and the short right renal vein; these structures separate the artery from the head of the pancreas and common bile duct and from the second part of the duodenum. Each artery enters the hilum of the kidney by dividing into three branches, two in front and one high up behind the pelvis of the ureter (Fig. 5.50).

Each renal artery gives off a *suprarenal branch* to the gland and a *ureteric branch* to the upper end of the ureter.

The Gonadal Arteries. These have a similar origin and course in both sexes. Both *testicular* and *ovarian arteries* arise from near the front of the aorta, below the renal arteries but well above the origin of the inferior mesenteric artery. They slope steeply downwards over the psoas muscle (the right artery first crossing the inferior vena cava), crossing the ureter and supplying its middle portion, and being themselves crossed by the colic vessels and the peritoneum of the floor of the infracolic compartment. They reach the pelvic brim about halfway between the sacro-iliac joint and the inguinal ligament, after which their course is different in the two sexes. In the male the *testicular artery* runs along the pelvic brim above the external iliac artery and enters the deep inguinal ring and passes in the spermatic cord to the testis. In the female the *ovarian artery* crosses the pelvic brim and runs down the lateral wall of the pelvis to enter the infundibulo-pelvic fold of the peritoneum and passes to the ovary and uterine tube (PLATES 18, 20, Fig. 5.63, p. 334).

The **common iliac arteries** are asymmetrical. They commence to the left of the midline, on the body of L4, and each passes to the front of the sacro-iliac joint where the bifurcation into external and internal iliac arteries takes place. Thus the right artery is a centimetre longer than the left. Each common iliac artery is crossed at its bifurcation over the sacro-iliac joint by the ureter, and the left artery is crossed somewhat higher up by the inferior mesenteric (superior rectal) artery, vein and lymphatics. Each is also crossed by the hypogastric nerves (presacral nerves, lumbar splanchnics) and is covered by the parietal peritoneum of the posterior abdominal wall.

The **external iliac artery** runs along the pelvic brim on the psoas muscle, and passes beneath the inguinal ligament to enter the femoral sheath. Its two branches are given off just above the inguinal ligament. The *inferior epigastric artery* (p. 258) supplies the rectus muscle. The *deep circumflex iliac artery* runs above the inguinal ligament to the anastomosis at the anterior superior iliac spine (Fig. 5.8, p. 261).

The **external iliac vein** enters the abdomen on the medial side of the artery. Each vein runs along the pelvic brim behind the artery. Over the sacro-iliac joint each is joined by the internal iliac vein. The **common iliac veins** slope up behind the arteries; they unite behind the commencement of the right common iliac artery (PLATE 20).

The **inferior vena cava** (PLATE 18) has a longer course than the aorta in the abdomen. It commences, at a slightly lower level than the bifurcation of the aorta, by the confluence of the right and left common iliac veins behind the commencement of the right common iliac artery. It runs, on the right of the aorta, upwards beyond the aortic orifice of the diaphragm, from which level it lies on the right crus behind the bare area of the liver, and extends to the central tendon of the diaphragm, which it pierces on a level with the body of Th. 8 vertebra (four vertebræ higher than the commencement of the abdominal aorta). It lies on the bodies of the lumbar vertebræ and the right lumbar sympathetic trunk and crosses the right renal artery. Above this level it is separated from the right crus of the diaphragm by the contiguous edges of the right suprarenal gland and the right cœliac ganglion, and the right phrenic artery.

At its commencement below, its tributaries lie posterior to the aorta, while higher up it and its tributaries lie on a more anterior plane, a result of its development (p. 309).

In the infracolic compartment the inferior vena cava lies behind the peritoneum of the posterior abdominal wall; it is crossed by the mesentery of the small intestine and above this by the third part of the duodenum. In the supracolic compartment it lies at first behind the portal vein and the head of the pancreas and common bile duct, then behind the peritoneum of the posterior abdominal wall (here it forms the posterior wall of the epiploic foramen) and, above this, behind the bare area of the liver, into which it excavates a deep groove.

Due to its compound embryological origin, its *tributaries* are not identical with the branches of the abdominal aorta. In particular there are no tributaries corresponding with the three single ventral aortic branches to the gut. The blood from the alimentary canal, pancreas and spleen is collected by the portal venous system and only after passing through the liver does it eventually reach the inferior vena cava via the hepatic veins (p. 299). The derivatives of the paramesonephric ridge (suprarenal glands and gonads) have an asymmetrical venous drainage. On the right side their veins open into the inferior vena cava, but on the left they drain into the left renal vein. This adult asymmetry is, however, embryologically symmetrical since on each side the veins drain into a persisting segment of the subcardinal vein (p. 309).

The *median sacral veins* (companions to the artery) drain into the left common iliac vein in front of the body of L5.

The third and fourth **lumbar veins** drain into the inferior vena cava, those from the left side passing behind the aorta and all four passing behind the lumbar sympathetic trunks. (The upper two lumbar veins join the azygos system and the fifth joins the ilio-lumbar vein.)

The **gonadal veins** accompany the arteries (testicular or ovarian) and are usually paired. As they run up on the psoas muscle the two venæ comitantes usually unite. On the right the vein enters the inferior vena cava an inch or so below the renal vein; the left vein enters the left renal vein (p. 309).

The **renal veins** lie in front of the arteries and join the inferior vena cava at right angles, at the level of L2. Each emerges from the hilum of the kidney as five or six tributaries, in front of the branches of the renal artery. The *left vein* is much longer than the right (3 inches [7 cm] and 1 inch [2·5 cm] respectively). It receives the *left suprarenal vein* and the *left gonadal vein* and passes in front of the aorta to reach the vena cava. The renal veins are overlaid by the pancreas.

The **right suprarenal vein** is a short vessel that enters the inferior vena cava behind the bare area of the liver, where the right suprarenal gland lies in contact with both liver and vena cava.

The **hepatic veins** enter the inferior vena cava from the right and left sides of the liver. There are three veins, right, middle and left, all of about equal size. They open directly from the caval groove in the bare area of the liver and have no course outside the liver. They convey a mixture of hepatic artery and portal vein blood. They are indispensable to the support of the liver (p. 300).

The Lymph Nodes of the Abdomen

A series of outlying nodes drain lymph from the alimentary canal, liver, spleen and pancreas (p. 285). From these nodes lymphatics pass back along the cœliac, superior and inferior mesenteric arteries to groups situated around the origins from the aorta of these three gut arteries. These are the **pre-aortic lymph nodes.** Similarly, lymphatics pass back along the paired branches of the aorta, both visceral and somatic. They drain into nodes placed alongside the aorta at the origins of the paired arteries; these are the **para-aortic lymph nodes.** The lymph drainage of any viscus follows its artery back to the aorta. Lymph from the pelvic viscera is drained through nodes along the internal iliac vessels. The lower limb drains by the deep inguinal nodes through the femoral canal into nodes along the external iliac vessels. Both internal and external iliac nodes drain up into the para-aortic nodes.

The nodes lie, in front of and at the sides of the aorta, in masses that cannot readily be demonstrated separately by dissection, but clinically their described groups are accurately demarcated from each other.

The main groups drain up by efferent vessels which enter the **cisterna (receptaculum) chyli.** This is a thin-walled sac, 2 to 3 inches (5 to 7·5 cm) long, lying in front of the bodies of L1 and 2 between the aorta and azygos vein, and it leads directly into the thoracic duct. It receives gastro-intestinal trunks from the cœliac and superior mesenteric groups of pre-aortic nodes. It also receives lumbar trunks from the para-aortic nodes and lymphatics which descend from the posterior ends of the lower intercostal spaces.

The lymph drained from the central lacteals of the villi of the small intestine contains, after absorption of a meal, an emulsion of fat. The droplets could not pass through the cortex of a lymph node. Consequently, most of the lymph of the lacteals is conveyed direct to the cisterna by an intestinal lymph trunk with no interposed lymph nodes along its path. Forty per cent of the lymph from the lacteals enters directly, in the intestinal wall, into small venules and is thus conveyed directly to the liver.

THE AUTONOMIC NERVES

The abdomen receives both sympathetic and parasympathetic nerves. The sympathetic supply is twofold, by the ganglionated lumbar trunk and the cœliac plexus. The cœliac plexus, formed from splanchnic nerves that come from the thorax, is wholly visceral; it supplies all the abdominal organs, including the gonads. The ganglionated lumbar trunk supplies somatic branches for the lower abdominal wall and the lower limb, but its visceral branches supply only the pelvic organs. The parasympathetic supply is provided by the vagus from above and the pelvic parasympathetics from below; it is wholly visceral. The vagus joins the cœliac plexus and the pelvic parasympathetics join the inferior hypogastric plexus. Thus it is better to study the sympathetic plexuses first.

The Sympathetic System

The **lumbar sympathetic trunk** brings preganglionic fibres descending from the lower thoracic trunk, and it receives a further input of preganglionic fibres (white rami) from the first and second lumbar nerves. Its ganglia give off the regular somatic and visceral branches (p. 33), and the trunk passes down across the pelvic brim to become the sacral trunk (p. 343).

The lumbar sympathetic trunk enters the abdomen by passing behind the medial arcuate ligament on the front of the psoas muscle. Passing in front of the psoas fascia the trunk lies on the vertebral bodies just touching the medial margin of the psoas muscle. As elsewhere it lies in front of the segmental vessels; the lumbar arteries and veins pass behind the trunk to reach the fibrous arches in the origin of the psoas muscle. The common iliac vessels lie in front of the trunks at the pelvic brim; they are not segmental vessels in series with the lumbar vessels but, in the quadrupedal position, are merely branches of the aorta and vena cava passing ventrally to the pelvis and hind limb, so the sympathetic trunks naturally pass dorsal to them.

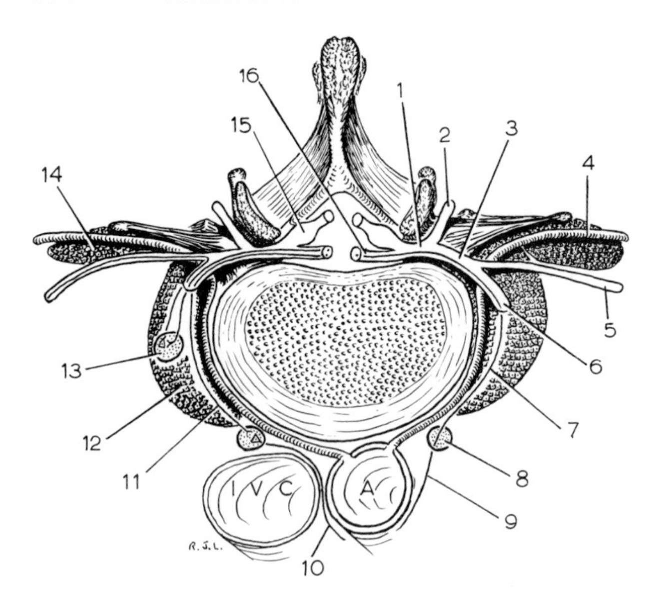

FIG. 5.45. THE SYMPATHETIC RAMI AT L1 OR L2 (diagrammatic). An accessory ganglion in psoas major (13) relays directly back to L1 or L2 segmental nerve. The remainder of L1 and L2 outflow passes into the sympathetic trunk for local or distant relay. 1. Mixed spinal nerve. 2. Posterior primary ramus. 3. Anterior primary ramus with preganglionic fibre from cell body in spinal cord. 4. Lumbar artery. 5. Branch of lumbar plexus (e.g. ilio-hypogastric or ilio-inguinal nerve). 6. Root of lumbar plexus. 7. Preganglionic fibre. 8. Lumbar sympathetic ganglion. 9. Hypogastric nerve (white). 10. Hypogastric nerve (grey). 11. Preganglionic fibre (cell body in lateral grey horn of spinal cord). 12. Psoas major. 13. Accessory ganglion in psoas major. 14. Quadratus lumborum. 15. Dorsal root ganglion. 16. Anterior nerve root.

The left lumbar trunk lies along the left margin of the aorta, while the right trunk lies behind the inferior vena cava, a result of the asymmetry of the two vessels; the trunks themselves are symmetrically placed in their whole extent.

The **lumbar ganglia** are conventionally four in number, but fusion of ganglia may reduce them. White rami communicantes are received from the first two lumbar nerves and small **accessory ganglia** lie at this level *within the substance of the psoas muscle* (Fig. 5.45). Here relay occurs and the grey rami return to the first two (especially the first) lumbar nerves within the substance of the muscle. In the operation of lumbar sympathectomy the accessory ganglia remain intact, hence the skin area supplied by L1 and L2 will not be completely denervated. The other white rami (for the sympathetic trunk itself) pass from the upper two lumbar nerves behind the fibrous arches in psoas. They join the trunk and run down to relay in the lumbar and sacral ganglia, for the somatic and visceral branches therefrom.

Branches of the Lumbar Trunk. *Somatic branches* pass from the lumbar ganglia to all five lumbar nerves, for distribution to the body wall and lower limb. These are grey rami, and they accompany the lumbar vessels behind the fibrous arches in psoas. Those for L1 and 2 are very sparse, since almost all the relay at these levels takes place in the accessory ganglia in psoas.

Visceral branches arise from all lumbar ganglia. They join loosely and pass down in front of the common iliac vessels. These are the *hypogastric nerves (lumbar splanchnics, presacral nerves)*. They contain a mixture of pre- and post-ganglionic fibres and are joined by similar fibres from the aortic plexus. They unite in front of the body of the fifth lumbar vertebra to make the small *superior hypogastric plexus*, which divides into right and left branches to pass down to the *inferior hypogastric (pelvic) plexuses*. There are no ganglia in the superior hypogastric plexus, and the preganglionic white fibres in the hypogastric nerves pass through to relay in ganglia in the inferior hypogastric plexuses. The hypogastric nerves pass caudally to supply pelvic viscera. They give no branches to the abdominal viscera, which are supplied by the cœliac plexus.

The **cœliac plexus** lies around the origin of the cœliac artery above the upper border of the pancreas. The greater and lesser splanchnic nerves pierce the crura of the diaphragm and enter the **cœliac ganglia.** These are two semilunar masses, right and left, which overlap the edges of the aorta and encroach laterally on the crura of the diaphragm. Either or both ganglia may be separated into two or more ganglionic masses. The splanchnic nerves are almost all preganglionic (white) and they relay in the cœliac ganglia. The least splanchnic nerve relays in a small *renal ganglion* behind the renal artery. This is merely an offshoot of the main cœliac ganglion itself. Separated masses of ganglion may lie on the aorta at the superior and even the inferior mesenteric artery origins. Such irregularity, of separation or of fusion, is common in autonomic ganglia. The cœliac ganglia are partially separated in the specimen illustrated in PLATE 20.

From the cœliac ganglia the mass of postganglionic fibres build up a rich network on the aorta, called the cœliac plexus. The fibres supply all the abdominal viscera, which they reach by streaming along the visceral branches of the aorta. Thus the midline gut arteries (cœliac, superior mesenteric, and inferior mesenteric) carry sympathetic fibres to the foregut and its derivatives, to the midgut, and to the hindgut down to the lower

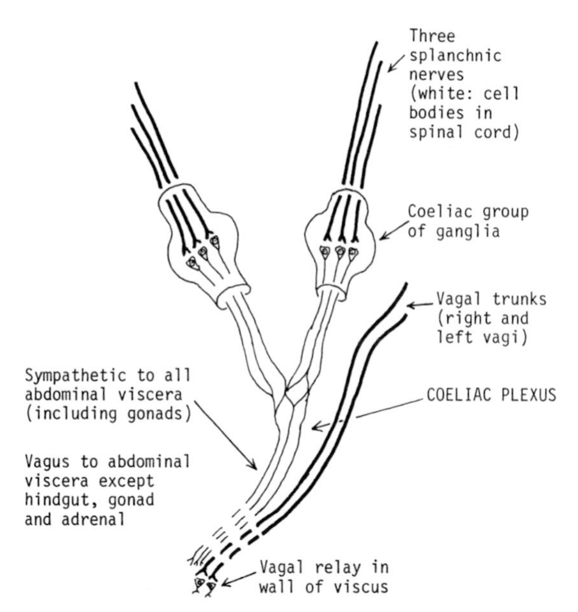

FIG. 5.46. THE CONSTITUENTS OF THE CŒLIAC PLEXUS. The plexus lies distal to the sympathetic relays in the cœliac group of ganglia (see PLATE 20). White fibres are shown in heavy line, grey fibres in slender line.

rectum. Similarly the paired visceral arteries carry fibres from the cœliac plexus. Those passing to the kidney pick up the branches of the renal ganglion to form the renal plexus behind the renal artery. Testis and ovary are likewise supplied by a sympathetic plexus that accompanies each gonadal artery. Sometimes the post-ganglionic fibres are not all used up at the inferior mesenteric artery, in which case they stream down over the aorta to enter the superior hypogastric plexus.

The sympathetic fibres are vaso-motor, motor to sphincters (e.g., ileo-colic), inhibitory to peristalsis, and *sensory* to all the viscera supplied.

The suprarenal has a second supply. White fibres from the lesser (or greater) splanchnic nerve pass without relay to the cells of the suprarenal medulla (these cells share a common origin from neural crest ectoderm with the cell bodies of sympathetic ganglia). These preganglionic fibres cause the suprarenal medulla to pour forth adrenalin. The vaso-motor supply to the suprarenal gland reaches it by postganglionic fibres from regular relay in the cœliac ganglion.

The Parasympathetic System

Both vagi, intermixed in the right vagal trunks, go to the cœliac plexus. Without relay they accompany the postganglionic sympathetic fibres of the plexus, but not to the whole territory. To the gut they pass along the foregut and midgut arteries only as far as the transverse colon—no mean feat, however, for a cranial nerve! They enter the renal plexus and pass into the kidney, for a purpose which is obscure. They do not supply the gonads or suprarenals, as far as is known.

The pelvic parasympathetics send parasympathetic fibres up from each side to the superior hypogastric plexus. Here the fibres of the two sides join into a single nerve that runs up, separate from the hypogastric nerves, to reach the inferior mesenteric artery an inch from its origin. This is the parasympathetic supply of the hindgut from splenic flexure to rectum.

The vagus is motor and secreto-motor to the gut and its glands down to the transverse colon; the pelvic parasympathetics fill this function from the splenic flexure to the rectum.

THE KIDNEYS

The kidney possesses a capsule which gives the fresh organ a glistening appearance. All surfaces are usually smooth and convex though traces of lobulation, normal in the fœtus, are often seen. Thick rounded lips of kidney substance bound the hilum, from which the pelvis emerges behind the vessels to pass down into the ureter.

The kidneys lie high up on the posterior abdominal wall behind the peritoneum, largely under cover of the costal margin. At best only their lower poles can be palpated in the normally built individual. Each kidney lies obliquely, with its long axis parallel with the lateral border of psoas major. On its vascular pedicle it lies well back in the paravertebral gutter, so that the hilum faces somewhat forwards as well as medially. As a result of this slight 'rotation' of the kidney an antero-posterior radiograph gives a somewhat foreshortened picture of the width of the kidney. The normal kidney measures rather more than 4 × 2 × 1 inches (say 12 × 6 × 3 cm) and weighs rather more than 4 oz (say 130 g). The hilum of the right kidney lies just below, and of the left just above the transpyloric plane about 2 inches (5 cm) from the midline. The bulk of the right lobe of the liver accounts for the lower position of the right kidney. The upper pole of the left kidney may overlie the eleventh rib in a radiograph, that of the right kidney seldom ascends so high, though it must be remembered that each kidney moves in a vertical range of almost 1 inch (2 cm) during full respiratory excursion of the diaphragm.

The **structure** of the kidney is displayed when the organ is split open longitudinally. A dark reddish and vascular cortex lies beneath the capsule and extends towards the pelvis between a number of darker striated areas, triangular in outline, the pyramids of the medulla. The apices of several pyramids open together into a renal papilla, each of which projects into a **minor calyx** of the pelvis. The cortex contains glomeruli and convoluted tubules, the medulla contains all the collecting tubules, the parallel arrangement of which lends the striated appearance to the pyramids.

The histological (and functional) unit of the kidney is the **nephron.** Each human kidney comprises about one million nephrons. The **glomerulus** consists of a tuft of capillaries clasped in Bowman's capsule; the incoming arteriole is of larger bore than the outgoing vessel. The efferent arteriole breaks up into capillaries that enmesh the convoluted tubules. A **proximal convoluted tubule** in the cortex leads from Bowman's capsule to the **loop of Henle,** which lies in the medullary pyramid. From the loop a **distal convoluted tubule** returns to the cortex and makes contact with its own Bowman's capsule at the **macula densa;** here lie the large gland-like cells of the juxtaglomerular apparatus possibly concerned in the production of renin.

The **relations** of the kidneys are roughly symmetrical. Posteriorly the relations are the same, comprising mostly the diaphragm and quadratus lumborum muscles, with overlap medially on to the psoas and laterally on to the transversus abdominis muscles. The upper pole lies on those fibres of the diaphragm which arise from the lateral and medial arcuate ligaments. Thus the **posterior recess of the pleura** lies posteriorly, a point of importance in posterior approaches to the kidney (Fig. 5.47). The subcostal vein, artery and nerve, on emerging beneath the lateral arcuate ligament, lie behind the posterior surface of the kidney, as do the ilio-hypogastric and ilio-inguinal nerves. The upper lumbar arteries and veins lie behind the quadratus lumborum and thus are more distant from the kidney.

The hilum of the kidney lies over the psoas muscle and the convexity of the lateral border lies on the aponeurosis of origin of the transversus abdominis. The suprarenal glands lie somewhat asymmetrically (PLATE 18). The right gland, pyramidal in shape, surmounts the upper pole of the right kidney, behind the inferior vena cava and the bare area of the liver, while the left gland, crescentic in shape, is applied to the medial border of the left kidney above its hilum, behind the peritoneum of the posterior wall of the lesser sac.

The anterior relations of the two kidneys are more symmetrical than appears at first sight and may be studied simultaneously with advantage (Fig. 5.48). On each side the peritoneum of the posterior abdominal wall lies in contact with certain areas of the kidney, while intervening structures force it away from the kidney in other areas. The hilum is separated from the peritoneum, on the right side by the second part of the duodenum and on the left side by the tail of the pancreas. The lateral part of the lower pole is separated from peritoneum by the hepatic and splenic flexures of

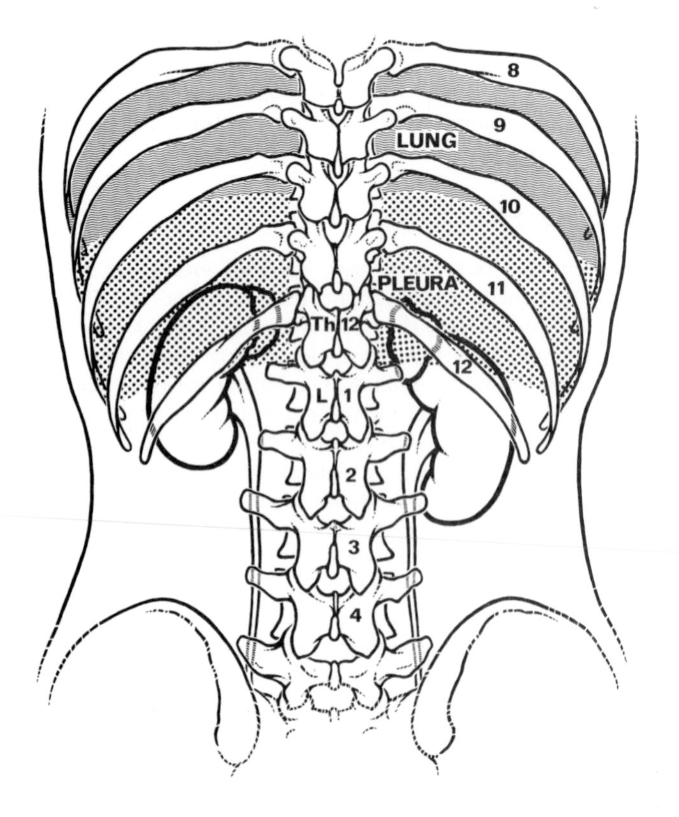

FIG. 5.47. THE RELATIONSHIP OF THE PLEURAL SAC TO THE UPPER POLE OF THE KIDNEY (posterior view). The ureter lies just medial to the tips of the lumbar transverse processes.

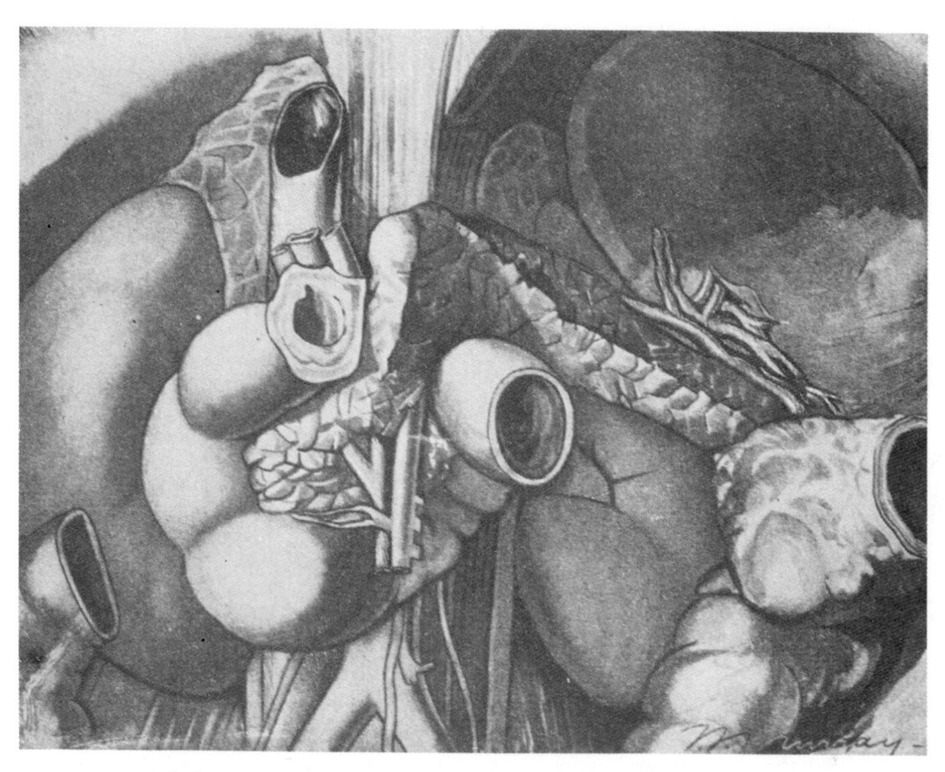

FIG. 5.48. THE RETROPERITONEAL VISCERA OF THE UPPER ABDOMEN, AND SPLEEN.

the colon on the right and left sides respectively. The medial part of the lower pole, on each side, lies in contact with peritoneum which separates it from coils of jejunum; here, between peritoneum and kidney, is an artery, the ascending branch of the right colic and of the upper left colic arteries respectively. The upper halves of each kidney, up to the superior pole, lie in contact with peritoneum. On the right kidney is the peritoneum of the hepato-renal pouch, in contact with the under surface of the liver. This is in the greater sac. The left kidney is in contact with both stomach and spleen and the lieno-renal ligament passes forwards from it along a line of attachment which separates these two areas. That is, the medial part of the upper pole is in the omental bursa, forming part of the stomach bed.

The **perinephric fat** is, at body temperature, of rather more solid consistency than the general body fat. It is in the shape of an inverted cone, filling the funnel-shaped hollow of the supra-iliac part of the paravertebral gutter, and it plays a part in retaining the kidney in position. The development of nephroptosis ('floating kidney') after severe loss of weight is thus explained.

The **perinephric fascia** surrounds the perinephric fat and separates the kidney from the suprarenal gland. It is no very obvious membrane in the living, but appears more convincingly in the coagulated dissecting-room cadaver. In truth it is little more than a vague condensation of the areolar tissue between the parietal peritoneum and the posterior abdominal wall, but certain of its attachments are worthy of note, since they serve to restrain the extension of a perinephric abscess (Fig. 5.49). At the hilum of the kidney the fascia is firmly attached to the renal vessels and the ureter, a further factor in stabilizing the kidney and in discouraging spread of pus across the midline. It ascends as a dome between the upper pole of the kidney and the suprarenal, and explains why in nephrectomy the latter gland is not usually displaced (or even seen). It is usually described as deficient below when traced downwards, but it is far better to regard it as merging below into the areolar tissue which connects the peritoneum to the posterior abdominal wall. Pus in the perinephric space does not track downwards, and injections into the space do not flow downwards.

Similar remarks apply to an alleged layer of fascia passing between the anterior surfaces of the dome-shaped perinephric fasciæ in front of the aorta and inferior vena cava; this 'layer' is no more than the areolar tissue that attaches the parietal peritoneum to all the structures on the posterior abdominal wall.

Blood Supply. The blood supply of the kidneys is furnished by the wide-bored renal arteries, with a blood flow in excess of 1 litre per minute (for the body

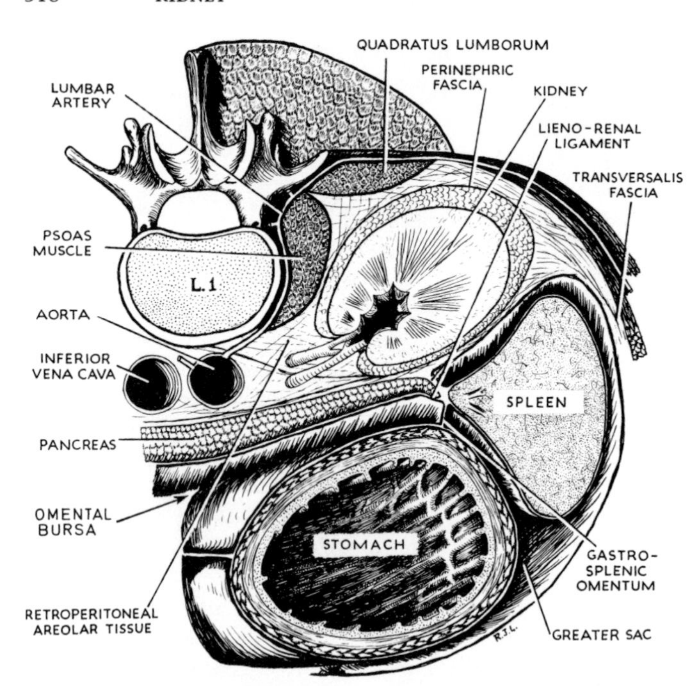

FIG. 5.49. A HORIZONTAL SECTION THROUGH THE LEFT KIDNEY, SPLEEN AND STOMACH.

economy, not merely for the arterial supply of the renal substance). They leave the abdominal aorta at right angles and lie behind the pancreas and renal veins. Each divides, usually, into three branches which enter the hilum of the kidney two in front, and one behind the renal pelvis. The posterior branch passes to the *upper pole* (a posterior branch, passing behind the ureter to the *lower pole*, is often associated with hydronephrosis). After further division the arteries enter the kidney substance. The anterior and posterior halves of the arterial supply are somewhat separate (Brodel), hence the kidney is usually split longitudinally rather than transversely at operation. A well-planned transverse incision, however, is almost as bloodless as a longitudinal splitting of the organ, for the fields of the three main branches are separate.

Distribution. The two upper branches of the renal artery supply anterior and posterior halves of the upper part of the kidney. The lower branch supplies both anterior and posterior halves of the lower pole (Fig. 5.50). Although infarction of the kidney is common, the branches of the renal artery are not end arteries in the ordinary sense. There is a separate artery for each unit of kidney substance, or renule. Its branches anastomose with each other within the renule, but do not anastomose with adjacent renules, so that infarction can only affect one or more complete renules. A *renule* is the counterpart of the discrete lobule into which the kidney is divided in many species (e.g., bears, cetacea, etc.). In man and most mammals the renules fuse into a single kidney, but they retain their separate blood supply. A human renule consists of the two or three medullary pyramids (with their corresponding cortical substance) that open by one renal papilla into a minor calyx of the pelvis. A renal infarct is thus pyramidal with its base on the convex surface of the kidney and its apex at a renal papilla.

Contrary to common belief, arterial arcades do not exist, though some branches of the arteries run in the

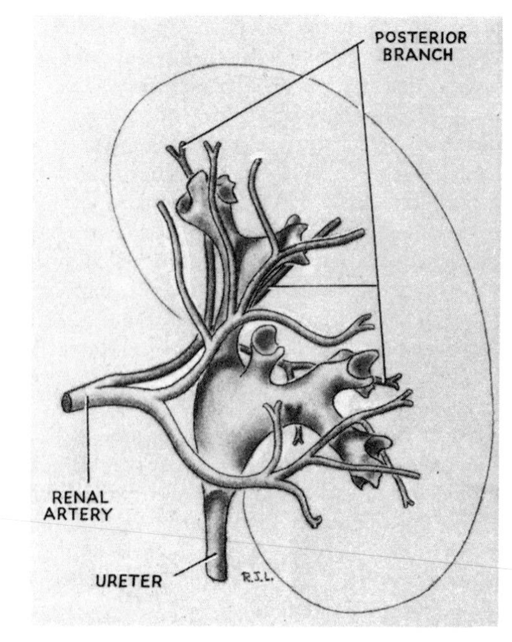

FIG. 5.50. A CAST OF THE RENAL PELVIS AND THE RENAL ARTERY (left side viewed from in front). Drawn from Specimen S.289A in R.C.S. Museum.

boundary zone across the bases of the pyramids. These branches supply straight vessels into the pyramids, giving them their typical striated appearance.

The *veins* of the kidney are very profuse, and they tend to form venous arcades along the bases of the medullary pyramids. The venous arcades in the boundary zone collect blood from cortex and medulla and empty into the tributaries of the renal vein; five or six veins lie with the renal arteries and unite in or just beyond the hilum to form the wide-bored renal vein. A few capsular veins drain to the lumbar veins.

The portal system of the kidney differs from that of the liver, for the first set of capillaries in the glomerulus merely loses glomerular filtrate, and the blood is still oxygenated as it passes into the second set of capillaries around the convoluted tubules. Arterio-venous shunts occur in the kidney, as in so many other parts of the body. They by-pass the capillary bed of glomerulus and tubule.

The lymphatics of the kidney drain to para-aortic nodes at the level of origin of the renal arteries (L2). The surface of the upper pole may drain through the diaphragm into nodes in the posterior mediastinum.

Abnormal Renal Arteries. These are of two types. First, an abnormal branch may arise from a normally situated renal artery; a branch passing behind the *lower* part of the renal pelvis constitutes abnormality, for it may be associated with hydronephrosis.

Second, there is quite commonly an accessory branch from the aorta. This is usually from below the renal artery, and it passes to the lower pole. Occasionally there is an aortic branch from above the renal artery passing to the upper pole. These aortic branches usually fail to enter the hilum, and pierce the convexity of the kidney surface. They are abnormally persisting fœtal arteries, which grow in sequence from the aorta to supply the kidney as it ascends the posterior abdominal wall. When they are present, the kidney usually shows some persistence of its fœtal lobulation.

The **nerve supply** of the kidney is derived from both parts of the autonomic system. The *sympathetic* connector cells lie in the spinal cord from Th. 12 to L1 segments and they send preganglionic fibres to all three splanchnic nerves. The excitor cells are in the cœliac ganglia and, for the lowest splanchnic nerve, in the renal ganglion in the hilum of the kidney. They are vasomotor in function. The afferent nerves from the kidney are, segmentally, Th. 10 to 12. The *parasympathetic* supply is derived from the vagus, but its function is unknown. The renal plexus is a loose network, most of it behind the renal artery.

The development of the kidney runs parallel with the comparative anatomy of the vertebrate excretory system and the former is made more intelligible by an acquaintance with the latter.

Comparative Anatomy. Three separate excretory organs appear in the vertebrate scale. They are, successively, the pronephros, the mesonephros and the metanephros. The first two consist of excretory tubules arranged segmentally and they empty into the same duct (the Wolffian duct). The third consists of a mass of tubules having no segmental arrangement and it drains into a new duct that develops specifically for the purpose (the ureter).

The **pronephros,** a mass of segmentally arranged tubules in the cervical region, empties into a longitudinal duct which conveys the urine to the cloaca. This is the pronephric duct. The pronephros is a very primitive formation, appearing well developed in the vertebrate scale only in embryo fish. In subsequent development it is replaced by a mass of tubules, segmentally arranged, and more caudally placed. This is the **mesonephros** and it constitutes the excretory organ of all fish and amphibia.

The only real interest of the evanescent pronephros is that its duct persists and receives the efferent tubules of the mesonephros. The duct changes its name to **mesonephric (Wolffian) duct** and it carries urine from the mesonephros to the cloaca.

In vertebrate forms higher than amphibia a mass of excretory tubules develops caudal to the mesonephros. This is the **metanephros.** From the caudal end of the mesonephric duct an outgrowth buds, and grows up to meet the tubules of the metanephros. This is the **ureter.** The mesonephros and its duct disappear in female reptiles, birds and mammals, but the mesonephric duct is retained in males; some mesonephric tubules persist as the vasa efferentia of the testis, opening into the mesonephric duct, which persists as the ductus deferens (Fig. 5.14, p. 266).

Progression up the vertebrate scale thus shows a caudal drift of the excretory organ and the appearance of a new excretory duct, with persistence of the original (pronephric, later mesonephric) duct in the male only, as the excretory duct of the testis. These changes occur likewise in the developing human embryo.

Lobulation. There is great variation in the amount of lobulation of the mammalian kidney, and this bears no relation to the animal's position in the vertebrate scale. At one extreme the multiple lobules (called *renules*) fail to fuse with each other, and each opens separately into a minor calyx of the pelvis (e.g., bears, cetacea). At the other extreme there is complete fusion of the lobules, and a single medulla is capped by cortex (e.g., the unipyramidal kidney of the cat). The human kidney lies between these extremes, for the cortex fuses into a continuous mass, while the medulla of each lobule remains separate as a pyramid opening into a minor calyx of the pelvis.

Development of the Kidney. The early embryo

is distinguished by a longitudinal ridge that projects into the body cavity (cœlom) on each side of the dorsal midline (Fig. 1.47, p. 39). This is a mass of mesoderm called the **intermediate cell mass.** It is drained by the subcardinal vein. From its lateral side develop the pronephros, mesonephros and metanephros in that order from cranial to caudal, while its medial side gives rise to the cortex of the adrenal and the gonad.

The pronephros is very evanescent, but its duct persists. Mesonephric tubules develop and open into the pronephric duct which is henceforth called the mesonephric (Wolffian) duct. Caudal to the mesonephros the intermediate cell mass gives rise to a number (one million) of excretory tubules whose lobulated mass is called the **metanephrogenic cap.** The latter induces a bud to grow from the caudal end of the mesonephric duct. This bud, the ureter, grows up and divides into the calyces of the pelvis (major and minor) and the collecting tubules of the medullary pyramids, into which the distal convoluted tubules of the metanephros come to drain. The manner of their meeting and joining is unknown; failure to join is the probable cause of congenital polycystic disease of the kidney.

The metanephros, like the mesonephros, develops from mesoderm. The mesonephric duct is commonly thought to arise likewise, but this is not certain, and it may be that the duct is, in fact, ectodermal in origin.

The definitive kidney (metanephros) develops in the pelvis and is supplied from the median sacral artery. It subsequently migrates to its adult position, gaining successively new arteries of supply from the common iliac or internal iliac arteries, and from the abdominal aorta. Deep grooves give the fœtal and neonatal kidney a prominently lobulated appearance.

Abnormalities of the kidneys are common. The fœtal lobulations may persist in greater or lesser degree. Failure to ascend may occur, and persistence of one of the fœtal arteries (an 'aberrant' renal artery) is not unusual. Should such an artery pass behind the ureter there may be an accompanying hydronephrosis, though there may be no causative association between the two conditions. Fusion of the lower poles may occur ('horseshoe kidney') in which case the ureters pass ventral (anterior) to the isthmus of kidney substance. Polycystic disease has already been mentioned. Finally, one kidney may be congenitally absent, a fact which should always be recalled before considering nephrectomy.

The **renal pelvis** consists of the funnel-shaped upper expansion of the ureter. It lies within the hilum of the kidney and is dilated in its upper and lower extremities into the **major calyces.** A dozen or more minor calyces receive the renal papillæ (Fig. 5.50). The pelvis, like the ureter, is lined with transitional epithelium and its wall is composed of smooth muscle and fibrous tissue. The capacity of the average pelvis is under 5 ml.

The Ureter

The ureter is 10 English inches long; but English makers do not calibrate their ureteric catheters in inches. It is worth remembering that an inch measures 2·5 cm, so that the catheter will not be inserted beyond the 25 cm mark from the ureteric orifice in the bladder. Its points of narrowest calibre are at the pelvi-ureteric junction, at the halfway mark where it crosses the pelvic brim, and at its termination in the bladder mucosa. It is a tube of smooth muscle, circular and longitudinal. The muscle lies in three layers, an innermost and outermost circular and an intermediate longitudinal layer. The 'circular' layers are in fact close-wound spirals and the 'longitudinal' layer is a very gradual spiral of muscle (cf. the muscle layers of the wall of the alimentary canal). The outer circular layer is not present at all levels. Unlike the discrete layers of the alimentary canal, the muscle coats of the ureter intermingle to some extent. The ureter possesses a mucous membrane of lax areolar tissue, as is to be expected in a dilatable tube. It is lined with transitional epithelium. Its lumen is large compared with the thickness of the muscle wall (contrast these features with those of the ductus deferens, p. 266, and note the urine-proof transitional epithelium of the ureter and the tall ciliated columnar epithelium of the ductus; thus may the two be easily distinguished). (Figs. 5.51, 5.52.)

The ureter passes down on the psoas muscle and crosses the genito-femoral nerve (PLATE 20, p. *x*). It leaves the psoas muscle at the bifurcation of the common iliac artery, over the sacro-iliac joint, and passes into the pelvis (p. 331). It is plastered to the posterior abdominal wall by the peritoneum with its colic vessels, and the gonadal vessels lie between it and the latter. It adheres to the peritoneum of the posterior abdominal wall when that membrane is stripped up from the psoas fascia.

Its **surface markings** are of use in palpating it for tenderness and in identifying X-ray shadows. On the anterior abdominal wall it can be marked from the tip of the ninth costal cartilage to the bifurcation of the common iliac artery. This latter point is found by joining the point of bifurcation of the abdominal aorta (2 cm below and to the left of the umbilicus) to a point midway between the anterior superior iliac spine and the symphysis pubis. The common iliac artery bifurcates one-third of the way down this line.

More important is the line of projection of the ureter on an X-ray film. It lies medial to the tips of the transverse processes of the lumbar vertebræ and crosses the pelvic brim at the sacro-iliac joint (Fig. 5.47). From here its pelvic shadow passes to the ischial spine and thence, foreshortened, to the pubic tubercle.

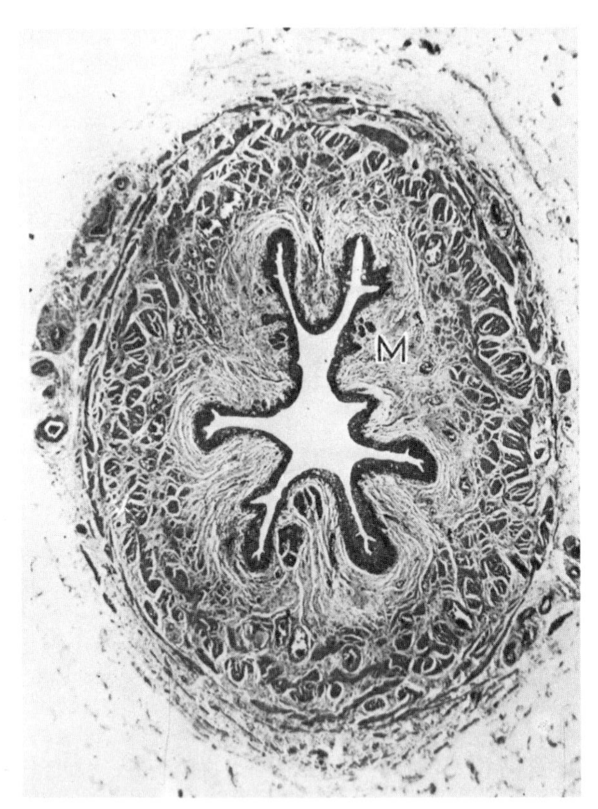

FIG. 5.51. THE URETER IN CROSS SECTION. CONTRAST ITS WIDE LUMEN AND LOOSE MUCOUS MEMBRANE (**M**) WITH THE NARROW LUMEN AND DENSE THIN MUCOUS MEMBRANE OF THE DUCTUS DEFERENS (VAS).

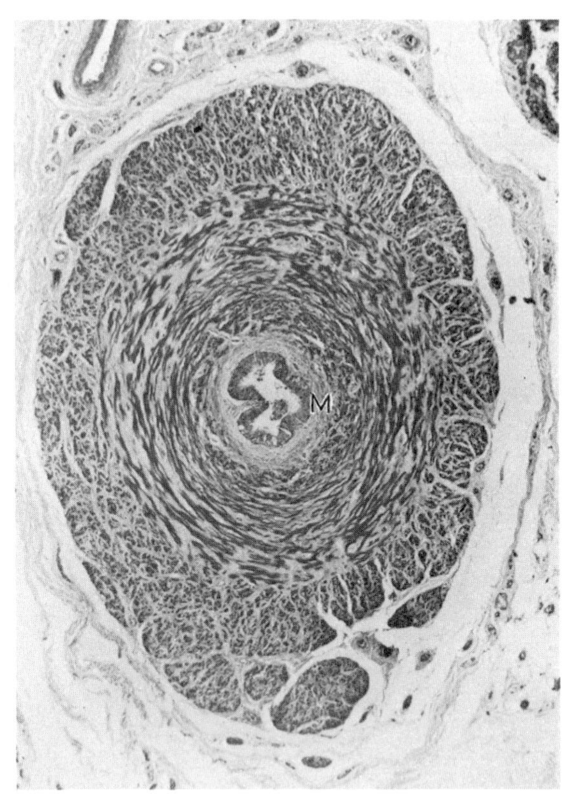

FIG. 5.52. THE DUCTUS DEFERENS (VAS) IN CROSS SECTION. **M.** Mucous membrane. Here the thick muscle wall is in two layers, an inner circular and an outer longitudinal.

The **blood supply** of the ureter is derived from four sources. The upper end is supplied by the ureteric branch of the renal artery and the lower end by a ureteric branch from either the inferior or the superior vesical artery. The middle reaches of the ureter are supplied by branches from the gonadal artery, and, in many cases, by branches from the common iliac artery. All these vessels make a fairly good anastomosis with each other. The ureter can be displaced forwards, with the peritoneum, so dividing the branches from the common iliac artery, with no danger to its subsequent blood supply. In transplanting the lower end of the ureter into the colon, however, the best possible blood supply is needed to combat infection, and the ureter should therefore not be displaced from the common iliac artery. The arteries of the ureter supply the adventitia, i.e., the loose areolar tissue around the muscle wall. The peritoneum of the infracolic compartment is supplied by the colic arteries. Thus in the retroperitoneal areolar tissue there is some meeting of terminal capillaries of the two systems, but the colic arteries themselves supply no blood to the ureter.

The veins of the ureter drain into the renal, gonadal and internal iliac veins. The lymphatics run back alongside the arteries; the abdominal portion of the ureter drains into para-aortic nodes below the renal arteries, the pelvic portion into nodes on the side wall of the pelvis alongside the internal iliac arteries.

The **nerve supply** of the ureter is derived from sympathetic connector cells in the spinal cord from L1 to L2. Synapse occurs in the cœliac and renal ganglia and postganglionic fibres reach the tube along the ureteric arteries. Nothing is known of a parasympathetic supply.

Development of the Ureter. The ureter is of mesodermal origin; it is derived by a process of budding from the caudal end of the mesonephric duct. Its upper end divides into two (the major calyces of the renal pelvis) and further subdivisions produce the collecting tubules. Low division of the ureteric bud produces double ureter (p. 331). If the mesonephric duct is, in fact, ectodermal then the ureter which buds from it is likewise ectodermal.

The Suprarenals (Adrenals)

These glands lie one alongside the upper part of each kidney. They are somewhat asymmetrical.

The **right suprarenal gland** is pyramidal in shape

and surmounts the upper pole of the right kidney. It lies between the inferior vena cava and the right crus of the diaphragm, its right border projecting to the

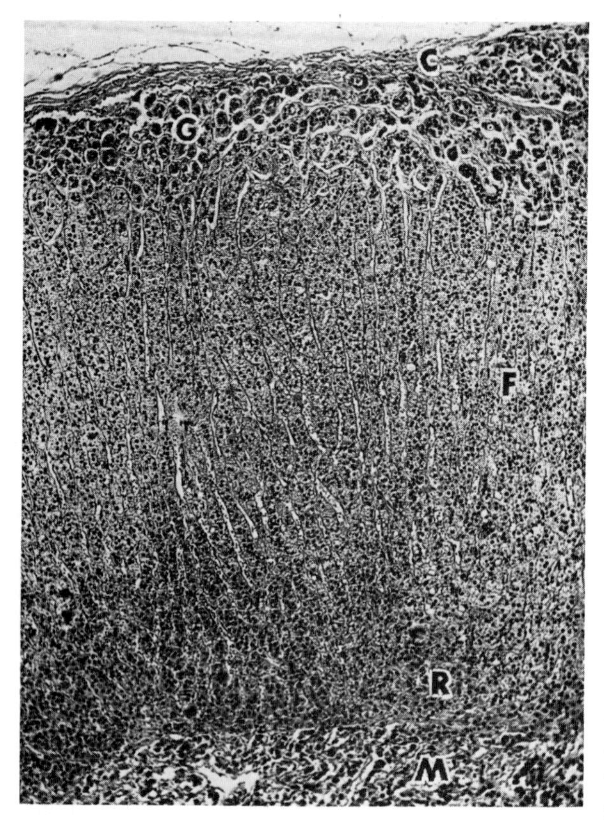

FIG. 5.53. THE SUPRARENAL GLAND. The broad zona fasciculata of the cortex is characteristic. **C**—capsule. **G**—zona glomerulosa. **F**—zona fasciculata. **R**—zona reticularis. **M**—medulla. (Photomicrograph × 60.)

right of the vena cava and coming into contact with the bare area of the liver. Like the left gland, it has three sources of arterial supply. Its own proper branch from the aorta is reinforced by adrenal branches of the phrenic and renal arteries. It drains by one vein directly into the inferior vena cava where that vessel lies behind the bare area of the liver.

The **left suprarenal gland** is crescentic in shape and drapes the medial border of the left kidney above the hilum. Its lower pole is covered in front by the tail of the pancreas, the rest of the gland is covered with peritoneum of the omental bursa and forms part of the stomach bed. It lies on the left crus of the diaphragm (PLATES 14, p. *vii* and 20, p. *x*).

Like the right gland it has three sources of arterial supply, directly from the aorta and by branches from the phrenic and renal arteries. It drains by a single vein into the left renal vein (cf. the gonad); this is in embryological symmetry with the right side, since in each case the suprarenal vein drains into a persisting segment of the subcardinal vein of the embryo.

Microscopic Anatomy. To the naked eye a section across the suprarenal resembles a sandwich. Two layers of cortex (the bread) enclose a much thinner layer of medulla (the meat) between them. In places there is no medulla, and the two layers of cortex then meet each other. The **cortex** is characterized by a poorly-stained *zona fasciculata*, consisting of parallel rows of large pale cells lying at right angles to the surface. The cells are distended with steroid secretion. The zona fasciculata is the intermediate, and the widest, of three zones in the cortex. On the surface are small clumps of better-stained cells, the *zona glomerulosa*, and in the depths smaller well-stained cells are arranged in a network, the *zona reticularis*. The **medulla** is surprisingly small in volume. It consists of larger well-stained cells. Eosinophilic granules of pre-adrenalin lie in the cytoplasm. These granules are positive to the *chromaffin reaction*; they are stained golden-brown by chromium salts. Large venous spaces are numerous in the medulla.

Development. The suprarenal gland is derived from two sources. The medulla is derived by migration of cells from the neural crest and is ectodermal in origin while the cortex is derived *in situ* from the mesoderm of the intermediate cell mass (p. 38).

THE PELVIS

The Bony Pelvis. The individual features of the os innominatum (p. 190) and the sacrum (p. 473) are considered separately. When articulated the bones enclose a cavity; from the brim of the cavity the ala of each ilium projects up to form the iliac fossa, part of the posterior abdominal wall. The **pelvic brim** is formed in continuity by the pubic crest, pectineal line of the pubis, arcuate line of the ilium, ala and promontory of the sacrum. The plane of the brim is oblique, lying at 60 degrees with the horizontal (Fig. 5.55). From the brim the pelvic cavity projects back to the buttocks.

Sex differences are striking and easily recognized; they are due to the two facts that the female pelvis is broader than that of the male for easier passage of the fœtal head and that the female bones, including the head of the femur, are more slender than those of the male. In the **male** pelvis the sturdy bones make an acute subpubic angle, pointed like a Gothic arch, while in the **female** the slender bones make a wide subpubic

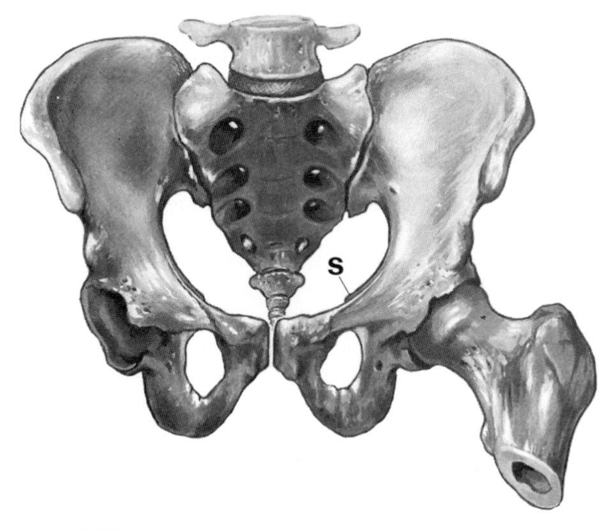

FIG. 5.54. THE MALE PELVIS VIEWED FROM IN FRONT. An imaginary horizontal plane through the top of the symphysis pubis traverses the tip of the coccyx, the ischial spine (**S**), the centre of the acetabulum and femoral head and the tip of the greater trochanter.

angle, rounded like a Roman arch. The outline of the pelvic brim differs. In the **male** the sacral promontory indents the outline, and the brim is widest towards the back (a 'heart-shaped' outline) while in the **female** there is less indentation of the outline by the sacral promontory and the brim is widest further forwards (a 'transversely oval' outline). Sex differences in the os innominatum (p. 196) and in the sacrum (p. 475) are equally obvious in the articulated pelvis.

Position of the Pelvis. Hold an articulated pelvis in the position it occupies in the erect individual, and note the degree of tilting of the os innominatum and sacrum. The anterior superior iliac spines and the upper margin of the symphysis pubis lie in the same vertical plane. Note an important **horizontal plane**—the upper border of the symphysis pubis, the spine of the ischium, the tip of the coccyx, the head of the femur and the apex of the greater trochanter lie in the one plane (Fig. 5.54). This plane passes through the pelvic cavity at a level with the tip of the examining finger of the clinician. The ovaries in the female and the seminal vesicles in the male lie in this plane.

THE PELVIC WALLS

Continental anatomists aptly name the pelvis the basin, and in shape it resembles a pudding basin lying on its side. The rim of the basin is the pelvic brim (promontory of sacrum and ilio-pectineal line). The side of the basin resting on the table constitutes the pelvic floor (levator ani) the rest of the basin represents the pelvic walls (sacrum and os innominatum, piriformis and obturator internus muscles, Fig. 5.56).

The side wall of the pelvis is formed by the os innominatum, clad with the obturator internus muscle

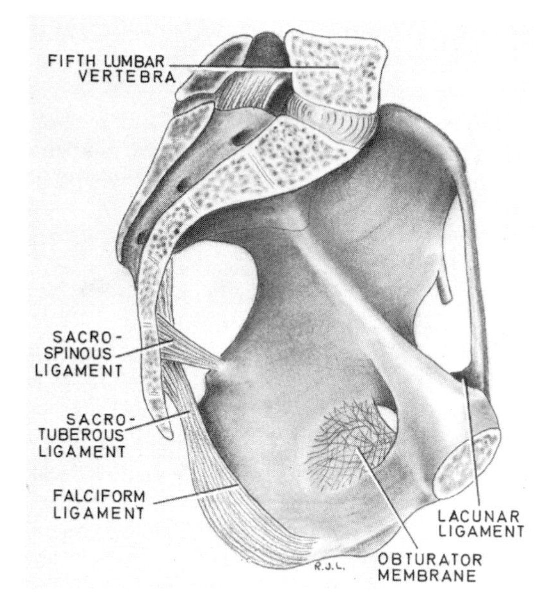

FIG. 5.55. THE LEFT HALF OF THE PELVIS VIEWED FROM WITHIN. (cf. Fig. 3.66, p. 195). Drawn from Specimen S 90 in R.C.S. Museum.

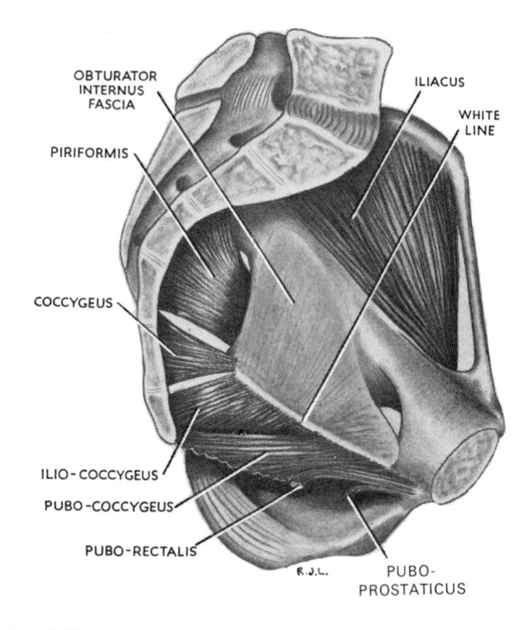

FIG. 5.56. THE LEFT HALF OF THE LEVATOR ANI MUSCLE. cf. Fig. 5.55.

and its fascia, the curved posterior wall is formed by the sacrum. The great sciatic notch between the two is occupied by the piriformis muscle.

The obstetrical term 'false pelvis' refers to part of the abdominal cavity, the iliac fossæ. It is part of the general abdominal cavity. The term has no meaning in descriptive anatomy. The bony pelvis as part of the skeleton is one thing, the pelvic cavity as an extension of the abdominal cavity is another thing. Neither is false.

Obturator Internus Muscle. Examine the os innominatum. The large obturator foramen contains in life a felted mass of fibrous tissue called the obturator membrane, with a gap above that converts the obturator notch into a canal for the obturator nerve and vessels (Fig. 3.66, p. 195). The muscle arises from the whole membrane and from the bony margins of the obturator foramen (Fig. 5.67, p. 340). The origin extends posteriorly as high as the pelvic brim and across the flat surface of the ischium to the margin of the great sciatic notch. On the ischial tuberosity the origin extends down to the falciform ridge. From this wide origin the muscle fibres converge fan-wise towards the lesser sciatic notch. Above the notch is a curved bare area of bone, with a large bursa lying on it beneath the muscle. Tendinous fibres develop on the muscle surface where it bears on the lesser sciatic notch and the bone often shows low ridges and grooves where the tendon takes a right-angled turn to pass into the buttock (Fig. 3.66, p. 195). It is inserted, with the gemelli, into the medial surface of the greater trochanter (Fig. 3.27, p. 146). Its nerve supply and action are considered in the buttock (p. 148).

Note the extent of the *upper margin* of the muscle. From the pelvic brim at the sacro-iliac joint the line slopes downwards along the side wall of the pelvis until anteriorly it lies below the obturator canal. Above this line lies bare bone; below it the muscle is covered with a strong membrane. This is attached to bone at the margins of the muscle down to the falciform edge of the sacro-tuberous ligament on the ischial tuberosity. The white line for origin of levator ani slopes across the obturator internus fascia (the pelvic cavity is above this line, the ischio-rectal fossa below it). The posterior surface of the body of the pubis and of the pubic symphysis is bare of both muscle and fascia (Fig. 5.56).

Piriformis Muscle. The muscle arises from the middle three pieces of the sacrum and from the adjoining lateral mass. Its origin extends medially between the anterior sacral foramina, so that the emerging sacral nerves and the sacral plexus lie upon the muscle. It passes laterally through the great sciatic notch, which it virtually fills, and crosses the buttock to the upper border of the greater trochanter. The pelvic surface of the muscle and the sacral plexus are covered with a strong membrane of pelvic fascia attached to the periosteum of the sacrum at the margins of the muscle; elsewhere the sacrum presents bare bone. The nerve supply and action of piriformis are considered in the buttock (p. 147).

THE PELVIC FLOOR

The pelvic floor consists of a gutter-shaped sheet of muscle, the pelvic diaphragm, slung around the midline body effluents (urethra and anal canal and, in the female, the vagina).

The muscles of the pelvic floor are called coccygeus and levator ani, but it is better to regard them as one morphological entity, ischio-coccygeus, ilio-coccygeus and pubo-coccygeus from behind forwards. They arise in continuity from the spine of the ischium, from the 'white line' over the obturator fascia, and from the body of the pubis, and are inserted into the coccyx and the ano-coccygeal raphé (Fig. 5.57). From their origin the muscle fibres slope downwards and backwards to the midline; the pelvic floor so produced is a gutter that slopes downwards and forwards. Appreciation of this point clarifies the whole of obstetrical mechanics—the lowest part of the fœtus is the first to meet this sloping gutter and during delivery is mechanically rotated to the front.

The **coccygeus muscle** is best thought of as ischio-coccygeus. It arises from the tip of the ischial spine, alongside the posterior margin of the obturator internus muscle. Its fibres fan out to be inserted into the side of the coccyx and the lowest piece of the sacrum; it lies edge to edge with the lower border of piriformis. In the tailed animals this muscle is the 'agitator caudæ'. Man has no tail to wag and the muscle is degenerating. Its gluteal surface is, indeed, not muscle, but fibrous tissue, and is none other than the sacro-spinous (small sciatic) ligament (Figs. 5.56 and 5.55).

The **levator ani muscle** consists of two chief parts, ilio-coccygeus and pubo-coccygeus, often separated by a triangular gap. Their fibres arise in continuity, from the ischial spine to the body of the pubis, across the obturator fascia. Here is a thickening called the **white line** or **arcus tendineus.** It is densely adherent to the obturator fascia and is usually described as a thickening thereof. This is not really correct; it is the property of levator ani, not of the obturator fascia. The levator ani originally arose from the pelvic brim (its present origin in most mammals) and in man has migrated down the side wall of the pelvis, bringing the tendinous arch with it; in rare cases the arch hangs free between ischial spine and pubis, with levator ani swinging from it (Fig. 5.56).

The **ilio-coccygeus muscle** arises from the posterior

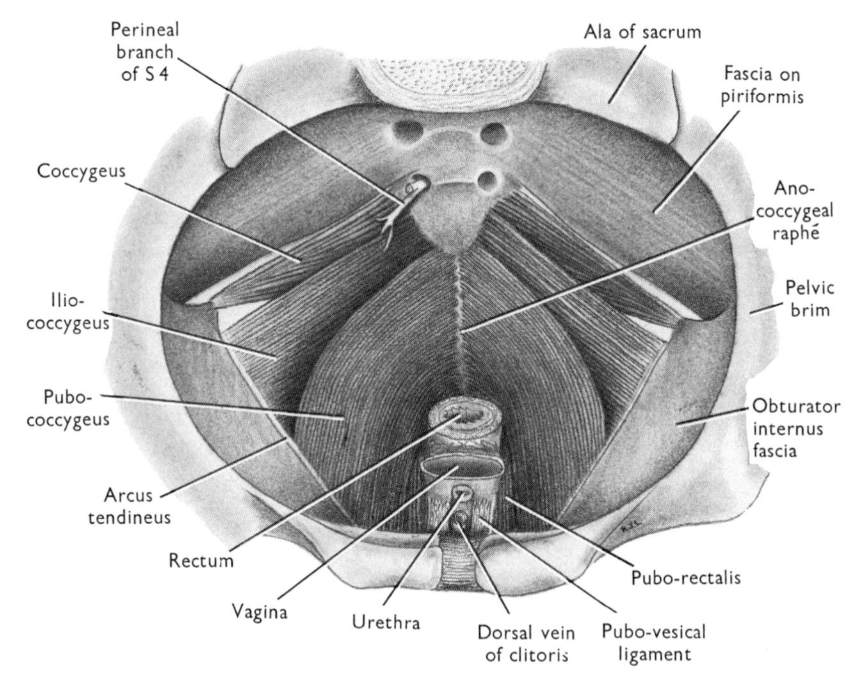

FIG. 5.57. THE PELVIC FLOOR SEEN FROM ABOVE (FEMALE).

half of the white line and, crossing the pelvic surface of coccygeus its fibres are inserted into the side of the coccyx and the **ano-coccygeal raphé.** This latter extends from the tip of the coccyx to the junction of rectum and anal canal. It consists of an interdigitation of the fibres of the levator ani muscles of right and left sides, and is easily elongated. During defæcation, and especially during the second stage of labour, it is passively stretched.

Note that the ilio-coccygeus does not arise from the ilium; its name derives from its former origin on the iliac bone at the pelvic brim.

The **pubo-coccygeus muscle** is that part of the levator ani which arises from the anterior half of the white line and from the posterior surface of the body of the pubis on a level with the *lower* border of the symphysis pubis. There is often a triangular gap between the adjacent borders of this muscle and ilio-coccygeus. The pubo-coccygeus forms a flat muscle whose fibres are in different functional sets. The bulk of its posterior fibres (i.e., those arising from the white line) sweep backwards in a flat sheet on the pelvic surface of the ilio-coccygeus and are inserted into the tip of the coccyx and the ano-coccygeal raphé. These constitute the pubo-coccygeus muscle proper. Fibres arising more anteriorly, from the periosteum of the body of the pubis, swing more medially and more inferiorly around the ano-rectal junction and join with fibres of the opposite side and *with the posterior fibres of the profundus part of the external anal sphincter.* No raphé exists here, and the muscles form a U-shaped sling which holds the ano-rectal junction angled forwards; this part of the muscle is called the **pubo-rectalis** (Fig. 5.71, p. 347). More medially still, a U-shaped sling of fibres passes behind the prostate into the perineal body; the muscle was named, not very aptly, **levator prostatæ.** It is now named **pubo-prostaticus.** In the female a similar muscular sling passes behind the vagina into the perineal body; it is named **pubo-vaginalis,** or **sphincter vaginæ** (Fig. 5.76, p. 354). Do not confuse this 'sphincter vaginæ' of the pelvic floor with the sphincter of the introitus (the bulbo-spongiosus muscle). The pubo-prostatic and pubo-vaginal slings are of fibres that interdigitate widely, between the pelvic floor and the skin of the perineum, bound together by fibrous tissue and constituting the perineal body (p. 348).

The whole pelvic floor is thus seen to consist of a gutter-shaped muscular diaphragm, the fibres of which are arranged in a series of U-shaped loops sloping progressively downwards towards the midline. The fibres of the U-shaped loops are inserted progressively into the coccyx and the ano-coccygeal raphé from behind forwards, and when traced in this direction the anterior fibres overlap the upper or pelvic surface of the posterior fibres of the raphé.

There is a gap anteriorly between the medial edges of the pubo-prostatic or pubo-vaginal muscles. This is almost completely filled by the **pubo-prostatic ligaments** in the male and the **pubo-vesical ligaments**

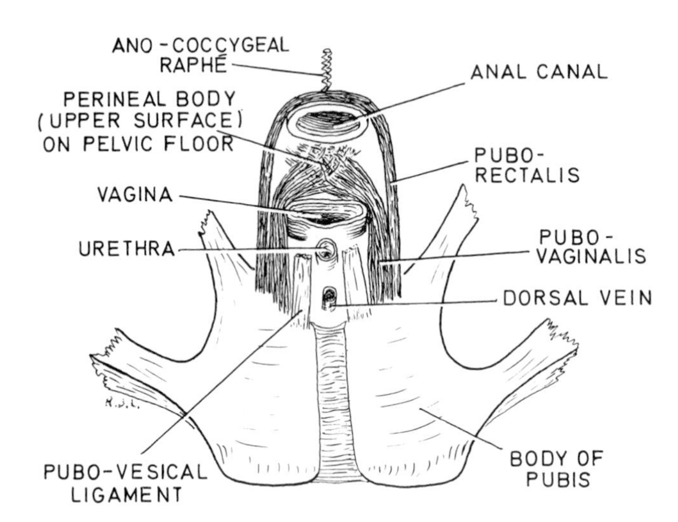

FIG. 5.58. THE PELVIC EFFLUENTS (female). The pubo-vesical ligament is attached to the bladder neck (in the male this is the pubo-prostatic ligament, attached to the apex of the prostatic fascia). The pubo-vaginalis muscle decussates behind the vagina into the perineal body (in the male this is the pubo-prostaticus). The dorsal vein is the deep dorsal vein of the clitoris (penis). The pubo-rectal sling is identical in the two sexes, and behind this the ano-coccygeal raphé receives the pubo-coccygeus muscles (see Fig. 5.57 and cf. Fig. 5.71).

in the female. In the midline between these ligaments lies the deep dorsal vein of the penis or clitoris (Fig. 5.57).

The muscles of the pelvic floor are innervated as follows. The perineal branch of the fourth sacral nerve passes from the pelvis to the perineum through the pelvic floor between the coccygeus and the ilio-coccygeus, and, according to the rule (p. 8), supplies both from its proximal part, i.e., on their upper or pelvic surfaces. The major part of the muscle is supplied, on its under or perineal surface, by branches from the inferior rectal (hæmorrhoidal) and, further forward, by the deep branch of the perineal nerve (p. 353).

The actions of the muscles of the pelvic floor are best analysed under the headings of static or postural tonus and contractile tonus. In the erect posture the weight of the viscera (and this comprises chiefly the mobile intestines and their liquid contents) falls on the bodies of the pubic bones and, further back, on the fibro-muscular perineal body. Very little weight is supported on the more posterior and higher ano-coccygeal raphé. Even when from advanced age or other cause the levatores ani are atonic it is very unusual to find any undue prolapse of the pelvic floor.

Abdominal pressure is increased by contraction of the diaphragm and abdominal wall. This may be momentary, as in coughing and sneezing, more prolonged as in yawning, micturition and defæcation and muscular efforts such as lifting, etc., and is most prolonged and intense in the second stage of labour. Increased intra-abdominal pressure is counteracted by an appropriate contraction of the levator ani and coccygeus muscles, wholly or in part. This assists in maintaining continence of the bladder and rectum in muscular effort, coughing, etc. In micturition, defæcation and parturition the appropriate pelvic effluent is open, but contraction of other fibres of the levator and coccygeus muscles resists increased intra-abdominal pressure and prevents evisceration through the pelvic outlet. The rôle of the pelvic floor in the second stage of labour has already been mentioned, and the better functioning of these muscles following prenatal exercises is an important adjunct to easier childbirth.

THE PELVIC FASCIA

The fascia of the pelvis is usually described under the headings parietal and visceral. Its arrangement is essentially simple, and this simplicity is best appreciated by considering the pelvic fascia under the headings of the pelvic wall, the pelvic floor and the pelvic viscera. Two principles govern the arrangements of the fascia. The first is that over non-expansile parts the fascia is a strong membrane, while over expansile or mobile parts no membrane exists, the fascia consisting here of a loosely felted areolar tissue. The second principle is that fascia does not extend over bare bone.

The **fascia of the pelvic wall** is a strong membrane which covers the muscles (obturator internus and piriformis) and is firmly attached to the periosteum at their margins. Elsewhere the bone of the pelvic wall is bare of fascia. An exception is the *fascia of Waldeyer*, which sweeps downwards from its attachment in the hollow of the sacrum to the ampulla of the rectum. The spinal nerves lie external to the fascia of the pelvic wall, the vessels lie internal to it. The sacral plexus lies behind the pelvic fascia, between it and the piriformis muscle, and its branches to the buttock do not, therefore, pierce the fascia. The vessels of the buttock (superior and inferior gluteal) on the other hand have to pierce this fascia to establish continuity between pelvis and buttock.

The **fascia of the pelvic floor,** though usually included under the term parietal pelvic fascia, bears no possible resemblance to that of the pelvic wall. Here is no inexpansile membrane; indeed such a structure would nullify the necessary mobility of the pelvic diaphragm. The surface of the levator ani is covered with no more than the epimysium (loose areolar tissue) that distinguishes muscle surfaces anywhere in the body.

Between the pelvic floor and the pelvic peritoneum lie the pelvic viscera. The extraperitoneal space between these viscera is composed of loose areolar tissue, which forms 'dead space' for distension of bladder and rectum, and vagina too. The space allows for ready compression during the passage of the fœtus in childbirth. Through this loose tissue the infection in pelvic cellulitis travels widely and fast. In the dead-space tissue are found so-called ligaments which are of two types.

First, condensations of areolar tissue surround the branches of the iliac vessels and the branches of the hypogastric plexuses to the viscera. Some of these are very strong. The lateral ligaments of the uterus and bladder and the fascia of Waldeyer are examples of ligaments that form around neuro-vascular bundles.

Second, certain ligaments exist in their own right, independently of neuro-vascular bundles. The pubo-prostatic and pubo-vesical ligaments are examples. Further examples are the round ligaments of the uterus and the utero-sacral ligaments; both contain very much smooth muscle mingled with their fibrous tissue.

The **fascia of the pelvic viscera** is loose or dense in conformity with the distensibility of the organ. The non-distensible prostate is surrounded by a tough membrane of fascia, the highly distensible bladder and rectum have no membrane around their muscle walls, only a loose and cellular tissue invests them.

THE RECTUM

The Latin word 'rectus' means straight, as if ruled. The rectum is misnamed, for it is curved in conformity with the hollow of the sacrum and, in addition, has a secondary bulge to the left of the midline.

The rectum is continuous with the sigmoid (pelvic) colon and there is no change of structure at the junction. The distinction is merely a matter of peritoneal attachments; where there is a mesocolon the gut is called sigmoid, where there is no mesentery it is called rectum (cf. the duodenum, retroperitoneal, and the jejunum on a mesentery). There is nothing vague about where the rectum ends, however; this is where its muscle coats are replaced by the sphincters of the anal canal. This is the ano-rectal junction, and it is slung in the U-loop of pubo-rectalis.

The three tæniæ of the large intestine, having broadened out over the sigmoid colon, come together over the rectum to invest it in a complete outer layer of longitudinal muscle. The rectum commences in the hollow of the sacrum at the level of the third piece and curves forward over the coccyx and ano-coccygeal raphé to pass through the pelvic floor into the anal canal just behind the perineal body. The ano-rectal junction lies at the pelvic floor $1\frac{1}{2}$ inches (3 cm) above the cutaneous margin of the anus, at about 2 inches (5 cm) from the tip of the coccyx in the position of rest. The anal canal runs from the termination of the rectum to the anal orifice (p. 344); it is in the perineum.

The rectum possesses no mesentery, but the pelvic peritoneum of the posterior wall of the pouch of Douglas is draped over its upper part. Thus the upper third is clad in front and on both sides, the middle third in front only and the lower third not at all by peritoneum.

The upper part of the rectum is usually empty, but the lower part, resting on the pelvic floor (the ano-coccygeal raphé), is distended into the **ampulla,** which contains resting flatus and fæces. The upper and lower ends of the rectum lie in the midline, but the ampulla is convex to the left. Three lateral curves are thus produced, each being marked on the interior of the ampulla by a horizontal shelf; these are named 'valves' (the **rectal valves** of Houston). They lie, *two on the left and one on the right between them,* across half the circumference of the rectal lumen and are produced by the circular muscle of the gut and are not confined merely to the mucous membrane as is the case with the circular folds of the duodenum and jejunum. The valves vary in position but are always present, and they appear early in development. Their purpose is not quite clear, but they may be concerned in the separation of flatus from the fæcal mass, holding up the fæces while allowing the flatus to pass.

The **blood supply** of the rectum is derived principally from the artery of the hindgut, the inferior mesenteric artery, whose **superior rectal** branch supplies the mucous membrane of the gut as far down as the muco-cutaneous junction of the anal canal (Hilton's line). The muscle wall of the rectum receives a reinforcement from the **middle rectal** branches of the internal iliac artery. Thirdly, some small twigs from the **median sacral** artery reach the back of the rectum in the hollow of the sacrum; these are in series with the three main gut arteries, for the median sacral artery is the atrophic continuation of the abdominal aorta.

The **superior rectal artery** is a direct continuation of the trunk of the inferior mesenteric. After the lowest sigmoid artery has entered the pelvic mesocolon the vessel, touching but not crossing the medial side of the

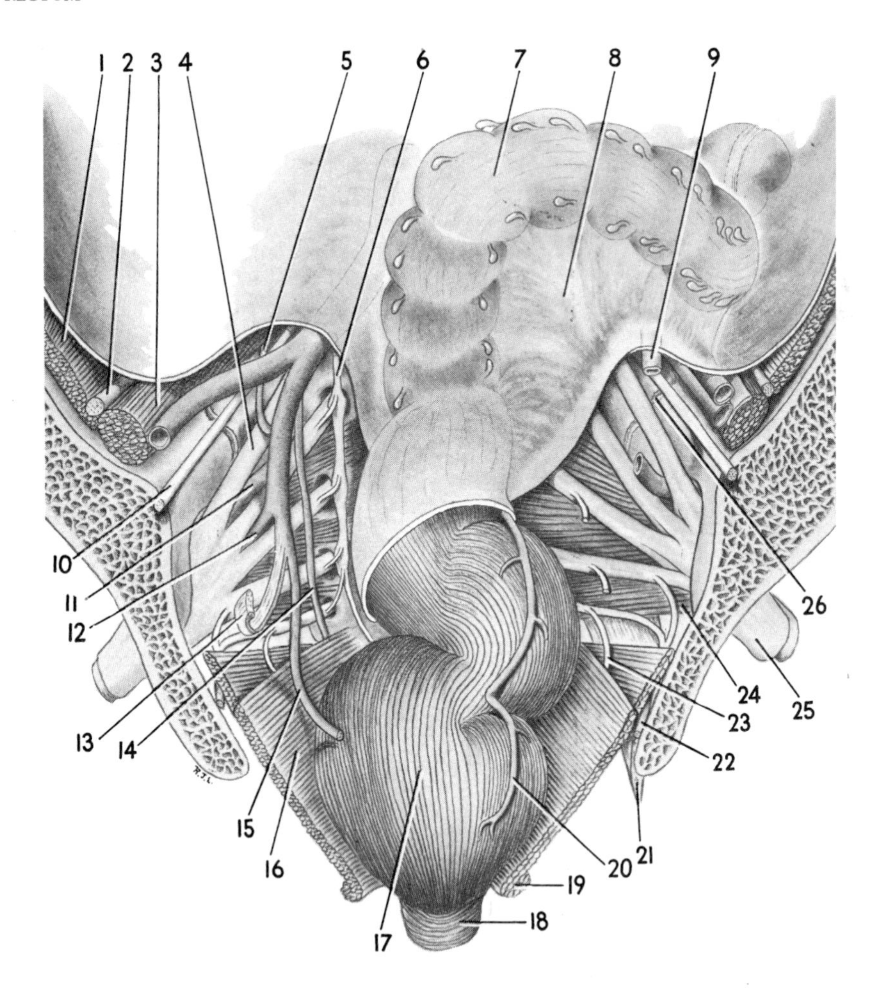

FIG. 5.59. THE RECTUM FROM IN FRONT. This coronal section of the pelvis is the posterior half of that in Fig. 5.63. 1. Iliacus. 2. Femoral nerve. 3. Psoas. 4. Lumbo-sacral trunk. 5. Ilio-lumbar artery. 6. Sacral sympathetic trunk. 7. Pelvic (sigmoid) colon. 8. Pelvic (sigmoid) mesocolon. 9. Ureter. 10. Obturator nerve. 11. Superior gluteal artery. 12. Inferior gluteal artery. 13. Superior vesical artery with obliterated umbilical artery (medial umbilical ligament). 14. Lateral sacral artery. 15. Uterine artery. 16. Levator ani. 17. Ampulla of rectum. 18. Anal canal. 19. Pubo-rectalis. 20. Branch of superior rectal artery, unusually far forward. 21. Sacro-tuberous ligament (cut). 22. Pudendal nerve. 23. Perineal branch of S4. 24. Piriformis. 25. Sciatic nerve. 26. Sacro-iliac joint.

left ureter, crosses the bifurcation of the left common iliac vessels and descends in the base of the inferior limb of the pelvic mesocolon to the commencement of the rectum. The artery here divides into right and left branches. The right one divides into anterior and posterior branches. These three main branches sink into the muscle walls in the line of the three primary hæmorrhoids (4, 7 and 11 o'clock around the clockface at the anal margin when the patient is viewed in the lithotomy position).

The muscle wall of the ampulla and the upper part of the anal canal receives branches from the middle rectal artery, but the latter supply little blood to the mucous membrane and anastomose but little with the branches of the superior rectal artery.

The **venous return** of the rectum follows the arterial supply, with the distinctive difference that there is a *very free anastomosis* between the tributaries of the venous systems. A submucous plexus of veins in the rectum and anal canal drains through the muscle wall of the ampulla into a plexus that surrounds the ampulla. This is the *external rectal plexus.* It drains to two destinations; upwards to the portal system by the superior rectal vein and across to the internal iliac vein by a plexus of middle rectal veins. This plexus (and not the superior rectal vein) communicates freely with the venous plexuses in the base of the broad ligament of the female and with the vesical plexus in the male. The superior rectal vein is formed by union of tributaries corresponding with the three branches of the superior rectal artery (two on the right and one on the left). It runs up in the base of the inferior limb of the pelvic mesocolon, receives the sigmoid veins from the pelvic colon and, crossing the pelvic brim, becomes the inferior mesenteric vein. Portal obstruction produces varicosities (hæmorrhoids) at the junctional zone between superior and middle rectal veins on the ampulla. The submucous veins of the ampulla, and those of the anal canal, thereby become obstructed and varicose.

The **lymph drainage** of the rectum is via lymphatics which run back with the three arterial sources of supply (superior rectal, middle rectal and median sacral arteries). Lymph follicles in the mucous membrane provide the first filter. Thence vessels pierce the wall of the rectum and travel to nodes (i) in the hollow of the sacrum along the *median sacral artery*, (ii) on the side wall of the pelvis along the *middle rectal artery*, and (iii) along the *inferior mesenteric artery* and upwards to the pre-aortic nodes at the origin of the latter vessel. All these nodes must be removed in the radical extirpation of malignant disease of the rectum.

The **nerve supply** of the rectum is derived from both parts of the autonomic system. The *sympathetic* supply is derived by branches directly from the hypogastric plexuses and by fibres which accompany the inferior mesenteric and superior rectal arteries from the cœliac plexus. The *parasympathetic* supply is from S2 and 3 (or S3 and 4) by the nervi erigentes, which pass through the hypogastric plexuses. These pelvic parasympathetics (nervi erigentes) are motor to detrusor muscles. In addition they register crude sensation (and pain), and they have the additional ability to distinguish flatus from fæces. Loss of the rectal mucosa and its surgical replacement from above results in incontinence even though the sphincters be preserved with an intact nerve supply, for the sensory side of the reflex arc is lost. Mucosa from above the rectum (sensory supply from the sympathetic) cannot distinguish flatus from fæces.

The **fascia of the rectum** consists of loose areolar tissue surrounding the rectal venous plexus. Posteriorly a sheet of fascia, more membranous in character, suspends the lower part of the ampulla to the hollow of the sacrum; it encloses the superior rectal vessels (artery, veins and lymphatics). It is known as the *fascia of Waldeyer*. Laterally, just above the pelvic floor, the middle rectal artery and the branches of the pelvic plexuses are enclosed in a slight condensation of areolar tissue that is known to surgeons as the *lateral ligament* of the rectum. The fascia of Waldeyer, the lateral ligaments, the pelvic peritoneum and the vessels and most of all the pelvic floor combine to hold the rectum stable in its position.

Anteriorly some muscle fibres leave the lower part of the ampulla and pass forwards towards the apex of the prostate and the commencement of the membranous urethra. They form the *recto-urethralis muscle*, a surgical landmark in operations in this region.

THE BLADDER

The bladder is made of smooth muscle arranged in whorls and spirals—the detrusor muscle. It is adapted for mass contraction, not peristalsis. The muscle is lined by a loose and readily distensible mucous membrane, surfaced by transitional epithelium (p. 17).

The form and size of the bladder are the same in both sexes, but the attachments of its base and the posterior relations of the viscus require separate consideration in male and female. In each the trigone is fixed and the fundus varies in size above this in accordance with the volume of the contained urine. The distended bladder is globular (ovoid) in both sexes, while the empty bladder is flattened from above downwards by the pressure of the overlying intestines.

The **fundus** lies behind the bodies of the pubic bones; during distension it rises over the upper border of the pubis like the sun rising above the horizon. The fundus is held in position by the tone of its muscular walls on the fixed base of the trigone, as a toy balloon is fixed in space when its rubber teat is clasped firmly.

The bladder is separated from the pubic bones by a space lying between the pelvic floor and the pelvic peritoneum. This is the **retropubic space** (or **cave of Retzius**); it is occupied by very tenuous areolar tissue. The upper surface of the fundus has a serous coat, for here the pelvic peritoneum is firmly adherent behind the attachment of the median umbilical ligament. Elsewhere the bladder is surrounded by loose and tenuous tissue. The rising bladder strips peritoneum from behind the rectus abdominis, for the fascia transversalis is here loose and tenuous. Thus may the *distended* bladder be approached by knife or needle above the symphysis without opening the peritoneal cavity.

The trigone or base of the bladder is, unlike the fundus, relatively indistensible and immobile. It is held in position by the lateral ligaments of the bladder and by being fixed, in the male to the immobile prostate, and in the female to the cervix uteri and anterior vaginal fornix. It is more mobile, as can be readily understood, in the female (Fig. 5.76, p. 354).

The **interior of the bladder** can be studied at autopsy, at operation or by cystoscopy. The cavity varies in capacity with the contained urine (from zero to a litre or more), and the appearance of its walls depends upon the state of distension of the organ. When collapsed the mucous membrane is thick and thrown into folds, when distended it is thin and smooth. The trabeculæ of the muscle fibres can be seen through the mucous membrane. These remarks do not apply to the trigone, which varies but little with the state of distension of the organ.

The **trigone** (Fig. 5.60) is a triangular area lying between the internal urethral orifice and the orifices of the ureters. It is smooth-walled and the mucous membrane is rather firmly adherent to the underlying muscle. The ureteric orifices are connected by a transverse ridge, prominent when viewed through the cystoscope, called the *interureteric bar*. The orifices of the ureters lie at the ends of the bar; they are usually in the shape of an oblique slit, but considerable variations exist. The ureters pierce the muscle and mucosal walls very obliquely; more than any sphincteric muscular action the valve-like flap of mucosa thus produced is the important factor in preventing reflux of urine when intravesical pressure rises. The ureteric orifices are closed by this pressure, except to open rhymically in response to ureteric peristalsis each time a jet of urine is injected into the bladder (four or five times a minute normally).

In the male the trigone overlies the middle lobe of the prostate which, after the first flush of youth, may project above the internal urethral orifice as a rounded elevation called the **uvula vesicæ** (Fig. 5.60).

Structure of the Bladder. The smooth-muscle wall is composed of fibres running in whorls and spirals. Sections of the bladder wall show the fibres cut in many directions, longitudinal, latitudinal and oblique; both externally, and internally beneath the mucous membrane, the fibres produce a trabeculated appearance (the *detrusor muscle*). At the internal urethral orifice circular fibres provide an internal sphincter of smooth muscle which, though of obvious capability, is rather unconvincing anatomically.

The mucous membrane is thick and lax and lined with transitional (urine-proof) epithelium. It has no glands; mucus in the shed urine has come from the urethra. It possesses no muscularis mucosæ; it is therefore incorrect to use the term 'submucous' in the bladder (see p. 16).

Relations of the Bladder. In front lies the retropubic space, above are the pelvic peritoneum and coils of intestine (ileum and/or sigmoid colon), behind is the uterus in the female, or the seminal vesicles and rectum in the male (Fig. 5.76, p. 354, and PLATE 21).

The **median umbilical ligament** is a variably developed fibrous remnant (it is often absent). It is the remains of the urachus. When well developed it constitutes a midline fibrous cord joining the fundus of the bladder to the umbilical scar and it raises a low ridge of peritoneum on the posterior aspect of the anterior abdominal wall.

The **medial umbilical ligaments** are the two obliterated umbilical arteries of the fœtus. These two arteries remain patent as far from their origin from the internal iliac arteries as the superior vesical branch. From this situation the fibrous cords pass from the sides of the bladder obliquely upwards to the navel. Each raises a ridge of peritoneum on the posterior aspect of the anterior abdominal wall (Fig. 5.5, p. 258).

The **blood supply** of the bladder is derived from the superior and inferior vesical arteries and by a few small twigs from the pubic branch of the inferior epigastric artery (in 30 per cent of people this branch constitutes the abnormal obturator artery).

The veins of the bladder form a plexus, the **vesical plexus,** at the base of the bladder. In the female this plexus communicates with the veins at the base of the broad ligament and drains, on each side, across the pelvic floor into the internal iliac veins. In the male the vesical plexus communicates with the prostatic plexus and the middle rectal veins and drains across the pelvic floor into the internal iliac veins.

The **lymph drainage** of the bladder follows back along the arteries. Most of the lymphatics run with the superior and inferior vesical arteries to nodes on the side wall of the pelvis alongside the internal iliac artery, but a few from the fundus run with the pubic artery to nodes alongside the external iliac vessels.

The **nerve supply** of the bladder is derived from both parts of the autonomic system. The *sympathetic nerves* are derived from the pelvic plexuses and the *parasympathetic* via the pelvic plexuses from the nervi erigentes (S2 and 3 or S3 and 4). The sympathetic fibres to the detrusor muscle are inhibitory, those to the internal sphincter are motor and reach the pelvic plexuses by way of the hypogastric nerves (lumbar splanchnics, or presacral nerves) and are divided in the operation of presacral neurectomy. The parasympathetic nervi erigentes innervate the detrusor muscle. Afferent fibres of normal bladder distension and of pain pass back chiefly in the nervi erigentes. Presacral neurectomy, if it relieves bladder pain, does so not by dividing pain pathways, but by paralysing the vesical sphincter and thus preventing it from going into painful reflex spasm.

The **fascia of the bladder** is very loose and tenuous over the fundus, as would be expected in such a distensible viscus. Behind the trigone it is firm and tough. In the female this fascia serves to anchor the trigone to the cervix uteri and anterior fornix, while in the male it constitutes the upper limit of the recto-vesical fascia (of Denonvilliers). From the base of the bladder thickenings of areolar tissue pass laterally across the pelvic floor. They are the **lateral ligaments of the bladder** and enclose the vesical veins and inferior vesical artery and lymphatics, together with the nerves of the bladder.

The Ureter in the Pelvis

The ureter crosses the pelvic brim at the bifurcation of the common iliac vessels over the sacro-iliac joint. It passes downwards in contact with the peritoneum of the pouch of Douglas, towards the spine of the ischium. It then runs forward above the pelvic floor to pass obliquely into the base of the bladder; in this part of its course it lies in the loose areolar tissue of the pelvic fascia some distance below the pelvic peritoneum. In each sex only one structure of note passes above it, between it and the pelvic peritoneum—in the female the uterine artery (Fig. 5.63) and in the male the vas (PLATE 21).

The pelvic part of the ureter constitutes half its total length. The blood supply, etc., are considered on p. 321.

Development of the Bladder

The embryonic cloaca is divided by the growth of a transverse septum into a ventral uro-genital sinus and a dorsal extension of the hindgut which is destined to become the rectum and upper part of the anal canal. The uro-genital sinus is continued cranialwards into the urachus, whose cavity extends as a blind diverticulum into the allantoic stalk (umbilical cord). This cavity may persist as a patent urachus, in which case urine drains from the umbilicus, but normally it shrivels and persists as a fibrous remnant called the median umbilical ligament.

The mesonephric duct opens into the ventral part of the cloaca which afterwards becomes the uro-genital sinus. The ureter grows as a bud from the caudal end of this duct; the mesonephric duct distal to this point is called the common excretory duct.

The upper part of the uro-genital sinus enlarges to become the bladder. The common excretory duct dilates and becomes incorporated into its walls. The terminal part of the ureter shares in this progressive incorporation into the bladder wall, so that the mesonephric ducts and ureters finally open as four separate orifices. These are the ejaculatory ducts in the prostatic urethra and the ureteric orifices at the angles of the trigone. The mucous membrane between these four points is thus derived from the mesonephric duct. In the female the mesonephric ducts later atrophy and vanish, but the female trigone is developed in precisely similar manner.

The epithelium of the bladder is thus of endodermal origin, that of the trigone is probably mesodermal, though histologically and functionally they are identical (transitional epithelium).

Abnormalities. *Ectopia vesicæ* is a condition in which the symphysis pubis and the endodermal part of the bladder wall are lacking. The bladder trigone is present and joined to surrounding skin; urine leaks from the ureteric orifices. It is presumably due to a very early breakdown of the cloacal membrane and failure of the somatopleure to meet and close over the gap.

Double ureter is common; it is due to premature division of the ureteric bud. The two ureters may branch from a single distal segment or they may open separately into the bladder. In the latter case, as dilatation of the distal ureteric segments incorporates them into the bladder wall, rotation about each other results in the lower ureteric orifice draining the upper pole of the kidney and the upper ureteric orifice draining the lower pole of the kidney.

THE PROSTATE

This is a male organ and has no counterpart in the female. It is roughly the size and shape of a chestnut and it lies, apex below, between the bladder and the pelvic floor, clasped by muscular fibres that are called pubo-prostaticus (levator prostatæ). It is perforated by the urethra, though most of its substance lies behind and lateral to that passage. Between the alveoli the gland possesses a fibro-muscular stroma that is in direct continuity with the muscle of the bladder wall. It is perforated by the ejaculatory ducts, which open separately into the prostatic portion of the urethra. Its own ducts, some 30 to 40 in number, open separately by minute orifices into the prostatic urethra. The surface of the prostate is covered by a thin condensation of fibrous tissue that constitutes the capsule. The prostate is really a bilateral structure, right and left halves failing to fuse completely, leaving a posterior midline groove.

The **prostatic urethra** extends from the internal urinary meatus to the apex of the prostate and is the widest part of the urethra. It is characterized posteriorly by a midline longitudinal ridge, the **urethral crest** or **verumontanum.** In the centre of this crest is a minute depression, an embryological remnant resulting from union of the caudal ends of the paramesonephric (Mullerian) ducts; it is thus the homologue of the female uterus. It is called the **uterus** (or **utriculus**) **masculinus.** Alongside it, on the urethral crest, the ejaculatory ducts open. The prostatic ducts open on the crest and in the sulcus on each side (Fig. 5.60).

The **prostatic fascia** is a dense and tough membrane which completely invests the gland. It is a condensation of the areolar tissue of the pelvic fascia. Posteriorly, between the prostate and rectum, the prostatic fascia has a different origin, being derived from the pelvic peritoneum. In the fœtus the recto-vesical pouch, between the bladder and rectum, extended down to the pelvic floor. Fusion of anterior and posterior layers of the pouch made it more shallow, and the fused layers persist as a membrane between the pouch and the pelvic

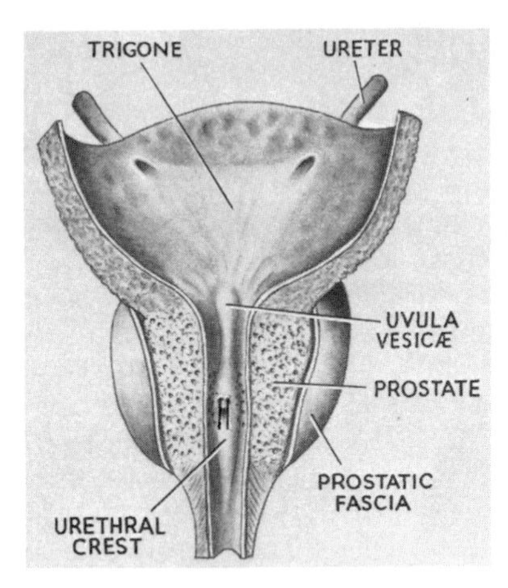

FIG. 5.60. THE PROSTATIC URETHRA OPENED FROM IN FRONT. A bristle lies in each ejaculatory duct; the utriculus masculinus lies between them. Drawn from a specimen.

floor. It covers the seminal vesicles and, below them, becomes incorporated into the fascial investment of the prostate. It is called the **recto-vesical fascia,** or **fascia of Denonvilliers.** The anterior wall of the rectum is freely mobile over it. The prostatic fascia is separated from the fibrous-tissue capsule of the gland by a narrow space which contains the prostatic plexus of veins.

The **blood supply** of the prostate is derived from the inferior vesical and middle rectal branches of the internal iliac artery. Whatever the variation in origin a large prostatic artery enters the prostate laterally at the junction with the bladder (in the specimen illustrated in PLATE 21 the prostatic artery on the right side branches from the obturator artery). The emerging veins accompany this main artery. Ligation of these vessels greatly reduces the bleeding that otherwise accompanies prostatectomy. The prostatic veins drain into the prostatic plexus. The **prostatic plexus** consists of a mass of thin-walled veins lying between the prostatic fascia and the fibrous capsule of the gland. It receives the deep dorsal vein of the penis and the veins of the base of the bladder and, after communicating with the rectal plexus, drains across the pelvic floor into the internal iliac veins. The **lymphatics** of the prostate pass across the pelvic floor to nodes on the side wall of the pelvis alongside the internal iliac vessels.

Structure of the Prostate. Lying in a stroma of fibrous tissue and smooth muscle fibres are the alveoli of the gland, large cavities lined with columnar epithelium. By early adult life infolding of the epithelium is common and cystic dilatation of many of the alveoli makes them visible to the naked eye in histological sections (Fig. 5.61). Hyperplasia of the epithelium (two layers or papillomatous ingrowths) is common. Corpora amylacea are often seen.

Applied Anatomy of Prostatic Neoplasm. The prostate consists of a continuum of glandular tissue and stroma; most of the gland lies behind the urethra. A midline groove posteriorly divides the gland descriptively into two 'lateral' lobes. This groove is readily palpable on rectal examination with the finger; the rectal wall can be made to slide over the prostatic fascia. In advanced carcinoma the groove is obliterated and the rectal wall becomes fixed by neoplastic induration to the prostatic fascia. The pyramidal portion of gland between the bladder and the ejaculatory ducts is known to surgeons as the 'median lobe'. In minor degrees of prostatic hypertrophy it projects into the bladder as a rounded elevation known as the uvula vesicæ. Further enlargement obstructs the outflow of urine and results in incomplete emptying of the bladder, with the train of symptoms of frequency of micturition, straining during the act, and a diminished stream.

'Hypertrophy' of the prostate is usually not a generalized hypertrophy of the whole gland, but the growth of an adenoma within it. The surrounding prostate becomes condensed around the adenoma. 'Prostatectomy',

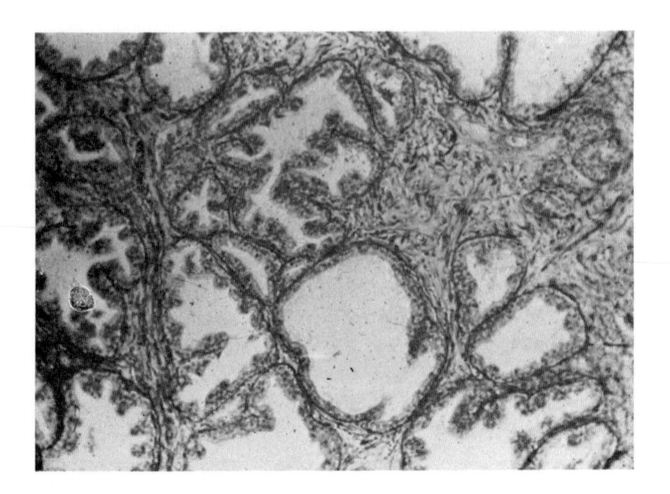

FIG. 5.61. HUMAN PROSTATE (aged about 40 years). Infolding of the epithelium is typical, as are the large alveoli. The fibrous stroma contains a good deal of smooth muscle. (Photomicrograph.)

whether transvesical or retropubic, amounts to shelling out of the adenoma from the peripheral condensed prostatic tissue. The prostatic plexus of veins is thus not entered. Complete removal of the prostate from within its prostatic fascia involves opening up the prostatic plexus.

The Ductus Deferens. The former name of **vas deferens** still lingers on, especially in common parlance (p. 266). This thick-walled muscular tube enters the abdomen at the deep inguinal ring and passes along the side wall and floor of the pelvis to reach the back of the bladder. In its course *no other structure intervenes between it and the peritoneum.* After hooking around the interfoveolar ligament and inferior epigastric artery at the deep inguinal ring it crosses the external iliac vessels (PLATE 21).

It passes over the obturator nerve and vessels, lies on the obturator fascia and descends below the peritoneum. It curves medially and forwards, *crosses above the ureter* and approaches its opposite fellow. The two ducts now turn downwards side by side and each dilates in fusiform manner. This dilatation is the **ampulla,** the storehouse of spermatozoa. The ampullæ lie vertically between the seminal vesicles; at their lower ends each loses its thick muscle wall and joins with the outlet of the seminal vesicle to form the **ejaculatory duct.** Each ejaculatory duct passes obliquely through the prostate to open on the side of the urethral crest (Fig. 5.62).

The artery to the ductus deferens, a branch of the inferior (or superior) vesical artery, accompanies the duct to the lower pole of the epididymis.

The Seminal Vesicles. These are a pair of thin-walled, elongated and lobulated sacs, each consisting of a blind tube folded on itself. They are applied to the base of the bladder above the prostate and are covered posteriorly by the recto-vesical fascia of Denonvilliers. Each joins the lower end of the ampulla of the ductus deferens behind the prostate to form the ejaculatory duct. They produce the seminal fluid and contain no spermatozoa. The secretion imparts motility to the spermatozoa. The two vesicles lie, with their long axes oblique, lateral to the ampullæ (Fig. 5.62).

The seminal vesicles are supplied by branches of the vesical arteries. Each develops as an outgrowth from the mesonephric duct. Their muscle walls are innervated via the hypogastric nerves and plexuses. Motor fibres (sympathetic) lie in the branch from the first lumbar ganglion—their division produces sterility, for the paralysed vesicles cannot expel their secretion. It is noteworthy that ejaculation is a sympathetic phenomenon while the erection that normally precedes it is parasympathetic (the nervi erigentes).

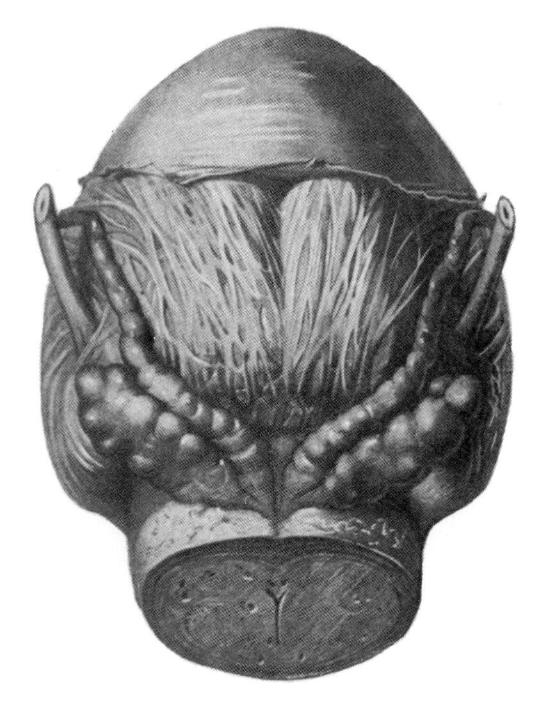

FIG. 5.62. POSTERIOR VIEW OF THE EMPTY BLADDER AND PART OF THE PROSTATE. THE VAS (DUCTUS DEFERENS) HOOKS OVER THE TERMINAL PART OF THE URETER AND THEN DILATES INTO THE AMPULLA. DISTAL TO THE AMPULLA IT JOINS THE DUCT OF THE SEMINAL VESICLE TO FORM THE EJACULATORY DUCT. THE TWO EJACULATORY DUCTS ARE SEEN AT THE URETHRAL CREST ON THE CUT SURFACE OF THE PROSTATE. THE SEMINAL VESICLE LIES LATERAL TO THE AMPULLA.

Microscopic Anatomy. The wall is made of smooth muscle. The sacculations are lined with a delicate mucous membrane that is projected into high folds joining each other in a honeycomb pattern. The epithelium, tall columnar, stains deeply.

THE UTERUS

This is a muscular organ whose function is to provide a nidus for the developing embryo. In the virginal state it is the shape of a flattened pear. Its size is about 3 × 2 × 1 inches (say 8 × 5 × 3 cm). It possesses a fundus, body and cervix. It receives the uterine tubes, and the cervix protrudes and opens into the vault of the vagina.

The **fundus** is the part above the entrance of the tubes. It is convex and measures about 2 inches (5 cm) from side to side and about an inch (3 cm) thick. It possesses a serous coat of pelvic peritoneum which continues downwards over the front and back of the body (Fig. 5.65).

The **body** of the uterus tapers downwards from the fundus and is flattened antero-posteriorly. The upper angles, at the junction of fundus and body, receive

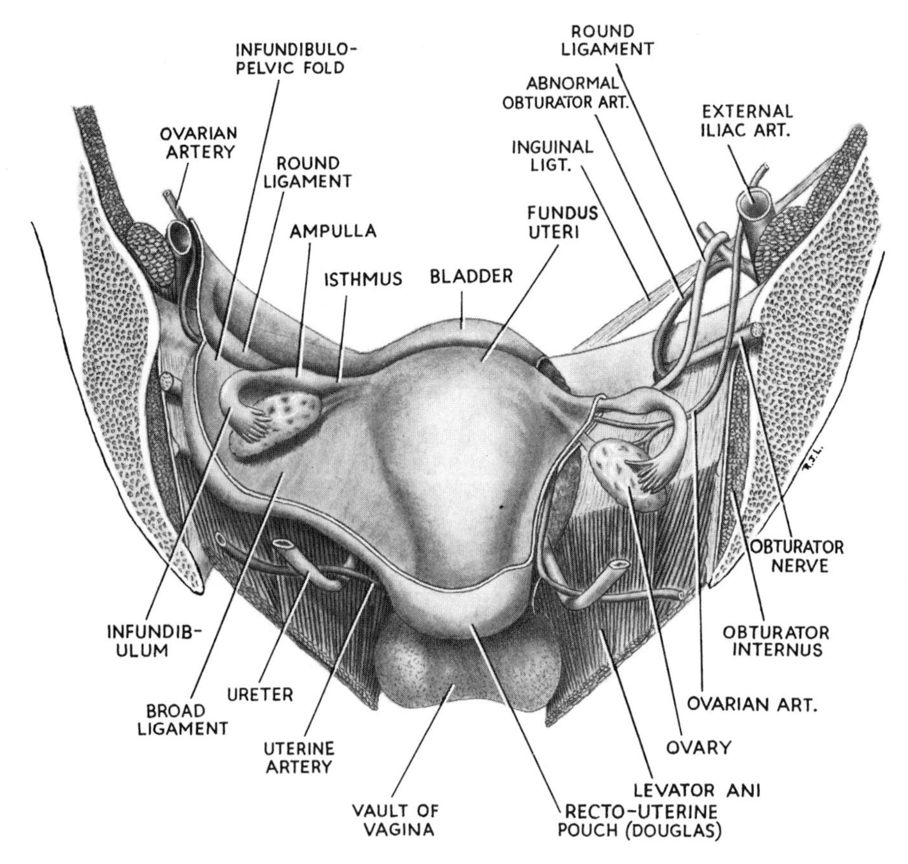

FIG. 5.63. THE UTERUS VIEWED FROM BEHIND AFTER SPLITTING THE PELVIS. The broad ligament and parietal peritoneum have been removed from the right side. Drawn from a dissection.

the uterine tubes. They are called the *cornua*. The cavity of the uterus occupies the body. A narrow slit in the virgin, it enlarges by growth of the uterine walls during pregnancy to accommodate the fœtus.

The **cervix** or neck of the uterus tapers below the body and is clasped by the vault of the vagina, into which it protrudes. The deep sulcus which surrounds the protruding cervix is known as the fornix of the vagina; it is deepest posteriorly. The canal of the cervix opens from the cavity of the body at the isthmus and opens into the vagina at the (external) os. The os is a circular dimple in the virgin, but usually a transverse slit after childbirth.

The internal os is in the supravaginal part of the cervix, now called the *isthmus uteri*. The isthmus dilates and is taken up into the uterus as the organ enlarges during pregnancy. This expansion of the upper cervix, or isthmus, constitutes the 'lower uterine segment' of the obstetrician. The lower part of the canal of the cervix dilates during childbirth to provide passage for the fœtus.

Anteriorly the lower part of the body and the supravaginal part of the cervix are firmly attached by fibrous tissue to the base of the bladder, but posteriorly the uterus lies free and is invested in the pelvic peritoneum (of the pouch of Douglas) right down to the cervix and the posterior fornix of the vagina (Fig. 5.76, p. 354).

The normal **position of the uterus** is one of anteversion and slight anteflexion. That is to say, the fundus and upper part of the body are bent slightly forward, while the organ thus flexed leans forwards as a whole from the vagina. It is maintained in this position by the support of the pelvic floor and the presence of certain ligaments, especially the round and utero-sacral ligaments.

The **blood supply** of the uterus is by the *uterine artery*, a branch of the internal iliac artery. It passes medially across the pelvic floor, in the base of the broad ligament, *above the ureter*, to reach the side of the organ at the supravaginal part of the cervix. Giving a branch to the cervix and vagina the vessel turns upwards between the leaves of the broad ligament to run alongside the uterus as far as the entrance of the tube, where it anastomoses end on with the tubal branch of the ovarian artery. In its course it freely gives off branches which penetrate the walls of the uterus. The *veins* of the uterus course below the artery at the lower edge of the broad ligament where they form a wide plexus across the pelvic floor. This plexus communicates with the vesical and rectal plexuses and drains into the internal iliac vein.

The *lymphatics* pass with each artery to the internal iliac group of lymph nodes. Carcinoma of the fundus may occasionally spread along the round ligament to superficial inguinal nodes, but this does not mean that lymph from the uterus normally drains along this pathway (cf. carcinoma of pylorus spreading to nodes in the porta hepatis, and see p. 18).

The nerves of the uterus are branches from the inferior hypogastric plexuses. Little is known of motor pathways to this organ, whose muscle is so sensitive to hormonal influence. There is evidence that the sympathetic supply (vaso-constrictor) may also be motor to the uterine muscle. Sensory pathways are better understood. Pain from the cervix is carried by the nervi erigentes (S2, 3) while that from the body of the uterus (labour pains) travels up the hypogastric nerves to the lower thoracic segments.

Peritoneal Attachments. The uterus is draped by an adherent fold of pelvic peritoneum which is continued laterally on each side to be attached to the side wall of the pelvis; this reflexion is the broad ligament. The front of the uterus possesses a serous coat as far down as the attachment of the base of the bladder, at the level of the isthmus, while the back of the organ is completely clad with very adherent peritoneum. At the sides of the organ the anterior and posterior serous coats pass laterally as the anterior and posterior leaves of the broad ligament (Figs. 5.63 and 5.65).

Fascia and Ligaments of the Uterus. The lower part of the front of the uterus is adherent to the bladder, as already mentioned, by condensed fibrous tissue. The extraperitoneal areolar tissue (pelvic fascia) is condensed in other places to form named ligaments. These 'ligaments' are composed histologically of a high proportion of smooth muscle fibres. The most important ligaments are the round ligament (of the body) and the lateral and utero-sacral ligaments (of the cervix).

The **round ligament** extends from the junction of uterus and tube to the deep inguinal ring. It lies in the anterior leaf of the broad ligament, below the uterine tube. It is continuous with the 'ligament of the ovary' (p. 337) and the two represent, in continuity, the gubernaculum, the counterpart of the gubernaculum testis of the male. The round ligament passes through the inguinal canal and is attached at its distal extremity to

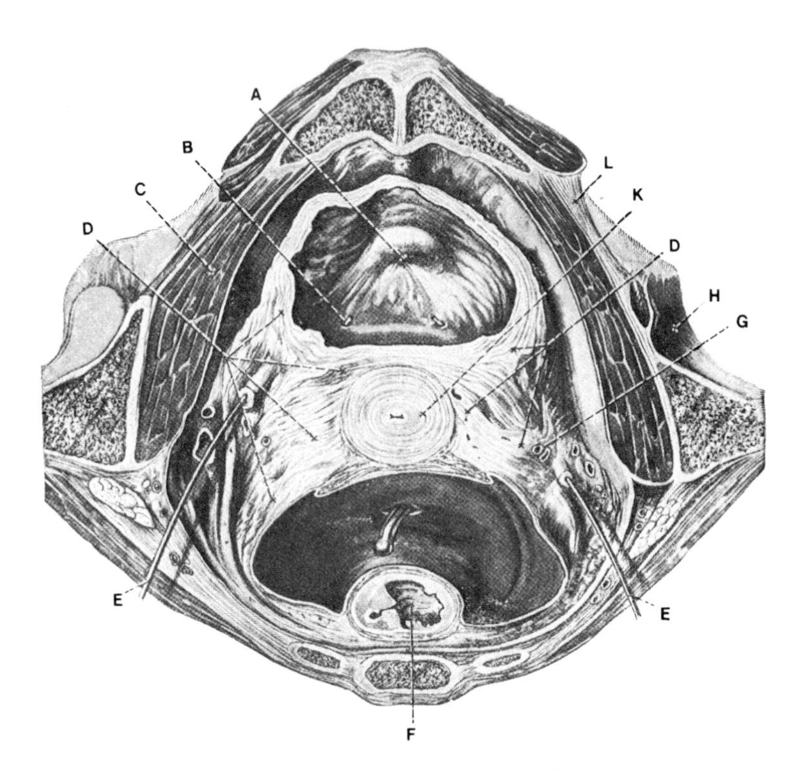

FIG. 5.64. SECTION THROUGH THE CERVIX UTERI PARALLEL WITH THE PELVIC BRIM. A. Internal urinary meatus. B. Ureteric orifice. C. Obturator internus. D. Pelvic fascia. E. Probe in ureter. F. Rectum (in front of this the points of a pair of forceps have entered the pouch of Douglas through the posterior fornix). G. Uterine artery and vein. H. Acetabulum. K. Cervix uteri. Behind this the utero-sacral folds are seen. L. Obturator membrane.

the fibro-fatty tissue of the labium majus of the vulva. It is supplied by the ovarian artery in the broad ligament and by a branch from the inferior epigastric artery in the inguinal canal (Fig. 5.63).

It is comprised largely of smooth muscle. Its function is to hold the fundus forward in anteversion, especially when forces tend to push the uterus back (e.g., distension of the bladder, gravity during recumbence).

The **utero-sacral ligaments** extend backwards from the cervix. They lie below the pelvic peritoneum, embrace the pouch of Douglas and the rectum, and are attached to the front of the lower part of the sacrum (Fig. 5.64). They consist of bundles of smooth muscle connected and surrounded by the fibrous tissue of the pelvic fascia. They serve to keep the cervix braced backwards against the forward pull of the round ligaments on the fundus and so maintain the body of the uterus in anteversion.

The **lateral ligaments** (of Mackenrodt) consist of thickenings of connective tissue around the uterine arteries and the uterine venous plexuses in the base of the broad ligaments. They extend from the cervix laterally to the side wall of the pelvis. They impart lateral stability to the cervix uteri (Fig. 5.64).

It has already been mentioned that the base of the bladder is firmly attached to the front of the uterus and cervix. The connective tissue binding them together extends inferiorly to bind the urethra to the anterior wall of the vagina. Posteriorly the vagina and anal canal are separated, but bound together, by the fibromuscular mass of the perineal body. This mass helps to support the uterus; when torn, or stretched into atrophy after parturition, the posterior vaginal wall prolapses, and this condition is often followed by prolapse or retroversion of the uterus (Fig. 5.76, p. 354).

The Uterine Tubes. Each tube has a lateral free extremity, expanded and open, surrounded by a number of small finger-like processes called *fimbriæ*. One, longer than the others, is attached by its tip to the ovary; this is the *ovarian fimbria*. The fimbriated end of the tube, or *infundibulum*, lies coiled behind the broad ligament. The remainder of the tube lies between the two layers of the upper margin of the broad liga-

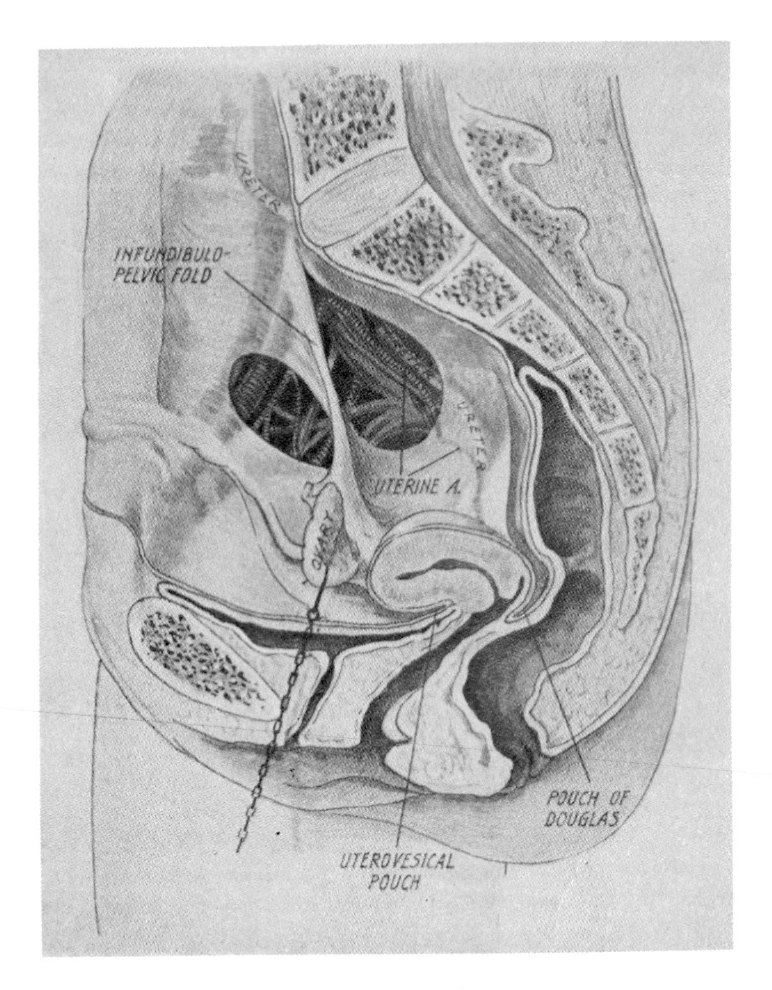

FIG. 5.65. THE RIGHT HALF OF THE FEMALE PELVIS. Windows are cut in the parietal peritoneum and the ovary is displaced forwards.

ment. Laterally this part is somewhat tortuous and dilated in spindle shape, the *ampulla.* The portion of tube adjoining the uterine wall is straight and narrow; it is known as the *isthmus* (Fig. 5.63). The *intramural part* of the tube is surrounded by its own muscular wall up to its opening into the cavity of the uterus.

The tube, formed of smooth muscle, is lined by a mucous membrane that is thrown into complicated folds. The surface epithelium is of the ciliated columnar type. In the isthmus the folds flatten and the cilia tend to disappear.

The tube is supplied by the tubal branch of the ovarian artery. The tubal artery runs below the tube, between the layers of the broad ligament, and anastomoses end on with the uterine artery. The veins of the tube join the pampiniform plexus and the lymphatics pass with the ovarian artery to para-aortic nodes just above the level of the umbilicus (opposite L2 vertebra).

The Broad Ligament. This is broad, but is not strictly speaking a ligament, since it consists of no more than a double fold of peritoneum lying lateral to the uterus. It is square in shape. The medial edge, attached to the side wall of the uterus, flows over anterior and posterior surfaces of that organ as its serous coat. The lateral edge is attached to the side wall of the pelvis, whence its two layers pass forwards and backwards to line the pelvic cavity. The inferior edge approaches the pelvic floor, whence its two layers pass forwards and backwards. Its superior border is free. The uterine tube lies in the medial three-fourths of this edge. The lateral quarter of the upper edge is called the **infundibulo-pelvic fold** and contains the ovarian artery, veins (pampiniform plexus) and lymphatics (Fig. 5.63).

The anterior leaf of the broad ligament is bulged forwards by the round ligament of the uterus just below the uterine tube. The posterior leaf of the broad ligament is attached to the ovary; this portion is known as the mesovarium. Between the two layers is a mass of areolar tissue in which lie the uterine and ovarian arteries, veins and lymphatics, the round ligament and the ligament of the ovary and the vestigial remnants comprising the epoöphoron and the paroöphoron (p. 338).

Development of the Uterus and Tubes. The paramesonephric (Mullerian) ducts fuse at their caudal ends to make the uterus and upper part of the vagina. Their cranial ends persist as the uterine tubes. Incomplete fusion results in a median septum in the uterus or in a bicornuate uterus. The latter is the usual state in most mammals, where the multiple pregnancies occupy both uterine horns.

THE OVARY

The ovary is ovoid in shape, smaller than the testis. It is firm to the touch, being composed of rather dense fibrous tissue in which the ova are embedded. The ovary projects into the pelvic cavity, attached to the posterior leaf of the broad ligament by a double fold of peritoneum which is called the mesovarium. The **mesovarium** is attached equatorially around the ovary, but does not invest the surface of the gland, which is covered with low columnar epithelium. The ovary often lies flush within the posterior leaf of the broad ligament, in which case there is no mesovarium, but traction on the ovary will pull up the peritoneum into a temporary 'mesovarium'.

The ovary lies on the peritoneum of the side wall of the pelvis in the angle between the internal and external iliac vessels, on the obturator nerve. Pain in the ovary is often referred along the cutaneous distribution of this nerve (the inner side of the thigh down to the knee). The parietal peritoneum against which the ovary lies is supplied by the obturator nerve. The obturator internus muscle and its fascia separate the ovary from the thinnest part of the hip bone, the concavity of the acetabulum. The ovary in its normal position can just be reached through the vagina by the tip of the examining finger. It is overlaid by the coils of sigmoid colon and ileum that occupy the pouch of Douglas.

The ovary lies with its long axis oblique, its upper pole medial and its lower pole lateral (Fig. 5.63). It is slung to the upper angle of the uterus by the **ligament of the ovary.** This is a mass of smooth muscle and fibrous tissue lying between the two layers of the broad ligament. It attaches the ovary to the upper angle of the uterus, whence it continues on as the round ligament. This continuous ligament is the remnant of the gubernaculum.

The ovary is supplied by the ovarian artery, a branch of the abdominal aorta just below the renal artery. The vessel runs down behind the peritoneum of the infracolic compartment and the colic vessels, crossing the ureter obliquely, on the psoas muscle. It crosses the brim of the pelvis and enters the infundibulo-pelvic fold at the lateral extremity of the broad ligament. It gives off a branch to the uterine tube which runs medially between the layers of the broad ligament and anastomoses with the uterine artery, and it ends by entering the ovary (Fig. 5.63).

The ovarian veins form a plexus in the mesovarium and the infundibulo-pelvic fold (the **pampiniform plexus,** cf. the testis). The plexus drains into a pair of ovarian veins which accompany the ovarian artery. They usually combine as a single trunk before their termination. That on the right joins the inferior vena cava, that on the left the left renal vein. This is embryologically symmetrical, since on each side the segment of vein concerned is a persistent part of the subcardinal vein of the embryo (p. 309).

The lymphatics of the ovary drain to para-aortic nodes alongside the origin of the ovarian artery, just above the level of the umbilicus (L2).

Structure of the Ovary. The ovary consists of a fibrous stroma covered with a cubical epithelium known as the germinal epithelium (note that it has no serous coat). At birth each ovary contains (such is the prodigality of nature) some 3 million ova. The origin of the ova is uncertain, but present opinion is that they are derived in the very early embryo by endodermal cells of the yolk sac that migrate into the developing gonad. Each ovum is surrounded by a single layer of cells, called *follicular* or *thecal cells.* The theca cells are derived from the surface (germinal) epithelium. Less than one in 15,000 ova is destined to be shed.

Maturation of the Ovum. The layer of theca cells divides into two layers which become separated in part by accumulating fluid. Each layer of cells divides into many layers and the fluid increases in amount (Fig. 5.66). This is the **Graafian follicle.** The cells covering the ovum are known as the *cumulus* or *discus proligerus,* those lining the follicle as the *stratum granulosum.* Once a month a follicle ruptures and extrudes its contained ovum. The liquor folliculi escapes and hæmorrhage occurs into the collapsed follicle. The blood clots; into the clot grow cells from the stratum granulosum which become distended with lipoid of a yellow colour. The mass of cells is visible to the naked eye and is called the **corpus luteum.** The corpus luteum persists for a week if pregnancy has not occurred, for nine months if it has. Then the luteal cells atrophy and become replaced by white fibrous tissue (i.e., a scar). This is called the **corpus albicans.**

Note that these are all changes in the theca cells, not in the ovum itself. Before fertilization the ovum must undergo meiosis. In man this reduction division takes place at about the time of rupture of the follicle: a large 'secondary oöcyte' and a small polar body result. Each divides again, the oöcyte into the mature ovum and another polar body, while the first polar body splits into two. The three polar bodies disappear; their rôle is unknown.

Not all the ova, even those in follicles, are destined to be shed. Many of the ova and follicles are arrested and atrophy, a process known as **atresia.** An atretic ovum is pyknotic. An atretic follicle undergoes coagulation of the liquor folliculi and gradual reduction of the granulosa cells to a single layer. All or most of the above stages, including atresia, can be made out by examining a section of a mature ovary.

Descent of the Ovary. The ovary develops from the paramesonephric ridge of the intermediate cell mass in the same way as the testis. Its site of origin lies in the peritoneum of the posterior abdominal wall. It descends, preceded by the gubernaculum. The gubernaculum proceeds through the inguinal canal, as in the male, and becomes attached to the labium majus. The ovary does not follow its gubernaculum so far, and its descent is arrested in the pelvis. The gubernaculum persists as the ligament of the ovary and the round ligament of the uterus. The ovary is supplied by its own branch from the dorsal aorta and drains into the subcardinal vein. As with the testis, artery and vein persist, becoming merely elongated as the gonad descends to its adult position.

Embryological Remnants. The mesonephric tubules and mesonephric duct normally disappear in the female. Should they persist their tubules are to be found between the layers of the broad ligament. The epoöphoron consists of a number of tubules joining at right angles a persistent part of the mesonephric duct. It lies between ovary and tube. The mesonephric duct may persist as a tube opening into the lateral

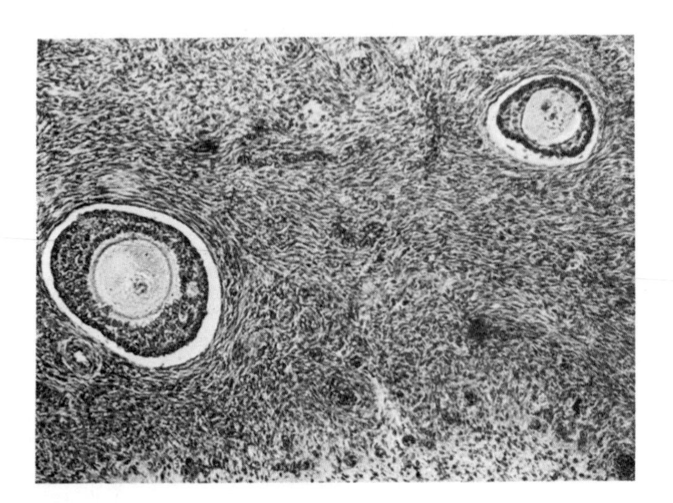

FIG. 5.66. HUMAN OVARY. The ovary is made of dense fibrous tissue. Two maturing ova are seen.

fornix of the vagina or even at the vestibule of the vulva alongside the vaginal orifice. The paroöphoron lies nearer the base of the broad ligament. It consists of a number of minute tubules, blind at each end, the homologue of the paradidymis in the male (p. 267). Distension of such a tubule produces a parovarian cyst, characterized by its thin wall and crystal-clear fluid content.

THE PELVIC PERITONEUM

The pelvic peritoneum is draped over the pelvic viscera and invests them in part with a serous coat. Between the viscera it hangs in dependent folds or pouches. These differ somewhat in the two sexes.

Nerve Supply. As elsewhere, the visceral layer has no nerve supply. The parietal peritoneum, on the pelvic walls, is supplied by the obturator nerve (L3).

Pelvic Peritoneum in the Male

From the margin of the pelvic brim the peritoneum descends across the pelvic walls to line the cavity, but nowhere does it descend far enough to reach the pelvic floor. Anteriorly in the midline it descends but little below the upper margin of the symphysis pubis, and then only when the bladder is empty. It passes from the lower part of the anterior abdominal wall and is reflected in a shallow fold to the fundus of the bladder, where it is densely adherent. The fold roofs in the retropubic space (cave of Retzius). If the bladder rises in distension above the symphysis pubis this fold of peritoneum is stripped upwards from the anterior abdominal wall. A trocar can then be introduced through the anterior abdominal wall into the bladder below the fold; in this way extraperitoneal drainage of the bladder may be carried out as a relief measure in cases of urethral obstruction. From the back of the bladder the peritoneum descends in a fold before ascending over the rectum and the hollow of the sacrum. This fold is the **recto-vesical pouch.** In the early fœtus it reaches the pelvic floor between bladder and rectum, but the depths of the fold later become obliterated by fusion of peritoneal surfaces. The fused surfaces persist as a fibrous septum between bladder and prostate in front and the rectum behind—the recto-vesical fascia of Denonvilliers. The fascia is attached above to the recto-vesical fold and below to the pelvic floor (perineal body).

The midline visceral peritoneum, including its folds, passes laterally in a continuous sheet to the side wall of the pelvis and becomes continuous with the parietal peritoneum of the abdominal cavity at the pelvic brim.

The pelvic peritoneal cavity is occupied by coils of the sigmoid colon and the lower part of the ileum, which lie in front of the rectum and above the bladder.

Pelvic Peritoneum in the Female

The basic arrangement of the peritoneum is similar in the female, the only difference being the presence of the uterus and the broad ligaments. From the back of the bladder the peritoneum ascends over the front of the uterus, making a fold called the **utero-vesical pouch** between the two viscera. The visceral peritoneum of the uterus is firmly attached to the back of that organ *and to the posterior fornix* of the vagina, whence it is reflected up over the rectum and sacrum. This deep pouch of peritoneum is the **recto-uterine pouch** (the **pouch of Douglas**). It is opened by incision through the posterior fornix, and this method is employed in draining an intraperitoneal pelvic abscess in the female (Figs. 5.63, 5.64, 5.65).

Each side of the uterus is attached to the side wall of the pelvis by the broad ligament, which extends down towards the pelvic floor (p. 337).

THE PELVIC VESSELS

The pelvic walls and viscera are supplied by branches of the internal iliac artery and drain into tributaries of the internal iliac veins. Just as the commencement of the inferior vena cava lies to the right of the aorta and dorsal to the right common iliac artery, so the internal iliac veins tend to lie to the right and dorsal to their companion arteries. Arteries and veins lie within the parietal pelvic fascia and only their branches that pass out of the pelvis (except the obturator vessels) need to pierce this membranous structure.

The common iliac artery bifurcates at the pelvic brim opposite the sacro-iliac joint (Fig. 5.67). From this point the **internal iliac artery** passes downwards and soon divides into a small posterior and a larger anterior division. The posterior division breaks up into three branches all of which are parietal. The anterior division breaks up likewise into three parietal branches and has in addition, three visceral branches. They are named as follows:

Branches of the Posterior Division (all parietal)

(1) Ilio-lumbar artery
(2) Superior gluteal artery
(3) Lateral sacral artery

Branches of the Anterior Division

A. Parietal
  (1) Inferior gluteal artery
  (2) Internal pudendal artery
  (3) Obturator artery

B. Visceral
  (1) Superior vesical artery
  (2) Inferior vesical artery
  (3) Middle rectal arteries (in the female largely replaced by the *uterine artery*).

Branches of the Posterior Division. The **ilio-lumbar artery** (Fig. 5.59) passes upwards out of the pelvis in front of the lumbo-sacral trunk and behind the obturator nerve. It passes laterally deep to the psoas muscle. Its *lumbar branch* is really the fifth lumbar segmental artery. It passes laterally to supply the psoas and quadratus lumborum muscles and, by its posterior branch, the erector spinæ muscle. This vessel gives a spinal branch into the foramen between L5 and the sacrum.

The *iliac branch* supplies the iliac fossa, i.e., the iliacus muscle and the iliac bone. It extends to the anastomosis around the anterior superior iliac spine (deep and superficial circumflex iliac arteries, ascending branch of lateral femoral circumflex and the upper branch of the deep division of the superior gluteal artery).

The **superior gluteal artery** passes backwards by piercing the pelvic fascia. It passes above S1 in the sacral plexus (the inferior gluteal artery passes below S1) and leaves the pelvis through the greater sciatic foramen above the upper border of piriformis. Its course and distribution in the buttock are considered on p. 147.

The **lateral sacral artery** runs down lateral to the anterior sacral foramina, i.e., in front of the roots of the sacral plexus. In the pelvis it supplies the roots and the piriformis muscle. *Spinal branches* enter the anterior sacral foramina, supply the spinal meninges and the roots of the spinal nerves and pass through the posterior sacral foramina to supply the muscles over the back of the sacrum. The artery takes over the segmental supply from the lumbar arteries and, as might be expected, it is often multiple; the commonest variant is for an upper sacral artery to supply the first two sacral segments and a lower sacral artery to supply the remaining segments in the pelvis (Fig. 5.67).

Branches of the Anterior Division. The **inferior gluteal artery** passes backwards by piercing the parietal

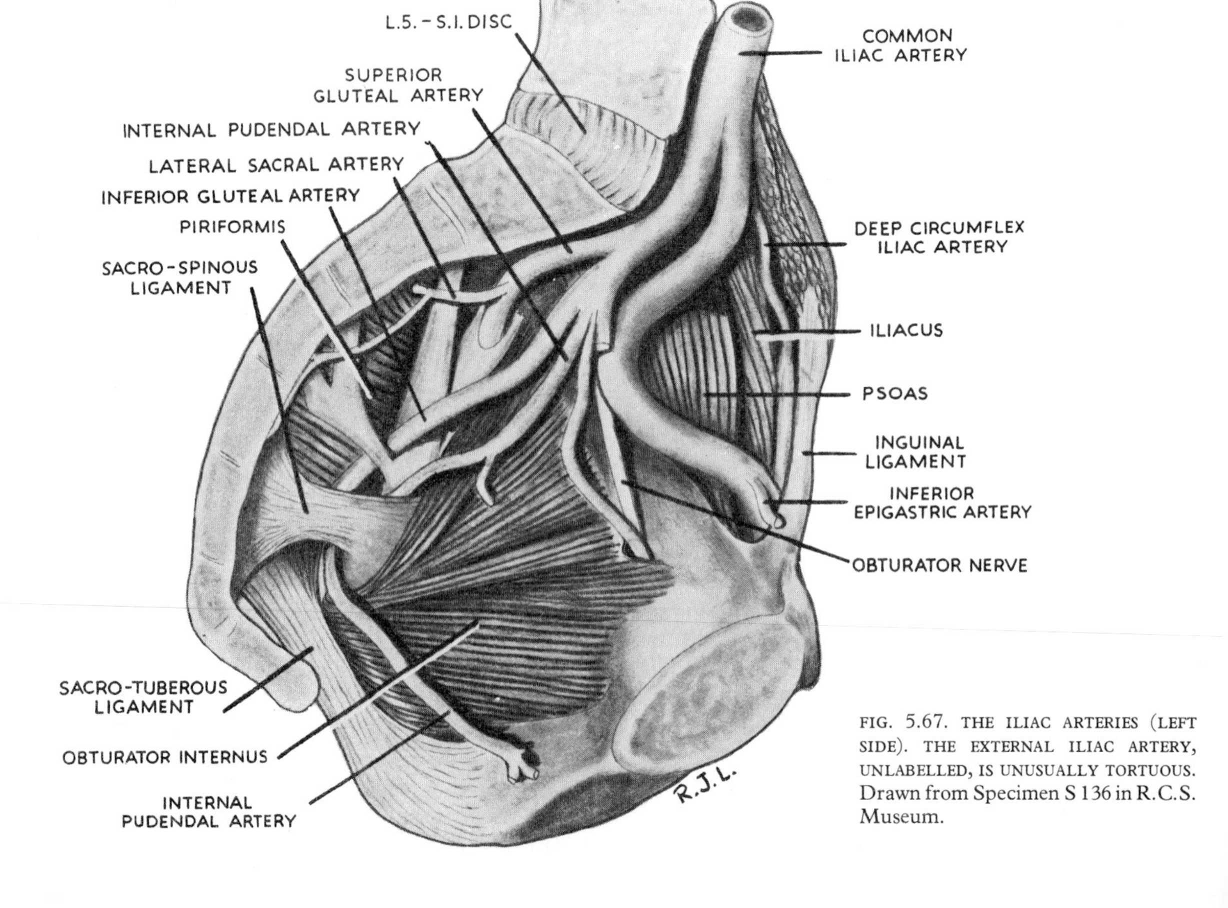

FIG. 5.67. THE ILIAC ARTERIES (LEFT SIDE). THE EXTERNAL ILIAC ARTERY, UNLABELLED, IS UNUSUALLY TORTUOUS. Drawn from Specimen S 136 in R.C.S. Museum.

pelvic fascia, then passes below the S1 nerve root of the sacral plexus (which separates it from the superior gluteal artery). It leaves the pelvis through the greater sciatic foramen below the piriformis muscle; its course in the buttock is considered on p. 147.

The **internal pudendal artery** (Fig. 5.67) pierces the parietal pelvic fascia and passes out of the pelvis through the greater sciatic foramen below the piriformis muscle. It is distributed in the perineum to the anal triangle (p. 348) and the external genitalia (p. 353). It is the artery of the perineum.

The **obturator artery** passes along the side wall of the pelvis below the nerve. The nerve, artery and vein pass through the obturator foramen into the thigh. The artery gives off a small branch to the periosteum on the back of the pubic bone, which anastomoses with the pubic branch of the inferior epigastric artery. In over a third of cases this anastomotic connexion opens up and no obturator artery arises from the internal iliac. Such replacement by the branch from the inferior epigastric artery is named the **abnormal obturator artery** (Fig. 5.63, right side). In its passage from the inferior epigastric artery to the obturator foramen it passes down to one or other side of the upper end of the femoral canal. It usually passes laterally, lying alongside the femoral (i.e., external iliac) vein. When lying medially, alongside the edge of the lacunar ligament, it is vulnerable to injury or division if the ligament is divided in freeing an incarcerated femoral hernia at operation. The abnormal artery is present in a third of cases; when present it lies lateral or medial to the neck of a femoral hernia in the proportion of 10 to 1. That is to say, it is found on the free edge of the lacunar ligament in about 3 per cent (1 in 30) of individuals (the same incidence as the thyroidea ima artery) (Fig. 5.63).

The **superior vesical artery** represents the canalized proximal end of the umbilical artery of the embryo. A relic of the large umbilical artery persists as a fibrous band attached alongside the artery and continued beyond the vascular termination as the medial umbilical fold (Fig. 5.59). The superior vesical artery supplies the fundus of the bladder, and the terminations of vas and ureter if these are not supplied by the inferior vesical artery.

The **inferior vesical artery** supplies the trigone and lower part of the bladder; usually it gives a branch to the terminal part of the ureter and provides the artery to the vas.

The **middle rectal artery** is badly named. It is characterized by three features: (1) It is often absent, especially in the female; (2) very little of its blood goes to the rectum, and that only to the muscle coats thereof; and (3) most of its blood goes to the prostate.

It branches from the anterior division of the internal iliac artery as a single vessel, but immediately breaks up into a leash of vessels, a few of which supply the muscle coats of the rectum while most of them pass forwards into the prostate. The main prostatic artery is variable in its origin (middle rectal, obturator) but constant in its entry to the lateral part of the prostate at the junction with the base of the bladder (PLATE 21). The prostatic branches, absent in the female, are replaced by the uterine and vaginal arteries.

The **uterine artery** is a large vessel which crosses the pelvis in the base of the broad ligament. It passes *above the ureter*. At the cervix it turns upwards closely applied to the muscle thereof and runs alongside the uterus in the broad ligament. At the entrance of the uterine tube it anastomoses end on with the tubal branch of the ovarian artery (Fig. 5.63).

The **vaginal artery,** usually a separate branch of the internal iliac artery, nevertheless often branches from the uterine artery. It supplies the very vascular walls of the upper part of the vagina.

The Veins of the Pelvis

The rectal plexus drains partly into the superior rectal veins and so by the inferior mesenteric vein into the portal system. But part of the rectal plexus drains into the internal iliac veins. This part communicates with the uterine venous plexus or with the prostatic-vesical venous plexus as these pass into the internal iliac veins. The *internal vertebral plexus* (p. 493) drains through the anterior sacral foramina into the lateral sacral veins, and so into the same internal iliac veins. There are no valves in this system. Sudden increase in pelvic pressure (as in coughing) may be momentarily more than the inferior vena cava can accommodate, and this drives blood backwards up the internal vertebral plexus, into posterior intercostal veins and by azygos veins into the superior vena cava, by-passing the diaphragm. It is conceivable that emboli from disease of the pelvic viscera can find their way by occasional reflux blood flow into the vertebræ. In this way secondary carcinomatous deposits may appear in the vertebræ from primary growths in any of the pelvic viscera.

THE NERVES OF THE PELVIS

The nerves to be seen in the pelvis are the obturator, which passes along the side wall of the pelvis to reach the thigh, the sacral plexus and its branches in the pelvis, the sacral part of the ganglionated trunk of the sympathetic and the hypogastric plexuses which provide the autonomic supply of the pelvic viscera.

The **obturator nerve** is a branch of the lumbar plexus formed within the substance of the psoas muscle from the anterior divisions of the second, third and fourth lumbar nerves (anterior primary rami). It is the nerve of the adductor compartment of the thigh, which it reaches by piercing the medial border of psoas and passing straight along the side wall of the pelvis to the obturator foramen. It crosses the pelvic brim medial to the sacro-iliac joint (i.e., on the ala of the sacrum) and runs forward between the internal iliac vessels and the fascia on the obturator internus muscle. It appears in the angle between the internal and external iliac vessels, in which part of its course it is separated from the normally situated ovary only by the parietal peritoneum lining the pelvic wall. Pain from the ovary is frequently referred along the nerve to the skin on the medial side of the thigh. This may be less an irritation of the main nerve trunk than irritation or inflammation of the parietal peritoneum, which is here supplied by the obturator nerve.

Obturator artery and vein converge to the obturator foramen, in which the nerve lies highest, against the pubic bone (Fig. 5.67) with the artery and vein beneath it. The nerve divides in the obturator foramen into anterior and posterior divisions; the former passes anterior to the upper border of obturator externus, while the posterior division, first giving off a branch to supply the obturator externus, pierces the muscle. The distribution in the thigh is considered on p. 144.

The **accessory obturator nerve** is incorrectly named. Its only characteristic in common with the obturator nerve is that it leaves the medial border of the psoas muscle. Its more important bony relation is shared with the femoral nerve; like the femoral nerve it passes over, not under, the pubic ramus and, like the femoral nerve, it is derived from *posterior* and not anterior divisions of the nerves of the lumbar plexus. It should have been named the 'accessory femoral' nerve. It is formed in the substance of the psoas from the posterior (not anterior) divisions of the third and fourth lumbar nerves and it supplies the pectineus muscle. It is present in only one-third of individuals.

The Sacral Plexus

Not all the lumbar nerves are used up in the formation of the lumbar plexus. Much of L4 and all of L5 enter the sacral plexus. After L4 has given off its branches to the lumbar plexus it emerges from the medial border of psoas and joins the anterior primary ramus of L5 to form the **lumbo-sacral trunk.** This large nerve passes over the ala of the sacrum and crosses the pelvic brim, separated from the obturator nerve by the ilio-lumbar artery and veins. It descends to join the anterior primary rami of the upper four sacral nerves in the formation of the sacral plexus (Figs. 5.59 and 5.78, p. 358).

The **sacral plexus** is a broad triangular structure formed by the junction of the nerves lateral to the anterior sacral foramina. It rests upon the piriformis muscle and is covered anteriorly by the strong membrane of parietal pelvic fascia which invests that muscle. Anterior to the fascia the lateral sacral arteries and veins lie in front of the sacral nerves. At a higher level the common iliac vessels lie over the lumbo-sacral trunk. S1 separates the superior and inferior gluteal arteries. The ureter, in front of the iliac vessels (Fig. 5.59), crosses the upper part of the plexus and in front of all are the parietal pelvic peritoneum and pelvic viscera.

The sacral nerves give off certain branches and then divide, as does the lumbo-sacral trunk, into anterior and posterior divisions which thereupon branch and re-unite to form the nerves for the supply of flexor and extensor compartments of the lower limb. The branches of the sacral plexus will be considered under these three headings. (A summary of their distribution is given on p. 357.)

Branches from the Sacral Nerves. These are six, three from behind and three from in front of the anterior primary rami. They are: twigs to piriformis (S1, 2), the perforating cutaneous nerve (S2, 3) and the posterior cutaneous nerve of the thigh (S2, 3) branching from behind, and the pelvic parasympathetics (S2, 3), the pudendal nerve (S2, 3, 4) and the perineal branch of S4 from in front.

The **piriformis** is supplied by separate twigs which pass backwards from S1 and 2.

The **perforating cutaneous nerve** arises from the posterior surfaces of S2 and 3. It pierces the sacro-tuberous ligament and the lower border of gluteus maximus and supplies the skin of the buttock over the area where the two buttocks are just losing contact.

The **posterior cutaneous nerve of the thigh** is formed by branches that pass backwards from S2 and 3, with a small contribution from S1. The nerve passes laterally to leave the lower border of piriformis behind the sciatic nerve, which separates it from the ischium. It thus enters the buttock (p. 148).

The **pelvic parasympathetics (nervi erigentes)** arise by several rootlets from the anterior surfaces of S2 and 3 (or 3 and 4). They pass forward into the inferior hypogastric plexuses where they mix with the sympathetic nerves and are distributed to the derivatives of the cloaca (p. 344).

The **pudendal nerve** (S2, 3, 4) arises from the anterior surfaces of its parent nerves. The three twigs unite to form a nerve which passes back between the piriformis and coccygeus muscles, medial to the pudendal vessels. In the buttock it appears between the piriformis and the sacro-spinous ligament, and curls around the latter to run forward into the ischio-rectal fossa (p. 148).

The **perineal branch of S4** runs forward on the

coccygeus and enters the perineum by passing between that muscle and ilio-coccygeus, first supplying each on its pelvic surface. It is distributed to the external anal sphincter and peri-anal skin (p. 349).

Branches from the Anterior Divisions. These nerves are destined for the flexor compartment of the lower limb. They are the tibial part of the sciatic nerve and the nerves to obturator internus and quadratus femoris.

The *tibial* (medial popliteal) part of the **sciatic nerve** is a big branch from the anterior divisions; it is formed by union of branches from all five anterior divisions (L4, 5, S1, 2, 3). It usually joins the extensor compartment nerve (the common peroneal) in the pelvis and the sciatic nerve so formed leaves the pelvis below the lower border of piriformis, lying on the ischium in the greater sciatic notch, lateral to the ischial spine. Its course in the buttock is considered on p. 148.

The **nerve to obturator internus** (L5, S1, 2) also supplies the superior gemellus. It leaves the pelvis, lateral to the pudendal vessels, below the piriformis (p. 148).

The **nerve to quadratus femoris** (L4, 5, S1) also supplies the inferior gemellus and the hip joint. It leaves the pelvis in front of the sciatic nerve, which holds it down on the ischium (p. 148).

Branches from the Posterior Divisions. These are the nerves of the extensor compartment of the lower limb. They are the common peroneal part of the sciatic nerve and the superior and inferior gluteal nerves.

The *common peroneal* (lateral popliteal) part of the **sciatic nerve** is formed by union of branches from the posterior divisions of L4, 5, S1, 2. There is no S3 in the extensor compartment. It usually joins the tibial (medial popliteal) part to form a combined nerve (p. 148), but not infrequently it fails to do so. In these cases the nerve pierces the lower border of piriformis (Fig. 3.28, p. 147).

The **superior gluteal nerve** is formed from the posterior divisions of L4, 5 and S1. It passes out of the pelvis above the piriformis muscle (p. 147).

The **inferior gluteal nerve** is formed from the posterior divisions of L5 and S1 and 2. It passes below the lower border of piriformis into the buttock (p. 147).

The **coccygeal plexus** consists of a minor mingling of a branch from S4 with S5 and the coccygeal nerve. Branches supply the coccygeus muscle and the postanal skin near the tip of the coccyx.

The Sacral Sympathetic Trunks

The trunks cross the pelvic brim behind the common iliac vessels and lie in the concavity of the sacrum medial to the anterior sacral foramina. Each has characteristically four ganglia. The trunks lie parallel with the lateral margin of the sacrum and converge at the front of the coccyx to unite at a small swelling, the ganglion impar (Fig. 5.59, p. 328).

Somatic branches are given off to all the sacral nerves (lower limb and perineum), and visceral branches leave the upper ganglia to mingle with the inferior hypogastric plexus (pelvic viscera).

The Inferior Hypogastric Plexus (Pelvic Plexus)

It is more concise to keep the old name of 'hypogastric plexus' for the small plexus on the sacral promontory (no ganglia here) and the old name 'pelvic plexus' for the large plexus alongside the pouch of Douglas (ganglia for sympathetic relay are here). The new names of

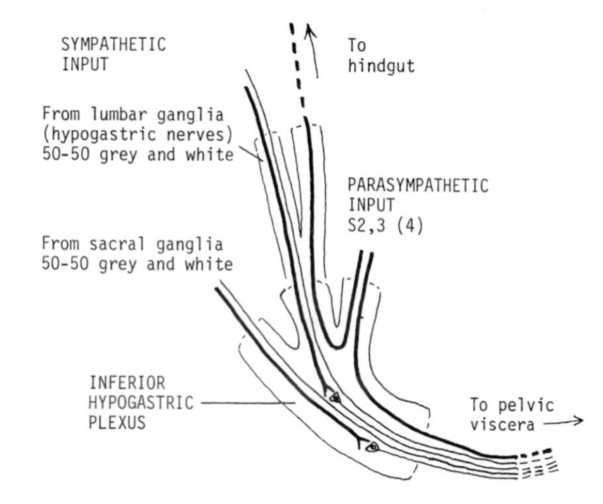

FIG. 5.68. THE CONSTITUENTS OF THE INFERIOR HYPOGASTRIC PLEXUS. White fibres are shown in heavy line, grey fibres in slender line. Cell bodies of white fibres are in the spinal cord. Cell bodies of grey fibres are in the sympathetic ganglia in the lumbar and sacral chain.

'superior hypogastric plexus' and 'inferior hypogastric plexus' can be confusing. The plexuses are shown in the dissection illustrated in PLATE 20, p. *x*, but the pelvic plexus (see the pointer) is seen edge on, much foreshortened.

The inferior hypogastric plexus is an autonomic plexus on the side wall of the pelvis, lateral to the rectum and within the parietal pelvic fascia. Its sympathetic components are derived from the superior hypogastric plexus (p. 314) and by branches from the upper sacral ganglia of the sympathetic trunk. Its parasympathetic components are carried by branches from the second and third (or third and fourth) sacral nerves; these are the pelvic parasympathetic nerves (nervi erigentes). The plan of its constituent parts is indicated in Fig. 5.68.

The plexus is a coarse, flat meshwork, enlarged in places by ganglia; it measures nearly 2 inches (5 cm) in antero-posterior and about one inch (2 cm) in vertical dimension. About half the fibres in the hypogastric nerves are medullated (preganglionic) and they relay in the ganglia of the inferior hypogastric plexus (Fig. 5.68). The remaining sympathetic fibres and all the parasympathetic (nervi erigentes) fibres pass through without relay. The parasympathetic motor and secreto-motor fibres relay in the walls of the viscera.

The visceral branches of the inferior hypogastric plexuses run in leashes of nerves bound up in condensed bundles of fibrous tissue. In their passage through the cellular fibrous tissue above the pelvic floor they are accompanied by visceral branches of the internal iliac artery and vein; these neuro-vascular bundles in their fibrous tissue condensations produce certain of the named 'ligaments' of the pelvic viscera (e.g., lateral ligaments of the bladder, of the cervix, of the rectum). While it is well established that sympathetic vaso-constrictor fibres accompany all arteries to the pelvic viscera, much ignorance still exists as to the exact course of certain of the motor and sensory fibres to these organs. In general it appears that the muscles of the bladder (detrusor muscle) and rectum are innervated by the nervi erigentes, the smooth muscle of the internal sphincter of the bladder through the hypogastric (presacral) nerves and the smooth muscle of the internal sphincter of the anal canal by branches from the sacral ganglia which pass through the pelvic plexuses. The course of the afferent fibres is less well understood. Normal sensations of distension of bladder and rectum probably pass through the nervi erigentes; pain fibres probably take the same course, though some may pass in the hypogastric nerves as well. Pain fibres from the body of the uterus appear to lie mainly in the hypogastric nerves, the cell bodies being in the posterior root ganglia of Th. 11 and 12 spinal nerves. Pain fibres from the cervix travel in the nervi erigentes (parasympathetic nerves).

As well as being motor to the smooth muscle of the pelvic viscera the pelvic parasympathetic nerves (nervi erigentes) supply the colon below the splenic flexure. These branches run up from the inferior hypogastric plexus and leave the superior hypogastric plexus by a separate trunk which joins the inferior mesenteric artery, by which they are distributed to the descending and pelvic colon.

Thus in the pelvis the two parts of the autonomic system conform to their *motor* functions higher up. The pelvic parasympathetics are motor to the emptying muscle of the bladder, and of the gut from splenic flexure to rectum. They are secreto-motor to the gut. The sympathetics are motor to the smooth-muscle sphincters of the bladder and anal canal; they are motor, too, to the seminal vesicles. Recent work indicates that the sympathetic supplies motor fibres to the uterine muscle.

The *sensory* supply to abdominal viscera and gonads is sympathetic, and this includes the descending colon. But there is a change of sensory supply to the derivatives of the cloaca. Bladder, rectal ampulla and anal canal receive sensory fibres from the pelvic parasympathetics. So, too, do the cervix uteri and upper vagina, but the body of the uterus receives its sensory supply from the sympathetic. Thus pain from the prostate or rectum, mediated by the pelvic parasympathetics to S2 and 3 segments, may be referred to these dermatomes (posterior cutaneous nerve of thigh) and so be mistaken for sciatica.

THE PERINEUM

The perineum consists of that part of the pelvic outlet caudal to the pelvic diaphragm (levator ani and coccygeus). A line joining the *anterior* parts of the ischial tuberosities divides the perineum into a large posterior anal triangle and a smaller anterior urogenital triangle (Fig. 5.69).

ANAL TRIANGLE

The anal triangle contains the anal canal and the ischio-rectal fossæ with their contents. Its sides are formed by the sacro-tuberous ligaments (covered by the lower border of gluteus maximus) and its base is formed by the line between the anterior parts of the ischial tuberosities. Its contents are the same in each sex.

Anal Canal

The anal canal is some 3 cm (over an inch) long. Like the rest of the gut it is a tube of muscle, but the fibres are all circular, consisting of the internal (smooth) and the external (striated) sphincters. These sphincters hold it continually closed except for the temporary passage of flatus and fæces. The junction of rectum and anal canal is at the pelvic floor, i.e., at the level where the pubo-rectalis muscle (p. 325) clasps the gut and angles it forwards. From this right-angled junction with the rectum the anal canal passes downwards and somewhat backwards to the skin of the perineum.

The Lining of the Anal Canal. The canal is lined with mucous membrane in its upper two-thirds and skin in its lower one-third. The junction of the two is abrupt; it is at Hilton's white line (Fig. 5.72). This line is the site of attachment of the fascia derived

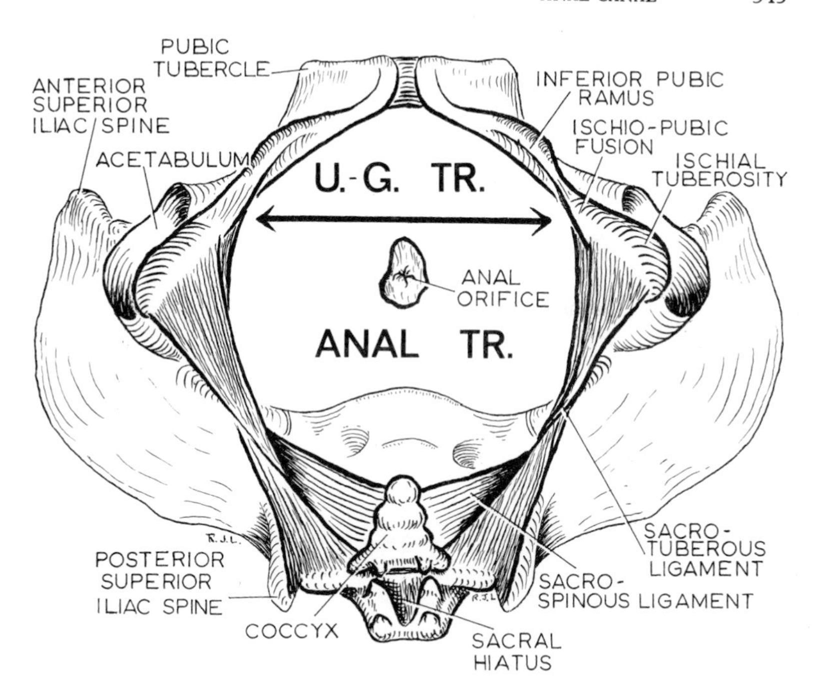

FIG. 5.69. THE 'TRIANGLES' OF THE PELVIC OUTLET. The dividing line joins the ischio-pubic fusions. The (anterior) uro-genital triangle houses the external genitalia. The (posterior) anal 'triangle' is really a pentagon; it contains the anal canal and the ischio-rectal fossæ. The position of the anal orifice is shown; it is anterior to the line joining the convexities of the ischial tuberosities (cf. Fig. 5.74).

from the longitudinal muscle coat of the rectum (p. 347) and it is a watershed dividing upper and lower zones of arterial supply and venous and lymphatic return. It separates also two zones of different nerve supply. It is not 'white' on inspection, but was so described by Hilton on account of its relative avascularity. The part above Hilton's line (mucous membrane) is derived from the endoderm of the cloaca, while the part below the line (skin) is derived from the ectoderm of the anal pit or proctodæum.

The **cloacal part,** is lined with typical large-gut mucous membrane containing mucous crypts and covered with columnar epithelium (goblet cells). There are mucous glands in the submucosa. Occasionally in the lower part, between the annulus hæmorrhoidalis (*v. inf.*) and Hilton's line, there are no mucous crypts and the epithelium is in several layers. This is the zone of so-called 'transitional' epithelium; it is not, however, the urine-proof transitional epithelium of the urinary passages, but a thinned out epidermis extending up from the anal part. It is a modified squamous stratified epithelium. This zone is called the *pecten* by clinicians.

The **anal-pit part** is lined with thin hairless skin. It is surfaced with squamous stratified epithelium.

The *arterial supply* of the cloacal part is derived from the superior rectal artery (the artery of the hind gut) as far as the white line, where the hind gut ends. The lower third (skin) below Hilton's line, is supplied by the inferior rectal artery (which also supplies the sphincters). These arteries do not anastomose with each other.

The *veins* of the upper part (mucous membrane) drain upwards into the submucous plexus of the ampulla of the rectum (*v. inf.*). The veins of the skin part, below Hilton's line, drain downwards into the inferior rectal vein or into tributaries of the saphenous vein. They do not communicate with the veins of the cloacal part. Hilton's line is a venous watershed.

The *lymphatics* of the cloacal part pass upwards from lymphatic follicles in the mucous membrane to join those of the rectum (p. 329). Those of the skin part

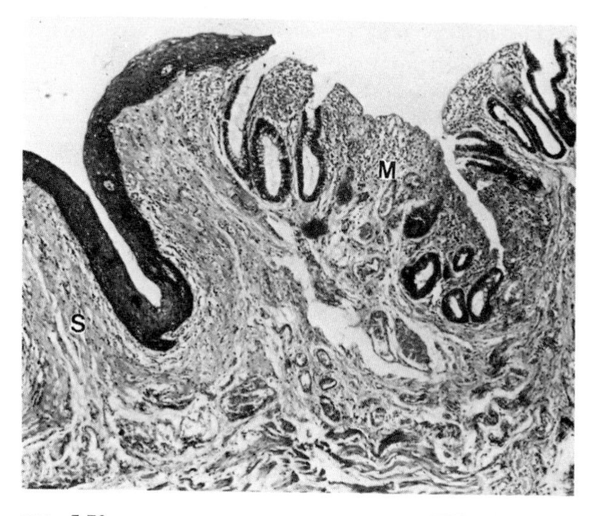

FIG. 5.70. THE ANO-CUTANEOUS JUNCTION (Hilton's Line). Note the abrupt change from columnar epithelium (goblet cells) to squamous stratified epithelium (human). The columnar epithelium has been shed from the surface and persists only in the mucous crypts (a common artefact). S. Skin. M. Mucous membrane.

drain with the rest of the perineum into the medial group of superficial inguinal nodes.

The *nerves* of the cloacal part are autonomic, from the inferior hypogastric plexuses, and this part of the mucous membrane is relatively insensitive to touch though it registers pressure (e.g., of the examining finger or a rectal tube) and it can *distinguish between fæces and flatus*. The nerves of the skin part are somatic, from the inferior rectal nerve and this skin is highly sensitive (e.g., to the pain of a fissure in ano).

Veins of the Anal Canal. The mucous membrane of the upper two-thirds contains a rich plexus of veins. These drain upwards by vertical channels which, when full of blood, raise ridges in the mucous membrane known as *anal columns*. Three of these veins, situated at 4, 7 and 11 o'clock when the patient is viewed in the lithotomy position, are apt to become varicose as the three 'primary' hæmorrhoids. The veins pass up in the submucosa to join the submucous plexus in the ampulla of the rectum, whence the blood is diverted to either the portal or systemic systems (p. 328).

The anal columns are joined by cross channels of anastomosing veins which raise small mucosal folds known as anal '*valves*'. These anastomosing cross channels form a venous ring known as the *annulus hæmorrhoidalis*, or *zona hæmorrhoidalis* (Fig. 5.72). The situation of this dentate ring of valves is variable. It often lies low in the canal, close to the mucocutaneous junction (Hilton's line), but wherever it may be it is always in the large bowel (cloacal) part of the anal canal. It never lies at the junctional part, and it is wrong to imagine that it represents the site of the cloacal membrane of the embryo.

The anal columns vary in prominence according to the amount of their contained blood. The anal valves remain constant irrespective of the amount of blood in the annulus. The tiny curved ridges of mucous membrane produce pockets above them; these may become infected.

The Anal Sphincters. The anal canal is always closed except for the passage of flatus or fæces. It is held closed by a sphincteric *tube* of muscle. The tube is 3 cm (over an inch) long, lined as described above. The muscle is in two distinct sphincteric entities, internal and external. Each occupies two-thirds of the canal, so that they overlap at the middle third. The internal sphincter is of smooth muscle and lies around the upper two-thirds of the canal. The external sphincter is of striated muscle; it clasps the lower part of the internal sphincter and surrounds the lower two-thirds of the anal canal.

Internal Sphincter. This occupies the upper two-thirds of the anal canal, i.e., down to Hilton's line. It is the thickened lower end of the inner circular muscle of the rectum, with which it is continuous. It is made of smooth muscle. It is innervated from the inferior hypogastric plexuses; sympathetic stimulation contracts the muscle, parasympathetic stimulation relaxes it. It is under a voluntary control similar to that of the smooth muscle of the bladder. The sphincter is relatively weak, and is not competent when acting alone; at least some of the deepest part of the external sphincter is essential for complete continence of flatus and fæces. The internal sphincter is surrounded by a loose and distensible fibrous sheath, which is the downward continuation of the longitudinal muscle coat of the rectum (*v. inf.*).

External Sphincter. This tube of striated muscle surrounds the lower two-thirds of the anal canal. It consists of three parts (each a 'ring' of muscle) lying adjacent to each other in series. The three rings lie superficial, middle and deep, but, unfortunately, they are not so named. The middle 'ring' is, very confusingly, named 'superficialis'; the superficial ring is named 'subcutaneous', and the two are separated by a fascial septum. This fascia is a downward prolongation of the longitudinal muscle of the rectum. It invests the internal sphincter, separating it from the surrounding external sphincter, and turns inwards to become attached to Hilton's white line. From the same level a sheet of this peri-anal fascia passes outwards to the pudendal canal and separates the peri-anal space from the ischio-rectal space (p. 347). It likewise separates the 'subcutaneous' from the 'superficial' part of the external sphincter (Fig. 5.72).

Corrugator Cutis Ani Muscle. This small muscle consists of thin slips of smooth muscle fibres which radiate out from Hilton's white line to be attached to the peri-anal skin. It is part of the panniculus carnosus (p. 3) and has nothing to do with the external sphincter. Its contraction shrinks and therefore cleans the peri-anal skin in many quadrupeds; it is an almost functionless vestige in man.

Subcutaneous External Sphincter. This is a thick ring of muscle, not attached to bone. It lies immediately beneath the skin and corrugator fibres, and is separated from the superficial external sphincter by the fascial attachment to the white line mentioned above. It is easily palpated by the examining fingertip. It is supplied by the inferior rectal nerve (S3, 4).

Superficial External Sphincter. This is the *middle* of the three parts of the external sphincter. It is an elliptical muscle, attached to the tip of the coccyx posteriorly (Fig. 5.74) and to the perineal body anteriorly. It is supplied by the perineal branch of S4.

Deep External Sphincter. This is again an annular muscle, not attached to bone. It encircles the lower part of the internal sphincter. Posteriorly it blends with the embracing loop of the pubo-rectalis muscle (p. 325), but anteriorly it forms a complete ring which

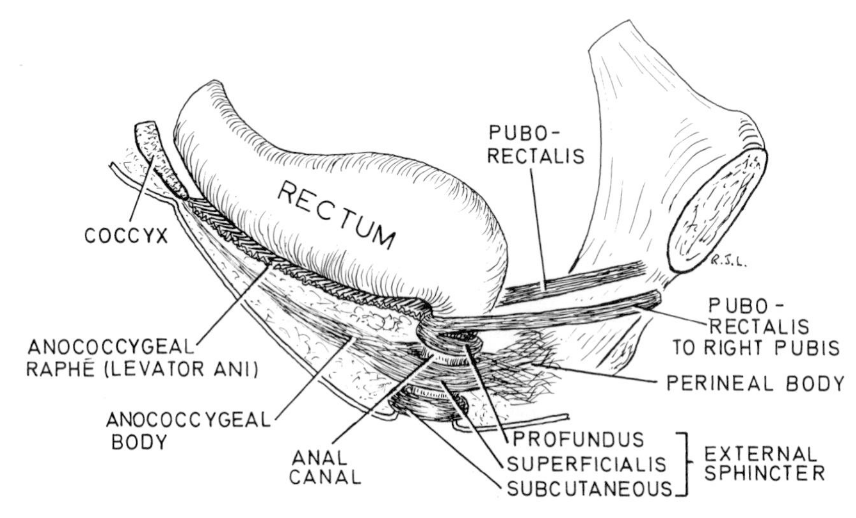

FIG. 5.71 THE PUBO-RECTAL SLING AND EXTERNAL ANAL SPHINCTER (lateral view).

separates from the pubo-rectalis and fills in the space between the two halves of that muscle in front of the recto-anal junction (Fig. 5.71).

It is essential to the continence of flatus and fæces. Posteriorly the whole ano-rectal ring (pubo-rectalis and profundus sphincter) functions, but anteriorly there is only the profundus to maintain continence; hence the danger of anterior lacerations of the anal canal (e.g., obstetric). The profundus part of the external sphincter is supplied by the inferior rectal nerve (S3, 4).

The peri-anal fascia surrounding the internal sphincter is expansile to accommodate the passage of a fæcal mass, but it cannot be stretched longitudinally. Many of its fibres (not illustrated in Fig. 5.72) traverse the subcutaneous sphincter to become attached to peri-anal skin. Thus can the longitudinal muscle fibres of the *rectum* retract the *anal canal* proximally over an extruding fæcal mass.

The Ischio-rectal Fossa

This is a wedge-shaped space filling in the lateral part of the anal triangle, and extending forwards into the uro-genital triangle (p. 352). It is filled with soft fat which forms 'dead-space' into which the anal canal can expand during defæcation. Its lateral wall is formed by the fascia over the lower part of obturator internus, the falciform margin of the sacro-tuberous ligament, and the tuber ischii. Medially the two fossæ are separated by the perineal body, the anal canal and the ano-coccygeal body, and they are roofed in by the downward sloping levator ani muscles of the pelvic floor (Fig. 5.72).

The junction of the rectum and anal canal is slung in the pubo-rectalis muscle; the rectum is wholly in the pelvic cavity. The name ischio-*rectal* fossa is a misnomer, and the inferior 'rectal' vessels and nerves supply no rectum but only the anal canal, peri-anal space and skin.

Ano-coccygeal Body. The fibres of the ilio-coccygeus and of the pubo-coccygeus interdigitate in front of the coccyx in a midline raphé which extends from the tip of the coccyx to the ano-rectal junction. Raphé and skin diverge from each other as they pass forwards to the upper and lower ends of the anal canal. In the midline space between them is a fibro-muscular mass of tissue called the ano-coccygeal body, which separates the two ischio-rectal fossæ behind the anal canal (Fig. 5.74). Fibres of the intermediate (superficialis) part of the external sphincter traverse the ano-coccygeal body to become attached to the tip of the coccyx. Lateral to this the sacro-tuberous ligament limits the fossa.

Peri-anal Space. A prolongation from the longitudinal muscle of the rectum passes downwards as a fascial membrane which splits and is inserted into Hilton's white line internally. An outer prolongation reaches the side wall of the pelvis, at the pudendal canal. This delicate membrane, the *peri-anal fascia*, separates the depths of the ischio-rectal fossa from a shallow subcutaneous *peri-anal space* (Fig. 5.72). The fat in the latter space is contained in small loculi which are separated by fibrous septa almost complete (contrast the large loculi in the ischio-rectal space separated by very incomplete septa). Hence infection in this space gives rise to considerable tension as swelling occurs, with consequent great pain. The peri-anal space, in other words, consists of ordinary subcutaneous fat, and the peri-anal fascia separates it from the fat in the ischio-rectal space.

The Ischio-rectal Space. The fat in this space is arranged in large loculi which are but incompletely separated by delicate septa. Infection in this space (by

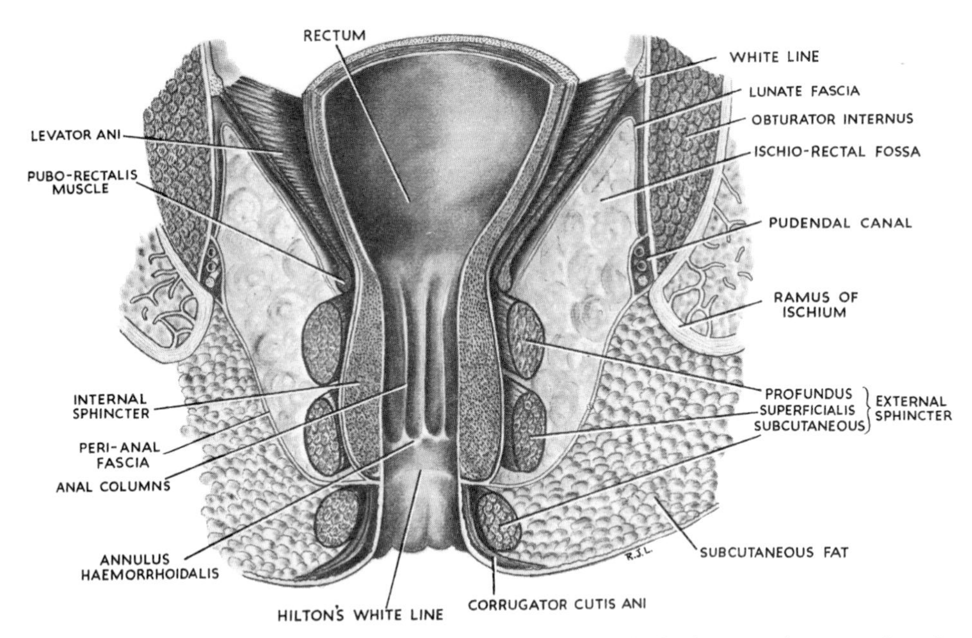

FIG. 5.72. THE ANAL CANAL AND ISCHIO-RECTAL FOSSÆ IN CORONAL SECTION. Author's preparation, somewhat schematic.

far the greater part of the whole ischio-rectal fossa) can cause swelling without tension and with, therefore, a minimum or absence of pain. The space allows dilatation of the anal canal during defæcation. Although the rectum lies above the pelvic floor it can dilate by pressing the sloping levator ani into the space (Fig. 5.72 is drawn with the rectum and anal canal somewhat distended into the space). Similarly the vagina can dilate into the space, especially during parturition, when the passage of the fœtal head obliterates the space.

Pudendal Canal. The sacro-tuberous ligament is attached to the medial half of the lower part of the tuber ischii. Its upper edge is prolonged forward on the medial surface of the ischium as the *falciform ligament*. Above the falciform ligament is the dense fascia on the obturator internus. Here lies a fibrous canal containing the internal pudendal vessels and nerve. This canal, the pudendal canal, is formed from the lateral prolongation of the delicate peri-anal fascia, which splits and thickens to enclose the pudendal neuro-vascular bundle. The pudendal canal connects the lesser sciatic foramen to the posterior edge of the perineal membrane (PLATE 13). In former times the pudendal canal was euphoniously called Alcock's canal.

Lunate Fascia. Arching over the upper margin of the ischio-rectal fat, separating it from the areolar tissue on the lower surface of the levator ani muscle, is the lunate fascia. It commences laterally at the pudendal canal and fades out medially over the profundus division of the external sphincter ani muscle (Fig. 5.72). At its anterior extremity it is prolonged forwards to fuse with the areolar tissue on the lower surface of levator ani muscle. This situation is anterior to the posterior end of the perineal membrane; thus each ischio-rectal fossa possesses a forward prolongation into the uro-genital triangle on the lateral aspect of the membranous urethra (male) (Fig. 5.74) or vagina (female) along the inferior pubic ramus.

The Perineal Body

Lying in front of the anal canal is a fibro-muscular mass of tissue fixed to, and forming part of, the pelvic floor. It is composed chiefly of the interdigitating fibres of pubo-prostaticus, but is incremented by both transversus perinei muscles and the 'superficial' part of the external anal sphincter. It extends from the level of the pelvic floor to the skin of the perineum, plugging the space between right and left ischio-rectal fossæ (Fig. 5.74). It is indispensable to the support of the pelvic viscera (see p. 353).

Nerves and Vessels of the Ischio-rectal Fossa. The **pudendal nerve** and **internal pudendal vessels** leave the pelvis through the greater sciatic foramen, passing beneath the lower border of piriformis muscle to reach the buttock. Their course in the buttock is short. They turn and enter the lesser sciatic foramen, the vessels passing over the tip of the spine of the ischium, the nerve more medially over the sacro-spinous ligament. They enter the pudendal canal, which is in continuity with the lesser sciatic foramen, and run forwards in it to supply the perineum (PLATE 13).

Note that the nerve to obturator internus, which has crossed the base of the ischial spine, is now above this level and, moreover, is beneath the fascia over the

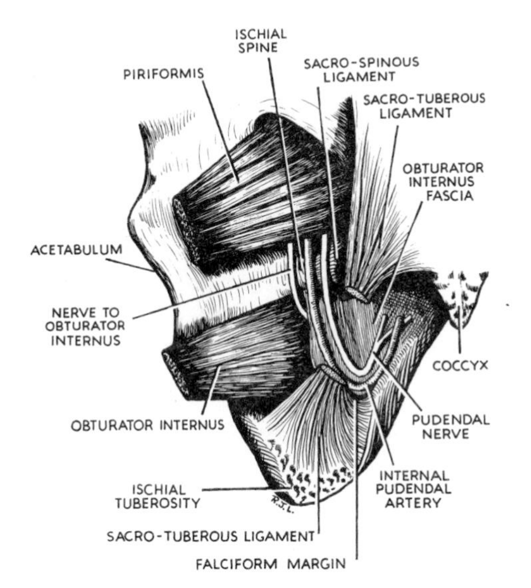

FIG. 5.73. POSTERIOR VIEW, LOOKING FORWARDS INTO THE LEFT ISCHIO-RECTAL FOSSA. THE SACRO-TUBEROUS LIGAMENT IS DIVIDED TO EXPOSE THE PUDENDAL CANAL.

muscle, and sinks into the fleshy fibres to innervate them (Fig. 5.73). From the neuro-vascular bundle in the pudendal canal the **inferior rectal artery and vein** and the **inferior rectal nerve** are given off almost at once. Together they arch up obliquely forwards over the lunate fascia and then curve downwards and medially (PLATE 13, p. *vi*). They break up into multiple branches to supply the external sphincter and peri-anal skin. They also give branches of supply to the middle part of levator ani. Note that the inferior rectal vessels and nerves in their oblique passage between the ischial tuberosity and the anal canal *arch convexly upwards*, against the side wall and roof of the fossa. They do not penetrate the ischio-rectal fat and are safe from injury during incision into the fossa (Fig. 5.74 and PLATE 13).

Do not be misled by the name inferior 'rectal' nerve and vessels. If their former names of inferior 'hæmorrhoidal' were unfortunate, their present names are wrong. They supply no part of the rectum.

Towards the anterior part of the anal triangle the pudendal nerve divides into the **dorsal nerve of the penis** and the **perineal nerve;** these now lie above and below the perineal membrane (PLATE 13). The vessels and dorsal nerve of the penis enter the deep perineal pouch; the perineal nerve, first giving a branch into the deep pouch, enters the superficial pouch. The *perineal branch of S4* enters the posterior part of the fossa between coccygeus and ilio-coccygeus muscles. It supplies the intermediate part of the external sphincter called superficialis, gives a sympathetic branch of supply to the dartos muscle, and ends by supplying the peri-anal skin. It runs in company with the inferior rectal nerve in the space between the lunate fascia and lower surface of levator ani.

Cutaneous Nerves. The skin of the anal triangle is supplied by the inferior rectal nerve (S3, 4), the perineal branch of S4, and some twigs from the coccygeal plexus (S5).

URO-GENITAL TRIANGLE

This triangle is contained between the ischio-pubic rami and the line passing between the anterior parts of the ischial tuberosities. Its contents differ in the two sexes.

The Male Triangle

The **perineal membrane** (*O.T.* **triangular ligament**) is an unyielding sheet of fibrous tissue which forms the basis upon which the penis and penile musculature are fixed; below it the scrotum depends, above it the membranous urethra lies, surrounded by the sphincter urethræ, below the apex of the prostate. It is pierced by the urethra and by foramina for nerves and vessels. It is attached to the ischio-pubic rami from the subpubic angle back to the level of the anterior part of the ischial tuberosities, along a ridge which lies on the inner part of the medial surface of each ramus. Its antero-posterior extent is almost 1½ inches (3·5 cm). The fascia of Colles is attached to its posterior margin.

The **superficial perineal pouch** is a name given to the space enclosed between the perineal membrane and the fascia of Colles. It contains the testes and their spermatic cords as well as the penis and the muscles of the corpora. In a word it constitutes the external genitalia.

The Root of the Penis (Fig. 5.74). The urethra pierces the perineal membrane in the midline an inch (2·5 cm) below the subpubic angle. It is surrounded by the **bulb** of the penis, which is firmly attached to the inferior surface of the membrane. The bulb consists of cavernous tissue enclosed in a fibrous membrane. It is continued anteriorly as the corpus spongiosum. Posteriorly it expands in bulbous shape, with a midline notch. At the sides the **crura** of the corpora cavernosa arise from the angle between the perineal membrane and the everted margin of the pubic ramus. They are conical in shape with their apices pointing posteriorly; anteriorly they are continued into the corpora cavernosa which become connected to the corpus spongiosum at the subpubic angle, from which point the penis becomes free. The bulb of the penis transmits the ducts of the bulbo-urethral glands (after they have pierced the perineal membrane) to the urethra.

The **blood supply** to the root of the penis reaches the bulb via the arteries to the bulb, and the crura via the deep artery of the penis. These arteries pierce the perineal membrane from above.

The Penile Musculature. The bulb and each crus of the penis are provided with a penile muscle; in addition there is a transverse muscle along the margin of the perineal membrane. This last is the *transversus perinei superficialis*, which arises from the ischial ramus just posterior to the attachment of the perineal membrane and is inserted into the perineal body.

Ischio-cavernosus. This muscle arises from the posterior part of the perineal membrane and from the ramus of the ischium. The fibres spiral forwards over the crus and are inserted into the upper surface of the commencement of the corpus cavernosum. Their function is to aid in the support of, and to move slightly, the erect organ.

Bulbo-spongiosus. This is the current name for what was appropriately named *compressor urethræ* or *accelerator urinæ* in former days. It arises from the perineal body and, in front of that, from a median raphé. Its posterior fibres are directed forwards and laterally over the bulb of the penis to be inserted into the perineal membrane. The fibres arising from the raphé are inserted into a dorsal fibrous expansion on the penis; the more posterior of these fibres clasp the corpus spongiosum, the more anterior clasp the whole penis. The older names indicate the action of the muscle; it empties the urethra.

Nerve Supply of Penile Musculature. The three superficial perineal muscles are supplied by the perineal branch of the pudendal nerve (S2, 3).

The **penis** consists of a corpus spongiosum, perforated by the urethra and continuous distally with the glans, and the two corpora cavernosa. All three corpora are firmly anchored posteriorly to the perineal membrane (p. 349). Each of the three is enclosed in a tough fibrous membrane, the *tunica albuginea* of the corpus (Fig. 5.75); that of the corpus spongiosum enlarges distally to enclose the glans. The fibrous sheaths of the corpora are fused together; between the corpora cavernosa the fibrous tissue forms a septum with vertical strands like a comb (the septum pectiniforme). In some mammals a bone, the os penis, lies here. The fused fibrous sheaths are attached to the under surface of the symphysis

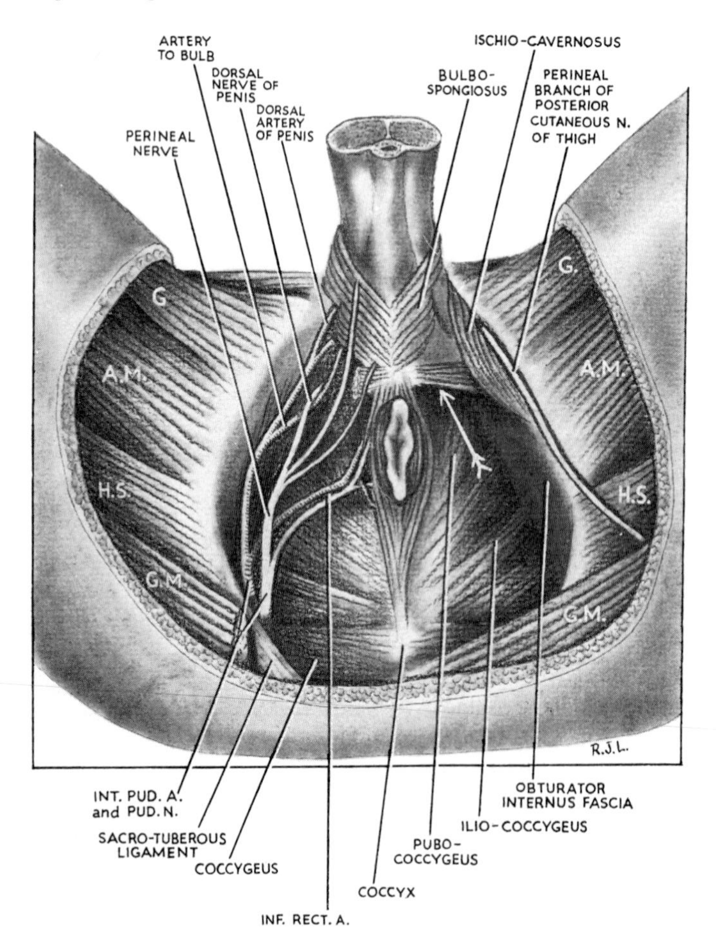

FIG. 5.74. THE MALE PERINEUM. The arrow points to the prolongation from the left ischio-rectal fossa into the deep perineal pouch. Anterior to the arrow head lies the deep transversus perinei muscle. This is inserted into the perineal body, just anterior to the anal orifice. The right half of the perineal membrane has been removed to open up the deep pouch. G. Gracilis. A.M. Adductor magnus. H.S. Hamstrings. G.M. Gluteus maximus. Drawn from Specimen S 353 in R.C.S. Museum.

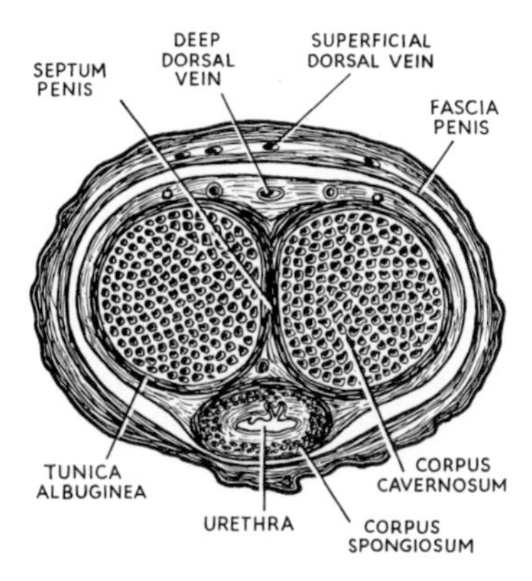

FIG. 5.75. A CROSS SECTION OF THE BODY OF THE PENIS. Somewhat diagrammatic, but drawn to scale from a specimen.

pubis by a triangular sheet of fibrous tissue called the *suspensory ligament*.

The three corpora thus fused together are loosely surrounded by the **fascia penis,** a cylindrical prolongation of the fascia of Colles (p. 352), beneath which lie the midline deep dorsal vein, a dorsal artery on each side and laterally the dorsal nerve. The skin is hairless and prolonged forwards in a fold, the prepuce, which invests the corona and some or all of the glans. Beneath the skin in the midline is the superficial dorsal vein accompanied by lymphatics from the skin and the anterior part of the urethra. The vein divides right and left into the superficial external pudendal veins, and the lymphatics pass to the medial group of superficial inguinal nodes. Some lymphatics from the glans are said to pass direct to the node (of Cloquet) in the femoral canal.

Blood Supply of the Penis. The artery to the bulb supplies the corpus spongiosum and the glans. The deep artery supplies the corpus cavernosum and the dorsal artery supplies the skin, fascia and glans. Note the anastomosis, via the continuity of corpus spongiosum and glans, between the artery to the bulb and the dorsal artery. The deep artery is separate; it supplies the corpus cavernosum only, and this is a closed vascular system whose sole function is erection. Venous return from the corpora is partly by way of veins that accompany the arteries and join the internal pudendal veins but mostly by the deep dorsal vein which pierces the suspensory ligament, passes above the perineal membrane and enters the prostatic venous plexus. The superficial dorsal vein, draining to the superficial external pudendal and saphenous veins, drains the dorsal skin of the penis.

Mechanism of Erection. Erectile tissue, in the corpora of the penis as elsewhere in the body, consists of fibrous saccules into which arterioles open directly. The small arteries are corkscrew-shaped (*helicine arteries*) to allow of their elongation in erection. When the arterioles open, the fibrous saccules become tightly distended with arterial blood. Erect tissue is red and warm. Venous obstruction, which makes the parts blue and cold, plays no part in erection. The corpora, whether erect or flaccid, drain freely by way of the deep dorsal vein of the penis. This vein is held permanently open at its passage above the perineal membrane (cf. the inferior vena cava in the diaphragm and the jugular bulb in the jugular foramen). Stimuli resulting in erection of the external genitalia, in either sex, are mediated by the parasympathetics (nervi erigentes). Ejaculation, on the other hand, is initiated by the sympathetic system (hypogastric nerves).

The **penile urethra** consists of the **bulb** (attached with the corpus spongiosum to the perineal membrane) and the **free penile part.** It is lined with transitional epithelium throughout, except at its dilated anterior part in the glans, the **fossa navicularis;** here is stratified squamous epithelium. The mucous membrane contains numerous glands whose ducts open in a proximal direction (against the stream of urine). These are the urethral glands (of Littré). The empty urethra is horizontal in cross-section; the meatus is a vertical slit—hence the spiral stream of urine, which delays separation of the stream into discrete droplets.

Lymph drainage. The penile urethra drains to the inguinal nodes, both superficial and deep. Lymph vessels from the glans penis are said to run directly to the node (of Cloquet) in the femoral canal.

Nerve Supply. The penile urethra is supplied in its entire length by branches of the perineal nerve which pierce the bulb of the corpus spongiosum.

Development. The prostatic urethra down to the ejaculatory ducts is developed, with the bladder trigone, by incorporation of the mesonephric ducts (mesodermal). From here to the perineal membrane (cf. female urethra) it is developed from the uro-genital sinus part of the cloaca (endodermal). The penile part is enclosed by the external genital swellings, into which a prolongation of the uro-genital sinus grows (cf. labia minora of the female). The glans penis is perforated by an ectodermal tube which joins the endodermal penile part.

The **scrotum** is a pouch of skin containing the testes and spermatic cords. The subcutaneous tissue has no fat, but contains a part of the panniculus carnosus, the **dartos muscle** which sends a sheet into the midline fibrous septum of the scrotum. The rugosity of the skin is due to contraction of the dartos. The dartos is unstriped muscle, and is supplied by sympathetic fibres that are carried in the perineal branch of S4. Deep to dartos is the layer of the superficial fascia (Colles' fascia) attached behind to the posterior edge of the perineal membrane, at the sides to the ischio-pubic rami and bodies of the pubic bones, and in front continuous with Scarpa's fascia. The blood supply of the skin is from superficial and deep external pudendal arteries (from the femoral). Posteriorly there are some branches from the internal pudendal artery. Venous drainage is by external pudendal veins, superficial and deep, to the great saphenous vein (Fig. 3.10, p. 133). Lymph drainage is to the medial group of superficial inguinal nodes.

Nerve Supply of the Scrotum. The anterior axial line (p. 26) crosses the scrotum. The anterior one-third of the scrotal skin is supplied by the ilio-inguinal nerve (L1). The posterior two-thirds is supplied by scrotal branches of the perineal nerve (S3), reinforced laterally by the perineal branch of the posterior cutaneous nerve of the thigh (S2).

The **fascia of Colles** extends downwards as the continuation of the fascia of Scarpa. Its edges are attached to the front of the pubic bone, to the pubic ramus and to the posterior margin of the perineal membrane, thus closing in the subfascial space which lies beneath it in continuity with the space beneath the fascia of Scarpa on the anterior abdominal wall. From its marginal attachments in the urogenital triangle the fascia of Colles is projected into a bulbous scrotal expansion and a cylindrical penile expansion, the distal end of the latter being attached around the corona of the glans penis. Rupture of the penile urethra permits extravasation of urine beneath the fascia of Colles whence, if untreated, the collection distends the tissues of scrotum and penis and then passes upwards over the anterior abdominal wall beneath the fascia of Scarpa. It may reach the submammary space and the axilla, but never extends to the back, for no fascia of Scarpa exists beyond the midaxillary lines, and the subfascial space is obliterated there.

The Deep Perineal Pouch

This is the name given to the space between the perineal membrane and the levator ani muscles of the pelvic floor. It is bounded above by the areolar tissue (epimysium) on the under surface of the levator ani muscles; *no definitely formed membrane exists here.* It contains the membranous urethra surrounded by the sphincter urethræ muscle and the bulbo-urethral glands of Cowper. The deep transversus perinei muscle and perineal body bound it posteriorly. Laterally, between the roof (levator ani) and floor (perineal membrane) the dorsal penile nerve and internal pudendal artery lie against the pubic ramus in a small pyramidal prolongation of the ischio-rectal fossa (Fig. 5.74 and PLATE 13).

The **membranous urethra** passes down from the apex of the prostate at the pelvic floor to pierce the perineal membrane an inch (2·5 cm) behind the symphysis pubis. It is ½ inch (1·5 cm) in length and consists of a tube lined with transitional epithelium. The walls of the tube are of soft fibrous tissue containing a fair amount of non-striated muscle, mostly circular in disposition. It is surrounded by the sphincter urethræ (striated) muscle, by which it is readily compressed into occlusion of its lumen.

The **bulbo-urethral glands** (Cowper's glands) are two small glands, one lying on each side of the urethra just above the perineal membrane. The single duct of each pierces the perineal membrane and opens into the roof of the bulbous (attached) part of the penile urethra. Their function is unknown.

The **sphincter urethræ muscle** is often called by clinicians the 'external sphincter' of the bladder for the very good reason that it is capable of maintaining continence of urine after the internal sphincter has been destroyed or paralysed by disease or operation. It is a broad band of loosely packed muscle fibres which surround the membranous urethra spindle fashion. Most of the fibres are attached laterally to the pubic ramus. They consist of U-shaped loops behind and in front of the urethra; some, however, freely encircle the urethra while others pass backwards to mingle with the fibro-muscular tissue of the perineal body. The muscle is innervated by the perineal branch of the pudendal nerve (S2, 3). The action of the circular fibres is to close as a sphincter on the urethra; that of the curved anterior and posterior fibres which are attached to bone is to compress the membranous urethra between their interlacing U-shaped loops.

The interlacing fibres of the sphincter urethræ were formerly known as the 'deep transversus perinei' and the epimysium on the deep surface of this muscle, coagulated in the fixed cadaver, was called the 'superior fascia of the uro-genital diaphragm'. It does not exist in the living. Neither is there really a uro-genital 'diaphragm'. Only the sphincter urethræ lies deep to the perineal membrane. The deep transverse perineal muscle lies in the perineal body *behind* the uro-genital triangle (Fig. 5.74).

The **deep transversus perinei muscle** consists of fibres which are attached laterally to the ischio-pubic ramus and medially to the perineal body; very few fibres pass directly from bone to bone (Fig. 5.74). It is supplied by the perineal branch of the pudendal nerve. Its action is probably to brace the perineum somewhat against downward pressure from the pelvic floor above it; but it is a weak muscle.

The **ischio-rectal fossa** (p. 347) extends forwards above the deep transversus perinei and the perineal membrane as a pointed process on each side along the pubic rami. These processes do not communicate with each other; they are separated by the midline structures, membranous urethra and the sphincter urethræ. The arteries and nerves passing to the penis lie in the forward prolongation of the ischio-rectal fossa (Fig. 5.74 and PLATE 13).

The **internal pudendal artery,** having given off the inferior rectal artery in the anal triangle, passes forwards along the ischio-pubic ramus above the perineal membrane; its branches to the penis pierce the membrane to reach their destination. The **artery to the bulb** pierces the perineal membrane alongside the urethra and the duct of Cowper's gland and enters the corpus spongiosum. It gives branches to the cavernous tissue of this corpus and passes forwards to supply the glans penis. It has no anastomosis with the arteries of the corpora cavernosa.

The internal pudendal artery perforates the anterior angle of the perineal membrane and immediately divides into (*a*) the **deep artery of the penis** which enters the crus and supplies, by *helicine arteries*, the erectile cavernous tissue of the corpus cavernosum, and (*b*) the **dorsal artery of the penis.** This latter vessel passes along the medial surface of the crus against the perineal membrane to reach the dorsum of the penis. The two dorsal arteries pierce the suspensory ligament and run forward, alongside the median deep dorsal vein, with the dorsal nerves lying laterally, between the fascia of the penis and the fibrous sheaths of the corpora cavernosa (Fig. 5.75). The arteries pass to the glans, where they anastomose with the terminal branches of the arteries to the bulb.

The **deep dorsal vein of the penis** drains most of the blood from the corpora. It runs proximally in the midline (Fig. 5.75) and pierces the suspensory ligament and runs upwards in the midline gap between the symphysis pubis and perineal membrane. It passes, in the midline, through the deep perineal pouch and enters the pelvis by passing up between the two puboprostatic ligaments. It joins the prostatic plexus.

The **dorsal nerve of the penis** is the continuation of the pudendal nerve. It runs forward in the ischiorectal fossa and pierces the anterior angle of the perineal membrane on the lateral side of the internal pudendal artery. It supplies the skin of the penis and glans and gives branches to the corpus cavernosum. It has no branches in the deep perineal pouch.

The **perineal nerve** (p. 349) passes into the superficial pouch to supply the penile musculature, the penile urethra and scrotal skin. Before passing superficial to the perineal membrane, however, it gives a branch to the deep pouch which is motor to sphincter urethræ and the anterior fibres of levator ani and sensory to the membranous urethra (PLATE 13).

The Female Urogenital Triangle

All the male formations and structures are present in the female, but modified greatly for functional reasons. The essential difference is the failure in the female of midline fusion of the genital folds. The male scrotum is represented by the labia majora of the vulva and the corpus spongiosum of the male urethra is represented by the labia minora and bulb of the vestibule.

The **perineal membrane** (triangular ligament) of the male is represented in the female by only a narrow shelf of membrane attached along the pubic rami. It gives attachment to the crura of the clitoris, each of which is covered by an ischio-cavernosus muscle, as in the male. Medial to each crus, attached to the margin of the shelf of perineal membrane, is a mass of erectile tissue. This is the **bulb of the vestibule,** separated into two halves by the orifices of vagina and urethra. They join together in front of the urethral orifice and pass forwards to the glans of the clitoris. The glans is enclosed by fusion together of the anterior ends of the labia minora, each of which splits to form a dorsal prepuce and a ventral frenulum to the glans.

The erectile tissue of the bulb is covered by the **bulbo-spongiosus muscle,** whose fibres extend from the perineal body around vagina and urethra to the clitoris; they form a perineal sphincter for the vagina in addition to its pelvic sphincter (pubo-vaginalis part of levator ani).

The **vestibular glands** (glands of Bartholin) lie at the orifice of the vagina, one behind the posterior end of each half of the bulb of the vestibule, deep to the bulbo-spongiosus muscle. Each opens by a single duct into the vaginal orifice. The glands are homologous with Cowper's glands in the male.

The deep perineal pouch is, in the female, perforated by both the urethra and vagina (Fig. 5.76). Alongside these canals is a forward projection of the ischio-rectal fossa containing the dorsal nerve of the clitoris and the deep and dorsal arteries, as in the male.

The **perineal body** in the female lies below the pelvic floor between vagina and anal canal; it is a mass of fibrous tissue into which mingle many muscle fibres from the bulbo-spongiosus and transversus perinei muscles, as well as from the pubo-vaginalis and external anal sphincter. Lacking the rigid support of the complete perineal membrane of the male, the perineal body is more mobile in the female. It helps to support the levator ani above it; its integrity is indispensable to the stability of the pelvic organs. It is liable to laceration during parturition; its obstetrical importance is indicated by its obstetrical name—simply, '*the perineum*'. (In anatomy the term perineum refers to the whole pelvic outlet below the pelvic floor.)

The **vagina** is a fibrous tube lined with stratified squamous epithelium. When collapsed it is thrown into transverse folds and its anterior and posterior walls lie in contact. Above the pelvic floor it projects into the pelvic cavity. The cervix uteri projects downwards, surrounded by the fornices. The posterior fornix is in

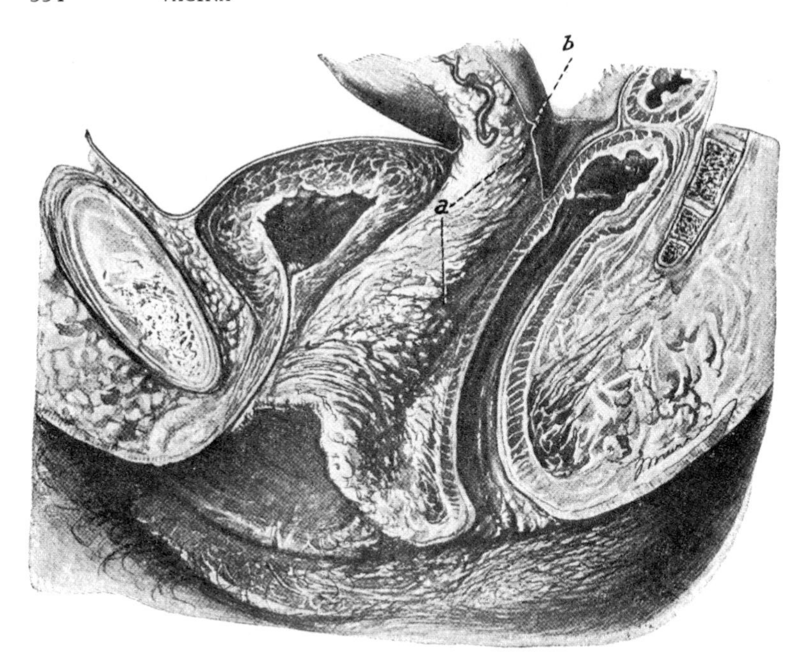

FIG. 5.76. THE FEMALE PELVIS IN SAGITTAL SECTION, LEAVING THE VAGINA INTACT. a. Levator ani (sphincter vaginæ). b. Pouch of Douglas.

contact with the peritoneum of the utero-rectal pouch (Fig. 5.65, p. 336). The long axes of vagina and uterus lie at right angles to each other. The upper part of the vagina is clasped by the pelvic floor—fibres that loop around behind it. The U-shaped sling so formed is named pubo-vaginalis (sphincter vaginæ). A perineal sphincter (bulbo-spongiosus) surrounds its outlet. There are no glands in the vagina (except Bartholin's glands); the mucous membrane is kept moist by mucus secreted from the cervix uteri.

The **female urethra** is $1\frac{1}{2}$ inches (3·5 cm) long. Its mucous membrane is lined with transitional epithelium throughout. There are a few poorly-developed, pit-like glands, said to be homologous with the prostate, but bearing no resemblance to the structure of that gland. The wall of the urethra consists of smooth muscle, with a good deal of fibrous tissue. The muscle is almost entirely longitudinal, with a few circular fibres forming a thin outer layer.

The urethra is surrounded by the *sphincter urethræ* as in the male. Curved loops of muscle fibres pass anterior and posterior to the urethra (the former are usually the better developed); they are attached to the pubic rami and perineal membrane. A few circular fibres surround the urethra. As in the male the muscle is supplied by the perineal nerve.

Stress incontinence results from malfunction of the sphincter urethræ. The smooth muscle of the bladder neck is not strong enough to withstand much pressure (coughing, sneezing, skipping etc.). A few drops of urine escaping into the proximal urethra stimulate the quick-acting and much stronger striped muscle of the sphincter urethræ into reflex contraction. Should the sphincter urethræ fail, the urine escapes to the exterior (stress incontinence). The anterior U-shaped fibres demand a fixture against which to compress the urethra, and surgical relief provides such a firm structure (usually postoperative scar tissue) behind the urethra.

Cutaneous Nerve Supply of the Urogenital Triangle

The ilio-inguinal nerve supplies the anterior third of the scrotum (labium majus) down to the anterior axial line. The skin of the penis (clitoris) is supplied by the dorsal nerve. The posterior two-thirds of the scrotum (labium majus) is supplied laterally by the perineal branch of the posterior cutaneous nerve of the thigh and medially (labium minus in the female) by scrotal (or labial) branches of the perineal branch of the pudendal nerve. The mucous membrane of the penile urethra (labia minora) is supplied by the perineal branch of the pudendal nerve.

The segmental supply of the penis is derived from L1 on the dorsal aspect of the root, while distal to the axial line the penile skin and adjoining scrotum are supplied by S3, which is overlapped further back by S2 (Fig. 1.32, p. 27).

THE JOINTS OF THE PELVIS

The **sacro-iliac joint** is a synovial joint formed between the cartilage-covered auricular surfaces of the ilium (Fig. 3.66, p. 195) and ala (costal element) of the sacrum. The hip bone is strongly buttressed for weight bearing between the auricular surface and the acetabulum (bears weight in standing) and between the auricular surface and ischial tuberosity (bears weight in sitting). The hyaline cartilage and underlying bone of the auricular surfaces are undulating and irregular enough to discourage movement at this joint. The articular margins give attachment to the capsule and its lining of synovial membrane. In the adult, especially the male, the joint cavity is obliterated in places by fibrous bands which pass from one articular surface to the other. Ligamentous bands, very strong posteriorly and weak anteriorly, surround the capsule.

The **anterior sacro-iliac ligament** is a flat band which joins the bones above and below the pelvic brim; stronger in the female, it indents a pre-auricular groove on the female ilium just below the pelvic brim.

The **posterior sacro-iliac ligaments** are attached behind to the contiguous surfaces of sacrum and ilium. The shortest fibres lie deepest between the two bones, the longest fibres lie, naturally, more superficially. These ligaments are very strong. They are attached into deep pits on the posterior surface of the lateral mass of the sacrum (Fig. 6.81, p. 474).

Accessory ligaments give added stability to the joint. The fifth lumbar vertebra is attached to the sacrum by the intervertebral disc and other ligaments (p. 473). Hence the *ilio-lumbar ligaments*, acting through this vertebra, assist in strengthening the bond between ilium and sacrum.

The *sacro-tuberous* and *sacro-spinous ligaments* (Fig. 5.55, p. 323) are joined strongly to the sacrum (and coccyx) and to the ischium and help to stabilize the sacrum on the os innominatum—in particular they oppose forward tilting of the sacral promontory.

The **sacro-tuberous ligament** is a flat band of great strength. It is attached to the posterior border of the ilium between the posterior superior and posterior inferior iliac spines, to the transverse tubercles of the sacrum below the auricular surface, and to the upper part of the coccyx. From this wide area the ligament slopes down to the medial surface of the ischial tuberosity. The upper edge of the ischial attachment is prolonged forwards and attached to a curved ridge of bone. This prolongation is the *falciform process*; it lies just below the pudendal canal. The sacro-tuberous ligament is narrower in the middle than at either end. Its gluteal surface gives origin to gluteus maximus. The ligament is said to be the phylogenetically degenerated tendon of origin of the long head of biceps femoris (Fig. 5.55, p. 323).

The **sacro-spinous ligament** lies on the pelvic aspect of the sacro-tuberous ligament. It has a broad base which is attached to the side of the lower part of the sacrum and the upper part of the coccyx. It narrows as it passes laterally, where its apex is attached to the spine of the ischium. The coccygeus muscle lies on the pelvic surface of the ligament. The ligament is the phylogenetically degenerated posterior surface of the coccygeus muscle.

The sacro-tuberous and sacro-spinous ligaments enclose the lesser sciatic foramen, whose lateral part is occupied by the emerging obturator internus muscle and whose medial part leads forwards into the pudendal canal above the falciform process of the sacro-tuberous ligament (Fig. 5.73).

Stability. The sacro-iliac articulation depends entirely upon ligaments. The two joint surfaces lie in diverging planes; the weight of the fifth lumbar vertebra tends to push the sacrum down towards the symphysis. There is no bony factor in stability; true the sacrum is wedge-shaped, but in the reverse direction of a keystone. Opposing any simple gliding movement of the joint surfaces are the posterior sacro-iliac ligaments, and the ilio-lumbar ligament acting through the fifth lumbar vertebra, while opposing forward rotation of the sacral promontory around the joint are the sacro-tuberous and sacro-spinous ligaments. While the ligaments are intact the bony surfaces so held in apposition are irregular enough to discourage gliding and rotation, but this bony factor is entirely dependent upon the integrity of the ligaments.

Note that the bony surfaces are not weight-bearing. The body weight is suspended by the sacro-iliac ligaments, which sling the sacrum below the iliac bones. Body weight tends to separate, not compress, the cartilage-covered articular surfaces.

The sacro-iliac ligaments soften towards the later months of pregnancy and permit some slight rotation of the sacrum during parturition.

The **sacro-coccygeal joint** is produced through the intermediary of an intervertebral disc of fibro-cartilage. The cornua of the coccyx and sacrum are joined by ligaments. A good deal of movement is possible in this articulation, but only flexion-extension. There is no side to side movement.

The **symphysis pubis,** as its name implies, is a secondary cartilaginous joint. The bony surfaces of the pubes are each covered with a thin plate of hyaline cartilage and the two sides are joined by a broad mass of transversely running fibres. Centrally a tissue-fluid space may develop, but it is never lined with synovial membrane. Ligamentous fibres reinforce the symphysis above, and especially below, the bony margins of the joint.

BRANCHES OF THE LUMBAR AND SACRAL PLEXUSES

The two plexuses have already been described, and the situations of their branches may be found in the descriptions of the appropriate regions. It is convenient, however, to summarize the whole course of each branch, and this is done in the following condensed review.

Lumbar Plexus

After the anterior primary rami of the upper four lumbar nerves have supplied psoas and quadratus lumborum segmentally, they form the plexus in the substance of psoas major. The plexus innervates part of the lower abdominal wall, but is chiefly concerned in supplying skin and muscle 'borrowed from the trunk' by the lower limb. In this way it reinforces the sacral plexus, which is the true plexus of the lower limb (p. 24 *et seq.*).

Branches of the Lumbar Plexus

- L1—Ilio-hypogastric and ilio-inguinal
- L1, 2—Genito-femoral ('femoro-genital')
- L2, 3 (posterior divisions)—Lateral cutaneous nerve of thigh
- L2, 3, 4 (posterior divisions)—Femoral
- L2, 3, 4 (anterior divisions)—Obturator

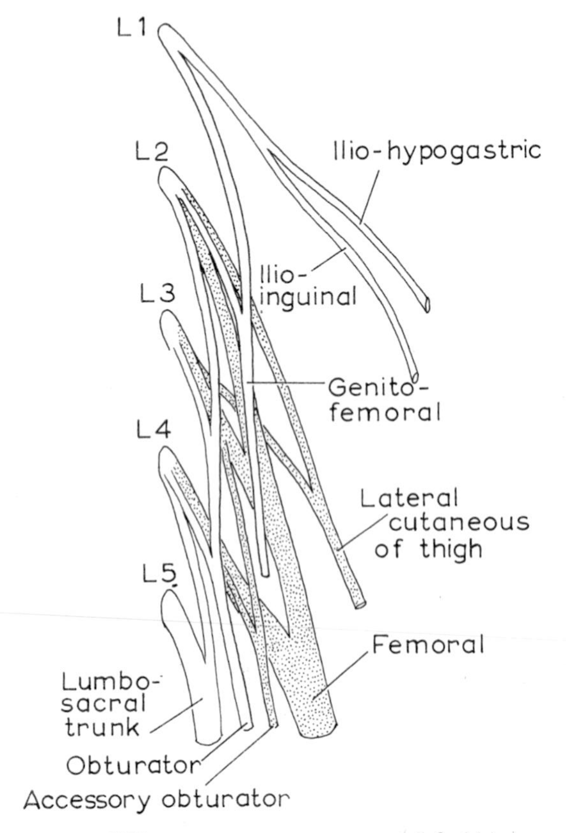

FIG. 5.77. PLAN OF THE LUMBAR PLEXUS (left side).

Ilio-hypogastric and Ilio-inguinal Nerves. These are only by convention included in the lumbar plexus. They are really just the first lumbar segmental body-wall nerve and its collateral branch, in series with the thoracic nerves (see p. 22).

Genito-femoral Nerve. (L1, 2—best remembered segmentally as the 'femoro-genital' nerve.) L1 is the femoral part, and supplies an area of skin below the middle of the inguinal ligament. L2 is the genital part, and supplies that part of the abdominal wall herniated into the scrotum for the descent of the testis (i.e., the spermatic cord). It is sensory to tunica vaginalis and the spermatic fasciæ, motor to cremaster muscle. It supplies neither the testis nor the scrotal skin. The genital branch is very small in the female, supplying only the fascia on the round ligament in the inguinal canal. The genito-femoral nerve is described on p. 311.

Lateral Cutaneous Nerve of Thigh. (L2, 3—posterior divisions, mostly L2). The nerve is wholly sensory, to the iliac fascia and peritoneum of the iliac fossa, and to the lateral side of the thigh down to the knee. It emerges from the lateral border of psoas (occasionally as a branch of the femoral nerve) and its course is described on pp. 131 and 310.

Femoral Nerve. (L2, 3, 4—posterior divisions.) The nerve issues from the lateral border of psoas and crosses the iliac fossa in the gutter between psoas and iliacus, deep to the iliac fascia. It *supplies iliacus* here, and then passes beneath the inguinal ligament lateral to the femoral sheath. Lying on iliacus, it breaks up at once into 9 or 10 branches. The lateral femoral circumflex artery runs through this leash, dividing the branches into 'superficial' and 'deep'. There are four superficial branches (two cutaneous and two muscular). The deep branches supply quadriceps femoris (a branch for each vastus and two for rectus femoris) and include one cutaneous branch, the saphenous nerve.

Superficial Branches. The *nerve to pectineus*, often double, runs behind the femoral sheath to reach the muscle. The *nerve to sartorius* often pierces the muscle and continues on as an intermediate cutaneous nerve. The *intermediate cutaneous nerve* of the thigh, often piercing sartorius, supplies skin and fascia lata over the front of the thigh down to the knee. The *medial cutaneous nerve* of the thigh supplies the upper medial side of the thigh, and an anterior branch reaches the front of the knee; but the lower medial side of the thigh is supplied by the obturator nerve.

Deep branches. The nerve to *rectus femoris* is usually double, and the upper branch supplies also the *hip joint* (Hilton's law, p. 15). The nerve to *vastus lateralis* runs down with the descending branch of the lateral femoral circumflex artery between rectus femoris and vastus intermedius. The nerve to *vastus intermedius* sinks into the anterior surface of that muscle. The nerve to *vastus medialis* enters the upper part of the subsartorial canal and sinks into the muscle. It is a very large nerve for, although the nerves to the two vasti supply also a few fibres to the knee joint, this nerve carries most of the femoral branches to the knee.

The *saphenous nerve* gradually crosses the femoral artery in the subsartorial canal, gives some twigs to the subsartorial plexus, and runs on to emerge below the posterior border of sartorius. Here its *infrapatellar branch* pierces sartorius to run into the patellar plexus. The saphenous nerve, now cutaneous, supplies skin and periosteum over the subcutaneous surface of the tibia. It runs with the great saphenous vein in front of the medial malleolus and ends on the medial side of the foot just short of the big toe.

Obturator Nerve. (L2, 3, 4—anterior divisions.) Coming out of the medial side of psoas the nerve lies on the ala of the sacrum lateral to the lumbo-sacral trunk. It slants down to the side wall of the pelvis between the origin of the internal iliac artery and the ilium. From the angle between external and internal iliac vessels it runs straight to the obturator foramen, supplying the parietal peritoneum of the side wall of the pelvis (in the female the ovary lies here). In the obturator canal it splits into anterior and posterior divisions.

The *posterior division* supplies obturator externus, then pierces the upper border of that muscle and runs into the thigh deep to adductor brevis. It runs down on adductor magnus, whose pubic part it supplies (the ischial part of adductor magnus is supplied by the sciatic nerve). A slender branch accompanies the femoral artery into the popliteal fossa to supply the knee joint.

The *anterior division* passes over obturator externus and, emerging into the thigh, it supplies the hip joint. It runs down over adductor brevis, deep to pectineus and adductor longus. It supplies these two adductors and often helps the femoral nerve to supply pectineus. It supplies also the gracilis. It supplies the lower medial side of the thigh by a cutaneous branch which runs through the subsartorial plexus.

Accessory Obturator Nerve. See p. 342.

Sacral Plexus

This is a flat, triangular formation on the front of piriformis muscle. It is formed of the lumbo-sacral trunk (L4, 5) and the upper four sacral nerves. Its constituent nerves divide into anterior and posterior divisions. Its branches total a round dozen, six from the nerves before they divide, and three each from the anterior and posterior divisions.

The six branches from the main nerves all come from sacral segments and all have the initial 'P'. They are:

Nerves to Piriformis (S1, 2)	from behind
Perforating Cutaneous Nerve (S2, 3)	
Posterior Cutaneous Nerve of Thigh (S2, 3)	
Pelvic Parasympathetics (S2, 3)	from in front
Pudendal Nerve (S2, 3, 4)	
Perineal Branch of S4	

Piriformis. The muscle is supplied segmentally by twigs which pass back from the upper sacral nerves directly into the muscle.

Perforating Cutaneous Nerve. (S2, 3). This perforates the sacro-tuberous ligament and the gluteus maximus fibres that arise there. The nerve supplies a small area of skin on the lower medial side of the buttock.

Posterior Cutaneous Nerve of the Thigh. (S2, 3—the tiny S1 component is negligible.) This cutaneous nerve has a wide distribution. It runs down below piriformis on the sciatic nerve. At the lower border of gluteus maximus it becomes cutaneous, but is unique in that it remains beneath the deep fascia. It runs down the posterior midline beneath the fascia as far as the lower ends of the gastrocnemius bellies. It supplies a strip of deep fascia and skin, between anterior and posterior axial lines, from the buttock to the mid-calf by a series of *perforating branches*, each of which pierces the deep fascia separately.

Branches. A *gluteal branch* winds around gluteus maximus to supply skin over the convexity of the buttock. The *perineal branch* winds around the ham-strings and gracilis origins and pierces the fascia lata at the medial convexity of the upper thigh. It supplies the lateral part of the posterior two-thirds of the scrotum (labium majus).

Pelvic Parasympathetics. (S2, 3 or S3, 4.) These are the *nervi erigentes*. They pass through the pelvic plexuses to supply the derivatives of the cloaca. They are sensory (both normal distension and pain) to the bladder, cervix uteri, anal canal and lower rectum (note that referred pain from these segments is felt along the posterior cutaneous nerve of the thigh). They are motor to the bladder (inhibitory to the internal sphincter). In addition they are motor to the descending colon and rectum.

Pudendal Nerve. (S2, 3, 4.) The nerve runs down over the front of the plexus and curls around the coccygeus muscle (whose gluteal surface is the sacro-spinous ligament) and enters the pudendal canal. It is the nerve of the pelvic floor and perineum (PLATE 13).

Branches. The *inferior rectal nerve* arches over the lunate fascia beneath levator ani, and its branches run alongside the anal canal to reach peri-anal skin. It supplies the under surface of levator ani, and is motor to the profundus and subcutaneous parts of the external anal sphincter.

The *perineal nerve* is a terminal branch of the pudendal. It runs forward superficial to the perineal membrane and breaks up to supply skin of the posterior two-thirds of the scrotum (labium majus) and the mucous membrane of the urethra (labia minora). It is motor to the muscles of the uro-genital triangle, namely, ischio-cavernosus, bulbo-spongiosus, superficial and deep transversus perinei and the sphincter urethræ, and it supplies also the anterior part of levator ani. Sphincter urethræ and levator ani are supplied by a separate branch that runs above the perineal membrane into the deep perineal pouch; this branch is also sensory to the membranous (or female) urethra.

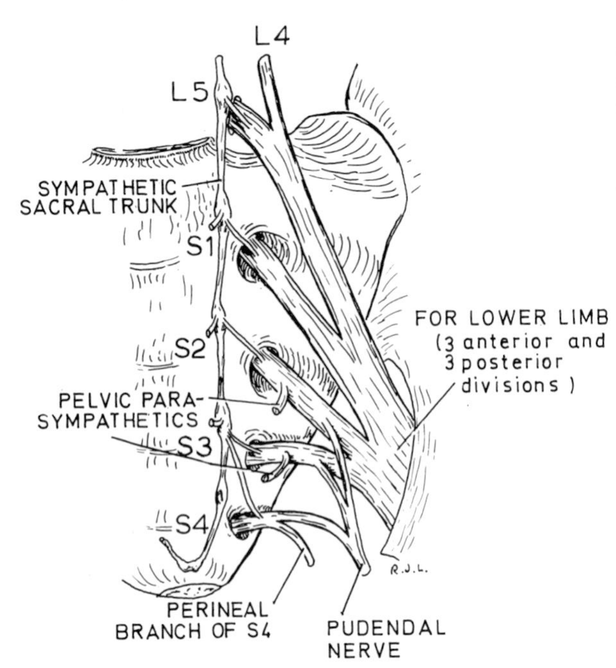

FIG. 5.78. THE SACRAL PLEXUS (left side). The branches from its anterior surface (three P's) are shown. S4 may give off a pelvic parasympathetic, otherwise it is for the perineum and does not enter the lower limb.

The *dorsal nerve* of the penis (clitoris) is the other terminal branch of the pudendal nerve. It runs in the forward prolongation of the ischio-rectal fossa into the deep perineal pouch, deep to the perineal membrane, which it pierces just below the symphysis pubis. It runs forward to supply the skin of the whole penis (clitoris) except on the dorsum just at the root (ilio-inguinal nerve).

Perineal Branch of S4. Supplying coccygeus and ilio-coccygeus on their pelvic surfaces, the nerve slips between them into the roof of the ischio-rectal fossa. It supplies the intermediate part of the external anal sphincter known as 'superficialis' and reaches the skin of the anal margin. Sympathetic fibres that hitch-hike along it reach the dartos muscle.

Anterior Divisions of the Sacral Nerves

Three nerves stem from this part of the sacral plexus.

The **nerve to obturator internus** (L5, S1, 2—same as the inferior gluteal nerve) passes below piriformis, curls around the base of the ischial spine and sinks into obturator internus. It supplies also the superior gemellus.

The **nerve to quadratus femoris** (L4, 5, S1—same as the superior gluteal nerve) lies on the ischium deep to the sciatic nerve and, continuing in contact with the bone, runs down deep to obturator internus to sink into the deep surface of quadratus femoris. It supplies also the inferior gemellus and, exemplifying Hilton's Law, gives a branch to the hip joint.

The **tibial (medial popliteal) component** of the sciatic nerve is formed from all five divisions (L4, 5, S1, 2, 3).

Posterior Divisions of the Sacral Nerves

Three nerves arise from these divisions.

The **superior gluteal nerve** (L4, 5, S1) passes back around the great sciatic notch above piriformis, runs in the plane between glutei medius and minimus, supplies both, and ends in the tensor fasciæ latæ. It supplies no skin.

The **inferior gluteal nerve** (L5, S1, 2) passes back below piriformis and sinks into the deep surface of gluteus maximus, rather on its medial side. It supplies no skin.

The **common peroneal** (lateral popliteal) **component** of the sciatic nerve is formed from only four posterior divisions (L4, 5, S1, 2). It joins the tibial in the pelvis and the sciatic nerve so formed lies on the ischium at the lower border of piriformis. If fusion fails to occur, the common peroneal nerve pierces piriformis to reach the buttock.

The Sciatic Nerve

This is the nerve of the lower limb. The adductor and extensor compartments of the thigh and the skin over the tibia, muscle and skin 'borrowed from the trunk' are supplied from the lumbar plexus by the obturator and femoral nerves. The main trunk of the sciatic nerve supplies the hamstring compartment and then separates into the tibial and common peroneal nerves. The tibial nerve supplies the flexor part of the calf and the sole of the foot around to the toenails. The common peroneal nerve supplies the extensor and peroneal compartments of the leg and the dorsum of the foot.

The sciatic nerve is formed at the lower margin of piriformis by union of the tibial and common peroneal components and emerges into the buttock lying on the ischium. The nerve to quadratus femoris is deep to it and the posterior cutaneous nerve of the thigh lies superficial. The sciatic nerve, midway between the greater trochanter and the ischial tuberosity, passes vertically downwards into the hamstring compartment. It lies on obturator internus and gemelli, quadratus femoris and then on adductor magnus. It is overlaid by the long head of biceps. It lies a finger's breadth lateral to the flat tendon of semimembranosus. It divides, usually at the upper angle of the popliteal fossa, into the tibial and common peroneal nerves.

Branches. Motor branches supply all three hamstrings and the ischial fibres of adductor magnus. These last, as well as those for the long head of biceps and the two 'semi-' muscles, are from the tibial component, but the branch to the short head of biceps comes from the common personeal component (p. 25).

The Common Peroneal Nerve. This enters the apex of the popliteal fossa and runs alongside the biceps tendon just beneath the deep fascia. It lies on the fat of the fossa and crosses plantaris, the lateral head of gastrocnemius, the popliteus tendon inside the knee-joint capsule and the fibular origin of soleus. It then sinks into the upper fibres of peroneus longus and divides into two terminal branches, the deep peroneal and superficial peroneal nerves. It can be palpated in almost the whole of its course, but especially where it lies on the neck of the fibula, for here it can be rolled on the bone.

Branches. The common peroneal nerve itself supplies no muscles, for it is the nerve of the extensor compartment, and the popliteal fossa through which it runs contains only flexor compartment muscles.

Two articular branches, the *upper* and *lower lateral genicular nerves*, supply the knee joint.

Two cutaneous branches should be noted. The *lateral sural nerve* joins the sural nerve below the gastrocnemius heads and runs to the lateral side of the little toe. The *lateral cutaneous nerve of the calf* pierces the deep fascia as the nerve lies on soleus and supplies skin and deep fascia over the upper half of the peroneal compartment. The third and last cutaneous branch, called *recurrent genicular*, is unimportant. It supplies the superior tibio-fibular joint, a few fibres of tibialis anterior, and a little skin over the ligamentum patellæ.

The Deep Peroneal Nerve. Formed in the substance of peroneus longus, the nerve spirals down over the fibula deep to extensor digitorum longus and reaches the interosseous membrane. It runs down lateral to the vessels, crosses the lower end of the tibia and the dorsum of the foot, and ends by supplying the skin of the first interdigital cleft.

Branches. It supplies the muscles of the extensor compartment of the leg: extensor digitorum longus, tibialis anterior, extensor hallucis longus and peroneus tertius. On the dorsum of the foot it gives a lateral branch which passes deep to extensor digitorum brevis and supplies that muscle. It is beneath the deep fascia, and supplies periosteum and ligaments of the tibia and the dorsum of the foot.

The Superficial Peroneal Nerve. Formed in the substance of peroneus longus the nerve runs down in the muscle, emerging from its anterior border about a third of the way down the leg. It supplies peroneus longus and brevis, then perforates the fascia over them about halfway down the leg. It supplies the skin over the peronei and extensor muscles in the lower half of the leg. Above the ankle it divides into a medial and a lateral branch which diverge only narrowly. They supply skin and deep fascia on the dorsum of the foot (the deep peroneal nerve supplies the underlying bones). The medial branch breaks up to supply the medial side of the big toe and the second interdigital cleft, while the lateral branch breaks up to supply the third and fourth clefts.

The Tibial Nerve. The nerve enters the apex of the popliteal fossa and, in the midline of the limb, passes vertically down deep to the heads of gastrocnemius behind the knee joint and across the popliteus muscle, to run beneath the fibrous arch in soleus.

Branches. Three *genicular nerves*, upper and lower medial and a middle, accompany the arteries and supply the knee joint. *Muscular branches* supply the muscles of the fossa: plantaris, both heads of gastrocnemius, soleus and popliteus. The last-named branch recurves around the lower border of the muscle to enter its deep (i.e. anterior) surface. A single cutaneous branch, the *sural nerve*, lies in the groove between the two heads of gastrocnemius and pierces the deep fascia halfway down the leg. Here it is joined by the lateral sural nerve. The sural nerve runs down alongside the small saphenous vein behind the lateral malleolus and ends on the lateral side of the little toe.

The Tibial Nerve in the Calf. From the fibrous arch the nerve runs down with the posterior tibial vessels beneath the soleus muscle. The neuro-vascular bundle lies in the groove between the bellies of flexor hallucis longus and flexor digitorum longus. Behind the medial malleolus, beneath the flexor retinaculum, the nerve divides, distal and superficial to the vessels, into its terminal medial and lateral plantar branches.

Branches. Muscular branches supply soleus, tibialis posterior and the flexors hallucis and digitorum longus. It supplies periosteum on the flexor surfaces of tibia and fibula. *Medial calcanean* branches pierce the flexor retinaculum and supply the weight-bearing skin of the heel.

The **medial** and **lateral plantar nerves** are described on pp. 180 and 181.

Section 6. The Head and Neck

GENERAL TOPOGRAPHY OF THE NECK

The first thoracic vertebra lies at the highest part of the sloping thoracic operculum. From its upper border rises the cervical spinal column, gently convex forwards, and supporting the skull. A mass of extensor musculature lies behind the vertebræ. It is supplied segmentally by *posterior* primary rami and supports the cervical spine and head. A much smaller amount of prevertebral flexor musculature lies in front of the vertebræ and, more laterally, is attached to the thoracic operculum and the scapula. It is supplied segmentally by *anterior* primary rami. This musculature comprises longus colli, rectus capitis anterior and rectus capitis lateralis, longus capitis, scalenus anterior, scalenus medius, scalenus posterior and levator scapulæ. The whole mass, projecting but little in front of the cervical spine, lies flat from side to side behind the pharynx, and from here curves away posteriorly; it is covered over by the prevertebral fascia, which thus forms a vertical sheet passing from side to side across the central part of the neck. The function of the prevertebral fascia is to provide a basis upon which the mobile viscera of the neck may move freely. Projecting down from the base of the skull is the face, which thus lies in front of the upper part of the prevertebral fascia. The hard palate lies on a level with the anterior arch of the atlas, the lower border of the mandible lies between C2 and 3 vertebræ. Suspended from the back of the face is the pharynx, which extends below to the level of the cricoid cartilage (C6) and then continues on as the œsophagus.

In front of the sheet of prevertebral fascia lie the viscera of the neck—pharynx and œsophagus, larynx and trachea in front of these and, on each side, the carotid sheaths, lying one each side of the pharynx. Lying above the larynx is the hyoid bone. It is connected to the mandible by the mylo-hyoid muscle, which forms the upper limit of the anterior part of the neck. The mylo-hyoid muscle separates the mouth from the neck. The hyoid bone and larynx are suspended by muscles from the skull. Inferiorly they are connected by muscles to the sternum and the scapula; beneath these muscles the thyroid gland, enclosed in the pretracheal fascia, lies alongside the respiratory canal. The anterior part of the neck extends no higher than the mandible, being limited by the mylo-hyoid muscle; above this level the face extends to the base of the skull. Further back the neck extends as high as the base of the skull; on each side of the pharynx is a carotid sheath, with the cervical sympathetic trunk behind it. Emerging into the neck are the ninth, tenth, eleventh and twelfth cranial nerves. Finally surrounding the whole neck is a collar of fascia, the investing layer of deep cervical fascia, which contains trapezius and sterno-mastoid muscles.

The Fasciæ of the Neck

As in other parts of the body, so in the neck, the term fascia is applied in anatomy to widely differing structures. On the one hand, it may apply to a very real membrane, visible and demonstrable as such, a membrane that may be incised and sutured (e.g., investing layer, prevertebral fascia, pretracheal fascia). On the other hand, the term may be applied to a loosely knit areolar tissue which is no more than that lying on the surface of any muscle in the body and quite unworthy of a name. Especially is this true of the fibrous tissue on the surface of muscles that can be distorted appreciably—here there is no true membrane. The cheeks could not be puffed out nor the pharynx dilated over a large bolus if the buccinator and pharyngeal constrictors were covered with an inexpansile membrane like the prevertebral fascia. There is no such thing, for example, as a bucco-pharyngeal fascia in the sense that there is a prevertebral fascia, and much confusion has resulted from failure to appreciate this point. Tissue spaces exist between the muscles of the neck as they do between the muscles in a limb. There is no need in either case to describe the muscle as being covered with a named 'fascia'. The fasciæ of the neck demonstrable as membranes existing in their own right, so to speak, comprise only the investing layer, the prevertebral fascia, the pretracheal fascia, and the carotid sheath. All other fascia that has been described has no anatomical reality as a separate structure and none of it is worth a name to itself.

The Investing Layer of Deep Cervical Fascia

This fascia, comparable in every way to the deep fascia that underlies the subcutaneous fat in the limbs and elsewhere, surrounds the neck like a collar. It splits around the sterno-mastoid and trapezius muscles with a layer adherent to the superficial and deep surfaces of each of them. Posteriorly it meets the ligamentum nuchæ, from which the cervical part of the trapezius muscle arises. It is attached to the hyoid bone, hence the chin and Adam's apple of man in profile; there is no 'dewlap' as in cattle. Its upper and lower attachments should be studied with care and understood with precision. Examine the base of a skull. The attachments of trapezius and sterno-mastoid extend from the external occipital protuberance in a curve along the superior nuchal line to the tip of the mastoid process. The investing layer enclosing both muscles is attached to the skull over the whole extent of this line; in the front of the neck it is attached to the lower border of the mandible, from chin to angle on each side (PLATE 37, p. *xx*).

Between the angle of the mandible and the tip of the mastoid process it is slightly complicated in that it here splits into two layers that diverge from each other like the covers of a half-open book. The parotid gland lies

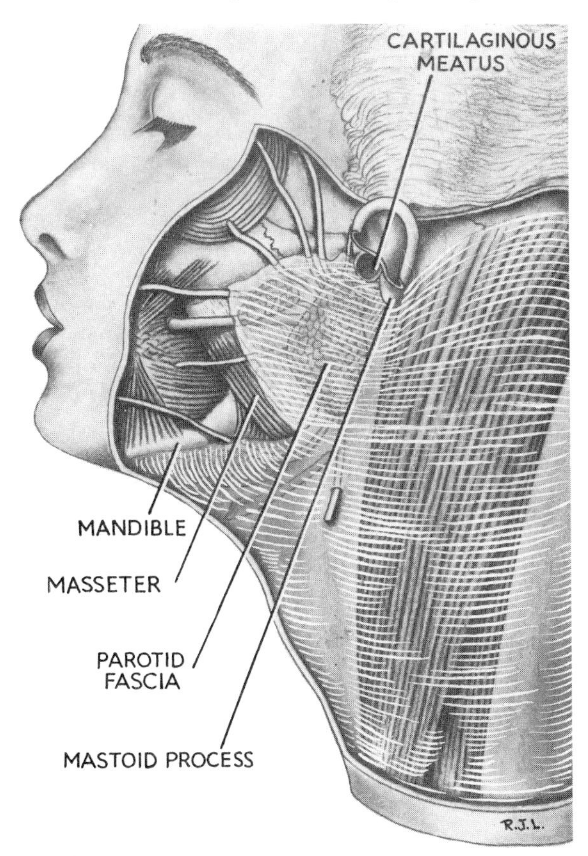

FIG. 6.1. THE UPPER ATTACHMENT OF THE INVESTING LAYER OF DEEP CERVICAL FASCIA. THE BRANCHES OF THE FACIAL NERVE PIERCE THE PAROTID FASCIA. For the arrangement of the veins see PLATE 37.

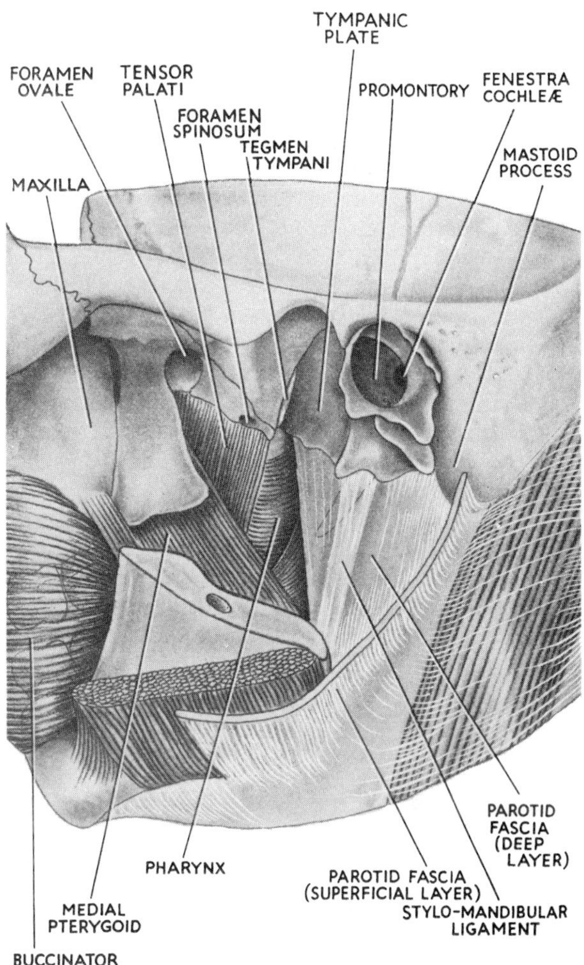

FIG. 6.2. THE DEEP LAYER OF THE PAROTID FASCIA, EXPOSED BY REMOVAL OF THE PAROTID GLAND, RAMUS OF MANDIBLE AND LATERAL PTERYGOID MUSCLE.

between these two layers, which may thus be spoken of as the **parotid fascia.** The upward attachment of the *superficial layer* extends from the tip of the mastoid process across the cartilaginous part of the external acoustic meatus to the lower border of the zygomatic process of the temporal bone (Fig. 6.1). It follows the lower border beyond the temporo-zygomatic suture. There is thus a 'free border' of the superficial layer of the parotid fascia extending from the zygomatic bone to the angle of the mandible. It is not really free, since the parotid fascia passes forwards beyond the parotid gland and blends with that epimysium on the surface of the masseter so unnecessarily named the masseteric fascia. The *deep layer* of the parotid fascia is attached along the base of the skull from the tip of the mastoid process to the lower border of the tympanic plate, as far medially as the carotid foramen (Fig. 6.2). Here it blends with the carotid sheath. The part of it that extends between the styloid process and the angle of the mandible is usually rather thicker than the rest and is called the **stylo-mandibular ligament.** As in the case of the superficial layer, so in the deep layer there is a 'free' anterior border. This thins out and becomes adherent to the deep surface of the parotid gland, with whose histological capsule it finally blends (Fig. 6.2).

The lower attachment of the investing layer is to the pectoral girdle. It is attached to the spine of the scapula and the lateral flat part of the clavicle with the trapezius muscle, and to the clavicle and sternum with the sterno-mastoid. In the intervals between these muscles, it is attached to both clavicles and to the jugular notch by two layers into which it splits a short distance above them. The layers are attached to the anterior and posterior borders of the jugular notch, enclosing between them the *suprasternal space* in which the anterior jugular veins angle laterally to pass behind the sterno-mastoid muscles. Of the two layers that adhere to the middle third of the clavicle (between sterno-mastoid and trapezius) the deeper splits around the inferior belly of the omo-hyoid, forming a fascial sling which binds the muscle down to the clavicle (Fig. 2.3, p. 55). The two layers represent, above the clavicle, the two layers of clavi-pectoral fascia that enclose the subclavius muscle. They are pierced by the external jugular vein (PLATE 37).

The relations of the veins of the face and neck to the investing layer are dealt with on p. 385.

The Prevertebral Fascia (Fig. 6.9, p. 375)

This is a firm, tough membrane that lies in front of the prevertebral muscles. It extends from the base of the skull in front of longus capitis and rectus capitis lateralis downwards to the lower limit of the longus colli muscle (body of Th. 3 vertebra). It extends sideways across the scalenus anterior, scalenus medius and levator scapulæ muscles, getting thinner further out and finally fading out of existence well under cover of the anterior border of trapezius. In the posterior triangle of the neck it covers the muscles that floor the triangle, and, since it crosses in front of the anterior tubercles of the cervical transverse processes, all the cervical nerve roots (and thus the cervical plexus and trunks of the brachial plexus) lie deep to it. The lymph nodes of the posterior triangle and the accessory nerve lie superficial to it. It has also the third part of the subclavian artery (not the vein) deep to it; it becomes prolonged over the artery below the clavicle as the axillary sheath. It does not invest the subclavian or axillary vein; these lie in loose areolar tissue anterior to it, free to dilate during times of increased venous return from the upper limb. The fascia is pierced by the four cutaneous branches of the cervical plexus (*q.v.*). The purpose of the prevertebral fascia is to provide a fixed basis on which the pharynx, œsophagus and carotid sheaths can glide during neck movements and swallowing (Fig. 6.9, p. 375), undisturbed by any movements of the prevertebral muscles.

The Pretracheal Fascia

Like the other fasciæ of the neck, this forms a layer between sliding surfaces. It lies deep to the infrahyoid strap muscles (sterno-thyroid, sterno-hyoid and omohyoid) so that its upward attachment is limited by the respective attachments of those muscles, namely, the hyoid bone at the midline and the oblique line of the thyroid cartilage more laterally. It splits to enclose the thyroid gland, to which it is not adherent except between the isthmus and second, third and fourth rings of the trachea. It is pierced by the thyroid vessels. Laterally, it fuses with the front of the carotid sheath on the deep surface of the sterno-mastoid (PLATE 23, p. *xii*) and inferiorly it passes behind the brachiocephalic veins to blend with the adventitia of the arch of the aorta (PLATE 44). No structure lies between the trachea and the pretracheal fascia, which thus provides a slippery surface for up and down gliding of the trachea during swallowing and neck movements.

The Carotid Sheath

This is not a fascia in the sense of a demonstrable membranous layer, but consists of a dense feltwork of areolar tissue that surrounds the carotid arteries (common and internal) and invests the vagus nerve. It is thin and almost non-existent where it overlies the internal jugular vein; this is to be expected, for the vein must be free to dilate during increased blood-flow. The carotid sheath is attached to the base of the skull at the margins of the carotid foramen, and is continued downwards along the vessels to the aortic arch. In front the lower part of the sheath is firmly attached to the deep surface of the sterno-mastoid and along this line the pretracheal fascia blends with it. Behind the carotid sheath there is a minimum of loose areolar tissue between it and the prevertebral fascia—not enough to limit the spread of infection. In this tenuous tissue the cervical sympathetic trunk lies, attached to the front of the prevertebral fascia (Fig. 6.9, p. 375).

The Tissue Spaces of the Neck

Behind the prevertebral fascia is a closed space from which the only escape can be made by a perforation in the fascia. For example, an abscess from a cervical vertebra can lift the prevertebral fascia as far as the third thoracic body, but can extend no lower unless the fascia gives way. Immediately in front of the prevertebral fascia is a space that extends from the base of the skull to the diaphragm (Fig. 4.9, p. 222). Its upper part is the retropharyngeal space; and below this the space extends behind the œsophagus through the superior into the posterior mediastinum. In the neck it passes freely behind the carotid sheath into the posterior triangle and pus from the retropharyngeal space frequently points there. In such cases pathological walling off has prevented spread of pus through the superior into the posterior mediastinum; no anatomical structure hinders its descent. The space behind the pharynx and larynx likewise extends laterally *behind* the carotid sheath into the posterior triangle, and below into the superior and posterior mediastinal spaces. In front of the pretracheal fascia, however, the space is limited above and laterally by the attachments of the fascia, but extends freely below through the front of the superior mediastinum to the pre-pericardial space, the anterior mediastinum (Fig. 4.9, p. 222).

THE POSTERIOR TRIANGLE

This is an area enclosed between the sterno-mastoid and trapezius muscles. Its apex lies high up at the *back* of the skull on the superior nuchal line. Its base is *in front* at the root of the neck and consists of the part of the clavicle lying between the two muscles, generally the middle third of the bone. The triangle is a spiral and therefore illustrations tend to distort its true appearance. Its roof is formed by the investing layer of deep cervical fascia described above. Its floor consists of the prevertebral fascia. Beneath this fascia may be seen, from above downwards, portions of the following muscles: splenius, levator scapulæ, scalenus posterior, scalenus medius, and scalenus anterior, while in the lateral angle just beyond the border of the first rib, part of the first digitation of serratus anterior is visible, especially if the shoulder be depressed. The prevertebral fascia plasters down upon these muscles the subclavian artery, the three trunks of the brachial plexus and the loops of the cervical plexus. In operations on the posterior triangle all these structures are safe provided the prevertebral fascia is left intact (Fig. 6.9, p. 375).

Contents of the Triangle. Lying between the floor and roof are the lymph nodes of the posterior triangle, and the accessory nerve. The cutaneous branches of the cervical plexus pass straight through to pierce the deep fascia at the posterior border of sterno-mastoid. The transverse cervical and suprascapular arteries, with their veins, cross the lower part of the triangle just above the clavicle. The nodes are most numerous just above the clavicle and are generally known as the **supraclavicular lymph nodes.** They are really outlying members of the postero-inferior group of deep cervical lymph nodes (p. 443). Two or three small nodes lie at the apex of the triangle in the subcutaneous tissue; known as **occipital nodes,** they become enlarged in German measles and scalp infections. The **accessory nerve** emerges beneath the posterior border of sterno-mastoid at the junction of its upper and middle thirds, and passes almost vertically downwards on levator scapulæ to disappear beneath the anterior border of trapezius at the junction of its middle and lower thirds. It is thus particularly liable to injury in operations for removal of nodes from the posterior triangle (Fig. 6.4, PLATE 25, p. *xiii*). The *surface marking* of the accessory nerve can best be demonstrated by palpating the lateral mass of the atlas (the nerve lies on the jugular vein here). With the subject viewed in *accurate lateral profile* a vertical line running down from this point and out towards the shoulder marks the course of the nerve, as seen in Fig. 6.4.

The Floor of the Posterior Triangle

At the apex, between sterno-mastoid and trapezius, the upper border of splenius capitis (p. 467) is often low enough to expose a little of semispinalis capitis (p. 466). Here the occipital artery and great occipital nerve (*posterior* primary ramus of C2) emerge and pass up to the scalp (PLATE 33, p. *xvii*).

Levator scapulæ is described with the muscles of the pectoral girdle (p. 58). It is usually split longitudinally into two ribbons of muscle. The accessory nerve lies upon it, separated by the prevertebral fascia. The transverse cervical artery divides at its anterior border, the ascending branch running up over the muscle, the descending branch running down deep to it, in company with the dorsal scapular nerve.

Scalenus posterior, scalenus medius and scalenus anterior are described with the root of the neck (p. 376).

The Cervical Plexus

The cervical plexus is formed by simple loops between the anterior primary rami of the upper four cervical nerves, after each has received a grey ramus from the superior cervical ganglion. It lies in series with the brachial plexus, on the scalenus medius, plastered down beneath the prevertebral fascia. It is covered by the upper part of sterno-mastoid, and does not lie actually in the posterior triangle.

Muscular branches are given off segmentally to the prevertebral muscles (longus capitis, longus colli and the scalenes). Other muscular branches are:

(*a*) A loop from C1 to the hypoglossal nerve, by which the fibres are carried to its meningeal branch and the superior ramus of the ansa cervicalis, the former 'descendens hypoglossi' (infrahyoid muscles) and the nerves to thyro-hyoid and genio-hyoid.

(*b*) Branches from C2 and 3 to the sterno-mastoid, and from C3 and 4 to the trapezius. These fibres are proprioceptive, all the motor fibres to the muscles being in the accessory nerve (Fig. 6.4, PLATE 25, p. *xiii*).

(*c*) The **inferior ramus of the ansa cervicalis** (formerly called 'descendens cervicalis') is formed by union of a branch each from C2 and C3. The nerve spirals around the lateral side of the internal jugular vein and descends to join the superior ramus (C1) at the ansa (p. 370).

(*d*) **The Phrenic Nerve.** This is formed from C4 and runs down vertically over the obliquity of the scalenus anterior muscle, passing from lateral to medial borders, beneath the prevertebral fascia, lateral to the ascending cervical branch of the inferior thyroid artery. It receives unimportant contributions from C3 and C5. It passes behind the subclavian vein into the mediastinum (p. 226). It may be joined below the vein by a branch (the **accessory phrenic nerve**) from the nerve to subclavius. It is not uncommon for the phrenic nerve to descend in front of the subclavian vein (Fig. 6.3). Rarely it may penetrate the vein. This is understood by recalling that the vein forms by a coalescence of a rich venous plexus in the embryo.

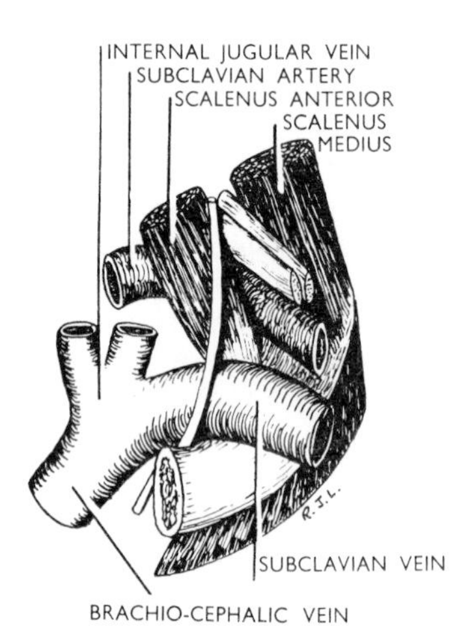

FIG. 6.3. THE LEFT PHRENIC NERVE (NOT LABELLED) PASSING IN FRONT OF THE SUBCLAVIAN VEIN. Drawn from Specimen S 173 B in R.C.S. Museum.

Cutaneous branches of the plexus supply the front and sides of the neck.

The Cutaneous Nerves of the Neck
(Fig. 6.4, PLATE 25)

A strip of skin over the extensor muscles of the neck and extending over the back of the skull as high as the vertex is supplied segmentally by posterior primary rami of cervical nerves (p. 21). The rest of the neck skin is supplied by anterior primary rami of C2, 3 and 4 from the cervical plexus. C1 has no cutaneous branch and from C5 downwards the dermatome concerned is projected peripherally to clothe the upper limb. Four nerves emerge from the posterior border of the sterno-mastoid muscle just below the accessory nerve and radiate like the spokes of a wheel.

The **lesser occipital nerve** (C2) is a slender branch that hooks around the accessory nerve and runs up along the posterior border of sterno-mastoid to supply the posterior part of the neck below the superior nuchal line (i.e., over the upper part of sterno-mastoid). It may overlap to the tip of the auricle.

The **great auricular nerve** (C2 and 3) is much more important. A large trunk passing vertically upwards over the sterno-mastoid, it is distributed to an area of skin on the face over the parotid gland and to the parotid fascia as well as to the auricle. It supplies the skin of the auricle over the whole of its cranial surface and on the lower part of its lateral surface below the external acoustic meatus. Branches passing deep to the parotid gland (PLATE 25) supply the deep layer of the parotid fascia. The great auricular nerve contains mostly C2 fibres (Fig. 1.36, p. 31).

The **transverse cervical** (anterior cutaneous nerve of the neck) (C2 and 3) emerges as a single trunk behind the posterior border of sterno-mastoid, and quickly breaks up into a number of slender twigs that innervate the skin in the midline of the neck from chin to sternum. Interruption of the main trunk thus produces a very elongated area of anæsthesia. (*Note:* The transverse cervical nerve must not be confused with the transverse cervical artery, which is just above the clavicle—see PLATE 25.)

The **supraclavicular nerves** emerge in common with the other three at the posterior border of sterno-mastoid. They contain fibres from C3 and 4 (overwhelmingly C4). They are distributed in three main groups. The *suprasternal nerves* are the most medial and supply the skin as far down as the sternal angle of Louis, and the sterno-clavicular joint. The *supra-clavicular nerves* proper pass anterior to (occasionally through) the clavicle and supply skin as far down as the anterior axial line. The *supra-acromial nerves* are the most lateral of this group, and pass not only across the acromion to the skin half-way down the deltoid muscle, but also over the posterior aspect of the shoulder to supply skin as far down as the spine of the scapula (posterior axial line: Fig. 1.25, p. 23).

Dermatomes of the Neck (see p. 30)

The cylindrical part of the neck is supplied by C3. Above this, the expanded part of the neck (from hyoid bone to chin, across the parotid gland and much of the ear and scalp behind this) is supplied by C2. Below

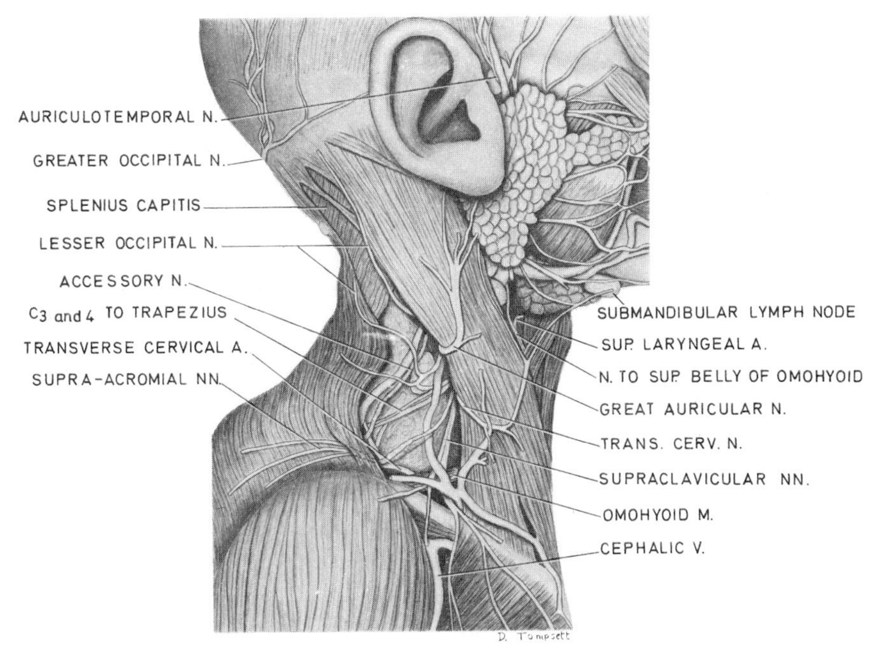

FIG. 6.4. A SUPERFICIAL DISSECTION OF THE RIGHT SIDE OF THE NECK AND FACE (see PLATE 25 for full colour). THE PRE-VERTEBRAL FASCIA IS INTACT IN THE LOWER PART OF THE POSTERIOR TRIANGLE. Illustration of Specimen S.350 in R.C.S. Museum.

the cylinder of the neck a wide area, extending across the chest to the anterior axial line (p. 26), over the tip of the shoulder half-way down the deltoid muscle, and posteriorly down to the spine of the scapula, an area that supports the weight of a cape, is supplied by C4. There is much overlap across the boundary lines shown in Figs. 1.36, p. 31 and 6.16, p. 384.

Sterno-mastoid Muscle. This muscle is usually not adequately described. The French anatomists describe it with great precision. It consists of four parts. The two heads of origin below spring from manubrium and clavicle respectively; that from the manubrium is a rounded tendon, that from the clavicle a thick fleshy mass. A triangular interval exists between the two above the sterno-clavicular joint. Behind this lies the termination of the internal jugular vein. The manubrial tendon is attached to the front of the bone below the jugular notch (it can be seen, unlabelled, in Fig. 6.5, PLATE 26). From the manubrial tendon a broad fleshy belly develops that passes upwards to be inserted into a curved line extending from the tip of the mastoid process to the superior nuchal line of the occiput. Thus one can speak of two parts of the muscle here, namely, *sterno-mastoid* and *sterno-occipitalis*, though the two parts are intimately blended. The posterior fibres of the clavicular head are inserted into the superior nuchal line medial to and in line with the sterno-occipitalis fibres. This part of the muscle may be called *cleido-occipitalis*, and it forms with the other two a continuous sheet on the surface of the muscle. The fourth part of the muscle, the *cleido-mastoid*, lies, on the other hand, deep to these (PLATE 25); in the interval between it and the other three the accessory nerve passes from the side of the atlas into the posterior triangle. According to the rule (p. 8), the accessory nerve supplies the muscle by a branch which leaves it proximal to its point of entry. Branches from C2 and 3 also enter the muscle, but these are only proprioceptive. It is probable that the cell bodies of the motor fibres in the accessory lie symmetrically in the cord with the proprioceptive fibres, namely, in the second and third segments. The sterno-mastoid muscle lies enclosed within a sheath of the investing layer of deep cervical fascia, which splits to surround it. The attachment more anteriorly of this fascia to the hyoid bone draws the muscle forward in a gentle convexity that is very noticeable when the head is rotated.

The blood supply of the sterno-mastoid muscle is derived from branches of the occipital and superior thyroid arteries.

Action of Sterno-mastoid. Contraction of one muscle produces the 'wry neck' position, namely, the ear approaching the tip of the shoulder and the chin rotating to the opposite side. If the opposite abducting muscles contract (scalenus medius, semispinalis cervicis, etc.) the neck is held vertical and simple rotation of the head results. Sterno-mastoid is the principal rotator of the atlanto-axial joint, and in turning the face towards the opposite side its contraction can easily be seen and felt.

The action of both muscles contracting simultaneously is more complex. The result depends in part upon the previous position of the head, and above all upon what other groups of muscles are contracting at the same time. The posterior fibres pass up behind the atlanto-occipital joint, which they therefore extend. The most anterior fibres, on the other hand, lie just in front of the joint, and this part of the muscle flexes the skull on the atlas.

The chief purpose of the muscle, however, starting with the head is the normal position (face vertical, gaze horizontal) is to protract the head; that is, to move the face forwards while keeping it vertical and maintaining a horizontal gaze, as in peering over someone's shoulder in a crowd. Note that this action is a combination of flexion of the cervical spine and extension of the atlanto-occipital joint simultaneously. It cannot be over-emphasized, however, that the sterno-mastoids seldom contract alone, and their action must always be considered in association with other groups. They are essentially antagonists to the extensor muscles behind the vertebræ, but since they are not attached to the vertebral column, but only to the movable head, they of necessity exert a powerful action on the atlanto-occipital joints in addition to flexing the cervical spine.

Sterno-mastoid Bed

The muscle separates the posterior and anterior triangles of the neck. When the triangles have been studied there remains a wide strip of territory deep to the muscle. In general one can say that deep to the upper half of sterno-mastoid lies the cervical plexus (p. 363) and deep to its lower part lies the common carotid artery and the carotid sheath (p. 369) over the scalenus anterior muscle (p. 376).

THE ANTERIOR TRIANGLE OF THE NECK

The descriptive subdivisions of this area into several triangles are unnecessary and, apart from the submandibular triangle, they are not dealt with separately in this book. The general area bounded by the lower border of the mandible, and lying between the two sterno-mastoid muscles, is dealt with as a whole. It contains the viscera of the neck.

Beneath the investing layer of deep cervical fascia is a group of longitudinal muscles that extends from the mandible to the sternum, and which is supplied segmentally by the anterior primary rami of the upper three cervical nerves. Above the hyoid bone is the genio-hyoid muscle. It lies on the mouth side of the mylo-hyoid, which prevents its being seen from the neck, but since the mylo-hyoid is a special muscle of mastication that has migrated below genio-hyoid, its description is not given here (p. 371). Below the hyoid are the sterno-hyoid and omo-hyoid muscles, lying side by side in the same plane, and more deeply a wider sheet of muscle that is attached to the thyroid cartilage and receives two names, the thyro-hyoid and sterno-thyroid muscles. All these muscles are in the same plane of the body wall musculature as rectus abdominis. One should picture the rectus sheet as extending from chin to symphysis pubis, supplied segmentally from C1 to Th. 12 spinal nerves, but interrupted by the intervention of the thoracic cage. Remnants of the longitudinal strip sometimes lie over pectoralis major, forming the rectus sternalis muscle (Fig. 2.2, p. 54).

Genio-hyoid Muscle. A slender flat ribbon of muscle extends from each inferior genial tubercle to the upper border of the body of the hyoid bone. The two muscles lie side by side between the mylo-hyoid muscle and the base of the tongue (genio-glossus). The nerve supply is by a branch from the hypoglossal nerve, but it consists exclusively of fibres from C1 that 'hitch-hike' along the hypoglossal nerve (PLATES 39 and 27, pp. *xxi* and *xiv*).

Sterno-hyoid Muscle. A flat strap of muscle extends from the lower border of the hyoid bone to the back of the sterno-clavicular joint and adjoining parts of manubrium and clavicle. The two muscles lie edge to edge at the hyoid bone, but diverge from each other below; the Adam's apple projects through the space between them (PLATE 26). The muscle is supplied by a branch from the ansa cervicalis which enters the lower end of the muscle and runs up in its substance to supply it segmentally (C1, 2 and 3 from above down). Tendinous intersections, occasionally present, indicate the segmental origin of the muscle from paravertebral myotomes (cf. rectus abdominis).

Omo-hyoid Muscle. This flat strap of muscle lies edge to edge with the sterno-hyoid at its attachment to the lateral part of the inferior border of the hyoid bone. As it descends it diverges somewhat from the sterno-hyoid and, passing beneath the sterno-mastoid muscle, it comes to lie over the carotid sheath. Where it lies on the internal jugular vein, the muscle fibres are replaced by a flat tendon that slides readily on this structure. A change of direction now occurs, and the inferior belly passes almost horizontally just above the level of the clavicle to pass back to its attachment to the transverse scapular ligament and a little of the upper border of the scapula. The intermediate tendon and supraclavicular portion of the muscle are bound down to the clavicle in a fascial sling derived from the deep lamina of the investing layer of deep cervical fascia Fig. 2.3, p. 55). The muscle is supplied segmentally (C1, 2, 3 from above down) by the ansa cervicalis. The *development* of the muscle is interesting. At first attached in the fœtus to the medial end of the clavicle it is a longitudinal strip of muscle like the rest of the infrahyoid musculature. By a process of migration, however, it moves along the clavicle and finally reaches its adult attachment to the scapula. The process of migration may be arrested at any point. When the head is rotated to one side, the opposite omo-hyoid is straightened, and can exert a more direct downward pull on the hyoid bone (Fig. 6.54, p. 443).

Thyro-hyoid Muscle. This is a broader muscle, flat like a strap, that lies under cover of the upper ends of sterno-hyoid and omo-hyoid. It arises beneath these two muscles from the greater horn of the hyoid bone, and is inserted into the oblique line of the thyroid cartilage, end to end with the sterno-thyroid. It is supplied by a branch of the hypoglossal nerve, but the fibres are all 'hitch-hiking' from C1, they have no cell bodies in the hypoglossal nucleus (cf. genio-hyoid) (PLATE 39).

Sterno-thyroid Muscle. Broader than sterno-hyoid and lying deep to it, this muscle is attached lower down than sterno-hyoid on the posterior surface of the manubrium; so low, in fact, that it is below the sterno-clavicular joint and has no attachment to the clavicle. It extends laterally from the manubrium to the first costal cartilage. Its upper attachment is to the oblique line of the thyroid cartilage. The muscle is supplied by the ansa cervicalis (C2 and 3). Its C1 part is represented by the thyro-hyoid muscle (PLATE 26).

Action of the Infrahyoid Muscles. The infrahyoid muscles are all depressors of the larynx. The sterno-thyroid acts directly on the thyroid cartilage, the others act indirectly via the hyoid bone. Depression of the larynx increases the volume of the resonating chambers during phonation and thus affects the quality of the voice. Probably a much more significant role of the infrahyoid muscles is to oppose the elevators of the larynx (mylo-hyoid, palato-pharyngeus, stylo-pharyngeus, salpingo-pharyngeus), 'paying out rope' during contraction of the elevators; descent of the larynx after elevation is due to elastic recoil of the trachea. The infrahyoid muscles prevent ascent of the hyoid bone when the digastric is in action (see p. 446). A contracting thyro-hyoid approximates the thyroid cartilage and hyoid bone. Which moves the more depends on what other muscles are in action at the time.

The Thyroid Gland

(PLATES 44 and 28, pp. *xxiii* and *xv*)

The gland consists of two symmetrical lobes united in front of the second, third and fourth tracheal rings by an isthmus of gland tissue. Each **lobe** is pear-shaped, consisting of a narrow upper pole and a broader lower pole. It lies under cover of the sterno-thyroid and sterno-hyoid muscles, to the *side* of the larynx and trachea. The upper pole lies tucked away beneath the upper end of the sterno-thyroid muscle, between it and the ala of the thyroid cartilage. The lower pole extends along the side of the trachea as low as the sixth tracheal ring. The gland possesses its own delicate histological capsule, or fascia propria. It lies free within an envelope of pretracheal fascia. The **isthmus** joins the anterior

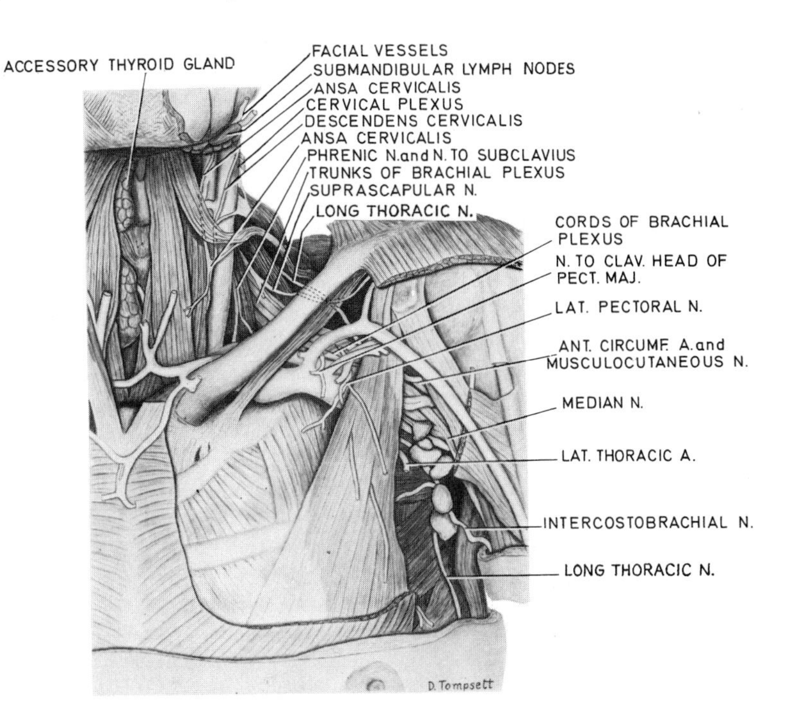

FIG. 6.5. THE FRONT OF THE NECK AND THE LEFT AXILLA (see PLATE 26 for full colour). Illustration of Specimen S.350 in R.C.S. Museum.

surfaces of the lobes, towards their lower poles (PLATE 28). The posterior surface of the isthmus is firmly adherent to the second, third and fourth rings of the trachea, and the pretracheal fascia is here fixed between them. This fixation and the investment of the whole gland by pretracheal fascia are responsible for the gland moving up and down with the larynx during swallowing.

Pyramidal Lobe. A small portion of gland substance often projects upwards from the isthmus, generally to the left of the midline. It is named the pyramidal lobe and represents a development of glandular tissue from the caudal end of the thyro-glossal duct. It is attached to the inferior border of the hyoid bone by fibrous tissue—muscle fibres sometimes present in it are named *levator glandulæ thyroideæ* and are innervated by a branch of the external laryngeal nerve (PLATE 39).

Accessory Thyroid Glands. Separate masses of thyroid tissue are not uncommonly found near the hyoid bone (PLATE 26), in the superior mediastinum, or beneath the sterno-mastoid muscle.

Blood Supply. The **superior thyroid artery,** the first branch from the anterior aspect of the external carotid artery, after giving off its sterno-mastoid and superior laryngeal branches, pierces the pretracheal fascia as a single vessel to reach the summit of the upper pole. It divides on the gland into an anterior branch that runs down to the isthmus and a posterior branch that runs down the back of the lobe and anastomoses with an ascending branch of the inferior thyroid artery from the lower pole. The **inferior thyroid artery,** on the other hand, divides outside the pretracheal fascia into four or five branches that pierce the fascia separately to reach the lower pole of the gland. The recurrent laryngeal nerve lies normally behind these branches, but it is common for it to pass between them before they pierce the pretracheal fascia. The nerve always lies behind the pretracheal fascia, and if this structure remains intact during thyroidectomy the nerve will not have been divided. It is close behind the fascia, however, and may be bruised or caught in a ligature; hence the advisability of ligating the inferior thyroid artery well lateral to the gland before it begins to divide into its terminal branches. (Note that the inferior thyroid artery gives off œsophageal and inferior laryngeal branches before its terminal distribution into the thyroid gland). The **thyroidea ima artery** (PLATE 28) enters the lower part of the isthmus in 3 per cent of individuals. It springs from the brachio-cephalic trunk or direct from the arch of the aorta.

The **venous return** from the upper pole follows the superior thyroid artery. This vein, the *superior thyroid vein,* enters either the internal jugular or common facial vein in about equal proportions. The *middle thyroid vein,* short and wide, is usually present. It passes from the middle of the lobe, directly into the internal jugular vein. From the isthmus and lower poles the *inferior thyroid veins* form a plexus that lies in the petracheal fascia, in front of the cervical part of the trachea. The plexus drains into the brachio-cephalic veins, most of it into the left one (PLATE 44).

The **lymphatics** follow the arteries. From the upper pole they enter the antero-superior group of deep cervical lymph nodes. From the lower pole they pass with the inferior thyroid artery back to its point of origin from the subclavian behind the carotid sheath into the postero-inferior group. A few pass downwards into pretracheal nodes, following the course of the thyroidea ima artery.

Nerve Supply of the Thyroid Gland. The bulk of the sympathetic supply is derived from the middle cervical ganglion and enters the gland on the inferior thyroid artery; some fibres from the superior cervical ganglion travel with the superior thyroid artery. The sympathetic fibres are vaso-constrictor. Vagus nerve filaments are traceable to the gland; their purpose is unknown.

Microscopic Anatomy. A section shows the typical colloid vesicles, whose appearance depends on the state of activity of the gland. In the 'resting' state the vesicles are uniformly distended with structureless colloid. They are rounded in outline, and lined with a

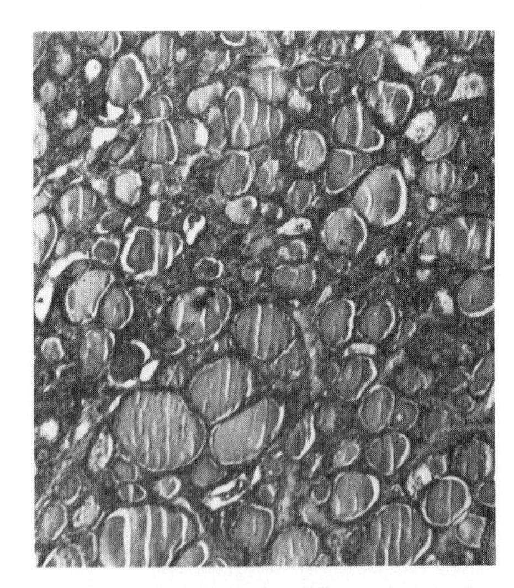

FIG. 6.6. THE THYROID GLAND. (Photomicrograph ×130.) This is the 'resting' stage; the vesicles are distended with stored secretion. The parallel fractures in the colloid are a common artefact.

well-stained layer of low columnar epithelium. Clumps of small round cells lie between the vesicles; they may produce calcitonin. In the more active state the amount of colloid is less, and the vesicles are uniformly smaller and crinkled in outline. The lining columnar epithelium is much taller. This appearance is well seen in the fœtal gland.

Development of the Thyroid Gland. The gland develops as a proliferation of cells from the caudal end of the thyro-glossal duct (p. 42). A contribution from the fourth pouch probably exists. There is some evidence that the calcitonin-producing cells between the vesicles may develop from the fourth and even the fifth (ultimo-branchial body) pouch.

Parathyroid Glands (PLATE 42, p. *xxiii*)

These small masses of gland tissue lie characteristically behind the lateral lobe of the thyroid gland, one at the upper pole, the other at the lower pole, near the anastomosing artery that joins superior and inferior thyroid arteries. The gland behind the upper pole is called parathyroid IV because it develops from the dorsal diverticulum of the fourth pharyngeal pouch; that behind the lower pole, parathyroid III, develops from the third pouch, but is displaced caudally by descent of the thymus (p. 41). All four parathyroid glands lie on or within the thyroid gland, inside the pretracheal fascia. Aberrant nodules of parathyroid tissue, outside the pretracheal fascia, are common, and have been discovered as low down as the superior mediastinum. The glands have a somewhat glairy appearance and are pinkish-buff in colour.

Microscopic Anatomy. In section the gland is homogeneous and very vascular, two points that aid identification. The gland is a mass of small round cells, something like lymphocytes in appearance. But there is no arrangement into follicles, as in lymphatic tissue, nor is there a cortex and medulla, as in the thymic lobule. The cells are arranged in irregular spiral columns, but these, of course, cannot be seen in a section, and appear only as small clumps of cells.

The Trachea

The trachea is in continuity with the larynx. It is attached to the lower margin of the cricoid cartilage. Its patency as an airway is maintained by the presence in its wall of a series of C-shaped cartilages joined together by a fibro-elastic membrane. The gaps in the C-shaped rings lie posteriorly and this part of the trachea is closed by a sheet of unstriped muscle, the *trachealis muscle*. The trachea lies in the midline of the neck, in contact with the front of the œsophagus. In the groove between trachea and œsophagus runs the recurrent laryngeal nerve. To the side of the trachea is the carotid sheath. The isthmus of the thyroid gland is adherent to the second, third and fourth tracheal rings and the lobes of the gland lie against the lateral side of the trachea as far down as the sixth ring.

The trachea is supplied with blood by the inferior thyroid artery and its veins drain to the brachio-cephalic (innominate) veins by the inferior thyroid plexus. The lymphatics of the cervical part drain to the postero-inferior group of deep cervical nodes. The nerve supply is from the recurrent laryngeal branch of the vagus. Sympathetic fibres come in from the middle cervical ganglion on the artery. The vagus supplies sensory, secreto-motor and motor (trachealis muscle) fibres, while the sympathetic fibres are vaso-constrictor.

The thoracic part of the trachea and its functional anatomy are dealt with on p. 225.

The Œsophagus

The œsophagus commences in continuity with the crico-pharyngeus muscle at the level of the lower border of the cricoid cartilage. Its inner muscular coat breaks free from the lower border of this muscle to form a continuous circular coat. The outer longitudinal coat is attached to the midline ridge of the lamina of the cricoid cartilage and to the arytenoid cartilages. The fibres spiral down from this origin to the back of the œsophagus, whence the longitudinal coat is continuous (Fig. 6.34, p. 419).

The cervical part of the œsophagus is supplied by œsophageal branches of the inferior thyroid artery, and its veins drain into the brachio-cephalic (innominate) veins. The lymphatics drain back along the inferior thyroid artery to the postero-inferior group of deep cervical lymph nodes. The cervical part of the œsophagus is innervated by the recurrent laryngeal nerves (PLATE 42). These provide sensory, motor and secreto-motor fibres. The sympathetic supply comes along the inferior thyroid arteries from cell bodies in the middle cervical ganglion.

Note that the cervical and superior mediastinal parts of the œsophagus have the same neuro-vascular supply as the trachea, lower larynx and hypo-pharynx.

The thoracic part of the œsophagus is dealt with on p. 238.

The **carotid sheath** contains the common carotid artery, internal jugular vein and vagus nerve. It is a thickly matted fibrous tissue investment over the artery and nerve, extending up above the bifurcation on the internal carotid artery to the base of the skull (p. 393) and extending down through the superior mediastinum to the arch of the aorta. It is free posteriorly to slide over the prevertebral fascia, but its lower part is connected anteriorly by fibrous tissue to the fascia on the deep surface of the sterno-mastoid. Pus tracking laterally from around the pharynx thus passes behind, not in front of the sheath, to point in the posterior triangle. The sheath is really non-existent over the internal jugular vein, which is thus free to dilate enormously for increased blood-flow. The 'dead space' around the vein contains the inferior deep cervical lymph nodes in anterior and posterior groups, the latter extending behind the sterno-mastoid into the posterior triangle, forming a supraclavicular group (p. 363).

In the *lower part* of its course the carotid sheath extends from the sterno-clavicular joint vertically upwards to the level of the upper border of the lamina of the thyroid cartilage (C3 vertebra), where the **common carotid artery** bifurcates. The terminal portion of the artery is often dilated into the carotid sinus, continuous with that on the commencement of the internal carotid artery (p. 393). The common carotid artery arises on the left side from the arch of the aorta, where it lies in front of the subclavian artery up to the sterno-clavicular joint. Here the two arteries diverge. On the right the brachio-cephalic trunk bifurcates behind the sterno-clavicular joint into common carotid and subclavian arteries. The common carotid gives off no branches proximal to its bifurcation.

The *internal jugular vein* receives the *middle thyroid vein.* This short wide vessel passes from halfway down the lateral border of the gland horizontally across the common carotid artery (PLATE 44, p. *xxiii*).

The **ansa cervicalis** lies on the front of the internal jugular vein and gives branches to the infrahyoid muscles. It is formed by union of superior and inferior rami. The *superior ramus* (descendens hypoglossi) is a branch of the hypoglossal nerve (p. 374) given off where the nerve loops just below the posterior belly of the digastric muscle, on the occipital, external carotid and lingual arteries. It runs down on the front of the internal jugular vein. It contains only C1 fibres, which have hitch-hiked along the hypoglossal nerve.

The *inferior ramus* (descendens cervicalis) is formed by union of a branch each from C2 and C3 in the cervical plexus. The single nerve so formed spirals from behind around the internal jugular vein and runs down to join the superior ramus at a variable level. Sometimes a wide loop is formed over the lower part of the vein and the branches arise from the loop. Sometimes the two nerves join, Y-shaped, high up and the branches are given off from the stem of the Y. In either case they are distributed to the infrahyoid muscles (sterno-hyoid, sterno-thyroid and omo-hyoid) segmentally, C1, C2 and C3 from above down (PLATES 26 and 35, pp. *xiv* and *xix*).

The *upper part* of the carotid sheath lies at the back of the infratemporal fossa and has different contents. Its attachments and relations are dealt with on p. 393.

THE SUPRAHYOID REGION

The hyoid bone is connected to the mandible by a thin sheet of muscle, the mylo-hyoid, which forms the floor of the mouth, and arises in uninterrupted continuity from the whole length of the mylo-hyoid line, from below the genial tubercles at the midline to the posterior surface of the third molar on each side. Superficially (i.e., below it) lies the anterior belly of the digastric muscle. The nerve to the mylo-hyoid and the submental branch of the facial artery run between the two. Finally lying upon it, half hidden under the mandible, are the submandibular salivary gland and lymph nodes. These structures are covered in by the investing layer of deep cervical fascia, which is attached to the hyoid bone and the inferior border of the mandible. In the subcutaneous tissue lie the platysma muscle, the anterior jugular veins and the submental lymph nodes. The skin of this region is supplied by the upper division of the transverse cervical nerve (C2).

The Platysma. This part of the panniculus carnosus forms a broad flat sheet that extends from the deep fascia beneath the breast upwards to the inferior border of the mandible (PLATE 31). The two muscles are separated below, but converge above towards the midline and actually overlap just beneath the chin. The lateral border of each muscle slants up across the parotid gland and converges into a triangle whose apex is attached to the modiolus, constituting the *risorius muscle*. The platysma is supplied by the cervical branch of the facial nerve. Its purpose is obscure; it elevates the skin over the upper part of the chest if the jaw be fixed; *per contra* it may be a minor factor in opening the jaw, but this is very improbable except in extreme dyspnœa (test it on yourself).

Anterior Jugular Veins. These veins commence beneath the chin and pass downwards, side by side beneath the platysma, to the suprasternal region. Here they pierce the deep fascia and come to lie in the suprasternal space, where they are often connected by a short anastomotic vein. Each now angles laterally and passes deep to the sterno-mastoid muscle to open into the external jugular vein after the latter has pierced the deep fascia (PLATE 26).

Submental Lymph Nodes. A few small nodes lie beneath the chin, some superficial, others deep to the investing layer of deep fascia. They drain, *across the midline*, a wedge of tissue in the floor of the mouth opposite the premaxilla of the upper jaw, namely tip of tongue, floor of mouth, gums and lips opposite the four lower incisor teeth. In their turn they drain into the submandibular group of lymph nodes or directly into the jugulo-omo-hyoid node (Fig. 6.55, p. 443).

The Submandibular Fossa

This space lies deep to the investing layer of deep fascia and the inferior border of the mandible. The investing layer of the deep fascia is attached to the inferior border of the mandible and to the whole length of the hyoid bone up to the tip of the greater cornu.

It is a demonstrable membrane that spans the fossa and supports the submandibular gland. Deep to this the fossa extends to the mylo-hyoid muscle; the space thus projects upwards under cover of the mandible as high as the mylo-hyoid line. No membrane surfaces the mylo-hyoid muscle deep to the gland; only the loose and *distensible* epimysium lies here. The deep fascia does not split to enclose the submandibular gland (Fig. 6.29, p. 412). Swelling in the space is limited by the attachments of the mylo-hyoid muscle and the deep fascia; distension of the space is seen in Ludwig's angina.

The Mylo-hyoid Muscle. This thin sheet of muscle forms the 'diaphragm' of the floor of the mouth. It arises by two halves in continuity from the whole length of the mylo-hyoid line of the mandible as far back as the posterior surface of the third molar tooth.

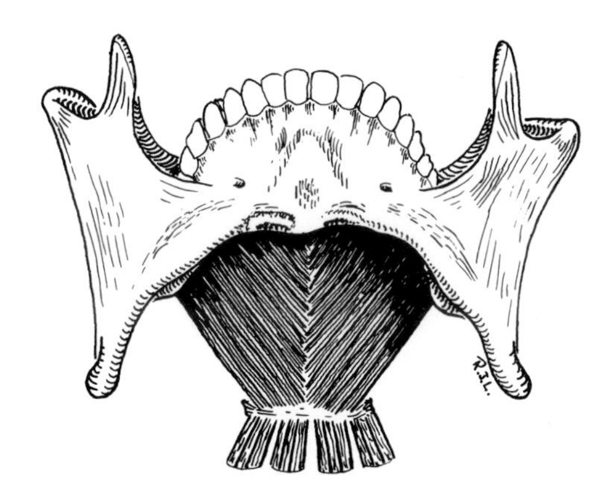

FIG. 6.7. The right and left halves of the mylo-hyoid muscle interdigitate in a midline raphé which extends from the chin to the hyoid bone.

The two halves of the muscle slope downwards towards each other. The posterior quarter of each muscle is inserted into the anterior surface of the body of the hyoid bone. In front of this the anterior three-quarters of each muscle interdigitate in a midline raphé which extends from the chin to the hyoid bone (Fig. 6.7).

The *nerve supply* is derived from the motor root of the trigeminal nerve by way of the mylo-hyoid branch of the inferior alveolar nerve.

Its *action* is to form a mobile but stable floor to the mouth and support the weight and thrust of the tongue. The two halves of the muscle together form a gutter. Contraction of the fibres makes the gutter more shallow, thus elevating the tongue and hyoid bone as, for example, in swallowing or in protruding the tongue. The action of the mylo-hyoid is discussed with mandibular movements on p. 447.

The Nerve to Mylo-hyoid. This nerve branches from the inferior alveolar nerve at the mandibular foramen (p. 392) and pierces the spheno-mandibular ligament at its attachment to the inferior margin of the foramen. It is accompanied by small branches of the inferior alveolar artery and vein; the vessels indent the mandible, forming the mylo-hyoid groove. The groove marks the upper limit of the attachment of the medial pterygoid muscle (Fig. 8.15, p. 572).

The small artery supplies the periosteum and some fibres of the medial pterygoid; the nerve, however, passes forwards on the neck surface of the mylo-hyoid muscle. It ends between the anterior belly of the digastric and the mylo-hyoid muscle, supplying both. It is accompanied by the *submental branch of the facial artery*. The submental artery gives off a branch which perforates the mylo-hyoid muscle to supply the anterior part of the floor of the mouth, including the sublingual gland. *Submental veins* run back to the common facial vein. Lymphatics pierce the anterior part of mylo-hyoid to drain tip of tongue and floor of mouth into the submental lymph nodes (Fig. 6.55, p. 443).

The Submandibular Gland (PLATE 35, p. *xix* and Fig. 6.29, p. 412).

This salivary gland, two thirds serous in man, lies in the submandibular fossa, partly under cover of the mandible. Its contact with the bone is responsible for the smooth elongated concavity that lies below the mylo-hyoid line opposite the bicuspid and molar teeth. Of the size of a walnut, it becomes narrow posteriorly and curves around the free posterior border of the mylo-hyoid muscle, and a small deep part of the gland thus comes to lie in the floor of the mouth, between the mandible and the side of the tongue (i.e., between the mylo-hyoid and hyo-glossus muscles). The main duct, within the gland, thus curves around the posterior border of the mylo-hyoid muscle, but does not become free until it leaves the anterior end of the deep, or mouth part of the gland. The facial artery, curving downwards over the posterior belly of the digastric muscle, lies in a deep groove (or even a tunnel) in the gland, between it and the mandible, before it arches upwards around the inferior border of the mandible to reach the face at the front of the masseter muscle. The common facial vein, sloping downwards to the internal jugular vein, grooves the surface of the posterior part of the gland (PLATE 35). The mandibular branch of the facial nerve (p. 382) crosses the gland here (PLATE 35, p. *xix*).

Blood Supply. It is supplied from the facial artery and its veins drain into the common facial vein. Its lymph passes to the submandibular lymph nodes.

Nerve Supply. Secreto-motor fibres to the gland have their cell bodies in the submandibular ganglion or in small ganglionic masses on the surface of the gland itself. The preganglionic fibres pass from cell bodies in the superior salivatory nucleus in the pons by way of the nervus intermedius and travel with the facial nerve as far as the stylo-mastoid canal. They leave the facial nerve in the chorda tympani (in company with taste fibres to the anterior two-thirds of the tongue), and passing across the lateral wall of the middle ear leave the skull through the petro-tympanic fissure. The chorda tympani, passing downwards, often grooves the medial side of the spine of the sphenoid. It joins the lingual nerve from behind, 2 cm below the base of the skull.

Development. A groove in the floor of the mouth becomes converted into a tunnel whose blind end proliferates to form the secreting acini. Its origin is almost certainly ectodermal (p. 39).

The Submandibular Lymph Nodes. Half a dozen lymph nodes lie on the surface of the submandibular salivary gland. They drain the submental lymph nodes, the lateral parts of the lower lips, all the upper lip and external nose, and the anterior two-thirds of the tongue (except the extreme tip) of their own side (PLATE 35). They receive also from the anterior half of the nasal walls and the paranasal sinuses that drain there (frontal, anterior and middle ethmoidal, maxillary sinus).

The Side of the Neck

The **digastric muscle** arises as a fleshy belly from the digastric notch on the medial surface of the mastoid process. A triangular belly tapers down to the intermediate tendon, which is held beneath a fibrous sling attached near the lesser cornu of the hyoid bone. The tendon is lubricated by a synovial sheath within the fibrous sling. The bifurcated tendon of insertion of the stylo-hyoid muscle plays no part in holding down the tendon. The anterior belly lies on the inferior surface of the mylo-hyoid muscle; it connects the intermediate tendon to the digastric fossa on the lower border of the mandible. The posterior belly is supplied by the facial nerve, the anterior belly by the mylo-hyoid nerve (trigeminal). The muscle is formed by fusion of derivatives of the second and first pharyngeal arches (p. 40). Its action is to depress and retract the chin; it is, with the lateral pterygoid, the opener of the mouth (Fig. 6.56, p. 446).

The **external carotid artery** commences at the bifurcation of the common carotid, level with the upper border of the thyroid lamina. At first somewhat medial to, it slopes upwards in front of, the internal carotid artery, passes deep to the posterior belly of the digastric and the stylo-hyoid muscle, above which it pierces the deep lamina of the parotid fascia and enters the gland. It divides behind the neck of the mandible into maxillary and superficial temporal arteries (PLATES 34 and 36).

Relations. At its commencement it lies against the side wall of the pharynx; its place is soon taken by the internal carotid artery, which then lies against the pharynx all the way up to the base of the skull. As the external carotid artery lies in the parotid gland it is separated from the internal carotid by the deep part of the gland and the 'pharyngeal' structures: stylopharyngeus muscle, glosso-pharyngeal nerve and pharyngeal branch of the vagus. At the commencement of the artery the internal jugular vein lies lateral, but higher up it is posterior and deep to the artery. The common facial vein crosses the artery, with the hypoglossal nerve lying between. Except at its commencement the vessel lies in front of the anterior border of sterno-mastoid.

Before it enters the parotid gland the external carotid artery gives off *six branches*, three from in front, two from behind and one deep (medial). The three from in front are the superior thyroid, lingual and facial, which arise below, at, and above the tip of the greater cornu of the hyoid bone and diverge widely. The two from behind are the occipital and posterior auricular, which pass up along the lower and upper borders of the posterior belly of the digastric muscle. The branch from the medial side is the ascending pharyngeal, which ascends to the base of the skull on the side wall of the pharynx, alongside the internal carotid artery.

The **superior thyroid artery** arises at the commencement of the external carotid. It slopes almost vertically downwards, with the vein, to the upper pole of the thyroid gland (p. 368). Close to it, alongside the larynx, lies the external laryngeal nerve (see PLATE 34, p. *xviii*). The main branch is the *superior laryngeal artery* (PLATE 36) which pierces the thyro-hyoid membrane with the internal laryngeal nerve. It also gives a branch to the sterno-mastoid muscle.

The **lingual artery** arises from the front of the external carotid at the tip of the greater cornu of the hyoid bone. It forms a short upward loop, then passes forwards along the upper border of the greater cornu, deep to the hyo-glossus muscle (p. 413). It is accompanied by a deep lingual vein which passes back to the internal jugular vein. It is crossed by the hypoglossal nerve and its companion vein, the latter opening into the common facial vein. The common facial vein overlies the hypoglossal nerve and the loop of the lingual artery on its way to the internal jugular vein (Fig. 6.8).

The **facial artery** arises from the front of the external carotid above the tip of the greater cornu of the hyoid bone, and runs upwards on the superior constrictor, deep to the digastric and stylo-hyoid muscles, then deep to the submandibular salivary gland. It indents the surface of the gland. As the artery lies on the superior constrictor muscle it gives off a branch to the tonsil

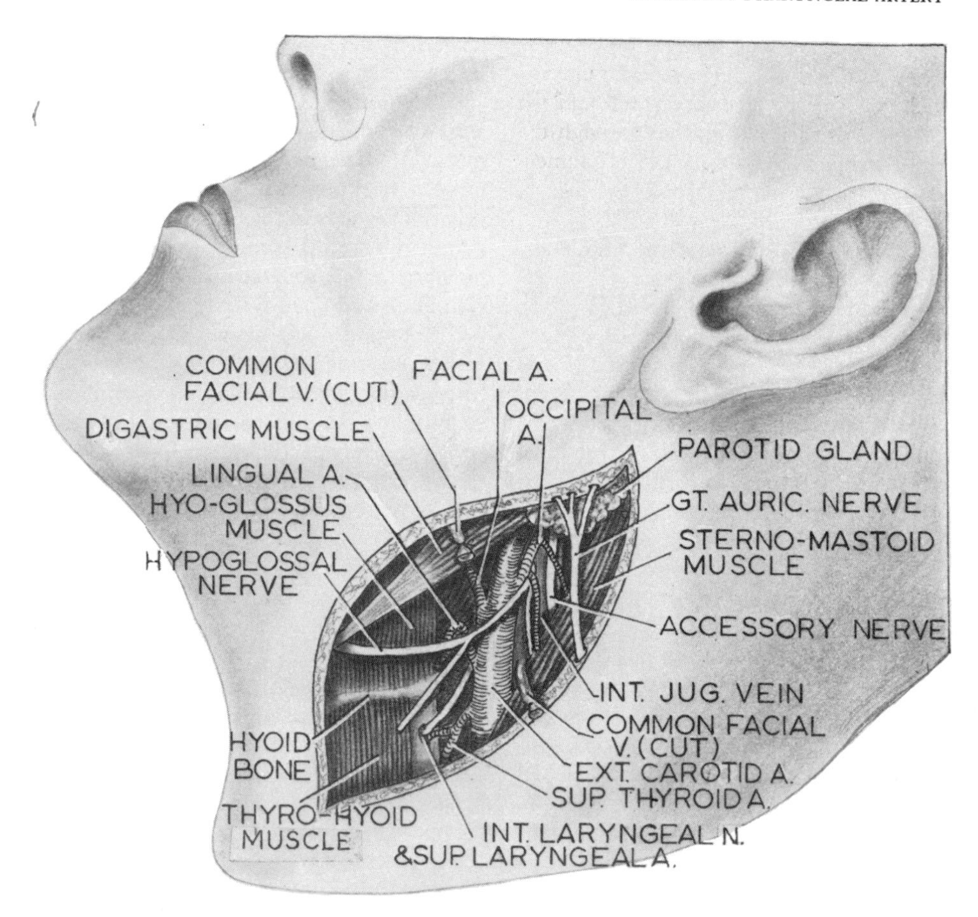

FIG. 6.8. AN EXPOSURE OF THE LEFT SIDE OF THE NECK. The common facial vein has been divided between ligatures. The three branches from the front of the external carotid artery are exposed. The occipital artery, from the back of the external carotid, gives off two sterno-mastoid branches. The upper branch marks the accessory nerve, lying on the internal jugular vein. The lower branch holds down the hypoglossal nerve, which here gives off the superior ramus of the ansa cervicalis.

and soft palate. The facial artery then makes an S-bend, curling down over the submandibular gland (Fig. 6.29, p. 412) and up over the mandible. Just before it crosses the inferior border of the mandible it gives off a sizable branch, the **submental artery** (PLATE 30), which accompanies the mylo-hyoid nerve into the submandibular fossa. There it supplies the anterior belly of the digastric and the mylo-hyoid muscle, and sends perforating branches through the floor of the mouth to the sublingual gland.

The **occipital artery** arises from the back of the external carotid on a level with the facial artery. It courses along the lower border of the digastric muscle, passing deeper and grooving the base of the skull at the occipito-mastoid suture deep to the digastric notch on the mastoid process. It passes back through the apex of the posterior triangle to supply the back of the scalp (p. 386). The artery gives off two branches to the sterno-mastoid. The upper branch is a guide to the accessory nerve in front of the upper border of the muscle. The artery at its origin is crossed by the hypoglossal nerve, which hooks around it from behind; here the nerve is held down by the lower sterno-mastoid branch of the artery (Fig. 6.8).

The **posterior auricular artery** arises above the level of the digastric muscle, sometimes within the substance of the parotid gland. It runs up superficial to the styloid process across the upper border of the digastric, and crosses the surface of the mastoid process. It supplies the skin over the mastoid process; it is cut in incisions for mastoid operations. Auricular branches supply the pinna of the ear. Its *stylo-mastoid branch* enters the stylo-mastoid foramen, supplies the facial nerve and gives off the *stapedial artery* to the stapedius muscle, of interest as being the remnant of the artery of the second pharyngeal arch (p. 43).

The **ascending pharyngeal artery** arises just above the commencement of the external carotid, from its deep aspect. It runs up along the side wall of the pharynx in front of the prevertebral fascia, deep to the internal carotid artery. It supplies the pharyngeal wall and the soft palate and sends meningeal branches through the foramina nearby (foramen lacerum, jugular foramen, hypoglossal canal).

The Thyro-glosso-facial Confluence of Veins

Many large veins converge towards the tip of the greater cornu of the hyoid bone to empty into the internal jugular vein. They come from face, tongue and thyroid gland and *their pattern, like most venous patterns, is very variable*. They overlie the hypoglossal nerve, which itself here overlies three arteries (occipital, external carotid, and loop of lingual). The conventional pattern usually described is given here, but it must be emphasized that one or more variants are almost always found.

The **common facial vein** crosses the hypoglossal nerve and external carotid artery to empty into the internal jugular vein beneath the sterno-mastoid muscle. It is formed on the surface of the submandibular gland by the confluence of the (anterior) facial vein and the anterior branch of the retromandibular vein. It receives the submental vein and the vena comitans nervi hypoglossi. The deep lingual vein, running with the lingual artery, empties directly into the internal jugular vein, just above this level. The superior thyroid vein empties into the internal jugular just below the entrance of the common facial vein.

Another way of putting it would be to say that all the veins mentioned above drain into the common facial vein with the proviso that any one of them (especially the superior thyroid vein) may drain directly into the internal jugular vein (PLATE 30, p. *xvi*).

The **jugulo-digastric node** (see PLATE 35) lies anterior to the internal jugular vein and sterno-mastoid muscle, just below the posterior belly of the digastric muscle. It is merely a named member of the antero-superior group of deep cervical lymph nodes. It receives lymphatics from the tonsil and neighbouring mucous membrane, and from the posterior members of the sub-mandibular group of lymph nodes (p. 443).

The **hypoglossal nerve** is described in the upper part of the carotid sheath on p. 396. It emerges from between the internal carotid artery and internal jugular vein deep to the digastric muscle. Hooking around the occipital artery it curves forwards over the external carotid artery and the loop of the lingual artery. As it crosses these three arteries it descends a little below the lower border of the digastric muscle, just behind the intermediate tendon. Here it gives off the superior ramus of the ansa cervicalis (descendens hypoglossi) (C1 fibres). *This is where to look for it in a dissection of the neck* (Fig. 6.8). It passes forwards just above the greater cornu of the hyoid bone, on the surface of the hyo-glossus, which muscle here separates it from the lingual artery. It gives off a branch to the thyro-hyoid muscle (C1 fibres) and passes above the mylo-hyoid muscle to enter the floor of the mouth (PLATES 34 and 39, pp. *xviii* and *xxi*).

The *superior ramus* of the ansa cervicalis (descendens hypoglossi) passes vertically downwards over the front of the internal jugular vein where it is joined by the inferior ramus (descendens cervicalis) to form the ansa cervicalis (p. 370).

THE PREVERTEBRAL REGION

The Prevertebral Muscles of the Neck
(Fig. 6.9)

Some relatively weak flexor muscles extend in front of the vertebral column from skull to superior mediastinum. They are covered anteriorly by a strong fibrous membrane, the prevertebral fascia, upon whose slippery surface the pharynx and œsophagus, together with the carotid sheaths, glide freely during neck movements or swallowing.

The **rectus capitis anterior** extends from the lateral mass of the atlas to the front of the foramen magnum.

The **rectus capitis lateralis** lies edge to edge with the former muscle; it extends from the lateral mass of the atlas to the jugular process of the occipital bone. The **anterior primary ramus of C1,** passing forwards lateral to the atlanto-occipital joint, supplies each muscle and then passes between them to sink into the overlying longus capitis muscle. It gives a branch to the hypoglossal nerve, which is distributed in the meningeal branch, the superior ramus of the ansa cervicalis and the branches to the thyro-hyoid and genio-hyoid muscles.

The **longus capitis** muscle is attached by four slender tendons, in line with those of scalenus anterior, to the anterior tubercles of the four 'typical' cervical vertebræ (C3, 4, 5 and 6). The ribbon-shaped muscle so formed is inserted into the basi-occiput, the two muscles lying side by side in front of the foramen magnum alongside the pharyngeal tubercle. The muscles bulge slightly into the upper part of the naso-pharynx. They are supplied segmentally by anterior primary rami of the upper four cervical nerves. Their action is to flex the skull and upper neck; they are weak because their action is assisted by gravity and by the powerful sterno-mastoids.

The **longus colli** muscle extends from the anterior tubercle of the atlas into the superior mediastinum. It consists of upper, lower and central fibres. The *upper fibres* connect the anterior tubercle of the atlas with the anterior tubercles of the transverse processes of the third, fourth and fifth cervical vertebræ. The *lower fibres* connect the bodies of the upper three thoracic vertebræ with the anterior tubercles on the transverse processes of the fifth and sixth cervical vertebræ. The lateral border of this lower part makes with the medial border of scalenus anterior a pyramidal space whose apex is the carotid tubercle (of Chassaignac) and whose base is the first part of the subclavian artery. In this space lie the vertebral artery and cervical sympathetic trunk with the stellate ganglion (Fig. 6.11). The *central fibres* of longus colli connect the bodies of the second, third and fourth cervical vertebræ with the remaining cervical and the upper three thoracic vertebræ (Fig. 6.9).

The longus colli is supplied segmentally by the anterior primary rami of the spinal nerves. It is a flexor of the neck, weak because it is aided by gravity and the strong sterno-mastoids.

The *prevertebral fascia* is considered on p. 362.

The Cervical Sympathetic Trunk
(Fig. 6.9)

The cervical trunk ascends from the thorax across the neck of the first rib, medial to the highest intercostal vein. It runs up medial to the vertebral artery and lies in front of the prevertebral fascia, to which it

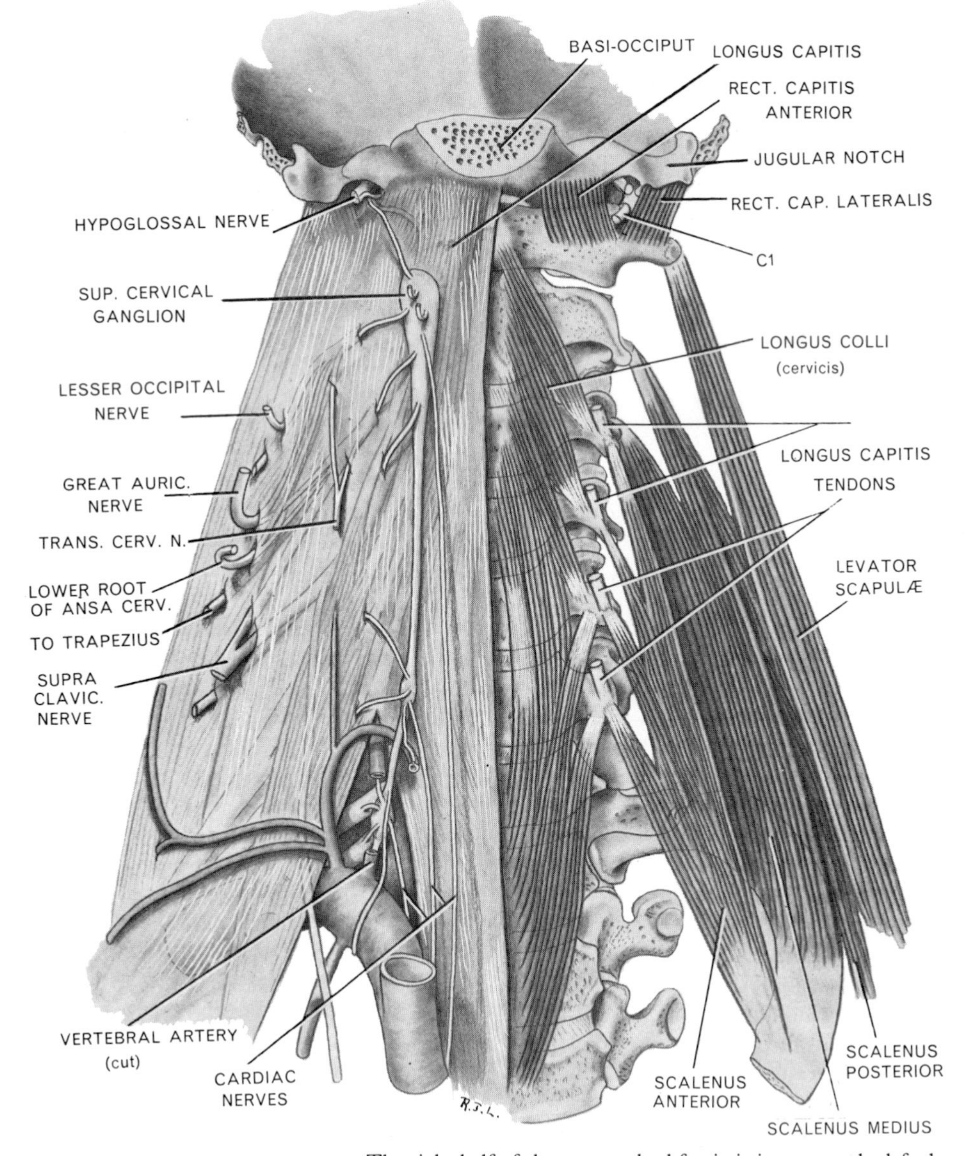

FIG. 6.9. THE PREVERTEBRAL REGION OF THE NECK. The right half of the prevertebral fascia is intact; on the left the prevertebral muscles are exposed.

is attached by loose areolar tissue. It ends at the superior cervical ganglion. The trunk lies behind the carotid sheath, just medial to the vagus nerve.

The **superior cervical ganglion** is over an inch (3 cm) long and lies in front of the lateral mass of the atlas and axis. The **middle cervical ganglion** is a small, inconstant ganglion lying on the trunk medial to the carotid tubercle (sixth cervical vertebra). The **inferior cervical ganglion** lies behind the commencement of the vertebral artery. A small mass when separate, it is more often fused with the first thoracic ganglion to form the **stellate ganglion,** a mass which may measure as much as 1 cm by 0·5 cm; it replaces the trunk in front of the neck of the first rib (PLATE 11, p. *v*). The middle ganglion is connected to the inferior ganglion (or stellate ganglion when present) by a part of the trunk which passes in front of the subclavian artery. This is the *ansa subclavia* (Fig. 6.10). More often, as in Fig. 6.10, it lies lower, and joins the inferior cervical to the first or second thoracic ganglion.

No white rami enter the trunk from the cervical nerves: all the fibres ascend from the thorax. As elsewhere, the branches of the ganglia are somatic and visceral in their distribution.

Somatic branches pass as grey rami to all eight cervical nerves. The superior ganglion gives grey rami to the first four (i.e., to the cervical plexus), the middle ganglion to the next two (5 and 6) and the inferior ganglion to the last two (7 and 8) anterior primary rami (i.e., to the brachial plexus for distribution to the upper limb).

Visceral branches include a branch from each ganglion to the cardiac plexuses. The branch from the upper left ganglion runs down to the superficial cardiac plexus, the other five ganglionic branches all pass to the deep cardiac plexus. All six cardiac branches pass discretely through the neck and superior mediastinum.

Vascular branches 'hitch-hike' their way along arteries. The *superior ganglion* gives branches to the internal carotid and external carotid arteries. The carotid 'plexus' passes into the skull with the internal carotid artery as one or two separate nerve bundles. As well as accompanying all branches of the artery the fibres are distributed to the pterygo-palatine ganglion and the eyeball, the latter including the motor supply of the dilator pupillæ muscle of the iris. The plexus on the external carotid artery accompanies all branches of the vessel and in addition supplies sympathetic fibres to the pharyngeal plexus and the submandibular and otic ganglia. Thus the branches to the carotid arteries are both somatic and visceral in their distribution.

The *middle cervical ganglion* gives branches to the subclavian and inferior thyroid arteries—the latter supply the lower larynx, the trachea, the hypopharynx and the upper œsophagus. The branches to the subclavian artery are for only its local branches. The branches along the inferior thyroid artery are visceral only (not somatic).

The *inferior cervical ganglion* gives a large branch to the vertebral artery (the vertebral 'plexus') which runs with all the branches of that vessel. It supplies only the branches of the vertebral artery (no viscera).

Note that only the three cervical ganglia, and no others, give off direct branches to arteries.

THE ROOT OF THE NECK

This should be studied in the dissecting room or the anatomical museum. The root of the neck lies above the apex of the lung; to understand the former is to appreciate the relations of the latter. The key to the root of the neck is the scalenus anterior muscle (Fig. 6.10). Study its relations precisely; they comprise almost all that needs to be known of the part. Thence pass to the thoracic operculum and note the structures that pass through it between the apices of the lungs.

Scalenus Anterior Muscle. This flat muscle arises from the anterior tubercles of the four 'typical' cervical vertebræ, the lowest and largest being the carotid tubercle of the sixth. At the anterior tubercles of the sixth, fifth, fourth and third vertebræ four slender tendons of origin lie end to end with those of longus capitis. The muscle slopes forwards and laterally as it passes down to end in a narrow tendon attached to the scalene tubercle on the inner border of the first rib. It belongs to the group of prevertebral flexor muscles of the neck, and with them is covered by a prolongation of prevertebral fascia, the foundation upon which the viscera of the neck gain their mobility. The *nerve supply* is derived segmentally by separate branches from the anterior primary rami of C5 and 6. Its *action* is to assist other muscles in stabilizing the neck. Actively contracting it bends the neck forwards and laterally or, if the neck is fixed, elevates the first rib as an accessory muscle of inspiration. It is, however, a very slender muscle.

Anterior Relations (Fig. 6.11). The *phrenic nerve* passes vertically down across the obliquity of the muscle, plastered thereto by the prevertebral fascia. The nerve leaves the medial border of the muscle low down and crosses the subclavian artery and its internal thoracic branch behind the subclavian vein. Lying on the suprapleural membrane it passes medial to the apex of the lung, in front of the vagus nerve, to enter the superior mediastinum. The nerve not infrequently lies in front of the subclavian vein (Fig. 6.3, p. 364). The *ascending cervical artery* is a branch of the inferior

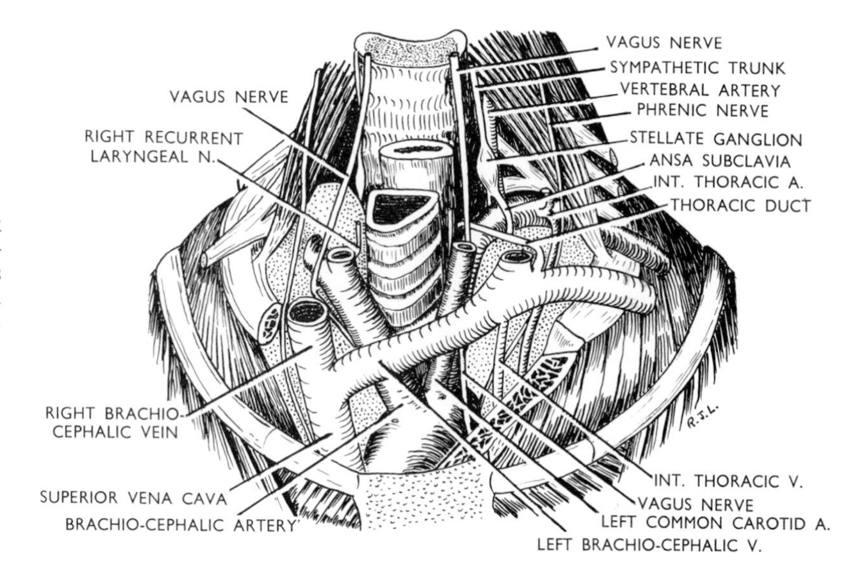

FIG. 6.10. THE ROOT OF THE NECK AND SUPERIOR MEDIASTINUM EXPOSED BY REMOVAL OF THE CLAVICLES AND MANUBRIUM STERNI. Drawn from Specimen S 233 B in R.C.S. Museum.

thyroid artery or the thyro-cervical trunk. It runs up on the prevertebral fascia medial to the phrenic nerve. The uninjected artery can easily be mistaken for the nerve in a dissecting room specimen (Fig. 6.11).

In front of the prevertebral fascia the *transverse cervical* and *suprascapular arteries* lie between the scalenus anterior and the carotid sheath (internal jugular vein). The *vagus nerve* in the carotid sheath passes down in front of the subclavian artery, on the right side giving off its *recurrent laryngeal* branch. The latter

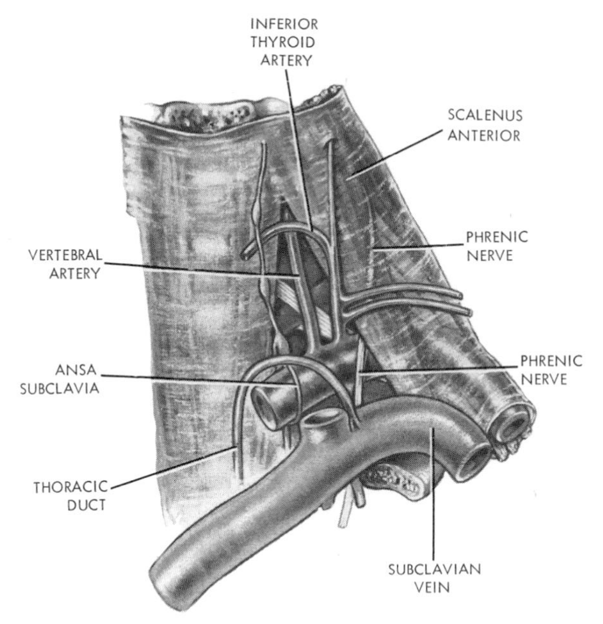

FIG. 6.11. THE RELATIONSHIP OF THE PREVERTEBRAL FASCIA TO THE SCALENE MUSCLES AND THE STRUCTURES OVER THE APEX OF THE LEFT LUNG.

hooks around the artery and passes upwards. The vagus nerve inclines posteriorly and runs on the medial surface of the apex of the lung to enter the superior mediastinum. The *internal jugular vein* is surrounded by the inferior *deep cervical lymph nodes.* One of them, behind the vein and above the inferior belly of the omo-hyoid muscle, is named the *jugulo-omo-hyoid node.* It receives all the lymph from the tongue (p. 444); it lies beneath the posterior border of the sterno-mastoid (Fig. 6.54, p. 443). The *subclavian vein* lies in a groove on the first rib and, due to the slope of the rib, lies at a lower level than the insertion of scalenus anterior. It joins the internal jugular vein at the medial border of the scalenus anterior muscle; the thoracic duct on the left and the right lymph duct on the right enter the angle of confluence of the two veins. In front of the carotid sheath runs the intermediate tendon of the omo-hyoid. All the structures enumerated above are covered in front by the lower part of the sterno-mastoid muscle (Fig. 6.5, PLATE 26, p. *xiv*).

Medial Relations. The edge of *longus colli muscle* runs up to the anterior tubercle on the transverse process of the sixth vertebra. It makes, with the medial edge of scalenus anterior, a pyramidal space; the base of the space is formed by the subclavian artery, and the neck of the first rib, between which lies the suprapleural membrane. Note that the apex of the pyramidal space is the carotid (Chassaignac's) tubercle, so named because the *common carotid artery* lies on it and can be there compressed by an assailant. The common carotid artery, medial to the vein, lies deep to sterno-mastoid immediately in front of the pyramidal space.

The space contains the stellate ganglion and the vertebral artery and vertebral vein (or veins). The inferior thyroid artery arches medially in a bold curve whose upper convexity lies in front of the apex of the pyramidal space. At a lower level, and further forward, the *thoracic duct* (or right lymph duct) makes a similar convexity as it arches over the lung apex and subclavian artery to enter the confluence of the subclavian and internal jugular veins (Fig. 6.11). The *first part* of the **subclavian artery** arches over the suprapleural membrane and impresses a groove upon the apex of the lung; it disappears behind the scalenus anterior muscle. It has three branches. The *vertebral artery* is the first branch; it arises from the upper convexity of the subclavian and passes up to disappear, at the apex of the pyramidal space, into the foramen transversarium of the sixth cervical vertebra. The sympathetic 'plexus' on the vertebral artery is often a single nerve that runs up behind the vessel. Medial to it lie the middle and inferior cervical sympathetic ganglia, the latter more often than not fused with the first thoracic ganglion to form the *stellate ganglion*. These ganglia lie on a more posterior plane; a connecting loop between middle and inferior cervical ganglia passes in front of the subclavian artery, forming the *ansa subclavia*. The *thyro-cervical axis* arises lateral to the vertebral artery from the upper surface of the subclavian. It divides immediately into transverse cervical, suprascapular and inferior thyroid arteries, which have already been noted. The *internal thoracic artery* arises from the lower surface of the subclavian and passes downwards over the lung apex, crossed by or crossing the phrenic nerve (PLATE 28, p. *xv*).

The *vertebral vein* emerges from the foramen transversarium of C6 and runs forward across the apex of the lung and beneath the subclavian artery to empty into the brachio-cephalic vein (Fig. 4.32, p. 248). It may be accompanied by a companion vein that passes through the foramen transversarium of C7.

Posterior Relations. The scalenus anterior is separated from the scalenus medius by the subclavian artery and the anterior primary rami of the lower cervical and first thoracic nerves. The last-named occupies the 'subclavian' groove, the subclavian artery usually lying somewhat further above the rib. The *second part* of the **subclavian artery** is that part which lies behind scalenus anterior. Here the fourth and last branch of the artery arises; it is named the *costo-cervical trunk* (PLATE 11). It passes back across the suprapleural membrane towards the neck of the first rib and there divides into a descending branch, the *superior intercostal artery*, which enters the thorax across the neck of the first rib and an ascending branch, the *deep cervical artery*, which passes beneath the transverse process of the seventh cervical vertebra and runs upwards behind the transverse processes, in front of semispinalis capitis, to anastomose with the occipital artery.

Lateral Relations. The *trunks* of the brachial plexus and the *third part* of the **subclavian artery** emerge from the lateral border of the scalenus anterior. They lie behind the prevertebral fascia, which plasters them down upon the floor of the posterior triangle and is projected along them into the axilla as the *axillary sheath*. No such sheath surrounds the axillary and subclavian veins; the latter lies in front of the prevertebral fascia and is free to dilate to accommodate an increased blood-flow. The *prevertebral fascia* is attached to bone at the edge of longus colli and is projected down from above over the scalenus anterior, being attached to the medial border thereof. There is no fascial roof across the pyramidal space between the two muscles and the subclavian artery by passing in behind scalenus anterior passes behind the lateral expansion of the prevertebral fascia (Fig. 6.11). Occasionally, the transverse cervical or suprascapular artery or both arise from the third part of the subclavian artery (that is, behind the prevertebral fascia), in which case they pass laterally between the trunks of the brachial plexus.

The **scalenus medius** arises from the posterior tubercles and costo-transverse lamellæ of all the cervical vertebræ and is inserted into the quadrangular area between the neck and subclavian groove of the first rib (Fig. 4.35, p. 252). The muscle is supplied segmentally by separate branches of the anterior primary rami of C3 to 8 inclusive. The muscle assists others in stabilizing the neck. Actively contracting it is an abductor of the neck towards its own side or an accessory muscle of inspiration.

Scalenus Posterior. The muscle arises from the posterior tubercles of the lower cervical vertebræ, passes across the outer border of the first rib deep to the upper digitation of serratus anterior, and is inserted into the second rib. Supplied segmentally by the lower four or five cervical nerves (anterior primary rami), the muscle is, like the other scalenes, an elevator of the ribs (an accessory muscle of inspiration) and a lateral flexor of the neck.

The root of the neck is floored in by the suprapleural membrane to which the dome of the parietal pleura is attached. Many of the structures studied above form relations of the apex of the lung (Fig. 4.32, p. 248).

The midline structures in the root of the neck consist of trachea and œsophagus (p. 369), the thoracic duct (p. 240) and the **recurrent laryngeal nerves.** These nerves run up in the groove between trachea and œsophagus, usually behind but not infrequently between the branches of the inferior thyroid artery near the lower pole of the thyroid gland. They give branches to the cervical parts of œsophagus and trachea and to the crico-pharyngeus, and enter the lower part of the pharynx below the inferior border of the crico-pharyngeus muscle (PLATE 42 and Fig. 6.34, p. 419).

THE FACE

Muscles of the Face

Embryologically the muscles of 'facial expression' are developed from the mesoderm of the second pharyngeal arch, from which they migrate widely to their adult positions. They are supplied by the nerve of the second arch, the seventh cranial (facial) nerve. They are, *morphologically*, specialized members of the panniculus carnosus (p. 3). The essential point here is that the panniculus carnosus is in places attached to the dermis, which it therefore wrinkles or dimples. There is no deep fascia on the face.

Functionally the muscles are differentiated to form groups around the orifices. The orifices of orbit, nose and mouth are guarded by eyelids, nostrils and lips and there is a *sphincter* and an opposing *dilator* arrangement peculiar to each. The purpose of the facial muscles is to control these orifices. The varying expressions so produced on the face are side effects; man has educated himself to be particularly sensitive to minor changes of expression in the faces of his fellows. Some of the muscles supplied by the facial nerve are incapable of affecting the expression of the face; moreover, certain facial expressions full of meaning are produced by muscles (e.g., levator palpebræ superioris, ocular muscles, tongue) not supplied by the facial nerve. The muscles of 'facial expression' are best understood and remembered by appreciating their functional arrangement around the orifices on the face, and it is strongly recommended that the muscles be studied in these functional groups.

The Muscles of the Eyelids

The palpebral fissure is surrounded by a sphincter, the orbicularis oculi, and has a dilator mechanism consisting of levator palpebræ superioris and occipito-frontalis.

Orbicularis Oculi. The muscle is rightly described in two parts, the palpebral part, confined to the lids, and the orbital part, extending beyond the bony orbital margins on to the face. The **palpebral part** of orbicularis oculi consists of fibres that arise from the medial palpebral ligament and arch across both lids, anterior to the tarsal plates, and are inserted into the lateral palpebral raphé. Some of the lower fibres are attached medially to the posterior lacrimal crest and the lacrimal sac itself. Contraction of the palpebral fibres closes the lids gently without burying the eyelashes, and stretches the lacrimal sac (p. 431). This movement of 'blinking' does not diminish the volume of the conjunctival sac, and even when the eye is brimful of tears none are spilt when the eyelids close. The **orbital part,** much the larger, arises from the anterior lacrimal crest and the frontal process of the maxilla, whence the fibres circumscribe the orbital margin in a series of concentric loops. The muscle lies flat over the forehead and cheek (frontal and zygomatic bones). When it contracts it lowers the eyebrow, to shade the eye from a light that is too bright from above. Orbital and palpebral parts contracting together close the eyelids forcibly so that the eyelashes are buried and only their tips are visible. This movement ('screwing up the eyes') diminishes the volume of the conjunctival sac, and causes an eye that is brimful of tears to spill out over the cheek.

Levator Palpebræ Superioris. This muscle is the opponent of the sphincter of the palpebral fissure, i.e., of the palpebral fibres of orbicularis oculi. It is dealt with under the orbital muscles (p. 432).

Occipito-frontalis. The muscle is considered on p. 386. Contraction of both bellies elevates the eyebrows and throws the frontal skin into horizontal wrinkles; in a proportion of individuals fixation of the muscle to the scalp moves the latter when the muscle contracts. Generally occipitalis anchors the scalp and only the frontalis fibres shorten. The occipito-frontalis is the opponent of the orbital part and levator palpebræ superioris is the opponent of the palpebral part of orbicularis oculi. Both combine to elevate the lid and eyebrow to clear the field of vision when looking fully upwards.

The Muscles of the Nostrils

The sphincter muscle of the nostril is compressor naris (PLATE 32), which embraces the alar cartilages, and its opponent is dilator naris, which is inserted into the lateral part of the ala. Each arises from the maxilla. The nose, in addition, can be elevated somewhat by procerus (PLATE 32) and levator alæ nasi and depressed by the depressor septi, much weaker because its action is assisted by elastic recoil of the nasal cartilages.

The Muscles of the Lips and Cheeks

The sphincter is the orbicularis oris; the dilator mechanism consists of the remainder of the facial muscles, which radiate outwards from the lips like the spokes of a wheel.

Orbicularis Oris. The muscle consists of fibres proper to itself, attached near the midline to upper and lower jaws, and fibres that are added to these from the dilator muscles. The intrinsic fibres are attached to bone near the midline and well away from the alveolar margin. The *incisive* and *mental slips* curve around the angle of the mouth in a loop on either side (Fig. 6.12). They are the deepest of all the orbicularis fibres and the mucous membrane of the lips is firmly attached to them. They are only a thin sheet. The bulk of the orbicularis muscle is formed of extrinsic

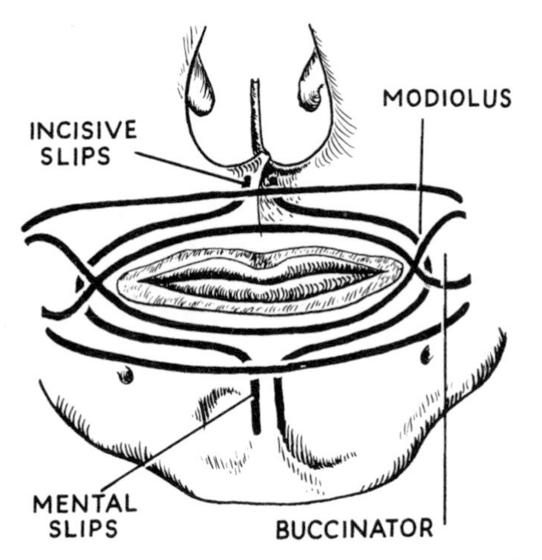

FIG. 6.12. THE FORMATION OF ORBICULARIS ORIS BY INTRINSIC AND EXTRINSIC FIBRES.

fibres; most of these come from the buccinator. The fibres of buccinator converge towards the modiolus. At the modiolus they form a chiasma; the uppermost and lowermost fibres pass straight on into their respective lips, while the middle fibres decussate, the upper fibres of buccinator passing into the lower lip, the lower into the upper lip (Fig. 6.13). Contraction of the orbicularis oris causes a narrowing of the mouth, the lips becoming pursed up into the smallest possible circle (the whistling expression).

Buccinator. The muscle arises from both jaws opposite the molar teeth, and from the pterygo-mandibular raphé. Examine the vestibule of the mouth opposite the molars with tongue and finger. The buccinator arises just beyond the bucco-gingival fold; note its distance from the gum margin. Examine a skull. Mark out the origin of buccinator with great

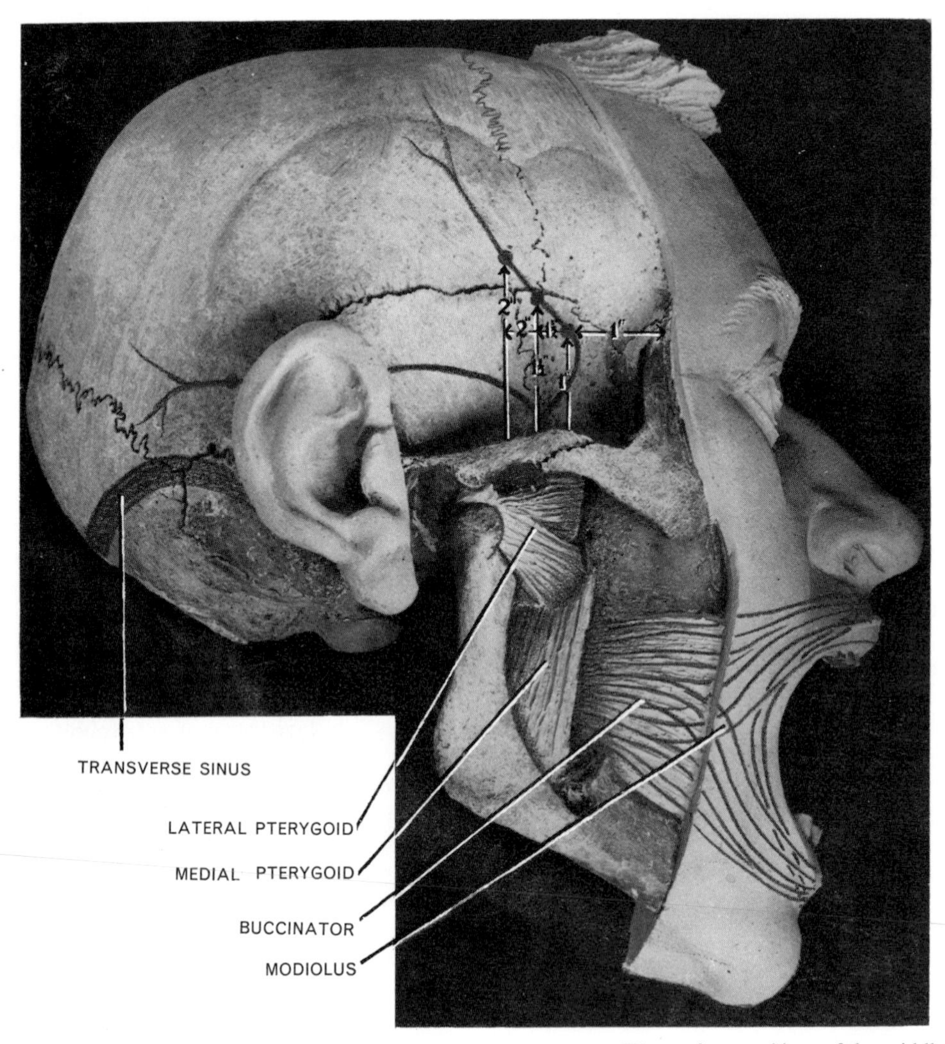

FIG. 6.13. THE DECUSSATION OF THE FIBRES OF THE BUCCINATOR AT THE MODIOLUS. The surface markings of the middle meningeal artery and the transverse sinus are outlined.

precision. Commencing at the anterior border of the first molar of the upper jaw, well away from the alveolar margin, the line passes horizontally backwards, skirting the root of the zygomatic process, and curves downwards to the tuberosity of the maxilla. Here there is a gap in the bony origin, the muscle arising from a fibrous band that extends from the tip of the hamulus to the nearest part of the tuberosity of the maxilla. This band is called the pterygo-maxillary ligament, not to be confused with the pterygo-mandibular *raphé*.

Through the gap above this band passes the tendon of tensor palati as it hooks around the base of the hamulus (Fig. 6.14). From the tip of the hamulus the pterygo-mandibular raphé extends to the mandible just above the posterior end of the mylo-hyoid line. The buccinator arises from the whole length of the raphé, along which it interdigitates with the fibres of the superior constrictor; the raphé is passively elongated when the mouth is open. The mandibular attachment of the raphé is separated by a narrow interval from the posterior attachment of the mylo-hyoid muscle to the mylo-hyoid ridge; here the lingual nerve rests on the mandible, which, indeed, it often grooves (Fig. 6.30, p. 413). The line of attachment of buccinator should now be traced behind the third molar (across the retromolar fossa) to the external oblique line of the mandible; it now runs down just above this line as far forwards as the anterior border of the first molar (Fig. 6.13). The muscle converges on the modiolus, where its fibres of origin from the raphé decussate; the maxillary and mandibular fibres pass medially without decussation into the upper and lower lips respectively. The muscle is pierced by the parotid duct opposite the *third* upper molar tooth, and by the buccal branch of the mandibular nerve, which supplies it with proprioceptive fibres.

The buccinator is the muscle of the cheek pouch and is lined by adherent mucous membrane. The nerve supply is by the buccal branch of the facial nerve. Its action is essentially that of an accessory muscle of mastication; it is indispensable to the return of the bolus from the cheek pouch to the grinding mill of the molars. It is not in any sense a muscle of 'facial expression'. When the cheeks are puffed out the muscle is relaxed. Only in forcible expulsion of air from the mouth, as in blowing a trumpet, are the elongated fibres of the puffed out cheeks under contraction. Otherwise contraction of the muscle obliterates the cavity of the vestibule and pulls the closed lips tightly back against the teeth.

The **molar mucous glands,** four or five nodules, each the size of a split pea, lie on its surface, and their ducts pierce the buccinator to open on the mucous membrane of the cheek (PLATE 34).

The Dilator Muscles of the Lips. Radiating from the orbicularis oris like the spokes of a wheel is a series of dilator muscles, some inserted into the lips, some into the modiolus. All contracting together open the lips into the widest possible circle, an action that is usually accompanied by simultaneous opening of the jaws. Upper and lower lips have flat sheets of elevator and depressor muscles. Other muscles converge towards the angle of the mouth, where their decussating fibres build up, with the chiasma in the buccinator fibres, a

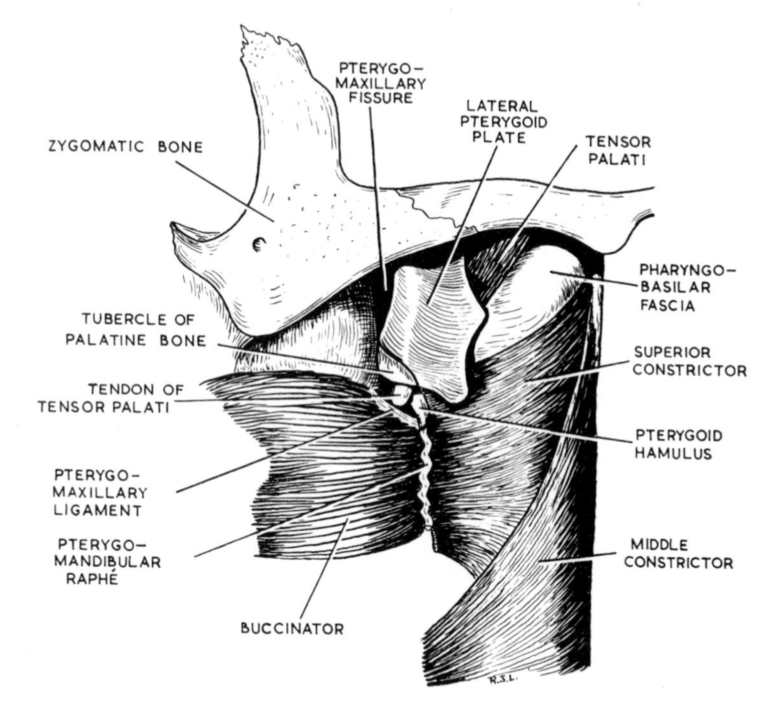

FIG. 6.14. THE ATTACHMENTS OF THE PTERYGOID HAMULUS.

knot of muscle known as the **modiolus.** The intersecting fibres cannot slip through the modiolus, for at their crossing they are bound together by fibrous tissue. Thus the modiolus (and the angle of the mouth) can be moved about. The modiolus lies just lateral to the angle of the mouth, opposite the second upper premolar tooth. Its position and movements are of great importance in prosthetic dentistry. The individual dilator muscles are of little practical importance; only their chief features are summarized here (PLATE 31).

Traced from the nose laterally a continuous sheet of elevator fibres comprises two separate muscles. The *levator labii superioris alœque nasi* arises from the frontal process of the maxilla and is inserted into the alar cartilage and the upper lip; it elevates both. The *levator labii superioris* arises from the inferior orbital margin and is inserted into the remainder of the upper lip, which it elevates. The muscle overlies the exit of the infra-orbital nerve. From the canine fossa below the infra-orbital margin arises the *levator anguli oris*; the infra-orbital nerve lies sandwiched between it and the overlying levator labii superioris. The fibres of this muscle, deep to the superficial sheet of muscle, converge to the modiolus and pass through it to become superficial. They merge into the fibres of depressor anguli oris. *Zygomaticus minor* from the zygomatico-maxillary suture and *zygomaticus major* further out on the surface of the zygomatic bone converge to the modiolus. *Risorius* is a variable muscle, an upward extension from the platysma. It converges on the modiolus, with a gap above and below it exposing the facial artery and its companion vein. The *depressor anguli oris* arises from the oblique line of the mandible. It lies superficial; its fibres pass through the modiolus to the deeper stratum (levator anguli oris). The *depressor labii inferioris* arises deep to the former muscle. It is quadrangular in shape; its fibres are inserted into the lower lip. *Mentalis* is a muscle that arises from the symphysis menti near the midline of the mandible. Its fibres pass downwards through the depressor labii inferioris to reach the skin. It is an elevator of the skin of the chin (which it sometimes dimples) and thus of the centre of the lower lip and its contraction may disturb a lower denture.

Nerve Supply of the Face Muscles. All the muscles thus far described receive their motor supply from the facial nerve. This nerve contains no sensory fibres on the face and proprioceptive impulses from the facial muscles are conveyed centrally by the trigeminal nerve, whose cutaneous branches intermingle freely on the face with branches of the facial nerve. A muscle supplied by the facial nerve receives its *proprioceptive innervation* from branches of the sensory nerve supplying the *skin over the muscle.* On the face itself the trigeminal nerve supplies all the muscles. Platysma is supplied by the transverse cervical nerve and occipitalis by the lesser occipital nerve; these two muscles lie beyond the cutaneous distribution of the trigeminal nerve.

Extracranial Course of the Facial Nerve

The nerve emerges from the base of the skull through the stylo-mastoid foramen. It immediately gives off the **posterior auricular nerve** which passes upwards behind the ear to supply the occipital belly of occipitofrontalis. A **muscular branch** is next given off which divides to supply the posterior belly of the digastric and the stylo-hyoid muscles. The nerve now approaches the postero-medial surface of the parotid gland. Just before entering the gland it divides into upper and lower branches, the *temporo-zygomatic* and *cervico-facial* branches. Within the substance of the parotid gland each divides and rejoins to divide again and finally emerge from the parotid gland in five main groups of branches. This plexiform arrangement, the **pes anserinus,** lies in the gland superficial to the retromandibular vein and the external carotid artery. Place the heel of the hand over the parotid gland, thumb on the temple, little finger down the neck. The five digits indicate the five divisions of the nerve.

1. **Temporal Branch.** This emerges from the upper border of the gland, crosses the zygomatic arch, and supplies auricularis anterior and superior, and part of frontalis. It is unimportant.

2. **Zygomatic Branches.** They consist of upper and lower parts, which proceed above and below the eye. The upper branches supply frontalis and the upper half of orbicularis oculi. They cross the zygomatic arch and may be divided in incisions for operations on the temporal fossa, or injured in fractures of the zygomatic arch. The lower branches supply the lower half of orbicularis oculi and the muscles below the orbit. Two or three branches pass to both upper and lower lids (Fig. 6.15, PLATE 31). Paralysis prevents blinking, the precorneal film of tears is no longer spread, and the dry cornea ulcerates.

3. **Buccal Branch.** This supplies the buccinator and the muscle fibres of the upper lip (i.e., orbicularis oris and the lower fibres of the elevators). Paralysis prevents emptying of the cheek pouch; the bolus lodges there and cannot be returned to the molar teeth. Chewing has to be performed on the other side.

4. **Mandibular Branch.** This supplies the muscles of the lower lip. It is important to remember that this nerve emerges from the lower border of the parotid gland and usually passes *into the neck* below the angle of the mandible. It crosses the inferior border of the bone to reach the face at the anterior border of the masseter muscle. At this point it lies on the facial artery and (anterior) facial vein. A small lymph node lies here

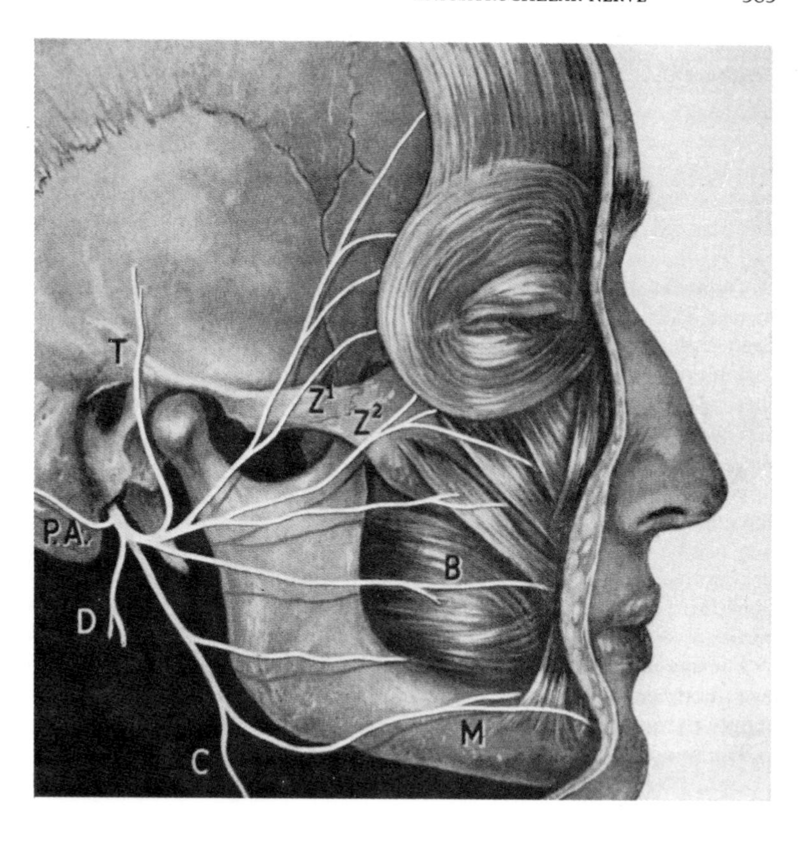

FIG. 6.15. THE EXTRACRANIAL DISTRIBUTION OF THE FACIAL NERVE. **P.A.** Posterior auricular. **D.** Branch to posterior belly of digastric and stylo-hyoid. **T.** Temporal branch. **Z^1.** Upper zygomatic branches. **Z^2.** Lower zygomatic branches. **B.** Buccal branch. **M.** Mandibular branch. **C.** Cervical branch.

and an incision of an abscess in the node has resulted in permanent paralysis of the lower lip. The nerve is in danger when an incision is made along the lower border of the mandible (PLATE 30 and Fig. 6.54, p. 443).

5. **Cervical Branch.** This nerve passes vertically downwards from the lower border of the parotid gland behind the mandible and supplies the platysma muscle. The mandibular branch occasionally arises from it below the parotid gland, a fact which apparently originated the erroneous though common statement that the muscles of the lower lip are supplied by the cervical branch of the facial nerve (Fig. 6.15 and PLATE 31).

Sensory Nerve Supply of the Face

The skin of the face is supplied in three zones by branches of the three divisions of the trigeminal nerve. The three zones meet at the margins of the eyelids and the angles of the mouth; the lines of junction of the zones curve upwards, evidence of the displacement of skin necessary to cover the large cranium of man (p. 30).

The muscles of the face are supplied with proprioceptive fibres by these same cutaneous nerves, which make multiple junctions with the branches of the seventh nerve on the face. The proprioceptive impulses from any face muscle are carried by the same nerves that supply the skin over the muscle.

The fifth nerve has three divisions. The number of named cutaneous branches are five, three and three (Fig. 6.16).

Cutaneous Branches of the Ophthalmic Division

(five in number)

The **lacrimal nerve** supplies a small area of skin and conjunctiva over the lateral part of the upper lid.

Towards the medial end of the upper margin of the orbit the **supra-orbital nerve** indents the bone into a notch or a foramen. The nerve passes up, breaking into half a dozen or more branches which radiate out and supply the forehead and scalp up to the vertex (PLATE 32).

The **supratrochlear nerve** passes up on the medial side of the supra-orbital nerve and divides to supply the middle of the forehead up to the hairline. Above this the supra-orbital nerves meet in the midline up to the vertex.

The **infratrochlear nerve** supplies skin on the medial part of the upper lid and, passing *above* the medial palpebral ligament, descends along the side of the external nose. It supplies no skin of the lower lid, which is second division territory. It supplies skin over the nasal bone (the 'bridge' of the nose).

These four branches of the ophthalmic division supply upper lid conjunctiva.

The **external nasal nerve** supplies the middle of the external nose down to the tip. It emerges between the nasal bone and the upper nasal cartilage (PLATE 32).

Cutaneous Branches of the Maxillary Division

(three in number)

These are the infra-orbital and the zygomatico-facial and zygomatico-temporal nerves.

The **infra-orbital nerve** emerges through its foramen and lies between levator labii superioris and the deeper placed levator anguli oris. It is a large nerve that immediately breaks up into a tuft of branches; these radiate away from the foramen to supply *palpebral branches* to the lower eyelid and cheek, *nasal branches* to the side and ala of the nose, *labial branches* to the upper lip and the labial gum from the midline to include the gum over the second premolar tooth. Many twigs join nearby branches of the facial nerve for the proprioceptive supply of adjacent face muscles. These junctions form a small network known as the *infra-orbital plexus*.

The **zygomatico-facial nerve** emerges from one or two foramina in the zygomatic bone; its branches supply the overlying skin.

The **zygomatico-temporal nerve** emerges *in the temporal fossa* through a foramen in the temporal (posterior) surface of the zygomatic bone. It makes its way to the surface and supplies a small area of skin over the front of the temple on a level with the upper eyelid (i.e., the 'hairless' part of the temple).

Cutaneous Branches of the Mandibular Division

(three in number)

The **auriculo-temporal nerve** passes around the neck of the mandible and ascends over the posterior root of the zygoma behind the superficial temporal vessels. The *auricular* part of the nerve supplies the external acoustic meatus and the surface of the tympanic membrane and the skin of the auricle above this level. The *temporal* part of the nerve supplies the hairy skin over the temple, that part in which grey hairs usually first appear (p. 31).

The **long buccal nerve** gives off cutaneous twigs before it pierces the buccinator muscle. They supply a 'thumb-print' area over the cheek just below the zygomatic bone, between the areas of the infra-orbital nerve and the great auricular nerve. (N.B. The *great auricular nerve* supplies an area of skin over the angle of the mandible and parotid gland, and the parotid fascia. The skin is neck skin (C2), which in the embryo is

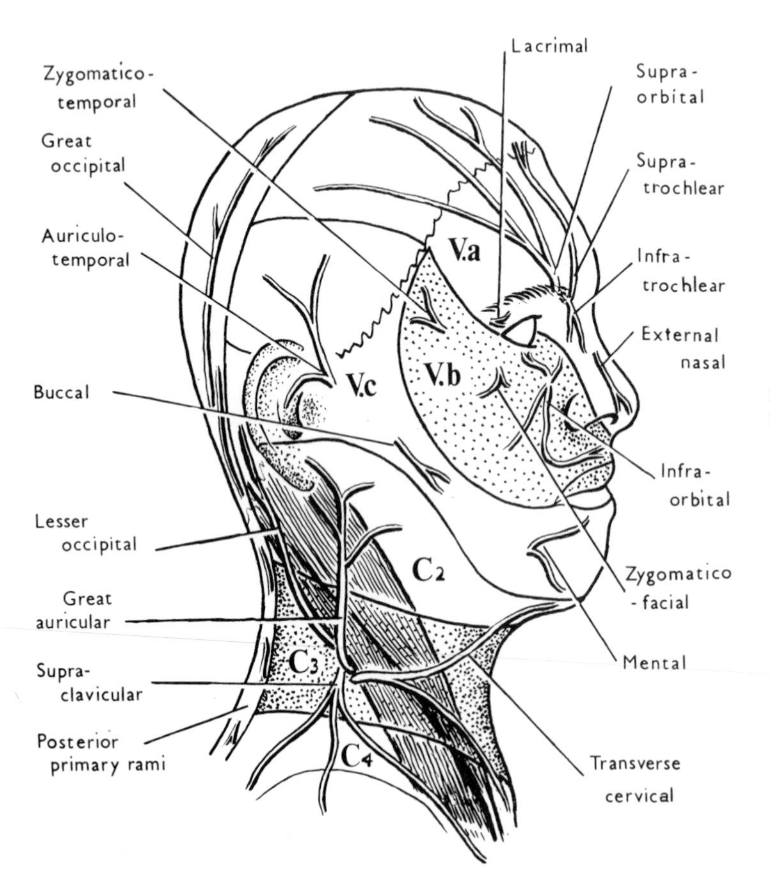

FIG. 6.16. THE DERMATOMES AND CUTANEOUS NERVES OF THE HEAD AND NECK.

drawn up to close the gap left by the beard skin. The latter migrates up over the temple, helping to cover the large cranium of man).

The **mental nerve** is a cutaneous branch of the inferior alveolar nerve. Like the infra-orbital nerve it breaks up into a tuft of branches that radiate away from the mental foramen to supply the skin and mucous membrane of the lower lip and labial gum from the midline to the bicuspid teeth (PLATE 30).

Blood Supply of the Face

The **facial artery** is the highest of the three branches that issue from the anterior aspect of the external carotid artery. It passes upwards on the side wall of the pharynx and hooks downwards over the digastric muscle to come into contact with the upper surface of the submandibular gland (p. 371). It then hooks upwards over the inferior border of the mandible at the anterior border of the masseter muscle. It pursues a tortuous course towards the medial angle of the eye, lying deep to the sheet of dilator muscles that radiate out from the lips. It is visible only in the interval above and below the zygomatic and risorius muscles. Its labial branches are important; one for each lip, the **labial artery** divides into two, each of which runs across the lip beneath the red margin, one in front and the other behind. They anastomose end to end at the midline. The severed artery spurts from both ends. The temple is supplied by the **superficial temporal artery,** a terminal branch of the external carotid. At its beginning it gives off the **transverse facial artery,** which runs across the cheek just above the parotid duct. The forehead is supplied from the orbit by the **supra-orbital** and **supratrochlear** branches of the ophthalmic artery. In the scalp these anastomose freely with the superficial temporal artery, establishing a free communication between internal and external carotid systems.

The **venous return** from the face is normally entirely superficial (PLATE 37). From the forehead the supra-orbital and supratrochlear veins pass to the medial canthus, where they unite to form the *angular vein.* This accompanies the facial artery as the (anterior) **facial vein** to a point just below the border of the mandible. Here in the neck it pierces the investing layer of deep fascia and is joined by the anterior branch of the retromandibular vein to form the common facial vein. Blood from the temple is collected into the tributaries of the superficial temporal vein. The latter is joined by the maxillary veins from the pterygoid plexus to form the **retromandibular vein.** This passes downwards in the substance of the parotid gland and on emerging from its lower border divides into anterior and posterior branches. The anterior branch joins with the (anterior) facial vein to form the **common facial vein,** which empties into the internal jugular. It often receives the superior thyroid vein and the vena comitans nervi hypoglossi. The posterior branch pierces the investing layer of deep cervical fascia and is joined by the posterior auricular vein to form the **external jugular vein.** This courses down in the subcutaneous tissue over the sterno-mastoid muscle and pierces the investing layer of deep cervical fascia a finger's breadth above the midpoint of the clavicle, to empty into the subclavian vein (PLATE 37).

Deep Venous Anastomoses (PLATE 37, p. *xx*)

The (anterior) facial vein communicates with the cavernous sinus. At the medial canthus there is a communication with the **ophthalmic veins,** which drain directly into the sinus. Blood from the forehead normally flows via the (anterior) facial vein; if the latter is blocked by pressure or thrombosis blood above the obstruction will flow through the orbit into the cavernous sinus. Hence the 'danger area' of infection of the upper lip and nearby cheek. A further communication is the **deep facial vein.** This passes, in front of the masseter muscle, between the (anterior) facial vein and the pterygoid plexus. The plexus receives a vein from the cavernous sinus through the foramen ovale or, if it be present, the foramen of Vesalius, a small hole medial to the foramen ovale (see p. 487). The danger area of the face lies between the angular and deep facial veins.

Lymph drainage of face (see page 443). The middle of the lower lip drains bilaterally into the submental nodes. The area in front of the facial artery (external nose, upper lip, remainder of lower lip) drains to the submandibular nodes, and the rest of the face drains to the pre-auricular (parotid) group.

THE SCALP

The scalp may be defined as the hair-bearing area extending from the top of the neck muscles at the back of the head to the forehead and eyebrows at the front of the head and extending down over the temples to the ears and zygomatic arches.

The skin is thick. Many of the fibres of the occipitalis and frontalis muscles are inserted into it; elsewhere it is tethered by fibrous bands in the subcutaneous tissue to these muscles and the intervening aponeurosis.

The **blood supply** of the scalp is derived from the *external carotid artery* by the occipital, posterior auricular and superficial temporal arteries, and from the *internal carotid artery* by the supratrochlear and supra-orbital arteries (PLATE 32). *All these arteries anastomose* *very freely with each other*. The junction of forehead and temple, above the outer end of the eyebrow, is the area where the external and internal carotid arteries anastomose most freely with each other. The arteries of the scalp are attached to the deepest layer of the dermis, and in a scalp wound it is difficult to catch them cleanly in a hæmostat. From the richly-anastomosing skin arteries very few branches cross the subaponeurotic space to the underlying bones. Scalping does not cause necrosis of the bones of the vault, most of whose blood comes from the middle meningeal artery.

The *occipital artery* emerges from the apex of the posterior triangle and runs with the great occipital nerve to supply the back of the scalp up to the vertex. The *posterior auricular artery* runs with the lesser occipital nerve to supply the scalp behind the ear.

The *superficial temporal artery* is a terminal branch of the external carotid artery. Running up behind the temporo-mandibular joint it crosses the posterior root of the zygomatic arch and branches out widely into the skin that overlies the temporal fascia. One branch, called the *middle temporal artery*, pierces the temporal fascia and runs up vertically deep to the muscle. Its companion vein produces a vertical groove seen on the squamous bone of almost every skull examined (Fig. 6.2, p. 361).

The *supra-orbital artery* runs with the supra-orbital nerve and the *supratrochlear artery* runs with the supratrochlear nerve. They are the terminal branches of the ophthalmic artery. The supra-orbital is the larger and supplies the front of the scalp up to the vertex. Its anastomosis with the superficial temporal artery connects the internal and external carotid systems.

The *veins* of the scalp run back with the arteries. In forehead, temple and occipital regions they receive diploic veins from frontal, parietal and occipital bones.

The supra-orbital and supratrochlear veins drain by the angular vein into the (anterior) facial vein. The superficial temporal veins run into the retromandibular vein, and occipital veins reach the plexus around the semispinalis capitis muscle. The posterior auricular vein drains the scalp behind the ear, and receives also the mastoid emissary vein from the sigmoid sinus. Infection here can be dangerous or fatal, from retrograde thrombosis of cerebellar and medullary veins (PLATE 37).

The *lymph drainage* of the scalp is to preauricular and occipital lymph nodes (see p. 444).

The **nerves of the scalp** run with the arteries. Posteriorly the great occipital and third occipital nerves (*posterior* primary rami of C2 and C3 respectively) extend to the vertex and the posterior scalp respectively. The lesser occipital (*anterior* primary ramus of C2) supplies skin behind the ear. The temple is supplied by the auriculo-temporal and the zygomatico-temporal nerves, and the forehead and front of the scalp by the supratrochlear and supra-orbital nerves (PLATES 33, 32, 31).

The **occipito-frontalis muscle** consists of occipital and frontal bellies, separated by an aponeurosis into which both are inserted. The *occipitalis* arises from the highest nuchal line and passes forward into the aponeurosis. The **galea aponeurotica** lies over the vertex between occipitalis and frontalis bellies. It fades out laterally by blending with the temporal fascia just above the zygomatic arch. The *frontalis muscle* arises from the front of the aponeurosis and is inserted into the upper part of orbicularis oculi and the overlying skin of the eyebrow. The skin of the scalp is firmly bound down to these muscles and to the galea aponeurotica (epicranial aponeurosis).

The occipitalis muscle is supplied by the posterior auricular and frontalis by the superior zygomatic branches of the facial nerve. Occipitalis pulls the scalp back in certain individuals. Usually it merely anchors the aponeurosis while contraction of the frontalis elevates the eyebrows and produces horizontal wrinkles in the skin of the forehead. The scalp muscles are the opponents of the orbital part of orbicularis oculi; thus the scalp may be regarded as *an upward prolongation of the face* that extends back to the occiput.

The **subaponeurotic space** extends beneath these muscles over the vault of the skull. It is limited behind by the attachments of occipitalis to the highest nuchal line and at the sides by the blending of the aponeurosis with the temporal fascia. In front the space extends down beneath the orbicularis oculi into the eyelids—bleeding anywhere beneath the aponeurosis may appear as a 'black eye' by the blood tracking down through the space.

The **pericranium** is merely the periosteum of the bones of the vault. It is rather loosely attached to the bone and can be stripped up by a subperiosteal hæmatoma. Such a hæmatoma outlines the bone, for the pericranium is very firmly attached at the sutures.

The Temporal Fossa and Zygomatic Arch

Examine a skull. There are two temporal lines, superior and inferior. They diverge from a common origin at the border of the zygomatic process of the frontal bone, and sweep boldly up in a convexity that takes them back behind the ear. The area of bone between them is polished.

The **temporal fascia** is attached to the superior temporal line and passes down to the upper border of the zygomatic arch. It is a rugged membrane; the superficial temporal vessels and auriculo-temporal nerve lie upon it, and it is perforated by the middle temporal artery and vein (PLATE 31).

The **temporalis muscle** arises from the temporal fossa over the whole area between the inferior temporal line and the infratemporal crest. The muscle is small in the newborn and rises up the side of the skull as the individual grows.

The large fan-shaped muscle converges towards the coronoid process of the mandible, and its muscle fibres become tendinous. Very little of the tendon is inserted beyond the border of the coronoid process into the outer plate of the mandible: the main insertion is into the bevelled surface on the *inner plate of the bone*, at the posterior border of the coronoid process and at the anterior border down the ascending ramus as far as the attachment of the buccinator in the retromolar fossa (Figs. 6.17, 8.15, p. 572).

The blood supply of the muscle is derived from the temporal branches of the maxillary artery, reinforced by the middle temporal artery (p. 386).

The muscle is supplied by the two deep temporal nerves, sometimes reinforced by the middle temporal nerve; they are branches of the anterior division of the mandibular nerve. Its upper and anterior fibres elevate the mandible (close the jaws) its posterior fibres, sweeping over the root of the zygoma, retract the mandible. They are the *only* fibres to do so; no other muscle retracts the condyle.

The **zygomatic arch** is formed by processes of the squamous temporal and zygomatic bones, which meet at a suture sloping downwards and backwards. The arch is completed anteriorly by the zygomatic process of the maxilla (p. 555).

Nerves crossing the arch are vulnerable in incisions or in fractures. The auriculo-temporal nerve crosses well back, just in front of the ear (PLATE 25, p. *xiii*). The zygomatic branches of the facial nerve cross the arch, to the muscles of the upper and lower eyelids (Fig. 6.15).

THE PREAURICULAR REGION

The **masseter muscle** arises from the zygomatic arch. Its commonly described two heads are, in fact, three, though they are fused together on the arch. At their insertions on the mandible the three sets are fused anteriorly, but diverge from each other posteriorly for the passage of the masseteric nerve and the masseteric artery. The *superficial part* of the muscle is the largest. It arises from the anterior two-thirds of the lower border of the zygomatic arch as far forward as the zygomatic process of the maxilla. Its fibres slope down at 45 degrees and are inserted into a wide area: from the angle of the mandible forwards along the lower border, and upwards to include the lower part of the ascending ramus. The upper part of the muscle is covered in aponeurotic fibres (to slide on the parotid duct and the accessory parotid gland), the lower half is fleshy (PLATE 30). The *intermediate part* of the muscle arises from the middle third of the arch and the *deep part* from the deep surface of the arch in continuity. The fibres pass vertically downwards to be inserted into the ramus of the mandible. Their insertions fuse with each other and with the superficial fibres at the anterior border of the ramus, but are quite separate posteriorly. The masseteric nerve runs down between the deep and intermediate parts, and a branch from the superficial temporal or transverse facial artery runs forwards between the superficial and intermediate parts. Nerve and artery thus incompletely divide the muscle into

FIG. 6.17. THE THREE PARTS OF INSERTION OF THE MASSETER.

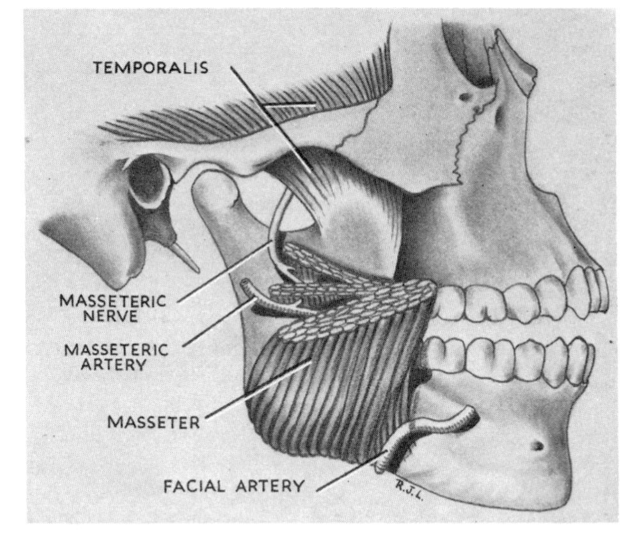

three parts (Fig. 6.17). The artery is the chief source of supply to the muscle; the vessel accompanying the masseteric nerve is very small.

The muscle is supplied by the masseteric branch from the anterior division of the mandibular nerve (p. 392). Its **action** is to close the jaws by elevating and drawing forwards the angle of the mandible.

The Parotid Gland. This salivary gland is usually serous, in some animals wholly so; a few scattered mucous acini are found in the human gland. It is wedge-shaped and lies, base outwards, behind the posterior border of the mandible. Its base extends forwards over the masseter muscle; this part is usually described as the *superficial lobe*. The deep part, the tapered edge of the wedge, is usually called the *deep lobe*. The upper limit of the gland reaches the glenoid fossa behind the mandibular joint. This part is often referred to as the *glenoid lobe*. The gland is not separated into recognizable lobes, however, and these names refer but vaguely to different parts of the wedge-shaped whole.

The gland lies firmly within its sheath, the *parotid fascia*, two diverging layers derived from an upward extension of the investing layer of deep cervical fascia (p. 361). Moderate swelling of the gland produces a considerable tension within the unyielding fascia, with consequent discomfort or pain.

The **superficial lobe,** so-called, lies wedged between the mastoid process and the posterior border of the mandible and extends forwards over the masseter for a variable distance (PLATE 31). Its anterior border is convex forwards; from it emerge the parotid duct and the five divisions of the branches of the facial nerve (p. 382). This part of the gland has anterior and posterior relations that are similar, namely, bone sandwiched between two muscles. In front is the mandible sandwiched between masseter and medial pterygoid, behind lies the mastoid process sandwiched between sterno-mastoid and digastric muscles.

The **glenoid lobe,** so-called, lies behind the mandibular joint and is in contact with both osseous and cartilaginous parts of the external acoustic meatus. The auriculo-temporal nerve is in contact with this part of the gland, to which it gives off its secreto-motor branch from the otic ganglion.

The **deep lobe,** so-called, is the narrowed edge of the wedge-shaped gland, lying in contact with the internal jugular vein below the glenoid lobe.

Beneath the parotid fascia **lymph nodes** lie on the surface and within the substance of the gland. Other members of this **pre-auricular group** lie superficial to the parotid fascia, just beneath the skin of the face. The parotid gland contains the pes anserinus of the **facial nerve** within the substance of its superficial lobe. Immediately deep to the nerve lies the **retromandibular vein** and, deepest of all, the **external carotid artery.** The gland encloses all three. The superficial situation of the nerve makes incision into the gland very hazardous—the face can be paralysed before serious bleeding is encountered (PLATE 30, p. *xvi*).

The **parotid duct** passes forwards across the masseter and turns around its anterior border to pierce the buccinator. It lies in the line between the intertragic notch of the auricle and the midpoint of the philtrum and is palpable (feel it on yourself as it passes over the clenched masseter muscle). The duct opens on the mucous membrane of the cheek opposite the second upper molar tooth; it pierces the buccinator further back and runs forwards beneath the mucous membrane to its orifice—the valvular flap of mucous membrane so produced prevents inflation of the gland when intra-oral pressure is raised.

An **accessory gland** usually lies on the masseter between the duct and the zygomatic arch. Several ducts open from it into the parotid duct. It and the duct lie on the tendinous part of the surface of the masseter muscle (PLATE 31).

The **deep relations** of the gland are best appreciated by looking at what remains when the gland is removed (Fig. 6.19). The bed of the parotid consists of the posterior belly of the digastric and the styloid process and stylo-hyoid muscle and deep to them the internal jugular vein, crossed by the accessory nerve as it lies on the lateral mass of the atlas. Further forward is the internal carotid artery, with the 'pharyngeal' structures lying on it (p. 393). It is not in direct contact with the gland.

Blood Supply. Branches from the external carotid artery supply the gland. Venous return is to the retromandibular vein. Lymph drains to the nodes within the parotid sheath and thence with the external carotid artery to nodes of the antero-superior group of deep cervical lymph nodes.

Nerve Supply. Secreto-motor fibres arise from cell bodies in the otic ganglion and reach the gland by 'hitch-hiking' along the auriculo-temporal nerve. The preganglionic fibres arise from cell bodies in the inferior salivatory nucleus in the medulla, and travel by way of the glosso-pharyngeal nerve, its tympanic branch, the tympanic plexus and the lesser (superficial) petrosal nerve to the otic ganglion. The sympathetic fibres reach the gland from the superior cervical ganglion by way of the plexus on the external carotid and middle meningeal arteries. The gland itself receives sensory fibres from the auriculo-temporal nerve, but the parotid fascia receives its sensory innervation from the great auricular nerve (C2).

Development. A groove that appears in the cheek (ectoderm) becomes converted into a tunnel from the blind end of which cells proliferate to form the gland.

THE INFRATEMPORAL FOSSA

This is a space lying beneath the base of the skull between the side wall of the pharynx and the ascending ramus of the mandible. It is bounded in front by the posterior wall of the maxilla, and behind by the styloid apparatus and the carotid sheath, which lie in front of the prevertebral fascia. It has no anatomical floor and continues downwards through the tissue spaces of the neck, alongside the pharynx and œsophagus. The space is continuous through the superior into the posterior mediastinum.

Contents of the Fossa

The infratemporal fossa is occupied by the two pterygoid muscles, medial and lateral, around and between which pass the branches of the mandibular nerve, the maxillary artery and the pterygoid venous plexus. Examine a skull and mandible. The attachments of the pterygoid muscles should be studied carefully. Each has two heads of origin.

The Lateral Pterygoid Muscle (PLATE 35). The two heads of this muscle arise from the infratemporal surface of the skull and from the lateral pterygoid plate respectively. The temporal fossa on the side of the skull ends under cover of the zygomatic arch, at a series of tubercles (the infratemporal crest). This line forms the junction between the side wall and base of the skull. On the base of the skull, medial to the crest, the smooth area of bone consists of the greater wing of the sphenoid. In front of the eminentia articularis a small area of the squamous part of the temporal bone completes the infratemporal surface of the skull. The whole of this area gives origin to the *upper head* of the muscle. The *lower head* arises from the lateral surface of the lateral pterygoid plate. The two heads, lying edge to edge, converge and fuse into a short thick tendon that is inserted into the pterygoid pit (beneath the medial end of the mandibular condyle). The upper fibres of the tendon, lying above this pit, pass back into the articular disc of the mandibular joint, and into the anterior part of the capsule.

When the muscle contracts it draws condyle and disc forwards from the glenoid fossa down the slope of the eminentia articularis (p. 446). It is indispensable to *active* opening of the mouth.

The Medial Pterygoid Muscle (PLATE 34). The great bulk of this muscle arises from the deep (medial) surface of the lateral pterygoid plate and the depths of the fossa between the two plates (Fig. 6.31, p. 414). (Note that the medial pterygoid plate has the pharynx attached to its posterior border and thus forms the

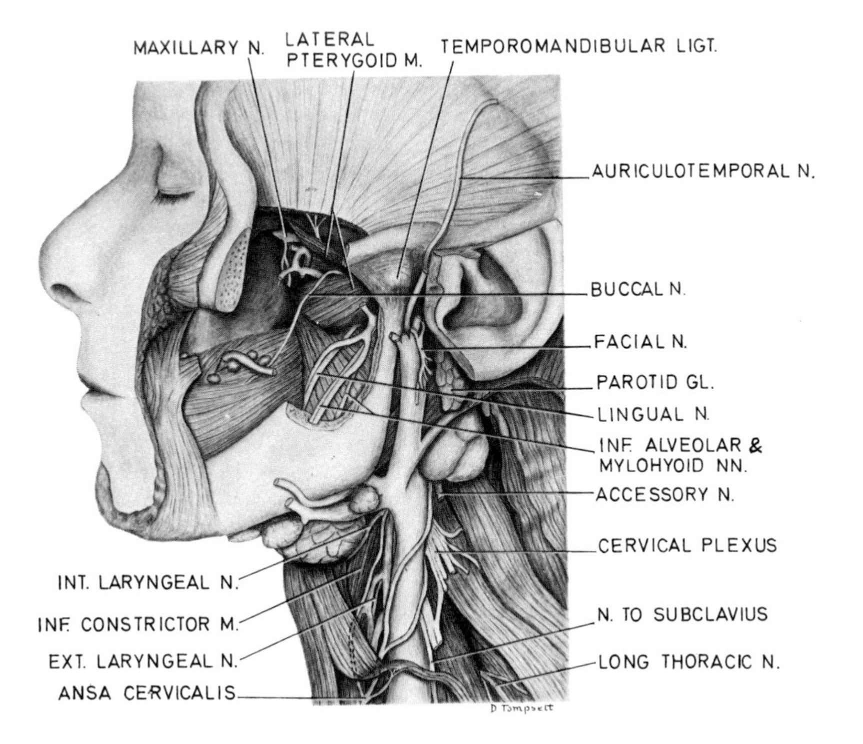

FIG. 6.18. THE LEFT INFRATEMPORAL FOSSA AND THE SIDE OF THE NECK (see PLATE 35 for full colour). Illustration of Specimen S.350 in R.C.S. Museum.

posterior part of the lateral wall of the nose, while the lateral pterygoid plate lies sandwiched between the two pterygoid muscles.)

The bulky muscle diverges down from the lateral pterygoid muscle at nearly a right angle from their common origin beside the lateral pterygoid plate. A small band of muscle, the *inferior head*, arises from the tuberosity of the maxilla and the tubercle of the palatine bone (identify this on the skull) and, passing over the lower margin of the lateral pterygoid muscle, fuses with the main muscle mass. In this way the two heads, very unequal in size, embrace the lower edge of the lateral pterygoid muscle. The broad band of muscle passes down and back at 45 degrees (and *laterally*) to reach the angle of the mandible. It is inserted from the border of the angle over the rough area seen on the bone as far as the groove for the mylo-hyoid vessels and nerve. The nerve runs in the narrow angle between the muscle and the mandible (PLATE 35, Figs. 6.31, p. 414, 8.15, p. 572).

The pull of the muscle on the angle of the mandible is upwards and forwards and *medially* (i.e., it closes the mouth and moves the mandible towards the opposite side). It is the great chewing muscle for the molar teeth and is particularly developed in the chewing mammals, especially the ruminants, which chew and re-chew the cud (p. 446).

The pterygoid muscles are supplied by the mandibular division of the trigeminal nerve; the medial directly from the main trunk, the lateral from the anterior division thereof.

The **maxillary artery** is, with the superficial temporal artery, a terminal division of the external carotid. It enters the infratemporal fossa by winding around deep to the neck of the mandible and passing forwards between the neck of the mandible and the spheno-mandibular ligament. Here the auriculo-temporal nerve lies above it, and the maxillary veins below it. It usually runs deep to the lower head and passes forward between the two heads of the lateral pterygoid muscle (Fig. 6.18) but, like vascular patterns elsewhere, variation is common and the artery may pass below (or, more rarely, above) the muscle. In any case it passes deeply into the pterygo-maxillary fissure and so into the pterygo-palatine fossa (identify these on the skull).

It is described conventionally in three parts, before, on and beyond the lateral pterygoid muscle and this is useful, since *five branches come from each part*. From first and third parts the five branches all enter foramina in bones, from the second part none go through foramina in bones.

The five ('bony') branches from the **first part** are the inferior alveolar, middle meningeal, accessory meningeal and, to complete the five, two branches to the ear.

The **inferior alveolar** (dental) **artery** passes downwards and forwards (vein behind it) towards the inferior alveolar nerve, which it meets at the mandibular foramen, in which all three lie. It passes forwards in the mandible, supplying the pulps of the mandibular teeth and the body of the mandible. Its *mental branch* emerges from the mental foramen and supplies the nearby muscles and skin.

The **middle meningeal artery** passes vertically upwards to the foramen spinosum. It is embraced by the two roots of the auriculo-temporal nerve. It supplies the bones of the skull (p. 482). From the sympathetic plexus on the artery a branch peels off to enter the otic ganglion.

The **accessory meningeal artery** passes upwards through the foramen ovale and supplies the dura mater of the floor of the middle fossa and of the trigeminal (Meckel's) cave. It is the chief source of blood supply to the trigeminal ganglion.

The remaining two arteries pass up through the squamo-tympanic fissure superficial and deep to the tympanic membrane. The **deep auricular artery** is the superficial of the two and supplies the external acoustic meatus. The deeper of the two is the **anterior tympanic artery** which, in the middle ear, joins the circular anastomosis around the tympanic membrane.

The **second part** of the maxillary artery gives off branches to the pterygoid muscles and branches to the temporalis muscle. Further branches accompany the lingual and long buccal nerves. Together they can be summarized as five in number to the soft parts.

The **third part** of the maxillary artery divides in the pterygo-palatine fossa into five branches which accompany the five branches of the pterygo-palatine ganglion (p. 397). The artery then passes forwards through the inferior orbital fissure, along the floor of the orbit and infra-orbital canal to emerge in the cheek with the infra-orbital nerve.

The **pterygoid plexus** is a network of very small veins that lie around and *within* the lateral pterygoid muscle. It is frequently not demonstrable in the dissected cadaver (because it lies *within* the muscle); sometimes it appears as a knot of veins on the lateral surface of the lower head of the muscle. In the living it is often very full, and can easily be punctured by the needle delivering an anæsthetic solution in the region of the posterior superior alveolar nerves. The veins draining into the pterygoid plexus correspond with the branches of the maxillary artery, but they do not return all the arterial blood, much of which returns from the periphery of the area by other routes (facial veins, pharyngeal veins, diploic veins). On the other hand the pterygoid plexus receives the drainage of the inferior ophthalmic veins via the inferior orbital fissure (blood from the *internal* carotid artery). The pterygoid plexus drains into a pair of large, but very short, **maxillary veins** which lie deep to the neck of the mandible. They run back and join the superficial temporal vein to form the retromandibular vein (PLATE 37).

In addition to its input and outflow the pterygoid plexus has two very important *connecting veins*. The *deep facial vein* communicates, across the buccinator and in front of the masseter muscle, with the (anterior) facial vein. A small vein reaches it from the *cavernous sinus* through the foramen ovale or, when present, through the foramen of Vesalius, a small foramen on the medial side of the foramen ovale. Thus infection from the face can spread, via the plexus, to the cavernous sinus.

The rôle of the deep facial vein is to provide an alternative pathway, via the (anterior) facial vein, for drainage of the pterygoid plexus if the maxillary vein be temporarily occluded by local pressures from outside or from the mandible in certain movements, or in the reverse way for drainage of the angular vein if the lower part of the anterior facial vein be occluded by pressure.

The *inferior ophthalmic veins*, running with the infra-orbital artery, drain partly into the cavernous sinus and partly into the pterygoid plexus, providing a further pathway of communication between the two.

The rôle of the pterygoid plexus is to act as a 'peripheral heart', aiding venous return by the pumping action of the lateral pterygoid muscle. The plexus is valved and sucks blood from incompressible parts (face bones, orbit) and pumps it back into the maxillary veins. It pumps each time the mouth is actively opened (talking or chewing or both) but not when the mandible droops by gravity (e.g., in sleep). The prolonged and forcible contraction of the lateral pterygoid muscle to open the mouth in yawning is accompanied by a like contraction of the diaphragm to aid venous return from the abdomen, and often also by 'stretching' of the limbs to empty them of stagnant venous blood. Thus yawning is a purposive reflex triggered off by venous stagnation, and is contagious only to those whose veins are likewise stagnant. Compare the reflex emptying of a *full* bladder encouraged by the sound of falling water; the *empty* bladder is unaffected by the sound of Niagara itself (this can be confirmed).

The **spheno-mandibular ligament** is a flat band of tough fibrous tissue extending from a narrow attachment on the spine of the sphenoid. It broadens as it passes downwards to be attached to the lingula and inferior margin of the mandibular foramen. It is the perichondrium of Meckel's cartilage (Fig. 1.49, p. 40). Between it and the neck of the mandible pass the auriculo-temporal nerve and the maxillary artery and vein. Between it and the ramus of the mandible the inferior alveolar vessels and nerve converge to the mandibular foramen. It is pierced by the **mylo-hyoid nerve** which, branching from the inferior alveolar nerve, lies in the groove on the mandible at the margin of attachment of the medial pterygoid muscle. The *mylo-hyoid artery*, a similar branch of the inferior dental artery, is very small and ends in the medial pterygoid muscle. Its place is taken below the jaw, alongside the mylo-hyoid nerve, by the submental branch of the facial artery. The small vein accompanying the mylo-hyoid artery is actually responsible for making the mylo-hyoid groove on the mandible (Fig. 6.30A, p. 413).

The tensor palati muscle (p. 418) lies deep to the lateral pterygoid muscle against the side wall of the pharynx (Fig. 6.14). Between the two muscles the mandibular nerve enters the infratemporal fossa.

The Mandibular Nerve

The mandibular branch from the trigeminal ganglion lies in the dura mater of the middle cranial fossa lateral to the cavernous sinus. With the motor root of V it enters the foramen ovale, where the two join and emerge as the mandibular nerve (cf. spinal nerves in intervertebral foramina). The nerve lies deep to the upper (infratemporal) head of the lateral pterygoid muscle, on the tensor palati muscle, from which it is separated by the otic ganglion. This point is 4 cm deep to the eminentia articularis, through the mandibular notch. After a short course the nerve divides into a small anterior (mainly motor) and a large posterior (mainly sensory) branch.

Branches from the Main Trunk. (*a*) *Meningeal.* A recurrent nerve, the *nervus spinosus*, re-enters the middle cranial fossa via the foramen spinosum, or more usually the foramen ovale, supplying the posterior half of the middle cranial fossa, and the mastoid antrum and air cells through the petro-squamous suture. It gives twigs to the cartilaginous part of the Eustachian tube before entering the skull.

(*b*) *Motor.* The nerve to the medial pterygoid muscle runs forwards to the muscle, and gives off the motor root to the otic ganglion. This root passes near or through the ganglion without synapse and its fibres supply the two tensor muscles, tensor palati and tensor tympani. Do not confuse this miscalled motor root with the *secreto*-motor root of the ganglion, which is the lesser petrosal nerve.

Branches from the Anterior Division. This division is motor, except for one branch (the long buccal nerve).

(*a*) *Temporal branches* to the muscle pass above the upper border of the lateral pterygoid muscle. Two in number, *anterior* and *posterior*, they are sometimes joined by a third, the *middle*, which comes out, with the long buccal nerve, between the two heads of the lateral pterygoid muscle.

(*b*) The *masseteric nerve*, passing above the upper border of the lateral pterygoid, emerges through the mandibular notch to enter the deep surface of the masseter (Fig. 6.17). It gives an articular branch to the mandibular joint (Hilton's Law).

(*c*) *Pterygoid branches*, one to each head of the lateral pterygoid muscle, run with the buccal nerve and enter the deep surface of the lateral pterygoid muscle.

(*d*) The **buccal nerve,** wholly sensory, contains all the fibres of common sensation in the anterior division of the mandibular nerve. It emerges between the two heads of the lateral pterygoid muscle and courses downwards and forwards in a fascial tunnel on the deep surface of the temporal muscle. It runs on the buccinator, giving branches to the skin over the cheek, then pierces the buccinator (giving proprioceptive fibres to it) and supplies the mucous membrane of the cheek and the gum of the lower jaw opposite the lower molars and second premolar (i.e., up to the mental foramen) (PLATES 34, 35, p. *xix*).

Branches from the Posterior Division. This division is sensory except the motor fibres which are distributed via the mylo-hyoid nerve. There are three branches.

(*a*) **Auriculo-temporal Nerve.** This branch is derived by two roots from the posterior division; they embrace the middle meningeal artery. The nerve passes backwards between the neck of the mandible and the spheno-mandibular ligament, lying above the maxillary vessels, deep to the glenoid lobe of the parotid gland. It gives a branch to the mandibular joint, and ascends over the posterior root of the zygoma behind the superficial temporal vessels (PLATE 39). The *auricular* part innervates the skin of the tragus and upper part of the pinna, the external acoustic meatus and the outer surface of the tympanic membrane. The *temporal* part is distributed to the skin of the temple over that part which first turns grey (this area is really skin borrowed from the beard, pulled up towards the vertex as the skull enlarges to accommodate the brain; its former site over the angle of the jaw is replaced by the great auricular—C2). The auriculo-temporal nerve also supplies the parotid gland, and carries to it the postganglionic secreto-motor fibres from the otic ganglion (PLATE 39). Note that the nerve does not supply the parotid fascia, whose deep and superficial laminæ are supplied by the great auricular nerve.

(*b*) **Inferior Alveolar Nerve.** This large branch emerges below the lower head of the lateral pterygoid and curves down on the medial pterygoid muscle (Fig. 6.18, PLATE 35). The nerve lies anterior to its vessels between the spheno-mandibular ligament and the ramus of the mandible, and enters the mandibular foramen. The *mylo-hyoid nerve* leaves the inferior alveolar nerve at the mandibular foramen. It pierces the spheno-mandibular ligament, lies on the mandible above the insertion of the medial pterygoid muscle (Fig. 6.30A, p. 413), and runs forward on the superficial (cervical) surface of the mylo-hyoid, supplying it and the anterior belly of the digastric (PLATE 30, p. *xvi*).

The inferior alveolar nerve runs with its vessels in the mandibular canal. It supplies the three molar and two premolar teeth. Then it divides into the mental nerve (p. 385) and the *incisive nerve*. This nerve supplies the pulps and periodontal membranes of the canine and both incisors, with some overlap into the opposite central incisor.

(*c*) **Lingual Nerve.** This large branch appears below the lateral pterygoid muscle on the side wall of the pharynx and passes forwards and downwards between the medial pterygoid muscle and the mandible. It then comes *into contact with the mandible*, sometimes leaving a groove below and medial to the third molar, just above the posterior end of the mylo-hyoid ridge. This groove separates the attachments of the pterygo-mandibular raphé above and mylo-hyoid muscle below (Fig. 8.15, p. 572). The nerve thus enters the mouth on the surface of the mylo-hyoid, beneath the mucous membrane of the floor of the mouth. The **chorda tympani nerve** emerges through the petro-tympanic fissure, grooves the spine of the sphenoid, and joins the lingual nerve 2 cm below the base of the skull and is distributed with it to the anterior two-thirds of the tongue. It carries all the parasympathetic secreto-motor fibres to the submandibular ganglion and all the taste fibres from the anterior two-thirds of the tongue (PLATE 39).

The Otic Ganglion. This small body lies between the tensor palati muscle and the mandibular nerve, just below the foramen ovale. It is about 2–3 mm in diameter. It is a relay station for parasympathetic secreto-motor fibres (IX) to the parotid gland; the lesser (superficial) petrosal nerve brings these fibres. A branch from the nerve to the medial pterygoid muscle passes through the ganglion to the tensor muscles (tympani and palati). The connexions of the otic ganglion are given on p. 37 (PLATE 40, p. *xxii*).

Anterior Wall of the Infratemporal Fossa. Examine a skull. Confirm that the anterior boundary of the infratemporal fossa is formed by the posterior wall of the maxilla. Note that this wall is traceable medially into a slit between the maxilla and the lateral pterygoid plate. The slit is the pterygo-maxillary fissure and it leads into the pterygo-palatine fossa. The posterior wall of the maxilla is traceable upwards into a slit between the maxilla and the greater wing of the sphenoid; the slit is the inferior orbital fissure (PLATE 35).

Through the pterygo-maxillary fissure emerge the **posterior superior alveolar nerves.** They are branches of the maxillary nerve, given off in the pterygo-palatine fossa. Two nerves pierce the posterior wall of the maxilla separately; their foramina can usually be seen in the dried skull. They are distributed to the molar teeth and the mucous membrane of the antrum. A third nerve lying with them does not pierce the bone but runs along the alveolar margin of the maxilla as far forward as the first molar. It supplies the gum of the vestibule alongside the three molar teeth (the adjacent mucous membrane that lines the buccinator is supplied by the buccal nerve—V c). Branches of the maxillary artery accompany these nerves, and veins drain the blood back into the pterygoid plexus.

Communicating veins from the inferior ophthalmic veins emerge through the inferior orbital fissure and cross the anterior part of the infratemporal fossa to reach the pterygoid plexus.

The mucous membrane of the vestibule of the mouth is reflected at the lateral part of the anterior wall, and the posterior superior alveolar nerves can be blocked by an injection here. Around the tip of the injecting needle lie the vessels mentioned above. They or the pterygoid plexus itself can be punctured during such an injection, with a consequent hæmatoma of some size.

Posterior Part of the Infratemporal Fossa

In contact with the prevertebral fascia the carotid sheath lies on the side wall of the pharynx. Lateral to it the three muscles of the styloid apparatus slope down.

The **carotid sheath** extends from the base of the skull to the arch of the aorta (p. 369). In its upper part it is attached to the margins of the carotid foramen in the petrous bone and to the inferior border of the tympanic plate where it blends with the deep layer of the parotid fascia (p. 362). It contains here the internal

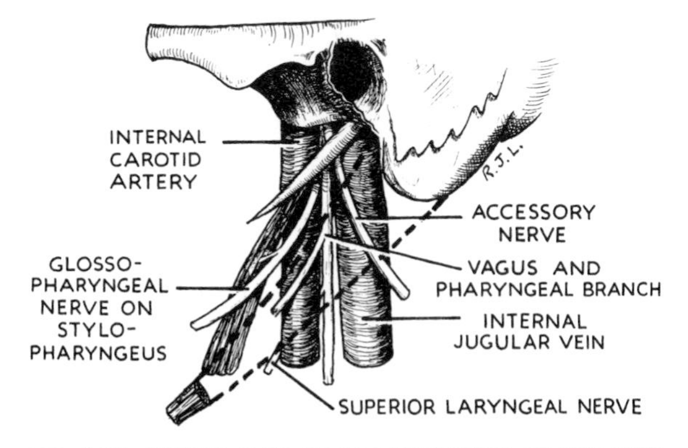

FIG. 6.19. LATERAL VIEW OF THE STRUCTURES IN THE LEFT CAROTID SHEATH AT THE BASE OF THE SKULL. THE POSTERIOR BELLY OF DIGASTRIC IS SHOWN IN DOTTED OUTLINE. The hypoglossal nerve (not shown) hooks around the vagus and appears between the artery and vein below the lower border of the digastric. Cf. PLATE 39.

carotid artery. The internal jugular vein and IX, X, XI and XII cranial nerves lie close by. Medial to it lies the pharynx; laterally the deep lobe of the parotid gland touches the sheath and above this the styloid process and its three muscles slope downwards and forwards. Anteriorly is the infratemporal fossa, of which the carotid sheath may be said to form the posterior boundary. Behind the carotid sheath lies the cervical sympathetic trunk on the prevertebral fascia (PLATE 39).

Examine a skull. The carotid foramen lies immediately in front of the jugular foramen. The latter lies deep to the external acoustic meatus. The internal jugular vein lies behind the artery at the base of the skull, but slopes as it descends, and at a lower level lies lateral to the common carotid artery as the vessels lie on scalenus anterior. At all levels the *vagus nerve lies deep in the groove between the two*, bound in the carotid sheath. The passage through the jugular foramen of IX and XI should be studied (p. 492). They emerge at the base of the skull between artery and vein and immediately *curve away from each other superficial to the vessels* (Fig. 6.19). The hypoglossal nerve (XII) emerges from the hypoglossal canal (anterior condylar foramen) medial to the sheath. It spirals down around the ganglion of the vagus and passes through the carotid sheath between artery and vein, and throughout its course to the tongue it has *arteries deep and veins superficial to it.*

The internal and external carotid arteries are separated from each other at the base of the skull. The external carotid, indeed, lies buried in the parotid gland. Apart from the deep lobe of the parotid gland the *only structures intervening between internal and external carotid arteries are pharyngeal in destination* and in name. They are the glosso-*pharyngeal* nerve, the stylo-*pharyngeus* muscle and the *pharyngeal* branch of the vagus (Fig. 6.19). Other nerves passing forwards into the neck pass deep to the internal carotid artery (superior laryngeal branch of the vagus) or superficial to the external carotid artery (hypoglossal nerve, facial nerve) (PLATE 39).

The **internal carotid artery** arises at the bifurcation of the common carotid, usually at the level of the upper border of the thyroid cartilage, level with C3 vertebra. At its commencement it bulges into the **carotid sinus;** here the arterial wall is thin and has a rich nerve supply (IX). The nerves mediate blood pressure impulses to the vital centres. The internal carotid artery is really external (i.e., lateral) to the external carotid at its origin, but soon slopes up posteriorly to occupy a medial and deeper level. It has no branches and passes straight up to the carotid foramen. It carries with it a rich contribution of sympathetic nerves derived from the superior cervical ganglion. This is known as the **carotid plexus,** though it is usually in the form of two or three single nerves rather than a network. The artery curves forwards in the petrous bone and then curves upwards into the foramen lacerum in the middle cranial fossa (PLATE 41). From the carotid foramen to its termination at the anterior perforated substance (pp. 485, 519) the internal carotid artery has half a dozen bends in its course well seen in a lateral arteriogram. It may be that these bends damp down the pulsations and give a more regular stream of blood for delivery to the brain.

The **carotid body** is a small yellowish-grey structure lying behind the bifurcation of the common carotid artery. Its cells are chemo-receptors and are innervated by the glosso-pharyngeal nerve.

The **internal jugular vein** emerges from the jugular bulb at the posterior compartment of the jugular foramen. At first behind the internal carotid artery, it lies on the lateral mass of the atlas, crossed by the accessory nerve. It receives the inferior petrosal sinus just beneath the base of the skull; the sinus passes back lateral (or sometimes medial) to the vagus nerve. The vein passes down to gain the lateral side of the internal carotid artery and of the common carotid artery; in the lower part of their course the vessels are overlaid by the sloping sterno-mastoid muscle. The termination of the vein lies beneath the triangular interval between sternal and clavicular heads of the muscle. It joins the subclavian vein to form the brachio-cephalic vein behind the sterno-clavicular joint.

The **tributaries** of the internal jugular vein below the inferior petrosal sinus are several veins from the walls of the pharynx (on which the veins form the pharyngeal plexus) and, at the level of the hyoid bone, the common facial vein with the lingual and superior thyroid veins. At a lower level the short middle thyroid vein enters.

The carotid sheath itself is really a firm feltwork of fibres binding the vagus nerve and the internal jugular vein to the internal carotid artery. It surrounds the artery and vagus nerve, but does not *surround* the jugular vein, which can expand into a loose 'dead-space' zone that lies lateral to it. In this dead space lie the deep cervical lymph nodes (p. 443).

The Cranial Nerves of the Carotid Sheath

The last four cranial nerves pass through the upper part of the 'sheath'.

The **glosso-pharyngeal nerve** emerges from the anterior part of the jugular foramen on the lateral side of the inferior petrosal sinus (p. 492). It makes a deep notch in the inferior border of the petrous bone and here its *petrous ganglion* bulges the nerve. The ganglion contains the cell bodies of *all* sensory fibres in the nerve. The nerve passes down on the internal carotid artery and curves forward around the lateral side of stylo-pharyngeus muscle (PLATE 39). Passing now parallel with the lower border of stylo-glossus, below the lower border of the superior constrictor muscle, it passes high up behind the posterior border of the hyo-glossus and so reaches the tongue (p. 414) (PLATE 36).

Its branches are as follows:

Meningeal branches run back through the jugular foramen and supply the inferior surface of the tentorium cerebelli.

The **tympanic branch** (Jacobson's nerve) leaves the nerve at the petrous ganglion and passes through the ridge of petrous bone between the carotid and jugular foramina to supply the middle ear and bony part of the auditory tube with sensory fibres. In this branch are also parasympathetic fibres from the inferior salivatory nucleus. They run through the tympanic plexus on the promontory, leave the middle ear in the lesser (superficial) petrosal nerve (p. 451) and so pass to relay in the otic ganglion (PLATE 40, p. *xxii*) for the secreto-motor supply of the parotid gland and the other small glands of the vestibule of the mouth.

The **motor branch** to stylo-pharyngeus is given off as the nerve spirals around the posterior border of that muscle. The cell bodies of the motor axons lie in the nucleus ambiguus.

The **sinu-carotid nerve** supplies the carotid sinus and carotid body (baro-receptors and chemo-receptors) from which the fibres pass centrally to the vasomotor and other 'vital centres' in the brain stem.

The **pharyngeal branches** join the pharyngeal plexus on the middle constrictor muscle. They pierce the muscle and supply the mucous membrane of the oro-pharynx with common sensation and (a few) taste fibres and with parasympathetic fibres from the inferior salivatory nucleus to the mucous and serous glands of the oro-pharynx (these fibres relay in small ganglia in the mucous membrane of the pharynx).

The **lingual** distribution is to the posterior one-third of the tongue (p. 414). It supplies sensory fibres (common sensation and taste) and secreto-motor fibres to the glands of the posterior third of the tongue. These last relay in small ganglia in the mucous membrane of the posterior one-third of the tongue.

Embryology. The glosso-pharyngeal is the nerve of the third pharyngeal arch and supplies the derivatives of that arch. The endoderm of the third arch provides the mucous membrane of the posterior third of the tongue (including the vallate papillæ) and the mucous membrane of the oro-pharynx; this area is supplied by IX with common sensation and taste. The endoderm of the *second pouch* provides part of the middle ear—the pretrematic branch of IX is the tympanic (Jacobson's) nerve. The mesoderm of the arch produces the stylo-pharyngeus muscle, supplied by IX. The third arch artery persists as the internal carotid artery; IX supplies its dilatation, the carotid sinus, and the adjacent carotid body, developed from third arch mesoderm.

Neuro-physiology. IX is a very mixed nerve. Its *afferent fibres* come from tongue, pharynx, tympanum, carotid sinus and carotid body, and comprise common sensation, taste and the special pressure reception and chemo-reception for the vital centres. Their cell bodies are in the petrous ganglion.

Its *efferent fibres* are both branchial motor and parasympathetic. The solitary muscle supplied is stylopharyngeus (cell bodies in nucleus ambiguus). The parasympathetic supply comes from cell bodies in the inferior salivatory nucleus. They are secreto-motor to the oro-pharynx (relay in tiny ganglia in mucosa of posterior third of tongue and pharyngeal wall) and to the parotid gland and the other small glands that open into the vestibule of the mouth (relay in the otic ganglion).

The **vagus nerve** emerges through the middle compartment of the jugular foramen, in which a small enlargement constitutes the *superior ganglion.* Just below the base of the skull the larger **inferior ganglion** dilates the trunk. The ganglia contain cell bodies of the afferent fibres of the vagus; the superior ganglion for the unimportant meningeal and auricular branches, the *inferior ganglion for all the sensory fibres that matter.* Below the inferior ganglion the vagus receives a large branch from the accessory nerve; this is its complement of nucleus ambiguus fibres, which XI has carried for it through the jugular foramen. These nucleus ambiguus fibres supply all the *striated* muscle of the viscera (pharynx, soft palate, œsophagus, larynx).

The nerve runs straight down the neck, contained in the back of the carotid sheath, between carotid artery and jugular vein. In the root of the neck it passes in front of the subclavian artery and so enters the mediastinum (p. 226) to supply thoracic and abdominal viscera.

Its branches in the neck are as follows:

Meningeal branches pass up from the superior ganglion to supply the dura mater of the posterior fossa below the tentorium.

The **auricular branch** runs laterally between the tympanic plate and the mastoid part of the temporal bone. It supplies the postero-inferior quadrant of the outer surface of the tympanic membrane and a small adjacent area of skin of the external acoustic meatus and a little of the corresponding skin behind the auricle.

The **pharyngeal branch** of the vagus slopes forward across the internal carotid artery parallel with and below the glosso-pharyngeal nerve and joins the pharyngeal plexus on the middle constrictor muscle (p. 416). Its fibres are derived from the accessory nerve by its branch to the inferior ganglion of the vagus; their cell bodies lie in the nucleus ambiguus in the medulla. The fibres supply the constrictor muscles of the pharynx and the muscles of the soft palate (except tensor palati, V c).

The **superior laryngeal nerve** slopes downwards on the side wall of the pharynx deep to the internal carotid artery. At about the level of the hyoid bone it divides into a large *internal laryngeal nerve* which pierces the thyro-hyoid membrane to reach the piriform fossa (p. 416) and a small *external laryngeal nerve* which runs close to the superior thyroid vessels and passes down outside the larynx to supply the crico-thyroid and crico-pharyngeus muscles.

The **cervical cardiac branches** are two on each side. On the right they pass down behind the subclavian artery to the deep cardiac plexus. On the left the upper nerve passes alongside the trachea to the deep cardiac plexus, the lower nerve crosses the arch of the aorta to the superficial cardiac plexus. (Note that the superficial cardiac plexus receives only the upper sympathetic and lower vagal branch from the left side; all the other cardiac branches of sympathetic and vagus go to the deep plexus.)

The **recurrent laryngeal nerve** hooks around the ligamentum arteriosum on the left side and around the subclavian artery on the right side (PLATE 42, p. *xxiii*). Thence it runs up alongside the trachea to pass under the lower border of the inferior constrictor (crico-pharyngeus muscle). The nerve has cardiac branches, and also supplies the trachea, œsophagus and crico-pharyngeus muscle before entering the hypopharynx.

Neuro-physiology. The vagus supplies all striped muscle at the upper end of the food and air passages (i.e., the constrictors and the muscles of the soft palate and larynx) by its pharyngeal and laryngeal branches. But these motor fibres are not truly vagal. Their cell bodies lie in the nucleus ambiguus in the medulla. The axons leave the medulla as the cranial root of the accessory nerve. Below the jugular foramen all these axons pass as a branch to the inferior ganglion of the vagus and so travel via the vagus and its branches to the pharynx and larynx.

The true vagus is a mixed nerve. Its *afferent fibres* come from larynx and air passages including the lungs, from pharynx and alimentary canal down to the cardiac orifice, from the ascending aorta and the aortic bodies. It supplies the epiglottis and laryngeal part of the pharynx with taste and the rest of the larynx and trachea with common tactile sensibility, the aorta with baro-receptor and the aortic bodies with chemo-receptor fibres which pass to the vital centres. It supplies the stretch receptors of the lung. Its *efferent fibres* are branchial motor (XI from the nucleus ambiguus) to the pharynx and larynx muscles. The true vagal efferents are parasympathetic; motor to the smooth muscle of the alimentary canal and gall-bladder, secreto-motor to the pancreas and the œsophageal and gastric glands, motor to the smooth muscle and secretomotor to the glands of the bronchial tree, and depressor to the heart. The parasympathetic efferent cell bodies are in the vagal trigone of the medulla, and their axons relay in the walls of the viscera.

The **accessory nerve** is formed in the posterior cranial fossa by union of cranial and cervical roots. The nerve occupies the middle compartment in the jugular foramen, just lateral to the vagus. All the fibres of its cranial root leave the nerve in a branch which joins the vagus. The nerve, now consisting of cervical fibres only, slopes down on the internal jugular vein where the latter lies on the lateral mass of the atlas. It passes deep to the styloid process and posterior belly of the digastric and gives off a branch to the sterno-mastoid muscle. It pierces the sterno-mastoid near the entrance of the motor branch; here it is crossed by the upper sterno-mastoid branch of the occipital artery. The nerve passes through the muscle across the posterior triangle to the trapezius (p. 363).

Neuro-physiology. The *cranial root* of the nerve is an accessory part of the vagus. The cell bodies lie in the nucleus ambiguus and their fibres leave XI just below the base of the skull to join X, by which they supply the striped muscles of the pharynx, palate, œsophagus and larynx. This is analagous to the motor root of a spinal nerve joining the sensory root just distal to the ganglion. The only mystery of the accessory nerve lies in the curious pathway of its spinal fibres. The *cervical root* supplies the sterno-mastoid and trapezius muscles. The unique exit of these fibres from the spinal cord and peculiar course into the cranium and out to the neck are not understood. Their chief segments of origin are C2 and 3 for the sterno-mastoid and C3 and 4 for the trapezius (there is some overlap into C1 and C5 segments). Instead of emerging in the anterior roots of these cervical nerves in the ordinary way the motor axons emerge on the lateral side of the cord (p. 535). The proprioceptive fibres from the muscles reach the spinal cord in the ordinary way, through C2 and 3 for the sterno-mastoid and through C3 and 4 for the trapezius.

The **hypoglossal nerve** emerges from the hypoglossal canal by the anterior condylar foramen. It picks up *a substantial branch from the anterior primary ramus of C1* and then spirals behind the inferior ganglion of the vagus to emerge between the internal carotid artery and internal jugular vein. It lies on the carotid sheath deep to the styloid muscles and the posterior belly of the digastric, then curls forward just beneath the tendon of the digastric, across the arteries and deep to the veins (p. 374) to pass to the tongue (PLATE 36).

Neuro-physiology. All the branches of the hypoglossal nerve are formed by hitch-hiking C1 fibres, from the small *meningeal branch* which enters the posterior fossa through the hypoglossal canal to the branch to the ansa and the branches to thyro-hyoid and genio-hyoid muscles. The true hypoglossal fibres supply only tongue muscles. Their cell bodies are in the medulla. The tongue muscle is derived from suboccipital myotomes, which in their migration through the neck pass between the carotid arteries (external and internal) lying deep, and the jugular vein lying superficial, dragging their nerve behind them. There are no sensory fibres in the hypoglossal nerve. In this the nerve resembles III, IV, VI, VII and XI. The muscles of the orbit and face have a proprioceptive supply (by V to its mesencephalic nucleus) and sterno-mastoid and trapezius, supplied by XI, have a proprioceptive supply from C2, 3 and 4. The muscles of the tongue have been shown to contain spindles, but their innervation remains uncertain. It may be via the lingual branch of the mandibular nerve.

The Styloid Apparatus

The styloid process is a part of the temporal bone that ossifies in cartilage. From its tip the *stylo-hyoid ligament* passes to the lesser cornu of the hyoid bone. Both process and ligament are remnants of the second pharyngeal arch cartilage (p. 41); the unossified cartilage disappears, its perichondrium persisting as the ligament. Hence the styloid process is very variable in length, the ligament varying likewise, inversely with it.

Three muscles diverge from the styloid process. The stylo-pharyngeus is deepest and arises highest; it passes almost vertically downwards to the larynx. The stylo-hyoid arises from behind high up and the stylo-glossus from in front low down; they diverge to the lower and upper borders of the side of the tongue. Each of the three muscles has a different nerve supply. They all act significantly during the act of swallowing (p. 447). The styloid apparatus lies lateral to the carotid sheath and helps the latter to close in the infratemporal fossa posteriorly. Its upper part lies in the bed of the parotid gland (PLATE 36).

The **stylo-pharyngeus** muscle arises from the deep aspect of the styloid process high up. It slopes down across the internal carotid artery, in front of which it crosses the lower border of the superior constrictor and passes down inside the middle constrictor. Here it lies behind the palato-pharyngeus and is inserted into the thyroid cartilage and the side wall of the pharynx (p. 429). The glosso-pharyngeal nerve, hooking around the posterior border of the muscle where it crosses the internal carotid artery, supplies the muscle, which is an elevator of the larynx and pharynx (PLATE 39).

The **stylo-glossus** muscle arises from the front of the styloid process and the upper part of the stylo-hyoid ligament. It passes forwards below the superior constrictor and is inserted into the side of the tongue, where it interdigitates with the upper fibres of hyo-glossus. The glosso-pharyngeal nerve curves parallel with its lower border on a slightly deeper plane. The stylo-glossus muscle is supplied by the hypoglossal nerve. It is a retractor of the side of the tongue; it is used in swallowing (PLATE 39).

The **stylo-hyoid** muscle arises from the back of the styloid process high up. It slopes down along the upper border of the digastric muscle. Its lower end bifurcates around the intermediate tendon of the digastric muscle and is inserted by two slips into the base of the greater cornu of the hyoid bone. It is supplied by the facial nerve. It is a retractor and elevator of the hyoid bone, used in swallowing (PLATE 34).

The **stylo-mandibular ligament** is a thickening in the deep lamina of the parotid fascia (p. 362). It extends from the vaginal process of the tympanic plate to the angle of the mandible (Fig. 6.2, p. 361).

The **external carotid artery** passes between the muscles of the styloid apparatus. It runs up deep to digastric and stylo-hyoid, but superficial to stylo-pharyngeus (PLATE 36), to enter the parotid gland. The retromandibular vein on the other hand, passes down from the parotid gland superficial to stylo-hyoid and digastric (PLATE 35).

THE PTERYGO-PALATINE FOSSA

This fossa is concerned with the blood and nerve supply to the upper jaw; the maxillary vessels, the maxillary nerve and the pterygo-palatine ganglion occupy the fossa. The pterygo-palatine ganglion sends branches into the nose and down to the palate, the maxillary nerve sends branches to the upper teeth, floor of the orbit and skin of the face. Branches of the maxillary vessels accompany all these nerves.

Osteology. Examine a skull. In the infratemporal fossa the maxilla and tubercle of the palatine bone fuse with the lower part of the lateral pterygoid plate (Fig. 6.14, p. 381); they separate above at the pterygo-maxillary fissure, which leads into the pterygo-palatine fossa. On the lateral wall of the nose the maxilla is separated from the medial pterygoid plate by the vertical process of the palatine bone. The vertical process articulates with the maxilla; between the two lies the greater palatine canal which opens below at a foramen on the hard palate. The greater palatine canal opens above into the fossa, for here the vertical plate of the palatine bone bifurcates; one limb remains attached to the maxilla, the other limb passes back to articulate with the sphenoid (Fig. 8.13, p. 569). Between the two lies the spheno-palatine foramen, leading from the fossa into the lateral wall of the nose (Fig. 6.24).

The pterygo-palatine fossa is thus seen to be bounded posteriorly by the sphenoid bone (the root of pterygoid process and greater wing containing the foramen rotundum) medially by the palatine bone (with its notch bounding the spheno-palatine foramen) and anteriorly by the posterior wall of the maxilla, below the apex of the floor of the orbit (p. 570).

The passages leading from the fossa should be identified on the skull with the aid of a piece of flexible fine wire or a fine bristle. The wire passed into the pterygo-maxillary fissure passes across the fossa through the spheno-palatine foramen, the last-named closed in life by the mucous membrane of the lateral wall of the nose. The wire passed through the foramen rotundum from the middle cranial fossa can be seen to enter the pterygo-palatine fossa and pass through the inferior orbital fissure into the infra-orbital groove; this is the course of the maxillary nerve. The wire passed through the pterygoid canal from the foramen lacerum can be seen to enter the fossa; this is the course of the nerve of the pterygoid canal. The wire passed into the greater or lesser palatine foramen enters the fossa along the canal which lodges the nerves of the palate. The wire passed through the palatino-vaginal canal in the roof of the nose passes forward into the fossa; this is the course of the pharyngeal branch of the pterygo-palatine ganglion (Fig. 8.5, p. 560).

Contents of the Fossa

The **maxillary nerve,** giving a meningeal branch to the front of the middle cranial fossa, passes through the foramen rotundum in the greater wing of the sphenoid bone into the pterygo-palatine fossa. Deviating laterally, it divides at the inferior orbital fissure into two branches, the infra-orbital and zygomatic, both of which enter the orbit and are distributed to the skin of the face (p. 384).

In the fossa the pterygo-palatine ganglion is suspended by two short thick branches from the maxillary nerve and, a little further forward, the three *posterior superior alveolar nerves* are given off. They pass through the pterygo-maxillary fissure to the posterior wall of the maxilla (PLATE 34).

The **pterygo-palatine** (spheno-palatine) **ganglion** (PLATE 41) is a relay station between the superior salivatory nucleus in the pons and the lacrimal gland and mucous and serous glands of the palate, nose and paranasal sinuses. It is the ganglion of hay fever ('running nose and eyes'). Its connexions are summarized on p. 37.

The autonomic root is the *nerve of the pterygoid canal* (Vidian nerve). This nerve is formed in the foramen lacerum by union of the greater (superficial) petrosal nerve (p. 491), containing parasympathetic secreto-motor fibres, with the deep petrosal nerve, containing sympathetic vaso-constrictor fibres. The latter is a branch given off from the carotid plexus in the foramen lacerum. The combined nerve passes forward in the pterygoid canal and joins the ganglion.

The branches from the maxillary nerve to the ganglion are sensory and, like the sympathetic fibres in the deep petrosal nerve, they pass through the ganglion without relay. The only cell bodies in the ganglion are parasympathetic (secreto-motor). The five branches of the pterygo-palatine ganglion are distributed to the nose and palate. *Every branch carries a mixture of all three kinds of fibres : sensory, secreto-motor and sympathetic.*

The secreto-motor fibres to the lacrimal gland leave their cell bodies in the ganglion and join the maxillary nerve, pass in its zygomatic branch into the orbit, join the lacrimal branch of the ophthalmic nerve and so reach the lacrimal gland.

The five branches of the pterygo-palatine ganglion are distributed as follows:

The **naso-palatine nerve** (long spheno-palatine) passes through the spheno-palatine foramen, crosses the roof of the nose, and is distributed to the septum and incisive gum of the hard palate (p. 401).

The **posterior superior lateral nasal nerves** (short spheno-palatine) pass through the spheno-palatine foramen and turn forward to supply the postero-superior quadrant of the lateral wall of the nose.

The **anterior palatine nerve** (greater palatine) passes down through the greater palatine canal and at the greater palatine foramen turns forward to supply the mucous membrane of the hard palate (PLATE 41). Its nasal branches supply the postero-inferior quadrant of the lateral wall of the nose (p. 401). It gives branches to the medial wall of the maxillary sinus.

The **middle and posterior palatine nerves,** two in number, were formerly called simply the lesser palatine nerves. They pass down behind the anterior palatine nerve and emerge through the lesser palatine foramina (medial and lateral) behind the crest of the palatine bone. They pass back to the soft palate and the mucous membrane of the supratonsillar recess.

The **pharyngeal nerve** passes back through the palatino-vaginal canal, emerges at the roof of the nose and supplies the mucous membrane of the naso-pharynx as far down as Passavant's muscle (p. 416).

The **maxillary artery** passes through the pterygo-maxillary fissure and enters the pterygo-palatine fossa, where it gives off five branches that pass with the five branches of the pterygo-palatine ganglion.

The main trunk of the artery passes through the inferior orbital fissure and accompanies the infra-orbital nerve along the floor of the orbit and, almost spent, emerges on the face through the infra-orbital foramen.

Veins run back with these arteries and, passing through the fossa, emerge at the pterygo-maxillary fissure to drain into the pterygoid plexus.

THE NOSE

The nose is for breathing; the design of its cavity results in warming and moistening the inspired air, and in cleaning it too. Since odours are air-borne, the olfactory receptors are placed in the nose. The floor of the nose is the hard palate. Hence chewing can go on in the mouth cavity without interfering with breathing; the flap-valve of the soft palate meanwhile shuts off the mouth cavity from the airway through the oro-pharynx. Breathing is arrested during swallowing; the soft palate is elevated and shuts off the nose (i.e., the naso-pharynx) from the foodway through the oro-pharynx. Thus the oro-pharynx is the crossroads of airway and foodway; collisions between air stream and food are avoided by the control mechanism of the soft palate acting as the policeman on point duty.

The nasal cavity extends from the external nostrils to the posterior end of the nasal septum, a midline partition which divides the cavity into two halves. Posteriorly it opens into the naso-pharynx (p. 417). The respiratory mucous membrane of the nasal cavity is very vascular and red in colour. In the roof is an area of olfactory mucous membrane which is less vascular and appears yellow in colour. The lateral wall of the nose is increased in area by the projection of the nasal conchæ; on this wall open the paranasal sinuses and the tear duct.

Respiratory mucous membrane lines the respiratory passages from the vestibule of the nose to the walls of the naso-pharynx including all the paranasal sinuses, and from the lower larynx to bronchioles of a diameter of 1 mm. The corium is lax and cellular, containing small groups of serous glands and mucous glands with their serous crescents. Over the inferior conchæ the mucous membrane contains masses of erectile tissue in which arterio-venous shunts are numerous. The surface epithelium is columnar. The cells are tall, and obliquity of histological section may give a false impression of two or more layers of cells (whence the confusing name 'pseudo-stratified') and the surface cells are ciliated. Very many of these ciliated columnar cells are goblet cells. A surface film of mucus traps particulate matter. The surface is self-cleansing, for the cilia beat the film of mucus from the nose back into the naso-pharynx (and from the trachea up towards the larynx). The watery secretion of the serous glands evaporates to

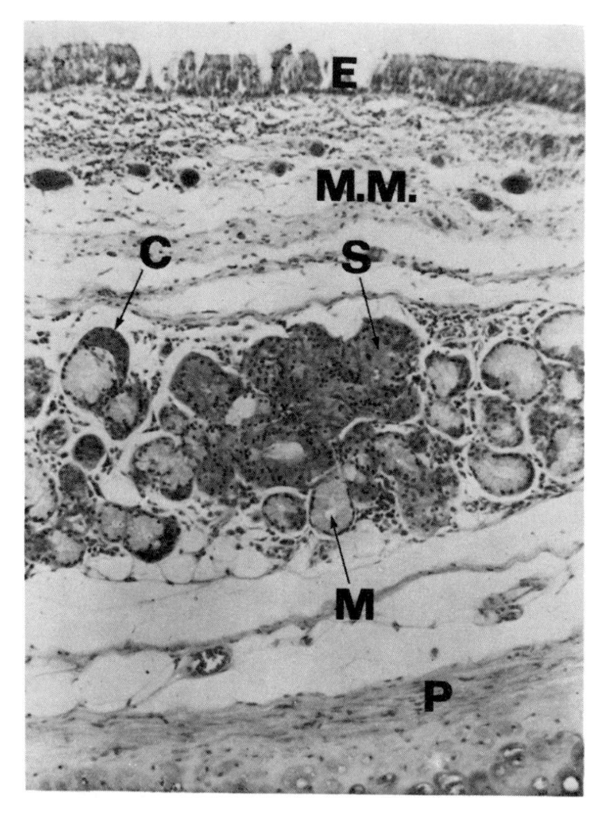

FIG. 6.20. RESPIRATORY MUCOUS MEMBRANE. The mucous membrane (**M.M.**) is thick and loose-woven. It contains serous glands (**S**) and mucous glands (**M**). The mucous glands have serous crescents (**C**). The mucous membrane is surfaced by a single layer (**E**) of ciliated columnar epithelium. **P** denotes perichondrium on hyaline cartilage. (Human trachea, photomicrograph × 350.)

moisten the inspired air. In the nasal cavity the mucous membrane is redder (much more vascular than elsewhere) to warm the inspired air.

Olfactory mucous membrane extends from the cribriform plate downwards on the lateral wall to the middle concha and downwards on the septum over a corresponding area. It is less vascular than respiratory mucous membrane. The epithelium is non-ciliated columnar; the cells contain a yellow lipoid pigment. This is *neuro-epithelium.* In nature olfaction seems to occur by the intermediary of pigmented cells (not always yellow as in man) which transmute the physical stimulus into a nerve impulse. The olfactory cells (bipolar *nerve cells*) are elongated, and hair-tufts project from their surface. The basal parts of the cells are thin and pass inwards to form a rich plexus of non-myelinated nerve fibres in the corium. From this plexus some twenty *olfactory nerves* are formed; these nerves pierce the cribriform plate and pass to the olfactory bulb. The corium of the olfactory area contains numerous serous glands; olfactory epithelium must be moist to function, for substances can be smelt only when they are in solution.

The **external nose** projects from the face; its skeleton is largely cartilaginous. Examine a skull. The nasal orifice is bounded above by the nasal bones and elsewhere by the two maxillæ (Fig. 6.50, p. 434). The external nose consists of the nasal bones (the 'bridge' of the nose) and the upper and lower nasal cartilages, supported in the midline by the cartilaginous part of the nasal septum. A small alar cartilage forms the lateral boundary of the nostril itself; this is made of elastic cartilage, and is moved by the compressor and dilator naris muscles. The other cartilages are of the hyaline variety. The whole is covered by adherent skin that contains many sebaceous glands; its upper part is lined with respiratory mucous membrane. The skin extends into the *vestibule* within the nostrils and here has a variable crop of stiff hairs. The muco-cutaneous junction lies beyond the hair-bearing area.

The skin is supplied by the external nasal nerve, which notches the nasal bone and passes down on the upper and lower nasal cartilage to the tip of the nose, by the infratrochlear nerve above this and, around the nostrils, by nasal branches of the infra-orbital nerve (PLATE 32).

The Cavity of the Nose

The floor of the nose is the roof of the mouth—the hard palate. The lateral wall of the nose is, above, the medial wall of the orbit (the ethmoidal air cells intervening) and, below, the medial wall of the maxillary antrum. The floor of the antrum lies at a lower level than the floor of the nasal cavity.

The lateral wall and septum are close together at the narrow roof and further apart at the broader floor. In cross-section the nasal cavity is pear-shaped, but three conchæ project into it with increasing prominence from above down, so that the distances between conchæ and septum are equal (Fig. 6.21).

The **lateral wall of the nose** should be studied in the intact specimen and in the dried skull (Fig. 6.24). In the living it is covered with mucous membrane that is very vascular and adherent to the periosteum of the underlying bone.

The shape of the lateral wall is roughly *semicircular,* that is, highest halfway along, at the cribriform plate of the ethmoid. From the vestibule the roof curves up to this level; behind the cribriform plate the roof curves down over the body of the sphenoid into the nasopharynx (Fig. 6.22). The semicircular lateral wall is not vertical, but slopes up from the broad nasal floor to the narrow roof.

There are three *nasal conchæ (turbinate bones)* projecting downwards like scrolls from the lateral wall. The

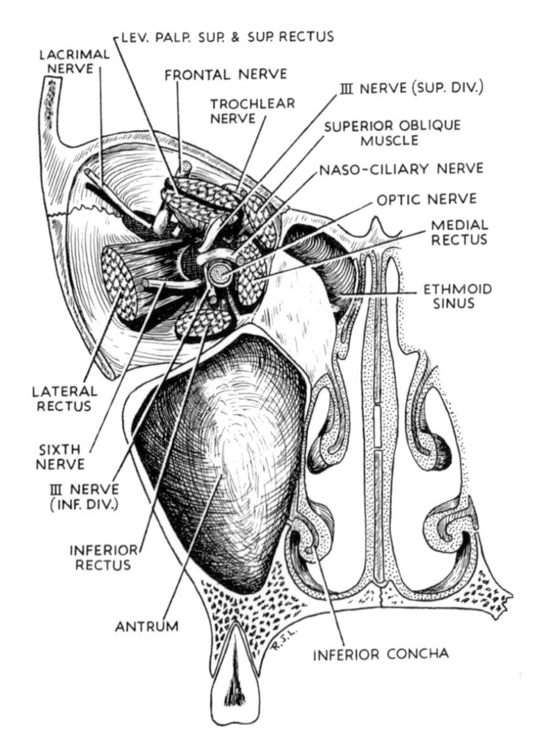

FIG. 6.21. CORONAL SECTION THROUGH THE SECOND UPPER PREMOLAR TOOTH. Drawn from Specimens M 137. 3, and M 138. 3, in R.C.S. Museum.

lowest is the longest. The middle and upper are joined anteriorly, but diverge away from each other posteriorly. Beneath the free inferior border of each concha is a meatus, called superior, middle and inferior respectively. Above the superior concha is the spheno-ethmoidal recess. A single opening exists in this recess, in the superior meatus, and in the inferior meatus—the middle meatus receives *all the other openings* into the lateral wall (Fig. 6.22).

The **inferior concha** is the longest and broadest of the three conchæ. It is covered with mucous membrane that contains large vascular spaces, erectile tissue that controls the calibre of the nasal cavity. It can swell and 'block the nose' instantaneously. It overhangs the **inferior meatus,** which receives the naso-lacrimal duct, draining excess tears from the eye. The duct opens 2 cm behind the nostril.

The **spheno-ethmoidal recess** lies above and behind the superior concha. It receives the ostium of the sphenoidal air sinus. Strictly speaking the sphenoidal ostium is in the narrow roof rather than in the lateral wall (Fig. 6.24).

The **superior concha** is small. It extends posteriorly from its junction with the middle concha. Its lower edge is free and overlies the **superior meatus,** into which drain the posterior ethmoidal air cells.

The **middle concha** is midway in size and position between superior and inferior. It extends horizontally back from its junction with the superior concha. It overhangs the middle meatus, which can be seen only when the concha is displaced (Fig. 6.22).

The **middle meatus** presents a convex bulge beneath the concha. This is the **bulla** of the ethmoid, produced by the bulging middle ethmoidal air cells, whose ostium opens on the bulla, usually high up. Beneath the bulla is a semicircular slit, the **hiatus semilunaris,** into which open the remaining paranasal sinuses. The frontal sinus opens through the *infundibulum* into the anterior end of the hiatus semilunaris. The anterior ethmoidal cells open through an ostium nearby in the hiatus or, quite frequently, directly into the infundibulum of the frontal sinus. The maxillary antrum opens near the posterior end of the hiatus semilunaris, often by a double ostium.

The **blood supply** and **nerve supply** of the lateral

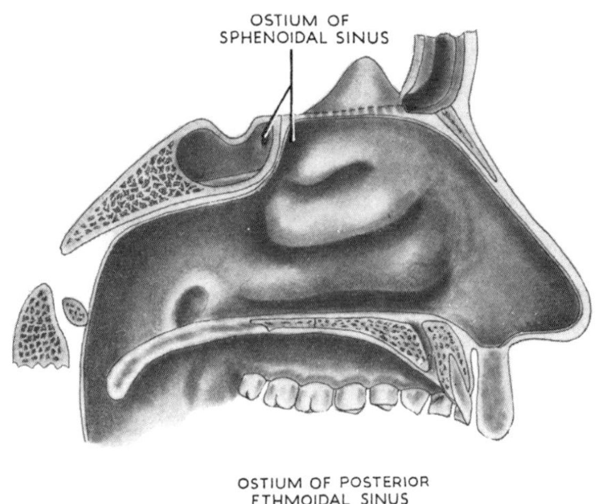

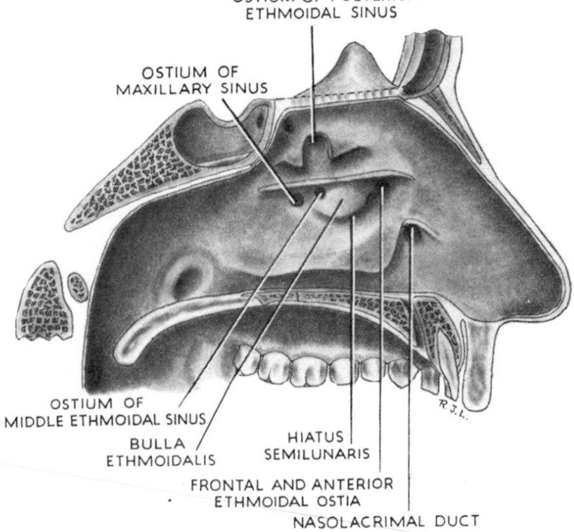

FIG. 6.22. THE LATERAL WALL OF THE NOSE AND NASO-PHARYNX. IN THE LOWER FIGURE THE CONCHÆ ARE DIVIDED TO SHOW THE OSTIA. Compare Fig. 6.24.

wall practically coincide. A minor difference is that nerves tend to pass out of the nose while arteries are not quite adequate and are reinforced by vessels that pass into the nose. If the lateral wall be bisected by a vertical and a horizontal line the four quadrants so produced receive separate supplies.

The *postero-superior quadrant* is supplied by the posterior superior lateral nasal vessels and nerves from the maxillary artery and the pterygo-palatine ganglion. Both reach the lateral wall by passing through the spheno-palatine foramen (see osteology of lateral wall, p. 402).

The *postero-inferior quadrant* is supplied by multiple branches of the greater palatine artery and anterior palatine nerve, which pierce the perpendicular plate of the palatine bone and pass forwards into the nasal mucous membrane.

The *antero-superior quadrant* is supplied by the *anterior ethmoidal nerve*, which passes down through the nasal slit in the cribriform plate alongside the crista galli of the ethmoid bone. The nerve gives off lateral branches (those under discussion) and medial or septal branches, and passes out to the surface between the nasal bone and upper nasal cartilage, where it is called the external nasal nerve. The blood supply of this quadrant is from the anterior ethmoidal artery assisted by the posterior ethmoidal and some branches that enter the nose from the facial artery (PLATE 41).

The *antero-inferior quadrant* is supplied by the anterior superior alveolar nerve on its way to its termination in the tiny septal branch. The arteries, on the other hand, are assisted by entering branches from the facial and perforating branches from the greater palatine arteries.

The **venous drainage** from the central parts of the lateral wall is via veins accompanying the arteries; they pass back to the pterygoid plexus. Posteriorly the veins reach the pharyngeal plexus, while anteriorly they pass out to the (anterior) facial vein. The **lymphatic drainage** follows the veins rather than the arteries. From the front half of the nose the lymphatics pass out across the face to the submandibular lymph nodes. From the back half of the nose and naso-pharynx they pass to retropharyngeal and antero-superior deep cervical lymph nodes.

The **nasal septum** consists of two bones and a septal cartilage. The cartilage is anterior and extends forwards to give shape and prominence to the external nose. It is almost always deviated from the midline. The vomer and vertical plate of the ethmoid compose the back part of the septum; the latter frequently shares in the deviation of the septal cartilage.

The mucous membrane of the septum is firmly bound to the perichondrium of the septal cartilage; this 'muco-perichrondrium' is readily stripped from the cartilage, as in the operation of submucous resection of the septum (the term is inaccurate; it is actually a subperichondrial resection).

The *nerve supply* and *blood supply* of the septum are similar. The long spheno-palatine artery and the naso-palatine nerve enter the spheno-palatine foramen and pass medially across the roof of the nose to the upper part of the posterior border of the septum. They pass forwards together in the mucous membrane of the septum, sloping down to the incisive canal. The nerves pass through the canal to the hard palate; it is usually said that they lie in the midline of the canal with the left nerve in front of the right nerve. This is difficult to remember, and possesses the further disadvantage of being untrue. Each nerve keeps to its own side of the canal. Blood flows upwards in the canal. That is to say, the long spheno-palatine arteries fall short of the canal, and on the anterior part of the septum each anastomoses with an ascending branch of the greater palatine artery.

The naso-palatine nerve and spheno-palatine artery supply the postero-inferior part of the septum. The antero-superior part of the septum is innervated by the septal branches of the anterior ethmoidal nerve. The anterior ethmoidal artery is assisted by branches that enter the anterior nares from the facial artery. Note that whereas the lateral wall is supplied in quadrants, the septum is supplied in halves (antero-superior and postero-inferior).

The *venous drainage* of the septum is towards the face from the front half and with the spheno-palatine artery from the back half, the vein flowing into the pterygoid plexus. *Lymph drainage* from the front half of the septum is into the submandibular lymph nodes and from the posterior part is across the floor of the nose to retropharyngeal and antero-superior deep cervical lymph nodes.

Sneezing. This protective reflex expels irritant material trapped in the nose. First there is a reflex secretion (no use to sneeze through a dry nose), then this secretion itself (cf. coryza) initiates the sneeze reflex. The excess secretion with its trapped particles is blasted out of the nostrils. The soft palate controls the volume of the nasal blast, the rest of the compressed air escaping harmlessly through the mouth (see p. 422).

Osteology of the Nose

Examine the longitudinal section of a skull made alongside the nasal septum.

The **septum** consists essentially of vomer and ethmoid. The vomer is fixed to the rostrum of the sphenoid by its alæ. It forms the posterior border of the septum and, slotted into a grooved ridge on the hard palate, extends like a ploughshare forwards beyond the incisive canal. It is grooved on each side by the long spheno-palatine vessels (Fig. 6.23).

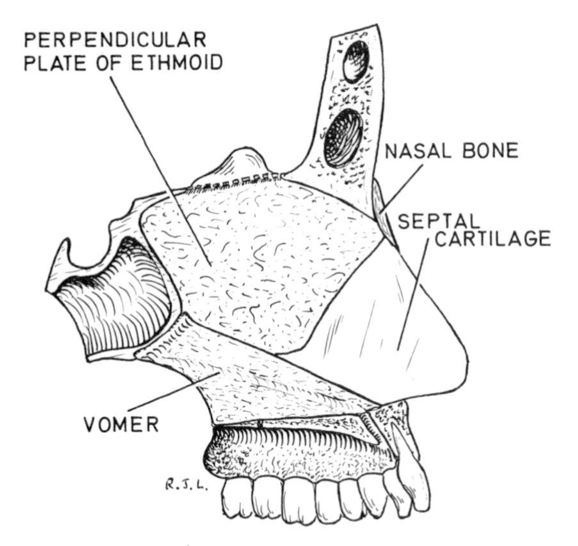

FIG. 6.23. THE SKELETON OF THE NASAL SEPTUM. It consists essentially of two bones (vomer, ethmoid) and one plate of hyaline cartilage.

The perpendicular plate of the ethmoid forms a suture with the upper border of the vomer. It completes the posterior part of the bony septum, but falls short of the anterior extremity of the vomer. The septal cartilage fills the angle between the two bones.

The **lateral wall** of the nose can only be studied adequately if the delicate bones have not been broken away. The frontal process of the maxilla and the nasal bones are in front. Behind them, study the lateral wall by building it up on the maxilla (Fig. 6.24). The maxilla (carrying all the upper teeth) is stabilized laterally by the flying buttress of bone called the zygomatic arch (Fig. 6.17, p. 387) and medially by the firm articulation of the palatal process with the opposite one.

The maxillary sinus opens on the medial wall of the maxilla by a large gap (Fig. 8.11, p. 568) which is made much smaller by the encroachment of neighbouring bones. The *palatine bone* sends a vertical plate across the posterior part of the maxillary hiatus. Between the two bones lies the greater palatine canal (Fig. 8.11, p. 568). The upper part of the vertical plate of the palatine bone is divided by a deep notch into two flat processes which articulate with the maxilla in the apex of the floor of the orbit and with the body of the sphenoid bone (Fig. 8.13, p. 569). Thus is enclosed the *spheno-palatine foramen* which opens from the pterygo-palatine fossa through the lateral wall of the nose just above the posterior end of the middle turbinate bone and just beneath the downward-sloping roof of the nasal cavity (Fig. 6.24). The *inferior concha* is a separate bone. Its vertical plate covers the lower part of the maxillary hiatus; the anterior part of the concha articulates with the conchal crest of the maxilla (Fig. 8.11, p. 568), the posterior part of the concha articulates with the conchal crest of the palatine bone (Fig. 8.12, p. 569). The *lacrimal bone* articulates with the frontal process of the maxilla and with the base of the inferior concha; the tear duct is enclosed between them. The *ethmoid bone* articulates by its lateral mass (labyrinth) at the upper border of the maxillary hiatus. Below this level the uncinate process curves back beneath the bulla to articulate with the base of the inferior turbinate bone. Between bulla and uncinate process lies the hiatus semilunaris. The *medial pterygoid plate* lies edge to edge behind the vertical plate of the palatine bone and completes the lateral wall of the nose.

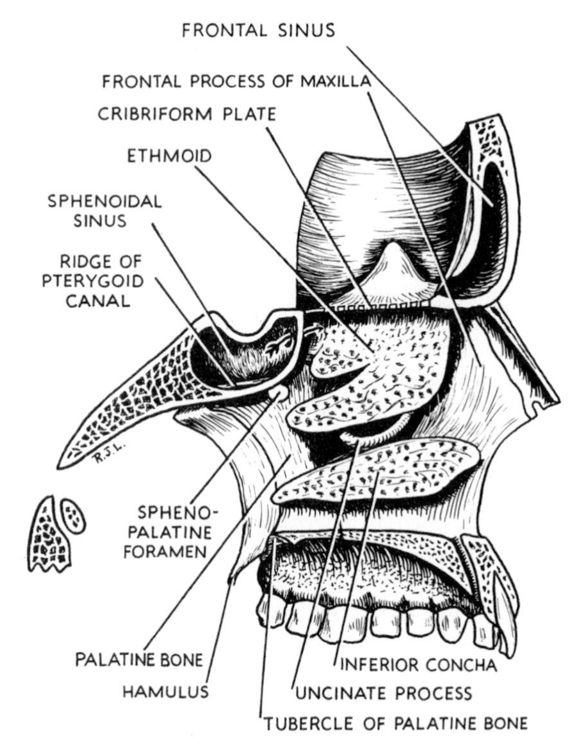

FIG. 6.24. THE BONES OF THE LATERAL WALL OF THE NOSE. AN ARROW PASSES THROUGH THE OSTIUM OF THE SPHENOIDAL AIR SINUS. (Cf. Fig. 8.11, p. 568.)

The auditory (Eustachian) tube opens along the upper part, the pharynx is attached along the lower part of the posterior border of the medial pterygoid plate down to the hamulus.

The bones of the lateral wall are considered individually on p. 567 *et seq.*

Anteriorly the framework of the nose (external nose) is completed down to the nostrils by an upper and lower hyaline cartilage and, at the nostril margin, by the elastic alar cartilage. These are bound to each other by fibrous tissue.

Development of the Nose

(see p. 45)

From each upper jaw region a flange of mesoderm grows medially above the tongue. The two flanges meet to form the palate (joining from before backwards). A midline flange grows downwards from the forebrain capsule (base of cranium) simultaneously, and all three meet and fuse to enclose the nasal cavities. Thus is formed the **nasal capsule,** in mesenchyme. The mesenchyme chondrifies at the sixth week and the nasal capsule then consists of one continuum of hyaline cartilage. The upper part of the nasal capsule, like the neighbouring base of the cranium, ossifies in cartilage, but in the lower part ossification *in membrane* proceeds from various centres outside the perichondrium. At birth the vomer is complete, the ethmoidal plate is still cartilage; the latter commences to ossify after one year. The original cartilage of the septum remains sandwiched for a time between two laminæ of membrane bone; it finally atrophies and disappears.

The anterior portion of the cartilaginous nasal septum is never enclosed by bone. The hyaline cartilage grows and persists as the cartilaginous part of the septum and the cartilages of the external nose. The side walls and floor of the nose (palate) are ossified at birth.

THE PARANASAL SINUSES

Certain of the bones that form the boundaries of the nasal cavities are hollowed out. The cavities so produced are lined with respiratory mucous membrane, and they communicate by relatively tiny apertures with the nasal cavity. They are known as paranasal sinuses. Their function is quite unknown. That they lighten the face bones is obvious enough, but a pair of spectacles and a pipe weigh down the head as much as sinuses full of cancellous bone would do. An elephant carries a pair of heavy tusks or a man on its head—it makes little difference to its neck muscles if its face bones are full of air or of bone marrow. When the sinuses are blocked or full of fluid the resonance of the voice is impaired, but it is difficult to imagine that the real purpose of such complicated structures is mere voice production. They are well developed in many silent animals. It may be that they serve as insulators to prevent incoming cold air from cooling down the surrounding parts, while the inspired air itself is being warmed and humidified by the conchæ.

All the sinuses are lined with respiratory mucous membrane. The glands produce a film of mucus which is moved by the cilia in spiral fashion towards the ostium. Gravity plays no part in draining a *normal* sinus (cf. trachea and bronchi).

There are four (bilateral) sinuses. The maxillary and ethmoidal sinuses are beyond the lateral wall of the nose. The frontal and sphenoidal sinuses abut at the midline, separated by a thin bony septum that is almost always off centre, causing asymmetry of these sinuses.

Development of the Sinuses. The frontal sinus is absent and the remainder are rudimentary at birth. They continue to enlarge throughout life, but not at a constant rate—there are spurts of enlargement. From birth to adult life 'growth' of a sinus is due to enlargement of the bone that encloses it; in old age 'growth' of a sinus is due to resorption of surrounding cancellous bone.

Sinus	State at Birth	Rapid Enlargement	
Maxillary	Rudimentary	6–7 years	Post-puberty
Ethmoidal	Rudimentary	,,	,,
Sphenoidal	Rudimentary	,,	,,
Frontal	Absent (appears in 2nd year)	,,	,,

The early enlargement of all the sinuses coincides with the eruption of the second dentition. At this period the baby-face elongates and takes on more adult contours (see p. 49).

The Maxillary Sinus (Antrum of Highmore). This is a space contained within the body of the maxilla, but the hiatus in the latter is closed by the palatine bone, inferior concha, and uncinate process (*v. sup.*). The roof of the sinus is the floor of the orbit. The floor of the sinus is the alveolar part (tooth bearing area) of the maxilla; it lies at a lower level than the floor of the nose. The sinus is pyramidal in shape, the base at the lateral wall of the nose and the apex in the zygomatic process of the maxilla. Anterior and posterior walls are the corresponding walls of the maxilla. Certain ridges appear within the cavity; a constant one is at the junction of roof and anterior wall, produced by the downward passage of the infra-orbital nerve (Fig. 8.11, p. 568).

The maxillary sinus is present at birth, but is no more than a shallow slit, slightly overgrown into a short *cul-de-sac* anteriorly and posteriorly. It excavates the lateral wall of the nose, beneath the middle concha, and lies just beneath the medial side of the floor of the orbit. The body of the neonatal maxilla lateral to this is full of developing teeth.

Relationship of Teeth. The maxillary canine raises a ridge on the facial surface of the maxilla but not on the wall of the sinus, whose cavity lies behind this. The posterior five teeth lie below the floor of the sinus. The roots of the first molar and second premolar lie nearest to the down-curving floor of the sinus, but they do not project through the floor until, occasionally with advancing age, excavation of the floor may expose them. The roots are usually enclosed in a thin layer of compact bone; when this is absent the apex of the root is in contact with the mucous membrane of the antrum. Extraction of such a tooth must leave an antro-oral fistula by rupture of the mucous membrane. These fistulæ mostly heal spontaneously.

The **ostium** of the antrum is high up and well back on its nasal wall. It is 3–4 mm in diameter. A second smaller ostium often lies posteriorly. It opens at the posterior end of the hiatus semilunaris in the middle meatus of the lateral wall of the nose (Fig. 6.22 and PLATE 40).

The **mucous membrane** is respiratory in type (p. 398). The cilia wash the mucus in spiral tracks towards the ostium. The *blood supply* of the mucous membrane is by small arteries that pierce the bone, mostly from the facial, maxillary, infra-orbital and greater palatine arteries, and veins accompany these vessels to the (anterior) facial vein and to the pterygoid plexus. Lymph drainge is for the most part via the infra-orbital foramen or the ostium; in either case the lymphatics flow to the submandibular lymph nodes.

The *nerve supply* of the mucous membrane is by the superior alveolar nerves (posterior, middle and anterior), the anterior palatine nerve, and the infra-orbital nerve; all are branches of the second (maxillary) division of the trigeminal nerve.

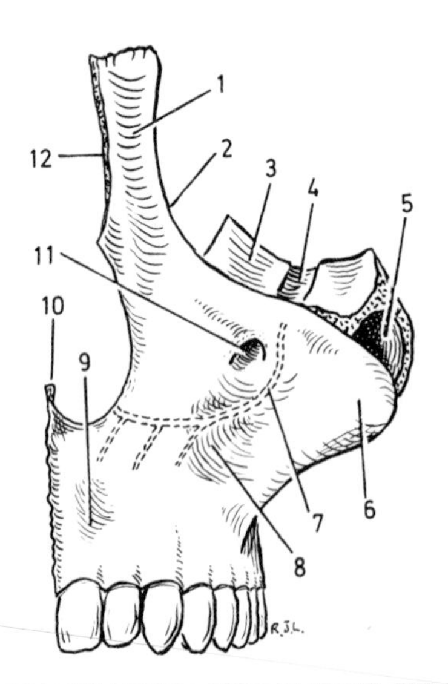

FIG. 6.25. THE LEFT MAXILLA FROM IN FRONT. THE COURSE OF THE ANTERIOR SUPERIOR ALVEOLAR NERVE IS SHOWN IN BROKEN LINE. **1.** Frontal process. **2.** Anterior lacrimal crest. **3.** Orbital plate. **4.** Infra-orbital groove. **5.** Excavation of maxillary sinus into zygomatic bone (cf. Fig. 8.11, p. 568). **6.** Zygomatic process. **7.** Course of anterior superior alveolar nerve. **8.** Canine fossa. **9.** Incisive fossa. **10.** Anterior nasal spine. **11.** Infra-orbital foramen. **12.** Articulation with nasal bone.

The *posterior superior alveolar nerves*, usually two in number, pierce the posterior wall of the maxilla separately and pass forwards in the bone above the apices of the molar teeth, which they supply. Minute branches pierce the bone to supply the mucous membrane of the antrum.

The *middle superior alveolar nerve* leaves the infra-orbital nerve on the floor of the orbit and runs down in the lateral wall of the maxilla to supply the premolar teeth, and the overlying mucous membrane of the antrum. It forms loops in the bone with the posterior and anterior superior alveolar nerves. It is sometimes absent, in which case its supply to the teeth is usually taken over by the anterior superior alveolar nerve.

The *anterior superior alveolar nerve* leaves the infra-orbital nerve in the infra-orbital canal in the roof of the antrum. It passes laterally, then curves medially below the infra-orbital foramen, in the anterior wall of the maxilla (Fig. 6.25). It supplies the pulps of the canine and incisors, the antero-inferior quadrant of the lateral wall of the nose, the floor of the nose and a small area of the septum anteriorly. Branches innervate the mucous membrane of the anterior wall of the antrum.

The *anterior palatine nerve* in its canal gives off minute branches which perforate the maxilla to supply the posterior part of the medial wall of the antrum.

The *infra-orbital nerve* gives perforating branches that supply the roof of the antrum.

The **ethmoidal sinuses** lie between the orbit and the nose, in the lateral mass (*labyrinth*) of the ethmoid. Examine a skull. The *ethmoid bone* consists of two lateral masses, connected above by the cribriform plate. From the cribriform plate superiorly the crista galli projects up into the anterior cranial fossa, and inferiorly the perpendicular plate projects down in the midline to articulate with the vomer in the bony part of the nasal septum. The *lateral mass* of the ethmoid is walled in by very thin bone; the superior and middle conchæ project from its nasal wall. The cavity is lined by mucous membrane of the ethmoid sinuses. The lateral mass possesses no roof; it is closed in by the orbital plate of the frontal bone, which projects medially to articulate with the cribriform plate. The ethmoidal nerves and vessels pass between the roof of frontal bone and the ethmoid bone itself. The lateral mass is divided by two bony septa into three cavities, anterior, middle and posterior (Fig. 6.49, p. 433). These drain by separate ostia into the lateral wall of the nose. Each cavity contains incomplete bony septa, making its interior contours very irregular. Thus an ethmoidal sinus is really an intercommunicating group of ethmoidal air cells (cf. mastoid air cells) (Fig. 6.49, p. 433). The ethmoidal sinuses are properly developed at birth but very small.

The *posterior ethmoidal sinus* is an irregular cavity in the back of the lateral mass. Its bony wall is completed posteriorly by fusion of the orbital process of the palatine bone and the sphenoidal concha (*v. inf.*). It is roofed in by the orbital plate of the frontal bone, between which and the lateral wall of the labyrinth (the pars papyracea, at the medial wall of the orbit) lies the posterior ethmoidal foramen. Here pass the posterior ethmoidal nerve and vessels. The nerve supplies the posterior ethmoidal sinus and the sphenoidal sinus. The artery and vein do likewise and, unlike the nerve, overrun the labyrinth to supply an area on the lateral wall of the nose. The sinus opens by a small ostium beneath the superior concha (in the superior meatus). Its lymph drains into retropharyngeal lymph nodes.

The *middle ethmoidal sinus* projects as a convexity into the lateral wall of the nose, under cover of the middle concha. This is the *bulla* of the ethmoid. An ostium on its upper margin communicates with the middle meatus (Fig. 6.22). The sinus is supplied by the anterior ethmoidal nerve and vessels. Its lymph drains across the face into the submandibular lymph nodes.

The *anterior ethmoidal sinus* occupies the anterior part of the labyrinth. Roofed in by the frontal bone, its bony walls are completed by the lacrimal bone. It opens into the anterior part of the hiatus semilunaris in the middle meatus. A separate septum in the anterior part encloses a funnel-shaped canal (the *infundibulum*) which connects the frontal sinus above with the hiatus semilunaris below. The anterior ethmoidal sinus may open into the infundibulum directly, instead of separately into the hiatus semilunaris. The anterior ethmoidal sinus is supplied by the anterior ethmoidal nerve and vessels. Its lymph drains into the submandibular lymph nodes.

The **sphenoidal sinus** is present at birth, but is only rudimentary. The *sphenoidal conchæ* are two small crescentic flakes of bone which project from the under surface of the body of the sphenoid. The sinus is subsequently formed by excavation of the body of the sphenoid. At the same time the sphenoidal concha grows forward, fusing with the posterior part of the labyrinth of the ethmoid, and closes in the anterior wall of the sinus. An antero-posterior ridge on the floor, when present, is made by the bony wall of the pterygoid (Vidian) canal (Fig. 6.24). When small the sinus lies in front of the pituitary fossa; as it enlarges it lies beneath the fossa, and in old age atrophy of the cancellous bone results in posterior extension of the sinus into the basi-occiput. The sinus opens on each side through its anterior wall into the spheno-ethmoidal recess. It is supplied by the posterior ethmoidal nerve. Its blood supply is from the maxillary artery and its lymph drainage is into the retropharyngeal lymph nodes.

The **frontal sinus** is undeveloped at birth; it appears during the second year. It consists of an excavation into the diploë of the frontal bone. It extends upwards above the medial end of the eyebrow and *backwards into the medial part of the roof of the orbit* (Fig. 6.49, p. 433). The two sinuses are unequal in extent and are separated by a bony septum that lies off the midline. The frontal bone articulates with the ethmoid labyrinth (Fig. 8.7, p. 565). The margins of the frontal sinus fit the upper end of the infundibulum. It is thus, by way of the ethmoid labyrinth, that the frontal sinus opens into the hiatus semilunaris of the middle meatus.

The sinus is supplied from the supra-orbital nerve by branches that pierce the frontal bone. Its blood supply is by the supra-orbital artery and its lymphatics drain across the face to the submandibular lymph nodes. Note the different drainage of the overlying eyebrow skin, which drains to pre-auricular lymph nodes (p. 444).

THE MOUTH

The mouth is for eating and talking through, and its structure is adapted accordingly. It serves also as an emergency airway in dyspnœa, but its structure has nothing to do with this function—it merely provides a bigger air-hole than the narrow nostrils. The mobile lips (p. 379) are indispensable to articulate speech (except in ventriloquists). They are prehensile, too, for grasping food or sucking in liquid. The cheek pouch of the vestibule prevents chewed food from spilling to the ground and the buccinator returns it to the molar teeth for rechewing.

The tongue is for grasping food, for moving it during mastication, and for swallowing it. The delicate movements of the tongue turn laryngeal noise into articulate speech. In addition its skin is highly sensitive, even more than finger-tips. It detects a fish bone and spits it out if nobody is looking, or feels a hot potato and spits it out even if people *are* looking. It is better than a finger-tip, for it possesses also the sense of taste, to accept or reject what is in the mouth. The tongue *is* the mouth; all the rest is accessory.

The roof of the mouth is the hard palate; breathing goes on during mastication, with the flap-valve soft palate standing sentinel behind.

The mouth extends from the lips to the anterior pillars of the fauces. It is enclosed by the lips and cheeks; the slit-like space between lips and cheeks on the one hand and the teeth and gums on the other hand is known as the *vestibule*. The space inside the teeth and gums is the mouth cavity proper. It is roofed in by the palate, and the floor is largely occupied by the tongue.

The mucous membrane of the mouth is adherent to the deeper structures; on lips and cheek to the face muscles, on tongue to the muscles thereof, on the hard palate to the periosteum of the bone. It is therefore seldom caught between the teeth when chewing. It is covered with stratified squamous epithelium and is supplied by the trigeminal nerve—above by the second (maxillary) division and below by the third (mandibular) division.

The lips and cheeks are covered with hairy skin. Their substance consists of the facial muscles in cadaverous individuals, added to by fat in the plump-faced. The groove between the buccinator and the anterior border of the masseter contains an encapsuled mass of fat, the *suctorial pad*, in the newborn. This pad often persists into adult life and does not shrivel even when, for any cause, the individual loses the rest of his body fat. It is alleged to prevent indrawing of the cheeks during sucking.

The mucous membrane (the red border) of the margins of the lips is highly sensitive and is represented by a large area in the sensory cortex. It is the main exploratory sensory area in babies, before they learn to use their hands for stereognosis.

The Vestibule

This is a closed space, lips and cheeks lying in contact with the teeth and gums. It leads, by the space behind the molar teeth, into the cavity of the mouth and in the rest position, with the teeth slightly parted, it communicates all around, between the teeth, with the mouth cavity.

On the mucous membrane of the cheek is a low papilla opposite the second upper molar tooth; it is the opening of the parotid duct. Nearby are the tiny openings of the ducts of the *molar mucous glands*. These are four or five small mucous glands lying on the surface of the buccinator near the parotid duct; their ducts pierce the buccinator to reach the mucous membrane of the vestibule. There are many other mucous glands (*buccal* and *labial*) scattered in the vestibule, especially in the lower lip. Unlike the molar glands they all lie internal to the face muscles, in the mucous membrane of the vestibule. Many of them are visible ('sago granules'). They are palpable to the examining finger or to the individual's own tongue. The orifices of their tiny ducts are too small to see.

The Gums. The mucous membrane of the gums is firmly attached to the underlying alveolar bone and, in a normal mouth, to the necks of the teeth. From the body of maxilla and mandible *alveolar bone* grows up to clasp the necks of the erupting teeth. After loss of a tooth the alveolar bone becomes resorbed. It is cancellous bone, covered with a thin plate of compact bone, and, like the mucous membrane of the mouth, has a high resistance to pathogenic organisms. The bacteria of the mouth in contact with bare bone elsewhere in the body would produce a gross osteomyelitis; saliva in contact with bare alveolar bone after tooth extraction produces an osteomyelitis of the jaw with the utmost rarity.

Nerve Supply of the Vestibule. The maxillary gums are supplied from the midline as far as the second premolar gum by branches of the infra-orbital nerve, which also supplies the adjacent mucosa of the upper lip. The vestibular gum over the upper molars is supplied by one of the posterior superior alveolar nerves. The overlying mucosa of the cheek is supplied, from the upper reflexion down to the lower sulcus, *including the vestibular gum over the lower molars*, by the long buccal nerve. Over the premolar teeth to the midline the lower sulcus is supplied, on both gums and lip, by branches of the mental nerve.

The Teeth

Structure. A tooth consists of a crown, which projects beyond the gum, and a root, which is fixed beneath the gum. The cavity in the interior of the tooth contains the pulp. The pulp cavity of a tooth is completely surrounded by dentine, from the crown of the tooth to the apex of the root. The projecting part of the tooth is capped with enamel and the buried part is surrounded with cementum, so that nowhere does the dentine appear on the surface of the tooth.

The **pulp** is a loose fibrous tissue containing blood vessels and lymphatics and nerves; all of these gain access to the pulp through a foramen in the apex of the root. The pulp is covered with a single layer of tall columnar cells which lie in contact with the inner surface of the dentine. They are known as *odontoblasts*, and throughout life they retain the power to produce dentine within the pulp cavity if the surface of the dentine is breached.

Dentine is a calcified material containing much organic matter, in the same proportions as bone. It contains spiral tubules which radiate out from the pulp cavity. Each tubule is occupied by a slender protoplasmic process from one of the odontoblasts.

Cementum is rather like bone; it exists as calcified lamellæ. Near the apex of the root it contains lacunæ and canaliculi occupied by cells with long branching processes, known as cementoblasts. Like enamel and dentine it has no blood supply. It lacks a nerve supply, too.

Enamel is the hardest animal substance in existence. It is 97 per cent calcified material, and exists in the form of crystalline prisms lying roughly at right angles to the surface of the tooth.

The **periodontal membrane** holds the cementum to the bony walls of the tooth socket. It consists of collagen fibres passing obliquely from the alveolar bone towards the apex of the tooth; the fibres 'sling' the tooth in position against pressure on its occlusal surface. It is really the modified periosteum of the alveolar bone. The membrane is radiolucent; in a radiograph of the tooth it shows as a clear interval between tooth and bone shadows.

Tooth Form. The shape of a tooth is adapted to its function. The human adult has from the midline two incisors, one canine, two premolars and three molars; that is, eight teeth in each half-jaw, or thirty-two teeth in all.

Roots. The *upper molars* have three roots each, two on the convexity and one on the concavity of the curve of the alveolar margin. The *lower molars* have two roots each, one anterior and one posterior. All the other teeth have but a single root—though a bifid root is not uncommon, especially in the first upper premolar.

Crowns. The incisor crowns are chisel-shaped, adapted for biting. Upper and lower incisors do not meet edge to edge, but by a sliding overlap, like the blades of a pair of scissors. The canine crowns are pyramidal or conical, much more rugged than the incisor crowns. They are the 'holding' teeth, well developed in carnivores. The premolar teeth have two cusps (lingual and buccal) whence their name of bicuspid teeth. It is of interest to note the definition of a **premolar tooth.** It is a *permanent* tooth occupying the former site of a deciduous molar. ('Deciduous premolar' is a meaningless term.) Upper molars have four, lower molars five, cusps on their crowns.

Nerve Supply of the Teeth. The pulp and periodontal membrane share the same nerve, which does *not* supply the overlying gum.

In the *upper jaw* the molars are supplied by two posterior superior alveolar nerves (a third, misnamed, supplies only the molar gum and no teeth). The anterior buccal root of the first molar is supplied by the middle superior alveolar nerve, which supplies also the two premolars. The canine and incisor teeth are supplied by the anterior superior alveolar nerve.

In the *lower jaw* the three molars and two premolars are supplied by the main trunk of the inferior alveolar nerve, whose terminal incisor branch supplies the canine and both incisors, overlapping to the opposite central incisor.

Tooth Position. The teeth of the upper jaw lie in a continuous curve, like a horseshoe. In the alveolar bone the outer plate is thinner than the palatal plate. In the lower jaw the curve of the anterior teeth straightens out in the molar region. In the alveolar bone the labial plate is thinner than the lingual plate over incisors, canines and premolars, but in the posterior molar region the lingual plate of the mandible is thinner than the buccal plate.

The attachment of the mylo-hyoid muscle is below the apices of most of the mandibular teeth—an apical abscess thus points in the mouth. In the second and third molars the apices lie below the mylo-hyoid line—an apical abscess bursting through the inner plate points in the neck.

The upper teeth make a *larger curve than the lower*. The upper incisors lie in front of the lowers in the closed position. The upper canine lies just behind the lower, in front of the first premolar, to their outer (buccal) side. The palatal cusps of the upper premolars and molars lie in the groove between lingual and buccal cusps of their opposite members, and each upper tooth articulates with its opposite member and the tooth behind that.

The Deciduous Teeth. The deciduous, or milk, teeth begin to erupt at about the sixth month and are completely erupted at the end of the second year. They consist of five teeth in each half-jaw, twenty in all. There are two incisors, one canine and two molars. They are shed as the permanent teeth erupt.

Development and Eruption of Teeth. Teeth are derived by budding of the epithelium (ectoderm) lining the mouth. The buds of ectoderm produce only the enamel; they evoke a reaction in the surrounding mesoderm, which differentiates to produce the dentine and cementum.

In the mouth cavity (stomodeum) of the five-week embryo [$\frac{1}{2}$ inch (12 mm) long] an ingrowth of ectoderm occurs over the site of the future gums. A curved sheet of ectoderm grows into the adjacent mesoderm, tilting medially. This is the *primary dental lamina.* From its outer surface a series of buds grow into the mesoderm—one for each deciduous tooth. At a later stage a similar series of buds grow (more medially) from the depths of the primary dental lamina—one bud for each permanent tooth. These epithelial buds are the *tooth germs* and when they are well formed the primary dental lamina becomes absorbed. Remnants of this epithelium may later grow into cysts or tumours.

The tooth germs grow away from the mouth surface into a wineglass-shaped mass, attached by its stalk to the primary dental lamina. This mass is the *enamel organ.* Its surface epithelium becomes columnar; the cells lining the concavity are called *ameloblasts* (Fig. 6.26). They produce enamel. The epithelium of the rest of the enamel organ takes on a different appearance; its cells develop long branching processes and fluid separates them. From this appearance the mass is known as the *stellate reticulum.* After the ameloblasts commence to secrete enamel the stellate reticulum undergoes atrophy. Beyond the stellate reticulum the enamel organ is prolonged as an epithelial sheath around the root.

The mesoderm within the cavity of the wineglass-shaped enamel organ is evoked to differentiate into the *dental papilla.* Its surface cells become columnar; they are known as *odontoblasts* since they produce dentine.

The ameloblasts (ectodermal) lining the concavity of the enamel organ and the odontoblasts (mesodermal) covering the convexity of the dental papilla lie in contact. Though the ameloblasts were there first, so to speak, it is the odontoblasts that become active first. They produce dentine and, later, the ameloblasts produce enamel. As the two substances accumulate, the secreting cells are pushed further away from the amelodentinal junction (Fig. 6.27). Around the root the dentine is contained by the epithelial sheath prolonged down from the enamel organ.

The mesoderm of the dental papilla persists as the pulp of the tooth, surrounded by the dentine it has secreted.

Cementum is produced in the mesoderm outside the dentine of the root. The process is comparable to membranous ossification. As the cementum is formed the epithelial sheath around the dentine is absorbed and the cementum becomes firmly bound to the dentine of the root.

The crown of the tooth is fully formed before eruption, but the root is only one-third formed. At this stage the tooth lies within a fibrous tissue condensation within the bone of the jaw. This is the *dental follicle*, and in the permanent teeth it communicates by a tiny orifice with the surface of the bone. The fibrous tissue in this orifice is known as the *gubernaculum*.

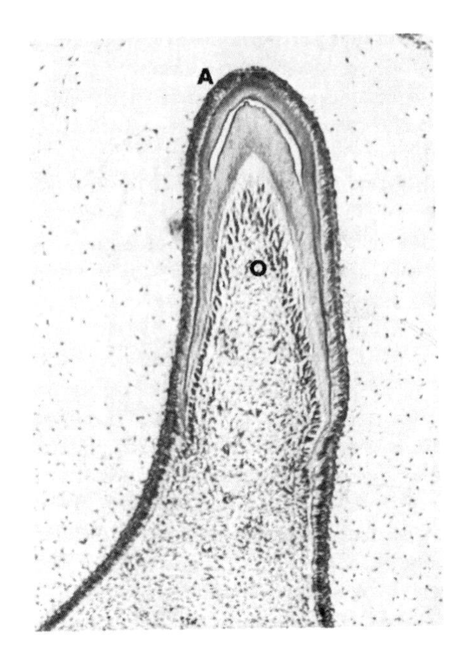

FIG. 6.27. A DEVELOPING TOOTH. (Photomicrograph, high power.) **A.** Ameloblasts. **O.** Odontoblasts. Same specimen as Fig. 6.26.

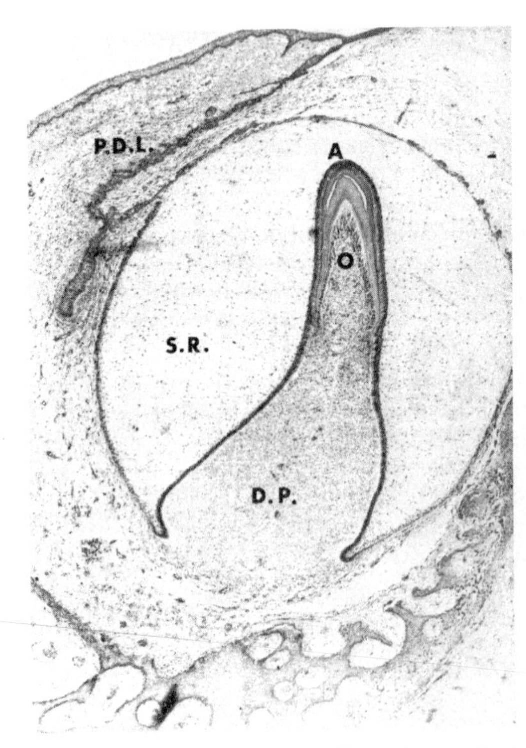

FIG. 6.26. A DEVELOPING TOOTH. (Photomicrograph, low power.) **P.D.L.** Primary dental lamina. **S.R.** Stellate reticulum of the ectodermal enamel organ. **A.** Ameloblasts. **O.** Odontoblasts differentiated from the mesodermal dental papilla, **D.P.**

The tooth erupts by a combination of elongation of the root and absorption of the overlying bone. The elongating root is ensheathed in an upgrowth of alveolar bone.

The normal times of eruption are:

Deciduous Teeth

6 months	Lower central incisors
7 months	Upper central incisors
8 months	Upper lateral incisors
9 months	Lower lateral incisors
1 year	First molars
18 months	Canines
2 years	Second molars

Permanent Teeth

6 years	First permanent molars
7 years	Central incisors
8 years	Lateral incisors
9 years	First premolars
10 years	Second premolars
11 years	Canines
12 years	Second permanent molars
18 years plus	Third permanent molars (wisdom teeth).

Summary of Eruption of Permanent Teeth. A lower tooth precedes its opposite number in the upper jaw.

The first permanent molar (the six-year molar) erupts before any deciduous teeth have been shed. The second permanent molar does not erupt until twelve years of age. In the intervening period the five deciduous teeth in each half-jaw are replaced. The order of replacement is first the incisors, central and lateral, then the milk molars, first and second and, last of all, the long-rooted canine.

The Cavity of the Mouth

The *palate* is the roof of the mouth. Between the teeth it lies on a basis of bone, the hard palate. Behind the teeth and hard palate the soft palate projects down.

The **hard palate** is made up of premaxilla, maxilla and palatine bones. Examine a skull. On the hard palate lies a midline foramen just behind the incisor teeth; it is the incisive foramen and it leads obliquely backwards up into the nose. In a young skull an irregular suture passes from the foramen to the alveolar bone between lateral incisor and canine teeth, there to be lost. This is the outline of the premaxilla, the bone that carries the incisor teeth and appears on the face in all mammals but man. In man it is overlaid on the face by the maxilla; the suture between maxilla and premaxilla runs across the sockets of the incisor teeth (Fig. 1.57, p. 45). The main mass of the hard palate is made by the palatal processes of the maxillæ; posteriorly the horizontal plates of the palatine bones complete the bony shelf. The greater palatine foramen lies between the palatine bone and the maxilla; the lesser palatine foramina perforate the palatine bone itself (Fig. 8.5, p. 560).

The **mucous membrane** of the front of the hard palate is strongly united with the periosteum and the two cannot be stripped apart. Together they can be readily stripped from the bone; these combined layers are known as **muco-periosteum.** The attachment of the periosteum to the bone is secured by multiple fibrous-tissue pegs (Sharpey's fibres) that leave a finely pitted bone surface on the dried skull. This fixation of the mucous membrane is for mastication; the moving bolus does not displace the mucous membrane. There are transverse masticatory ridges in this part of the muco-periosteum. Over the horizontal plate of the palatine bone mucous membrane and periosteum are separated by a mass of mucous gland tissue; no large Sharpey's fibres are needed here, and the bone surface is smoothly polished. From between these areas the mucous membrane curves down to the under surface of the soft palate.

The *blood supply* of the hard palate is provided by the greater palatine artery, which emerges from the greater palatine foramen and passes around the palate (lateral to the nerve) to enter the incisive foramen and pass up into the nose. Veins accompany the artery back to the pterygoid plexus. Other veins pass back to the supratonsillar region and join the pharyngeal plexus. Lymph return is alongside these latter veins to retro-pharyngeal and deep cervical lymph nodes.

The *nerve supply* of the hard palate, including the palatal gums, is by the anterior palatine nerve (a branch of the maxillary nerve via the pterygo-palatine ganglion) as far forward as the incisive foramen. The anterior part of the palate, behind the incisor teeth (the area of the premaxilla), is supplied by the two naso-palatine nerves, from the same source.

Development of the Palate (see p. 45).

Soft Palate. The soft palate is in the pharynx, not in the mouth (see p. 418).

The Tongue

The **mucous membrane** on the dorsum of the tongue consists of two parts, an anterior two-thirds and a posterior one-third; they have different developmental origins and nerve supplies and different surface appearances. The anterior two-thirds lies in the mouth; the posterior one-third is in the oro-pharynx. Each surface has a different function; the mouth part grips for chewing and the oro-pharynx part is slippery for swallowing.

The **anterior two-thirds** of the tongue is covered by a thick fibrous mucous membrane into which the underlying muscles are inserted. The surface epithelium is of the stratified squamous variety and is projected into **papillæ.** In man these are of two types, filiform (conical) and fungiform (mushroom-shaped). The former give rise to the velveted appearance (the 'fur' of the tongue). Fungiform papillæ are just visible as discrete pink spots, more numerous towards the edges of the tongue. They bear taste buds.

The mucous membrane of the under surface of the tongue and the floor of the mouth is thin and smooth; it projects as a midline flange, the frenulum, beneath the tip of the tongue.

There are no glands on the dorsum of the anterior two-thirds of the tongue, but on the under surface behind the tip there is a mucous gland, the *anterior lingual gland*, on each side of the midline. From each gland half a dozen tiny ducts open on the under surface of the tongue.

The **posterior third** of the tongue is not in the mouth; none the less it is studied here. It is bounded by the **vallate papillæ**. They number about a dozen and are arranged in the form of a V, apex back. There are no papillæ behind the vallate papillæ on the posterior third of the tongue. The smooth mucous membrane has a nodular appearance from the presence of underlying masses of mucous and serous glands and aggregations of lymphatic tissue. The latter constitute the 'lingual tonsil', part of Waldeyer's ring (p. 417).

The mouth part (anterior two-thirds) is prehensile;

the filiform papillæ make a non-slip surface for moving a bolus. The posterior one-third is in the oro-pharynx and is not used for mastication. The vallate papillæ lie just behind the mouth; they are not in actual contact with the bolus that is being chewed, but the nearby mouth saliva reaches them and thus transmits the flavour of the bolus. Behind the vallate papillæ the posterior one-third, like the soft palate above it, is coated with mucus from its multiple glands and makes a smooth and slippery surface for swallowing.

Just behind the apex of the V of vallate papillæ lies a minute pit, the *foramen cæcum*, a remnant of the thyro-glossal duct. A shallow groove behind the vallate papillæ, the *sulcus terminalis*, is incorrectly stated to be the line of union of the two embryological areas of mucous membrane. The vallate papillæ, however, supplied essentially by the glosso-pharyngeal nerve, belong to the posterior third of the tongue; the line of union is anterior to them, and does not show on the tongue. The posterior third of the tongue slopes down to the epiglottis. A midline flange of mucous membrane is raised up between the two, the glosso-epiglottic fold. From each side of the epiglottis a similar mucosal fold extends laterally to the wall of the pharynx, the pharyngo-epiglottic fold. It is now called the 'lateral glosso-epiglottic fold' but this is a misnomer, for it is not attached to the tongue. Its framework is fibrous tissue joining the epiglottis to the greater cornu of the hyoid bone (Fig. 6.41B, p. 424). These three folds enclose the two valleculæ, which are shallow oval pits below the posterior end of the tongue.

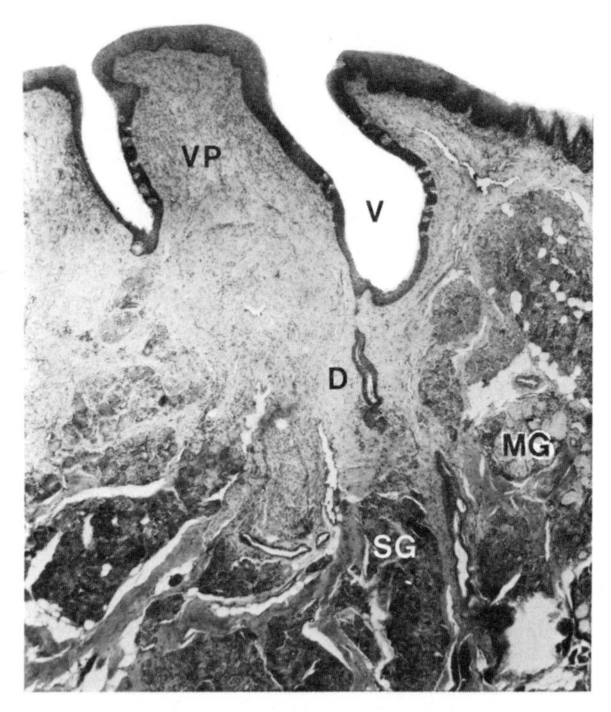

FIG. 6.28. A HUMAN VALLATE PAPILLA. TASTE BUDS ARE ON EACH WALL OF THE SLIT. SEROUS GLANDS OPEN INTO THE SLIT. **V.P.** Vallate papilla. **V.** Valla, taste buds on both walls. **D.** Duct of serous gland. **S.G.** Serous gland. **M.G.** Mucous gland.

Microscopic Anatomy. The substance of the tongue is composed of straight fibres of striated muscle lying for the most part in three planes at right angles to each other. Mucous glands with their crescents and serous glands are abundant in the posterior third. Lymphatic follicles occur here; there are none in the mouth (i.e., in the anterior two-thirds). The surface is covered by a stratified squamous epithelium which, in the mouth part, is projected into papillæ.

The *filiform papillæ* are by no means thread-like, but project as truncated cones whose flat tops are covered with keratin. The appearance of the protruded tongue depends on the state of the keratin. When there is little keratin (as after chewing roughage) the tongue is a 'healthy' pink. When the keratin is thick (as from lack of chewing in illness, starvation) it becomes white by maceration (cf. 'washerwoman's fingers'). Keratin dries after mouth breathing and is then brown in colour. The *fungiform papillæ* are so scanty as to be seldom seen in random sections of the tongue; they have a slender base and bulbous tip which is not keratinized. A *vallate papilla* consists of a cylindrical mass about 2 mm in diameter surrounded by a deep slit, both walls of which contain taste buds. The ducts of serous glands open into the base of the slit, to wash it out ready to taste the next arrival (Fig. 6.28).

Taste buds lie as spherical masses entirely within the epithelium; they are seldom seen except in the walls of vallate papillæ, where they abound. A taste bud is made up of slender, spindle-shaped, pale cells. It opens on the surface by a tiny gustatory pore, through which projects a tuft of minute hairs.

The **anterior pillars of the fauces** (nowadays called **palato-glossal arches**) are ridges of mucous membrane raised up by the palato-glossus muscles. They extend from the under surface of the front of the soft palate to the sides of the tongue at the vallate papillæ. The whole constitutes the **oro-pharyngeal isthmus.** In front of it is the mouth, behind it is the pharynx; and it is narrower than either. It is closed by depression of the palate and elevation of the dorsum of the tongue, and narrowed by contraction of the palato-glossus muscles.

The Muscles of the Tongue. The tongue is a mass of striated muscle, containing glands and lymphatic tissue and fat. There is a midline fibrous septum dividing the tongue into symmetrical halves.

The main bulk of the tongue is made up of the genio-glossus muscle, with some vertical, longitudinal and transverse intrinsic muscles fibres. The whole muscular mass thus constituted is connected by extrinsic muscles to the hyoid bone (hyo-glossus) the palate (palato-glossus) and the styloid process (stylo-glossus).

The tongue rests on the floor of the mouth (mylo-hyoid muscle and the hyoid bone). This highly mobile shelf greatly enhances the mobility of the tongue.

The **genio-glossus muscle** accounts for most of the bulk of the tongue. It arises from the superior genial tubercle, whence the fibres radiate widely to be inserted into the mucous membrane of the dorsum of the tongue from tip to base. The lowest fibres, passing through to the base, are attached to the body of the hyoid bone.

Intrinsic muscles lie wholly within the tongue and are not attached to bone. They are arranged in three planes at right angles to each other.

The *superior longitudinal fibres* form a broad band from side to side beneath the mucous membrane, into which most of them are inserted from the back of the tongue to the tip.

The *inferior longitudinal fibres* lie along the sides of the genio-glossus in the lower part of the tongue, medial to the hyo-glossus.

The *transverse fibres* arise from the median *fibrous septum* in the tongue and radiate out through the other muscle fibres to reach the mucous membrane of the sides of the tongue.

The *vertical fibres* arise from the mucous membrane of the dorsum near the midline and radiate down towards the sides at the lower part of the tongue.

The Extrinsic Muscles. The genio-glossus muscle has already been described.

The **hyo-glossus muscle** arises from the length of the greater horn of the hyoid bone and from the body of that bone lateral to genio-hyoid (PLATE 27, p. *xiv*). It extends as a quadrilateral sheet on the side of the tongue; its upper border, interdigitating at right angles with the fibres of stylo-glossus, is attached to the side of the tongue (PLATE 39, p. *xxi*). This muscle is the key to the floor of the mouth; its relations are therefore considered under that heading (p. 413).

Palato-glossus descends from the under surface of the palatal aponeurosis to the side of the tongue (p. 421).

Stylo-glossus is a muscle of the styloid apparatus (p. 396).

Movements of the Tongue. The tongue is used in sucking, in prehension of food, in chewing, in swallowing and in speaking. It is also used in toilet (licking the lips in man, licking the fur in animals) and occasionally in gesture. Its moist 'fur' makes an excellent damping pad for licking.

The intrinsic muscles alter the *shape* of the tongue, the extrinsic muscles stabilize the organ and by their contraction alter its *position*, as well as its shape.

Alteration of Shape. The transversus muscle narrows the tongue and consequently heaps up the dorsum into a side to side convexity. With simultaneous contraction of the vertical muscle this convexity is flattened and, since the total *volume* of the tongue remains constant, the organ becomes elongated and pointed at the tip. If now the lowest fibres of genio-glossus contract the back of the tongue is drawn forwards, i.e., the pointed tongue is extruded.

Contraction of the longitudinal fibres shortens the tongue, which then becomes convex from front to back. Contraction of the vertical fibres produces a midline groove with consequent heaping up of the sides of the tongue, as in the first stage of swallowing.

The essential point is that the tongue volume remains constant. Therefore shortening of some fibres can only occur if others passively elongate; this results in change of shape.

Alteration of Position. It has already been pointed out that the lowest fibres of genio-glossus draw the tongue forward. Stylo-glossus opposes this movement and retracts the organ. Hyo-glossus draws the sides of the tongue downwards.

The position of the tongue is altered by the mylo-hyoid muscle, on which the tongue rests. The mobile floor of the mouth can be elongated or shortened, raised or lowered, thus still further altering the position of an already mobile organ (p. 447).

The movements of the tongue in the *first (voluntary) stage of swallowing* illustrate its mobility. Contraction of the vertical intrinsic muscle makes a longitudinal groove on the dorsum; the heaped-up tip and edges are in contact with the hard palate and teeth. The liquid or moist bolus is thus imprisoned in the groove. Contraction of the mylo-hyoid now raises the floor of the mouth, compressing the tongue against the hard palate. The vertical intrinsic fibres relax from before backwards; pressure from the contracting mylo-hyoid muscle obliterates the groove in the same sequence, forcing the bolus backwards (Fig. 6.46, p. 427). It is more difficult to swallow a single bolus with the mouth open; but it is still possible. While drinking from a vessel the lips and teeth remain open and the tongue spills no liquid during swallowing. The tongue is mobile enough to perform almost any movement.

Blood Supply of the Tongue. The *lingual artery* supplies the tongue. It runs above the greater cornu of the hyoid bone, deep to hyo-glossus and passes forward to the tip. Beneath hyo-glossus it gives off *dorsales linguæ* branches into the posterior part of the dorsum. At the anterior border of the hyo-glossus it gives a branch to the sublingual gland and adjacent floor of the mouth.

The venous return from the tip is by a large vein visible on each side of the midline on the under surface of the tongue. These are the *ranine veins.* Each runs back superficial to the hyo-glossus alongside the hypoglossal nerve; the *vena comitans nervi hypoglossi* empties into the common facial vein. From the depths of the tongue and from the dorsales linguæ veins blood is collected into the *lingual vein* which runs alongside the lingual artery and empties into the internal jugular vein (p. 374).

Lymph Drainage of the Tongue (p. 444).

Nerve Supply of the Tongue. The anterior two-thirds is supplied by the lingual nerve, whose trigeminal component mediates common sensibility (cell bodies in the trigeminal ganglion) and whose chorda tympani component mediates taste (cell bodies in geniculate ganglion). The parasympathetic secreto-motor fibres to the anterior lingual gland run in the chorda tympani from the superior salivatory nucleus; they relay in the submandibular ganglion.

The posterior one-third of the tongue, *including the vallate papillæ,* is supplied by the glosso-pharyngeal nerve. This has fibres of common sensibility and taste, and carries parasympathetic secreto-motor fibres to the glands. These last relay in small *lingual ganglia* in the mucosa.

Vaso-constrictor sympathetic fibres travel with the lingual artery from the external carotid plexus; their cell bodies are in the superior cervical ganglion.

Motor Supply. All the muscles of the tongue, intrinsic and extrinsic, are supplied by the hypoglossal nerve (except palato-glossus). The motor cell bodies lie in the hypoglossal triangle in the medullary part of the floor of the fourth ventricle. The pathway of proprioceptive impulses from the tongue muscles is not yet known; it is probably the lingual nerve.

Development of the Tongue. The mucous membrane and glands of the tongue are derived from the floor of the pharynx. The bucco-pharyngeal membrane breaks down so early that it is not possible to say where ectoderm and endoderm meet on the floor of the mouth. There is some evidence that the anterior two-thirds of the tongue may be covered in mucous membrane of ectodermal origin. However this may be, the anterior two-thirds derives from the first pharyngeal arch and the posterior one-third from the third arch. The muscles are derived from suboccipital myotomes and migrate into the tongue (p. 42).

The Floor of the Mouth

The floor of the mouth is the mylo-hyoid muscle; above this muscle lies the mouth, below it lies the neck.

The mylo-hyoid muscle (p. 371) slopes down from the mylo-hyoid line of the mandible into a midline raphé; only its posterior fibres are inserted into the body of the hyoid bone. The arrangement resembles that of the levator ani in the pelvic floor. An important difference is that the raphé of the pelvic floor is attached behind to a fixed bone (the tip of the coccyx) whereas the mylo-hyoid raphé is attached behind to the mobile hyoid bone. The hyoid bone can be moved up and down, forwards and backwards; thus the floor of the mouth is much more mobile than the pelvic floor. This contributes a greater mobility to the already very mobile tongue.

The hyoid bone is slung between mandible and styloid process by the genio-hyoid and stylo-hyoid muscles; their reciprocal contractions determine the antero-posterior position of the hyoid bone. Each of these

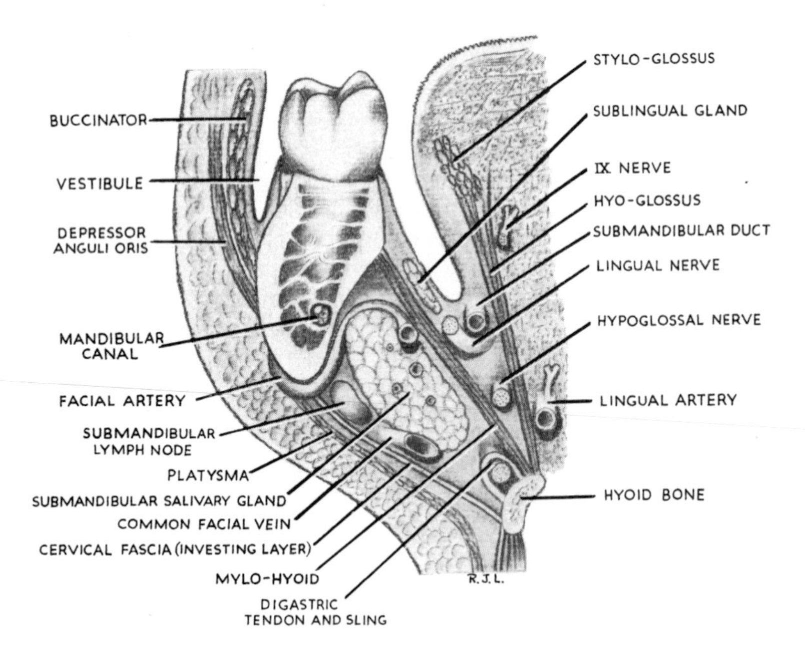

FIG. 6.29. CORONAL SECTION THROUGH THE LOWER JAW JUST BEHIND THE FIRST MOLAR TOOTH. Left side, viewed from behind (semidiagrammatic).

muscles slopes *down* to the hyoid, as does the mylo-hyoid muscle; these three muscles elevate the hyoid bone. The infrahyoid muscles depress the hyoid bone; the reciprocal contractions of these groups determine the vertical level of the hyoid bone.

Lying on the mylo-hyoid muscle is the mass of the tongue. Between the two are the genio-hyoid muscles near the anterior midline and, laterally, the hyo-glossus muscle (PLATE 39).

The **genio-hyoid muscle** arises from the inferior genial tubercle and passes back as a flat strap to the upper border of the body of the hyoid bone. The two muscles lie side by side, above the mylo-hyoid and below the genio-glossus. They are supplied by a branch from the hypoglossal nerve, but the motor fibres are derived from the first cervical nerve (p. 396). The muscle pulls the hyoid bone forwards and upwards towards the chin; it is opposed by stylo-hyoid and the infrahyoid strap muscles.

The **hyo-glossus muscle** lies vertically alongside the tongue. Knowledge of its relations, superficial and deep, gives the course of the structures of the floor of the mouth. Superficial (lateral) to the muscle lie three structures, the lingual nerve, submandibular duct and hypoglossal nerve. Deep (medial) to the muscle lie three structures, the glosso-pharyngeal nerve, the stylo-hyoid ligament and the lingual artery (PLATE 36).

Hyo-glossus and mylo-hyoid pass upwards from the hyoid bone, the former to the side of the tongue the latter to the mylo-hyoid line of the mandible. The structures in the angular interval between them (Fig. 6.29) should be studied in a dissection.

The **submandibular duct** passes forwards from the deep lobe of the gland (p. 371). It opens on the sub-lingual papilla, a low elevation at the side of the frenulum of the tongue (PLATE 34).

The **hypoglossal nerve** passes forwards on the lower border of hyo-glossus, just above the greater cornu of the hyoid bone. At the anterior border of hyo-glossus the round nerve trunk becomes flattened and breaks up into a number of branches that radiate into the muscles of the tongue. The genio-hyoid muscle receives a branch containing C1 fibres only (PLATE 36, p. *xix*). There is always a communication between the hypoglossal and lingual nerves, and perhaps this carries proprioceptive impulses from the muscles. The nerve is accompanied by a companion vein that drains the front of the tongue into the common facial vein.

The **lingual nerve** enters the mouth from outside the pharynx by passing below the inferior border of the superior constrictor at its attachment to the mandible, at the lower end of the pterygo-mandibular raphé. The nerve lies against the lingual plate of the mandible at the posterior border of the third molar tooth, and runs forward on the upper surface of the mylo-hyoid muscle (Fig. 6.30). It gives off a gingival branch that supplies all the lingual gum and mucous membrane of the adjoining sulcus to the midline. The lingual nerve then dips under the submandibular duct and runs forward on the surface of the hyo-glossus above the level of the duct (PLATE 36). It is thus distributed to the mucous membrane of the anterior two-thirds of the tongue.

The **submandibular ganglion** hangs suspended from the lingual nerve, on the surface of hyo-glossus below the submandibular duct. It is a relay station for the parasympathetic secreto-motor fibres in the chorda tympani. Post-ganglionic fibres supply the sub-lingual gland (the fibres to the submandibular salivary gland mostly relay in cell bodies in the hilum of the gland). For the connexions of the ganglion see p. 37.

Three structures lie deep to the hyo-glossus muscle. The **lingual artery** passes deep to the posterior border of the muscle and runs forward above the greater cornu of the hyoid bone (level with the hypoglossal nerve on the other side of the muscle) (Fig. 6.29). Dorsales linguæ arteries are given off deep to the muscle and the lingual artery passes forward from beneath the anterior

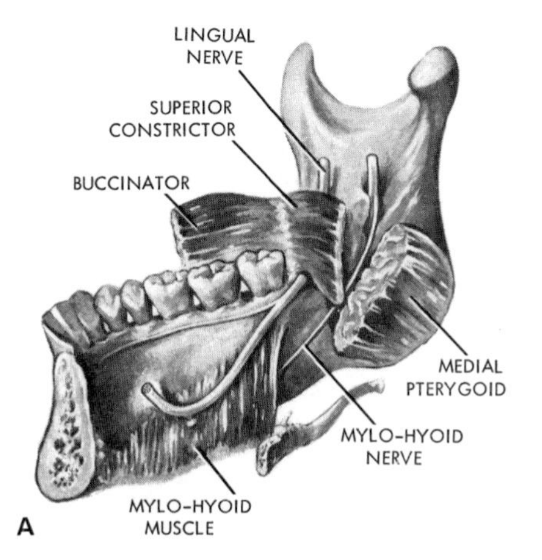

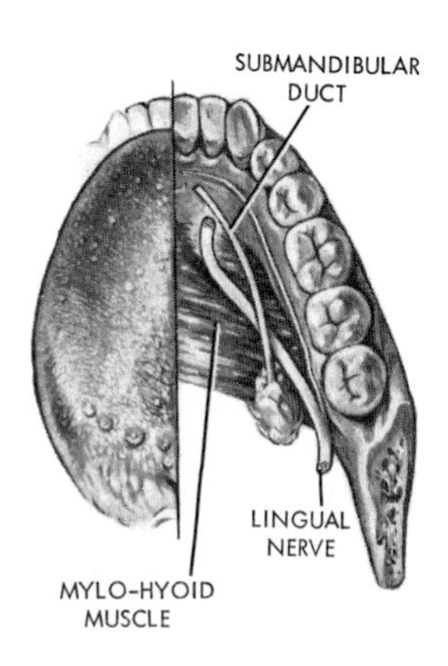

FIG. 6.30. THE PASSAGE OF THE LINGUAL NERVE FROM WITHOUT THE PHARYNX TO WITHIN THE MOUTH. A. The right nerve viewed from within the mouth. It passes under the free lower border of superior constrictor, which interdigitates with buccinator at the pterygo-mandibular raphé. B. The right nerve viewed from above.

border of the hyo-glossus to reach the front of the tongue (PLATE 34, p. *xviii*).

The **glosso-pharyngeal nerve** disappears beneath the upper part of hyo-glossus. It is much thinner than the hypoglossal nerve. It is distributed, like the dorsales linguæ arteries, to the mucous membrane of the posterior third of the tongue, including the vallate papillæ (PLATE 39, p. *xxi*).

The **stylo-hyoid ligament** runs down from the tip of the styloid process to the lesser cornu of the hyoid bone. From the angle between the ligament and the greater cornu the middle constrictor fibres arise and emerge from beneath the posterior border of hyo-glossus to fan out around the pharynx (Fig. 6.32).

The **sublingual gland** lies, in front of the anterior border of hyo-glossus, between the mylo-hyoid muscle and the side of the tongue (genio-glossus). It makes a smooth depression in the mandible alongside the mid-line. It lies below the termination of the submandibular duct. It is a mucous gland, twice the size of an almond kernel. Of its fifteen or so ducts, half open directly into the submandibular duct, the remainder separately on the sublingual papilla.

It is supplied by the lingual artery and by branches of the submental artery which pierce the mylo-hyoid muscle to reach the gland. The venous return is by corresponding veins. It is innervated from the sub-mandibular ganglion (p. 37).

THE PHARYNX

The pharynx is applied to the *back of the face* in the same way as a respirator is applied to the front of the face. The 'face piece' continues below into a tube, the œsophagus. The open part of the face piece, or pharynx, is applied below the base of the skull to the back of the nose, mouth, and larynx, the œsophageal tube hanging free below this level. Both pharynx and œsophagus are in contact posteriorly with the pre-vertebral fascia, which provides a foundation upon which pharynx and œsophagus can freely slide during swallowing and movements of the neck. The 'dead space' between the pharynx and the prevertebral fascia not only allows for free mobility of the pharynx and œsophagus, but also permits the extension of infection from one side to the other of the neck. It continues below into the posterior mediastinum.

The Wall of the Pharynx

The wall of the pharynx is thin. It contains three curved sheets of muscle, the superior, middle and inferior constrictors. They overlap posteriorly, being telescoped into each other like three stacked cups. But the muscle does not extend up to the base of the skull; here the immobile wall of the naso-pharynx consists of a rigid membrane, the pharyngo-basilar fascia. This extends down to the level of the soft palate, making a fourth 'cup' stacked inside the other three. The pharyngo-basilar fascia makes the wall of the naso-pharynx. Its stiffness keeps it always open for breathing, and no food enters it. Attached to the back of the nose it cannot hang straight down, for this would obstruct breathing, so it is pulled back well behind the back of the nose and slung to the skull base.

The Pharyngo-basilar Fascia. This non-expansile sheet of fascia is attached to the base of the skull and to the medial pterygoid plates (i.e., to the back of the nose). It is reinforced posteriorly by a midline thickening which may be called the *pharyngeal ligament* (Fig. 6.31). After making the general survey of the base of the skull (please read p. 557) the attachment of the pharyngo-basilar fascia can be studied (Fig. 8.4, p. 559).

Start from the pharyngeal tubercle. The attachment

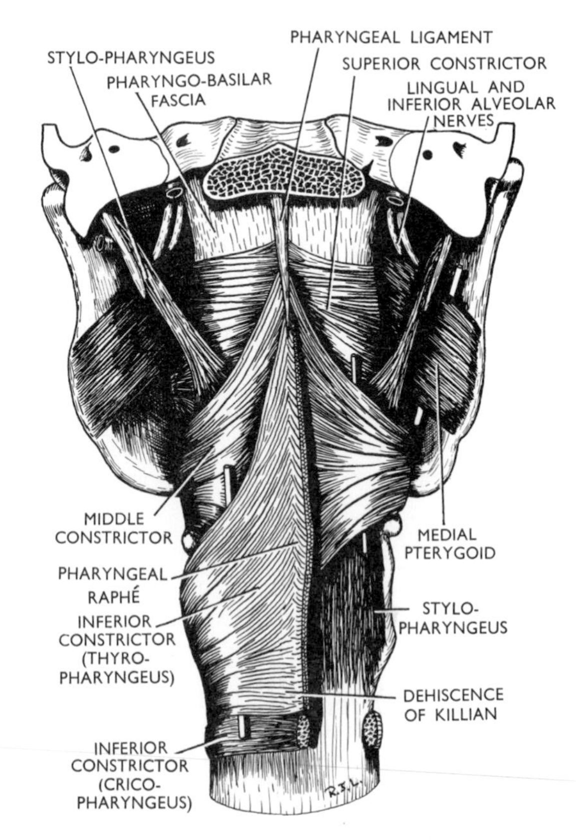

FIG. 6.31. POSTERIOR VIEW OF THE PHARYNX. ON THE LEFT SIDE A ROD LIES BENEATH THE THYRO-PHARYNGEUS. ON THE RIGHT SIDE RODS LIE BENEATH THE MIDDLE CONSTRICTOR AND THE MEDIAL PTERYGOID MUSCLE. Drawn from Specimen S 200 in R.C.S. Museum.

passes laterally, convex forwards over the longus capitis muscle, to the petrous part of the temporal bone just in front of the carotid foramen. From this point its attachment is to the cartilaginous part of the Eustachian tube, not to the skull. Below the orifice of the Eustachian tube it is attached to the sharp posterior border of the medial pterygoid plate, down to the hamulus. Suspended from the base of the skull, and sweeping around from one medial pterygoid plate to the other, reinforced posteriorly by the pharyngeal ligament, the pharyngo-basilar fascia makes a rigid wall that holds the naso-pharynx permanently open for breathing. The lower edge of the pharyngo-basilar fascia lies at Passavant's muscle, level with the hard palate, inside the superior constrictor muscle. Passavant's muscle (p. 420) acts like a purse-string on the lower free margin of the pharyngo-basilar fascia. Below this a soft and distensible mucous membrane lines the oro-pharynx.

Note that the quadrangular area at the apex of the petrous bone in front of the carotid foramen lies within a lateral recess of the pharynx. The levator palati muscle arises here; it is *entirely intrapharyngeal*, being covered medially by mucous membrane. Note too that the cartilaginous part of the Eustachian tube enters the naso-pharynx above the pharyngo-basilar fascia, which is firmly attached to it (Fig. 8.4, p. 559).

Superior Constrictor Muscle. Examine a skull. Note that the upper third of the posterior border of the medial pterygoid plate is rounded and blunt to the touch. It is rather concave, for it lodges the free end of the tubal (Eustachian) cartilage. This border now gives way to a spur, below which the posterior border of the medial pterygoid plate is sharp to the examining finger. The superior constrictor fibres arise below the tube, in continuity from the spur along the whole of the sharp two-thirds of the posterior border, down to and including the tip of the hamulus. The muscle fibres lie outside the pharyngo-basilar fascia as they share this attachment to the edge of the medial pterygoid plate. The pharyngo-basilar fascia ends at the hamulus, but the origin of the superior constrictor muscle continues down along the pterygo-mandibular raphé (interdigitating here with the buccinator) to the mandible just above the posterior end of the mylo-hyoid line at the level of the posterior border of the last molar tooth (Figs. 6.14, p. 381 and 8.15, p. 572).

Insertion. From this origin the muscle sweeps around the pharynx, its fibres diverging upwards and downwards as they approach the posterior midline of the pharynx. The uppermost fibres slope up to be inserted into the pharyngeal tubercle of the basi-occiput, the middle fibres pass to the median ligament in the 'naso-pharynx' and, below the level of Passavant's muscle, to the median raphé. The lowest fibres extend as far down as the level of the vocal folds, lying within the middle constrictor. Note that there is a space between the upper border of the superior constrictor and the base of the skull. Through here passes the cartilaginous part of the pharyngo-tympanic tube and the rest of the space is closed by the firm pharyngo-basilar fascia (Fig. 6.31).

There is a gap laterally between the superior and middle constrictors (Fig. 6.32). This is plugged by the back of the tongue (PLATE 39) and traversed by structures that pass from outside the pharynx to inside the mouth.

Middle Constrictor Muscle. The stylo-hyoid ligament is attached to the lesser cornu of the hyoid bone deep to the hyo-glossus muscle. From the angle between it and the greater cornu arises the middle constrictor, its fibres diverging widely as they sweep around the pharynx to end in the median raphé. The uppermost fibres reach the pharyngeal ligament and enclose the superior constrictor; the lower arch down as far as the level of the vocal folds. They enclose the superior constrictor and are themselves enclosed within the inferior constrictor (PLATE 39 and Fig. 6.32).

The gap between the middle and inferior constrictors is closed by the thyro-hyoid membrane (p. 428), which joins the hyoid bone to the thyroid cartilage (Fig. 6.32) and walls in the laryngeal part of the pharynx at the piriform fossa (p. 418).

Inferior Constrictor Muscle. This muscle arises from the oblique line on the lamina of the thyroid

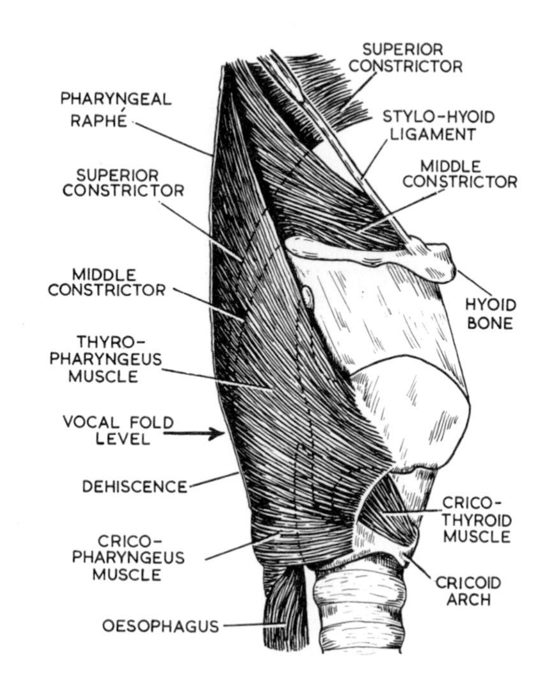

FIG. 6.32. THE PHARYNGEAL CONTRICTORS. Elevation during swallowing must telescope the muscles into greater mutual overlap, but the dehiscence remains the same.

cartilage, and from the side of the arch of the cricoid cartilage. The two parts are best considered separately, and additional nomenclature seems here justified. The **thyro-pharyngeus** part of the inferior constrictor arises from the oblique line of the thyroid cartilage and in continuity below this from a fibrous arch that spans the crico-thyroid muscle. It encloses the middle and superior constrictors as its fibres curve around to the midline raphé. The fibres diverge widely, sweeping up to the pharyngeal ligament at the level of Passavant's muscle. The lower fibres are horizontal, edge to edge with crico-pharyngeus. The **crico-pharyngeus muscle,** rounded and thicker than the flat sheets of the other constrictors, extends uninterruptedly from one side of the cricoid arch to the other, around the pharynx. There is no raphé here. The muscle acts as a sphincter at the lower extent of the pharynx, and is continuous with the circular muscular coat of the œsophagus. It is always closed, except for momentary relaxation during deglutition. It has a different nerve supply from the other constrictors. The closure of the crico-pharyngeus prevents air from being sucked into the upper œsophagus when intrathoracic pressure falls; air is sucked only into the permanently open trachea.

The outer surface of the pharynx is covered by the delicate epimysium of the pharyngeal constrictors. There is no fascia, in the sense of a properly developed membrane, surrounding this; the peristalsis of swallowing would be impeded or prevented if such existed. The so-called *bucco-pharyngeal fascia* is unworthy of mention except to condemn such misleading nomenclature. Similarly it is a misconception to imagine a pharyngo-basilar fascia inside the pharynx below the palate. Swallowing would be impeded by such an arrangement.

The Dehiscence of Killian. Note the identity of insertion of superior and middle constrictors; they interdigitate in the midline raphé down to the level of the vocal folds. Overlapped by thyro-pharyngeus, the three lie together from the pharyngeal ligament down to this level. Below this the posterior wall is only the single sheet of thyro-pharyngeus; this is the dehiscence (Fig. 6.32). Killian observed that a pharyngeal diverticulum (the so-called 'œsophageal' diverticulum) pouches through the dehiscence, always above and *never below* the crico-pharyngeus muscle. Such diverticula are normal in some species (e.g., the pig).

Motor Supply of the Pharyngeal Constrictors. *All the striped muscle of the pharynx derives its innervation from cell bodies in the nucleus ambiguus.* The motor axons leave the brain stem in the cranial root of the accessory nerve and pass to the vagus. Most leave it in the pharyngeal branch to the pharyngeal plexus, whence the constrictor muscles are innervated. The *crico-pharyngeus* part of the inferior constrictor, however, receives its innervation from the nucleus ambiguus by way of the recurrent laryngeal nerve and the external laryngeal branch of the superior laryngeal nerve. It has been suggested, not very convincingly, that this alteration in the nervous pathways from the nucleus ambiguus to the muscles may result in neuro-muscular incoordination during swallowing. If the crico-pharyngeus failed to relax in front of the advancing wave of pharyngeal peristalsis the pressure above the muscle would be exerted upon the weakest part of the pharyngeal wall, the dehiscence of Killian. Certainly this is the part through which a pharyngeal diverticulum almost invariably herniates.

The Pharyngeal Plexus. Lying on the lateral wall of the pharynx, for the most part over the middle constrictor, is a nerve plexus from which most of the pharynx is supplied. It derives its branches from the vagus (pharyngeal branch), the glosso-pharyngeal, and the cervical sympathetic. The glosso-pharyngeal supplies the mucous membrane of the oro-pharynx with sensory nerves and sends secreto-motor fibres to its mucous glands. The pharyngeal branch of the vagus carries nothing but accessory nerve fibres from the nucleus ambiguus for the motor supply of the three constrictors. The sympathetic supplies vaso-constrictor fibres to the wall of the pharynx.

Sensory Supply of the Pharynx. The mucosa of the naso-pharynx is supplied from the maxillary nerve through the pterygo-palatine ganglion, whose pharyngeal branch reaches the naso-pharynx via the palatino-vaginal canal. The mucosa of the oro-pharynx, including the valleculæ, is supplied from the glosso-pharyngeal nerve and that of the laryngeal part of the pharynx including the piriform fossa by the vagus via the internal and recurrent laryngeal nerves.

Blood Supply of the Pharynx. Many arteries take blood to the pharynx; the ascending pharyngeal artery, branches from the lingual, facial, and from the superior and inferior laryngeal arteries. Venous blood is collected into the pharyngeal plexus, whence it is returned from the upper part via the pterygoid plexus. Further down the pharynx drains into the internal jugular vein, while from the lowest part the veins find their way with the inferior thyroid veins into the brachiocephalic veins. Tissue fluid is carried to the retropharyngeal lymph nodes, and via these or directly, into the deep cervical nodes, upper and lower groups.

The Interior of the Pharynx

Just as the blank and windowless wall of a tall building gives no impression of the floors and furnishings within, so inspection of the outside of the pharynx (Fig. 6.31) gives no hint of the interior formations.

The pharynx is described as consisting of naso-pharynx, oro-pharynx and laryngeal pharynx, lying behind the nose, mouth and larynx respectively. A word of explanation is here required. The term 'naso-pharynx' is misleading; *functionally* this cavity is no part of the pharynx, but is purely respiratory. No aliment normally enters it. It consists of that part of the cavity above Passavant's muscle. It is lined with ciliated columnar respiratory epithelium, the rest of the pharynx with squamous stratified alimentary epithelium which continues down the whole length of the œsophagus. Its mucous membrane is supplied by the trigeminal nerve ('pharyngeal' branch of the pterygo-palatine ganglion); the rest of the pharynx is supplied by the glosso-pharyngeal and vagus nerves.

The interior of the naso-pharynx and oro-pharynx is understood better after the soft palate has been studied, and the laryngeal part of the pharynx cannot be properly understood until the larynx has been studied.

Nasal Part of the Pharynx. The wall of the naso-pharynx is the rigid pharyngo-basilar fascia, open in front for breathing. The rigid fascia is further stiffened behind by the midline pharyngeal ligament. Thus is the airway maintained patent at all times. Inside this wall are structures that project the lining mucous membrane into folds and ridges. Muscular ridges running vertically are produced by the longus capitis and salpingo-pharyngeus muscles. A midline gutter lies between the longus capitis ridges, and the lateral recess of the pharynx lies behind the salpingo-pharyngeus ridge.

The *lateral recess (fossa of Rosenmuller)* is a narrow slit quite unlike the large bulge in the pharyngo-basilar fascia visible from outside. The recess in the fascia is almost filled by levator palati. A catheter missing the tubal orifice and introduced into the fossa of Rosenmuller may perforate the pharyngo-basilar fascia and enter the internal carotid artery, which lies here against the wall of the pharynx.

The opening of the Eustachian tube lies above the soft palate in the lateral wall. It is in the shape of an inverted **J**, the long limb lying posteriorly and being continued downwards as the salpingo-pharyngeal fold, produced by the underlying salpingo-pharyngeus muscle (Figs. 6.22, p. 400 and 6.34). The opening shows prominent rounded lips, formed by the trumpet-shaped medial end of the tube, which is here covered with lymphatic tissue, the **tubal tonsil.** Further collections of lymphatic tissue lie posteriorly above Passavant's muscle, the **naso-pharyngeal tonsil,** or adenoid. These collections of lymphatic tissue constitute, with the palatine and lingual tonsils, the so-called **Waldeyer's ring.**

The floor of the naso-pharynx is the soft palate, a mobile floor like a trap-door. When elevated it comes in contact with Passavant's ridge (p. 420) and this marks the line of junction between naso-pharynx and oro-pharynx.

The calibre of the naso-pharynx is not large; there is no reason for its diameter to exceed that of the nostrils. The cavity equals in volume and shape the terminal segment of its owner's thumb.

Oral Part of the Pharynx. The wall of the oro-pharynx is formed posteriorly by all three constrictors. It closes completely behind a swallowed bolus, otherwise is open for breathing. Anteriorly there is a mobile wall, namely the posterior third of the tongue. Projecting ridges (the two faucial pillars) lie laterally. They are produced by palato-glossus and palato-pharyngeus muscles and the palatine tonsil lies between them. The oro-pharynx is separated from the mouth by the anterior pillar of the fauces and from the laryngeal part of the pharynx by the epiglottis and pharyngo-epiglottic ('lateral glosso-epiglottic') folds.

The **palatine tonsil** is a large collection of lymphatic tissue which projects into the oro-pharynx from between the anterior and posterior faucial pillars. It is covered with mucous membrane which in many places is invaginated into deep crypts whose mouths are visible to the naked eye. The base of the tonsil is covered with a layer of fibrous tissue which lies upon the palato-pharyngeus muscle within the embracing fibres of the superior constrictor muscle. This fibrous tissue is a tiny flap prolonged down from the pharyngo-basilar fascia. The tonsil is supplied with blood by tonsillar branches of the ascending palatine branch of the facial artery. Its tissue fluid drains into the jugulo-digastric node. The microscopic anatomy of the tonsil is considered on p. 20.

The function of the tonsils is not properly known. They can scarcely protect against all the bacteria that lie in the bolus which rushes past them in the second stage of swallowing. But bacteria trapped in the crypts may well induce the production of antibodies by the lymphatic follicles.

The **valleculæ** lie between the epiglottis and the posterior border of the tongue. They are shallow pits separated by the median glosso-epiglottic fold and limited inferiorly by the pharyngo-epiglottic ('lateral glosso-epiglottic') folds of mucous membrane.

Laryngeal Part of Pharynx. The posterior wall of the laryngeal part is formed by the three overlapping constrictors down to the level of the vocal folds (upper border of cricoid lamina). Below this (i.e., behind the cricoid lamina) is the dehiscence of Killian and finally the crico-pharyngeus sphincter. From each side of the epiglottis a fold of mucous membrane extends laterally to the side wall of the pharynx. This is the pharyngo-epiglottic fold, nowadays misnamed the lateral glosso-epiglottic fold. It separates the oro-pharynx from the laryngeal part of the pharynx. Below the fold is the **piriform fossa** (PLATE 42). The piriform fossa is bounded medially by the quadrate membrane of the

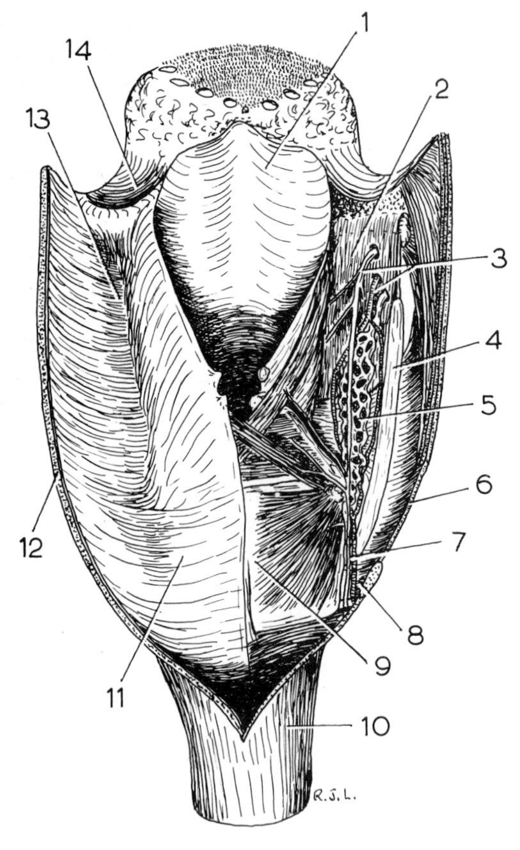

FIG. 6.33. THE LARYNGEAL PART OF THE PHARYNX. THE MUCOUS MEMBRANE IS REMOVED ON THE RIGHT SIDE TO SHOW THE ANASTOMOSES OF SUPERIOR AND INFERIOR LARYNGEAL VESSELS AND THE INOSCULATION OF INTERNAL AND RECURRENT LARYNGEAL NERVES IN THE PHARYNX. THERE IS NO SUCH OVERLAP IN THE LARYNX; THE VOCAL FOLDS ARE A COMPLETE 'WATERSHED'. 1. Epiglottis. 2. Thyro-hyoid membrane. 3. Internal laryngeal nerve and superior laryngeal vessels in the piriform fossa. 4. Thyroid cartilage (posterior border). 5. Submucous venous plexus. 6. Dehiscence of Killian. 7. Recurrent laryngeal nerve and inferior laryngeal vessels. 8. Crico-pharyngeus muscle. 9. Cricoid lamina and posterior crico-arytenoid muscle. 10. Œsophagus. 11. Hypopharynx. 12. Pharyngeal wall (constrictors). 13. Piriform fossa. 14. Vallecula (oro-pharynx, not laryngo-pharynx).

larynx, below the ary-epiglottic fold. The lateral wall of the fossa, beneath the mucous membrane, consists of thyro-hyoid membrane above and the lamina of the thyroid cartilage below.

The piriform fossæ, broad above and narrow below, lie beside the aperture of the larynx. Below the aperture the arytenoids and the lamina of the cricoid cartilage are draped over with mucous membrane. The lower part of the pharynx **(hypopharynx)** thus possesses an anterior wall. It is flat, and the posterior wall lies against it, obliterating the piriform fossæ as the hypopharynx tapers off, wedge-shaped, into the clasp of the crico-pharyngeus muscle. Note that the posterior wall of the hypopharynx is, in fact, the dehiscence of Killian.

The thyroid-hyoid membrane is perforated by the superior laryngeal vessels and the internal laryngeal nerve. Note that when these vessels and nerve perforate the thyro-hyoid membrane they are not yet in the larynx, but lie beneath the mucous membrane of the piriform fossa. Similarly the inferior laryngeal vessels and recurrent laryngeal nerve, passing beneath the crico-pharyngeus part of the inferior constrictor, enter first the pharynx. The superior and inferior vessels anastomose and the nerves communicate in the mucous membrane of the pharynx, which they supply (Fig. 6.33). After having given off these communicating branches the vessels and nerves enter the larynx. In the larynx they have no anastomoses; the vocal folds are a complete watershed for vessels and nerves.

THE SOFT PALATE

The soft palate consists functionally of an aponeurosis that is acted upon by several muscles to produce alterations in its shape and position, but it is to be observed that anatomically much of its bulk is accounted for by the large volume of mucous and serous glands present in its oral mucous membrane. The **palatine aponeurosis** consists of the flattened-out tendon of the tensor palati muscle.

(*Note on names.* The tensor and elevator of the palatine aponeurosis are now called 'tensor veli palatini' and 'levator veli palatini', but the older names are retained here because they are simpler and more euphonious, and there is no danger of confusion.)

Tensor Palati Muscle. This muscle arises from the base of the skull and the lateral side of the cartilaginous part of the pharyngo-tympanic tube, that is to say, outside the pharynx. Its origin extends from the scaphoid fossa at the upper end of the medial pterygoid plate along the greater wing to the spine of the sphenoid. From this oblique origin of about 2 cm in extent the flat muscle converges towards the base of the hamulus where it becomes a flat tendon. The tendon passes above the fibrous arch in the origin of the buccinator (Fig. 6.14, p. 381), bends at a right angle around the hamulus and so gets inside the pharynx. The tendon now broadens out into a wide triangular aponeurosis. The anterior border of this triangular aponeurosis is attached to the crest of the palatine bone, the medial border blends with that of the opposite side. The postero-lateral border blends with the side wall of the pharynx in front, but hangs free behind, forming the edge of the soft palate with the dependent uvula in the midline. The **uvula** is a mass of mucous glandular tissue; the musculus uvulæ can be ignored. Note that each tensor palati consists of two triangles converging on the base of the hamulus, the vertical one of muscle, the other of aponeurosis. The aponeurosis is not flat, but concave towards the mouth; when tensed by contraction of the muscle it is flattened and therefore depressed somewhat. The main action of the tensor palati is so to tense the palatine aponeurosis that other muscles may elevate and depress it without altering its shape. When the tensor palati contracts (e.g., in swallowing) it pulls upon the cartilage of the pharyngotympanic tube, opens the tube, and permits equalization of air pressure between the middle ear and nose (Fig. 6.34).

Nerve Supply. From the motor division of the trigeminal via the nerve to the medial pterygoid muscle.

Levator Palati Muscle. Arising from the quadrate area at the apex of the petrous bone anterior to the carotid foramen the muscle forms a rounded belly that is inserted into the nasal surface of the palatine aponeurosis between the two heads of the palato-pharyngeus muscle. Some fibres arise in addition from the medial part of the tubal cartilage. The two muscles in passing

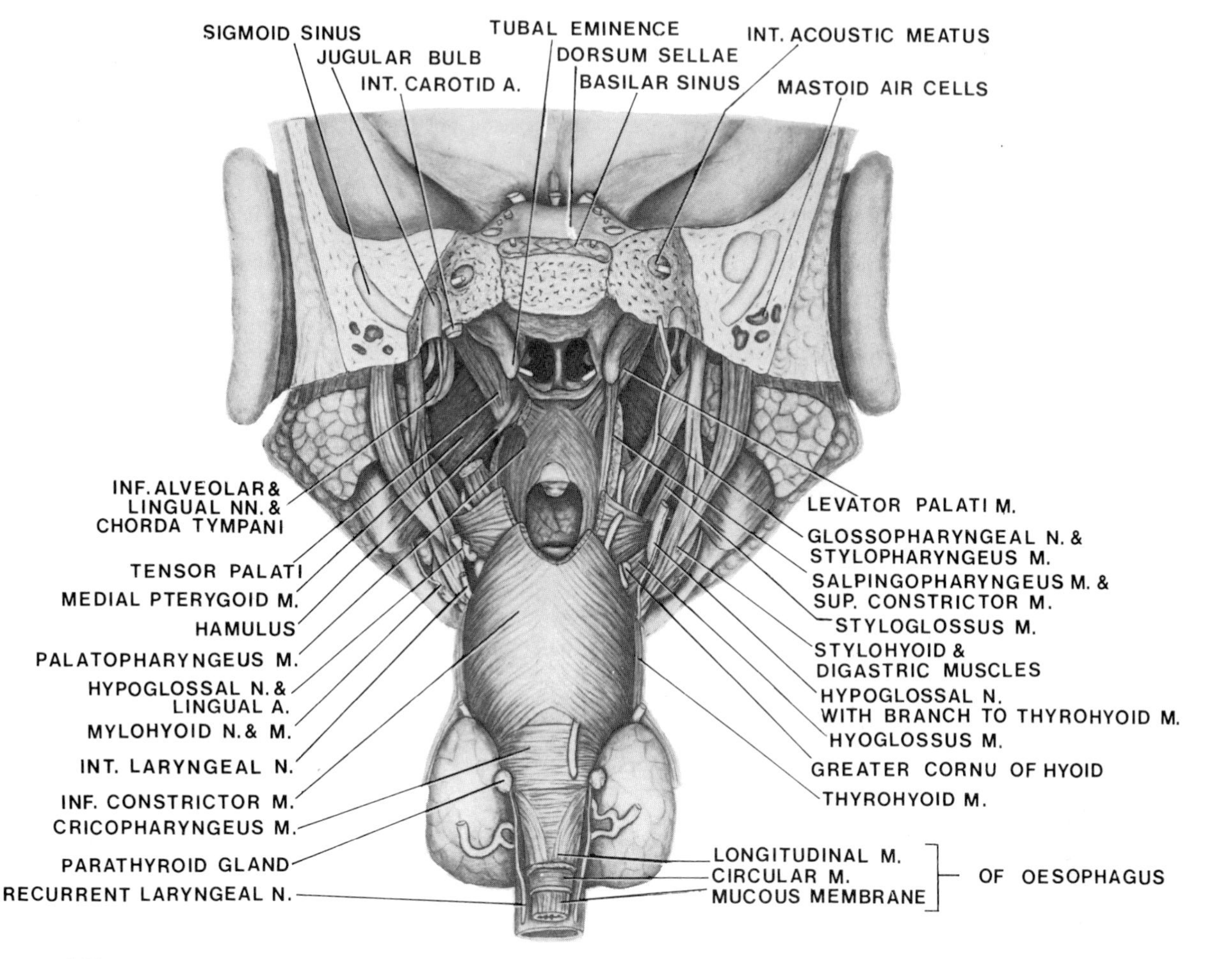

FIG. 6.34. THE NASO-PHARYNX OPENED TO VIEW THE SOFT PALATE FROM BEHIND. Illustration of Specimen S 198 in R.C.S. Museum.

down to the palate are directed forwards and medially, together forming a V-shaped sling (Fig. 6.36). Their contraction pulls the palate upwards and backwards. If the tensor be relaxed two dimples appear on the oral surface of the palate as the levators contract; much more usually the tone of the tensor stiffens the palatine aponeurosis and the soft palate is raised without alteration in shape. In either case the naso-pharynx is shut off from the oro-pharynx by the action of the levators. The soft palate comes into contact with the posterior wall of the pharynx at Passavant's ridge on a level with the anterior arch of the atlas vertebra. Contraction of the levator palati muscle opens the cartilaginous tube and equalizes air pressure between the middle ear and the nose.

Nerve Supply. By way of the pharyngeal plexus from the pharyngeal branch of the vagus, the fibres having come from the accessory nerve; their cell bodies are in the nucleus ambiguus.

Palato-pharyngeus Muscle. The muscle arises from two heads, the one fixed to bone, the other movable on the palatine aponeurosis. The two heads embrace the insertion of the levator palati on the upper surface of the palatine aponeurosis. The *anterior head* arises from the posterior border of the hard palate (i.e., the horizontal plate of the palatine bone). At this level the palatine aponeurosis lies just on its oral surface, and passes further forwards to reach the crest of the palatine bone (Fig. 6.35). The *posterior head* arises further back on the upper surface of the aponeurosis (Fig. 6.34). The two heads arch downwards over the lateral edge of the aponeurosis, join, and form a muscle that passes downwards beneath the mucous membrane

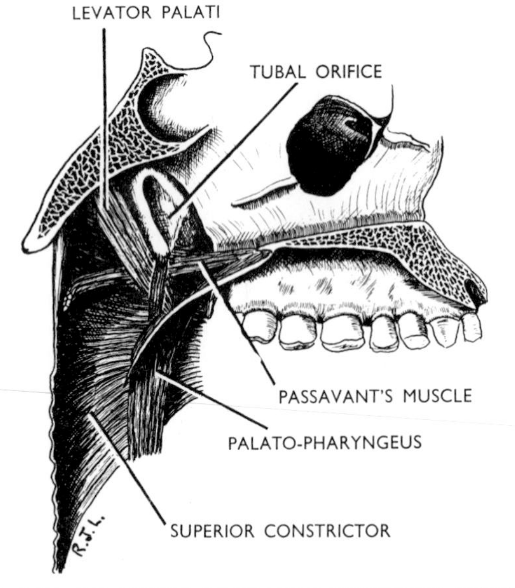

FIG. 6.35. THE MUSCLES OF THE NASO-PHARYNGEAL ISTHMUS. The left side of the soft palate viewed from within.

of the lateral wall of the pharynx just behind and lateral to the tonsil. The upper part of the muscle raises a ridge of mucous membrane that constitutes the posterior pillar of the fauces; the lower part is inserted chiefly into the posterior border of the thyroid lamina and cornua; the muscle ought rightly to be named palato-laryngeus. Some of the anterior fibres are inserted into the upper border of the thyroid lamina just in front of the superior cornu. Some of the posterior ones merge with the surrounding fibres of the inferior constrictor.

Action of Palato-pharyngeus. It is an elevator of the larynx and pharynx. It arches the relaxed palate, making it more concave on its oral surface. The posterior head, attached only to the palatine aponeurosis, depresses as a whole the tensed palate.

Nerve Supply. From the pharyngeal plexus (pharyngeal branch of vagus, which consists of accessory nerve fibres with their cell bodies in the nucleus ambiguus).

Comparative Anatomy of Palato-pharyngeus Muscle. In keen-scented mammals the epiglottis rides above the level of the soft palate and the larynx is intranarial. It is supported in this elevated position by stylo-pharyngeus and salpingo-pharyngeus muscles and held in a sphincter, the palato-pharyngeus muscle, that clasps the laryngeal inlet. In man the larynx is never intranarial: at its much lower level it has pulled down the sphincter to produce the palato-pharyngeus muscle as described above. Some fibres remain, however, to form a sphincter at the level of the hard palate. They are the laterally attached fibres of palato-pharyngeus, arising from the lateral aspect of the posterior border of the hard palate, in continuity with the anterior head of palato-pharyngeus, and encircling the pharynx inside the fibres of the superior constrictor (Fig. 6.36). They constitute **Passavant's muscle** and their contraction raises a ridge, **Passavant's ridge,** within the pharynx at the level of the hard palate and anterior arch of the atlas vertebra. It is against this ridge that the soft palate is elevated by the levator palati. The two levators constitute a V-shaped sling that hangs downwards and forwards; Passavant's muscle is a U-shaped sling that passes horizontally backwards to clasp the lower parts of the levators. Contraction of both slings shortens them, and causes them to interlock, shutting off the nose from the oro-pharynx. Passavant's muscle is greatly hypertrophied in cleft palate, a result of the effort to close the naso-pharyngeal isthmus when swallowing.

The *ridge* exists only when Passavant's muscle contracts at the moment of elevation of the soft palate. When the soft palate hangs down Passavant's muscle is relaxed and there is no ridge to obstruct the airway. Passavant's muscle is a purse-string at the inferior free margin of the pharyngo-basilar fascia.

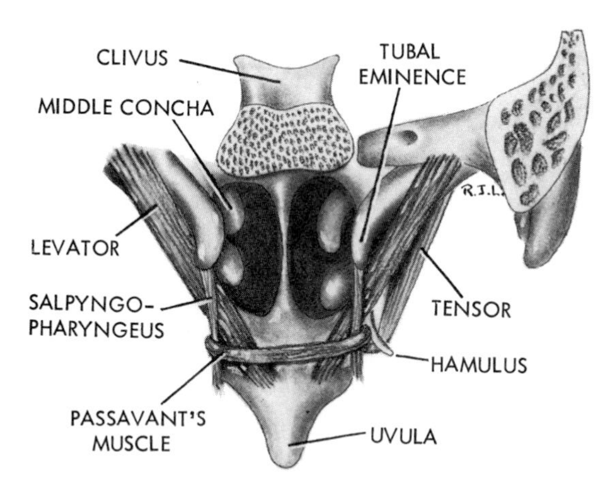

FIG. 6.36. THE NASO-PHARYNGEAL ISTHMUS, POSTERIOR VIEW. Modified from Specimen S 190 in R.C.S. Museum.

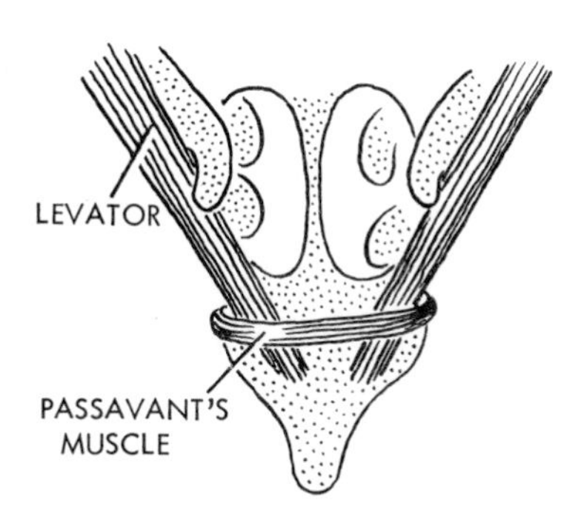

FIG. 6.37. THE ESSENCE OF PALATAL ELEVATION. (Cf. Fig. 6.36.) The **V** of the levators is imprisoned in the **U** of Passavant's muscle. The elevated soft palate thus seals the naso-pharyngeal isthmus.

Palato-glossus Muscle. This muscle arises from the under surface of the palatine aponeurosis, and passes downwards to interdigitate with the stylo-glossus. The muscle raises a ridge in front of the tonsil, the anterior pillar of the fauces, which marks the junction between mouth and pharynx, between mucous membrane supplied by the trigeminal (mouth) and glosso-pharyngeal (pharynx) nerves. Its action is sphincteric at the oro-pharyngeal isthmus; when it contracts it raises the tongue, and narrows the transverse diameter of the isthmus. It is innervated from the pharyngeal plexus (accessory fibres with cell bodies in the nucleus ambiguus).

Structure of the Soft Palate. The soft palate is covered with 'wear and tear' epithelium (squamous stratified) on its oral surface and on the posterior part of its nasal surface, where it comes into contact with Passavant's ridge. Its oral epithelium contains a few scattered taste buds. The anterior part of its nasal surface is covered with respiratory mucous membrane (a loose corium containing small mucous glands and covered with ciliated columnar epithelium with many goblet cells). The thickness of the palate and uvula is occupied by a large mass of mucous and serous glands below the palatine aponeurosis; the ducts of these glands open by orifices scattered over the oral epithelium. Scattered lymphatic follicles are found in the oral mucous membrane.

Blood Supply of the Soft Palate. Lesser palatine branches of the maxillary artery, the ascending palatine branch of the facial artery, and palatine branches of the ascending pharyngeal artery anastomose freely in the soft palate. Most of the venous blood is drained laterally through the wall of the pharynx into the pharyngeal venous plexus and the pterygoid plexus. Tissue fluid is drained by lymphatics that empty into the retro-pharyngeal and antero-superior deep cervical lymph nodes.

Nerve Supply of the Soft Palate. The **sensory** supply of the soft palate is essentially from the maxillary division of the trigeminal nerve, though on its oral surface there is an overlap of glosso-pharyngeal fibres that encroach on its lateral border from the lateral wall of the pharynx. Maxillary nerve fibres are derived from the middle and posterior (lesser) palatine nerves which pass through the pterygo-palatine ganglion without relay, their cell bodies lying in the trigeminal ganglion. Taste fibres in the greater (superficial) petrosal nerve supply the few taste buds on the oral surface of the soft palate (cell bodies in geniculate ganglion, central processes in nervus intermedius and so to tractus solitarius in the pons).

Secreto-motor fibres in the middle and posterior (lesser) palatine nerves are from cell bodies in the pterygo-palatine ganglion. These are activated from the superior salivatory nucleus in the pons by way of the nervus intermedius and the greater (superficial) petrosal nerve.

Muscles of the soft palate, except tensor palati, are innervated from the pharyngeal plexus; the motor cell bodies lie in the nucleus ambiguus, the motor axons passing in the cranial root of the accessory nerve, to the vagus, and thence by its pharyngeal branch to the pharyngeal plexus. The tensor palati is supplied by the motor root of the trigeminal, whose cell bodies are in the pons. The motor axons reach the tensor palati via the nerve to the medial pterygoid muscle and, without relay, through the otic ganglion.

Movements and Function. The soft palate is basically a flap-valve that can shut off the oro-pharynx from the mouth (e.g., during chewing, so that breathing is unimpeded) or from the naso-pharynx (e.g., in swallowing to prevent food entering the naso-pharynx, or in coughing when the whole blast escapes through the mouth, or when blowing the bugle, etc.). When elevated it locks into Passavant's ridge, which exists only at this moment and disappears as the soft palate comes down. The plentiful mucous glands make its oral surface slippery, like the posterior third of the tongue, for swallowing. Short of full upward or downward closure, its position is determined by the opposing pulls of levator palati and the posterior head of palato-pharyngeus; thus can the resonating chambers be altered to modify the quality of the voice. In sneezing it is held firm to resist being blown inside-out and its position determines how much of the blast is directed through the nose. Too violent an explosion might damage the conchæ; the excess escapes harmlessly through the mouth.

If there were no crossing of air and food pathways there would be no soft palate. The crossing is beneficial in several ways. The nose must be narrow for warming, moistening and cleaning the gentle stream of air in tranquil breathing. It is inadequate for increased pulmonary ventilation; the open mouth is needed then. Coughing would be very ineffective if sputum had to be expelled through the narrow nose instead of into the pharynx or mouth, to be expectorated or swallowed. And you cannot talk through your nose (try it with closed lips).

THE LARYNX

Comparative Anatomy. The essential reason for the existence of a larynx is not for phonation, but to provide a protective sphincter at the inlet of the air passages. The larynx first appears in the lung fish as a simple muscular sphincter surrounding the opening of the air passage in the floor of the pharynx. A functional improvement is found in the addition of dilator fibres which radiate outwards from the sphincter. Next a bar of cartilage appears on each side of the larynx; to it the dilator fibres become attached. A later modification is the division of each cartilaginous bar into a cranial and a caudal half. In the mammals the cranial halves appear as the arytenoid cartilages; the caudal cartilages fuse to form a ring, the cricoid. A protective shield, the thyroid cartilage, is developed anteriorly. Evidence of the essential function of the larynx is provided by the birds, in whom the rima glottidis in the floor of the mouth shuts to close the air inlet, but is silent; phonation is from a dilatation, the *syrinx*, at the lower end of the trachea just above its bifurcation.

Modifications for Olfaction and Deglutition. Keen-scented animals, in order not to interrupt respiration during swallowing, have developed an intranarial larynx. The epiglottis has evolved as a flap to protrude above the soft palate in these animals. The inlet of the larynx, protruding into the naso-pharynx, is suspended there by the elevators of the larynx, and clasped by the sphincter muscle of the naso-pharyngeal isthmus (the palato-pharyngeus). The epiglottis is a modification for olfaction in these animals (but not in man).

In animals that possess an intranarial larynx and also have to swallow large amounts of liquid food (e.g., ruminants, toothed whales, etc.) there are developed folds of mucous membrane that form side-walls rising above the glottis. A lateral gutter thus passes along each side of the epiglottis and laryngeal aperture—a food channel for deglutition during uninterrupted respiration. The ary-epiglottic folds are a modification for deglutition.

The Function of the Larynx. Obviously the airway must be permanently open, to permit continuous breathing. Functional needs, however, require temporary closure and the larynx provides *two separate mechanisms* that bring this about.

At the inlet is a sphincter which closes during swallowing; this is adequate to guard the inlet against entry of swallowed material, but added security is provided by the folding back of the upper part of the epiglottis, like a lid, over the tightly-shut inlet. Closure of the vocal folds is not required for swallowing—the sphincter of the inlet and the epiglottis are enough for this. Indeed, should a foreign body pass the sphincter of the inlet it is seldom blocked by the vocal folds but passes between them into the trachea.

Below the inlet, halfway down the larynx, lie the vocal folds (vocal cords). These too must be open for breathing. They are temporarily closed for three separate reasons: (1) during phonation, (2) prior to the explosion of a cough or sneeze, (3) for certain muscular efforts (*v. inf.*). The vocal folds, in addition to the movements of opening-closing, possess also a second movement, that of lengthening-shortening, with concomitant change of tension. Change of tension results in change of pitch and this movement is normally used only when the cords are *closed for phonation.*

The ary-epiglottic sphincter of the inlet is valvular, and it is not capable of withstanding pressure from below. The vocal folds provide a vastly stronger seal than this, able to withstand great pressure from above or from below.

The Skeleton of the Larynx

The skeletal framework of the larynx consists of cartilages and membranes.

The Cricoid Cartilage (Fig. 6.38). This is the

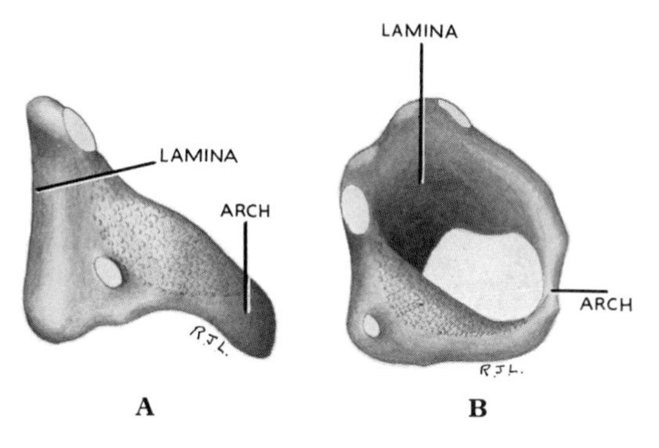

FIG. 6.38. THE CRICOID CARTILAGE. A. In lateral profile viewed from the right. B. Viewed from above and somewhat in front from the right side.

foundation of the larynx; to it the two arytenoids and the thyroid cartilage are articulated by synovial joints. It is the only complete cartilaginous ring in the whole of the air passages. The anterior part of the ring is known as the **arch;** posteriorly it is projected upwards into a quadrangular flat part, the **lamina.** Near the junction of arch and lamina is an articular facet for the inferior horn of the thyroid cartilage. Note that the lamina has *sloping* shoulders, which carry articular facets for the arytenoids. A vertical ridge in the midline of the lamina produces a shallow concavity on each side for the origin of the posterior crico-arytenoid muscle. The ridge itself gives attachment to the longitudinal muscle of the œsophagus.

Arytenoid Cartilages (Fig. 6.41). These two small cartilages, articulating with the upper border of the lamina of the cricoid, serve for attachment of the vocal folds and of many muscles. Each is in shape a gracefully curved pyramid, with a forward projection, the *vocal process*, attached to the vocal fold, and a lateral projection, the *muscular process*, for the reception of the crico-arytenoid muscles (posterior and lateral). The superior process of the arytenoid articulates with a small cartilage, the corniculate cartilage, to which is attached the ary-epiglottic fold. Each arytenoid cartilage

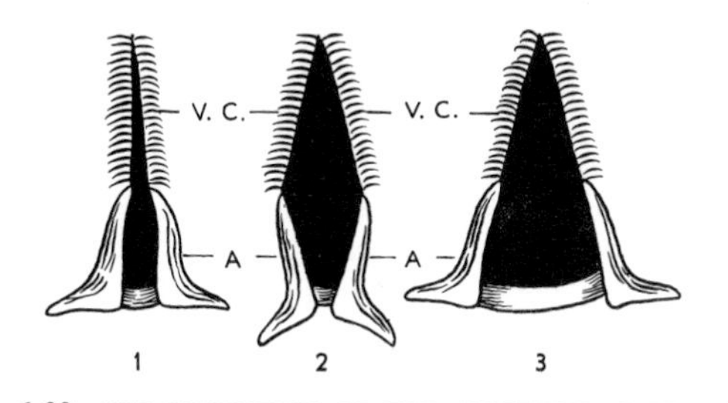

FIG. 6.39. THE MOVEMENTS OF THE ARYTENOID CARTILAGES. 1. Vocal folds adducted. 2. Rotation of the arytenoids produces a diamond-shaped opening (not human). 3. Lateral excursion of the arytenoids produces a V-shaped opening. V.C. Vocal cord. A. Arytenoid cartilage.

articulates with a sloping shoulder on the upper border of the cricoid lamina. The capsular ligament of this synovial joint is lax, allowing both rotary and gliding movements. The ratio of gliding to rotary movement varies widely in different species. Gliding of the arytenoids opens the rima glottidis in the shape of a V; rotation opens the rima in the shape of a diamond (Fig. 6.39). *Note that when the arytenoids are pulled downwards they slide apart from each other along the sloping shoulders of the cricoid lamina.* In man the cylindrical articulating surfaces permit a greater range of gliding than of rotary movement, and the open human glottis resembles a V and not a diamond (Fig. 6.39, 3).

The **thyroid cartilage** (Fig. 6.40) consists of two conjoined laminæ whose posterior borders are free and projected upwards and downwards as the superior and inferior cornua. Each inferior cornu articulates with the cricoid at a synovial joint. Movement between the thyroid and cricoid occurs around an axis that passes

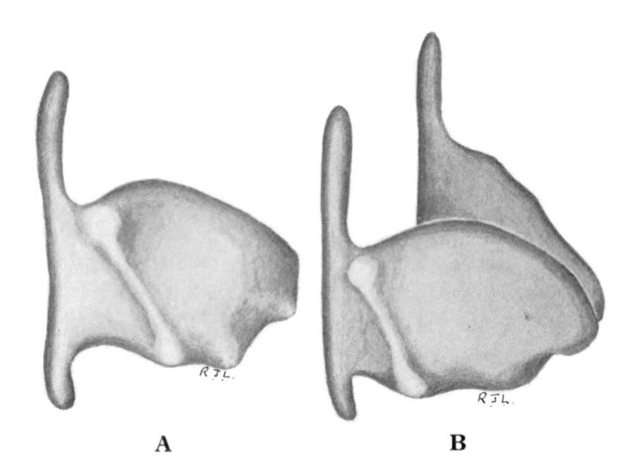

FIG. 6.40. THE THYROID CARTILAGE. A. In lateral profile from the right. B. Viewed from above and somewhat in front from the right side.

transversely between the two joints (Fig. 6.44). The outer surface of each lamina possesses an oblique ridge bounded above and below by a tubercle. The point of junction of the two laminæ anteriorly constitutes the 'Adam's apple' or prominence of the larynx. A deep midline notch (Fig. 6.40B) lies above the prominence.

The Epiglottis (Fig. 6.41). This slightly curled, leaf-shaped cartilage is prolonged below into a slender process attached in the midline below the notch in the upper border of the thyroid cartilage. The epiglottic cartilage leans back from its attached stalk to overhang the vestibule of the larynx. The prominence on the posterior surface below the apex, the 'cushion' of the epiglottis, is produced by the shape of the cartilage, enhanced by an overlying collection of mucous glands.

For the glosso-epiglottic and pharyngo-epiglottic folds, see p. 410.

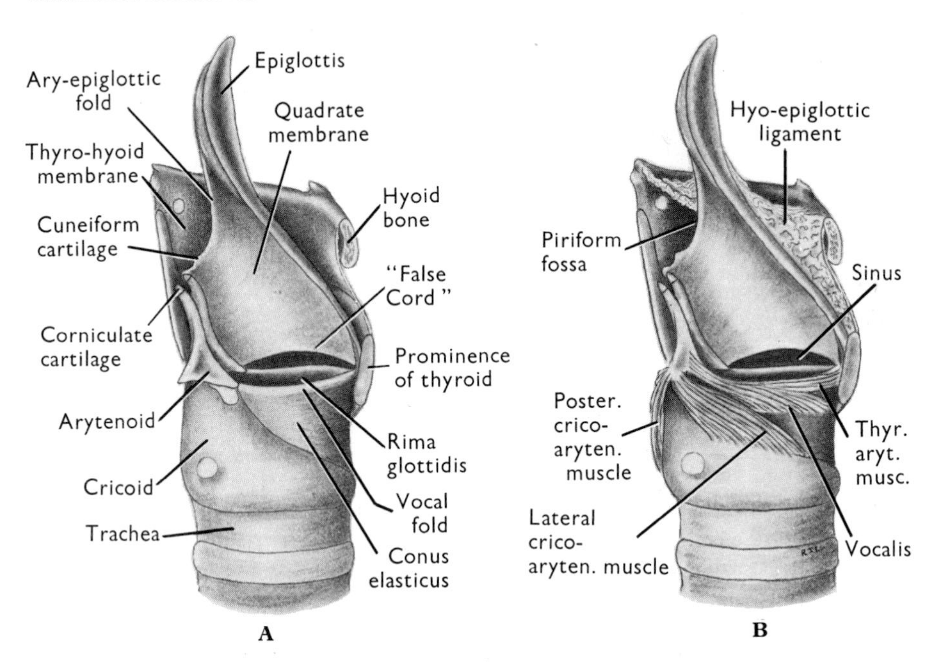

FIG. 6.41. THE SKELETON OF THE LARYNX. A. The interior of the larynx viewed from the right side. The right quadrate membrane and the right halves of the thyroid cartilage and hyoid bone have been removed. B. The muscles that are attached to the arytenoid cartilage seen in the same view. The hyo-epiglottic ligament extends laterally to form the pharyngo-epiglottic ('lateral glosso-epiglottic') fold.

The Laryngeal Cartilages. The epiglottis and the corniculate cartilages are composed of yellow elastic cartilage (which never calcifies). The remainder of the cartilages of the larynx are composed of hyaline cartilage (which in old age often calcifies and even ossifies).

The Conus Elasticus (Fig. 6.41). Projecting upwards from the arch of the cricoid is a fibro-*elastic* membrane which is attached anteriorly in the midline to the lower border of the thyroid notch. This part is very thick and strong and constitutes the **crico-thyroid ligament.** The upper portion of the conus elasticus is hidden beneath the laminæ of the thyroid cartilage. Its upper border is unattached on each side, constituting the **vocal folds** or **vocal cords,** whose anterior ends are attached to the thyroid cartilage at the crico-thyroid ligament and whose posterior ends are attached to the vocal process of each arytenoid cartilage. The slit between these free edges is known as the *rima glottidis.* The rima glottidis bisects the thyroid cartilage. The conus elasticus is elastic, but it is not a cone; its apex is not a point but a long slit (the rima glottidis). Its former name was *crico-vocal membrane.*

The Quadrate Membrane (Fig. 6.41). Extending between the arytenoid cartilages and the epiglottis is a similar fibro-elastic membrane. Its anterior border is attached to the side of the lower half of the epiglottis. Its posterior border, much shorter, is attached between the vocal process of the arytenoid and the corniculate cartilage. Its lower border is free, constituting the **vestibular band,** formerly known as the 'false cord'. Its upper border, much longer than the false cord, constitutes the **ary-epiglottic fold,** at the inlet of the larynx. The ary-epiglottic fold contains a tiny cartilage, the cuneiform cartilage.

In shape the quadrate membrane resembles the mainsail of a boat. The two ary-epiglottic folds make together an oval *aperture of the larynx* that lies in a vertical plane, with the upper half of the epiglottis projecting free above. The shape of the larynx, from the *vertical* aperture between the ary-epiglottic folds down to the *horizontal* lower border of the cricoid cartilage, can be compared with an old-fashioned ventilator on a ship's deck.

The Intrinsic Muscles of the Larynx

These consist of two separate groups: those which control the aperture of the inlet, and those which move the vocal folds.

The Sphincter of the Inlet (Ary-epiglottic Muscle) (Fig. 6.42). The free upper edge of the quadrate membrane, the ary-epiglottic fold, contains muscle fibres which connect the side of the epiglottis to the muscular process and posterior surface of the *opposite* arytenoid cartilage. The two muscles thus cross each other behind the transversely running fibres of the interarytenoid muscle. In crossing the apex of the arytenoid cartilage many of the fibres of this muscle are attached to the corniculate cartilage, thus giving rise to the erroneous description of the lower part of this muscle as an oblique interarytenoid muscle. It is more accurate to conceive of this muscle as a complete sphincter of the inlet. Its contraction opposes the arytenoids to each other and draws the epiglottis down to bring its lower half into contact with the arytenoids. It should be noted that the free edge of the ary-epiglottic fold sometimes contains no muscle fibres and that in these cases the oblique interarytenoid fibres are always well developed. In such cases the epiglottis is approximated to the arytenoids by the following muscle.

The **thyro-epiglottic muscle** arises from the upper border of the lamina of the thyroid cartilage. Its fibres lie outside the quadrate membrane, on which they run to be inserted into the side of the epiglottis.

The aperture of the inlet thus closed forms an effective valvular protection from above (i.e., against swallowed material) even without the epiglottic lid that normally is an added refinement. It can be readily opened by air pressure from below, such as an inescapable cough during swallowing.

Muscles of the Vocal Folds. These have easy names, determined by the cartilages to which they are attached. They work in opposing groups, and are best studied in their functional opposing pairs. The only movements they impart to the vocal folds are (*a*) opening and closing and (*b*) lengthening and shortening.

Opening-closing (Abduction-adduction) of the Vocal Folds. These movements are produced by movements of the crico-arytenoid joints. Opening (abduction) is produced by the posterior crico-arytenoid muscle, which has two separate actions. Each is opposed by a separate adductor muscle, the lateral crico-arytenoid and the interarytenoid muscle.

The Posterior Crico-arytenoid Muscle (Fig. 6.42). This is the most important single muscle in the larynx and perhaps in the whole body. It is *the only dilator muscle* of the rima glottidis. It arises from the concavity on each side of the lamina of the cricoid cartilage, whence its fibres converge on the muscular process of the arytenoid cartilage. Its upper fibres are almost horizontal, its lateral fibres almost vertical. The action of its oblique intermediate fibres can be resolved by the parallelogram of forces into a horizontal and vertical component. The *horizontal action* is a movement of the muscular processes of the arytenoids towards each other. This rotates the arytenoids, each about its own axis, so that their vocal processes are separated and the glottis opens with a diamond-shaped aperture. The *action of the vertical component* is to draw the arytenoids downwards. Since they articulate with the cricoid upon sloping shoulders this downward movement causes a bodily separation of the arytenoids from each other without rotation, the glottis opening with a V-shaped aperture, apex forwards. The horizontal or rotary movement and the vertical or bodily separation movement occur simultaneously. In man the latter constitutes the greater proportion of the total arytenoid movement. A subsidiary but indispensable action of the posterior crico-arytenoid muscle is to brace back the arytenoids on their cricoid articulation, and prevent the vocal processes from tilting downwards. The horizontal and vertical components of the posterior crico-arytenoid muscle are each opposed by a separate muscle.

The Lateral Crico-arytenoid Muscle (Fig. 6.43). This muscle arises from the upper border of the cricoid arch. Its fibres pass backwards and upwards beneath the thyroid lamina to attach themselves to the muscular process of the arytenoid. The action of the muscle, in drawing the muscular process forwards, is to cause the vocal processes to approximate each other by

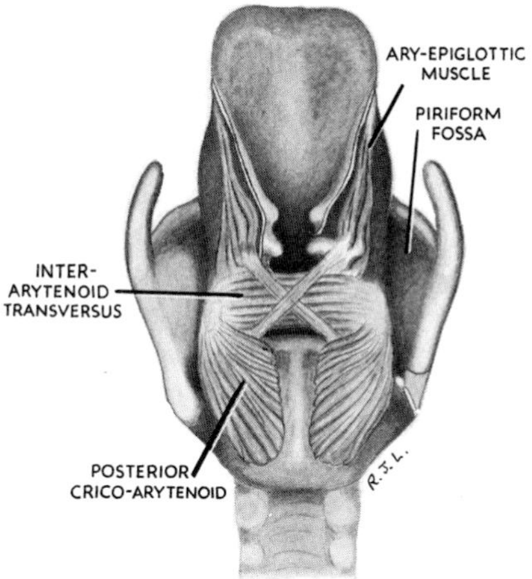

FIG. 6.42. THE LARYNX FROM BEHIND AFTER REMOVAL OF THE MUCOUS MEMBRANE.

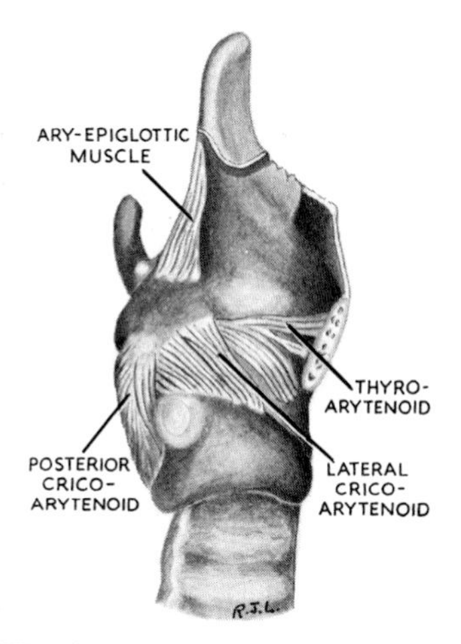

FIG. 6.43. THE RIGHT QUADRATE MEMBRANE AND THE MUSCLES OF THE LARYNX EXPOSED BY REMOVAL OF THE RIGHT HALF OF THE THYROID CARTILAGE. Drawn from a dissection.

rotation of the arytenoids. Thus it opposes the horizontal component of the posterior crico-arytenoid muscle. A subsidiary action is to assist the vertical component of the posterior crico-arytenoid to draw the arytenoid downwards on the shoulder of the cricoid lamina. Thus it is in part, and at times, a dilating muscle of the larynx, but this action can *never be possessed as a separate function*, but only by way of assisting the vertical component of the posterior crico-arytenoid muscle (Fig. 6.41B).

The Interarytenoid Muscle. This consists of a strong mass of transverse fibres which connect the posterior and part of the medial surfaces of the arytenoid cartilages to each other. Contraction of this muscle draws the arytenoid cartilages upwards along the sloping shoulders of the cricoid lamina, approximating them without rotation. The muscle is simply an opponent of the vertical action of the posterior crico-arytenoid muscle (Fig. 6.42).

Cadaveric Position. When all muscles are out of action (death, or complete loss of the recurrent laryngeal nerve) the human vocal fold approaches the midline, in about half-abduction. This is enough to cause vibration and stridor if the air flow is substantially increased for any reason.

Partial Palsy Position. With incomplete lesions of the recurrent laryngeal nerve the abductor muscle (posterior crico-arytenoid) is more vulnerable than the adductor (interarytenoid), and the partially paralysed vocal fold lies in, or even slightly across, the midline.

Lengthening-shortening of the Vocal Folds. These movements are produced by movements of the crico-thyroid joints.

Crico-thyroid Muscle (Fig. 6.44). This is a triangular muscle which diverges from the arch of the cricoid backwards to fan out towards its attachment to the inferior horn and lower border of the thyroid lamina. Its contraction causes the arch of the cricoid and the Adam's apple to approach each other. Usually the thyroid cartilage is tilted downwards towards the relatively fixed cricoid arch. The action of this muscle on the vocal fold is identical whether the thyroid moves on a fixed cricoid or vice versa. It is more readily understood if the thyroid cartilage is assumed to be fixed. Then, tilting upwards of the cricoid arch will tilt the lamina backwards; thus the vocal folds are *lengthened* (Fig. 6.45). Note especially that lengthening does not presuppose an increase in tension. The vocal folds consist of a felted membrane of fibro-elastic tissue whose strands run in all directions. Squares of this network are converted into diamonds by increased length of the vocal folds without a corresponding increase of tension.

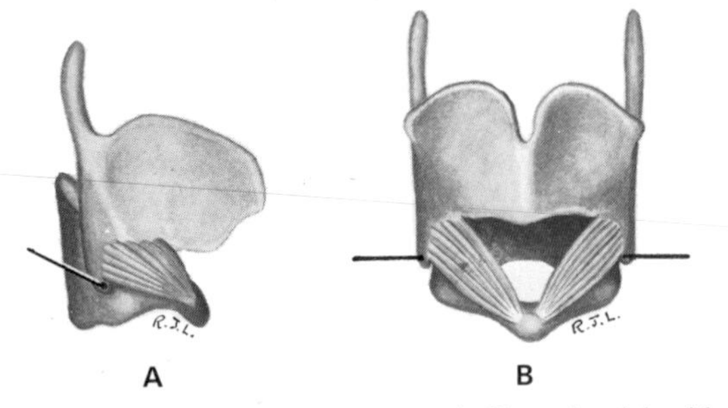

FIG. 6.44. THE CRICO-THYROID MUSCLE. A. From the right side. B. From in front. The pointer is directed to the axis of rotation between the cricoid and thyroid cartilages.

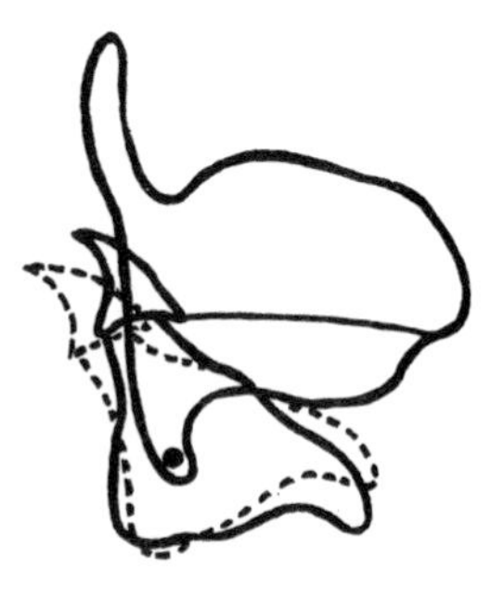

FIG. 6.45. THE ACTION OF THE CRICO-THYROID. Approximation of thyroid lamina and cricoid arch increases the distance between the arytenoids and the Adam's apple; the vocal cords are elongated.

The **thyro-arytenoid muscle** (Fig. 6.41) is the opponent of the crico-thyroid. It lies above the free edge of the crico-vocal membrane. It extends from the back of the Adam's apple of the thyroid cartilage to the vocal process and anterior surface of the arytenoid cartilage. Its contraction *shortens* (not relaxes) the vocal folds. Certain fibres of this muscle extend from the arytenoid cartilage to the crico-vocal membrane, falling short of the thyroid cartilage. They constitute the *vocalis muscle*. Their contraction pulls up portions of the crico-vocal membrane, thereby increasing the vertical depth of the opposing surfaces of the vocal folds. This thickening of the opposing surfaces of the vocal folds is assisted by the contraction of the main bulk of the thyro-arytenoid muscle. The tension of the fold is not a function of its length, but of the tonic contraction of the thyro-arytenoid muscle. The length of the folds having been determined by the opposing contractions of the crico-thyroid and thyro-arytenoid muscles, their tension is now determined by the isometric contractile tonus of the thyro-arytenoid muscle, counter-balanced by the requisite opposing tonus of the crico-thyroid muscle. It is apparent that the longer the vocal fold the more readily can the stretched thyro-arytenoid muscle contract.

It must be emphasized that these changes in length and tension control the pitch of the voice and occur normally *only when the cords are in contact for phonation.*

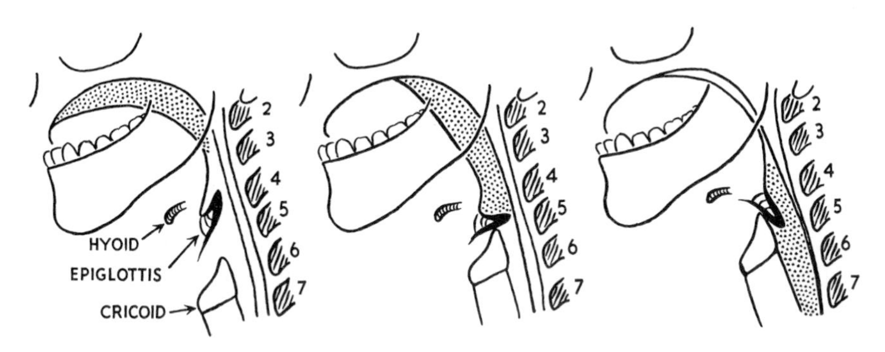

FIG. 6.46. THE STAGES OF SWALLOWING. (1) The bolus entering the oro-pharynx. (2) The bolus in the oro-pharynx. The larynx (see cricoid outline) has risen and the epiglottis is tilting. *The hyoid has not moved.* (3) The bolus entering the œsophagus. Hyoid and larynx have together risen and the epiglottis is inverted. Tracings from a cine-radiograph made by the late Dr Russell Reynolds.

Intrinsic Movements of the Larynx

1. *Sphincteric.* The inlet is protected during swallowing by the sphincteric action of the ary-epiglottic muscle. A second sphincter is provided by the rima glottidis itself, but this is very rarely needed. Not only is the larynx hauled up beneath the posteriorly bulging tongue, but the epiglottis itself is inverted by the passing bolus and closes like a lid over the laryngeal inlet (Fig. 6.46). Thus during swallowing the entry of a foreign body into the aperture of the larynx is a very rare event.

The sphincter of the inlet is used only during swallowing. It plays no part in phonation, coughing or the closure for muscle effort; in these the larynx is closed at the vocal folds.

2. *Phonation.* The stream of air emitted during phonation emerges as a series of discrete jets, as from a siren. This is not only a more effective means of sound production, but is very economical of expired air. At rest the vocal folds are separated, and have sharp edges. During phonation they are held together, and the crico-vocal membrane is pulled up by the vocalis muscle, so that the folds are in contact over a vertical extent of the order of 3 mm below their free edges (Fig. 6.47). It is this part of the larynx that is lined with thick squamous stratified epithelium (for 'wear and tear'); the lower part of the larynx is lined with the typical ciliated columnar epithelium of the respiratory mucous membrane. The apposed vocal folds are blown apart intermittently by the pressure of the expired air below them. The frequency of emission of the jets depends on the length and tension of the vocal folds, and determines the pitch.

The function of the vocal folds is to produce sound, varying only in intensity and pitch. The *quality* of the voice depends on the resonators above the larynx. These are altered by change of position of the soft palate and the tongue, thus altering the volume of naso-pharynx and oro-pharynx. Depression of the larynx (p. 429) increases the volume of the resonating chambers. *Articulation* depends on breaking up the sound into recognizable consonants and vowels by the use of tongue, teeth and lips.

Whispering. Here the vocal folds are separated, and vibrations are imparted to a constant stream of expired air. This is inefficient as a means of sound production, and is very wasteful of air.

3. *Coughing.* A cough or sneeze is an explosion of *compressed* air. The vocal folds are powerfully adducted, a strong expiratory contraction is made to build up the

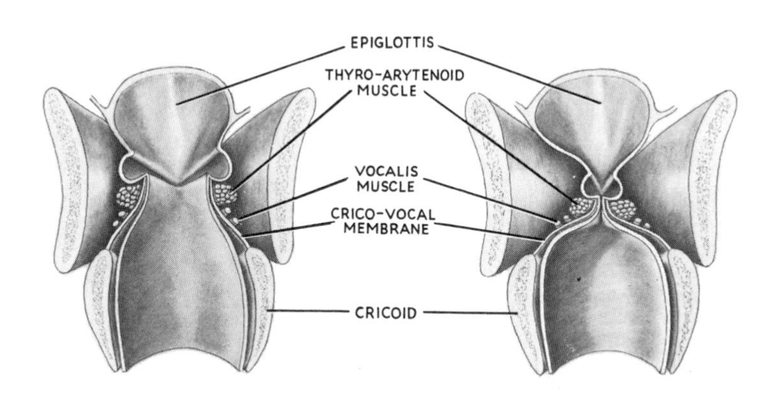

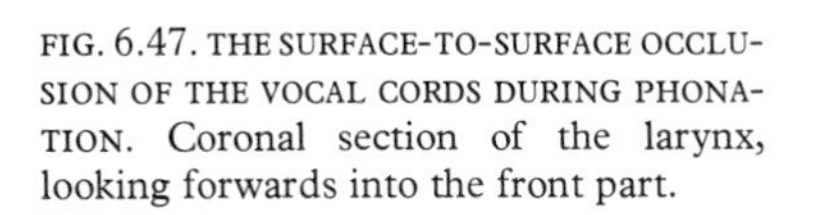
FIG. 6.47. THE SURFACE-TO-SURFACE OCCLUSION OF THE VOCAL CORDS DURING PHONATION. Coronal section of the larynx, looking forwards into the front part.

intrathoracic pressure (see p. 222), the folds are then suddenly abducted and the blast of compressed air explodes through the larynx (its expulsive force increased by the simultaneous 'choke-barrel' narrowing of the trachea).

4. *Muscular Efforts.* Abdominal straining is made more effective by adduction of the vocal folds. The diaphragm is weaker than the muscles of the anterior abdominal wall. To prevent loss of intra-abdominal pressure by upward displacement of the diaphragm the folds are closed after a deep breath and the diaphragm is forced up against a cushion of compressed air. This manœuvre is used for evacuation of pelvic effluents and also for the straining of heavy lifting. Escape of a jet of compressed air causes the characteristic grunt.

The act of 'holding the breath' by powerfully adducted vocal folds may be to prevent either inspiration or expiration. Air entry is prevented by the valve effect of the adducted vocal cords. This is used in powerful adducting movements of the upper limb, where the vocal folds in apposition prevent air entry, and so the adducting muscles do not waste their effort by elevating the ribs.

Summary of Vocal Fold Movements

1. The vocal folds are held open constantly (for breathing), and their length is never altered in this position.
2. They are closed intermittently for:
   (a) phonation
   (b) prior to a cough
   (c) abdominal straining.
3. Change of length occurs *only when they are closed for phonation.* When closed they are bunched up by vocalis to strengthen the seal.

The Epiglottis. This thin sheet of elastic cartilage (p. 423) provides the anterior wall of the vestibule of the larynx. But it has attachments also to the pharynx and, while not relevant to the larynx, it is convenient to note them now. The hyoid bone lies at a level that bisects the epiglottis, but the epiglottis leans back, away from the body of the hyoid. The two are connected by loose fibrous tissue with fat in its interstices—the *hyo-epiglottic ligament.* This is projected laterally to join the tip of the greater cornu to a point half-way up the side of the epiglottis. This lateral extension of the hyo-epiglottic ligament raises a ridge in the mucous membrane of the pharynx—a pharyngo-epiglottic fold. It has nothing to do with the tongue, in spite of its newly-dubbed name of lateral 'glosso'-epiglottic fold. It marks the junction between oro-pharynx (vallecula) above and the laryngeal part of the pharynx (piriform fossa) below (Fig. 6.41B).

The Thyro-hyoid Membrane (Fig. 6.41). This is a fibrous membrane connecting the whole length of the upper border of the thyroid laminæ and superior cornua to the body and greater cornua of the hyoid bone. Its free posterior border is thickened, is named the **thyro-hyoid ligament,** and is reinforced by a small unimportant cartilage. Note that the thyro-hyoid membrane passes up behind the body of the hyoid bone to be attached posteriorly to its upper border—a bursa lies between the membrane and the back of the hyoid bone. It is here that remnants of the thyro-glossal duct (p. 42) may persist, necessitating splitting or resection of part of the hyoid bone to give adequate surgical access (Fig. 1.51, p. 42).

The thyro-hyoid membrane forms the lateral wall of the piriform fossa and is perforated by the superior laryngeal vessels and internal laryngeal nerve. It is not part of the larynx, but it is convenient to study it here.

The Mucous Membrane of the Larynx

The interior of the larynx is draped with mucous membrane. It sweeps down from the epiglottis over the ary-epiglottic folds, wherein on each side two small bumps indicate the corniculate and cuneiform cartilages. The mucous membrane lines the *vestibule,* which extends from the aperture (i.e., the ary-epiglottic folds) down to the false cords. It is walled in by the two quadrate membranes and the lower half of the epiglottis. To give sharp edges to the false cords and the vocal folds the mucous membrane herniates between them. The open slit so produced is named the *sinus* of the larynx, and it leads into the herniated *saccule* of the larynx. This contains mucous glands, probably for lubrication of the vocal folds, which themselves contain no glands. The saccule lies under cover of the thyroid cartilage and does not extend beyond the thyroid lamina. Potentially it could, and in some primates an enormous sac herniates down behind the clavicle into the axilla. The reason for this is obscure (i.e., not known).

Microscopic Anatomy. The fibrous tissue of the corium is loosely woven in all parts except over the vocal folds. It permits of great swelling except at the rima glottidis. Hence the danger of œdema of the glottis; the swelling accumulates above the rima and suffocation results because the tissue fluid cannot disperse across the watershed of the vocal fold. The vestibule of the larynx is lined with a very low stratified squamous epithelium. The walls here come into gentle apposition during constriction of the aperture in swallowing. Greater wear and tear occurs below this, and the false cords and vocal folds are covered with very thick squamous stratified epithelium with well-marked papillæ in the dense corium. The lower half of the larynx is lined with ciliated columnar epithelium, and beneath this the mucous membrane contains both mucous and serous glands.

The mucous membrane over the *epiglottis* conforms to the functional needs. Over all its *laryngeal (posterior) surface* it contains a few scattered glands, and is surfaced by a low stratified squamous epithelium. The *anterior surface* of its free part (not in the larynx) faces the tongue. Its mucous membrane is very thick, and is packed with mixed glands, mostly mucous. The glands indent pits or even perforations in the elastic cartilage. Their mucus makes this inverting surface of the epiglottis slippery for swallowing. The surface epithelium consists of a very thick layer of squamous stratified ('wear and tear') cells.

Blood Supply of the Larynx. As in the innervation, so in the vascular pattern, the vocal folds form a dividing line between the upper and lower halves of the larynx. Above the vocal folds blood is brought to the larynx by the superior laryngeal branch of the superior thyroid artery. This vessel enters the piriform fossa with the internal laryngeal nerve by piercing the thyro-hyoid membrane (Fig. 6.33). The superior laryngeal veins accompany the artery and empty into the superior thyroid veins. Tissue fluid from the upper half of the larynx is returned by lymphatics that accompany the superior thyroid artery and drain into nodes of the antero-superior group of deep cervical lymph nodes.

The lower half of the larynx is supplied from the inferior laryngeal branch of the inferior thyroid artery; it accompanies the recurrent laryngeal nerve beneath the inferior constrictor of the pharynx. It supplies all of the larynx below the vocal folds. Venous return is by the inferior laryngeal veins to the inferior thyroid veins, which drain into the brachio-cephalic veins, chiefly the left. Tissue fluid is returned by lymphatics that accompany the inferior thyroid artery and empty into nodes of the postero-inferior group of deep cervical lymph nodes. A few lymphatics run with the veins to the pretracheal lymph nodes. These and the tracheo-bronchial group are commonly involved in the spread of malignant disease.

Nerve Supply of the Larynx. The *upper half* of the larynx is innervated with sensory and secreto-motor fibres by the internal laryngeal branch of the superior laryngeal branch of the vagus. The muscles are innervated by the recurrent laryngeal nerve by fibres derived from the accessory nerve; the cell bodies lie in the nucleus ambiguus. The *lower half* of the larynx is innervated with sensory and secreto-motor fibres by the recurrent laryngeal branch of the vagus, which innervates also all the intrinsic muscles of the larynx except the crico-thyroid. The crico-thyroid muscle is supplied by the external laryngeal branch of the superior laryngeal nerve, whose fibres are in like manner derived from the accessory nerve and whose cell bodies lie in the nucleus ambiguus.

The *sympathetic supply* comes in along the arteries. Above the cords the fibres run with the superior laryngeal artery from the superior cervical ganglion, while below the cords the inferior laryngeal artery brings fibres from the middle cervical ganglion.

Development of the Larynx (see p. 43). Knowledge is not precise, but it can be stated that the cartilages of the larynx are derived from the fourth and sixth pharyngeal arches, the epiglottis and ary-epiglottic folds from the furcula.

Extrinsic Muscles of the Larynx

The larynx moves upwards during the act of swallowing and afterwards returns to the position of rest. Two opposing groups of muscles exist, the one elevating and the other depressing the larynx. It is to be noted that an elevator of the hyoid bone must elevate the larynx by transmitting its pull through the thyro-hyoid membrane to the thyroid cartilage, and that the larynx cannot be depressed without an equal depression of the hyoid. Normally after elevation of the larynx the return to the position of rest is by elastic recoil of the trachea.

Elevators of the Larynx. The elevator of the hyoid bone is the mylo-hyoid muscle. The actions of the digastric, stylo-hyoid and genio-hyoid muscles are discussed on p. 446.

The '-pharyngeus' Muscles. The three muscles, stylo-pharyngeus, salpingo-pharyngeus and palato-pharyngeus are all badly named. In each case their essential insertion is into the larynx, though certain of their fibres fade out inside the lateral wall of the pharynx. Their attachments to the larynx are chiefly into the posterior border of the lamina and cornua of the thyroid cartilage, though some of their fibres encroach a little on the upper border of the thyroid lamina just in front of the superior cornu. They should rightly be named the stylo-laryngeus, salpingo-laryngeus and palato-laryngeus. They are the elevators of the larynx. They are described on pp. 396 and 420. It is understandable that most of their fibres should reach the larynx, which is elevated against the elastic stretch of the trachea, whereas the pharynx itself requires much less force for elevation because the longitudinal muscle of the œsophagus relaxes.

Depressors of the Larynx. These muscles are described elsewhere (p. 366). They consist of the infrahyoid 'strap' muscles sterno-hyoid, sterno-thyroid and omo-hyoid. Only one, the sterno-thyroid, is directly a depressor of the larynx, the others acting indirectly via the hyoid bone. Generally the elevated larynx is returned to its rest position by the elastic recoil of the trachea. Active depression of the larynx increases the capacity of the resonating chambers during phonation and, thus affects the quality of the voice.

THE ORBIT

Examine a skull. The **orbit** is a pyramidal space enclosed within bony walls. The **roof** is the orbital plate of the frontal bone, the floor of the anterior cranial fossa. It is very thin. In front, on the medial side, the frontal sinus obtrudes between the cranial and orbital plates of the frontal bone (Fig. 8.7, p. 565). At the extreme posterior end the lesser wing of the sphenoid completes the roof.

The **medial wall** extends in front from the anterior lacrimal crest on the frontal process of the maxilla, backwards across the lacrimal bone and the orbital plate of the ethmoid, to the body of the sphenoid and the optic foramen. The posterior lacrimal crest is a vertical ridge on the lacrimal bone. Between anterior and posterior lacrimal crests is the lacrimal fossa, a groove that leads down into the naso-lacrimal canal. At the junction of medial wall and roof two foramina lie between ethmoid and frontal bones; they are the anterior and posterior ethmoidal foramina. The medial walls lie antero-posterior, parallel with each other. They separate the orbits from the ethmoidal air cells and the lateral walls of the nose.

The **lateral wall** of the orbit is composed of the zygomatic bone and the greater wing of the sphenoid. Posteriorly there is a gap between lateral wall and roof (greater and lesser wings of the sphenoid), the superior orbital fissure, leading into the middle cranial fossa; a further gap diverges from the medial end of this fissure between lateral wall and floor. This is the inferior orbital fissure; it leads into the pterygo-palatine and infratemporal fossæ. The lateral wall separates the orbit from the temporal fossa in front and from the middle cranial fossa behind.

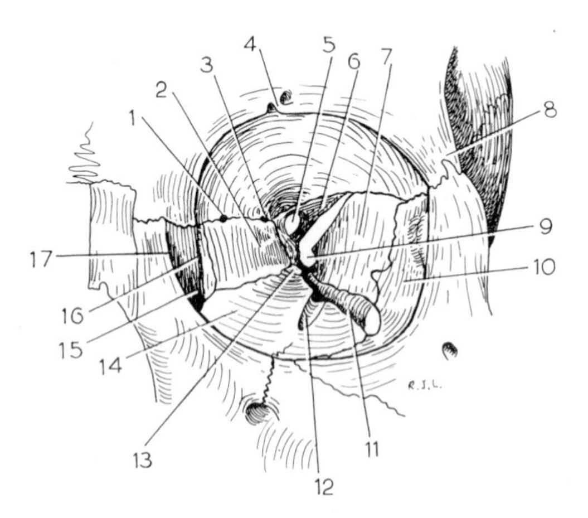

FIG. 6.48. THE BONY WALLS OF THE LEFT ORBIT, VIEWED ALONG THE ORBITAL AXIS, AT 25 DEGREES TO THE SAGITTAL PLANE. **1.** Anterior ethmoidal foramen. **2.** Orbital plate of ethmoid, articulating posteriorly with body of sphenoid. **3.** Posterior ethmoidal foramen. **4.** Supra-orbital notch and foramen. **5.** Optic canal. **6.** Lesser wing of sphenoid. **7.** Greater wing of sphenoid. **8.** Zygomatic process of frontal bone. **9.** Superior orbital fissure. **10.** Zygomatic bone. **11.** Inferior orbital fissure. **12.** Infra-orbital groove. **13.** Orbital process of palatine bone. **14.** Orbital plate of maxilla. **15.** Lacrimal fossa. **16.** Posterior lacrimal crest on lacrimal bone. **17.** Anterior lacrimal crest on frontal process of maxilla.

The **floor** consists of the orbital plate of the maxilla (grooved and canalized by the infra-orbital nerve) completed in front by the zygomatic bone laterally and completed behind by a tiny piece of bone at the apex, the orbital process of the palatine bone.

The **orbital margin** consists of four curved sides. It slopes downwards from the nose. The upper margin (frontal bone) is notched or canalized near its medial end for the passage of the supra-orbital nerve and artery. The lateral margin is formed by the conjoined processes of the frontal and zygomatic bones. The lower margin is formed half and half by zygomatic bone and maxilla, which meet at a suture. An irregular suture leads from this margin into the infra-orbital foramen, indicating the sinking of the infra-orbital nerve through the bone from its original passage around the infra-orbital margin (Fig. 6.50). The suture closes in later life and is not seen in elderly skulls. The *medial margin* of the orbit consists of two ridges which overlap. From the inferior margin the anterior lacrimal crest rises upwards and from the superior margin the posterior lacrimal crest runs down behind this.

The **eyelids** are covered in front with loose skin and behind with adherent conjunctiva. Their skeletal framework is the septum orbitale, thickened at the margins of the lids to form the tarsal plates. The orbicularis oculi muscle lies in front of the septum.

The **septum orbitale** is attached to the anterior lacrimal crest and to the margins of the orbit. It has a wide 'buttonhole' in it, the palpebral fissure between the lids. It is not of equal density throughout. It is greatly thickened at the upper margin of the buttonhole to form the crescent-shaped **superior tarsal plate** and similarly, though to a smaller extent, in the lower margin to form the **inferior tarsal plate.** From the medial end of the buttonhole to the anterior lacrimal crest the septum orbitale is thickened; this is the **medial palpebral ligament,** which anchors the tarsal plates to the anterior lacrimal crest. No such structure exists laterally. Here the orbital septum is thin and fuses with the raphé of the palpebral part of the orbicularis oculi muscle **(lateral palpebral raphé).**

In front of the tarsal plates and septum orbitale lie the curving fibres of the palpebral part of orbicularis oculi (p. 379). The angular vein and termination of the facial artery lie superficial to the orbicularis at the medial canthus. In front of the muscle fibres is thin skin overlying a loose fibrous subcutaneous tissue. The skin is thin and densely adherent to the margins of the palpebral fissure; here, just outside the margin, are the eyelashes, long hairs whose follicles are stabilized by being attached to the rigid tarsal plates.

The **tarsal plates** are crescent-shaped laminæ of very dense fibrous tissue, curved to the curve of the eyeball. They are rigid and retain their curve after being distorted. As well as anchoring the roots of the eyelashes they contain the **tarsal (Meibomian) glands.** These modified sebaceous glands secrete an oil that makes the lid margins waterproof and, covering the tears on the cornea, delays their evaporation. The palpebral conjunctiva is firmly adherent to the deep surface of each tarsal plate, so it is not wrinkled by eyelid movements.

At the medial end of each lid margin is a low elevation surmounted by a minute **punctum.** This opens into a **canaliculus,** a tiny canal which conveys excessive tears to the lacrimal sac.

The **blood supply** of the eyelids is derived from palpebral branches of the ophthalmic artery.

The Conjunctiva. This transparent membrane is attached to the sclera at the margins of the cornea, with which it blends. It is loosely attached over the anterior part of the sclera and thence reflected to the inner surfaces of the eyelids. It is firmly attached to the tarsal plates. It blends with the skin at the margins of the lids. A fold near the medial canthus, the plica semilunaris, is said to be a vestige of the third eyelid. It is nothing of the sort. Its slack allows lateral deviation of the cornea. Elsewhere the slack conjunctiva is tucked in as the fornices; the presence of the lacrimal canaliculi and sac prevents any infolding here at the medial canthus. At the medial canthus is a small pimple of modified skin, called the caruncle. Between plica and caruncle is the **lacus lacrimalis,** bounded by hairless lid margins; in the lacus excessive tears accumulate before passing into the canaliculi.

The conjunctiva is a firm membrane of fibrous tissue covered with wear-and-tear epithelium (squamous stratified) except for scattered islands mainly in the upper fornix. These are islands of columnar epithelium, all the cells of which are goblet cells, secreting mucus. Into the superior fornix of the conjunctiva the lacrimal gland opens laterally by fifteen or more separate little ducts. A flange from the tendon of levator palpebræ superioris is attached along the superior fornix.

Nerve supply. The palpebral and adjoining scleral surfaces of the conjunctiva are supplied by the nerves of the overlying skin (Fig. 6.16, p. 384). They are four branches of Va for the upper lid and the palpebral branches of Vb for the lower lid. Note that the cornea has a separate supply via the ciliary ganglion (p. 437).

Circulation of Tears. Tears are received into the lateral part of the superior conjunctival fornix and flow to the lacus lacrimalis. The closing eyelids brush excessive tears across, for the lateral canthus moves medially by elongation of the lateral palpebral raphé. Normally the volume of tears secreted equals that lost by evaporation, and no tears pass into the lacrimal sac. The sac is an emergency apparatus for removing *excessive* tears to clear the vision.

Excessive tears are discouraged from spilling over on to the cheeks by the waterproof lid margins, greased by the secretion of the Meibomian glands. The eye can be 'brimful of tears' before it spills. Opening and closing the lids (blinking) pumps the tears out of the conjunctival sac, and the movement is accelerated during times of increased tear production. Blinking, however, serves a much more important function, that of moistening the corneal epithelium. The upper lid acts like a windscreen wiper; it leaves a film of tears on the cornea, a film so thin that it would evaporate instantaneously in most climates. Evaporation is retarded by a film of mucus (from the tarsal conjunctiva) on the film of water and a film of oil (from the Meibomian glands) on the film of mucus. **Blinking** is a reflex act, set off by dryness of the cornea to replace the precorneal film, or by excessive tears to pump them into the nose.

Blinking is produced by the palpebral fibres (not the orbital fibres) of orbicularis oculi, and the movement does not diminish the volume of the conjunctival sac. 'Screwing up the eyes' by the orbital fibres compresses the conjunctival sac, and an eye brimful of tears is caused thereby to spill over down the cheek (see p. 379).

The lacrimal canaliculi lead from the puncta to the lacrimal sac; flap valves of mucous membrane prevent reflux of tears. The **lacrimal sac** lies in the lacrimal groove and some of the palpebral fibres of the orbicularis oculi are inserted into its walls. When the muscle contracts the lids are closed and the puncta turned inwards to dip into the lacus lacrimalis. Simultaneously the sac is drawn widely open, so that tears are sucked up through the canaliculi. When the muscle relaxes the lacrimal sac contracts by its own elasticity and pumps its contents down the naso-lacrimal duct. There is very much elastic tissue in the wall of the lacrimal sac.

The **naso-lacrimal duct,** 2 cm long, slopes downwards and *laterally*, in conformity with the pear-shaped nasal cavity, to open in the inferior meatus rather more than 2 cm behind the nostril. It is a wide-bored tube (goose-quill size). The mucous membrane is raised into several variable folds which act as valves to prevent air being blown up the duct into the lacrimal sac. The valves enforce one-way traffic in both canaliculi and naso-lacrimal duct. Blinking pumps the lacrimal sac like a rubber syringe, sucking excess fluid from the conjunctival sac and squirting it down into the nose.

Movements of the Eyelids. The lower lid

possesses very little mobility, the upper lid a great deal. The lids are closed gently by the palpebral fibres and forcibly when the orbital fibres of the orbicularis oculi join in (p. 379). The lids are opened in the ordinary way by the levator palpebræ superioris muscles. The levators hold the lids open while the orbital fibres of orbicularis are contracting to lower the eyebrows as a pair of sun visors.

The Contents of the Orbit

The orbit contains the eyeball, with the optic and other nerves that innervate it, the extrinsic muscles that move it and the vessels and nerves that supply them. Some nerves and vessels send branches through the orbit to the nose and face.

The eyeball does not lie symmetrically in the axis of the orbit. The two eyeballs face forwards parallel with each other. The medial walls of the orbits are parallel, but the lateral walls slope away to make a right angle with each other. The optic nerve and ocular muscles come from the apex of the orbit, at the back of the medial wall, and pass laterally, not straight forwards, to their ocular attachments.

Examine a skull. The superior orbital fissure is retort-shaped, with the broad end medially (Fig. 8.2, p. 556). A fibrous ring surrounds the 'bulb of the retort' and the optic foramen. It is attached to bone except where it bridges the 'neck of the retort'. From the ring the four recti arise; from the bone above the ring the levator palpebræ superioris and the superior oblique take origin. As these muscles pass forwards from the apex of the orbit they broaden out, to form a **cone of muscles** around the eyeball. Many nerves pass through the superior orbital fissure (Fig. 6.21, p. 400). Three pass through the lateral part, outside the fibrous ring, and they remain *outside the cone of muscles.* They are the lacrimal, frontal and trochlear nerves. The rest of the lateral part of the fissure is closed by the fibrous layer of the dura mater of the middle cranial fossa. The nerves that pass through the fibrous ring enter the cone of muscle; those destined for the eyeball or ocular muscles remain *inside the cone.* The only branches that pass out penetrate *through the cone beneath the superior oblique muscle*; they are the posterior and anterior ethmoidal and the infratrochlear nerves, all branches of the naso-ciliary nerve. If these points are grasped, the disposition of the nerves in the orbit becomes intelligible.

The Muscles of the Orbit

The **levator palpebræ superioris** (Figs. 6.21, p. 400, 6.49) arises at the apex of the orbit from the roof (lesser wing of the sphenoid). It is a flat muscle that broadens out as it passes forward beneath the roof of the orbit. The thick frontal nerve lies along its upper surface. At the anterior end the muscle broadens into a ribbon-like tendon that is projected on each side into a pair of crescentic horns. This broad tendon is inserted into the superior tarsal plate. In front, a flange penetrates the septum orbitale and palpebral muscle to become attached to the skin of the upper lid, while behind a weaker flange is attached to the superior fornix of the conjunctiva. The muscle elevates the upper lid, pulling skin and conjunctiva up with it. Its opponent is the palpebral part of orbicularis oculi. The levator is innervated from the superior division of the oculomotor (III) nerve. The branch either pierces the superior rectus or passes on its medial side to enter the lower surface of the levator. It carries sympathetic fibres for the smooth muscle part of the levator (p. 486).

A thin sheet of unstriated muscle lies beneath the main tendon; it passes into the upper margin of the tarsal plate. It is supplied by sympathetic fibres from cell bodies in the superior cervical ganglion. Stimulation results in widening of the palpebral fissure (this is not true exophthalmos). Division of the cervical sympathetic, by paralysis of the muscle, causes slight ptosis.

The **superior rectus** arises from the upper part of the fibrous ring and from the dural sheath of the optic nerve. It passes forward and laterally beneath the levator and pierces the fascia bulbi (Tenon's capsule); its tendon is inserted into the upper part of the sclera anterior to the coronal equator of the eyeball (Fig. 6.49). It is supplied on its ocular surface by the superior division of the oculomotor nerve (III) and its action is to rotate the eyeball so as to elevate the cornea and draw it nasally.

The **medial rectus** arises from the medial part of the fibrous ring and from the dural sheath of the optic nerve. It passes along the medial wall of the orbit below the superior oblique muscle and, piercing the fascia bulbi, its tendon is inserted into the medial surface of the sclera anterior to the coronal equator of the eyeball. The muscle is supplied on its ocular surface by the inferior division of the oculomotor (III) nerve; its action is to rotate the eye to make the cornea deviate horizontally towards the nasal side.

The **inferior rectus** arises from the lower part of the fibrous ring. It deviates laterally as it passes forwards to the eyeball. Its tendon pierces the fascia bulbi and is inserted, anterior to the coronal equator of the eyeball, into the inferior surface of the sclera. An expansion from the fascia bulbi passes above the inferior oblique into the inferior tarsal plate. The muscle is innervated on its upper surface from the inferior division of the oculomotor (III) nerve; its action is to depress the cornea and deviate it towards the nasal side, at the same time slightly depressing the lower lid.

The **lateral rectus** arises from the lateral convexity of the fibrous ring, in continuity across the superior orbital fissure from the bone above to the bone below.

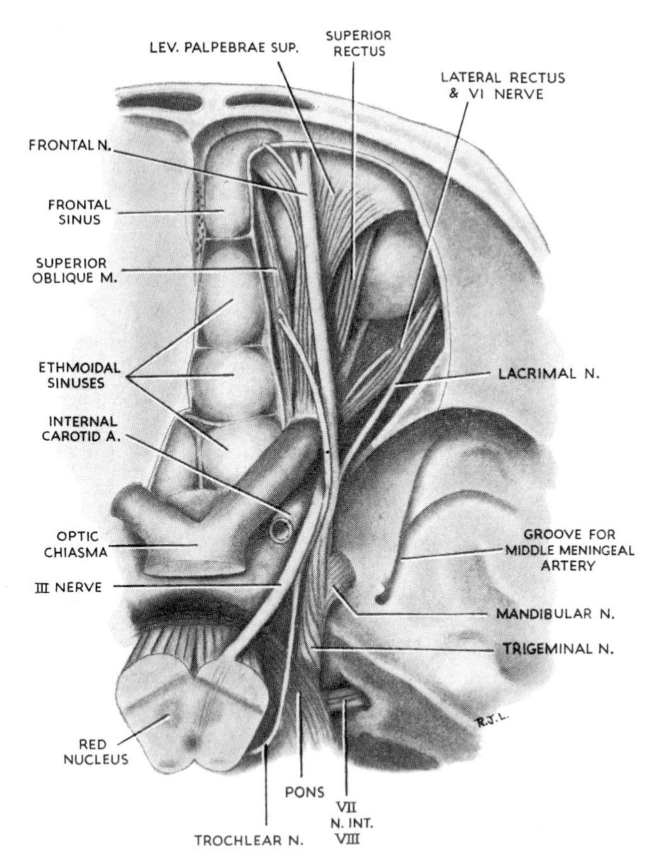

FIG. 6.49. THE ANTERIOR CRANIAL FOSSA AFTER REMOVAL OF THE ORBITAL PLATE OF THE FRONTAL BONE. THE ORBIT AND ETHMOIDAL SINUSES ARE EXPOSED. Drawn from Specimen M 138. 2, in R.C.S. Museum.

Its origin is C-shaped; upper and lower limbs of the C are usually referred to as the upper and lower heads of origin, but actually there is only one continuous curved origin. The muscle passes along the lateral wall of the orbit (Fig. 6.49) and, piercing the fascia bulbi, its tendon is inserted into the lateral surface of the sclera in front of the coronal equator of the eyeball. It is supplied by the abducent (VI) nerve on its ocular surface, and its action is to deviate the cornea horizontally towards the temporal side.

The **superior oblique** arises from a narrow tendon above and medial to the fibrous ring and its rounded belly passes forward above the medial rectus. It gives way to a slender tendon (Fig. 6.93, p. 486) which passes through the trochlea (where it is surrounded by a synovial sheath) and turns backwards and laterally to pierce the fascia bulbi and pass *under the superior rectus*. It is inserted into the postero-superior quadrant of the sclera (i.e., behind the coronal equator of the eyeball). The **trochlea** (pulley) is a loop of dense fibrous tissue slung to the under surface of the frontal bone (beneath the frontal sinus) a few millimetres behind the supero-medial angle of the orbital margin. The muscle is innervated on its upper surface by the trochlear (IV) nerve. Its action is to depress the cornea and turn it out towards the temporal side.

The **inferior oblique** (Fig. 6.50) arises from the floor of the orbit vertically beneath the trochlea. A small depression or rough area of bone on the orbital plate of the maxilla is occasionally seen, but usually the muscle leaves no mark. The muscle passes obliquely back *below the inferior rectus* and curls up to insinuate itself between the lateral rectus and the sclera. It is attached into the postero-inferior quadrant of the sclera (i.e., behind the coronal equator of the eyeball). The muscle is supplied on its upper surface by the inferior division of the oculomotor (III) nerve. Its action is to elevate the cornea and deviate it towards the temporal side.

The **fascia bulbi (Tenon's capsule)** is applied like a bursa to the back of the eyeball from the corneo-scleral junction to the attachment of the optic nerve. Its inner layer is thin and blended with the sclera. Its outer layer is pierced by the tendons of the four recti and two obliques, and is prolonged as a tubular investiture proximally along each tendon towards the muscle belly. This outer layer is not of equal density throughout. Over the lateral rectus the tubular prolongation is thickened where it is attached to *Whitnall's tubercle*, a palpable (if not visible) elevation on the orbital surface of the zygomatic bone just within the orbital margin (Fig. 8.14, p. 570). Over the medial rectus the tubular prolongation of fascia bulbi is thickened where it is attached to the posterior lacrimal crest on the lacrimal bone, at the horizontal level of Whitnall's tubercle. These two attachments are known as **check ligaments** (Fig. 6.50). Between them the inferior part of the fascia bulbi is thickened, and forms a sling, like a hammock, for the support of the eyeball. This thickened segment of the fascia bulbi is known as the **suspensory ligament** (of Lockwood). It supports the eyeball in the orbit.

Stability of the Eyeball

Vertical Stability. The eyeball does not rest on the floor of the orbit. It is suspended above this level by the suspensory ligament of Lockwood and, in fact, lies nearer the superior than the inferior orbital margin, as can be confirmed by simple palpation. The whole maxilla can be removed, with the side walls of the orbit

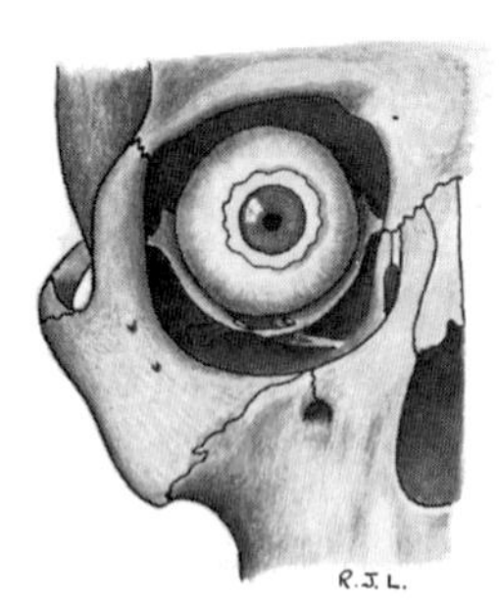

FIG. 6.50. THE EYEBALL SUSPENDED IN THE ORBIT BY THE LIGAMENT OF LOCKWOOD, WHICH IS PIERCED BY THE INFERIOR OBLIQUE MUSCLE.

up to the level of Whitnall's tubercle, without descent of the eyeball. Above this level removal of bone destroys the attachments of the suspensory ligament, the eyeball falls down, and intolerable diplopia results.

Antero-posterior Stability. The two pupils always converge towards the object being viewed. The eyeball is very mobile in its socket (fascia bulbi) rotating up or down or sideways, but with all these movements the eyeball does not pop in or out. In other words, it always rotates about a fixed centre, which is the geometrical centre of its own spheroid. Neither the resting tonus nor the pull of the contracting recti displaces the eyeball posteriorly.

The eyeball is prevented from posterior displacement by three factors: (1) the bony attachments of the recti; (2) the orbital fat; (3) the forward pull of the obliques.

The medial and lateral recti are attached to the bony margins of the orbit by the check ligaments and this attachment discourages posterior displacement of the eyeball.

The orbital fat lies within and without the cone of muscles and forms a cushion towards which the four recti can pull the eyeball back. If its volume increases (e.g., in hyperthyroidism) exophthalmos results, while if the volume decreases (e.g., in *extreme* dehydration) enophthalmos occurs.

The two obliques exert a forward pull on the eyeball. It is oblique in direction and may be thought to be a rather irrelevant factor. Nevertheless, it is known that the forward pull of the obliques can be considerable, as in the case of the stage comedian who by voluntary contraction of his obliques could so protrude the eyeballs that he could close his eyelids behind them, an astonishing performance that was brought to an untimely end by cellulitis of the orbit and meningitis following conjunctival ulceration.

Movements of the Eyeballs. The actions of the medial and lateral recti are simple. Each lies in a horizontal plane and rotates the eyeball so that the cornea looks medially and laterally respectively. There are no secondary movements concerned.

The actions of the superior and inferior recti and obliques are more complex. Of these four muscles *each is inserted beyond the coronal equator, and the line of pull passes medial to the axis of rotation of the eyeball.* Thus the superior rectus pulls the pupil up and in, the inferior oblique pulls it up and out; combined they produce a vertical excursion of the pupil upwards, for their medial and lateral components cancel each other out. Similarly, the inferior rectus pulls the pupil down and in, the superior oblique pulls it down and out; combined they move the pupil vertically down. Pure upward and downward excursion of the pupil is produced by one rectus acting with its opposite oblique, and by no other means.

The obliquity of pull of the superior and inferior recti and obliques produces, too, a certain amount of wheel rotation of the eye (12 o'clock on the cornea rotates clockwise or anticlockwise). This results in obliquity of the false image when one of these muscles is paralysed (Fig. 6.51).

The **results of paralysis** of individual muscles can now be understood. Muscular imbalance alters the axis of the affected eyeball and diplopia results. The diplopia increases as the gaze is carried in the direction of action of the paralysed muscle. The **real image** falls on the macula of the unaffected eye. The **false image** is mediated from some peripheral part of the retina of the paralysed eye. The results of paralysis of individual eye muscles are shown in Fig. 6.51.

The **lacrimal gland** is a serous gland with a larger orbital and a smaller palpebral part (PLATE 34). The *orbital lobe* lies in a shallow fossa on the lateral part of the roof of the orbit, supported by the lateral horn of the expanded tendon of levator palpebræ superioris (PLATE 32). It extends around the posterior margin of this horn and turns forward as the *palpebral lobe*, visible

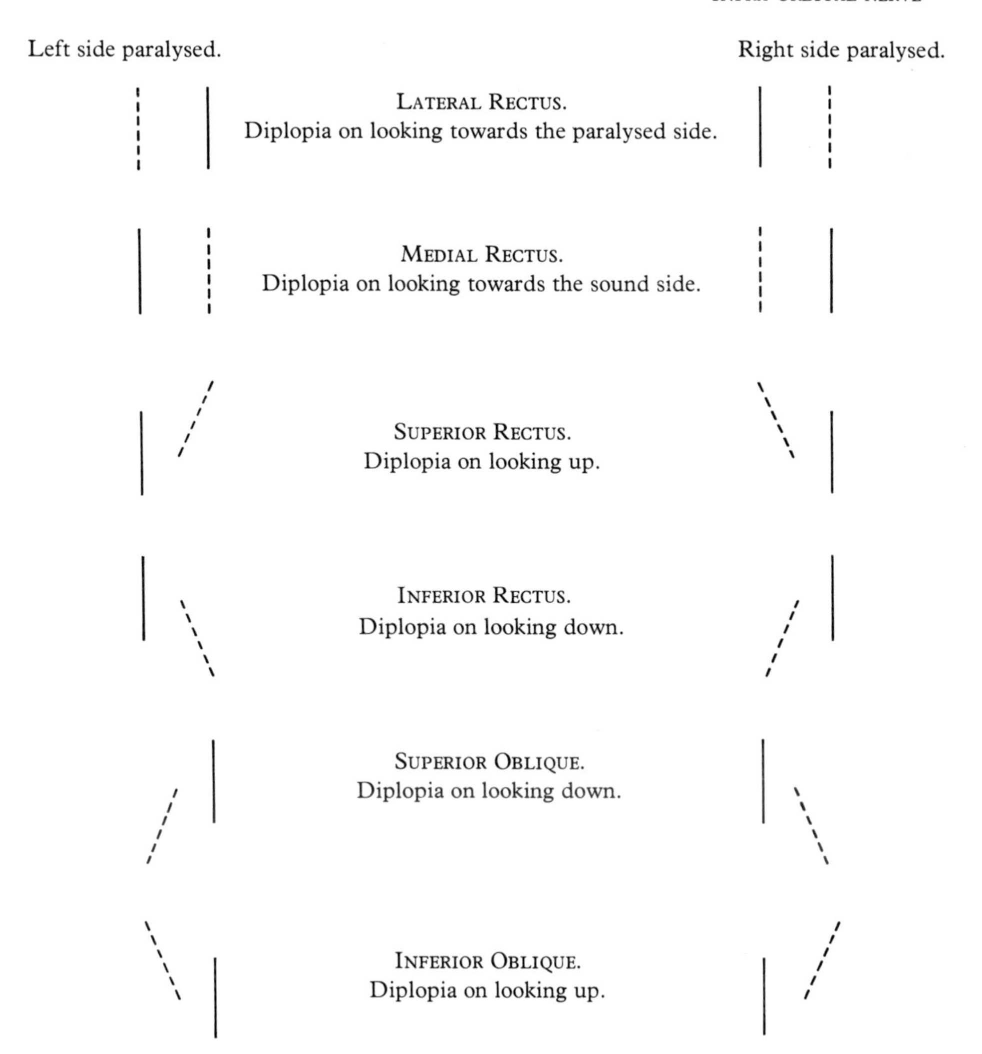

FIG. 6.51. THE RESULTS OF PARALYSIS OF THE EXTRINSIC MUSCLES OF THE EYEBALL. Real image solid, false image in dotted outline.

through the superior fornix of the conjunctiva. The dozen or more ducts lead from the palpebral part into the lateral extent of the superior fornix. Its secreto-motor fibres from the superior salivatory nucleus travel in the greater (superficial) petrosal nerve, and relay in the pterygo-palatine ganglion. The postganglionic fibres travel via the zygomatic and lacrimal nerves to the gland. Under normal conditions the lacrimal gland secretes just enough tears to replace those lost by evaporation. The lacrimal sac and naso-lacrimal duct are an emergency mechanism to pump away *excessive* tears; removal of the lacrimal sac does not produce a flow of tears down the cheek (epiphora) unless there is excessive secretion.

The Nerves of the Orbit

The infra-orbital and zygomatic branches of the maxillary nerve pass from the pterygo-palatine fossa through the inferior orbital fissure and run forward on the floor of the orbit. They are destined for the skin and, apart from some periosteal twigs, they give no branches to the orbital contents. They use the orbit only as a means of transit.

The **infra-orbital nerve,** accompanied by the infra-orbital artery, occupies the groove in the posterior part of the floor. The two enter the infra-orbital canal and proceed to the face, supplying the antrum and upper central teeth *en route* (p. 404). The **zygomatic nerve** passes along the lower side of the lateral wall and enters the zygomatic bone. Just before or just after this it divides into its two terminal divisions, the zygomatico-facial and zygomatico-temporal nerves, which supply the skin of the face. Along the lateral wall runs its communicating branch to the lacrimal nerve; this contains the secreto-motor fibres to the lacrimal gland.

The **optic nerve** comes from the chiasma (above

and medial) and the ophthalmic artery comes from the internal carotid artery at the roof of the cavernous sinus (below and lateral) and in this relationship the two pass through the optic foramen in the sphenoid bone.

The optic nerve, really an outdrawn cord of white matter of the brain, is surfaced by pia mater and lies loose within a tube of arachnoid and dura mater as far as the back of the eyeball. Its length in the orbit is 3 cm. It meets the sclera 3 mm medial to the posterior pole. From optic foramen to sclera the nerve curves laterally and downwards as it passes forwards. Across this obliquity the naso-ciliary nerve, ophthalmic artery and superior ophthalmic vein lie antero-posteriorly on top of the nerve. The ciliary ganglion lies to the lateral side of the nerve one-third of the way from optic foramen to eyeball.

The continuation of the subarachnoid space around the optic nerve up to the eyeball accounts for the appearance of papillœdema in cases of increased intracranial pressure.

Blood supply. The intracranial part of the optic nerve is supplied by the anterior cerebral artery. Here in the orbit the posterior 2 cm receives a branch at the optic foramen from the ophthalmic artery. The anterior 1 cm is supplied by the central artery. As elsewhere in the central nervous system, these perforating vessels are end arteries.

The **central artery** of the retina, the first branch of the ophthalmic artery, slants gradually through the meningeal coverings and enters the infero-medial part of the optic nerve half-way between optic foramen and eyeball. The anterior part of the optic nerve is surrounded by the short ciliary vessels and nerves.

The remaining nerves of the orbit are the ophthalmic (V), oculomotor (III), trochlear (IV) and abducent (VI) and they all enter via the superior orbital fissure. The ophthalmic and oculomotor have divided beforehand, so that seven nerves pass through the fissure. They all come from the dura mater on the lateral wall of the cavernous sinus, where they cross as indicated on p. 486. There is no logical reason to explain their particular disposition at the superior orbital fissure, and to aid his memory any student is justified in learning a mnemonic from a friend.

Nerves Outside the Cone of Muscles. Three nerves pass outside the fibrous ring and therefore remain outside the cone of the ocular muscles (see p. 432).

The **lacrimal nerve,** a terminal branch of the ophthalmic, is a slender filament which runs forward on the lateral wall of the orbit along the upper border of the lateral rectus muscle (Fig. 6.49). It picks up a branch from the zygomatic nerve (secreto-motor, p. 37) which it gives off to the lacrimal gland. It supplies periosteum and, piercing the septum orbitale, reaches the skin of the outer part of the upper lid. It is accompanied in the distal part of its course by the lacrimal branch of the ophthalmic artery. It supplies both surfaces of the conjunctiva in the upper fornix.

The **frontal nerve** is a large nerve which runs straight forward above the levator muscle in contact with the periosteum of the orbital roof (Fig. 6.49). It is a branch of the ophthalmic nerve. It supplies minute twigs to the periosteum and a branch to the mucous membrane of the *frontal sinus.* Just behind the orbital margin it divides into medial and lateral branches which pass to the forehead and redivide. These are the *supratrochlear* and *supra-orbital* nerves (PLATE 32, p. *xvii*). They are accompanied by corresponding branches of the ophthalmic artery.

The **trochlear (IV) nerve** has a simple course. Lying medial to the frontal nerve it passes forward across the origin of levator palpebræ superioris and sinks into the upper surface of the superior oblique muscle (Fig. 6.93, p. 486).

Nerves Inside the Cone of Muscles. Aided by his mnemonic the student will appreciate that the oculomotor nerve (III) enters the fibrous ring in two divisions, upper and lower, with the naso-ciliary nerve (Va) between them and the abducent nerve (VI) below all three. The optic nerve lies medial and superior to this group, with the ophthalmic artery intervening.

The **abducent (VI) nerve** has a simple course. It passes forward, diverging away from the optic nerve, and sinks into the ocular surface of the lateral rectus muscle, just below its middle (Fig. 6.49).

The **superior division** of the oculomotor nerve runs forwards above the optic nerve and supplies the overlying superior rectus and levator palpebræ muscles one-third of the way along the orbit (Fig. 6.21, p. 400). It carries sympathetic fibres from the cavernous plexus to the smooth-muscle part of the levator.

The **inferior division** of the oculomotor nerve is the larger of the two. It breaks up immediately into its three branches. The branch to the medial rectus passes below the optic nerve and sinks into the ocular surface of the muscle one-third of the way along the orbit. The branch to the inferior rectus runs along the upper surface of the muscle and enters it one third of the way along the orbit. The branch to the inferior oblique runs along the lateral edge of the inferior rectus and enters the ocular surface of the inferior oblique muscle. It gives the motor branch to the ciliary ganglion (Fig. 6.93, p. 486).

Summary of Motor Branches. Six long muscles are entered by their motor nerves one-third of the way along the orbit. The superior oblique is the only muscle supplied by the fourth nerve, and this is the only motor nerve outside the cone of muscles. The lateral rectus is the only muscle supplied by the sixth nerve. All the other muscles are supplied by the third nerve on their ocular surfaces. The superior division of III supplies levator palpebræ and the superior rectus, while the inferior division supplies medial rectus, inferior rectus and the inferior oblique.

The **naso-ciliary nerve** passes straight forwards alongside (lateral to) the ophthalmic artery, across the obliquity of the optic nerve, and divides into its terminal branches, the anterior ethmoidal and infratrochlear nerves. These leave the cone of muscle by passing below the superior oblique muscle (above the medial rectus). The **anterior ethmoidal nerve,** accompanied by a branch of the ophthalmic artery, enters the anterior ethmoidal foramen, lies under the roof of the labyrinth of the ethmoid, crosses the cribriform plate of the ethmoid and by the nasal slit enters the roof of the nose (PLATE 41). The **infratrochlear nerve,** accompanied by a branch of the ophthalmic artery, passes forward on the medial wall of the orbit just below the trochlea (Fig. 6.93, p. 486). It supplies nearby periosteum, the lacrimal sac, conjunctiva and, passing *above the medial palpebral ligament,* is distributed to skin of the upper lid and the bridge of the nose.

Within the cone of muscles the naso-ciliary nerve gives off several **branches.** The most distal of these is the **posterior ethmoidal nerve,** which, accompanied by a branch of the ophthalmic artery, leaves the cone of muscles by the same route (beneath the superior oblique muscle) and enters the posterior ethmoidal foramen for the supply of the posterior ethmoidal and sphenoidal air sinuses. The proximal branches of the naso-ciliary nerve remain within the cone of muscle.

The **sensory root of the ciliary ganglion** leaves the nerve at the fibrous ring and passes through the ciliary ganglion. The **long ciliary nerves,** two in number, pierce the posterior part of the sclera medial to the short ciliary nerves. These nerves carry sympathetic fibres, which the naso-ciliary nerve has picked up from the cavernous plexus. Their cell bodies are in the superior cervical ganglion; they are motor to the dilator pupillæ muscle. A few sensory fibres supply the cornea; their cell bodies are in the trigeminal ganglion.

The **ciliary ganglion** is a minute body (2 mm diameter) lying on the lateral side of the optic nerve just in front of the ophthalmic artery as that vessel spirals around the nerve (Fig. 6.93, p. 486). Three roots enter its posterior end (PLATE 29). The *sensory root* is a branch of the naso-ciliary nerve and passes through the ganglion without relay to supply cornea, iris and ciliary body. The *sympathetic root* is a branch from the cavernous plexus. It passes through the fibrous ring just below the naso-ciliary nerve, and passes through the ganglion without relay. It carries vasoconstrictor fibres to the vessels of the eyeball; its cell bodies are in the superior cervical ganglion. The *motor root* (parasympathetic) leaves the nerve to the inferior oblique and its fibres relay in the ganglion; their cell bodies are in the Edinger-Westphal nucleus. From the nerve cells in the ganglion the postganglionic fibres pass to the sphincter pupillæ and ciliary muscle (pupil constriction and accommodation).

The branches of the ciliary ganglion are the **short ciliary nerves,** a dozen or more in number. Each nerve contains fibres from all three roots of the ganglion (motor, sensory and sympathetic). The nerves pierce the back of the sclera around the attachment of the optic nerve and pass to their respective destinations within the eyeball.

The **ophthalmic artery** is a branch of the internal carotid given off as that vessel emerges from the roof of the cavernous sinus. It passes through the optic foramen, in which the optic nerve joins it from above. Both lie in a tubular prolongation of dura. In the orbit it pierces the dura and spirals around the lateral side of the optic nerve to pass forwards above the nerve, medial to the naso-ciliary nerve. Its branches accompany all the branches of the naso-ciliary, the frontal and the lacrimal nerves. Thus it supplies the ethmoidal air cells, part of the lateral wall of the nose, external nose, eyelids and forehead, in all of which places its branches anastomose with branches of the external carotid (maxillary, facial, superficial temporal) thus establishing connexions between internal and external carotid systems (p. 385).

The ophthalmic artery supplies all the muscles of the orbit from within the cone of muscle, supplies the lacrimal gland by branches from the artery accompanying the lacrimal nerve and supplies the eyeball. This last is by two sets of vessels. The **central artery** supplies the optic nerve and retina, while the **posterior ciliary arteries** pierce the sclera to enter the choroid coat of the eye. There is no anastomosis between the two sets of vessels. *The central artery is an end artery.* **Anterior ciliary arteries,** from the muscular branches to the recti, pierce the anterior part of the eyeball.

The Veins of the Orbit. Two veins drain the orbit, the superior and inferior ophthalmic veins. The **superior ophthalmic vein** commences above the medial palpebral ligament and passes back above the optic nerve with the ophthalmic artery. It receives tributaries that correspond in the main with the branches of the ophthalmic artery. It communicates at its commencement with the angular vein (PLATE 37). The **inferior ophthalmic vein** commences in front of the orbit (communicating over the inferior orbital margin with tributaries of the facial vein) and runs back within the cone of muscles. It communicates with the superior ophthalmic vein and drains through the inferior orbital fissure into the pterygoid plexus.

The **lymphatics** of the orbit drain through the eyelids and cheeks to the pre-auricular and parotid lymph nodes and so to the antero-superior group of deep cervical nodes (p. 443).

The **orbital fat** fills the spaces in the orbit. In particular it surrounds the optic nerve within the cone of muscles and makes a cushion to stabilize the eyeball against the backward pull of the recti. It remains very constant in volume during any changes in disposition of the general body fat and fluid.

THE EYEBALL

The eyeball contains the light-sensitive retina and, like a camera, it is provided with a lens system for focusing images (the refractive media) and with means of controlling the amount of light admitted (the iris diaphragm). Like a camera, its inside is black to prevent internal reflections.

The wall of the eyeball, enclosing the refractive media, is made up of three coats. The outer coat is fibrous and consists of the sclera and cornea; a vascular coat (the choroid, ciliary body and iris) intervenes between this and the innermost nervous coat (the retina). The sclera can be regarded as a cup-like expansion of the dural sheath of the optic nerve. The choroid, similarly, is an expansion of the arachnoid and pia, the retina being an expansion of the brain substance of the optic nerve.

The Fibrous Coat

The **sclera** is a dense feltwork of fibrous tissue comprising the posterior five-sixths of the eyeball. The interlacing fibres run mostly antero-posteriorly and coronally. The sclera is opaque (the 'white' of the eye). It is thinnest at the entrance of the optic nerve, whose perforating fibres give it a sieve-like appearance, the **lamina cribrosa.** If a sustained increase of intra-ocular pressure occurs (chronic glaucoma) the lamina cribrosa yields and bulges posteriorly ('cupping' of the disc).

The *sheath of dura mater* around the optic nerve blends with the sclera. The sclera receives the insertions of the ocular muscles. It is pierced obliquely by the ciliary nerves and arteries around the entrance of the optic nerve, and by the venæ vorticosæ (the choroid veins) just behind the coronal equator. The anterior ciliary arteries (from muscular branches to the recti) perforate the sclera near the corneo-scleral junction.

The sclera is almost avascular. The bulbar conjunctiva is attached to it by loose connective tissue (episclera) which is vascular; engorgement of its vessels produces a circumcorneal injection indicative of inflammation within the eyeball.

The **sinus venosus scleræ (canal of Schlemm)** circumscribes the corneo-scleral junction, at the periphery of the anterior chamber. It lies deeply and its endothelial lining is in contact with the mesothelium of the spaces of the ligamentum pectinatum. Thus aqueous humour drains into the canal. The canal is connected with anterior scleral veins, into which it drains. Normally the outflowing aqueous humour fills the canal, but if the veins are obstructed the canal fills with regurgitated blood (cf. the thoracic duct at its opening into the brachio-cephalic vein).

The **cornea** is one with the sclera, with the difference that its laminæ of fibrous tissue are transparent instead of opaque white. It bulges forward from the sclera, being the segment of a smaller sphere. It occupies the anterior one-sixth of the eyeball. It is completely avascular. The corium of the conjunctiva is attached to the sclera at the corneo-scleral junction, but the conjunctival epithelium is projected over the surface of the cornea. This is wear-and-tear epithelium (squamous stratified). Dryness of the epithelium sets off the blinking reflex, which re-lubricates the epithelium with tears (p. 431). The cornea is supplied mainly by the short ciliary nerves, via the sensory branch from the naso-ciliary nerve through the ciliary ganglion. The cell bodies are in the trigeminal ganglion. This main sensory supply is assisted by a few fibres that run in the long ciliary nerves.

Microscopic Anatomy. The squamous stratified epithelium on the anterior surface rests upon a thin, optically structureless membrane. This is **Bowman's membrane,** which is the homogeneous surface of the underlying substantia propria; it contains no elastic tissue. The **substantia propria** consists of lamellæ of dense fibrous tissue; in each lamella the fibres are parallel. The fibres of alternate lamellæ cross at right angles. **Corneal corpuscles** lie in the lamellæ; in section they are spindle-shaped, but viewed on the surface they are seen to be large, flat, stellate cells with long, branching, intercommunicating processes. The substantia propria is limited posteriorly by the **posterior elastic membrane (Descemet's membrane).** This is optically structureless, but is truly elastic. Peripherally in the angle of the anterior chamber it is continued as a spongework of fibres attached to sclera, ciliary muscle and front of iris (the **ligamentum pectinatum**). The posterior surface of Descemet's membrane is covered with a single layer of **pavement cells** (mesothelium) which is continued over the fibres of the ligamentum pectinatum and the anterior surface of the iris (Fig. 6.52).

The Vascular Coat (Uveal Tract)

The intermediate coat of the eyeball consists of a

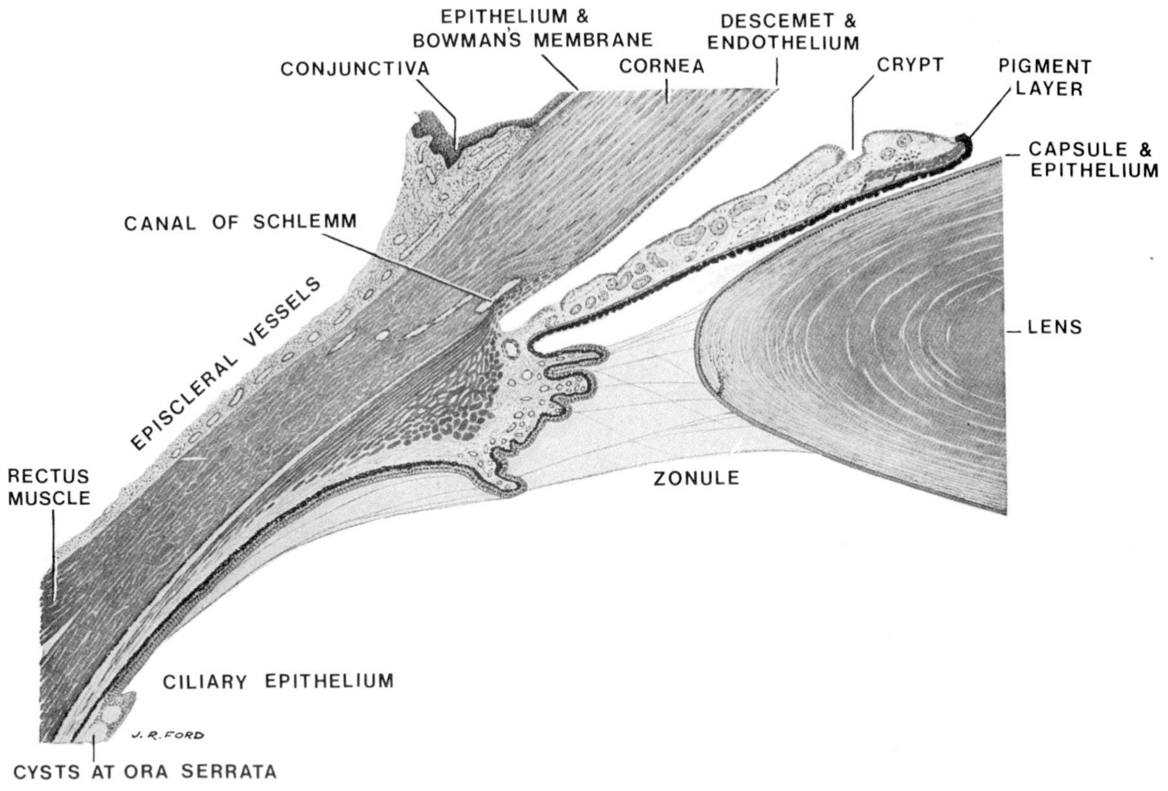

FIG. 6.52. SECTION THROUGH THE CORNEO-SCLERAL JUNCTION.

continuum of vascular tissue which is made up of the choroid, the ciliary body and iris.

The **choroid** is a thin, pigmented 'skin' that lines the posterior part of the sclera, from which it can readily be stripped up. It is brown in man, black in many animals. It ends anteriorly at the ora serrata, where it merges into the ciliary body. It is perforated by the optic nerve, to which it is firmly attached. It is separated from the sclera by the **suprachoroidal space,** across which the ciliary arteries, veins and nerves pass. A few loose connective-tissue fibres cross the space and hold the choroid in position. The retina is firmly united to the inner surface of the choroid.

Microscopic Anatomy. The choroid consists of masses of blood vessels sandwiched between two avascular membranes. The **suprachoroid lamina** consists of lamellæ of fibrous tissue between which are numerous pigmented cells. The **vascular layer** occupies the greatest thickness of the choroid. In the outer part are the larger veins and arteries, in the inner part lie the capillaries, the whole bound together by fibrous tissue containing pigment cells. The larger veins in the outer layer converge in spirals or whorls to four or five veins, called venæ vorticosæ, which pass symmetrically through the sclera just behind the equator. The **lamina vitrea** (membrane of Bruch) is a thin, structureless, transparent membrane which limits the choroid internally. The outer layer (pigmented epithelium) of the retina is attached very firmly to it.

The purpose of the rich capillary layer of the choroid is to nourish, by diffusion, the rods and cones of the retina. It is especially thick in the otherwise avascular macular region.

The **ciliary body** extends from the corneo-scleral junction backwards rather more than halfway to the equator (i.e., for about 6 mm). It lies as a flat ring applied to the inner surface of the sclera. The ring is thickest around its internal anterior circumference (at the corneo-scleral junction) whence it thins out to its external posterior circumference, here to blend at the ora serrata with the anterior margin of the choroid.

Being thicker in front and thinner behind, the ciliary body appears triangular in section. The two long sides of the triangle are in contact with the sclera externally (or anteriorly) and the vitreous internally (or posteriorly). The periphery of the iris is attached halfway along the short anterior side of the triangle. The scleral surface of the ciliary body contains the ciliary muscle. The vitreous surface of the ciliary body appears smooth and black peripherally, where it is continuous with the choroid at the ora serrata, but further forward this surface is projected into about seventy ridges which radiate for some 2 mm from the anterior margin. The ridges are the *ciliary processes*; they are white, but the grooves between them are black like the peripheral part of the vitreous surface. They lie in reciprocal grooves on the anterior surface of the vitreous. Their central ends are free and rounded.

The **ciliary muscle** consists of unstriped muscle; its function is to focus the lens for near vision. It has two sets of fibres, radial and circular. The *radial fibres* are attached to the periphery of Descemet's membrane and to a spur of sclera at the corneo-scleral junction. They radiate around the scleral surface of the ciliary body and are attached peripherally into the supra-choroid lamina. Their action is to draw this part of the choroid forward, relaxing the suspensory ligament of the lens and allowing the lens to bulge. The *circular fibres* lie within the anterior part of the radial fibres; their contraction diminishes the circumference of the attachment of the suspensory ligament, relaxing the latter and allowing the lens to bulge. Both parts of the ciliary muscle are supplied from the Edinger-Westphal part of the oculomotor (III) nucleus in the midbrain, by fibres which relay in the ciliary ganglion and enter the eyeball in the short ciliary nerves.

The muscle has no opponent—elastic recoil of the suprachoroid lamina tenses the suspensory ligament as the muscle relaxes, and this tension in the ligament flattens the lens.

Microscopic Anatomy. The scleral surface of the ciliary body is occupied by the radiating fibres of the ciliary muscle, which blend posteriorly with the supra-choroid lamina. Internal to the muscle is the vascular layer. In the periphery of the ciliary body (i.e., posteriorly) the ciliary processes consist of even more vascular tissue, for the most part veins. The lamina vitrea extends forwards from the choroid. Attached to its inner surface is a layer of highly pigmented cells continuous with those of the retina, and covered with a single layer of non-pigmented very adherent cells which here replace the nervous layer of the retina. These layers are called the **pars ciliaris retinæ.**

The **iris** is attached at its periphery to the middle of the anterior surface of the ciliary body; peripheral to this attachment the ciliary body itself and a narrow rim of sclera form the angle of the anterior chamber (Fig. 6.52). From its peripheral attachment the iris is pushed slightly forwards, in the form of a very low cone, by contact with the anterior convexity of the lens. The iris is perforated centrally by the pupil, the varying size of which controls the amount of light entering the eyeball. The size of the pupil is determined by the opposing contractions of a marginal sphincter and a radially disposed dilator muscle. Iris, ciliary body and choroid are continuous, and all three have a similar structure.

Microscopic Anatomy. The main bulk of the iris is made up of blood vessels and the loose connective tissue between them. The vascular pattern is derived from a major arterial circle behind the attached periphery of the iris. The two long ciliary arteries enter the back of the sclera and run forwards in the suprachoroidal space. In front of the ciliary body they join with the anterior ciliary arteries (from the muscular branches to the recti) to form the **major circle of the iris.** Vessels pass in the iris radially towards the pupil, at the margins of which they form the **minor circle of the iris.** The connective tissue contains a variable number of pigment cells, which determine the *colour of the iris.* If they are absent the iris is blue in colour by diffusion of light in front of the black posterior surface. As the pigment cells increase in number the iris colour becomes increasingly darker. The colour of the pigment, too, varies in different individuals.

The **sphincter pupillæ** is a well-developed circular band of smooth muscle lying posteriorly at the margin of the pupil. It is supplied, like the ciliary muscle, from the Edinger-Westphal part of the oculomotor (III) nucleus in the midbrain, by fibres that relay in the ciliary ganglion and pass into the eyeball in the short ciliary nerves. The **dilator pupillæ** is an ill-defined sheet of radial fibres (smooth muscle) extending over the back of the iris from the ciliary body to the pupil margin behind the sphincter. It is supplied by the cervical sympathetic. The connector cells lie in Th.1 segment of the spinal cord. They relay in the superior cervical ganglion, whence the postganglionic fibres pass in the carotid plexus and cavernous plexus, leaving the latter to travel with the naso-ciliary nerve, from which they branch off as the long ciliary nerves. Both sphincter and dilator muscles (but not the mesodermal ciliary muscle) are derived from neural ectoderm.

The front of the iris is covered with a single layer of flattened cells (mesothelium) continued from the back of the cornea. The back of the iris is covered by two layers of deeply pigmented cells which, continuous with those of the ciliary body (pars ciliaris retinæ) extend to the inner margin of the pupil.

The Nervous Coat

The **retina** is the delicate innermost membrane of the eyeball. It lies attached to the lamina vitrea of the choroid, and its inner surface is in contact with the vitreous. The light-sensitive area ends abruptly, halfway between equator and corneo-scleral junction, at a dentate line named the **ora serrata.** Forward of this a thin insensitive layer passes on in continuity as the pars ciliaris retinæ and the pigmented epithelium on the back of the iris. At the entrance of the optic nerve is a circular pale area, 1·5 mm in diameter; this is the **optic disc.** It overlies the lamina cribrosa of the sclera. The optic disc is excavated to a variable degree, producing the **physiological cup.** There are no rods or cones in the optic disc, hence it is insensitive to light—the **'blind spot'** (PLATE 43).

At the posterior pole of the eye (3 mm lateral to the optic disc) is a shallow depression of comparable size to the disc; it is completely free of blood vessels and is yellowish in colour. It is called the **macula lutea.** In the centre of the macula is a shallow pit, the **fovea centralis.** This is the thinnest part of the retina. It is the area of the most acute vision, for here the nerve cells of the retina are pushed aside and the rods disappear. Only cones lie in the fovea centralis.

The outer layer of the retina consists of a single layer of pigmented epithelial cells firmly attached to the lamina vitrea of the choroid. Next to this layer lie the light receptors (the rods and cones), forming a layer less firmly attached to the pigment cells, so that in detachment of the retina the pigment cells remain in position while the rods and cones, with the other layers of the retina, become displaced inwards from them.

The physiological arrangement of the nervous elements is very similar to that in any other sensory pathway (e.g., in the spinal cord). From the sense receptor, the first neurone has its cell body peripherally placed (in the retina this is the bipolar cell). It leads by synapse to the second neurone (in the retina this is the ganglion cell) whose axon passes to the thalamus (in this case the lateral geniculate body) whence, after relay, the third neurone leads through the internal capsule to the sensory cortex (Fig. 7.8, p. 503).

The light receptors are of two kinds, rods and cones. Only the **rods** contain visual purple; they are the more primitive type of light receptor. They do not register colour, but are sensitive to dim light (*scotopic vision*). The periphery of the retina contains rods only, the fovea centralis none at all. Eighty rods by their bipolar cells share one ganglion cell (i.e., one axon to the lateral geniculate body). They are of low threshold; eighty rods together 'whisper up the line' to the thalamus and cortex. The **cones** have a higher threshold (*photopic vision*) and they register colour. Cones alone occupy the fovea centralis. Beyond this they share equally with the rods, but they fall short of the periphery of the retina. Each cone is connected to a single ganglion cell—alone it 'shouts up the line' to the thalamus and cortex.

Microscopic Anatomy. A detailed account of the minute anatomy of the retina is beyond the scope of this book. For the student of general anatomy the following is a brief summary. Ten layers are described (Fig. 6.53).

1. The outermost is the layer of **pigment cells** already mentioned. Their processes project inwards between the rods and cones; the pigment is mobile in these processes.

2. The light receptors. The **rods** consist of two segments. The outer is slender and rod-like; it contains

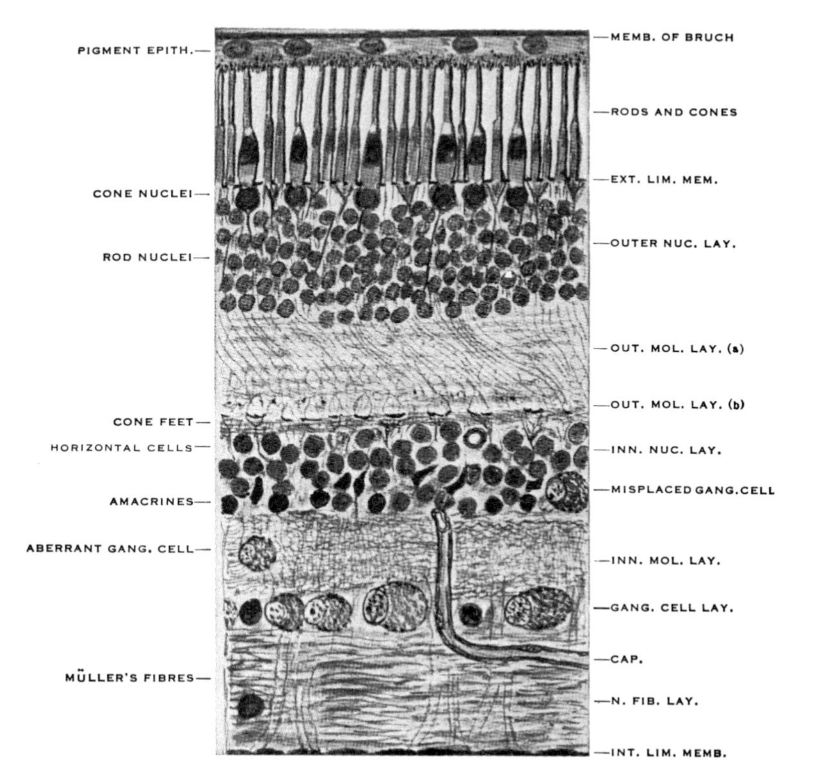

FIG. 6.53. THE LAYERS OF THE RETINA. Preparation by the late Dr Eugene Wolff.

the visual purple. The inner segment is slightly bulbous and transparent. The **cones** likewise consist of two segments, their shape varying in different parts of the retina. In general the outer segment is conical in shape; it contains no visual purple. The whole cone is squatter and more bulbous than the rod.

3. The **external limiting membrane** is a thin membrane perforated by the central processes of the rods and cones.

4. The **outer nuclear layer** contains the nuclei and central processes of the rods and cones. The central process of the rod has minute bulges (varicosities) upon it and it swells around its own nucleus, which may lie at any level in this layer. The central process of the cone is thicker and the nucleus, around which the process swells, lies against the external limiting membrane. These three outer layers of the retina are avascular, and nourished by diffusion from the capillaries of the choroid.

5. The **outer molecular layer** is acellular and consists of a zone in which the central processes of the rods and cones arborize with the dendrites of the bipolar cells.

6. The **inner nuclear layer** contains the cell bodies of the bipolar cells. It corresponds with the sensory ganglion of a spinal nerve, for the bipolar cells are the first neurone on the light pathway. This layer is nourished by capillaries of the retinal vessels; in obstruction of the central artery of the retina this and all the layers internal to it undergo avascular necrosis.

7. The **inner molecular layer** is acellular; it is the zone of arborization of the bipolar and ganglion cells.

8. The **ganglion cell layer** contains the large cell bodies of the second order neurones (comparable, for example, to the nuclei gracilis and cuneatus). Their axons lie in the next layer, whence they pass through the optic nerve, chiasma and tract to the lateral geniculate body. These cells form a single layer in the retina except around the macula lutea where they are several layers thick, having been 'pushed away' from the fovea, where there are none to obstruct the light passing to the cones.

9. The **nerve fibre layer** consists of the axons of the ganglion cells streaming towards the disc to enter the optic nerve. The retinal vessels lie for the most part in this layer, and the nerve fibres force a hæmorrhage here into the characteristic flame shape.

10. The **internal limiting membrane** is a thin hyaline membrane between retina and vitreous. Attached to it are the conjoined expanded bases of the *fibres of Muller*. These fibres penetrate the retina and bind its nervous elements together. Each fibre contains a nucleus in the bipolar cell (inner nuclear) layer. The fibres pass outwards to the outer limiting membrane.

Blood Supply of the Retina. The *central artery* passes through the lamina cribrosa within the optic nerve and in the optic disc divides into an upper and lower branch. Each gives off nasal and temporal branches. The upper and lower temporal branches curve up and down respectively to clear the macula lutea (PLATE 43). The branches of the central artery are end arteries. The central artery supplies the neurones (bipolar and ganglion cells) of the retina. The light receptors (rods and cones and their nuclei in the outer nuclear layer) are supplied by diffusion from the capillaries of the choroid. The *retinal veins* run with the branches of the central artery. The central vein leaves via the optic disc and emerges from the optic nerve and its coverings to join the superior ophthalmic vein. Perivascular lymph spaces exist in the retina, but it is doubtful if lymphatic vessels emerge from the eyeball.

Development of the Retina. The retina is developed from a hollow out-growth the *optic vesicle*, which protrudes from the cerebral vesicle. The optic vesicle becomes invaginated to form the *optic cup*, consisting of two layers of cells. The outer layer differentiates to form the pigment cell layer. The inner layer forms the remaining layers of the retina with the rods and cones outermost (next the pigment cells). The ganglion cells and their axons are innermost; light has therefore to pass through them to activate the receptors.

The Refracting Media

Most of the refraction of light takes place at the junction of air and corneal epithelium. Beyond the cornea light passes through the aqueous humour, the lens and vitreous to reach the retina.

The **vitreous body** is a colourless, jelly-like mass which occupies the posterior four-fifths of the eyeball. It is enclosed in the delicate homogeneous *hyaloid membrane*. It is indented in front by the posterior convexity of the lens and, beyond this, has radial furrows reciprocal with the ciliary processes. It is attached to the optic disc and just in front of the ora serrata; elsewhere it lies free, in contact with the retina.

The **lens** is a transparent bi-convex body enclosed in a transparent elastic *capsule*. It is 10 mm in diameter and its posterior surface, resting on the vitreous, is more highly convex than the anterior. The latter surface is in contact with the pupillary margin of the iris.

Microscopic Anatomy. The *lens capsule* is homogeneous. It is thickest anteriorly near the circumference. The *capsular epithelium* lies anteriorly, deep to the capsule. Centrally it is a single layer of cubical cells, but more peripherally the cells elongate to produce *fibres* which, by their accumulation, make up the lens substance.

The **suspensory ligament** is a series of delicate fibrils attached to the ciliary processes and the furrows between them and, further back, to the ora serrata.

The fibres pass centrally to attach themselves to the lens, mostly in front of, but a few behind, the circumference. In the rest position they hold the lens flattened under tension; when relaxed by contraction of the ciliary muscle, elasticity of the lens causes its anterior surface to bulge.

The **aqueous humour** lies between the back of the cornea and the front of the lens. The space is divided by the iris into anterior and posterior chambers, which communicate with each other through the pupil.

The **anterior chamber** is limited peripherally by the angle between the sclera (just beyond the corneoscleral junction) and the ciliary body (beyond the peripheral attachment of the iris). The angle contains the sponge-work of the fibres of the ligamentum pectinatum. The spaces of the sponge-work drain into the canal of Schlemm. Obliteration of the angle therefore prevents absorption of the aqueous, with consequent rise of intraocular tension. The anterior chamber is 3 mm deep centrally.

The **posterior chamber** is bounded in front by the iris and behind by the lens and its suspensory ligament. It is triangular in cross-section (pupil and lens in contact with each other forming the apex of the triangle). The base of this triangle is formed by the ciliary processes. Aqueous humour lies between the fibres of the suspensory ligament as far back as the anterior surface of the vitreous.

LYMPH DRAINAGE OF THE HEAD AND NECK

With the outstanding exception of the *tongue*, the lymph drainage of the head and neck tends to follow the general rule of superficial lymphatics accompanying veins and deep lymphatics accompanying arteries. All the lymph from the head and neck drains ultimately into the deep cervical lymph nodes, a chain of nodes surrounding the whole length of the internal jugular vein. Most of this lymph has already filtered through outlying nodes that are arranged in two 'circles', the 'inner circle' and the 'outer circle'. The outer circle is made up of superficial nodes from chin to occiput. The inner circle lies within it, surrounding the upper air and alimentary passages, comprising the pretracheal and paratracheal and retropharyngeal nodes. With few exceptions, lymph from a node in either of these circles drains into the nearest deep cervical lymph nodes. The internal jugular veins, surrounded by deep cervical nodes, lie vertically between the inner and outer circles.

Lymph from the deep cervical nodes is collected into the jugular lymph trunk. This joins the thoracic duct on the left side, but on the right side usually opens independently into the internal jugular or brachiocephalic vein.

Deep Cervical Lymph Nodes. These are found around the internal jugular vein from base of skull to root of neck. Though not arranged in demonstrable groups convenience of description requires that they be known as superior and inferior, anterior and posterior. Two nodes are important enough to know by name (Fig. 6.54). The **jugulo-digastric node** lies below

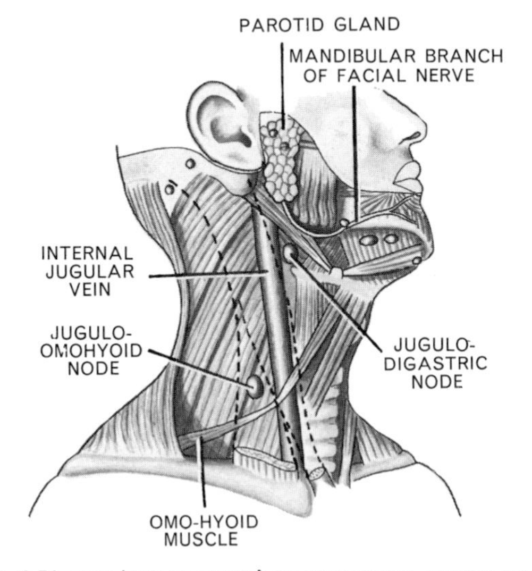

FIG. 6.54. THE 'OUTER CIRCLE' OF SUPERFICIAL LYMPH NODES AND THE TWO NAMED NODES OF THE DEEP CERVICAL CHAIN. A SMALL MANDIBULAR NODE LIES OVER THE MANDIBULAR BRANCH OF THE FACIAL NERVE. The sterno-mastoid muscle is indicated in dotted outline.

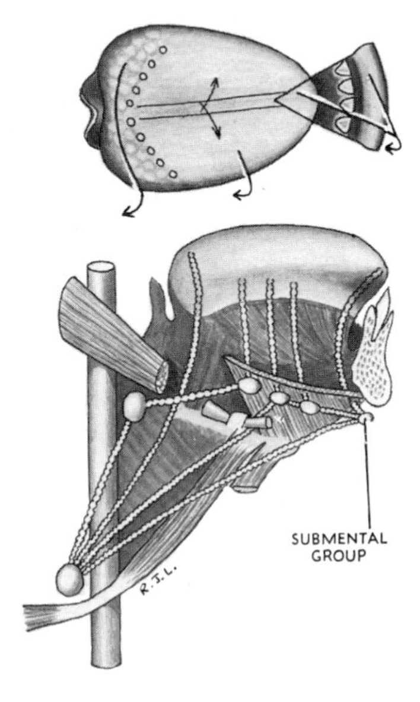

FIG. 6.55. SCHEME OF LYMPH DRAINAGE OF THE TONGUE. Cf. Fig. 6.54.

the posterior belly of the digastric, between the angle of the mandible and the anterior border of sterno-mastoid. It is a member of the antero-superior group. The **jugulo-omo-hyoid node** lies above the inferior belly of the omo-hyoid, behind the jugular vein just under cover of the posterior border of sterno-mastoid. It is a member of the postero-inferior group.

Other members of the postero-inferior group extend behind the border of sterno-mastoid into the posterior triangle; they are named the supraclavicular nodes.

The **'outer circle' of superficial nodes** (Fig. 6.54) extends from chin to occiput. It is made up of a number of named groups of superficial nodes.

Submental Nodes. Three or four small nodes lie just beneath the chin, within the thickness of the deep fascia or just superficial to this. They drain a wedge of tissue opposite the premaxilla of the upper jaw, namely tip of tongue, floor of mouth, lingual and labial gum opposite the incisor teeth and the same extent of the lower lip. This drainage is *bilateral* (Fig. 6.55). The submental nodes drain into the submandibular group, but a few efferents pass direct to the jugulo-omo-hyoid node (postero-inferior group).

Submandibular Nodes. Besides receiving lymph from the submental nodes, this group drains a wide area extending from the centre of the forehead, nose and nearby cheek, upper lip, and the anterior two-thirds of the tongue (excluding tip), floor of mouth and gums. They receive lymph from the upper teeth (via the infra-orbital foramen and face) and most of the lower teeth (via the mental foramen). They receive, too, the lymph from the anterior half of the nasal cavity and from the frontal and maxillary and the middle and anterior ethmoidal sinuses. The nodes lie beneath the deep fascia on the surface of the submandibular salivary gland. Most of them drain into the jugulo-omohyoid node; a few of the most posterior in the group drain into the jugulo-digastric node (Fig. 6.55).

Buccal and Mandibular Nodes. A small node often lies isolated on the buccinator muscle, another on the lower border of the mandible at the anterior border of the masseter. The latter node is near the mandibular branch of the facial nerve, which has been divided inadvertently in opening an abscess of this node. They drain part of the cheek and lower eyelid. Their efferents pass to the antero-superior group of deep cervical nodes.

Pre-auricular Nodes. These lie on and within the parotid gland; one or two are subcutaneous. They drain temple, vertex, eyelids and orbit and the external acoustic meatus. Some of the superficial members of this group extend down into the neck along the external jugular vein. The pre-auricular nodes drain into the upper group of the deep cervical lymph nodes.

Occipital Nodes. A few nodes lie at the apex of the posterior triangle and over the mastoid process in the subcutaneous tissue. They drain the posterior part of the scalp and auricle. Their efferents pass to the supraclavicular nodes (PLATE 33).

Lymph Drainage of the Face. The face drains into the three anterior groups of the outer circle in *three wedge-shaped blocks of tissue*. Centrally chin and tongue-tip drain into the *submental* nodes. A wedge of tissue above this drains into the *submandibular* nodes; the wedge extends from central forehead and frontal sinuses through the anterior half of the nose and maxillary sinus to the upper lip and lower part of the face, and includes the side of the tongue and floor of the mouth. Beyond the second wedge forehead, temple, orbital contents and cheek drain to the *pre-auricular* group.

The 'Inner Circle' of Lymph Nodes. Surrounding the larynx, trachea and pharynx is a collection of scattered nodes that extend from the pretracheal region around the retropharyngeal space.

The pretracheal nodes drain part of the lower larynx and trachea and thyroid isthmus, the retropharyngeal nodes (many on the *side* wall of the pharynx) drain the soft palate and posterior parts of hard palate and nose, as well as the pharynx itself. The nodes of the 'inner circle' drain to the nearest group of deep cervical nodes.

Lymph Drainage of the Tongue (Fig. 6.55). The lymph drainage of the tongue is irregular. It does not follow the inner and outer circle pattern of the rest of the head and neck, neither do the lymphatics follow arteries or veins.

The tip drains *bilaterally* to the submental nodes. The sides of the rest of the anterior two-thirds drain unilaterally through the floor of the mouth into the submandibular group; there is some overlap across the midline of the tongue. The posterior one-third of the tongue drains, *bilaterally*, into the jugulo-omo-hyoid node. Note that all lymph from the tongue ultimately reaches the jugulo-omo-hyoid node; from the tip by way of submental and submandibular nodes, from the sides by way of submandibular nodes, and from the posterior one-third direct.

THE TEMPORO-MANDIBULAR JOINT

The mandibular (temporo-mandibular) joint is a synovial joint between the condyle of the mandible and the under surface of the squamous part of the temporal bone. It is separated into upper and lower cavities by a fibrous disc within the joint. Both bone surfaces are covered with a layer of dense glistening fibrous tissue ('fibro-cartilage') identical with that of the disc. There is no hyaline cartilage in this joint. The articulating surfaces of both bones and of the disc are covered with synovial membrane in the newborn, but with use of the jaws this soon disappears and the membrane is then restricted to a narrow fringe lining the capsule.

The **capsule** is attached high up on the neck of the mandible around the articular margin of the condyle. It is more spacious above. It is attached anteriorly just in front of the transverse prominence of the eminentia articularis (Figs. 8.4, p. 559 and 8.5, p. 560), posteriorly to the squamo-tympanic fissure, and medially and laterally between these lines. It is lax, but strong.

The **disc** of the joint is attached around its periphery to the inside of the capsule. Anteriorly it is attached near the condyle, and this part of the disc moves forward with the condyle. Posteriorly it is attached nearer the temporal bone and therefore cannot move forward so freely; but the texture of this posterior part is less dense than elsewhere. The fibres are crinkled (like knitted socks) so that they can elongate and recoil a little. The disc seen in section has an undulating contour which moulds it to the incongruity of the bony surfaces; it is thinnest at the centre, thickest around its edges (PLATE 34).

The anterior margin of the disc and adjoining capsule receive the insertion of the upper fibres of the lateral pterygoid muscle.

The **temporo-mandibular ligament** is a stout band of fibrous tissue passing obliquely down and back from the lower border of the zygoma to the posterior border of the neck and ramus of the mandible. Its deep fibres blend with the capsule (PLATE 35). It tightens in retrusion of the head of the mandible. It tightens also in that protrusion of the head which accompanies opening of the jaw. The glenoid fossa is an ovoid concavity, and the ligament is most relaxed in the rest position of the mandible, but is tightened by movements away from the rest position. The head and neck of the mandible move in the glenoid fossa like a toggle switch.

The **spheno-mandibular ligament** is included as an accessory ligament of the joint because its lower attachment is at the axis of rotation of the mandible, and it remains constant in length and tension in all positions of the mandible (p. 391).

Stability of the Joint. The joint is much more stable with the teeth in occlusion than when the jaw is open.

In **occlusion** the teeth themselves stabilize the mandible on the maxilla and no strain is thrown on the joint when an upward blow is received on the mandible. In the occluded position apart from the stabilizing effect of the teeth, forward movement of the condyle is discouraged by the prominence of the eminentia articularis and by contraction of the posterior fibres of temporalis, while backward movement is prevented by the obliquity of the fibres of the temporo-mandibular ligament and by contraction of the lateral pterygoid muscle.

In the **open position** the joint is less stable. As well as rotating, the condyle lies forward on the slope of the eminentia articularis. Backward dislocation towards the tympanic plate is opposed by the obliquity of the fibres of the temporo-mandibular ligament and by contraction of the lateral pterygoid muscle. Forward dislocation is opposed by the slope of the articular eminence, by the tension of the temporo-mandibular ligament and by contraction of the masseter, temporalis and medial pterygoid muscles. In the open position there is little to prevent upward dislocation through the thin plate of bone roofing in the glenoid fossa; the injury is very rare because a blow on the drooping chin usually either closes or more fully opens the mouth of the recipient.

Forward dislocation is the commonest form of displacement. Reduction is prevented by spasm of the posterior deep fibres of the masseter (p. 387). These fibres by their spasm hold the *dislocated* jaw open, because the condyle is so far forward that they pass down *behind* its axis of rotation. The spasm must be overcome (with or without an anæsthetic) before the anterior dislocation can be reduced. Anterior dislocation readily occurs in the edentulous. It is easily reduced; the joint is less stable because the increased elevation of the edentulous mandible permanently elongates the temporo-mandibular ligament.

Movements of the Joint. In the rest position there is 3 or 4 mm separation of the teeth. Movements at the joint are made understandable when analysed into their component parts. These are three. The condyle is able to protract-retract, hinge open and shut, and move from side to side. Move your own!

1. *Protraction-retraction.* This movement occurs mainly in the upper compartment of the joint; condyle and disc move together. If one condyle protrudes more than the other the chin is slewed to the *opposite* side.

2. *Opening-closing.* Opening is either passive or active. *Do not confuse the two.*

(a) **Passive** (by gravity, as in sleep or anæsthesia). There is a hinge movement in the lower compartment of the joint, with its axis through the condyles. This swings the tongue back, narrows the oro-pharynx, and obstructs the airway. The individual may snore (like a student at a lecture) or choke.

(b) **Active** (by muscle action, as in chewing, talking). Here the hinge glides. *This is a compound, not an elemental, joint movement.* The resultant of the combined movements (hinge in the lower compartment, glide in the upper compartment) is of mandibular rotation around an axis passing through the two mandibular foramina (see below). The tongue does not swing back and there is no obstruction to the airway.

3. *Side-to-side movements*, with no protraction-retraction and no opening-closing. This movement occurs in the upper compartment; condyle and disc move together. It is rarely performed. The articular surfaces of the two condyles lie on the arc of a circle whose circumference intersects the foramen magnum (confirm this on a skull). Lateral movements thus produce a turning of the mandible around the centre of this circle, which lies in front of the chin. The chin and condyle move towards the *same* side.

The *muscles* moving the mandible (see also below) are as follows: Protraction is produced by the lateral pterygoid, retraction by the posterior fibres of temporalis. As the disc is pulled forward by the lateral pterygoid it flattens somewhat, and its posterior crimped fibres are pulled straight. Their recoil during retraction replaces the disc; there is no muscle to oppose the pull of lateral pterygoid *on the disc*. Lateral excursions are produced by unilateral contraction of lateral and of medial pterygoid muscles. Opening is produced by the lateral pterygoids and digastric muscles, closing by the masseters, medial pterygoids and temporal muscles.

The *nerve supply* of the mandibular joint is derived from the auriculo-temporal nerve, assisted by the nerve to the masseter (Hilton's Law).

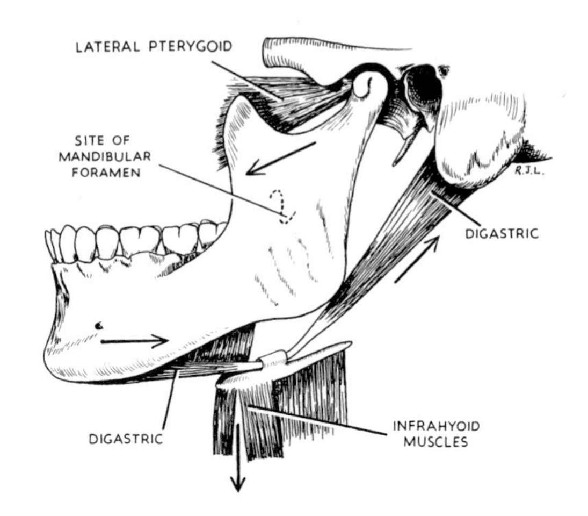

FIG. 6.56. OPENING THE MOUTH. THE MANDIBLE IS ROTATED BY THE DIGASTRIC AND LATERAL PTERYGOID MUSCLES. This illustrates *active* rotation; the axis traverses the mandibular foramina. The infrahyoid muscles tether the pulley for the digastric tendon.

Movements of the Mandible and the Floor of the Mouth

Some confusion exists concerning the identity of the muscles responsible for opening the mouth. The matter can be viewed in the best perspective by considering the rôle of all the muscles of the head and neck.

The skull is mobile on the neck in all positions of the cervical vertebræ. The mandible is mobile on the skull in all positions of the latter. The floor of the mouth is mobile on the mandible in all positions of the latter. There is *a separate set of muscles for each of these movements*, and confusion has arisen from failure to place certain muscles in their correct group.

Movements of the Head

The head, in the erect position, has its centre of gravity in front of the occipital condyles; unsupported it nods forward. It is held erect by the tonus of the postvertebral extensor muscles. Extension of the head and neck is produced by this powerful group. Flexion is produced by the prevertebral muscles and the sternomastoids. Any of these muscles acting unilaterally abducts and rotates the head and neck.

Movements of the Mandible

Active opening is a simple movement. The mandible, slung in muscle, rotates about an axis that passes through the mandibular formina. The condyles are pulled forwards by the lateral pterygoids and the chin is pulled down and back by the digastric muscles. Viewed from the side, the mandible rotates as does a ship's steering wheel when two opposite spokes are pulled upon in different directions. A ship's steering wheel is stabilized upon a central axle. The mandible is stabilized by being held in the grasp of the opposing muscles as they 'pay out rope' and by its condyles abutting against the temporal bones. Movements between the condyle and the temporal bone must of necessity be a combination of gliding and hingeing, and the disc provides a means whereby each movement may take place in a separate compartment of the joint.

The digastric muscle acts as a whole; its anterior belly could produce no mandibular movement were the posterior belly not to contract with equal force. In many mammals the muscle, though in two bellies, has no connexion with the hyoid bone; it passes forward (the occipito-mandibularis muscle) from behind the ear to the lower border of the mandible, on which it acts as a simple depressor. Its attachment to the hyoid bone in man and other mammals gives the mandibular attachment of the muscle a more efficient line of pull. The angulation so produced between the two bellies of the digastric gives the contracting muscle a secondary action, namely, elevation of the hyoid bone. This unwanted action is prevented by synergic contraction of the infrahyoid strap muscles. The muscle does not act as a prime mover to elevate the hyoid bone. Gravity assists the opening of the mouth to a very minor extent; in the ordinary rapid movements of chewing and talking it plays no part.

In closing the mouth the lateral pterygoids elongate

and 'pay out rope' against the posterior fibres of temporalis as these retract the upper part of the mandible. Similarly, the digastric 'pays out rope' against the remainder of temporalis, the masseters and the medial pterygoids as these muscles elevate the body of the mandible.

In chewing food the mouth does not close like a rat-trap, but the mandible approaches the maxilla with a slewing movement. This grinding between the molars is produced by some antero-posterior as well as lateral excursion of the mandible. Antero-posterior movement of one half of the mandible is produced by alternate actions of lateral pterygoid and posterior (i.e., *horizontal*) fibres of temporalis of the same side, alternately protruding and retruding that side of the mandible.

Lateral excursion of the closing (chewing) mandible is produced mainly by the medial pterygoid muscle on the chewing side, pulling the angle of the mandible upwards, forwards and *medially*. Picture this action by looking at Fig. 6.31, p. 414.

Movements of the Floor of the Mouth

The floor of the mouth is the mylo-hyoid muscle, whose midline raphé and posterior fibres are attached to the body of the hyoid bone. It is its own elevator; contraction of the muscle flattens out the angle between its two halves, and hyoid bone and raphé are drawn upwards.

Genio-hyoid and stylo-hyoid determine by their relative lengths the antero-posterior position of the hyoid bone. They lengthen or shorten the floor of the mouth, the mylo-hyoid raphé being passively lengthened or shortened, like a concertina. Both muscles slope down to the hyoid bone; hence their opposing contractions elevate the floor of the mouth.

The elevators of the floor (mylo-hyoid, genio-hyoid and stylo-hyoid) are opposed by the infra-hyoid strap muscles. The resultant of the opposing contractions of these muscles determines the vertical and antero-posterior position of the hyoid bone (and therefore the state of the floor of the mouth and the position of the larynx). In ordinary movements these muscles play no part in depressing the mandible. In very forced movements of depression they may act as accessory muscles to assist the digastric. In very forced depression even the platysma may join in the tug of war on the mandible —watch an exhausted runner nearing the finishing post.

Swallowing. The first or voluntary stage of swallowing consists in passing the bolus through the oro-pharyngeal isthmus, and is discussed with tongue movements (p. 411). The back of the tongue is *pushed back* by compression of the contracting mylo-hyoid on the closed mouth cavity, and it is *pulled back* and up by stylo-glossus after the bolus has passed (Fig. 6.46, p. 427).

The second or involuntary stage commences as the bolus enters the oro-pharynx, but the pharynx is *already elevated* before the bolus arrives.

The *larynx* is suspended from the hyoid bone by membrane and muscle and elevates with the floor of the mouth. In addition it possesses its own elevators. The medial muscle of the styloid apparatus (stylo-pharyngeus), assisted by salpingo-pharyngeus and palato-pharyngeus, draws up the larynx and pharynx. This elevation *precedes* elevation of the hyoid bone. Elevation of the hyoid bone in the first stage of swallowing is accompanied by further elevation of the larynx and pharynx. The laryngeal inlet, elevated beneath the protecting shelf of the overhanging epiglottis, is closed by contraction of the ary-epiglottic muscles, and respiration is suspended. The epiglottis itself is inverted by the passing bolus, and closes like a lid over the larynx. It recovers its normal position by its own elastic recoil; its framework consists of elastic cartilage (Fig. 6.46, p. 427).

The sequence of swallowing for the bolus is obviously first stage in the mouth, second stage in the pharynx. But the sequence of muscle contraction is not the same. The sequence is (1) larynx and pharynx move up to the hyoid bone (Fig. 6.46, 2), *then* (2) larynx and pharynx and hyoid move up together, *then* (3) larynx and pharynx and hyoid move down together, *then* (4) larynx and pharynx move down from the hyoid bone.

The *soft palate* is elevated against the projecting shelf of Passavant's ridge, sealing off the naso-pharynx. This opens the auditory tube.

With nose and larynx thus sealed off the bolus has nowhere to go but down the pharynx. The bolus is much hastened in its descent by the elevation of the pharynx to receive it, like a snake darting upwards open-mouthed for its prey, and by a swift-moving peristaltic wave of contraction that passes behind it down the pharyngeal constrictors and striped-muscle part of the œsophagus. Beyond this level it is safe from regurgitation, and the laryngeal inlet opens to resume respiration. The bolus proceeds on its leisurely way propelled by the slow worm-like contractions of smooth muscle.

THE EAR

The **external ear (auricle)** has a skeleton of resilient yellow elastic cartilage which is thrown into folds. The folds give the auricle its characteristic shape. The cartilage is covered on both surfaces with adherent hairy skin; it does not extend into the lobule of the ear. The lobule is a tag of skin containing soft fibro-fatty tissue; it is easily pierced for ear-rings. The cartilage of the auricle is prolonged inwards in tubular fashion as the cartilaginous part of the external acoustic meatus, whose attachment to bone stabilizes the auricle in position. *Intrinsic muscles* in the auricle act as sphincter and dilator mechanisms of the external meatus in lower animals, but they are vestigial and functionless in man. *Extrinsic muscles* move the auricle in lower animals, but in man are without function. They lie anterior, superior and posterior to the auricle and all are supplied by the facial (VII) nerve. The blood supply of the auricle is mainly from the posterior auricular artery, assisted by the superficial temporal. Lymph drainage is to the pre-auricular, mastoid (i.e., occipital) and external jugular nodes.

The skin of the auricle is supplied by the great auricular and the auriculo-temporal nerves. The *great auricular nerve* supplies the whole of the cranial surface and the lateral surface below the meatus with fibres from C2. The *auriculo-temporal nerve* supplies the outer surface of the tympanic membrane, the external acoustic meatus and the skin of the auricle above this level. The auricular branch of the vagus supplies the postero-inferior quadrant of the tympanic membrane and skin of the adjoining meatus and a small area of skin of the cranial surface near the mastoid. The facial nerve by branches from the tympanic plexus also supplies the cutaneous surface of the tympanic membrane and external meatus, and these areas show vesicles in cases of facial herpes. The lesser occipital nerve (C2) overlaps the great auricular nerve at the upper margin of the cranial surface.

The **external acoustic meatus** is a sinuous tube nearly 3 cm in length; it is straightened for introduction of an otoscope by pulling the auricle upwards and backwards. Due to the obliquity of the drum its antero-inferior wall is longest and its postero-superior wall shortest. Its outer third is cartilage, its inner two-thirds bone; in both zones the skin is firmly adherent.

The bony part is formed by the tympanic plate, C-shaped in cross-section, the gap in the C being applied to the under surface of the squamous and petrous bones. The cartilaginous portion is likewise C-shaped; the gap is filled with fibrous tissue attached to the periosteum of the squamous bone. Hairs and sebaceous glands abound in the cartilaginous part. Here also are the *ceruminous glands*, long coiled tubules like modified sweat glands, which secrete a yellowish-brown wax. The meatus is narrowest a few millimetres from the membrane; this portion is known as the isthmus.

The blood supply is derived from the posterior auricular and superficial temporal arteries and, nearer the membrane, by the deep auricular branch of the maxillary artery, which enters the meatus through the squamo-tympanic fissure. Lymph drainage is to the pre-auricular, mastoid and external jugular nodes. The nerve supply is from the auriculo-temporal nerve overlapped by the facial nerve and the auricular branch of the vagus.

The **tympanic membrane** is a thin fibrous membrane covered externally with a thin layer of squamous stratified epithelium and internally with low columnar epithelium. The fibres are a lateral radial set and a medial circumferential set. The membrane is circular, 1 cm in diameter, and lies obliquely at 55 degrees with the external acoustic meatus, facing downwards and forwards as well as laterally. It is concave towards the meatus. At the depth of the concavity is a small depression, the *umbo*, produced by the handle of the malleus. When the drum is illuminated for inspection, the concavity of the membrane produces a 'cone of light' radiating from the umbo over the antero-inferior quadrant. The handle of the malleus is firmly attached to the inner surface of the membrane. From the lateral process of the malleus two thickened fibrous folds (malleolar folds) diverge up to the margins of the tympanic plate; between them the small upper segment of the membrane is lax (membrana flaccida, Shrapnell's membrane). This part is crossed internally by the chorda tympani.

The tympanic membrane is thickened at its circumference and slotted into a groove in the tympanic plate at the site of the tympanic ring of the fœtus (p. 48). It is held tense by the inward pull of the tensor tympani muscle. Its tension is affected by difference of pressure in the tympanic cavity and external meatus in cases of Eustachian tube obstruction.

The blood supply is derived from the deep auricular branch of the maxillary on the outer side and from the stylo-mastoid artery (posterior auricular) on the inner side. The tympanic branch of the maxillary assists the latter vessel to form a circular anastomosis around the margins of the membrane.

The cuticular epithelium is supplied by the auriculo-temporal (with overlap from the trespassing facial and vagus nerves), the inner mucosal surface by the tympanic branch of the glosso-pharyngeal (IX) nerve.

The Middle Ear (Tympanic Cavity)

The middle ear is an air-containing cavity in the petrous bone. It is floored in by the scroll-like tympanic plate, which is C-shaped in cross-section. The margins of the rolled plate form the tympanic notch. The purpose of the middle ear is to transform air-borne vibrations from the tympanic membrane to liquid-borne vibrations in the internal ear, to which end it contains three ossicles. It is the intermediate portion of a blind diverticulum that leads out from the respiratory mucous membrane of the naso-pharynx. The diverticulum consists of the Eustachian tube, middle ear, and mastoid antrum and air cells. The diverticulum is formed by interposition of part of the second pharyngeal pouch into the first pharyngeal pouch and thus has a double nerve supply (p. 41).

The auditory tube is supplied by the pharyngeal branch of the pterygo-palatine ganglion (p. 398) at its ostium and by the meningeal branch of Vc along its cartilaginous part, and the mastoid antrum and air cells by the meningeal branch of Vc (p. 391). They are developed from the first pharyngeal pouch. The middle ear interrupts their continuity, bringing endoderm of the second pouch. Its mucosa is supplied from the tympanic plexus, mostly by the tympanic branch of the glosso-pharyngeal, assisted partly by the facial nerve.

The **tympanic cavity** is the shape of a biconcave lens, tilted to the plane of the tympanic membrane. It is about 15 mm in diameter. The *lateral wall* is largely occupied by the tympanic membrane, which extends upwards for 10 mm from the floor. Above the membrane the temporal bone is hollowed out into the *epitympanic recess.* The tympanic membrane bulges into the cavity to within a millimetre or two of the promontory. The *medial wall* is the bony lateral wall of the internal ear. Opposite the membrane is a convexity named the *promontory,* covered with fine grooves indented by the tympanic plexus. It is formed by the prominence of the basal turn of the cochlea. Behind and above the promontory is an oval window, the *fenestra vestibuli,* closed in life by the foot-piece of the stapes. Inside this window is the perilymph of the vestibule. At the bottom of the conical depression behind the promontory is a round window, the *fenestra cochleæ,* closed in life by a fibrous membrane known as the *secondary tympanic membrane.* Inside this window is the perilymph of the blind end of the cochlea (scala tympani). A ridge runs horizontally above the promontory and fenestra vestibuli; it overlies the *canal for the facial nerve.* The *processus cochleariformis* is a funnel-shaped projection from the posterior end of the canal for the tensor tympani. It faces laterally towards the tympanic membrane and lies just below the anterior end of the canal for the facial nerve (Fig. 6.58).

The *roof* of the tympanum is the tegmen tympani, a laminar projection of petrous bone that roofs in also the canal for the tensor tympani and the tympanic antrum. Above it the temporal lobe lies in the middle cranial fossa.

The *floor* is a thin plate of bone separating the cavity from the jugular fossa and the carotid canal. Between these two the tympanic branch of IX (Jacobson's nerve) enters.

The *anterior wall* is shortened by approximation of roof and floor. It is perforated by the openings of the two canals. The lower and larger of these is the bony part of the auditory tube, the upper and smaller is the canal for the tensor tympani muscle. The latter canal extends somewhat backwards before opening on the medial wall as the processus cochleariformis (Fig. 6.62, p. 455). The lower part of this wall is perforated by carotico-tympanic branches of the internal carotid artery. Sympathetic fibres from the carotid plexus run with them to join the tympanic plexus.

The *posterior wall* is deficient above, where there is an aperture, the *aditus,* which leads back into the tympanic antrum. The ridge of the canal for the facial nerve passes back along the medial wall of the aditus and above this ridge is the convex bulging of the lateral semicircular canal. Below the aditus is a shallow con-

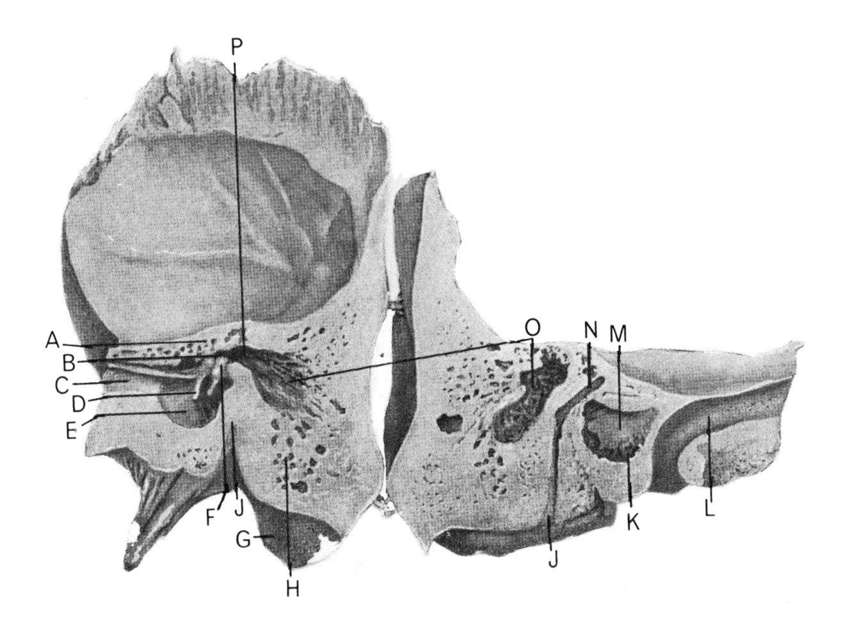

FIG. 6.57. THE RIGHT TEMPORAL BONE IN OBLIQUE SECTION ALONG THE AXIS OF THE PETROUS BONE, WITH THE MEDIAL PART TURNED BACK. The slight inequality of the two cut surfaces results from loss of bone in the saw cut. A. Tegmen tympani. B. Epitympanic recess. C. Eustachian tube. D. Handle of malleus. E. Tympanic membrane. F. Incus. G. Mastoid process. H. Mastoid air cells. J. Stylo-mastoid foramen. K. Medial wall of tympanic cavity. L. Carotid canal. M. Promontory. N. Canal for facial nerve. O. Antrum. P. Aditus.

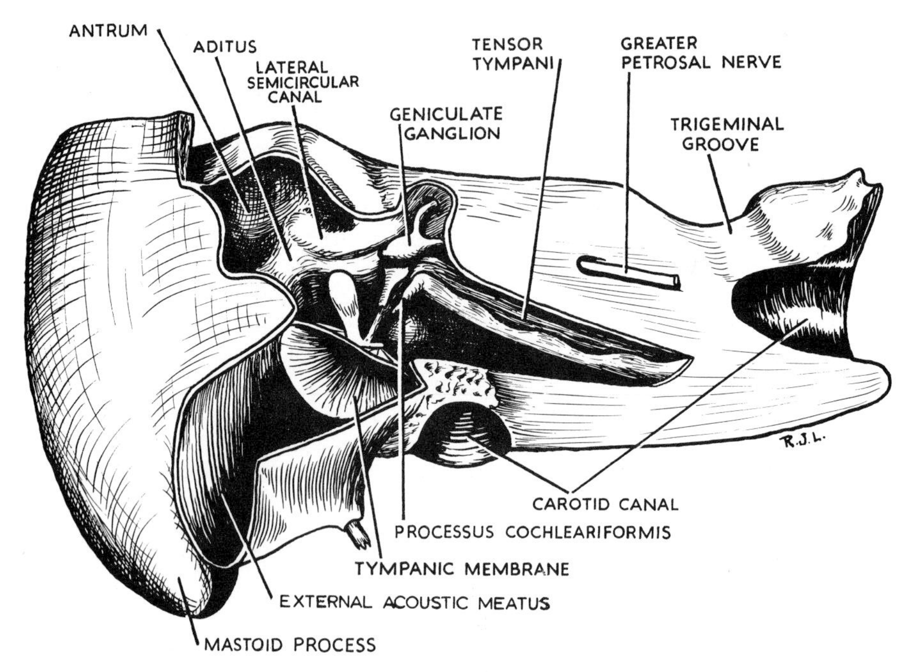

FIG. 6.58. THE RIGHT MIDDLE EAR AFTER REMOVAL OF THE TEGMEN TYMPANI. Viewed from the right and from slightly above. Drawn from a dissection.

cavity for the short process of the incus, and lower still there projects a hollow cone, the *pyramid*, whose apex is perforated by the tendon of stapedius. Close to the posterior margin of the tympanic membrane is a tiny canal in the posterior wall; this is the posterior canal for the chorda tympani.

The **ossicles** form by synovial joints a bony chain for transmission of vibrations from the tympanic membrane to the internal ear.

The **malleus** is shaped like a round-headed club. There is a constriction, the neck, between head and handle (Fig. 6.58). The convex *head* lies in the epitympanic recess. Its posterior surface has an articular facet for the incus. The narrow *neck* lies against the pars flaccida of the tympanic membrane. The chorda tympani crosses medial to the neck. The *handle* projects somewhat backwards down to the umbo: its upper end has a projection, the *lateral process*. The two form a lateral concavity moulded to the medial convexity of the tympanic membrane; the periosteum of lateral process and handle is firmly fixed to the fibrous layer of the membrane. The malleolar folds are attached to the apex of the lateral process. The *anterior process* is slender; it is embedded in the fibres of the anterior ligament.

The **incus** has a relatively large body and two slender processes (Fig. 6.59). The *body* is rounded and laterally compressed. It lies in the epitympanic recess and articulates anteriorly with the head of the malleus. The *short process* projects backwards to lie in a shallow fossa in the posterior wall just below the aditus. The *long process* projects down into the cavity of the middle ear, just behind and parallel with the handle of the malleus. Its tip hooks medially and is bulbous—the lentiform nodule—for articulation with the stapes.

The **stapes** has a small *head* showing a concave facet for articulation with the lentiform nodule. A narrower *neck* diverges into slender anterior and posterior *limbs*, the former shorter and less curved than the latter. The limbs are attached to the *base* (or footpiece); a delicate obturator membrane, clothed with mucosa, fills the space between them. The base is oval and fits into the fenestra vestibuli; the two are united by circumferential ligamentous fibres of elastic tissue.

The **joints of the ossicles** are synovial; the joint surfaces are coated with hyaline cartilage. The simple capsular ligaments contain an overwhelming preponderance of elastic fibres. Accessory ligaments stabilize the mobile ossicles. They too are overwhelmingly elastic in constitution. The malleus is attached by a superior

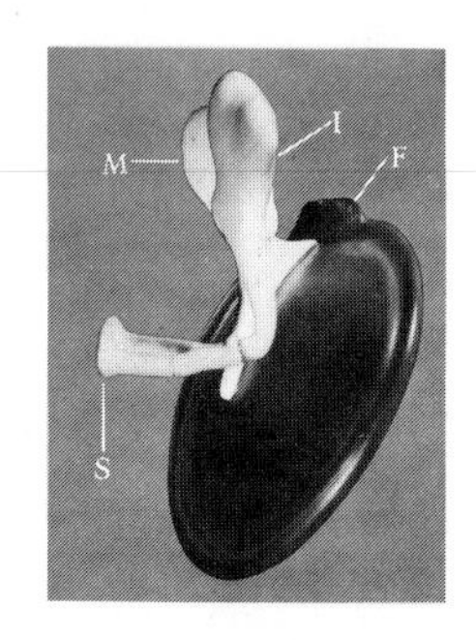

FIG. 6.59. POSTERIOR VIEW OF THE TYMPANIC MEMBRANE AND OSSICLES OF THE RIGHT SIDE. M. Head of malleus. I. Short process of incus. S. Footpiece of stapes. F. Pars flaccida. Photograph of Specimen S 231 in R.C.S. Museum, a scale model made by Dr D. H. Tompsett.

ligament passing from the head to the roof of the epitympanic recess. An anterior ligament passes from the anterior process through the squamo-tympanic fissure to the spine of the sphenoid (whence it enlarges downwards as the spheno-mandibular ligament). Another band passes from the anterior process to the anterior margin of the tympanic notch; this forms with a posterior ligament passing from the head to the posterior margin of the notch an 'axis-ligament' around which the malleus rotates.

The cartilage-covered tip of the short process of the incus is held by a ligament to its fossa below the aditus. The foot-piece of the stapes is united by elastic fibres around its circumference to the margins of the fenestra vestibuli; both bone margins are covered with hyaline cartilage.

The muscles of the ossicles are two, one for the malleus, developed with it from the first pharyngeal arch, and one for the stapes, developed with it from the second pharyngeal arch. They are thus innervated by mandibular and facial nerves respectively. Their function is to damp down over-vibration from low-pitched sound waves.

The **tensor tympani** arises from both cartilaginous and bony walls of the Eustachian tube. The slender muscle ends in a round tendon which bends around the processus cochleariformis, passes across the cavity of the middle ear and is inserted into the upper end of the handle of the malleus (Fig. 6.58). Its nerve of supply from the mandibular nerve runs in the nerve to the medial pterygoid, from which it passes near the otic ganglion to reach the muscle. The cell bodies lie in the motor nucleus of V (in the pons); there is no relay in the otic ganglion. Contraction of the muscle draws the handle of the malleus inwards, making the drum more highly concave and therefore more tense. It has no opponent—elastic recoil restores the *status quo* as the muscle relaxes.

The **stapedius** arises from the interior of the hollow pyramid. Its tendon emerges from the apex of the pyramid and is inserted into the back of the neck of the stapes. The muscle is supplied from the facial nerve by a branch given off in the facial (stylo-mastoid) canal. Its action is to retract the neck of the stapes, thus tilting the foot-piece in the fenestra vestibuli; the anterior margin of the foot-piece cants towards the tympanic cavity. It has no opponent; elastic recoil replaces the foot-piece.

Development of the Ossicles. The malleus and incus are developed from the proximal end of the first arch cartilage (Fig. 1.49, p. 40), the stapes from that of the second arch (Fig. 1.50, p. 41). Formed in hyaline cartilage, they all commence to ossify at the twelfth week, and at this stage the ossicles are already almost *full adult size.*

Nerves of the Middle Ear. The middle ear is supplied by branches of the **tympanic plexus.** The plexus lies on the promontory. It receives contributions from three sources:

1. The tympanic branch of the glosso-pharyngeal nerve (Jacobson's nerve) passes through the petrous bone between jugular and carotid foramina, and runs up to the promontory. It contains sensory fibres (cell bodies in the ganglion of the nerve) and preganglionic secreto-motor parasympathetic fibres (cell bodies in inferior salivatory nucleus).
2. Branches from the facial nerve at the geniculate ganglion pierce the temporal bone. A few of these are sensory (cell bodies in geniculate ganglion), but most are parasympathetic secreto-motor fibres coming by the nervus intermedius from the superior salivatory nucleus.
3. Sympathetic twigs from the carotid plexus run in on the carotico-tympanic arteries. They are vaso-constrictor, their cell bodies being in the superior cervical ganglion.

There are no synapses in the tympanic plexus. Branches of the plexus supply the mucous membrane of the tympanic cavity with common sensation. These are from the glosso-pharyngeal component, assisted in part by a few fibres from the facial nerve. These latter are embryological remnants from the nerve supply of the second pharyngeal pouch. They supply some of the mucosa, even extending beyond the drum to reach the skin of the external acoustic meatus. The occurrence of vesicles in the meatus is thus explained in cases of 'facial herpes'. In the same way a few sensory twigs of the facial nerve may supply mucous membrane above the tonsil (the other derivative of the second pharyngeal pouch, see p. 41).

The **lesser** (superficial) **petrosal** nerve is a branch of the tympanic plexus which collects all the parasympathetic (secreto-motor) fibres of the tympanic branches of VII and IX. It leaves the middle ear through a canal in the petrous bone and, running parallel with the greater (superficial) petrosal nerve, enters the middle cranial fossa. It reaches the otic ganglion, where it relays for the supply of the parotid gland (PLATE 40).

The **chorda tympani** crosses the middle ear. It is a mixed visceral nerve. It contains taste fibres from the tongue (cell bodies in the geniculate ganglion) and secreto-motor fibres for the salivary glands of the floor of the mouth (cell bodies in the superior salivatory nucleus in the pons). At about 6 mm above the stylo-mastoid foramen the chorda tympani leaves the facial nerve in the facial canal and pierces the posterior wall of the tympanic cavity lateral to the pyramid (PLATE 40). It runs forward over the pars flaccida of the tympanic membrane and the neck of the malleus, lying *just beneath the mucous membrane* throughout its course. It passes out of the front of the middle ear at the anterior margin of the tympanic notch, lying between the tympanic plate and that part of the tegmen tympani that walls in the canal for the tensor tympani. It emerges from the petro-tympanic fissure and joins the lingual nerve 2 cm below the base of the skull (p. 392).

The **mucous membrane** of the middle ear, continuous with that of the Eustachian tube and the mastoid antrum, covers all the structures enumerated above —the irregularities of the walls, the ossicles, their ligaments and their muscles. Its contours in the middle ear are thus very irregular, and certain pouches are produced, particularly in the epitympanic recess alongside the superior and lateral ligaments of the malleus. The mucous membrane is thin and adherent to the underlying structures. It is covered with columnar epithelium carrying stereocilia.

The *blood supply* of the middle ear comes from several branches of the external carotid artery. From the maxillary artery a *tympanic* branch enters the squamo-tympanic fissure and supplies the front of the tympanum and internal surface of the membrane. From the posterior auricular artery the *stylo-mastoid* branch enters the foramen of that name and supplies the back of the tympanum and internal surface of the membrane. One of its branches, the *stapedial artery*, supplies the muscle; it is of interest as being the remnant of the artery of the second pharyngeal arch of the embryo. From the middle meningeal artery a branch accompanies the greater (superficial) petrosal nerve, supplies the geniculate ganglion and passes through to the tympanum. A small branch of the ascending pharyngeal artery accompanies the tympanic branch of the glossopharyngeal nerve. From the *internal carotid artery* a small branch, the *carotico-tympanic artery*, perforates the wall of the carotid canal. Often this artery is broken up into several small vessels whose foramina are visible in the dried skull on the ridge between carotid and jugular foramina.

The *veins* leave the squamo-tympanic fissure to reach the pterygoid plexus. Others pass upwards through the petrous bone to the superior petrosal sinus; spreading thrombophlebitis may involve these veins after suppuration in the middle ear and so involve the veins of the temporal lobe (meningitis or cerebral abscess). The lymph vessels run down to the parotid and retropharyngeal lymph nodes.

The **facial nerve** passes through the petrous bone from the internal acoustic meatus to the stylo-mastoid foramen in three different directions, laterally, posteriorly and downwards in that order. Firstly it runs laterally, with the nervus intermedius. Here it lies above the vestibule, cochlea in front and semicircular canals behind. At the geniculate ganglion the two nerves unite. The greater (superficial) petrosal nerve passes forward from the ganglion through a canal in the petrous bone and emerges from a hiatus into the middle cranial fossa. (It runs forward beneath the trigeminal ganglion into the foramen lacerum and joins the deep petrosal nerve to form the nerve of the pterygoid canal.) The second part of the facial nerve passes backwards from the ganglion in a canal which raises a ridge on the medial wall of the tympanum above the promontory and fenestra vestibuli; here branches perforate the bone to enter the tympanic plexus. This part of the nerve lies below the prominence of the lateral semicircular canal (Fig. 6.58). The third part of the facial nerve passes down medial to the aditus and emerges from the stylomastoid foramen. The nerve to the stapedius and the chorda tympani leave the third part of the nerve.

The Auditory Tube

The auditory tube (pharyngo-tympanic tube, Eustachian tube) connects the naso-pharynx by way of the middle ear with the air cavities in the mastoid part of the temporal bone. Over 3 cm long, it slopes from the middle ear forwards and medially at 45 degrees and downwards at 30 degrees. Like the external acoustic meatus it has bony and cartilaginous parts, but the proportions are reversed.

The **bony part,** over 1 cm long, tapers down from the anterior wall of the middle ear to its orifice. This is the narrowest part of the tube, known as the *isthmus*; it lies beside the spine of the sphenoid. The bony part perforates the petrous part of the temporal bone (Fig. 8.9, p. 566). It is lined with muco-periosteum, a thin mucous membrane that is densely adherent to a thin periosteum. It is surfaced by columnar epithelium and, like the rest of the bony cavities, it contains no glands. It is kept moist by evaporation.

The **cartilaginous part,** over 2 cm long, joins the bony orifice at the isthmus and is lodged in the groove between the greater wing of the sphenoid and the apex of the petrous part of the temporal bone (Fig. 8.4, p. 559). It is made of elastic cartilage, in two flanges joined above. The posterior is longer than the anterior flange; thus a vertical section through the cartilage resembles an inverted **J** (long limb posterior). It enlarges from the isthmus like a trumpet, with its open end greatly expanded, particularly the long posterior limb which forms the *tubal eminence*. The margins of the flange are joined by a sheet of fibrous tissue. The cartilaginous tube thus completed is lined by respiratory mucous membrane (loose-woven fibrous tissue containing mucous and serous glands) and surfaced by ciliated columnar epithelium (see p. 398). The cilia beat towards the naso-pharynx, thus protecting the middle ear from air-borne particles, including bacteria. Loose longitudinal folds of mucous membrane gently occlude the cartilaginous tube in most individuals; the folds are momentarily parted during swallowing.

The opening of the tube is attached to the back of the medial pterygoid plate just below the skull base. The tubal elevation is made prominent, especially in the young, by lymphatic follicles in the mucous membrane. This is the tubal tonsil (p. 417). The posterior limb (the tubal eminence) is elongated by a vertical fold of mucous membrane, called the salpingo-pharyngeal fold, which is draped over the **salpingo-pharyngeus muscle.**

The muscle passes vertically down inside the pharynx (Fig. 6.34, p. 419). It is slender and is inserted into the posterior border of the thyroid cartilage and the side wall of the pharynx. It is supplied from the pharyngeal plexus. Its contraction must move the tubal elevation and so assist in opening the tube. It is too slender to have much effect in elevating the larynx and pharynx.

The pharyngo-basilar fascia is attached to the lower part of the tube; lateral to this the tensor palati muscle arises outside the pharynx, and medial to this the levator palati muscle arises inside the pharynx. Both are attached in part to the tube. Swallowing therefore, by pulling on the lower wall, opens the tube and allows equalization of air pressure on the two sides of the tympanic membrane. Air is slowly lost from the middle ear and mastoid cavities by absorption into the capillaries thereof.

The tube is supplied by the ascending pharyngeal and middle meningeal arteries. Its veins drain into the pharyngeal plexus and its lymphatics to retropharyngeal lymph nodes. Its nerve supply is from the pharyngeal branch of the pterygo-palatine ganglion (Vb) at the ostium, and the nervus spinosus (meningeal branch of Vc) for the cartilaginous part. The bony part is supplied by branches from the tympanic plexus (IX).

The Mastoid Air Cavities

The **tympanic (mastoid) antrum** lies behind the epitympanic recess in the petrous part of the temporal bone. It is connected to the recess by the aditus. It is spherical, but its size is very variable. It may be up to 1 cm in diameter. When large it is covered by a thin layer of bone, when small by a thick layer. Its medial wall lies 15 mm deep to the suprameatal triangle at the postero-superior margin of the external acoustic meatus. It is roofed in by the tegmen tympani (Fig. 6.58).

During the first year **mastoid air cells** burrow out into the thin plate of bone at the bottom of the sigmoid sinus. Lined with adherent muco-periosteum they pneumatize the mastoid process and may eventually extend into the petrous bone for a variable distance, even to the tip.

Antrum and air cells are supplied by the stylomastoid branch of the posterior auricular artery. The veins drain into the mastoid emissary vein, the posterior auricular vein and the sigmoid sinus. Spreading infection can involve these veins in the posterior cranial fossa, leading to sigmoid sinus thrombosis, meningitis, cerebellar or cerebral abscess. Lymph drainage is to the mastoid and upper deep cervical lymph nodes.

The nerve supply is from the nervus spinosus (meningeal branch of Vc), which passes from the middle cranial fossa through the squamo-petrous suture.

The Internal Ear

Buried within the stony hardness of the petrous bone is the internal ear, practically full adult size at birth. The hearing and balance parts form one continuous cavity, enclosed in membrane and containing endolymph. This is a completely closed system; it is supplied by the hearing and balance parts of the eighth nerve and is known as the membranous labyrinth. The delicate membranous labyrinth lies in a corresponding cavity in the ivory-like bone; this is the osseous labyrinth. It is much larger than the membranous labyrinth; the space between the two contains perilymph.

The membranous labyrinth consists of the spiral duct of the cochlea in front (hearing), the saccule and utricle between (organ of static balance) and the semicircular ducts behind (organ of kinetic balance). The bony cavities which contain these three parts of the membranous labyrinth are known as the cochlea, vestibule and semicircular canals respectively (Fig. 6.63).

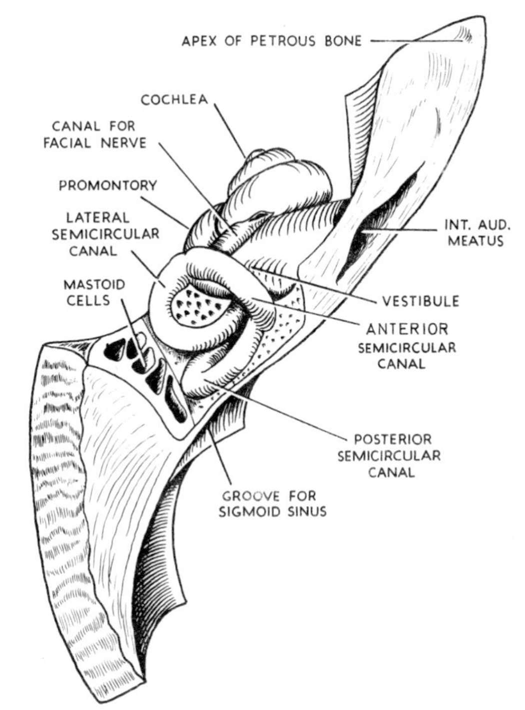

FIG. 6.60. THE LEFT OSSEOUS LABYRINTH IN POSITION IN THE TEMPORAL BONE. Viewed from above. Drawn from Specimen 152, D, in R.C.S. Museum.

The **osseous labyrinth** consists of a complicated series of cavities within the petrous bone, lined with periosteum and containing perilymph. The smaller membranous labyrinth lies in the perilymph. The bony cavity opens into the middle ear through the fenestra vestibuli (closed in life by the foot-piece of the stapes) and the fenestra cochleæ (closed in life by the secondary tympanic membrane) and it opens into the posterior cranial fossa through the aqueduct of the vestibule (closed in life by the ductus endolymphaticus) and the aqueduct of the cochlea, which opens into the sub-arachnoid space. Through the aqueduct of the cochlea the perilymph drains into the cerebrospinal fluid.

The **cochlea** is the cavity in the bone which surrounds the membranous duct of the cochlea. It is conical in shape, and consists of two and three-quarter spiral turns of a tapering cylindrical canal, like the spiral slide at a fun fair. The bony canal is of greatest calibre at the basal turn; this part projects laterally, producing the promontory on the medial wall of the middle ear (Fig. 6.60).

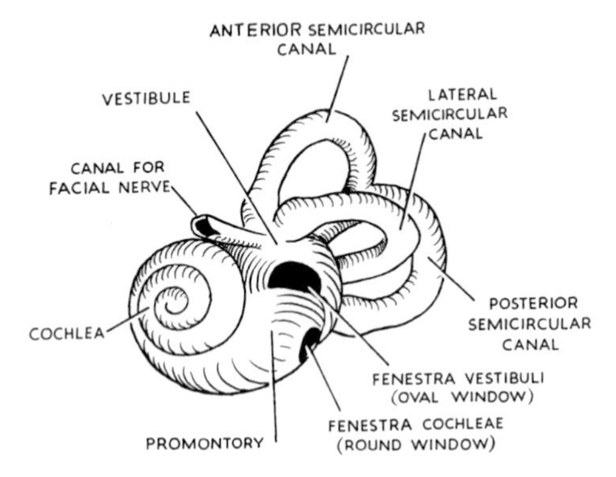

FIG. 6.61. THE LEFT OSSEOUS LABYRINTH. Viewed from the lateral side. Drawn from Specimen 152, C, in R.C.S. Museum.

The axial bony stem around which the canal spirals is known as the **modiolus.** The base of the modiolus lies at the fundus of the internal acoustic meatus and its apex lies across the long axis of the petrous bone, pointing towards the canal for the tensor tympani. The apex of the modiolus is overlaid by the blind extremity of the apical turn of the cochlea, which forms the dome, or *cupola*, of the cochlea (Fig. 6.62).

From the modiolus a shelf of bone projects into the canal, like a thread projecting from a screw. This is the **spiral lamina;** it ends at the apex of the modiolus, beneath the cupola, in a hook-like process, the hamulus. Its projection is widest in the basal and narrowest in the apical turn. The membranous part of the cochlea is known as the **duct of the cochlea.** It is attached to the spiral lamina and to the outer bony wall of the canal. It commences blindly at the hamulus and spirals in two and three-quarter turns to its basal extremity, where it is connected by a minute canal to the saccule (Fig. 6.63).

The bony canal of the cochlea is thus partitioned by the spiral lamina and the membranous duct of the cochlea, which contains endolymph. The canal on the apical side of the partition is the **scala vestibuli,** that on the basal side the **scala tympani;** they contain perilymph. At the cupola of the cochlea they communicate with each other around the blind apical extremity of the membranous duct of the cochlea. This point of communication is known as the *helicotrema.*

The basal turn of the cochlea sees the termination of the lamina spiralis. Here the scala tympani is sealed off into a blind end. There are two holes in this cul-de-sac. One leads laterally into the middle ear; it is the fenestra cochleæ, closed in life by the secondary tympanic membrane. The other is the beginning of a canal, the **aqueduct of the cochlea,** which leads down through the substance of the petrous bone and opens into the glosso-pharyngeal notch, below the internal acoustic meatus, in the anterior compartment of the jugular foramen (p. 492). The aqueduct of the cochlea is patent in life. The arachnoid mater is attached to the margin of its opening, so that perilymph draining down the aqueduct is received into the cerebro-spinal fluid in the subarachnoid space.

The *modiolus* is perforated spirally at its base in the internal acoustic meatus by the branches of the cochlear nerve. These run into the modiolus and fan out spirally towards the base of the lamina spiralis, where their bipolar cell bodies lie.

The cell bodies lie in a canal at the base of the spiral lamina and together are known as the **spiral ganglion.** The spiral ganglion is the counterpart of the posterior root ganglion of a spinal nerve (i.e., it contains the cell bodies of the first neurone of a sensory pathway). The spiral ganglion (cochlear nerve) connects the sound receptors in the organ of Corti with the cochlear nuclei in the brain stem.

The **vestibule** is a hollow in the bone which contains the membranous saccule and utricle. It is 3 mm wide and measures 6 mm long and 5 mm high. The scala vestibuli of the cochlea, containing the minute ductus reuniens, opens into the front of the vestibule and the five orifices of the semicircular canals open posteriorly (Fig. 6.62).

The *medial wall* abuts on the internal acoustic meatus. It is excavated into two concavities separated by an oblique curved ridge called the *vestibular crest.* The anterior inferior concavity is the *spherical recess*; it lodges the saccule. The floor is the inferior vestibular area of the internal acoustic meatus; it is perforated by over a dozen minute foramina for the nerves to the saccule from the inferior division of the vestibular nerve. The

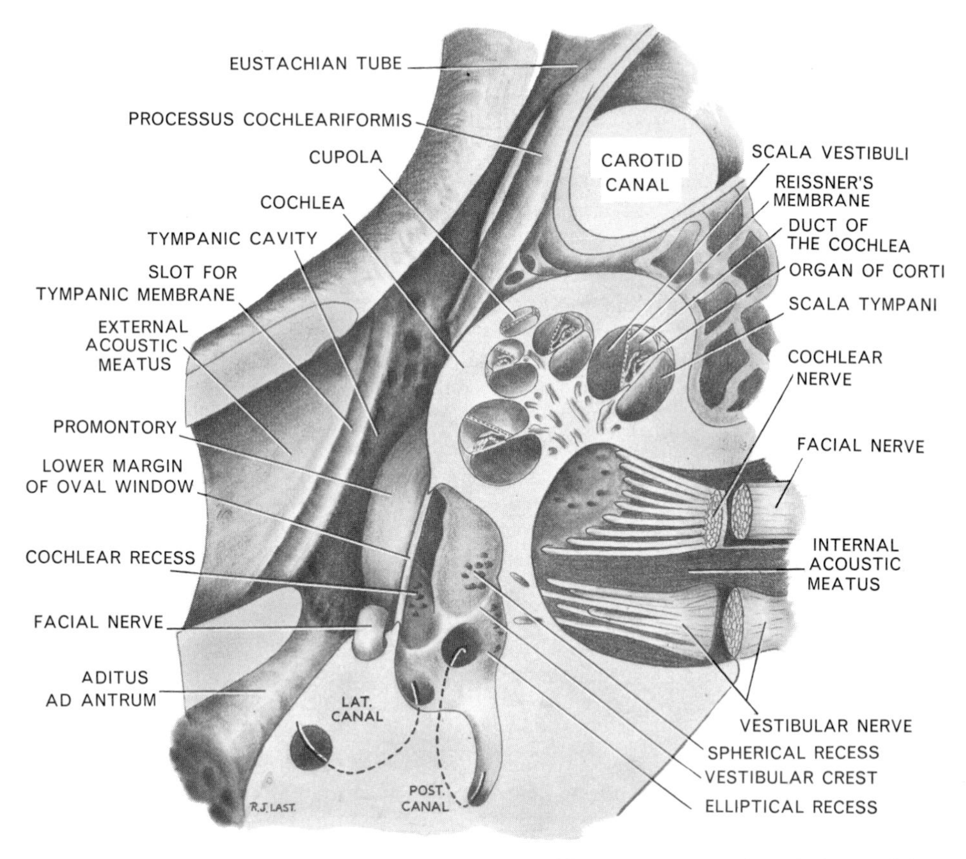

FIG. 6.62. A HORIZONTAL SECTION THROUGH THE LEFT TEMPORAL BONE. The section passes through the modiolus and the oval window. Author's preparation.

posterior superior concavity is the *elliptical recess*, which extends also on the roof of the vestibule. The openings of the anterior and lateral semicircular canals lie behind it. The elliptical recess lodges the utricle. Its floor is the superior vestibular area of the internal acoustic meatus; it is perforated by over a score of minute foramina for passage of nerves to the utricle and the anterior and lateral semicircular ducts. Just posterior to the basal termination of the lamina spiralis of the cochlea is a minute pit called the *cochlear recess* which transmits nerves to the basal end of the membranous duct of the cochlea. These nerves are in series with those passing through the tractus spiralis in the base of the modiolus to the remainder of the cochlear duct. Beneath the cochlear recess, at the lower margin of the elliptical recess, is the opening of the *aqueduct of the vestibule.* This canal is badly named, for no water is led along it. It is a minute canal nearly 1 cm long which opens in the posterior cranial fossa on the posterior surface of the petrous bone. It is plugged in life by the ductus endolymphaticus and a vein; no perilymph can escape through this canal.

The *lateral wall* of the vestibule abuts on the middle ear behind the promontory. Here is the kidney-shaped opening of the fenestra vestibuli, closed in life by the foot-piece of the stapes and its annular ligament (Fig. 6.62).

The **semicircular canals** lie in three planes at right angles to each other. Each is about two-thirds of a circle; in length along the curve they measure about 20 mm. Their calibre is 1 mm except at one end, where each is dilated to a calibre of 2 mm. The dilated part is named the *ampulla.*

The *superior semicircular canal* is placed in a vertical plane across the long axis of the petrous bone, convexity upwards, ampulla laterally. Its convexity produces the arcuate eminence on the upper surface of the petrous bone in the middle cranial fossa. It still lies highest of the three canals, in spite of its new name of *anterior* semicircular canal (Fig. 6.61).

The *posterior semicircular canal* is placed in a vertical plane in the long axis of the petrous bone, convexity backwards, ampulla below. The ampulla is innervated separately by a branch of VIII which pierces the foramen singulare in the internal acoustic meatus.

The *lateral semicircular canal* is placed 30 degrees off

the horizontal plane, convexity backwards, ampulla anteriorly. The ampulla bulges the medial wall of the aditus and epitympanic recess above the facial canal. The lateral semicircular canal lies horizontal if the head nods 30 degrees forwards.

The lateral semicircular canal opens by each end separately into the back of the vestibule. The anterior and posterior canals open separately at their ampullated ends, but their non-ampullated ends fuse into a common canal about 4 mm long. Thus only five openings connect the three canals with the cavity of the vestibule. The ampullated ends of anterior and lateral canals lie near each other high up, behind the elliptical recess (they share with the utricle the upper division of the vestibular nerve). The ampullated end of the posterior canal opens low down (it has its own branch from the inferior division of the vestibular nerve). The non-ampullated end of the lateral and the common opening of the anterior and posterior canals lie near each other low down towards the medial wall of the vestibule.

It should be noticed that anterior and posterior canals, lying across and along the axis of the petrous bone, are each at 45 degrees with the sagittal plane. Thus the posterior canal of one side lies parallel with the anterior canal of the opposite side.

The **membranous labyrinth** is a reduced replica of the hollow bony labyrinth. It consists of one continuous closed cavity containing endolymph. The origin of the endolymph is unknown. The membranous covering consists of three layers. The outer fibrous layer is vascular and in places adherent to the endosteum of the bony labyrinth. The intermediate layer is homogeneous like a basement membrane and the inner epithelial layer is elaborated in three places into receptors of sound, static balance and kinetic balance. It is supplied by the cochlear (hearing) and vestibular (balance) divisions of the eighth nerve. There is no known reason why hearing and balance are so intimately connected in the same peripheral structure and innervated by the same cranial nerve.

The **duct of the cochlea** is the spiral anterior part of the membranous labyrinth which contains the sound receptors (do not confuse with *aqueduct* of the cochlea, a canal for perilymph, p. 454). It is attached to the apical surface of the spiral lamina and to the outer bony walls of the cochlea. It commences at a blind extremity in the cupola of the cochlea and after two and three-quarter spiral turns ends in a bulbous extremity in the basal turn of the cochlea. A minute membranous canal connects this extremity with the saccule (Fig. 6.63).

Two membranes enclose the duct of the cochlea. The *basilar membrane* extends in the line of the lamina spiralis to the outer bony wall of the cochlea. It is widest at the apex (0·4 mm) and narrowest (0·2 mm) at the base of the cochlea. It is over 3 cm long. Throughout its length it supports the spiral organ of Corti. It contains stiff straight fibres (24,000 of them) passing from the edge of the lamina spiralis to the outer wall of the cochlea.

The *vestibular (Reissner's) membrane* is very delicate: it passes obliquely across the cochlea on the apical side of the basilar membrane. Between these membranes the duct of the cochlea is triangular in cross-section (Fig. 6.62).

The *spiral organ of Corti* contains the sound receptors (the inner and outer hair cells), supported by other cells and overlaid by the membrana tectoria. It extends throughout the whole length of the duct of the cochlea.

The *rods of Corti* lie in two rows, an outer and an inner. Each rod is supported by a foot plate on the basilar membrane; inner and outer rods incline towards each other and meet at their upper extremities. They enclose, above the basilar membrane, a triangle which is the cross-section of the tunnel of Corti. The rods of Corti support the *auditory hair cells.*

The *inner hair cells* are a single layer of some 4,000 columnar cells lying against the inner rods. Each cell has 100 acoustic hairlets projecting in a tuft from its surface. Internal to these cells the epithelium soon becomes cubical.

The *outer hair cells* are similar to the inner, but are more numerous. They lie in three or four rows, interspersed with supporting columnar cells (*the cells of Deiters*), against the outer row of rods of Corti. External to these cells the columnar epithelium gradually becomes lower and is continued over the outer wall of the duct of the cochlea as a cubical epithelium. This latter overlies a vascular connective tissue, the *stria vascularis.*

The *membrana tectoria* is a soft fibrillated lamina that projects from beneath the inner attachment of the vestibular membrane and rests like a pad upon the inner and outer hair cells.

The cochlear nerve supplies, by the dendrites of the cell bodies of the spiral ganglion, the inner and outer hair cells.

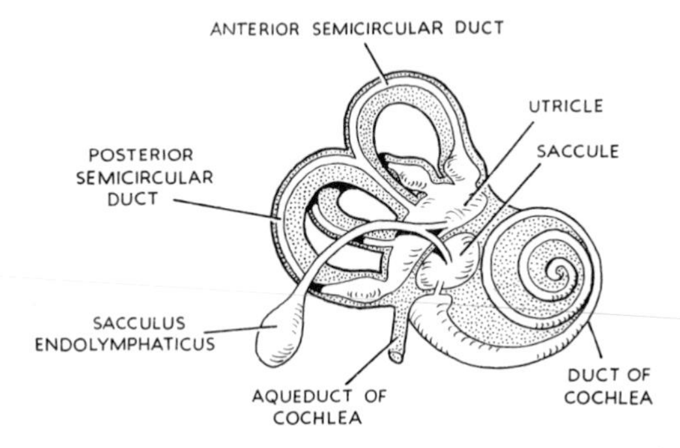

FIG. 6.63. THE LEFT MEMBRANOUS LABYRINTH. Viewed from the medial side. Compare Fig. 6.95, p. 491.

The mechanism of the organ of Corti is not fully understood. It is probable that sound vibrations are communicated from the perilymph to the endolymph through the delicate vestibular membrane and cause appropriate fibres in the basilar membrane to bulge. The overlying hair cells are thus stimulated.

The **saccule** lies low down in the front of the vestibule, connected to the basal part of the duct of the cochlea by a narrow membranous canal, the ductus reuniens. Its *medial wall*, lying in the spherical recess of the vestibule, receives the filaments of the lower division of the vestibular nerve. This wall is thickened, the *macula*, over an area 1·5 mm in diameter. The macula is lined by columnar cells from which hairlets project into an adherent plaque of a gelatinous material. Particles of calcium carbonate (otoliths) are embedded in the gelatinous material; they stimulate the hair cells and the vestibular nerve when they pull on the hairs, i.e., when the macula lies uppermost.

The **utricle** is a similar fibrous sac occupying the upper and back part of the vestibule. It receives the three semicircular ducts by five posterior openings. It has a *macula* in its *floor* similar to that of the saccule (*v.s.*), but rather larger.

The saccule and utricle lie with their adjacent walls in contact. From each leads a narrow membranous canal, the duct of the saccule and the duct of the utricle; these unite in Y-shaped manner to form the *ductus endolymphaticus*. The ductus endolymphaticus lies in the aqueduct of the vestibule and projects as a blind diverticulum, the *sacculus endolymphaticus*, beneath the fibrous dura mater of the posterior cranial fossa (p. 492).

The **semicircular ducts** are only a quarter the calibre of the bony canals except at the ampullæ, which they almost fill. Each membranous duct is adherent by its *convexity* to the wall of the bony canal in which it lies. The ducts open into five orifices in the back of the utricle inside the elliptical recess.

The kinetic balance receptors lie in each ampulla at the *crista*. This is a transverse crest on the medial surface of each ampulla surmounted by hair cells similar to those of the maculæ of saccule and utricle. The hairs are much longer and reach the middle of the lumen. They are embedded in an adherent gelatinous mass.

Blood Supply of the Labyrinth. The labyrinthine artery divides in the internal acoustic meatus into branches which accompany the cochlear and vestibular nerves to the labyrinth. Branches of the stylo-mastoid artery assist. The veins unite to form a labyrinthine vein which leaves the internal acoustic meatus and joins the inferior petrosal sinus. Various irregular veins penetrate the petrous bone independently to open into the superior petrosal sinus. A small vein lies in each aqueduct; that in the aqueduct of the cochlea joins the inferior petrosal sinus, that in the aqueduct of the vestibule joins the superior petrosal sinus.

Distribution of the Eighth Nerve. The cochlear division enters the front of the inferior part of the lamina cribrosa in spiral fashion to reach the organ of hearing (p. 491). The lower division of the vestibular nerve supplies the macula of the saccule and, through the foramen singulare, the ampulla of the posterior semicircular duct. The upper division of the vestibular nerve supplies the macula of the utricle and the ampullæ of of the anterior and lateral semicircular ducts and overlaps a little on to the macula of the saccule. The cell bodies of the cochlear fibres lie in the spiral ganglion in the modiolus. The cell bodies of the vestibular fibres lie in the vestibular ganglion in the depths of the internal acoustic meatus (Figs. 6.63 and 6.95, p. 491).

THE VERTEBRAL COLUMN

The backbone forms the central axis of the skeleton. It supports the skull and gives attachment, by way of the ribs, to the thoracic cage and, by way of this cage, to the pectoral girdle and upper limb. By way of the pelvic girdle it is strongly united to the lower limbs, which serve the double function of support and propulsion. The great strength of the backbone comes from the size and architecture of the bony elements, or vertebræ, and the ruggedness of the ligaments and muscles that hold them together. This great strength is combined with great flexibility; the backbone is flexible because it has so many joints so close together. Finally, the backbone contains in its cavity the spinal cord, to which it gives great protection from external violence.

The backbone, or vertebral column, is made up of five parts with individual vertebræ peculiar to each. These are the cervical, thoracic, lumbar, sacral and coccygeal parts.

In the fœtus in utero the column lies flexed in its whole extent, like the letter C. This anterior flexion is known as the *primary curvature* of the column, and it is retained throughout life in the thoracic, sacral and coccygeal parts. The neonatal backbone has no intrinsic curvature; it will take up any bend imposed from outside. Secondary extension of the column produces the *secondary curvatures* in the neck and lumbar region, the former associated with muscular support of the head and the latter with that of the trunk. In each part, neck and lumbar region, the secondary curvatures are of lordosis (i.e., forward convexity).

As the secondary curvatures develop in the neck and lumbar regions the vertebral column is opened out from its original C-shape, and elongated into a vertical column characterized by gentle sinuous bends. These bends give a certain resilience to the column, but the actual shock-absorbing factors in the spinal column are the intervertebral discs.

Before considering the functional anatomy of the column as a whole (p. 462) certain anatomical details of its several parts should be studied.

General Characteristics of Vertebræ

Examine a thoracic vertebra. It consists of a ventral *body* and a dorsal *neural arch*; they enclose between them the *spinal canal*. From the neural arch three processes diverge; in the posterior midline is the *spinous process* while, symmetrically on either side, the *transverse processes* project. That part of the neural arch between spinous process and transverse process is known as the *lamina*, that between transverse process and body is called the *pedicle*. The vertical extent of the pedicle is less than that of the body, to allow room for passage of the spinal nerve (Fig. 6.64). At the junction of lamina and pedicle (i.e., at the root of the transverse process) are found *articular processes*, above and below, which have cartilage-covered facets for the synovial joints between the neural arches. The direction of the facets conforms with and determines the nature of the movement possible between adjacent vertebræ.

Finally, *costal elements* (ribs) articulate with both body and transverse process of all vertebræ.

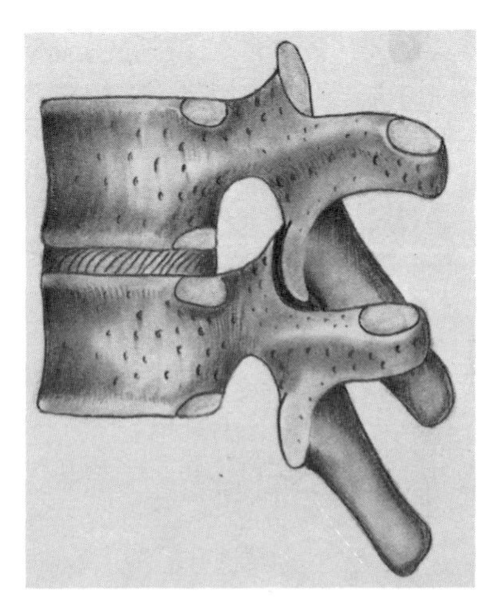

FIG. 6.64. TWO LOWER THORACIC VERTEBRÆ VIEWED FROM THE LEFT SIDE, TO SHOW THE BOUNDARIES OF THE INTERVERTEBRAL FORAMEN. (Cf. Fig. 6.84.)

Morphology. The development of a vertebra gives the clue to its morphology. It ossifies in three parts, the *centrum* and the right and left halves of the *neural arch*, and these are the three morphological parts of a vertebra. In the thoracic region costal elements develop separately as the ribs, articulating with the neural arches; but in all other parts of the vertebral column the costal elements become fused to the neural arches and incorporated as morphological parts of the vertebræ.

It should be particularly noted that the morphological *centrum* is not the same thing as the anatomical *body* of a vertebra. Part of the neural arch is incorporated into the body of the vertebra, and the neuro-central junction lies anterior to the costal facet on the body; that is, the costal facet on the body lies on the neural arch, and not on the centrum (Figs. 6.64 and 6.65). The term *neural arch* thus has two meanings. In descriptive anatomy it consists of the pedicles and laminæ with the processes that project from them. In morphological language it comprises also that part of the body which develops from the neural arch of the embryo, that part with which the costal element articulates.

Typical Vertebræ

Examine a mid-cervical, mid-thoracic and mid-lumbar vertebra (Fig. 6.65). Compare and contrast all three. For comparison note that each possesses a body and a neural arch, the latter consisting of pedicles and laminæ with the projecting articular processes and the spinous and transverse processes. By contrast note that the cervical vertebra has a foramen, the foramen transversarium, in the transverse process and that it has no costal facets. The thoracic vertebra has costal facets, smooth areas that are covered with hyaline cartilage in life. One costal facet lies towards the tip of the transverse process, the others lie as demi-facets on the side of the body at its upper and lower margins. The lumbar vertebra has neither a foramen transversarium nor costal facets.

These two features, foramen transversarium and costal facet, serve to distinguish cervical, thoracic and lumbar vertebræ in all mammals.

Costal Elements

Developed in association with each vertebra are costal elements. In the thoracic region they develop separately to form the ribs, which articulate with the vertebræ (neural arch) by synovial joints. Elsewhere the costal elements are vestigial and fuse with the neural arches to become incorporated into the adult vertebræ. It is desirable to distinguish the costal elements in the various vertebræ.

Cervical Vertebræ. The foramen in the transverse process is produced by the costal element. In the typical cervical vertebræ the costal element consists of the anterior tubercle, the costo-transverse bar, and the tip of the posterior tubercle (Fig. 6.65).

Cervical ribs are not uncommon, existing as either bony elements or fibrous tissue bands, passing down from the seventh cervical vertebra to the first rib. The subclavian artery and lowest root (Th. 1) of the brachial plexus become displaced upwards over such structures, and pressure upon them may cause severe symptoms, which usually appear about the age of puberty, when normally the neck elongates and the shoulders droop somewhat (Fig. 6.66). The pressure produced by a thin fibrous band may be more irritating than that due to a smooth bony rib. The presence of a fibrous band may be inferred if the anterior tubercle of the seventh cervical vertebra is enlarged. The patient whose radiograph is shown in Fig. 6.66 had symptoms on the right side which were due to a fibrous band, and the well-ossified cervical rib on the left side produced no symptoms. When a cervical rib is well developed the brachial plexus is more likely to be prefixed (i.e., its roots are C4, 5, 6, 7 and 8) thus preserving the normal nerve to rib relationship.

Lumbar Vertebræ. The so-called transverse processes of the lumbar vertebræ are in reality costal elements. The true morphological transverse process is contracted into a small mass of bone which is grooved by the posterior primary ramus of the spinal nerve. Above the groove lies the *mamillary process* (beside the superior articular facet) and below the groove is found the *accessory tubercle*; both belong to the true transverse element (Fig. 6.79, p. 472). The costal element may fail to fuse and remain with a synovial joint connecting it to the neural arch; this is a lumbar rib, symptomless and discovered by accident in X-ray examination.

Sacrum. The five sacral vertebræ are fused into a single bone and so, too, are the five costal elements. The latter produce the lateral mass of the sacrum, lying lateral to the transverse tubercles on the back of the sacrum and extending between the anterior sacral foramina on the front of the bone. The auricular surface (the sacro-iliac joint) lies wholly on the lateral mass. In morphological terms the pelvis articulates with ribs, not with the centra of vertebræ.

The Coccyx. This bone is itself so vestigial that no costal elements can be made out with any certainty. The matter is of no importance.

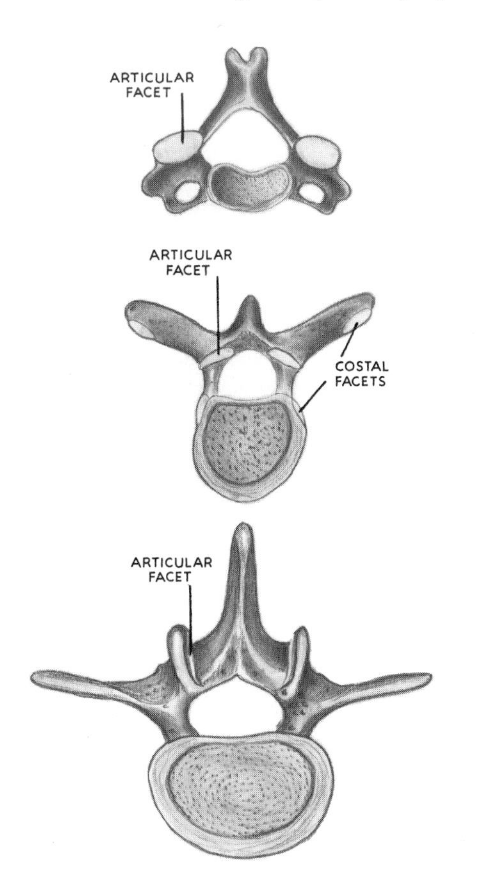

FIG. 6.65. THE ESSENTIAL CHARACTERISTICS OF CERVICAL, THORACIC AND LUMBAR VERTEBRÆ.

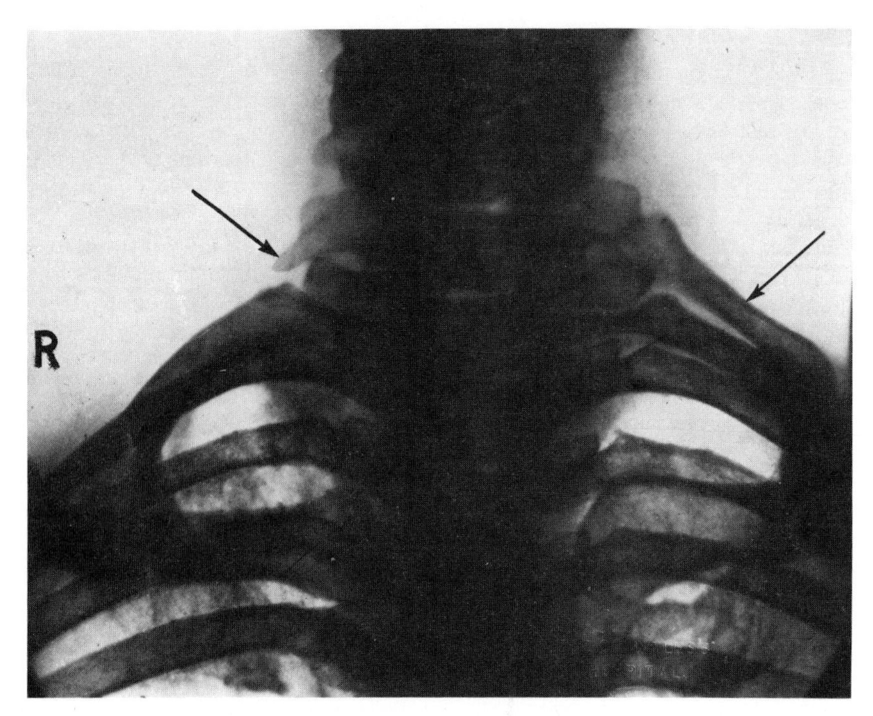

FIG. 6.66. A CERVICAL RIB ON THE LEFT SIDE. On the right side the anterior tubercle of the seventh cervical vertebra is greatly enlarged. A fibrous band passed from this tubercle to the first rib. Radiograph by courtesy of Dr. Seymour J. Reynolds.

Articulations between Vertebræ

Adjacent vertebræ are held together by strong ligaments; these ligaments allow greater range of movement between the neural arches than between the bodies. The vertebræ articulate between their bodies and between their neural arches. These joints are very different from each other, and require separate study.

Articulations between the Bodies. The bodies of adjacent vertebræ are held together by the strong intervertebral disc. An **intervertebral disc** is a secondary cartilaginous joint, or symphysis. The upper and lower surfaces of each vertebral body are covered completely by a thin plate of hyaline cartilage. These plates of cartilage are united by a peripheral ring of fibrous tissue, named the **annulus fibrosus.** The annulus fibrosus consists of concentric laminæ, the fibres of which lie at 45 degrees with the bodies of the vertebræ. Alternate layers of the annulus fibrosus contain fibres lying alternately in the 45 degrees slope at right angles to each other. By this means the annulus fibrosus is able to withstand strain in any direction. Inside the annulus fibrosus is a bubble of semi-liquid gelatinous substance known as the **nucleus pulposus.** It is derived from the embryonic notochord. (The notochord extended originally as far cranially as the sella turcica, but it disappears except in the nucleus pulposus of each intervertebral disc.) The nucleus pulposus in the embryo lies at the centre of the disc. Subsequent growth of the vertebral bodies and discs occurs in a ventral and lateral direction (the spinal cord prevents a corresponding growth dorsally). Thus in the adult the nucleus pulposus lies nearest to the back of the disc and if it herniates through the annulus fibrosus it will be most likely to do so posteriorly and press on the roots of a spinal nerve near the intervertebral foramen, or on the spinal cord itself.

Note that there is no dividing interface between the annulus fibrosus and the nucleus pulposus. Peripherally the oblique fibres of the annulus are extremely densely packed, with interchange of fibres between adjacent laminæ. Centrally the fibres become gradually less dense but they penetrate the nucleus pulposus. Similarly the nucleus pulposus is centrally a gel and becomes gradually less albuminous peripherally until it is replaced by ordinary tissue fluid.

The function of the nucleus pulposus is that its semi-liquid nature allows of ready distortion, so that the annulus fibrosus can alter its thickness at any part of its circumference to conform with the movements of the vertebral bodies. Liquids are incompressible—it is the annulus fibrosus, not the nucleus, that acts as a shock-absorber to minimize jarring of the skull. The annulus fibrosus is not compressed between adjacent vertebræ, for the incompressible nucleus pulposus transmits pressure in a centrifugal direction and the fibres of the annulus fibrosus are thereby put on the stretch. Should the fibres yield, the nucleus pulposus herniates them beyond the circumference of the vertebral bodies. The high osmotic pressure of the gel prevents water being extruded by the pressure of body weight or jarring movements. Thus the nucleus pulposus maintains a constant volume. If the fluid in the nucleus were lymph it would be so squeezed out by body weight that at the end of the day a man would be six inches shorter than when he arose from his bed.

Thus the discs unite the vertebral bodies and prevent their separation, yet at the same time keep them apart, thanks to the nucleus pulposus in each. The discs are very strong—much stronger than the vertebral bodies themselves. A vertebral body will fracture before a *normal* disc gives way.

In the typical cervical vertebræ there is an upturned lip at either side of the body, and a reciprocal bevel on the lower surface. Tiny neuro-central synovial joints are formed between these surfaces, at the lateral edges of the intervertebral discs.

The bodies of the vertebræ are further held together by longitudinal ligaments. The **anterior longitudinal ligament** extends from the anterior tubercle of the atlas to the front of the upper part of the sacrum. It is firmly united to the periosteum of the vertebral bodies, but is free over the intervertebral discs. It is a flat band, broadening gradually as it passes downwards.

The **posterior longitudinal ligament** extends from the back of the body of the axis (second cervical) vertebra to the sacral canal. It narrows gradually as it passes downwards. It has serrated margins. The serrations are broadest over the intervertebral discs, to which they are firmly united. The ligament narrows over the vertebral bodies, from which it is separated by the emerging basivertebral veins. The ligament is continued above the body of the axis as the membrana tectoria.

Articulations between the Neural Arches. Adjacent neural arches are articulated by synovial joints and ligaments. The synovial joints possess a simple capsule. The articular surfaces permit of gliding of one surface upon another. The direction of the joint surface determines the direction of the movements possible between adjacent vertebræ (see below).

Several ligaments attach adjacent neural arches to each other. The strongest are the ligamenta flava and the supraspinous ligaments (Fig. 6.67).

The **ligamenta flava** are yellow in colour from their high content of elastic fibres. They join the contiguous borders of adjacent laminæ. They are attached to the front of the upper lamina and to the back of the lower lamina—shallow grooves on the macerated vertebræ indicate their lines of attachment. They are stretched by flexion of the spine; in leaning forward their increasing elongation becomes an increasing antigravity support.

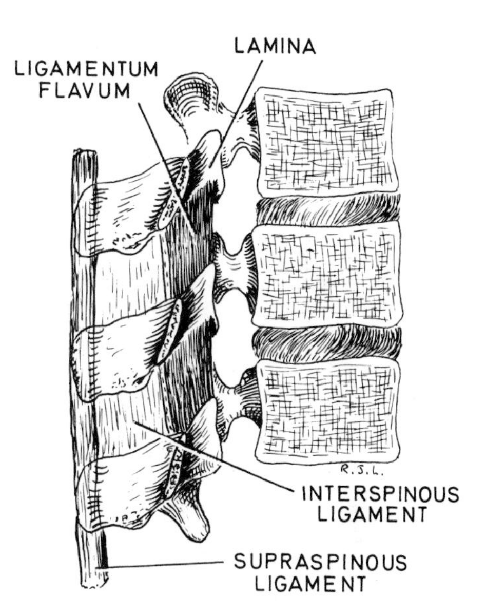

FIG. 6.67. THE VERTEBRAL COLUMN BISECTED. The imbrication of alternating laminæ and ligamenta flava is seen from within.

The **supraspinous ligaments** join the tips of adjacent spinous processes. They are strong bands of white fibrous tissue; they are lax in the extended spine. They are drawn taut by full flexion, and then support the spine (no action currents can be obtained from the erector spinæ muscles when the spine is fully flexed, as in touching the toes).

The **interspinous ligaments** are relatively weak sheets of fibrous tissue uniting spinous processes along their adjacent borders. They are well developed only in the lumbar region. They fuse with the supraspinous ligaments.

The **intertransverse ligaments** are similar weak sheets of fibrous tissue joining the transverse processes along their adjacent borders.

The Vertebral Column

In the normal erect posture the vertebral column supports the head and trunk on the pelvis. (The pelvis is supported by the lower limbs in standing and by its own ischial tuberosities in sitting.) This support is maintained by the bodies of the vertebræ and the intervertebral discs, which thus become progressively larger from above downwards. The curvatures of the spine are produced partly by the wedge-shape of the vertebral bodies, but mostly by the wedge-shape of the intervertebral discs. This is particularly noticeable in the lower part of the spine; the fifth lumbar vertebra is usually wedge-shaped and the disc between it and the sacrum is very thick anteriorly (Fig. 5.55, p. 323).

Note that the upper surface of the sacrum slopes steeply downwards; sliding of the fifth lumbar vertebra down this slope produces the condition known as spondylo-listhesis. This is normally prevented by the strength of the intervertebral disc and the direction of the articular facets between the neural arch of L5 and the first sacral arch (Fig. 6.81, p. 474). Bony locking at this synovial joint prevents forward displacement of the fifth lumbar vertebra if the neural arch is intact; in spondylo-listhesis the pedicle of L5 is deficient.

The **spinal canal** (p. 493) becomes progressively smaller from above downwards. It is closed anteriorly by the vertebral bodies, the intervertebral discs and the posterior longitudinal ligament and posteriorly by the laminæ and the ligamenta flava. Laterally it is occupied by the pedicles, which are narrower than the height of the vertebral bodies. Thus a series of **intervertebral foramina** is produced between adjacent pedicles. Each intervertebral foramen is bounded in front by the lower part of the body of a vertebra and its adjacent intervertebral disc, and bounded behind by the capsule of the synovial joint between adjacent neural arches (Figs. 6.64 and 6.84, p. 476). The intervertebral foramina lodge the spinal nerves and posterior root ganglia and give passage to the spinal arteries and veins.

Movements of the Vertebral Column

In general the movements of the spine are simple enough. Flexion and extension, and lateral flexion (or abduction) are possible in cervical, thoracic and lumbar regions, though in varying degree in the three parts. Rotation occurs, strangely enough, mainly in the thoracic region. Movements of the head occur at the specialized atlanto-occipital and atlanto-axial joints and are considered on p. 463. For the vertebral column as a whole the movements are best understood by considering each of the three regions separately. In each region movements occur around the nucleus pulposus as around a ball-bearing, and the direction of the movements is determined by the direction of the articular facets on the neural arches.

The Lumbar Region. Examine the five lumbar vertebræ in an articulated skeleton. The articular facets lie in an antero-posterior plane; they lock, and greatly limit rotation of the bodies on each other. Flexion and extension are free, and a good deal of abduction is possible; combination of these movements produces circumduction, but, to repeat, *practically no rotation is possible.*

The Thoracic Region. Examine the articulated thoracic vertebræ. The synovial joints between Th. 12 and L1 are of the lumbar type, but elsewhere the direction of the articular facets on the neural arches is quite different. On any one neural arch the upper facets face backwards and laterally; they lie on the circumference of a circle whose centre lies in the vertebral body. The lower facets are reciprocal. Thus rotation of the bodies on each other is possible, in spite of the splinting effect of the ribs (Fig. 6.65).

Flexion and extension occur, as well as abduction (lateral flexion). The thoracic spine is thus the most versatile region of all, though the *range* of movements is limited by the ribs. Flexion of the thoracic spine crowds the ribs together (expiratory), extension spreads the ribs (inspiratory), while abduction crowds them on the concave side and spreads them on the convex side. Abduction involves sliding of the neural arches downwards and backwards on the concave side and upwards and forwards on the convex side, thus producing passively a certain amount of simultaneous rotation. This is well seen in the condition of scoliosis, where the lateral rotation of the spinous processes towards the convex side exaggerates the apparent deformity.

The Cervical Region. The atlanto-occipital and atlanto-axial joints are specialized for head nodding and head rotation. They are considered below.

The articular facets of the other joints slope similarly to those in the thoracic region, but they do not lie on the circumference of a circle. Looking along them will confirm that they both lie in the same plane. While flexion and extension are free, *pure* rotation is obviously impossible. Abduction is not a simple movement. The neural arch of the abducted vertebra slides downwards (and therefore backwards) on the concave side and upwards (and therefore forwards) on the convex side, thus inevitably producing a concomitant rotation greater than that produced in abduction of the thoracic spine.

Special Vertebræ

The Atlas and Axis. The first and second cervical vertebræ are atypical in several respects, both structural and functional. Weight-bearing between them and the skull is not by way of the vertebral bodies, and their joints permit a much wider range of movement than elsewhere in the vertebral column.

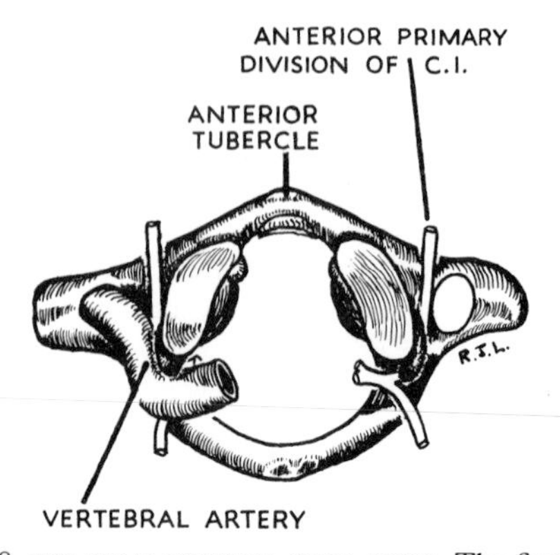

FIG. 6.68. THE ATLAS VERTEBRA FROM ABOVE. The first cervical nerve divides into anterior and posterior primary divisions just behind the atlanto-occipital joint (right side), where it lies in the groove beneath the vertebral artery (left side).

The **atlas** lacks its centrum, which is fused to the centrum of the axis vertebra, forming the dens (odontoid process). Each neural arch is thick and strong to constitute the powerful *lateral mass* of the atlas. The articular facets are in series with the small synovial joints on the bodies of the other cervical vertebræ, not with the articular facets on the neural arches. Thus the first and second cervical nerves send their anterior primary rami behind, and not in front of, the joints (Fig. 6.68). The neural arches of the atlas join posteriorly to make the posterior arch. The anterior arch of the atlas represents morphologically the ossified hypochordal bow; this latter is represented in the thorax by the intermediate strand of the triradiate ligament of the head of the rib (p. 212). The *articular facets* differ markedly. The upper ones are kidney-shaped and deeply concave for the occipital condyles, while the lower ones are circular and flat for the axis. Thus may the isolated atlas be orientated.

The **axis vertebra** is characterized by three main features, (i) the dens, (ii) a rugged lateral mass and (iii) the large spinous process (Fig. 6.69). The *dens (odontoid process)* is the centrum of the atlas, fused to the centrum of the second cervical vertebra. It bears no weight. The weight of the skull is transmitted through the lateral mass of the atlas to the lateral mass of the axis. The lower articulations of the axis are as for the ordinary cervical vertebræ—body to body with intervening disc and laterally placed synovial joints, and the ordinary articular facets on the neural arch. From the axis downwards the weight of the skull is supported by the vertebral bodies. The bifid spinous process is very large; this is due to powerful muscle attachments, the semispinalis cervicis from below and muscles of the suboccipital triangle above (Fig. 6.75).

The *atlanto-occipital joints* and the *atlanto-axial joints* are adapted to provide freedom of head movement, the former for nodding and lateral flexion, the latter for rotation.

The **atlanto-occipital joint** is a synovial joint between the convex occipital condyle and the concave facet on the lateral mass of the atlas. Both surfaces are covered with hyaline cartilage. The epiphyseal line between basi-occiput and jugular process crosses the occipital condyle along the line of the hypoglossal canal, into which it extends (Fig. 8.10, p. 567). The synovial cavity of the joint is contained in a lax but strong capsule, which is innervated by the first cervical nerve.

The anterior and posterior **atlanto-occipital membranes** are attached to the upper borders of the respective arches of the atlas and to the outer margins of the foramen magnum (Fig. 6.71). The anterior membrane completely closes the space between the two synovial joints, but the posterior membrane is deficient at each lateral extremity to allow passage for the vertebral artery and the first cervical nerve; the lateral margin of the membrane sometimes ossifies, converting the groove for the vertebral artery into a foramen. The membranes are innervated by the first cervical nerve (Fig. 6.96, p. 492).

The curved surfaces of the joint are well adapted for head flexion and extension, and allow also for a considerable amount of abduction (lateral flexion) of the skull on the atlas. In the ordinary erect position the centre of gravity of the skull lies in front of the joint and the head is maintained in position by the tonus of the extensor muscles, notably semispinalis capitis. It is flexed by relaxation of the extensors (i.e. by gravity) and, actively, by longus capitis and sterno-mastoids acting together. The effect of gravity is considerable [the head weighs 7 lb (over 3 kg)] and, of course, varies with position, being greatest when the neck is horizontal. Lateral flexion is produced by unilateral contraction of such muscles as sterno-mastoid, longissimus capitis, etc.

No rotation is possible at the atlanto-occipital joints. The articular surfaces are sections of a spheroid, like an egg lying on its side. Half buried in a reciprocal socket, the egg can be rolled to and fro in sagittal and coronal planes; but its oval shape prevents rotation around a vertical axis.

The **atlanto-axial joints** are synovial joints formed between the dens and the atlas and, one on each side, between the lateral masses of the two vertebræ (Fig. 6.69). In addition it must be noted that certain ligaments, bypassing the atlas, connect axis and occiput. Examine the joint surfaces of the lateral mass of axis and atlas; they are circular and flat, allowing for free gliding. They are covered in life with hyaline cartilage,

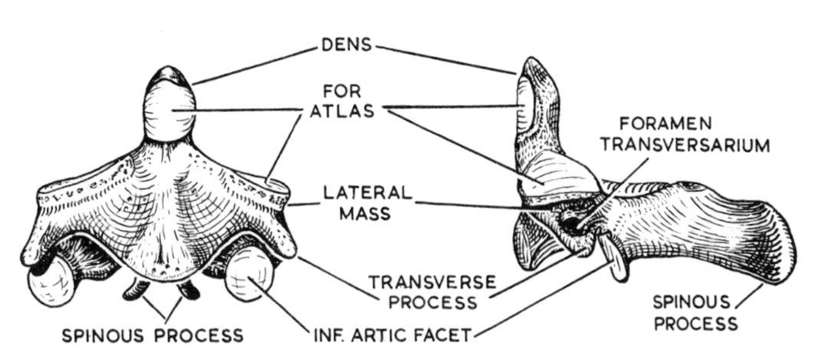

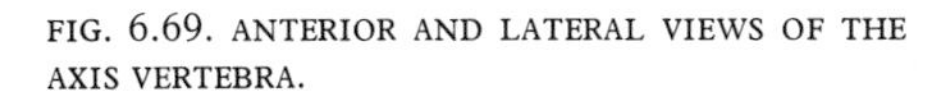
FIG. 6.69. ANTERIOR AND LATERAL VIEWS OF THE AXIS VERTEBRA.

and the synovial cavity is contained within a lax capsule supplied by the second cervical nerve.

The dens articulates with the back of the anterior arch of the atlas by a small synovial joint; the smooth facets seen on the dry bones are covered with hyaline cartilage in life. The dens is held in position by the transverse limb of the cruciform ligament; between the two is a relatively large synovial cavity, or bursa.

Accessory ligaments of these joints are the membrana tectoria, the cruciform ligament, and the apical and alar ligaments of the dens.

The **membrana tectoria** extends upwards in continuity with the posterior longitudinal ligament. It is attached to the back of the body of the axis and diverges upwards to become attached to the margin of the anterior half of the foramen magnum. It lies in front of the spinal dura mater, which is firmly attached to it.

The **cruciform ligament** lies in contact with the front of the membrana tectoria. It consists of a strong transverse band attached to the atlas and a vertical band which joins the body of the axis to the foramen magnum (Fig. 6.70). It holds the dens in position; rupture of this ligament allows the dens to dislocate backwards with fatal pressure on the medulla. It articulates with the back of the dens by a large synovial bursa which is surrounded by a loose fibrous capsule.

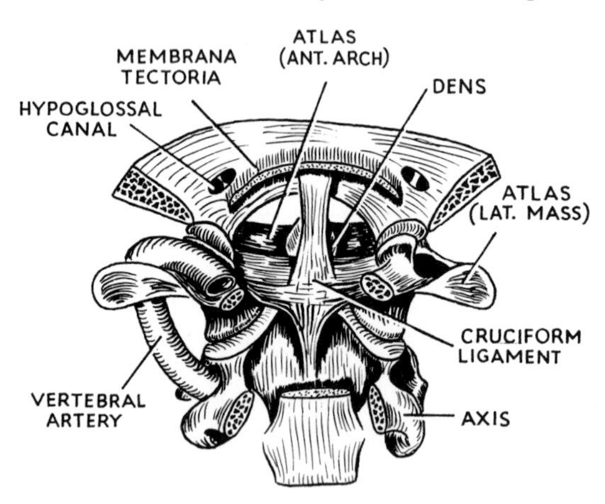

FIG. 6.70. THE CRUCIFORM (CRUCIATE) LIGAMENT EXPOSED FROM BEHIND BY DIVISION OF THE MEMBRANA TECTORIA.

The **apical ligament** lies in front of the upper limb of the cruciform ligament. It joins the apex of the dens to the anterior margin of the foramen magnum; it is a fibrous remnant of the notochord.

The **alar ligaments (check ligaments)** lie obliquely one on either side of the apical ligament. Attached to the sloping upper margin of the dens they diverge upwards to be attached to the margins of the foramen magnum. They limit rotation of the head.

Immediately in front of the alar and apical ligaments lies the anterior atlanto-occipital membrane.

Movements at the atlanto-axial joints are simply those of rotation about a vertical axis passing through the dens. The atlas rotates by its anterior arch and

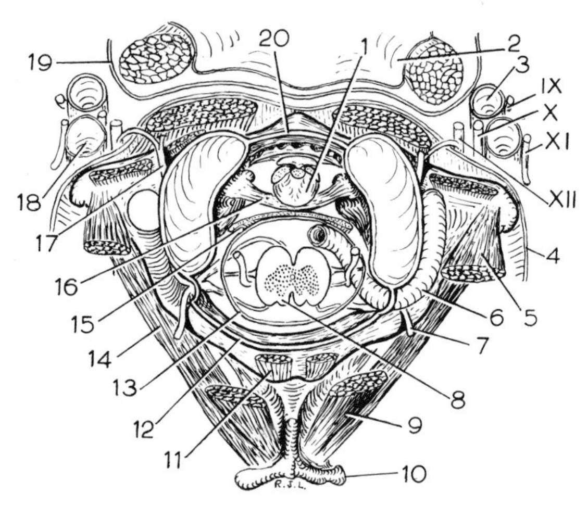

FIG. 6.71. THE ATLAS AND NEARBY STRUCTURES. This complex diagram is clarified if the atlanto-occipital membranes and joint capsules are picked out in colour (say green) and the vessels and nerves tinted. The chaos of such an assortment of structures becomes orderly if the *regions* are studied separately.
Outside the atlanto-occipital membranes and joints: In front, the prevertebral muscles and naso-pharynx, with the carotid sheath and last four cranial nerves alongside. Laterally the vertebral artery approaches the lateral mass of the atlas, and C1 emerges. Posteriorly is the suboccipital triangle covered by the extensor muscles.
Within the vertebral canal (i.e. within the atlanto-occipital membranes and joint capsules): In front lies the atlanto-odontoid joint with its ligaments. This compartment is closed behind by the membrana tectoria. The spinal dura mater (13) is adherent to the membrana tectoria (15) but lies free laterally and posteriorly. In this epidural space runs the vertebral artery, here giving off its meningeal branches to the posterior fossa. The roots of C1 emerge. The internal vertebral plexus of veins, communicating with the occipital venous sinus, lies here surrounded by loose fat.
Within the spinal dura mater (lined by arachnoid mater, 13) is the subarachnoid space. Here lies the medulla-spinal cord junction (8), stabilized by the ligamentum denticulatum. The accessory nerve runs up, with the vertebral artery, to the posterior fossa. The roots of C1 pierce the dura mater.
1. Dens (odontoid process) with its apical and alar ligaments. **2.** Mucous membrane of naso-pharynx over levator palati. **3.** Internal carotid artery. **IX, X, XI, XII.** Last four cranial nerves. **4.** Prevertebral fascia. **5.** Obliquus capitis superior. **6.** Vertebral artery. **7.** Posterior primary ramus of C1. **8.** Medulla-spinal cord junction (ligamentum denticulatum, accessory nerve and roots of C1 not labelled). **9.** Rectus capitis posterior major. **10.** Spinous process of axis. **11.** Rectus capitis posterior minor arising from posterior arch of atlas. **12.** Posterior atlanto-occipital membrane. **13.** Spinal dura mater (theca) lined with arachnoid mater. **14.** Obliquus capitis inferior. **15.** Membrana tectoria. **16.** Transverse band of cruciform (cruciate) ligament. **17.** Anterior primary ramus of C1, giving a branch to the hypoglossal nerve and ending in longus capitis. **18.** Internal jugular vein. **19.** Pharyngo-basilar fascia. **20.** Anterior atlanto-occipital membrane.

transverse limb of the cruciform ligament gliding around the dens and the lower flat facets on its lateral mass gliding on the facets on the lateral mass of the axis. The head rotates with the atlas; the curved surfaces of the atlanto-occipital joints allow of no independent rotation of occiput on atlas.

The muscles chiefly responsible for rotation are sterno-mastoid, splenius capitis and the inferior oblique, usually with other muscles acting in synergism to prevent lateral flexion.

Summary of Movements in the Vertebral Column

From the fixed sacrum up to the mobile skull a survey of the whole column can now be made. Flexion and extension are common to all parts. These movements are very free in the specialized atlanto-occipital joint and in the lumbar region, free in the cervical region, and restricted in the thoracic region. Lateral flexion (or abduction) is free in the lumbar region and the atlanto-occipital joints, less free in the neck and rather restricted in the thorax. Rotation is free in the specialized atlanto-axial joints. In the remainder of the vertebral column rotation is negligible except in the thoracic region. Rotation is almost absent in the lumbar region.

Blood Supply of the Vertebral Column

The vertebræ and the longitudinal muscles attached to them are supplied by segmental arteries. The ascending cervical, the intercostal and the lumbar arteries give multiple small branches to the vertebral bodies. The extensor muscles are supplied in the neck by the occipital, the deep cervical and the transverse cervical arteries. In thoracic and lumbar regions the muscles receive posterior branches of the intercostal, lumbar and lateral sacral arteries.

Venous drainage. The richly supplied red marrow of the vertebral body drains almost wholly by a pair of large basi-vertebral veins into the internal vertebral plexus (p. 493). Drainage of the neural arch and of the attached muscles is into the *external vertebral plexus.* The external vertebral plexus is intramuscular, and non-existent over the bare fronts of the vertebral bodies. The internal and external vertebral plexuses together drain into the regional segmental veins (vertebral, posterior intercostal, lumbar and lateral sacral veins). In the pelvis venous communication is thus established with the pelvic viscera, in the abdomen with the renal veins, in the thorax with the veins of the breast that enter the intercostal veins and in the neck with the inferior thyroid veins via the brachiocephalic veins. In this way, by reflux blood flow through these largely valveless veins, malignant disease possibly spreads from prostate, uterus, breast and thyroid gland to the bodies of the vertebræ.

The Muscles of the Vertebral Column

The movements of the vertebral column are produced by muscles widely separated from each other and, in almost all cases, are assisted by gravity. The action of paradox, whereby muscles 'pay out rope' against gravity, must always be borne in mind. Nevertheless for the sake of simplicity it is desirable to discuss the actions of the various muscle groups contracting as prime movers, and only after these actions are understood to interpret the functions of the muscles when combined with gravity.

Running along the whole length of the spine, from skull to sacrum, is a posterior mass of longitudinal extensor muscles known collectively as the erector spinæ. It is derived from the outer of the three layers of the body wall and is supplied segmentally by posterior primary rami, being the only muscle in the body so supplied (p. 211). At the front of the vertebral column are some flexor muscles derived from the innermost of the three layers of the body wall, constituting the prevertebral rectus. Longus capitis, longus colli and psoas belong to this group. They are supplied segmentally by anterior primary rami. The more usual muscles for flexing the spine are, however, the two rectus abdominis muscles, which act indirectly by way of the pelvis and thoracic cage. Rotators of the column are to be found in the abdominal flank muscles (the two obliques) which likewise act indirectly on the spine by way of the pelvis and the thoracic cage (p. 259).

The Erector Spinæ Muscle

This great mass of muscle lies behind the spine throughout its whole length, from the sacrum to the skull. In the neck it is specialized into certain separate muscles for particular head movements. It lies beneath the lumbar and thoraco-lumbar fascia and in the neck, where the fascia fades out, its superficial layer is the splenius muscle. These muscles and only these muscles are supplied by posterior primary rami of all the spinal nerves; the supply is segmental.

The erector spinæ mass of muscle consists of three layers, with three named muscles in each layer. The extent of the muscle fibres is not uniform, some are short and some are long, like a system of guy ropes. It is evident that the shortest fibres must lie deepest and so it is that the deepest layer of all consists of three groups of muscles that extend no further than adjacent vertebræ.

The Deepest Layer. None of these muscles are very strong. The **interspinales** join adjacent borders of the spinous processes, alongside the interspinous ligaments. The **intertransversales** join adjacent transverse processes; they are best developed in the upper part of the spinal column. They consist of a double sheet of thin muscle. The **rotatores** (Fig. 6.72) are

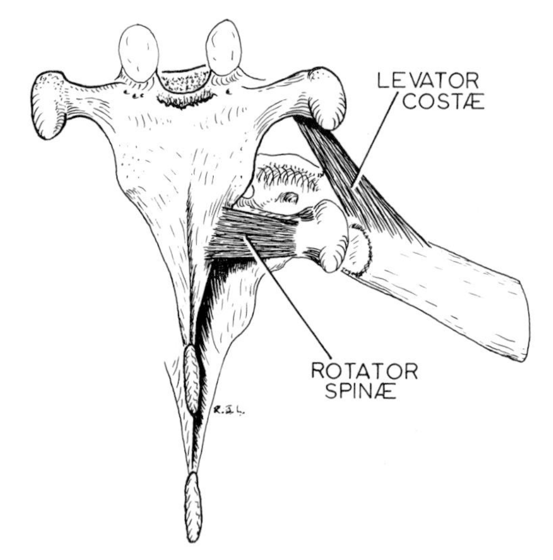

FIG. 6.72. THE ROTATOR SPINÆ AND LEVATOR COSTÆ MUSCLES. Drawn from Specimen S.242.2, in R.C.S. Museum.

interesting in that they are confined to the thoracic spine, the only region where pure rotation occurs. Each extends from the base of a transverse process to the root of the spinous process of the vertebra above; due to the declination of the spinous processes the course of the muscle fibres is horizontal, giving them the best possible leverage.

In the suboccipital triangle these muscles are specialized into the recti and obliques.

The Intermediate Layer. This is much more massive than the deepest layer. It consists of two sheets of oblique fibres that slope up to the spinous processes from the transverse processes. The deeper fibres constitute the multifidus spinæ and they are shorter (cross fewer vertebræ) than the fibres of the semispinalis which overlie this muscle. The levatores costarum make the third group of muscles in this layer.

The **multifidus** fibres slope upwards from the laminæ and mamillary processes to the spinous processes of vertebræ 2 or 3 above their level of origin. They commence at the upper part of the sacrum and extend to the upper part of the neck (C2).

Semispinalis lies on the surface of multifidus. Its fibres arise from the transverse processes and slope steeply upwards to the spinous processes. It extends from the lower thoracic region to the skull. **Semispinalis thoracis** extends from the transverse processes of the lower six thoracic vertebræ and each part is inserted into the spinous process of the eighth vertebra above it. **Semispinalis cervicis** arises in continuity at a high level; the uppermost part of the muscle is inserted into the concavity of the bifid spinous process of the axis. It is a powerful muscle (Fig. 6.74).

Semispinalis capitis is the most powerful part of this layer. It arises from the transverse processes of the upper six thoracic and lower four cervical vertebræ and is inserted into the occipital bone near the midline, between the superior and inferior nuchal lines. Its medial part is usually somewhat separated from the main mass; a tendinous intersection usually present denotes the segmental origin of the muscle. It covers the semispinalis cervicis and lies beneath splenius and trapezius. It can be seen in contraction in a thin neck, for the overlying muscles are but flat sheets (PLATE 33). It is the chief extensor of the head, and in the erect position or in leaning forward it is in isometric contraction as a muscle of posture. Like the soleus muscle in the calf it contains a large plexus of veins within as well as around it.

The ascending branch of the transverse cervical

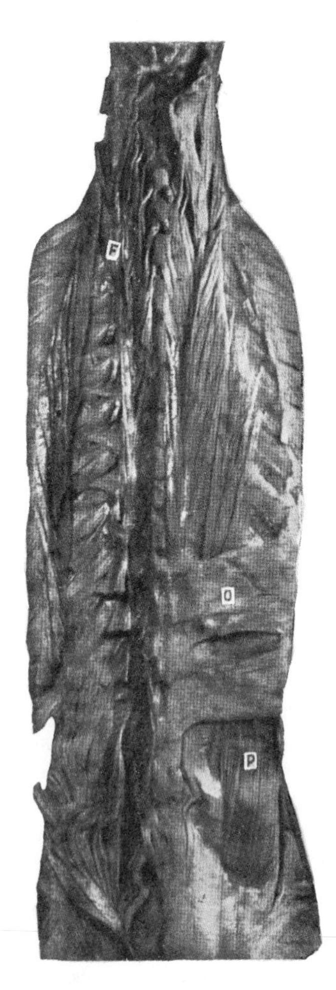

FIG. 6.73. PORTION OF THE ERECTOR SPINÆ MUSCLE. On the right 'P' is the sacro-spinalis muscle and 'O' part of latissimus dorsi. Above this the powerful mass of longissimus thoracis fills the gutter between ribs and vertebræ. On the left 'F' marks longissimus cervicis. Photograph of Specimen S 242 in R.C.S. Museum.

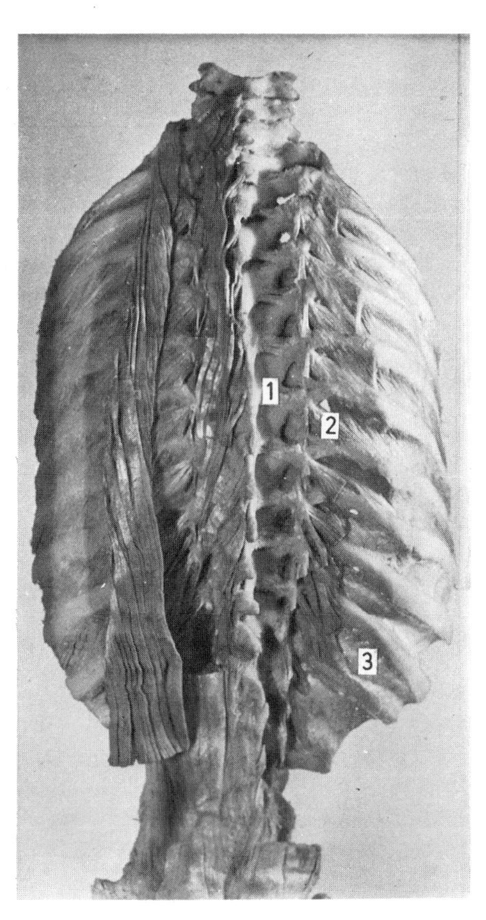

FIG. 6.74. PORTION OF THE ERECTOR SPINÆ MUSCLE. On the left ilio-costalis and semispinalis are shown. On the right 1, indicates the rotatores spinæ, 2, the levatores costarum, and 3, the internal intercostal membrane. Photograph of Specimen S 242. 2, in R.C.S. Museum.

artery passes up behind the muscle, the deep cervical artery (the ascending branch of the costo-cervical trunk) passes up in front of the muscle. Each anastomoses with a branch of the occipital artery. It is their corresponding veins that form the plexus already mentioned.

The **levatores costarum** are fan-shaped muscles spreading down from near the tip of each transverse process to be inserted into the upper border of the next rib, lateral to its tubercle. Incorrectly said to be supplied by anterior primary rami, the muscles are, in fact, supplied by posterior primary rami and are correctly included here in the erector spinæ group (Fig. 6.72).

The Superficial Layer. The superficial layer is the most powerful of all. It consists of three parts lying side by side, medial, intermediate and lateral. It commences below, deep to the lumbar fascia, on the back of the sacrum and the inner side of the iliac crest; this part is named *sacro-spinalis*. The thick mass of fibres diverges up and divides into two main bundles. The lateral part of the muscle is called **ilio-costalis.** It is inserted by shining tendons into the angles of the lower six ribs (Fig. 6.74). From these attachments new muscle bundles arise and each runs up to be attached to the angle of the sixth rib above. This lateral band is the **costalis.** From the attachments of costalis new fibres arise to run upwards to be inserted into the transverse processes of the lower four cervical vertebræ. These constitute the **costo-cervicalis.** It should be noted that this mass of ilio-costo-cervicalis forms the most lateral part of the superficial layer of the erector spinæ. Fibres from below are attached into ribs as far lateral as possible, beneath the thoraco-lumbar fascia, just within the angle of each rib. These fibres are replaced at their insertions by new fibres which arise by tendons on their medial side to pass upwards.

The intermediate part of the massive bundle arising from the sacrum and iliac crest passes up to be inserted into the gutter between transverse processes and ribs. This is the **longissimus thoracis.** At its insertion it is replaced by new fibres on the medial side that pass up to the transverse processes of the lower cervical vertebræ. This is the **longissimus cervicis.** From these insertions new bundles arise and pass upwards as **longissimus capitis;** this muscle is inserted into the mastoid process deep to the splenius capitis. It overlies the lateral border of semispinalis capitis and is covered over by the splenius (Fig. 6.73).

The third (medial) part of the superficial layer is not very strong. It is the **spinalis** muscle. Its fibres run alongside the spinous processes and the longitudinal mass of muscle is divided into *spinalis thoracis* and *spinalis cervicis.* Neither is very strong, and they should not be confused with the interspinales of the deepest layer of the erector spinæ.

In the neck, the underlying extensor muscles are bound down by the **splenius.** This muscle arises from the upper six thoracic spinous processes and the supraspinous ligaments between them, and from the ligamentum nuchæ. The fibres slope upwards and laterally. They lie deep to trapezius and sterno-mastoid (PLATE 33) and are inserted into the superior nuchal line and mastoid process (deep to the sterno-mastoid) and into the transverse processes of the upper three or four cervical vertebræ (deep to levator scapulæ). The whole muscle is like a bandage that holds down the deeper extensor muscles at the back of the neck. It and all the muscles deep to it are supplied by posterior primary rami.

The Back of the Neck

The back of the neck consists of muscles connecting the skull to the spine and pectoral girdle and, beneath these muscles, of two parts which lie above and below the prominent spine of the second cervical vertebra. In the midline the ligamentum nuchæ separates the two sides and the flat sheet of oblique muscle fibres named the splenius separates the deep extensor mass (supplied by posterior primary rami) from the superficial muscles supplied by the accessory nerve (PLATE 33).

The superficial muscles are the trapezius and sternomastoid, which are attached edge to edge along the superior nuchal line of the occipital bone. From this attachment they diverge to the pectoral girdle, enclosing within their borders the apex of the posterior triangle of the neck. Above the attachment of these muscles (the superior nuchal line) lies the scalp, while below this level the fleshy mass of the neck is seen; the junction of the two marks the attachment of the tentorium cerebelli and the line of the transverse sinus within the cranium. The muscles of the back of the neck cover the posterior cranial fossa and the cerebellum; above their attachment is the supratentorial compartment containing the occipital lobes of the cerebral hemispheres (PLATE 34).

The **ligamentum nuchæ** is a strong band of elastic tissue well developed in those mammals with protuberant necks and heads. In man, with his vertical neck, it consists of no more than a midline intermuscular septum of fibrous tissue. Triangular in shape, its three borders are attached to the external occipital crest, the bifid spines of the cervical vertebræ and the investing layer of deep cervical fascia which encloses the trapezius muscle (PLATE 23).

Beneath the trapezius and sterno-mastoid lies the splenius muscle. Beneath the splenius lie the semispinalis capitis and longissimus capitis muscles. When all these are removed the deeper structures of the back of the neck are seen to be divided into upper and lower portions by the prominent backward projection of the massive spinous process of the second cervical (axis) vertebra. Below this level the semispinalis cervicis is seen converging almost vertically upwards to the internal surfaces of the bifid axial spine. Above the spine lie the right and left suboccipital triangles.

The **suboccipital triangle** is bounded by the rectus capitis posterior major and the superior and inferior oblique muscles. Its floor contains the posterior arch of the atlas and the posterior atlanto-occipital membrane. Across the floor runs the vertebral artery, and through the floor emerges the suboccipital (first cervical) nerve. Across the roof run the greater occipital (C2) nerve and the occipital artery (Fig. 6.75).

The **rectus capitis posterior major muscle** is badly named. It is not vertical, but oblique. It arises from the outer surface of the bifid spinous process of the second cervical vertebra and extends obliquely upwards and outwards to be attached to the lateral

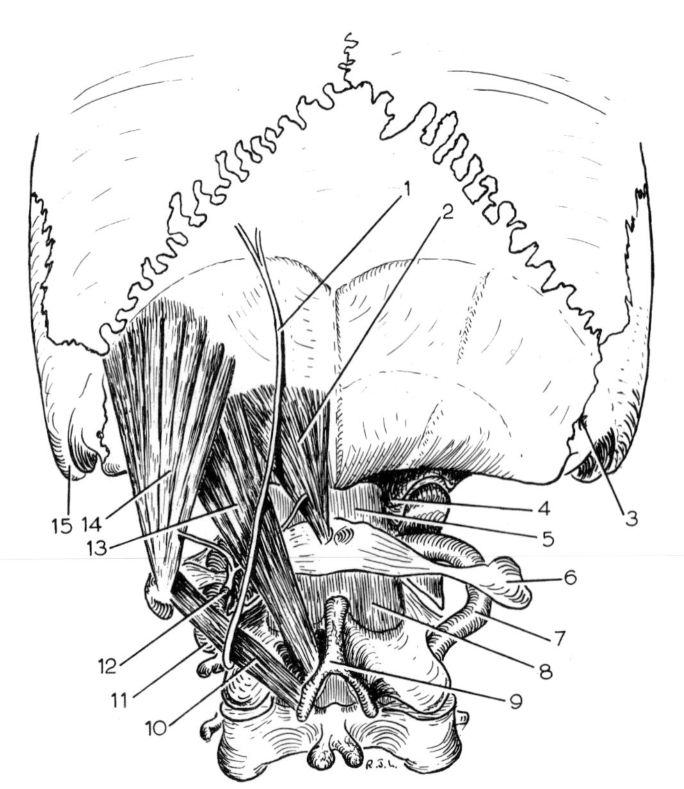

FIG. 6.75. THE SUBOCCIPITAL REGION AND THE SUBOCCIPITAL TRIANGLE. **1.** Great occipital nerve. **2.** Rectus capitis posterior minor. **3.** Mastoid emissary foramen. **4.** Occipital condyle. **5.** Posterior atlanto-occipital membrane. **6.** Transverse process of atlas. **7.** Vertebral artery. **8.** Ligamentum flavum (C1–C2). **9.** Spinous process of axis. **10.** Inferior oblique muscle. **11.** Vertebral artery. **12.** Posterior primary ramus of C1. **13.** Rectus capitis posterior major. **14.** Superior oblique muscle. **15.** Mastoid process.

part of the area below the inferior nuchal line. It is supplied by the posterior primary ramus of C1, and its action is to extend the head and rotate it (with the atlas) back towards its own side.

The **inferior oblique muscle (obliquus capitis inferior)** is attached between the outer surface of the bifid spine of the axis (below rectus capitis posterior major) and the back of the lateral mass of the atlas (Fig. 6.71). It is supplied by the posterior primary ramus of C1, and its action is to rotate the atlas (and the skull with it) back towards its own side.

The **superior oblique muscle (obliquus capitis superior)** extends from the back of the lateral mass of the atlas to the lateral part of the occipital bone between superior and inferior nuchal lines. It is supplied by the posterior primary ramus of C1. Attached to atlas and skull it can only move one on the other—the movement is of lateral flexion of the skull combined with slight extension.

The **rectus capitis posterior minor** is the only muscle attached to the posterior arch of the atlas. It arises from a small fossa near the midline and passes vertically upwards to be inserted into the medial part of the area below the inferior nuchal line. It is supplied by the posterior primary ramus of C1 and its action is to extend the head.

The actions of the small muscles as stated, and those of the recti capitis anterior et lateralis (p. 374) are of use less as prime movers than as synergists. They adjust and stabilize the skull by cancelling out any unwanted secondary effects of such powerful prime movers as sterno-mastoid, splenius, semispinalis, longissimus, etc.

The second part of the **vertebral artery** ascends through the foramina transversaria of the upper six cervical vertebræ, anterior to the emerging spinal nerves. It gives a *spinal branch* into each intervertebral foramen. Its course from the sixth to the second cervical vertebra is vertical. Between the foramina transversaria of the axis and atlas it passes laterally, and here has a pronounced posterior convexity. This must be to allow for taking up slack during rotation of the atlas on the axis (Fig. 6.96, p. 492).

The **vertebral veins** exist only in the neck, outside the skull. Some of the blood from the muscles of the suboccipital triangle is collected in a plexus of veins that descends alongside the vertebral artery, within and without the foramina transversaria. From this plexus two vertebral veins usually emerge, one from the sixth with the vertebral artery and one alone through the foramen transversarium of the seventh cervical vertebra. The vertebral plexus of veins receives the segmental drainage of the cervical part of the spinal cord along the nerve roots, as well as drainage from the internal vertebral plexus (see p. 493). The two vertebral veins join the brachio-cephalic vein in the root of the neck, on the apex of the lung (Fig. 4.32, p. 248).

The vertebral artery on emerging from the foramen transversarium of the atlas passes medially behind the atlanto-occipital joint. Here it lies in the floor of the suboccipital triangle before piercing the lateral angle of the posterior atlanto-occipital membrane. It deeply grooves the posterior arch of the atlas. In this groove, between artery and bone, lies the **suboccipital (first cervical) nerve.** The posterior primary ramus of this nerve passes backwards to supply the two recti, the two obliques and the upper fibres of semispinalis capitis. The anterior primary ramus winds around the lateral side of the atlanto-occipital joint and passes forwards (Fig. 6.68) to join the cervical plexus. Neither branch reaches the skin.

The posterior primary ramus of C2 (the **greater occipital nerve**) emerges below the posterior arch of the atlas. It curls around the lower border of the inferior oblique muscle and passes upwards across the roof of the suboccipital triangle (Fig. 6.75). It pierces semi-spinalis capitis (first supplying it) and extends up to supply the skin of the scalp up to the vertex (PLATE 33).

The **occipital artery** is a large vessel that passes back along the occipito-mastoid suture of the skull deep to the digastric and longissimus capitis muscles. It runs across the upper part of the roof of the suboccipital triangle and passes to the scalp on the lateral border of semispinalis capitis. It gives off descending branches that lie both superficial and deep to the semispinalis capitis, where they anastomose with ascending branches of the transverse cervical and costo-cervical arteries. The companion veins form a rich plexus around and within semispinalis capitis.

OSTEOLOGY OF THE VERTEBRÆ

A simple review of the principal features of vertebræ is made on p. 458, with enough detail to understand how vertebræ articulate into a column that works. The following accounts of vertebræ are intended *only* for those who must have more complete details of the individual bones.

Thoracic Vertebræ

It is better to begin with these, since the costal elements ossify separately as thoracic ribs. *Examine the articulated column*, allowing for possible distortion introduced by artificial discs. Note the backward convexity of the normal kyphosis; this results in the upper thoracic vertebræ leaning forwards, so their transverse processes are tilted up a little. This tilts the corresponding rib necks. The neck of the first rib, for example, slopes upwards from the head to the tubercle (PLATE 11). The concave facets on these upper transverse processes thus look somewhat downwards, while the flat facets on the lower transverse processes look somewhat upwards.

The spinous processes slope downwards with gradually increasing declivity as far as the 7th, below which they very gradually level out to become almost horizontal at the 12th. Thus the tips of the upper four spines lie opposite the tubercles of ribs one lower in the series, the next four spines (those of the 5th, 6th, 7th and 8th) opposite those two lower in the series, and the spines of the last thoracic vertebræ are opposite ribs only one lower in the series. The prominent spinous process of the 'vertebra prominens', seen and felt in the living at the upper end, is usually that of the 7th cervical, but it may be that of the 1st thoracic, so it is uncertain to count spinous processes downwards from this prominence (they should be counted up from the L4–5 interval, p. 196).

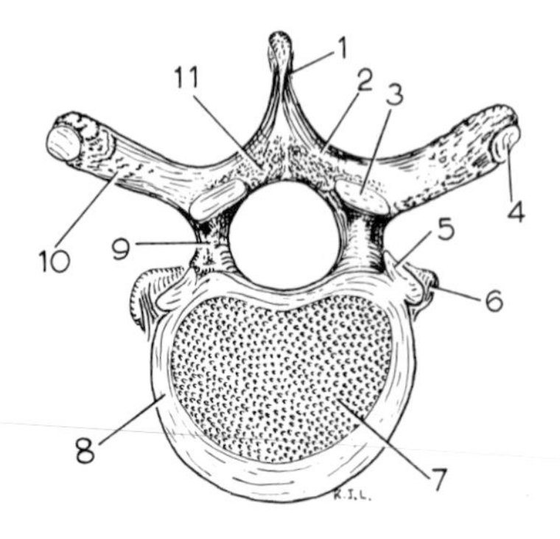

FIG. 6.76. A TYPICAL THORACIC VERTEBRA (SUPERIOR VIEW). **1.** Spinous process. **2.** Lamina. **3.** Articular process. **4.** Costal facet (for rib tubercle). **5.** Costal facet (for rib head). **6.** Process for inferior costal facet (for lower rib head). **7.** Cancellous bone of body. **8.** Epiphyseal ring of compact bone. **9.** Pedicle. **10.** Transverse process. **11.** Area for ligamentum flavum.

Only a narrow space separates adjacent laminæ, which are wide and imbricated. The transverse processes are all of corresponding length except the much shorter 12th and sometimes 11th. The bodies become more massive from above down; the middle four often show slight asymmetry, with a left-sided excavation produced by the aorta.

A Typical Thoracic Vertebra (Fig. 6.76). The essential characteristic is, of course, the presence of costal facets. On the body these consist of a pair of demi-facets. The upper of these is on the body at its junction with the upper border of the pedicle. It is semicircular in outline and lies vertical. The lower is smaller, and faces downwards from the lower border of the body. These two facets lie on the body but not on the centrum (in morphological language they are on the neural arch). Covered with hyaline cartilage, each makes a separate synovial joint with one of the facets of a rib head (p. 212).

The body is concave from above down around its circumference, and the surface is perforated by multiple vessels. The anterior convexity gives attachment to the anterior longitudinal ligament. From the 4th downwards the body shows a sharper curve around this anterior convexity, giving the upper and lower surfaces a heart-shaped outline. This outline is asymmetrical from a slight left-sided excavation produced by the descending aorta. The posterior surface of the body is concave from side to side, making with the laminæ a spinal canal that is almost circular in outline (in lumbar and cervical vertebræ the canal is more triangular in outline). Two large foramina separated by a strip of bone (looking like a pair of nostrils) open centrally on the back of the body. They are for the basi-vertebral veins. They are spanned by the narrow part of the posterior longitudinal ligament, which is attached only at the upper and lower borders of the body (and to the adjacent discs). The upper and lower surfaces of the body are similar. Each surface is enclosed in a heart-shaped ring of compact bone at the margin. This is the fused epiphyseal ring of the body, and it encloses a large central area in which cancellous bone reaches the surface. Both compact and cancellous areas are covered in life by a thin plate of hyaline cartilage, to which the fibres of the annulus fibrosus are attached (p. 460).

The *pedicle* projects back from the upper half of the body. Its upper border is level with the upper surface of the body. Thus the body itself takes no part in the formation of the intervertebral foramen above it (Fig. 6.64, p. 458). The upper border of the pedicle curves up to a superior articular process. The lower border curves down to the inferior articular process. The concave excavation in the lower border makes

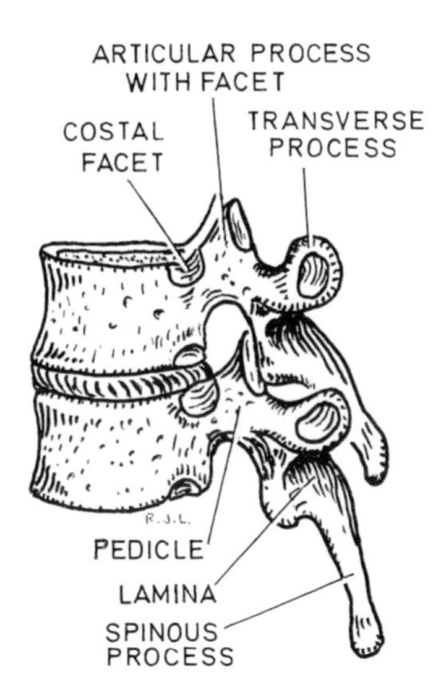

FIG. 6.77. TWO MID-THORACIC VERTEBRÆ VIEWED FROM THE LEFT SIDE. The intervertebral disc is in position. The intervertebral foramen is bounded in front by half the body of the upper vertebra and the intervertebral disc (cf. lumbar vertebræ and contrast cervical vertebræ, Fig. 6.84).

the upper border of the intervertebral foramen. This accommodates the thoracic nerve of the same number.

The flat *laminæ* slope down from the pedicles to unite in the midline and complete the neural arch. Their upper borders slope down to form a deep notch. Between the superior articular processes this notch is grooved from side to side, along the *posterior* surface of the lamina, for attachment of the ligamentum flavum which goes up to the vertebra above. The lower border of the lamina is likewise deeply grooved between the inferior articular processes, but this groove lies along the *anterior* surface of the lamina. It is for attachment of the ligamentum flavum coming up from the lamina below. Thus alternate laminæ and ligamenta flava are imbricated, like overlapping tiles of a roof.

The superior *articular process* projects up from the junction of pedicle and lamina. It carries an oval articular facet, set very steeply, facing backwards and slightly laterally; so the two articular surfaces lie on the arc of a circle, permitting rotation of adjacent bodies. The inferior articular facets are reciprocal, facing forward and somewhat medially. They project below the pedicle at the inferior angle of the lamina. Upper and lower facets, each cartilage-covered, form synovial joints, with capsule and synovial membrane attached to the articular margins. The ligamentum flavum extends from one capsule across to the other and fills the bony hiatus between adjacent laminæ. The only exit from the spinal canal is the line of intervertebral foramina.

The *spinous process* slopes down from the junction of the laminæ. It tapers from a broad base down to a tip that is expanded. Weak interspinous ligaments and interspinal muscles connect adjacent sharp borders of the processes, while the thick and strong supraspinous ligament connects their expanded tips. The thoracolumbar fascia is attached here, with trapezius overlying splenius and latissimus dorsi. The base of the spinous process receives the rotator spinæ muscle from the vertebra below, and more superficially the multifidus, semispinalis and spinalis muscles are inserted into each side as far as the tip.

The *transverse process* projects backwards as well as laterally from the junction of pedicle and lamina. Its anterior surface expands towards the tip, to carry the characteristic costal facet. The upper six costal facets are concave, the lower ones flat (the hyaline cartilage that coats them is not always of uniform thickness, so the facet on the dried bone may not correspond accurately with the living cartilaginous facet). The tip of the transverse process gives attachment to the lateral costo-transverse ligament. The root of the transverse process, adjoining the lamina, gives origin to a rotator spinæ muscle. Upper and lower borders carry weak intertransverse ligaments and muscles; the lower margin gives origin to a levator costæ muscle, and receives the superior costo-transverse ligament from the neck of the rib below. The posterior surface carries erector spinæ attachments (semispinalis, longissimus) and the anterior surface is covered by the attachment of the inferior costo-transverse ligament.

The First Thoracic Vertebra. The body is broad, not heart-shaped. Longus colli (cervicis) is attached to its anterior convexity over the anterior longitudinal ligament. On the body, at its junction with the pedicle, is a large facet for the single articular surface of the head of the first rib. There is a demi-facet for the second rib at the lower border of the body. The pedicle is attached below the upper margin of the body, as in a cervical vertebra, so the body takes part in the formation of the intervertebral foramen above it as well as below it. These are the foramina for exit of C8 and Th.1 spinal nerves.

The Eleventh Thoracic Vertebra. This carries a single costal facet, behind the body on the *upper* part of the broad pedicle (Fig. 6.78). The articular facets are typically thoracic. The transverse process is often stunted. At times these features are shown also by the 10th thoracic vertebra.

The Twelfth Thoracic Vertebra. The broad pedicle carries a single costal facet near its *lower* margin (Fig. 6.78). The transverse process is stunted and its base projects upwards into a rounded mamillary process behind the articular facet and downwards into a sharp accessory tubercle. The posterior primary ramus of Th.11 lies in the groove between them. The superior articular facet looks backwards like a typical thoracic

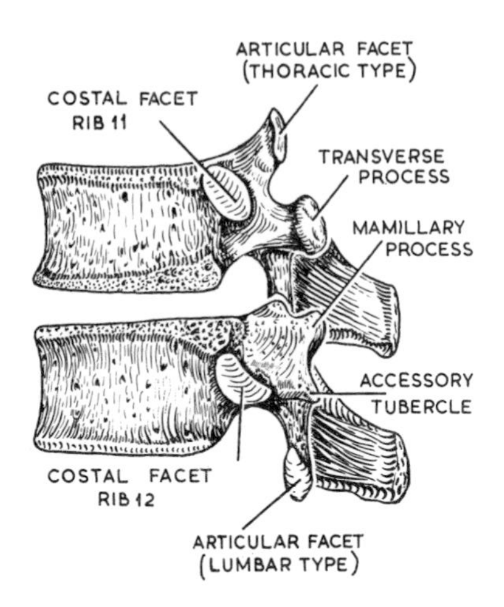

FIG. 6.78. THE 11TH AND 12TH THORACIC VERTEBRÆ VIEWED FROM THE LEFT SIDE.

facet. The inferior articular facet is lumbar in type, a vertical cylinder facing laterally. The upper limit of psoas major attachment is at the lower border of this vertebra, and psoas minor arises just above this.

Lumbar Vertebræ

The normal lordosis of the lumbar vertebræ and the angulation between the fifth and the sacrum (Fig. 5.55, p. 323) are best appreciated by examination of a fresh specimen; artificial discs used in the articulated skeleton are seldom thick and wedged enough. The *bodies* may be wedge-shaped, deep in front and shallow behind; this is especially so of the fifth. But as often as not they show no wedging, and if the five dry bones are piled on each other their spinous processes make a backward convexity. When separated by discs of normal thickness the bodies fan out enough to produce a backward concavity of the spines. The wedge-shaped discs themselves, not the bodies, produce the normal lordosis. The massive bodies increase in breadth from above down, and this is reflected posteriorly by a progressive widening between the articular processes. Thus in each of the upper three vertebræ the four processes make a rectangle set vertically (Fig. 6.80). In the fourth they are square and in the fifth they make a horizontal rectangle (though the best way to recognize the fifth is by a glance at its transverse process). The *transverse processes* are variable in length, but the fourth is usually the longest. In the upper four they are spatulate and set well back on the pedicle (Figs. 5.49, p. 318 and 6.65, p. 459). The transverse process of the fifth, however, is quite characteristic. Short, massive, pyramidal, its base is attached from the pedicle *well forwards on the lateral side of the body itself* (Fig. 6.88).

From this general survey of the lumbar spine, study the individual characteristics of a single lumbar vertebra (Fig. 6.79). The *body* shares with the smaller thoracic vertebræ the characteristics of being concave from above down, of having pedicles attached to its *upper* half, and of being perforated by a pair of basivertebral veins posteriorly. It differs from the thoracic vertebra in being kidney- not heart-shaped, and the posterior surface is flatter, less concave from side to side, so the spinal canal is not circular but somewhat triangular in cross section. Anterior and posterior longitudinal ligaments are attached as in the thoracic vertebræ. Psoas major is attached to the upper and lower borders, but not to the concavity of the body, where psoas spans the vertebra on a fibrous arch. Lumbar vessels and sympathetic rami curve around beneath the fibrous arch.

The *pedicles* enclose intervertebral foramina identical in formation with the thoracic foramina. The *laminæ* do not show such a downward slope as in the thoracic vertebræ, but the upper border is grooved posteriorly and the lower border is grooved anteriorly in identical manner, for attachment of the imbricated ligamenta flava.

The *spinous process* is roughly horizontal. The upper

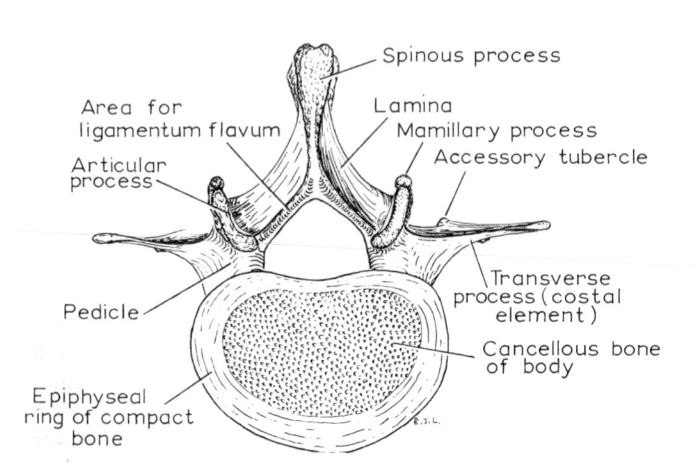

FIG. 6.79. A LUMBAR VERTEBRA. SUPERIOR VIEW.

border is straight but the lower border curves down, producing the so-called hatchet shape. The interspinous ligaments are attached to adjacent upper and lower borders. The posterior border is much thickened for the supraspinous ligament. Interspinales and multifidus muscles are attached to the spines.

The *transverse processes* have already been noticed. The upper four, spatulate, are characterized by a vertical ridge on the anterior surface (palpable if not readily visible). This is for attachment of the psoas fascia and the anterior lamella of the lumbar fascia. The medial and lateral arcuate ligaments are attached to the ridge on the first lumbar transverse process. Medial to the ridge psoas is attached, while lateral to the ridge quadratus lumborum receives partial insertion. The tip of the transverse process receives the middle lamella of the lumbar fascia (Fig. 1.24, p. 22). Adjacent borders carry intertransverse ligaments and muscles. The posterior surface receives the attachments of sacro-spinalis.

The *articular processes* are characteristic. The upper pair rise up and carry articular facets that face medially (the upper facets of the 4th and 5th face posteriorly as well as medially). The articular surfaces are cylindrical, being concave from front to back (Fig. 6.79). The lower pair of articular processes project down from the lateral angles of the laminæ and are mortised into the superior processes of the vertebra below. Each carries a reciprocal convex facet, a section of a vertical cylinder. It has already been explained that the transverse processes are fused ribs (costal elements), and now the true transverse element of the lumber vertebra can be seen. It consists of two elevations with a groove between them made by the medial branch of the posterior primary ramus of the overlying lumbar nerve (Fig. 6.80). The *mamillary process* is a 'breast-shaped' convexity projecting back from the superior articular process behind the margin of the articular facet. The *accessory tubercle* lies below this, at the root of the transverse process; it varies from a prominent sharp spike to complete absence. The mamillary process and accessory tubercle together represent the stunted transverse process of a thoracic vertebra, and they carry

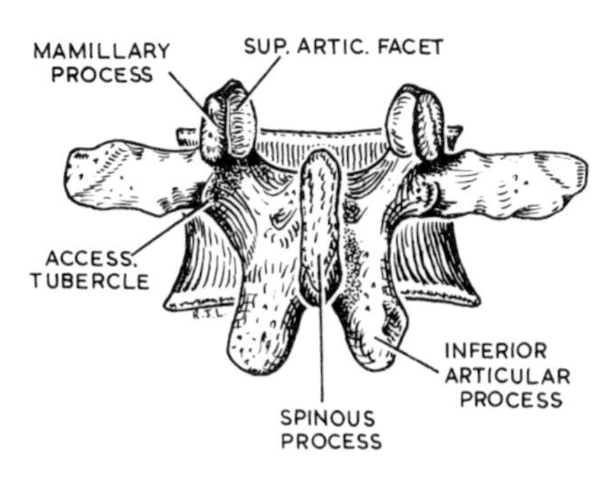

FIG. 6.80. POSTERIOR VIEW OF THIRD LUMBAR VERTEBRA.

similar muscle attachments (multifidus, longissimus) of the erector spinæ musculature. (Note the same arrangement in the 12th thoracic vertebra in Fig. 6.78.)

The Fifth Lumbar Vertebra. The characteristic pyramidal transverse process has been noted above (see Fig. 6.88). It carries the strong ilio-lumbar ligament, and quadratus lumborum arises from both. The anterior and middle lamellæ of the lumbar fascia are attached to the process and the ligament. The inferior articular processes face well forwards, and are received into backward-facing facets on the sacrum, and this locking prevents the fifth lumbar vertebra from sliding forwards down the slope of the first sacral vertebra. Furthermore, the adjacent bodies are strongly united by the intervertebral disc. Thus, although the sloping lumbosacral joint carries the whole body weight, it is extremely stable. A strongly contracting sacro-spinalis muscle acts as a supporting strap posteriorly.

The fifth lumbar vertebra may be fused on one or both sides to the first sacral vertebra, a condition known as 'sacralization'.

The Sacrum

Five progressively smaller vertebræ and their costal elements fuse to make this bone, which is triangular in outline and curved to a concavity towards the pelvis. It carries the whole body weight. It articulates with the ilium to make the upper posterior wall of the pelvic basin but the sacro-iliac joint so formed *is not weight-bearing*. The sacrum is slung on ligaments above and behind the joint, and they carry the body weight. Below the sacro-iliac joints the sacrum tapers off down to its apex. The anatomical position of the sacrum can be assessed by inspecting the articular facets for the fifth lumbar vertebra. These surfaces are cylindrical, curved from side to side but flat from above down; *the concave cylinder lies vertical*. The upper surface of the first sacral body is thus seen to slope down at 30 degrees or more, and from here the sacrum is directed backwards before curving down over the pelvic cavity (Fig. 5.67, p. 340).

Anterior Surface. (Pelvic surface.) This concave surface is smooth. In the midline five diminishing bodies are fused, with four ridges persisting to mark the lines of ossification. These transverse lines represent the intervertebral discs. On each side the four anterior sacral foramina are large but diminish from above down. The rounded bars of bone between adjacent foramina (costal elements) represent the heads and necks of ribs. The medial boundaries of the anterior sacral foramina are thus formed by the bodies of the sacral vertebræ, but the other three-fourths of their circumference is costal in origin. The rounded bar of bone above the first sacral foramen continues the arcuate line of the ilium to form the posterior part of the pelvic brim.

This continues medially to the prominent anterior lip of the first sacral body, known as the *promontory* of the sacrum.

The mass of bone lateral to the foramina is formed by fusion of the costal elements (shafts of ribs) with each other; it is known as the lateral mass. It is deeply indented by grooves for the anterior primary rami of the upper four sacral nerves, which pass laterally from the anterior sacral foramina. The piriformis muscle arises from the three ridges (costal elements) that separate the anterior foramina, and from the lateral mass nearby. Below the promontory peritoneum is draped over the upper two bodies, but the retroperitoneal rectum lies against the lower three bodies. The fascia of Waldeyer and the superior rectal vessels lie between the sacrum behind and the peritoneum and rectum in front. The superior hypogastric plexus extends on to the promontory, and below this the fascia of Waldeyer is attached. The midline of the sacral hollow contains the median sacral artery and vein, with some lymph nodes alongside them (p. 329). On each side the sacral sympathetic trunk crosses alongside the promontory to lie medial to the sacral foramina. Lateral to the foramina lies the sacral plexus on the piriformis muscle. The plexus and muscle are covered by a dense sheet of parietal pelvic fascia, on which the lateral sacral artery and vein lie lateral to the foramina (Fig. 5.59, p. 328).

Posterior Surface. This convex surface is irregular and rough. In the midline it is closed by fusion of adjacent laminæ. A gap above due to the caudal slope of the first sacral laminæ is closed by the ligamentum flavum attached to the laminæ of the fifth lumbar vertebra. A *hiatus* below, variable in its extent, indicates failure of fusion of the laminæ of the fifth and often of the fourth sacral vertebræ. This hiatus is closed by fibrous tissue. Adjacent spinous processes are fused with each other to produce a midline ridge that projects dorsally from the fused laminæ. It is called the *median sacral crest* (Fig. 6.81). The superior articular process, on the first sacral vertebra, carries a concave cylindrical facet for the synovial joint with the fifth lumbar vertebra. Below this, medial to the posterior foramina and at the lateral margin of the fused laminæ, is a line of irregular tubercles that represent fusion of adjacent articular processes of the sacral vertebræ. This low ridge forms the '*articular*' *crest*, and it is projected below, alongside the sacral hiatus, to end in the rounded *sacral cornu*, for articulation with the coccyx. Lateral to the superior articular process is a prominent boss of bone which is the transverse process of the first sacral vertebra. Below this the transverse processes are fused with each other, making a ridge lateral to the posterior foramina. This ridge is called the *lateral sacral crest*, and its lower part is in line with the ridge on the posterior end of the iliac crest. It is marked by bosses of bone that represent the tips of the fused transverse processes. Thus the posterior sacral foramina are enclosed wholly by the fused sacral vertebræ, for the fused costal elements lie lateral to the lateral sacral crest. The groove between the median and lateral sacral crests is filled by the erector spinæ muscle, and the posterior lamella of the lumbar fascia that covers it is attached to both crests.

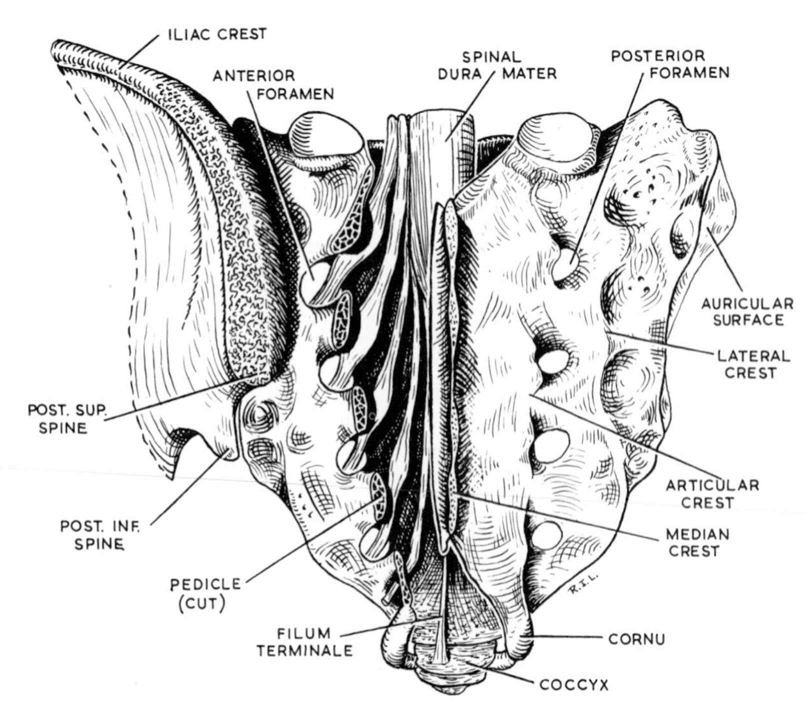

FIG. 6.81. THE SACRUM FROM BEHIND. The five pedicles and the laminæ have been cut through on the left side, to show the sheaths of dura mater around the nerve roots.

Between the lateral sacral crest, and the auricular surface the presence of three or four deep fossæ gives an undulating irregularity to the lateral mass. The whole of this area carries the attachments of the weight-bearing sacro-iliac ligaments. The *auricular surface* extends to the lateral border of the anterior surface and ala. Broad above and narrow below, the articular cartilage has an irregular surface from tubercles and depressions in the bone itself.

Below the weight-bearing postauricular surface the sacrum is more slender, and the lateral crest meets the lateral border of the sacrum at the level of the fourth sacral vertebra. This point is the apex of a small triangular area below the auricular surface. Part of the sacro-tuberous ligament with its attached gluteus maximus arises from the triangular area (Fig. 3.63, p. 190). Below this the lateral margin of the sacrum gives attachment to the sacro-spinous ligament and the coccygeus muscle. The apex of the sacrum is attached by an intervertebral disc to the body of the coccyx, and the cornua in some cases make tiny synovial joints with the cornua of the coccyx.

The *ala* of the sacrum projects laterally from the upper surface of the first sacral vertebra (Fig. 6.82). Its margin forms the brim of the pelvis. Laterally it gives attachment to the weak anterior sacro-iliac ligament, and iliacus arises from the ligament over the ala. The base of the ala is crossed by the sympathetic trunk. Lateral to this, on the ala, lie the lumbo-sacral trunk medially, the obturator nerve laterally, and the ilio-lumbar artery between them.

The **sacral canal,** triangular in cross section, curves with the sacrum. It is closed in front and behind. Pedicles project back from the upper half of each sacral body, making four intervertebral foramina which provide lateral exits as in the rest of the spinal canal. But these exits are blocked further out by the lateral mass, which must be circumvented via anterior or posterior sacral foramina. The fifth sacral foramen is formed behind by the cornua of the sacrum and coccyx (Fig. 6.81); emerging here are the insignificant fifth sacral and the coccygeal nerves. The sacral canal contains the dura mater, which extends down to the second vertebra. The dura mater supports the arachnoid and contains cerebro-spinal fluid and the roots of the sacral and coccygeal nerves. The dura mater is prolonged as tubular sheaths around the sacral nerve roots, which unite just distal to the spindle-shaped bulgings of the posterior root ganglia. Each of these lies just inside its own intervertebral foramen between the sacral pedicles (Fig. 6.81). The filum terminale, piercing the dura, runs down to blend with the periosteum on the back of the coccyx. The space around the dura mater and its prolongations is filled with loose fat and the internal vertebral plexus of veins.

Sex differences in the sacrum are pronounced. The most useful guide is the comparison of the width of the body of the first sacral vertebra with the width of the lateral mass, or ala. In the male the rugged vertebræ and narrower pelvis, in the female the slender vertebræ and broad pelvis, imprint their characteristics on the sacrum. The body is wider than the ala in the male, narrower than the ala in the female (Fig. 6.82). Other differences lie in the curvature of the bone. In the male the anterior surface is gently and uniformly concave, in the female it is flat above and turns forward more prominently below. The auricular surface occupies two and a half vertebræ in the male, only two in the female (who has more potential mobility at the sacro-iliac joint).

Clinical considerations. Fusion of the first and second sacral vertebræ may be incomplete and they may articulate on one or other side after the manner of lumbar vertebræ, producing unilateral 'lumbarization' of the first sacral vertebra.

Anæsthetic solutions put into the lower part of the sacral canal act on nerves after their issue from the dura mater. These nerves are S2, S3, S4, and the coccygeal plexus. The sacral nerves contain the pain fibres from the derivatives of the cloaca (bladder, vagina, rectum). Caudal anæsthesia is of use in midwifery, since pain fibres from the *cervix* are blocked and the perineum is likewise anæsthetized for passage of the fœtal head. Pain fibres from the *body* of the uterus enter the cord

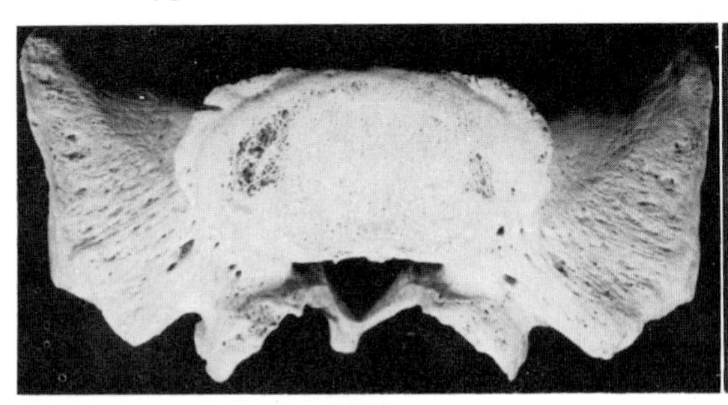

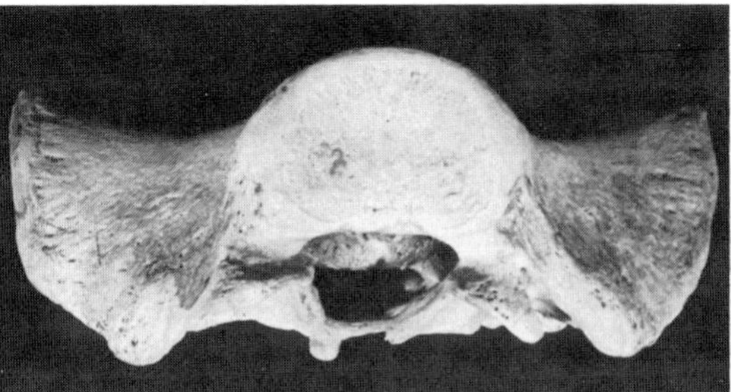

MALE FEMALE

FIG. 6.82. THE UPPER ASPECT OF THE SACRUM. In the male the body is broad and the alæ narrow. In the female the body is narrow and the alæ broad.

at Th.11 by way of the hypogastric nerves; these are unaffected by caudal anæsthesia. A needle introduced into the lower part of the sacral canal, through the inferior hiatus, must not be advanced beyond the third sacral vertebra for fear of piercing the spinal meninges and entering the subarachnoid space.

The Coccyx. This represents in man the multi-jointed tail of other vertebrates. It is contracted into four pieces fused together into a small triangular bone joined to the apex of the sacrum by an intervertebral disc and by tiny synovial joints at the lateral cornua. The upper surface of the coccyx is in the pelvic floor, the lower surface is in the buttock, beneath the skin of the natal cleft. The tip of the coccyx gives attachment to the ano-coccygeal raphé and to the intermediate part (called superficialis) of the external anal sphincter. Its borders give attachment to the ischio-coccygeus (coccygeus) muscle and sacro-spinous ligament, and to the overlapping posterior fibres of ilio-coccygeus and pubo-coccygeus.

Cervical Vertebræ

The cervical column makes a forward convexity. The widest of the transverse processes are those of the first vertebra, the atlas (Fig. 6.9, p. 375). The most massive of the spinous processes is that of the second vertebra, the axis; but this is not noticed in the living neck because it lies deep in the extensor muscles beneath the projecting occipital pole of the skull. The change of curvature between the cervical and thoracic parts of the column projects the 7th spinous process backwards. The prominence of the tip of the spinous process gives the name vertebra prominens to the seventh cervical vertebra—but it is fallacious to count on this, for the uppermost spinous process to be visible or readily palpable may be that of the first thoracic. The first and second cervical vertebræ are adapted to the functions of head nodding and rotation and the seventh has certain peculiar characteristics, so there are only four 'typical' cervical vertebræ, the 3rd, 4th, 5th and 6th.

A Typical Cervical Vertebra (Fig. 6.83). The broad kidney-shaped *body* is the same size as, or smaller than, the spinal canal (Fig. 6.65). On each side it is projected up into a lip, and its lower margin laterally is bevelled reciprocal with this. In the midline the lower margin is projected downwards a little. The flat upper and lower surfaces, each covered by a plate of hyaline cartilage, give attachment to the intervertebral disc. This flat area is the centrum of the vertebra. The upturned lips and bevelled margins are on the (morphological) neural arch. The hyaline cartilage on the centrum continues across these surfaces lateral to the disc attachments. A synovial joint is thus formed between each adjacent lip and bevel, and this is known as the *neuro-central synovial joint*. The basi-vertebral

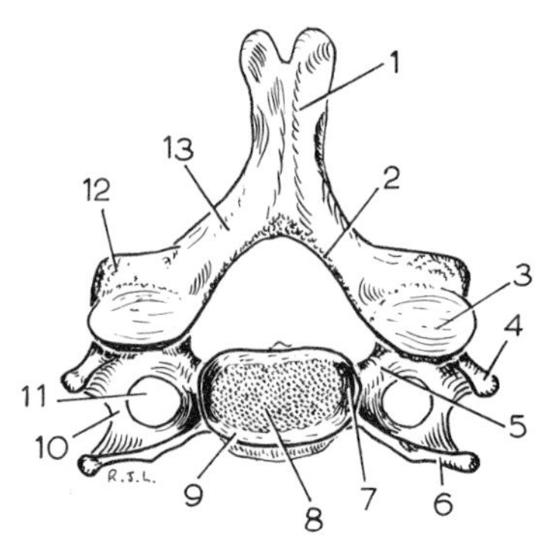

FIG. 6.83. THE SIXTH CERVICAL VERTEBRA (SUPERIOR VIEW). **1.** Spinous process. **2.** Area for ligamentum flavum. **3.** Articular facet. **4.** Posterior tubercle of transverse process. **5.** Pedicle. **6.** Anterior tubercle. **7.** Upturned lip of neuro-central junction. **8.** Cancellous bone of centrum. **9.** Epiphyseal ring of compact bone. **10.** Costo-transverse lamella. **11.** Foramen transversarium. **12.** Articular process. **13.** Lamina.

foramina on the posterior surface are wider apart than in thoracic and lumbar vertebræ. Basi-vertebral veins emerge from them, and the posterior longitudinal ligament spans the veins. The anterior surface of the body is concave from above down, and the anterior longitudinal ligament is firmly attached to this surface. Longus colli (cervicis) overlies the ligament.

The *pedicle* is attached below the upturned lip on the body. Thus an intervertebral foramen in the neck is bounded in front by *both* vertebral bodies and the synovial joint and disc between them (Figs. 6.84, 6.85). Contrast the foramina lower down, where the anterior boundary is the intervertebral disc and the body above it (Fig. 6.77, p. 471). Attached to the pedicle and body is the lateral projection of the transverse process, perforated by the typical foramen transversarium. The bar of bone that projects from the pedicle behind the foramen is the *true transverse element*, and it ends in the posterior tubercle, which is part of the costal element. The bar of bone that projects from the body in front of the foramen transversarium ends in the upturned anterior tubercle. The anterior tubercles enlarge progressively from the 3rd down to the 6th. The large 6th anterior tubercle is called the carotid tubercle (of

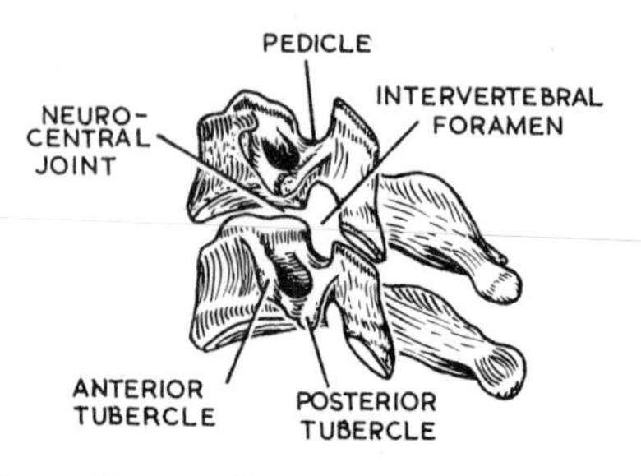

FIG. 6.84. THE 3RD AND 4TH CERVICAL VERTEBRÆ ARTICULATED TO SHOW THE BOUNDARIES OF THE INTERVERTEBRAL FORAMEN.

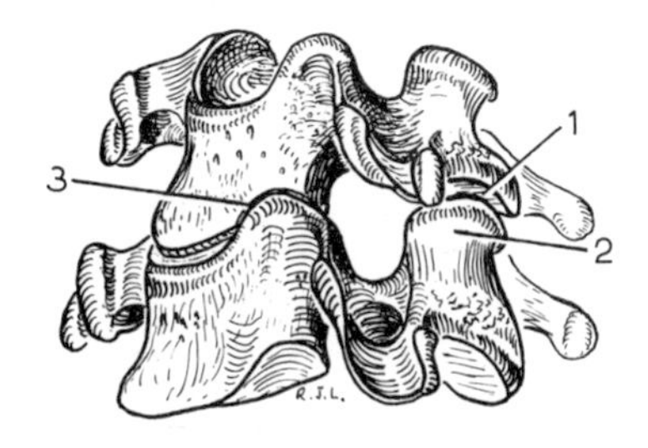

FIG. 6.85. THE C3–C4 INTERVERTEBRAL FORAMEN FROM THE ANTERO-LATERAL ASPECT. The foramen is bounded in front by the lower half of the upper vertebra and the disc, as in thoracic and lumbar foramina, but here in addition is the neuro-central synovial joint and part of the body of the lower vertebra. **1.** Synovial joint between neural arches. **2.** Articular process of C4. **3.** Synovial joint (neuro-central).

Chassaignac) because the common carotid artery can be compressed against it (a former method of producing unconsciousness, now superseded by modern anæsthesia). The anterior and posterior tubercles are joined by a down-curved costo-transverse lamella (Fig. 6.84). The anterior bar and tubercle, the lamella and the posterior tubercle are the costal element, a vestigial rib fused to the vertebra. The vertebral artery lies in the foramen transversarium, and the posterior root ganglion of the nerve of the same number lies behind it on the costo-transverse lamella (Fig. 6.96, p. 492). The anterior tubercle gives attachment to the tendons of longus capitis, scalenus anterior and longus colli (Fig. 6.9, p. 375). The posterior tubercle gives origin to scalenus medius. Levator scapulæ arises from the posterior tubercles down to the 4th, and scalenus posterior from those of the 5th and 6th (and 7th).

The laminæ enclose a relatively large spinal canal, somewhat triangular in cross section. Their borders are grooved for ligamenta flava (the back of the upper border and the front of the lower border, as in thoracic and lumbar vertebræ). At the junction of pedicle and lamina there are upper and lower *articular processes.* The two processes form a cylindrical column sliced off obliquely above and below for the articular facets. The upper facets face up and back, and usually both lie in the same plane. The lower facets face down and forward. The capsule and synovial membrane of these joints are attached to the articular margins.

The *spinous process* is usually bifid, and excavated inferiorly by a pair of concavities for semispinalis cervicis.

Atypical Cervical Vertebræ. The **vertebra prominens** (7th cervical) is atypical in that its long spinous process is not bifid like the others, and the foramen transversarium does not transmit the vertebral artery. The foramen is small, and contains the posterior vein when the vertebral vein is doubled. It is said to transmit the grey ramus from the inferior cervical ganglion to the anterior primary ramus of C7, but this is doubtful. The anterior tubercle is very small; to it are attached the scalenus pleuralis and the suprapleural membrane (PLATE 11).

The Atlas. The essential features are summarized on p. 462. To study further details a glance at the articular facets serves to distinguish upper from lower surfaces. The kidney-shaped facets for the occipital bone are deeply concave from front to back, and the two together make a more gentle concavity from side to side. In the 'hilum' of each kidney is a tubercle for attachment of the vital transverse ligament, which stabilizes the dens (odontoid process) of the axis.

The *anterior arch* is short and its posterior concave surface carries a concave cylindrical facet for the synovial joint with the dens. It is projected into a tubercle in front, for attachment of the anterior longitudinal ligament and longus colli (Fig. 6.68, p. 462). The upper border of the arch carries the anterior atlanto-occipital membrane between the capsules of the atlanto-occipital

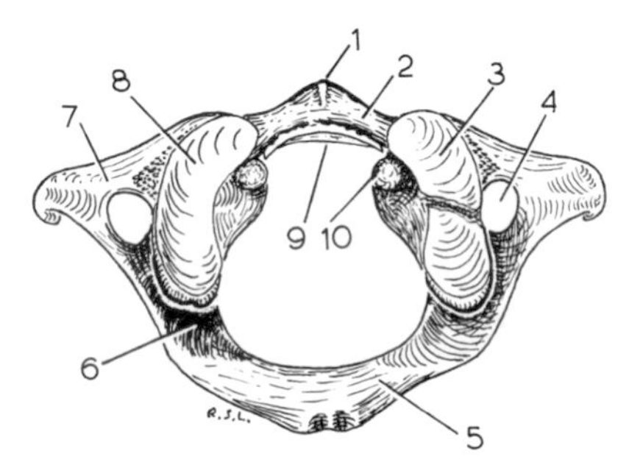

FIG. 6.86. THE ATLAS (SUPERIOR VIEW). **1.** Anterior tubercle. **2.** Anterior arch. **3.** Facet for occipital condyle (double, not unusual) on lateral mass. **4.** Foramen transversarium. **5.** Posterior arch. **6.** Groove for vertebral artery. **7.** Transverse process **8.** Facet for occipital condyle (single, normal). **9.** Facet for dens of axis. **10.** Tubercle for transverse band of cruciform (cruciate) ligament.

joints. The *posterior arch* is longer, and makes almost a semicircle. At the root of the arch the upper surface is grooved, below the projecting articular facet, by the vertebral artery and the first cervical nerve (p. 469). The groove is an unreliable guide for orientation of the disarticulated atlas, because it is often very shallow, and the under surface may show a similar groove for the posterior root ganglion of the 2nd cervical nerve. The upper border of the posterior arch gives attachment to the posterior atlanto-occipital membrane between the arterial grooves, while the lower border receives the ligamentum flavum from the axis. A pair of dimples at the posterior convexity of the arch (Fig. 6.86) give origin to the rectus capitis posterior minor muscles. The *lateral mass* carries the weight-bearing articular facets. In contrast to the deep concavity of the upper kidney-shaped surfaces the lower facets are circular and flat, or very gently concave. They articulate with the lateral mass of the axis. The lateral mass is projected into the transverse process, which is perforated by the foramen transversarium. There is usually no distinct formation of anterior and posterior tubercles, but all the bone anterior and lateral to the foramen transversarium is costal in origin.

The internal jugular vein, crossed by the accessory nerve, lies on the transverse process. Posteriorly the transverse process gives attachment to obliquus capitis superior and inferior, to splenius cervicis, and to levator scapulæ. Anteriorly the little recti capitis anterior and lateralis are attached to the lateral mass (Figs. 6.9, p. 375 and 6.75).

The Axis. The second cervical vertebra provides the pivot around which the atlas and head rotate. The dens (odontoid process) is quite characteristic, projecting up from the body between a pair of massive weight-bearing shoulders on the lateral mass (Fig. 6.69, p. 463). The weight is communicated from these shoulders through the body to the body of the 3rd vertebra; the articulations between the 2nd and 3rd vertebræ are typical of the rest of the column. The *dens (odontoid process)* carries anteriorly a cylindrical facet which makes a synovial joint with the anterior arch of the atlas. Its posterior surface is usually smooth, but no synovial joint exists here, only a bursa to lubricate the movements of the transverse ligament during head rotation. The apex of the dens carries the apical ligament (a remnant of the notochord) and the borders that slope down from the apex give attachment to the alar ligaments, one on either side. The *body* shows the bevel along the lateral margin of its lower surface typical of the cervical vertebræ, and in the midline the lower margin projects down a little. The lower surface articulates typically with the 3rd body by a disc and a pair of lateral neuro-central synovial joints.

The anterior longitudinal ligament is attached to the front of the body, with longus colli muscle and the prevertebral fascia overlying it. The posterior surface of the body gives attachment to the vertical limb of the cruciform ligament and, below it, to the membrana tectoria, which sweeps up from the attachment of the posterior longitudinal ligament. The upper surface of the body carries a pair of large facets. These slope down from the dens like shoulders, and each extends from the body well back onto the massive pedicle and well out on the lateral mass, where it overhangs the foramen transversarium (Fig. 6.69, p. 463). Each surface makes a synovial joint with the facet on the under surface of the lateral mass of the atlas. The costal element slopes steeply down from the body to end in a prominent transverse process. This represents the posterior tubercle of a typical cervical vertebra. The foramen transversarium is not vertical as in the other vertebræ, but is directed upwards and outwards to communicate a lateral bend to the vertebral artery (Fig. 6.96, p. 492). Scalenus medius, splenius cervicis and levator scapulæ are attached to the transverse process. An inferior articular process extends down from the junction of pedicle and lamina; its articular facet faces downwards and forwards as in typical cervical vertebræ (Fig. 6.69, p. 463). The *laminæ* are thick and rounded and project posteriorly into a massive *spinous process* (Fig. 6.87).

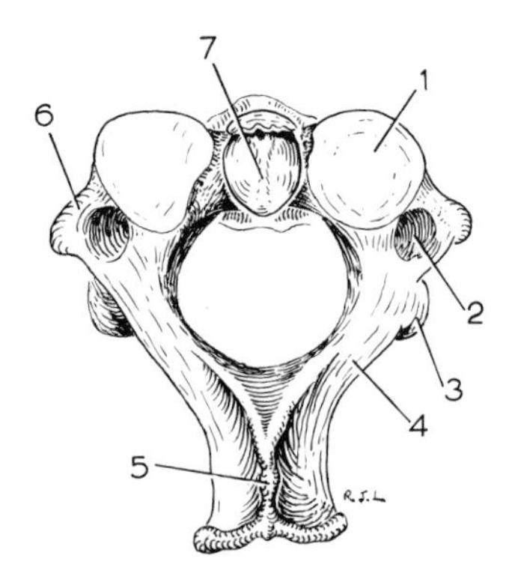

FIG. 6.87. THE AXIS, THE SECOND CERVICAL VERTEBRA (SUPERIOR VIEW). **1.** Weight-bearing facet on lateral mass for lateral mass of atlas. **2.** Foramen transversarium. **3.** Inferior articular process. **4.** Lamina. **5.** Spinous process. **6.** Transverse process. **7.** Dens (odontoid process).

The upper border of this is ridged, but the lower surface is grooved and ends in a wide bifurcation, so the tip of the spinous process resembles an inverted U (Fig. 6.75). The pair of strong semispinalis cervicis muscles are inserted into the concavity, while rectus capitis posterior major and the inferior oblique diverge widely from the outer surface on either side. The thick rounded upper and lower borders of the laminæ give attachment to the ligamenta flava. The posterior surface of the laminæ carries multifidus and longissimus attachments.

Movements of Vertebræ. These are discussed on p. 462.

Development of Vertebræ. The vertebræ develop from the sclerotome parts of the mesodermal somites (p. 38). The sclerotomes surround the notochord and neural tube in a sheath of mesoderm. A series of cartilaginous rings appears in the mesodermal sheath; each ring ossifies in three centres to form the centrum and the two halves of the neural arch of a vertebra. Each ring (and the vertebra to which it gives rise) is formed by fusion of adjacent halves (caudal and cranial) of the original somites. Thus the vertebræ lie not in segments of the body wall, but in the intersegmental planes.

Ossification of Vertebræ. All the vertebræ ossify in hyaline cartilage. The centre for the centrum is

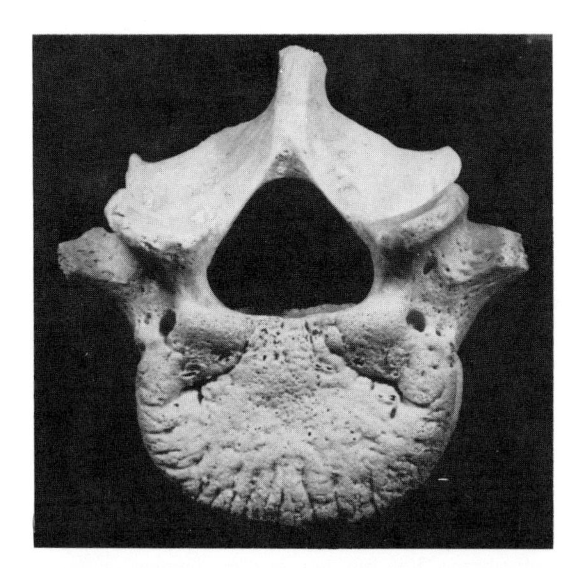

FIG. 6.88. THE UPPER ASPECT OF THE FIFTH LUMBAR VERTEBRA AT THE AGE OF TWENTY. The neuro-central junction has not completely ossified. The pyramidal transverse process fuses with the body of this vertebra, but not with the centrum. The radial grooves around the antero-lateral convexity of the body indicate that the epiphysis had not united. (Photograph.)

double, but the two areas rapidly fuse; failure of one half results in hemi-vertebra. The centrum and the two halves of the neural arch are ossified by the eighth week of fœtal life. By the third year all these parts are joined. The vertical cylindrical surface of the body is covered with compact bone, but the cancellous bone on the flat upper and lower surfaces, bevelled around its circumference, remains covered with a layer of hyaline cartilage. The epiphyses for the body appear as bony rings, upper and lower, soon after puberty. They are ridged and grooved reciprocally with the bevelled margin of the body. Fusion of the epiphyseal ring and body occurs in the early twenties. Thus the vertebra of an adolescent, when macerated and separated from its annular epiphysis, shows irregular radial grooves at the circumference of upper and lower surfaces (Fig. 6.88).

Soon after puberty secondary centres appear also at the tip of the spinous process (double in the bifid spines of the cervical vertebræ) and at the tips of the transverse processes of all the vertebræ, and in the mamillary processes of the 12th thoracic vertebra. These fuse in the early twenties. It is noteworthy that the costal elements of cervical and lumbar vertebræ do not have a separate bony centre, but ossify by direct extension from the neural arch. An occasional centre in the costal element of the 7th cervical or 1st lumbar vertebra may lead to the formation of a cervical or lumbar rib. On the other hand the weight-bearing costal elements of the sacrum have bony centres. These appear at 6 months of fœtal life and fuse with the neural arches at about 5 years. They fuse with each other and with the sacral bodies in the early twenties.

The *atlas* ossifies in the 7th week by a centre in each lateral mass. These extend around the posterior arch and unite at the 4th year. In the meantime a centre in the anterior arch has appeared at the first year. Its junction with the bone of the lateral mass cuts across the anterior part of the upper articular surface; these epiphyses fuse at the 7th year. This epiphyseal junction may permanently divide the articular surface (as in Fig. 6.86, right side).

The *axis* ossifies regularly in its centrum and neural arch, but the dens is exceptional. The dens is the detached centrum of the atlas vertebra, but only its lower part is ossified at birth, and it fuses with the centrum of the axis at 4 years. At this time its cartilaginous apex has begun to ossify, and this fuses at 12 years.

THE CRANIAL CAVITY

THE MENINGES

The interior of the cranium is lined with dura mater, the surface of the brain is covered with pia mater. Between the two, in contact with the dura mater, lies a membrane known as arachnoid mater.

A stranded whale dies because it is too heavy to breathe; its weight must be permanently suspended in water (remember the Principle of Archimedes). The soft brain is similarly suspended in water; without this support it flops flat (observe sheep's brains on a slab in the butcher's shop). The cerebro-spinal fluid in which the brain 'floats' is held in a thin bag of arachnoid mater. This delicate membrane is cerebrospinal fluid-proof, but not strong enough to withstand the hydrostatic pressure. It is supported everywhere by surface contact with dura mater. This is a tough sheet of thick fibrous tissue. In the spinal canal it lies free, but supports the arachnoid mater—two wet surfaces in contact, with only a film of tissue fluid between them. In the cranial cavity the fibrous dura mater is largely fused with the periosteum that lines the skull bones; but whatever its contour it supports the arachnoid mater, which everywhere lies in surface contact with it. The surface of the brain itself, closely covered with a vascular fibrous membrane called pia mater, fits snugly over its convexity against the arachnoid mater, but along its sulci and across its base it is out of contact with arachnoid mater. Here it 'floats' free in the cerebrospinal fluid.

General Arrangement

(Fig. 7.25, p. 518)

The **dura mater** is conventionally described as consisting of two layers, but it must be emphasized that this is really a false concept. The **outer layer** is none other than the ordinary periosteum which invests the surface of any bone; since it lies within the skull it is usually called endosteum, or the 'endosteal layer of the dura'. Blood vessels pass through it to supply the bone, but *no structure intervenes between it and the bare bone,* similar to the arrangement of periosteum elsewhere in the body. Around the margins of every foramen in the skull it lies in continuity with the periosteum on the outer surface of the cranial bones. It is not prolonged into the dura mater of the spinal canal nor is it ever invaginated by any cranial nerve. It is ordinary periosteum. *It never leaves the bone.*

The **inner layer** of the dura mater is very different. It consists of a dense, strong, fibrous membrane. It is the dura mater proper. In many places it lies in contact with the 'endosteal layer'. In such places of contact the two are very intimately fused together and cannot be separated from each other even by sharp dissection. Over the vault the fused layers are easily stripped away as a 'single' membrane from the bare bone, a fact which makes removal of the calvaria relatively easy. On the base of the skull the fused layers are so firmly attached in all three cranial fossæ that they can be stripped away from the bare bone only with some difficulty.

In other places the inner layer of the dura mater is separated from the 'endosteal layer'. In this way the venous sinuses of the dura mater are formed; they lie, lined with vascular endothelium, between the two layers. Folds of the inner layer project into the cranial cavity; one such fold roofs in the posterior cranial fossa, forming the tentorium cerebelli. Another fold forms the falx cerebri, lying in the midline between the two cerebral hemispheres. The function of these flanges, or septa, is to minimize rotary displacement of the brain. Concussion is caused more readily by rotary displacement of the brain than by a mass displacement of the head.

The **pia mater** invests the brain and spinal cord as periosteum invests bone. Like periosteum it contains blood vessels, and *nowhere does any structure intervene between pia mater and the underlying nervous tissue.* It invests the surface of the central nervous system to the depths of the deepest fissures and sulci. It is made of vascular fibrous tissue and can fairly easily be stripped away from the brain surface. It is prolonged out over the cranial nerves and spinal nerve roots to fuse with their epineurium, and it is invaginated into the substance of the brain by the entering cerebral arteries. The arteries lie loose in these sheaths of pia, surrounded by a narrow perivascular space containing cerebrospinal fluid. The pia mater of the spinal cord is projected laterally, on each side, to form the ligamentum denticulatum (p. 494).

The **arachnoid mater** consists of an impermeable delicate membrane that everywhere is supported by the inner surface of the fibrous layer of the dura mater. *Nothing save a thin film of tissue fluid (lymph) lies between the two* throughout the whole extent of the cranial and spinal cavities. That is to say, vessels and nerves pierce the dura mater and arachnoid mater both at the same place, and never run along between the two membranes.

The contours of the surface of the brain (covered with pia) and the fibrous layer of the dura (lined with arachnoid) do not correspond, especially over the base. Over convexities of the brain (e.g., the gyri on the supero-lateral surface of the cerebral hemisphere) pia and arachnoid not only touch, but often fuse, giving rise in such situations to the term 'pia-arachnoid'. Over the sulci on the cerebral hemispheres and over the irregularities along the base of the brain the pia and arachnoid (brain and dura) are separated. Such spaces between pia and arachnoid are filled with the circulating cerebrospinal fluid; they are large below, constituting the basal cisterns.

The membranous part of the arachnoid, in contact with the dura, is joined to the pia across the fluid-filled subarachnoid spaces by delicate strands of fibrous tissue that give a cobweb appearance in places, whence the name is derived (Fig. 7.25, p. 518).

The Dura Mater

Examine the interior of a dried skull and compare it with the wet specimen of the skull after removal of the brain. In the latter the bone is invested with the 'outer' or 'endosteal' layer of the dura mater (Fig. 6.89).

The fibrous layer is raised into two prominent flanges, the falx cerebri and tentorium cerebelli, which are connected to each other posteriorly. In other places the fibrous layer is separated from the 'endosteal layer' by venous sinuses and by the meningeal arteries (middle meningeal, etc.) and, in places, by the passage of nerves. The fibrous layer leaves the endosteal layer at the foramen magnum and is projected down the spinal canal, as far as the second sacral vertebra, as the spinal dura mater, or *theca.*

The fibrous layer is likewise evaginated around the cranial nerves and spinal nerve roots. In places these evaginations pass straight out, as around the spinal nerve roots; in other places they lie between the two layers of the dura, as in the trigeminal (Meckel's) cave.

The **tentorium cerebelli** is a flange of the fibrous layer which projects from the margins of the transverse sinuses and the margins of the superior petrosal sinuses. It is attached, from the apex of one petrous bone to the other, along the upper borders of the petrous

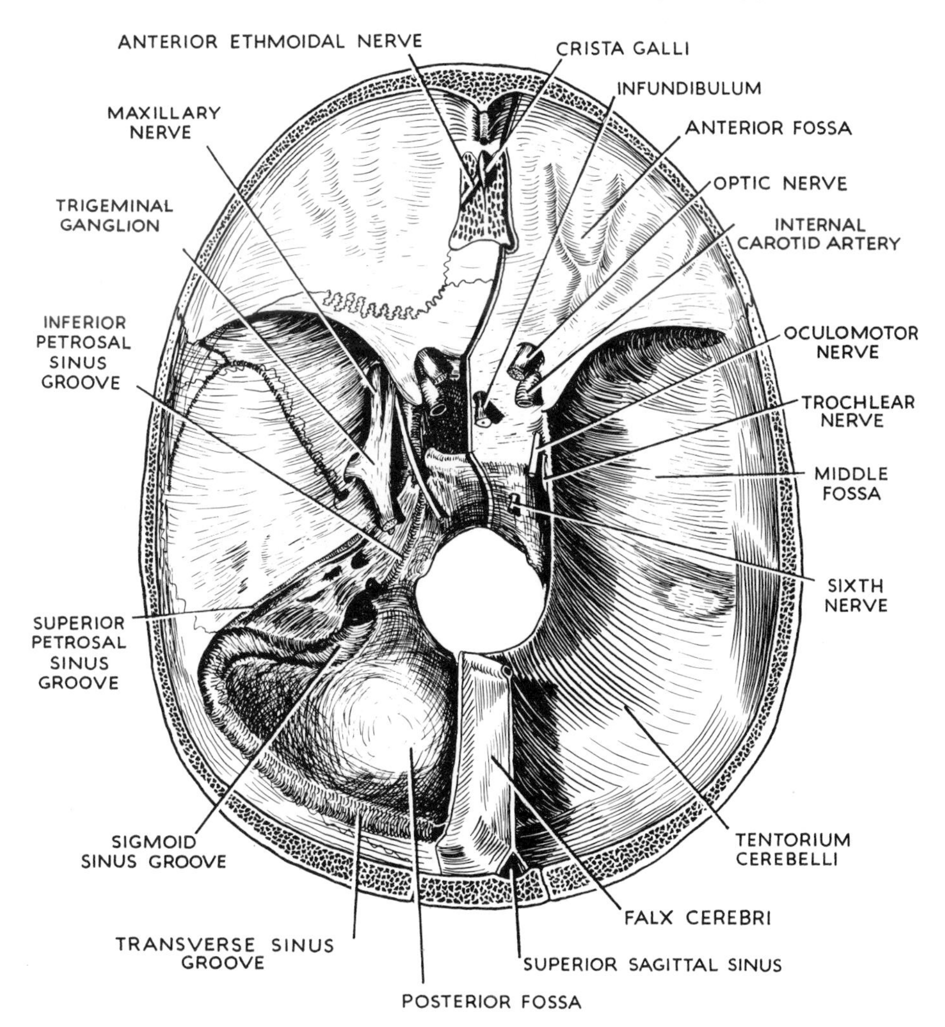

FIG. 6.89. THE CRANIAL FOSSÆ VIEWED FROM ABOVE. BOTH 'LAYERS' OF THE DURA MATER HAVE BEEN REMOVED ON THE LEFT SIDE. Drawn from a dissection.

temporal bones and horizontally along the inner surface of each side of the skull to the internal occipital protuberance. Its upper and lower layers are separated at their bony attachments by the superior petrosal and transverse sinuses, but elsewhere are intimately fused with each other. There is no venous sinus in the margin of the tentorium. The free margin is U-shaped and lies at a higher level than the bony attachment of the tentorium. The membrane slopes concavely upwards as it converges from the attached to the free margin, in conformity with the shape of the upper surface of the cerebellum and the under surface of the posterior part of the cerebral hemisphere.

The free concave margin is traceable forwards to the anterior clinoid processes. Over the superior petrosal sinus it overlies the attached margin, and from this point forwards to the anterior clinoid processes it lies as a ridge of dura mater on the roof of the cavernous sinus. To the medial side of the ridge is a concave triangular fossa which is pierced by the third and fourth nerves (Fig. 6.89).

The midline attachment of the falx suspends the tentorium, which falls down if the falx is divided (e.g., at autopsy). The straight sinus, which lies in the midline at the junction of the two, slopes more steeply than 45 degrees in the intact dura mater (PLATE 41).

The **falx cerebri** is a sickle-shaped flange of fibrous dura lying in the midline between the cerebral hemispheres. Its anterior margin is attached to the crista galli of the ethmoid bone and to the cavity of the foramen cæcum, into which it projects like a peg—an enlarged Sharpey's fibre. No vein passes through the foramen cæcum. The posterior margin is attached to the upper surface of the tentorium cerebelli in the midline, from the attached to the free margin of the tentorium; here its layers are separated for the passage of the straight sinus. Its convex upper border is attached alongside the midline to the whole length of the concave inner surface of the skull, from the foramen cæcum to the internal occipital protuberance. Its two layers are separated a short distance above the foramen cæcum to accommodate the superior sagittal sinus, which becomes progressively broader from this point to the internal occipital protuberance. The concave lower border of the falx cerebri is free and contains the inferior sagittal sinus within its two layers; this border lies just above the corpus callosum (Fig. 7.29, p. 521). Between superior and inferior sagittal sinuses the two layers of the falx are firmly united to form a strong inelastic membrane.

The **falx cerebelli** is a low elevation of the fibrous dura in the midline of the posterior cranial fossa, extending from the internal occipital protuberance along the internal occipital crest to the posterior margin of the foramen magnum. It lodges the small occipital sinus between its layers, and it projects a little into the sulcus between the cerebellar hemispheres.

Blood Supply of the Dura Mater. The fibrous layer of the dura mater requires very little blood to nourish it. The endosteal layer, on the other hand, is richly supplied, with the adjacent bone. In the supratentorial part it is supplied by the middle meningeal artery and, in the anterior cranial fossa, by meningeal branches of the ophthalmic and anterior ethmoidal arteries, and over the cavernous sinus by meningeal branches of the internal carotid artery and the accessory meningeal artery. The latter vessel enters the skull, from the maxillary artery, through the foramen ovale. All these arteries lie, like the venous sinuses, between the fibrous and endosteal layers of the dura. In the posterior fossa arteries run in the same plane, which they enter at the foramen magnum as meningeal

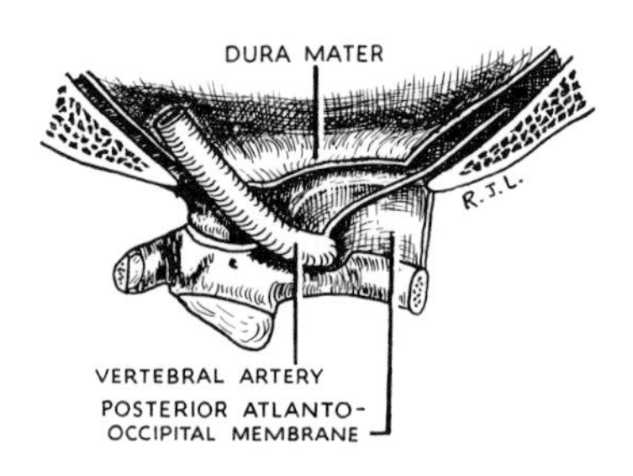

FIG. 6.90. THE RIGHT HALF OF THE FORAMEN MAGNUM VIEWED FROM WITHIN. The vertebral artery, having pierced the posterior atlanto-occipital membrane, gives off meningeal branches before piercing the dura mater to enter the subarachnoid space.

branches of the vertebral artery given off, of course, before this vessel pierces the spinal theca (Figs. 6.90, 6.96).

The **middle meningeal artery,** a branch of the maxillary, arises in the infratemporal fossa and passes upwards between the two roots of the auriculo-temporal nerve to enter the foramen spinosum. It thus enters the middle cranial fossa, accompanied by its own plexus of sympathetic nerves. It is accompanied throughout all its ramifications by veins which lie between it and bone and which are responsible for the grooves indented on the inner surface of the calvaria. It courses laterally on the floor of the middle cranial fossa and turns upwards and forwards on the greater wing of the sphenoid, where it divides into anterior and posterior branches (Fig. 6.13, p. 380).

The *anterior branch* courses up towards the pterion and then curves back to ascend towards the vertex, lying over the precentral gyrus. Hæmorrhage from the vessel thus causes pressure on the motor area. In the region of the pterion the artery frequently lies in a bony tunnel for a centimetre or more.

The *posterior branch* courses horizontally backwards, on a groove in the squamous part of the temporal bone, and ramifies over the posterior part of the skull. It lies along the superior temporal gyrus; hæmorrhage here may cause contralateral deafness (see p. 503).

The purpose of the middle meningeal artery is to supply the bones of the vault of the skull. These bones receive very little blood from the vessels of the scalp; the procedure of scalping produces no necrosis of the underlying bones. Only where the bones give attachment to muscles (temporal fossa and suboccipital region) is any substantial supply received from the exterior.

Much of the blood from the marrow is drained by large *diploic veins*, which emerge on the exterior. There are one frontal, two parietal, and one occipital on each side. Other diploic veins drain into the venous sinuses. The remaining blood drains into the middle meningeal veins.

The **middle meningeal veins** are sinuses in the dura mater and accompany the branches of the artery. They lie between the artery and the bone, grooving the latter. Some converge to two veins which leave the skull through the foramen spinosum and foramen ovale to join the pterygoid plexus. Most middle meningeal veins join the spheno-parietal sinus.

The **surface markings** of the middle meningeal artery are important in the matter of trephining for hæmorrhage. The foramen spinosum lies at a level just above the tubercle of the root of the zygoma (eminentia articularis). The point of division of the artery lies just above the midpoint of the zygomatic arch. The anterior branch can be mapped out in its possible bony tunnel by joining three points 1 inch (2·5 cm), 1½ inches (3·75 cm) and 2 inches (5 cm) respectively behind the zygomatic process of the frontal bone and above the zygomatic arch (Fig. 6.13, p. 380). The posterior branch runs backwards parallel with the upper border of the zygomatic arch and the supramastoid crest; it is usually exposed vertically above the mastoid process on a level horizontal with the upper margin of the orbit.

Nerve Supply of the Dura Mater. Most of the supratentorial part of the dura mater is supplied, rather surprisingly, from the ophthalmic division of the trigeminal nerve. The *tentorial nerves* course up and back from the anterior end of the cavernous sinus to supply the falx, the dura of the vault, and the upper surface of the tentorium cerebelli. The anterior cranial fossa receives some twigs from the anterior ethmoidal nerve, and from the maxillary nerve.

The middle fossa is supplied, in its anterior portion by a branch of the maxillary division of the trigeminal (the middle meningeal nerve), and in its posterior part by the meningeal branch of Vc (nervus spinosus).

The posterior fossa is supplied by meningeal branches of the ninth and tenth cranial nerves. These innervate the under surface of the tentorium cerebelli and the upper part of the bony fossa, but the dura around the foramen magnum is supplied by the upper three cervical nerves, indicating an upward migration of spinal dura to invest the outsize brain of man (cf. the upward migration of the skin to cover the skull and face, p. 31).

Venous Sinuses of the Dura Mater

All the venous sinuses, except the inferior sagittal and straight sinuses, lie between the fibrous dura and the endosteum. They receive all the blood from the brain. Except the inferior sagittal and straight sinuses they receive blood also from the adjacent bone. Several of them have important communicating branches with veins outside the skull.

The venous sinuses are held permanently open by the unyielding fibrous dura mater. Provided the head is above the heart they drain into the jugular veins by a siphonage effect (e.g. with the head nodding forwards the superior sagittal sinus flows *upwards* from the forehead against gravity).

The **superior sagittal sinus** lies between the two layers of the falx cerebri along the convexity of its attached margin. It commences just above the foramen cæcum and grows progressively larger as it passes back to the internal occipital protuberance. It grooves the bones along the midline of the vault of the skull. Three or four lakes of blood project laterally from it, between the fibrous dura and the endosteum; into these lakes the villi and the Pacchionian bodies (granulations) of the arachnoid project to return cerebro-spinal fluid to the blood stream (Fig. 7.25, p. 518).

The arachnoid granulations leave indentations on the adult skull. The superior sinus does not drain the frontal pole of the hemisphere, but receives veins from the upper and posterior parts of both medial and lateral surfaces of both hemispheres. These veins enter the sinus obliquely, against the flow of the blood stream. The superior sagittal sinus turns at the occipital protuberance, generally to the right, and becomes the transverse (lateral) sinus (Fig. 6.89).

The **inferior sagittal sinus** begins some little distance above the crista galli and lies between the folds of the free margin of the falx cerebri. It drains the lower parts of the medial surface of each hemisphere. At the attachment of falx cerebri and tentorium cerebelli it flows into the straight sinus (Fig. 7.29, p. 521).

The **straight sinus** lies between the folds of the fibrous dura at the junction of falx cerebri and tentorium cerebelli. It commences anteriorly by receiving the inferior sagittal sinus, the two basal veins (right and left) and the single great cerebral vein (of Galen). The straight sinus receives veins from the adjoining occipital lobes and from the upper surface of the cerebellum. It ends at the internal occipital protuberance by turning into the transverse (lateral) sinus, generally the left. The straight sinus slopes down steeply, at more than 45 degrees (PLATES 34 and 41).

The **transverse sinus** commences at the internal occipital protuberance and runs laterally between the two layers of the attached margin of the tentorium cerebelli. It courses horizontally forwards, grooving the occipital bone and the angle of the parietal bone. Reaching the junction of petrous and mastoid parts of the temporal bone it curves downwards, deeply grooving the inner surface of the mastoid bone, as the sigmoid sinus. One sinus is larger than the other, namely that which receives the superior sagittal sinus; this is usually the right. It should be noted that in clinical work the transverse and sigmoid sinuses are still usually referred to by their Old Terminology name of **lateral sinus.**

The two transverse sinuses communicate at their commencement at the internal occipital protuberance. Each receives tributaries from the nearby surfaces of cerebral and cerebellar hemispheres and, at its termination at the commencement of the sigmoid sinus, the superior petrosal sinus enters (Fig. 6.89).

The **surface marking** of the transverse sinus is lower than is commonly thought (Fig. 6.13, p. 380). It runs horizontally from the external occipital protuberance to the mastoid, that is, at the upper limit of the neck muscles where they join the skull, along the superior nuchal line.

The **sigmoid sinus** commences as the termination of the transverse sinus, deeply grooving the inner surface of the mastoid bone. It curves downwards and then curves forwards to the posterior margin of the jugular foramen. Here it expands into the *jugular bulb* which occupies the posterior and largest compartment in the jugular foramen, from which it emerges as the internal jugular vein. The sigmoid sinus is connected with the exterior in its upper part by the mastoid emissary vein which joins the posterior auricular vein and in its lower part by a vein which passes through the posterior condylar foramen (when present) to join the suboccipital plexus of veins.

It receives the superior petrosal sinus at its upper end and the occipital sinus at its lower end. Cerebellar veins drain to it, and it receives veins also from the mastoid air cells. Thrombophlebitis in these veins may lead to cerebellar abscess from mastoid infection.

The **occipital sinus** runs downwards from the beginning of the transverse sinus to the foramen magnum, skirts the margin of the foramen and drains into the sigmoid sinus. The two sinuses, lying along the attachment of the falx cerebelli, are often fused into a single trunk. Around the margins of the foramen magnum the sinuses communicate with the veins outside the spinal dura (the internal vertebral plexus). The occipital sinus receives tributaries from the cerebellum and medulla and drains the choroid plexus of the fourth ventricle.

The **basilar sinuses** consist of a network of veins, lying between the endosteal and fibrous layers of the dura mater, on the clivus. They connect the two inferior petrosal sinuses and receive veins from the lower part of the pons and from the front of the medulla. Thrombosis is thus fatal.

Note that no veins accompany the vertebral and basilar arteries; the vertebral vein itself commences *outside* the skull below the occipital bone (p. 469).

The Cavernous Sinus. The right and left cavernous sinuses are symmetrical, lying alongside the body of the sphenoid bone in the middle cranial fossa. Each contains the internal carotid artery and transmits some cranial nerves, each receives blood from three sources (orbit, vault bones, and cerebral hemisphere), each drains by three emerging veins (superior and inferior petrosal sinuses, and a small vein to the pterygoid plexus) and, finally, the two sinuses intercommunicate.

The cavernous sinus lies in a space between the periosteum of the body of the sphenoid ('endosteal' layer of the dura mater) and a fold of fibrous dura mater. The fold of fibrous dura mater commences medially, as the roof of the sinus, where the side wall of the pituitary fossa and the diaphragma sellæ meet (p. 489). The **roof** is attached to the anterior and middle clinoid processes of the sphenoid bone; it is perforated anteriorly by the emerging internal carotid artery, and is raised into a ridge laterally by the forward continuation of a fold of dura mater from the free edge of the tentorium cerebelli which extends to the anterior clinoid process. Just medial to this fold the third and fourth cranial nerves invaginate the roof. Medial to these nerves the roof is attached to the posterior clinoid process (Fig. 6.91). This part of the roof is continued medially as the diaphragma sellæ (Fig. 6.92). Forwards of the anterior clinoid process the anterior part of the roof is the endosteal layer on the under surface of the

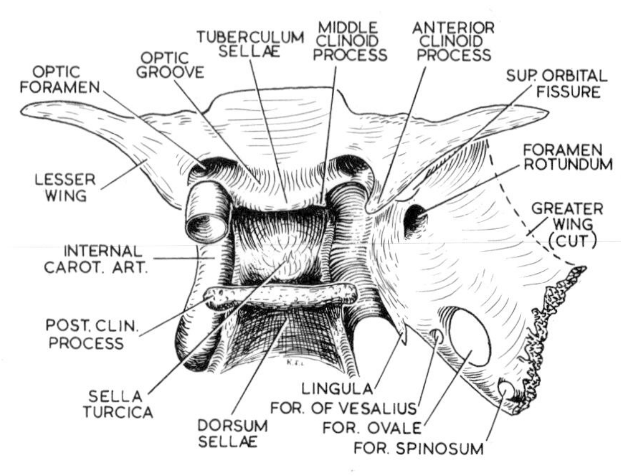

FIG. 6.91. THE SELLA TURCICA.

lesser wing of the sphenoid (Fig. 6.93). From the ridge on the roof the fibrous layer of the dura mater descends to the floor of the middle cranial fossa; this sheet of fibrous dura constitutes the **lateral wall** of the cavernous sinus.

In it the third, fourth and fifth cranial nerves run forward. The lateral wall, meeting the floor of the middle cranial fossa, is continued laterally as the fibrous layer of dura across the middle cranial fossa (Fig. 6.89). The anterior limit of the lateral wall is its vertical line of fusion with the endosteum of the greater wing of the sphenoid at the lateral margin of the foramen rotundum. Further back the lateral wall is attached medial to the foramen ovale and the trigeminal (Meckel's) cave. Thus the **floor** of the sinus is a narrow strip of endosteum along the base of the greater wing of the sphenoid.

The **medial wall** of the sinus is for the most part the endosteum on the body of the sphenoid. A single layer of fibrous dura mater completes the medial wall alongside the sella turcica, occluding the sinus from the pituitary fossa (Fig. 6.92).

The cavernous sinus differs from all the other venous sinuses of the dura mater by being intersected with numerous septa of fibrous tissue which divide the blood space into a series of tiny caves, like the corpora cavernosa of the penis, hence the name of cavernous sinus. The fibrous dura mater which covers in the posterior fossa passes vertically upwards to join the roof of the cavernous sinus, thus forming a very narrow **posterior wall.** The venous blood of the sinus leaves through this posterior wall, raising the fibrous layer of the dura mater from the underlying periosteum of the petrous bone to form the superior and inferior petrosal sinuses (Fig. 6.94).

The **anterior wall** of the cavernous sinus is narrow, and largely taken up by the entrance through it of the ophthalmic veins from the orbit. It plugs up the medial end of the superior orbital fissure.

The cavernous sinus thus extends from the *apex* of the orbit back to the *apex* of the petrous temporal bone. Each end is pointed, so that the narrow slit-like sinus is spindle-shaped in lateral view (PLATE 37).

The Relations of the Cavernous Sinus. Examine a dried skull and a museum or a dissecting-room specimen from which the brain has been removed (Fig. 6.89).

Medial to the sinus lies the body of the sphenoid bone and the fibrous lateral wall of the pituitary fossa. The sphenoidal air sinus lies towards the front of the pituitary fossa at a lower level; its extent is very variable and tends to increase posteriorly with advancing years. *Lateral* to the sinus lies the medial surface of the temporal lobe of the hemisphere and, postero-inferiorly, in the floor of the middle cranial fossa, the trigeminal cave (Fig. 6.93).

Superiorly the emerging internal carotid artery lies in contact with the forepart of the roof of the sinus; the artery here passes backwards a little before turning up towards the anterior perforated substance. Further back and somewhat above the roof lies the uncus. *Inferiorly* the sinus rests on the greater wing of the sphenoid bone. *Posteriorly* lies the cerebral peduncle at its junction with the upper border of the pons. *Anteriorly* lies the apex of the orbit.

The Contents of the Cavernous Sinus. Lying within the cavity of the cavernous sinus are the internal carotid artery and the sixth cranial nerve. The **internal carotid artery** curves upwards from the foramen

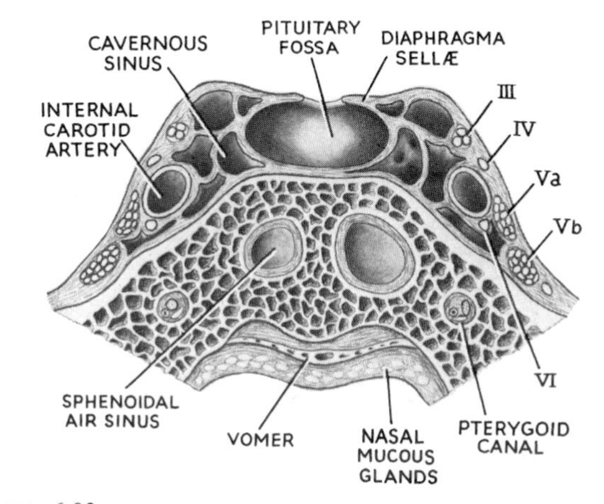

FIG. 6.92. THE PITUITARY FOSSA AND CAVERNOUS SINUSES IN CORONAL SECTION. Drawn to scale from a specimen.

lacerum to enter the posterior part of the sinus between the periosteum of the sphenoid bone ('endosteal layer' of dura) and the fibrous layer of the dura mater. It arches upwards and then forwards, deeply grooving the medial wall of the sinus, then curves upwards to pierce the roof of the sinus just medial to the anterior clinoid process (Fig. 6.91). The artery is accompanied by a plexus of sympathetic fibres from the superior cervical ganglion. The plexus is here named the **cavernous plexus;** it gives off branches which run forwards into the orbit (the long ciliary nerves via the nasociliary nerve to the dilator pupillæ muscle and sympathetic fibres to the ciliary ganglion and thence to the eyeball, and a branch to the oculo-motor nerve for the smooth-muscle part of levator palpebræ superioris).

The **sixth cranial nerve** pierces the dura mater of the clivus, which it invaginates for a short distance. It runs upwards between the fibrous dura mater and underlying periosteum ('endosteal layer' of dura mater) and generally enters the inferior petrosal sinus, whence it is carried across the groove at the apex of the petrous bone and so into the cavernous sinus. It runs forwards on the lateral side of the internal carotid artery (Fig.

6.94). Further forward it lies below the upturning artery (Fig. 6.92), and leaves the anterior wall of the sinus to enter the superior orbital fissure (Fig. 6.21, p. 400).

The Intradural Course of the Third, Fourth and Fifth Nerves. The third and fourth nerves enter the fibrous roof of the cavernous sinus and run forwards in its lateral wall; the fifth nerve ganglion, with its ophthalmic and maxillary divisions, lies in the lateral wall of the sinus. All these nerves lie on the medial side of the wall of the sinus, i.e., more within the sinus than within the middle fossa (Figs. 6.92, 6.93).

Some crossing of nerves occurs in the lateral wall of the cavernous sinus. The nerve relations at the points of crossing are in conformity with their relations as they enter the superior orbital fissure (p. 436). The third enters the fibrous ring at the medial end of the fissure, the fourth enters the fissure lateral to the fibrous ring; thus the third crosses medial to the fourth. The sixth lies more medially still, in the cavernous sinus itself.

The **oculomotor nerve** (p. 436) enters the roof anteriorly, medial to the ridge raised up by the forward continuation of the free margin of the tentorium cerebelli, and then passes forward in the lateral wall. It inclines downwards towards the medial end of the superior orbital fissure, and thus passes *medial* to the other nerves, namely, the fourth nerve and the branches of the ophthalmic nerve (Fig. 6.93). At the anterior end of the sinus it breaks into two, the superior and inferior divisions. In its course it picks up sympathetic fibres from the cavernous plexus; these are for the smooth-muscle part of levator palpebræ superioris.

The **trochlear nerve** (see pp. 436, 490) enters the roof of the cavernous sinus, behind the third nerve, alongside the ridge of dura mater (Fig. 6.89) and courses horizontally forwards in the lateral wall of the cavernous sinus and enters the superior orbital fissure at a higher level and lateral to the third nerve. The third nerve inclines downwards medial to the fourth as they lie in the lateral wall of the cavernous sinus.

The **trigeminal ganglion,** in its anterior part, lies forward of Meckel's cave (p. 490). Its **mandibular**

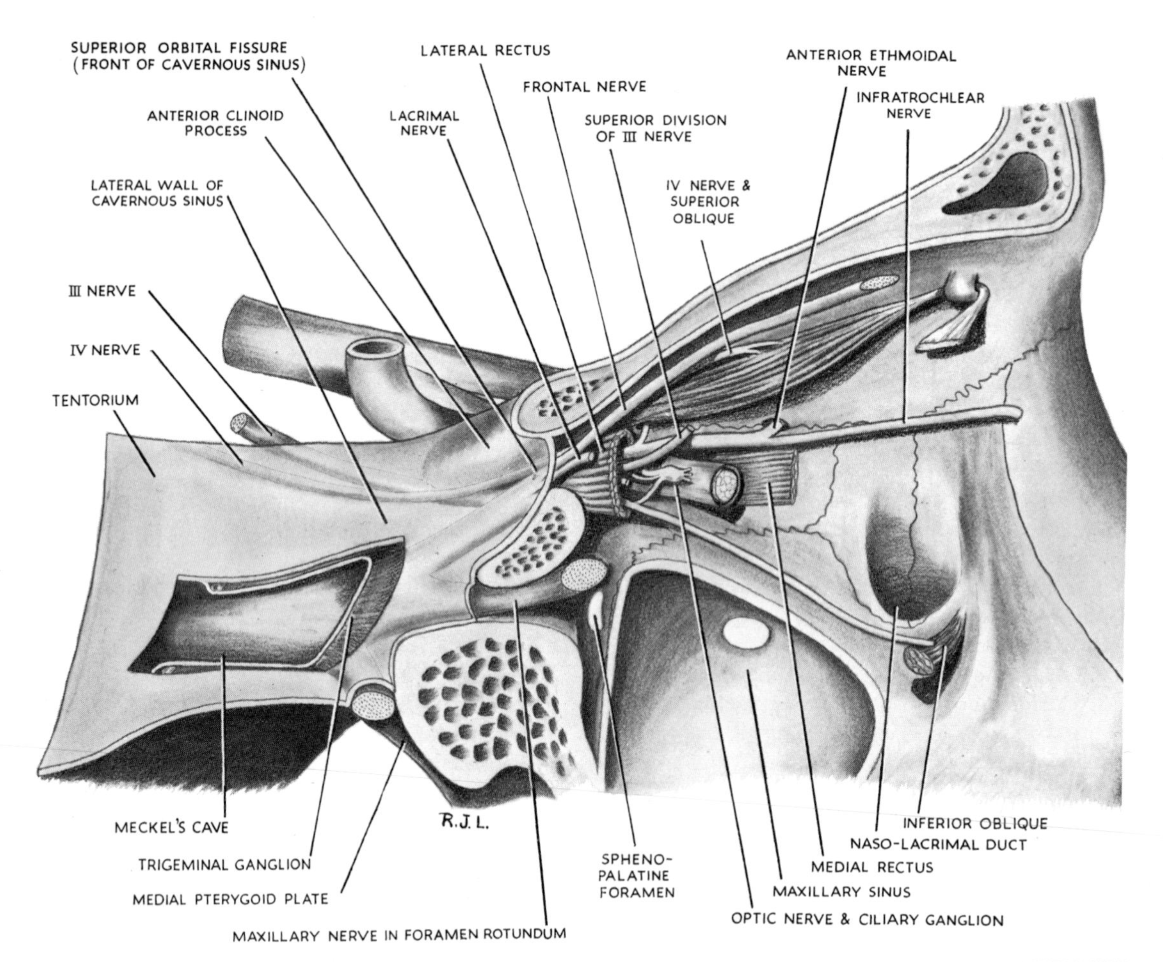

FIG. 6.93. A LONGITUDINAL SECTION THROUGH THE RIGHT ORBIT AND MIDDLE CRANIAL FOSSA, VIEWED FROM THE LATERAL SIDE.

division passes downwards to the foramen ovale and does not, strictly, come in contact with the lateral wall of the cavernous sinus. The **maxillary division** runs horizontally forwards on the medial side of the lateral wall and leaves the middle fossa through the foramen rotundum. The **ophthalmic division** divides into its three branches towards the anterior limit of the lateral wall. Two of these branches, the lacrimal and frontal, enter the orbit through the lateral part of the superior orbital fissure, the other (naso-ciliary nerve) slopes downwards to enter the fibrous ring at the medial end of the superior orbital fissure. All three branches are crossed medially by the downward-sloping third nerve. At the anterior end of the sinus the ophthalmic division gives off its tentorial branches to the dura mater (p. 483). In its course through the sinus the ophthalmic division picks up sympathetic fibres from the cavernous plexus (these eventually enter the long ciliary nerves to the dilator pupillæ muscle, see p. 437).

Summary of Nerve Relations. In its course through the cavernous sinus no nerve lies lateral to the fifth and its branches and no nerve lies medial to the sixth. The fourth runs parallel with the fifth at a higher level and the third slopes downwards crossing medial to both (Figs. 6.92, 6.93).

Afferent Veins of the Cavernous Sinus

1. The **superior ophthalmic vein** passes back directly into the anterior end of the sinus at the superior orbital fissure. The inferior ophthalmic veins also enter the front of the cavernous sinus, but before doing so they drain much of their blood through the inferior orbital fissure into the veins of the pterygoid plexus.
2. The **superficial middle cerebral vein** traverses the subarachnoid space and drains into the cavernous sinus by piercing its fibrous roof near the emerging carotid artery.
3. The **spheno-parietal sinus** drains blood from the skull bones over the temporal region of the vault (middle meningeal artery). It runs beneath the edge of the lesser wing of the sphenoid, lying between fibrous and endosteal layers of the dura mater, and enters the roof of the cavernous sinus.

Efferent Veins of the Cavernous Sinus

The **superior petrosal sinus** leaves the top of the posterior wall and, bridging the groove made by the underlying fifth nerve, runs back along the upper border of the petrous bone between the two layers at the attached margin of the tentorium cerebelli (Fig. 6.94). It enters the commencement of the sigmoid sinus, at the termination of the transverse sinus. It receives blood from the internal ear by several small veins that emerge from separate foramina in the petrous bone.

The **inferior petrosal sinus** is larger and empties the bulk of the blood from the cavernous sinus. It leaves the posterior wall of the cavernous sinus beneath the **petro-clinoid ligament.** This is a fibrous band stretched between the apex of the petrous part of the temporal bone and the side of the dorsum sellæ (Fig. 6.94). The sixth cranial nerve lies either within or alongside the sinus beneath this ligament, and grooves the apex of the petrous temporal bone. The inferior petrosal sinus runs down between the two layers of the

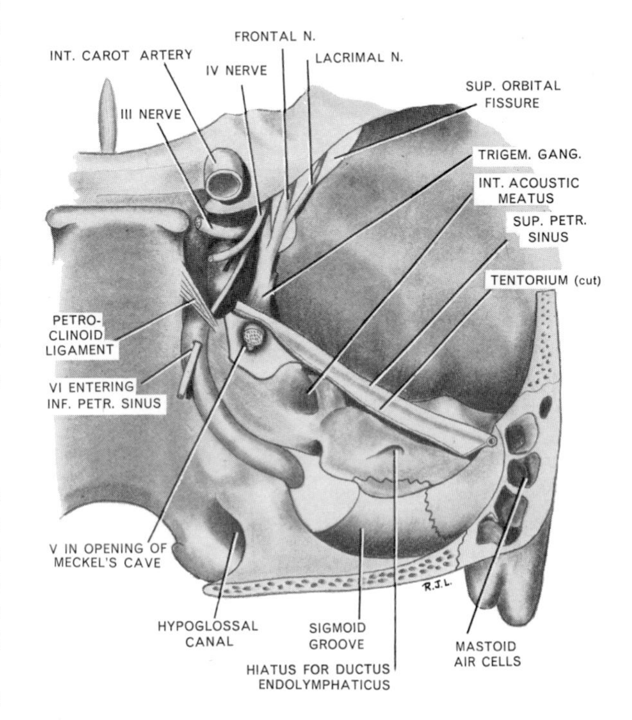

FIG. 6.94. THE POSTERIOR AND MIDDLE CRANIAL FOSSÆ VIEWED FROM BEHIND AFTER A CORONAL SECTION OF THE SKULL THROUGH THE MASTOID PROCESS. All dura mater is removed except the margin of the tentorium cerebelli attached to the superior petrosal sinus and a small piece of dura around the opening of Meckel's cave. The inferior petrosal sinus is in position. Diagrammatic.

dura along the suture between the apex of the petrous bone and the side of the clivus (occipital bone). It leaves a well-marked groove on the dried skull. It enters the anterior compartment of the jugular foramen medial to the ninth nerve and joins the internal jugular vein a short distance below the base of the skull.

The cavernous sinus drains also into the **pterygoid venous plexus.** The vein usually passes through the foramen ovale, but when the foramen of Vesalius is present it transmits a vein also. The foramen of Vesalius lies medial to the foramen ovale (PLATE 37 and Fig. 6.91, p. 484).

Communicating Veins of the Cavernous Sinus

The cavernous sinuses communicate with each other through an **intercavernous plexus** that lies between the fibrous dura mater and the endosteum of the pituitary fossa. A few communicating veins lie also in the diaphragma sellæ (Fig. 6.92).

Applied Anatomy of Cavernous Sinus Thrombosis. The cavernous sinus is in venous connexion with the skin of part of the face (the 'danger area') whence infection may produce thrombosis. By the superficial middle cerebral vein such thrombosis can spread to the hemisphere. The danger area of the face lies above the level of the deep facial vein; it comprises the upper lip and nose and medial part of the cheek. It is significant that it lies between the two veins of communication between face skin and cavernous sinus, namely, (1) the angular vein via superior ophthalmic vein directly to the cavernous sinus, and (2) deep facial vein via pterygoid plexus and the communicating veins in the foramen ovale and foramen of Vesalius (PLATE 37).

Thrombosis of the cavernous sinus produces sixth nerve palsy—not dangerous—and retrograde thrombosis of the inferior petrosal sinus and medullary veins—usually fatal.

The Arachnoid Mater (see p. 480)

The arachnoid is an impermeable membrane of fibrous tissue which contains the cerebro-spinal fluid. It lies everywhere closely supported by the fibrous layer of the dura mater in the cranium and spinal canal; the two are separated by a mere film of tissue fluid (lymph). Except at the aqueduct of the cochlea they are never blended together; their arrangement is similar to that of, for example, the visceral and parietal layers of the pleura, except that they do not become continuous with each other. In addition, the arachnoid has a network of fibrous strands which connect the membrane to the underlying pia mater on the cerebro-spinal nervous system (Fig. 7.25, p. 518).

Beneath the arachnoid is the **subarachnoid space.** In places the subarachnoid space is obliterated by fusion of pia mater with the arachnoid membrane; elsewhere the two are widely separated and the subarachnoid spaces so formed contain a number of delicate strands like spider-webs. In the spinal canal these spider-web strands are condensed into a delicate posterior midline lamina that forms an incomplete posterior median septum. The subarachnoid spaces provide a pathway for circulation and absorption of the cerebro-spinal fluid on its escape from the cerebral ventricles.

Structures connecting the surface of the brain with foramina necessarily pass through the subarachnoid space. Thus all the cranial nerves and the roots of the spinal nerves lie in the space, as well as all the arteries and veins of the brain and spinal cord. The space extends down to the termination of the spinal arachnoid and dura at the level of the second sacral vertebra.

In certain areas the arachnoid herniates through little holes in the dura mater into the venous sinuses. Such herniæ are called **arachnoid villi;** through their walls the cerebro-spinal fluid 'oozes' back into the blood. The arachnoid villi are most numerous in the superior sagittal sinus and its laterally projecting blood lakes. In the child the villi are discrete; as age progresses they become aggregated into visible clumps, called **arachnoid granulations (Pacchionian bodies).** These latter leave indentations on the inner table of the cranial vault alongside the superior sagittal sinus, at the site of the blood lakes (Fig. 7.25, p. 518). Only in the villi and granulations is the arachnoid mater pervious to cerebro-spinal fluid.

Between the base of the brain and the base of the skull several larger spaces exist as a result of the incongruities in the contours of bone and brain. These spaces are known as the **basal cisterns.** The most notable are the cisterna magna, cisterna pontis, cisterna interpeduncularis and the cisterna chiasmatica.

The **cerebello-medullary cistern (cisterna magna)** is the largest of the basal cisterns. It occupies the angle between the under surface of the cerebellum and the posterior surface of the medulla. Cerebro-spinal fluid flows into it from the midline aperture (foramen of Magendie) in the roof of the fourth ventricle. The lateral part of the cistern contains the vertebral artery and its posterior inferior cerebellar branch on each side (Fig. 6.96). The cisterna magna can be tapped in the midline by a needle passed through the posterior atlanto-occipital membrane and spinal dura.

The **cisterna pontis** lies between the clivus and the front of the pons and medulla. Cerebro-spinal fluid flows into it from the lateral apertures (foramina of Luschka) of the fourth ventricle. The cisterna pontis contains the basilar artery and its pontine and labyrinthine branches and the fifth to twelfth cranial nerves.

The **interpeduncular cistern** lies between the dorsum sellæ and the cerebral peduncles; it is roofed in by the floor of the third ventricle (mamillary bodies and posterior perforated substance). The floor of the cistern, on the dorsum sellæ, is formed by the arachnoid membrane passing across, in contact with the dura mater, between the right and left temporal lobes. The cistern contains the terminal branches of the basilar artery (including the posterior part of the circle of Willis), the stalk of the pituitary gland, and the third and fourth cranial nerves (Fig. 7.36, p. 528).

The **cisterna chiasmatica** lies above the optic chiasma, beneath the rostrum of the corpus callosum. It contains the anterior communicating artery and the intracranial part of the optic nerves.

THE CRANIAL FOSSÆ

The three fossæ within the base of the skull, anterior, middle and posterior, descend backwards like the steps of a staircase. They are described on p. 562.

The Anterior Cranial Fossa

This is described on p. 562. The floor of the fossa roofs in the orbits, ethmoidal sinuses and the nose. From the nose up to 20 olfactory nerves perforate the dura mater and arachnoid mater over the cribriform plate, and stream back in the subarachnoid space to enter the olfactory bulb, from which the olfactory tracts pass back to the inferior surface of the frontal lobe.

The Middle Cranial Fossa

This is described on p. 562. In life the sella turcica on the body of the sphenoid is roofed over by the diaphragma sellæ, while from the upper border of the petrous bone the tentorium cerebelli prolongs the middle fossa posteriorly as far back as the internal occipital protuberance in the midline. This supratentorial compartment lodges the posterior poles of the cerebral hemispheres (Fig. 6.89).

The middle fossa contains the pituitary gland and the trigeminal ganglion. The second, third, fourth, fifth and sixth cranial nerves leave the skull through foramina in the fossa.

The **pituitary fossa** occupies the sella turcica, which in life is lined with dura mater. This latter is prolonged up the sides of the fossa, hitched between middle and posterior clinoid processes, as a flange lying between the pituitary fossa and the cavernous sinus. The fossa is completed above by a horizontal sheet of dura mater, the diaphragma sellæ, perforated centrally to accommodate the stalk of the pituitary gland. The diaphragma sellæ flows out laterally to form the roof of the cavernous sinus (Fig. 6.92).

The cavity of the pituitary fossa is occupied by the gland, which makes a snug fit with the dura mater, so that the original subarachnoid space around the gland is usually obliterated and the pituitary stalk lies close against the margins of the aperture in the diaphragma sellæ.

The **pituitary gland (hypophysis cerebri)** is a composite structure consisting of two parts histologically very different from each other. The larger *anterior lobe* is mostly cells and is very vascular, the smaller *posterior lobe* is mostly fibres and is less vascular. The posterior lobe is connected with the stalk, a prolongation downwards and forwards from the tuber cinereum. The anterior lobe is adherent to the posterior lobe by a narrow zone of gland substance that is often divided by a cleft from the main mass of the anterior lobe—this narrow zone is often named the pars intermedia, but it is developmentally, structurally and functionally part of the anterior lobe.

Microscopic Anatomy. The posterior lobe (neurohypophysis) consists largely of unmyelinated nerve fibres whose cell bodies lie in hypothalamic nuclei. Scattered among these fibres and the neuroglia are a few cells named *pituicytes.* Their function is uncertain. The pars intermedia commonly contains large colloid vesicles resembling those of the thyroid gland. The anterior lobe (adeno-hypophysis) is very cellular, with a minimum amount of fibrous tissue forming a skeletal framework. The cells are of three kinds. About 50 per cent of them have pale agranular cytoplasm; they are called *chromophobe* cells. The remaining 50 per cent have either eosinophilic or basophilic granules in their cytoplasm. *Eosinophils* outnumber *basophils* by about 4 to 1 (PLATE 24).

Blood Supply. Branches from the internal carotid artery supply the stalk and the posterior lobe. From the capillaries here a portal system of vessels runs on to supply the anterior lobe. The veins drain into the intercavernous anastomosis.

Relations. On each side is a flange of dura mater separating the gland from the cavernous sinus. Below the pituitary fossa lies the body of the sphenoid bone, containing the sphenoidal air sinus; when small the sinus lies antero-inferior to the fossa, but when large the sinus extends back beneath the fossa, from which it is then separated by only a thin plate of bone (Fig. 6.24, p. 402).

As already stated, the pituitary gland is in theory surrounded by a subarachnoid space, but in practice it is usually found that pia and arachnoid have fused together on its surface and around the stalk, lying snugly against, but not fused with, the dura mater of the sella turcica.

The optic chiasma, at a higher level, usually lies towards the back of the diaphragma sellæ, so that a tumour of the pituitary gland, rising upwards usually passes in front of the chiasma, pressing on the medial sides of the optic nerves (fibres from nasal half of retina, hemianopia of temporal fields). The stalk of the pituitary gland slopes *forwards* down to the diaphragma sellæ.

Development of the Pituitary Gland. The gland is all ectodermal. The anterior lobe is formed from Rathke's pouch, the posterior lobe growing out as a diverticulum from the floor of the diencephalon (neuro-ectoderm).

The **internal carotid artery** emerges from the roof of the cavernous sinus medial to the anterior clinoid process and curves immediately backwards, lying on the roof of the sinus before curving upwards lateral to the optic chiasma. At the anterior perforated substance it divides into its terminal branches. Of these the anterior cerebral artery passes forward above the optic nerve. The anterior communicating artery, lying in the cisterna chiasmatica, lies vertically above the groove that joins the two optic foramina (PLATE 34).

The **ophthalmic artery** branches from the internal carotid immediately above the roof of the cavernous sinus. The carotid arteries come from below and laterally, the optic nerves come from above and medially (i.e., from the chiasma), and this is the relationship of nerve and artery in the optic foramen.

The **optic nerve** slopes forward, down and laterally from the chiasma to the optic canal. Clad only in pia mater, it receives its tube of arachnoid and dura mater at the optic canal. Its intracranial part, in the cisterna chiasmatica, is here supplied by branches of the anterior cerebral artery that run down from the chiasma.

The **oculomotor nerve** leaves the medial side of the crus of the cerebral peduncle (Fig. 7.35, p. 528). The nerve passes forwards between the posterior cerebral and superior cerebellar arteries, crosses the interpeduncular cistern and enters the roof of the cavernous sinus in the middle fossa, slightly indenting the dura mater thereof before running down in the lateral wall of the sinus (Fig. 6.89).

The **trochlear nerve** curls around the cerebral peduncle below the posterior cerebral artery and runs forward above the superior cerebellar artery, lateral to the third nerve in the interpeduncular cistern, just below the free margin of the tentorium cerebelli (i.e., in the posterior fossa). It enters the middle fossa just behind the third nerve and pierces the dura mater of the roof of the cavernous sinus (Fig. 6.89) at the margin of the tentorium cerebelli.

The **posterior communicating arteries,** joining the internal carotid and posterior cerebral arteries in the circle of Willis, lie in the interpeduncular cistern, above and lateral to the pituitary gland in the middle cranial fossa (Fig. 7.36, p. 528).

The **trigeminal ganglion** lies beneath the dura mater in the floor of the middle cranial fossa alongside the cavernous sinus. The fifth nerve leaves the pons in the posterior fossa and runs forwards to cross the upper border of the petrous bone, upon which it leaves a shallow groove some 5 mm wide. It passes beneath the superior petrosal sinus at this point (Fig. 6.94). The layer of fibrous dura mater joining the superior petrosal sinus to the petrous bone is evaginated around the sensory and motor roots of the fifth nerve and does not fuse with the pia mater until as far forwards as the middle of the trigeminal ganglion (Fig. 6.93). As elsewhere, the evagination of dura mater supports the contiguous arachnoid, so that sensory and motor roots of the fifth nerve *and the posterior half of the ganglion* are bathed in cerebro-spinal fluid. This fluid space, surrounding the nerve and posterior half of the ganglion, is known as *Meckel's cave.* It lies in the middle cranial fossa, but its mouth opens backwards into the posterior cranial fossa. The dura mater forming the walls of the cave is evaginated between the fibrous and endosteal layers of dura mater which elsewhere clothe the floor of the middle fossa. Thus above the cave lie two fibrous layers of dura fused and continuous with each other posteriorly around the superior petrosal sinus and tentorium, while below the cave lies the fibrous wall of the cave fused to the endosteal layer of the dura mater (Figs. 6.93, 6.94).

Applied Anatomy of the Trigeminal Ganglion. The sensory root and posterior half of the ganglion, bathed in the cerebro-spinal fluid in Meckel's cave, can be approached only across the subarachnoid space. The anterior half of the ganglion and the three divisions of the trigeminal nerve lie in front of Meckel's cave. The upper part of the ganglion, with the ophthalmic and maxillary divisions, here lies in the fibrous lateral wall of the cavernous sinus. The lower part of the ganglion and the mandibular division lie in the middle fossa fused between the two layers of dura mater. Extradural approach is therefore possible across the floor of the middle fossa by stripping the dura from the bone, avoiding entry into the subarachnoid space.

Transient *facial* palsy sometimes follows the latter approach. The explanation is thought to be that in stripping up the dura mater from the floor of the middle fossa tension is exerted on the greater (superficial) petrosal nerve and therefore on the geniculate ganglion. Subsequent œdema here causes pressure on the motor fibres of the facial nerve with paralysis of the muscles until the œdema subsides and the nerve recovers.

Blood Supply of the Trigeminal Ganglion. Ganglionic branches leave the internal carotid artery in the cavernous sinus and these are usually stated to be the blood supply to the ganglion. But a very appreciable amount of blood is brought to the ganglion by the accessory meningeal artery which enters the foramen ovale and runs up along the mandibular division to reach the ganglion, a similar arrangement to that in the spinal nerves.

Nerve Supply. The epineurium of the ganglion and the adjacent dura mater are supplied by the meningeal branch of V c (nervus spinosus).

Relations of the Trigeminal Ganglion. The posterior half of the ganglion lies, as already stated, in Meckel's cave beneath the fibrous floor of the middle cranial fossa, where it indents the petrous bone alongside the foramen lacerum. Here the greater (superficial) petrosal nerve lies beneath the cave, between fibrous and endosteal layers of dura mater. The anterior half of the ganglion lies, in its upper part, in the fibrous dura mater of the lateral wall of the cavernous sinus.

The **mandibular division** of the trigeminal nerve

passes laterally to descend through the foramen ovale, so that it never comes into contact with the lateral wall of the cavernous sinus. It is joined in the foramen ovale by the small motor root to emerge as the (mixed) mandibular nerve, a similar arrangement to that of the spinal nerves in the intervertebral foramina.

The **maxillary nerve** passes forwards within the fibrous layer to leave the skull through the foramen rotundum. The branches of the *ophthalmic division* pass forwards likewise, to leave the skull through the medial end of the superior orbital fissure. The *superior orbital fissure* is elsewhere plugged up by the fibrous layer of dura mater which passes across between the edges of the greater and lesser wings of the sphenoid bone.

The **greater** (superficial) **petrosal nerve** emerges from its hiatus in the petrous bone and runs obliquely forwards, between the two layers of the dura mater and beneath the trigeminal ganglion to the foramen lacerum. Here it is joined by the **deep petrosal nerve,** a branch from the carotid plexus of sympathetic nerves. The two join to form the **nerve of the pterygoid canal (Vidian nerve).** This nerve enters the posterior end of the pterygoid canal in the foramen lacerum and runs along the canal to join the pterygo-palatine ganglion.

The **lesser** (superficial) **petrosal nerve** leaves its hiatus in the petrous bone and runs forwards beneath the fibrous floor of the middle cranial fossa to emerge through the foramen ovale to join the otic ganglion. At times it leaves the skull through its own foramen posterior to the foramen ovale; this foramen is known as the **foramen innominatum.**

The middle meningeal artery is considered on p. 482.

The Posterior Cranial Fossa

This fossa lodges the convexities of the cerebellar hemispheres as well as the pons and medulla oblongata. In it the fifth to twelfth cranial nerves inclusive pierce the dura mater (Fig. 6.96).

The **trigeminal nerve** leaves the antero-lateral surface of the pons by two roots, a large sensory and a small motor. They lie close together. The motor root emerges somewhat above and medial to the sensory root, but spirals to enter the mouth of Meckel's cave below it (Fig. 6.49, p. 433). The sensory root itself shows a spiral arrangement of its fibres. At the junction with the pons the mandibular fibres lie superior, the ophthalmic inferior with the maxillary fibres between, but in Meckel's cave the mandibular fibres lie most laterally and the ophthalmic fibres most medially.

The **abducent nerve** leaves the brain stem near the ventral midline at the junction of pons and medulla and runs upwards through the cisterna pontis. It evaginates the dura mater of the clivus some distance above its origin from the brain stem, and runs thence upwards between the fibrous and endosteal layers of the dura mater to enter the inferior petrosal sinus, in which it passes over the groove at the apex of the petrous bone. It thereby enters the posterior end of the cavernous sinus. The relatively long intracranial course of this delicate nerve renders it particularly vulnerable to increase of intracranial pressure; paralysis of the lateral rectus is often an early sign in such cases.

The **facial** and **auditory** nerves, with the intervening **nervus intermedius,** leave the junction of pons and medulla and pass laterally and somewhat upwards to enter the internal acoustic meatus. The *labyrinthine artery* lies with them. This vessel is usually said to be a branch of the basilar artery, but in a high proportion of cases it arises from the anterior inferior cerebellar artery.

The **internal acoustic meatus** is a foramen directed laterally in the posterior surface of the obliquely set petrous bone. Its fundus consists of a plate of bone, the **lamina cribrosa,** divided by a horizontal crest into an upper and lower semicircle. The fibrous layer of the dura mater is everywhere fused with the endosteum within the meatus (Fig. 6.95).

If it be remembered that in the petrous bone the cochlea lies in front of the vestibule and that the facial nerve passes in its bony canal above the vestibule, the arrangement of the structures will become clear. Thus the facial nerve and nervus intermedius pierce the front of the upper part, the cochlear nerve the front of the lower part (by many branches in spiral arrangement). The vestibular nerve pierces the plate posteriorly, by upper and lower divisions that lie behind the facial nerve foramen, and the spiral cochlear foramina respectively. Each division of the vestibular nerve has a ganglion deep in the meatus. Behind the vestibular area is a single foramen (the *foramen singulare*) for the passage of the nerve to the posterior semicircular duct. The labyrinthine artery divides in the meatus and its branches accompany the nerves through the bony plate.

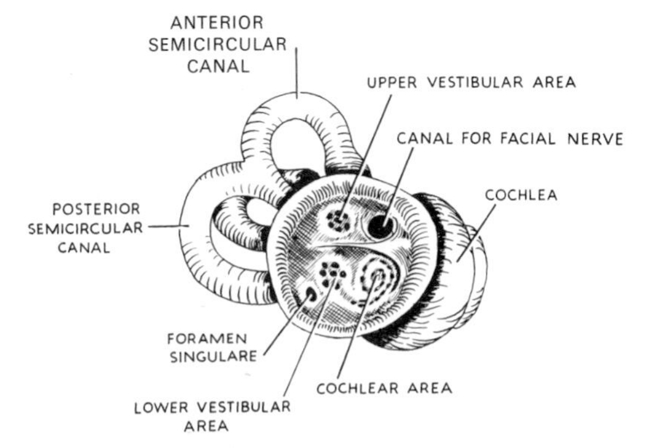

FIG. 6.95. THE LEFT INTERNAL ACOUSTIC MEATUS WITH THE OSSEOUS LABYRINTH. Drawn from Specimen 152, C, in R.C.S. Museum.

The **subarcuate fossa** lies lateral to the internal acoustic meatus, below the arcuate eminence. It is a very shallow fossa against which the flocculus of the cerebellum lies; in the neonatal skull this fossa is as large and as deep as the internal acoustic meatus and it lodges the flocculus.

Further laterally on the posterior surface of the petrous bone is the orifice of the aqueduct of the vestibule, a narrow slit overhung by a sharp scale of bone. The sacculus endolymphaticus depends from this slit beneath the fibrous layer of the dura (Fig. 6.94).

The **glossopharyngeal, vagus and accessory nerves** arise from a series of rootlets lying vertically between the olive and the inferior cerebellar peduncle. The three nerves pass laterally across the occipital bone in that order, behind the jugular tubercle. All three pass through the jugular foramen (Fig. 6.96).

The **cervical part of the accessory nerve** is seen entering the posterior fossa through the foramen magnum. It arises by a series of rootlets that emerge from the lateral surface of the upper five segments of the cervical cord *posterior* to the ligamentum denticulatum. These rootlets unite into a single trunk that passes forwards over the top of the ligamentum denticulatum lateral to the vertebral artery. It unites with the cranial root medial to the jugular foramen.

The **jugular foramen** is divided by two transverse septa of the fibrous dura mater into three compartments. These septa may ossify. The ninth nerve and inferior petrosal sinus share the anterior compartment, tenth and eleventh nerves lie in the middle compartment, while the large posterior compartment is occupied by the termination of the sigmoid sinus. Examine the jugular foramen in the dried skull. The inferior border of the petrous bone shows a deep notch immediately below the internal acoustic meatus. This notch is indented by the glosso-pharyngeal nerve; the aqueduct of the cochlea opens into the depths of the notch and by this means perilymph drains into the subarachnoid space (the arachnoid making a tubular projection into the opening of the aqueduct). The groove made by the inferior petrosal sinus will be seen to enter the jugular foramen medial to the glosso-pharyngeal notch of the petrous bone (Fig. 6.94).

In the central compartment of the jugular foramen the accessory nerve descends lateral to the vagus. It is worth noting the arrangement of the nerves as they pass down through the jugular foramen, since it explains their relationship at the base of the skull between the carotid sheath and the deep lobe of the parotid gland (p. 393). The essential point is that the ninth and eleventh nerves lie more laterally than the tenth in the foramen.

The **hypoglossal nerve** leaves the medulla by a vertical series of rootlets between the pyramid and the olive. The rootlets unite into two trunks which enter the hypoglossal canal separately, divided from each other by a septum of the fibrous dura mater which occasionally ossifies. The hypoglossal canal lies in the epiphyseal junction between the basi-occiput and the jugular process (ex-occiput) of the occipital bone (Fig. 8.10, p. 567).

The arteries in the posterior fossa comprise the two

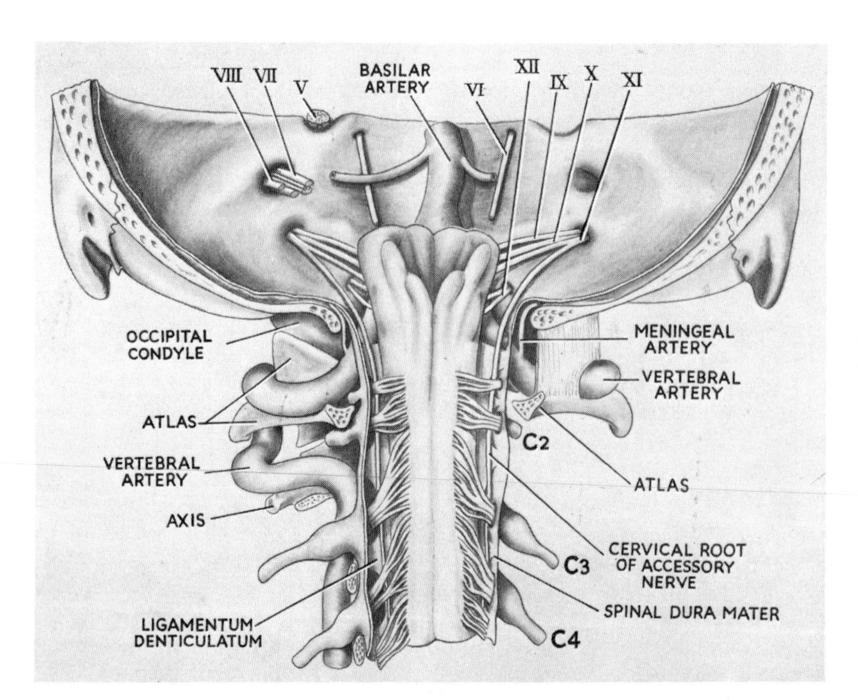

FIG. 6.96. THE POSTERIOR CRANIAL FOSSA AND SPINAL CANAL OPENED FROM BEHIND.

vertebral and the basilar arteries with their branches. Before piercing the spinal dura mater, but after piercing the posterior atlanto-occipital membrane the vertebral arteries give off meningeal branches which enter the posterior fossa between the fibrous and endosteal layers of the dura mater at the foramen magnum (Figs. 6.90, 6.96).

The **vertebral artery** then pierces the spinal dura mater and arachnoid, gives off two small posterior spinal arteries, and runs forward in front of the ligamentum denticulatum between the lower rootlets of the hypoglossal nerve and the upper rootlets of the first cervical nerve. It gives off the anterior spinal artery and the posterior inferior cerebellar artery and spirals up to meet its opposite fellow at the lower border of the pons to form the basilar artery. The posterior and anterior spinal arteries pass downwards through the foramen magnum on the spinal cord (p. 536). The posterior inferior cerebellar artery is perhaps the most tortuous artery in the body. Its coils insinuate themselves between the rootlets of XII, XI and X and the vessel is distributed to the cerebellum and medulla (p. 534).

The **basilar artery** runs up in front of the pons. It is not responsible for the ventral median groove in the pons; indeed, the artery is usually curved to one side of the midline (Fig. 6.96). It gives off the anterior inferior cerebellar artery and many pontine branches. The labyrinthine artery arises from the anterior inferior cerebellar or directly from the basilar trunk. The basilar artery ends at the upper border of the pons by branching on each side into superior cerebellar and posterior cerebral arteries (Fig. 7.36, p. 528).

THE SPINAL CANAL

The spinal canal (p. 462) is a smooth-walled tubular space whose only openings from top to bottom are the line of intervertebral foramina. It contains the spinal cord and the spinal meninges. The spinal cord is narrower than the spinal dura mater and the spinal dura mater is narrower than the spinal canal. Thus each can adapt itself without strain to the movements of the spinal column.

The spinal canal is lined with a layer of extradural fat in which lies the **internal vertebral plexus** of veins (Fig. 6.97). The plexus receives its tributaries mostly from the large **basi-vertebral veins** draining the active red marrow in the bodies of the vertebræ (p. 465). The plexus is largely a longitudinal line of veins on each side of the spinal canal, like a ladder whose rungs are made by each pair of basi-vertebral veins. The internal vertebral plexus sends its efferent veins (the *intervertebral veins*) through the intervertebral foramina to drain into the segmental veins. The basi-vertebral veins might have drained independently through their intervertebral foramina, without linking into this longitudinal extradural plexus. The internal vertebral plexus exists as a bypass of the diaphragm. It functions when the inferior vena cava cannot cope with a sudden flush of blood resulting from a sudden increase of intra-abdominal pressure (e.g., in coughing or abdominal straining). Thus pelvic and abdominal venous blood is momentarily squirted up the plexus into posterior intercostal veins and so above the diaphragm into the superior vena cava. There are no valves in these veins, and so communication exists between the bodies of the vertebræ and, particularly, the thyroid gland, breast and prostate (p. 341), significant perhaps in the spread of cancer.

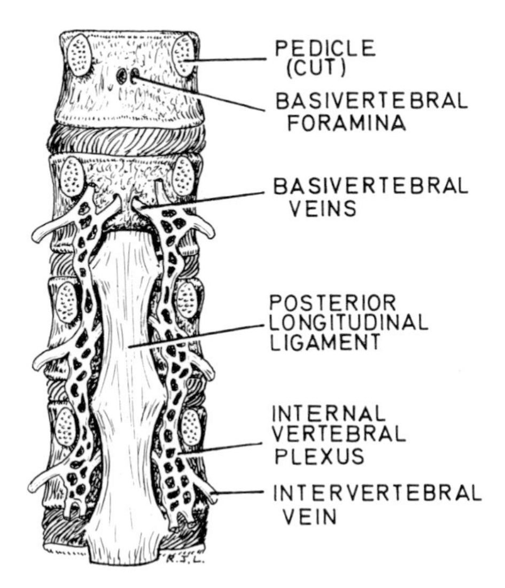

FIG. 6.97. THE INTERNAL VERTEBRAL PLEXUS. THE BASI-VERTEBRAL VEINS EMERGE FROM BENEATH THE POSTERIOR LONGITUDINAL LIGAMENT AND ENTER THE PLEXUS. THE PLEXUS DRAINS BY INTERVERTEBRAL VEINS INTO THE LOCAL SEGMENTAL VEINS.

The Spinal Meninges

The **spinal dura mater,** or **theca,** is a prolongation of the fibrous layer of the dura mater of the posterior cranial fossa. It extends downwards through the foramen magnum to the level of the second sacral vertebra. It is attached rather firmly to the membrana tectoria and to the posterior longitudinal ligament on the body of the axis (second cervical) vertebra, but elsewhere in the spinal canal it lies free of bony or ligamentous attachments. It is separated from the spinal canal by a layer of fat in which lies the internal vertebral plexus. The spinal dura mater is pierced segmentally by the anterior and posterior roots of the spinal nerves and is prolonged over these roots to form a series of lateral projections, one entering each intervertebral foramen. Thus the loose-fitting theca is stabilized within the spinal canal.

The **spinal arachnoid** is supported by the inner surface of the spinal dura; nothing but a thin film of lymph separates these two membranes. The arrangement is similar to that in the skull. Below the level of the spinal cord (i.e., over the cauda equina) the arachnoid is nothing but a delicate membrane that is supported by the dura mater, but over the spinal cord itself the spinal arachnoid sends many delicate web-like processes across the subarachnoid space to the pia mater on the cord. They are rather well developed in the posterior midline, where they form an incomplete *posterior median septum.*

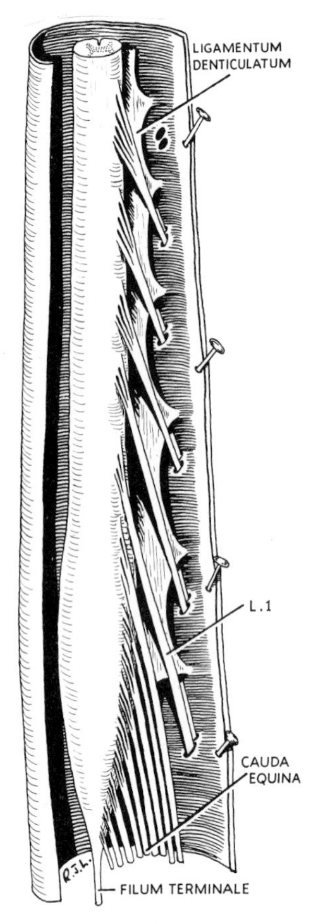

FIG. 6.98. THE LOWER END OF THE SPINAL CORD EXPOSED FROM BEHIND BY OPENING THE THECA. On the left side the nerve roots and the ligamentum denticulatum have been removed.

The **spinal pia mater,** as in the cranium, invests the surface of the central nervous system. It clothes the spinal cord and enters to line the anterior median sulcus. It is prolonged over the spinal nerve roots and blends with their epineurium. It is projected below the apex of the conus medullaris, whence it extends as the **filum terminale** to perforate the spinal theca at S2. The filum terminale lies centrally in the cauda equina. A lateral projection of pia mater on each side forms the **ligamentum denticulatum.** This forms a flange which crosses the subarachnoid space and, piercing the arachnoid, connects the side of the spinal cord to the dura mater. It is attached in an unbroken line along the spinal cord from the foramen magnum to the conus medullaris, but its lateral edge is connected to the spinal dura by a series of 'teeth', which are attached to the spaces between the issuing spinal nerves. The root of L1 lies at the lowest denticulation. The ligamentum denticulatum, the filum terminale and the attached nerve roots serve to stabilize the loose-fitting spinal cord within the spinal dura mater (Fig. 6.98).

The **spinal subarachnoid space** is relatively large. It communicates through the foramen magnum with the subarachnoid space of the posterior cranial fossa. Usually it is regarded as having no connexion with tissue spaces outside the spinal arachnoid, but certain experimental evidence suggests that some communication may exist. Knowledge is incomplete in this matter.

Below the level of the conus medullaris it contains only the cauda equina, and it is here that the operation of lumbar puncture is performed, usually through the fourth lumbar interspace. Spinal puncture performed above the level of the second lumbar spinous process endangers the spinal cord.

Section 7. The Central Nervous System

THE BRAIN

The brain consists of the cerebral hemispheres, the diencephalon, the brain stem (midbrain, pons and medulla) and the cerebellum. All are contained in the cranial cavity; the medulla is continued below into the spinal cord. The cerebral hemispheres and brain stem contain cavities (the ventricles) which continue into the minute central canal of the spinal cord. Cerebro-spinal fluid is produced by the choroid plexuses within the ventricles; its only exit is through foramina in the roof of the fourth ventricle, in the medulla.

The Cerebral Hemispheres

These extend throughout the length of the skull from forehead to occiput, above the anterior and middle cranial fossæ and, behind this, above the tentorium cerebelli. Their medial surfaces are flat and lie against the falx cerebri; below the falx the two hemispheres are joined by the corpus callosum. The under surface of the hemisphere is more irregular than the medial surface; the orbital surface of the frontal lobe is slightly concave from the impression of the anterior cranial

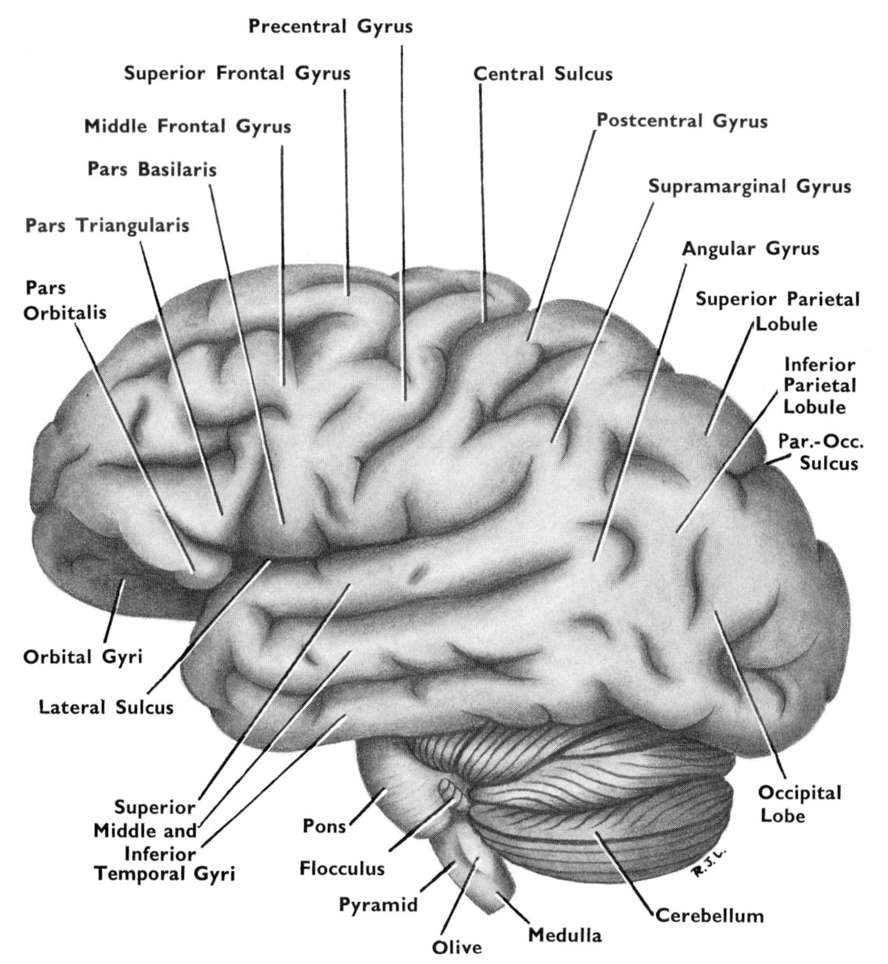

FIG. 7.1. THE BRAIN FROM THE LEFT SIDE, WITH THE GYRI SOMEWHAT SIMPLIFIED. Drawn from Specimen S 219 in R.C.S. Museum.

fossa, the temporal pole is boldly convex in conformity with the middle cranial fossa while the under surface of the occipital lobe slopes downwards and outwards to conform with the shape of the tentorium. The under surfaces of the two hemispheres are joined by the cerebral peduncles of the midbrain; anteriorly lie the structures of the floor of the third ventricle. The lateral surfaces of the hemispheres are boldly convex in conformity with the shape of the skull; the more complete term 'supero-lateral' is usually applied to this convex surface.

All surfaces of the cerebral hemisphere are covered with grey matter (the cells of the cerebral cortex) which is thrown into a complicated series of tortuous folds. The convex folds are termed gyri; the grooves between them are called sulci. All the gyri and sulci are named, but it adds nothing to the appreciation of cerebral anatomy to memorize all the names, the more especially as current ideas of cortical function assign many areas to the cortex that have little or no relation to the anatomical gyri and sulci.

Only the most important gyri and sulci will therefore be here described by name. Though the patterns of no two brains are identical, an underlying similarity exists, and the simplified pattern common to all brains should be appreciated by the student.

The Supero-lateral Surface of the Hemisphere (Fig. 7.1). A deep fissure that separates the frontal and temporal lobes on the under surface of the brain is continued to the lateral surface and passes backwards, above the temporal lobe. This is the **lateral sulcus** (fissure of Sylvius). When separated by the fingers the lateral sulcus is seen to branch upwards from its commencement in V-shaped manner. The areas of bulging cortex between these little sulci are subdivisions of the inferior frontal gyrus. Separately they are named the pars orbitalis, pars triangularis and pars basilaris (Fig. 7.1). Together they are called the **opercula** (of the frontal and parietal lobes), and with the temporal operculum they overlie a buried part of the cortex known as the insula.

An oblique sulcus passes up from just behind the opercula to indent the superior border of the hemisphere *just behind the midpoint.* It is *the only long sulcus to pass over on to the medial surface* of the hemisphere. This is the **central sulcus** (fissure of Rolando) and it separates frontal and parietal lobes. The precentral and postcentral gyri lie in front of and behind it; they contain the motor and sensory cortical areas. In front of the precentral gyrus the frontal lobe is divided by two horizontal sulci into three gyri, the superior, middle and inferior frontal gyri. A similar arrangement divides the temporal lobe into superior, middle and inferior temporal gyri.

The parietal lobe is divided by a transverse sulcus into superior and inferior parietal lobules. Into the latter project the lateral sulcus and the temporal sulci; the posterior ends of these sulci are closed by little gyri named the supramarginal, angular and posterior parietal gyri.

The occipital lobe is not marked anatomically from the parietal lobe on this surface of the hemisphere. The parieto-occipital sulcus on the medial surface extends across the superior border of the hemisphere and here divides parietal and occipital lobes; the line of division on the lateral surface of the hemisphere is an imaginary one. It extends down in a 45-degree slope to the inferior border where there is often a slight notch indented into the border by a fold of dura mater over the transverse venous sinus. Such a notch is seen in Fig. 7.2.

The **insula** is an area of cortex buried beneath the

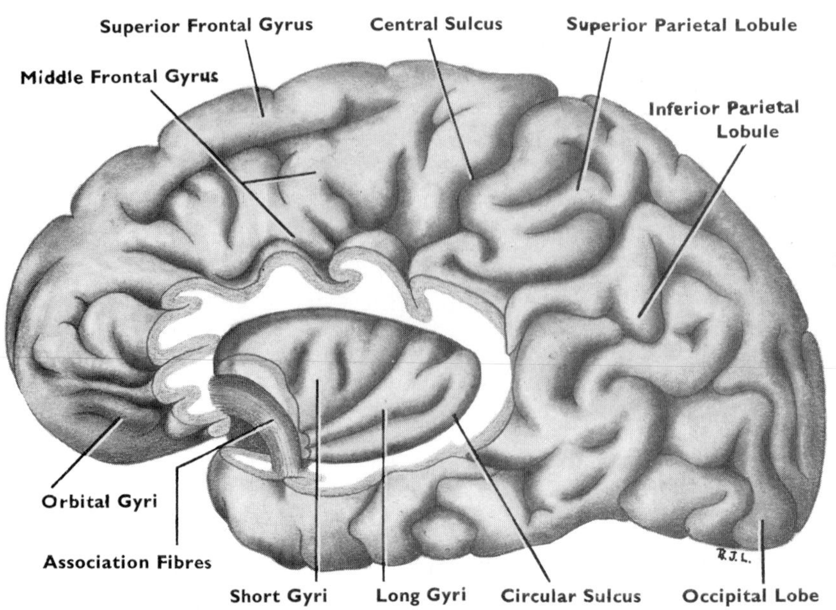

FIG. 7.2. THE LEFT INSULA EXPOSED BY EXCISION OF THE OPERCULA. Drawn from Specimen S 218.23 in R.C.S. Museum.

bulging opercula; when the latter are separated or removed it is exposed as a few long and short gyri almost completely surrounded by the *circular sulcus* (Fig. 7.2).

The Medial Surface of the Hemisphere. The two medial surfaces are flat and lie close together; they can be inspected only when their midline connexions are divided by sagittal section (Fig. 7.3). Such a section severs the corpus callosum and the roof and floor of the third ventricle, as well as the brain stem and cerebellum if these are still attached to the cerebral hemispheres. The wall of the third ventricle thus exposed beneath the corpus callosum is described on p. 514. The medial surface of the hemisphere above the corpus callosum shows a sulcus curving back. This is the sulcus cinguli; between it and the corpus callosum is the gyrus cinguli (Fig. 7.3). Above the sulcus cinguli the medial frontal gyrus extends, anteriorly, to the superior border of the hemisphere. Just behind the mid-point of the superior border the central sulcus turns on to the medial surface; it is enclosed in the paracentral lobule.

Now examine the posterior end of the hemisphere. The oblique parieto-occipital sulcus separates the parietal from the occipital lobe; it extends over the superior border to appear, as previously noted, on the supero-lateral surface. The medial surface of the occipital lobe is wedge-shaped and is named the cuneus. Between the parieto-occipital sulcus and the paracentral lobule is the precuneus.

The **cuneus** is limited inferiorly by a sulcus which runs forward from the occipital pole to the medial surface of the temporal lobe. The parieto-occipital sulcus runs into it. This sulcus is named calcarine in front of the junction and postcalcarine behind (Fig. 7.3).

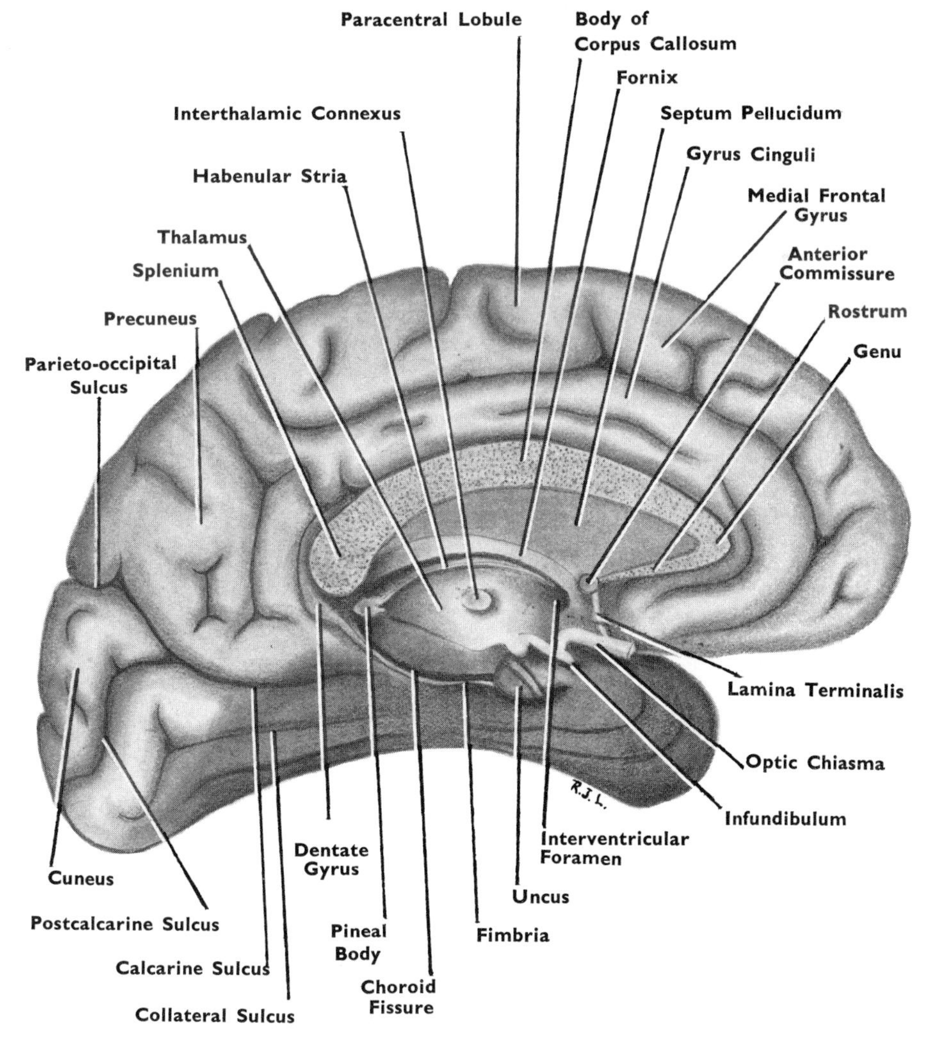

FIG. 7.3. THE MEDIAL SURFACE OF THE LEFT CEREBRAL HEMISPHERE. THE CEREBELLUM AND BRAIN STEM ARE REMOVED TO EXPOSE THE CHOROID FISSURE AND THE UNDER SURFACE OF THE HEMISPHERE. Drawn from Specimen S 218.25 in R.C.S. Museum.

The lingual gyrus lies between the calcarine and collateral sulci, at the border between medial and inferior surfaces of the occipital lobe.

The medial surface of the temporal lobe can be seen fully only when the cerebral peduncles are divided and the brain stem removed. It is best studied with the choroid fissure (see p. 516).

The parts on the medial surface immediately surrounding the corpus callosum and diencephalon are collectively referred to as the *limbic lobe* because of some functional continuity between them. The limbic 'lobe' has no reality as an anatomical formation in its own right; it is not a lobe. Its function is mainly integrative between afferent and efferent cortical pathways (p. 504).

The **inferior surface of the hemisphere** (Fig. 7.4) shows the orbital surfaces of the frontal lobes and the sloping inferior surface of the temporo-occipital part of the brain.

The orbital surface of the frontal lobe has a straight gyrus, the gyrus rectus, along its medial margin. Lying on the gyrus rectus is the olfactory bulb. The olfactory tract runs in the olfactory sulcus alongside the gyrus rectus. Lateral to the olfactory bulb and tract this surface is gently concave and is divided by an H-shaped sulcus into anterior, medial, posterior and lateral orbital gyri. The orbital gyri and sulci leave prominent impressions on the orbital plate of the frontal bone (Fig. 6.89, p. 481).

The temporal pole is boldly convex; the temporal lobe merges posteriorly with the occipital lobe and the continuous surface so formed is concave and oblique in conformity with the slope of the tentorium cerebelli, against which it lies. Hence much of the medial surface

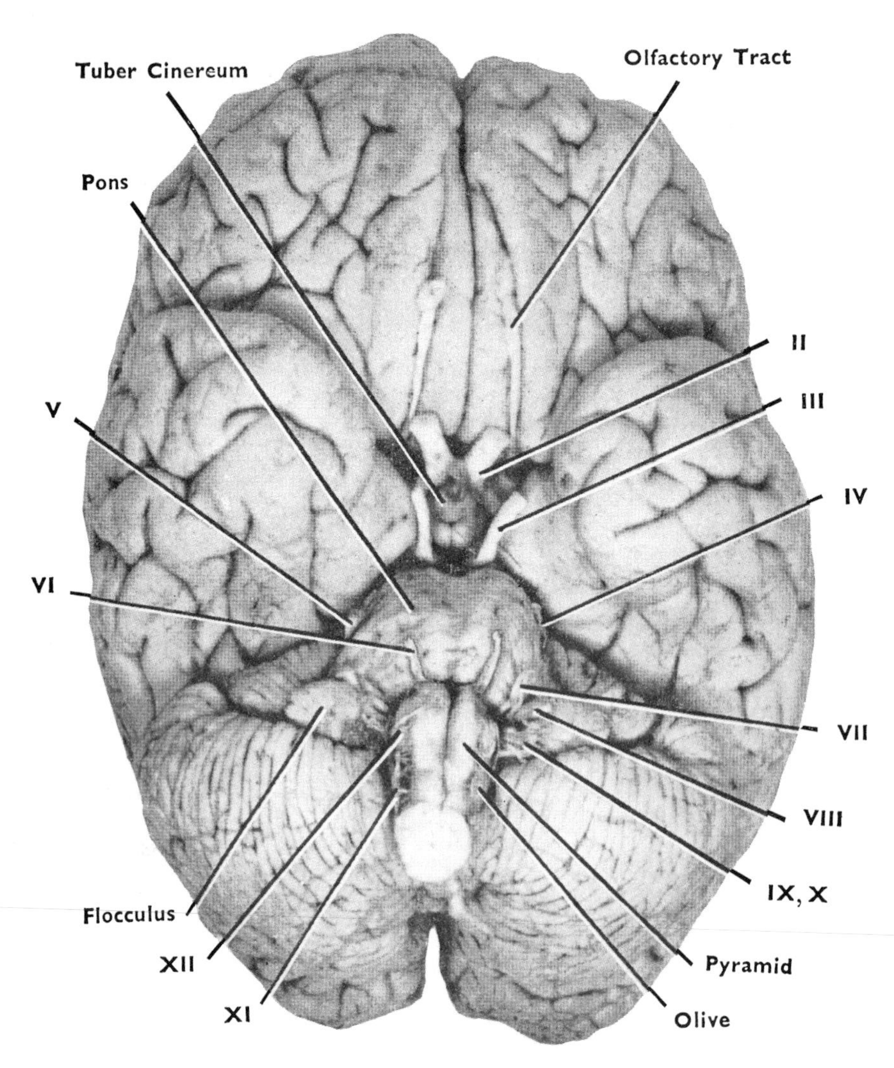

FIG. 7.4. THE INFERIOR SURFACE OF THE BRAIN, WITH THE ATTACHMENTS OF THE CRANIAL NERVES. The right cerebral hemisphere is set in advance of the left; such asymmetry is normal. (Photograph.)

of the temporal lobe can be seen from the inferior view; the tentorial surface under review is really infero-medial on the hemisphere. It is characterized by two long parallel sulci, the occipito-temporal sulcus laterally and the collateral sulcus medially (Fig. 7.3). They run antero-posteriorly between the temporal and occipital poles. Medial to the collateral sulcus is the hippocampal gyrus, confined to the temporal lobe and recurved anteriorly to form the uncus (Fig. 7.3).

Between the temporal poles the midline structures which form the floor of the third ventricle can be seen. They lie in front of the cerebral peduncles of the midbrain and are bounded in front by the **optic chiasma.** From the chiasma the optic tracts diverge around the cerebral peduncles, high up under cover of the temporal lobe. Behind the chiasma lies a rounded elevation, the **tuber cinereum,** from which the stalk of the pituitary gland depends in the intact brain. Behind the tuber cinereum are the rounded eminences of the **mamillary bodies** and, behind these, deep in the angle between the cerebral peduncles, the posterior perforated substance (Figs. 7.4, 7.35, p. 528).

On the inferior surface of the frontal lobe immediately lateral to the chiasma is the anterior perforated substance; the medial and lateral divisions of the olfactory tract can be seen diverging around this area (Fig. 7.35, p. 528).

Internal Structure of the Cerebral Hemisphere. The interior of the cerebrum is characterized by the presence within the white matter of large masses of grey matter which together are known as the basal ganglia and, also, by cavities which receive and transmit the cerebro-spinal fluid.

The **basal ganglia** consist essentially of the corpus striatum complex which is motor in function, and the thalamus which is sensory. Since they lie buried in the white matter they can be seen only in sections of the brain; but their shape is irregular, so that their cut appearance varies markedly with the level and direction of the particular section studied. A solid model of the basal ganglia gives a better idea of their shape (Figs. 7.5—posterior view, 7.6—lateral view, 7.23, p. 516—superior view).

The **thalamus** lies in the walls of both third and lateral ventricles, and posteriorly it projects below the splenium. It is shaped like a wedge lying on its side, sharp edge anteriorly and base posteriorly; the upper part of the base projects posteriorly—this projection is named the pulvinar. The thalamus contains nuclei of the sensory tracts; it projects fibres through the internal capsule to the cerebral cortex. One such nucleus, on the visual pathway, is visible as a swelling on the back of the thalamus below the pulvinar. It is called the lateral geniculate body; the optic tract passes around the cerebral peduncle to reach it. The thalamus will be better understood after the lateral and third ventricles

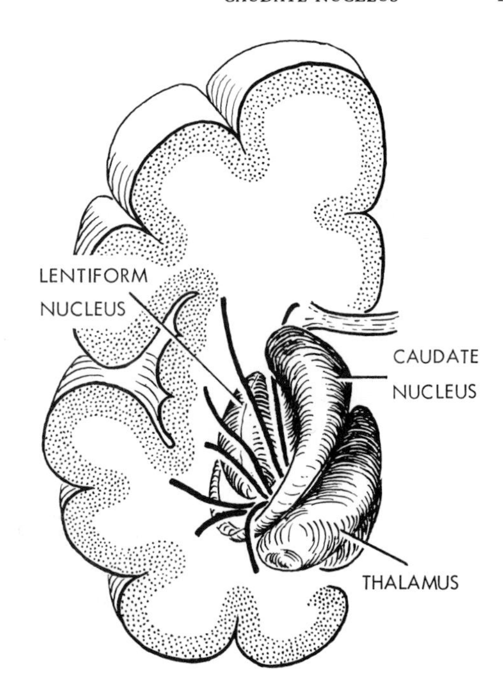

FIG. 7.5. THE BASAL GANGLIA OF THE LEFT HEMISPHERE VIEWED FROM BEHIND. The disposition of the fibres of the corona radiata and internal capsule is indicated.

have been studied, so its further description is postponed to p. 514.

The **corpus striatum** is usually described as consisting of the lentiform nucleus and the caudate nucleus. These nuclei are cut almost completely apart by the white fibres of the internal capsule. Some bands of connecting grey matter survive across the internal capsule (Fig. 7.15) and give the body its striped name. Two other grey masses, the amygdaloid nucleus and the claustrum, are conjoined, and it is better to include them also in the corpus striatum complex. All four are joined together in the base of the hemisphere, and at this place they are joined also to the grey matter of the cortex at the anterior perforated substance (Fig. 7.15). Here pass the striate arteries and veins. Above the anterior perforated substance the four nuclei diverge into the white matter like the petals of some strange orchid (Fig. 7.6).

The **lentiform nucleus** is the shape of a highly biconvex lens, completely buried in the hemisphere. It is oval in outline (Fig. 7.6). It has two parts, a large lateral *putamen*, curved and roughly quadrilateral in shape, and a small medial *globus pallidus*, bluntly conical. It is the putamen that is joined to the head of the caudate nucleus by bands of grey matter (Fig. 7.15).

The 'C-shaped' **caudate nucleus** is rather the shape of a highly curved comma (Fig. 7.6). A bulbous *head* tapers back to the *body* which, curving around the lateral part of the thalamus, curves sharply forward into a long,

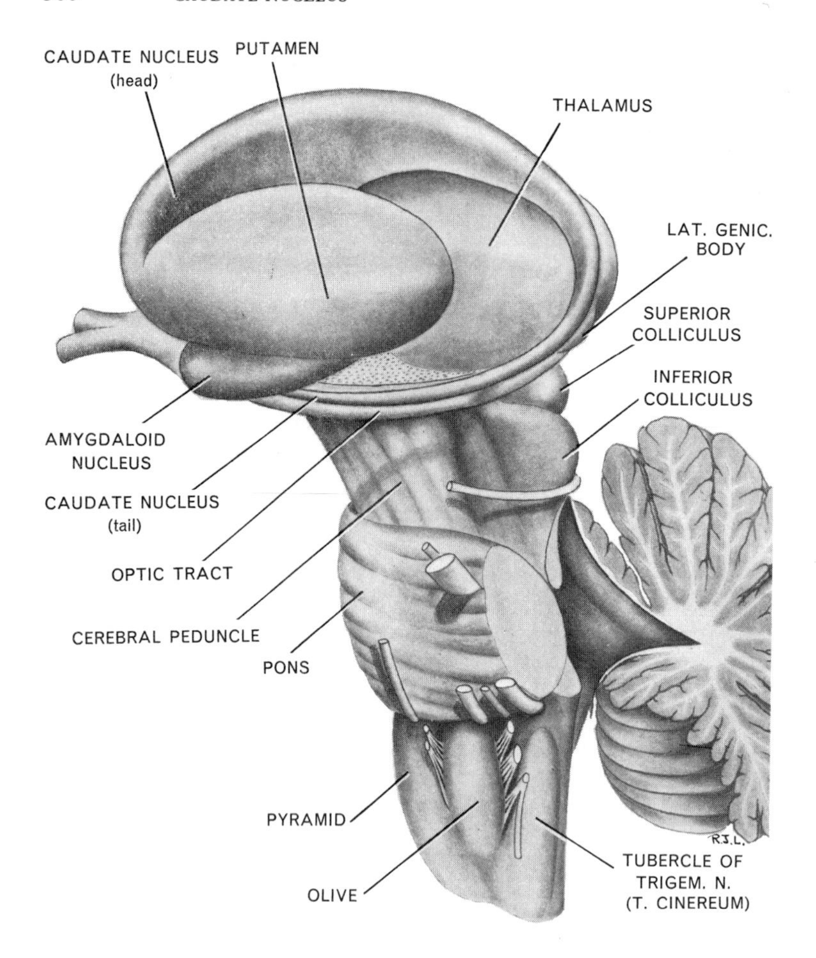

FIG. 7.6. THE BASAL GANGLIA OF THE LEFT HEMISPHERE. The crus cerebri is cut through and the internal capsule removed. The right half of the fourth ventricle is exposed by division of the left cerebellar peduncles, and sagittal division of the cerebellum. Drawn from specimens in R.C.S. Museum.

thin *tail*. The tail joins the amygdaloid nucleus. The caudate nucleus is curled snugly around the fibres of the internal capsule, like a hand holding a bunch of flowers (Fig. 7.5). The whole length of its outer convexity projects into the lateral ventricle.

The **amygdaloid body (amygdaloid nucleus)** is not a 'body' because it is fragmented into several scattered masses, and it is not really shaped like an almond (its representation in Fig. 7.6, is diagrammatic). The amygdaloid is joined to the lower convexity of the putamen, and its posterior part projects into the tip of the inferior horn of the lateral ventricle (Fig. 7.17).

The **claustrum** is a thin lamina, circular in outline, curved into a saucer shape. Its lower margin joins the putamen (Fig. 7.16), and the claustrum leans laterally above this (Fig. 7.23, p. 516). The white matter (largely association fibres) between claustrum and putamen is called the 'external capsule'. The claustrum lies just deep to the cortex of the insula.

The connexions of the corpus striatum are complicated and to a very large extent still unknown. The ganglia seem to be part of a complex system of feedback circuits between cortex and thalamus on the extrapyramidal pathway. Destructive lesions here, especially of the caudate nucleus, produce the muscular rigidity and tremor characteristic of paralysis agitans.

The **white matter** of the cerebral hemisphere is made up of fibres belonging to three main groups.

Commissural fibres join the two hemispheres. Most of them are gathered together in the corpus callosum; a few lie in the anterior and habenular commissures. They radiate widely and symmetrically through the white matter of the hemispheres.

Association fibres are confined to their own hemisphere, in which they connect different parts of the cortex.

Projection fibres are those which join the grey matter of the hemisphere with lower centres; they are both sensory (afferent) and motor (efferent) and in the base of the hemisphere they lie lateral to the thalamus and the head of the caudate nucleus, where they are known as the *internal capsule*. The lentiform nucleus lies lateral to the internal capsule and the tail of the caudate nucleus curls around, also lateral (Fig. 7.5). From the internal capsule the fibres radiate upwards and outwards in the shape of a curved fan to reach the cortex; this fan-shaped arrangement is termed the *corona radiata.* The fibres of the corpus callosum (i.e., tapetum) intersect it.

The **internal capsule** consists of afferent fibres passing up to the cortex from cell bodies in the thalamus, and of efferent fibres passing down from cell bodies in the cortex to the crus cerebri of the midbrain. It lies within the concavity of the C-shaped caudate nucleus, which separates it from the C-shaped concavity of the lateral ventricle. Its anterior and middle fibres, passing through the corpus striatum, almost separate the caudate nucleus from the lentiform nucleus; the two remain connected mainly at the anterior perforated substance, on the inferior surface of the frontal lobe (Fig. 7.5).

Within the concavity of the C-shaped caudate nucleus the white matter of the internal capsule is indented by the medial convexity of the lentiform nucleus (i.e., by the globus pallidus) so that a horizontal section through the internal capsule shows it to possess a bend called the genu, at the apex of the globus pallidus (Figs. 7.7, 7.9).

The **anterior limb** contains *fronto-pontine fibres* from cell bodies in the frontal cortex; these are the fibres that indent the corpus striatum between the head of the caudate nucleus and the anterior part of the lentiform nucleus. They pass down below the thalamus into the cerebral peduncle, where they occupy the medial one-fifth of the crus (basis pedunculi). They arborize around the pontine nuclei.

The **genu** and anterior two-thirds of the **posterior limb** of the internal capsule contain the cortico-spinal fibres that constitute the *pyramidal tracts.* The cell bodies lie in the motor cortex. The fibres pass below the thalamus in the cerebral peduncle; they occupy the middle three-fifths of the crus (basis pedunculi). They decussate in the medulla and arborize around lower motor neurones to the skeletal muscles. In the internal capsule the head fibres lie anteriorly, at the genu; behind these lie the arm, hand, trunk, leg and perineum fibres in that order (in the midbrain the head fibres lie medially, the perineum laterally, in the same order). It is in this part of the internal capsule that hæmorrhage or thrombosis of a thalamo-striate artery commonly occurs. The muscles of the opposite side of the body are thus paralysed; they become spastic and show increased muscle jerks, the signs of an upper motor neurone lesion. Fibres from the speech area (Broca's area) are interrupted in lesions of the left internal capsule; thus loss of speech accompanies hemiplegia of the *right side of the body.*

The head fibres in the pyramidal tract decussate to the motor nuclei of the brain stem; this part of the tract is called *cortico-bulbar.* The remainder, decussating mostly in the medulla, reach the anterior horn cells of the spinal cord; this part of the tract is called *cortico-spinal.*

Behind the pyramidal tracts the posterior limb of the internal capsule contains sensory fibres mediating impulses derived from the opposite half of the body; their cell bodies are in the thalamus and the fibres pass upwards through the corona radiata to the sensory cortex. Behind these tactile fibres are the *visual fibres.* Their cell bodies lie in the lateral geniculate body, whence the fibres pass back in the optic radiation to the occipital cortex. This part of the internal capsule is behind the lentiform nucleus, but it still lies within the concavity of the C-shaped caudate nucleus. Behind the visual fibres are the *auditory fibres* passing out from cell bodies in the medial geniculate body to the auditory centre at the posterior end of the superior temporal gyrus. Finally the most posterior fibres of all are the *temporo-pontine* fibres, which pass from cell bodies in the temporal cortex through the internal capsule and so beneath the thalamus into the cerebral peduncle, where they occupy the lateral one-fifth of the crus (basis pedunculi). They are accompanied by fibres from the occipital and parietal cortex; all arborize around the cells of the pontine nuclei.

The **corpus callosum** (Fig. 7.3) consists of a mass of commissural fibres each of which extends from cortex to cortex between symmetrical parts of the two hemispheres. It commences at the anterior commissure, at the upper end of the lamina terminalis of the diencephalon and, traced from here to its termination it becomes increasingly thicker. From the anterior commissure the mass passes upwards and forwards; this part is known as the *rostrum.* It now takes a sharp bend backwards; the bend is known as the *genu.* From here it is gently convex upwards (the *body* of the corpus callosum) and it ends posteriorly as a thick rounded free border, the *splenium.* The corpus callosum can be seen by separating the two hemispheres, and its cut surface is exposed in a midline sagittal section through the brain (Fig. 7.20).

The fibres of the corpus callosum extend to all parts of the cerebral cortex. In a horizontal section the fibres of the genu are seen arching forwards on each side to the frontal cortex; this appearance gives them the name *forceps minor.* Similarly, the fibres of the massive splenium curve backwards symmetrically to the occipital cortex, forming the *forceps major* (Fig. 7.7).

Between forceps minor and forceps major the fibres of the corpus callosum spread out to the cortex on the lateral surface of the hemisphere. They pass across the anterior horn and body of the lateral ventricle, for each of which they form the roof; this mass of fibres extending laterally deep into the hemisphere is called the *tapetum.* As it turns down into the temporal lobe the tapetum forms the lateral wall of the inferior horn of the lateral

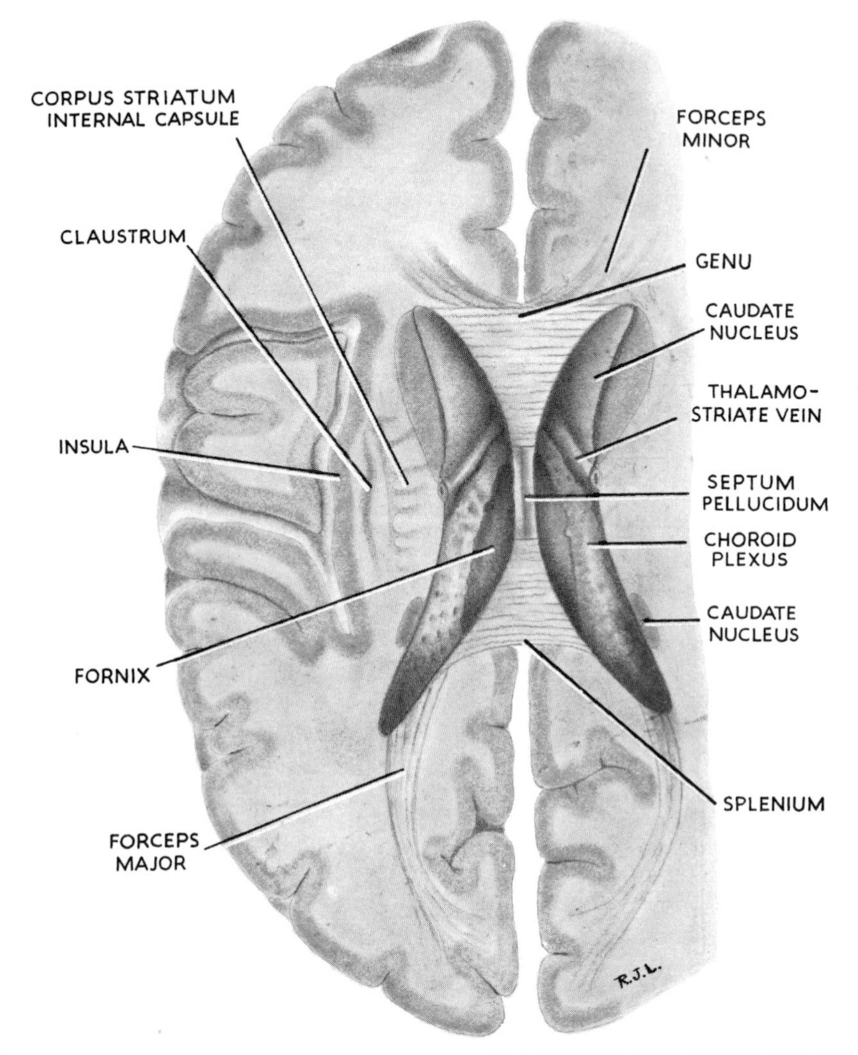

FIG. 7.7. A HORIZONTAL SECTION THROUGH THE GENU AND SPLENIUM OF THE CORPUS CALLOSUM, VIEWED FROM ABOVE. THE LATERAL VENTRICLE IS OPENED. (cf. Fig. 7.9).

ventricle. The transverse fibres of the tapetum intersect the vertical fibres of the corona radiata.

The part of the corpus callosum lying between the two hemispheres is covered with a thin film of grey matter (the indusium griseum) which is a remnant of the hippocampal formation (smell brain) of the archæpallium. Lying in the indusium griseum, on each side of the corpus callosum, is a pair of longitudinal ridges, the medial and lateral *striæ longitudinales* (Fig. 7.22).

Areas of the Cortex. Certain areas of the cortex have long been identified, many otheres are in the process of identification. Around the specific areas mentioned below are areas of cortex concerned with association.

The **motor cortex** lies in the precentral gyrus and the anterior wall of the central sulcus (area 4). Movements of various areas of the body are initiated here; the cells of the motor cortex send their axons down the pyramidal tracts. The opposite half of the body is represented upside down along the motor cortex. The face lies lowest, then the hand (a very large area), then arm, trunk and leg. The leg and perineum areas overlap the superior border and extend down on the medial surface of the hemisphere into the paracentral lobule. Below the face area on the left side (frontal and parietal opercula) lies the speech centre (Broca's area).

The **sensory cortex** for appreciation of kinæsthetic sensibility, touch, temperature, etc., from the opposite half of the body lies in the postcentral gyrus and the posterior wall of the central sulcus, in areas that correspond roughly with those of the motor cortex. In man there is a good deal of overlap of adjacent sensory areas.

Both motor and sensory cortex lie near the ascending part of the anterior branch of the middle meningeal artery and may be damaged by the pressure of a hæmorrhage from the vessel.

The blood supply of the motor and sensory cortex is from the middle cerebral artery up to within a finger's breadth of the superior border of the hemisphere (i.e., for areas face, arm and trunk), and above this level by the anterior cerebral artery (i.e., for areas leg and perineum) (Fig. 7.26, p. 519).

The **auditory cortex** lies in the upper part of the temporal operculum, where the anterior *transverse temporal gyrus* passes deeply inwards to the circular sulcus behind the insula. The auditory cortex is supplied by the middle cerebral artery. The posterior branch of the middle meningeal artery lies near it, and hæmorrhage from this vessel may injure the centre and cause some deafness (not entirely unilateral, since each cochlea registers in both auditory areas).

The **visual cortex** lies on the medial surface of the occipital lobe. The true centre lies in the depths of the calcarine and postcalcarine sulci, but association areas extend from the sulcus upwards on the cuneus and downwards on the lingual gyrus and posteriorly around the occipital pole on to the lateral surface of the occipital lobe. The true centre is characterized by a white line which bisects the grey matter of the cortex, the stria of Gennari. It is present in the inferior wall of the calcarine sulcus and both walls of the postcalcarine sulcus (Fig. 7.9).

Each occipital cortex receives from its own half of each retina; that is, it registers the *opposite visual field.* In each cortex the upper half receives from the upper half of each half-retina, the lower half from the lower half of each half-retina, i.e., the upper and lower *visual fields are crossed.* The macula registers at the posterior end of the sulcus and more peripheral parts of the retina progressively more anteriorly in the visual area.

The Visual Pathways. The peripheral nerves of ordinary sensation, with their cell bodies in posterior root ganglia, are represented in the visual pathway by the bipolar cells of the retina. These cells receive impulses from the rods and cones of the retina. The bipolar cells synapse with ganglion cells in the inner part of the retina (next the vitreous, one cell for each cone, one cell for eighty rods). These are homologous with the second neurone cell bodies in the central nervous system in the other sensory pathways. Their axons run on the surface of the retina and enter the disc and so pass to the optic nerve.

The **optic nerve** is not a nerve in the sense of the other cranial and spinal nerves; it is an elongated piece of white matter stretched out from the brain and enclosed in the meninges thereof as far forward as its attachment to the sclera. Histologically it is identical with white matter of the central nervous system. The essential point is that, like all white matter, it has no Schwann cells and, likewise, no power of regeneration when divided. In the orbit it is surrounded by a tube of dura mater and arachnoid, with cerebro-spinal fluid in the subarachnoid space. At the optic foramen the dura and arachnoid leave it and the nerve, still sheathed in pia mater, passes up to meet its fellow at the *optic chiasma*, which is attached to the anterior part of the floor of the third ventricle.

In the chiasma the nasal fibres of each optic nerve decussate and pass into the optic tract of the opposite side; the temporal fibres from each retina pass on to their own side (Fig. 7.8). Thus the right optic tract contains fibres from the right half of each retina, the left optic tract contains fibres from the left half of each retina and since there is no further decussation this holds true right back to the visual cortex.

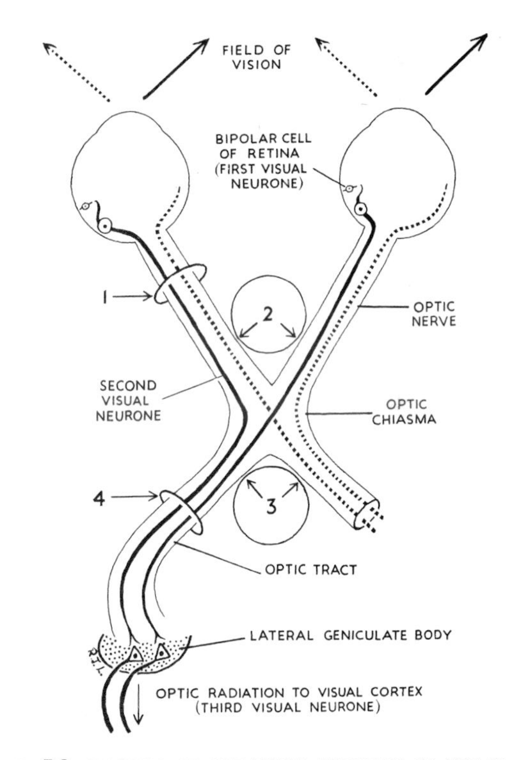

FIG. 7.8. DIAGRAM OF THE THREE NEURONES IN THE VISUAL PATHWAY. FIBRES FROM THE NASAL HALF OF EACH RETINA (SECOND VISUAL NEURONE) DECUSSATE IN THE OPTIC CHIASMA. A LESION AT 1 CAUSES MONOCULAR BLINDNESS, A LESION AT 2 OR 3 CAUSES BITEMPORAL HEMIANOPIA, A LESION AT 4 CAUSES CONTRALATERAL HEMIANOPIA.

Cortical pathways for common sensation consist of three neurones. They reach the opposite hemisphere by a *complete* decussation of the second order neurones. The visual pathway by the *half* decussation of its second order neurones at the chiasma achieves the same object.

There is *complete crossing of the visual fields.* One hemisphere registers common sensation from the opposite half of the body and also from the opposite half of the visible environment.

The **optic tract** passes from the chiasma around the cerebral peduncle, high up against the temporal lobe and, reaching the side of the thalamus, divides into two branches. The larger of these enters the lateral geniculate body, in which the fibres synapse. These are visual fibres. The smaller branch passes down medially, between the lateral and medial geniculate bodies, and synapses in the superior colliculus and the pretectal nuclei; these are fibres mediating light reflexes.

Blood supply. The optic tract is supplied chiefly by the anterior choroidal and posterior communicating arteries, the chiasma and intracranial part of the optic nerve by the anterior cerebral. In the orbit the nerve is supplied by the ophthalmic artery and, distally, by the central artery on its way to the retina.

The **lateral geniculate body** is a small rounded elevation below the pulvinar on the posterior surface of the thalamus. From the ganglion cells of the second neurone (in the retina) axons pass in the optic nerve, chiasma and optic tract to synapse with cells in **six layers** of the geniculate body. The fibres from the half-retina of the same side (i.e., temporal fibres) synapse at layers 2, 3 and 5, those from the half-retina of the opposite side (i.e., nasal fibres) synapse at layers 1, 4 and 6. From these layers the cell bodies send their axons through the optic radiation to the occipital cortex (p. 503).

The *superior brachium* is the name given to the small (medial) branch of the optic tract. It passes down on the thalamus to the midbrain, where it ends in the tectum. The fibres in the superior brachium arborize around cells in the superior colliculus. The cell bodies in the *superior colliculus* send their fibres, by tecto-bulbar and tecto-spinal tracts, to motor nuclei in the brain stem and spinal cord for the mediation of light reflexes (e.g., reflex blinking and turning away from a flash of bright light). The superior colliculi are united by the posterior commissure (at the entrance to the aqueduct, Fig. 7.20), and thus general body reflexes to light are usually bilateral. The special fibres concerned in the *pupillary light reflex* do not synapse in the colliculus, but pass bilaterally to *each* pretectal nucleus. The **pretectal nucleus** is a small group of cells lying in the tegmentum under the upper and lateral margin of the superior colliculus. It passes light impulses to each Edinger-Westphal nucleus and so to the sphincter pupillæ. A lesion here produces the Argyll-Robertson pupil; contraction to light is lost, but the pupil still contracts to accommodation and convergence. (The pathway of these latter reflexes is explained on p. 543.)

A third branch of the optic tract passes to the medial geniculate body. These are not light fibres, but commissural auditory fibres (between the two medial geniculate bodies) whose course in the white matter is via the optic tracts and chiasma (Gudden's commissure).

The **olfactory cortex** or rhinencephalon is proportionately much smaller in the brain of man than in the brains of the keener-scented mammals. The symptoms of lesions of the rhinencephalon in man are of limited diagnostic value.

The **olfactory nerves** pass from the olfactory mucous membrane (p. 399) through the perforations of the cribriform plate of the ethmoid bone. Their bipolar cell bodies lie in the olfactory mucous membrane. The central processes of these first order neurones gather themselves into twenty or so nerves, which pass through the cribriform plate ensheathed in dura, arachnoid and pia mater. The nerves pass back to the **olfactory bulb,** in which they synapse. The *olfactory tract* is an elongated extension of white matter of the brain (cf. the optic nerve). It lies in the olfactory sulcus to the side of the gyrus rectus on the inferior surface of the frontal lobe. It passes back to the anterior perforated substance and there divides (Fig. 7.36, p. 528).

The *lateral root* of the olfactory tract skirts the anterior perforated substance and passes up across the insula, then bends sharply back upon itself and enters the *uncus*, which is a recurvature at the front of the hippocampal gyrus. This part of the uncus is known as the *temporal prepiriform cortex*. The *medial root* of the olfactory tract sends some of its fibres by way of the anterior commissure to the opposite olfactory bulb (inhibitory), but most end in the *frontal prepiriform cortex* (anterior perforated substance) and a small part of the amygdaloid nucleus.

By the above two routes the second neurone fibres pass from the olfactory bulb to the uncus and adjoining areas of the anterior perforated substance, where they synapse around cell bodies in the grey matter. This area of grey matter is called the *piriform area* because it is large and pear-shaped in lower vertebrates; it bears no resemblance to a pear in the human brain. The arrangement is unique in that the second neurone reaches cells in the cerebral cortex without having relayed through the thalamus. This is the **olfactory centre** for conscious appreciation of smell, and no other cortex is apparently involved.

Further synapses connect the olfactory bulb with the hypothalamus and brain stem, as is the case with other sensory pathways (light, sound, taste, touch) for visceral and somatic effects, distinct from conscious appreciation.

'Limbic' Formations. Surrounding the corpus callosum and diencephalon the grey matter on the medial surface of the hemisphere is often known under this name ('limbic system', 'limbic lobe'). Most of the formations are easily seen on the surface of the brain.

On the medial surface of the frontal lobe alongside the lamina terminalis is the small paraterminal gyrus. This can be traced around the genu and over the body of the corpus callosum into the gyrus cinguli already noticed.

On the upper surface of the corpus callosum is a thin film of grey matter named the *indusium griseum*, beneath which lie the *medial* and *lateral longitudinal striæ*. These appear to be aberrant fibres of the fornix formation (v. inf.).

Behind the splenium of the corpus callosum is the dentate gyrus, whose tail can be traced along the upper part of the hippocampal gyrus to cross the uncus (Fig. 7.3).

The Hippocampus. This is part of the 'limbic system'. Just above the anterior part of the hippocampal gyrus (here known as the *subiculum*) lies the hippocampal sulcus (Figs. 7.17, 7.18), which is projected into the floor of the inferior horn of the lateral ventricle (Fig. 7.9). This projection is the hippocampus. On the ventricular surface of the hippocampus is a thin film of white matter called the *alveus*; its cell bodies are in the hippocampus and the subiculum. The fibres of the alveus thicken medially to form the *fimbria*. This breaks free from the hippocampus as the *crus* (posterior pillar) *of the fornix*.

This flat band curves up behind the thalamus to join its fellow in a partial decussation across the midline. This is called the *commissure of the fornix*. It is really a chiasma, and is in the nature of an association tract rather than a true commissure. The conjoined mass of white matter, lying beneath the corpus callosum, is known as the *body of the fornix*. From it the conjoined anterior columns arch down in front of the anterior poles of the thalami, forming the anterior margins of the interventricular foramina (Fig. 7.20).

The columns of the fornix pass both anterior and posterior to the anterior commissure. The anterior fibres pass mainly to so-called *septal nuclei* near the lamina terminalis (*not* in the septum pellucidum). The posterior fibres pass directly to the thalamus or into the mamillary body. From the mamillary body fibres pass in the lateral wall of the third ventricle as the mamillo-thalamic tract (bundle of Vicq d'Azyr) to the anterior pole of the thalamus. Here they relay and the thalamic neurones send their fibres through the internal capsule to the gyrus cinguli.

Much of the 'limbic lobe', notably the hippocampus and its fornix fibres relaying to the gyrus cinguli, was formerly thought to be rhinencephalic. This is no longer tenable, and the hippocampus and its connexions are now thought to be concerned in integrating afferent and efferent pathways of behavioural patterns, especially those of a repetitive nature that have been learned. Ablation of the hippocampus results in loss of memory for recent events, while memory for distant events is retained.

Taste. The cortex for conscious appreciation of taste appears to lie near the lower end of the postcentral gyrus (i.e., the parietal operculum) near the face area for tactile appreciation. Fibres reach this area through the internal capsule from the medial surface of the thalamus, after relay from the nucleus of the tractus solitarius (p. 526).

The Ventricles of the Brain

The central nervous system is hollow; it develops from a neural tube whose cavity persists. The cavity is lined throughout with ependyma. The brain formation requires cerebro-spinal fluid, whatever the reasons may be, and this fluid is produced within the cavity. The places where the cerebro-spinal fluid is produced are known as ventricles. In each ventricle the cavity comes to the surface without opening thereon; that is to say, the lining ependyma comes into contact with the surface pia mater, with no grey or white matter between. This is to allow the choroid plexuses to invaginate. These vascular fringes invaginate the whole length of each surface encroachment of the ventricle.

Each cerebral hemisphere possesses its cavity, the *lateral ventricle*, and this comes to the surface at a curved slit called the choroid fissure. The choroid plexus of the lateral ventricle is invaginated here. The diencephalon has a cavity, the *third ventricle*, that comes to the surface on its roof, and here are invaginated the two choroid plexuses of the third ventricle. The pons and medulla share a cavity which reaches the surface at the upper medulla, where the roof is invaginated by the right and left choroid plexuses of the *fourth ventricle*.

The choroid plexuses of the lateral ventricles are large and highly vascular; this pair secretes the bulk of the cerebro-spinal fluid. The choroid plexuses of the third and fourth ventricles are minute and they secrete only a small percentage of the total C.S.F. output. Each lateral ventricle opens into the third ventricle by the interventricular foramen. From the third ventricle the aqueduct opens below into the fourth ventricle, which is a cavity in both pons and medulla. Below the fourth ventricle the central canal extends as a tiny tube through the spinal cord into the upper end of the filum terminale. The only apertures in this system lie in the roof of the fourth ventricle, whence the cerebro-spinal fluid escapes into the subarachnoid space.

If the ventricles are filled with air they throw a shadow on an X-ray film. Use is made of this by direct injection of air (ventriculography) or by allowing air to rise through a lumbar puncture needle (encephalography). To interpret such radiographs knowledge of the shape of the ventricles is essential. The midline cavities (third ventricle, aqueduct and fourth ventricle) are symmetrical, but *asymmetry, especially posteriorly, is the rule in the lateral ventricles*. Look down on the

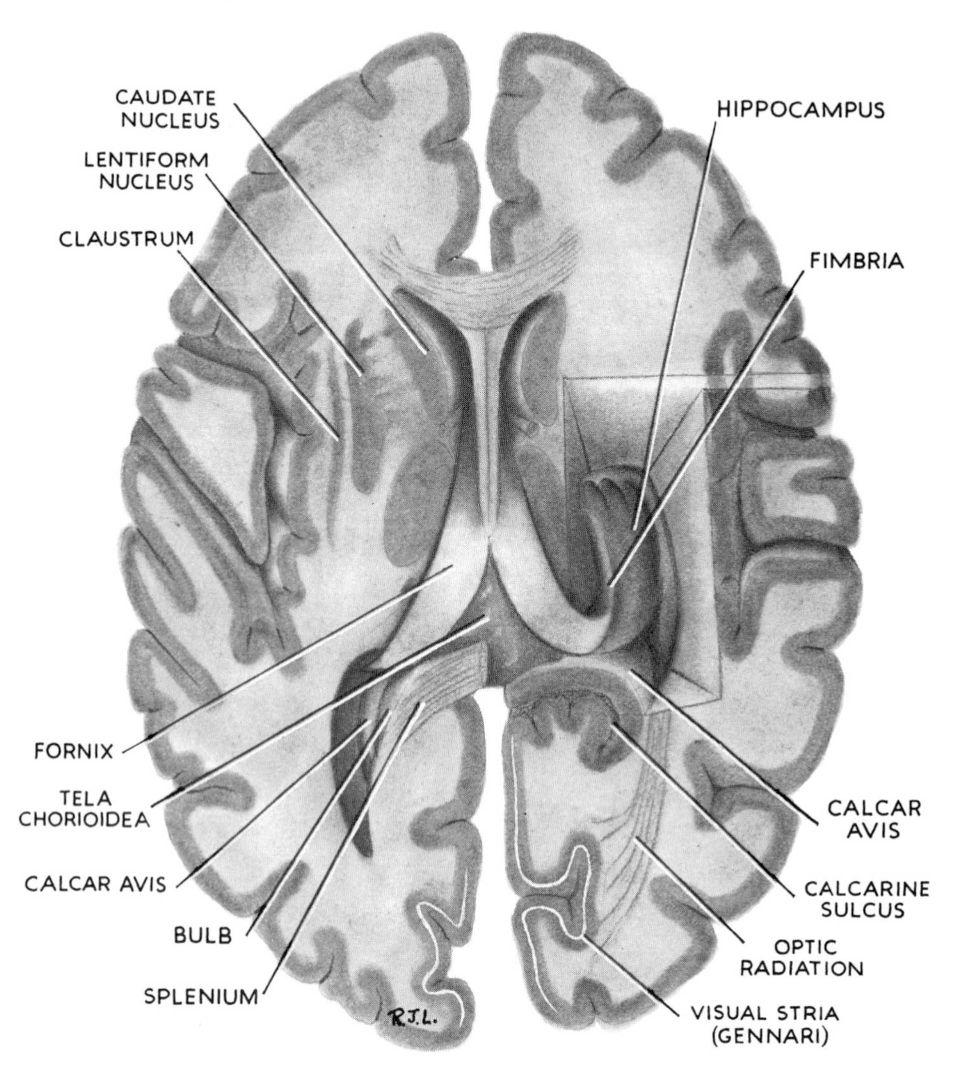

FIG. 7.9. A HORIZONTAL SECTION PASSING BETWEEN THE CORPUS CALLOSUM AND FORNIX. ON THE RIGHT SIDE THE OCCIPITAL AND TEMPORAL LOBES ARE CUT AT A LOWER LEVEL, AND THE INFERIOR HORN OF THE LATERAL VENTRICLE IS OPENED BY REMOVAL OF ITS ROOF. Drawn from a dissection.

vertex of any skull; it is asymmetrical in that one brow lies anterior to the other and the opposite occipital region lies more posteriorly than its fellow. The cerebral hemispheres are similarly asymmetrical and their contained lateral ventricles even more so (Figs. 7.4, 7.12).

The **lateral ventricle** is a C-shaped cavity, lined with ependyma, lying within the cerebral hemisphere. It does not lie entirely within the white matter of the hemisphere; indeed on its medial side it lies against the pia mater of the medial surface of the hemisphere, where pia mater and ependyma come into contact with each other. This line of contact is narrow, and curves around the top of the thalamus and the tail of the caudate nucleus, forming a C-shaped slit on the medial surface of the hemisphere known as the **choroid fissure** (Fig. 7.3). The choroid fissure should be regarded as the medial wall of the body and inferior horn. Into the choroid fissure the choroid plexus of the lateral ventricle passes, invaginating the pia mater and ependyma before it. The lips of the choroid fissure meet around the invaginated plexus, which thus lies hidden within the body and inferior horn of the ventricle.

In other places the cavity of the lateral ventricle lies further from the surface, so that grey matter at the bottom of a sulcus indents the cavity. Such sulci are the hippocampal, calcarine and collateral, which show as convexities within the cavity of the ventricle. Of the basal ganglia, only the caudate nucleus and thalamus project into the cavity. Elsewhere the walls of the

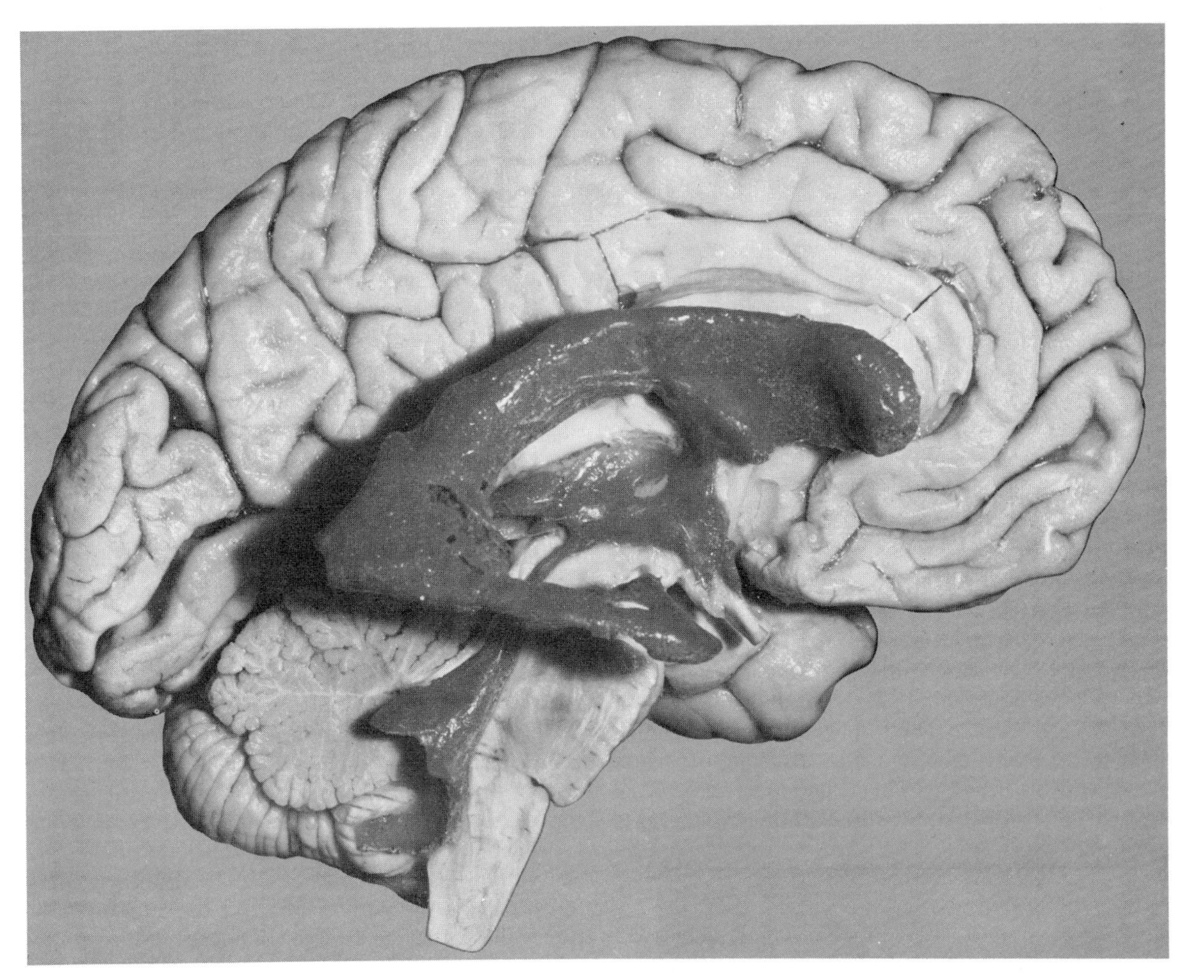

FIG. 7.10. A CAST OF THE CEREBRAL VENTRICLES FROM WHICH THE RIGHT HALF OF THE BRAIN HAS BEEN REMOVED. The lateral ventricle has no posterior horn but the trigone is grooved by the calcar avis and the bulb. In the third ventricle the supra-optic recess and the infundibulum are well shown. Above the aqueduct is the delicate pineal recess; the posterior commissure indents the cast between these two. A large suprapineal recess projects back above the pineal. Photograph of Specimen D 717.A in R.C.S. Museum.

cavity are formed by white matter of the cerebral hemisphere (tapetum).

The C-shaped cavity consists of named parts. On the upward convexity is the body, projected forwards into a blind extremity, the anterior horn. These parts are floored in by the caudate nucleus and the thalamus and the roof is the corpus callosum and the fornix. The blind extremity of the **anterior horn** is contained by the fibres of the corpus callosum that run laterally from the genu and rostrum (forceps minor). The bulbous head of the caudate nucleus lies in the floor, meeting the roof at an angle on the lateral side (Fig. 7.15), but separated from the roof medially by a thin partition between the fornix and corpus callosum called the septum pellucidum. Behind the anterior column of the fornix, between it and the anterior pole of the thalamus, is a small aperture, the **interventricular foramen of Monro,** which leads from the lateral into the third ventricle. The choroid plexus does not extend into the anterior horn, its anterior limit being the interventricular foramen (Figs. 7.7, 7.23).

The **body** of the lateral ventricle lies behind the level of the interventricular foramen. Its floor is the thalamus and body of the caudate nucleus, with the thalamo-striate groove between them. The stria semicircularis lies in the groove; it is a band of white matter running the whole length of the caudate nucleus, from the amygdaloid nucleus at the tail to the anterior perforated substance at the head. The roof of the body is the corpus callosum with, on the medial side, the crus and body of the fornix.

If the corpus callosum is removed in a dissection the

FIG. 7.11. A CAST OF THE CEREBRAL VENTRICLES VIEWED FROM THE RIGHT SIDE. Note, above the upper end of the aqueduct, the tiny slender spike of the pineal recess and the large suprapineal recess above this. The tip of the inferior horn is indented from above by the amygdaloid nucleus. Photograph of Specimen D 717.G in R.C.S. Museum.

fornix is seen lying upon the thalamus, with the choroid plexus projecting between the two. It *appears* to be a floor structure of the body of the lateral ventricle. None the less it is actually a roof structure—a roof can collapse upon a floor and this has, so to speak, happened to the fornix. Through the medial wall of the body of the ventricle, between roof (fornix) and floor (thalamus) the choroid plexus is invaginated, thrusting pia mater and ependyma before it. This is the upper part of the choroid fissure which is now seen to be limited anteriorly by the interventricular foramen (Figs. 7.9, 7.17, 7.23).

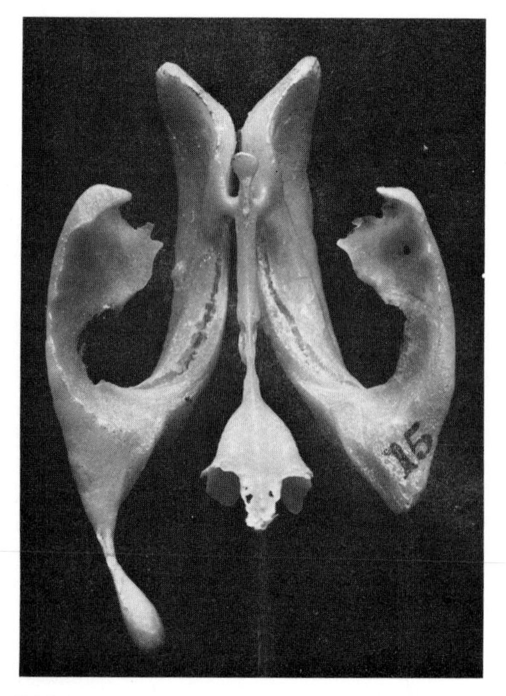

FIG. 7.12. A CAST OF THE CEREBRAL VENTRICLES VIEWED FROM BELOW. The tip of the inferior horn is indented from below by the pes hippocampi. Same specimen as Fig. 7.11.

From the body the cavity of the lateral ventricle arches downwards and then forwards into the temporal lobe.

The **posterior horn** is a projection from the backward convexity of the ventricle backwards towards the occipital pole. It is most variably developed, is often completely absent, and is seldom (12 per cent) well developed on both sides. When present it shows convexities on its walls. The floor is convex; it is the grey matter at the depths of the collateral sulcus; the convexity is called the *collateral trigone* and it extends forwards into the posterior part of the inferior horn. Above this on the medial wall is a bold projection called the *calcar avis*, produced by the grey matter at the depths of the calcarine sulcus (Fig. 7.9). The calcar avis is variable in its development and this accounts for the variation in the posterior horn. If ill-developed, the posterior horn is large. If well-developed its projection gives a bulbous tip to the posterior horn or even leaves a tear-drop cavity separated from the cavity of the posterior horn. In many cases the calcar avis is so well developed as completely to obliterate the posterior horn; in these cases it indents the posterior convexity of the lateral ventricle itself (Fig. 7.10). The development of the calcar avis and therefore of the posterior horn is seldom symmetrical in the two hemispheres.

Above the calcar avis is a convexity of white matter produced by the forceps major of the splenium—this is called the bulb of the posterior horn. The lateral wall of the posterior horn is formed by tapetal fibres and the optic radiation. When the posterior horn is absent the optic radiation comes into contact with the calcar avis, i.e., with its own cortex (Fig. 7.9). Put in other words, this is to say that the size of the posterior horn is determined by the manner in which the fibres of the optic radiation approach the visual cortex. A gradual approach leaves a big posterior horn, an abrupt approach obliterates the cavity.

The phylogenetic significance of the posterior horn caused much bitter controversy in the post-Darwinian era. Today we are still ignorant of its phylogenetic significance, if any. The choroid plexus never projects into its cavity.

The posterior convexity of the lateral ventricle, from which the posterior horn projects when present, is known as the trigone of the lateral ventricle. From this region the C-shaped cavity curves forwards into the temporal lobe, where it is known as the inferior horn.

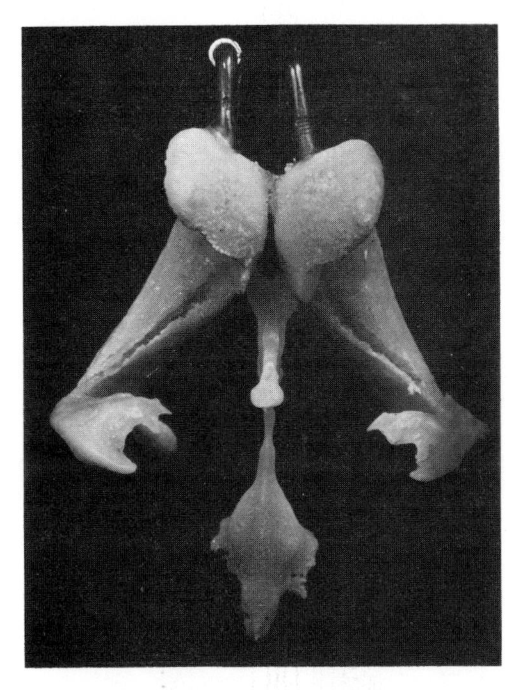

FIG. 7.13. A CAST OF THE CEREBRAL VENTRICLES VIEWED FROM IN FRONT. Photograph of the specimen illustrated in Fig. 7.11.

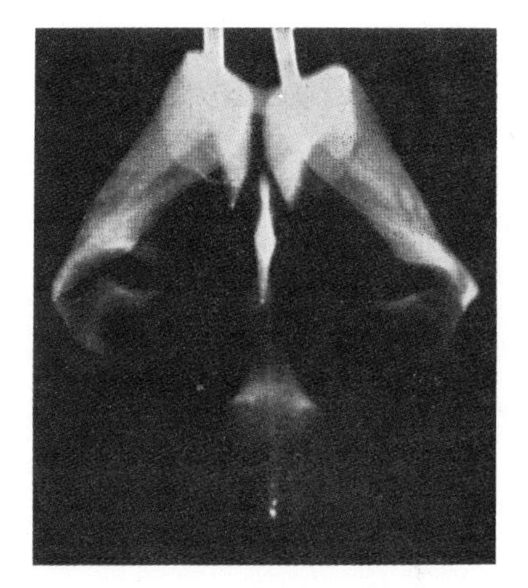

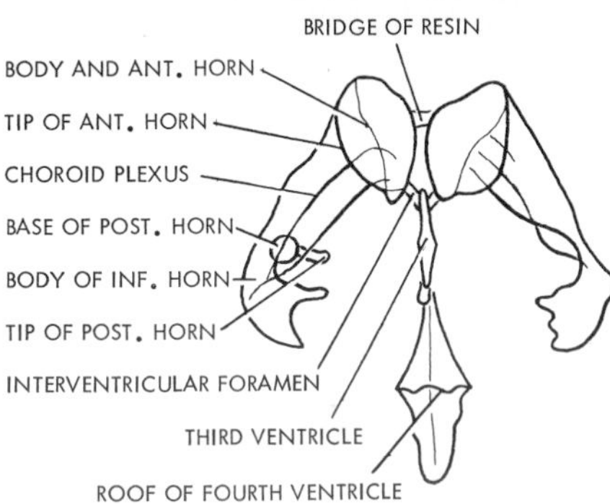

FIG. 7.14. AN ANTERO-POSTERIOR RADIOGRAM OF THE CAST ILLUSTRATED IN Fig. 7.13. The shadows can be interpreted by reference to the tracing.

The **inferior horn** shows on its floor the forward projection of the collateral trigone. Medial to this the floor shows a convexity of grey matter, called the **hippocampus,** which is seen to expand anteriorly into a grooved eminence known as the *pes hippocampi*. The hippocampus can be traced backwards towards the concavity of the trigone of the lateral ventricle as a tapering cylinder (an elongated cone) of grey matter (Fig. 7.9). It is the inward surface of the grey matter in the depths of the hippocampal sulcus on the medial surface of the hemisphere. The axons of the cell bodies in the hippocampus and subiculum form, from the pes backwards, a band of white matter which increases in thickness as the grey matter of the hippocampus diminishes. This white matter commences on the pes as a film called the *alveus*. The fibres pass backwards along the medial surface of the hippocampus, forming an ever-widening ribbon of white matter called the *fimbria*. The fimbria, leaving the hippocampus, ascends over the back of the thalamus as the **crus** (pillar) **of the fornix.** The two crura partially decussate in the midline to form the body of the fornix (p. 505). The fimbria forms the lower lip of the choroid fissure in the inferior horn just as its continuation the fornix forms the upper lip of the fissure in the body of the lateral ventricle; they lie in continuity around the convexity of the C. Similarly, within the concavity of the C the **caudate nucleus** lies in continuity. Its bulbous head in the anterior horn and its thinner body in the floor of the body of the ventricle are continued into the *roof* of the inferior horn as an ever-diminishing *tail of the caudate nucleus*. At the extremity of the tail of the caudate nucleus is an expansion of grey matter, the *amygdaloid nucleus*, which lies to the lateral side of the anterior perforated substance. It produces a shallow convexity on the roof at the tip of the inferior horn, just above the pes hippocampi (Fig. 7.11). These two projections often lie together, separated only by the

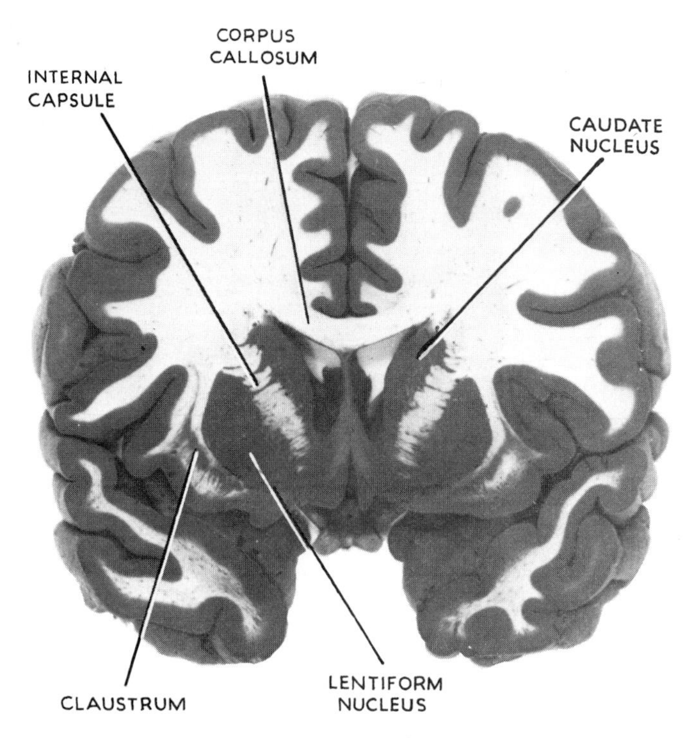

FIG. 7.15. CORONAL SECTION THROUGH THE OPTIC CHIASMA, VIEWED FROM IN FRONT. The section is anterior to the interventricular foramen (identify the position on Fig. 7.10). Only the anterior horns of the lateral ventricles are cut. They are separated by the septum pellucidum, which here joins the body and rostrum of the corpus callosum behind the genu (see Fig. 7.3). There is no fornix or thalamus in this section. Of the lentiform nucleus only the putamen is cut; it is joined to the head of the caudate nucleus by bands of grey matter that give the internal capsule its typical striated appearance here. The caudate nucleus, putamen and claustrum are conjoined with the cortex of the anterior perforated substance lateral to the optic chiasma. (Photograph.)

choroid plexus, and in a ventriculogram the injected air often fails to separate them. In such cases the X-ray of the tip of the inferior horn shows a fenestrated shadow. The inferior horn is closed laterally by the white matter of the tapetum (Fig. 7.17).

The choroid plexus of the lateral ventricle is considered on p. 520.

The C-shape of the lateral ventricle has certain implications. Structures around the convexity appear in the roof of the body and the floor of the inferior horn,

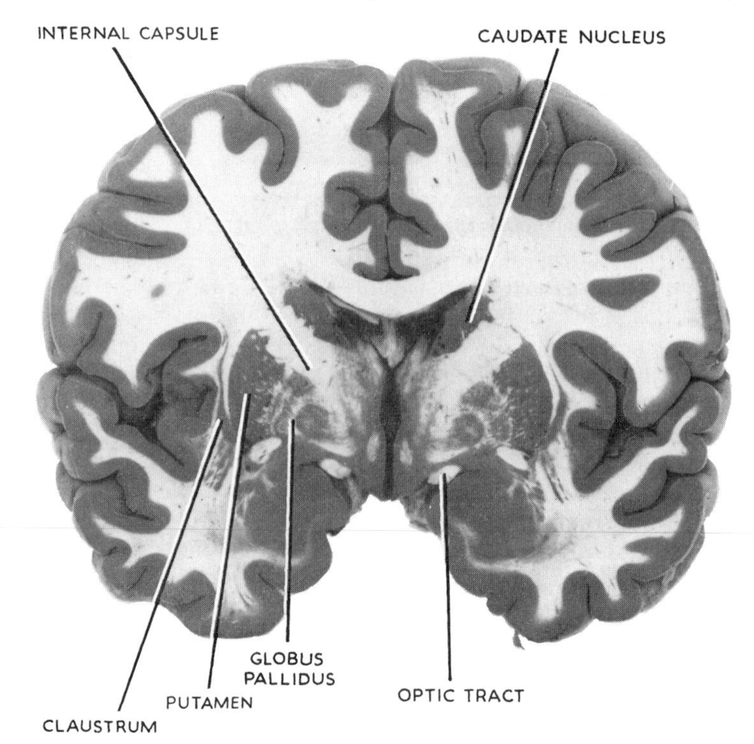

FIG. 7.16. CORONAL SECTION THROUGH THE CORPORA MAMILLARIA, VIEWED FROM IN FRONT. The section is just behind the interventricular foramen (identify the position in Fig. 7.10). The lateral ventricle (body) and caudate nucleus are smaller than in Fig. 7.15. The fornix appears in the midline above the narrow anterior pole of the thalamus. It is joined to the corpus callosum by a shallow septum pellucidum. Topographically they form the medial wall of the lateral ventricle, but developmentally the fornix is here a roof structure that has collapsed, leaving only the choroid fissure between it and the thalamus. The lentiform nucleus is here much bigger and the globus pallidus is seen. The claustrum lies just deep to the insula; it is still joined below to the putamen. Lateral to the optic tract is the amygdaloid, but it cannot be distinguished from its fusion with the lentiform nucleus. Below the optic tract is the slit-like tip of the inferior horn of the lateral ventricle, with the bulge of the uncus below it. (Photograph.)

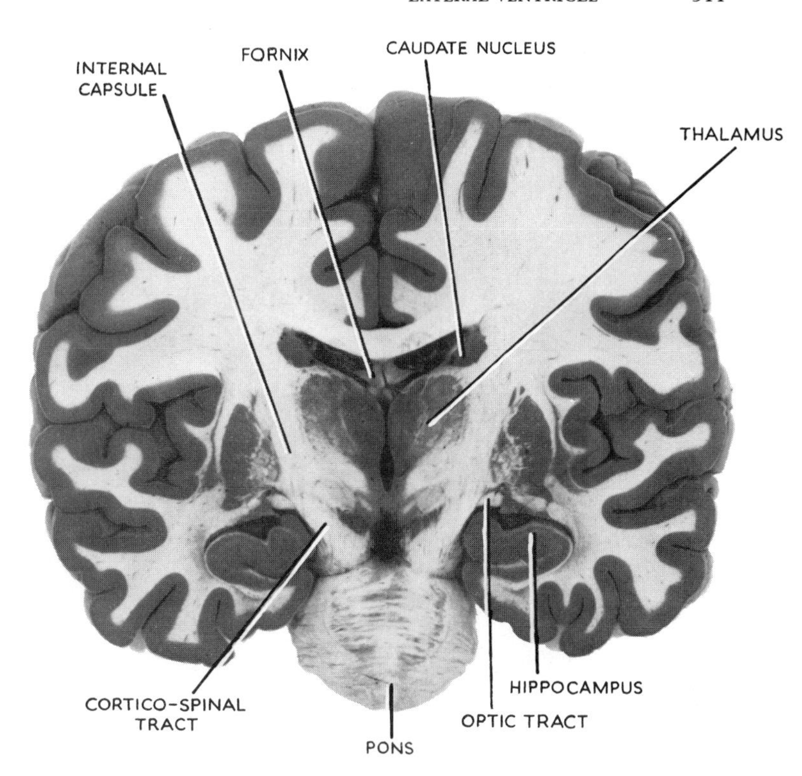

FIG. 7.17. CORONAL SECTION THROUGH THE POSTERIOR PERFORATED SUBSTANCE, VIEWED FROM IN FRONT. The caudate nucleus is smaller than in Fig. 7.16. The fornix touches the corpus callosum; the septum pellucidum has vanished. The lentiform nucleus is smaller here, and there are no bands of grey matter in the internal capsule. The choroid fissure continues between fornix and thalamus. The thalamus is bigger here, and its lateral surface is bevelled inferiorly by the internal capsule. The slit-like cavity of the third ventricle lies between the thalami. Lateral to the optic tract is the amygdaloid in the roof of the inferior horn of the lateral ventricle. The hippocampus, opposite in the floor, is made by the infolding of the hippocampal sulcus. The alveus shows as a delicate film of white matter on the hippocampus. (Photograph.)

structures in the concavity appear in the floor of the body and the roof of the inferior horn. Sections, be they vertical or horizontal, through the C-shaped formations must cut them in two places.

Contained within the concavity of the C-shaped lateral ventricle is the C-shaped caudate nucleus, and contained within the concavity of the caudate nucleus is the white matter of the internal capsule.

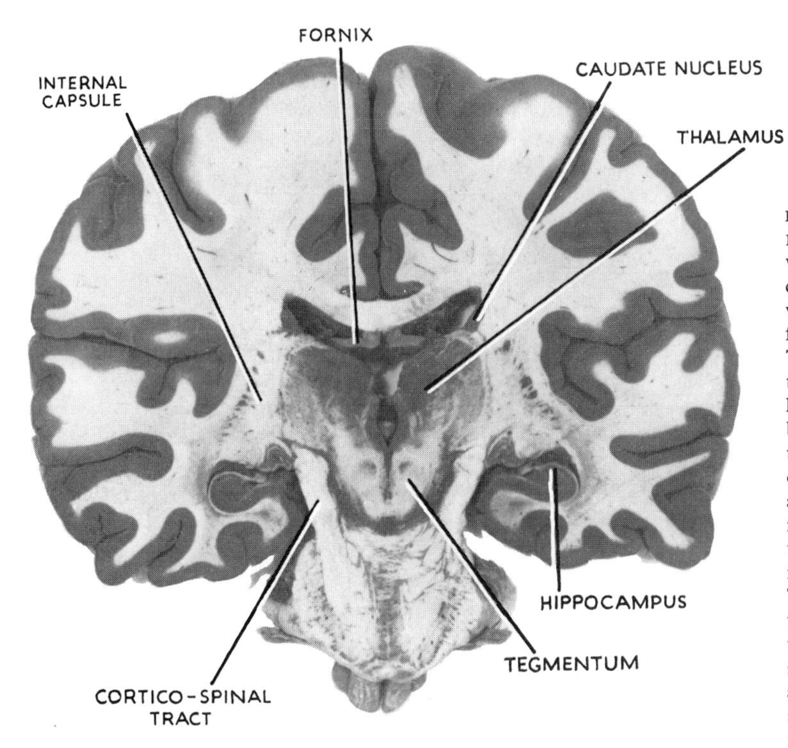

FIG. 7.18. CORONAL SECTION THROUGH THE POSTERIOR END OF THE THIRD VENTRICLE, VIEWED FROM IN FRONT. The body of the caudate nucleus is small. The fornix is wider (the crura just apart) and separated from the thalamus by the choroid fissure. The thalamus is continuous medially into the grey matter of the hypothalamus, and lateral to this the tegmentum of the mid-brain ascends to the lateral part of the thalamus ('neothalamus'). The inferior horn of the lateral ventricle shows well. The alveus on the hippocampus is thickening medially to form the fimbria. Above this the tail of the caudate nucleus hugs the internal capsule (i.e. cortico-spinal tract). The fimbria and tail of the caudate are the lips of the choroid fissure leading into the inferior horn. Note that a coronal section through the third ventricle is almost longitudinal through the brain stem. (Photograph.)

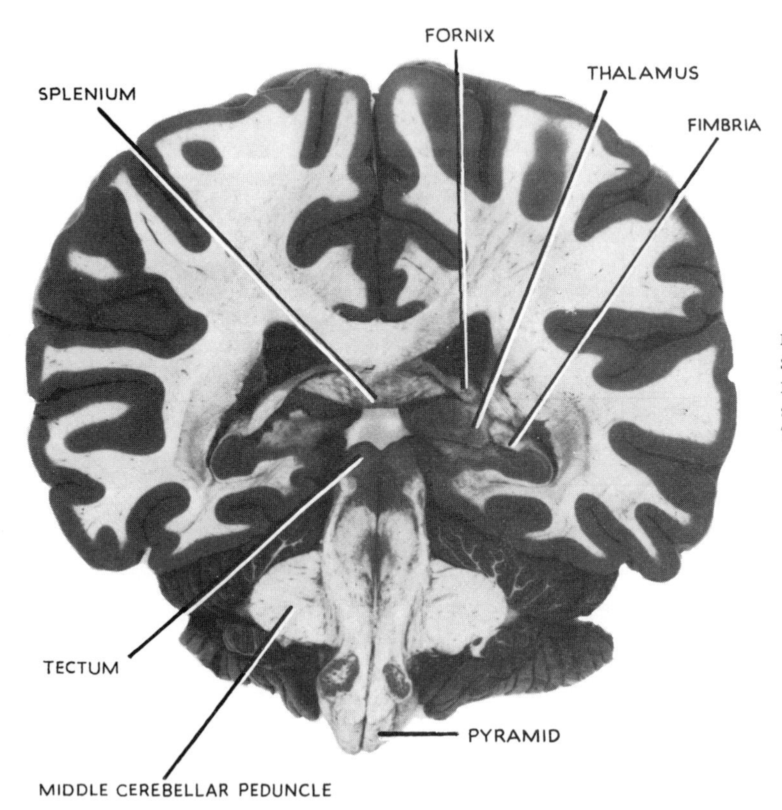

FIG. 7.19. CORONAL SECTION THROUGH THE SPLENIUM AND THE BRAIN STEM, VIEWED FROM IN FRONT. This and the previous sections should be compared with Figs. 7.3 and 7.6.

Several typical sections are illustrated in Figs. 7.15–7.19). These are coronal sections, each 1 cm thick, through the ventricles.

The Diencephalon and Third Ventricle

That part of the brain cranial to the midbrain is called the forebrain. Developed as a single tube (the fore-end of the neural tube) its cranial end is formed by a thin plate of grey matter named the lamina terminalis. Just to the caudal side of this lamina the side walls of the forebrain blow out two enormous balloons, or vesicles, which become the cerebral hemispheres, already described. The remainder of the forebrain, relatively unexpanded, becomes the **diencephalon,** still closed anteriorly by the lamina terminalis. The cavity within its substance is the third ventricle, into which the lateral ventricles of the cerebral hemispheres open through the interventricular foramina. The diencephalon, enclosing this cavity, has two side walls, a floor and a roof. The floor and roof converge towards each other posteriorly, where they join the midbrain; and the cavity of the third ventricle is continued through the midbrain as a narrow canal, the aqueduct (of Sylvius). The aqueduct leads through the midbrain into the cavity of the fourth ventricle (Fig. 7.20).

The anterior wall and floor of the diencephalon can be seen in the undissected brain. The anterior wall is the **lamina terminalis,** a thin sheet which extends between the two hemispheres from the rostrum of the corpus callosum to the top of the optic chiasma. It contains, in its upper part, the anterior commissure which joins the two piriform areas. The floor, seen from below, extends from the optic chiasma, tuber cinereum and infundibulum, and mamillary bodies to the posterior perforated substance, where the floor joins the tegmentum of the cerebral peduncles (Fig. 7.4). The anterior wall and floor can be seen cut through in a median sagittal section through the third ventricle; this section exposes the side wall of the diencephalon (Fig. 7.20).

In such a section a thin lamina is seen connecting the rostrum, genu and front of the body of the corpus callosum on the one hand to the anterior column of the fornix on the other; it is called the **septum pellucidum.** This is a part of the medial surface of the hemisphere cut off by the backward growth of the corpus callosum; its function is unknown. The two septa pellucida may be adherent; when they lie apart the closed cavity between them is called the cavum septi pellucidi; it has no connexion with the ventricular system and the term 'fifth ventricle' sometimes applied to it, should therefore be avoided. The cavum is lined with pia mater, not ependyma (Fig. 7.20).

The **third ventricle** (Fig. 7.20) is a slit-like space.

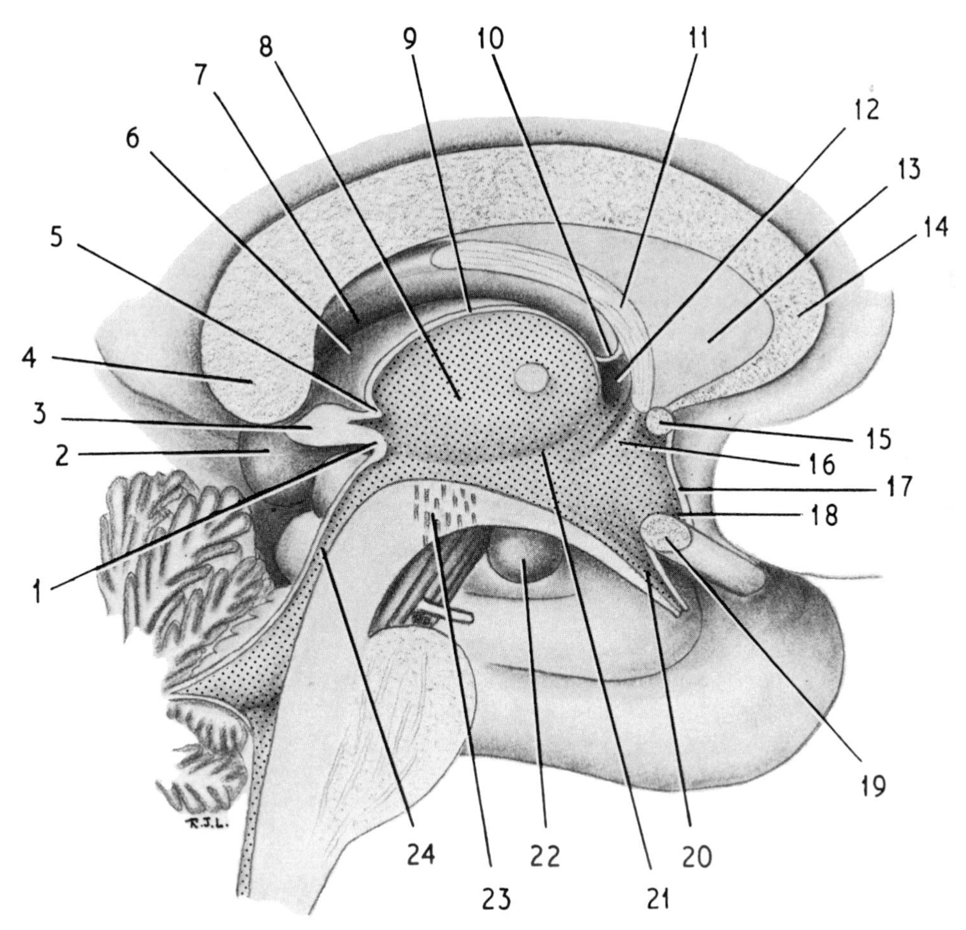

FIG. 7.20. THE THIRD VENTRICLE AND NEARBY FORMATIONS IN SAGITTAL SECTION. THE EPENDYMA IS STIPPLED. 1. Posterior commissure. 2. Pulvinar. 3. Pineal body. 4. Splenium. 5. Habenular commissure. 6. Upper surface of thalamus, on diencephalon, covered in pia mater (the lower layer of the tela chorioidea). 7. Lower surface of fornix, covered in pia mater (the upper layer of the tela chorioidea). The slit-like space between fornix and thalamus extends laterally into the choroid fissure of the cerebral hemisphere. Compare Fig. 7.23, which shows a superior view of the same place. 8. Thalamus. 9. Habenular stria. The pia mater is cut flush with the stria, thus removing the choroid plexus and the suprapineal recess. 10. Reflexion of pia mater from thalamus to fornix, at apex of tela chorioidea, closing the interventricular foramen. 11. Fornix ('body', or 'commissure', cut through). 12. Interventricular foramen. 13. Septum pellucidum. 14. Genu of corpus callosum. 15. Anterior commissure. 16. Anterior column of fornix. 17. Lamina terminalis. 18. Supra-optic recess. 19. Optic chiasma. 20. Infundibulum. 21. Hypothalamic sulcus. 22. Mamillary body. 23. Posterior perforated substance. 24. Aqueduct.

FIG. 7.21. THE OUTLINE OF THE CAVITY OF THE THIRD VENTRICLE. (For comparison with Fig. 7.20). From the pineal to the anterior column of the fornix pia mater alone roofs in the third ventricle and the interventricular foramen. It is lined, like all the ventricular surfaces, with ependyma.

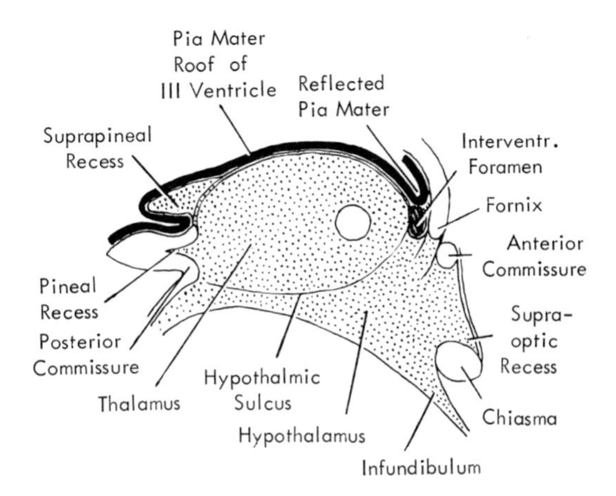

It lies in the sagittal plane (Fig. 7.17). Much of the lateral wall is occupied by the thalamus, a rounded mass of grey matter that bulges convexly into the ventricle. The two thalami often (60 per cent of cases) become gummed together; their union is called the *interthalamic adhesion*. It is not a commissure, there is no interchange of fibres between the two sides. The adhesion, when present, gives a fenestrated shadow in ventriculography of the third ventricle (Fig. 7.11). The *hypothalamic (subthalamic) groove* curves down from the interventricular foramen below the thalamus towards the aqueduct of the midbrain (Fig. 7.20). A sagittal X-ray projection shows the lateral bulge produced by the groove (Fig. 7.14, and cf. Fig. 7.17).

Below the hypothalamic groove the side wall slopes down to the floor. This region, including the floor, is known as the **hypothalamus** and it contains nuclei with marked physiological activity. In general they may be said to be cell stations for the autonomic nervous system comparable to the basal ganglia on the somatic side. Their axons pass by the spinal sympathetic tract to the connector cells in the lateral grey columns of the thoracic spinal cord. They also have direct connexions with the posterior lobe of the pituitary gland.

Behind the optic chiasma the *infundibulum* is the hollow within the tuber cinereum; from this projection the floor of the ventricle curves up smoothly, back to the aqueduct (Figs. 7.11, 7.20). The dependent corpora mamillaria and the posterior perforated substance make no mark inside the ventricle. From the indentation of the optic chiasma the anterior wall (lamina terminalis) passes up. A tiny angle between the lamina terminalis and the chiasma is known as the *supra-optic recess*. Attached behind the upper end of the lamina terminalis is the *anterior commissure*, a rounded cord that joins the two piriform areas. It is a commissure of the archæpallium (p. 518). Behind this the conjoined anterior columns of the fornix lie in contact before they diverge to sink down into the lateral wall of the ventricle. Behind each anterior column is an *interventricular foramen*, ependyma-lined and roofed in by bare pia mater sweeping from the under surface of the body of the fornix to the upper surface of the anterior pole of the thalamus. This is the anterior limit of the tela chorioidea (Figs. 7.20, 7.23).

The cavity of the third ventricle is lined with ependyma, continuous through the interventricular foramen with that lining the lateral ventricle and through the aqueduct with that lining the fourth ventricle. As in all ventricles, the lining of ependyma reaches the surface pia mater to allow for invagination of a choroid plexus. In the third ventricle this place is the narrow roof.

Attached to the upper curvature of the bulging thalamus is a thin band of white matter, the *habenular stria*. Right and left striæ join posteriorly in a U-shaped loop, the *habenular commissure*. Anteriorly the striæ pass from the thalamus below the interventricular foramen alongside the anterior column of the fornix and so to the piriform area. The habenular commissure is probably olfactory, like the anterior commissure. The drawing back into the U-shaped loop exposes the narrow *roof* of the third ventricle (Fig. 7.22).

The upper surface of each thalamus alongside the habenular stria is on the surface of the diencephalon; it is covered with pia mater. This pia mater is attached to the habenular striæ and sweeps across between them to roof in the third ventricle, back to the habenular commissure. This pial roof is lined with ependyma. It is variably slack towards its posterior end and bulges back as the suprapineal recess (Figs. 7.10, 7.21). The whole length of the roof is invaginated by the pair of (small) *choroid plexuses* of the third ventricle, which hang down as slender fringes inside the cavity.

The pial roof extends forward to the anterior pole of the thalamus, sweeps across to the anterior column of the fornix to close the interventricular foramen, thence no longer lined with ependyma it passes back on the under surface of the fornix as the upper layer of the tela chorioidea, out of contact with the third ventricle (Figs. 7.20, 7.23).

Attached to the habenular commissure the mysterious **pineal body** projects back, lying above the superior colliculi between the posterior parts of the thalami, just below the splenium (Fig. 7.37, p. 529). It is a soft conical body, less than half a centimetre long. It is noteworthy for the number of corpora amylacea it contains. These calcify, and to such an extent that after the age of forty years they normally throw a shadow in an X-ray photograph of the skull. Such calcified particles (known to the ancients) were called brain sand. Brain sand occurs also in the choroid plexuses.

The stalk of the pineal is attached also to the posterior commissure, which connects the two superior colliculi above the entrance to the aqueduct. Between the habenular and posterior commissures the pineal stalk is hollowed out a little; a minute spike shows in a ventriculogram above the aqueduct and below the large suprapineal recess (the pineal recess is seen in Fig. 7.11).

The **thalamus** is seen in horizontal and coronal sections buried in the cerebral hemisphere, and by its connexions with the sensory parts of the internal capsule it appears to be part of the hemisphere. This is not so. Actually it is part of the wall of the diencephalon, the part of the forebrain which does not expand into the cerebral hemispheres. However, as the cerebral hemispheres develop, new cells in the thalamus send fibres up to the sensory cortex and in this way the 'neothalamus' becomes incorporated in the pathways leading to the cortex. The 'neothalamus' lies on the lateral side. The medial side of the thalamus has connexions with the hypothalamus, while the anterior pole projects to the gyrus cinguli (p. 505).

The mass of grey matter making up the thalamus is roughly wedge-shaped. The medial walls of the two thalami lie parallel, near each other across the third ventricle, where in two-thirds of cases they are gummed together to form the *interthalamic adhesion (interthalamic connexus*, or *massa intermedia)*. This part of the medial surface is covered with the ependyma of the third ventricle. Behind this the medial surface diverges from the midline and expands into a large posterior convexity called the *pulvinar*; the lateral geniculate body bulges down from its lateral part (Fig. 7.37, p. 529). The medial geniculate body, a 'thalamic' nucleus which relays auditory impulses, is separated from the main mass of the thalamus and lies on the midbrain (p. 524).

The superior surface of the thalamus is also convex; it is triangular in outline, tapering forward from the large pulvinar to the small blunt anterior pole (Fig. 7.23). The superior surface and the posterior surface (pulvinar) of the thalamus are on the external surface of the diencephalon itself. They are covered in pia mater. An oblique strip along the lateral margin of the superior surface lies in the lateral ventricle. Covered with ependyma, this narrow strip belongs rather to the hemisphere than to the diencephalon. The body and tail of the caudate nucleus are in contact here with the lateral margin of the thalamus (Fig. 7.6).

The lateral surface of the thalamus is bevelled by the internal capsule (Fig. 7.17), whose descending fibres lie in contact. The ascending fibres of the internal capsule arise further back from several dozen nuclei in this lateral part ('neothalamus') and make the neothalamus all one, as it were, with the hemisphere.

The inferior surface of the thalamus is narrower than the superior surface. Medially it joins the hypothalamus down to the posterior perforated substance in the floor of the third ventricle. Here pass the thalamo-striate arteries and thalamic veins. Lateral to this, and posterior too, the lemnisci of the tegmentum enter the thalamus (Fig. 7.18) and attach it to the top of the midbrain.

All four surfaces (medial, inferior, lateral and superior) of the thalamus converge to the small blunt anterior pole. Covered in ependyma this lies at the interventricular foramen.

The pia mater over the thalamus originally invested the cylindrical tube of the diencephalon, as, for example, it still invests the mesencephalon or the spinal cord in the adult. Over the dorsal surface of the diencephalon, however, the corpus callosum and fornix extend back. Each of these is clothed in pia mater. Thus the pia mater is folded back on itself from the interventricular foramina. Behind this it extends over the roof and dorsal surface of the thalamus and continues down over the pineal body and tectum of the midbrain. From the interventricular foramina it likewise extends backwards

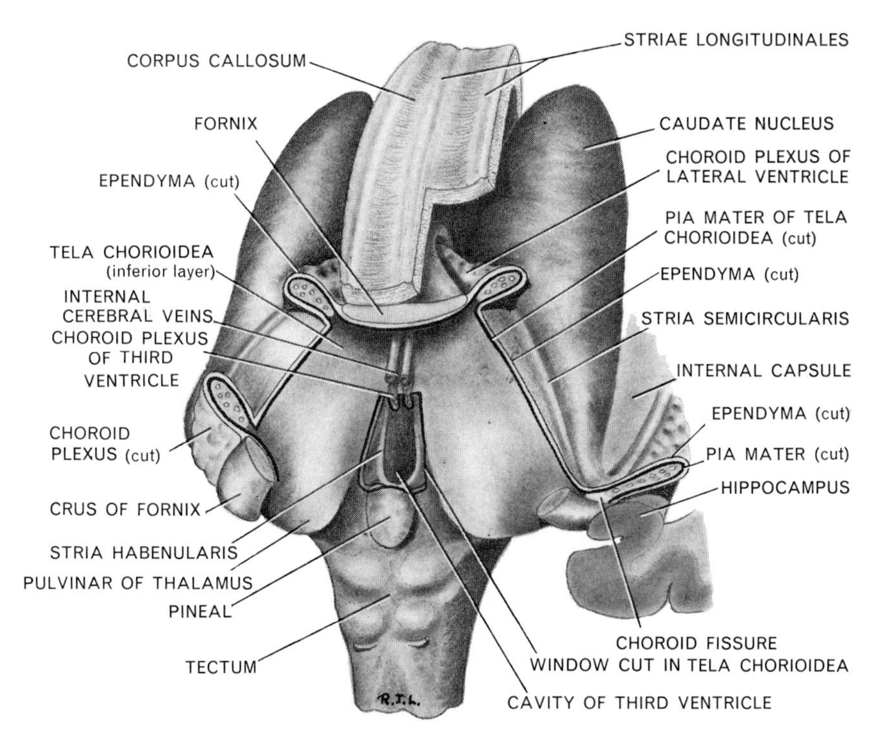

FIG. 7.22. THE LOWER LAYER OF THE TELA CHORIOIDEA EXPOSED BY REMOVAL OF THE BODY OF THE FORNIX AND THE POSTERIOR PART OF THE CORPUS CALLOSUM. Drawn from a dissection.

on the under surface of fornix and corpus callosum and is reflected around the splenium of the latter to the upper surface of the body of the corpus callosum. Upper and lower layers of pia mater, thus lying in contact, are no more than the pia mater investing the surface of any part of the central nervous system; they are only lying in contact because the fornix and corpus callosum project back over the dorsal surface of the diencephalon. These two layers of pia mater are known as the **tela chorioidea.** They are reflected, the upper on to the lower, between the interventricular foramina and are similarly reflected on to each other at the lateral extent of the tela. This follows the line of the body of the fornix, which passes obliquely across the upper surface of the thalamus (Fig. 7.23).

The layers are thus reflected on each other laterally along the lateral margins of the body of the fornix; the lower layer is attached to the upper surface of the thalamus as far laterally as this line. The tela chorioidea thus forms a pouch, triangular in outline, apex forward. The base is open, where the two layers of pia mater part, the lower passing down over the back of the midbrain, the upper passing forwards on the corpus callosum around the splenium. The lateral reflexion of pia mater

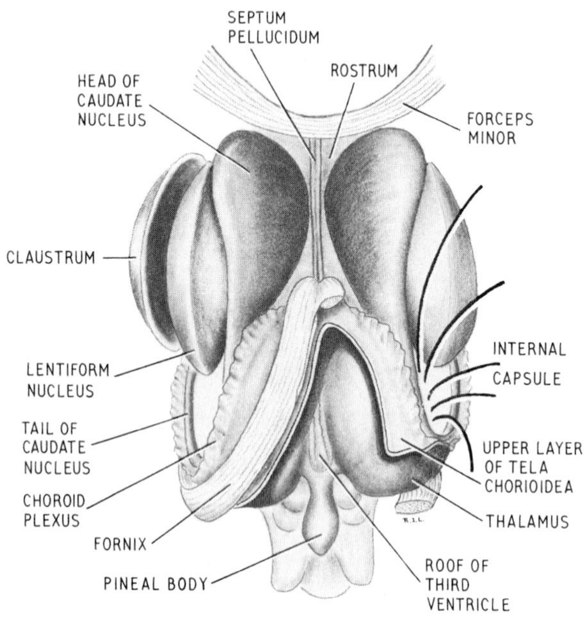

FIG. 7.23. THE BASAL GANGLIA AND THE TELA CHORIOIDEA (superior view). The upper layer of the tela, here partly cut away, clothes the under surface of the body of the fornix and the splenium. The lower layer of the tela clothes the thalami and roofs the third ventricle (for simplicity the suprapineal recess is omitted from the drawing). The two layers of the tela form a triangular pouch, open posteriorly across the base. Each side of the pouch lies at the choroid fissure, where it is 'pushed' into the body of the lateral ventricle by the choroid plexus. The apex of the pouch roofs the interventricular foramina (see Fig. 7.20).

along the edge of the tela, between the body of the fornix and the upper surface of the thalamus, comes into contact with the ependyma lining the lateral ventricle. This is the upper part of the C-shaped choroid fissure, and into it is invaginated the *choroid plexus of the lateral ventricle*, pushing the conjoined pia mater and ependyma in advance of it into the cavity of the lateral ventricle. At the posterior margin of the interventricular foramen the conjoined pia mater and ependyma of the medial wall of the lateral ventricle are traceable backwards into the conjoined pia mater and ependyma of the roof of the third ventricle. Here the conjoined pia mater and ependyma are invaginated by the *choroid plexuses of the third ventricle.* These are much smaller, but each is continuous, right and left, with the plexus of the lateral ventricle (Fig. 7.22).

The *internal cerebral veins* run back on the pia mater of the inferior lamina of the tela chorioidea (Fig. 7.30), on the roof of the third ventricle.

The **choroid fissure** is a C-shaped slit in the medial wall of the cerebral hemisphere, extending from the interventricular foramen around the thalamus and cerebral peduncle as far as the uncus of the temporal lobe (Fig. 7.3). Its convexity is contained by the body and crus (pillar) of the fornix, the fimbria and the hippocampus, its concavity is contained by the thalamus (upper and posterior surfaces) and the tail of the caudate nucleus. At the slit pia mater and ependyma come into contact with each other and both are invaginated into the lateral ventricle by the choroid plexus. The choroid fissure on the medial wall of the hemisphere was originally on the roof of the cerebral vesicle, and the arrangement is better understood by considering the development.

Development of the Forebrain

A neural groove on the ectodermal (dorsal) surface of the embryonic plate closes in to form the neural tube. The anterior end of the neural tube is destined to become the forebrain. The whole neural tube is hollow (as is any tube) and the cavity comes to the surface in two places, each on the dorsal surface of the tube. In these two places the ependyma lining the cavity of the tube lies in contact with the pia mater clothing the surface of the tube. The anterior of these two areas is destined to become the forebrain, the posterior the hindbrain (fourth ventricle).

The forebrain is limited by the lamina terminalis, which remains *in situ* as the adult lamina terminalis. Across its dorsal margin lies a commissure connecting the two sides; this is the anterior commissure and persists as such. The roof, as already stated, is devoid of nervous tissue; here pia mater and ependyma blend. The side wall contains a good deal of nervous tissue (nerve cells) and this is the primitive thalamus.

From near the front end of the hollow forebrain right and left hollow diverticula grow out laterally, like bubbles. The thalamus lies behind (caudal to) these **cerebral vesicles,** and this part of the forebrain remains relatively stable; it is called the diencephalon. The cavities of diencephalon and cerebral vesicles communicate at the site of the future interventricular foramen, in the side wall of the diencephalon (Fig. 7.24).

As the cerebral vesicle grows out it takes the thin roof of the forebrain with it. Into the L-shaped roof thus produced a choroid plexus, itself L-shaped, is invaginated. The plexus is fed by an artery at its lateral extremity (the anterior choroid artery) and it drains by a vein which runs back along the roof of the unexpanded diencephalon, to join with the vein of the opposite side.

The cerebral vesicles grow mightily, and become rotated, with their surfaces folded to accommodate themselves to the limits of the cranial cavity in which they develop. The thin roof-plate of the vesicle, with its invaginated choroid plexus, becomes compressed and distorted. The cerebral vesicles grow by multiplication of nerve cells in the grey matter on their surfaces. As new daughter cells appear their axons grow in two main ways, (*a*) to produce commissural and association fibres, and (*b*) to produce projection fibres. Both sets of fibres produce new complications. The commissural fibres form the corpus callosum and association fibres form the fornix. The former grows first forwards (the rostrum) then backwards (the genu, body and splenium) in conformity with the growth of the vesicles (cerebral hemispheres). The fornix is forced backwards by the backward growth of the corpus callosum, so that both come to overlie the roof of the unexpanded forebrain (diencephalon). The cerebral vesicles (cerebral hemispheres) grow caudally, then ventrally, and finally forward. The corpus callosum cannot do this without cutting through the brain stem; hence its fibres, sweeping away to the temporal lobe, become heaped up into a thick mass posteriorly, forming the splenium. The fornix, on the other hand, consists rather of association than commissural fibres, and it parts from its fellow and sweeps around the side of the diencephalon to pass forwards into the temporal pole (Fig. 7.24).

In addition to these commissural and association fibres, projection fibres pass down from the cortex of the developing cerebral vesicle into lower centres in the

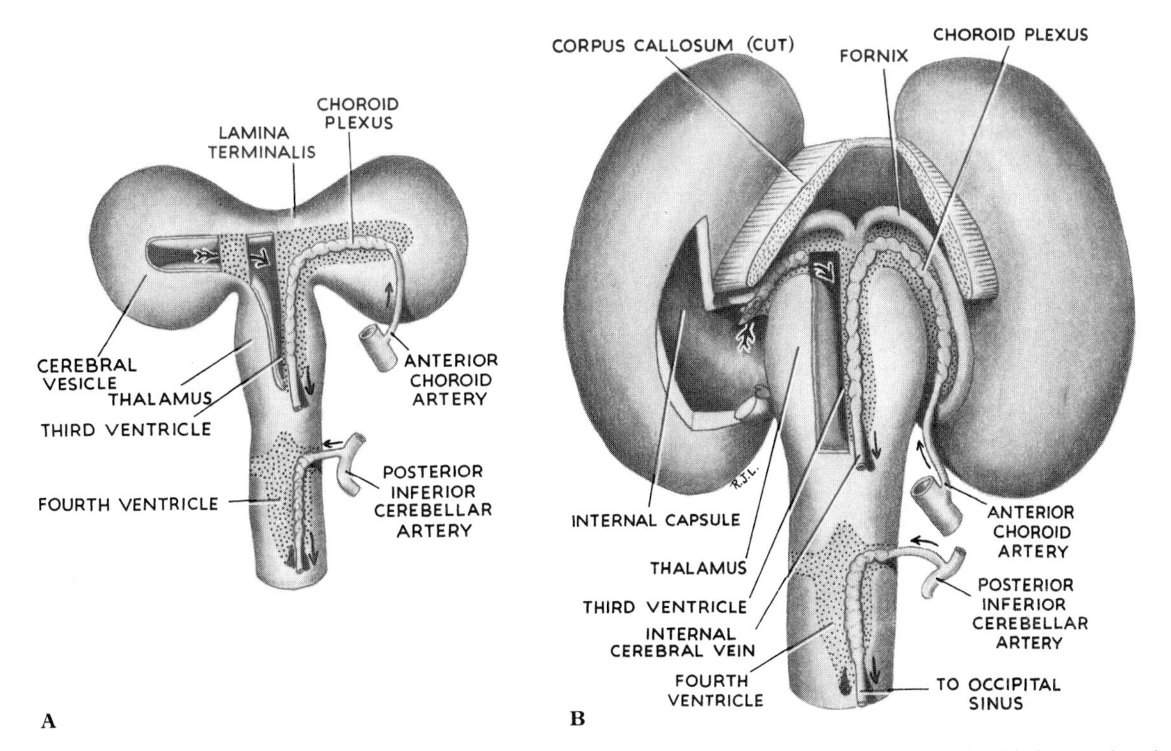

FIG. 7.24. DORSAL VIEW OF THE DEVELOPING CEREBRAL HEMISPHERES AND THE CHOROID FISSURE (Diagram). In (A) the ependymal roof-plate of the telencephalon is drawn out on the roof of the growing cerebral vesicle and is invaginated by the choroid plexus. The ependymal roof-plate of the early cerebral vesicle (A) comes to lie on the medial side of the growing hemisphere (B) and it curves around the developing basal ganglia and the fibres of the internal capsule, forming the choroid fissure. The roof-plate is partly removed on the left side, and an arrow passes from the lateral ventricle (in the hemisphere) through the interventricular foramen of Monro into the third ventricle. Small arrows indicate the direction of blood flow in the choroid plexuses of the third and fourth ventricles. The cerebellum is omitted in order to show the roof and choroid plexus of the fourth ventricle.

developing neural tube. These cortico-spinal fibres stream down alongside the thalamus. Meanwhile, ascending tracts are developing and extend up to new cells developing in the lateral wall of the thalamus. From these thalamic cells new fibres pass up to the cortex, establishing continuity between the lateral part of the thalamus and the cortex of the cerebral vesicle. The two bundles, of ascending and descending fibres, fuse together into a mass of white matter called the internal capsule. The ascending fibres of the internal capsule are connected with the thalamus; the descending fibres are closely applied to the thalamus; thus the thalamus, originally lying in the side wall of the diencephalon, becomes partly incorporated in the cerebral vesicle (hemisphere).

Meanwhile, cells have been dividing in the floor of the cerebral vesicle to produce the corpus striatum. Fibres of the internal capsule interrupt the continuity of this cell mass and almost completely separate the caudate from the lentiform nucleus. The fibres bulge the caudate nucleus (and lateral ventricle, in the floor of which it lies) convexly around their mass.

Meanwhile, the cerebral vesicle itself has been altering its position. Unequal growth on its surface forces the roof to the medial surface, and the curvature of the caudate nucleus and lateral ventricle around the fibres of the internal capsule is reflected in a similar curvature of the thin roof plate around the thalamus and internal capsule. Thus the original *dorsal* roof plate of the vesicle (choroid fissure) becomes *medial and curved*; but the anterior choroid artery still enteres the distal extremity of the fissure and the choroid plexus of the cerebral vesicle is still in continuity with the choroid plexus of the diencephalon (third ventricle) at the point of junction of the original roof plates (interventricular foramen of Monro) (Figs. 7.22, 7.24).

Morphology

Certain parts of the cortex, notably the insula and piriform area, can be regarded as remnants of the primitive brain (*archæpallium*). Anterior and posterior commissures unite the two halves. The anterior pole of the thalamus and the globus pallidus provide the basal ganglia. The archæpallium was predominantly a smell brain. The coeval cerebellum was mainly vestibular in its connexions (archæcerebellum).

The huge growth of hemisphere which buries the insula is the *neopallium*. Its commissure is the corpus callosum and it is associated with the appearance of the putamen and caudate nucleus and the lateral nuclei of the thalamus. With the neopallium appear also the neocerebellum and the red and olivary nuclei.

The smell world has been replaced by a new sight, hearing and touch world, and fine co-ordinated movements have become possible.

Blood Supply of the Cerebral Hemispheres

The cerebral hemispheres and the walls of the diencephalon are supplied from both the internal carotid and vertebral systems. The arteries are directed in essence to the grey matter, which needs more blood than the white matter. Superficial cortical arteries supply the grey matter on the surface, perforating arteries supply the grey matter of the basal ganglia. Both sets of arteries send branches to the adjacent white matter.

An artery that has entered the surface of the brain

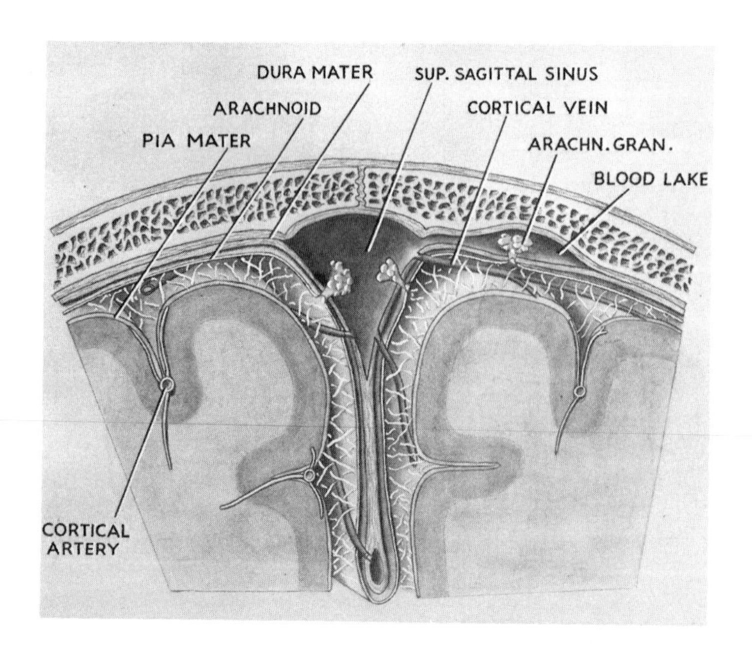

FIG. 7.25. THE MENINGES AND THE DISPOSITION OF THE CORTICAL ARTERIES AND VEINS AT THE VERTEX (Diagrammatic).

from either of these sets is always an end artery (i.e., it has no precapillary anastomosis with its fellows), and thus cerebral softening follows its obstruction. Entering arteries invaginate a tubular prolongation of pia mater around them, forming a *perivascular space* that extends to the finest branches of the vessel and invests the actual nerve cells.

The internal carotid and vertebral systems anastomose with each other around the optic chiasma and infundibulum of the pituitary stalk, forming the **circle of Willis** (the French call it, more accurately, the polygon of Willis). The communicating vessels are small and in many cases are inadequate to maintain an effective circulation if one internal carotid artery is suddenly blocked; in such cases hemiplegia of the other side of the body results. Yet the circle of Willis is so constantly present that it suggests there is some functional reason for its existence. A possible reason is that it equalizes the pressure (and the volume of blood flow) between the two sides of the brain. The circle of Willis is formed in the following way. The basilar artery from the vertebral system divides at the upper border of the pons into right and left *posterior cerebral arteries.* From each posterior cerebral a small *posterior communicating artery* runs forward through the interpeduncular cistern to join the internal carotid artery at the anterior perforated substance. Each internal carotid artery gives off an *anterior cerebral artery*; the circle of Willis is completed by the *anterior communicating artery*, a small vessel that unites the anterior cerebrals in the cisterna chiasmatica, below the rostrum of the corpus callosum. The only structures encircled by the circle of Willis are the optic chiasma and the pituitary stalk (Fig. 7.36, p. 528).

The **arterial supply of the cerebral cortex** is by three cerebral arteries, anterior, middle and posterior. The former two are branches of the internal carotid, the last is the terminal branch of the basilar artery (from the vertebrals). The branches of these three arteries anastomose across the frontiers of their respective territories, on the surface of the pia mater, but very sparsely and only by arterioles. Their *perforating* branches are invariably end arteries.

The **internal carotid artery** emerges from the roof of the cavernous sinus, gives off the ophthalmic artery, then curls back to lie on the front half of the roof. It then turns vertically upwards to the anterior perforated substance where it divides into two branches for the supply of the cortex. It here gives off also the striate arteries, the anterior choroid artery, and the posterior communicating artery (Fig. 7.36, p. 528).

The **middle cerebral artery** is the largest and most direct branch of the internal carotid and therefore most subject to embolism. It passes deep into the lateral fissure to supply the cortex of the insula and overlying

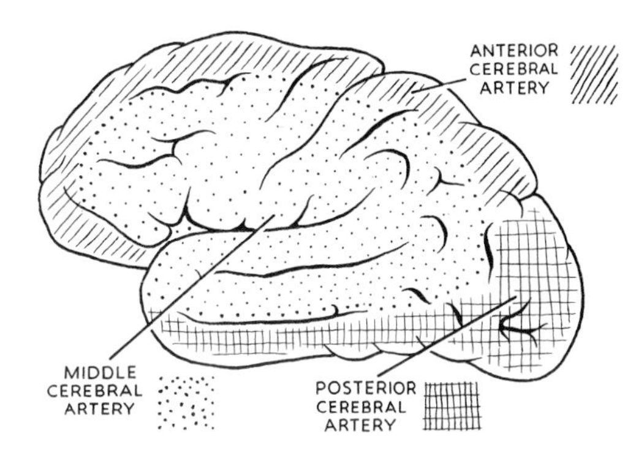

FIG. 7.26. THE AREAS OF ARTERIAL DISTRIBUTION TO THE LATERAL SURFACE OF THE CEREBRAL HEMISPHERE (cf. Fig. 7.1, p. 495).

opercula. It reaches the lateral surface of the hemisphere by passing in the lateral fissure, from which its branches emerge for the most part deep in the sulci and ramify over an area that falls short of the borders of the lateral surface by one gyrus or its equivalent breadth (Fig. 7.26). It does not reach the superior frontal gyrus or the inferior temporal gyrus. In its area of cortical distribution lie the motor and sensory areas for the opposite half of the body excluding leg and perineum (on the left side this includes the speech area of Broca) and the auditory centre. For striate arteries see below.

The **anterior cerebral artery** leaves the internal carotid artery at the anterior perforated substance and passes forwards above the optic nerve (the two arteries are here connected by the anterior communicating artery). It is distributed to the orbital surface of the frontal lobe and to the whole of the medial surface of the hemisphere above the corpus callosum as far back as the parieto-occipital sulcus. Its distribution extends over the superior border to meet the area supplied by the middle cerebral artery; it thus supplies the superior frontal gyrus and an equal zone behind this over the

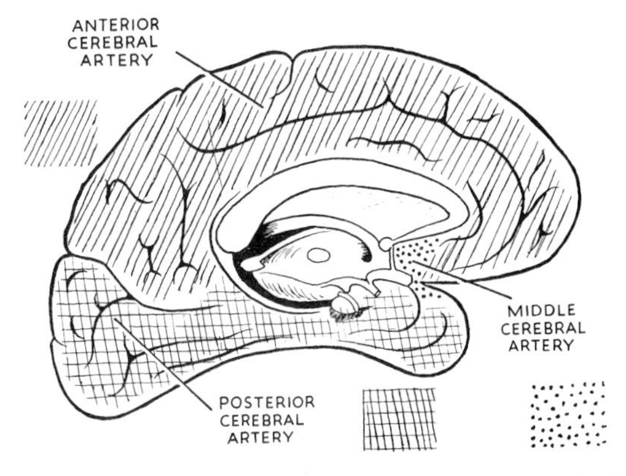

FIG. 7.27. THE AREAS OF ARTERIAL DISTRIBUTION TO THE MEDIAL AND INFERIOR SURFACES OF THE CEREBRAL HEMISPHERE (cf. Fig. 7.3, p. 497).

parietal lobe. The motor and sensory areas for the opposite leg and perineum lie in its territory (Figs. 7.26, 7.27).

The **posterior cerebral artery** curls back around the cerebral peduncle (supplying it and the optic tract), and passes back above the tentorium to supply the infero-medial surface of the temporal and occipital lobes. Its territory meets that of the anterior cerebral artery at the parieto-occipital sulcus. Its branches extend around the borders of the brain to supply the inferior temporal gyrus and a corresponding strip of cortex on the lateral surface of the occipital lobe. The visual area for the opposite field of vision lies wholly within its territory (PLATE 34 and Figs. 7.26, 7.36).

For thalamo-striate branches *v. inf.*

The *basal ganglia* are supplied by arteries that enter the perforated substances; branches from the internal carotid system for the corpus striatum and from the vertebral system (posterior cerebral artery) for the thalamus.

The *anterior perforated substance* receives several branches from the middle cerebral artery (or from the internal carotid). These are the *striate arteries.* They pass straight up into the corpus striatum where they supply the amygdaloid nuclei, the globus pallidus, putamen and head of the caudate nucleus (a few deep branches of the anterior cerebral reach the head of the caudate nucleus and some from the middle cerebral pierce the cortex of the insula to reach the putamen). The striate arteries also supply the anterior limb of the internal capsule.

The *posterior perforated substance* receives branches from both posterior cerebral arteries. These are the *thalamo-striate arteries.* They pass straight up into the thalamus and body of the caudate nucleus; they also give branches to the posterior part of the internal capsule; hæmorrhage, thrombosis or embolism here causes hemiplegia.

The **choroid plexus** of the lateral ventricle is supplied by the *anterior choroid artery*, a branch of the internal carotid (or middle cerebral). It enters just above the uncus at the inferior extremity of the choroid fissure (the tip of the inferior horn of the ventricle) and it is here that the choroid plexus begins. A 'booster' supply comes in through the choroid fissure behind the thalamus by a few *posterior choroid arteries* which branch from the posterior cerebral. The vein of the choroid plexus flows in the same direction as the artery and emerges through the superior extremity of the choroid fissure (i.e., the interventricular foramen).

Venous Drainage of the Hemispheres. The venous return *does not follow the arterial pattern.* Unlike the cortical arteries, which tend to travel deep in the sulci, the cortical veins tend to travel superficially, in the arachnoid mater (Fig. 7.28). They lie adherent to

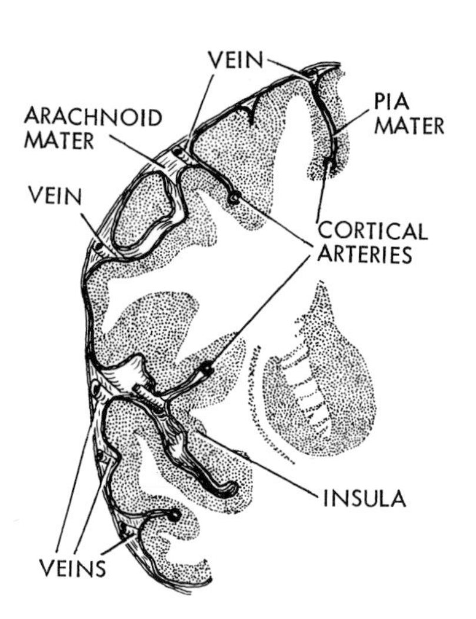

FIG. 7.28. CORTICAL ARTERIES LIE IN THE PIA MATER DEEP IN THE SULCI. CORTICAL VEINS LIE ADHERENT UNDER THE ARACHNOID MATER WHERE THIS BRIDGES THE SULCI. On the insula lies the deep middle cerebral vein (no arachnoid here). Over the insula the lateral sulcus is bridged by arachnoid mater; this carries the superficial middle cerebral vein. Coronal section, drawn to scale from a specimen.

the deep surface of the arachnoid mater that bridges each sulcus. In general blood flows into the nearest available venous sinus of the dura mater, generally entering obliquely against the blood stream. Only when there is no sinus near enough (anterior part of the hemisphere and lower parts of basal ganglia) is a venous pattern formed that resembles the arterial pattern.

The supero-lateral surface of the hemisphere drains above into the superior sagittal sinus and below into the transverse sinus, in each case by veins that enter against the direction of blood flow. The superior veins, if encountering a blood lake, pass on its cerebral surface beneath the arachnoid (the blood lakes are between the 'two layers' of the dura, Fig. 7.25).

Adherent to the deep surface of the arachnoid mater that bridges the lateral sulcus runs the *superficial middle cerebral vein*, draining the adjacent cortex and emptying into the cavernous sinus. At the posterior end of this vein are *superior and inferior anastomotic veins* which join the superior sagittal and transverse sinuses. The depths of the lateral sulcus and the surface of the insula are too far from a sinus in the dura mater. Blood from this region drains into the *deep middle cerebral vein* which runs along with the artery to the anterior perforated substance, where it joins the basal vein.

The medial and inferior surfaces of the hemisphere drain into the geometrically nearest venous sinus of the dura mater (the two sagittals and the straight sinus) except anteriorly, where there is no sinus present in the margins of the falx. Here the blood from the surface of the hemisphere is collected into the *anterior cerebral*

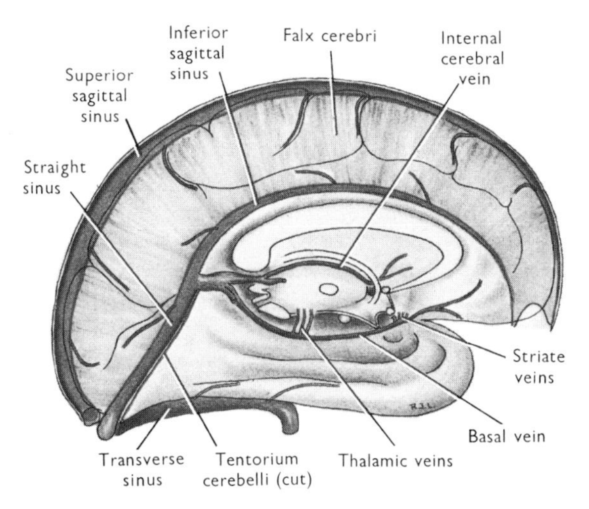

FIG. 7.29. THE VENOUS DRAINAGE OF THE CEREBRAL HEMISPHERE (MEDIAL ASPECT) AND OF THE BASAL GANGLIA. Note that the lower halves of corpus striatum and thalamus drain via the perforated substances into the basal vein. The upper halves drain into the internal cerebral vein. These veins then reach the straight sinus.

vein, which returns around the genu of the corpus callosum alongside the anterior cerebral artery. The anterior cerebral vein drains also the orbital surface of the frontal lobe.

At the anterior perforated substance striate veins emerge through the perforations. They drain the lower part of the corpus striatum and join the deep middle cerebral vein and the anterior cerebral vein; the veins from these three sources form the *basal vein*. This passes around the cerebral peduncle below the optic tract, with the fourth nerve and the posterior cerebral artery. It receives veins from the posterior perforated substance; these drain the lower part of the thalamus. Just below the splenium the two basal veins join the great cerebral vein to enter the straight sinus. Only the lower parts of the basal ganglia drain through the perforated substances into the basal vein. Their upper parts drain into the internal cerebral vein (Fig. 7.29).

The *internal cerebral vein* receives blood from three sources. It is formed at the interventricular foramen by the meeting of (*a*) the choroid vein, draining the choroid plexus of the lateral ventricle, (*b*) the thalamo-striate vein which lies in the thalamo-striate groove and receives blood from the upper parts of thalamus and body of caudate nucleus, and (*c*) the veins of the septum pellucidum which bring blood from the corpus callosum and adjacent cortex and the head of the caudate nucleus (Fig. 7.30).

The internal cerebral vein so formed runs back in the pia mater of the roof of the third ventricle (the tela chorioidea). It recevies the veins from the tiny choroid plexus of the third ventricle and then joins its fellow to make the *great cerebral vein* (of Galen) just beneath the splenium. This vein is joined by the two basal veins, and with the inferior sagittal sinus it enters the straight sinus.

Microscopic Anatomy of the Brain

The **cortex** is composed of layers of cells which vary in their characteristics in different regions. In general, motor cortex has many large pyramidal cells, sensory cortex has small round cells called granular cells. A fibrous tissue called neuroglia (ectodermal in origin) binds the cells together. The fibres of the neuroglia are produced by neuroglia cells, whose nuclei stain readily with ordinary stains.

In most parts of the cortex six layers of nerve cells can be made out. On the surface is a layer of nerve fibres with very few cells; this is the molecular layer.

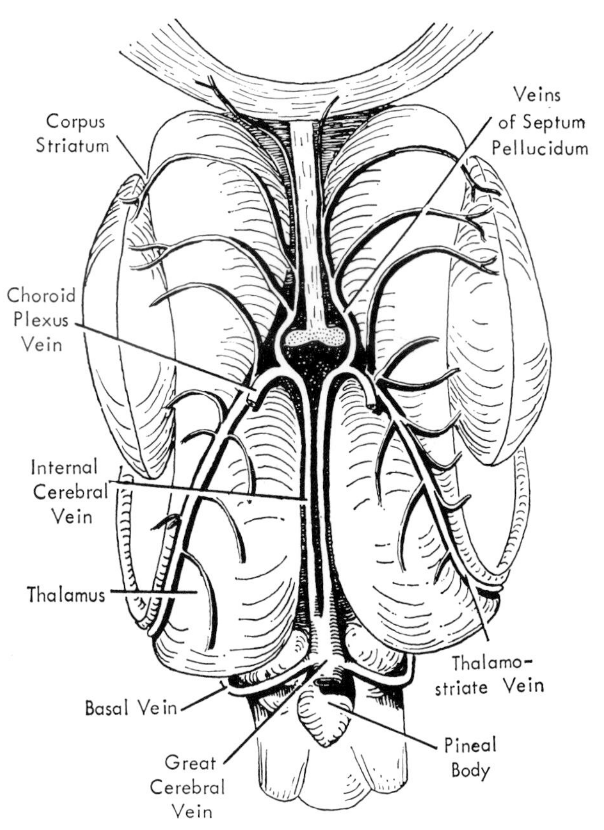

FIG. 7.30. THE VENOUS DRAINAGE OF THE BASAL GANGLIA. The upper part of the corpus striatum drains by the veins of the septum pellucidum. The tail of the caudate nucleus and the upper part of the thalamus drain by the thalamo-striate vein. These converge to the interventricular foramen, where they are joined by the choroid vein from the plexus of the lateral ventricle. Thus is formed the internal cerebral vein. The pair of internal cerebral veins along the roof of the third ventricle unite to form the great cerebral vein (of Galen). This receives each basal vein, which brings blood via the anterior and posterior perforated substances from the lower parts of the corpus striatum and thalamus.

The next four cell layers are an outer granular, outer pyramidal, inner granular and inner pyramidal in that order. The deepest layer of the cortex consists of fusiform cells lying at right angles to the surface.

Changes in the relative distribution of these layers are most pronounced in the known sensory and motor areas. The postcentral gyrus (touch), the superior temporal gyrus (hearing), and the calcarine sulcus (sight) are covered by cortex in which the two pyramidal layers are replaced by granular cells.

On the other hand, the cortex of the precentral gyrus (area 4) and the posterior part of the frontal lobe contains pyramidal cells in place of the granular layers. In area 4 are found certain very large pyramidal cells known as Betz cells; they are very similar to the cells found in the anterior grey columns of the spinal cord.

The **white matter** is composed of medullated nerve fibres bound together by the fibres of the neuroglia. This is well seen in a section of the optic nerve, where the rather solid looking mass of neuroglia enclosing the medullated nerve fibres is incompletely broken into segments by ingrowing septa from the pia mater. The nuclei of the neuroglia cells are plentiful, but there are no Schwann cells—the cut fibres will not regenerate.

The cross-section of a spinal or cranial nerve is very different (p. 20).

THE BRAIN STEM

The brain stem or bulb is a mass of nervous tissue connecting the cerebral hemispheres with the spinal cord. It extends in life from just above the aperture in the tentorium cerebelli to the foramen magnum, and the cerebellum projects from its dorsal surface. The brain stem lies very nearly vertical in the body. It consists of midbrain, pons and medulla oblongata, and each of these three structures looks very different when viewed from the dorsal and ventral surfaces.

The brain stem consists of fibres and cells. Most of the fibres in the brain stem ascend or descend longitudinally, as in the spinal cord, and most of the cells are aggregated into nuclei. These **nuclei** consist of three groups:

1. The *nuclei of the cranial nerves* III to XII.
2. Other *named nuclei* which are demonstrable, such as the colliculi, the red nucleus, the pontine nuclei and the olivary nucleus.
3. A number of *physiological centres* (in the medulla) in the region of the nuclei of the vagus. They constitute the cardiac, respiratory, vasomotor, etc., centres, and are often known collectively as the 'vital centres'. They are not demonstrable as visible aggregates of cells. Very great bodily disturbances result from minor lesions and immediate death follows major lesions of the area containing the vital centres. This seems adequate reason to call them 'vital', in spite of current physiological doubts.

The **levels of the nuclei** of the cranial nerves are as follows (Fig. 7.47, p. 541):

Those of III and IV, with the red nucleus, lie in the midbrain. The sensory nucleus of V lies in all three, midbrain, pons and medulla. [The mesencephalic nucleus is proprioceptive for the muscles of the orbit and face as well as for the muscles of mastication, the pontine nucleus is for ordinary cutaneous sensibility from the skin area supplied by the trigeminal, including the nose and mouth, while the medullary nucleus is for pain, with the three divisions of V lying upside down in it.] The motor nucleus of V lies in the pons. The nuclei of VI and VII lie likewise in the pons, while VIII overlaps the junction of pons and medulla and lies partly in each. IX, X, XI and XII nuclei lie in the medulla, though it is to be noted that XI possesses also a spinal nucleus in the upper five segments of the cervical cord, especially in cervical segments 2, 3 and 4 (the same levels as the sensory roots that supply its muscles, sterno-mastoid and trapezius). The nuclei of VI, VIII, IX, X and XII are indicated by visible projections on the floor of the fourth ventricle (p. 530). The situation in the brain stem of the nuclei at their respective levels is best understood by studying their development (p. 540).

In the following account of the three parts of the brain stem the same general plan is followed, namely, to describe the external appearances, both ventral and dorsal, since they differ so much, the site of the attached cranial nerve roots, and the situation, relations and blood supply of the part. Finally, the internal architecture, consisting of nuclei and the main fibre tracts and connexions, will be considered.

It is useful to remember that while the nuclei of *motor* cranial nerves contain the cell bodies of peripheral (i.e., lower motor) neurones, the nuclei of *sensory* cranial nerves (except the mesencephalic part of V) contain the cell bodies not of the nerves themselves but of second order neurones. The cell bodies of sensory nerves lie in the ganglia on the nerves.

The Midbrain (Mesencephalon)

(Figs. 7.35, 7.37)

This part of the brain stem lies between the lower part of the cerebral hemisphere (where it is connected with the internal capsule and the thalamus) and the upper part of the pons, into which it disappears.

External Appearance. *Ventrally* it consists of the crura of the two cerebral peduncles, which lie in V-shaped manner cranial to the pons, enclosing the posterior perforated substance of the diencephalon between them. They converge down towards the upper border of the pons from their point of emergence below the thalamus.

Dorsally the midbrain is roughly cylindrical. It shows four low rounded eminences, the colliculi (corpora quadrigemina) two superior and two inferior. The superior colliculi lie below the pineal body, behind the posterior ends of the thalami and the roof of the third ventricle, and are overlapped somewhat by the splenium of the corpus callosum. Lateral to each superior colliculus is the medial geniculate body. Below the inferior colliculi the superior cerebellar peduncles converge into the dorsal surface of the midbrain from the front of the cerebellum. The triangular space between them is closed by a thin sheet of white matter called the superior medullary velum. On it lies the superior vermis of the cerebellum. The superior medullary velum roofs in the pontine part of the fourth ventricle.

The third and fourth cranial nerves leave the brain stem at the midbrain, but the sites of their emergence are very different. The **third nerve** leaves through the medial surface of the basis pedunculi, on the ventral surface of the midbrain, and passes forwards between the posterior cerebral and superior cerebellar arteries in the interpeduncular cistern of cerebro-spinal fluid to reach the roof of the cavernous sinus. The **fourth nerve** leaves the *dorsal* surface of the midbrain, just below the inferior colliculus, that is, through the superior medullary velum. It is unique in that it is a motor nerve leaving the cerebro-spinal axis dorsally; all other motor nerves except the cervical roots of XI leave ventrally near the midline. Furthermore, unlike all other lower motor neurones, it crosses the midline between its nucleus and its point of emergence. The two fourth nerves decussate dorsal to the aqueduct of the midbrain. The reason or significance of this is unknown. The fourth nerve curls around the cerebral peduncle and passes forward, just lateral to the third nerve between the posterior cerebral and superior cerebellar arteries, lying just below the free edge of the tentorium cerebelli, to enter the roof of the cavernous sinus behind the third nerve (PLATE 34).

Situation and Relations of the Midbrain. The midbrain extends from a level just above the dorsum sellæ of the sphenoid to a line that joins the apices of the petrous parts of the temporal bones. In other words, most of it lies in the posterior cranial fossa. The aperture in the tentorium cerebelli lies on its dorsal surface at the level of the upper part of the superior colliculi.

The **basilar artery** divides at the caudal extremity of the midbrain, and the two terminal divisions on each side (posterior cerebral and superior cerebellar arteries) sweep around the cerebral peduncles, the superior cerebellar artery lying at the junction of the midbrain and pons. The third and fourth cranial nerves pass forwards between these two arteries, the fourth nerve having curled around the peduncle from behind. At this level the basal vein also curls around the peduncle on its way from the anterior perforated substance to the commencement of the straight sinus. Higher up the optic tract curls around the cerebral peduncle. The posterior communicating artery runs forward below the medial surface of the peduncle, in the interpeduncular cistern (Fig. 7.36). Dorsally the midbrain lies near the cerebellum; higher up is the tentorium cerebelli lying near the superior colliculi. The splenium of the corpus callosum holds the pineal body over the upper part of the midbrain, just above the superior colliculi. Above and behind this lies the inferior surface of the cerebral hemisphere.

Internal Structure of the Midbrain (Figs. 7.31, 7.32). The diencephalon and fourth ventricle have

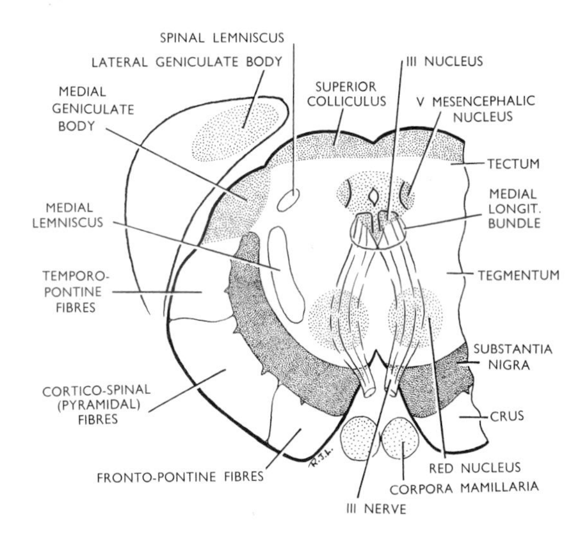

FIG. 7.31. CROSS SECTION OF THE MIDBRAIN THROUGH THE SUPERIOR COLLICULI.

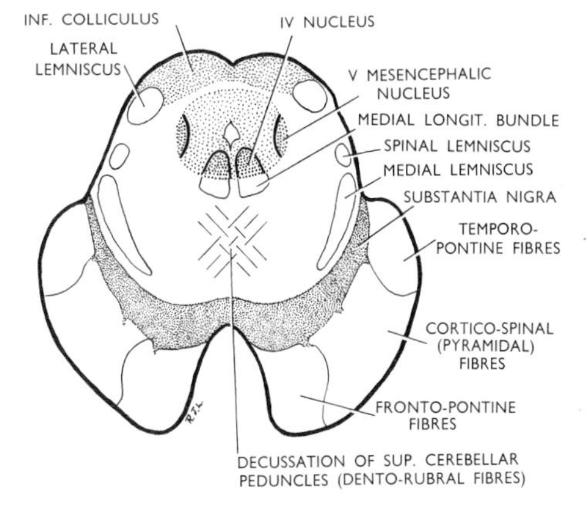

FIG. 7.32. CROSS SECTION OF THE MIDBRAIN THROUGH THE INFERIOR COLLICULI.

no dorsal nervous tissue, being roofed in by ependyma and pia mater only, but the midbrain that joins them is closed in around a central canal like that of the spinal cord. The canal is known as the *aqueduct of the brain (aqueduct of Sylvius)* and it conveys cerebrospinal fluid from the third to the fourth ventricle. It is surrounded by grey matter, as in the spinal cord. Dorsal to the canal is the *tectum* (i.e., the four colliculi) and ventral is the *cerebral peduncle.* This latter structure is cut by a transverse band of deeply pigmented grey matter known as the *substantia nigra,* into the *tegmentum,* lying between the aqueduct and the black substance, and the *crus (basis pedunculi).* It is the crura that are seen as converging rope-like bands on the ventral surface of the midbrain. The tegmentum has parallel sides as its fibres pass upwards to the thalamus. The tectum contains reflex centres for light and sound, the basis pedunculi contains descending (efferent) fibres from the cortex, the tegmentum contains ascending fibres approaching the thalamus (among them lies the red nucleus) while the grey matter around the aqueduct contains the nuclei of the third and fourth cranial nerves, and the mesencephalic nucleus of the fifth nerve.

The **crus cerebri** contains white matter only, and all the fibres are descending (efferent) from the cortex, having passed down through the internal capsule. The central three-fifths is occupied by the 'pyramidal tracts' (cortico-spinal fibres) of which only 5 per cent are the axons of Betz cells, the remaining 95 per cent being the axons of other motor cells in the precentral gyrus. Fronto-pontine fibres occupy the medial one-fifth and temporo-pontine fibres the lateral one-fifth of the crus cerebri; both sets of fibres arborize around the nuclei of the pons. The bundle named temporo-pontine contains also fibres from occipital and parietal cortex.

The **colliculi** consist of nuclei on the reflex pathways for light and sound, the superior a cell station for light reflexes and the inferior a cell station for sound reflexes. They receive from the retina and cochlea respectively and send fibres to the motor nuclei of the cranial and spinal nerves. They transmit impulses for the reflex rotary movements of eyes, head, body and limbs away from or towards light and sound stimuli. Fibres from the tectum to the nuclei of the cranial nerves form the tecto-bulbar tract, fibres from the tectum to the anterior horn cells of the spinal cord form the tecto-spinal tract.

Cranial to the superior colliculus (there is one on each side) lies a group of cells, the *pretectal nucleus* (p. 504). It is worthy of note, since it relays the pupillary light reflex (i.e., it receives from the retina and sends fibres to the underlying Edinger-Westphal nucleus on each side). A lesion here banishes the pupillary light reflex (the Argyll-Robertson pupil).

The central canal (aqueduct of the brain) is surrounded by grey matter, the ventral part of which contains the nuclei of cranial nerves III and IV. III lies at the level of the superior and IV at the level of the inferior colliculi (Fig. 7.47, p. 541). Just ventral to III is the red nucleus. The **substantia nigra** is made up of large nerve cells rich in melanin and iron content. Its purpose is not well understood, but it seems to be a pathway in the extrapyramidal system, receiving impulses from the lemnisci and superior colliculi as well as being connected with the globus pallidus and frontal cortex. It sends fibres to the red nucleus and pontine nuclei.

The **third nerve nucleus** lies close against the midline ventral to the aqueduct, in line with the other somatic motor nuclei (IV, VI and XII). The parasympathetic nucleus (**Edinger-Westphal**) lies near the midline in the cranial part of the nucleus; its axons run out with the third nerve and relay in the ciliary ganglion, from which postganglionic fibres innervate the sphincter pupillæ and ciliary muscles. The third nerve passes ventrally through the red nucleus to emerge from the brain stem on the medial side of the basis pedunculi. There is some interchange of fibres between the two sides (Fig. 7.31).

The **fourth nerve nucleus** lies caudal to the third nucleus, ventral to the aqueduct. The nerve proceeds dorsally and crosses the midline, where it decussates with its fellow dorsal to the aqueduct. It emerges through the superior medullary velum (Fig. 7.37).

The **red nucleus** lies in the tegmentum just ventral to the third nerve nucleus, the axons of which pass through it. It is easily made out in sections of the midbrain, being slightly larger than a full-sized pea. It receives the superior peduncle of the cerebellum, the fibres coming from the dentate nucleus in the opposite cerebellar hemisphere. It has many other connexions, especially with the cortex and globus pallidus. Its efferent fibres decussate at the level of the inferior colliculi and descend to the nuclei pontis and spinal cord. The rubro-spinal tract is part of the extrapyramidal system.

The Medial Geniculate Body. Though homologous with the thalamic nuclei, this collection of cells lies in the midbrain and is therefore considered here. It receives fibres from the cochlear nerve by way of the nuclei of that nerve and the lateral lemniscus, and relays them through the posterior limb of the internal capsule to the auditory cortex, situated at the medial part of the superior temporal gyrus (p. 503). It is a cell station on the pathway of conscious hearing (Fig. 7.37).

The **mesencephalic nucleus of V** lies in the central grey matter, lateral to the aqueduct, throughout the whole length of the midbrain. This long slender nucleus receives proprioceptive fibres from the muscles supplied by V (muscles of mastication) and from the muscles of the orbit and face and, perhaps, the muscles of the tongue (see p. 396).

It is unique in being a collection of first neurone cells buried in the central nervous system. All other first neurone cell bodies lie in the sensory ganglia of the cranial nerves or in the posterior root ganglia of the spinal nerves. The fibres pass straight through the trigeminal ganglion from the periphery to the mesencephalic nucleus.

Between the medial lemniscus and the central grey matter the tegmentum contains fragments of grey matter broken up by criss-cross bundles of white fibres. The 'network' appearance so produced gives it the name *Formatio reticularis*. It is traceable through the pons and medulla into the formatio reticularis of the upper spinal cord (p. 537). Great activity is ascribed to these **reticular nuclei** by present-day physiologists.

The **tracts of the midbrain** consist of descending tracts in the crus (basis pedunculi) and ascending tracts in the tegmentum. Of the latter the most important are the lemnisci (medial, spinal, trigeminal and lateral). Most are on their way to relay in the thalamus. The superior cerebellar peduncles enter the tegmentum and decussate on the way upwards to the red nuclei, from which the rubro-spinal tracts decussate and descend in the tegmentum. The *medial longitudinal bundle* lies immediately ventral to the grey matter around the aqueduct. It extends from the upper border of the midbrain to the lower border of the medulla. It links the vestibular nucleus with the motor nuclei of the cranial nerves. The *posterior commissure* joins right and left colliculi; tectal reflexes are often bilateral.

Blood Supply of the Midbrain. The midbrain is supplied by the posterior cerebral artery as the vessel curls around the basis pedunculi. Medial and lateral central branches enter the crus cerebri to supply the substantia nigra and red nucleus and, on the dorsal aspect, supply the colliculi, grey matter around the aqueduct, and the pineal body. The veins drain for the most part into the basal vein as it passes around the peduncle. From the colliculi some blood enters the great cerebral vein.

The Pons (Fig. 7.35)

Ventrally the pons is a broad transverse mass that extends prominently beyond the sides of midbrain and medulla before curving back to sink into the cerebellum. It has a midline groove, and close inspection of the surface shows that its fibres run transversely. These are the ponto-cerebellar fibres; they constitute the middle cerebellar peduncle. The bulge on each side of the groove is due to the bulk of the underlying cortico-spinal (pyramidal) tracts (Fig. 7.33).

The *dorsal surface* of the pons is concealed by the attached cerebellum. The central canal (the aqueduct) of the midbrain opens out at the upper border of the pons into the cavity of the fourth ventricle. The pontine part of the roof of the latter consists only of a thin sheet of white matter, the superior medullary velum, upon which the lingula of the superior vermis lies. The superior cerebellar peduncles give attachment to this velum.

Only one cranial nerve, the fifth, emerges from the pons, though VI, VII, the nervus intermedius and VIII are attached to the brain stem along the line of junction of pons and medulla. The **fifth nerve** is attached to the pons by two roots, a very large sensory and a small motor. The motor root lies ventral and cranial to the sensory root. Together they are attached to the supero-lateral part of the ventral surface of the pons, where the latter is commencing to curve back into the middle cerebellar peduncle. From this point the two roots of V pass forwards in the posterior cranial fossa (i.e., below the tentorium), to pass in the groove on the upper surface of the apex of the petrous bone into Meckel's cave and so to the trigeminal ganglion lying in the middle cranial fossa (Fig. 6.94, p. 487 and Fig. 7.35).

Situation and Relations of the Pons. The pons separates midbrain and medulla. It lies against the upper part of the clivus and grooves the basi-occiput above the jugular tubercle of that bone. It grooves the apex of the petrous bone in the posterior cranial fossa almost as far laterally as the internal acoustic meatus. Inspection of the posterior fossa of a skull shows at a glance the broad transverse groove in which the pons lies. It is separated from the underlying bone by the cisterna pontis, in which the basilar artery runs upwards. The artery may or may not occupy the midline groove in the pons; usually the artery has a gentle curve that carries it to one side of the groove. The superior cerebellar artery curls backwards around the upper margin of the pons. The labyrinthine artery lies ventral to the pons. The sixth nerve runs upwards across its ventral surface, while the seventh, nervus intermedius and eighth emerge more laterally between pons and medulla. More laterally the flocculus and choroid plexus of the fourth ventricle lie beside its lower border, in the cerebello-pontine angle (Fig. 7.35).

Internal Structure. The structure of the midbrain has been seen to be relatively simple. Dorsal to the aqueduct lies the tectum (the four colliculi) and ventral to it lie the ascending tracts (tegmentum) and still more ventrally the descending tracts (crus cerebri). The pons differs from this simple structure in that its central canal opens out in triangular shape. The dorsal wall of the pons consists of the ependyma of the dilated central canal (roof of the fourth ventricle) covered with a thin sheet of white matter called the superior medullary velum; this latter is bounded on each side by the superior cerebellar peduncles passing up to the mid-

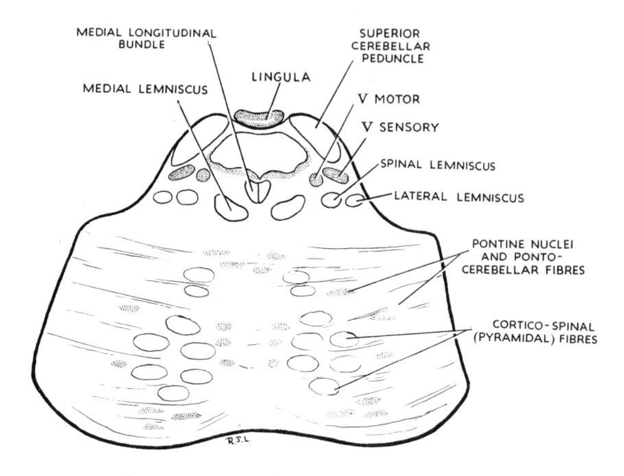

FIG. 7.33. CROSS SECTION OF THE UPPER PART OF THE PONS.

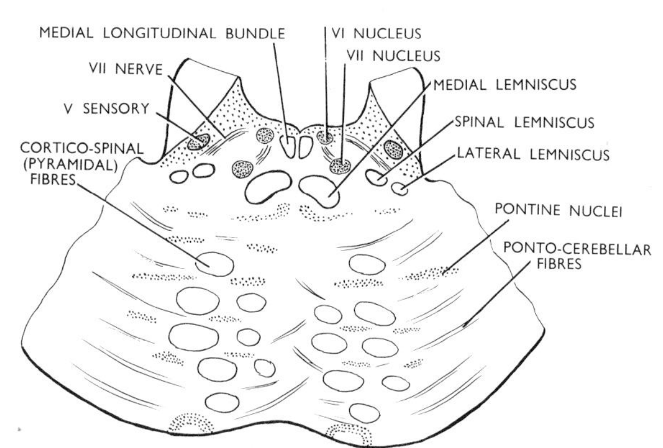

FIG. 7.34. CROSS SECTION OF THE LOWER PART OF THE PONS.

brain. Beneath the floor of the fourth ventricle (p. 531) the pons contains the nuclei of cranial nerves V (sensory and motor), VI and VII and part of VIII. Ascending tracts surround the nuclei.

More ventrally lie the descending tracts. The cortico-pontine tracts end here, arborizing around small groups of cells called the pontine nuclei. The axons of these cells decussate across the midline and form the middle cerebellar peduncle, through which they pass to the cortex of the cerebellar hemisphere. The cortico-spinal (pyramidal) tract, a single bundle in the internal capsule and crus cerebri of the midbrain, passes straight down through the pons, but becomes broken up into separate small bundles by the transversely running ponto-cerebellar fibres. The bundles of pyramidal fibres bulge the ponto-cerebellar fibres forward on each side, thus producing the midline ventral groove of the pons. The ventral part of the pons possesses this crossing of pyramidal and ponto-cerebellar fibres throughout the whole of its length, but dorsally the arrangement of nuclei and ascending tracts differs somewhat between the upper and lower parts of the pons.

Nuclei of the Pons. The nuclei of cranial nerves VI and VII lie in the pons, with the pontine (sensory) nucleus and the motor nucleus of V. Part of the vestibular nucleus encroaches into the pons. In addition, the pontine nuclei themselves form an important part of the grey matter of this part of the brain stem.

The **nuclei pontis** consist of masses of grey matter whose cells receive axons from the ipsilateral cortex. The fibres from the nuclei pontis decussate in the midline and pass transversely into the cortex of the opposite cerebellar hemisphere (neocerebellum). These transverse bundles interrupt the cortico-spinal (pyramidal) fibres and break them up into separate longitudinal bundles.

The central canal in the pons is opened out to form the cavity of the fourth ventricle. Here in the floor somatic motor cells lie near the midline, somatic afferents furthest from the midline, while visceral and branchial centres lie between the two, sensory lateral to motor (p. 540 and Fig. 7.46, p. 541).

The **pontine nucleus of V** lies beneath the superior cerebellar peduncle, with the spinal lemniscus medial to it. Dorsal to the latter, nearer the floor of the fourth ventricle, is the **motor nucleus of V** (Fig. 7.33).

The **sixth nerve nucleus** lies near the midline in the pontine part of the floor of the fourth ventricle, and is responsible for the facial colliculus. The **seventh nerve nucleus** lies deeper and further from the midline (Fig. 7.34). More superior still is the motor nucleus of V and just lateral to this lies the pontine sensory nucleus of V. A collection of cells alongside the seventh nucleus is known as the **superior salivatory nucleus.** It is the secreto-motor nucleus for the pterygo-palatine and submandibular ganglia; the axons of its cells pass out in the nervus intermedius. It is continuous below with the inferior salivatory nucleus of IX, situated in the medulla (Fig. 7.47, p. 541).

Lateral to the superior salivatory nucleus is a column of cells called the **nucleus of the tractus solitarius.** These are cell bodies of the second neurone on the pathway for taste, and they receive taste impulses from the nervus intermedius. The nucleus of the tractus solitarius of the pons continues below into that of the medulla (p. 530); it sends its fibres to the opposite thalamus to be relayed to the sensory cortex for taste (parietal operculum).

The nuclei of VIII lie at the junction of pons and medulla; vestibular and cochlear nuclei are separate. The vestibular fibres pass anterior to the inferior cerebellar peduncle, pass through the medulla and synapse in the **vestibular nucleus,** which forms an eminence at the lateral angle of the floor of the fourth ventricle in both pons and medulla. Fibres from the cell bodies of the vestibular nucleus pass into the adjacent inferior peduncle and so to the archæcerebellum (vermis and flocculus). Other fibres join the medial longitudinal bundle and connect with ocular and spinal nuclei. Vestibulo-spinal tracts descend to anterior horn cells in the spinal cord.

The **cochlear nuclei** are more complicated. Actually they lie in the upper medulla, but are more conveniently studied here. Fibres from the spiral ganglion of the cochlea pass transversely across the inferior cerebellar peduncle. Most pass ventrally to the ventral nucleus of the cochlea, just anterior to the inferior peduncle. A few pass dorsally, behind the inferior peduncle, and synapse in the dorsal cochlear nucleus. From this nucleus the central fibres run transversely through the inferior peduncle beneath the floor of the ventricle. Fibres from the ventral nucleus pass transversely in the substance of the brain stem to join the dorsal fibres across the midline. The transversely running fibres form the **corpus trapezoideum.** The two together turn upwards to form the **lateral lemniscus,** which terminates in the medial geniculate body and the inferior colliculus. Both the corpus trapezoideum and the lateral lemniscus contain cell stations which make connexions via the medial longitudinal bundle with the ocular nuclei.

Tracts and Connexions in the Pons. The main longitudinal fibres of the pons are as those of the midbrain and medulla. The medial longitudinal bundle runs uninterruptedly through it, lying near the midline just ventral to the floor of the fourth ventricle. The axons from the sixth nerve nucleus pass ventrally to emerge near the midline at the junction of pons and pyramid. The fibres from VII pass dorsally towards the floor of the fourth ventricle, curve around over VI nucleus (forming the facial colliculus) and then pass ventrally to emerge between pons and olive lateral to the sixth nerve.

The parasympathetic fibres from the superior salivatory nucleus leave the brain stem in the nervus intermedius, which emerges through the upper part of the medulla, and lies, indeed, nearer the eighth than to the seventh nerve. The nervus intermedius is a mixed nerve, for it contains also sensory fibres (cell bodies in the geniculate ganglion) which mediate taste from the mouth (tongue and palate) to the cell bodies in the nucleus of the tractus solitarius.

The fibres of the motor nucleus of V pass ventrally and laterally, accompanied by the sensory fibres, which latter have three destinations. Some pass upwards to the mesencephalic nucleus (where their cell bodies lie); these are proprioceptive. The others have their cell bodies in the trigeminal ganglion; some (normal sensation) are directed to the pontine nucleus, the remainder (pain and temperature extremes) descend to the spinal nucleus in the medulla. The medial lemniscus passes up near the midline, intersecting the corpus trapezoideum. Ventral to the corpus trapezoideum lie the cortico-spinal (pyramidal) tracts, intersected by the transversely running decussating fibres of the pontocerebellar pathways that constitute the middle cerebellar peduncle and which give to this part of the brain stem its name of pons.

The **blood supply** is from the pontine branches of the basilar artery; venous return is into the inferior petrosal sinuses and the basilar plexus.

The Medulla (Figs. 7.35, 7.37)

The medulla oblongata is the upward continuation of the spinal cord. It lies almost vertically and extends between the anterior arch of the atlas and a line joining the jugular tubercles of the occipital bone. It is embraced dorsally by the convex cerebellar hemispheres and lies in the deep groove between them called the vallecula.

External Appearance. The lower part of the medulla is cylindrical and resembles the spinal cord, with a midline groove ventrally and dorsally, but the upper part is conical and possesses a very different surface appearance.

Ventrally the upper part of the medulla is deeply grooved in the midline. On each side of the median groove is a bold cylindrical convexity called the *pyramid.* It contains the cortico-spinal (pyramidal) tracts. These tracts *decussate* towards the lower end of the medulla, near the anterior lip of the foramen magnum, and here the anterior median fissure is obliterated by the decussating fibres (Fig. 7.35).

Lateral to the pyramid in the upper (open) half of the medulla is an oval bulge called the *olive*—produced by the prominence of the underlying olivary nucleus. A few transverse ridges lie across the olive; these are the *anterior external arcuate fibres,* arising from pontine nuclei and passing into the inferior cerebellar peduncle. They are aberrant pontine fibres and, like the fibres of the middle peduncle, connect lower pontine nuclei (called *arcuate nuclei*) with the opposite neocerebellum. They probably contain also some crossed fibres from the gracile and cuneate nuclei.

Lateral to the olive, and forming the lateral surface of the medulla is the *inferior cerebellar peduncle* (restiform body). From the lower cylindrical part of the medulla the two inferior peduncles diverge up from each other to enter the lower part of the cerebellum medial to and below the middle peduncles.

Attached to the junction of medulla and pons are the cranial nerves VI, VII, nervus intermedius and VIII. The abducent nerve (VI) emerges between pyramid and pons. The facial nerve (VII) emerges between olive and pons, while the nervus intermedius and VIII are attached more laterally, at the juction of pons and inferior cerebellar peduncle.

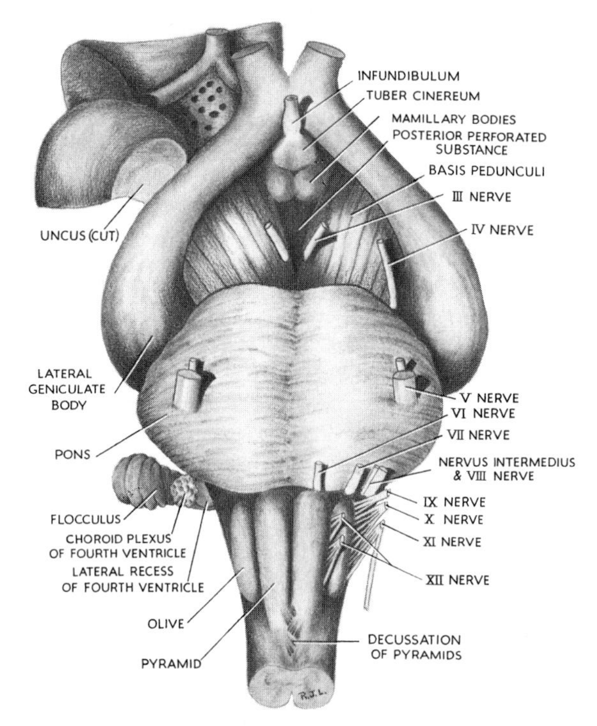

FIG. 7.35. VENTRAL ASPECT OF THE BRAIN STEM. The optic tract embraces the crus (basis pedunculi). Lateral to the optic chiasma the medial and lateral roots of the olfactory tract enclose the anterior perforated substance. Drawn from a specimen.

The rootlets of IX, X, XI and XII cranial nerves are attached to the upper part of the ventral surface of the medulla. XII is a somatic motor nucleus and its rootlets emerge, as would be expected, nearest the midline. They occupy the groove between pyramid and olive: the posterior inferior cerebellar artery passes below them. The other three are visceral nuclei and their rootlets are attached, therefore, more laterally. They lie in a vertical line between olive and inferior cerebellar peduncle, IX above X, with the posterior inferior cerebellar artery making its tortuous course between them. The cranial root of XI lies as a series of rootlets below X at the lower part of the medulla; the cervical root passes up through the foramen magnum to join it (Fig. 6.96, p. 492).

The *posterior surface* of the medulla presents a very different appearance. The lower portion is closed around a central canal, but the upper portion expands into the fourth ventricle, whose roof is a thin lamina of ependyma and pia mater, perforated for the escape of cerebrospinal fluid. The floor of the fourth ventricle is the expanded anterior wall of the central canal and is described on p. 531.

In the closed part of the medulla the posterior surface is formed by the upward continuation of the gracile and cuneate columns of the spinal cord, bounded laterally by the inferior cerebellar peduncles. The columns diverge from each other and each ends in a low elevation at the inferior boundary of the fourth ventricle. These are the *gracile* and *cuneate nuclei*. The fibres of the posterior columns relay here. Most pass in the medial lemniscus to the thalamus, but some of the proprioceptive fibres pass in the *posterior external arcuate fibres* through the nearby inferior peduncle to the cortex of the neocerebellum.

Towards the lateral surface of the inferior cerebellar peduncle is an elongated oval eminence, rather ill-defined, called the *tubercle of the trigeminal nerve*. It is produced by the spinal nucleus (gelatinous substance) and the overlying spinal tract of the trigeminal nerve (Fig. 7.6, p. 500 and Fig. 7.37).

Internal Structure of the Medulla. The internal structure of the lower extremity of the medulla resembles that of the adjoining spinal cord. Passing upwards through the medulla there is considerable rearrangement of the fibres in the closed part, while in the open part (fourth ventricle) the arrangement of fibres is very different and, in addition, the nuclei of the cranial nerves VIII to XII lie among the tracts.

The Closed Part. The lower part contains the decussation of the pyramids. Above this level the

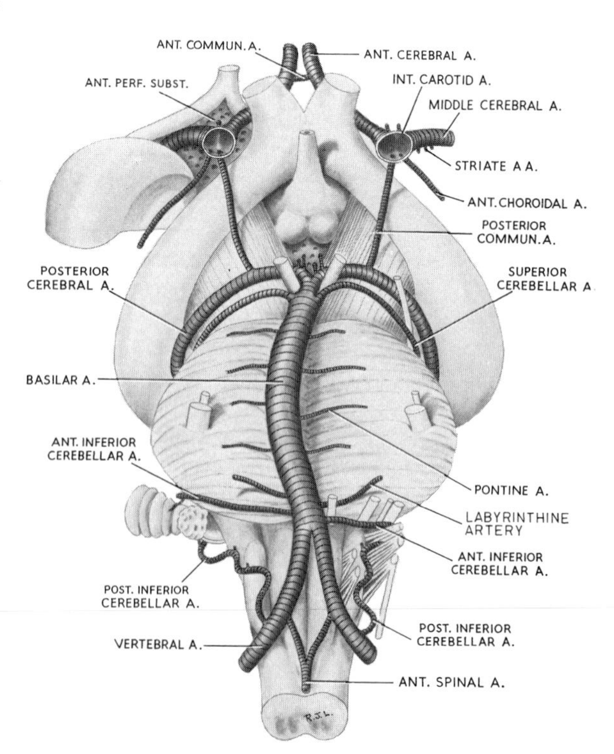

FIG. 7.36. THE CIRCLE OF WILLIS AND THE ARTERIES OF THE BRAIN STEM. The arterial circle (horizontal) lies at right angles to the basilar artery (vertical).

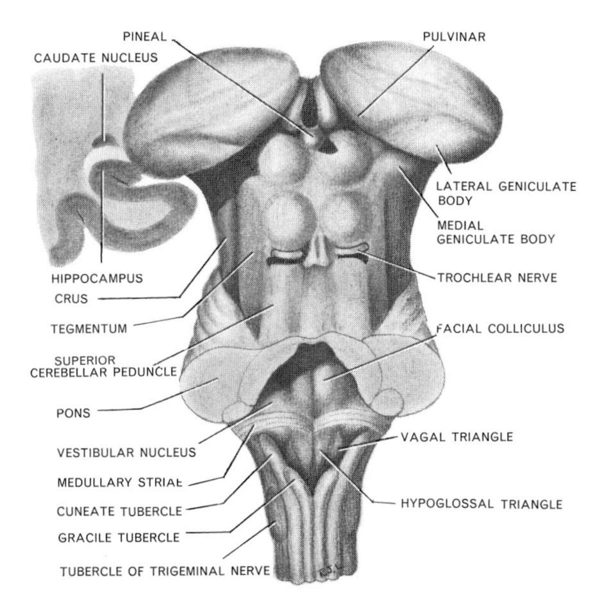

FIG. 7.37. DORSAL ASPECT OF THE BRAIN STEM AFTER REMOVAL OF THE CEREBELLUM AND THE ROOF OF THE FOURTH VENTRICLE. THE INFERIOR HORN OF THE LEFT LATERAL VENTRICLE IS CUT IN CORONAL SECTION (cf. Fig. 7.47). Drawn from a specimen.

pyramids descend from the pons alongside the median sulcus; below this level the crossed fibres pass back to enter the lateral white columns of the spinal cord, almost obliterating in their passage the anterior grey columns around the central canal. Posteriorly the grey matter around the central canal becomes increasingly more prominent above. Two posterior bulges occur in it, the nuclei gracilis and cuneatus (Fig. 7.38). They enlarge progressively when traced upwards, as they gradually replace the gracile and cuneate fasciculi. In the upper part of the closed portion of the medulla their central processes *decussate* ventrally, between the central canal and the pyramids, and pass upwards as the **medial lemnisci,** to the thalamus. The gracile and cuneate nuclei terminate at the inferior border of the fourth ventricle.

The side of the closed part is occupied by the

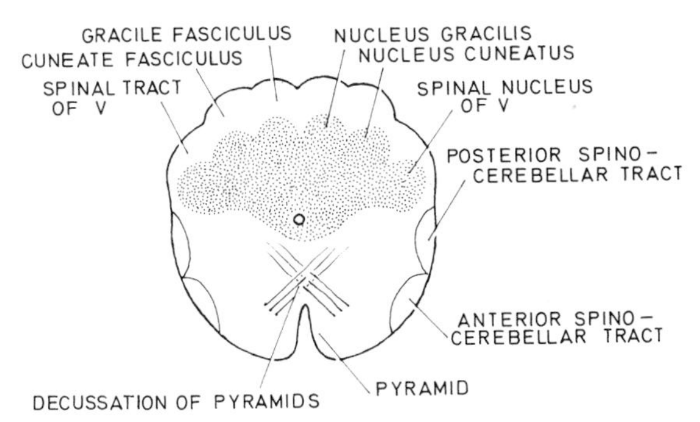

FIG. 7.38. CROSS SECTION THROUGH THE CLOSED PART OF THE MEDULLA.

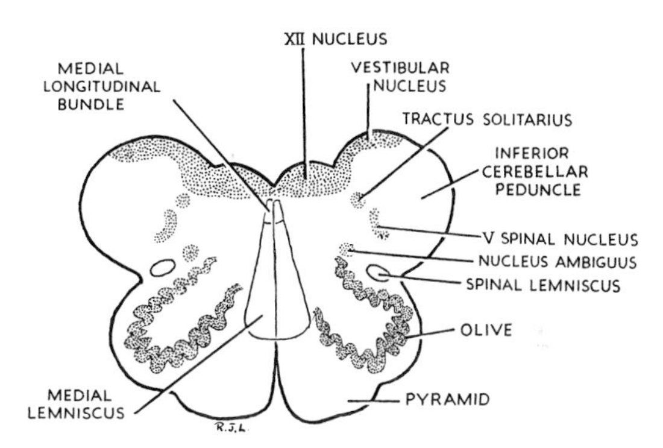

FIG. 7.39. CROSS SECTION THROUGH THE OPEN PART OF THE MEDULLA OBLONGATA (LOWER PART OF FOURTH VENTRICLE).

anterior and posterior spino-cerebellar tracts, with the spinal lemniscus deep to the former; this is similar to the arrangement in the spinal cord (Fig. 7.45, p. 539).

One cranial nerve nucleus appears in the closed part of the medulla. This is the spinal nucleus of V, in line with the substantia gelatinosa of the spinal cord, overlaid laterally by incoming white fibres, the spinal tract of V. It lies near the postero-lateral surface of the medulla. It contains cell stations on the pain pathway of V, with the three divisions of V represented upside down; i.e. ophthalmic division lowest.

The Open Part. The upper part of the medulla encloses the lower half of the fourth ventricle. Posteriorly it consists of merely the perforated roof of the ventricle (p. 530). All the substance of the medulla lies ventral to the floor, which is merely the expanded central canal. Cranial and other nuclei lie amongst the tracts. The pyramids bulge forwards on each side of the median ventral sulcus. Behind (dorsal to) the pyramids the medial lemnisci, each surmounted by the medial longitudinal bundle, lie side by side against the midline. Lateral to the pyramid lies the *olivary nucleus*, a crenated C-shaped lamina of grey matter, open towards the opposite inferior cerebellar peduncle. Its fibres leave the hilum and decussate across the midline to enter the opposite inferior peduncle and pass to the neocerebellum. Like the dentate nucleus of the cerebellum the olivary nucleus evolved with the neopallium and the neocerebellum. *Parolivary nuclei*, more ancient, lie medial to this nucleus; their fibres pass through the inferior cerebellar peduncle to the palæocerebellum.

The fibres of the posterior spino-cerebellar tract likewise enter the inferior cerebellar peduncle; they pass to the palæocerebellum. The anterior spino-cerebellar tract passes up uncrossed, on the surface, between the olive and the inferior peduncle; the spinal lemniscus continues up deep to it (Fig. 7.45, p. 539), with the spinal nucleus and tract of V more dorsal, as in the lower medulla. A new structure, the *nucleus ambiguus*, lies between them. It contains motor cell bodies for the striated muscles of the pharynx, larynx and soft palate. These nuclei and tracts are all supplied by the posterior inferior cerebellar artery (p. 534). The floor of the ventricle is occupied by the hypoglossal and vagal nuclei, with the vestibular nucleus more laterally (p. 531). The *hypoglossal nucleus* lies near the midline; its fibres, in a series of rootlets, pierce the medial limb of the olivary nucleus and emerge from the surface between olive and pyramid.

The *nucleus of the vagus* lies in the lower part of the vagal trigone. It is a mixed nucleus, containing motor cell bodies for smooth muscle, secreto-motor cell bodies for glands, cell bodies of second order sensory neurones and, in its inferior angle, vital centres (at least motor cell bodies for heart muscle). Terminology is confusing. Some name the whole mixture as the dorsal nucleus of the vagus, to distinguish it from the nucleus ambiguus. Others restrict the name *dorsal nucleus* to only those cell bodies which supply smooth muscle (respiratory and alimentary tracts) to distinguish them from the others. In this case the secreto-motor cell bodies (for glands of the respiratory and alimentary tracts) are named the *inferior salivatory nucleus*, and the sensory cell bodies of the second order neurones are named the *nucleus of the tractus solitarius*. Understanding of the reality matters more than names. Only a few fibres from the nucleus ambiguus enter the vagus; instead they run in the cranial root of XI and join the vagus below the skull.

The *glosso-pharyngeal nucleus* lies at the upper end of the vagal trigone. It has secreto-motor cell bodies which constitute the *inferior salivatory nucleus* (continuous with the pontine superior salivatory nucleus whose fibres leave in the nervus intermedius). By peripheral relay these fibres of IX supply glands of the oro-pharynx and the vestibule of the mouth. IX supplies only one muscle (striated) and carries the necessary fibres from the nucleus ambiguus to the stylo-pharyngeus muscle. IX supplies no smooth muscle. Sensory fibres enter (from cell bodies in the petrous ganglion) to synapse on the second order cell bodies of the *nucleus of the tractus solitarius*. This nucleus is in line with that of the pons, and here IX, like the nervus intermedius, mediates taste impulses. Unlike the nervus intermedius, IX has other sensory fibres than taste. Common sensation is relayed, probably, through the same nucleus while pain, perhaps, relays in the nearby spinal nucleus of V. The special impulses from the carotid sinus and carotid body are passed to the 'vital' centres.

Blood Supply of the Medulla. The medulla is supplied ventrally by the vertebral and basilar arteries, but laterally and dorsally the posterior inferior cerebellar artery provides the supply (see p. 534). The veins drain dorsally to the occipital sinus and ventrally into the basilar plexus of veins and the inferior petrosal sinus. The medullary veins communicate with the spinal veins.

The Fourth Ventricle

The substance of the midbrain surrounds the aqueduct and the substance of the lower medulla surrounds the central canal. Between the two, however, the substance of pons and upper medulla lies ventral and the central canal is expanded into a cavity known as the fourth ventricle, which is roofed in by little more than ependyma and pia mater (Fig. 7.6, p. 500).

The **roof** is updrawn into a tent shape (the ridge-pole of the tent lying transversely) and is covered by the cerebellum. The *upper part* of the roof lies over the pons. The ependyma here is covered with a thin sheet of white matter called the superior medullary velum, which is bounded by the superior cerebellar peduncles. The *lower part* of the roof lies over the medulla. The ependyma is covered in its upper part by the inferior medullary velum, a thin sheet of white matter from the base of the flocculus, but in the lower part ependyma and pia mater alone form the roof. The lower margin of the roof is attached to the margins of the gracile and cuneate tubercles and is perforated by a midline slit, the **median aperture** (foramen of Magendie), by which cerebro-spinal fluid escapes into the cisterna cerebello-medullaris.

The cavity is prolonged laterally as a narrow **lateral recess** around the inferior cerebellar peduncle; here the roof is attached to the margins of the medullary striæ. The narrow, tubular lateral recess has a patent extremity, the **lateral aperture** (foramen of Luschka), which opens anteriorly, just behind the eighth nerve, into the cisterna pontis (Fig. 7.35). Through these apertures the cerebro-spinal fluid escapes from the ventricular system into the subarachnoid space for absorption by the arachnoid villi. The extent of the lateral recess is indicated by the cast in Fig. 7.40.

The **choroid plexus** of the fourth ventricle is a small bilateral L-shaped structure which indents the medullary part of the roof. It commences at the lateral aperture by a branch of the posterior inferior cerebellar artery. Here it lies just below the flocculus. It indents the roof of the lateral recess, passes medially to meet its fellow, and the two turn down towards the median aperture. They drain back into the occipital sinus.

The **floor** of the fourth ventricle is diamond shaped (it is once more known as the *rhomboid fossa*). The upper boundaries are the superior cerebellar peduncles, the lower are formed by the gracile and cuneate nuclei and, above them, by the inferior cerebellar peduncles. A deep midline groove runs from the aperture of the aqueduct of the midbrain above to the commencement of the central canal below. On each side of the groove the floor is symmetrical. At its widest part the floor is crossed transversely by glistening white fibres; these are the *medullary striæ* (Fig. 7.37). They lie between pontine and medullary parts of the floor. At the lateral angle

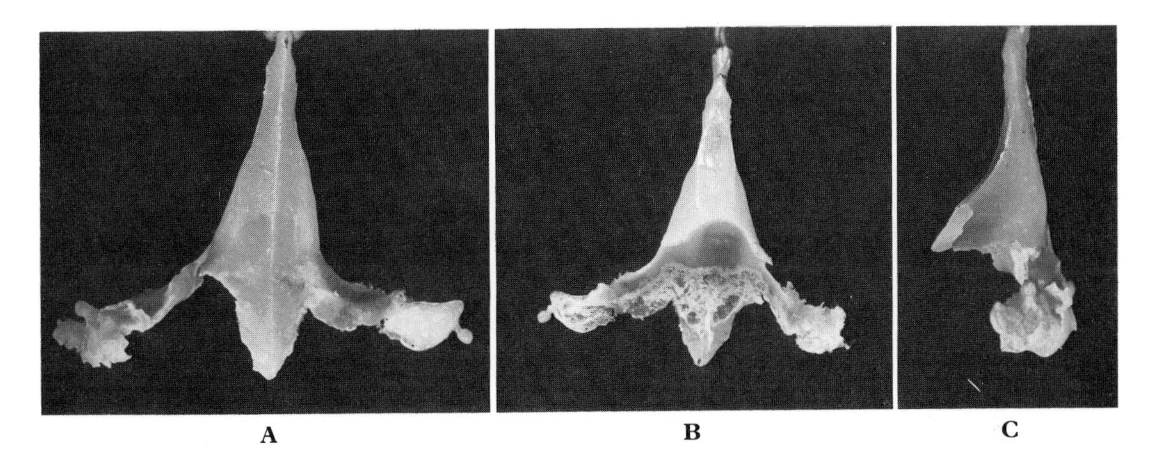

FIG. 7.40. A CAST OF THE FOURTH VENTRICLE. A. VENTRAL ASPECT. B. DORSAL ASPECT. C. RIGHT LATERAL ASPECT. (Photograph)

of the floor beneath the medullary striæ is a rounded prominence which extends into both pontine and medullary parts of the floor. This is the vestibular nucleus (Fig. 7.37).

The **pontine part** of the floor is characterized by a rounded eminence near the midline; this is called the *facial colliculus.* The recurving fibres of the seventh nerve lie beneath the ependyma, but the prominence is actually caused by the underlying sixth nerve nucleus. The seventh nerve nucleus lies deeply and makes no impression on the floor of the fourth ventricle. Between the facial colliculus and the vestibular eminence is a groove known as the superior fovea which leads up to the opening of the aqueduct. In the upper part of the fovea is a small area known as the locus cæruleus, the bluish colour being caused by the underlying substantia ferruginea, a collection of pigmented cells whose function is unknown.

The **medullary part** of the floor is smaller than the pontine part. Its lateral angle is occupied by the lower part of the vestibular area. From the inferior angle a groove (the inferior fovea) passes up to the medullary striæ to meet the edge of the vestibular eminence. The groove divides the floor into two triangles. The medial triangle, apex down, is the *hypoglossal triangle.* Beneath it lies the twelfth nerve nucleus. The lateral triangle, apex upwards, lies between the hypoglossal and vestibular nuclei. It overlies the dorsal nucleus of the vagus and the nucleus of IX and is known as the *vagal triangle.*

There is nothing else worthy of note in the fourth ventricle.

The Nuclei of the Cranial Nerves

The cranial nerve nuclei have been noted in the preceding account of the internal structure of the brain stem, but perhaps they may be there confused with other named nuclei. It is convenient to list them separately (see Fig. 7.47, p. 541).

Motor nuclei to *striated muscles* send their fibres *direct* to the muscles. Nuclei for *smooth muscle* and for *glands* effect their distribution by *peripheral relay* in a ganglion cell.

Sensory nuclei are *not* the actual nuclei of the cranial nerves, whose cell bodies lie in the ganglion on the nerve itself. Sensory nuclei are aggregates of cell bodies of the *second* sensory neurones. Their central processes go to the usual three sensory destinations: (1) to motor nuclei for reflex effects, (2) to the cerebellum and (3) to the opposite thalamus for relay to the sensory cortex. There is a solitary exception: the midbrain part of the trigeminal nucleus contains the cell bodies of the *first* neurone on the proprioceptive pathway from muscles of the orbit, face and (probably) tongue. Their central processes are directed chiefly to the cerebellum.

Oculomotor Nucleus. Motor. Somatic, thus near the midline. Lies in the floor of the aqueduct of the midbrain, level with the superior colliculi. The parasympathetic (Edinger-Westphal) part lies cranial to the somatic part.

Trochlear Nucleus. Motor. Somatic, thus near the midline. Lies in the floor of the aqueduct, level with the inferior colliculi.

Trigeminal Nucleus. Two parts. *Motor.* Branchial (first pharyngeal arch), thus off centre. In upper pons, deep to floor of fourth ventricle. *Sensory.* Continuous through whole brain stem and extends into upper spinal cord. Mesencephalic (midbrain) part is in the grey matter lateral to the aqueduct. Pontine part very lateral in pons, ventral to superior cerebellar peduncle: it relays agreeable sensations (light touch, etc.). Medullary part in the inferior cerebellar peduncle extends through lower medulla into spinal cord, the whole long nucleus being named the 'spinal' nucleus. It relays the three divisions of the trigeminal nerve upside-down. Transmits nasty sensations (temperature extremes and pain).

Abducent Nucleus. Motor. Somatic, thus near the midline. In the pons, deep to 'facial' colliculus in floor of fourth ventricle.

Facial Nucleus. Motor. Branchial (second pharyngeal arch), thus off centre. In pons, deep and lateral to 'facial' colliculus.

Nervus Intermedius Nucleus. In lower pons, lateral to facial nucleus, deep to floor of fourth ventricle. *Secreto-motor* part in superior salivatory nucleus. *Sensory* part is nucleus of tractus solitarius, lateral to this. [N.B. This is a cranial nerve in its own right. It is not the 'sensory nucleus of the facial nerve', because (*a*) most of it is secreto-*motor* and (*b*) its sensory part is for taste only, and one does not taste with the face.]

Eighth Nerve Nucleus. In medulla are *cochlear nuclei* buried ventrally and dorsally in the inferior cerebellar peduncle. In both pons and medulla, *vestibular nucleus* makes a bulge in lateral angle of floor of fourth ventricle.

Glosso-pharyngeal Nucleus. In upper medulla at apex of vagal trigone. *Secreto-motor* in inferior salivatory nucleus. *Sensory* nucleus is nucleus of tractus solitarius lateral to this; pain probably relays in spinal nucleus of trigeminal. *Motor* (to stylo-pharyngeus only) cell bodies are in the nucleus ambiguus.

Vagal Nucleus. In medulla, caudal to IX in vagal trigone. Inferior salivatory nucleus and nucleus of tractus solitarius as for IX but the cells are intermingled. To these are added motor cell bodies for smooth muscle; all are intermingled. To these are added, at caudal end, the 'vital centres', denied by present-day physiologists but still vital, for a lesion here is immediately fatal. At least cardiac cell bodies intermingle here.

Accessory Nucleus. Nucleus ambiguus in inferior cerebellar peduncle.

Hypoglossal Nucleus. Motor. Somatic, thus near the midline. In medulla, in hypoglossal trigone in floor of fourth ventricle.

THE CEREBELLUM

The cerebellum occupies the posterior cranial fossa, where it lies posterior to the brain stem. It consists of two hemispheres united in the midline by a portion of cerebellar substance known as the vermis. Three peduncles connect each hemisphere to the three parts of the brain stem. The superior peduncle enters the midbrain, the middle peduncle consists of the transverse fibres of the pons and the inferior peduncle arises from the medulla. The ventral surface of the vermis lies upon the superior medullary velum above and the roof of the medullary part of the fourth ventricle below.

The *superior surface* of the cerebellum is bounded posteriorly by a convex border that lies below the attached margin of the tentorium cerebelli. From this border the superior surface slopes concavely upwards, in conformity with the shape of the tentorium, to the highest part of the cerebellum, which lies at the aperture in the tentorium. The *postero-inferior surfaces* are boldly convex below the posterior border. Known to the ancients by the appropriate name of 'nates', they occupy the concavities in the occipital bone. Between them lies a deep groove, the vallecula, which lodges the three parts of the inferior vermis.

External Appearance and Parts

The surface of the cerebellum is indented by fine slit-like sulci, between which lie more or less parallel cerebellar *folia.* In the main the folia and sulci lie transversely from side to side across the whole extent of the cerebellum. Several transverse fissures pass deeply into the substance; the folia extend into their depths.

A well-marked groove indents the convex posterior border. This is the deepest sulcus of all; it extends from side to side and around towards the front, where its margins embrace the middle peduncle (Figs. 7.4, p. 498, 7.19, p. 512). It is known as the *horizontal fissure*; it has no known functional significance. Anterior to it, on the superior surface, is a much shallower groove known as the **fissura prima.** The primary fissure is significant, for it separates palæocerebellum and neocerebellum (Fig. 7.1, p. 495).

The ancients gave many fanciful names to different parts of the cerebellar surface. The parts of the vermis so named chance to have functional and morphological significance; but all the named parts of the hemispheres are functionally meaningless and can be ignored.

The **hemispheres** consist, then, of a small *anterior lobe* on the superior surface in front of the primary fissure and a pair of large *posterior lobes* comprising the rest of the hemispheres behind the primary fissure. This division possesses not only the advantage of simplicity, but is of morphological and functional significance. The anterior lobe is part of the palæocerebellum, the posterior lobes comprise the neocerebellum.

The **vermis** consists of superior and inferior parts, separated from each other by the neocerebellar posterior lobes, which meet in the midline behind the primary fissure.

The *lingula* is that part of the superior vermis which lies in contact with the superior medullary velum. On the superior surface of the anterior lobe there is no line of demarcation between vermis and cerebellar hemispheres.

In the vallecula the inferior vermis consists of three small lobules which retain their ancient names of pyramid, uvula and nodule. The *nodule* lies highest, on the roof of the fourth ventricle. Projecting laterally from each side of the nodule is a slender band of white matter whose bulbous extremity, capped with grey matter, can be seen from in front, lying in the angle between cerebellum and pons. This is the *flocculus*; the choroid plexus of the fourth ventricle projects just below it (Fig. 7.35).

The *uvula* and *pyramid*, larger than the nodule, occupy the remainder of the vallecula. Projecting laterally is a slender lobule, the *paraflocculus*, attached mainly to the pyramid. It lies beneath the flocculus. Slender in man, it is relatively enormous in the marine mammals (it may be concerned with preserving rotary stability about the long axis of the body).

Morphology of the Cerebellum

Study of the history of investigations into the functions of the cerebellum brings to light a series of theories, each one replacing former ideas. Much remains to be known; current ideas, based on morphological data, provide the most convincing theories so far advanced. The cerebellum is imagined to have three distinct morphological parts, evolved sequentially, and *possessing different functions.*

The *archæcerebellum* has vestibular connexions only. It is represented in mammals by the lingula, the uvula and the flocculo-nodular lobule. Lesions of this part produce vestibular symptoms; disturbances of equilibrium with no alteration to spinal reflexes. Fine movements are well performed, there is no nystagmus, no tremor and no alteration of muscle reflexes. The condition is known as trunk ataxia—the victim walks as if drunk.

The *palæocerebellum*, superimposed dorsally, divides the archæcerebellum into superior and inferior parts. It has spinal connexions (spino-cerebellar tracts), and is concerned with postural and righting reflexes. It is represented in mammals by the anterior lobe, the pyramid and the paraflocculus. Lesions of this part cause great disturbance of postural mechanisms with increased muscle reflexes.

The *neocerebellum*, superimposed dorsally upon the palæocerebellum, separates the anterior lobe from the pyramid and paraflocculus. It comprises the whole of the cerebellar hemispheres behind the primary fissure (i.e., it is synonymous with the posterior lobe, as distinct from the inferior vermis). It appears with the neopallium and is best developed in man. It has cerebro-pontine connexions from the pontine nuclei via the middle peduncle. It is concerned in feed-back circuits with the basal ganglia and the cerebral cortex. It functions in the control of the synergic background of muscle tone in the performance of accurate voluntary movements (p. 8). Lesions of the neocerebellum lead to hypotonia, diminished or pendulum muscle jerks, intention tremor, adiadokokinesia, nystagmus, etc.

Internal Structure (Fig. 7.41)

Like the cerebrum, the cerebellum is surfaced with a cortex of grey matter, with the white matter internal. Each hemisphere contains nuclei of grey matter near the roof of the fourth ventricle. The **dentate nucleus** is a large crenated crescent, resembling the olive in the medulla, open towards the superior peduncle. Its main connexions are from the neocerebellum, and its efferent fibres leave the hilum and pass to the contralateral red nucleus and thalamus. Three small masses lie medial to the hilum of the dentate nucleus, called the nuclei *emboliformis*, *globosus* and *fastigii* (in that order, E.G.F. from lateral to medial). From their proximity to the fourth ventricle these nuclei are known collectively as the **roof nuclei** of the cerebellum. The nucleus fastigii belongs to the archæcerebellum, the other two to the palæocerebellum.

Microscopic Anatomy

The cerebellar cortex is very characteristic and is of *identical appearance in all areas.* Two equally thick cortical layers sandwich a single layer of Purkinje cells between them. The *molecular layer* of the cortex lies on the surface; it consists almost entirely of fibres, with a few cells scattered among them. These are called *basket cells* because their axons arborize in basket shape around the Purkinje cell bodies. The *granular layer* lies deep to the molecular layer and consists overwhelmingly of

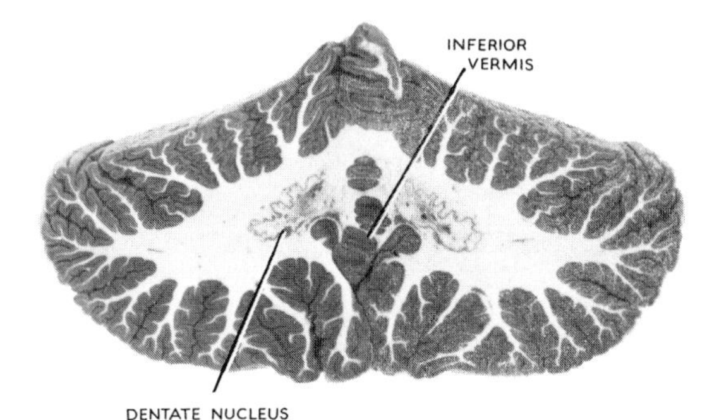

FIG. 7.41. A CORONAL SECTION OF THE CEREBELLUM, VIEWED FROM IN FRONT. (Photograph.)

small round granular cells tightly packed together (in low power appearance somewhat resembling the lymphocytes in lymphatic tissue). The layer of *Purkinje cells* lies between the two and consists of very large flask-shaped cells lying separately at intervals.

The microscopic connexions of the cells are relatively simple. Incoming (afferent) fibres are of only two kinds. Both activate the Purkinje cells, one (the climbing fibre) directly, the other (the moss fibre) through the intermediary of granular and basket cells. The Purkinje axons (efferent) synapse in the dentate nucleus.

The *Purkinje dendrites* form an elaborate branching pattern in the molecular layer. All the branches lie in one plane, like an ornamental tree or the cross-beams on a telegraph pole. The plane of the dendrites lies at right angles to the long axis of a cerebellar folium. Adjacent dendrites lie crosswise along a folium, like telegraph poles along a roadside. The *Purkinje axons* pass through the granular layer, and relay in the dentate nucleus. Thence fibres run in the superior peduncle to the red nucleus, thalamus and cerebral cortex of the opposite side.

The *climbing fibres* (the terminals of pontine and vestibular fibres) pass from the white matter through the granular layer and weave around the Purkinje dendrites in the molecular layer.

The *moss fibres* (terminals of spino- and olivo-cerebellar fibres) arborize as a tuft around a granular cell. The axon of the granular cell passes out into the molecular layer and bifurcates in T-shaped manner. Some of these fibres arborize around the few basket cells, the majority arborize directly with the Purkinje dendrites. The axon of the basket cell divides and arborizes around the cell bodies of some 500 Purkinje cells.

Note that a climbing fibre directly activates one Purkinje cell, while a moss fibre indirectly activates many hundred Purkinje cells.

Cerebellar Connexions

The superior and middle peduncles are simple; the inferior peduncle contains a great mixture of fibres.

The *superior peduncle* contains a preponderance of efferent fibres, passing from the dentate nucleus to the red nucleus, thalamus and cortex of the opposite side. Its one afferent tract of note is the anterior spino-cerebellar tract, passing to the anterior lobe (palæocerebellum).

The *middle peduncle* contains fibres from the pontine nuclei of the opposite side; they are afferent to the neocerebellum.

The *inferior peduncle* is predominantly afferent. The only efferent tract of note is the cerebello-vestibular tract, from the roof nuclei to the vestibular nucleus of the same side. The afferent fibres to the vermis (archæcerebellum) form the vestibulo-cerebellar tract, from the vestibular nucleus of the same side. Those to the anterior lobe (palæocerebellum) comprise the posterior spino-cerebellar tract and the parolivo-cerebellar tract. Those to the posterior lobe (neocerebellum) form the olivo-cerebellar tract (from the olive of the opposite side), the *posterior external arcuate fibres* from the gracile and cuneate nuclei of both sides, and the anterior external arcuate fibres from pontine nuclei.

Blood Supply of the Cerebellum

Two arteries supply the large convex under surface and one artery supplies the small upper surface of each cerebellar hemisphere. They all anastomose with each other on the cerebellar surface, but their perforating branches into the cerebellum are, as elsewhere in the C.N.S., end arteries.

The **posterior inferior cerebellar artery** is one of the most tortuous arteries in the body. It is the largest branch of the vertebral artery. It arises ventrally, near the lower end of the olive, and spirals back around the medulla below the hypoglossal rootlets and then between the rootlets of the glosso-pharyngeal and vagus nerves. It supplies the choroid plexus of the fourth ventricle and is distributed to the vallecula and the back of the cerebellar hemispheres. It supplies, in passing, the adjacent part of the medulla, which contains among other tracts the nucleus ambiguus, the spinal tract of the fifth, and the spinal lemniscus. *Thrombosis* of the posterior inferior cerebellar artery thus gives a distinct clinical picture. There is unilateral 'bulbar palsy' (loss of nucleus ambiguus). Paralysis of the semi-abducted vocal fold of that side causes dysphonia. Paralysis of the soft palate and pharyngeal muscles of that side causes dysphagia—swallowed fluids gush out through the nose. Loss of the uncrossed spinal tract of V and of the crossed spinal lemniscus results in loss of pain and thermal sensibility of the same side of the face and of the opposite half of the body.

The *anterior inferior cerebellar artery* arises from the basilar artery at the lower part of the pons and passes back on the inferior surface of the cerebellar hemisphere, supplying this surface and the adjacent flocculus. It commonly gives rise to the labyrinthine artery.

The *superior cerebellar artery* arises near the termination of the basilar artery and passes laterally to wind around the cerebral peduncle below the fourth nerve. It is distributed over the superior surface of the cerebellum.

Venous drainage is from the surface of the cerebellum into the nearest available venous sinus of the dura mater. Thus the superior and posterior surfaces drain into the straight and transverse sinuses, inferior surfaces into the inferior petrosal, sigmoid and occipital sinuses. The superior vermis drains anteriorly into the great cerebral vein at its entrance into the straight sinus.

THE SPINAL CORD

The spinal cord is a cylinder, somewhat flattened from front to back, whose lower end tapers into a cone. Ventrally it possesses a deep midline groove, the *anterior median sulcus*, and dorsally it shows a shallow sulcus, from which a *posterior median septum* of neuroglia extends into its substance. The posterior median septum within the spinal cord is attached to the incomplete posterior median septum of arachnoid in the subarachnoid space.

In the fœtus the spinal cord extends to the lower limit of the spinal dura mater at the level of the second sacral vertebra. The spinal dura remains attached at this level throughout life, but the spinal cord becomes relatively shorter, which is to say that the bony spinal column and the dura mater grow more rapidly than the spinal cord. Thus at birth the conus medullaris lies opposite the third lumbar vertebra and does not reach its permanent level opposite L1 or L2 until the age of twenty years. The spinal nerve roots, especially those of the lumbar and sacral segments, thus come to slope more and more steeply downwards.

The spinal cord possesses two symmetrical enlargements which occupy the segments of the limb plexuses. That for the brachial plexus is known as the **cervical enlargement** and that for the lumbo-sacral plexus as the **lumbar enlargement.** They occupy, *in the cord*, the segmental levels of the plexuses concerned (C5 to Th. 1 for the cervical enlargment and L2 to S3 for the lumbar enlargement), but their levels measured by vertebræ are, of course, quite different. Thus the cervical enlargment lies roughly corresponding to the vertebræ (C3 to Th. 1), but the lumbar enlargment extends only from Th. 9 to L1. Both cervical and lumbar enlargements are due to the greatly increased mass of motor cells in the anterior columns of grey matter in these situations (*v. inf.*).

The Spinal Nerve Roots. No spinal nerves lie inside the spinal theca; indeed, no nerve lies, strictly speaking, within the spinal canal. The anterior and posterior *roots* of the spinal nerves unite within the intervertebral foramina. Within the subarachnoid space the anterior and posterior nerve roots are attached to the spinal cord each by a series of rootlets. Each anterior root is formed by three or four rootlets which emerge irregularly along the antero-lateral surface of the spinal cord. Each posterior root is formed by several rootlets, attached vertically to the postero-lateral surface of the cord. A short distance from the cord the rootlets are found combined into a single root (Fig. 6.96, p. 492). The anterior and posterior roots pass from the cord to their appropriate intervertebral foramina, where each evaginates the dura mater separately before uniting to form the mixed spinal nerve. The **ganglion** on the posterior nerve root lies, in the intervertebral foramen, within the little tubular evagination of dura mater immediately proximal to the point of union of anterior and posterior nerve roots. For all levels from C1 to L1 the anterior and posterior nerve roots pass in front and behind the ligamentum denticulatum, and evaginate the dura mater between the denticulations. Below L1 the anterior and posterior nerve roots pass almost vertically downwards through the subarachnoid space and form, with the centrally disposed filum terminale of pia mater, the **cauda equina.** This contains, as explained above, only anterior and posterior nerve roots; the posterior root ganglia and spinal nerves occupy evaginations of dura mater within the intervertebral foramina (Fig. 6.98, p. 494).

The posterior root ganglia of the cervical nerves lie lateral to the intervertebral foramina, in contact with the vertebral artery (Fig. 6.96, p. 492).

In conformity with the shortness of the spinal cord, the lower a nerve root the more steeply it slopes down to its intervertebral foramen. The upper cervical roots are horizontal, the lumbar and sacral roots almost vertical. The upper few thoracic roots slope down to their point of evagination of the spinal dura only to become kinked upwards at an angle to reach their foramen (Fig. 7.42).

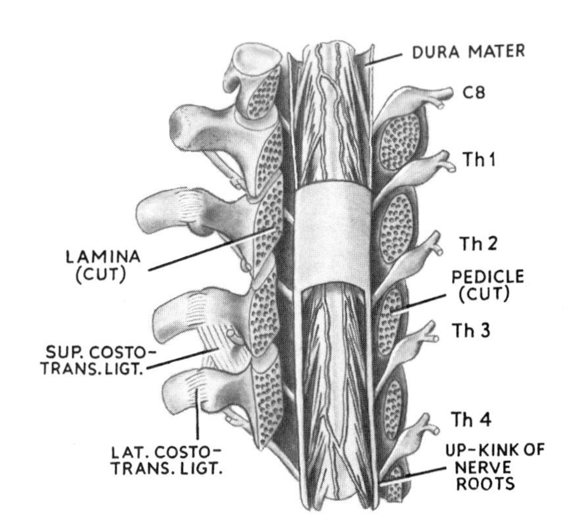

FIG. 7.42. THE ACUTE ANGULATION OF THE UPPER THORACIC NERVE ROOTS AT THEIR PASSAGE THROUGH THE SPINAL THECA.

The roots of the **cervical part of the accessory (XI) nerve** emerge from the lateral surface of the upper five segments of the cord, *behind the ligamentum denticulatum.* They unite into a single trunk which passes upwards through the foramen magnum into the cranium (Fig. 6.96, p. 492).

Blood Supply of the Spinal Cord

The spinal cord is supplied by anterior and posterior spinal arteries which descend from the level of the foramen magnum. The anterior spinal artery is the larger.

The **posterior spinal arteries** consist of one or two vessels on each side which branch from the posterior inferior cerebellar or vertebral artery at the foramen magnum. They supply the posterior columns, both grey and white, of the spinal cord. They receive a 'booster' supply from spinal branches that enter Th. 1 and Th. 11 intervertebral foramina and are reinforced at the conus medullaris by anastomosis with the anterior spinal artery.

The **anterior spinal artery** is a midline vessel that lies on the anterior median fissure. It is formed at the foramen magnum by union of two arteries, one from each vertebral artery, and divides at the conus medullaris into two branches which pass upwards to reinforce the weaker posterior spinal arteries. It supplies the whole cord anterior to the posterior grey columns, namely, the lateral columns and the anterior grey and white columns. Spinal branches of the trunk arteries run along the anterior nerve roots from their entrance through the intervertebral foramina; those at Th. 1 and Th. 11 are especially large and are known as the arteries of Adamkiewicz.

The Arteries of Adamkiewicz. Spinal branches from the first and eleventh intercostal arteries are especially large. They pass along the nerve roots to the spinal cord and reinforce the anterior and posterior spinal arteries, especially the former. The spinal arteries at the foramen magnum are not themselves able to supply the whole length of the cord. The nerve root arteries at Th. 1 and Th. 11 provide an indispensable booster input (the radicular arteries at all other levels supply only the nerve roots, and have no effective anastomosis with the spinal arteries).

The artery at Th. 11 supplies the spinal cord upwards and downwards; its damage (as from fracture of the spine) may result in softening of several segments of the cord above the level of Th. 11. The artery of Th. 1 supplies the cord *only downwards from this level.* It communicates with the anterior spinal artery in a kind of valvular manner peculiar to itself, and its blood will not flow upwards along the anterior spinal artery. Consequently if the anterior spinal artery is interrupted anywhere between the foramen magnum and C8 the greatest ischæmia results at C8, furthest from the point of interruption; below this level the artery of Adamkiewicz maintains the blood supply of the cord. Symptoms of a lesion at C8 may thus be produced by a vascular interruption from any cause at any level from C1 to C8 itself.

Spinal Veins. As is usual in the body, even when arterial input is by end arteries, the emerging veins anastomose freely. The spinal veins form loose-knit plexuses anteriorly and posteriorly. On each side the posterior spinal veins are double, straddling the posterior nerve roots. Both anterior and posterior spinal veins drain along the nerve roots through the intervertebral foramina and so into the segmental veins (vertebral veins in the neck, azygos veins in the thorax, lumbar veins in the lumbar region, lateral sacral veins in the sacral region). At the foramen magnum they communicate with the veins of the medulla.

Internal Structure of the Spinal Cord

The spinal cord consists of a central mass of grey matter, in the form of a fluted column surrounding the central canal, enclosed in a cylindrical mass of white matter. It is almost divided into two halves by the anterior median fissure and the posterior median septum. The latter extends forward as far as the grey matter of the grey commissure, which contains the central canal. The anterior fissure does not completely separate the white matter—a narrow white commissure lies anterior to the grey (Fig. 7.43).

The Grey Matter. The fluted column is like an H-girder. In section it shows an anterior and a posterior grey horn. The former falls short of the surface of the cord, while the latter is connected to the surface by the gelatinous substance. Elsewhere than in the enlargements (e.g., in the thoracic region) the anterior and posterior columns lie in line and the grey matter is truly H-shaped (Fig. 7.43). In the enlargements the *anterior grey column* is much broadened, so that its lateral extent possesses actually a *posterior border* (Fig. 7.44). The medial part of the anterior column is concerned with the innervation of the longitudinal flexor and extensor muscles of the trunk—the lateral extensions in the enlargements contain the motor cells for the muscles of the limbs.

Between the limb enlargments (actually from Th. 1 to L2) there is a *lateral column* of grey matter midway between anterior and posterior grey columns (Fig. 7.43). This slight lateral projection, or *lateral horn*, contains groups of small cells which are connector cells of the sympathetic part of the autonomic nervous system (Fig. 1.38, p. 32). Their axons pass out in the anterior (motor) roots and enter the mixed spinal nerves from Th. 1 to L2 which they leave as the white rami communicantes passing to the ganglionated sympathetic trunks.

The *posterior columns* of grey matter contain nerve cells which are for the most part connector cells (internuncial neurones) of spinal reflex arcs. The *substantia gelatinosa* at the tip of the posterior column, together with the overlying grey matter on its surface (large cells here) is concerned with the transmission of pain im-

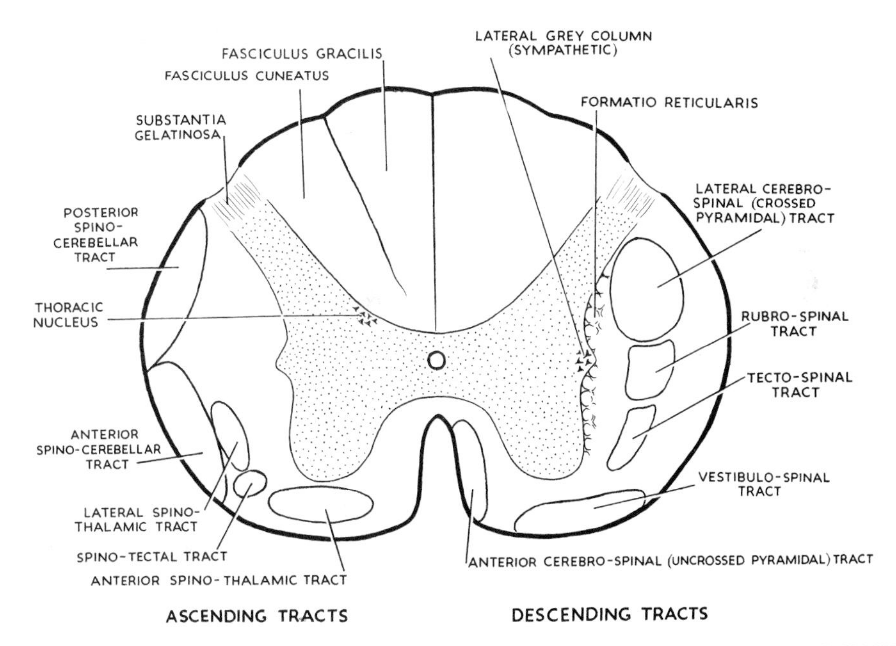

FIG. 7.43. CROSS SECTION OF THE SPINAL CORD TO SHOW THE MAIN ASCENDING AND DESCENDING TRACTS.

pulses. One group of cells, the *thoracic nucleus* (Clarke's column), lies near the root of the posterior column; these cells are the second neurones on the spino-cerebellar pathway (Fig. 7.43).

The *grey commissure* connects the grey matter in the right and left halves of the cord. It is the limb of the H seen in cross section. Centrally it contains the *central canal* of the spinal cord. This tiny tube lined with ependyma (a columnar epithelium with stereocilia) is the downward continuation of the cavity of the fourth ventricle. It extends, surrounded by a little grey matter, into the upper few centimetres of the filum terminale.

The White Matter. The white matter of the spinal cord contains three kinds of fibres, ascending, descending, and intersegmental (or connecting). The white matter is divided into three main columns by the fluted grey matter and the attached nerve roots (Fig. 7.43).

The posterior white columns lie between the posterior grey horn and the posterior median septum. The lateral white columns lie between the anterior and posterior grey columns, and the anterior white columns lie between the anterior grey columns and the anterior median fissure. The anterior white columns are joined by the white commissure. The posterior white columns contain ascending (afferent or sensory) fibres, the anterior white columns contain descending (efferent or motor) fibres while the lateral white columns contain a mixture of the two, but mostly descending (efferent or motor).

In addition, intersegmental connecting fibres run up and down between various segments of the cord. Interspersed with a certain amount of grey matter they produce the appearance of a network under low magnification. This is known as the *formatio reticularis*; it lies lateral to the grey matter and is best marked in the cervical region of the cord. It is the downward continuation of a similar formation throughout the brain stem (p. 525).

The **posterior white columns** are wholly occupied by ascending fibres whose cell bodies lie in the posterior root ganglia of the spinal nerves. They convey 'normal' sensations, that is to say, ordinary cutaneous sensibility, moderate degrees of temperature variation (warmth and coolness) and proprioceptive impulses from joints, ligaments, tendons and muscles. The fibres from the lowest parts of the body (segmentally speaking) lie nearest the midline and from these levels upwards incoming fibres are placed progressively to the lateral side. In this way fibres from the perineum and lower limb form a slender column (the fasciculus gracilis or column of Goll) alongside the midline. Above the lumbar enlargment the incoming fibres from trunk and upper limb form a new column lateral to this. The column is pointed below and this wedge-shape gives it the name of fasciculus cuneatus (column of Burdach). The two columns end at the posterior border of the fourth ventricle by each arborizing around a group of cells that form the nucleus gracilis and nucleus cuneatus. (From these nuclei the medial lemniscus, having crossed the midline in the sensory decussation, passes to the thalamus for relay to the sensory cortex.)

The **anterior white columns** contain uncrossed pyramidal fibres (cell bodies in area 4 of the precentral gyrus) and the vestibulo-spinal fibres (cell bodies in the vestibular nucleus in the floor of the fourth ventricle).

These descending fibres arborize around motor cells of the anterior grey columns, the pyramidal fibres having previously decussated in the white commissure.

Near the antero-lateral surface of the cord, crossed by the anterior nerve roots, is the *anterior spino-thalamic tract*, carrying crude touch and pressure impulses from the opposite side of the body. These are second order neurones, their cell bodies lying in and on the gelatinous substance; from these cell bodies the axons *cross obliquely* in the grey commissure to the antero-lateral surface of the cord. They pass to the thalamus and relay to the cortex.

The **lateral white columns** contain both ascending and descending fibres (most of the latter are crossed). In addition, intersegmental fibres lie medially, in the formatio reticularis alongside the crescent of grey matter. In general, the descending fibres lie deeply, the ascending fibres lie on the surface of the lateral column; but there is much intermingling of adjacent tracts (Fig. 7.43).

The **descending fibres** consist of the *crossed pyramidal tract* (cortico-spinal tract) lying well posteriorly, against the posterior grey horn, and the *rubro-spinal*, *tecto-spinal* and *vestibulo-spinal tracts* lying in alphabetical order in front of the pyramidal tracts, the last named reaching the surface of the cord at the line of emergence of the anterior nerve rootlets and intermingling with the anterior spino-thalamic tract. The human rubro-spinal tract is said to diminish rapidly at the upper part of the cord. This is not to say that the red nucleus has no connexions with the motor cells of the lower spinal segments. On the contrary, it is connected to them by intersegmental relay through the formatio reticularis.

The *spinal sympathetic tract* passes from cell bodies in the hypothalamic nuclei, and runs down the cord on the medial side of the cortico-spinal tract, to synapse with the connector cells in the lateral grey columns.

The **ascending fibres** consist of *spino-cerebellar tracts*, anterior and posterior, occupying almost the entire surface of this part of the cord and, just deep to the anterior tract, at the antero-lateral part of the cord, the *lateral spino-thalamic tract* mediating pain and temperature extremes from the opposite side of the body.

The **appearance in cross-section** of the cord is different in the cervical, thoracic and lumbar regions (Fig. 7.44). The higher the level of section the greater is the amount of white matter, for both ascending and descending tracts leave the cord in each spinal nerve. The amount of grey matter is greatest in the anterior horns at the limb enlargments; the shape of the 'H' in cross section serves to distinguish the regions of the cord from each other.

In the *cervical region* below C3 the anterior grey horn

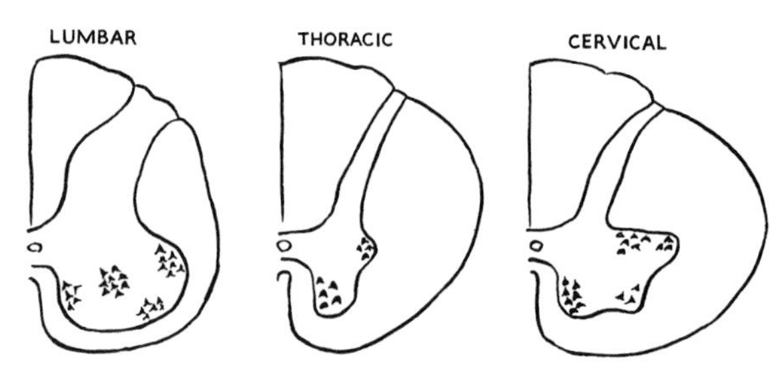

FIG. 7.44. THE SPINAL CORD IN CROSS SECTION IN THE LUMBAR, THORACIC AND CERVICAL REGIONS.

extends far laterally into the white matter. At C4 lie the motor cells for the diaphragm (phrenic nerve), while below this level the lateral extension of the anterior horn contains the motor cells for the upper limb muscles (brachial plexus). The posterior grey horns are slender. There is much white matter.

In the *thoracic region* the anterior grey horn is slender and in line with the posterior grey horn; the grey matter is truly H-shaped in cross-section. The *lateral horn* projects slightly from the grey matter.

In the *lumbar enlargement* there is more grey matter than white matter in the cross-section, and both posterior and anterior grey horns have bulging convex outlines.

	Lumbar	*Thoracic*	*Cervical*
Posterior horn	Bulbous	Slender	Slender
Lateral horn	Absent	Present	Absent
Anterior horn	Bulbous	Slender	Broad

Summary of the Projection Fibres

The situation in the cord of the principal descending, ascending and intersegmental tracts has been indicated in the preceding pages. To know the situation of a nerve fibre is, however, meaningless unless the situation of its cell body is also known; you should train yourself to think in whole neurones. For this reason a brief account of the course and destination of the principal pathways is given here.

Afferent impulses are conveyed to the cerebral cortex by three neurones. The cell body of the first neurone lies in the posterior root ganglion of a spinal nerve (or the sensory ganglion of a cranial nerve). The cell body of the second neurone lies in the grey matter of the spinal cord (or brain stem) and its central process goes to the opposite thalamus. The cell body of the third neurone lies in the thalamus, whence its axon passes through the internal capsule to the cortex. (There is an odd exception; the cell bodies of the first proprioceptive neurone of the trigeminal nerve do not lie in the trigeminal ganglion, but are buried in the brain stem as the mesencephalic nucleus of the trigeminal nerve.)

Efferent impulses from the *motor cortex* are conveyed to the striated muscles by two neurones, the upper and lower motor neurones. The upper motor neurones decussate and pass to the motor nuclei of the cranial nerves and the anterior horn cells of the spinal nerves, whence the lower motor neurones pass into the motor nerves of the muscles concerned. Efferent impulses from *other cortical areas* (the extrapyramidal system) are mediated by several neurones. The cell bodies of the intermediate neurones lie in the brain stem (tectum, red nucleus, vestibular nucleus, olive, etc.) and are acted upon by the basal ganglia and cerebellum as well as the cerebral cortex.

Fibres descending from the cortex, both pyramidal and extrapyramidal, cross to motor nuclei lying near the ventral midline. The motor roots from these nuclei leave the brain stem and cord near the ventral midline. (There is one odd exception; the fourth cranial nerve after leaving its nucleus, decussates and emerges posteriorly from the brain stem.)

Ascending Tracts. All incoming fibres are destined for nerve cells in either (*a*) the cortex of the opposite cerebral hemisphere, via thalamic relay, for consciousness of sensation, (*b*) the cerebellum, for balanced muscle actions, or (*c*) the brain stem or spinal cord for reflex actions.

(*a*) Fibres destined for the opposite thalamus and **cortex** convey sensations which run two different courses in their ascent to the thalamus. They are:

1. Kinæsthetic (joint and muscle sense, spatial appreciation, vibration sense), light discriminative touch and moderate temperature variation (warmth and coolness). These forms of sensibility are conveyed by large, well-myelinated fibres that enter on the medial side of the posterior nerve rootlets and turn up in the *posterior white columns*, where the lowest fibres to enter lie most medially. They relay in the medulla in the gracile and cuneate nuclei. From these nuclei some second neurone fibres pass to the cerebellum, but the vast majority, decussating in the upper medulla, ascend in the *medial lemniscus* through the pons and tegmentum of the midbrain to the opposite *thalamus*. After relay in the lateral part of the thalamus the axon of the third neurone passes through the *internal capsule* to the *post-central gyrus*.

2. Crude touch, temperature extremes and pain. These forms of sensibility are conveyed by smaller, less myelinated fibres. Pain fibres are almost non-myelinated and lie laterally in the posterior nerve rootlets. These fibres relay in the dorsal grey matter at the tip of the posterior grey horn, whence the axons of the second neurones slope obliquely upwards, decussate in the grey matter in front of and behind the central canal and pass into the anterior and lateral *spino-thalamic tracts*. The lateral spino-thalamic tract conveys pain and extreme heat and cold.

As these fibres decussate they lie in different obliquities, the pain fibres least of all, the temperature fibres next and the crude touch fibres most oblique of all. Thus clinical investigation of pain in cord disease gives more accurate segmental localization than investigation of temperature and crude touch. The proximity of these spino-thalamic fibres to the central canal renders them vulnerable in syringomyelia (where loss of pain and heat sensibility may result in burning of skin which has normal sensation to light touch).

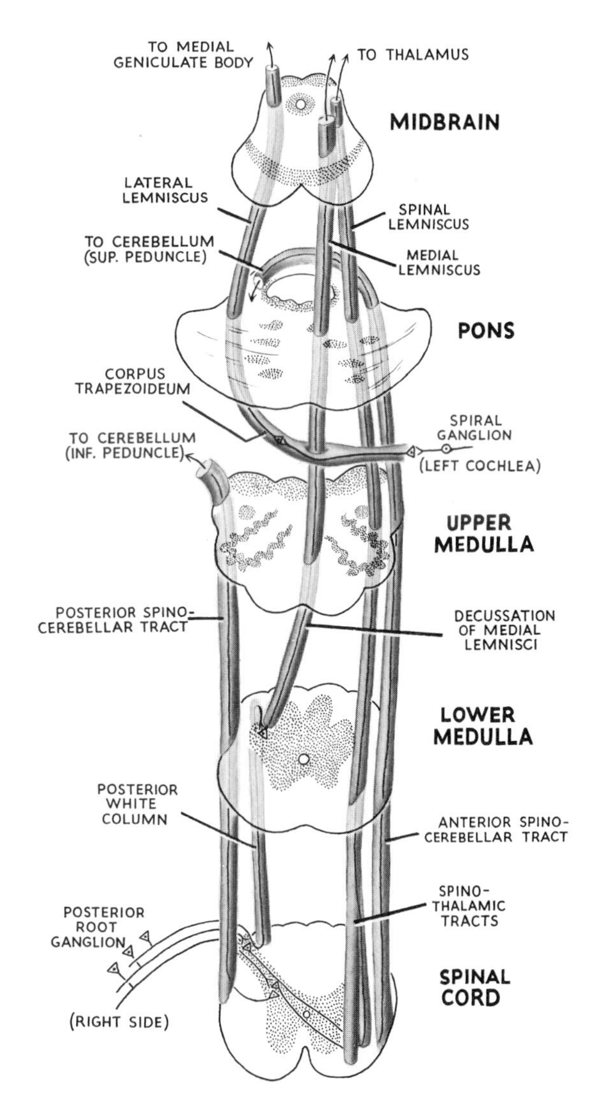

FIG. 7.45. DIAGRAM (not to scale) OF THE CHIEF ASCENDING PATHWAYS. THE SENSORY PATHWAYS FROM THE RIGHT SIDE AND THE AUDITORY PATHWAY (much simplified) FROM THE LEFT SIDE ARE SHOWN. The basic pattern of cortical pathways is that the cell body of the first neurone lies in a peripheral ganglion, and that the central process of the *second* neurone crosses the midline to ascend to the thalamus for relay to the cortex. Cerebellar pathways, on the other hand, are ipsilateral.

The anterior and lateral spino-thalamic tracts pass upwards in the lateral part of the medulla (they are thus blocked in thrombosis of the posterior inferior cerebellar artery, which supplies this part). They are known together as the *spinal lemniscus*; they join the medial lemniscus and pass with it to relay in the *thalamus*, whence third neurone fibres pass through the *internal capsule* to the *postcentral gyrus*.

(*b*) Fibres destined for the **cerebellum** (carrying proprioceptive impulses) pass in small number in the posterior columns, relay in the cuneate and gracile nuclei and enter the inferior cerebellar peduncle. The vast majority of the spino-cerebellar fibres, however, take a quite different course. Passing into the posterior horn of grey matter they synapse in the *thoracic nucleus* (Clarke's column), which lies in the base of the posterior horn (Fig. 7.43). From these cell bodies the axons of the second neurones enter the dorsal (direct or uncrossed) spino-cerebellar tract of their own side or cross to the anterior spino-cerebellar tract of the opposite side. The anterior spino-cerebellar tract passes into the superior cerebellar peduncle and (probably) crosses to the opposite cerebellar hemisphere, the posterior (dorsal) spino-cerebellar tract passes into the inferior cerebellar peduncle and does not cross (Fig. 7.45). Both tracts are received into the *anterior lobe* of the cerebellum (p. 532), i.e., the palæocerebellum.

(*c*) Fibres destined for the **brain stem** pass upwards in spino-tectal and spino-vestibular and other tracts, while those for the **spinal cord** relay in the posterior horn of grey matter and by internuncial or intersegmental neurones are connected to anterior horn cells of their own or other segments for the purpose of spinal reflexes.

Descending Tracts. Fibres descend to the motor nuclei of the brain stem and of the spinal cord from the cerebral cortex in two main systems, the pyramidal and the extrapyramidal. The pyramidal fibres pass directly to the cell bodies of the lower motor neurones, while the extrapyramidal fibres are interrupted by relays *en route* (see p. 8).

The **pyramidal fibres** run from the precentral gyrus and the front of the central sulcus (area 4, the motor area) through the corona radiata of the cerebral hemisphere to the internal capsule, beneath the thalamus, into the crus cerebri of the midbrain, through the pons (here intersected by the transversely running ponto-cerebellar fibres of the middle cerebellar peduncle) into the pyramids of the upper medulla. Most of them decussate in the lower medulla; these fibres form the crossed pyramidal tract, which lies in the posterior part of the lateral column of white matter in the spinal cord. The uncrossed pyramidal fibres descend in the anterior column of white matter; they ultimately cross in the anterior (white) commissure to arborize around anterior horn cells of the opposite side.

The **extrapyramidal fibres** run from the frontal and temporal cortex through the internal capsule and relay at different levels. Most of them synapse with the pontine nuclei and so pass to the cerebellum. From the cerebellum fibres pass to the red, reticular, vestibular and olivary nuclei. From the cortex other fibres pass to the basal ganglia (corpus striatum and thalamus) and thence to the red and reticular nuclei. These brain stem nuclei are also activated by (reflex) sensory pathways. The tectal nuclei are activated reflexly by light and sound.

From all these brain stem nuclei, fibres of the extrapyramidal system descend to synapse with the lower motor neurones. The descending tracts lie in the lateral columns of white matter in the spinal cord, anterior to the crossed pyramidal tracts.

DEVELOPMENT OF THE NERVOUS SYSTEM

The nervous system is entirely *ectodermal* in origin. A **neural groove** forms in the antero-posterior midline of the dorsal surface of the embryonic plate. This is the amniotic, or ectodermal, surface of the embryo. The groove folds into the **neural tube,** which becomes depressed below the surface. Some of the ectodermal cells are left isolated between neural tube and overlying ectoderm; they form the **neural crest** and are destined to migrate and become the cell bodies in sensory ganglia or in autonomic ganglia (Fig. 7.46). The cells of the suprarenal medulla are also derived from the neural crest.

The cranial end of the neural tube becomes dilated into vesicles and its walls thicken by proliferation of cells; the cerebral hemispheres, brain stem, and cerebellum are so developed. More caudally the neural tube enlarges in a simple manner by proliferation of cells, to form the spinal cord. In all regions these proliferating cells arrange themselves reguarly in *functional groups*.

Spinal Cord. The central canal, relatively very large at first, is not rounded in cross-section, but is projected laterally into a groove on the inner wall of the spinal cord known as the **sulcus limitans.** The inner wall of the spinal cord is separated by the sulcus limitans into a dorsal part, the alar lamina and a ventral part, the basal lamina. The **alar lamina** contains sensory (afferent) cells and the **basal lamina** contains motor (efferent) cells. In each lamina the cells are of two kinds; near the sulcus limitans lie the autonomic (**visceral**) cells, while further away lie the body wall and limb (**somatic**) cells (Fig. 7.46).

Brain Stem. A similar arrangement holds in the

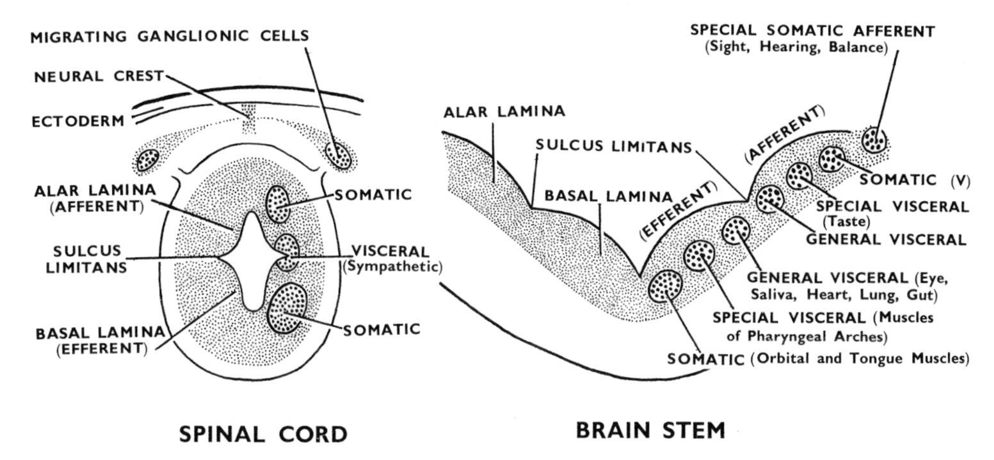

FIG. 7.46. DIAGRAMMATIC CROSS SECTIONS TO SHOW THE AFFERENT AND EFFERENT CELL GROUPS IN THE DEVELOPING SPINAL CORD AND THE SIMILAR ARRANGEMENT IN THE OPEN PART OF THE BRAIN STEM (FOURTH VENTRICLE).

brain stem as in the spinal cord. But here a third type of cell appears in each lamina, namely the special **branchial** afferent and efferent cells of cranial nerves supplying the derivatives of the pharyngeal arches (p. 40). These branchial cells lie between the autonomic and somatic cells of each lamina. They are the central cell stations of the nerves of the pharyngeal arches (V, VII, IX and X).

In the **fourth ventricle** the central canal is opened out, and basal and alar laminæ lie roughly in the same plane, but still separated by the upward continuation of the sulcus limitans. The dorsal afferent and ventral

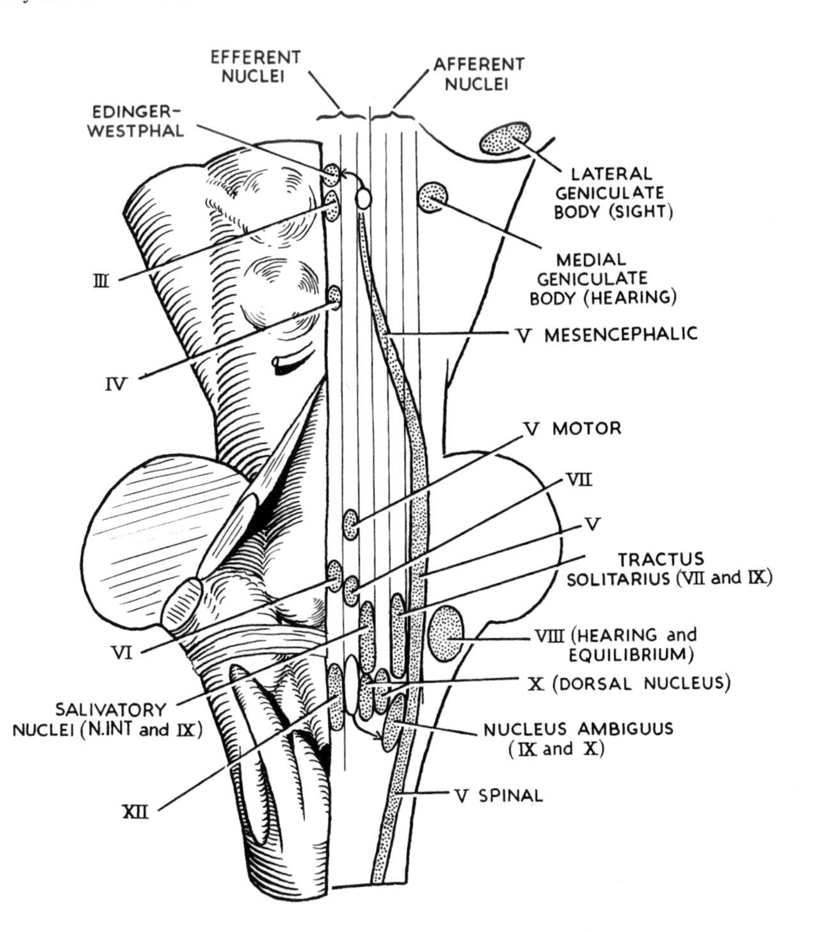

FIG. 7.47. THE SITES OF THE NUCLEI OF THE CRANIAL NERVES. The nuclei occupy afferent and efferent columns as shown in Fig. 7.46. Dorsal view of brain stem (diagrammatic).

efferent cells here become lateral and medial respectively. The order of the cell groups is basically the same, namely three longitudinal columns of nuclei in each part, motor and sensory, of the brain stem (Fig. 7.46), but migration of certain cell groups (neurobiotaxis) in the developing brain stem alters, in places, the relatively simple basic arrangement.

Motor Nuclei of the Brain Stem. The motor nuclei are arranged according to the type of muscle they supply. The ordinary skeletal muscle of the head (somatic muscle) consists of the muscles of the orbit and the muscles of the tongue. In line with the motor cell columns of the spinal cord the brain stem nuclei of these muscles (III, IV, VI and XII) lie near the midline ventral to the 'central canal' (i.e., ventral to the aqueduct or floor of the fourth ventricle as the case may be).

Developed from the region of the embryonic pharynx are the striated muscles of mastication, of the face, and of the pharynx and larynx. Their motor nuclei (branchial) lie slightly more laterally. They comprise the motor nucleus of V, the nucleus of VII and the nucleus ambiguus (Fig. 7.47).

More laterally still lie the autonomic or visceral (splanchnic) motor nuclei, represented by the Edinger-Westphal (parasympathetic) part of III, the secreto-motor nuclei of the nervus intermedius and IX (the salivatory nucleus) and the dorsal nucleus of X (motor to the visceral muscles of thorax and abdomen, i.e., heart, lungs, alimentary canal).

The above basic arrangement is altered somewhat by medial migration of the Edinger-Westphal nucleus cranial to the somatic part of III, and by lateral migration of the nucleus ambiguus (Fig. 7.47), but in general the three lines of nuclei retain their embryonic relationships.

Afferent Nuclei of the Brain Stem. These follow the general plan outlined above, but an extra column of sensory nuclei is imposed upon the three already mentioned. Special sensations are mediated by nuclei which lie most lateral of all; these are sight, hearing and balance. They form the **fourth column,** lateral to the three basic columns already described. The lateral and medial geniculate bodies and the cochlear and vestibular nuclei lie most lateral of all at their respective levels. Next medially lie the ordinary somatic nuclei of skin sensibility; these are the sensory nuclei of V in pons and medulla, in line with the posterior grey column of the spinal cord. More medially still are the branchial afferent nuclei. They subserve taste and are represented by the nucleus of the tractus solitarius. Most medially of all in the sensory column lie the nuclei for reception of visceral sensibility, and they are represented by the dorsal nucleus of IX and X.

SUMMARY OF CRANIAL NERVES

The local features of the cranial nerves have been noted in the regions already studied, but it is useful to summarize each nerve. The following account merely brings into continuity the features already seen in the various regions.

Olfactory Pathways (First cranial nerve)

From the neuro-epithelium in the roof of the nose trace the two neurones to the olfactory cortex. The first neurones are scattered bipolar cells in the neuro-epithelium. Their central processes gather themselves into nearly twenty olfactory nerves that pass through foramina in the cribriform plate and here pierce the dura mater and arachnoid mater. They enter the olfactory bulb to synapse on cell bodies of the second neurone. The central processes of these cell bodies pass in the olfactory tract to the region of the anterior perforated substance (prepiriform cortex and a small part of the amygdaloid nucleus, p. 504). The arrangement is unique and 'primitive'—the second order neurone directly activates the conscious cortex, bypassing the thalamus. Other olfactory pathways, by polysynaptic junctions, activate hypothalamic and brainstem nuclei (as is the case with all sensory pathways) for visceral and somatic effects, distinct from conscious appreciation.

Light Pathways (Second cranial nerve)

From the neuro-epithelium in the retina trace the neurones centrally. The rods and cones, near the choroidal surface of the retina, activate the bipolar cells of the retina; these are the first sensory neurones. Their central processes synapse on 'ganglion' cells that lie scattered on the vitreous surface of the retina. The central processes of these second neurones converge to the optic disc and enter the dura mater and arachnoid mater there. They pass in the optic nerve up to the chiasma where the nasal fibres from each retina decussate (this is a *complete* functional decussation). From the chiasma the optic tract passes around the midbrain (cerebral peduncle) to three destinations, (1) the thalamus for relay to the visual cortex, (2) the pretectal nuclei for pupil constriction to light and (3) the superior colliculus for body reflexes to light.

The **visual pathway** is that part of the optic tract which reaches the lateral geniculate body on the thalamus (Fig. 7.8, p. 503). Here the temporal fibres relay on layers 2, 3 and 5 while the crossed nasal fibres do so on layers 1, 4 and 6. From the lateral geniculate body fibres of the third neurones pass through the posterior limb of the internal capsule and backwards by the optic radiation to the visual cortex on the medial surface of the occipital pole. The visual cortex lies deep in the postcalcarine sulcus, with the macula represented posteriorly. More peripheral parts of the retina register more anteriorly.

Accommodation Reflex. The eye automatically focuses to near vision, and this change in lens curvature is accompanied by pupil constriction to sharpen the focus. Both these reflexes are cortical; that is, they accompany conscious *seeing*. The near object is seen clearly even though it may not be understood (like writing in a foreign—or one's own—language). From the visual cortex an association bundle reaches the posterior end of the middle frontal gyrus. Cell bodies in this cortex send their efferent fibres via the internal capsule down to the Edinger-Westphal nucleus, whence by ciliary ganglion relay they activate the ciliary muscle and the sphincter pupillæ. Note that the pupil constriction which accompanies accommodation to near vision has nothing to do with the amount of light reaching the retina. It reduces spherical aberration by preventing light traversing the periphery of the lens.

Pupillary Light Reflex. Another part of the optic tract (second light neurone) bypasses the thalamus and reaches *each* pretectal nucleus. The cell bodies here send their axons to *each* Edinger-Westphal nucleus. Thus the pupillary light reflex is double-crossed, explaining the consensual pupil reaction to light. Each Edinger-Westphal nucleus, by ciliary ganglion relay, activates the sphincter pupillæ. The Argyll-Robertson pupil (which constricts on accommodation but not to light) can be explained by a lesion in the pair of pretectal nuclei; but there is still doubt about this.

General Light Reflexes. A third part of the optic tract passes in the superior brachium to cell bodies in the superior colliculus. These relay through tecto-bulbar tracts to motor nuclei of the brain stem to move the eyes, eyelids, etc., and by tecto-spinal tracts to spinal anterior horn cells for neck, trunk and limb movements. Thus reflex movements of any part of the body may result. These reflex responses are often bilateral (blinking, throwing up the hands, jumping at a flash of light, etc.) and this is explained by the connexion of the two superior colliculi by the posterior commissure at the entrance to the aqueduct.

Cranial Nerves III to XII

The first two cranial nerves, summarized above, are not really nerves, but rather outdrawn parts of the central nervous system. Only the peripherally scattered bipolar nerve cells correspond with an ordinary nerve. They are sensory, so in each case it is logical to follow the neurones from the localized periphery to their central connexions. However, for the widely distributed sensory nerves of the brain stem this is not practicable, and it is easier to trace V, IX and X from their brain-stem nuclei to the periphery, just as is done with the motor nerves.

The origin, course and distribution of these cranial nerves stand in increasing clinical importance. Knowledge of the nucleus matters least, of the course matters much more, and of the distribution is most important of all (what is lost if the nerve is lost?).

A scheme that takes note of specific landmarks is desirable, and the following summaries of the cranial nerves are based on this uniform pattern:

1. Origin, i.e., the nucleus. Its position and the nature of its cell bodies. (The cranial nuclei are listed together on p. 531.)
2. Transit of brain stem—any special features.
3. Point of attachment to surface of brain stem (Fig. 7.35, p. 528).
4. Course through the basal cistern of C.S.F.
5. Point of simultaneous perforation of arachnoid and dura mater (the ganglion of sensory nerves usually lies just beyond this).
6. Point of bony exit from skull, with details of its course between dura mater and bone (not applicable to the last four pairs, which pierce dura mater and bone in the same place).
7. Extracranial course and distribution. Here it is best to trace the nerve to its terminal destination and *afterwards* recount the details of its branches in their proper order (not applicable to VII, which radiates from the parotid gland).

Oculomotor Nerve (III)

The nucleus lies in the midbrain. It is near the midline, in the grey matter in the floor of the aqueduct, level with the superior colliculi. It is somatic (motor to striated ocular muscles) and visceral (motor to smooth muscle in the eyeball) and this latter, the Edinger-Westphal nucleus, lies cranial to the somatic nucleus. The fibres undergo partial decussation, which probably facilitates bilateral eye movements (looking up or down and convergence). They traverse the brain stem with a lateral convexity, passing through the red nucleus and the substantia nigra (Fig. 7.31, p. 523).

The nerve is attached to the medial surface of the crus (basis pedunculi) near the midline, just above the pons. The nerve is straight and has a gentle downward slope. It passes forwards between the posterior cerebral and superior cerebellar branches of the basilar artery, and traverses the interpeduncular cistern below the floor of the third ventricle (Fig. 7.36, p. 528). Lying below the optic tract it pierces arachnoid and dura mater at the roof of the cavernous sinus just behind the upward bend of the internal carotid artery and just in front of IV (Fig. 6.49, p. 433).

Still straight and sloping downwards it is attached to the lateral wall of the cavernous sinus above IV. It slants down medial to IV and V a (Fig. 6.93, p. 486). It picks up some sympathetic fibres that peel off the cavernous plexus—these are for the smooth-muscle part of levator palpebræ superioris. At the anterior pole of the cavernous sinus it splits into a superior and an inferior division and immediately enters the tendinous ring at the medial end of the superior orbital fissure. This takes it inside the cone of ocular muscles, with the naso-ciliary nerve between its divisions and VI below it (Fig. 6.21, p. 400).

Superior Division. This passes up to enter the ocular surface of superior rectus one-third of the way along. It supplies this muscle and passes on to levator palpebræ superioris. The sympathetic fibres supply the smooth-muscle part of levator (their loss accounts for the partial ptosis characteristic of Horner's syndrome).

Inferior Division. This breaks up quickly into three branches. One supplies medial rectus and another inferior rectus; each of these enters the ocular surface of its muscle one-third of the way along. The third branch, longer, passes forwards into the ocular surface of inferior oblique (Fig. 6.93, p. 486). All the Edinger-Westphal fibres are in this branch. They leave it to enter the ciliary ganglion, where they relay. The cell bodies of the ciliary ganglion supply, through the short ciliary nerves, the sphincter pupillæ (for pupil constriction) and the ciliary muscle (for accommodation).

Trochlear Nerve (IV)

The nucleus is in the midbrain. It is somatic (for the superior oblique muscle) and lies near the midline in the floor of the aqueduct, level with the inferior colliculi (Fig. 7.32, p. 523). The fibres *decussate* completely, dorsal to the aqueduct. This is unique, but the significance is not understood.

The nerve emerges dorsally, near the midline below the inferior colliculus, through the apex of the superior medullary velum (Fig. 7.37, p. 529). Thence it passes around the midbrain just below the free edge of the tentorium cerebelli and runs between the posterior cerebral and superior cerebellar arteries (Fig. 7.36, p. 528). Clinging to the under surface of the free edge of the tentorium cerebelli it is directed thereby to the roof of the cavernous sinus. Here it pierces the arachnoid and dura mater behind III. Running forwards in the lateral wall of the cavernous sinus below III it is then crossed medially by III and so becomes the uppermost nerve in the anterior end of the sinus. It enters the superior orbital fissure lateral to the tendinous ring and passes over levator palpebræ superioris to enter the orbital surface of the superior oblique one-third of the way along the muscle (Fig. 6.49, p. 433).

Trigeminal Nerve (V)

This is both motor and sensory. The **motor nucleus**

is in the upper pons (Fig. 7.33, p. 526). It is branchial, for the muscles of the first pharyngeal arch. It is thus off centre, and lies deep beneath the floor of the fourth ventricle. The **sensory nucleus** extends in continuity throughout the whole length of the brain stem and descends into the upper 2 or 3 segments of the spinal cord. The *mesencephalic part* extends through the whole length of the midbrain, wherein it lies lateral to the aqueduct, in the central grey matter. It is unique in that its cell bodies are those of the first order neurones; their peripheral processes have passed straight through the trigeminal ganglion. They mediate proprioceptive impulses from the fifth nerve muscles and also from those supplied by III, IV, VI, VII and (probably) XII. The *pontine part* is somatic and thus lies far lateral, ventral to the superior cerebellar peduncle. It consists of second order cell bodies which mediate nice sensations (light touch, minor changes of agreeable temperature, etc.) from the whole trigeminal territory via the first order neurones whose cell bodies are in the trigeminal ganglion. The *spinal nucleus* is the name given to the column of second order cell bodies that extends along the whole medulla, from the inferior cerebellar peduncle in the open part, through the closed part and into the substantia gelatinosa on the posterior horns of the upper two or three segments of the cervical cord. It mediates nasty sensations (crude touch and deep pressure, great extremes of temperature, and *pain*). The three divisions of V register on this nucleus upside-down (i.e., V c in the upper medulla, V b in the closed medulla and V a in the upper cervical cord).

The fibres that reach the pontine and spinal nuclei come from cell bodies in the trigeminal ganglion. Together with the direct fibres that reach the mesencephalic nucleus they form a single, large, sensory root attached to the pons, well lateral and just above centre. The motor root emerges separately, slightly cranial and medial to its companion (Fig. 7.6, p. 500). Together they pass, below the tentorium cerebelli, to the mouth of the trigeminal (Meckel's) cave (Fig. 6.94, p. 487). This is a tubular prolongation of arachnoid-lined fibrous dura mater around the sensory and motor roots, and it crosses the upper border of the petrous bone near its apex. The dural sheath containing the two nerve roots passes forwards, peeling apart the two layers of dura that floor the middle cranial fossa, just lateral to where the same two layers peel apart to enclose the cavernous sinus. The sensory root then expands into the large, flat, crescentic *trigeminal ganglion*; the motor root remains separate. The dural sheath obliterates the subarachnoid space by fusing with the pia mater halfway along the ganglion; this is the anterior extremity of Meckel's cave (Fig. 6.93, p. 486). The posterior half of the ganglion and both roots are thus bathed in cerebro-spinal fluid. The anterior half of the ganglion, beyond the subarachnoid space, gives off its three sensory divisions, ophthalmic, maxillary and mandibular. The first two pass forwards in the lateral wall of the cavernous sinus; they are wholly sensory. The mandibular division, likewise sensory, passes straight down from the lower part of the ganglion to the foramen ovale; here it is joined by the motor root.

Ophthalmic Division (V a)

This is the nerve of the fronto-nasal process (p. 45 and Fig. 1.56, p. 44). Leaving the upper part of the ganglion the ophthalmic division runs forward in the lateral wall of the cavernous sinus below IV. Here it picks up sympathetic fibres from the cavernous plexus; these are for the dilator pupillæ muscle. At the anterior end of the cavernous sinus it gives off meningeal branches, the *tentorial nerves*, which supply all the supratentorial dura mater except that in the bony floor of the middle cranial fossa. Finally it breaks into three branches that pass through the superior orbital fissure. These are the lacrimal, frontal and naso-ciliary nerves (Fig. 6.93, p. 486; Fig. 6.21, p. 400).

Lacrimal Nerve. Passing just lateral to the tendinous ring, the nerve proceeds along the upper part of the lateral wall of the orbit, there picking up a secreto-motor branch from the zygomatic nerve which it gives to the lacrimal gland. It is sensory to a fingertip area of skin at the lateral end of the upper eyelid and to both palpebral and ocular surfaces of the corresponding conjunctiva (Fig. 6.16, p. 384).

Frontal Nerve. This leaves the cavernous sinus to traverse the superior orbital fissure just lateral to the tendinous ring, bunched together between the lacrimal and trochlear nerves. A large nerve, it runs forward in contact with orbital periosteum above the levator palpebræ superioris (Fig. 6.49, p. 433). Just behind the superior orbital margin it divides into a large supra-orbital and a small supratrochlear branch (PLATE 32). The *supra-orbital nerve* supplies the frontal sinus, notches or perforates the orbital margin, supplies the upper eyelid (skin and both surfaces of conjunctiva), all the forehead except a central strip, and the frontal scalp up to the vertex. The *supratrochlear nerve* supplies the upper lid and conjunctiva and a narrow strip of forehead skin alongside the midline. It scarcely extends into the scalp, where the two supra-orbital nerves meet each other.

Naso-ciliary Nerve. This is sensory to the whole eyeball, to the paranasal sinuses along the medial wall of the orbit, to some mucous membrane of the nasal cavity and to the skin of the external nose. It carries hitch-hiking sympathetic fibres for the dilator pupillæ muscle.

The nerve leaves the cavernous sinus and enters the tendinous ring between the two divisions of III. It passes straight forwards into the cone of muscles above the optic nerve and divides into two terminal branches. These are the infratrochlear and anterior ethmoidal nerves (Fig. 6.93, p. 486). With the penultimate branch (the posterior ethmoidal) they leave the cone of muscles above the medial rectus and below the superior oblique. They are the only nerves to traverse the cone of muscles, and all three pass through the same intermuscular slit.

The branches of the naso-ciliary nerve are as follows:

1. The *sensory root* of the ciliary ganglion. This passes through the ciliary ganglion and via the short ciliary nerves is sensory to the whole eyeball including the cornea (not conjunctiva).

2. The *long ciliary nerves*. Two branches of the naso-ciliary nerve enter the sclera independently. They carry sympathetic fibres, picked up by V a in the cavernous sinus, to the dilator pupillæ muscle.

3. The *posterior ethmoidal nerve* enters the posterior ethmoidal foramen and supplies the posterior ethmoidal air cells and the adjacent sphenoidal sinus. It does not reach the nasal cavity.

4. The *infratrochlear nerve* passes forward just below the trochlea of the superior oblique tendon, supplies skin and conjunctiva at the medial end of the *upper* lid and ends on the skin over the bridge of the nose (nasal bone: PLATE 32).

5. The *anterior ethmoidal nerve* passes into the anterior ethmoidal foramen below the frontal bone. Running obliquely forwards it is in the roof of the middle and anterior ethmoidal air cells, and supplies both. It passes on to the cribriform plate, between the two layers of dura mater (Fig. 6.89, p. 481). It descends through the nasal slit alongside the front of the crista galli into the roof of the nose. It straddles the nose, supplying the antero-superior quadrant of the lateral wall and the antero-superior half of the nasal septum. It then continues, under the name of *external nasal nerve*, to notch the nasal bone and supply the skin over the external nasal cartilages, down to the tip.

Maxillary Division (V b)

This is the nerve of the maxillary process that differentiates from the first pharyngeal arch (p. 40). Leaving the middle part of the trigeminal ganglion the nerve runs forward in the lateral wall of the cavernous sinus below V a. The lateral wall here fuses with the endosteal layer of dura mater at the lateral margin of the foramen rotundum, and so V b is directed through the foramen into the upper part of the pterygo-palatine fossa (Fig. 6.93, p. 486). It has a short course, below the roof of the fossa, to the inferior orbital fissure where it breaks into its two terminal branches, the zygomatic and infra-orbital nerves.

The branches of the maxillary division are meningeal, ganglionic and posterior superior alveolar nerves. The meningeal branch supplies the dura mater of the anterior half of the middle cranial fossa; it is named the *middle meningeal nerve*.

The **ganglionic branches** are two short nerves that suspend the pterygo-palatine (spheno-palatine) ganglion (PLATE 40). They pass through the ganglion into its branches, where they mingle with the postganglionic fibres of the greater petrosal nerve and the sympathetic fibres of the deep petrosal nerve. By way of the ganglion V b has the following five branches:

1. The *naso-palatine nerve* (long spheno-palatine) enters the spheno-palatine foramen, crosses the roof of the nose and slopes down along the nasal septum, supplying its postero-inferior half. It goes through the incisive canal into the hard palate and supplies the gum behind the two incisor teeth.

2. The *posterior superior lateral nasal nerves* (short spheno-palatine) enter the spheno-palatine foramen and supply the postero-superior quadrant of the lateral wall of the nose.

3. The *anterior palatine nerve* (greater palatine) runs down in the greater palatine canal, between the vertical plate of the palatine bone and the body of the maxilla (PLATE 41). Multiple branches supply the postero-inferior quadrant of the lateral wall of the nose and the adjacent floor of the nose, others supply the maxillary sinus nearby. The nerve emerges from the greater palatine foramen and supplies all the hard palate except the incisor gum.

4. The *middle* and *posterior palatine nerves* (lesser palatine) descend through the lesser palatine foramina in the palatine bone and pass back to supply the mucous membrane on both surfaces of the soft palate.

5. The *pharyngeal branch* passes back through the palatino-vaginal canal to supply the mucous membrane of the naso-pharynx down to the level of Passavant's muscle.

The **posterior superior alveolar nerves** (posterior superior dental) are three in number and they emerge through the pterygo-maxillary fissure. Two enter the posterior wall of the maxilla above the tuberosity. They supply mucous membrane of the maxillary sinus and the three molar teeth (except the anterior buccal root of the first molar). The third stays outside the maxilla, pierces the buccinator, and supplies the gum of the vestibule alongside the three molar teeth.

The maxillary division picks up postganglionic fibres from the pterygo-palatine ganglion for the secreto-motor supply of the lacrimal gland.

The **zygomatic nerve** is a terminal branch of the maxillary division. It enters the inferior orbital fissure and runs along the lower part of the lateral wall of the orbit. It carries the secreto-motor fibres for the lacrimal gland. These leave the zygomatic nerve and go up the lateral wall of the orbit to join the lacrimal nerve. The zygomatic nerve enters the zygomatic bone and there divides into two branches.

The *zygomatico-facial nerve* perforates the facial surface of the zygomatic bone and supplies the skin over the bone. The *zygomatico-temporal nerve* perforates the temporal surface of the zygomatic bone, pierces the temporalis fascia, and supplies skin above the zygomatic arch (the 'hairless' skin of the temple Fig. 6.16, p. 384).

The **infra-orbital nerve** passes forward along the floor of the orbit, sinks into a groove, then enters a canal and emerges on the face through the infra-orbital foramen. It supplies multiple small branches through the orbital plate of the maxilla to the roof of the maxillary sinus. In the infra-orbital groove it gives off the *middle superior alveolar nerve.* This runs down, supplying adjacent mucosa of the maxillary sinus, to supply the two premolar teeth and the anterior buccal root of the first molar. The infra-orbital nerve gives off, in the infra-orbital canal, the *anterior superior alveolar nerve.* This goes lateral, then turns inferior, to the infra-orbital canal, supplies the maxillary sinus, the canine and the two incisors, and reaches the anterior inferior quadrant of the lateral wall of the nose (including the naso-lacrimal duct) and the adjacent floor of the nose. It ends on the nasal septum (Fig. 6.25, p. 404).

Emerging on the face the infra-orbital nerve lies between levator labii superioris and levator anguli oris (PLATE 32). It has many communications with local branches of the facial nerve; these are for the proprioceptive supply of the nearby facial muscles. It is distributed in three groups of branches (Fig. 6.16, p. 384). The *palpebral branches* supply the skin of the lower lid and both surfaces of the conjunctiva. The *nasal branches* supply a small strip of skin along the external nose. The *labial branches* supply the skin and mucous membrane of the whole upper lip and also the adjacent gum: that is, from the midline to include the gum of the second premolar tooth.

Mandibular Division (V c)

This is the nerve of the first (mandibular) pharyngeal arch. It is a very short nerve. Leaving the inferior part of the ganglion the division passes down to the foramen ovale (Fig. 6.89, p. 481). It is accompanied by the small motor root of V which passes beneath the ganglion and joins the sensory root at the foramen ovale. The mixed nerve then passes from the foramen ovale into the infratemporal fossa, between the upper head of the lateral pterygoid muscle and tensor palati (which lies on the side wall of the naso-pharynx). After 4 or 5 mm the nerve divides into a 'cat of nine tails'. These branches consist of an anterior group (all motor except one) and a posterior group (all sensory except one small branch). There are two branches from the short trunk before it breaks up.

1. A **meningeal branch (nervus spinosus)** passes up through the foramen ovale (sometimes through the foramen spinosum); here it supplies the cartilaginous part of the Eustachian tube. In the middle fossa it supplies the dura mater in the posterior half thereof, then passes between the squamous and petrous parts of the temporal bone to supply the mastoid antrum and the mastoid air cells that extend from it. It is the nerve of the first pharyngeal pouch.

2. The **nerve to the medial pterygoid** sinks into the deep surface of the muscle. It has a branch that passes close to the otic ganglion and supplies the two tensor muscles, tensor palati and tensor tympani.

The anterior leash of branches consists of the following six:

1. Two **nerves to the lateral pterygoid,** one to each head.

2. Two **deep temporal nerves.** These pass above the upper head of the lateral pterygoid, turn above the infratemporal crest, and sink into the deep surface of temporalis.

3. The **nerve to masseter** likewise passes above the upper head of the lateral pterygoid, proceeds laterally behind temporalis and through the mandibular notch to sink into masseter. It gives a branch to the mandibular joint (Hilton's law).

4. The **buccal nerve** (long buccal) is the only sensory branch of the anterior group and is the only nerve to pass between the two heads of the lateral pterygoid. It carries sometimes a *middle temporal nerve* that enters the deep surface of temporalis. The buccal nerve passes down, deep to temporalis, on the lower head of lateral pterygoid. It reaches the buccinator (PLATE 35), and gives off here a cutaneous branch which supplies a thumb-print area of skin over the soft cheek immediately below the zygomatic bone (Fig. 6.16, p. 384). The buccal nerve then pierces the buccinator and supplies the mucous membrane adherent to the deep surface of the muscle and ends by supplying the vestibular gum of the three mandibular molar teeth. The buccal nerve carries secreto-motor fibres from the otic ganglion; they are for the molar and buccal glands.

The posterior branches are three in number, and are all sensory except the mylo-hyoid branch of the inferior alveolar nerve.

1. The **auriculo-temporal nerve** has two roots that pass back around the middle meningeal artery. The nerve picks up postganglionic secreto-motor fibres from the otic ganglion; these are for the parotid gland. It passes back deep to the neck of the mandible, and gives the major sensory supply to the mandibular joint. Curving around the neck of the mandible it supplies the parotid gland, with sensory fibres and the

secreto-motor fibres it picked up from the otic ganglion. The nerve now divides into its terminal branches. The *auricular branch* supplies the external acoustic meatus and the external surface of the auricle above this. The *temporal branch* runs up over the root of the zygomatic process of the temporal bone, behind the superficial temporal vessels (PLATE 35), and supplies the hairy skin of the scalp (the hair that first turns grey).

2. The **inferior alveolar nerve** (inferior dental) passes down deep to the lower head of the lateral pterygoid, and lies on the medial pterygoid between the mandible and the spheno-mandibular ligament. It enters the mandibular foramen in front of the inferior alveolar artery and vein. Here it gives off the *nerve to mylo-hyoid* (PLATE 35). This is the only motor part of the posterior branches of V c; it pierces the spheno-mandibular ligament, lies in the gutter between the ramus and the medial pterygoid, then lies on the mylo-hyoid groove made by its accompanying small vessels (Fig. 6.30, p. 413). The mylo-hyoid nerve passes forwards down into the neck (i.e., below mylo-hyoid muscle) and runs between mylo-hyoid and the anterior belly of the digastric, supplying both these muscles. It is accompanied here by the submental branches of the facial artery and vein.

The inferior alveolar nerve runs forwards in the mandibular canal and supplies the posterior five teeth (3 molars and 2 premolars). Then it divides into its two terminal branches. The *incisive branch* goes on to supply the remaining three teeth (canine and both incisors) and overlaps to the opposite central incisor. The *mental nerve* passes from the mental foramen (PLATE 32) to supply the lower lip (both surfaces) and the adjacent gum, which is from the midline to include the second premolar gum. It carries a few fibres from the otic ganglion to the labial glands of the lower lip.

3. The **lingual nerve** is joined by the chorda tympani about 2 cm below the base of the skull, deep to the lower border of the lateral pterygoid muscle. It curves down on the medial pterygoid in front of the inferior alveolar nerve (PLATE 35). It then passes under the free lower border of the superior constrictor and goes forward above the mylo-hyoid muscle (i.e., in the mouth). It grooves the lingual plate of the mandible just below the last molar tooth, and here gives off a *gingival branch* that supplies the lingual gum to the midline (Fig. 6.30, p. 413). Dipping now below the submandibular duct (PLATE 36) it ascends on hyo-glossus to the anterior two-thirds of the tongue, which it supplies with common sensation and taste, the latter mediated by the chorda tympani fibres. The secreto-motor fibres of the chorda tympani (nervus intermedius) are given off to the submandibular ganglia, where they relay to the salivary glands in the floor of the mouth cavity. The lingual nerve and no other supplies all the mucous membrane of the floor of the mouth.

Abducent Nerve (VI)

The nucleus is in the lower pons (Fig. 7.34, p. 526). It is somatic and lies near the midline of the fourth ventricle, where it produces the bulge known as the 'facial' colliculus (Fig. 7.37). The nerve emerges at the lower border of the pons, above the pyramid of the medulla (Fig. 7.35). It enters the cisterna pontis and turns upwards, between the anterior inferior cerebellar artery and the pons, to pierce the arachnoid and dura mater on the clivus, above the level of the jugular tubercle (Fig. 6.89, p. 481). Running up now between the two layers of the dura mater it enters the inferior petrosal sinus at the apex of the petrous temporal bone (Fig. 6.94, p. 487). Here it bends forwards under the petro-clinoid ligament to enter the cavernous sinus. It passes straight forwards in the sinus, lateral to the S-bend of the internal carotid artery, and reaches the medial end of the superior orbital fissure. It enters the tendinous ring below the inferior division of III (Fig. 6.21, p. 400). In the orbit it passes within the cone of muscles to enter the ocular surface of the lateral rectus one-third of the way along the muscle (Fig. 6.49, p. 433).

Facial Nerve (VII)

The nucleus lies in the lower pons. It is branchial, motor to all the muscles derived from the second pharyngeal arch. Thus it lies off centre, lateral to the 'facial' colliculus and deep beneath the floor of the fourth ventricle. A few sensory fibres survive from the embryo, supplying skin of the external acoustic meatus and some mucous membrane at the supratonsillar recess. Their nucleus, for what it is worth, is in the nucleus of the tractus solitarius.

The fibres of VII approach the floor of the fourth ventricle and make a 'knee-bend' on the surface of the abducent nucleus at the facial colliculus (Fig. 7.34, p. 526), but it is the abducent nucleus deep to them that is responsible for the little hillock. The facial nerve traverses the pons and emerges at its lower border, above the olive (Fig. 7.35, p. 528). It passes laterally in the cerebello-pontine angle, through the cisterna pontis, and enters the internal acoustic meatus. It pierces the antero-superior quadrant of the fundus of the meatus (Fig. 6.95, p. 491) and is here accompanied by the nervus intermedius. These nerves run transversely in a canal in the petrous bone above the vestibule of the internal ear. A tube of arachnoid-lined dura mater encloses them for a short distance but fuses with them proximal to the genu. Near the middle ear the nerve makes a sharp posterior bend (the genu, or *geniculum*). Here the nervus intermedius is distended by the geniculate ganglion, and here it joins the facial nerve.

The combined nerve runs back in the medial wall of the middle ear, above the promontory and just below the bulge of the lateral semicircular canal (PLATE 40). It now curves downwards behind the middle ear, deep to the aditus ad antrum, and passes vertically down the stylo-mastoid canal. Having shed all the hitch-hiking nervus intermedius fibres it emerges from the stylo-mastoid foramen, now a purely motor nerve.

Branches in the Petrous Bone. Four branches are given off, but two of them are pure nervus intermedius branches. They are the greater (superficial) petrosal nerve and the chorda tympani, and are described where they belong, with the nervus intermedius (*v. inf.*). The facial nerve in the middle ear gives branches to the tympanic plexus. Some of these **tympanic branches** are nervus intermedius fibres for the lesser petrosal nerve and otic ganglion, but some are true facial nerve fibres and are sensory. They traverse the tympanic membrane and trespass into auriculo-temporal territory on the skin of the external acoustic meatus and pinna. Their chief interest is clinical. Their cell bodies lie with those of the nervus intermedius in the geniculate ganglion, whence herpes causes the appearance of vesicles in the skin area.

In the stylo-mastoid canal the facial nerve gives a **branch to stapedius.**

Extracranial Course of the Facial Nerve. This is described on p. 382. The face muscles supplied by the facial nerve receive their proprioceptive supply from the cutaneous nerve of the overlying skin.

Nervus Intermedius

This cranial nerve has no number. Its recognition as a separate cranial nerve is long overdue. To call it the 'sensory root of the facial nerve' is philosophical, and stems from the effort to conform to an old-fashioned belief that there are but 12 pairs of cranial nerves. This pair makes 13! The facial is a branchial nerve, but the nervus intermedius is a visceral nerve, secreto-motor and sensory. Its secreto-motor distribution extends from the floor of the mouth to the lacrimal gland in the orbit, including the palate, nasal cavity, naso-pharynx and the paranasal sinuses. Its sensory fibres mediate only taste (from the mouth, not from the face). It hitch-hikes for only a short distance with VII and, in fact, afterwards travels for greater distances with branches of V.

Its nucleus lies in the lower pons (Fig. 7.47, p. 541). The motor (i.e., secreto-motor) nucleus is the superior salivatory nucleus, which lies in the visceral efferent column lateral to the branchial column (i.e., lateral to the nucleus of VII). The sensory nucleus is the nucleus of the tractus solitarius, and this lies further lateral; it is acted on by central processes of the taste cell bodies that form the geniculate ganglion.

The nervus intermedius passes through the pons and emerges at its lower border, between the pons and the inferior cerebellar peduncle, near the eighth nerve (Fig. 7.35, p. 528). It passes, together with VII, into the antero-superior quadrant of the internal acoustic meatus. Passing laterally in the petrous bone, above the vestibule of the internal ear, it shares a common tube of arachnoid and dura mater with VII. The dura mater fuses, the nerve expands into the geniculate ganglion, and here the two nerves join. At once the greater (superficial) petrosal nerve is given off. It is 100 per cent nervus intermedius, and has nothing to do with VII. Further branches are given off to the tympanic plexus. Finally the remainder of the nervus intermedius fibres leave VII as the chorda tympani, and VII runs on, purely motor now, to the stylo-mastoid foramen. The details of the three branches of the nervus intermedius, all given off from its short journey with VII, follow.

Greater (superficial) **Petrosal Nerve.** This is almost entirely secreto-motor for the palate and above, but a few taste fibres supply the scattered taste buds on the oral surface of the palate. The cell bodies of the taste fibres are in the geniculate ganglion. The greater petrosal nerve leaves the geniculate ganglion and travels forwards and medially at a 45 degrees slant through the petrous bone (PLATE 40). It emerges and slants forwards in a groove on the petrous bone, between the two layers of the dura mater. Here in the middle cranial fossa it may be pulled on in extradural operations and so cause small hæmorrhage or œdema at the geniculate ganglion with consequent pressure on VII and a temporary facial paresis. The nerve passes beneath the trigeminal ganglion in Meckel's cave and reaches the foramen lacerum. Here it is joined by the deep petrosal nerve (PLATE 41), which peels off from the sympathetic plexus on the internal carotid artery (the cell bodies of the deep petrosal nerve are in the superior cervical ganglion). The two join and pass forwards through the pterygoid canal. The nerve of the pterygoid canal emerges into the pterygo-palatine fossa and enters the pterygo-palatine ganglion. Here the nervus intermedius (secreto-motor) fibres relay. The taste fibres and the sympathetic fibres pass straight through the ganglion, together with the sensory fibres of V b. Joined by postganglionic secreto-motor fibres they innervate the five territories of the ganglion (nasal septum, lateral nasal wall, paranasal sinuses, hard and soft palates, and the naso-pharynx). Lacrimatory post-ganglionic fibres join the maxillary nerve and enter the orbit in its zygomatic branch. They pass up the lateral wall of the orbit to join the lacrimal nerve (V a) and so reach the lacrimal gland.

Tympanic Branches. These join the tympanic plexus, where they are gathered by IX into the lesser (superficial) petrosal nerve and reach the otic ganglion for relay to the parotid and the small glands of the oral vestibule. They are entirely secreto-motor.

Chorda Tympani. This is secreto-motor for the glands in the floor of the mouth cavity and taste for the mouth part of the tongue (i.e., its anterior two-thirds). It leaves the facial nerve in the stylo-mastoid canal, 6 mm above the foramen, and passes through the posterior wall of the middle ear (PLATE 40). It is draped over by the mucous membrane as it passes across the pars flaccida of the tympanic membrane and the neck of the malleus. It leaves through the anterior wall of the middle ear, passes through the petrous bone, and emerges at the medial end of the petro-tympanic fissure. Passing down deep to the spine of the sphenoid, which it grooves, the nerve slopes forwards to join the lingual nerve just above the lower border of the lateral pterygoid muscle. By the lingual nerve its taste fibres are taken to the mouth part (anterior two-thirds) of the tongue. Its secreto-motor fibres, by relay in the submandibular ganglia, pass to the glands in the floor of the mouth cavity.

Stato-acoustic (Auditory) Nerve (VIII)

The nuclei are in the medulla and encroach on the pons. The nerve is wholly sensory, special to sound reception and balance; thus the nuclei lie lateral. It is logical and convenient to trace the nerve from the periphery to its brain stem nuclei, as in studying the first and second cranial nerves (the diffuse peripheral distribution of the sensory nerves V, IX and X makes it imperative to trace them in the reverse direction, from the nucleus to the periphery). There are two quite distinct parts of the eighth nerve. From fishes to man sound reception and balance are combined in the one sensory organ and in a single cranial nerve; nobody knows why this should be, or what connexion (if any) sound reception has with balance. It is a mystery.

Cochlear Nerve. The neuro-epithelium for sound reception consists of the hair cells of the organ of Corti. The first sensory neurone is bipolar. Its cell body lies in the base of the bony spiral lamina. The cell bodies thus lying in a spiral line are together referred to as the *spiral ganglion*. Their central processes run along the modiolus of the cochlea and join into many small nerves that pierce dura and arachnoid mater at the base of the modiolus in a spiral pattern (Fig. 6.62, p. 455). This is at the antero-inferior quadrant of the internal acoustic meatus (Fig. 6.95, p. 491). They join together in the subarachnoid space and enter the cisterna pontis joined with the vestibular part. The nervus intermedius and VII lie in front (Fig. 6.96, p. 492). Together these three pass through the cerebello-pontine angle in front of the flocculus of the cerebellum and the lateral aperture of the fourth ventricle (Fig. 7.35, p. 528). The eighth nerve (cochlear and vestibular parts bound together) enters the inferior cerebellar peduncle at the lower border of the pons.

The cochlear fibres relay on **cochlear nuclei** in the inferior cerebellar peduncle. These second order cell bodies are in two groups, ventral and dorsal. Their central processes pass across to the lower pons and turn up as they cross the medial lemniscus, here forming the trapezoid body. They ascend in a tract called the lateral lemniscus. Some fibres do not cross but run up in the lateral lemniscus of their own side; cochlear representation is largely bilateral. The lateral lemniscus ascends through the tegmentum of the mid-brain and relays in the *inferior colliculus (for reflex effects of sound)*. Tecto-bulbar and tecto-spinal tracts activate motor nuclei for head and neck and for body and limb reflex movements (cf. light reflex relay in the superior colliculus). The remainder of the lateral lemniscus relays in the *medial geniculate body (for hearing)*. This body, though anatomically in the midbrain, is actually a displaced thalamic nucleus. Here third order cell bodies send their processes far back in the posterior limb of the internal capsule to the auditory cortex. This lies in the transverse gyrus that runs deeply towards the insula from the posterior end of the superior temporal gyrus.

Vestibular Nerve. The neuro-epithelium consists of the hair cells in the maculæ of utricle and saccule (for static balance) and the ampullæ of the semi-circular ducts (for kinetic balance). The first neurones emerge from the fundus of the internal acoustic meatus, above and below the horizontal crest (Fig. 6.95, p. 491). From the postero-superior quadrant of the fundus emerge the nerve from anterior and lateral semicircular ducts and the utricle into which they open (Fig. 6.63, p. 456). Through the postero-inferior quadrant comes the lower division of the vestibular nerve, from the saccule. Alongside it, through the foramen singulare, emerges the nerve from the ampulla of the posterior semicircular duct. Having pierced the fibrous dura and arachnoid mater the upper and lower divisions lie in the internal acoustic meatus and are here distended into *vestibular ganglia* by the cell bodies of these first order neurones (Fig. 6.62, p. 455). The central processes combine into the vestibular nerve. This joins the cochlear nerve and passes into the cisterna pontis. The vestibular fibres relay in the **vestibular nucleus,** in the lateral angle of the fourth ventricle (Fig. 7.37, p. 529), in both pons and medulla. The nucleus sends its fibres to the archæcerebellum and, by the medial longitudinal bundle, to the motor nuclei of the brain stem. Vestibulo-spinal tracts descend to anterior horn cells. There appear to be cortical con-nexions from the vestibular nucleus, by way of bilateral thalamic relay (i.e., medial geniculate body), to the region of the auditory centre.

Glosso-pharyngeal Nerve (IX)

This is the nerve of the third pharyngeal arch. The nucleus is in the medulla (Fig. 7.47, p. 541). It is visceral as well as branchial, and mixed sensory and motor (i.e., secreto-motor). The two nuclei are in line with, and in continuity with, those of the nervus intermedius. They lie at the upper angle of the vagal trigone in the floor of the fourth ventricle (Fig. 7.37, p. 529). The inferior salivatory nucleus contains secreto-motor cell bodies, and lateral to it is the sensory nucleus of the tractus solitarius, which receives from the first order cell bodies of the petrous ganglion. These mediate ordinary sensation, taste and sinu-carotid impulses. A third origin is in the nucleus ambiguus; this is for the striated muscle stylo-pharyngeus.

From the three nuclei the fibres reach the surface of the medulla between olive and inferior cerebellar peduncle (Fig. 7.35, p. 528). In a series of rootlets which join to make a single nerve it emerges in the cisterna pontis and runs laterally behind the jugular tubercle of the occipital bone and enters the anterior compartment of the jugular foramen (Fig. 6.96, p. 492). Here it lies lateral to the inferior petrosal sinus; together they are separated from X and XI by a septum of fibrous dura mater. The glosso-pharyngeal nerve deeply notches the inferior border of the petrous bone, just below the internal acoustic meatus. Perforating the arachnoid and dura mater the nerve is distended into its elongated *petrous ganglion*. (Turn to p. 394.)

Vagus Nerve (X)

The nucleus lies in the medulla, in the vagal trigone in the floor of the fourth ventricle, caudal to IX (Fig. 7.37, p. 529). It is similar to IX, but in addition has motor cell bodies for the *smooth* muscle of the respiratory and alimentary systems (IX supplies no smooth muscle). Further, its caudal part contains scattered cell bodies of the 'vital centres', or at least those of the heart.

A group of secreto-motor cell bodies continues down from the inferior salivatory nucleus of IX. The nucleus of the tractus solitarius similarly continues down from IX into the vagal nucleus. In this narrower part of the fourth ventricle they are naturally not so widely separated as in the nucleus of IX, and are further intermingled with the smooth-muscle motor cell bodies. These smooth-muscle motor cell bodies are often called the 'dorsal nucleus' of the vagus to distinguish them from the secreto-motor (inferior salivatory nucleus) sensory (tractus solitarius) and cardiac cell bodies. A different nomenclature includes the whole conglomerate of cell bodies in the name 'dorsal nucleus', to distinguish these from the nucleus ambiguus. These two different meanings of the name 'dorsal nucleus' will confuse the unwary. The vagus supplies also the *striated* muscle of the viscera (soft palate, pharynx, œsophagus, larynx). The motor cell bodies lie in the nucleus ambiguus, but almost all the fibres run from the medulla in the cranial root of the accessory nerve (XI) and join the vagus outside the skull.

From the vagal nucleus the fibres leave the surface of the medulla in a series of rootlets below IX in the sulcus between olive and inferior cerebellar peduncle (Fig. 7.35, p. 528). These unite into a single nerve that crosses, below IX and above XI, the basi-occiput behind the jugular tubercle (Fig. 6.96, p. 492). With XI it enters the transverse slit of the middle compartment of the jugular foramen. Here it lies medial to XI, and perforates the arachnoid and dura mater. A small superior ganglion lies just above the long *ganglion nodosum* (*inferior ganglion*), which distends the vagus just below the skull base. The superior ganglion has cell bodies for the unimportant meningeal and auricular branches. The inferior ganglion lodges the cell bodies of all the other sensory fibres in the vagus nerve. At the inferior ganglion the accessory nerve gives *all* its nucleus ambiguus fibres to the vagus. This is similar to the motor root of a spinal nerve joining the sensory root just distal to the sensory ganglion. The nucleus ambiguus fibres are wholly motor; they are paid off in the appropriate branches of the vagus to *striated* muscle.

In the neck the vagus lies vertical, like a plumb line, in the carotid sheath, adherent to the internal carotid and then to the common carotid artery, always deep in the gutter between the artery and the internal jugular vein.

The branches in the neck are described on p. 395. The branches in the thorax are described on p. 226.

The vagus fibres not used up on the œsophagus leave the plexus as the vagal trunks, each a mixture of right and left vagi. The pattern as they pass on the œsophagus into the abdomen is two on the right and one on the left, but any of these may be double or multiple. Rotation of the stomach carries the left vagal trunk to its anterior surface; some of these nerves apparently run on to the porta hepatis (gall-bladder). Similarly the right vagal trunks pass to the posterior wall of the stomach, but most of them go to the cœliac plexus. They pass through this plexus to join the postganglionic sympathetic fibres. The vagal fibres are distributed along the cœliac artery to the liver, gall-bladder, and pancreas, along the superior mesenteric artery to the small gut and large gut as far as the splenic flexure, and along the renal artery to the kidney (renal function unknown). The branches to the gut and its derivatives are motor to smooth muscle and secreto-motor to glands, but not sensory. The sensory supply of viscera of the abdomen is sympathetic, but the vagus is sensory to the thoracic viscera (œsophagus, trachea and lungs).

Accessory Nerve (XI)

The nucleus is in the medulla. It is motor (both visceral and branchial) and gives a direct supply to the *striated* muscle of the viscera of the neck (soft palate, pharynx and œsophagus, larynx). The striated laryngeal muscles are branchial, developing from the fourth and sixth pharyngeal arches. The nucleus is named the *nucleus ambiguus*; it lies in the inferior cerebellar peduncle medial to the spinal nucleus of V (Fig. 7.39, p. 529).

The fibres leave the medulla by a series of rootlets in vertical line below X, between the olive and the inferior cerebellar peduncle (Fig. 7.35, p. 528). They unite into a single nerve called the **cranial root.** This is joined by a spinal root coming up through the foramen magnum into the cisterna pontis (Fig. 6.96, p. 492).

The **spinal root** forms from cell bodies in the anterior horn cells of the upper five segments of the cervical cord (mainly in 2, 3 and 4). The fibres have an aberrant course. They emerge as a series of five roots from the *lateral* surface of the spinal cord *behind* the ligamentum denticulatum. Joining in series they form a single nerve that passes over the upper tooth of the denticulate ligament to join the cranial root. These two roots have nothing to do with each other and are bound for different destinations. They merely 'join hands' to pass through the jugular foramen together, then part company.

The accessory nerve lies caudal to X and IX on the basi-occiput behind the jugular tubercle (Fig. 6.96, p. 492). It passes through the lateral part of the slit-like middle compartment of the jugular foramen, piercing arachnoid and dura mater here. The vagus lies medial to it in the jugular foramen, just in front of the jugular bulb. Outside the skull the accessory gives all its cranial (i.e., nucleus ambiguus) fibres to the vagus, and these are distributed by X in its branches to the striated muscle of the soft palate, pharynx and œsophagus, and larynx. The accessory nerve in the neck, wholly motor and wholly spinal now, lies on the internal jugular vein over the lateral mass of the atlas (Fig. 6.19, p. 393), passes into sterno-mastoid which it supplies, and then descends within the posterior triangle over levator scapulæ to supply trapezius by passing into its deep surface (PLATE 25).

The nucleus ambiguus fibres joining the vagus conform with the regular pattern of mixed (spinal) nerves. The curious pathway of the spinal fibres in the accessory nerve constitutes a mystery.

Hypoglossal Nerve (XII)

The nucleus is in the medulla. It is motor and somatic, and thus lies against the midline, in the hypoglossal trigone in the floor of the fourth ventricle (Fig. 7.37, p. 529). The fibres pass ventrally through the olivary nucleus and emerge from the surface of the medulla as a vertical line of rootlets between pyramid and olive (Fig. 7.35, p. 528). These join into two roots that enter the hypoglossal (anterior condylar) canal in the occipital bone (Fig. 6.96, p. 492). They are here separated by a flange of fibrous dura mater that sometimes ossifies. They pass forwards above the atlanto-occipital joint and join in the canal, emerging as a single nerve. They pierce the arachnoid and dura mater in the canal.

Examine a skull; note that the exit is deep (i.e., medial and posterior) to the jugular foramen where IX, X and XI emerge. The hypoglossal nerve from this deep position spirals down and forwards with a lateral convexity that brings it gradually more superficial. It spirals down behind the inferior ganglion (nodosum) of X, then descends behind and around the internal and then the external carotid arteries (PLATE 36). At this level, below the posterior belly of digastric, it lies on three arteries (occipital, external carotid and lingual) and is very near the skin, covered only by the common facial vein beneath the investing layer of deep cervical fascia (Fig. 6.8, p. 373). This spiral course curves outside the arteries but deep to the veins; it is the pathway of migration of the suboccipital myotomes that form the muscles of the tongue. The hypoglossal nerve now passes forward on hyo-glossus deep to mylo-hyoid (i.e., in the mouth). Running above the greater horn of the hyoid bone it splits up to supply the intrinsic muscles of the tongue and the extrinsic muscles genio-glossus, hyo-glossus and stylo-glossus (i.e., all the tongue muscle except only palato-glossus).

Branches. The branches of the hypoglossal nerve are entirely C1 fibres that join XII at its exit from the skull (Fig. 6.9, p. 375). XII itself has no supply outside the tongue. The *meningeal branch* (C1) supplies a ring of dura mater in the posterior fossa beyond the margin of the foramen magnum.

The **descending branch,** nowadays called the *upper root of the ansa cervicalis*, has hitherto been known as descendens hypoglossi. It branches off as XII curves down between the internal carotid artery and the internal jugular vein. It is joined at a variable level by a branch that combines from C2 and 3 in the cervical plexus. This nerve is called the *lower root of the ansa cervicalis*; it used to be known as descendens cervicalis. Together these nerves make the **ansa cervicalis** (ansa hypoglossi). The ansa lies on the internal jugular vein under cover of sterno-mastoid (PLATE 26, p. *xiv*); its branches supply segmentally sterno-hyoid (C1, 2, 3), omo-hyoid (C1, 2, 3) and sterno-thyroid (C2, 3).

The hypoglossal nerve next gives a slender **branch to thyro-hyoid** (C1). This comes off as the nerve lies on the lingual artery (PLATE 39). The last branch is the **nerve to genio-hyoid** (C1). It is given off in the mouth, above mylo-hyoid, and contains the last of the C1 fibres that travel along the hypoglossal nerve (PLATE 36).

Section 8. Osteology of the Skull

THE INTACT SKULL

The bones of the skull articulate to form the cranium and the face. The cranium encloses the brain and the face hangs down below it. A study of the *intact skull* is much more fruitful than undue contemplation of each separate bone. Examination of the disarticulated skull bones displays certain features not visible in the intact skull, but the chief justification for studying individual bones is to understand their manner of articulation as they combine to build up the skull.

External Features of the Skull

Superior View. The most striking feature of the convex upper surface is its asymmetry. One side is set in front of the other, and this is reflected internally by a similar asymmetry of the cerebral hemispheres and their contained lateral ventricles. Anteriorly the frontal bone articulates with the pair of parietal bones at the **coronal suture**; the original two halves of the frontal bone occasionally fail to fuse, leaving a midline *metopic suture*. The midline meeting place of the bones is the **bregma,** the site of the anterior fontanelle (Fig. 1.58, p. 47). The coronal suture is straight for an inch or more (3 cm) lateral to the bregma (this is the line of closure of the anterior fontanelle) and then becomes highly tortuous as it curves transversely down to the lateral surface of the skull. Behind the bregma the parietal bones articulate in the midline **sagittal suture.** The anterior inch (3 cm) of this is straight (the line of closure of the anterior fontanelle) and then comes a tortuous part for two inches (5 cm), followed by a straight part. Alongside this a foramen often perforates each parietal bone; an emissary vein leaves the superior sagittal sinus through it. A suggestion that the foramen marks the site of the problematical pineal eye of an unstated human ancestor is purely speculative. Thence the sagittal suture, tortuous again, curves down to the **lambda,** at the apex of the occipital bone. The centre of the parietal bone is a low prominence, the parietal eminence, and this lies on the profile of the skull from this view.

Posterior View. The lambda is the midline point where the sagittal suture meets the tortuous sutures between the occipital and parietal bones. It lies above the posterior pole of the skull. Along these sutures accessory bones are commonly found. They are known as *sutural bones*, and they are sometimes seen along other borders of the parietal bone, including the pterion. They must be distinguished from the occasional *interparietal bone*, a large triangular bone at the apex of the occipital. This is not an accessory bone but exists as a separate entity when the cartilaginous and membranous parts of the occipital bone fail to fuse (see p. 567).

Some $2\frac{1}{2}$ inches (6 cm) below the lambda the occipital bone is projected into the external occipital protuberance, from which a low ridge crosses towards the base of the mastoid process. This is the *superior nuchal line* (Fig. 8.3); gently convex upwards it lies at the junction of neck and scalp. It is the surface marking of the internal attachment of the tentorium cerebelli, which straddles the transverse sinus at the roof of the posterior fossa. Trapezius and sterno-mastoid are attached side by side here (trapezius at the medial one-third and sterno-mastoid along the lateral two-thirds of the superior nuchal line). Splenius capitis is inserted into the lateral third of the line deep to sterno-mastoid (PLATE 33). Above this is the *highest nuchal line*, which gives origin to occipitalis and the galea aponeurotica; beneath this part of the scalp the bare bone covers the occipital pole of the cerebral hemisphere. Below the superior nuchal line the bone that covers the cerebellar hemispheres gives attachment to the neck muscles (p. 468).

The mastoid part of the temporal bone articulates with the parietal and occipital. The mastoid process projects down from the side, its deep surface channelled into the digastric notch (Fig. 6.75, p. 468). It is best studied in the lateral and inferior views. The suture between the mastoid and occipital bones is commonly perforated by a mastoid emissary foramen, carrying a vein from the sigmoid sinus to the posterior auricular vein (Fig. 8.4).

Lateral View

The Cranium. Note the sutures between the cranial bones. The **coronal** suture curves down along the posterior border of the frontal bone; it is tortuous above and straight below the inferior temporal line (Fig. 6.13, p. 380). As it leaves the parietal bone the coronal suture curves forward where the frontal bone articulates with the greater wing of the sphenoid. The upper border of the greater wing articulates with the inferior angle of the parietal bone; this suture meets the coronal suture at the **pterion.** The squamous part of the temporal bone articulates in front with the greater wing of the sphenoid and then arches up and back to overlap the parietal bone. The mastoid part of the temporal bone continues this suture with the parietal until posteriorly it meets the occipital bone. The meeting place of these sutures is called the *asterion.*

The mastoid part of the temporal bone is projected down as the blunt **mastoid process.** To its lateral surface is attached the sterno-mastoid, with splenius and longissimus capitis lying deep to it. The lateral surface of the mastoid process is built up by a downgrowth of bone from the squamous part of the temporal, and incomplete fusion of the two results commonly in an irregular groove just short of the borders of the mastoid process (Fig. 6.19, p. 393). In front of this is the external acoustic meatus, to be studied later (p. 560). Above the meatus is a horizontal ridge called the suprameatal crest; this is projected forwards as the upper border of the zygomatic process of the temporal bone. The zygomatic arch is continued by the zygomatic bone, which articulates with the maxilla and frontal bone as a pair of anterior pillars of the arch (Fig. 8.1).

The frontal process of the zygomatic bone meets the frontal bone at a suture which is palpable in the living and is used as a landmark for the pterion (p. 482). The zygomatic bone has a sharp posterior border which continues up across the suture as a ridge on the frontal bone. This ridge arches up and back and diverges into the superior and inferior **temporal lines.** The superior line is traceable around its convexity down towards the mastoid process. The temporalis fascia is attached to the superior line and to the polished strip of bone between the two lines and sweeps down to be attached to the upper border of the zygomatic arch. The inferior temporal line curves around from the zygomatic process of the frontal bone across the frontal and parietal bones (dipping down as it crosses the coronal suture, Fig. 6.13, p. 380) as far as the mastoid part of the temporal bone. Here it runs forward into the suprameatal crest (PLATE 37).

The **temporal fossa** is enclosed by the inferior temporal line. In the fossa the greater wing of the sphenoid articulates behind with the squamous temporal to form a convexity which is the lateral wall of the middle cranial fossa, and it articulates in front with the zygomatic bone to form a deep concavity which is the lateral wall of the orbit. The upper border of the greater wing articulates with the parietal and frontal bones. Sutural bones are commonly found here. The lower limit of the temporal surface of the greater wing is marked by a line of sharp tubercles, the infratemporal crest; this is level with the upper border of the zygomatic arch. On the squamous bone a vertical groove lies above the external acoustic meatus (Fig. 6.2, p. 361). This lodges a deep vein and artery of the superficial temporal vessels. Temporalis arises from the whole surface of the temporal fossa. The posterior pillar of the zygomatic arch is

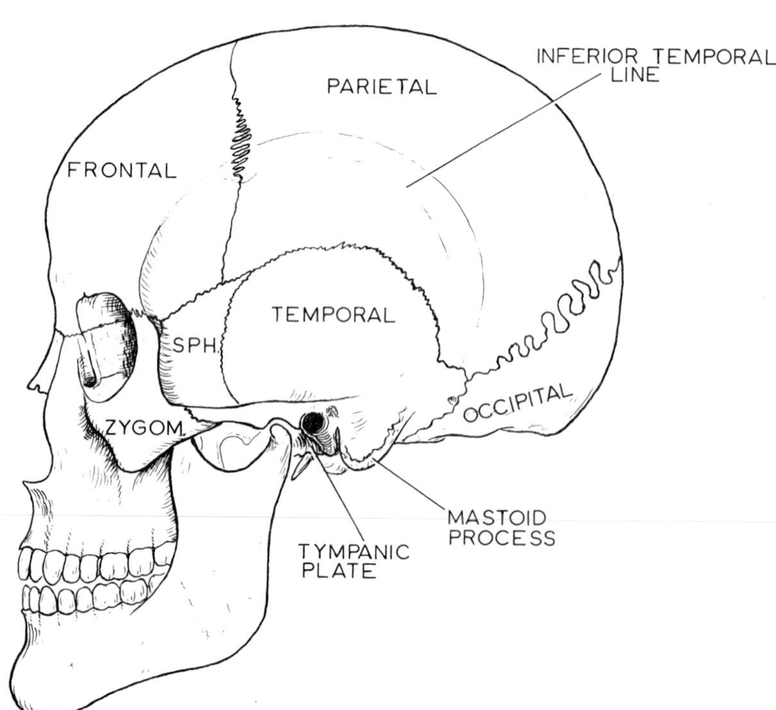

FIG. 8.1. THE SKULL IN LATERAL VIEW.

grooved to lodge the posterior fibres of the muscle, which glide to and fro across it during mandibular movement. The anterior wall of the temporal fossa is formed by the concave surface of the sygomatic bone, which is here perforated by the zygomatico-temporal nerve.

The Face. The zygomatic process of the temporal bone is hollowed below by the glenoid fossa in front of the external acoustic meatus; in front of this is the eminentia articularis. Note the position of the mandible (Fig. 6.17, p. 387). Its condyle lies in the glenoid fossa and the coronoid process lies under cover of the temporo-zygomatic suture. Remove the mandible and reserve it for future study.

The **zygomatic arch** is composed of three bones. The zygomatic process of the temporal bone meets the zygomatic bone at an oblique suture (Fig. 6.13, p. 380), and the zygomatic process of the maxilla forms the anterior pillar of the lower border of the arch. (N.B. The term *zygoma* is synonymous with the zygomatic process of the temporal bone only, and does not include the whole arch.) The superficial head of masseter arises from the lower border of the arch as far forwards as the maxilla, and the deep head arises from the concave medial surface of the arch. The temporo-mandibular ligament is attached to the lateral surface of the eminentia articularis deep to the parotid fascia, whose superficial layer is attached to the lower border of the arch in front of this (Fig. 6.1, p. 361). The temporalis fascia is attached to the upper border of the arch. The arch is crossed in front of the external acoustic meatus by the auriculo-temporal nerve and the superficial temporal vessels. Further forward the arch is crossed by the upper and lower zygomatic branches of the facial nerve (PLATE 31, p. *xvi*). The nerves are vulnerable on this subcutaneous bone, and their division will endanger vision from inability to close the lids with resulting desiccation and ulceration of the cornea. Deep to the arch the temporalis slides freely up and down with movements of the mandible.

The zygomatic bone forms the bony prominence of the cheek; it is perforated by a foramen, sometimes double, for the zygomatico-facial nerve. When no foramen is visible the zygomatico-facial nerve has perforated the surface by multiple small filaments. Zygomaticus major arises from the surface of the zygomatic bone, and zygomaticus minor from the zygomatico-maxillary suture.

Below the arch the upper jaw can be studied. The posterior convexity of the maxilla nears the lower part of the lateral pterygoid plate; inferiorly the tubercle of the palatine bone is wedged between them (PLATE 36, p. *xix*). The maxilla and pterygoid plate are separated above this by the pterygo-maxillary fissure. In its depths lies the pterygo-palatine fossa, and the spheno-palatine foramen can be seen here, opening into the lateral wall of the nose. The fossa is not fully visible in the intact skull (it is described on p. 397).

The tuberosity of the maxilla is a prominent boss of bone that projects above the *posterior* surface of the last molar tooth. The buccinator arises from a linear strip on the body of the maxilla. The line extends from level with the anterior border of the first molar tooth, skirting the base of the zygomatic process of the maxilla and keeping well above the alveolar bone until it turns down posteriorly to the tuberosity. From the tuberosity a fibrous band for the origin of buccinator passes to the tip of the hamulus. It is called the pterygo-maxillary ligament (Fig. 6.14, p. 381). Below the line of buccinator is the vestibule of the mouth, above it are the soft tissues of the face. From the tuberosity of the maxilla and the tubercle of the palatine bone the small inferior head of medial pterygoid arises (Fig. 6.2, p. 361) overlapping the inferior head of lateral pterygoid, which arises from the whole surface of the lateral pterygoid plate. The posterior convexity of the maxilla above the tuberosity shows two or more foramina for the posterior superior alveolar nerves and vessels (PLATE 35).

The zygomatic process of the maxilla forms the anterior pillar of the zygomatic arch. It is palpable through the cheek, or better still in the vestibule (feel your own!). It is hollowed by the apex of the maxillary sinus. The anterior pillar of the zygomatic arch is buttressed higher up, for here the zygomatic bone articulates with the zygomatic process of the frontal bone.

The anterior nasal aperture, bounded below by the sharp projection of the anterior nasal spine, is on the profile of the skull, but the face bones visible from this aspect are best studied from in front. Note in this lateral view, however, the ridge formed by the root of the canine tooth. Between the canine ridge and the zygomatic process is the canine fossa. Between the canine ridge and the midline is the incisive fossa (note the confusion of current terminology, which uses the term 'incisive fossa' also to denote the small pit on the hard palate into which the incisive foramen opens, see p. 409).

Anterior View

The Cranium. The frontal bone (sometimes bisected by a metopic suture) curves down to make the upper margins of the orbits. Medially it goes down to meet the frontal process of each maxilla, between which it articulates with the nasal bones. Laterally it projects down as a zygomatic process to make a suture with the zygomatic bone at the lateral margin of the orbit. Postpone the study of the orbital cavity (p. 430) and note that the frontal bone occupies the upper third of the anterior view of the skull, the maxillæ and mandible making the other two thirds (Fig. 1.58, p. 47).

The Face. The nasal bones curve downwards and forwards from their articulation with the frontal bone.

Each articulates with the frontal process of the maxilla, and they arch forward to meet in a midline suture. The lower border of each is notched by the external nasal nerve. These free borders make with the two maxillæ a pear-shaped anterior nasal aperture (Fig. 8.2). The bony septum and the conchæ of the lateral wall are visible in the nasal cavity, but their study is best postponed. The two maxillæ meet in a midline intermaxillary suture, and are projected forward as the anterior nasal spine at the lower margin of the nasal aperture (Fig. 6.12, p. 380 and Fig. 6.25, p. 404). The canine root makes a ridge on the anterior surface of the maxilla. Between this and the intermaxillary suture lies the *incisive fossa*. The incisive slips of orbicularis oris arise high in this fossa, near the midline. Lateral to the canine ridge is the *canine fossa*, from which levator anguli oris arises. Above this the anterior surface of the maxilla is perforated by the infra-orbital foramen. Levator labii superioris alæque nasi arises from the frontal process of the maxilla, and levator labii superioris arises from the lower margin of the orbit (i.e., from maxilla and zygomatic bone) and hangs down like a curtain over the infra-orbital foramen and its issuing nerve. Note that the attachments of buccinator and levator anguli oris, horizontal around the body of the maxilla, bisect the maxillary sinus, which lies with its lower half deep to the vestibule and its upper half deep to the soft tissues of the face. Nearer the midline the incisive slips of orbicularis oris arise from the same horizontal level; the mucous membrane of the vestibule is reflected from the maxilla to line all these muscles. As before, the study of the mandible will be deferred, but while it is in position note that the supra-orbital notch, infra-orbital foramen and mental foramen lie all three in a vertical line. The osteology of the orbital walls and margin is summarised on p. 430.

Inferior View

The Cranium. Discard the mandible and look at the inferior surface of the skull. The occipital, temporals and part of the sphenoid bones make the base of the cranium behind the face bones (Fig. 8.3).

The area behind the foramen magnum is simple. From the mastoid process to the external occipital protuberance the **superior nuchal line** lies in a curve *concentric with the foramen magnum.* Halfway between them the **inferior nuchal line** is concentric with both. The **external occipital crest,** in the midline between the external occipital protuberance and the foramen magnum, bisects this area. A rather vague line, radiating back and outwards from the foramen magnum, further bisects each half. Thus four areas are demarcated in each half (Fig. 8.3). There are two alongside the foramen magnum and they receive the recti. The medial area receives rectus capitis posterior minor and the lateral area receives rectus capitis posterior major (Fig. 6.75, p. 468). Between superior and inferior nuchal lines the medial area receives semispinalis capitis and the lateral area gives attachment to the superior oblique muscle of the head.

The area in front of the foramen magnum is more difficult. It is a complicated mixture of processes and

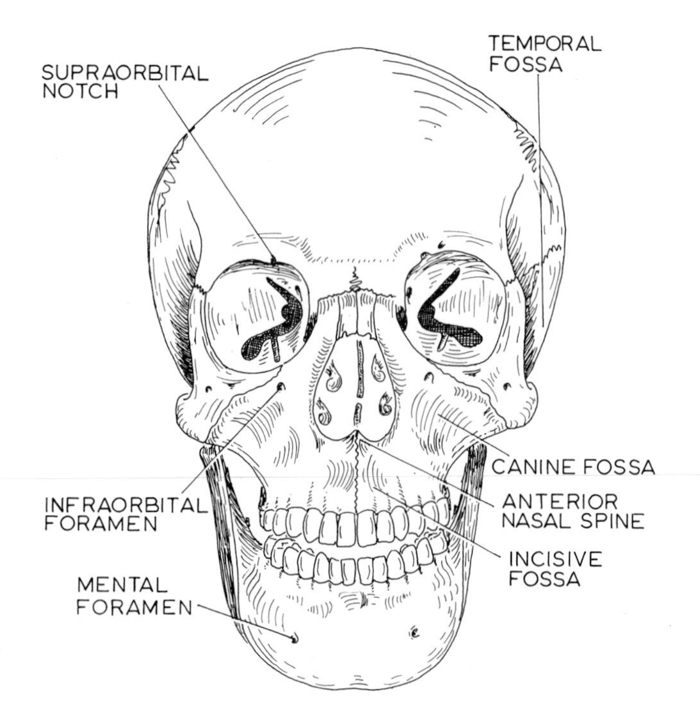

FIG. 8.2. OUTLINE OF THE SKULL IN ANTERIOR VIEW.

foramina, but do not be dismayed. A general survey will clarify the picture, and the following 'one-third: two-thirds' summary will prove useful (Fig. 8.3).

General Survey. Join the tips of the mastoid processes. One-third of the foramen magnum lies in front and two-thirds behind this line. The occipital condyles have the reverse proportions; two-thirds of the condyle lies in front of the line. Now identify the pterygoid tubercle by tracing the posterior border of the medial pterygoid plate from the hamulus up to the base of the skull. Here the pterygoid tubercle projects back towards the foramen lacerum. Note that the back of the nose (between the medial pterygoid plates) is the *same width as the diameter of the foramen magnum*, and is set forward at such a distance that a line joining the tip of the mastoid process to the pterygoid tubercle slants in at 45 degrees—right and left lines lie at right angles to each other (Fig. 8.3). This 45 degree slant along the petrous part of the temporal bone imprints itself on many structures at the base of the skull. *The line is divided into thirds* by the *styloid process* and *spine of the sphenoid* so the stylo-mastoid foramen and the foramen spinosum can be located. The stylo-mastoid foramen lies behind the base of the styloid process, and the foramen spinosum perforates the base of the spine of the sphenoid. The spine of the sphenoid overlies the opening in the petrous bone of the bony part of the Eustachian tube. The cartilaginous part of the tube, slanting at 45 degrees, lies in the slit between the greater wing of the sphenoid and the apex of the petrous bone (Fig. 8.4). The foramen ovale perforates the greater wing along a 45 degree slant anterior and medial to the foramen spinosum. Lateral to the pterygoid tubercle is the scaphoid fossa; from here to the spine of the sphenoid tensor palati arises from the greater wing along the same 45 degree slant.

One third of the way from the anterior margin of the foramen magnum to the back of the nasal septum (vomer) the *pharyngeal tubercle* projects from the basi-occiput (Fig. 8.3). On either side a ridge curves laterally, marking the attachment of the prevertebral fascia and the pharyngo-basilar fascia. The attachment of the pharynx can now be marked out on the base of the skull (p. 414). Note that the apex of the petrous bone with its attached levator palati muscle lies wholly in the pharynx (Fig. 8.4).

From this general survey more particular attention can now be given to some special points.

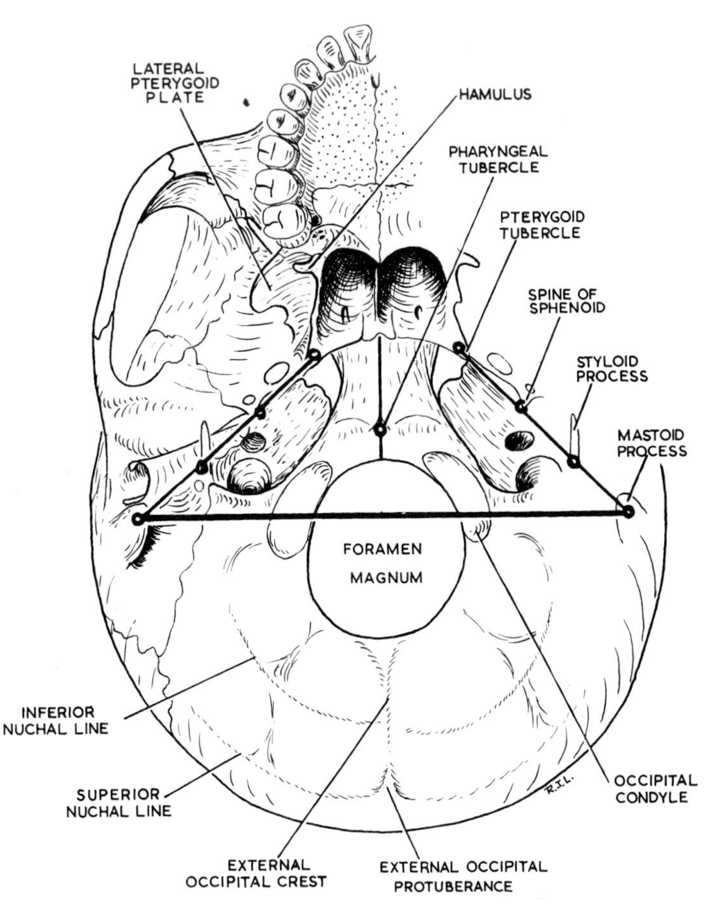

FIG. 8.3. THE BASE OF THE SKULL. SURVEY OF LANDMARKS.

The Occipital Bone. The **foramen magnum** is oval in shape, its long diameter being antero-posterior. Its transverse diameter is rather less than one third of the distance between the mastoid processes. The fibrous dura mater is attached to the margins of the foramen as it sweeps down from the posterior cranial fossa. Within the tube of dura mater the lower medulla with the spinal arteries and veins, the vertebral arteries and the cervical roots of the accessory nerves traverse the foramen in the subarachnoid space (Fig. 6.96, p. 492). Meningeal branches of the vertebral artery and communicating veins from the occipital sinuses to the internal vertebral plexus lie outside the fibrous dura, between it and the periosteum of the foramen magnum laterally and posteriorly. Anteriorly the margin of the foramen gives attachment to the ligaments sweeping up from the axis. Adherent to dura mater is the membrana tectoria and in front of this is the vertical limb of the cruciform ligament; in front again are the apical and the pair of alar ligaments of the odontoid process. All are attached, in front of the dura mater, to the anterior margin of the foramen magnum. The alar ligaments are attached each to a triangular area medial to the anterior pole of the occipital condyle. These triangular areas are limited anteriorly by a ridge that joins the anterior poles of the occipital condyles. The anterior atlanto-occipital membrane is attached to the ridge. The posterior atlanto-occipital membrane is attached to the posterior margin of the foramen magnum; observe that this is a curve of more than half the foramen. Both membranes are attached right up to the capsules of the atlanto-occipital joints. Tangential with the posterior margin of the foramen magnum is the line of fusion between the squamous and jugular parts of the occipital bone (it fuses at the 2nd year, see p. 567).

The Occipital Condyles. These convex surfaces, covered with hyaline cartilage, lie at the front half of the foramen magnum. Their posterior poles are separated by the diameter of the foramen, but their anterior poles are much closer together. The two convexities make a ball-and-socket joint with the atlas. But the antero-posterior curve is more pronounced than the combined side to side curvature; the ball is oval-shaped, like an egg lying on its side, and thus permits nodding and some abduction but *no rotation*. The capsule and synovial membrane are attached to the articular margins. The epiphyseal cartilage between basi-occiput and ex-occiput crosses this joint. Fusion (at the 6th year) is sometimes incomplete and the condyle may then possess two separate articular surfaces. Behind the condyle is a shallow fossa floored by thin bone. This is commonly perforated by the *posterior condylar canal*, carrying a vein from the bottom of the sigmoid sinus to the suboccipital venous plexus. Deep to the summit of the occipital condyle at the site of the original epiphysis the bone is perforated by the *hypoglossal canal*. The XIIth nerve enters this as two roots, separated by a flange of fibrous dura mater. The flange may ossify but does not extend to the anterior aperture of the canal, for here the nerve issues as a single trunk, medial to the jugular foramen. This foramen, lateral to the canal, is bounded by the jugular notch in the lateral projection of the occipital bone called the **jugular process** (Fig. 8.4). The jugular process articulates laterally with the mastoid part of the temporal bone; the occipital artery grooves the suture between them. The jugular process, at the posterior margin of the jugular foramen, gives attachment to rectus capitis lateralis, and the prevertebral fascia is attached to the edge of the bone, behind the internal jugular vein (Fig. 6.9, p. 375).

The **basi-occiput** extends forward from the foramen magnum and fuses (at 25 years) with the basi-sphenoid just behind the nose. In front of the pharyngeal tubercle the bone forms the roof of the naso-pharynx, whose mucous membrane is attached to the periosteum. Alongside the pharyngeal tubercle the ridges, convex forwards, for the prevertebral fascia and pharyngo-basilar fascia have already been noted (Fig. 8.4). Behind each is the insertion of longus capitis, with rectus capitis anterior behind this muscle, immediately in front of the occipital condyle and medial to the hypoglossal canal (Fig. 6.9, p. 375).

The *squamous part* of the occipital bone has already been studied (pp. 553, 556).

The Temporal Bone. There are four parts of this bone, separately ossified and afterwards fused. The *petro-mastoid* part, set at a slant of 45 degrees, forms a substantial part of the skull base alongside the occipital bone. The *squamous* part, undulating from the concavity of the glenoid fossa into the convexity of the eminentia articularis, makes a small part of the skull base lateral to this, but most of the squamous part is in the temporal fossa on the side wall of the skull. There is an anterior angle between the squamous and petrous parts, and here the greater wing of the sphenoid is slotted in. The *tympanic plate*, rolled up like a scroll, lies below the petrous and squamous parts, and behind it the *styloid process* projects like a nail from the surface of the petrous bone. Study these four parts of the temporal bone in turn (Fig. 8.4).

The mastoid process is grooved, sometimes deeply, by the digastric notch for the origin of the posterior belly of the digastric. Medial to this notch a groove for the occipital artery indents the bone along the temporo-occipital suture. The base of the **styloid process** lies one-third of the way from the tip of the mastoid process to the pterygoid tubercle. The length of the process is very variable. The stylo-pharyngeus arises high up medially, the stylo-hyoid high up posteriorly, and the stylo-glossus low down in front. The stylo-hyoid

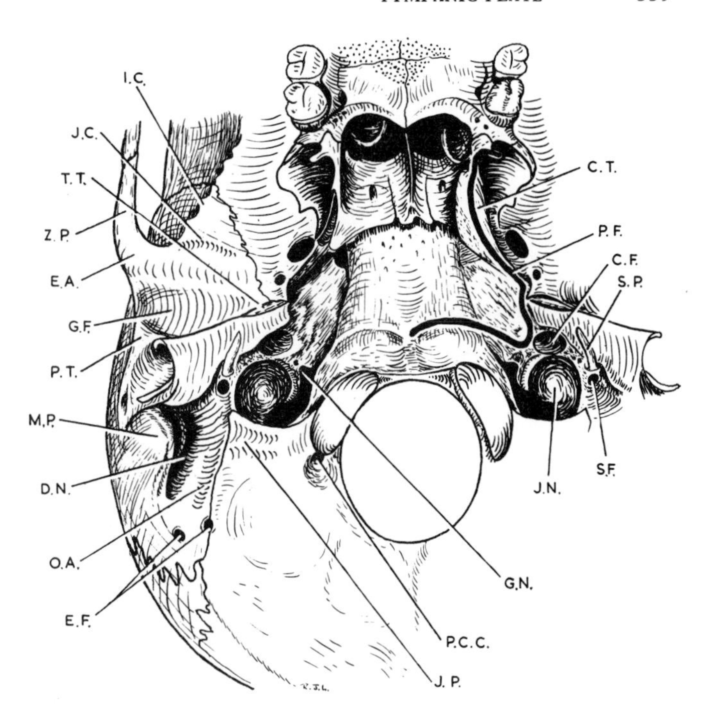

FIG. 8.4. INFERIOR VIEW OF RIGHT TEMPORAL BONE. THE LEFT CARTILAGINOUS AUDITORY TUBE IS SHOWN, WITH THE ATTACHMENT OF THE LEFT HALF OF THE PHARYNGO-BASILAR FASCIA. THE LEFT FORAMEN LACERUM IS CLOSED BY FIBROUS TISSUE. **C.T.** Cartilaginous part of auditory tube. **P.F.** Attachment of pharyngo-basilar fascia. **C.F.** Carotid foramen. **S.P.** Styloid process. **S.F.** Stylo-mastoid foramen. **J.N.** Jugular notch in temporal bone. **G.N.** Glosso-pharyngeal notch. **P.C.C.** Posterior condylar foramen. **J.P.** Jugular process of occipital bone. **E.F.** Mastoid emissary foramina. **O.A.** Occipital artery groove. **D.N.** Digastric notch. **M.P.** Mastoid process. **P.T.** Postglenoid tubercle. **G.F.** Glenoid fossa. **E.A.** Eminentia articularis. **Z.P.** Zygomatic process of temporal bone. **T.T.** Tegmen tympani. **J.C.** Attachment of joint capsule. **I.C.** Infratemporal crest.

ligament passes on from its tip. Behind its base is the stylo-mastoid foramen, transmitting the facial nerve and the stylo-mastoid branch of the posterior auricular artery. Medial to the styloid process the petrous bone is deeply hollowed out to form the jugular notch. The shallower jugular notch in the occipital bone lies behind, and the two notches form the jugular foramen, which here lodges the jugular bulb at the beginning of the internal jugular vein. Antero-medially each bone has a smaller notch for the anterior part of the jugular foramen. Here emerge the Xth and XIth cranial nerves with, in front of them, the IXth nerve and the inferior petrosal sinus. The IXth nerve itself makes a deep notch in the petrous bone (better seen from the posterior cranial fossa) and here lies its sensory ganglion (Fig. 8.4). In the floor of the notch is a foramen which is the exit of the aqueduct of the cochlea.

Anterior to the jugular notch the **petrous bone** is perforated by the carotid foramen. The internal carotid artery enters here and turns forward into the bone. The carotid sheath is attached to the margins of the carotid foramen. The ridge of bone between the jugular notch and the carotid foramen is perforated by multiple small foramina. One of these (unidentifiable) transmits the tympanic branch of the IXth nerve; the others carry carotico-tympanic branches from the internal carotid artery and sympathetic filaments from the carotid plexus. Antero-lateral to the carotid foramen, at the margin of the petrous bone, is the opening of the *bony part of the auditory tube*, overlapped somewhat by the spine of the sphenoid. Introduce into the opening a bristle from the housewife's broom, and note that it passes back into the middle ear. The apex of the petrous bone, rectangular from this view, passes forward at 45 degrees to lodge itself between the basi-occiput and greater wing of the sphenoid. No sutures are formed here, and a wide slit often lies between the petrous bone and the greater wing. The cartilaginous part of the auditory tube is lodged in the slit, on a 45 degree antero-medial slant (Fig. 8.4). Levator palati arises from the rectangular area at the apex of the petrous and from the cartilaginous tube. The pharyngo-basilar fascia is attached behind levator palati, anterior to the carotid foramen, and the mucous membrane of the naso-pharynx is attached in front of the muscle. The extreme tip of the petrous bone, perforated by the internal carotid artery as it turns up into the middle cranial fossa, varies greatly in shape. This irregularity forms a boundary of the *foramen lacerum* between the basi-occiput and the greater wing of the sphenoid. The foramen lacerum is completely closed here in life by dense fibrous tissue that extends across from the periosteum of the adjacent bones (Fig. 8.4). Apart from a few minute vessels nothing passes through the foramen lacerum from the cranial cavity to the exterior.

The Tympanic Plate (Fig. 8.1). This scroll-like bone is projected into a sharp lower border which lies in the characteristic 45 degree antero-medial obliquity. It makes a ripple where it crosses in front of the styloid process (Fig. 6.2, p. 361). This small wave in the bone is dignified by the name of vaginal process. The deep layer of the parotid fascia is attached to the sharp border as far medially as the carotid foramen. The parotid fascia is thickened between the vaginal process and angle of the mandible to receive the name of stylo-mandibular ligament. The thickening might unambiguously have been called the vagino-mandibular ligament.

The tympanic plate is C-shaped in section, open above. Laterally it is lodged below the squamous and mastoid parts, and here forms the bony part of the **external acoustic meatus.** More medially it lies below an excavation in the petrous part and here forms the bony wall of the middle ear. Now make the postponed inspection of the lateral margin of the tympanic plate. The bone is everted somewhat, and to its margins is attached the cartilaginous part of the external acoustic meatus. Posteriorly the plate lies against the mastoid process, and the trespassing auricular branch of the vagus emerges from its imprisonment between the two bones. Anteriorly the plate rests against a flange of squamous bone appropriately named the postglenoid tubercle (Fig. 6.2, p. 361—unlabelled). Note now that the squamous bone carries a sharp crest in the gap between the margins of the tympanic plate. This is called the *suprameatal spur*, and it completes the bony ring of the external acoustic meatus. A horizontal upper tangent and a vertical posterior tangent of the meatus enclose with the suprameatal spur a concavity named the *suprameatal triangle*. This is the mark for the tympanic antrum, which lies 15 mm deep to the surface. In the depths of the meatus can be seen the promontory (Fig. 6.2, p. 361), sometimes etched by fine lines made by the tympanic plexus. The round and oval windows can be seen too, for there is no tympanic membrane to obstruct this view of the medial wall of the middle ear.

Return now to the inferior view of the skull base. The tympanic plate and squamous bone meet at a transverse *squamo-tympanic fissure* deep in the glenoid fossa. Here is attached the capsule of the mandibular joint. In the depths of this fissure, well to the medial end, behind the spine of the sphenoid, is a thin flange of bone. This is the projecting margin of the *tegmen tympani*, part of the petrous bone that has grown down from the roof of the middle ear (see p. 49). It divides the fissure narrowly into a petro-squamous in front, for the capsule of the jaw joint, and the *petro-tympanic fissure* behind (Fig. 8.4). Deep in the petro-tympanic fissure the chorda tympani emerges and passes down in a groove on the *medial* surface of the spine of the sphenoid. Inspect once again the opening of the bony part of the auditory tube and note that the flange from the tegmen tympani forms its lateral boundary. The flange encloses, in fact, the canal for the tensor tympani muscle (Fig. 8.9).

The **squamous part** of the temporal bone is hollowed by the glenoid fossa. The gutter is not precisely transverse, in conformity with the obliquity of the condyle of the mandible (p. 446). The capsule of the joint is attached to the medial and lateral margins of the glenoid fossa. In front of the glenoid fossa is the convexity of the *eminentia articularis*. The lateral extremity of this transverse ridge projects at the lower margin of the zygomatic process. Anterior to the convexity of the eminentia is a curved line which marks the attachment of the capsule of the mandibular joint (Fig. 8.5). A small triangular area of squamous bone in front of this forms part of the infratemporal surface of the skull.

The Sphenoid Bone. Only the greater wing enters into the base of the cranium here. It forms a suture with the squamous temporal. Together these make the infratemporal surface, the size of a thumb print, and the pad of the thumb rests comfortably here. The upper head of lateral pterygoid arises from this surface. Anteriorly the surface ends at the inferior orbital fissure behind the maxilla. Medially the greater wing ends in a straight edge alongside the 45 degree obliquity of the apex of the petrous bone. The posterior end of this edge projects

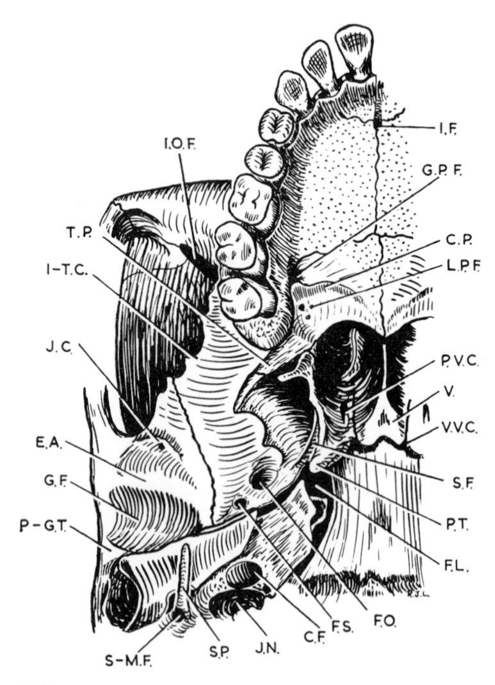

FIG. 8.5. THE INFRATEMPORAL SURFACE OF THE SKULL. **I.F.** Incisive foramen. **G.P.F.** Greater palatine foramen. **C.P.** Crest of palatine bone. **L.P.F.** Lesser palatine foramina. **P.V.C.** Palatino-vaginal canal. **V.** Vomer. **V.V.C.** Vomero-vaginal canal. **S.F.** Scaphoid fossa. **P.T.** Pterygoid tubercle. **F.L.** Foramen lacerum. **F.O.** Foramen ovale. **F.S.** Foramen spinosum. **C.F.** Carotid foramen. **J.N.** Jugular notch. **S.P.** Styloid process. **S.-M.F.** Stylo-mastoid foramen. **P.-G.T.** Postglenoid tubercle. **G.F.** Glenoid fossa. **E.A.** Eminentia articularis. **J.C.** Joint capsule attachment. **I.-T.C.** Infra-temporal crest. **T.P.** Tubercle of palatine bone. **I.O.F.** Inferior orbital fissure.

down as the spine of the sphenoid, whose base is perforated by the foramen spinosum for the middle meningeal artery. The spheno-mandibular ligament is attached to the tip of the spine. Forwards and medially is the much larger foramen ovale for the mandibular nerve (Fig. 8.5). Medial to the foramen ovale may be a small *foramen of Vesalius* for a vein joining the cavernous sinus to the pterygoid plexus. Behind the foramen ovale, less commonly, may be the small foramen innominatum for the lesser (superficial) petrosal nerve. When these foramina do not exist the vein and the nerve traverse the foramen ovale itself. Tensor palati arises from the edge of the greater wing (between the scaphoid fossa and the spine) and from the cartilaginous tube alongside it.

The Face. The inferior view exposes the hard palate and the back of the nose, with the pterygoid plates (Fig. 8.5). The medial and lateral pterygoid plates project down from a common pterygoid process of the sphenoid at the base of the skull.

The **medial pterygoid plate** forms the posterior boundary of the nose. Its lower end is projected laterally as a delicate spur, called the *pterygoid hamulus* (Fig. 8.3). To its tip is attached a fibrous band (the pterygo-maxillary ligament) which passes across to the tuberosity of the maxilla. This band gives origin to the buccinator, and it converts the curved hamulus into a fibro-osseous canal for the tensor palati (Fig. 6.14, p. 381). Attached to the tip of the hamulus is the pterygo-mandibular raphé. The posterior border of the medial pterygoid plate is projected into a spur, below which the pharyngo-basilar fascia and the superior pharyngeal constrictor are attached down to the hamulus. The expanded free end of the cartilaginous tube is slotted against the border of the medial pterygoid plate above the spur. The upper end of the medial plate projects back into the foramen lacerum as the pterygoid tubercle already seen (Fig. 8.5). From the spur a slender ridge runs up to the base of the skull towards the lateral pterygoid plate; a deeply concave fossa is thus enclosed lateral to the pterygoid tubercle. This is the **scaphoid fossa,** and from it the anterior fibres of tensor palati arise. The pterygoid tubercle projects beyond the opening of the **pterygoid canal** in the foramen lacerum. A bristle inserted here will traverse the canal and enter the pterygo-palatine fossa. This access from below is closed in life by a dense fibrous tissue that floors the foramen lacerum, but in the dried skull it is easier to enter the canal from below. In some skulls the opening of the canal is actually visible from below (see p. 563). A glance at a disarticulated sphenoid bone will clarify this point (Fig. 8.8).

In the roof of the nose the medial pterygoid plate extends medially below the body of the sphenoid to articulate with the vomer. This medial extension of the root of the medial pterygoid plate, ensheathing the body of the sphenoid, is known as the **vaginal process.** A vomero-vaginal suture is thus formed, opening posteriorly as the vomero-vaginal canal. Nothing goes through this canal, which does not deserve a name. In the lateral wall of the nose the vertical plate of the palatine bone articulates with the medial pterygoid plate (p. 402), and the articulation between the palatine bone and vaginal process of the medial pterygoid plate encloses the **palatino-vaginal canal.** This opens posteriorly in the vault of the nose half a centimetre in front of its posterior limit (Fig. 8.5, **P.V.C.**). Pass a bristle through the palatino-vaginal canal; it enters the pterygo-palatine fossa, as can be seen by a glance at the lateral surface of the skull, looking into the pterygo-maxillary fissure. The pharyngeal branch of the pterygo-palatine ganglion runs back through the palatino-vaginal canal and supplies the back of the roof of the nose and thence the naso-pharynx. It is accompanied by a branch from the third part of the maxillary artery.

The **lateral pterygoid plate** extends back and laterally into the space of the infratemporal fossa. It is a bony flange whose only purpose is to give attachments to the pterygoid muscles. The whole of its lateral surface gives origin to the lower head of the lateral pterygoid muscle, while the whole of its medial surface gives origin to the upper head of the medial pterygoid. The lateral plate is attached to the base of the skull at the anterior pole of the foramen ovale; it springs from a common stem with the medial plate.

The adjacent lower borders of the two plates are excavated into a triangular notch (Fig. 8.8). This is filled by the triangular extremity of the tubercle of the palatine bone (Fig. 8.12).

The Hard Palate (see p. 409). The hard palate is arched more by the downward projecting alveolar processes of the maxillæ than by any upward concavity of the palatal processes. The incisive foramen and the greater and lesser palatine foramina have already been studied. The sharp crest of the palatine bone forms the posterior border of the greater palatine notch, which fits a similar notch in the body of the maxilla to make the greater palatine foramen. The crest curves back towards the midline, where it becomes less prominent and fades out in the backward projection of the bone at the midline posterior nasal spine. Behind the crest the palatine bone is perforated by two or three lesser palatine foramina. Here the palatine bone forms its *tubercle*, which projects laterally behind the tuberosity of the maxilla and then expands into a small *pyramidal process* which occupies the notch between the lower borders of the pterygoid plates of the sphenoid (Fig. 8.13).

The *teeth* (p. 406) are carried in the alveolar process of the maxilla, which projects behind the palatal process to articulate \ :th the horizontal plate of the palatine bone (Fig. 8.5). Two incisors, one canine, two premolars (bicuspids) and three molars make up the permanent complement of each maxilla, matching those in each half of the mandible (see p. 406).

The Nasal Walls

The osteology of the nasal septum and lateral wall of the nose has been dealt with on p. 401. Study these in a split skull.

Internal Features of the Skull

The sutures are much less tortuous inside than outside and they ossify sooner; that is to say, each suture ossifies slowly from within out, and this process begins at about 40 years.

The Calvaria. The inner surface of the vault shows a midline groove, widening as it is traced back, for the superior sagittal sinus. Pits in and lateral to the groove are the indentations of arachnoid granulations, those outside the sinus being in the lateral blood lakes. The grooves for the anterior and posterior branches of the middle meningeal vessels run back and up across the side of the skull and reach the vault. Their surface markings can now be checked (p. 482).

The *base* of the skull is in three levels, like the steps of a staircase, and these are called the anterior, middle and posterior **cranial fossæ**. The anterior fossa lodges the frontal lobe and its floor is level with the upper margin of the orbit. The middle fossa lodges the temporal lobe and its floor is level with the upper border of the zygomatic arch. The posterior fossa lodges the brain stem and cerebellum and the attachment of its roof (the tentorium) lies behind the suprameatal crest along the superior nuchal line, at the upper limit of the soft neck muscles.

The **anterior fossa** lies highest of the three. Its posterior boundary is a sharp concavity made by the lesser wing of the sphenoid (Fig. 6.89, p. 481). Laterally the lesser wing meets the frontal bone and greater wing, and so the posterior border is continued to the side wall of the skull at the pterion. Here the bone is commonly tunnelled for a short distance by the anterior branch of the middle meningeal artery running up and back. The medial end of the lesser wing is projected back as the anterior clinoid process. In front of this the base of the lesser wing is perforated by the optic canal. The two optic foramina are joined by the optic groove, but by common consent the groove is accepted as being in the middle fossa. Thus in the anterior fossa the bone in front of the optic groove is the body of the sphenoid and this articulates with the cribriform plate (in many Asiatic skulls the two halves of the frontal bone fuse in the midline between the cribriform plate and the body of the sphenoid).

The orbital plate of the frontal bone is the largest contributor to the anterior fossa. Over the orbit it is convex and ridged in roughly H-shape in conformity with the orbital surface of the frontal lobe of the cerebral hemisphere. The frontal sinus invades a variable area at the antero-medial part of the roof of the orbit. Medially the upward convexity is replaced by a flatter part which roofs in the ethmoidal sinuses and articulates with the cribriform plate. Anteriorly the groove for the sagittal sinus is traceable down as a midline crest for the falx cerebri, and behind the lower end of the crest is the foramen cæcum, which is plugged by the fibrous tissue of the falx, making a huge Sharpey's fibre. The midline of the cribriform plate is projected up as a sharp triangle called the crista galli, for attachment of the falx. Alongside the anterior end of the crista galli is the elongated nasal slit. Pass a bristle through the anterior ethmoidal foramen in the medial wall of the orbit; it will pass obliquely forwards between frontal and ethmoid bones and lie over the cribriform plate. This is the oblique course of the anterior ethmoidal nerve and vessels (Fig. 6.89, p. 481), which, beneath the fibrous dura mater, lie on the cribriform plate and enter the roof of the nose by way of the nasal slit. The remaining perforations, small and round, in the cribriform plate are for the olfactory nerves. At the postero-lateral angle of the cribriform plate a shallow fossa indents the frontal bone; this lodges the olfactory bulb.

The **middle cranial fossa** is butterfly-shaped. The small 'body' of the butterfly is the body of the sphenoid between the clinoid processes, while the 'wings' of the butterfly expand hugely into a concavity that extends out to the lateral wall of the skull and back to the upper border of the petrous bone. The body of the sphenoid is centrally hollowed out into the sella turcica ('Turkish saddle') with a hump in front called the tuberculum sellæ (Fig. 8.6). At the back a transverse flange projects up called the dorsum sellæ ('the back of the saddle'), and the upper border of the dorsum sellæ ends at each

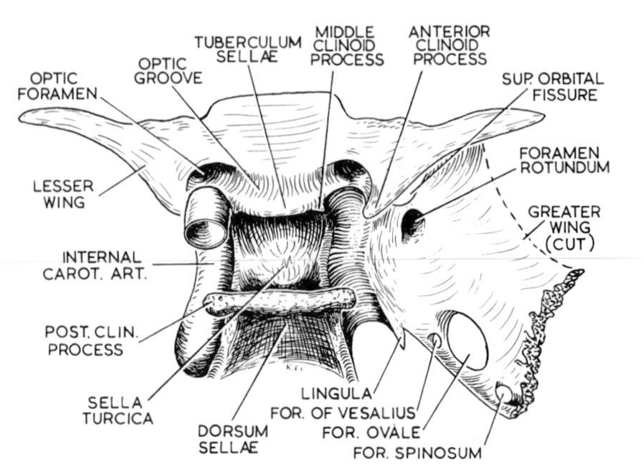

FIG. 8.6. THE SELLA TURCICA.

side as the posterior clinoid process (a bed-post on the back of the saddle—such is anatomical nomenclature even after three changes in the last half-century).

A sheet of fibrous dura mater sweeps across from the tuberculum sellæ to the anterior and posterior clinoid processes to roof the pituitary fossa and the cavernous sinuses. The diaphragma sellæ (p. 489) is perforated centrally for the pituitary stalk, and the roof of each cavernous sinus is pierced anteriorly by the internal carotid artery, which grooves the medial edge of the anterior clinoid process. Medial to the carotid groove is an elevation called the *middle clinoid process* (Fig. 8.6). It lies at the lateral end of the tuberculum sellæ. To it and to the posterior clinoid process is attached a flange of dura mater that descends vertically between the cavernous sinus and the pituitary fossa, and sweeps medially to floor the sella turcica. The tip of the anterior clinoid process gives attachment to a ridge of dura mater along the lateral border of the roof of the cavernous sinus; the ridge of dura passes lateral to the posterior clinoid process and is continued posteriorly as the free edge of the tentorium cerebelli (Fig. 6.89, p. 481).

In front of the tuberculum sellæ and carotid grooves the fibrous dura mater floors the optic groove and is projected into the optic canals around the optic nerves as they pass into the orbits. The optic chiasma lies at a higher level well behind the optic groove. Vertically above the optic groove is a space in which lies the anterior communicating artery and above that is the rostrum; this space is the cisterna chiasmatica.

The side of the body of the sphenoid is grooved by the S-shaped internal carotid artery from the foramen lacerum to the anterior clinoid process (Fig. 6.94, p. 487). Lateral to this the floor of the middle fossa is made by the greater wing of the sphenoid and the temporal bone. The greater wing here borders on the apex of the petrous bone and the side of the body of the sphenoid bone, enclosing the irregular foramen lacerum. The internal carotid artery emerges from the apex of the petrous bone and then deeply grooves the sphenoid. The lateral ridge of the groove in the sphenoid is often prominent, and this projection is called the *lingula* (Fig. 8.6). Beneath the lingula is the *opening of the pterygoid canal*, but it is more difficult to introduce a bristle into the foramen under the lingula than from the base of the dried skull beneath the pterygoid tubercle (p. 561). This is, however, the course of the greater (superficial) petrosal nerve, for the foramen lacerum is floored in by dense fibrous tissue. The nerve passes below the trigeminal ganglion and the internal carotid artery, above the fibrous floor of the foramen lacerum, to join the deep petrosal nerve at the pterygoid canal. Lateral to the foramen lacerum the apex of the petrous bone is grooved by the trigeminal ganglion (Fig. 6.89, p. 481).

At the tip of the apex of the petrous bone is a sharp spike, medial to which is a narrow groove for the inferior petrosal sinus and the VIth nerve. To this spike is attached the petro-clinoid ligament; it extends from the spike to halfway up the side of the dorsum sellæ (Fig. 6.94, p. 487). Lateral to the spike is a broad shallow groove made by the sensory root of the Vth nerve (the small motor root is deep to it). The trigeminal groove passes forward into the fossa for the trigeminal ganglion. Lateral to this the posterior border of the middle fossa is grooved by the superior petrosal sinus, and to the lips of this narrow groove the tentorium cerebelli is attached, straddling the sinus.

The posterior border of the middle fossa is the upper border of the petrous bone. At its junction with the side wall of the skull it is grooved by the inferior temporal gyrus (Fig. 6.94, p. 487). Medial to the groove is a prominence called the *arcuate eminence*, which is made by the underlying anterior semicircular canal. Medial to the arcuate eminence the petrous bone passes forwards and medially at 45 degrees. Its upper surface is perforated and grooved by the greater petrosal nerve; the groove passes obliquely into the foramen lacerum, and the nerve lies beneath the trigeminal ganglion here. Parallel and antero-lateral to this is a small groove made by the lesser petrosal nerve; this groove is directed towards the foramen ovale.

In front of the petrous part of the temporal bone is the greater wing of the sphenoid and a small part of the squamous temporal; this area has already been seen from below as the thumb-print sized roof of the infratemporal fossa (p. 560). The greater wing is perforated by the small foramen spinosum and, antero-medial to this, by the much larger foramen ovale. From the foramen spinosum a groove for the middle meningeal artery leads forward; this soon splits into anterior and posterior branches whose courses are marked by the grooves already noted. The grooves are made by the middle meningeal veins.

The *foramen ovale* perforates the greater wing in front of the trigeminal fossa at the apex of the petrous temporal bone. The foramen ovale transmits the mandibular nerve, whose sensory and motor roots join in the foramen. A vein from the cavernous sinus passes down to the pterygoid plexus. The accessory meningeal artery runs up to supply the trigeminal ganglion, and the meningeal branch of V c (nervus spinosus) commonly runs with the artery instead of going through the foramen spinosum. Just medial to the foramen ovale the fibrous dura mater peels away from the 'endosteal layer' to stretch up as the lateral wall of the cavernous sinus. In front of the foramen ovale is the *foramen rotundum*, which opens forwards into the pterygo-palatine fossa. It transmits the maxillary nerve. The fibrous dura mater of the lateral wall of the cavernous sinus blends with the periosteum of the greater wing at the lateral margin of the foramen rotundum.

The greater wing with its deeply concave forward projection into the lateral wall of the orbit fails to meet the lesser wing; the slit between them is the superior orbital fissure. Lateral to the line of the foramen rotundum this is closed by fibrous dura mater. Medial to the line of the foramen rotundum the medial end of the superior orbital fissure is open for the anterior end of the cavernous sinus and the nerves that run along it into the orbit (p. 436). Further laterally the greater wing, very thin, makes the lateral wall of the middle fossa, which is the floor of the temporal fossa on the side of the cranium, behind the orbit. Behind this the lateral wall of the middle cranial fossa is made by the squamous part of the temporal bone. At its junction with the floor (petrous bone) it is perforated by the meningeal branch of V c (nervus spinosus) on its way to the mastoid antrum and air cells.

The anterior end of the middle fossa is roofed by the projecting lesser wing of the sphenoid. From the pterion the spheno-parietal sinus runs below the margin of the lesser wing (i.e. in the roof of the middle fossa) to the anterior clinoid process, where it enters the roof of the cavernous sinus.

The **posterior cranial fossa,** deeply concave, lies above the foramen magnum. Anteriorly, its upper limit is the upper border of the petrous temporal bone. Behind this at the same horizontal level is a wide groove on the inner surface of the skull which extends to the midline. Here the two grooves meet at the *internal occipital protuberance*, which lies opposite the external occipital protuberance. The grooves are made by the transverse sinuses. Above the internal occipital protuberance (i.e., above the posterior fossa) is the groove made by the superior sagittal sinus. Here in the supratentorial part of the occipital bone is a concavity on either side for the occipital pole of the cerebral hemisphere. At the internal occipital protuberance the sagittal groove turns to one side (usually the right) into the transverse groove along the roof of the posterior fossa. The other transverse groove (usually the left) is narrower; it begins at the internal occipital protuberance by the inflow of the straight sinus. The transverse groove, straddled by the tentorium cerebelli, runs forward along the occipital bone to the inferior angle of the parietal and so to the mastoid part of the temporal. Here at the roof of the posterior fossa the groove for the superior petrosal sinus joins it. The *sigmoid groove* indents the cranial surface of the mastoid bone, and a mastoid emissary canal often runs posteriorly from this part (p. 484). Lower down the sigmoid groove indents the jugular process of the occipital bone, and here an emissary canal (the posterior condylar canal) runs back to open behind the occipital condyle. The jugular foramen (p. 492) is formed between the deep jugular notch of the petrous bone and the shallow jugular notch of the occipital bone. Right and left sides are seldom of equal size; the larger jugular foramen is almost always that which receives the superior sagittal sinus via the transverse and sigmoid sinuses, and this is more commonly the right one.

Running up from the foramen magnum to the dorsum sellæ is a broad groove known as the *clivus*. Just above the foramen magnum each border of the groove shows a rounded prominence called the *jugular tubercle*. This is the line of fusion of the basi-occiput and the ex-occiput (p. 567). It lies above the occipital condyle, and between them the bone is perforated obliquely (almost transversely) by the hypoglossal canal (Fig. 6.94, p. 487). The hypoglossal nerve enters here as two roots (Fig. 6.96, p. 492). They are separated by a flange of dura mater that often ossifies. Between the jugular tubercles the clivus is occupied by the medulla, and the groove behind the jugular tubercle lodges the IXth, Xth and XIth nerves on their way to the jugular foramen. Above the jugular tubercles the groove of the clivus broadens to indent the apex of each petrous temporal almost as far as the internal acoustic meatus; this broad groove lodges the pons. A glance into the posterior fossa will show the imprint of pons and medulla, with the jugular tubercle lying in the angle between them. Opposite the upper borders of the petrous bones the basi-occiput and basi-sphenoid meet; the cartilaginous epiphysis between them ossifies at 25 years. The groove of the inferior petrosal sinus indents the adjacent margins of the basi-occiput and temporal bones. The osteology of the petrous temporal bone in the posterior fossa is considered on p. 491. Note especially the internal acoustic meatus, the glosso-pharyngeal notch in the jugular foramen, and the opening of the aqueduct of the vestibule.

The internal occipital crest runs down in the midline from the internal occipital protuberance; to it is attached the narrow falx cerebelli over the occipital sinus. To either side of the internal occipital crest the deep concavities of the occipital bone lodge the cerebellar hemispheres; the external surface of this bone is covered by the insertions of the extensor muscles of the back of the neck.

Ossification of the Skull

In the anterior fossa only the ethmoid ossifies in cartilage. The floors of the middle and posterior fossæ ossify in cartilage. The rest of the skull, cranium above and face below, ossifies in membrane. Details of ossification are noted in the following descriptions of individual bones.

SEPARATE SKULL BONES

Most of the essential features of each bone have been seen in the study of the intact skull, but the new views obtained by handling the separate bones complete the survey, and the *additional* features thus displayed are now to be described. Each bone should be orientated for study alongside an intact skull.

The Frontal Bone. Inspect the disarticulated bone from below. There is a deep *ethmoid notch* between the concave orbital roofs. Alongside the notch the lower surface of the orbital plate shows fossæ indented by the upper limits of the ethmoid sinuses, and anterior to these the frontal sinus opens (Fig. 8.7). Note that the rough mark for the trochlea in the roof of the orbit lies below the floor of the frontal sinus. The *foramen cæcum* in the anterior cranial fossa, at the lower limit of the frontal crest, is now seen to be a blind pit; it does not perforate the frontal bone.

Trace the edge to edge articulation of adjacent skull bones. The nasal bone and frontal process of the maxilla articulate side by side over a wide area of rough bone which projects down medial to the superior orbital margin. Behind this is the area of ethmoid sinus indentation, with the oblique groove for the anterior ethmoidal nerve and the transverse groove for the posterior ethmoidal nerve crossing between the fossæ (Fig. 8.7). This area fits like a lid over the lateral mass of the ethmoid. The cribriform plate of the ethmoid fills the ethmoid notch. The lesser wing of the sphenoid articulates along the posterior border of the orbital roof. Lateral to this is a broad rough area for the greater wing of the sphenoid. The lateral margin of the orbit is completed in front by the zygomatic bone. The posterior border of the *squamous part* of the frontal bone, pitted by the diploë, articulates above the greater wing of the sphenoid with the parietal bone. The meeting place of the three bones is the pterion.

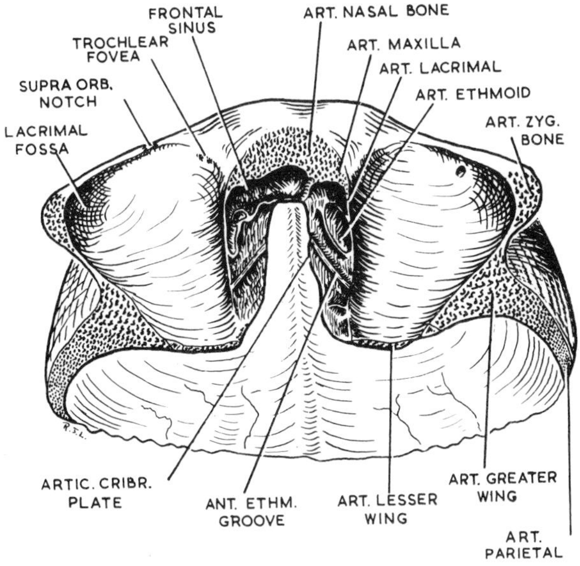

FIG. 8.7. INFERIOR VIEW OF THE FRONTAL BONE.

Ossification is wholly in membrane. A centre appears above each orbital margin and the bone develops in two halves. The metopic suture between them unites in the second year.

The Sphenoid Bone. Take the opportunity of noting certain features not visible in the intact skull. The greater wing and pterygoid process project from a common junction with the lower part of the body. The **pterygoid canal** perforates the root of the pterygoid process. It is almost as large as the foramen rotundum in the nearby greater wing. Its posterior aperture can now be confirmed as lying between the lingula and the pterygoid tubercle. Its anterior opening lies medial to the foramen rotundum (Fig. 8.8). The smooth area of the pterygoid process below these foramina is the posterior wall of the *pterygo-palatine fossa*. The rough area

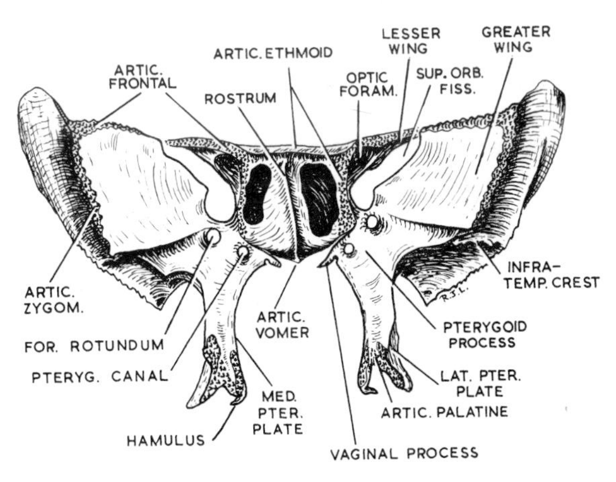

FIG. 8.8. ANTERIOR VIEW OF THE SPHENOID BONE.

below this opens into a notch between the lower ends of the pterygoid plates; the tubercle of the palatine bone articulates here. The vertical plate of the palatine bone articulates with the medial pterygoid plate to form the medial wall of the pterygo-palatine fossa. The *vaginal process* of the medial pterygoid plate is grooved at the base of its lower surface (Fig. 8.8); the palatine bone extends up to cover this groove and convert it into the palatino-vaginal canal. The body of the sphenoid shows an inferior ridge for the ala of the vomer, which fits under the vaginal process. The front of the body of the sphenoid is projected into the prominent *rostrum*, which articulates with the perpendicular plate of the ethmoid at the upper part of the nasal septum far back. The rostrum is projected back into the body as a septum between the two sphenoidal air sinuses; the septum is always deviated. The sinuses can be inspected; they excavate the bone to a very variable extent, seen better in a sagittal section of the skull.

Trace the articulations of adjacent skull bones. Posteriorly the body (here called basi-sphenoid) articulates with the basi-occiput at a cartilaginous epiphysis which ossifies at the 25th year. The straight postero-medial border of the greater wing lies near the apex of the petrous bone at the foramen lacerum. It is grooved for the cartilaginous part of the auditory tube. The curving postero-lateral border articulates with the squamous temporal bone by a bevelled suture that slants between the two bones. In the infratemporal fossa at the base of the skull the greater wing overlies the squama, in the temporal fossa on the side of the skull the squamous bone overlaps the greater wing. The upper border of the greater wing articulates with the parietal and frontal; the frontal articulation extends medially to bridge the superior orbital fissure between greater and lesser wings. The optic foramen perforates the base of the lesser wing; medial to this the ethmoid labyrinth articulates. The anterior border of the greater wing articulates with the zygomatic bone. The articulations with the vomer and palatine bone have been noted.

Ossification. The floors of the middle and posterior cranial fossæ ossify in cartilage. Bone above and below this (cranium and face) ossifies in membrane. Accordingly the body, lesser wing, and base of the greater wing ossify in cartilage. The rest of the greater wing and the pterygoid plates ossify in membrane. Centres appear at the end of the 2nd month, and it is unpractical to attempt to memorize them.

The Temporal Bone. The features not visible in the intact skull should be noted. The petrous part is perforated by the carotid canal, in which the internal carotid artery bends through a right angle. The aperture of exit lies well to the lateral side of the apex, in the foramen lacerum. Behind this, on the lateral surface, the petrous bone is perforated by the bony part of the auditory tube; pass a bristle through it into the middle ear. A ridge above the tube floors in the canal for the tensor tympani muscle, which canal is roofed in by an extension of petrous bone from the tegmen tympani (Fig. 8.9). The flange curves down between squamous and tympanic parts to reach the base of the skull, as already noted. The deep jugular notch lodges the jugular bulb. It and the bend in the carotid canal lie very close to the floor and anterior wall of the middle ear. Note the opening of the aqueduct of the cochlea in the depths of the glosso-pharyngeal notch. The internal acoustic meatus can be studied (p. 491), and behind this the opening of the aqueduct of the vestibule can be seen

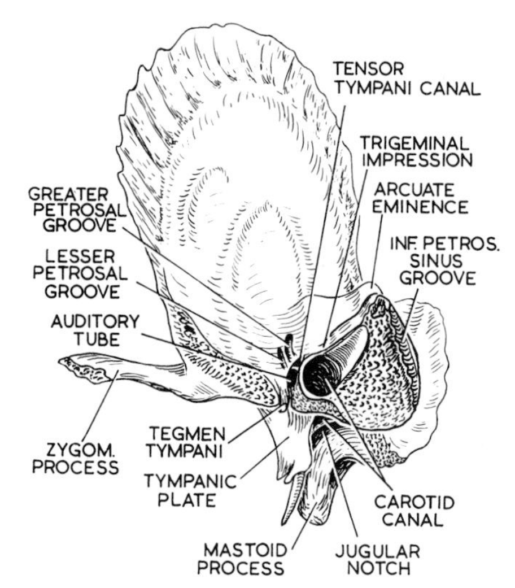

FIG. 8.9. THE RIGHT TEMPORAL BONE, LOOKING DIRECTLY AT THE APEX, TO SEE THE ORIFICE OF THE BONY PART OF THE AUDITORY (EUSTACHIAN) TUBE.

under a flange of bone that hangs down like a tiny curtain (Fig. 6.94, p. 487). The middle ear (p. 448) and labyrinth (p. 453) have already been studied.

Trace the articulations of adjacent skull bones. The apex of the petrous part is lodged between basi-occiput and greater wing of sphenoid. The bevelled and serrated edge of the squamous part articulates in front with the greater wing and above with the parietal bone, which it widely overlaps. Posteriorly the mastoid part articulates with the parietal and occipital. The suture at the tip of the zygomatic process is very oblique.

Ossification. The petro-mastoid part, in the skull base, ossifies in cartilage at the middle of pregnancy. The styloid process (not skull, but 2nd pharyngeal arch) begins to ossify at the end of pregnancy. The membrane parts of the bone (squamous and tympanic ring) ossify at the 8th week. Subsequent growth is described on p. 48.

The Occipital Bone. Examination of the disarticulated bone shows no features not already seen in the intact skull. Note that the jugular process, which forms the posterior border of the jugular foramen, is grooved in the posterior fossa by the sigmoid sinus and from this groove the posterior condylar canal (when present) passes back above the occipital condyle.

Trace the articulations of adjacent skull bones. The basi-occiput articulates, on the clivus, with the basi-sphenoid by a cartilaginous epiphysis that ossifies at the 25th year. Along each border of the basi-occiput is a groove for the inferior petrosal sinus; here the apical part of the petrous temporal articulates. Below the groove the two bones part to enclose the jugular foramen. Behind the foramen the mastoid part of the temporal bone articulates, and above this the parietal bone meets the occipital up to the lambda.

Ossification. According to the pattern (the base of the skull ossifies in cartilage), only the apical part of the squama, above the highest nuchal line, ossifies in membrane. There are four other centres. All centres appear at the end of the 2nd month. The basi-occiput (one centre) joins each jugular part, called *ex-occiput* (one centre), by a cartilaginous epiphysis across the occipital condyle and jugular tubercle at the hypoglossal canal. The squama is a combination of cartilaginous and membrane bones which join very quickly; failure of union results in a separate 'interparietal bone' below the lambda. The squamous bone, so formed, meets the pair of ex-occiputs along a cartilaginous junction that forms a posterior tangent to the foramen magnum (Fig. 8.10).

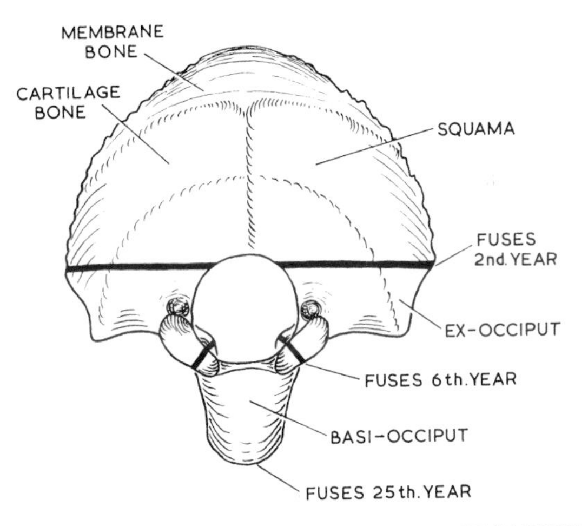

FIG. 8.10. THE FOUR PARTS OF THE OCCIPITAL BONE AT BIRTH.

Thus at birth the occipital bone is in four parts; basi-occiput, a pair of ex-occiputs, and the squama. The squama fuses with the pair of ex-occiputs at the end of the 2nd year, when the milk dentition is complete. The pair of ex-occiputs fuse with the basi-occiput (across the condyles) at the 6th year, when the permanent dentition begins to erupt. The basi-occiput fuses with the basi-sphenoid at the 25th year, when the permanent dentition is complete. Thus the backward elongation of the palate to accommodate the teeth is matched by a compensating growth of the skull base to keep the naso-pharynx patent.

The Parietal Bone. The isolated bone exposes the diploë along the anterior, superior and posterior margins. The inferior margin is sharply bevelled where it is overlapped superficially by the squamous temporal. Thus can the isolated bone be placed in the anatomical position. To distinguish right from left the groove in the postero-inferior angle made by the sigmoid sinus may help, but much more useful is to glance at the grooves made by the middle meningeal vessels. The anterior branch grooves the bone just behind the anterior border, while branches of the posterior division make grooves that pass up and *back* towards the posterior border. The articulations with adjacent bones can be checked in the intact skull. The upper border articulates with the other parietal at the sagittal suture. The anterior border articulates with the frontal at the coronal suture. The antero-inferior angle articulates with the sphenoid, and the bevelled inferior border is overlapped by the squamous part of the temporal. The postero-inferior angle meets the mastoid part of the temporal, and the posterior border articulates with the occipital.

Ossification. The bone ossifies in membrane before the 8th week; the centre appears at the site of the future parietal eminence.

The Ethmoid. The ethmoid consists of a pair of lateral masses complicated enough to have earned the name of labyrinth. They are joined above by the cribriform plate, in the floor of the anterior cranial fossa. From the midline of the cribriform plate the crista galli projects up and the perpendicular plate projects down.

Each **labyrinth** is a thin-walled box, approximately rectangular. Anteriorly the wall is deficient and is completed by incorporation of part of the lacrimal bone. A similar posterior deficiency is made good by the sphenoidal concha and the orbital process of the palatine bone. The labyrinth has no roof; it is overlapped by the orbital plate of the frontal bone, which articulates edge to edge with the cribriform plate. The cavity of the labyrinth is occupied by air cells, whose walls are undulating flakes of paper-thin bone. Two vertical bony partitions, one in front of the other, separate the air cells into anterior, middle, and posterior groups. A third partition near the anterior end of the labyrinth makes a funnel-shaped cavity, the **infundibulum,** beneath that part of the roofing frontal bone which contains the frontal sinus. The infundibulum is open at the lower border of the labyrinth; its lining mucous membrane, known as the *fronto-nasal duct,* thus leads from the frontal sinus through the ethmoid to the middle meatus of the nose. The infundibulum may be regarded as a specialized anterior ethmoidal air cell. The lateral wall of the labyrinth, paper thin, is called the orbital plate, or pars papyracea. It can be seen in the medial wall of the orbit. The medial surface of the labyrinth is irregular. It can be inspected in a sagittal section of the skull. The superior and middle conchæ spring from a common stem and diverge from each other posteriorly. The middle concha is the larger of the two, and it projects anteriorly to articulate with the frontal process of the maxilla and posteriorly to articulate with the perpendicular plate of the palatine just below the spheno-palatine foramen.

The middle ethmoidal air cells project as the bulla, under cover of the overhanging middle concha. The lower border of the labyrinth articulates with the upper edge of the maxillary hiatus. The uncinate process curves down in front of the bulla, across the maxillary hiatus, to articulate with the inferior concha. The anterior ethmoidal air cells open just behind the infundibulum; if the bone is incomplete the infundibulum itself serves for them as well as for the frontal sinus. The middle ethmoidal air cells open on the bulla and the posterior cells open below the superior concha. The crista galli (p. 562), cribriform plate (p. 562) and perpendicular plate (p. 402) have already been studied.

The articulations of the labyrinth have been noted. The perpendicular plate, in the nasal septum, articulates anteriorly with the frontal and nasal bones, inferiorly with the septal cartilage, and posteriorly with the vomer and rostrum of the sphenoid (Fig. 6.23, p. 402).

Ossification. The ethmoid ossifies in the cartilage of the nasal capsule. The labyrinth begins to ossify at the 5th month and the process is complete before birth. The cribriform plate, crista galli and perpendicular plate are still cartilaginous at birth and begin to ossify about the first year.

The Maxilla. The features to be noted in the disarticulated bone are on its medial and posterior surfaces. All the other features have been inspected in the intact skull. Through the *hiatus* in its nasal wall the thin-walled body can be seen to contain the sinus, which tapers to its apex in the zygomatic process. Sometimes the sinus has excavated through the maxilla into the zygomatic bone (Fig. 8.11). Note now that the floor of the sinus dips into the alveolar process and lies lower than the level of the palatine process. The roof and anterior wall of the sinus show a ridge projecting down from their junction; this is made by the infra-orbital nerve in its canal (Fig. 8.11). The outlines of the anterior and middle superior alveolar canals can usually be seen through the egg-shell thin bone by transillumination. The upper border of the maxillary hiatus is at the floor of the orbit; often this thin edge of bone is here and there prised apart into two layers by excavation from the adjacent ethmoid air cells. The upper and anterior edges of the hiatus meet at a tongue-like process that protrudes forwards over the *lacrimal groove* (Fig. 8.11). This groove, wide and deep, is overlapped from in front by the base of the frontal process. The groove is directed down and somewhat back, and its concave floor makes a convex elevation on the medial wall of the sinus in front of the hiatus. The frontal process shows a roughened low ridge across its base. This is the *conchal crest*, for articulation with the inferior concha. Above this is a less prominent ridge for the anterior end of the middle concha of the ethmoid.

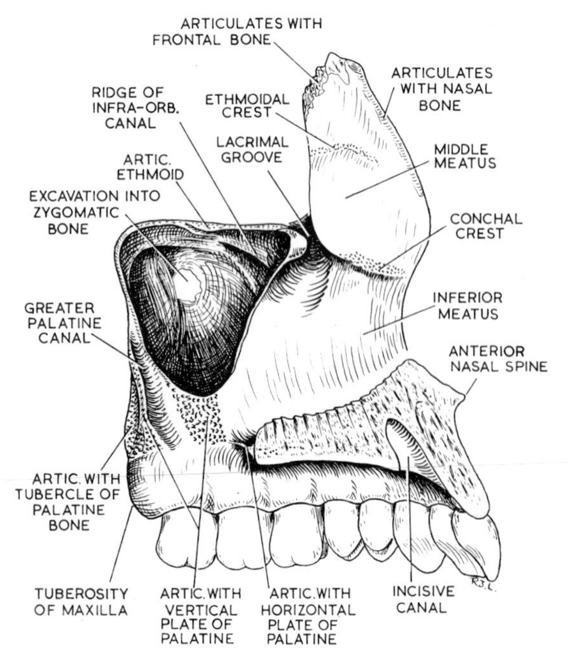

FIG. 8.11. THE LEFT MAXILLA (MEDIAL VIEW).

Inspect the roughened areas that articulate with the palatine bone. Below the hiatus is an approximately square area for the perpendicular plate of the palatine. Behind this, and behind the posterior edge of the hiatus, is the sloping *greater palatine groove*. Behind the groove is an elongated rough strip for the perpendicular plate of the palatine, which bridges the groove and converts it into the greater palatine canal (for the anterior palatine nerve and vessels). Above the tuberosity on the posterior surface of the maxilla is a triangular area of rough bone for the tubercle (pyramidal process) of the palatine bone.

The palatine process articulates with that of the opposite maxilla. The articulating surface is roughened by vertical ridges and grooves, and is projected up into a *nasal crest* for articulation with the vomer. Anteriorly it is grooved by the large *incisive canal*, which slopes down and forwards. The canal transmits the nasopalatine nerves, and the greater palatine arteries run upwards through it to the septum.

Trace the articulations with adjacent bones. The anterior margin of the frontal process has a narrow groove for the nasal bone. The apex of the frontal process is expanded and very rough for articulation with the frontal bone. The posterior border of the frontal process articulates with the lacrimal bone. The lacrimal bridges the lacrimal groove to articulate with the tongue-like process at the upper angle of the hiatus. The lower edge of the lacrimal bone articulates with the inferior concha, which bridges the lacrimal groove and completes its conversion into the naso-lacrimal canal. The lateral mass of the ethmoid articulates with the upper edge of the maxillary hiatus. The orbital process of the palatine bone articulates posteriorly between ethmoid and maxilla and lies at the apex of the floor of the orbit. The remainder of the articulation with the palatine bone (vertical and horizontal plates and pyramidal process) has already been noted.

The Palatine Bone. This delicate bone is the key to the understanding of the pterygo-palatine fossa and some of the canals that lead therefrom. Good specimens are not common, and while it is true that much of the bone can be seen in the intact skull every effort should be made to secure a disarticulated bone for separate study. Basically the shape is simple; a flat plate of bone is bent at a right angle in the form of a small horizontal plate (the back of the hard palate) and a larger perpendicular plate (in the lateral wall of the nose) and a pyramidal process (formerly called the **tubercle**) that projects posteriorly from the angle of junction of the plates. The base of the **pyramidal process** is at the junction of the plates, and from here the process tapers laterally and downwards to be slotted between the maxilla and the pterygoid plates of the sphenoid (Fig. 6.14, p. 381). It plugs the gap between the diverging lower ends of the medial and lateral pterygoid plates (Fig. 8.8) and it likewise articulates with a rough surface just above the tuberosity of the maxilla (Fig. 8.11). In a young skull, before ossification of these sutures, a separate triangular surface can be seen between the tuberosity of the maxilla and the lateral pterygoid plate (PLATE 36). The lower head of the medial pterygoid muscle arises here. Similarly there is a bare area of the pyramidal process facing posteriorly between the lower ends of the pterygoid plates (Fig. 8.12) and the deep head of the medial pterygoid muscle extends down to arise from this surface.

The **horizontal plate,** articulating with the palatine process of the maxilla, meets its twin in the midline of

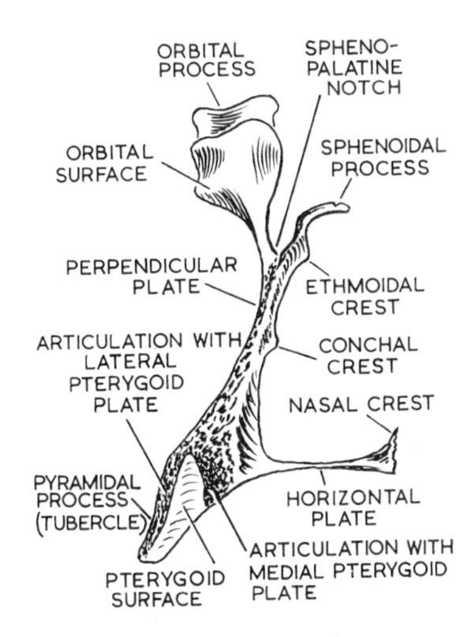

FIG. 8.12. THE LEFT PALATINE BONE (POSTERIOR VIEW).

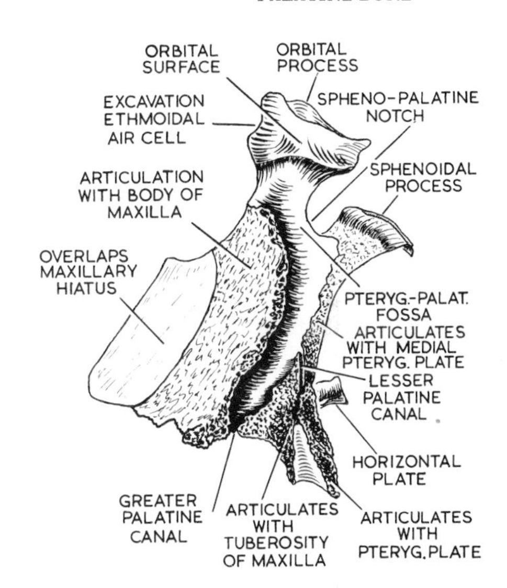

FIG. 8.13. THE LEFT PALATINE BONE (LATERAL VIEW).

the hard palate, and this border of the bone is marked by an upraised *nasal crest* for articulation with the vomer (cf. the nasal crest of the palatine process of the maxilla). The posterior border of the horizontal plate is concave and projects back to the midline of the hard palate as the posterior nasal spine. The inferior surface is marked by a **crest.** This begins at the base of the pyramidal process, and is prominent here, at the posterior boundary of the greater palatine foramen. The crest curves back to the posterior nasal spine, becoming gradually less prominent. The palatine aponeurosis is attached to the crest, and the upper head of palato-pharyngeus arises from the smooth area between the crest and the posterior border. More laterally Passavant's muscle arises in continuity from the same area. The base of the pyramidal process, behind the crest, is perforated by one or two *lesser palatine foramina.* The greater and lesser palatine foramina give exit to their respective nerves and vessels.

The **perpendicular plate** articulates with the body of the maxilla and with the medial pterygoid plate; it spans the gap between these two bones, and thus forms the medial wall of the pterygo-palatine fossa. It is projected up into two processes, called orbital and sphenoidal, with a deeply rounded spheno-palatine notch between them. The *lateral surface* can be inspected only in a carefully disarticulated bone; it is largely occupied by its articulation with the maxilla (Fig. 8.13). If the maxilla has already been studied it will easily be appreciated that the perpendicular plate of the palatine not only overlaps the posterior part of the maxillary hiatus (Fig. 8.11) but is applied to the body of the maxilla below and behind this. The posterior part of the maxillary surface shows the gradually deepening **greater palatine groove,** almost closed below by the approaching bony ridges that border it (Fig. 8.13). The greater palatine groove fits over that of the maxilla to form the **greater palatine canal.** From the lower part of the greater palatine groove one or two **lesser palatine canals** perforate the base of the pyramidal process. At the upper end of the greater palatine groove, below the spheno-palatine notch, is a smooth area which constitutes the medial wall of the *pterygo-palatine fossa.* The strip of bone on the pyramidal process that articulates with the medial pterygoid plate is continued up along the posterior border of the perpendicular plate to merge into the sphenoidal process.

The **sphenoidal process** (Fig. 8.12) articulates below the body of the sphenoid. It is a curled plate of thin bone that overlaps the vaginal process of the medial pterygoid plate. The palatino-vaginal canal lies between them and opens at the posterior border of the sphenoidal process. The canal transmits the pharyngeal branch of the pterygo-palatine ganglion and a corresponding branch of the maxillary artery.

The **orbital process** is roughly pyramidal in shape, attached by its apex to the perpendicular plate. It lies against the maxilla and extends below the posterior end of the ethmoid labyrinth to form a triangular apex to the floor of the orbit. It is usually excavated by extension into it of a posterior ethmoidal air cell. The lower part of the ethmoid with which the orbital process articulates is the original sphenoidal concha (see p. 405). This articulation converts the spheno-palatine notch into the **spheno-palatine foramen,** which is a communication between the pterygo-palatine fossa and the nose. It transmits the lateral nasal and naso-palatine nerves and vessels, and is closed over in life by the mucous membrane of the lateral wall of the nose.

The *medial surface* of the perpendicular plate can be seen in the intact skull. It forms part of the lateral wall of the nose. Halfway up is a conchal crest for articulation with the inferior concha. Just below the spheno-palatine foramen is an ethmoidal crest for articulation with the posterior end of the middle concha (Fig. 6.24, p. 402).

The articulations with adjacent bones have been noted in the above survey. The **pterygo-palatine fossa** will now be understood. It has been seen that its medial wall is the perpendicular plate of the palatine. The anterior wall is the body of the maxilla above the tubercle, and the posterior wall is the root of the pterygoid process and greater wing of the sphenoid. The roof of the fossa is the body of the sphenoid.

Ossification. In common with the face bones the palatine ossifies in membrane. A centre appears at the 7th week, in the pyramidal process, and bone spreads from this over the surface of the cartilaginous nasal capsule (p. 45).

The Zygomatic Bone. Inspection of the intact skull shows all the features except the articular surfaces.

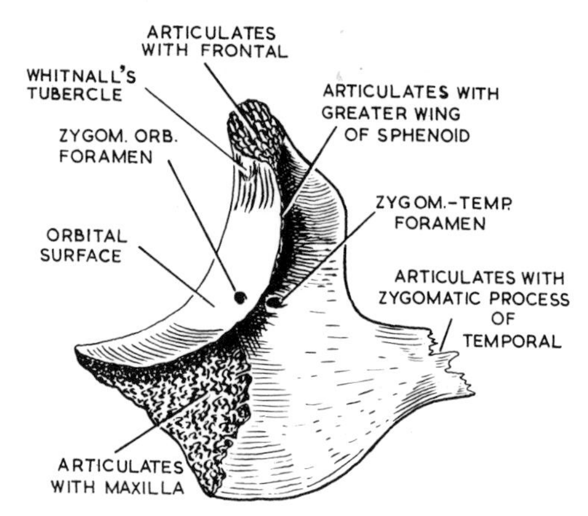

FIG. 8.14. THE RIGHT ZYGOMATIC BONE (MEDIAL ASPECT).

The maxillary surface is the most extensive of these, and it may be excavated by extension of the maxillary sinus. *Ossification*: The bone ossifies in membrane at the 8th week. There are three centres; they fuse at mid-term. Failure to fuse results in three or two parts to the bone. It is very often in two parts in the Japanese.

The Nasal Bone. The disarticulated bone shows a broad upper end with an extensive rough surface for articulation with the frontal bone. In the midline the two nasal bones articulate with each other and with the perpendicular plate of the ethmoid. The lateral border articulates with the frontal process of the maxilla. The inferior border is free at the bony margin of the nose, but in life it articulates with the external nasal cartilage. This border of the bone is notched by the external nasal nerve, which grooves the concave nasal surface of the bone. The nasal bone *ossifies* in membrane at the 8th week of pregnancy.

The Lacrimal Bone. This delicate bone lies in the medial wall of the orbit, between the frontal process of the maxilla and the labyrinth of the ethmoid. It overlaps the anterior ethmoidal air cells and lies edge to edge with the orbital plate of the labyrinth. It extends up to articulate with the frontal bone. Its orbital surface is projected into a vertical crest called the *posterior lacrimal crest.* In front of the crest lies the fossa for the lacrimal sac. The lower end of the posterior lacrimal crest is projected forwards as the *hamulus*; this spans the lateral margin of the lacrimal fossa to articulate with the maxilla at the upper end of the naso-lacrimal canal (Fig. 6.93, p. 486). The medial wall of this canal is continued by the *descending process* of the lacrimal bone, which articulates with the inferior concha in the lateral wall of the nose. *Ossification* occurs in membrane soon after the 8th week.

The Inferior Concha. This scroll-like bone is thicker than the delicate ethmoidal conchæ. At the middle of its curved upper border a vertical flange projects down to overlap the lower part of the hiatus in the maxilla. In front of this the sharp upper border articulates with the conchal crest on the frontal process of the maxilla, spanning the lacrimal groove. The lacrimal bone articulates with it here, and the two together form the medial bony wall of the naso-lacrimal canal. The uncinate process of the ethmoid articulates with the upper border of the concha across the maxillary hiatus. Behind the hiatus the upper border of the concha articulates with the conchal crest on the perpendicular plate of the palatine bone. *Ossification* is in the cartilage of the nasal capsule; the centre appears at about mid-term.

The Vomer is a flat plate in the form of a ploughshare. It rests on the nasal crests of the maxillæ and palatine bones, and its upper border articulates in front with the septal cartilage and behind with the perpendicular plate of the ethmoid bone (Fig. 6.23, p. 402). Its oblique posterior border is free at the posterior limit of the nasal septum. Above this the upper border is expanded into a pair of **alæ** which are slotted against the sphenoid, between the body of the bone and the vaginal processes. Each surface of the bone shows an oblique groove made by the naso-palatine vessels and nerve. The vomer is commonly deviated from the midline; here and there it may be separated into two laminæ.

Ossification is in membrane at the 8th week. A centre appears on either side of the cartilaginous septum and two plates are formed, with a layer of cartilage between them (see p. 45).

The Mandible

The mandible consists of a body which carries the teeth (deep to this part lies the cavity of the mouth) and a ramus which is for insertion of jaw-moving muscles (deep to this part lies the infratemporal fossa).

The **body** of the mandible is projected up around the teeth as alveolar bone which forms the walls of the tooth sockets. The alveolar bone is covered by muco-periosteum to form the inner and outer gums. The cavity of the tooth socket gives attachment to the periodontal membrane; loss of this fibrous tissue in the dried skeleton commonly allows the teeth to rattle in the bone. Healthy teeth will not fall out of the dried mandible, for the alveolar bone is constricted somewhat about their necks. After loss of a tooth the living alveolar bone atrophies and the bottom of the socket fills up with new bone; thus a glance at a gap will tell whether the tooth was lost before or after death.

Outer Surface. On the outer surface of the body the sharp anterior border of the ramus extends forward as the external oblique line. The buccinator is attached along this line as far as the anterior border of the first molar tooth. In front of this depressor labii inferioris and depressor anguli oris arise from a line just below the mental foramen. Nearer the midline, just above the mental protuberance, mentalis and the mental slips of orbicularis oris arise from this level. Below the gums the mucous membrane of the vestibule extends down to this line of muscle origin, whence it is reflected to line the cheek and lip. The **mental foramen** lies halfway between upper and lower borders (see p. 49 for changes with age). Its position varies with respect to the teeth; it is usually between the two premolars. It faces backwards and slightly upwards.

The **lower border** of the body shows a shallow oval depression near the midline (Fig. 8.15). Named digastric fossa, it receives the anterior belly of digastric. The lower border of the mandible gives attachment to the investing layer of deep cervical fascia from the midline back to the angle (PLATE 37); the attachment is interrupted at the anterior border of masseter for the passage of the facial vessels and the mandibular branch of the facial nerve, which lies on them here (PLATE 30).

The lateral surface of the **ramus** gives insertion to masseter from the angle forwards along the lower border to the body behind the external oblique line as far as the second molar tooth (Fig. 6.17, p. 387). Oblique ridges in the area of insertion of the muscle may roughen the region of the angle (Fig. 8.15). Above this the masseter is attached over the ramus almost as high as the mandibular notch (p. 387). The posterior border of the ramus is projected up as the neck, which expands into the head of the bone. The mandibular notch curves down between the coronoid process and the head. The coronoid process, sharp-bordered, has a slightly concave lateral surface. The temporalis is attached here by tendinous fibres that overlap the margins somewhat (Fig. 6.17, p. 387).

Inner Surface. The inner surface of the body is characterized by the **mylo-hyoid ridge** which forms a prominent obliquity below the molar teeth and fades out as it passes forward below the genial tubercles. The attachment of mylo-hyoid along this line divides the mouth from the neck. The muscle extends from *level with the posterior border of the last molar tooth* to the midline between the genial tubercles and the digastric fossæ; it may leave no mark here near the midline. The **genial tubercles** form four or less sharp projections low down in the midline; genio-glossus arises from the upper and genio-hyoid from the lower tubercles. Below the alveolar bone of the gums the mucous membrane of the mouth cavity lines the mandible down to the mylo-hyoid ridge and the genial tubercles. Above the anterior end of the mylo-hyoid line is a smooth concavity, the *sublingual fossa*, which lodges the sublingual gland, and here the mucous membrane loses contact with the mandible to be draped over the upper part of the gland. Below the prominence of the mylo-hyoid ridge (i.e., in

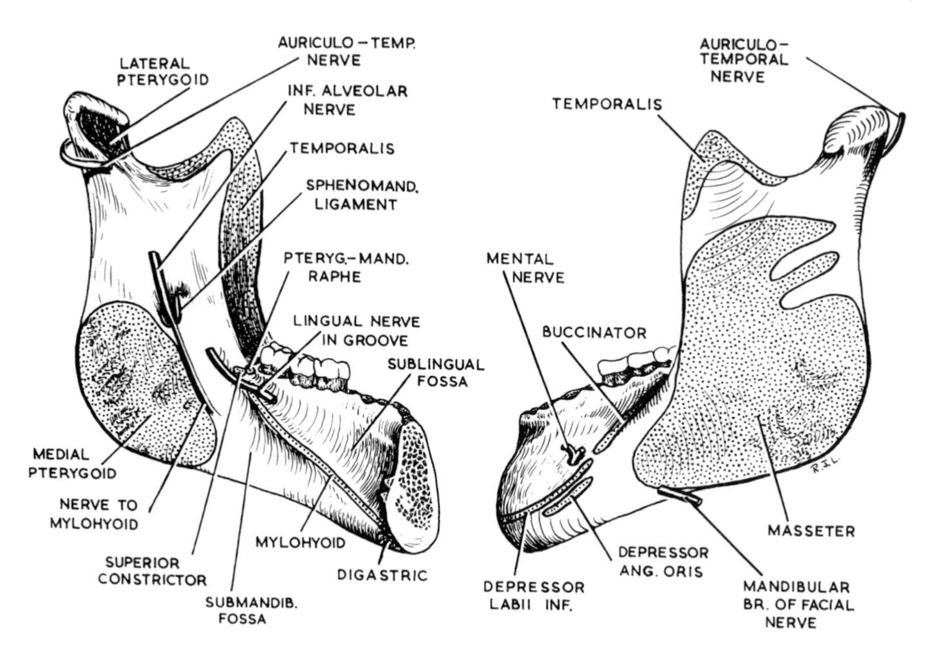

FIG. 8.15. THE LEFT HALF OF THE MANDIBLE (MEDIAL AND LATERAL VIEWS), SHOWING ADJACENT NERVES AND MUSCLE ATTACHMENTS.

the neck) the elongated smooth *submandibular fossa* lodges the superficial lobe of the submandibular gland. Above the mylo-hyoid ridge the medial surface of the mandible is grooved below the last molar tooth by the lingual nerve, and above this groove the pterygo-mandibular raphé and superior constrictor are attached level with the posterior border of the last molar tooth (Fig. 6.30A, p. 413). Buccinator origin crosses the upper border of the mandible in the retromolar fossa.

The medial surface of the **ramus** is characterized by the **lingula,** a sharp tongue of bone pointing towards the head of the mandible. It lies in front of the **mandibular foramen,** halfway between anterior and posterior borders of the ramus, and level with the occlusal surfaces of the teeth. The spheno-mandibular ligament is attached to the lingula and to the lower margin of the foramen behind it. The ligament is pierced by the mylo-hyoid vessels and nerve and these lie in the **mylo-hyoid groove,** a narrow sulcus that runs down from the mandibular foramen to die out at the posterior end of the submandibular fossa. Between the mylo-hyoid groove and the angle of the mandible the medial pterygoid muscle is inserted, and irregular bony ridges lie in this area for attachment of fibrous septa inside the muscle. The mylo-hyoid ridge, behind the last molar tooth, is traceable upwards to the lingula and on to the head of the mandible, this bare bone forming a mechanical strut in the mandible. In front of this the anterior border of the ramus is bevelled from the tip of the coronoid process down to the retromolar fossa on the upper border of the body. The posterior edge of the coronoid process is likewise bevelled down to the mandibular notch. Temporalis is inserted into these bevelled surfaces; the suprameatal fibres of the muscle, sliding smoothly over the posterior pillar of the zygomatic arch, run forwards to their insertion along the posterior border of the coronoid process and are the only retractors of the head of the mandible. The vertical and anterior fibres of temporalis are inserted into the bevelled surface along the anterior border as far down as the buccinator at the retromolar fossa.

The **neck** of the mandible expands transversely into the head, which carries the articular surface. The sharp border of the mandibular notch curves up to the lateral end of the head, but the posterior border of the ramus curves up medially to the medial end of the head. Between these diverging borders the anterior surface of the neck is hollowed into the *pterygoid pit* for the insertion of lateral pterygoid. The posterior triangular area below the articular surface is smooth for the attachment of the temporo-mandibular ligament. The **articular surface** is bevelled, with anterior and posterior sloping surfaces, and the transverse crest between them is directed medially and slightly backwards. The two condyles lie on the arc of a large circle which cuts across the front of the foramen magnum. The articular surface is covered by dense fibrous tissue identical in structure with the disc (the substance is called fibro-cartilage, but there is no cartilage in it). The capsule and synovial membrane are attached to the articular margins.

The *growth* of the mandible is considered on p. 49 and its *movements* are discussed on p. 446.

The Hyoid Bone (PLATE 27)

The hyoid bone lies free, suspended in muscle, and so is very mobile. The floor of the mouth and the tongue are attached to it above, and the larynx below, while behind are attached the epiglottis and the pharynx. Furthermore, it provides an adjustable pulley for the digastric muscle. It is palpable in the living (feel your own). A specimen is not usually present in students' sets of bones—obtain one for handling and inspection before reading any further, The attachments of the hyoid bone should not be studied until the anatomy of the floor of the mouth (p. 412) and larynx (p. 422) has been mastered.

The hyoid bone has a **body,** which is a curved sheet of bone convex forward and concave behind. On each side a **greater horn** projects back as a long slender process. At the junction of body and greater horn is the **lesser horn,** projecting up as a spike of bone (or hyaline cartilage) that does not always fuse but may remain permanently connected by fibrous tissue. Thus a U-shaped mass is produced, of a size that can be gauged by noting that it can be fitted inside the body of the mandible just above the mylo-hyoid line.

At rest the body lies just below the mandible, level with the last molar tooth, and the greater horn projects back as far as the angle of the mandible (Fig. 6.56, p. 446) at the level of the 3rd cervical vertebra. This is the level of the lateral glosso-epiglottic (pharyngo-epiglottic) folds within the pharynx, bisecting the epiglottis. Thus the hyoid marks on the surface the junction of oro-pharynx and laryngo-pharynx (Fig. 6.32, p. 415).

The thyro-hyoid membrane is attached to the sharp *upper* border of the body and greater horns, and hangs down free inside the curve of the bone. Behind the concave posterior surface of the body is a bursa between bone and membrane, and here the thyro-glossal duct made an upward kink behind the bone (Fig. 1.51, p. 42). From the upper surface of the body and from the thyro-hyoid membrane a rather fatty mass of loose fibrous tissue passes back to the epiglottis; it is named the *hyo-epiglottic ligament.* From the side of the epiglottis a flange of fibrous tissue passes across to the greater horn; covered with mucosa this constitutes the *pharyngo-epiglottic fold* now called the *lateral glosso-epiglottic fold* (Fig. 6.41, p. 424). Attached to the lower border of the body is an upward extension of the pretracheal fascia (PLATE 44). In front of this sterno-hyoid and omo-hyoid are attached to a pair of concave fossæ lying on the lower part of the anterior surface. Below omo-hyoid is the linear attachment of thyro-hyoid, extending back to the lower border of the greater horn. Above these muscles the investing layer of deep cervical fascia has a linear attachment to the bone, which is thus subcutaneous and palpable along this line. Just above this line is the linear attachment of mylo-hyoid to the body of the bone, and above this is a deep pit for the insertion of the genio-hyoid muscle. A few fibres of genio-glossus may be fixed to the upper border of the body and to the lesser horn (chondro-glossus) but they are insignificant. Fixation of the tongue to the hyoid bone is not attained by these fibres, but by hyo-glossus along the greater horn.

The lesser horn gives attachment to the stylo-hyoid ligament, and the middle constrictor arises from it and from the whole length of the greater horn, thus anchoring the hyoid bone to the pharynx. Lateral to the constrictor hyo-glossus arises from the whole length of the greater horn and the lateral part of the body alongside genio-hyoid, thus anchoring the hyoid to the tongue. The base of the greater horn has a lateral boss of bone. Here is attached the fibrous sling through which the intermediate tendon of digastric glides freely, and this is straddled by the insertion of the split tendon of stylo-hyoid.

The *movements* of the hyoid bone are discussed on p. 446 and the *development* is mentioned on p. 41.

BIOGRAPHICAL NOTES

by
Jessie Dobson, B.A., M.Sc.

Hunterian Trustee, Royal College of Surgeons of England.

ADAMKIEWICZ, Albert (1850–1921). Professor of Pathology in the University of Krakau. Arteries—arterial supply of the spinal cord. 'Die Blutgefässe des menschlichen Rückenmarkes.' *Sitzungsberichte der Kaiserliche Akademie der Wissenschaften zu Wien.* Bd. **85.** Abt. III. 1882. p. 101. (PAGE 536.)

ALCOCK, Benjamin. (Dates of birth and death not known.) In 1849 was appointed Professor of Anatomy in Queen's College, Cork, but was called upon to resign in 1853 in consequence of disputes about the working of the Anatomy Acts. Went to America in 1855. Canal—for the internal pudendal vessels in the ischio-rectal fossa. 'Iliac Arteries.' *The Cyclopædia of Anatomy and Physiology.* Edited by Robert B. Todd. Vol. II, 1836. p. 835. (PAGE 348.)

ARGYLL-ROBERTSON, Douglas Moray Cooper Lamb (1837–1909). Ophthalmic surgeon to Edinburgh Royal Infirmary from 1870 to 1897. Pupil—that does not react to light falling on the retina but will contract with accommodation for near objects. 'Four Cases of Spinal Myosis.' *Edinburgh Medical Journal.* Vol. **15,** 1869. p. 491. (PAGE 504.)

AUERBACH, Leopold (1828–1897). Professor of Neuropathology in Breslau. Plexus—plexus myentericus. 'Ueber einen Plexus myentericus, einen bisher unbekannten ganglionervösen Apparat im Darm der Wirbelthiere.' Breslau. 1862. (PAGE 287.)

BARTHOLIN, Caspar (Secundus) (1655–1738). Succeeded his father, Thomas Bartholin, as Professor of Medicine, Anatomy and Physics at Copenhagen. Glands—glandulæ vestibularis majores. 'De ovariis mulierum.' Rome, 1677. Nuremberg, 1679. (PAGE 353.)

BETZ, Vladimir Aleksandrovich (1834–1894). Professor of Anatomy in Kiev from 1868–1889. Cells—giant pyramidal cells of the motor cortex. 'Ueber die feinere Struktur der Gehirnrinde des Menschen.' *Centralblatt für medicinischen Wissenschaften.* Berlin. Bd. **12.** pp. 578 and 595. 1874; Bd. **19.** p. 193. 1881. (PAGE 522.)

BIGELOW, Henry Jacob (1818–1890). Professor of Surgery at Harvard, U.S.A., from 1849 to 1882. Ligament—ilio-femoral ligament or 'Y'-shaped ligament of the hip joint. 'The Mechanism of Dislocation and Fracture of the Hip.' Philadelphia, 1869. p. 17. (PAGE 150.)

BOWMAN, Sir William Paget (1816–1892). Professor of Anatomy and Physiology at King's College, London, from 1848 to 1856. Leading ophthalmic surgeon in England. Capsule—surrounding the glomerulus in the kidney. 'On the Structure and Use of the Malpighian Bodies of the Kidney, with Observations on the Circulation through that Gland.' *Philosophical Transactions of the Royal Society.* Vol. **132,** 1842. pp. 57–80. (PAGE 316.) Membrane—anterior 'elastic' membrane of the cornea. 'Lectures on the parts concerned in the Operations on the Eye and on the Structure of the Retina', delivered at the Royal London Ophthalmic Hospital, Moorfields. June, 1847. (PAGE 438.)

BROCA, Pierre Paul (1824–1880). Professor of Clinical Surgery and Director of the Anthropological Laboratories in Paris. Area—speech centre. 'Sur le volume et la forme du cerveau.' *Bulletin de la Société Anthropologique de Paris.* Vol. **2.** 1861. pp. 139–233. 'Perte de la parole; ramollissement chronique et destruction partielle du lobe antérieur gauche du cerveau.' *Ibid.* Vol. **2.** 1861. pp. 235–238. 'Remarques sur le siège de la faculté du langage articule, suivies d'une observation d'aphémie (perte de la parole).' *Bulletin de la Société Anatomique de Paris.* Vol. **36.** 1861. pp. 330–357. (PAGE 502.)

BRODEL, Max (1870–1941). Associate Professor and Director of the Institute of Art as Applied to Medicine, Baltimore. Bloodless line—on the kidney. The line of division between the areas supplied by the anterior and posterior branches of the renal artery. 'The Interesting Blood Vessels of the Kidney and their Significance in Nephrotomy.' Proceedings of the Association of American Anatomists. 1900. *Johns Hopkins Hospital Bulletin.* Vol. **12.** 1901. pp. 10–13. (PAGE 318.)

BRUCH, Carl Wilhelm Ludwig (1819–1884). Professor of Anatomy in Basle and, later, Giessen. Membrane—of the choroid—lamina vitrea. 'Untersuchungen zur Kenntniss des körnigen Pigments der Wirbelthiere.' Zurich. 1844. p. 3. (PAGE 439.)

BRUNNER, Johann Konrad (1653–1727). Professor of Anatomy at Heidelberg and, later, Strassburg. Glands—glandulæ duodenales. 'Descriptio de glandulis in duodeno intestino detectis.' Heidelberg. 1687. (PAGE 288.)

BURDACH, Karl Friedrich (1776–1847). Professor of Anatomy and Physiology in Königsberg. Column—postero-lateral column of the spinal cord. 'Vom Baue und Leben des Gehirns.' Leipzig. Bd. **I.** 1819. p. 134. (PAGE 537.)

CAMPER, Petrus (1722–1789). Professor of Medicine, Anatomy, Surgery and Botany in Gröningen from 1763 to 1773. Fascia—superficial layer of the superficial fascia of the abdomen. 'Icones herniarum.' Frankfurt a/M. 1801. p. 11. (PAGE 210.)

CHASSAIGNAC, Charles Marie Edouard (1805–1879). Surgeon in Paris. Tubercle—tuberculum caroticum on the sixth cervical vertebra. 'Quelques points d'Anatomie, de Physiologie et de Pathologie de la colonne vertébrale.' *Archives générales de Médecine.* Paris. Vol. **IV.** 1834. p. 458. (PAGE 477.)

CLARKE, Jacob Augustus Lockhart (1817–1880). Physician to the Hospital for Epilepsy and Paralysis, London. F.R.S., 1854. Column—nucleus dorsalis of the spinal cord—posterior vesicular column. 'Researches into the Structure of the Spinal Cord.' *Philosophical Transactions of the Royal Society.* Vol. **141.** 1851. p. 611. 'On the Anatomy of the Spinal Cord.' *Archives of Medicine.* London. Vol. I. 1857–1859. p. 206. (PAGE 537.)

CLOQUET, Jules Germain (1790–1883). Professor of Anatomy and Surgery in Paris. Node—lymphatic node in the femoral canal. 'Recherches anatomiques sur les Hernies de l'abdomen.' Paris. 1817. p. 68. (PAGE 138.)

COLLES, Abraham (1773–1843). Professor of Anatomy and Surgery in Dublin from 1804 to 1836. Fracture—of the lower extremity of the radius. 'On the Fracture of the Carpal Extremity of the Radius.' *Edinburgh Medical and Surgical Journal.* Vol. **10.** 1814. p. 182. (PAGES 91, 125.) Fascia—perineal fascia and deep layer of the superficial fascia of the abdomen. 'A Treatise on Surgical Anatomy.' Dublin. 1811. p. 174. (PAGE 352.)

COOPER, Sir Astley Paston (1768–1841). Surgeon to St. Thomas's and Guy's Hospitals, London. Ligament—upper part of the pectineal fascia. 'The Anatomy and Surgical Treatment of Internal and Congenital Hernia.' London. 1804. (PAGE 138.) Ligaments—skin attachments of the breast. 'Anatomy and Diseases of the Breast.' 1840. p. 49. (PAGE 66.)

CORTI, Alfonso (Marquis) (1822–1888). An eminent histologist who worked with Hyrtl, Johannes Muller, Kolliker, Gegenbaur and others, but held no academic post. Born in Sardinia. Organ—in the cochlea. 'Recherches sur l'organe de l'ouie des mammifères.' *Zeitschrift für wissenschaftlichen Zoologie.* Vol. **III.** 1851. p. 109. (PAGE 456.)

COWPER, William (1666–1709). London surgeon. F.R.S., 1698. Glands—glandulæ bulbo-urethrales. 'An Account of Two Glands and their Excretory Ducts Lately Discovered in Human Bodies.' *Philosophical Transactions of the Royal Society.* Vol. **21.** 1697. p. 364. (PAGE 352.)

CUVIER, Georges Leopold Chrétien Frédéric Dagobert (Baron) (1769–1832). The most eminent naturalist of his day, particularly noted as a zoologist and palæontologist. Professor of Natural History, Paris. Duct—junctional termination of the cardinal veins in the embryo. 'Leçons d'Anatomie Comparée.' Tom. **IV.** 1805. Paris. (PAGE 308.)

DARWIN, Charles Robert (1809–1882). Naturalist. Spent five years on a voyage of exploration on board H.M.S. *Beagle.* Private means enabled him to devote the whole of his time to his favourite studies. 'The Origin of Species by Means of Natural Selection.' 1850. 'The Descent of Man.' 1871. (PAGES 4, 509.)

DEITERS, Otto Friedrich Karl (1834–1863). Professor of Anatomy and Histology at the University of Bonn. Cells—outer hair cells in the organ of Corti. 'Untersuchungen über die Lamina Spiralis Membranacea.' Bonn. 1860. p. 59. (PAGE 456.)

DENONVILLIERS, Charles Pierre (1808–1872). Professor of Anatomy and Surgery in Paris. Fascia—prostato-peritoneal aponeurosis. 'L'anatomie du perinée.' *Bulletin de la Société Anatomique de Paris.* Vol. **12.** 1836. p. 106. (PAGE 332.)

DESCEMET, Jean (1732–1810). Professor of Anatomy and Surgery in Paris. Membrane—posterior membrane of the cornea. 'An sola lens crystallina cataractæ sedes.' Paris. 1758. (PAGE 438.)

DOUGLAS, James (1675–1742). Anatomist and 'man-midwife' of London. Physician to the Queen. F.R.S., 1706. Pouch—recto-uterine peritoneal pouch. Semicircular fold—rectus sheath. 'A Description of the Peritoneum and of that Part of the Membrana Cellularis which lies on its Outside.' London. 1730. (PAGES 339, 257.)

DUPUYTREN, Guillaume (Baron) (1777–1835). Professor of Surgery in Paris. Contracture—of the palmar fascia. 'De la rétraction des doigts par suite d'une affection de l'aponévrose palmaire.' *Journal universel et hebdomadaire de médecine et de chirurgie pratiques et des institutions médicales.* Paris. **V.** 1831. pp. 352–365. 'Leçons orales de Clinique chirurgicale faites à l'Hôtel-Dieu de Paris.' Paris. 1832. Vol. **I.** Retraction permanente des doigts. pp. 3 and 23. (PAGE 96.)

EDINGER, Ludwig (1855–1918). Anatomist and neurologist of Frankfurt-am-Main. **WESTPHAL, Karl**

Friedrich Otto (1833–1890). Professor of Psychiatry in Berlin. Edinger-Westphal nucleus—of the third cranial nerve. 'Ueber den Verlauf der central Hirnnervenbahnen mit Demonstration von Präparaten.' *Archiv. für Psychiatrie und Nervenkrankheiten*. Bd. **XVI.** 1885. p. 859. 'Ueber einen merkwürdigen Fall von periodischer Lähmung aller vier Extremitäten mit gleichzeitigen Erlöschen der elektrischen Erregbarkeit während der Lähmung.' *Berliner klinische Wochenschrift*. Bd. **22.** 1885. pp. 489 and 509. (PAGE 524.)

EUSTACHIO (EUSTACHI, EUSTACHIUS), Bartolomeo (1513–1574). Professor of Anatomy in Rome and Physician to the Pope. Tube—tuba auditiva. 'De auditus organis.' Venice. 1562. (PAGE 452.)

FALLOPPIO (FALLOPPIUS), Gabriele (1523–1563). Professor of Anatomy and Surgery in Padua. Tube—tuba uterina. 'Observationes anatomicæ.' Venice. 1561. (PAGE 267.)

FALLOT, Etienne Louis Arthur (1850–1911). French physician. Tetralogy—four commonly associated congenital heart defects. 'Contribution à l'Anatomie pathologique de la maladie bleue (Cyanose cardiaque).' *Marseille Médicale*. Vol **25.** pp. 77–93, etc. 1888. (PAGE 238.)

FOERSTER, Otfried (1873–1941). Neurologist at the Psychiatric Clinic in Breslau. Dermatomes—the area of skin which is supplied by the fibres of a certain spinal root. 'The Dermatomes in Man.' *Brain*. Vol. **56.** 1933. pp. 1–39. (PAGE 27.)

GALEN, ? Claudius (130–200). Physician in Rome. For two years physician to Marcus Aurelius in Venice. Great Vein—vena cerebri magna. Collected works, ascribed to Galen. (PAGE 521.)

GENNARI, Francesco (1750–). Physician and anatomist of Parma. Stria—cortical lamination line characteristic of the area striata of the occipital visual cortex. 'De peculiari structura cerebri.' Parma. 1782. pp. 72–75. (PAGE 503.)

GIANNUZZI, Giuseppe. Crescents—serous crescents on mucous alveoli. 'Von den Folgen des beschleunigten Blutstroms für die Absonderung des Speichels.' *Berichte d. kon. Sachs. Gesellsch. der wiss. Sitz*. Vol. **17.** 1865. p. 68. (PAGE 399.)

GIMBERNAT, Manuel Louise Antonio don (1734–1816). Professor of Anatomy in Barcelona from 1762 to 1774. Surgeon to King Charles III of Spain. Ligament—ligamentum lacunare. First demonstrated in 1768. 'Nuevo método de operar en la hernia crural.' Madrid. 1793. p. 28. (PAGE 255.)

GIRALDÈS, Joachim Albin Cardozo Cazado (1808–1875). Professor of Surgery in Paris. Died as the result of a wound inflicted while conducting an autopsy. Organ—the paradidymis. 'Note sur un nouvel organe glanduleux situé dans le cordon spermatique.' *Compte Rendu des Séances de la Société de Biologie*. 1859. pp. 123–124. (PAGE 267.)

GOLL, Friedrich (1829–1903). Professor of Anatomy in Zurich. Column—fasciculus gracilis—posterior column of the spinal cord. 'Beiträge zur feineren Anatomie des menschlichen Rückenmarks.' Zurich. 1860. p. 9. (PAGE 537.)

GRAAF, Regnier de (1641–1673). Anatomist and physician of Delft. Follicles—folliculus oöphorus vesiculosus. 'De mulierum organis generatione.' Leyden. 1672. (PAGE 338.)

GUDDEN, Bernhard Aloys von (1824–1886). Professor of Psychiatry in Zurich. Commissure—fibre tract between the medial geniculate bodies and the inferior colliculi of opposite sides, running in proximity with the optic tracts. 'Experimentaluntersuchungen über das peripherische und centrale Nervensystem.' *Archiv. für Psychiatrie*. Bd. **II.** 1870. p. 693. 'Ueber die Kreuzung der Nervenfasern im Chiasma nervorum opticorum.' *Archiv für Ophthalmologie*. Berlin. Bd. **XXV.** 1879. p. 1. (PAGE 504.)

HASSALL, Arthur Hill (1817–1894). Physician and botanist. Practised in London and, later, the Isle of Wight. Corpuscles—concentric corpuscles of the thymus. 'The Microscopic Anatomy of the Human Body in Health and Disease.' London. 1846. p. 478. (PAGE 227.)

HAVERS, Clopton (1657–1702). London physician. F.R.S., 1685. Lamellæ—bony layers surrounding the spaces or 'canals' in the compact tissue of bone. Fat pad—extrasynovial pads or fringes of synovial membrane consisting of intra-articular fat. 'Osteologia Nova.' London. 1691. (PAGES 11, 15.)

HEAD, Henry (1861–1940). Neurologist. Physician to London Hospital. F.R.S., 1899. Dermatomes—areas of hyperalgesia of skin, associated with diseases of viscera. 'On Disturbances of Sensation with Especial Reference to the Pain of Visceral Disease.' *Brain*. Vol. **XVI.** 1893. pp. 1–133. (PAGE 27.)

HENLE, Friedrich Gustav Jakob (1809–1885). Professor of Anatomy in Göttingen from 1852 to 1885. Loop—the looped portion of the uriniferous tubules of the kidney. 'Handbuch der systematischen Anatomie des Menschen.' Bd. **II.** Brunswick. 1866. pp. 303–308. (PAGE 316.)

HIGHMORE, Nathaniel (1613–1685). Physician of Sherborne, Dorsetshire. Antrum—sinus maxillaris. 'Corporis humani disquisitio anatomica.' La Hague. 1651. (PAGE 403.)

HILTON, John (1805–1878). Surgeon at Guy's Hospital, London, from 1849 to 1871. Joint Innervation—reflex control of muscles activating joints. 'Lectures on Rest and Pain.' 6th Edition. London. 1950. pp. 166–167.

White Line—linear interval between the external and internal sphincters. *Ibid.* p. 286. (PAGES 15, 344.)

HIS, Wilhelm (1831–1904). Professor of Anatomy and Physiology in Basle and, later, Leipzig. Copula—bond uniting the ventral ends of the third pharyngeal arches. 'Anatomie menschlicher Embryonen.' Leipzig. 1880. pp. 53–54. (PAGE 42.)

HIS, Wilhelm (1863–1934). Professor of Anatomy successively at Leipzig. Basle, Göttingen and Berlin. Bundle—atrio-ventricular bundle. 'Die Tätigkeit des embryonal Herzens.' Leipzig. 1893. p. 23. Also: *Centralblatt für Physiologie*. Vol. **IX.** 1895. p. 469. (PAGE 237.)

HOUSTON, John (1802–1845). Lecturer in Surgery in Dublin and physician to the City Hospital. Valves—sphincter ani tertius. 'Observations on the Mucous Membrane of the Rectum.' *Dublin Hospital Reports*. Vol. **V.** 1830. p. 158. (PAGE 327.)

HUMPHRY, Sir George Murray (1820–1896). Professor of Anatomy in Cambridge from 1866 to 1883, when he was appointed Professor of Surgery. Ligament—associated with the posterior cruciate ligament in the knee joint. 'A Treatise on the Human Skeleton.' Cambridge. 1858. p. 546. (PAGE 163.)

HUNTER, John (1728–1793). London surgeon and anatomist. Founder of the Hunterian Museum now in the custody of the Royal College of Surgeons of England. Canal—subsartorial canal. 'An account of Mr. Hunter's method of performing the Operation for the Popliteal Aneurism. Communicated in a letter to Dr. Simmons by Mr. Everard Home.' *London Medical Journal*. Vol. **7.** 1786. pp. 391–406. Vol. **8.** 1787. pp. 126–136. (PAGE 142.)

HUNTER, William (1718–1783). Anatomist and surgeon of London. Brother of John Hunter. Circulus vasculosus—vascular anastomoses around an articulation. 'Of the Structure and Diseases of Articulating Cartilages.' *Philosophical Transactions of the Royal Society*. Vol. **42.** 1743. No. 470. pp. 514–521. (PAGE 15.)

HUSCHKE, Emil (1797–1858). Professor of Anatomy in the University of Jena. Foramen—in the tympanic plate. 'Schädel, Hirn und Seele des Menschen.' Jena. 1854. (PAGE 48.)

JACOBSON, Ludwig Levin (1783–1843). Anatomist and Physician in Copenhagen. Nerve—tympanic branch of the glosso-pharyngeal nerve. 'Description anatomique d'une anastomose entre le nerf pharyngo-glossien, le trifacial et le trisplanchnique.' Read at a meeting of the Faculty of Medicine in Paris on July 22nd, 1813, and later published by Professor Breschet. Also published in the *Proceedings of the Royal Society of Medicine of Copenhagen*. Vol. **V.** 1818, with the title 'Supplementa ad otoiatriam'. pp. 293–303. (PAGE 394.)

KILLIAN, Gustav (1860–1921). Director of the Rhinolaryngological Clinic in Freiburg and, later, Berlin. Dehiscence—of the lowest fibres of thyro-pharyngeus immediately above the crico-pharyngeus muscle. 'La bouche de l'œsophage.' *Ann. Mal. Oreil. Larynx*. Paris. 1908. (PAGE 416.)

KUPFFER, Karl Wilhelm von (1829–1902). Professor of Anatomy in Kiel (1867), Königsberg (1875) and Munich (1880). Cells—'stellate cells' in the lining of blood channels in the liver. 'Ueber Sternzellen der Leber.' *Archiv für mikroskopische Anatomie*. Bonn. 1876. pp. 353–358. (PAGE 300.)

LANGER, Karl (1819–1887). Professor of Anatomy in Vienna. Lines—cleavage lines of the skin due to the disposition of the fibrous tissue of the dermis. 'Zur Anatomie und Physiologie der Haut: Über die Spaltbarkeit der Cutis.' *Sitzungsberichte der kaiserlichen Akademie der Wissenschaften*. Bd. **44.** 1862. p. 20. (PAGE 2.)

LANGERHANS, Paul (1847–1888). Professor of Pathological Anatomy in Freiburg. Islets—clumps of cells lying in the interalveolar tissue of the pancreas. 'Beiträge zur mikroskopischen Anatomie der Bauchspeicheldrüse.' Berlin. 1869. (PAGE 303.)

LECOMTE, O. (no dates known). Physician to the French Army. Anconeus—the pronator of the ulna. 'Du mouvement de rotation de la main.' *Archives générales de Médecine*. Vol. **24.** 1874. pp. 129–149. (PAGE 88.)

LIEBERKÜHN, Johann Nathanael (1711–1756). Physician and anatomist of Berlin, noted for his technique of injecting. Crypts—glandulæ intestinales. 'Dissertatio anatomica de Fabrica et Actione Villorum Intestinorum tenuium Hominis.' Leyden. 1745. **X.** (PAGE 288.)

LISTER, Joseph (Lord) (1827–1912). Professor of Surgery in Glasgow (1860). Edinburgh (1869) and King's College, London (1877). Tubercle—the prominence on the posterior surface of the lower end of the radius adjacent to the groove for the tendon of the extensor pollicis longus. 'On Excision of the Wrist.' *Lancet*, *i*, 1865, pp. 308, 335, 362. Lister does not actually refer to the tubercle in his writings, but probably referred to it in his lectures on the above subject. (PAGE 117.)

LITTRÉ, Alexis (1658–1726). Surgeon and anatomist of Paris. Glands—in the mucous membrane of the penile urethra. 'Description de l'urèthre de l'homme.' *Histoire de l'Académie des Sciences*. Année 1700, avec les Mémoires de Mathematique et de Physique pour la même Année. Paris. 1719. p. 312. (PAGE 351.)

LOCKWOOD, Charles Barrett (1856–1914). Surgeon to St. Bartholomew's Hospital, London. Suspensory ligament—of the globe of the eye. 'The Anatomy of the Muscles, Ligaments and Fasciæ of the Orbit, including an account of the Capsule of Tenon, the Check Ligaments of the Recti and of the Suspensory Ligament of the

Eye.' *Journal of Anatomy and Physiology*. Vol. **XX.** 1886. pp. 1–25. (PAGE 434.)

LOUIS, Pierre Charles Alexandre (1787–1872), a physician of Paris, is usually cited as being the first to note the presence of the 'angle', but no reference to the fact is to be found in his writings. American authorities usually give Antoine Louis (1723–1792), surgeon and physiologist of Paris, but it is not mentioned in his works. Angle—Angulus Ludovici—the angle formed by the manubrium and the body of the sternum. (PAGE 222.)

LUDWIG, Wilhelm Friedrich (1790–1865). Professor of Surgery and Midwifery in Tubingen. Court physician. Angina—swelling of the submandibular region combined with inflammatory œdema of the mouth. 'Ueber eine in neuerer Zeit wiederholt hier vorgekommene Form von Halsentzündung.' *Medicinisches Correspondenzblatt des Württembergischen ärztlichen Vereins*. Stuttgart. 1836. pp. 21–25. (PAGE 371.)

LUSCHKA, Hubert (1820–1875). Professor of Anatomy in Tubingen from 1849 to 1875. Foramen—in the lateral recesses of the fourth cerebral ventricle. 'Die Adergeflechte des menschlichen Gehirns.' Berlin. 1855. Also: 'Die Anatomie des Menschen.' Bd. **III.** Abt. II. Tubingen. 1867. p. 190. (PAGE 530.)

MACKENRODT, Alwin (1859–1925). Professor of Gynæcology in Berlin. Also known as a Pathologist. Ligament—ligamentum transversum colli of the uterus. 'Ueber die Ursachen der normalen und pathologischen Lagen der Uterus.' *Archiv. für Gynækologie*. Bd. **XLVIII.** 1895. p. 393. (PAGE 336.)

MAGENDIE, François (1783–1855). Professor of Pathology and Physiology in Paris and physician to the Hôtel Dieu. Foramen—apertura mediana ventriculi quarti. 'Mémoire physiologique sur le cerveau.' *Journal de Physiologie expérimentale et pathologique*. Paris. 1828. Vol. **VIII.** p. 222. (PAGE 530.)

MALPIGHI, Marcello (1628–1694). Professor of Medicine in Bologna. Has been named the 'founder of microscopical anatomy.' Corpuscles—splenic corpuscles. 'De Liene', in 'Exercitationibus de Viscerum Structura'. London. 1669. p. 12. (PAGE 305.)

MARSHALL, John (1818–1891). Fullerian Professor of Physiology at the Royal Institute; Professor of Anatomy at the Royal Academy; Professor of Surgery at University College, London. Vein—vena obliqua atrii sinistri. 'On the Development of the Great Anterior Veins in Man and Mammalia.' *Philosophical Transactions of the Royal Society*. 1850. p. 147. (PAGE 238.)

MECKEL, Johann Friedrich (1724–1774). Professor of Anatomy, Botany and Gynæcology in Berlin. Cave—dural space in which the trigeminal ganglion is lodged. 'Tractatus de quinto pare nervorum cerebri.' Göttingen. 1748. 'De ganglio secundi rami quinti paris nervorum cerebri nuper detecto.' Berlin. 1749. (PAGE 490.)

MECKEL, Johann Friedrich (1781–1833). Professor of Anatomy and Surgery in Halle. Grandson of the preceding. Cartilage—of the first branchial arch. 'Handbuch der menschlichen Anatomie.' Bd. **IV.** 1820. p. 47. (PAGE 40.)

MEIBOM, Heinrich (1638–1700). Professor of Medicine, History and Poetry in Helmstadt. Glands—sebaceous follicles of the eyelids. 'De vasis palpebrarum novis epistola.' Helmstadt. 1666. (PAGE 431.)

MEISSNER, Georg (1829–1905). Professor of Anatomy and Physiology in Basle and, later, Professor of Physiology in Göttingen. Plexus—plexus submucosus of the alimentary tract. 'Beiträge zur Anatomie und Physiologie der Haut.' Leipzig. 1853. (PAGE 287.)

MONRO, Alexander (1733–1817) (Secundus). Succeeded his father, Alexander Monro (Primus) as Professor of Anatomy in Edinburgh. Interventricular foramen. First noted in 1753. 'Observations on the Structure and Functions of the Nervous System.' Edinburgh. 1783. 'Of the Communication of the Ventricles of the Brain with each other, in Man and Quadrupeds.' Edinburgh. 1797. (PAGE 507.)

MORISON, James Rutherford (1853–1939). Surgeon to the Royal Infirmary, Newcastle-on-Tyne. Emeritus Professor of Surgery in the University of Durham. Kidney pouch—hepato-renal pouch. 'The Anatomy of the Right Hypochondrium relating especially to Operations for Gall-stones.' *British Medical Journal*. 1894, *ii*, p. 968. (PAGE 269.)

MULLER, Heinrich (1820–1864). Professor of Anatomy in Wurzburg. Fibres—neuroglial fibres of the retina. 'Anatomisch-physiologische Untersuchungen über die Retina des Menschen und der Wirbelthiere.' Leipzig. 1856. p. 68. (PAGE 442.)

MULLER, Johannes Peter (1801–1858). Professor of Anatomy and Physiology in Berlin. Duct—the primordial female genital duct or oviduct. 'Bildungsgeschichte der Genitalien aus anatomischen Untersuchungen an Embryonen des Menschen und der Thiere.' Dusseldorf. 1830. p. 60. (PAGES 267, 337.)

ODDI, Ruggero (no dates known). Physiologist of Perugia. Sphincter—sphincteric fibres around the termination of the common bile duct. 'D'une disposition à sphincter spéciale de l'ouverture du canal cholédoque.' *Archives Italiennes de Biologie*. Vol. **8.** 1887. pp. 317–322. (PAGE 302.)

PACCHIONI, Antoine (1665–1726). Professor of Anatomy in Rome and, later, Tivoli. Bodies—granula arachnoidea. 'Dissertatio epistolaris ad Lucam Schroeckium de glandulis conglobatis duræ meningis humanæ, indeque ortis lymphaticis ad piam matrem productis.' Rome. 1705. (PAGE 488.)

PACINI, Filippo (1812–1883). Professor of Anatomy in Florence from 1847 to 1883. Corpuscles—end organs of sensory nerves. 'Nuovi organi scoperti nel corpo umano.' Pistoja. 1840. (PAGE 303.)

PARONA, Francesco (end of 19th century). Chief Surgeon to Novara Hospital, Italy. Space—the deep forearm (intermuscular) space. 'Dell'oncotomia negli ascessi profondi diffusi dell'avambraccio.' *Annali universali di medicina e chirurgia.* **LXII.** Vol. 237. 1876. pp. 408–414. (PAGE 82.)

PASSAVANT, Philipp Gustav (1815–1893). Surgeon of Frankfurt. Bar—the projecting ridge on the posterior wall of the pharynx produced by contraction of the upper fibres of M. palato-pharyngeus. 'Ueber die Verschliessung des Schlundes beim Sprechen.' *Virchow's Archiv.* **XLVI.** 1869. (PAGE 420.)

PETIT, Jean Louis (1664–1750). Began to learn anatomy at the age of 7 and when he was 12 was demonstrator of anatomy for Littré. When he was 16 he was appointed Surgeon to La Charité Hospital in Paris and finally became Director of the Academy of Surgery. F.R.S. 1729. Triangle—lumbar triangle. 'Traité des maladies chirurgicales.' Paris. 1705. (PAGE 256.)

PEYER, Johann Conrad (1653–1712). Professor of Logic, Rhetoric and Medicine in Schaffhausen. Patches—aggregated lymphoid follicles in the lower ileum. 'Exercitatio anatomico-medica de glandulis intestorum, earumque usu et adfectionibus.' Schaffhausen. 1677. (PAGE 285.)

POUPART, François (1661–1709). Surgeon to the Hôtel Dieu, Paris. Ligament—ligamentum inguinale. 'Histoire de l'Académie Royal des Sciences.' Paris. 1705. p. 51 (PAGE 255.)

PURKINJE, Johannes Evangelista (1787–1869). Professor of Physiology in Breslau and, later, Prague. Fibres—muscular fibres beneath the endocardium. 'Mikroskopisch-neurologische Beobachtungen.' *Archiv für Anatomie, Physiologie und wissenschaftliche Medicin (Müller's Archiv).* Berlin. 1845. pp. 281–295. (PAGE 237.) Cells—of the cerebellar cortex, with many dendrites. 'Ueber die gangliösen Korperchen in verscheidenen Theilen des Gehirns.' *Berichte über d. Versamml. d. deutsch. Naturf.* Prague. 1837. p. 179. (PAGE 533.)

RANVIER, Louis Antoine (1835–1922). Physician and histologist in Paris. Nodes—interruptions of the medullary sheaths of nerves. 'Traité technique d'Histologie.' Paris. 1875. pp. 721–737. (PAGE 20.)

RATHKE, Martin Heinrich (1793–1860). Professor of Zoology and Anatomy in Königsberg. Pouch—a depression in the roof of the embryonic mouth in front of the bucco-pharyngeal membrane. 'Ueber die Entstehung der Glandula pituitaria.' *Archiv für Anatomie, Physiologie und wissenschaftliche Medicin (Müller's Archiv).* 1838. pp. 482–485. 'Entwickelungsgeschichte der Wirbelthiere.' Leipzig. 1861. (PAGE 39.)

REISSNER, Ernst (1824–1878). Professor of Anatomy in Dorpat and, later, Breslau. Membrane—membrana vestibularis of the cochlea. 'De auris internæ formatione.' Dorpat. 1851. (PAGE 456.)

RETZIUS, Andreas Adolf (1796–1860). Professor of Anatomy and Physiology in the Carolin Institute, Stockholm, from 1840 to 1860. Frondiform ligament—the deep attachment of the extensor retinaculum in the sinus tarsi that acts as a sling for the extensor tendons. 'Bemerkungen über ein schleuderförmiges Band in dem Sinus tarsi des Menschen und mehrerer Thiere.' *Archiv für Anatomie, Physiologie und wissenschaftliche Medicin (Müller's Archiv).* 1841. p. 497. (PAGE 169.) Cave—cavum prevesicale. 'Ueber das Ligamentum pelvioprostaticum oder den Apparat, durch welchen die Harnblase, die Prostata und die Harnröhre an der untern Beckenöffnung befestigt sind.' *Ibid.* 1849. pp. 188–189. (PAGE 329.)

ROLANDO, Luigi (1773–1831). First Professor of Practical Medicine at Sassari (Sardinia) and later Professor of Anatomy at Turin. Fissure—sulcus centralis. François Leuret named this structure eponymously because Rolando had called his attention to it some years previously. See: LEURET, François, and GRATIOLET. Louis Pierre: 'Anatomie comparée du système nerveux considerée dans ses rapports avec l'intelligence.' Paris. 1839–1857. (PAGE 496.)

ROSENMULLER, Johann Christian (1771–1820). Professor of Anatomy and Surgery in Leipzig from 1802 to 1820. Fossa—recessus pharyngeus. 'Handbuch der Anatomie.' Leipzig. 1808. (PAGE 417.)

SCARPA, Antonio (1747–1832). Professor of Anatomy in Pavia. F.R.S., 1791. Fascia—a fibrous layer of superficial fascia of the abdomen. 'Supplément au traité pratique des hernies.' Paris. 1823. p. 1. Trans. by C. P. Ollivier. (PAGES 133, 210.) Triangle—the femoral triangle. 'Sull'ernie. Memorie anatomico-chirurgiche.' Milan. 1809. (PAGE 136.)

SCHLEMM, Friedrich (1795–1858). Professor of Anatomy in Berlin from 1833 to 1858. Canal—at the junction of the cornea and the sclera. 'Theoretisch-praktisches Handbuch der Chirurgie.' Edited by Johann Nepomuk Rust. Bd. **III.** 1830. Bulbus Oculi. p. 333. (PAGE 438.)

SCHWANN, Theodor (1810–1882). Professor of Comparative Anatomy and Physiology in Liege. Cells—of the neurilemma. 'Mikroskopische Untersuchungen über die Uebereinstimmung in der Struktur und dem Wachstum der Thiere und Pflanzen.' Berlin. 1839. (PAGE 20.)

SHARPEY, William (1802–1880). Professor of Anatomy at University College, London, from 1836 to 1874 in succession to Jones Quain. Fibres—connective tissue fibres between periosteum and bone. 'Elements of Anatomy', by Jones Quain. 5th Edition, edited by Richard Quain and William Sharpey. p. clix. (PAGE 11.)

SHERRINGTON, Sir Charles Scott (1857–1952). Professor of Physiology in Liverpool from 1895 to 1913 and in Oxford from 1913 to 1936. F.R.S., 1893. Dermatomes—the 'segmental skin-field'. 'Experiments in Examination of the Peripheral Distribution of the Fibres of the Posterior Roots of some Spinal Nerves.' *Proceedings of the Royal Society*. Vol. **52.** 1892. pp. 333–337. *Philosophical Transactions of the Royal Society*. 1893, Vol. **184.** p. 641; and 1898, Vol. **190.** p. 45. (PAGE 27.)

SHRAPNELL, Henry Jones (–1834). Surgeon to S. Gloucestershire Regiment; married Edward Jenner's ward and resided for a time in London. Membrane—membrana flaccida of the membrana tympani. 'On the Form and Structure of the Membrane of the Tympanum.' *London Medical Gazette*. 1832. p. 120. (PAGE 448.)

STRUTHERS, Sir John (1823–1899). Professor of Anatomy in Aberdeen from 1863 to 1889. Ligament—of the humerus, passing to the medial condyle. 'On a Peculiarity of the Humerus and Humeral Artery.' *Monthly Journal of Medical Science*. October, 1848. p. 264. Reprinted in *Anatomical and Physiological Observations*. 1854. p. 208. (PAGE 73.)

SYLVIUS, François de la Boe (1614–1672). Professor of Practical Medicine in Leyden. Aqueduct—aqueductus cerebri. Fissure—the lateral cerebral fissure. 'Disputationum medicarum decas, primarias corporis humani functiones naturales ex anatomicis, practicis et chymicis experimentis deductas complectens; IV. De spiritum animalium in cerebro cerebelloque confectione, per nervos distributione atque usu vario.' Leyden. 1660. See also: 'Notæ de cerebro'. Edited by Caspar Bartholin. 1641. (PAGES 524, 496.)

TENON, Jacques Rene (1724–1816). Professor of Pathology in the Academy of Sciences, Paris, and Chief Surgeon at the Salpetrière. Capsule—fascia bulbi. 'Observations anatomiques sur quelques parties de l'Oeil.' *Mémoires sur l'anatomie, la pathologie et la chirurgie*. Paris. 1806. (PAGE 433.)

TREITZ, Wenzel (1819–1872). Professor of Pathological Anatomy in Krakau and, later, Professor of Pathology in Prague. Muscle—musculus suspensorius duodeni. 'Ueber einen neuen Muskel am Duodenum des Menschen, über elastische Sehnen, und einige andere anatomische Verhältnisse.' *Vierteljahrschrift für die praktische Heilkunde*. Vol. **37.** 1853. p. 113. (PAGE 293.)

TREVES, Sir Frederick (1853–1923). Surgeon to the London Hospital. Operated on King Edward VII for appendicitis. Bloodless fold—of the appendix. 'The Anatomy of the Intestinal Canal and Peritoneum in Man.' London. 1885. pp. 47–48. (PAGE 294.)

VATER, Abraham (1684–1751). Professor of Anatomy, Botany, Pathology and Therapeutics in Wittenberg. Ampulla—of the bile duct. 'Dissertatio anatomica qua novum bilis diverticulum circa orificium ductus choledochi ut et valvulosam colli vesicæ felleæ constructionem ad disceptandum proponit.' Wittenberg. 1720. (PAGE 302.)

VESALIUS, Andreas (1514–1564). Professor of Anatomy at Padua and, later, Bologna and Pisa. Physician to Charles V and Philip II of Spain. Foramen—immediately medial to the foramen ovale. 'De humani corporis fabrica.' Basle. 1543. (PAGES 385, 561)

VICQ D'AZYR, Felix (1748–1794). Physician and Comparative Anatomist of Paris. Physician to the Queen. Bundle—fasciculus mamillo-thalamicus. 'Sur la structure du cerveau, du cervelet, de la moelle alongée, de la moelle épinière; et sur l'origine des nerfs de l'homme et des animaux.' *Histoire de l'Académie des Sciences*. 1781. pp. 495–622. (PAGE 505.)

VIDUS VIDIUS (Guido Guidi) (1500–1569). Physician to Francis I of France and from 1548 Professor of Medicine at the University of Pisa. Nerve—of the pterygoid canal. 'De anatome corporis humani.' 1611. Edited by his nephew. (PAGE 491.)

VOLKMANN, Alfred Wilhelm (1800–1877). Professor of Physiology and Anatomy in Dorpat and, later, Halle. Canals—in bone, carrying blood vessels from the periosteum. 'Ueber die näheren Bestandstheile der menschlichen Knochen.' *Bericht ueber die Verhandlungen der königlichen Sachsischen Gesellschaft der Wissenschaften zu Leipzig (Math.-Phys. Kl.)*. Vol. **25.** 1873. p. 275. (PAGE 11.)

WALDEYER, Heinrich Wilhelm Gottfried (1836–1921). Professor of Pathological Anatomy in Breslau and, later, Berlin. Ring—of adenoidal tissue. 'Ueber den lymphatischen Apparat des Pharynx.' *Deutsche medicinische Wochenschrift*. Bd. **10.** 1884. p. 313. (PAGE 417.) Fascia—of the rectum. 'Lehrbuch der topographisch-chirurgischen Anatomie.' Joessel and Waldeyer. Vol. **II.** 1899. pp. 552–555. (PAGE 329.)

WHARTON, Thomas (1616–1673). Physician to St. Thomas's Hospital where he remained on duty during the Great Plague. Jelly—embryonic connective tissue in the umbilical cord. 'Adenographica sive glandularum totius corporis descriptio.' London. 1656. (PAGE 14.)

WHITNALL, Samuel Ernest (1876–1950). Professor of Anatomy at McGill University, Montreal (1919–1934) and at Bristol (1935–1941). Tubercle—on the zygomatic bone. 'On a Tubercle on the Malar Bone and on the Lateral Attachments of the Tarsal Plates.' *Journal of Anatomy*. Vol. **45.** 1911. pp. 426–432. (PAGE 433.)

WILLIS, Thomas (1621–1675). Physician to James II and one of the founders of the Royal Society. Circle—arterial circle at the base of the brain. 'Cerebri anatome, cui accessit nervorum descriptio et usus.' London. 1664. (PAGE 519.)

WINSLOW, Jacob Benignus (1669–1760). At the age of 74 was appointed Professor of Anatomy, Physic and Surgery in Paris and until his death was considered one of the best anatomical teachers in Europe. Foramen—foramen epiploicum. Ligament—oblique popliteal ligament of the knee joint. 'Exposition anatomique de la structure du corps humain.' Paris. 1732. pp. 352–365. (PAGES 274, 161.)

WOLFF, Kaspar Friedrich (1733–1794). Professor of Anatomy and Physiology at St. Petersburg. One of the founders of modern embryology. Duct—ureter primordalis. 'Theoria generationis.' Halle. 1759. (PAGES 267, 319.)

Index